Pierpaolo Degano   Roberto Gorrieri
Alberto Marchetti-Spaccamela   (Eds.)

Automata, Languages and Programming

24th International Colloquium, ICALP'97
Bologna, Italy, July 7-11, 1997
Proceedings

 Springer

Series Editors

Gerhard Goor, Karlsruhe University, Germany
Juris Hartmanis, Cornell University, NY, USA
Jan van Leeuwen, Utrecht University, The Netherlands

Volume Editors

Pierpaolo Degano
University of Pisa, Department of Computer Science
Corso Italia 40, I-56125 Pisa, Italy
E-mail: degano@di.unipi.it

Roberto Gorrieri
University of Bologna, Department of Computer Science
Mura Anteo Zamboni 7, I-40126 Bologna, Italy
E-mail: gorrieri@cs.unibo.it

Alberto Marchetti-Spaccamela
University of Rome, Department of Computer Science
Via Salaria 113, I-00198 Rome, Italy
E-mail: alberto@dis.uniroma1.it

Cataloging-in-Publication data applied for

Die Deutsche Bibliothek - CIP-Einheitsaufnahme

Automata, languages and programming : 24rd international colloquium ; proceedings / ICALP '97, Bologna, Italy, July 7 - 11, 1997. Pierpaolo Degano ... (ed.). - Berlin ; Heidelberg ; New York ; Barcelona ; Budapest ; Hong Kong ; London ; Milan ; Paris ; Santa Clara ; Singapore ; Tokyo : Springer, 1997
  (Lecture notes in computer science ; Vol. 1256)
  ISBN 3-540-63165-8

CR Subject Classification (1991): F, E, G.2, D.1, D.3, C.2

ISSN 0302-9743
ISBN 3-540-63165-8 Springer-Verlag Berlin Heidelberg New York

This work is subject to copyright. All rights are reserved, whether the whole or part of the material is concerned, specifically the rights of translation, reprinting, re-use of illustrations, recitation, broadcasting, reproduction on microfilms or in any other way, and storage in data banks. Duplication of this publication or parts thereof is permitted only under the provisions of the German Copyright Law of September 9, 1965, in its current version, and permission for use must always be obtained from Springer-Verlag. Violations are liable for prosecution under the German Copyright Law.

© Springer-Verlag Berlin Heidelberg 1997
Printed in Germany

Typesetting: Camera-ready by author
SPIN 10550031   06/3142 – 5 4 3 2 1 0   Printed on acid-free paper

Lecture Notes in Computer Science 1256
Edited by G. Goos, J. Hartmanis and J. van Leeuwen

Advisory Board: W. Brauer D. Gries J. Stoer

Springer
*Berlin
Heidelberg
New York
Barcelona
Budapest
Hong Kong
London
Milan
Paris
Santa Clara
Singapore
Tokyo*

Foreword

The International Colloquium on Algorithms, Languages and Programming (ICALP) is the annual conference of the *European Association for Theoretical Computer Science* (EATCS). The conference aims at enabling computer scientists to exchange theoretical ideas and results, as well as at stimulating cooperation between the theoretical and the practical community in computer science.

The main topics of ICALP '97 included computability, automata, formal languages, new computing paradigms, term rewriting, analysis and design of algorithms, computational geometry, computational complexity, symbolic and algebraic computation, cryptography and security, data types and data structures, theory of data base and knowledge bases, semantics of programming languages, program specification and verification, foundations of logic programming, parallel and distributed computation, theory of concurrency, theory of robotics, theory of logical design and layout.

ICALP '97 was held in Bologna, Italy, July 7–11, 1997. Previous colloquia took place in Paderborn (1996), Szeged (1995), Jerusalem (1994), Lund (1993), Wien (1992), Madrid (1991), Warwick (1990), Stresa (1989), Tampere (1988), Karlsruhe (1987), Rennes (1986), Nafplion (1985), Antwerpen (1984), Barcelona (1983), Aarhus (1982), Haifa (1981), Amsterdam (1980), Graz (1979), Udine (1978), Turku (1977), Edinburgh (1976), Saarbrücken (1974), and Paris (1972). The next ICALP will be held in Aalborg, Denmark, July 13–17, 1998.

ICALP '97 came in conjunction with the 25th anniversary of EATCS. The celebration of the association and of its founders included a historical perspective on the achievements of the community in the last 25 years with a talk by M.Nivat, the first EATCS President, and a discussion on the new challenges that EATCS will face in the future.

ICALP '97 was organised differently than before and accommodated further events, to react positively to the new challenges that the theoretical science community faces in the information technology society. Indeed, our community has developed and now utilizes several approaches and different methodologies that require increased specialization. As a consequence, there is a growing number of specialized conferences and workshops, and it is difficult for researchers to follow the recent developments on specialized research topics. ICALP '97 was a first step towards having a conference offering a single unifying environment while leaving room for specialization. In such an event, the computer science community interested in the development of formal methods and methodologies can stress the relationships that exist among different branches. The new organization of ICALP '97 can be summarized as follows.

- **Invited talks** There were more invited presentations than usual. The eight talks presented the main developments occurring in a specific area and the promising new trends.
- **Plenary and parallel sessions** Some papers were presented in plenary sessions. Parallel sessions were organised for the other submitted papers, according to the two tracks of the Journal of Theoretical Computer Science;

this reflects the main division in research topics within the community, while making evident its unifying aspects.

Satellite workshops Seven satellite workshop were held immediately before or after the main conference. Their specific topics were often at the interface between theoretical computer science and other information technology research areas.

Policy of research funding A panel discussion was held, with panelists including experts responsible for governmental and industrial research and development agencies in Europe and the U.S.

The Program Committee selected 73 papers out of 197 submissions, 183 of which were in electronic format. Their authors are from 30 countries from all over the world. Each submission has been sent to four Program Committee members, assisted by their own referees.

The selection meeting took place in Bologna, March 15–16, 1997. To permit a deeper evaluation of the papers, the Program Committee split in two parts for a preliminary discussion, according to the division mentioned above. Then, all the papers were evaluated again and all the decisions were taken altogether.

We would like to warmly thank all the Program Committee members and their referees for their invaluable contribution.

We are deeply indebted with all the members of the Organizing Committee for all their time and efforts. A special *"grazie"* to Vladimiro Sassone for his excellent automatic system that supported us through all the preparation of the colloquium, from receiving submissions and referees' reports to the preparation of the selection meeting and of the proceedings. *"Grazie"* also to Chiara Bodei for her precious help.

Finally, we gratefully acknowledge support from the UE - DG III, UNESCO Venice Office, Italian National Council of Research (Comitati 01, 07, 12), GNIM-CNR, IEI-CNR, the Universities of Bologna, Pisa, and Roma "La Sapienza", the Regione Emilia-Romagna, TELECOM Italia, and the United States Air Force European Office of Aerospace Research and Development.

April 1997

Pierpaolo Degano, Roberto Gorrieri, Alberto Marchetti-Spaccamela

Invited Lecturers

R. Milner, Cambridge M.O. Rabin, Jerusalem and Harvard
C. Papadimitriou, Berkeley D.S. Scott, Pittsburgh
K.R. Apt, Amsterdam K. Mehlhorn, Saarbrücken
R.J. Lipton, Princeton D. Perrin, Marne-la-Vallée

Program Committee

P. Degano, Pisa (Co-Chair) • M. Cosnard, Lyon
W. Drabent, Warsaw & Linköping J. Diaz, Barcelona •
R. Gorrieri, Bologna• A. Fiat, Tel Aviv•
G. Gottlob, Wien• J. van Leeuwen, Utrecht •
J.W. Klop, Amsterdam• A. Marchetti-Spaccamela, Roma (Co-Chair) •
T. Maibaum, London B. Monien, Paderborn •
J. Meseguer, Menlo Park T. Nishizeki, Sendai •
F. Nielson, Aarhus • J.E. Pin, Paris •
A. Pnueli, Rehovot B. Rovan, Bratislava•
P.S. Thiagarajan, Madras R. Tamassia, Providence •
P. Wegner, Providence E. Ukkonen, Helsinki •

Members marked with a • participated in the selection meeting

Organizing Committee

A. Asperti M. Bernardo
C. Bodei N. Busi
P. Ciancarini R. Davoli
R. Focardi R. Gorrieri (Chair)
S. Leonardi A. Masini
M. Roccetti P. Salomoni
V. Sassone R. Segala
G. Zavattaro

International Publicity Board

R. van Glabbeek, Stanford K. Larsen, Aalborg
J. Karhumaki, Turku T. Y. Nishida, Toyama
J. J. M. M. Rutten, Amsterdam B. Steffen, Passau

List of Referees

K. Aardal
C. Alvarez
T. Asano
Y. Azar
R. Barbuti
A. Berarducci
J. Berstel
S. Bezrukov
C. Bodei
D.J.B. Bosscher
M. Bousquet-Melou
O. Burkart
O. Carton
C. Choffrut
D. Clark
C. Coquand
P. Cousot
M. Crochemore
O. Danvy
M. Delorme
S. Domas
S. Dziembowski
E. Fachini
R. Feldmann
M. Flammini
A. Frangioni
G. Gallo
P. Gianni
B. Gramlich
D. Guijarro
T. Hagerup
M. Reichhardt Hansen
N. Heintze
H. Herbelin
R.N. Horspool
T. Ibaraki
K. Iwama
N. Jones
R. Kaivola
J. Karkkainen
C. Kenyon
J. Kivinen
E. Koehler
D. Kozen
O. Kupferman
T. Wah Lam

G. Aguzzi
K. Amano
E. Asarin
A. Bäumker
G. Barthe
P. Berenbrink
P. Berthome
J. Biccaregui
H.L. Bodlaender
V. Bouchitte
F.J. Brandenburg
N. Busi
R. Casas
S. Christensen
B. Codenotti
D. Corneil
N. Creignou
M. Dam
P. Darondeau
G. Delzanno
C.H. Cabral Duarte
A. Edalat
A. Fantechi
J. Fernandez
R. Focardi
R. Freivalds
F. Galvez
S. Gnesi
W.O.D. Griffioen
Y. Guo
S. Halevi
R. Hariharan
L. Hendren
J. Honkala
J. Hromkovic
L. Ilie
S. Iwata
Jose
G. Kant
H. Karloff
V. Keranen
R. Klasing
B. Konikowska
F. Kroeger
M. Kwiatkowska
C. Laneve

P. Alimonti
H.R. Andersen
J. Aspnes
S. van Bakel
G. Di Battista
M.T. de Berg
B. Berthomieu
G.M. Bierman
R. Bol
G. Boudol
V. Bruyere
C. Butz
G.L. Cattani
M. Chrobak
A. Compagnoni
F. Corradini
P. Crescenzi
P. Damaschke
R. Davoli
M. Devillers
P. Duris
T. Eiter
Farach
P. Ferragina
P. Fraigniaud
J. Gabarro
A. Garg
M. Grabowski
J.F. Groote
P. Gvozdjak
M.M. Halldorsson
T. Harju
F. Henglein
F. Honsell
T. Hsu
P. Inverardi
J. Jablonowski
K. Kühnle
J. Karhumki
T. Karvi
S. Khuller
H.C.M. Kleijn
U. Koppenhagen
W. Kuich
M.R. Lagana'
K. G. Larsen

N. Alon
C. Arapis
J. Autebert
J. L Balcazar
D. Beauquier
M. Bernardo
A. Bertoni
L. Boasson
M.L. Bonet
D. Boulanger
A. Bucciarelli
A. Carpi
A. Černý
P. Chrzastowski-Wachtel
A. Condon
B. Courcelle
S. Crespi-Reghizzi
D. R. Dams
M. Debbabi
L. Devroye
P. Dybjer
T. Erlebach
Y. Feldman
P. Flajolet
P. G. Franciosa
M. Gabbrielli
R. Gavalda
E. Graedel
S. Guerrini
M. Habib
C. Hankin
R. Heckmann
T. Henzinger
H.J. Hoogeboom
H. Hulgaard
G.F. Italiano
B. Jacobs
S. Kahrs
L. Kari
N. Katoh
T. Kimbrel
M. Knor
I. Korec
K. Narayan Kumar
J. Lagergren
M. Latteux

T. Lengyel	S. Leonardi	T. Lettmann	F. Levi
J. Lilius	A. Lingas	L. Lisovik	G. Longo
U. Lorenz	A. de Luca	B. Luttik	H.H. Luvengreen
Y. Métivier	I. Mackie	A. Maggiolo-Schettini	K. Makino
P. Malacaria	E. Malesinska	H. Mannila	J. Marcinkowski
L. Margara	T. Margaria	C. Martinez	S. Martini
A. Masini	A. Mateescu	G. Mauri	E. Mayordomo
R. Mayr	J. Mazoyer	K. Meer	P. Mellies
S. Melzer	K. Meyer	A. Kashem Mia	C. Michaux
A. Middeldorp	D. Miller	E. Moggi	F. Monin
U. Montanari	T. Moorhouse	A.W. Mostowski	M. Mukund
H. Nagamochi	P. Narbel	M. Nesi	R. De Nicola
P. Niebert	M. Nielsen	D. Niwinski	O. Nurmi
M. Nyknen	S. Ohta	T. Okamoto	D. Olejar
V. van Oostrom	F. Orava	P. Orponen	Ostrovsky
M. Oyamaguchi	P. Paczkowski	C. Palamidessi	P. Pananagden
P. Panangaden	G. Pani	D. Pardubska	D. Pedreschi
A. Peron	P. Persiano	I. Phillips	A. Pietracaprina
G.M. Pinna	M. Pistore	T. Plachetka	W. Plandowski
R. Platek	S. Prasad	V. Pratt	R. Preis
C. Priami	H. Przymusinska	P. Pudlak	P. Quaglia
F. van Raamsdonk	Y. Rabani	A. Rabinovich	M. Saidur Rahman
Y. S. Ramakrishna	A. Rauzy	J. Rehof	R. Rehrmann
A. Rensink	M. Riedel	M. Roettger	J.M.T. Romijn
K.H. Rose	A. Rosen	M. Rosendahl	G. Rosolini
V. Rottmann	P. Ružička	A. Rubio	J. Rutten
O. Sýkora	D. Sacca'	K. Salomaa	G. Salzer
D. Sands	D. Sangiorgi	M. Santha	U. Schmid
M. Schmidt-Schauss	U. Schoening	U. Schroeder	A. Schubert
P. Seebold	R. Segala	H. Seidl	A. Sen
G. Senizergues	M. Serna	E. Shahar	A. Shinohara
M. Vajteršic	M. Siegel	J. Petit i Silvestre	E. Singerman
J. Skurczynski	S. Skyum	A. Slobodova	S.A. Smolka
X. Song	D. Sotteau	Spinrad	P. Spirakis
M. Srebrny	L. Staiger	I. Stark	B. Steffen
M. Steinby	J. Stern	C. Stirling	W. Streicher
M. Sudan	K. Sunesen	S. D. Swierstra	R. Szelepcsenyi
V. Tannen	M. Tatsuta	G. Tel	B. Thomsen
S. Tison	J. Tiuryn	I.G. Tollis	Y. Toyama
L. Trevisan	J. Tromp	J. M. Troya	S. Tschoeke
F. Turini	R. Uehara	W. Unger	P. Urzyczyn
S. Varricchio	H. Veith	M. Veldhorst	B. Victor
E. de Vink	L. Vismara	P. Vitanyi	B. Voecking
W. Vogler	A. Voronkov	F. de Vries	I. Vrt'o
K. W. Wagner	I. Walukiewicz	Warnow	M. Westermann
F. Winkler	U. Wolter	M. Yamashita	H. Zantema
G. Zavattaro	X. Zhou	W. Zielonka	

Table of Contents

Invited Papers

Graphical Calculi for Interaction
R. Milner 1

NP-Completeness: A Retrospective
C. H. Papadimitriou 2

The LEDA Platform for Combinatorial and Geometric Computing
K. Mehlhorn, S. Näher, C. Uhrig 7

The Wadge-Wegner Hierarchy of ω-rational Sets
O. Carton, D. Perrin 17

From Chaotic Iteration to Constraint Propagation
K. R. Apt 36

DNA^{2}DNA Computations: A Potential "Killer App"?
L. F. Landweber, R. J. Lipton 56

Session 1: Formal Languages I

Tilings and Quasiperiodicity
B. Durand 65

Enumerative Sequences of Leaves in Rational Trees
F. Bassino, M.-P. Béal, D. Perrin 76

A Completion Algorithm for Codes with Bounded Synchronization Delay
V. Bruyère 87

The Expressibility of Languages and Relations by Word Equations
J. Karhumäki, W. Plandowski, F. Mignosi 98

Finite Loops Recognize Exactly the Regular Open Languages
M. Beaudry, F. Lemieux, D. Thérien 110

Session 2: Computability

An Abstract Data Type for Real Numbers
P. Di Gianantonio 121

Recursive Computational Depth
J. I. Lathrop, J. H. Lutz 132

Some Bounds on the Computational Power of Piecewise Constant Derivative Systems
O. Bournez 143

Monadic Simultaneous Rigid E-Unification and Related Problems
Y. Gurevich, A. Voronkov 154

Computability on the Probability Measures on the Borel Sets of the Unit Interval
K. Weihrauch 166

Session 3: Computational Complexity

Worst-Case Hardness Suffices for Derandomization: A New Method for Hardness-Randomness Trade-offs
A. A. Andreev, A. E. F. Clementi, J. D. P. Rolim 177

Results on Resource-Bounded Measure
H. Buhrman, S. Fenner, L. Fortnow 188

Randomization and Nondeterminism are Incomparable for Ordered Read-Once Branching Programs
F. Ablayev 195

Checking Properties of Polynomials
B. Codenotti, F. Ergün, P. S. Gemmell, S. Ravi Kumar 203

Exact Analysis of Dodgson Elections: Lewis Carroll's 1876 Voting System is Complete for Parallel Access to NP
E. Hemaspaandra, L. A. Hemaspaandra, J. Rothe 214

Session 4: Semantics I

Game Theoretic Analysis of Call-by-value Computation
K. Honda, N. Yoshida 225

On Modular Properties of Higher Order Extensional λ-Calculi
R. Di Cosmo, N. Ghani 237

On Explicit Substitutions and Names
E. Ritter, V. de Paiva 248

On the Dynamics of Sharing Graphs
A. Asperti, C. Laneve 259

Session 5: Algorithms I

Minimizing Diameters of Dynamic Trees
S. Alstrup, J. Holm, K. de Lichtenberg, M. Thorup 270

Improving Spanning Trees by Upgrading Nodes
S. O. Krumke, M. V. Marathe, H. Noltemeier, R. Ravi, S.S. Ravi, R. Sundaram, H.-C. Wirth 281

Dynamic Algorithms for Graphs of Bounded Treewidth
T. Hagerup 292

Session 6: Calculi for Concurrency I

The Name Discipline of Uniform Receptiveness
D. Sangiorgi 303

On Confluence in the π-calculus
A. Philippou, D. Walker 314

A Proof Theoretical Approach to Communication
Y. Fu 325

Session 7: Formal Languages II

Solving Trace Equations Using Lexicographical Normal Forms
V. Diekert, Y. Matiyasevich, A. Muscholl 336

Star-Free Picture Expressions are Strictly Weaker than First-Order Logic
T. Wilke 347

Session 8: Calculi for Concurrency II

An Algebra-Based Method to Associate Rewards with EMPA Terms
M. Bernardo 358

A Semantics Preserving Actor Translation
I. A. Mason, C. L. Talcott 369

Session 9: Algorithms II

Periodic and Non-periodic Min-Max Equations
U. Schwiegelshohn, L. Thiele 379

Efficient Parallel Graph Algorithms for Coarse Grained Multicomputers and BSP
E. Cáceres, F. Dehne, A. Ferreira, P. Flocchini, I. Rieping, A. Roncato,
N. Santoro, S. W. Song 390

Upper Bound on the Communication Complexity of Private Information Retrieval
A. Ambainis 401

Session 10: Logic and Verification

Computation Paths Logic: An Expressive, yet Elementary, Process Logic
D. Harel, E. Singerman 408

Model Checking the Full Modal Mu-Calculus for Infinite Sequential Processes
O. Burkart, B. Steffen 419

Symbolic Model Checking for Probabilistic Processes
C. Baier, E. M. Clarke, V. Hartonas-Garmhausen, M. Kwiatkowska,
M. Ryan 430

Session 11: Analysis of Algorithms

On the Concentration of the Height of Binary Search Trees
J. M. Robson 441

An Improved Master Theorem for Divide-and-Conquer Recurrences
S. Roura 449

Session 12: Process Equivalences

Bisimulation for Probabilistic Transition Systems: A Coalgebraic Approach
E. P. de Vink, J. J. M. M. Rutten 460

Distributed Processes and Location Failures
J. Riely, M. Hennessy 471

Basic Observables for Processes
M. Boreale, R. De Nicola, R. Pugliese 482

Session 13: Routing Algorithms

Constrained Bipartite Edge Coloring with Applications to Wavelength Routing
C. Kaklamanis, P. Persiano, T. Erlebach, K. Jansen 493

Colouring Paths in Directed Symmetric Trees with Applications to WDM Routing
L. Gargano, P. Hell, S. Perennes 505

On-Line Routing in All-Optical Networks
Y. Bartal, S. Leonardi 516

A Complete Characterization of the Path Layout Construction Problem for ATM Networks with Given Hop Count and Load
T. Eilam, M. Flammini, S. Zaks 527

Session 14: Petri Nets and Process Theory

Efficiency of Asynchronous Systems and Read Arcs in Petri Nets
W. Vogler 538

Bisimulation Equivalence is Decidable for One-Counter Processes
P. Jančar 549

Symbolic Reachability Analysis of FIFO-Channel Systems with Nonregular Sets of Configurations
A. Bouajjani, P. Habermehl 560

Axiomatizations for the Perpetual Loop in Process Algebra
W. Fokkink 571

Discrete-Time Control for Rectangular Hybrid Automata
T. A. Henzinger, P. W. Kopke 582

Session 15: Algorithms III

Maintaining Minimum Spanning Trees in Dynamic Graphs
M. Rauch Henzinger, V. King 594

Efficient Splitting and Merging Algorithms for Order Decomposable Problems
R. Grossi, G. F. Italiano 605

Efficient Array Partitioning
S. Khanna, S. Muthukrishnan, S. Skiena..................................616

Constructive Linear Time Algorithms for Branchwidth
H. L. Bodlaender, D. M. Thilikos...627

Session 16: Rewriting

The Word Matching Problem is Undecidable for Finite Special String-Rewriting Systems that are Confluent
P. Narendran, F. Otto..638

The Geometry of Orthogonal Reduction Spaces
Z. Khasidashvili, J. Glauert..649

The Theory of Vaccines
M. Marchiori..660

Session 17: Formal Languages III

The Equivalence Problem for Deterministic Pushdown Automata is Decidable
G. Sénizergues..671

On Recognizable and Rational Formal Power Series in Partially Commuting Variables
M. Droste, P. Gastin..682

On a Conjecture of J. Shallit
J. Cassaigne..693

Session 18: Cryptography

On Characterizations of Escrow Encryption Schemes
Y. Frankel, M. Yung..705

Randomness-Efficient Non-Interactive Zero Knowledge
A. De Santis, G. Di Crescenzo, P. Persiano...............................716

Session 19: Algorithms IV

Approximation Results for the Optimum Cost Chromatic Partition Problem
K. Jansen..727

The Minimum Color Sum of Bipartite Graphs
A. Bar-Noy, G. Kortsarz..738

A Primal-Dual Approach to Approximation of Node-Deletion Problems for Matroidal Properties
T. Fujito...749

Independent Sets in Asteroidal Triple-Free Graphs
H. Broersma, T. Kloks, D. Kratsch, H. Müller............................760

Session 20: Semantics II and Automata

Refining and Compressing Abstract Domains
R. Giacobazzi, F. Ranzato 771

Labelled Reductions, Runtime Errors, and Operational Subsumption
L. Dami 782

A Complete and Efficiently Computable Topological Classification of
D-dimensional Linear Cellular Automata over Z_m
G. Manzini, L. Margara 794

Recognizability Equals Definability for Partial k-Paths
V. Kabanets 805

Session 21: Biocomputing

Molecular Computing, Bounded Nondeterminism, and Efficient Recursion
R. Beigel, B. Fu 816

Constructing Big Trees from Short Sequences
P. L. Erdős, M. A. Steel, L. A. Székely, T. J. Warnow 827

Session 22: Logic Programming

Termination of Constraint Logic Programs
S. Ruggieri 838

The Expressive Power of Unique Total Stable Model Semantics
F. Buccafurri, S. Greco, D. Saccà 849

Author Index 861

Graphical Calculi for Interaction

Robin Milner

University of Cambridge, UK

Recently there has been great interest in operational models of interactive systems, and more recently especially in those which capture to some extent the elusive notion of *mobility*. The π-calculus [1] is one such model, and has had some success both in application and in prompting research in abstract models of interaction. But it can hardly claim to be canonical, and indeed nor can any of the other operational models.

We might consider that the quest for a canonical model of interaction is no more likely to succeed than that for a canonical model of computation. (In the latter case, we have to be content with many models – Turing machines, register machines, ... – and with translating between them.) Nonetheless, it would be timid not to seek aspects which are common to many, or even most, models of interactive behaviour.

In around 1992 I started from the π-calculus and tried to separate what seemed ad hoc from what seemed more essential. The exact communication discipline of the π-calculus fell into the ad hoc category; the rest – naming, restriction, parallel composition – have greater claim to be universal. This was the origin of *action calculi* [2]. To present the π-calculus as an action calculus, one starts from the common basis of action calculi and merely adds two or three so-called "controls" – for message-passing and replication. It turns out that the λ-calculus, the object calculus of Abadi and Cardelli, and many recent calculi can be similarly set up – and combined with each other – in the action-calculus framework. Considerable progress has been made, for example in [3], in the uniform treatment of models of action calculi.

In the conference lecture I shall emphasize one feature of action calculi: their graphical presentation. Several examples will be given – including some recent advances in calculi for representing *locality* – showing that this graphical element is exactly what all action calculi have in common. These examples motivate further development (which is certainly needed) in the general theory of action calculi and their models.

References

1. Milner, R., Parrow, J. and Walker, D., A calculus of mobile processes, Parts I and II. Information and Computation **100** (1992) 1–77.
2. Milner, R., Calculi for interaction. Acta Informatica **33** (1996) 707–737.
3. Milner, R., Mifsud, A. and Power, J., Control structures. Proc. IEEE Symposium on Logics in Computation, LICS (1995).

NP-Completeness: A Retrospective

Christos H. Papadimitriou[*]

University of California, Berkeley, USA

Abstract. For a quarter of a century now, NP-completeness has been computer science's favorite paradigm, fad, punching bag, buzzword, alibi, and intellectual export. This paper is a fragmentary commentary on its origins, its nature, its impact, and on the attributes that have made it so pervasive and contagious.

1. A keyword search in Melvyl, the University of California's on-line library, reveals that about 6,000 papers each year have the term "NP-complete" on their title, abstract, or list of keywords. This is more than each of the terms "compiler," "database," "expert," "neural network," and "operating system." Even more surprising is the diversity of the disciplines with papers referring to "NP-completeness:" They range from statistics and artificial life to automatic control and nuclear engineering. What is the nature and extent of the impact of NP-completeness on theoretical computer science, computer science in general, computing practice, as well as other domains of the natural sciences, applied science, and mathematics? And why did NP-completeness become such a pervasive and influential concept?

2. One of the reasons of the immense impact of NP-completeness has to be the appeal and elegance of the class P, that is, of the thesis that "polynomial worst-case time" is a plausible and productive mathematical surrogate of the empirical concept of "practically solvable computational problem." But, obviously, NP-completeness also draws on the importance of NP, as it rests on the widely conjectured contradistinction between these two classes. In this regard, it is crucial that NP captures vast domains of computational, scientific, and mathematical endeavor, and seems to roughly delimit what mathematicians and scientists had been aspiring to compute feasibly. True, there are domains, such as strategic analysis and counting, which have been within our computational ambitions, and still seem to lie outside NP; but they are the exceptions rather than the rule. NP-completeness has thus become a valuable intermediary between the abstraction of computational models and the reality of computational problems, grounding complexity theory to computational practice.

3. Also crucial for the success of NP-completeness has been its surprising ubiquity and effectiveness as a classification tool, and the scarcity of problems in

---

[*] christos@cs.berkeley.edu. Partially supported by the National Science Foundation. A version of this talk was given at a meeting in the Fall of 1995 celebrating the 60th birthday of Richard M. Karp, to whom this paper is also affectionately dedicated.

NP that resist classification as either polynomial-time solvable or NP-complete. (Ladner's result on intermediate degrees between P and NP-completeness [12] had been known almost as soon as NP-completeness was introduced, and thus theoretically the world could be full of mysterious intermediate problems.) In several occasions, extremely broad classes of computational problems in NP have been dichotomized with surprising accuracy into polynomially solvable and NP-complete, see [21, 22] for two early examples.

4. The founders of NP-completeness [2, 10, 13] appear to have anticipated its broad applicability and classification power. Leonid Levin [13] wrote in 1973: *"The method described here clearly provides a means for readily obtaining results of [this type] for the majority of important sequential search problems."* In Karp's paper [10] twenty one problems were proved NP-complete, showing beyond any doubt the surprisingly broad applicability of the method. Significantly, Karp seems annoyed and surprised that three other problems (linear programming, primality, and graph isomorphism) resisted at the time such classification. Primality and graph isomorphism were also mentioned by Cook [2]. Knuth was sufficiently convinced about the importance and broad applicability of the new concept to take early and deliberate action on the terminological front [11].

5. NP-completeness has had tremendous impact even in areas where, in some sense, it should not have. It is now common knowledge among computer scientists that NP-completeness is largely irrelevant to public-key cryptography, since in that area one needs sophisticated *cryptographic assumptions* that go beyond NP-completeness and worst-case polynomial-time computation [19]; furthermore, cryptographic protocols based on NP-complete problems have been ill-fated. Fortunately, the founders of modern cryptography did not know this. Diffie and Hellman base their famous pronouncement *"We stand today on the brink of a revolution in cryptography"* [3] on two facts: (1) Very fast hardware and software, and (2) novel techniques for proving problems hard (they cite Karp's paper [10]).

6. NP-completeness has also exhibited a great amount of versatility, adapting to contexts and computational aspects beyond its original scope of worst-case analysis of exact algorithms for decision and optimization problems. For example, it was used early on to show that certain optimization problems cannot be approximated satisfactorily [20], and indeed in a most ingenious and comprehensive way more recently [1]. By showing that even less ambitious goals than worst-case polynomial exact solution are unattainable, NP-completeness is thus a most useful tool for repeatedly pruning unpromising research directions and thus redirecting research to new ones (in a manner reminiscent of the struggle between Hercules and the monster Hydra [16]).

7. Let me illustrate this versatility of NP-completeness by a technical interlude on an aspect of efficient computation that has interested me recently, namely, *output polynomial time*. Certain computational problems require an output $f(x)$ on input x that is in the worst case exponential in the input. For such problems, one would like to have algorithms that are polynomial in $|x|$ *and* $|f(x)|$. The class of problems thus solvable can be called *output polynomial time*. One can use NP-completeness to prove that certain functions are not in output-polynomial time, unless P=NP. For example, consider the function MIN which maps a regular expression to the minimum-state equivalent *deterministic* finite-state automaton. MIN can be computed by first designing a nondeterministic automaton M, then an equivalent deterministic automaton M', and next minimizing the states of M' to obtain the final output; the problem is, of course, that the intermediate result M' could be exponential in *both* the input and the output. It is rather straightforward to use "traditional" NP-completeness techniques to show the following:

Theorem 1. *Unless P=NP, MIN is not in output polynomial time.*

In fact, we cannot even compute in output-polynomial time a deterministic automaton that has *at most polynomially more states than the minimum* —unless, of course, P=NP.

8. Often the required output $f(x)$ is a set $\{y_1, \ldots, y_k\}$ of strings that are related to x via an NP mapping; for example, if G is a graph, let AMIS(G) be the set of all maxim*al* independent sets of G. AMIS is known to be in output-polynomial time (see [9] for an exposition and strengthening of this result, and an early discussion of output polynomial time). For such problems we have an elegant alternative definition of output polynomial time. A function $f : \Sigma^* \mapsto 2^{\Sigma^*}$ is in output polynomial time if the following problem is solvable in polynomial time: Given x and $y \subseteq \Sigma^*$, either decide that $y = f(x)$, or find a string in $y \oplus f(x)$. It is easy to see that, if such an algorithm exists, then its iteration starting with $S = \emptyset$ gives an output polynomial time algorithm for f; and vice-versa, if an output polynomial time algorithm exists for f, it can be used to produce an element of $y \oplus f(x)$. For example, AMIS is in output polynomial time; its generalization to hypergraphs is open, but was recently shown to be in output $n^{c \log n}$ time [6]; see [5] for an extensive discussion of the hypergraph generalization of AMIS. One can use again "traditional" NP-completeness to show that the following generalization is *not* in output polynomial time, unless P=NP: Given a monotone circuit, compute the set of all minimal (with respect to the set of true inputs) satisfying truth assignments.

9. But, sometimes, "traditional" NP-completeness techniques do not seem to suffice to bring out the intractability of a problem, because this problem belongs to a class or computational mode that appears to be "between" P and NP. In such cases NP-completeness has acted as an open-ended research paradigm, spawning variants that are appropriate for the computational context being studied; examples are classes that capture local search [8], the parity argument [14], logarithmic nondeterminism [18], the related concept of fixed-parameter tractability [4], and approximability [17].

10. Complexity classes introduced this way, as abstractions of natural computational problems of mysteriously intermediate complexity, are in some precise sense well-motivated, indeed necessary; they are *discovered*, not *invented*, as they have always existed by dint of their natural complete problems. The only way to make them go away is to *collapse them with P or NP* —as occasionally happens, recall [17] and its brilliant follow-up [1].

11. NP-completeness is of course a valuable tool for demonstrating the difficulty of computational problems. However, NP-completeness is often used "allegorically;" a problem is shown NP-complete that is not, strictly speaking, a natural computational problem, but an artificial problem created to capture a mathematical concept. NP-completeness in this context suggests that a problem, area, or approach is *mathematically nasty.*. Because, if we believe that efficient algorithms are the natural outflow of the mathematical structure of a problem (a view shared by all computer scientists, with the possible exception of researchers in "metaphor-based" algorithmic paradigms such as neural nets, in which algorithmic behavior is thought to be "emergent"), then, contrapositively, complexity must be the manifestation of mathematical poverty, lack of structure. See [7] for an early example of such a use of NP-completeness in the theory of relational databases.

12. Beyond mathematics, NP-completeness (and complexity in general) can also be applied "allegorically" in other disciplines. It can be used as a metaphor for chaos in dynamical systems, for unbounded rationality in game theory, for unfairness in economics, for integrity of electoral systems in political science, for cognitive implausibility in artificial intelligence, for genetic indeterminism in genetics, and so on (see [16] for references).

13. NP-completeness is thus an important "intellectual export" of computer science to other disciplines. And it does fill a void in the interdisciplinary intellectual trade: It seems to me that the concept of lower bounds —and negative results in general— is particular to computer science, and has no well-developed counterpart in other disciplines. True, one sees isolated results in other sciences (such as Heisenberg's uncertainty principle in quantum mechanics, Arrow's impossibility theorem in economics, and Carnot's theorem in thermodynamics) which are arguably negative; however, nowhere else in science does one find such a comprehensive methodology for obtaining negative results (with the exception of complexity's own precursor mathematical logic, with its many incompleteness, undecidability, and inexpressibility results). NP-completeness is therefore valuable for another reason: It is one of the few precious features which give our science its special character, which set it apart from the other sciences (see [15] for another development of this argument).

14. In science, successful ideas are those that are pervasive and invasive, are invitingly elegant and methodical, are open to extensions and variants, and capture an objective necessity, answer a widespread but diffuse sense of dissatisfaction in the scientific community (in the case of NP-completeness, the widespread feeling among computer scientists in the 1960s that automata theory, the previous great paradigm, had run its course as a useful abstraction of computation). Thinking about the nature and history of NP-completeness could give us useful hints about computer science's next great paradigm, which, for all I know, has started being articulated somewhere else in this volume.

References

1. S. Arora, C. Lund, R. Motwani, M. Sudan, and M. Szegedy, "Proof verification and hardness of approximation problems." *Proc. 33rd FOCS* (1992) pp. 14–23.
2. S. A. Cook "The complexity of theorem-proving procedures," *Proc. 3rd STOC*, (1971), pp. 151–158.
3. W. Diffie and M. E. Hellman "New directions in cryptography," *IEEE Trans. Inform. Theory, 22*, pp. 644–654, 1976.
4. R. G. Downey and M. R. Fellows "Fixed-parameter tractability and completeness I: Basic results," *SIAM Journal on Computing, 24*, 4, pp. 873-921, 1995.
5. T. Eiter, G. Gottlob "Identifying the minimal transversals of a hypergraph and related problems" *SIAM Journal on Computing, 24*, 6, pp. 1278-1304, 1995.
6. M. Fredman and L. Khachiyan "On the complexity of dualization of monotone disjunctive normal forms" *Journal of Algorithms, 21*, 3, pp. 618–628, 1996.
7. P. Honeyman, R. E. Ladner, M. Yannakakis, "Testing the universal instance assumption," *Information Processing Letters, 12*, pp. 14–19, 1980.
8. D. S. Johnson, C. H. Papadimitriou, M. Yannakakis "How Easy is Local Search?" *J.CSS*, 1988 (special issue for the 1985 FOCS Conference).
9. D. S. Johnson, C. H. Papadimitriou, M. Yannakakis "On Generating All Maximal Independent Sets", *Information Processing Letters* 1988.
10. R. M. Karp "Reducibility among combinatorial problems," pp. 85–103 in *Complexity of Computer Computations*, R. E. Miller and J. W. Thatcher (eds), 1972.
11. D. E. Knuth "A terminological proposal," *SIGACT News, 6*, 1, pp. 12–18, 1974.
12. R. E. Ladner "On the structure of polynomial time reducibility," *J.ACM, 22*, pp. 155–171, 1975.
13. L. Levin "Universal sorting problems," *Pr. Inf. Transm., 9*, p¿ 265–266, 1973.
14. C. H. Papadimitriou "On the Complexity of the Parity Argument and other Inefficient Proofs of Existence" *JCSS, 48*, 3, 498–532, 1994.
15. C. H. Papadimitriou "Database metatheory: asking the big queries," *Proc. 1995 PODS Conf.*, reprinted in *SIGACT News*, spring 1996.
16. C. H. Papadimitriou "The complexity of knowledge representation," *Proc. 1996 Computational Complexity Symposium*.
17. C. H. Papadimitriou, M. Yannakakis "Optimization, approximation, and complexity classes" *Proc. 1988 STOC*, and *J.CSS,*, 1991.
18. C. H. Papadimitriou, M. Yannakakis "On limited nondeterminism and the complexity of the Vapnic-Chervonenkis dimension," special issue of *J.CSS* 1996 (special issue for the 1993 Structures Conf.).
19. R. L. Rivest "Cryptography," pp. 717–755 in *Handbook of Theoretical Computer Science*, J. van Leeuwen (ed), The MIT Press/Elsevier, 1990.
20. S. Sahni, T. Gonzalez "P-complete approximation problems," *J.ACM, 23*, pp. 555–565, 1976.
21. T. J. Schaeffer "The complexity of satisfiability problems," *Proc. 10th STOC*, (1978), pp. 216–226.
22. M. Yannakakis "Node- and edge-deletion problems," *Proc. 10th STOC*, (1978), pp. 253–264.

The LEDA Platform
for
Combinatorial and Geometric Computing

Kurt Mehlhorn* and Stefan Näher** and Christian Uhrig***

Abstract. We give an overview of the LEDA platform for combinatorial and geometric computing and an account of its development. We discuss our motivation for building LEDA and to what extent we have reached our goals. We also discuss some recent theoretical developments. This paper contains no new technical material. It is intended as a guide to existing publications about the system. We refer the reader also to our web-pages for more information.

1 What is LEDA?

LEDA [MN95, MNU96] aims at being a comprehensive software platform for combinatorial and geometric computing. It provides a sizable collection of data types and algorithms. This collection includes most of the data types and algorithms described in the text books of the area ([AHU83, Meh84, Tar83, CLR90, O'R94, Woo93, Sed91, Kin90, van88, NH93]). In particular, it includes stacks, queues, lists, sets, dictionaries, ordered sequences, partitions, priority queues, directed, undirected, and planar graphs, lines, points, planes, and polygons, and many algorithms in graph and network theory and computational geometry, e.g., shortest paths, matchings, maximum flow, min cost flow, planarity testing, spanning trees, biconnected and strongly connected components, segment intersection, convex hulls, Delaunay triangulations, and Voronoi diagrams. LEDA supports applications in a broad range of areas. It has already been used in such diverse areas as code optimization, VLSI design, graph drawing, graphics, robot motion planning, traffic scheduling, machine learning and computational biology.

We discuss different aspects of the LEDA system.

Ease of Use: The library is easy to use. In fact, only a small fraction of our users are algorithms experts and many of our users are not even computer scientists. For these users the broad scope of the library, its ease of use, and the correctness and efficiency of the algorithms in the library are crucial.

---

* Max-Planck-Institut für Informatik, Im Stadtwald, 66123 Saarbrücken, www.mpi-sb.mpg.de/~mehlhorn
** Martin-Luther-Universität Halle-Wittenberg, FB Mathematik und Informatik, Weinbergweg 17, 060099 Halle, www.informatik.uni-halle.de/~naeher
*** LEDA Software GmbH 66123 Saarbrücken, www.mpi-sb.mpg.de/LEDA/leda.html

The LEDA manual [MNU96] gives precise and readable specifications for the data types and algorithms mentioned above. The specifications are short (typically not more than a page), general (so as to allow several implementations) and abstract (so as to hide all details of the implementation).

Extendibility: Combinatorial and geometric computing is a diverse area and hence it is impossible for a library to provide ready-made solutions for all application problems. For this reason it is important that LEDA is easily extendible (see also section 4.4) and can be used as a platform for further software development. In many cases LEDA programs are very close to the typical text book presentation of the underlying algorithms. The goal is the equation

$$\text{Algorithm} + \text{LEDA} = \text{Program}.$$

We give an example. Dijkstra's shortest path algorithm takes a directed graph $G = (V, E)$, a node $s \in V$, called the source, and a non-negative cost function on the edges $cost : E \to R_{>0}$. It computes for each node $v \in V$ the distance from s. A typical text book presentation of the algorithm is as follows.

```
set dist(s) to 0.
set dist(v) to infinity for v different from s.
declare all nodes unreached.
while there is an unreached node
{ let u be an unreached node with minimal dist-value.             (*)
    declare u reached.
    forall edges e = (u,v) out of u
        set dist(v) = min( dist(v), dist(u) + cost(e) )
}
```

The text book presentation will then continue to discuss the implementation of line (*). It will state that the pairs $\{(v, dist(v)); v \text{ unreached}\}$ should be stored in a priority queue, e.g., a Fibonacci heap, because this will allow the selection of an unreached node with minimal distance value in logarithmic time. It will probably refer to some other chapter of the book for a discussion of priority queues.

We now give the corresponding LEDA program; it is very similar to the presentation above.

```
#include <LEDA/graph.h>
#include <LEDA/node_pq.h>
void DIJKSTRA(const graph &G, node s, const edge_array<double>& cost,
                              node_array<double>& dist)
{ node_pq<double> PQ(G);
  node v;
  edge e;

  forall_nodes(v,G)
```

```
    { if (v == s) dist[v] = 0; else dist[v] = MAXDOUBLE;
      PQ.insert(v,dist[v]);
    }
    while ( !PQ.empty() )
    { node u = PQ.del_min();
      forall_adj_edges(e,u)
         { v = target(e);
           double c = dist[u] + cost[e];
           if ( c < dist[v] )
           {  PQ.decrease_inf(v,c);   dist[v] = c;  }
         }
    }
}
```

We start by including the graph and the node priority queue data type. We use *edge_array*s and *node_array*s (arrays indexed by edges and nodes respectively) for the functions *cost* and *dist*. We declare a priority queue PQ for the nodes of graph G. It stores pairs $(v, dist[v])$ and is empty initially. The **forall_nodes**-loop initializes *dist* and PQ. In the main loop we repeatedly select a pair $(u, dist[u])$ with minimal distance value and then scan through all adjacent edges to update distance values of neighboring vertices.

Correctness: We try to make sure that the programs in LEDA are correct. We start from correct algorithms, we document our implementations carefully (at least recently), we test them extensively, and we have developed program checkers (see subsection 4.1) for some of them. We want to emphasize that many of the algorithms in LEDA are quite intricate and therefore non-trivial to implement. In the combinatorial domain it is frequently possible to obtain a correct implementation by sacrificing efficiency, e.g., by using linear search in the realization of a dictionary. In the geometric domain it is usually difficult to obtain a correct implementation even if efficiency plays no role. This is due to the so-called degeneracy and precision problem [MN94]. The geometric algorithms in LEDA use exact arithmetic and are therefore free from failures due to rounding errors. Moreover, they can handle all degenerate cases.

Efficiency: LEDA contains the most efficient realizations known for its types. For many data types the user may even choose between different implementations, e.g., for dictionaries he may choose between ab-trees, $BB[\alpha]$-trees, dynamic perfect hashing, and skip lists. The declarations

```
    dictionary<string,int> D1;
    dictionary<string,int,skip_list> D2;
```

declare *D1* as a dictionary from *string* to *int* with the default implementation and select the skip list implementation for *D2*.

Availability and Usage: LEDA is realized in C++ and runs on many different platforms (Unix, Windows95, Windows NT, OS/2) with many different compilers.

LEDA is now used at more than 1500 academic sites. Academic use is free, see http://www.mpi-sb.mpg.de/LEDA/leda.html. A commercial version of LEDA is marketed LEDA Software GmbH. There are license holders in the telecommunication industry (ATR (Japan), Comptel (Finland), E-Plus (Germany), France Télécom (France), MCI (USA)), in the graphics industry (Aristo Technologies (USA), Cadabra (Canada), Compass Design (USA), Fuji (Japan), Mentor Graphics (USA), MUS (Germany)), in the automotive industrie (Daimler Benz (Germany), Ford (USA), Honda (Japan)), in the computer industry (DEC (USA), IBM (USA), Siemens AG (Germany), Silicon Graphics (USA), SUN (USA)), and other industries (Chevron (USA), CFP (Germany), Dolphin (The Netherlands), Howmedica (Germany), Lufthansa (Germany), Neovista (USA), Prediction (USA), Sony (Japan), VTT (Finland)).

History: We started the project in the fall of 1988. We spent the first 6 months on specifications and on selecting our implementation language. Our test cases were priority queues, dictionaries, partitions, and algorithms for shortest paths and minimum spanning trees. We came up with the item concept as an abstraction of the notion "pointer into a data structure". It worked successfully for the three data types mentioned above and we are now using it for most data types in LEDA. Concurrently with searching for the correct specifications we investigated several languages for their suitability as our implementation platform. We looked at Smalltalk, Modula, Ada, Eiffel, and C++. We wanted a language that supported abstract data types and type parameters (polymorphism) and that was widely available. We wrote sample programs in each language. Based on our experiences we selected C++ because of its flexibility, expressive power, and availability. We are even more convinced now that our choice was the right one.

A first publication about LEDA appeared in MFCS 1989 (Lecture Note in Computer Science, Volume 379) and ICALP 1990 (Lecture Notes in Computer Science, Volume 443). Stefan Näher became the head of the LEDA project and he is the main designer and implementer of LEDA.

In the second half of 1989 and during 1990 Stefan Näher implemented a first version of the combinatorial part (= data structures and graph algorithms) of LEDA (Version 1.0). Version 2.0 allowed to use arbitrary data types (not only pointer and simple types) as actual type parameters of parameterized data types. It included a first implementation of the two-dimensional geometry library (libP) and an interface to the X-Window system for graphical input and output (data type window). Version 3.0 switched to the template mechanism to realize parameterized data types (macro substitution was used before), introduced implementation parameters that allow to choose between different implementations, extended the LEDA memory management system to user-defined classes, and further improved the efficiency of many data types and algorithms. Version 3.1 provided a more efficient graph data type and contained new data types (arbitrary precision number types and basic geometric objects) used for robust implementations of geometric algorithms and Versions 3.2 and 3.3 contained more geometry and new tools for documentation and manual production.

LEDA Software GmbH was founded in early 1995.

2 Why did we build LEDA?

We had four main reasons:

1. We had always felt that a significant fraction of the research done in the algorithms area was eminently practical. However, only a small part of it was actually used. We frequently heard from our former students that the effort needed to implement an advanced data structure or algorithm is too large to be cost-effective. We concluded that *algorithms research must include implementation if the field wants to have maximum impact*.
2. Even within our own research group we found different implementations of the same balanced tree data structure. Thus there was constant reinvention of the wheel even within our own tight group.
3. Many of our students had implemented algorithms for their master's thesis. Work invested by these students was usually lost after the students graduated. We had no depository for implementations.
4. The specifications of advanced data types which we gave in class and which we found in text books, including the one written by one of the authors, were incomplete and not sufficiently abstract. They contained phrases of the form: "Given a pointer to a node in the heap its key can be decreased in constant amortized time". This implied that a user of a data structure had to have knowledge of its implementation. As a consequence combining implementations was a non-trivial task. A case in point is the shortest path problem in graphs. We taught priority queues in the early weeks of an algorithm course and Dijkstra's algorithm for the shortest path problem in later weeks. Our students found it difficult to combine the programs.

The goal of the LEDA project is to overcome these shortcomings by creating a platform for combinatorial and geometric computing. The LEDA library should contain the major findings of the algorithms community in a form that makes them directly accessible to non-experts having only a limited knowledge in the area. In this way we hoped to reduce the gap between research and application.

3 Did we achieve our goals?

We believe that we have reached the last goal and have at least partially reached the first three goals.

LEDA was first distributed in the summer of 1990. Its user community has grown ever since. LEDA is now used at more than 1500 academic and industrial sites in over 50 different countries world-wide. Industrial use started in 1994.

Many users of LEDA are outside computer science and only a small fraction of our users are from the algorithms community. We therefore believe that we have reached our first two goals. The impact of algorithms research has increased and there is considerable use of LEDA and hence reuse of implementations. However, the gap between algorithms research and algorithms use is still quite large. In particular, many of the non-expert users of LEDA complain that a tutorial is missing. We hope that the forthcoming LEDAbook [MN] will help.

We have also partially achieved our third goal. We now do have a depository for our students work and we have just introduced the concept of LEDA extension packages (LEPs) that will allow a wider community to contribute. We come back to LEPs in section 4.4.

We have achieved our last goal. The specifications of our data types are sufficiently abstract and precise so as to allow their combination without any knowledge of implementation. We have seen an example in section 1. Many of our specifications are based on the so-called *item concept* which gives an abstract treatment of pointers into a data structure. Different components of LEDA can be combined without knowledge of the implementation.

The project also had a number of positive side-effects which we did not foresee. Firstly, LEDA's wide use gives us tremendous satisfaction[4]. Secondly, our experiences with the system suggested many difficult and well motivated problems for theoretical algorithms research. We will discuss program checking, running time prediction, and theoretical issues in the implementation of geometric algorithms below. *The system has changed the way we do algorithms research.*

4 Recent developments

A strength of the LEDA project is its strong theoretical underpinning. *We believe that only our strong theoretical background allowed us to build LEDA.* In the last two years we paid particular attention to program checking, running time prediction, and the correct implementation of geometric programs.

4.1 Program checking

Programming is a notoriously error-prone task; this is even true when programming is interpreted in a narrow sense: going from a (correct) algorithm to a program. The standard way to guard against coding errors is program testing. The program is exercised on inputs for which the output is known by other means, typically as the output of an alternative program for the same task. Program testing has severe limitations:

– It is usually only done during the testing phase of a program. Also, it is difficult to determine the "correct" suite of test inputs.

---

[4] We stated above that algorithms research must include implementation to have maximal impact. We might add: without implementation algorithm research is less rewarding.

- Even if appropriate test inputs are known it is usually difficult to determine the correct outputs for these inputs: alternative programs may have different input and output conventions or may be too inefficient to solve the test cases.

Given that program verification, i.e., formal proof of correctness of an implementation, will not be available on a practical scale for some years to come, *program checking* has been proposed as an extension to testing [BK89, BLR90]. The cited papers explored program checking in the area of algebraic, numerical, and combinatorial computing. In [MNS+96, MM95, HMN96] we discuss program checkers for planarity testing and a variety of geometric tasks. We have also added program checkers to some of the LEDA programs, e.g., the planarity test provides a planar drawing for a planar graph and a Kuratowski subgraph for a non-planar graph. A user of the planarity algorithm has thus the possibility to verify that the output of the algorithm is correct.

4.2 Running Time Prediction

Big-O analysis of algorithms is concerned with the asymptotic analysis of algorithms, i.e., with the behavior of algorithms for large inputs. It does not allow the prediction of actual running times of real programs on real machines and therefore its predictive value is limited.

- An algorithm with running time $O(n)$ is faster than an algorithm with running time $O(n^2)$ for sufficiently large n. Is $n = 10^6$ large enough? Asymptotic analysis of algorithms is of little help to answer this question. It is however true that a well-trained algorithms person who knows program and analysis can make a fairly good guess.
- For a user of LEDA statements of asymptotic running times are almost meaningless as he/she has no way to estimate the constants involved. After all, the purpose of LEDA is to hide the implementations from our users.

The two items above clearly indicate that we need more than asymptotic analysis in order to have a theory with predictive value. *The ultimate goal of analysis of algorithms must be a theory that allows to predict the actual running time of an actual program on an actual machine* with reasonable precision (say within a factor of two). We must aim for the following scenario: When a program is installed on a particular machine a certain number of well-chosen tests are executed in order to learn about machine parameters relevant for the execution of the program. This knowledge about the machine is combined with the analysis of the algorithm to predict running time on specific inputs. In the context of an algorithms library one could even hope to replace statements about asymptotic execution times by statements about actual execution times during installation of the library. In [FM97] we show for a small number of programs (Fibonacci heaps, Dijkstra's shortest path algorithm, and a maximum weight matching algorithm) that running time prediction within a factor of less than two and a wide range of machines is feasible.

4.3 Implementation of geometric algorithms

Geometric algorithms are frequently formulated under two unrealistic assumptions: computers are assumed to use exact real arithmetic (in the sense of mathematics) and inputs are assumed to be in general position. The naive use of floating point arithmetic as an approximation to exact real arithmetic very rarely leads to correct implementations. In a sequence of papers [BMS94a, See94, MN94, BMS94b, FGK+96, BRMS97] we investigated the degeneracy and precision issues and extended LEDA based on our theoretical work. LEDA now provides exact geometric kernels for two-dimensional and higher dimensional computational geometry [MMN+97] and also correct implementations for basic geometric tasks, e.g., two-dimensional convex hulls, Delaunay diagrams, Voronoi diagrams, point location, line segment intersection, and higher-dimensional convex hulls and Delaunay diagrams.

4.4 LEDA Extension Packages

LEDA extension packages are a new feature of the LEDA project structure. Up to two years ago, most of LEDA has been developed by a small group of persons under the tight supervision of Stefan Näher; no code went into the system that was not thoroughly understood by either Stefan Näher or Christian Uhrig. The growing numbers of contributors and the fact that Stefan Näher has new responsibilities as a professor has forced us to a change of the project structure. We decided to split LEDA into a core system (the actual LEDA version) and to shift enhancements into additional software packages.

LEDA extension packages (LEPs) extend LEDA into particular application domains and areas of algorithmics not covered by the core system. LEDA extension packages satisfy requirements, which guarantee compatibility with the LEDA philosophy. LEPs have a LEDA-style documentation, they are implemented as platform independent as possible and the installation process allows a close integration into the LEDA core library.

Currently, there are no released LEPs available, but there are several LEP under construction: PQ-trees (coordinated by Sebastian Leipert, Koeln), dynamic graph algorithms (coordinated by David Alberts, Halle), the homogeneous planar CGAL geokernel (coordinated by Stefan Schirra, Saarbrücken), a homogeneous d-dimensional geokernel (coordinated by Michael Seel, Saarbrücken), and a library for graph drawing (DFG-project Automatisches Graphenzeichnen).

References

[AHU83] A.V. Aho, J.E. Hopcroft, and J.D. Ullman. *Data structures and algorithms.* Addison-Wesley, 1983.

[BK89] M. Blum and S. Kannan. Programs That Check Their Work. In *Proc. of the 21th Annual ACM Symp. on Theory of Computing*, 1989.

[BLR90] M. Blum, M. Luby, and R. Rubinfeld. Self-testing/correcting with applications to numerical problems. In *Proc. 22nd Annual ACM Symp. on Theory of Computing*, pages 73–83, 1990.

[BMS94a]   Ch. Burnikel, K. Mehlhorn, and S. Schirra. On degeneracy in geometric computations. In *Proc. SODA 94*, pages 16–23, 1994.

[BMS94b]   Ch. Burnikel, K. Mehlhorn, and St. Schirra. How to compute the Voronoi diagram of line segments: Theoretical and experimental results. In Springer-Verlag Berlin/New York, editor, *LNCS*, volume 855 of *Proceedings of ESA'94*, pages 227–239, 1994.

[BRMS97]   Ch. Burnikel, R.Fleischer, K. Mehlhorn, and S. Schirra. A strong and easily computable separation bound for arithmetic expressions involving square roots. In *Proc. SODA 97*, pages 702–709, 1997.

[CLR90]   T.H. Cormen, C.E. Leiserson, and R.L. Rivest. *Introduction to Algorithms*. MIT Press/McGraw-Hill Book Company, 1990.

[FGK+96]   A. Fabri, G.-J. Giezeman, L. Kettner, S. Schirra, and S. Schönherr. The CGAL Kernel: A basis for geometric computation. In *Workshop on Applied Computational Geometry (WACG96)*, LNCS, 1996.

[FM97]   Ulrich Finkler and Kurt Mehlhorn. Runtime prediction of real programs on real machines. In *Proceedings 8th ACM-SIAM Symposium on Discrete Algorithms (SODA '97)*, January 1997.

[HMN96]   C. Hundack, K. Mehlhorn, and S. Näher. A Simple Linear Time Algorithm for Identifying Kuratowski Subgraphs of Non-Planar Graphs. Manuscript, 1996.

[Kin90]   J.H. Kingston. *Algorithms and Data Structures*. Addison-Wesley Publishing Company, 1990.

[Meh84]   K. Mehlhorn. *Data structures and algorithms 1,2, and 3*. Springer, 1984.

[MM95]   K. Mehlhorn and P. Mutzel. On the Embedding Phase of the Hopcroft and Tarjan Planarity Testing Algorithm. *Algorithmica*, 16(2):233–242, 1995.

[MMN+97]   K. Mehlhorn, Müller, S. Näher, S. Schirra, M. Seel, C. Uhrig, and J. Ziegler. A computational basis for higher-dimensional computational geometry and its applications. In *Proceedings of the Symp. on Computational Geometry*, 1997. http://www.mpi-sb.mpg.de/~seel.

[MN]   K. Mehlhorn and S. Näher. The LEDA Platform for Combinatorial and Geometric Computing. Cambridge University Press, forthcoming. Draft versions of some chapters are available at http://www.mpi-sb.mpg.de/~mehlhorn.

[MN94]   K. Mehlhorn and S. Näher. The implementation of geometric algorithms. In *13th World Computer Congress IFIP94*, volume 1, pages 223–231. Elsevier Science B.V. North-Holland, Amsterdam, 1994.

[MN95]   K. Mehlhorn and S. Näher. LEDA: A platform for combinatorial and geometric computing. *Communications of the ACM*, 38(1):96–102, 1995.

[MNS+96]   K. Mehlhorn, S. Näher, T. Schilz, S. Schirra, M. Seel, R. Seidel, and Ch. Uhrig. Checking Geometric Programs or Verification of Geometric Structures. In *Proc. of the 12th Annual Symposium on Computational Geometry*, pages 159–165, 1996.

[MNU96]   Kurt Mehlhorn, S. Näher, and Ch. Uhrig. The LEDA User Manual (Version R 3.4). Technical report, Max-Planck-Institut für Informatik, 1996. http://www.mpi-sb.mpg.de/LEDA/leda.html.

[NH93]   J. Nievergelt and K.H. Hinrichs. *Algorithms and Data Structures*. Prentice Hall Inc., 1993.

[O'R94]   J. O'Rourke. *Computational Geometry in C*. Cambridge University Press, 1994.

[Sed91]   R. Sedgewick. *Algorithms*. Addison-Wesley Publishing Company, 1991.

[See94]   Michael Seel. Eine Implementierung abstrakter Voronoidiagramme. Master's thesis, Max-Planck-Institut für Informatik, 1994.
[Tar83]   R.E. Tarjan. Data structures and network algorithms. In *CBMS-NSF Regional Conference Series in Applied Mathematics*, volume 44, 1983.
[van88]   C.J. van Wyk. *Data Structures and C programs*. Addison-Wesley Publishing Company, 1988.
[Woo93]   D. Wood. *Data Structures, Algorithms, and Performance*. Addison-Wesley Publishing Company, 1993.

The Wadge-Wagner Hierarchy of ω-Rational Sets

Olivier Carton and Dominique Perrin

Institut Gaspard Monge
Université de Marne-la-Vallée
93166 Noisy le Grand
France

Abstract. We present a unified treatment of the hierarchy defined by Klaus Wagner for ω-rational sets and also introduced in the more general framework of descriptive set theory by William W. Wadge. We show that this hierarchy can be defined by syntactic invariants, using the concept of an ω-semigroup.

1 Introduction

The idea of a Muller automaton was introduced by David Muller as a variant of usual finite automata, well suited for the recognition of infinite sequences. It was later proved by McNaughton that any recognizable set of ω-words can be recognized by a deterministic Muller automaton.

Klaus Wagner has introduced in 1979 [22] two concepts defined on Muller automata: chains and superchains. Together with an operation on automata called derivation, he has proved that the maximal lengths of chains and superchains (and the ones obtained on the derived automata) are enough to characterize the classes of recognizable ω-sets up to to the inverse image under a continuous function. This classification has also been investigated independently by W. Wadge. He has studied the reduction by a continuous function in abstract topological spaces, as a refinement of the classical Borel hierarchy. His results are based on a particular class of games, now called Wadge games. His classification itself is known as the Wadge hierarchy [10]. The connections between both theories were first discovered by Pierre Simonnet [19]. The Wagner hierarchy has been partially rediscovered several times [2, 9]. The interest in the classification of ω-rational sets was revived by the studies concerning the logic of distributed processing [15].

Since then Thomas Wilke [24] has shown how one could use, in the case of infinite words, algebraic methods allowing to replace finite automata by finite semigroups. This has lead to the notion of an ω-semigroup introduced in [17]. This approach has the advantage to make easier the definition of a variety along the line of Eilenberg's theory.

Another direction was investigated by Jean-Eric Pin in [18]. He has shown that the notion of ordered semigroup could be used to define families of recognizable sets that are not closed under complementation. This is especially interesting in the case of infinite words since very natural families like the open sets are not closed under complementation.

We would like to show here how Klaus Wagner's ideas fit into the present framework using ω-semigroups. In particular, we shall see that the definition of chains and superchains can be formulated in ω-semigroups, providing a clear explanation of the fact that they do not depend on the particular automaton used to recognize a given set but on the set itself. We shall show how the classes of the Wagner hierarchy are defined in topological terms. We will also investigate the link between Wagner's notions and that of ordered semigroups.

The work presented here is based on results obtained, in great part, in the first author doctoral thesis [4]. Part of it was presented at a conference held in Porto [6]. Those concerning the equivalence of the various definitions of chains and superchains will appear soon in [7]. The ones concerning the hierarchy itself will be published in a second paper [5].

2 Preliminaries

We assume a familiarity with the basic concepts of ω-rational sets and automata. For an introduction, we refer the reader to [21] or [16]. A word about notation. The alphabet is usually denoted by the symbol A. The set A^* (resp. A^+) is the set of finite words (resp. nonempty finite words) on the alphabet A. The set of (one-sided) infinite words on A is denoted by A^ω. We consider A^ω as a topological space with the usual Cantor topology.

We shall deal often with classes of sets. Since the sets considered are subsets of the topological space A^ω, a class of sets is really a mapping assigning to each alphabet A a set of subsets of A^ω. The *dual* class of a class Γ is formed of the complements (within each A^ω) of the sets in Γ. It is denoted by $\bar{\Gamma}$. We say that Γ is *ambiguous* if $\Gamma = \bar{\Gamma}$.

We shall use ordinals to index classes of sets. The symbol ω will thus be used in two ways, either to denote an ordinal in expressions like $\omega + 1$ or to denote an ω-rational set like $(a^*b)^\omega$. We hope that it will not bring confusion.

We now recall the definition of ω-semigroups and Wilke algebras. For a more detailed presentation, we refer the reader to [17]. We assume some familiarity with the basic notions of semigroup theory. We use the notation of [8] for all undefined notions in semigroup theory. We use the traditional notation S^1 to denote the semigroup obtained by adding an new neutral element 1 to S.

An ω-semigroup is a pair $S = (S_+, S_\omega)$ where S_+ is a semigroup and S_ω is a set with two operations in addition to the semigroup operation of S_+: A left action of S_+ on S_ω:

$$(s, u) \mapsto s.u$$

and an infinite product

$$\pi : S_+ \times S_+ \times S_+ \times \ldots \to S_\omega$$

These operations must satisfy the following axioms:

1. The action of S_+ on S_ω is associative: for $s, t \in S_+$ and $u \in S_\omega$

$$s.(t.u) = (st).u$$

2. The infinite product is ω-associative, in the sense that for any sequence $(s_n)_{n \geq 0}$ of elements of S_+ and any strictly increasing sequence $(n_i)_{i \geq 0}$ of integers with $n_0 = 0$, one has

$$\pi(s_0, s_1, s_2, \ldots) = \pi(t_0, t_1, t_2 \ldots)$$

with $t_i = s_{n_i} \ldots s_{n_{i+1}-1}$

3. The left action is compatible with the infinite product: for elements s and $(s_n)_{n \geq 0}$ of S_+, one has

$$s.\pi(s_0, s_1, s_2, \ldots) = \pi(s, s_0, s_1, s_2, \ldots)$$

The associativity of the operations allows one to denote all operations by mere concatenation, with su instead of $s.u$ and $s_1 s_2 \ldots$ instead of $\pi(s_1, s_2, \ldots)$.

An ω-semigroup morphism from $S = (S_+, S_\omega)$ into $S' = (S'_+, S'_\omega)$ is a pair $(\varphi_+, \varphi_\omega)$ where φ_+ is semigroup morphism from S_+ into S'_+ and φ_ω is a function from S_ω into S'_ω which is compatible with the ω-semigroup structure, i.e., the left action and the infinite product.

Thus an ω-semigroup is not an algebra in the usual sense since one of its operations has infinitely many arguments.

The concepts of rational expression and of ω-rational expressions extend to ω-semigroups in the following way. Let S be a semigroup and X be a subset of S. We denote by X^+ the subsemigroup generated by X in S. We denote by X^* the subset of S^1 defined by $X^* = \{1\} + X^+$. In this way, for any $s \in S$ and $X \subset S$, both subsets sX^* and X^*s are defined as subsets of S. Let now $S = (S_+, S_\omega)$ be an ω-semigroup. For $X, Y \subset S_+$, we denote by XY^ω the set

$$XY^\omega = \{xy_1 y_2 \ldots \mid x \in X, y_i \in Y\}$$

We further introduce a variant of ω-semigroups which is an algebra in the usual sense since all its operations have finite arity and is well suited to describe finite ω-semigroups. This concept is due to Wilke [23, 24].

A *Wilke algebra* is a pair $S = (S_+, S_\omega)$ where S_+ is a semigroup and S_ω is a set with two operations: A left action of S_+ on S_ω and a unary operation from S_+ into S_ω denoted

$$t \mapsto t^\omega$$

The operation ω must satisfy the following axioms:

$$(t^n)^\omega = t^\omega$$
$$s(ts)^\omega = (st)^\omega$$

for all $s, t \in T$ and $n \geq 1$.

A Wilke algebra morphism is a pair of functions compatible with the Wilke algebra structure.

A well-known version of Ramsey theorem says that if we define a coloring $\varphi : A^+ \to S$ of all words using only a finite number of colors, then each ω-word has a factorization:

$$x = v_0 v_1 v_2 \cdots$$

with all blocks except those involving the first one of the same color, *i.e.*, such that $\varphi(v_i v_{i+1} \ldots v_{i+k}) = \varphi(v_j v_{j+1} \ldots v_{j+l})$ for all $i,j \geq 1$, $k,l \geq 0$. This formulation holds even if the set of colors is a finite set without a multiplicative structure. In the case where S is a finite semigroup and φ a semigroup morphism, the result implies that for any ω-word x there is a pair of an element $s \in S$ and an idempotent $e = e^2 \in S$ such that $s = se$ and $x \in \varphi^{-1}(s)\varphi^{-1}(e)^\omega$.

The following result is essentially a consequence of Ramsey theorem. It shows that a finite ω-semigroup and a finite Wilke algebra are essentially the same thing.

Theorem 1. *For any finite Wilke algebra $S = (S_+, S_\omega)$, there is a unique infinite product from S_+ into S_ω making S an ω-semigroup such that $s^\omega = sss\ldots$ for all s in S_+.*

For a proof, see [24] or [17]. In the sequel, we shall not distinguish between finite Wilke algebras and finite ω-semigroups.

We say that a morphism $\varphi : A^\infty \to S$ from A^∞ onto an ω-semigroup $S = (S_+, S_\omega)$ *recognizes* an ω-set $X \subset A^\omega$ if $X = \varphi^{-1}(P)$ for some $P \subset S_\omega$.

The following result extends the classical concept of recognition by a finite semigroup for a rational set to ω-rational sets. The theorem can really be credited to Büchi since he had the original idea of introducing congruences of finite index to define rational ω-sets. For a proof, see [17].

Theorem 2. *A set $X \subset A^\omega$ is ω-rational iff there exists an ω-semigroup morphism from $A^\infty = (A^+, A^\omega)$ onto a finite ω-semigroup $S = (S_+, S_\omega)$ recognizing X.*

The notion of an ω-semigroup has been extended by Nicolas Bedon to countable ordinals in the sense that ω-words a replaced by words indexed by a countable ordinal [3]. This generalization has the advantage to give a more uniform structure: the operations are defined everywhere.

3 Chains and superchains

In this section, we introduce the notions of chains and superchains in automata and in ω-semigroups.

3.1 Chains and superchains in Muller automata

We recall that a Muller automaton is a deterministic finite automaton $\mathcal{A} = (Q, E, i, \mathcal{T})$ where Q is the state set, $E \subset Q \times A \times Q$ is the set of transitions and $i \in Q$ is the initial state. The table $\mathcal{T} \subset 2^Q$ is the set of accepting subsets of Q. We moreover suppose a Muller automaton to be *complete*: for each state $q \in Q$ and each symbol $a \in A$, there is a transition from q labeled by a. A set $R \subset Q$ is called *positive* if $R \in \mathcal{T}$ and *negative* otherwise.

A subset T of Q is said to be *admissible* if there is a cycle c in $\mathcal{A}$, accessible from the initial state i, such that the set of states encountered on c is exactly T. We say that T is the *content* of c.

Let $\mathcal{A} = (Q, E, i, \mathcal{T})$ be a complete Muller automaton. An $\mathcal{A}$-*chain* of length m is an increasing sequence

$$R_0 \subset R_1 \subset \cdots \subset R_m$$

of $m+1$ admissible subsets of Q such that, for $0 \leq i \leq m$, the R_i are alternately in $\mathcal{T}$ and outside $\mathcal{T}$.

We say that the chain is *positive* if $R_0 \in \mathcal{T}$ and *negative* if $R_0 \notin \mathcal{T}$. We denote by $m^+(\mathcal{A})$ (resp. $m^-(X)$) the maximal length of positive (resp. negative) $\mathcal{A}$-chains and we let $m(\mathcal{A}) = \max(m^+(\mathcal{A}), m^-(\mathcal{A}))$. It is obvious by the definition that $m(\mathcal{A})$ is finite for any finite Muller automaton $\mathcal{A}$. One indeed has the inequality $m(\mathcal{A}) \leq \text{card}(Q)$.

Wilke and Yoo have shown in [25] that $m(\mathcal{A})$ can be computed in polynomial time. This contrasts with the fact the computation of $m(X)$ for an ω-rational set X given by deterministic Rabin (or Streett) automata is NP-complete [11].

Example 1. Consider the set $X = (a^*b)^\omega$ of ω-words over $\{a,b\}$ which have an infinite number of symbols b. This set X is recognized by the automaton $\mathcal{A}_1$ represented in Figure 1 with $\mathcal{T} = \{\{2\}, \{1, 2\}\}$. The sequence $(\{1\}, \{1, 2\})$ is a negative chain of length 1. There are no positive chains of length 1 and thus $m = m^- = 1$.

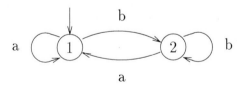

Fig. 1. Automaton $\mathcal{A}_1$.

An $\mathcal{A}$-*superchain* of length n is a sequence

$$C_0, C_1, \ldots, C_n$$

of $n+1$ $\mathcal{A}$-chains of length $m(\mathcal{A})$ such that:

(i) Each C_i is accessible from C_{i+1} for $1 \leq i \leq n$, *i.e.*, there exists a path from some state in C_{i-1} to some state in C_i.
(ii) The $\mathcal{A}$-chains C_i are alternately positive and negative.

We say that the superchain is *positive* if C_0 is positive and *negative* otherwise. We denote by $n^+(\mathcal{A})$ (resp. $n^-(\mathcal{A})$) the maximal length of positive (resp. negative) superchains and $n(\mathcal{A}) = \max(n^+(\mathcal{A}), n^-(\mathcal{A}))$. We let $n^+(\mathcal{A}) = -1$ (resp. $n^-(\mathcal{A}) = -1$) if the set of positive (resp. negative) superchains is empty. It is obvious by definition that $n(\mathcal{A})$ is finite for any finite Muller automaton $\mathcal{A}$. One indeed has the inequality $n(\mathcal{A}) \le \operatorname{card}(Q)$.

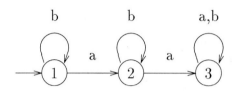

Fig. 2. Automaton $\mathcal{A}_2$.

Example 2. Consider the set $X = b^*ab^\omega$. It is recognized by the Muller automaton $\mathcal{A}_2$ of Figure 2 with $\mathcal{T} = \{\{2\}\}$. All chains are of length 0 and $m = m^+ = m^- = 0$. The sequence $(\{1\}, \{2\}, \{3\})$ is a negative superchain of length 2. One has $n = n^- = 2$ and $n^+ = 1$.

3.2 Chains and superchains in ω-semigroups

Let $S = (S_+, S_\omega)$ be an ω-semigroup and let X be a subset of S_ω. Let $C = (Y, Z)$ be a pair where Y is a non empty subset of S_+ and $Z = z_0, z_1, \ldots, z_m$ is a sequence of $m + 1$ elements of S_+. Let

$$Z_i = z_0 + z_1 + \ldots + z_i$$
$$W_i = Y Z_m^* (Z_i^* z_i)^\omega \qquad (1)$$

for $0 \le i \le m$.

We say that the pair C is an *X-chain* iff the sets W_i are alternately included in X and disjoint from X.

The number m is called the *length* of the chain C. It is important to observe that m is the number of alternations in the sequence $W_0, \ldots, W_m$ rather than the length of the sequence Z in the usual sense which would be $m + 1$.

We distinguish, among chains, positive and negative ones according to the nature of the first element. A *positive* chain is one such that $W_0 \subset X$ and a *negative* one such that $W_0 \cap X = \emptyset$. Two positive (resp. negative) chains are said to be of the *same sign*.

We denote by $m^+(X)$ (resp. $m^-(X)$) the maximal length of the positive (resp. negative) X-chains and $m(X) = \max(m^+(X), m^-(X))$. We set $m^+(X) = -1$ (resp. $m^-(X) = -1$) if the set of positive (resp. negative) chains is empty and

$m^+(X) = m^-(X) = \infty$ if the lengths of X-chains are unbounded. We shall see that $m(X)$ is always finite for an ω-rational set X.

We now come to the definition of a superchain in an ω-semigroup.

Let $S = (S_+, S_\omega)$ be an ω-semigroup and let X be a subset of S_ω. An X-*superchain* of length n is a sequence

$$C_0, C_1, \ldots, C_n$$

of $n+1$ X-chains $C_i = (Y_i, Z_i)$, all of maximal length $m = m(X)$ such that, with $Z_i = z_{i0}, z_{i1}, \ldots, z_{im}$, we have:

(i) Each C_i is accessible from C_{i-1} for $1 \le i \le n$, i.e., there is an element $u_i \in S_+$ such that $Y_{i-1} Z_{i-1}^* u_i \subset Y_i$.
(ii) The chains C_i are alternately positive and negative.

We say that the superchain is *positive* if C_0 is positive and *negative* otherwise. We denote by $n^+(X)$ (resp. $n^-(X)$) the maximal length of positive (resp. negative) superchains and $n(X) = \max(n^+(X), n^-(X))$. We let $n^+(X) = -1$ (resp. $n^-(X) = -1$) if the set of positive (resp. negative) superchains is empty. We shall see that $n(X)$ is also finite if X is ω-rational.

3.3 Correspondence between the definitions

We now come to the fact that the definitions of a chain in automata and in ω-semigroups are in correspondence. This has two main consequences: first it shows that the integers $m(X)$ are finite and computable for any ω-regular set X. Second, it shows that the integers $m(\mathcal{A})$ do not depend on the automaton but only on the set recognized. We have the following theorem.

Theorem 3. *Let $X \subset A^\omega$ be an ω-rational set recognized by a complete Muller automaton $\mathcal{A} = (Q, E, i, \mathcal{T})$. The following equalities hold:*

$$m^+(X) = m^+(\mathcal{A}) \quad and \quad m^-(X) = m^-(\mathcal{A}).$$

Let $\varphi : S \to S'$ be a morphism from an ω-semigroup $S = (S_+, S_\omega)$ onto an ω-semigroup $S' = (S'_+, S'_\omega)$. Let $X \subset S_\omega$ and $X' \subset S'_\omega$ be such that $X = \varphi^{-1}(X')$.

The image (Y', Z') of an X-chain (Y, Z) is an X'-chain of the same length and sign and each X'-chain is the image of an X-chain of the same length and sign.

Thus chains can be computed in any ω-semigroup recognizing X, in particular in a finite ω-semigroup when X is ω-rational. We will see in Section 6 that chains in finite ω-semigroups can be defined differently.

We now come to the fact that the definitions of a superchain in automata and in ω-semigroups are also in correspondence. As in the case of chains, this has two main consequences: first it shows that the integers $n(X)$ are finite and computable for any ω-regular set X. Second, it shows that the integers $n(\mathcal{A})$ do not depend on the automaton but only on the set recognized. We have the following theorem.

Theorem 4. Let $X \subset A^\omega$ be an ω-rational set recognized by a complete Muller automaton $\mathcal{A} = (Q, E, i, \mathcal{T})$. The following equalities hold:
$$n^+(X) = n^+(\mathcal{A}) \quad \text{and} \quad n^-(X) = n^-(\mathcal{A}).$$

4   Wagner's hierarchy

To a Muller automaton $\mathcal{A}$, one associates another Muller automaton called the *derived* automaton and denoted $\partial\mathcal{A}$. It is nonempty only when $n^+ = n^-$. It is then obtained from $\mathcal{A}$ by the following transformation:

1. All states that belong to a maximal positive superchain are collapsed into a single state q_+ and the set $\{q_+\}$ is positive.
2. All states that belong to a negative superchain are collapsed into a single state a q_- and the set $\{q_-\}$ is negative.

It was shown by Klaus Wagner that the set recognized by $\mathcal{A}$ only depends on the set X recognized by $\mathcal{A}$ and not on the particular Muller automaton used to recognize X. It can therefore be denoted ∂X.

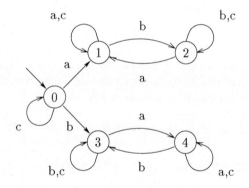

Fig. 3. Automaton $\mathcal{A}_3$.

Example 3. Consider the automaton $\mathcal{A}_3$ of Figure 3 with $\mathcal{T} = \{\{1\}, \{1,2\}, \{3\}\}$. We have for $\mathcal{A}_3$ $m = m^+ = m^- = 1$ and $n = n^+ = n^- = 0$. The derived automaton $\mathcal{A}_4 = \partial\mathcal{A}_3$ is represented in Figure 4. We then have for $\mathcal{A}_4$ $m = m^+ = m^- = 0$, $n^+ = 0$ and $n = n^- = 1$. Since $n^+ \neq n^-$, we have $\partial\mathcal{A}_4 = \emptyset$.

We associate to an ω-rational set X two ordinals denoted $\gamma(X)$ and $\mu(X)$ which are defined as follows. The ordinal $\gamma(X)$ is

$$\gamma(X) = \begin{cases} n(X) & \text{if } m(X) = 0 \\ \omega^{m(X)}(n(X)+1) & \text{otherwise} \end{cases}$$

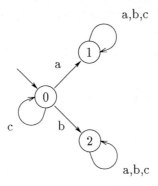

Fig. 4. The derived automaton $\mathcal{A}_4 = \partial \mathcal{A}_3$.

For example, we have for the sets X_1 and X_2 recognized by the automata $\mathcal{A}_1$ and $\mathcal{A}_2$ of the previous examples,

$$\gamma(X_1) = \omega \quad \text{and} \quad \gamma(X_2) = 2.$$

The ordinal $\mu(X)$ is then defined by

$$\mu(X) = \gamma(X) + \mu(\partial X).$$

The ordinal $\mu(X)$ is an arbitrary ordinal $< \omega^\omega$ and moreover, since $m(\partial X) < m(X)$ as soon as $m(X) \geq 1$, the decomposition given by the definition of $\gamma(X)$ produces the Cantor normal form of the ordinal $\gamma(X)$.

Both ordinals $\gamma(X)$ and $\mu(X)$ can be computed from any Muller automaton $\mathcal{A}$ recognizing the ω-set X since the integers $m(\mathcal{A})$ and $n(\mathcal{A})$ only depend on the set recognized by $\mathcal{A}$.

For example, we have for the ω-set X_3 recognized by the automaton $\mathcal{A}_3$ given above

$$\mu(X_3) = \omega + 1$$

We finally associate to an ω-rational set X an information called its *sign* and denoted $\text{sign}(X)$. It is an element of the three elements set $\{\sigma, \delta, \pi\}$ defined as follows. We first have

$$\text{sign}(X) = \begin{cases} \sigma & \text{if } n^- > n^+ \\ \pi & \text{if } n^- < n^+ \\ \delta & \text{if } n^- = n^+ \text{ and } m = 0 \\ \text{sign}(\partial X) & \text{otherwise} \end{cases}$$

It is clear that $\text{sign}(X) = \sigma$ iff $\text{sign}(A^\omega - X) = \pi$ and that $\text{sign}(X) = \delta$ iff $\text{sign}(A^\omega - X) = \delta$.

We introduce a preorder on the set $R(A)$ of ω-rational sets defined by lexicographically ordering the pair $(\mu(X), \text{sign}(X))$ with the convention that $\delta > \sigma$

and $\delta > \pi$ (σ and π being incomparable). The equivalence classes associated with the preorder are denoted

$$\Sigma_\alpha = \{X \in R(A) \mid \mu(X) = \alpha, \text{sign}(X) = \sigma\}$$
$$\Delta_\alpha = \{X \in R(A) \mid \mu(X) = \alpha, \text{sign}(X) = \delta\}$$
$$\Pi_\alpha = \{X \in R(A) \mid \mu(X) = \alpha, \text{sign}(X) = \pi\}$$

For any ordinal $\alpha < \omega^\omega$, the classes Σ_α and Π_α are dual of one another and the class Δ_α is ambiguous. The order defined on ω-rational sets by Wagner's theorem has the familiar shape given by Figure 5.

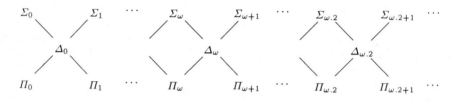

Fig. 5. The Wagner hierarchy

It may be useful for a reader used to Wagner's notation to realize that the correspondence between Wagner's notation and ours is the following. Our class Σ_n is his class C_0^n, our Δ_n his E_0^n and our Π_n his D_0^n. For $m \geq 1$ our class $\Sigma_{\omega^m.n}$ is Wagner's C_m^{n-1}, our $\Pi_{\omega^m.n}$ is his D_m^{n-1} and our Δ_m^n his E_m^{n-1}. Moreover if the normal form of the ordinal α is

$$\alpha = \omega^{m_k}.n_k + \ldots + \omega^{m_1}.n_1$$

then Σ_α is denoted in Wagner's notation

$$E_{m_k}^{n_k} \ldots E_{m_2}^{n_2} C_{m_1}^{n_1+1}$$

The idea of using ordinals instead of sequences of pairs of integers was suggested by Jean-Pierre Resseyre (oral communication).

The order thus defined happens to completely characterize another order called the *Wadge order* and defined in general as follows. Let E, F be topological spaces and let $X \subset E, Y \subset F$. We say that X *reduces* to Y, written $X < Y$ if there exists a continuous function $f : E \to F$ such that $X = f^{-1}(Y)$.

We can now state Wagner's main theorem.

Theorem 5. (K. Wagner) *Given ω-rational sets X, Y, one has the equivalence:*

$$X < Y \iff \gamma(X) < \gamma(Y).$$

The statement implies that for $X \in R(A)$, $Y \in R(B)$, one has $X < Y$ iff there exists a function $f : A^\omega \to B^\omega$ such that $X = f^{-1}(Y)$ and which is not only continuous but also rational. This is actually the content of the theorem of Büchi-Landweber (see [21]).

The main theorem due to Wadge is the following: in a topological space like A^ω, the order given by the reduction by a continuous function is a well ordering [10]. Thus the classes of the associated equivalence can be indexed by ordinals. When restricted to ω-rational classes, the order type of the hierarchy is ω^ω.

5 Topological classes

We shall give here a description in topological terms of the classes of the hierarchy. It allows one to prove Wagner's theorem in one direction since the topological characterization gives a definition of the classes invariant under the inverse of a continuous function. It is convenient to denote, for an ordinal $\alpha < \omega^\omega$

$$\Sigma_{\leq \alpha} = \bigcup_{\beta \leq \alpha} \Sigma_\beta$$

and correspondingly for $\Pi_{\leq \alpha}$ and $\Delta_{\leq \alpha}$.

We shall see that the classes of the Wagner hierarchy can be described using differences, separated unions and biseparated unions, starting from simple topological sets. We first describe the simple classes which happen to be classical classes of the Borel hierarchy.

5.1 Simple classes

The first kind is the class of open sets. We shall denote here by G the class of open sets, rational or not (and not by Σ_1 as it is sometimes done in topology). The following statement uses a special form of Büchi automata called *weak*: a path is successful if it contains at least one terminal state.

Theorem 6. *The following conditions are equivalent for an ω-rational set X.*

(i) $X \in \Sigma_{\leq 1}$.
(ii) X *is open.*
(iii) $X < a^* b(a+b)^\omega$
(iv) X *is recognizable by a weak deterministic Büchi automaton.*

Condition (i) can be formulated as follows: for all $x, y, z, t \in A^+$

$$xy^\omega \in X \Rightarrow xy^* zt^\omega \cap X \neq \emptyset$$

which precisely expresses that $m(X) = 0$ and $n^+ \leq 0$. We shall see later that this condition can formulated using an inequality in ordered ω-semigroups.

The second class is the class of sets which are countable intersections of open sets. We denote this class by G_δ (and not by Π_2 as it is done sometimes in topology, since it would contradict our use of this notation). Similarly, we denote by F_σ the class of countable unions of open sets. The following result is originally due to H. Landweber [12].

Theorem 7. *The following conditions are equivalent for an ω-rational set X.*

(i) $X \in \Sigma_{\leq \omega}$.
(ii) $X \in G_\delta$.
(iii) $X < (a^*b)^\omega$
(iv) X is recognizable by a deterministic Büchi automaton.

The equivalence between (ii) and (iii) is a general fact of descriptive set theory, independent of the hypothesis that X is ω-rational. A convenient way to prove the implications is (i) $\Rightarrow$ (iv) $\Rightarrow$ (iii) $\Rightarrow$ (ii) $\Rightarrow$ (i). The first one is proved using a well-known construction building a deterministic Büchi automaton from a Muller automaton satisfying $m^+ \leq 0$. The last one can be done by reformulating condition (i) as follows: for all $x, y, z \in A^+$

$$x(y+z)^* y^\omega \subset X \Rightarrow x(y^*z)^\omega \cap X \neq \emptyset$$

which expresses precisely that $m^+(X) \leq 0$.

5.2 Boolean combinations of open sets

In order to describe the boolean combinations of open sets, we introduce the notion of a difference of sets. Let Γ be a class of sets. We denote by $D_n(\Gamma)$ the class of sets X of the form

$$X = X_1 - X_2 + \ldots \pm X_n$$

where the sets X_i satisfy $X_i \in \Gamma$ and $X_1 \supset X_2 \supset \ldots \supset X_n$. Such an expression of X is called a *difference* of *length n*. According to a theorem of Hausdorff, if Γ is closed under finite unions and intersections and contains the empty set, the union of all the classes $D_n(\Gamma)$ for $n \geq 1$ is the boolean closure of Γ. This means that any set in the boolean closure of Γ is equal to a difference of sets of Γ. The classes $D_n(\Gamma)$ define a hierarchy within the boolean closure of Γ. As we shall see, it turns out that, when Γ is the class $\Sigma_{\leq 1}$ of ω-rational open sets or when Γ is the class $\Sigma_{\leq \omega}$ of ω-rational G_δ sets, the classes $D_n(\Gamma)$ coincide with classes of the Wagner hierarchy.

We consider here the classes Σ_n, *i.e.*, the classes of sets X such that $\gamma(X) < \omega$ or equivalently such that $m(X) = 0$. It is actually equivalent to assume on a connected Muller automaton $\mathcal{A} = (Q, E, i, \mathcal{T})$ that $m(\mathcal{A}) = 0$ or that each strongly connected component R of $\mathcal{A}$ is saturated in the sense that $S \in \mathcal{T}$ for all admissible sets $S \subset R$ or for none of them. Such an automaton is clearly equivalent to one of the following kind, that we propose to call a *weak Muller automaton*. It is a finite automaton $\mathcal{A} = (Q, E, i, \mathcal{T})$ with a definition of a successful path given by the following rule: a path γ is successful if the set of states met along γ is in $\mathcal{T}$.

The following result is originally due to Staiger and Wagner [20]. It means that an ω-rational set X belongs to the class $\Sigma_{\leq n}$ iff it is equal to a difference of length n of ω-rational open sets.

Theorem 8. *One has for all $n < \omega$*
$$\Sigma_{\leq n} = D_n(\Sigma_{\leq 1})$$

Moreover,
$$\Sigma_{\leq \omega} \cap \Pi_{\leq \omega} = \bigcup_{n < \omega} \Sigma_n$$
and coincides with the boolean closure of the family of rational open sets.

In the second equality, the inclusion from right to left is obvious since each Σ_n is contained in $\Sigma_{\leq \omega}$ and in $\Pi_{\leq \omega}$. The converse is also evident since a set $X \in \Sigma_{\leq \omega} \cap \Pi_{\leq \omega}$ satisfies $m^+(X) \leq 0$ and $m^-(X) \leq 0$ and therefore $m(X) = 0$.

Theorem 8 is really a counterpart for rational sets of a theorem of Hausdorff according to which, one has in a topological space such as A^ω
$$F_\sigma \cap G_\delta = \bigcup_{\alpha < \omega_1} D_\alpha(G)$$
where the union is on all countable ordinals (see [10] for example).

5.3 Separated classes and boolean combinations of G_δ-sets

In this section, we describe the classes $\Sigma_{\leq \alpha}$ for $\alpha = \omega^m.n$. We first consider the case of $\alpha = \omega^m$. The following result is originally due to K. Wagner [22].

Proposition 9. *For all $m < \omega$, we have the equality*
$$\Sigma_{\leq \omega^m} = D_m(\Sigma_{\leq \omega})$$

We now introduce the notion of a separated union. Let $X_1, X_2, Y \subset A^\omega$ be three ω-sets. Suppose furthermore that the three sets satisfy $X_1 \cap Y = \emptyset$ and $X_2 \subset Y$. Following a notation borrowed to Alain Louveau [14], let us denote by $\text{Sep}(Y, X_1, X_2)$ the union
$$X = X_1 + X_2$$
The picture is shown in Figure 6.

We say that X is the *separated union* of X_1 and X_2 or that X is the union of X_1 and X_2 separated by Y (we actually exchange X_1 and X_2 in the notation of [14]). We also define, for two classes Γ, Δ of ω-sets, a new class $\text{Sep}(\Gamma, \Delta)$ as the class of all sets of the form $X = \text{Sep}(Y, X_1, X_2)$ for $Y \in \Gamma$, $X_1 \in \Delta$ and $X_2 \in \bar{\Delta}$.

The following result gives a topological description of the classes $\Sigma_{\leq \omega^m.n}$. It is analogous to a statement given in [22].

Theorem 10. *For each $m \geq 1$ and $n \geq 2$, one has*
$$\Sigma_{\leq \omega^m.n} = \text{Sep}(D_{n-1}(G), \Sigma_{\leq \omega^m})$$
and dually
$$\Pi_{\leq \omega^m.n} = \text{Sep}(D_{n-1}(G), \Pi_{\leq \omega^m}).$$

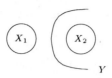

Fig. 6. Separated union of X_1 and X_2.

5.4 Biseparated classes

We now relate the definition of the set ∂X with the topological structure of X. We borrow again a notation from Alain Louveau [14] and introduce the notion of biseparated union. Let X_1, X_2, Y_1, Y_2 and Z be five ω-sets satisfying $X_1 \subset Y_1$, $X_2 \subset Y_2$, $Y_1 \cap Y_2 = \emptyset$, $Z \cap Y_1 = \emptyset$ and $Z \cap Y_2 = \emptyset$. Let us denote $\mathrm{Bisep}(Y_1, Y_2, X_1, X_2, Z)$ the union

$$X = X_1 + X_2 + Z$$

The picture is shown in Figure 7. We say that X is the *biseparated union* of X_1, X_2 and Z.

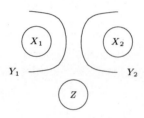

Fig. 7. Biseparated union of X_1, X_2 and Z.

If Φ, Γ, Δ are three classes of ω-sets, we denote by

$$\mathrm{Bisep}(\Phi, \Gamma, \Delta)$$

the class of sets $X = \mathrm{Bisep}(Y_1, Y_2, X_1, X_2, Z)$ with $Y_1, Y_2 \in \Phi$, $X_1 \in \Gamma$, $X_2 \in \bar{\Gamma}$ and $Z \in \Delta$.

The following result expresses that the elements of the class $\Sigma_{\omega^m.n+\beta}$ are the unions of sets of the same kind (but with opposite signs) separated by disjoint open sets plus some set of lower class of the same class.

Theorem 11. *For all $m \geq 1$ and $n \geq 1$ and $\beta < \omega^m$, one has*

$$\Sigma_{\omega^m.n+\beta} = \mathrm{Bisep}(G, \Sigma_{\omega^m.n}, \Sigma_\beta)$$
$$\Delta_{\omega^m.n+\beta} = \mathrm{Bisep}(G, \Sigma_{\omega^m.n}, \Delta_\beta)$$
$$\Pi_{\omega^m.n+\beta} = \mathrm{Bisep}(G, \Sigma_{\omega^m.n}, \Pi_\beta).$$

6 Finite ω-semigroups

The definition of chains an superchains in finite ω-semigroups uses the Green's relations $\mathcal{H}$ and $\mathcal{R}$ defined as follows. For elements s, t of a semigroup S, one has $s \geq_\mathcal{R} t$ if $s = t$ or $t \in sS$ and $s \geq_\mathcal{H} t$ if $s = t$ or $t \in sS$ and $t \in Ss$. The relation $\geq_\mathcal{R}$ is preorder and the restriction of $\geq_\mathcal{H}$ to the idempotents also.

In the case of a subset X of a finite ω-semigroup $S = (S_+, S_\omega)$, the definition of a chain relative to X can be used in the following form. It is a sequence $(s, e_0, e_1, \ldots, e_m)$ of elements of S_+ such that:

(i) For $0 \leq i \leq m$, the pair (s, e_i) is *linked*, i.e., $se_i = s$ and $e_i^2 = e_i$.
(ii) The sequence of idempotents $e_0, e_1, \ldots, e_m$ is decreasing for the $\mathcal{H}$ order.
(iii) The elements se_i^ω are alternately in X and outside of X.

We have again the notion of a *positive* or *negative* chain according to $se_0^\omega \in X$ or not. The definition of a chain in a finite ω-semigroup coincides with the definition of a chain we gave in a general ω-semigroup in the following sense. To any chain for the former definition, can be associated another chain for the latter one with the same length and same sign, and vice versa. The integers $m^+(X)$ and $m^-(X)$ do not depend on the definition of a chain considered.

The notion of a superchain is also adapted to the case of a finite ω-semigroup to be defined as a sequence $u_0, u_1, \ldots, u_n$ of chains $u_i = (s_i, e_{i0}, e_{i1}, \ldots, e_{im})$ of length m such that:

(i) The sequence s_i is decreasing for the $\mathcal{R}$ order, i.e.

$$s_0 \geq_\mathcal{R} s_1 \geq_\mathcal{R} \cdots \geq_\mathcal{R} s_n.$$

(ii) The chains u_i are alternately positive and negative.

As for chains, the definition of a superchain in a finite ω-semigroup is equivalent to the definition of superchain we gave in a general ω-semigroup.

The definition of chains and superchains on finite ω-semigroups allows one to give a characterization of the classes of Wagner's hierarchy. It would be interesting to extend these ideas to classes defined for finite words.

7 Ordered ω-semigroups

An *ordered ω-semigroup* is an ω-semigroup $S = (S_+, S_\omega)$ with a partial order on each of the sets S_+ and S_ω which are compatible with all operations: for all $s, t, u, v \in S$

$$s \leq t \Rightarrow usv \leq utv,$$

$$s \leq t, u \leq v \Rightarrow su^\omega \leq tv^\omega$$

A morphism $\varphi : S \to T$ of ordered ω-semigroup is a morphism of ω-semigroups which is also compatible with the orders: for all $s, t \in S$, $s, t \in S$ and $s \leq t$ imply $\varphi(s) \leq \varphi(t)$.

It has been shown by Jean-Eric Pin [18] that any ω-rational set has a finite *syntactic ordered ω-semigroup*. The context of finite a word v with respect to an ω-set $X \subset A^\omega$ is the the pair of sets $C(u) = (C_1(u), C_2(u))$ where $C_1(u)$ and $C_2(u)$ are respectively defined by

$$C_1(u) = \{(v, x) \in A^* \times A^\omega \mid vux \in X\}$$
$$C_2(u) = \{(v, w) \in A^* \times A^* \mid v(uw)^\omega \in X\}.$$

In the same way, the context of an ω-word x with respect to the ω-set $X \subset A^\omega$ is the set

$$C(x) = \{u \in A^* \mid ux \in X\}.$$

It is well known that if $S = (S_+, S_\omega)$ is the syntactic ω-semigroup of X, the elements of S_+ (resp. S_ω) correspond to contexts of finite words (resp. ω-words). More precisely two finite words u and u' (resp. two ω-words x and x') have the same image in the syntactic ω-semigroup iff they have the same context. This allow one to define the context of an element of S. Contexts could also have been directly defined in S with respect to the image P of X in S_ω. An order can be defined in S by

$$s \leq t \text{ iff } C(s) \subset C(t)$$

This order is compatible with the operation of S. The ω-semigroup S equipped with this order is then an ordered ω-semigroup. It is in fact the syntactic ordered ω-semigroup of X.

In a finite semigroup, we denote the unique idempotent which is a power of s by s^π instead of the usual notation s^ω since the symbol ω has another meaning here.

The following statement gives a characterization of open sets alternative to Theorem 6.

Theorem 12. *An ω-rational set X is open iff its syntactic ordered ω-semigroup satisfies the following identity*

$$x^\omega \leq x^\pi y z^\omega$$

The following result gives a syntactic characterization of the class $\Sigma_{\leq \omega}$

Theorem 13. *An ω-rational set X is in $\Sigma_{\leq \omega}$ iff its syntactic ordered ω-semigroup satisfies the following identity*

$$(x^\pi y)^\pi x^\omega \leq (x^\pi y)^\omega$$

As a consequence, we obtain the following syntactic characterization, due to Thomas Wilke [24], of the sets in $\Sigma_{\leq \omega} \cap \Pi_{\leq \omega}$, which are also the boolean combinations of open sets by Theorem 7.

Theorem 14. *An ω-rational set is a boolean combination of open sets iff its syntactic ω-semigroup satisfies the identity*

$$(x^\pi y)^\pi x^\omega = (x^\pi y)^\omega$$

Actually, the identity given in [24] is the identity

$$(x^\pi y^\pi)^\pi x^\omega = (x^\pi y^\pi)^\pi y^\omega$$

which can be shown to be equivalent to the previous one.

Conclusion

It would be interesting to investigate further on in several directions including the followings ones.

7.1 A syntactic definition of the derivative

Klaus Wagner has introduced the notion of the derivative ∂X of an ω-rational set X. It is defined using a Muller automaton recognizing X. We do not know how to define the derivative in a finite ω-semigroup in such a way that ∂X can be computed in the syntactic ω-semigroup of X.

7.2 Biinfinite words

The theory of ω-rational sets can be developed for sets of two-sided infinite words [16]. Such sets have also been considered in symbolic dynamics [13]. A *symbolic dynamical system* is by definition a set of biinfinite words which is topologically closed and invariant under the shift. Let S and T be two symbolic dynamical systems. A *morphism* from S into T is a function $f : S \to T$ which is continuous and commutes with the shifts of S and T. As a particular case of symbolic dynamical systems, a *sofic system* is defined by a set of forbidden blocks recognized by a finite automaton. As a still more restricted class, a *system of finite type* is a set of biinfinite words defined by a finite set of forbidden blocks. If X, Y are symbolic dynamical systems, it is natural to say that $X \subset A^Z$ reduces to $Y \subset B^Z$, denoted $X < Y$, if there exists a morphism f from A^Z to B^Z such that $X = f^{-1}(Y)$. One thus obtains a hierarchy of subsets of A^Z analogous to the Wadge-Wagner hierarchy. The three classes defined previously are precisely preserved by inverse morphisms. It would be interesting to know the Wadge-Wagner classes of symbolic dynamical systems.

7.3 Finite words

It is an open problem to define a hierarchy for finite words analogous to Wagner's one. An objective for such a classification could be to obtain a refinement of the characterization of some well known classes. For instance, the classes of locally testable sets is the boolean closure of the class of strictly locally testable ones. The latter are finite unions of sets of the form $UA^* \cap A^*V \setminus A^*WA^*$ where U, V and W are finite sets of words. If S denotes the family of strictly locally testable sets, the family $D_n(S)$ of differences of length n of elements of S defines a hierarchy within locally testable sets.

It is possible to define Muller automata on finite words. Let indeed $\mathcal{A} = (Q, E, \mathcal{T})$ be a finite automaton where $\mathcal{T}$ is a subset of $Q \times 2^Q \times Q$. A finite path $\gamma : i \xrightarrow{w} t$ in this automaton is successful if the set R of states met along the path is such that $(i, R, t) \in \mathcal{T}$. The usual definition of locally testable sets actually uses such automata: they are the sets recognized when the underlying automaton is the standard local automaton.

A full parallel with Wagner hierarchy requires a choice of a topology on finite words. A possibility would be to consider the profinite topology associated to a pseudo-variety of semigroups [1].

References

1. Jorge Almeida. *Finite Semigroups and Universal Algebra*. World Scientific, 1994.
2. Rana Barua. The Hausdorff-Kuratowski hierarchy of ω-regular languages and a hierarchy of Muller automata. *Theoretical Computer Science*, 96:345–360, 1992.
3. Nicolas Bedon. Automata, semigroups and recognizability of words on ordinals. IGM report 96-5, to appear in International Journal of Algebra and Computation.
4. Olivier Carton. *Mots infinis, ω-semigroupes et Topologie*. Thèse, Université Paris 7, 1993. Report LITP-TH 93-08.
5. Olivier Carton and Dominique Perrin. The Wagner hierarchy of ω-rational sets. To appear in International journal of algebra and computation.
6. Olivier Carton and Dominique Perrin. Chains and superchains in ω-semigroups. In Jorge Almeida, Grancinda Gomes, and Pedro Silva, editors, *Semigroups, Automata and Languages*, pages 17–28. World Scientific, 1994.
7. Olivier Carton and Dominique Perrin. Chains and superchains for ω-rational sets, automata and semigroups. *International journal of algebra and computation*, 1997. to appear.
8. John M. Howie. *Fundamentals of Semigroup Theory*. Oxford University Press, 1995.
9. Micheal Kaminski. A classification of ω-regular languages. *Theoretical Computer Science*, 36:217–229, 1985.
10. Alexander S. Kechris. *Classical Descriptive Set Theory*, volume 156 of *Graduate texts in mathematics*. 1995.
11. Sriram C. Krishnan, Anuj Puri, and Robert K. Brayton. Structural complexity of ω-languages. In *STACS '95*, volume 900 of *Lecture Notes in Computer Science*, pages 143–156, Berlin, 1995. Springer-Verlag.
12. Lawrence H. Landweber. Decision problems for ω-automata. *Mathematical Systems Theory*, 3:376–384, 1969.
13. Douglas Lind and Brian Marcus. *An Introduction to Symbolic Dynamics and Coding*. Cambridge University Press, 1995.
14. Alain Louveau. Some results in the Wadge hierarchy of Borel sets. In A.S. Kechris et al., editor, *Cabal Seminar 79-81*, volume 1019 of *Lecture Notes in Math.*, pages 28–55. Springer-Verlag, 1981.
15. Zohar Manna and Amir Pnueli. A hierarchy of temporal properties. In *Principles of Distributed Computing*, pages 377–408, 1990.
16. Dominique Perrin and Jean-Eric Pin. Infinite words. Version 1.4, Report LITP 97.04 (http://litp.ibp.fr/~jep/Resumes/MotsInfinis.html).

17. Dominique Perrin and Jean-Eric Pin. Semigroups and automata on infinite words. In J. Fountain and V. A. R. Gould, editors, *NATO Advanced Study Institute Semigroups, Formal Languages and Groups*, pages 49–72. Kluwer academic publishers, 1995.
18. Jean-Eric Pin. A variety theorem without complementation. *Russian Mathematics (Iz. VUZ)*, 39:80–90, 1995.
19. Pierre Simonnet. *Automates et Théorie Descriptive*. Thèse, Université Paris 7, 1992.
20. Ludwig Staiger and Klaus Wagner. Automatentheoretische und automatenfreie Charakterisierungen topologischer Klassen regulärer Folgenmengen. *Elektron. Informationsverarb. Kybernet.*, 10:379–392, 1974.
21. Wolfgang Thomas. Automata on infinite objects. In J. van Leeuwen, editor, *Handbook of Theoretical Computer Science*, volume B, chapter 4. Elsevier, 1990.
22. Klaus Wagner. On ω-regular sets. *Information and Control*, 43:123–177, 1979.
23. Thomas Wilke. An Eilenberg theorem for ∞-languages. In *ICALP '91*, volume 510 of *Lecture Notes in Computer Science*, pages 588–599, Berlin, 1991. Springer-Verlag.
24. Thomas Wilke. An algebraic theory for regular languages of finite and infinite words. *Int. J. Alg. Comput.*, 3(4):447–489, 1993.
25. Thomas Wilke and Haiseung Yoo. Computing the Rabin index of a regular language of infinite words. *To appear in International Journal of Algebra and Computation*, 1997.

From Chaotic Iteration to Constraint Propagation

Krzysztof R. Apt

CWI
P.O. Box 94079, 1009 AB Amsterdam, The Netherlands
and
Dept. of Mathematics, Computer Science, Physics & Astronomy
University of Amsterdam, The Netherlands

> *"Don't express your ideas too clearly.* Most people think little of what they understand, and venerate what they do not."
>
> (The Art of Worldly Wisdom, Baltasar Gracián, 1647.)

Abstract. We show how the constraint propagation process can be naturally explained by means of chaotic iteration.

1 Introduction

1.1 Motivation

Over the last ten years constraint programming emerged as an interesting and viable approach to programming. In this approach the programming process is limited to a generation of requirements ("constraints") and a solution of these requirements by means of general and domain specific methods. The techniques useful for finding solutions to sets of constraints were studied for some twenty years in the field of Constraint Satisfaction. One of the most important of them is *constraint propagation*, the elusive process or reducing a constraint satisfaction problem to another one that is equivalent but "simpler".

The algorithms that achieve such a reduction usually aim at reaching some "local consistency", which denotes some property approximating in some loose sense "global consistency", so the consistency of the whole constraint satisfaction problem. (In fact, most of the notions of local consistency are neither implied by nor imply global consistency.)

For some constraint satisfaction problems such an enforcement of local consistency is already sufficient for finding a solution or for determining that none exists. In some other cases this process substantially reduces the size of the search space which makes it possible to solve the original problem more efficiently by means of some search algorithm.

The aim of this paper is to show that the constraint propagation algorithms can be naturally explained by means of *chaotic iteration*, a basic technique used for computing limits of iterations of finite sets of functions that originated from numerical analysis (see Chazan and Miranker (1969)) and was adapted for computer science needs by Cousot and Cousot (1977). In fact, several constraint propagation algorithms proposed in the literature turn out to be instances of generic chaotic iteration algorithms studied here.

Moreover, by characterizing a given notion of a local consistency as a common fixed point of a finite set of monotonic and inflationary functions we can automatically generate an algorithm achieving this notion of consistency by "feeding" these functions into a generic chaotic iteration algorithm.

1.2 Preliminaries

Definition 1. Consider a sequence of domains $\mathcal{D} := D_1, \ldots, D_n$.

- By a *scheme* (on n) we mean a sequence of different elements from $[1..n]$.
- We say that C is a *constraint (on $\mathcal{D}$) with scheme* $i_1, \ldots, i_l$ if $C \subseteq D_{i_1} \times \cdots \times D_{i_l}$.
- Let $\mathbf{s} := s_1, \ldots, s_k$ be a sequence of schemes. We say that a sequence of constraints $C_1, \ldots, C_k$ on $\mathcal{D}$ is an $\mathbf{s}$-*sequence* if each C_i is with scheme s_i.
- By a *Constraint Satisfaction Problem* $\langle \mathcal{D}; \mathcal{C} \rangle$, in short CSP, we mean a sequence of domains $\mathcal{D}$ together with an $\mathbf{s}$-sequence of constraints $\mathcal{C}$ on $\mathcal{D}$. We call then $\mathbf{s}$ the *scheme* of $\langle \mathcal{D}; \mathcal{C} \rangle$. □

Given an n-tuple $d := d_1, \ldots, d_n$ in $D_1 \times \ldots \times D_n$ and a scheme $s := i_1, \ldots, i_l$ on n we denote by $d[s]$ the tuple $d_{i_1}, \ldots, d_{i_l}$. In particular, for $j \in [1..n]$ $d[j]$ is the j-th element of d. By a *solution* to a CSP $\langle \mathcal{D}; \mathcal{C} \rangle$, where $\mathcal{D} := D_1, \ldots, D_n$, we mean an n-tuple $d \in D_1 \times \ldots \times D_n$ such that for each constraint C in $\mathcal{C}$ with scheme s we have $d[s] \in C$.

Consider now a sequence of schemes $s_1, \ldots, s_k$. By its *union*, written as $\langle s_1, \ldots, s_k \rangle$ we mean the scheme obtained from the sequences $s_1, \ldots, s_k$ by removing from each s_i the elements present in some s_j, where $j < i$, and by concatenating the resulting sequences. For example, $\langle (3,7,2), (4,3,7,5), (3,5,8) \rangle = (3,7,2,4,5,8)$. Recall that for an $s_1, \ldots, s_k$-sequence of constraints $C_1, \ldots, C_k$ their *join*, written as $C_1 \bowtie \cdots \bowtie C_k$, is defined as the constraint with scheme $\langle s_1, \ldots, s_k \rangle$ and such that

$$d \in C_1 \bowtie \cdots \bowtie C_k \text{ iff } d[s_i] \in C_i \text{ for } i \in [1..k].$$

Further, given a constraint C and a subsequence s of its scheme, we denote by $\Pi_s(C)$ the constraint with scheme s defined by

$$\Pi_s(C) := \{d[s] \mid d \in C\},$$

and call it *the projection of C on s*. In particular, for a constraint C with scheme s and an element j of s, $\Pi_j(C) = \{a \mid \exists d \in C \ a = d[j]\}$.

Given a CSP $\langle \mathcal{D}; \mathcal{C} \rangle$ we denote by $Sol(\langle \mathcal{D}; \mathcal{C} \rangle)$ the set of all solutions to it. If the domains are clear from the context we drop the reference to $\mathcal{D}$ and just write $Sol(\mathcal{C})$. The following observation is useful.

Note 2. *Consider a CSP $\langle \mathcal{D}; \mathcal{C} \rangle$ with $\mathcal{D} := D_1, \ldots, D_n$ and $\mathcal{C} := C_1, \ldots, C_k$ and with scheme* **s**.

(i) $$Sol(\langle \mathcal{D}; \mathcal{C} \rangle) = C_1 \bowtie \cdots \bowtie C_k \bowtie_{i \in I} D_i,$$

where $I := \{i \in [1..n] \mid i \text{ does not appear in } \mathbf{s}\}$.

(ii) *For every* **s**-*subsequence* **C** *of* $\mathcal{C}$ *and* $d \in Sol(\langle \mathcal{D}; \mathcal{C} \rangle)$ *we have* $d[\langle \mathbf{s} \rangle] \in Sol(\mathbf{C})$.
□

Finally, we call two CSP's *equivalent* if they have the same set of solutions. Note that we do not insist that these CSP's have the same sequence of domains or the same scheme.

2 Chaotic Iterations

As already mentioned in the introduction, one of the corner stones of constraint programming is *constraint propagation*. In general, two basic approaches fall under this name:

- reduce the domains while maintaining equivalence;
- reduce the constraints while maintaining equivalence.

In what follows we study these two processes in full generality.

2.1 Chaotic Iterations on Simple Domains

In general, chaotic iterations are defined for functions that are projections on individual components of a specific function with several arguments. In our approach we study a more elementary situation in which the functions are unrelated but satisfy certain properties. These functions are defined on specific partial orders. We need the following concepts.

Definition 3. We call a partial order $(D, \sqsubseteq)$ an $\sqcup$-*po* if

- D contains the least element, denoted by $\bot$,
- for every increasing sequence

$$d_0 \sqsubseteq d_1 \sqsubseteq d_2 \ldots$$

of elements from D, the least upper bound of the set

$$\{d_0, d_1, d_2, \ldots\},$$

denoted by $\bigsqcup_{n=0}^{\infty} d_n$ and called the *limit* of $d_0, d_1, \ldots$, exists,
- for all $a, b \in D$ the least upper bound of the set $\{a, b\}$, denoted by $a \sqcup b$, exists.

Further, we say that

- an increasing sequence $d_0 \sqsubseteq d_1 \sqsubseteq d_2 \ldots$ *eventually stabilizes* at d if for some $j \geq 0$ we have $d_i = d$ for $i \geq j$,
- a partial order satisfies the *finite chain property* if every increasing sequence of its elements eventually stabilizes. □

Definition 4. Consider a set D, an element $d \in D$ and a set of functions $F := \{f_1, \ldots, f_k\}$ on D.

- By a *run* (of the functions $f_1, \ldots, f_k$) we mean an infinite sequence of numbers from $[1..k]$.
- A run $i_1, i_2, \ldots$ is called *fair* if every $i \in [1..k]$ appears in it infinitely often.
- By an *iteration of F associated with a run* $i_1, i_2, \ldots$ and starting with d we mean an infinite sequence of values $d_0, d_1, \ldots$ defined inductively by

$$d_0 := d,$$

$$d_j := f_{i_j}(d_{j-1}).$$

When d is the least element of D in some partial order clear from the context, we drop the reference to d and talk about an *iteration of F*.
- An iteration of F is called *chaotic* if it is associated with a fair run. □

Definition 5. Consider a partial order $(D, \sqsubseteq)$. A function f on D is called

- *inflationary* if $x \sqsubseteq f(x)$ for all x,
- *monotonic* if $x \sqsubseteq y$ implies $f(x) \sqsubseteq f(y)$ for all x, y,
- *idempotent* if $f(f(x)) = f(x)$ for all x.

□

The following observation can be easily distilled from a more general result due to Cousot and Cousot (1977). To keep the paper self-contained we provide a direct proof.

Theorem 6 (Chaotic Iteration). *Consider an $\sqcup$-po $(D, \sqsubseteq)$ and a set of functions $F := \{f_1, \ldots, f_k\}$ on D. Suppose that all functions in F are inflationary and monotonic. Then the limit of every chaotic iteration of F exists and coincides with*

$$\bigsqcup_{j=0}^{\infty} f \uparrow j,$$

where the function f on D is defined by:

$$f(x) := \bigsqcup_{i=1}^{k} f_i(x)$$

and $f \uparrow j$ is an abbreviation for $f^j(\bot)$, the j-th fold iteration of f started at $\bot$.

Proof. First notice that f is inflationary, so $\bigsqcup_{j=0}^{\infty} f \uparrow j$ exists. Fix a chaotic iteration $d_0, d_1, \ldots$ of F associated with a fair run $i_1, i_2, \ldots$. Since all functions f_i are inflationary, $\bigsqcup_{j=0}^{\infty} d_j$ exists. The result follows directly from the following two claims.

Claim 1 $\forall j \, \exists m \, f \uparrow j \sqsubseteq d_m$.

Proof. We proceed by induction on j.

Base. $j = 0$. As $f \uparrow 0 = \bot = d_0$, the claim is obvious.

Induction step. Assume that for some $j \geq 0$ we have $f \uparrow j \sqsubseteq d_m$ for some $m \geq 0$. Since

$$f \uparrow (j+1) = f(f \uparrow j) = \bigsqcup_{i=1}^{k} f_i(f \uparrow j),$$

it suffices to prove
$$\forall i \in [1..k] \, \exists m_i \, f_i(f \uparrow j) \sqsubseteq d_{m_i}. \qquad (1)$$

Indeed, we have then by the fact that $d_l \sqsubseteq d_{l+1}$ for $l \geq 0$

$$\bigsqcup_{i=1}^{k} f_i(f \uparrow j) \sqsubseteq \bigsqcup_{i=1}^{k} d_{m_i} \sqsubseteq d_{m'}$$

where $m' := max\{m_i \mid i \in [1..k]\}$.

So fix $i \in [1..k]$. By fairness of the considered run $i_1, i_2, \ldots$, for some $m_i > m$ we have $i_{m_i} = i$. Then $d_{m_i} = f_i(d_{m_i-1})$. Now $d_m \sqsubseteq d_{m_i-1}$, so by the monotonicity of f_i we have

$$f_i(f \uparrow j) \sqsubseteq f_i(d_m) \sqsubseteq f_i(d_{m_i-1}) = d_{m_i}.$$

This proves (1). □

Claim 2 $\forall m \, d_m \sqsubseteq f \uparrow m$.

Proof. The proof is by a straightforward induction on m. Indeed, for $m = 0$ we have $d_0 = \bot = f \uparrow 0$, so the induction base holds.

To prove the induction step suppose that for some $m \geq 0$ we have $d_m \sqsubseteq f \uparrow m$. For some $i \in [1..k]$ we have $d_{m+1} = f_i(d_m)$, so by the monotonicity of f we get

$$d_{m+1} = f_i(d_m) \sqsubseteq f(d_m) \sqsubseteq f(f \uparrow m) = f \uparrow (m+1).$$

□
□

In many situations some chaotic iteration studied in the Chaotic Iteration Theorem 6 eventually stabilizes. This is for example the case when $(D, \sqsubseteq)$ satisfies the finite chain property. In such cases the limit of every chaotic iteration can be characterized in an alternative way.

Corollary 7 (Chaotic Iteration). *Suppose that under the assumptions of the Chaotic Iteration Theorem 6 some chaotic iteration of F eventually stabilizes. Then every chaotic iteration of F eventually stabilizes at the least fixed point of f.*

Proof. It suffices to note that if some chaotic iteration $d_0, d_1 \ldots$ of F eventually stabilizes at some d_m then by Claims 1 and 2 $f \uparrow m = d_m$, so

$$\bigsqcup_{j=0}^{\infty} f \uparrow j = f \uparrow m. \tag{2}$$

Then, again by Claims 1 and 2, every chaotic iteration of F stabilizes at $f \uparrow m$ and it is easy to see that by virtue of (2) $f \uparrow m$ is the least fixed point of f. □

2.2 Chaotic Iterations on Compound Domains

Not much more can be deduced about the process of the chaotic iteration unless the structure of the domain D is further known. So assume now that $(D, \sqsubseteq)$ is the Cartesian product of the $\sqcup$-po's $(D_i, \sqsubseteq_i)$, for $i \in [1..n]$, defined in the expected way. It is straightforward to check that $(D, \sqsubseteq)$ is then an $\sqcup$-po, as well. In what follows we consider a modification of the situation studied in the Chaotic Iteration Theorem 6 in which each function f_i affects only certain components of D.

Consider the partial orders $(D_i, \sqsubseteq_i)$, for $i \in [1..n]$ and a scheme $s := i_1, \ldots, i_l$ on n. Then by $(D_s, \sqsubseteq_s)$ we mean the Cartesian product of the partial orders $(D_{i_j}, \sqsubseteq_{i_j})$, for $j \in [1..l]$.

Given a function f on D_s we say that f is *with scheme* s. Instead of defining iterations for the case of the functions with schemes, we rather reduce the situation to the one studied in the previous subsection. To this end we canonically extend each function f on D_s to a function f^+ on D as follows. Suppose that $s = i_1, \ldots, i_l$ and

$$f(d_{i_1}, \ldots, d_{i_l}) = (e'_{i_1}, \ldots, e'_{i_l}).$$

Let for $j \in [1..n]$

$$e_j := \begin{cases} e'_j & \text{if } j \text{ is an element of } s, \\ d_j & \text{otherwise.} \end{cases}$$

Then we set

$$f^+(d_1, \ldots, d_n) := (e_1, \ldots, e_n).$$

Suppose now that $(D, \sqsubseteq)$ is the Cartesian product of the $\sqcup$-po's $(D_i, \sqsubseteq_i)$, for $i \in [1..n]$, and $F := \{f_1, \ldots, f_k\}$ is a set of functions with schemes that are all inflationary and monotonic. Then the following algorithm can be used to compute the limit of the chaotic iterations of $F^+ := \{f_1^+, \ldots, f_k^+\}$. We say here that a function f *depends on* i if i is an element of its scheme.

Generic Chaotic Iteration Algorithm (CI)

$d := \underbrace{(\bot, \ldots, \bot)}_{n \text{ times}};$

$d' := d;$
$G := F;$
while $G \neq \emptyset$ **do**
    choose $g \in G$; suppose g is with scheme s;
    $G := G - \{g\};$
    $d'[s] := g(d[s]);$
    if $d[s] \neq d'[s]$ **then**
        $G := G \cup \{f \in F \mid f \text{ depends on some } i \text{ in } s \text{ such that } d[i] \neq d'[i]\};$
        $d[s] := d'[s]$
    fi
od

The following observation will be useful in the proof of correctness of this algorithm.

Note 8. *Consider the partial orders $(D_i, \sqsubseteq_i)$, for $i \in [1..n]$, a scheme s on n and a function f with scheme s. Then*

(i) f is inflationary iff f^+ is,
(ii) f is monotonic iff f^+ is.

$\square$

The following result summarizes the properties of the CI algorithm.

Theorem 9.

(i) Every terminating execution of the CI algorithm computes in d the least fixed point of the function f on D defined by

$$f(x) := \bigsqcup_{i=1}^{k} f_i^+(x).$$

(ii) If all $(D_i, \sqsubseteq_i)$, where $i \in [1..n]$, satisfy the finite chain property, then every execution of the CI algorithm terminates.

Proof. It is simpler to reason about a modified, but equivalent, algorithm in which the assignments $d'[s] := g(d[s])$ and $d[s] := d'[s]$ are respectively replaced by $d' := g^+(d)$ and $d := d'$ and the test $d[s] \neq d'[s]$ by $d \neq d'$.
(i) Note that the formula

$$I := \forall f \in F - G \; f^+(d) = d$$

is an invariant of the **while** loop of the modified algorithm. Thus upon its termination

$$(G = \emptyset) \wedge I$$

holds, that is
$$\forall f \in F \; f^+(d) = d.$$
Consequently, some chaotic iteration of F^+ eventually stabilizes at d. Hence d is the least fixpoint of the function f defined in item (i) because the Chaotic Iteration Corollary 7 is applicable here by virtue of Note 8(i) and (ii).

(ii) Consider the lexicographic order of the partial orders $(D, \sqsupseteq)$ and $(N, \leq)$, defined on the elements of $D \times N$ by
$$(d_1, n_1) \leq_{lex} (d_2, n_2) \text{ iff } d_1 \sqsupset d_2 \text{ or } (d_1 = d_2 \text{ and } n_1 \leq n_2).$$
We use here the inverse order $\sqsupseteq$ and N denotes the set of natural numbers.

By Note 8(i) all functions f_i^+ are inflationary, so with each **while** loop iteration of the modified algorithm the pair
$$(d, \text{card } G)$$
strictly decreases in this order $\leq_{lex}$. Howver, in general the lexicographic order $(D \times N, \leq_{lex})$ is not well-founded and in fact termination is not guaranteed. But assume now additionally that each partial order $(D_i, \sqsubseteq_i)$ satisfies the finite chain property. Then so does their Cartesian product $(D, \sqsubseteq)$. This means that $(D, \sqsupseteq)$ is well-founded and consequently so is $(D \times N, \leq_{lex})$ which implies termination. □

When all considered functions f_i are also idempotent, we can reverse the order of the two assignments to G, that is to put the assignment $G := G - \{g\}$ after the **if-then-fi** statement, because after applying an idempotent function there is no use in applying it immediately again. Let us denote by CII the algorithm resulting from this movement of the assignment $G := G - \{g\}$.

More specialized versions of the CI and CII algorithms can be obtained by representing G as a queue. To this end we use the operation **enqueue**(F, Q) which for a set F and a queue Q enqueues in an arbitrary order all the elements of F in Q, denote the empty queue by **empty**, and the head and the tail of a non-empty queue Q respectively by **head**(Q) and **tail**(Q). The following algorithm is then a counterpart of the CI algorithm.

GENERIC CHAOTIC ITERATION ALGORITHM WITH A QUEUE (CIQ)

$d := \underbrace{(\bot, \ldots, \bot)}_{n \text{ times}};$

$d' := d;$
$Q := \textbf{empty};$
enqueue$(F, Q);$
while $Q \neq \textbf{empty}$ **do**
    $g := \textbf{head}(Q);$ suppose g is with scheme $s;$
    $Q := \textbf{tail}(Q);$
    $d'[s] := g(d[s]);$

```
if d[s] ≠ d'[s] then
        enqueue({f ∈ F | f depends on some i in s such that d[i] ≠ d'[i]}, Q);
        d[s] := d'[s]
    fi
od
```

Denote by CIIQ the modification of the CIQ algorithm that is appropriate for the idempotent functions, so the one in which the assignment $Q := \mathbf{tail}(Q)$ is performed after the **if-then-fi** statement.

It is easy to see that the claims of Theorem 9 also hold for the CII, CIQ and CIIQ algorithms. A natural question arises whether for the specialized versions CIQ and CIIQ some additional properties can be established. The answer is positive. Namely, for these two algorithms the following result holds which shows that the nondeterminism present in these algorithms has no bearing on their termination.

Theorem 10. *If some execution of the CIQ algorithm terminates, then all the executions of the CIQ algorithm terminate.*

Proof. We first establish the following observation.

Claim 1 *If some chaotic iteration of F^+ eventually stabilizes, then all the executions of the CIQ algorithm terminate.*

Proof. We prove the contrapositive. Consider an infinite execution of the CIQ algorithm algorithm. Let $i_1, i_2, \ldots$ be the run associated with it and $\xi := d_0, d_1, \ldots$ the iteration of F^+ associated with this run. By the structure of this algorithm

$$\xi \text{ does not stabilize.} \qquad (3)$$

Let A be the set of the elements of $[1..k]$ that appear finitely often in the run $i_1, i_2, \ldots$. For some $m \geq 0$ we have $i_j \notin A$ for $j > m$. This means by the structure of this algorithm that after m iterations of the **while** loop no function f_i for $i \in A$ is ever present in the queue Q.

By virtue of the invariant I used in the proof of Theorem 9 we then have $f_i^+(d_j) = d_j$ for $i \in A$ and $j \geq m$. This allows us to transform the iteration ξ to a chaotic one by repeating each element d_j for $j \geq m$ *card A* times.

Assume now that a chaotic iteration of F^+ eventually stabilizes. Then by the Chaotic Iteration Corollary 7 the just constructed chaotic iteration stabilizes, as well. So the original iteration ξ also stabilizes which contradicts (3). □

Construct now a chaotic iteration of F^+ the initial prefix of which corresponds with a terminating execution of the CIQ algorithm. By virtue of the invariant I this iteration eventually stabilizes. This concludes the proof thanks to Claim 1. □

An analogous result holds for the CIIQ algorithm. On the other hand, it is easy to see that this result does not hold for the CI and CII algorithms.

3 Constraint Propagation

Let us return now to the study of CSP's. We show here how the results of the previous section can be used to explain the constraint propagation process.

3.1 Domain Reduction

In this subsection we study the domain reduction process. First we associate with each CSP an $\sqcup$-po that "focuses" on the domain reduction.

Consider a CSP $\mathcal{P} := \langle D_1, \ldots, D_n; \mathcal{C} \rangle$. Let for $X, Y \subseteq D_i$

$$X \sqsubseteq_i Y \text{ iff } X \supseteq Y.$$

Then for $i \in [1..n]$ $(\mathcal{P}(D_i), \sqsubseteq_i)$ is an $\sqcup$-po with $\bot_i = D_i$ and $X \sqcup_i Y = X \cap Y$. Consequently, the Cartesian product $(DO, \sqsubseteq)$ of $(\mathcal{P}(D_i), \sqsubseteq_i)$, where $i \in [1..n]$, is also an $\sqcup$-po. We call $(DO, \sqsubseteq)$ *the domain $\sqcup$-po associated with* $\mathcal{P}$.

As in in Subsection 2.2, for a scheme $s := i_1, \ldots, i_l$ we denote by $(DO_s, \sqsubseteq_s)$ the Cartesian product of the partial orders $(\mathcal{P}(D_{i_j}), \sqsubseteq_{i_j})$, where $j \in [1..l]$.

Note that $DO_s = \mathcal{P}(D_{i_1}) \times \cdots \times \mathcal{P}(D_{i_l})$. Because we want now to use constraints in our analysis and constraint are sets of tuples, we identify DO_s with the set

$$\{X_1 \times \cdots \times X_l \mid X_j \subseteq D_{i_j} \text{ for } j \in [1..l]\}.$$

In this way we can write the elements of DO_s as Cartesian products $X_1 \times \cdots \times X_l$, so as (specific) sets of l-tuples, instead of as $(X_1, \ldots, X_l)$, and similarly with DO.

Note that because of the use of the inverse subset order $\supseteq$ we have for $X_1 \times \cdots \times X_l \in DO_s$ and $Y_1 \times \cdots \times Y_l \in DO_s$

$$X_1 \times \cdots \times X_l \sqsubseteq_s Y_1 \times \cdots \times Y_l \text{ iff } X_1 \times \cdots \times X_l \supseteq Y_1 \times \cdots \times Y_l$$
$$(\text{iff } X_i \supseteq Y_i \text{ for } i \in [1..l]),$$

$$(X_1 \times \cdots \times X_l) \sqcup_s (Y_1 \times \cdots \times Y_l) = (X_1 \times \cdots \times X_l) \cap (Y_1 \times \cdots \times Y_l)$$
$$(= (X_1 \cap Y_1) \times \cdots \times (X_l \cap Y_l)).$$

Moreover, $D_1 \times \cdots \times D_n$ is the least element of DO.

So far we have defined an $\sqcup$-po associated with a CSP. Next, we introduce functions by means of which chaotic iterations will be generated. These functions are associated with constraints. Constraints are arbitrary sets of k-tuples for some k, while the $\sqsubseteq_s$ order and the $\sqcup_s$ operation are defined only on Cartesian products. So to define these functions we use the set theoretic counterparts $\supseteq$ and $\cap$ of $\sqsubseteq_s$ and $\sqcup_s$ which are defined on arbitrary sets.

Definition 11. Consider a sequence of domains $D_1, \ldots, D_n$ and a scheme s on n. By a *domain reduction function* for a constraint C with scheme s we mean a function f on DO_s such that for all $\mathbf{D} \in DO_s$

- $\mathbf{D} \supseteq f(\mathbf{D})$,
- $C \cap \mathbf{D} = C \cap f(\mathbf{D})$. $\square$

The first condition states that f reduces the "current" domains associated with the constraint C (so no solution to C is "gained"), while the second condition states that during this domain reduction process no solution to C is "lost". In particular, the second condition implies that if $C \subseteq \mathbf{D}$ then $C \subseteq f(\mathbf{D})$.

Note that for the partial order $(DO_s, \sqsubseteq_s)$ a function f on DO_s is inflationary iff $\mathbf{D} \supseteq f(\mathbf{D})$ and f is monotonic iff it is monotonic w.r.t. the set inclusion.

Example 1. As a simple example of a domain reduction functions consider a binary constraint $C \subseteq D_1 \times D_2$. Define now the functions f_1 and f_2 on $DO_{1,2} := \mathcal{P}(D_1) \times \mathcal{P}(D_2)$ as follows:

$$f_1(X \times Y) := X' \times Y,$$

where $X' = \{a \in X \mid \exists b \in Y \ (a, b) \in C\}$, and

$$f_2(X \times Y) := X \times Y',$$

where $Y' = \{b \in Y \mid \exists a \in X \ (a, b) \in C\}$. It is straightforward to check that f_1 and f_2 are indeed domain reduction functions. Further, these functions are monotonic w.r.t. the set inclusion and idempotent. □

Take now a CSP $\mathcal{P} := \langle D_1, \ldots, D_n; \mathcal{C} \rangle$ and a sequence of domains $D'_1, \ldots, D'_n$ such that $D'_i \subseteq D_i$ for $i \in [1..n]$. Consider a CSP $\mathcal{P}'$ obtained from $\mathcal{P}$ by replacing each domain D'_i by D_i and by restricting each constraint in $\mathcal{C}$ to these new domains. We say then that $\mathcal{P}'$ *is determined by* $\mathcal{P}$ *and* $D'_1 \times \ldots \times D'_n$.

Consider now a CSP $\mathcal{P} := \langle D_1, \ldots, D_n; \mathcal{C} \rangle$ and a domain reduction function f for a constraint C of $\mathcal{C}$. Suppose that

$$f^+(D_1 \times \cdots \times D_n) = D'_1 \times \cdots \times D'_n,$$

where f^+ is the canonic extension of f to DO defined in Subsection 2.2. We now define $f(\mathcal{P})$ to be the CSP determined by $\mathcal{P}$ and $D'_1 \times \ldots \times D'_n$. The following observation holds.

Lemma 12. *Consider a CSP $\mathcal{P}$ and a domain reduction function f. Then $\mathcal{P}$ and $f(\mathcal{P})$ are equivalent.*

Proof. Suppose that $D_1, \ldots, D_n$ are the domains of $\mathcal{P}$ and assume that f is a domain reduction function for C with scheme $i_1, \ldots, i_l$. Let

$$f(D_{i_1} \times \cdots \times D_{i_l}) = D'_{i_1} \times \cdots \times D'_{i_l}.$$

Take now a solution d to $\mathcal{P}$. Then $d[i_1, \ldots, i_l] \in C$, so by the definition of f also $d[i_1, \ldots, i_l] \in D'_{i_1} \times \cdots \times D'_{i_l}$. So d is also a solution to $f(\mathcal{P})$. The converse implication holds by the definition of a domain reduction function. □

When dealing with a specific CSP we have in general several domain reduction functions. To study their interaction we can use the Chaotic Iteration Theorem 6 in conjunction with the above Note. After translating the relevant notions into set theoretic terms we get the following direct consequence of these results. (In this translation DO_s corresponds to D_s and DO to D.)

Theorem 13 (Domain Reduction). *Consider a CSP $\mathcal{P} := \langle D_1, \ldots, D_n; \mathcal{C} \rangle$. Let $F := \{f_1, \ldots, f_k\}$, where each f_i is a domain reduction function for some constraint in $\mathcal{C}$. Suppose that all functions f_i are monotonic w.r.t. the set inclusion. Then*

- *the limit of every chaotic iteration of $F^+ := \{f_1^+, \ldots, f_k^+\}$ exists;*
- *this limit coincides with*

$$\bigcap_{j=0}^{\infty} f^j(D_1 \times \cdots \times D_n),$$

where the function f on DO is defined by:

$$f(\mathbf{D}) := \bigcap_{i=1}^{k} f_i^+(\mathbf{D}),$$

- *the CSP determined by $\mathcal{P}$ and this limit is equivalent to $\mathcal{P}$.* □

Informally, this theorem states that the order of the applications of the domain reduction functions does not matter, as long as none of them is indefinitely neglected.

Consider now a CSP $\mathcal{P}$ and suppose that the domain ⊔-po associated with it satisfies the finite chain property. Then we can use the `CI`, `CII`, `CIQ` and `CIIQ` algorithms to compute the limits of the chaotic iterations considered in the above Theorem. We shall explain in Subsection 4.1 how by instantiating these algorithms with specific domain reduction functions we obtain specific algorithms considered in the literature. In each case, by virtue of Theorem 9 and its reformulations for the `CII`, `CIQ` and `CIIQ` algorithms, we can conclude that these algorithms compute the greatest common fixpoint w.r.t. the set inclusion of the functions from F^+.

3.2 Constraint Reduction

We now study the constraint reduction process. As in the previous subsection we begin by associating with each CSP an ⊔-po that "focuses" on the constraint reduction.

Consider a CSP $\mathcal{P} := \langle \mathcal{D}; C_1, \ldots, C_k \rangle$. Let for $X, Y \subseteq C_i$

$$X \sqsubseteq_i Y \text{ iff } X \supseteq Y.$$

Let now $(CO, \sqsubseteq)$ be the Cartesian product of the ⊔-po's $(\mathcal{P}(C_i), \sqsubseteq_i)$, where $i \in [1..n]$. We call $(CO, \sqsubseteq)$ *the constraint ⊔-po associated with* $\mathcal{P}$.

Following the notation of the previous subsection, for a scheme $s := i_1, \ldots, i_l$ on k we denote by $(CO_s, \sqsubseteq_s)$ the Cartesian product of the partial orders $(\mathcal{P}(C_{i_j}), \sqsubseteq_{i_j})$, where $j \in [1..l]$, and identify CO_s with the set

$$\{X_1 \times \cdots \times X_l \mid X_j \subseteq C_{i_j} \text{ for } j \in [1..l]\},$$

and similarly with CO.

Next, we define functions that will be used to generate chaotic iterations.

Definition 14. Consider a CSP $\langle \mathcal{D}; C_1, \ldots, C_k \rangle$ and a scheme s on k. By a *constraint reduction function with scheme s* we mean a function g on CO_s such that for all $\mathbf{C} \in CO_s$

- $\mathbf{C} \supseteq g(\mathbf{C})$,
- $Sol(\mathbf{C}) = Sol(g(\mathbf{C}))$. □

$\mathbf{C}$ is here a Cartesian product of some constraints and in the second condition and in the example below we identified it with the sequence of these constraints, and similarly with $g(\mathbf{C})$. The first condition states that g reduces the constraints C_i, where i is an element of s, while the second condition states that during this constraint reduction process no solution to $\mathbf{C}$ is lost.

Example 2. As an example of a constraint reduction function consider the following function g on some CO_s:

$$g(C \times \mathbf{C}) := C' \times \mathbf{C},$$

where $C' = \Pi_t(Sol(C, \mathbf{C}))$ and t is the scheme of C. To see that g is indeed a constraint reduction function, first note that by the definition of Sol we have $C' \subseteq C$, so $C \times \mathbf{C} \supseteq g(C \times \mathbf{C})$. Next, note that for $d \in Sol(C, \mathbf{C})$ we have $d[t] \in \Pi_t(Sol(C, \mathbf{C}))$, so $d \in Sol(C', \mathbf{C})$. This implies that $Sol(C, \mathbf{C}) = Sol(g(C, \mathbf{C}))$.

Note also that g is monotonic w.r.t. the set inclusion and idempotent. □

Example 3. As another example that is of importance for the discussion in Subsection 4.1 consider a CSP $\langle D_1, \ldots, D_n; \mathcal{C} \rangle$ of binary constraints such that for each scheme i, j on n there is exactly one constraint, which we denote by $C_{i,j}$.

Define now for each scheme k, l, m on n the following function $g_{k,l}^m$ on CO_s, where s is the triple corresponding to the positions of the constraints $C_{k,l}, C_{k,m}$ and $C_{m,l}$ in $\mathcal{C}$:

$$g_{k,l}^m(X_{k,l} \times X_{k,m} \times X_{m,l}) := (X_{k,l} \cap \Pi_{k,l}(X_{k,m} \bowtie X_{m,l})) \times X_{k,m} \times X_{m,l}.$$

To prove that the functions $g_{k,l}^m$ are constraint reduction functions it suffices to note that by simple properties of the $\bowtie$ operation and by Note 2(i) we have

$$X_{k,l} \cap \Pi_{k,l}(X_{k,m} \bowtie X_{m,l}) = \Pi_{k,l}(X_{k,l} \bowtie X_{k,m} \bowtie X_{m,l})$$
$$= \Pi_{k,l}(Sol(X_{k,l}, X_{k,m}, X_{m,l})),$$

so these functions are special cases of the functions defined in Example 2. □

Take now a CSP $\mathcal{P} := \langle \mathcal{D}; C_1, \ldots, C_k \rangle$ and a sequence of constraints $C'_1, \ldots, C'_k$ such that $C'_i \subseteq C_i$ for $i \in [1..k]$. Let $\mathcal{P}' := \langle \mathcal{D}; C'_1, \ldots, C'_k \rangle$. We say then that $\mathcal{P}'$ is *determined by* $\mathcal{P}$ and $C'_1 \times \ldots \times C'_k$.

Consider now a CSP $\mathcal{P} := \langle \mathcal{D}; C_1, \ldots, C_k \rangle$ and a constraint reduction function g with scheme s. Suppose that

$$g^+(C_1 \times \cdots \times C_k) = C'_1 \times \cdots \times C'_k,$$

where g^+ is the canonic extension of g to CO defined in Subsection 2.2. We now define
$$g(\mathcal{P}) := \langle \mathcal{D}; C'_1, \ldots, C'_k \rangle.$$
We have the following observation.

Lemma 15. *Consider a CSP $\mathcal{P}$ and a constraint reduction function g. Then $\mathcal{P}$ and $g(\mathcal{P})$ are equivalent.*

Proof. Suppose that s is the scheme of the function g and let $\mathbf{C}$ be an element of CO_s. $\mathbf{C}$ is a Cartesian product of some constraints. As before we identify it with the sequence of these constraints. For some sequence of schemes $\mathbf{s}$, $\mathbf{C}$ is the s-sequence of the constraints of $\mathcal{P}$.

Let now d be a solution to $\mathcal{P}$. Then by Note 2(ii) we have $d[\langle \mathbf{s} \rangle] \in Sol(\mathbf{C})$, so by the definition of g also $d[\langle \mathbf{s} \rangle] \in Sol(g(\mathbf{C}))$. Hence for every constraint C' in $g(\mathbf{C})$ with scheme s' we have $d[s'] \in C'$ since $d[\langle \mathbf{s} \rangle][s'] = d[s']$. So d is a solution to $g(\mathcal{P})$. The converse implication holds by the definition of a constraint reduction function. □

As in the case of the domain reduction we can now apply the results of Section 2 to study the outcome of the constraint reduction process. To this end it suffices to translate the relevant notions into set theoretic terms. (In this translation CO_s corresponds to D_s and CO to D.) We get then the following counterpart of the Domain Reduction Theorem 13.

Theorem 16 (Constraint Reduction). *Consider a CSP $\mathcal{P} := \langle \mathcal{D}; C_1, \ldots, C_k \rangle$. Let $F := \{g_1, \ldots, g_k\}$, where each g_i is a constraint reduction function. Suppose that all functions g_i are monotonic w.r.t. the set inclusion. Then*

- *the limit of every chaotic iteration of $F^+ := \{g_1^+, \ldots, g_k^+\}$ exists;*
- *this limit coincides with*

$$\bigcap_{j=0}^{\infty} g^j(C_1 \times \cdots \times C_k),$$

where the function g on CO is defined by:

$$g(\mathbf{C}) := \bigcap_{i=1}^{k} g_i^+(\mathbf{C}),$$

- *the CSP determined by $\mathcal{P}$ and this limit is equivalent to $\mathcal{P}$.* □

When the constraint ⊔-po associated with a CSP $\mathcal{P}$ satisfied the finite chain property, we can use the algorithms discussed in Subsection 2.2 to compute the limits of the chaotic iterations considered in the above Theorem. We return to this issue in Subsection 4.1. Also here, as in the previous subsection, we can conclude by virtue of Theorem 9 that these algorithms compute the greatest common fixpoint w.r.t. the set inclusion of the functions from F^+. So the limit of the constraint propagation process could be added to the collection of important greatest fixpoints presented in Barwise and Moss (1996).

Next, we show how specific provably correct algorithms for achieving a local consistency notion can be automatically derived. As it is difficult to define local consistency formally, we illustrate the idea on an example.

Example 4. We consider here the notion of relational consistency proposed recently in Dechter and van Beek (1997).

To define it need to introduce some auxiliary concepts first. Consider a CSP $\langle D_1, \ldots, D_n; \mathcal{C} \rangle$. Take a scheme $t := i_1, \ldots, i_l$ on n. We call $d \in D_{i_1} \times \cdots \times D_{i_l}$ a tuple of *type t* and say that d is *consistent* if for every subsequence s of t and a constraint $C \in \mathcal{C}$ with scheme s we have $d[s] \in C$.

A CSP $\mathcal{P}$ is called *relationally m-consistent* if for any **s**-sequence $C_1, \ldots, C_m$ of different constraints of $\mathcal{P}$ and a subsequence t of $\langle \mathbf{s} \rangle$, every consistent tuple of type t belongs to $\Pi_t(C_1 \bowtie \cdots \bowtie C_m)$.

As the first step we characterize this notion as a common fixed point of a finite set of monotonic and inflationary functions.

Consider a CSP $\mathcal{P} := \langle D_1, \ldots, D_n; C_1, \ldots, C_k \rangle$. Assume for simplicity that for every scheme s on n there is a unique constraint with scheme s. Each CSP is trivially equivalent with such a CSP — it suffices to replace for each scheme s the set of constraints with scheme s by their intersection and to introduce "universal constraints" for the schemes without a constraint.

Consider now a scheme $i_1, \ldots, i_m$ on k. Let **s** be such that $C_{i_1}, \ldots, C_{i_m}$ is an **s**-sequence of constraints and let t be a subsequence of $\langle \mathbf{s} \rangle$. Further, let C_{i_0} be the constraint of $\mathcal{P}$ with scheme t. Put $s := \langle (i_0), (i_1, \ldots, i_m) \rangle$. (Note that if i_0 does not appear in $i_1, \ldots, i_m$ then $s = i_0, i_1, \ldots, i_m$ and otherwise s is the permutation of $i_1, \ldots, i_m$ obtained by transposing i_0 with the first element.)

Define now a function g_s on CO_s by

$$g_s(C \times \mathbf{C}) := (C \cap \Pi_t(\bowtie \mathbf{C})) \times \mathbf{C}.$$

It is easy to see that if for each function g_s of the above form we have

$$g_s^+(C_1 \times \cdots \times C_k) = C_1 \times \cdots \times C_k,$$

then $\mathcal{P}$ is relationally m-consistent. (The converse implication is in general not true). Note that the functions g_s are inflationary and monotonic w.r.t. the inverse subset order $\supseteq$ and also idempotent.

Consequently, by virtue of Theorem 9 reformulated for the CII algorithm, we can now use the CII algorithm to achieve relational m-consistency for a CSP with finite domains by "feeding" into this algorithm the above defined functions. The obtained algorithm improves upon the (authors' terminology) brute force algorithm proposed in Dechter and van Beek (1997) since the useless constraint modifications are avoided.

As in Example 3, by simple properties of the $\bowtie$ operation and by Note 2(i) we have

$$C \cap \Pi_t(\bowtie \mathbf{C}) = \Pi_t(C \bowtie (\bowtie \mathbf{C})) = \Pi_t(sol(C, \mathbf{C})).$$

Hence, by virtue of Example 2, the functions g_s are all constraint reduction functions. Consequently, by the Constraint Reduction Theorem 16 we conclude that the CSP computed by the just discussed algorithm is equivalent to the original one. □

It is perhaps worthwhile to note that the domain reduction process can be seen as a special case of the constraint reduction process. To this end it suffices to introduce unary constraints each of which coincides with a different domain of the given CSP and replace the reduction of the domains by the reduction of these unary constraints followed by the restriction of the other constraints to these reduced unary constraints. So the domain reduction functions can be seen as special cases of the constraint reduction functions.

We decided to consider the domain reduction process separately, because, as we shall see in the next section, it has been extensively studied, especially in the context of CSP's with binary constraints and of interval arithmetic. Consequently, it is useful to analyze it directly, without any introduction of new constraints.

4 Concluding Remarks

4.1 Related Work

It is illuminating see how the attempts of finding general principles behind the constraint propagation algorithms repeatedly reoccur in the literature on constraint satisfaction problems spanning the last twenty years.

As already stated in the introduction, the aim of the constraint propagation algorithms is most often to achieve some form of local consistency. As a result these algorithms are usually called in the literature "consistency algorithms" or "consistency enforcing algorithms".

To start with, in Mackworth (1977) a unified framework was proposed to explain the so-called arc- and path-consistency algorithms. Also the arc-consistency algorithm AC-3 and the path-consistency algorithm PC-2 were proposed and the latter algorithm was obtained from the former one by pursuing the analogy between both notions of consistency.

The AC-3 consistency algorithm can be obtained by instantiating the CII algorithm with the domain reduction functions defined in Example 1, whereas the PC-2 algorithm can be obtained by instantiating this algorithm with the domain reduction functions defined in Example 3.

In Dechter and Pearl (1988) the notions of arc- and path-consistency were modified to directional arc- and path-consistency, versions that take into account some total order $<_d$ of the domain indices, and the algorithms for achieving these forms of consistency were presented. These algorithms can be obtained as instances of the CIQ algorithm as follows.

For the case of directional arc-consistency the queue in this algorithm should be instantiated with the set of the domain reduction functions f_1 of Example 1 for the constraints the scheme of which is consistent with the $<_d$ order. These functions should be ordered in such a way that the domain reduction functions for the constraint with the $<_d$-large second index appear earlier. This order has the effect that the **enqueue** operation within the **if-then-fi** statement has always the empty set as the first argument, so it can be deleted. Consequently, the algorithm can be rewritten as a simple **for** loop that processes the selected domain reduction functions f_1 in the appropriate order.

For the case of directional path-consistency the constraint reduction functions $g_{k,l}^m$ should be used only for $k, l <_d m$ and the queue in the CIQ algorithm should be initialized in such a way that the functions $g_{k,l}^m$ with the $<_d$-large m index appear earlier. As in the case of directional arc-consistency this algorithm can be rewritten as a simple **for** loop.

In Montanari and Rossi (1991) a general study of constraint propagation was undertaken by defining the notion of a relaxation rule and by proposing a general relaxation algorithm. The notion of a relaxation rule coincides with our notion of a constraint propagation function instantiated with the functions defined in Example 2 and the general relaxation algorithm is the corresponding instance of our CI algorithm.

In Montanari and Rossi (1991) it was also shown that the notions of arc-consistency and path-consistency can be defined by means of relaxation rules and that as a result arc-consistency and path-consistency algorithms can be obtained by instantiating with these rules their general relaxation algorithm.

Van Hentenryck, Deville and Teng (1992) presented a generic arc consistency algorithm, called AC-5, that can be specialized to the known arc-consistency algorithms AC-3 and AC-4 and also to new arc-consistency algorithms for specific classes of constraints.

In Benhamou, McAllester and Hentenryck (1994) and Benhamou and Older (1997) specific functions, called narrowing functions, were associated with constraints in the context of interval arithmetic for reals and some properties of them were established that in our terminology mean that these are idempotent domain reduction functions. As a consequence the algorithms proposed in these papers, called respectively a fixpoint algorithm and a narrowing algorithm, become respectively the instances of our CIIQ algorithm and CII algorithm.

The importance of fairness for the study of constraint propagation was noticed in Montanari and Rossi (1991), while the relevance of the chaotic iteration was independently noticed in Fages, Fowler and Sola (1996) and van Emden (1996). In the latter paper the generic chaotic iteration algorithm CII was formulated and proved correct for the domain reduction functions defined in Benhamou and Older (1997) and it was shown that the limit of the constraint propagation process for these functions is their greatest common fixpoint.

The idea that the meaning of a constraint is a function (on a constraint store) with some algebraic properties was put forward in Saraswat, Rinard and Panangaden (1991), where the properties of being inflationary (called there extensive), monotonic and idempotent were singled out.

It is unrealistic to expect that all constraint propagation algorithms presented in the literature can be expressed as direct instances of the algorithms discussed in this paper. For example the `AC-4` algorithm of Mohr and Henderson (1986) associates with each domain element some information concerning its links with the elements of other domains. As a result this algorithm operates on some "enhancement" of the original domains.

We noted, however, that even in this case the analysis here provided can be used to explain this algorithm. To this end one needs to reason about the translation of the original CSP to a CSP defined on the enhanced domains. This analysis allows us to reduce the proof of the correctness of this algorithm to the proof that specific functions are monotonic domain reduction functions.

4.2 Idempotence

In each of the above papers the (often implicitly) considered semantic, domain or constraint reduction functions are idempotent, so we now comment on the relevance of this assumption.

To start with, in our study Apt (1997) of linear constraints on finite integer intervals we found that natural domain reduction functions are not idempotent. Secondly, as noticed in Older and Vellino (1993), another paper on constraints for interval arithmetic on reals, we can always replace each non-idempotent inflationary function f by

$$f^*(x) := \bigsqcup_{i=1}^{\infty} f^i(x).$$

The following is now straightforward to check.

Note 17. *Consider an $\sqcup$-po $(D, \sqsubseteq)$ and a function f on D.*

- *If f is inflationary, then so is f^*.*
- *If f is monotonic, then so f^*.*
- *If f is inflationary and $(D, \sqsubseteq)$ has the finite chain property, then f^* is idempotent.*
- *If f is idempotent, then $f^* = f$.*
- *Suppose that $(D, \sqsubseteq)$ has the finite chain property. Let $F := \{f_1, \ldots, f_k\}$ be a set of inflationary, monotonic functions on D and let $F^* := \{f_1^*, \ldots, f_k^*\}$. Then the limits of all chaotic iterations of F and of F^* exist and always coincide.* □

Consequently, under the conditions of the last item, every chaotic iteration of F^* can be modeled by a chaotic iteration of F, though not conversely. In fact, the use of F^* instead of F can lead to a more limited number of chaotic iterations. This may mean that in some specific algorithms some more efficient chaotic iterations of F cannot be realized when using F^*.

4.3 Semi-chaotic Iterations

The results of this paper can be slightly strengthened by considering the following generalization of the chaotic iterations.

Definition 18. Consider a set of functions $F := \{f_1, \ldots, f_k\}$ on a domain D.

- We say that an element $i \in [1..k]$ is *eventually irrelevant for an iteration* $d_0, d_1, \ldots$ of F if $\exists m \geq 0 \, \forall j \geq m \; f_i(d_j) = d_j$.
- An iteration of F is called *semi-chaotic* if every $i \in [1..k]$ that appears finitely often in its run is eventually irrelevant for this iteration. □

So every chaotic iteration is semi-chaotic but not conversely. Now, in all the results of this paper chaotic iterations can be replaced by semi-chaotic iterations. The reason is that, as shown in the proof of Theorem 10, every semi-chaotic iteration ξ can be transformed into a chaotic iteration ξ' with the same limit and such that ξ eventually stabilizes at some d iff ξ' does. The proof of Theorem 10 also shows that every infinite execution of the `CIQ` algorithm is associated with a semi-chaotic iteration of F^+.

However, the property of being a semi-chaotic iteration cannot be determined from the run only. So, for simplicity, we decided to limit our exposition to chaotic iterations.

Acknowledgements

This work was prompted by our study of van Emden (1996). Rina Dechter helped us to clarify (most of) our initial confusion about constraint propagation. Discussions with Eric Monfroy helped us to better articulate various points put forward here. Nissim Francez provided us with helpful comments.

References

Apt, K. (1997). A proof theoretic view of constraint programming, *Technical report*, CWI, Amsterdam. In preparation.

Barwise, J. and Moss, L. (1996). *Vicious Circles: on the mathematics of circular phenomena*, CSLI–Lecture Notes, Center for the Study of Language and Information, Stanford, California.

Benhamou, F. and Older, W. (1997). Applying interval arithmetic to real, integer and Boolean constraints, *Journal of Logic Programming*. Technical report 1994. To appear.

Benhamou, F., McAllester, D. and Hentenryck, P. V. (1994). CLP(intervals) revisited, *in* M. Bruynooghe (ed.), *Proceedings of the 1994 International Logic Programming Symposium*, pp. 124–138.

Chazan, D. and Miranker, W. (1969). Chaotic relaxation, *Linear Algebra and its Applications* **2**: 199–222.

Cousot, P. and Cousot, R. (1977). Automatic synthesis of optimal invariant assertions: mathematical foundations, *ACM Symposium on Artificial Intelligence and Programming Languages*, SIGPLAN Notices 12 (8), pp. 1–12.

Dechter, R. and Pearl, J. (1988). Network-based heuristics for constraint-satisfaction problems, *Artificial Intelligence* **34**(1): 1–38.

Dechter, R. and van Beek, P. (1997). Local and global relational consistency, *Theoretical Computer Science* **173**(1): 283–308.

Fages, F., Fowler, J. and Sola, T. (1996). Experiments in reactive constraint logic programming, *Technical report*, DMI - LIENS CNRS, Ecole Normale Supérieure. Submitted for publication.

Mackworth, A. (1977). Consistency in networks of relations, *Artificial Intelligence* **8**(1): 99–118.

Mohr, R. and Henderson, T. (1986). Arc-consistency and path-consistency revisited, *Artificial Intelligence* **28**: 225–233.

Montanari, U. and Rossi, F. (1991). Constraint relaxation may be perfect, *Artificial Intelligence* **48**: 143–170.

Older, W. and Vellino, A. (1993). Constraint arithmetic on real intervals, *in* F. Benhamou and A. Colmerauer (eds), *Constraint Logic Programming: Selected Research*, MIT Press, pp. 175–195.

Saraswat, V., Rinard, M. and Panangaden, P. (1991). Semantic foundations of concurrent constraint programming, *Proceedings of the Eighteenth Annual ACM Symposium on Principles of Programming Languages (POPL'91)*, pp. 333–352.

van Emden, M. H. (1996). Value constraints in the CLP scheme, *Technical Report CS-R9603*, CWI, Amsterdam. To appear in the Constraints journal.

Van Hentenryck, P., Deville, Y. and Teng, C. (1992). A generic arc-consistency algorithm and its specializations, *Artificial Intelligence* **57**(2–3): 291–321.

DNA^{2}DNA Computations: A Potential "Killer App"?

Laura F. Landweber[1]

Department of Ecology and Evolutionary Biology

Princeton University

Richard J. Lipton[2]

Department of Computer Science

Princeton University

and

Bellcore Research

1. Introduction

Ever since Adleman's seminal paper [1] there has been a flood of ideas on how one could use DNA to compute. Lipton was the first to show that DNA could be used to solve more than just a variation of the famous travelling salesman problem [12]. Since then there have been many other papers on using DNA to solve various computational problems. [3,5,4,6,7,15]

At the top level all these papers are similar: they all attempt to use DNA computation to solve some large search problem. Since a liter of water can hold 10^{22} bases of DNA, there is the possibility that one can outperform electronic machines.

However, this is currently problematic. There are several reasons for this. First, electronic machines are very fast; moreover, they are getting faster every day. Second, there are many models of how to do DNA computations. Yet, it is unclear if any of these models will be practical. The problem is mainly that DNA technology is not perfect. DNA operations are not error free.

Finally, there is the lack of a *killer app*. A killer app is an application that fits the DNA model; cannot be solved by the current or even future electronic machines; and is *important*. The latter is critical: to be a killer app the problem must be one for which people are willing to "pay money" for solutions. To date there are no viable candidates for the killer app.

We propose a new way to use DNA computations. This way allows us to use DNA computations to solve important and potentially killer applications. The potential applications include:

(1) DNA sequencing;
(2) DNA fingerprinting;
(3) DNA mutation detection or population screening;
(4) Other fundamental operations on DNA.

The key new idea is to use DNA computation to operate on *unknown* pieces of DNA. This is a fundamental change in the way that we use DNA computation. We call these DNA^2DNA computations: DNA *to* DNA computations. This idea was first proposed in [8] and called "analog" DNA computations there.

The key idea is the following. Suppose that one has a test tube that contains multiple copies of some unknown strand X of DNA. By *unknown* we mean that we do not known

---

[1] a Burroughs Wellcome Fund New Investigator in Molecular Parasitology.
[2] Supported in part by NSF CCR-9633103 and AFOSR F49620-97-0190.

the sequence of the strand. Suppose further that we wish to compute some property of X, i.e. for some function $f()$ we wish to obtain the value of $f(X)$. The current way to do this is: (i) sequence the strand X in the laboratory; (ii) then, determine the value of $f(X)$ on a PC. The difficulty with this method is that it requires the sequencing of the strand X.

Our new idea is to avoid the expensive step of sequencing the strand X. In particular, we plan to operate as follows: We will add to the test tube certain *known* strands of DNA and use these to perform a DNA computation on X. The result of this computation will be the answer $f(X)$.

The advantage of this method is that it avoids the sequencing step. Our hope is that this direct method of computing with unknown strands of DNA could be the key to finding "killer app's".

There is one huge advantage to our approach: since the problems we are solving are not digital, there is no way that electronic machines can compete. It's not that DNA based computation is faster, but that there is no way for electronic computers to do the the problems at all. One way to say this dramatically is that there is no place on a PC to "pour" in the unknown test tube of DNA. Without input, the problem cannot be solved at all on a PC.

Our method is based on a new transformation that allows us to "encode" an unknown piece of DNA. All of the DNA computations to date use special redundant codes. It is critical that the DNA be redundantly encoded. Without such a coding the computations cannot be performed. Indeed the main contributions of [1,12] were the construction of methods for creating and managing such codes.

Of course naturally occuring DNA is *not* coded in this redundant manner. This is a major roadblock: without codes the methods of DNA computation do not apply. However, we propose a method that allows us to transform DNA. This transformation causes the DNA to be *re-coded* into any redundant code that we choose.

There are many advantages to this re-coding. Mainly, it is now possible to apply all of the "tricks" of DNA computation to problems that involve unknown DNA. Since the DNA is coded the way that we choose we can operate on it much more freely. For example, one important application of this method is the following: (Note, the *exact* theorem statements are in section 3.)

Theorem: *Suppose that X and Y are unknown strands in distinct test tubes. Then, it is possible to check whether or not $X = Y$ in $O(\log(n))$ bio-steps where both strands are at most length n.*

Note, we mean that we test *exactly* whether or not X and Y are equal: the method will discover if they differ in even *one* base. Further, this is only a simple example of a more general type of theorem:

Theorem: *Suppose that $X^{(1)}, \ldots, X^{(k)}$ are unknown strands of length at most n that are in distinct test tubes. Then, in $O(\log(n))$ bio-steps we can compute the value of $F(X^{(1)}, \ldots, X^{(k)})$ where $F()$ is an NC^1 function.*

It is important to point out that our results avoid one of the key difficulties that face "classic" DNA computations. By "classic" we mean DNA computations that attempt to do purely digitial problems. The advantage is that our results are much more error tolerant. The reason is that in classic DNA computations there is often "one" strand that the experimenter seeks to find. In our new type of DNA computations, there are many many copies. Thus, small error rates or partial rates of completion for some of the operations should not be a problem.

We prove these results by combining our re-coding methods with a generalization of the pretty simulation method of Ogihara and Ray [13]. Other methods could be used but their method is perfect for our needs. Note, in [2] there is a criticism of [13] for using an unreaslistic model. We feel that this criticism is interesting but misses the essential point. They feel that the cost of the pour operation is not correctly included in [13]. The answer seems to be two-fold: First, even if the methods are linear in "pour" it's so fast that essentially the time is still logarithmic. Second, one can imagine using robots so that the pours can actually be done all at once.

2. Model

In this section we introduce our model of DNA computations. It is related to, but fundamentally different from, the models used in papers on classic DNA computations [3,5,4,6,7,15]. The key point is that in DNA²DNA computations the operations need not work perfectly. For example, we will only require assumptions about how selective DNA is when single strands anneal/ligate together. This is a major advantage of DNA²DNA computations. Of course the hope is that this weakening in the required models will make DNA²DNA computations really work in the laboratory. (Note, we are just beginning experiments in Laura Landweber's laboratory at Princeton University that we hope will show that this is correct.)

All our computations are described in terms of operations that are performed on *test tubes*. The state of a test tube is, thus, a critical concept. At any time a test tube will contain a multi-set of different pieces of DNA. Some pieces will be single strands, some double strands and others more complex structures. Clearly, in order to describe mathematically such a state, we need to supply the following information:

(*1*) The types of pieces of DNA that are in the test tube;
(*2*) The total number of pieces that are in the test tube;
(*3*) The number of pieces of each type that are in the test tube.

We will use string terminolgy to describe single strands of DNA. More precisely, we will identify strings S over the alphabet $\{A, T, C, G\}$ with the single strand of DNA of the form:

$$5' - S_1, \ldots, S_n - 3'.$$

Also by the *Watson-Crick* complement of S we will mean the string that is the reverse of S with each element changed into its complement, i.e. "A" with "T" and "C" with "G".

Suppose that a test tube T only contains single strands of DNA: note, this is an important special case. Clearly, its mathematical definition requires that we supply the following:

(*1*) A collection of that correspond to the single strands in T, i.e. $S^{(1)}, \ldots, S^{(k)}$;
(*2*) A integer M that is the total number of strands in T;
(*3*) A collection of frequencies $q_1, \ldots, q_k$ so that the ith strand $S^{(i)}$ occurs $q_i M$ times where $q_1 + \ldots + q_k = 1$.

One of the key insights about DNA²DNA computation is that we can simplify this definition: *we do not need to supply M*. That is we need not worry about the exact number of strands that are in the test tube. We need only to keep track of the frequency of each strand.

This is an important point about the difference between some classic DNA computations and DNA²DNA computations. In classic computations the number of types of strands k is the same order of magnitude as the total number of strands M. This is because in classic computations each strand is performing a separate computation: we need to have both k and M as large as possible.

On the other hand, in DNA²DNA computations k will often be relatively small. For example, $k = 1,000$ and $M = 10^{15}$ are quite reasonable parameters. Since M/k is so large we can essentially ignore the exact value of M. Of course it is critical for all DNA computations that there be enough material available to make the operations feasible. Note, if in some situation M became too small, then a standard "trick" is to use PCR to increase the number of total strands and thus restore M to a large enough value.

In summary, for the rest of the paper we will only supply the frequencies of each piece not the total number of pieces in describing a test tube. A common situation is the following:

Say that a test tube T contains $S^{(1)}, \ldots, S^{(k)}$ in *equal amounts* provided T contains the same number of copies of each the given single strands of DNA.

Now let us turn to consider the class of operations that we require: (Each is a *bio-step* in our computations.)

(*1*) *Cut.* This operation cuts or cleaves double strands of DNA at a certain pattern. This is done by using a restriction enzyme.

(*2*) *Gel Separate.* This operation uses denaturing polyacrylamide gel electrophoresis to separate DNA molecules by length.

(*3*) *Anneal.* This operation allows single strands to form double strands based on Watson-Crick pairing, i.e. "A" with "T" and "C" with "G".

As stated earlier we do *not* assume that each operation works perfectly. Let us now discuss the exact error model that we assume. Let $\tau > 0$ be a fixed small constant: we expect that it will be smaller than 10^{-3}. We will use τ to bound the error rate of all the operations that we perform. Note, we really have a collection of τ's: one for each operation. However, to avoid statements that are overly complex we will lump all the error rates together. Of course, one can in principle unravel this and get the exact dependence on each error rate, if one needs finer resolution.

Now let us turn to discuss the error rates of each type of operation. A cut can fail in two basic ways. First, a pattern that should be cut may not be cut. Second, some place that does not match the pattern may be incorrectly cut. We assume that at least $1/2$ of the correct sites are cut; we assume that at most τ of the incorrect ones are cut. Note, the action of most restriction enzymes are usually stated in terms of how long they take to cut $1/2$ of the population. One can increase this amount by either adding more enzyme or increasing the time of incubation.

Next let us discuss the separation of DNA by length. As in other papers we will arrange things so that no separation is ever required to separate strands that are too close in length. Further, we will arrange it so that the lengths are quite short. Gel methods work best for very short lengths. For lengths below several hundred one can tell i from $i+1$. We will assume that at least $1/2$ of the strands of the given length are correctly extracted; we also assume that at most τ strands of the wrong length are also extracted. Note, this means that we do *not* assume that strands are not lost in performing the gel. As long as approximately $1/2$ of the correct strands are not lost the operation fits our model.

Finally, we must discuss the error model used for annealing. This is the most complex. There are two cases. The first case is the "far-apart" case. In this case the single strands either exactly Watson-Crick bond or are such that they agree in at most $1/4$ of their positions. Furthermore we assume that the length of the match is above a fixed threshold. In this case we assume that at least $1/2$ of the correct pairs form; we assume that effectively *none* of the incorrect ones form. Note, that we are implicitly asssuming that there are enough of the DNA strands for these reactions to actually take place. However, we have already stated that there will always be "enough" material.

The second case is the "near" case. In this case, the correct and incorrect strands agree in more than $1/4$ of their positions. Now we can no longer assume that incorrect pairs will not form. For example, if two strands α and β are Watson-Crick complements except for one position, then they will likely bind each other. This is even more likely if the one place they differ is at the end. In this case they will bind almost as well as a perfectly matched pair. Thus, in this case we cannot assume that the rates of formation are vastly different for the correct and the incorrect case.

In this case we make the following weak assumption. We assume only that the rate or probability for a perfect match is strictly bigger than that of a partial match. This is themodynamically reasonable: More matches will be better. We make no assumption about the exact difference. However, it is important to make a small assumption that the gap is at least $\delta > 0$ for a fixed small value of δ.

In summary, the error model is as follows:

Operation		Correct	Incorrect
cut		1/2	τ
gel separate		1/2	τ
anneal	far-apart	1/2	0
anneal	near	p_1	p_2

where all that is claimed is that in each case, p_1 is strictly larger than p_2 by an amount that is at least δ. This is analogous to a selection coefficient. The big surprise, perhaps, is that we can assume so little about annealing accuracy in the near case. It is unclear that such a weak assumption is enough to get any results. However, it turns out that it is enough: how is the subject of the next section.

3. Re-Coding of Unknown DNA

In this section we will show how to re-code an unknown strand of DNA X by one that is coded as we wish. Suppose that we have a test tube that contains X. We must show how to create a new test tube that contains a *re-coded* version of the strand X. We will do this in two stages. Since our operations are only approximate we will *not* be able to do this exactly. Rather, we will be able to construct a test tube that "approximates" the desired one.

Definition: Let test tube T contain the single strands $S^{(1)}, \ldots, S^{(k)}$ with frequencies $q_1, \ldots, q_k$ and let T' contain $S^{(1)}, \ldots, S^{(k)}$ with frequencies $q'_1, \ldots, q'_k$. Then, say that T ϵ-*approximates* T' provided $\sum_{i=1}^{k} |q_i - q'_i| < \epsilon$.
Next a string definition:

Definition: A string α is *in* the string S provided α occurs as a consecutive substring of S, i.e. that for some i,

$$\alpha = S_i, \ldots, S_{i+l}$$

where l is the length of the string α. A string α is *in* the strand X provided α is in X.

There are two "tricks" that allow us to improve the quality of our basic operations. The first is that we can repeat a length separation multiple times. Clearly, less "correct" DNA is selected but also less "incorrect" DNA gets by. For example, the following is useful:

Lemma 1: *Suppose that test tube T contains equal amounts of $S^{(1)}, \ldots, S^{(k)}$ where each string is a different length. Then, for any $\epsilon > 0$ and each i we can in $\log_{1/\tau}(1/\epsilon)$ bio-steps construct a test tube that is an ϵ-approximation to the test tube that only contains $S^{(i)}$.*

We do not know in general how to get a similar lemma for cuts. Repeating a cut, for example, will cause more correct material to be cut but will also cut more incorrect material. However, there is a very important case where we can essentially do this. Suppose that we have a test tube T and we plan to first apply a cut step and then a separation step. If all the pieces from the cut have the same length then we can apply Lemma 1 after the cut. The effect of this will be that incorrectly cut material will be filtered out. In a sense we have made the cut appear to have an error rate of ϵ rather than τ.

Theorem 1: *Suppose that a test tube contains the unknown strand X. Also let $l = O(\log(n))$ and let $f()$ be a function that is defined on length l strings. Then, for any $\epsilon > 0$ in $O(\log(n) + \log(1/\epsilon))$ bio-steps we can create a test tube T' that is an ϵ-approximation to the test tube that contains in equal amounts the strands that correspond to $f(\alpha)$ where α ranges over the length l consecutive substrings of X.*

Thus, we can go from a test tube that contains one string to another that contains re-coded versions of all the consecutive pieces of X. This is not enough but is an important first step. We describe the complete method elsewhere [8]. (Here we will only sketch the proofs of the thorems. The full proofs will be in the final paper.)

Proof of Theorem 1: Recall that T is a test tube that contains many copies of the strand X. Our plan is to add to this test tube additional pieces of DNA. These will be from a set we call the *probe* set. In particular, assume that we have already created the following probe set in another test tube. The probe set will contain for each string α of length l, the strand that corresponds to the Watson-Crick complement of $\alpha f(\alpha)$. We then will anchor the strands of the test tube T to a surface of another test tube T'. Then, add the probe set and allow them to anneal. Now, wash off the excess. Next elute the bound probes from the solid support. Then, allow them to re-attach. Repeat these steps, i.e. perform a molecular selection procedure and call this collection of probes T'.

The result is that T' will contain those probes that survived the repeated washing steps. We claim that these will be almost totally the correct ones, i.e. a probe $\alpha f(\alpha)$ that survives will have with high probability α in X.

Let us calculate the survival probability in the correct and the incorrect case. In the correct case the probability that a probe survives is p_1^m where m is the number of iterated cycles of selection by binding; in the incorrect case it is p_2^m. Here $p_1 > p_2 + \delta$.

Note, this assumes that only the α part of the probes are available for bonding. We can easily arrange this in a number of ways. The simplest is to add additional material that block the $f(\alpha)$ part of the probes. We assume that this is done. See [8] for details.

Since p_2 is bounded below p_1 for $m = O(\log(n) + \log(\epsilon))$ we will expect that the fraction of incorrect probes that survive is at most ϵ.

Finally, we can arrange the probes so that we can cut away the $f(\alpha)$ part. Then, provided we have arranged that the length of the $f(\alpha)$ is much larger than l, a separation yields the desired test tube of DNA. In order to the error rate low we use Lemma 1 to repeat the separation. ∎

We plan now to use Theorem 1 to allow us to re-code the whole of the unknown X. Note, however, that already the transformation is quite useful. For example, in [8] we show how it can potentially be used to increase the power of "DNA chips". These chips attempt to sequence unknown DNA via hybridization.

Our plan is to use Theorem 1 to build a special test tube of DNA. It will contain pieces for each l consecutive substring of X. Moreover, these pieces will be able to anneal together to form the encoding of X. The key is that this method will only work on "reasonable" X's. The problem is that Theorem 1 only allows us to work with short parts of X. So that we need X to have the property that it's determined by it's short pieces. If X is not, then this approach cannot succeed.

Definition: Say that a string X is *l-determined* provided that X is uniquely reconstructible from its l long subsequences.

Note, if X is random then certainly for l about $2\log(n)$ all the l pieces are likely to be unique; in this case X is trivially l-determined. However, subsequences can be repeated and X can still be l-determined. This notion is already in use in DNA chips [16]. The method of sequencing via hybridization only can work for sequences that are l-determined for a small value of l. Also, we do not require that it is easy to find X from its pieces, only that it is possible.

Let us fix two functions $h_L(\beta, i)$ and $h_R(\beta, i)$ where β is a string of length $l-1$ and i is from 1 to n. These functions are the "re-coding" or "hashing" functions that we will use. We assume that they hash values so that distinct values in their range agree in at most $1/4$ of their positions. Thus, one can think of them as assigning a hash to a length $l-1$ string and an index i.

Theorem 2: *Suppose that T is a test tube with unknown DNA X that is l-determined for $l = O(\log(n))$ where n is its length. Suppose also that $h_L()$ and $h_R()$ are hash functions as above. Then, for any $\epsilon > 0$ in $O(\log(n) + \log(\epsilon))$ bio-steps we can form the test tube T' that is an ϵ-approximation to the test tube that contains in equal amounts the strands of the form:*

$$\cdots h_L(\beta[i], i) h_R(\beta[i], i+1) \cdots$$

where $\beta[i]$ is the ith substring of X of length $l-1$.

Proof of Theorem 2: Again we plan only to sketch the proof. The basic idea is as follows. Suppose that $\beta[i]$ is the $l-1$ long substring of X starting at index i. Then, we plan to put into a test tube the following pieces of DNA for each index i in the range 1 to n:

(*1*) if i is odd, $h_L(\beta[i], i) h_R(\beta[i+1], i+1)$;
(*2*) if i is even, the Watson-Crick complement of $h_L(\beta[i], i) h_R(\beta[i+1], i+1)$.
We do this by appealing to Theorem 1.

If Theorem 1 were perfect, then because we are in the far-apart case only the correct strands would form. The key is that as long as X is l-determined one can prove that no other strand will form that is of the correct length. Then, we could finish up the proof by using Lemma 1 to perform a length separation.

However, Theorem 1 only creates an approximation to the test that contains the pieces according to (1) and (2). Thus, we need to take into account the fact that ϵ of the test tube is incorrect.

Consider how the pieces anneal and ligate together to form the one correct long strand of the correct length. If there are errors sometimes incorrect pieces will asemble. Call these "miracle steps". The point is that these occur but the frequency is at most ϵ. Thus, the expected fraction of ways to assemble 2 correct pieces together with a miracle step is at most $\binom{n}{2}\epsilon$. In general with l miracle steps it is $\binom{n}{l}\epsilon^l$. An easy calculation, then shows that the fraction of incorrect strands allowing miracle steps is at most $O(n\epsilon)$. Thus, for ϵ small enough this will prove the theorem. ∎

Essentially, the proof of this theorem uses the same method Adleman used in his original paper [1]. The main difference is that we do not place the pieces explicitly into the test tube. Rather they are generated by the action of the molecular selection step of Theorem 1. Another key difference is that we allow errors.

4. Applications

Once DNA is re-coded the full power of DNA computation can be used to solve many interesting problems. In particular, we can now show that any reasonable computation can be efficiently performed on unknown DNA.

Theorem 3: *Suppose that $X^{(1)}, \ldots, X^{(k)}$ are unknown strands of length at most n that are in distinct test tubes. Moreover, assume that they are $O(\log(n))$-determined. Then, in $O(\log(n))$ bio-steps we can compute the value of $F(X^{(1)}, \ldots, X^{(k)})$ where $F()$ is an NC^1 function.*

Proof of Theorem 3: We will only sketch the main ideas. Our plan is to use Theorem 2 to construct a test tube that contains the information required by [13] to compute F. Think of the input as kn bits and select a coding method as required by [13]. Then, by Theorem 2 we can replace each of the test tubes by an approximation to one that contains $X^{(j)}$ as one long strand. Then, we cut this into n pieces: one for each of its bits. We pour equal amounts of these k test tubes together. Then, we perform the operations as in [13] to compute the value of F. A key point is that while we only have an approximation to the test tube, [13] is sufficiently robust that it will still compute correctly. ∎

Note, we needed a new operation here: the ability to take test tubes $T_1, \ldots, T_k$ and create a new test tube T that contains equal amounts of material from each test tube. We claim that this is a reasonable operation that can be done again with error rate at most τ.

Corollary 4: *Suppose that X and Y are unknown strands in distinct test tubes. Moreover assume that they are $O(\log(n))$-determined. Then, it is possible to check whether or not $X = Y$ in $O(\log(n))$ bio-steps where both are at most length n.*

5. Conclusions

Before discussing whether or not these results are practical, there is a generalization that should be mentioned. The main one is the case of *partially unknown* DNA. In many interesting situations the DNA in a test tube is not unknown. Rather we know that it is a equal to a known X_0 except for perhaps a few bases. This occurs in the case of mutation detection, for example. In this case the same theorems of section 4 can apply. However, now the probe set can be dramatically reduced in size. The full details of this will be in the final paper.

There are a number of issues that must be solved before we can claim that these methods are practical. We view them as the start of a new direction for DNA computation. We believe that they should be viewed as an "existence" proof. That is our results are not going directly into the laboratory. However, the idea of re-coding unknown DNA and then directly computing with it, DNA^2DNA computations, are potentially important.

Some of the practical issues are the following:
 (*1*) How can we build the probe sets?
 (*2*) Can we weaken the assumption on annealing in the far-apart case?
 (*3*) Can we weaken the assumption that the DNA is l-determined?
Clearly, one cannot use DNA synthesis machines directly to build large probe sets. At least two interesting methods seem possible. For one, we may be able to use the same technology that is used to create DNA chips. The micro-robotic methods used to create these chips might be useful for generating our probe sets. For another, we may be able to exploit the structure of the probe sets. The probe sets we need are very regular sets. Indeed the following seems to be an important open problem: *Given a set of strings what is the cost in bio-steps to create a test that contains only those strings?*

Second, an important question is how far can we weaken the assumption of how far-apart strands anneal and ligate? One of the most important questions for much of DNA computation is to better model annealing and ligation. Obviously, the more realistic we are in modelling how strands mis-pair, the more practical our results will be. In particular, can we prove Theorem 2 in the case where incorrect annealing/ligations occur but with a low probability?

Finally, there is the problem of the assumption we made that the DNA is l-determined. As stated earlier, random or even approximately random strings are $l = O(\log(n))$-determined. However, there are two problems with this. First, the size of l may be logarithmic but too large for practice. Second, real DNA is not random. Can the methods of Theorem 2 be improved to handle real DNA? We are currently investigating these questions.

Acknowledgement: We thank Chris Dunworth for a number of comments about this paper. We also like to thank Dan Boneh for his help in creating he original idea of "re-coding" DNA.

References

[1] Leonard M. Adleman. "Molecular Computation of Solution to Combinatorial Problems". Science, 266:1021–1024, 1994.

[2] M. Amos, A. Gibbons, P. Dunne. "The Complexity and Viability of DNA Computations", CTAG-97001.

[3] S. Roweis, E. Winfree, R. Burgoyne, N. Chelyapov, M. Goodman, P. Rothemund, L. Adleman, "A Sticker Based Architecture for DNA Computation", draft 1996.

[4] E. Bach, A. Condon, E. Glaser and C. Tanguay, DNA Models and Algorithms for NP-Complete Problems, Proceedings of the 11th Annual IEEE Conference on Computational Complexity, 1996.

[5] D. Beaver, Molecular Computing. Penn State University Tech Report CSE-95-001.

[6] D. Boneh, C. Dunworth, R. Lipton, "Breaking DES Using a Molecular Computer", Princeton CS Tech-Report number CS-TR-489-95.

[7] D. Boneh, C. Dunworth, R. Lipton, J. Sgall, "On the Computational Power of DNA", CS-TR-499-95.

[8] D. Boneh, L. Landweber, R. Lipton. "Analog DNA Computations", unpublished manuscript, July 1996.

[9] D. Boneh and R. Lipton, "Making DNA Computers Error Resistant", Princeton CS Tech-Report CS-TR-491-95.

[10] R. Dramanac, "DNA sequence determinationn by hybridizationn: a strategy for efficient large-scale sequencing", Science, 263, 596-596, 1994.

[11] R. Karp, C. Kenyon, and O. Waarts. "Error-resilent DNA computation". In Proceedings of the 7th ACM-SIAM Symposium on Discrete Algorithms, pages 458-467. ACM Press/SIAM, 1996.

[12] Richard J. Lipton. "DNA Solution of Hard Computational Problems". Science, 268:542-545, 1995.

[13] M. Ogihara and A. Ray. "Simulating Boolean Circuits on a DNA Computer", University of Rochester Technical Report 631, August 1996.

[14] C. Papadimitrou, private communication, 1994.

[15] J. Reif, "Parallel Molecular Computation: Models and Simulations". Seventh Annual ACM Symposium on Parallel Algorithms and Architectures, 1995.

[16] M. Waterman, *Introduction to Computational Biology,* Chapman Hall, 1995.

Tilings and Quasiperiodicity

Bruno Durand

LIP, ENS-Lyon CNRS, 46 Allée d'Italie, 69364 Lyon Cedex 07, France.

Abstract. Quasiperiodic tilings are those tilings in which finite patterns appear regularly in the plane. This property is a generalization of the periodicity; it was introduced for representing quasicrystals and it is also motivated by the study of quasiperiodic words. We prove that if a tile set can tile the plane, then it can tile the plane quasiperiodically —a surprising result that does not hold for periodicity. In order to compare the regularity of quasiperiodic tilings, we introduce and study a quasiperiodicity function and prove that it is bounded by $x \mapsto x + c$ if and only if the considered tiling is periodic. At last, we prove that if a tile set can be used to form a quasiperiodic tiling which is *not* periodic, then it can form an uncountable number of tilings.

1 Introduction

Matching rules in tilings are local constraints. Thus, tile sets have been used to model atomic positions in materials defined by short-range interactions. A traditional approach is then to focus on the periodicity or quasiperiodicity properties of tilings that can be formed. This study has been revived by quasicrystals (see [7] for an overview on the subject and pertinent references such as [9]). A relation between the quasiperiodicity property and the notion of self-similarity is established in [5].

In another hand, tilings can be considered as 2-dimensional infinite words with a local constraint. For 1-dimensional structures, an overview of results concerning infinite words can be found in [12]; bi-infinite words are studied in [10, 11], and the problem of quasiperiodicity is strongly related to the study of Sturmian words (see for instance [14] — references within).

We present in this paper three main results. First, a tile set that can tile the plane can always be used to form a quasiperiodic tiling of the plane. It is surprising because the same property for periodic tilings was conjectured by Wang in 1961 (see [15]) and was proved false by his student Berger in 1966 (see [1]). Furthermore it has been proved that there exists a tile set that can tile the plane but although possible tilings are non-recursive.

To prove this first result (in Section 3), we introduce a preorder between tilings of the plane that we call an *extraction* preorder. We show that quasiperiodic tilings are exactly the minimal elements of this preorder.

We introduce also a function to measure the regularity of a quasiperiodic tiling (Section 4). We prove that a quasiperiodic tiling is periodic if and only if this function is of the form $x \mapsto x + c$. We present some open problems in this field.

Our third main result (in Section 5) is that if a tile set can be used to form a strictly quasiperiodic tiling of the plane (*i.e.* non-periodic), then it can be used to form an uncountable number of different tilings. A corollary of this result is that if a tile set is aperiodic (*i.e.* cannot be used to form a periodic tiling) then it can form an uncountable number of different tilings. To prove this we are inspired by Dolbilin in [4] to introduce a tree representation for tile sets.

In our last Section, we present a topological approach of tilings. This approach allows us to give another point of view on our results of Section 3. We have not proved yet any new result using topology but we think that this approach may be fruitful.

Due to the page limit, some proofs are ommited.

2 Preliminaries

A tile is a square with color sides. Colors belong to a finite set C. A tile set τ is a subset of C^4. All tiles have the same (unit) size. A *configuration* is a mapping from the plane $\mathbb{Z}^2$ into the tile set. We call *pattern* a partial function of finite domain from $\mathbb{Z}^2$ into the tile set. We say that a pattern *appears* in a configuration, if the configuration is an extension of the image of this pattern by a shift. A *tiling of the plane* is a configuration in which all pairs of adjacent sides have the same color. Notice that it is not allowed to turn tiles.

The *tiling problem* consists of a tile set as input, and the question is whether it can be used to tile the plane. It was formulated by Wang in 1961 [15] for some logical purposes: a tile set can be reduced into some formula such that the formula is satisfiable if and only if the tile set can tile the plane. This tiling problem was conjectured decidable but was proved undecidable by Berger [1] in 1966; a simplified proof was given in 1971 by Robinson [13] (see also [2] for the consequences in logics —Hilbert's well-known *Entscheidungsproblem*).

A *periodic configuration* is formed by the juxtaposition of copies of the same rectangle. In other terms a periodic configuration should be periodic with respect to both axes. Thus, a periodic tiling is a periodic configuration which is also a tiling. This definition is justified by the following result of Wang: if a tile set can form a tiling which is periodic in only one direction, then it can form a tiling which is periodic. This property was one of the reasons why Wang conjectured that the tiling problem was decidable. The other reason was that he did not know any *aperiodic* tile sets, *i.e.* tile sets that can tile the plane but cannot form any periodic tiling. If such aperiodic tile sets did not exist, then one could decide the tiling problem by the following algorithm: try to tile a square of size n; if you cannot, then halt and answer "no", else if you can tile the square periodically, then halt and answer "yes", else add 1 to n and restart the same process. This algorithm does not halt if and only if the considered tile set is aperiodic. In the proof of Berger's theorem an aperiodic tile set is constructed with more than 20000 tiles, and in Robinson's simplified proof an aperiodic tile set containing approximatively 50 tiles is constructed. The smallest known aperiodic tile set contains 13 tiles and is due to Čulik and Kari ([3, 8]).

The *periodic tiling problem* consists of a tile set as input, and the question is whether it can be used to tile periodically the plane. It has been proved undecidable by Gurevich and his student Koriakov in 1972 [6]. They furthermore proved that you cannot recursively separate tile sets that cannot tile the plane from tile sets that can tile the plane periodically. This result is very important because in the previously mentioned reduction, tile sets that can tile the plane periodically correspond to formula having a finite model. Such reductions are called *conservative* in [2]. This book contains an appendix by ourself devoted to the proof of all these undecidability results.

It is often convenient to use other notions of tiles sets that differ slightly from above:

- one can use arrows on tiles; a tiling is considered as valid if and only if all pairs of adjacent sides have the same color, and if, for each arrow of the plane, its head points out on the tail of an arrow in the adjacent cell;
- one can replace squares by polygons of the plane and ask that two adjacent polygons neither overlap nor create holes;
- one can put a color not only on the sides of the squares put also on their corners; four corners in contact should have the same color;
- one could just assign a state (out of a finite set) to each considered cell and fix a neighborhood. The matching condition is replaced by a relation between states that should be verified in the neighborhood of each cell.

It is folklore that all these notions are equivalent: there exist transformations of tile set from one notion into another that preserve existence of valid tilings, periodicity or non-periodicity, quasiperiodicity, etc.

We could have considered tilings of the continuous plane $\mathbb{R}^2$ by polygons such as in the well known Penrose tilings. This notion of tilings is not equivalent to Wang tiles because the centers of the considered polygons may not have rational coordinates. Anyways, for these tilings, our theorems 6 and 13 are still valid if one consider that two tilings (or patterns) are equal if they can be superimposed using translations and rotations. The needed changes in the proofs are straightforward. Our study of the regularity of quasiperiodic tilings (Section 4) is slightly changed in this case: the size on a pattern — and thus quasiperiodicity functions — should be define up to a multiplicative constant.

3 Extraction and quasiperiodicity

Before defining quasiperiodic tilings, we introduce a partial preorder relation between configurations. We call this preorder the "extraction" preorder and prove that it has good properties with respect to the notion of tilings and —later— those notions of periodicity and quasiperiodicity.

3.1 Extraction

Definition 1. Let us consider two configurations c_1 and c_2. We say that c_1 is *extracted* from c_2 if and only if any pattern that appears in c_1 also appears in c_2. We denote this relation by $c_1 \prec c_2$.

Note that if $c_1 \prec c_2$, and if c_2 is a tiling of the plane, then c_1 is also a tiling of the plane. In other terms, a configuration which is extracted from a tiling is also a tiling: if c_1 had a defect, then this defect should also appear in c_2.

Let us now define what can be called a *diagonal extraction process*. We use it in order to prove the following proposition.

Proposition 2. *Assume that a sequence of patterns $(M_i)_{i \in \mathbb{N}}$ is given, that their domains increase $(dom(M_i) \subset dom(M_{i+1}))$, and that they cover the whole plane $(\bigcup_{i \in \mathbb{N}} dom(M_i) = \mathbb{Z}^2)$. Then there exist a configuration d such that any pattern that appears in d also appears in an infinite number of M_i's. If all the M_i's have been chosen in a configuration c, then the obtained configuration d is extracted from c ($d \prec c$). Furthermore, if c is a tiling, then d is also a tiling.*

Note that this diagonal extraction process is not effective; it is not an algorithmic procedure.

3.2 Quasiperiodicity

Definition 3. A *quasiperiodic* configuration is a configuration c with the following property: for all pattern M that appears in c, there exists an integer n such that M appears in all $n \times n$ squares in c.

A periodic configuration is also quasiperiodic. We call *strictly quasiperiodic* those configurations that are quasiperiodic but not periodic. The quasiperiodicity is a regularity property: a patterns that appears somewhere in a quasiperiodic configuration must appear regularly.

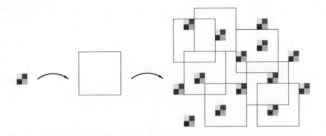

Fig. 1. Quasiperiodic configurations

If a tiling (resp. a configuration) is not quasiperiodic, then there exists a pattern in this tiling that can be associated to an infinite number of growing squares that belong to the tiling, and in which the pattern does not appear. In the sequel, we call such a pattern *critical* for the considered tiling (resp. for the configuration).

Lemma 4. *If pattern M is critical for a configuration c, then there exist at least one configuration c_M extracted from c in which the pattern M does not appear.*

Proof. Consider an infinite sequence of patterns in where M does not appear. Then make a diagonal extraction process to obtain a new configuration in which M does not appear. As this configuration is obtained by a diagonal extraction, then it is extracted from c.

Note that M is not critical in c_M since it does not appear in it.

Proposition 5. *Quasiperiodic tilings (resp. configurations) are exactly the minimal elements for the extraction preorder. More formally, c is quasiperiodic if and only if $\forall d \quad d \prec c \implies c \prec d$.*

Proof. Consider a quasiperiodic configuration c. Assume that $d \prec c$; let us prove that $c \prec d$ i.e. that any pattern of c can be found in d. Let us consider a pattern M in c; it can be found in all sufficiently large squares of c because of the quasiperiodicity hypothesis. Let us consider a square of the same size in d. As $d \prec c$, it appears somewhere in c and thus contains M. Hence M appears in d. The converse is straightforward using Lemma 4.

Theorem 6. *If a tile set admits a tiling, then it admits a quasiperiodic tiling.*

Before proving this theorem, we need to explain why the quasiperiodicity property is compatible with the extraction preorder.

Lemma 7. *If a pattern M is critical for a configuration c, and if c is extracted from a configuration d, then M is also critical for d.*

Proof. Consider d such that $c \prec d$. If M is critical for c, then it appears in c thus in d. Furthermore, the infinite family of rectangles in which M does not appear can be found in c hence in d. Thus M is critical for d.

Proof of Theorem 6. Remember that a tile set is given, that it can be used to tile the plane, and that our goal is to prove that one can form a quasiperiodic tiling of the plane with it.

Let t be a tiling of the plane using this tile set. Assume that it is not quasiperiodic. It contains some critical patterns. Among them, we consider the smallest pattern M_1: it is not difficult to define a total ordering of patterns; first order them by the size of their domain (more precisely by the size of the smallest square that contains their domain) and then by alphabetic order. Also note that the set of all patterns is countable. Using Lemma 4, we can construct a tiling $t_{M_1} \prec t$ in which M_1 does not appear. Because of Lemma 7, all critical patterns of t_{M_1} (if any) are also critical for t.

If t_{M_1} is not quasiperiodic, then we repeat this process: we choose the smallest critical pattern M_2 in t_{M_1} and obtain $t_{M_2} \prec t_{M_1} \prec t$.

If after a finite number of steps of this process, we obtain a quasiperiodic tiling, then the theorem is proved. Else, we obtain an infinite sequence of tilings (t_{M_k}) such that $\ldots t_{M_k} \prec \ldots t_{M_1} \prec t$. Let us consider now a 1×1-square in t_{M_1}, a 2×2-square in t_{M_2}, a $k \times k$-square in $t_{M_k}\ldots$ With a diagonal extraction process, we obtain a tiling d which is extracted from all the t_{M_k}'s: $d \prec \ldots t_{M_k} \prec \ldots t_{M_1} \prec t$. If this tiling had a critical pattern, then this pattern should be critical for all the t_{M_k}'s. But in the pattern ordering, one of the patterns M_k is greater than this critical patterns which contradicts our choice of the smallest possible pattern. Hence d is quasiperiodic and $d \prec c$.

Note that this proof is not constructive and uses the axiom of choice.

4 Quasiperiodicity functions

In this section, we introduce a *quasiperiodicity function* in order to measure the regularity of a quasiperiodic tiling.

Let us consider a quasiperiodic configuration. Coming back to the definition of quasiperiodicity (Definition 3 and Figure 1), it is natural to consider the function that maps a pattern to the smallest integer n such that the pattern appears in all squares of size $n \times n$. This function is not defined on those patterns that do not appear in the tiling. Since in the sequel we are only interested in upper bounds, we can restrict this function to square patterns —other patterns can be included in larger squares. Thus we can consider the maximum of this function on all patterns of size x: we map x to the minimal size of squares n in which one can find all those patterns of size x that appear in the tiling. We call it the *quasiperiodicity function* of the tiling.

Intuitively, if this function grows slowly to infinity, then the quasiperiodic tiling is rather regular, but if it grows fast, then the regularity is weak. Using this function we can characterize which quasiperiodic tilings are periodic:

Theorem 8. *A quasiperiodic tiling is periodic if and only if its quasiperiodicity function is bounded by $x \mapsto x + c$ where c is a constant.*

Proof. Let us consider a periodic tiling of period α. It is not difficult to prove that its quasiperiodicity function is bounded by $x \mapsto x + \alpha$. Such a situation is illustrated by Figure 2.

Let us consider now a quasiperiodic tiling of function bounded by $x \mapsto x+c$. Let us consider a pattern P_1 of size x_1 much larger that c. Let us consider a window of size $x_1 + c$ such that its left border is just 1 cell to the right of the left border of P_1 (see Figure 3). A copy of P_1 must appear in this window and overlaps P_1. Note that there are at most c^2 possible positions for this copy — it is essential in the rest of the proof.

Now let us consider a pattern P_2 of size $x_2 > x_1$ containing P_1. Let us consider a window of size $x_2 + c$ such that its left border is just 1 cell to the right

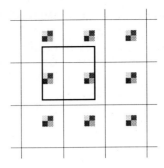

Fig. 2. The quasiperiodicity function of a periodic tiling

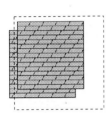 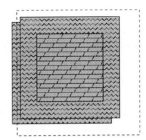

Fig. 3. The converse

of the left border of P_2. We find another copy of P_2 and there are at most c^2 possible translations from P_2 to this copy. Note that such a translation is also valid for P_1 since P_1 is embedded in P_2.

By iteration, we prove that there exist a common translation vector for all the $(P_i)_{i \in \mathbb{N}}$. Thus the tiling is periodic in at least one dimension. We use the same reasoning to find another periodicity vector: the difference is that instead of shifting the window to the right, we shift it in a direction which is orthogonal to the first periodicity vector. Note that this vector may not point exactly to the right: we can just say that it more or less points to the right.

Quasiperiodicity functions of the form $x \mapsto cx$ are not difficult to obtain. An example is any of the quasiperiodic tilings that can be formed using Robinson's aperiodic tile set (see [13] or [2] for the definition). Furthermore all these tilings have exactly the same multiplicative constant. For Penrose tilings, the study is a little more complicated since Penrose tiles cannot be placed on the vertices of $\mathbb{Z}^2$. There are several ways to measure sizes of patterns: we can consider the distance in $\mathbb{R}^2$ or the number of tiles included in it. These distances lead to different quasiperiodicity functions but each of them is bounded by a multiplicative non-zero constant times the other one. Anyway, *all* quasiperiodicity functions of quasiperiodic Penrose tilings are of the form $x \mapsto cx$.

Some important questions are still open: what are all quasiperiodicity functions that can be observed in quasiperiodic tilings? By "observed" we mean that *all* quasiperiodic tilings that can be formed with the considered tile set should be of the desired form. Otherwise it is not difficult to construct such tilings with a trivial tile set. Our last open problems are the following: is it possible to observe non-recursive quasiperiodicity functions? If a quasiperiodicity function is non-recursive and grows faster than any recursive function then the quasiperiodic tiling is regular but this regularity cannot be measured...

5 Counting

In this section, we introduce two structures: trees associated to tile sets, and trees associated to tilings of the plane; we are inspired by [4] to introduce them. We then combine these structures with the quasiperiodicity notion in order to prove the main result of this section (Theorem 13).

5.1 Trees

In the rest of this section, we only consider *valid* square patterns (the matching condition for the edges of the tiles is true inside the pattern). We call n-pattern any $2n \times 2n$ square pattern and we say that a $(n+1)$-pattern extends a n-pattern if the n-pattern *is the center* of the $(n+1)$-pattern. In other words, the $(n+1)$-pattern is obtained from the n-pattern by putting tiles around its border. Note that it is not always possible to do this because the matching condition must be true in this new border. A unique 0-pattern exists for all tile sets: it is the pattern with the empty domain.

Definition 9. The *tree associated to a tile set* τ is the tree $\mathcal{A}_\tau$ such that the vertices of $\mathcal{A}_\tau$ are n-patterns formed using the tiles of τ; the root is the 0-pattern; the children of a n-pattern node are those $(n+1)$-patterns that extend the n-pattern.

The tree such defined can be finite or infinite. All nodes are of finite degree but these degrees may not be bounded. Note that an infinite path in $\mathcal{A}_\tau$ corresponds to a tiling of the plane with τ. Conversely, let us consider a tiling of the plane with τ, and all n-patterns centered in the cell $(0,0)$. These patterns correspond to an infinite path in $\mathcal{A}_\tau$.

If the height of $\mathcal{A}_\tau$ is not bounded, using König's infinity lemma one can claim that it contains an infinite path and thus that it is possible to tile the plane (see also Proposition 2).

Definition 10. Let c be a valid tiling for tile set τ. The *tree associated to the tiling* c is the tree $\mathcal{A}_c$ such that $\mathcal{A}_c$ is the restriction of $\mathcal{A}_\tau$ containing all n-patterns of $\mathcal{A}_\tau$ that can be found in c.

All branches of the tree $\mathcal{A}_c$ are infinite since a pattern that appears somewhere in c can always be extended. Any infinite path of $\mathcal{A}_c$ corresponds to a tiling that can be extracted from c. Thus we obtain the following proposition:

Proposition 11. *Let c and d be two tilings of the plane with τ. $\mathcal{A}_c \subset \mathcal{A}_d$ if and only if $c \prec d$. If c is quasiperiodic, and if there exist d such that $\mathcal{A}_d \subset \mathcal{A}_c$, then $\mathcal{A}_d = \mathcal{A}_c$.*

Another interpretation of the previous proposition is the following: one can restrict $\mathcal{A}_c$ into some $\mathcal{A}_d$ if and only if c is not quasiperiodic.

5.2 Periodicity and quasiperiodicity

Let us now explain the difference between the tree associated to a periodic tiling and the tree associated to a strictly quasiperiodic one. Let us call *chain* of a tree an infinite path in the tree in which every node has exactly one child; the starting node of the chain is a node of the tree (usually not the root).

Proposition 12. *If c is quasiperiodic and if $\mathcal{A}_c$ contains a chain, then c is periodic.*

Proof. Consider the starting node of the chain —more precisely, the pattern M that is associated to it. There is no branching on this node and below hence if the pattern M appears in c centered in $(0,0)$ and in (i,j), then (i,j) is a periodicity vector of c. As c is quasiperiodic, the pattern c appears in all sufficiently large regions of c. Hence we can find 2 periodicity vectors for c of different directions; c is periodic.

Now we can present our main theorem. Its proof is easy with the help of the previous properties.

Theorem 13. *If a tile set can be used to form a strictly quasiperiodic tiling of the plane, then it can form an uncountable number of different tilings.*

First remark that this result is unchanged if we consider that two tilings that can be superimposed are equal. In this case, one can transform one of the tilings into the other by a translation. The set of translations is countable hence the theorem is still valid.

Proof. Let c be a strictly quasiperiodic tiling of the plane. $\mathcal{A}_c$ does not contain any chain otherwise c would be periodic (Proposition 11 et 12). Thus $\mathcal{A}_c$ contains an uncountable number of infinite paths. We can associate to each of these paths a tiling if we consider that all the patterns of the path are centered in the origin. Two different paths are associated to two different tilings thus the number of different tilings that can be formed is not countable.

Note that the uncountable set of tilings that is obtained in this proof consist of quasiperiodic tilings that can be mutually extracted; all these tilings can be obtained from c by extraction.

A corollary of this result is that one cannot separate quasiperiodic tilings with any computing device (computing devices usually belong to countable sets).

6 Topology

We present in this section another approach to tiling problems. This approach is based on the topological properties of the set of configurations.

Let us endow a tile set τ with the discrete topology for which all subsets are open. A configuration is a mapping of the plane $\mathbb{Z}^2$ into the tile set. Thus, the set of all configurations $\tau^{\mathbb{Z}^2}$ is a countable product of sets that we endow with the *product topology:* an open subset of $\tau^{\mathbb{Z}^2}$ is a union of finite intersections of sets of the form $\mathcal{O}_{i,a} = \{c \in \tau^{\mathbb{Z}^2},\ c(i) = a\}$.

In this topological approach, the notion of patterns is very natural since they correspond with basic open sets. More precisely, we can define a *basic open set* associated to a pattern as the set of all configurations equal to the pattern on its domain:

$$\mathcal{O}_p = \left\{c \in \tau^{\mathbb{Z}^2},\ c\big|_{\text{domain}(p)} = p\right\}.$$

Note that $\mathcal{O}_p$'s (and $\mathcal{O}_{i,a}$'s which are special $\mathcal{O}_p$'s) are both open and closed: their complements are finite union of the $\mathcal{O}_{p'}$ where $\text{domain}(p) = \text{domain}(p')$ and $p \neq p'$. Any open set $\mathcal{U}$ can be written as a union of basic open sets:
$$\mathcal{U} = \bigcup_{p \text{ pattern}} \mathcal{O}_p.$$

Proposition 14. $\tau^{\mathbb{Z}^2}$ *is a compact metric space.*

We shall use very often in the rest of this section the compactness of $\tau^{\mathbb{Z}^2}$ and more precisely the compactness of the set of tilings that can be formed using τ. Let us denote by T_τ this particular subset of configurations (which can be empty).

Proposition 15. *Let τ be a tile set. The subset T_τ of $\tau^{\mathbb{Z}^2}$ consisting of tilings of the plane by τ is compact.*

Furthermore, our process of diagonal extraction (Proposition 2) can be seen as a consequence of the compactness and of the shift invariance of T_τ.

Now let us interpret our relation of extraction (see Definition 1) in topological terms. To do that, let us consider the horizontal and vertical shifts σ_h and σ_v. Let us define $\Gamma(c)$ as the topological closure of the set of all images of c by any shift. It is natural to construct such a set since we tend to consider that two configurations that can be superimposed are the same. In the following formal definition, the topological closure is denoted by an over line:

$$\Gamma(c) = \overline{\bigcup_{i,j \in \mathbb{Z}} \{\sigma_h^i \circ \sigma_v^j(c)\}}.$$

Proposition 16. *Our relation of extraction corresponds exactly to the inclusion of our sets $\Gamma(c)$. More formally the following properties are equivalent:*
(a) $c_1 \prec c_2$,
(b) $c_1 \in \Gamma(c_2)$,
(c) $\Gamma(c_1) \subset \Gamma(c_2)$.

Note that if $\Gamma(c_1) \subset \Gamma(c_2)$ and if c_2 is a tiling then c_1 is also a tiling. It can be interpreted as a *monotonicity* property of tilings.

Let us come back now to the quasiperiodicity. We obtain from Section 3.2 that q is quasiperiodic if and only if $\Gamma(q)$ is minimal for the inclusion relation among all $\Gamma(c)$'s. Let us come back to our Theorem 6: "if a tile set can tile the plane, then it can be used to form a quasiperiodic tiling of the plane". In our context, it corresponds to the existence of a minimal $\Gamma(q)$ among all $\Gamma(c)$'s corresponding to tiling. Assume that it is possible to tile the plane; then using the monotonicity property of tilings and Zorn's lemma, we obtain the existence of a quasiperiodic tiling.

We do not know how to prove our combinatorial theorem (Theorem 13 of Section 5), or to interpret quasiperiodicity functions of Section 4 using only topological arguments.

References

1. R. Berger. The undecidability of the domino problem. *Memoirs of the American Mathematical Society*, 66, 1966.
2. E. Börger, E. Grädel, and Y. Gurevich. *The classical decision problem*. Springer-Verlag, 1996.
3. K. Čulik. An aperiodic set of 13 Wang tiles. *Discrete Mathematics*, 160:245–251, 1996.
4. N. Dolbilin. The countability of a tiling family and the periodicity of a tiling. *Discrete and Computational Geometry*, 13:405–414, 1995.
5. B. Durand. Self-similarity viewed as a local property via tile sets. In *MFCS'96*, number 1113 in Lecture Notes in Computer Science, pages 312–323. Springer Verlag, 1996.
6. Y. Gurevich and I. Koriakov. A remark on Berger's paper on the domino problem. *Siberian Journal of Mathematics*, 13:459–463, 1972. (in Russian).
7. K. Ingersent. *Matching rules for quasicrystalline tilings*, pages 185–212. World Scientific, 1991.
8. J. Kari. A small aperiodic set of Wang tiles. *Discrete Mathematics*, 160:259–264, 1996.
9. L. S. Levitov. *Commun. Math. Phys.*, 119(627), 1988.
10. M. Nivat and D. Perrin. Automata on infinite words. volume 192 of *Lecture Notes in Computer Science*. Springer, 1985.
11. M. Nivat and D. Perrin. Ensembles reconnaissables de mots biinfinis. *Canadian Journal of Mathematics*, 38:513–537, 1986.
12. J-E. Pin and D. Perrin. Mots infinis. (to appear) LITP repport 9340, 1993.
13. R.M. Robinson. Undecidability and nonperiodicity for tilings of the plane. *Inventiones Mathematicae*, 12:177–209, 1971.
14. P. Séébold. On the conjugation of standard morphisms. In *MFCS'96*, volume 113 of *Lecture Notes in Computer Science*, pages 506–516, 1996.
15. H. Wang. Proving theorems by pattern recognition II. *Bell System Technical Journal*, 40:1–41, 1961.

Enumerative Sequences of Leaves in Rational Trees

Frédérique Bassino[1] and Marie-Pierre Béal[2] and Dominique Perrin[1]

[1] Institut Gaspard Monge, Université de Marne-la-Vallée
[2] Institut Gaspard Monge, Université Paris 7 et CNRS

http://www-igm.univ-mlv.fr/~{bassino,beal,perrin}

Abstract. We prove that any IN-rational sequence $s = (s_n)_{n \geq 1}$ of nonnegative integers satisfying the Kraft strict inequality $\sum_{n \geq 1} s_n k^{-n} < 1$ is the enumerative sequence of leaves by height of a rational k-ary tree. Particular cases of this result had been previously proven. We give some partial results in the equality case.

1 Introduction

This paper is a study of problems linked with coding and symbolic dynamics. The results can be considered as an extension of the old results of Huffman, Kraft, McMillan and Shannon on source coding. We actually prove results on rational sequences of integers that can be realized as the enumerative sequence of leaves in a rational tree.

Let s be an IN-rational sequence of nonnegative numbers, that is a sequence $s = (s_n)_{n \geq 1}$ such that s_n is the number of paths of length n going from an initial state to a final state in a finite multigraph or a finite automaton. We say that s satisfies the Kraft inequality for a positive integer k if $\sum_{n \geq 1} s_n k^{-n} \leq 1$.

A rational tree is a tree which has only a finite number of non-isomorphic subtrees. If s is the enumerative sequence of leaves of a rational k-ary tree, then s satisfies Kraft's inequality for the integer k.

In this paper, we study the converse of the above property. Consider for example the series $s(z) = \frac{3z^2}{1-z^2}$. We have $s(1/2) = 1$ and we can obtain s as the enumerative sequence of the tree of the figure below associated with the prefix code $X = (aa)^*(ab + ba + bb)$ on the binary alphabet $\{a, b\}$. We dont know however if the same can be done for the series $s(z) = z^2(\frac{1}{1-z^2} + \frac{2}{1-2z^3})$.

Fig. 1. Tree associated to $3z^2(z^2)^*$

Known constructions allow one to obtain a sequence s satisfying Kraft's inequality as the enumerative sequence of leaves of a k-ary tree, or as the enumerative sequence of leaves of a (perhaps not k-ary) rational tree. These two constructions lead in a natural way to the problem of building a tree both rational and k-ary. This question was already considered in [9], where it was conjectured that any ℕ-rational sequence satisfying Kraft's inequality is the enumerative sequence of leaves of a k-ary rational tree.

In this paper, we prove this conjecture in the case where the sequence satisfies Kraft's inequality with a strict inequality, and we give some partial results in the equality case. For example, we state the following weaker property for such a sequence: If s is an ℕ-rational sequence of nonnegative numbers satisfying Kraft's equality, then there is a positive integer m such that $ms = \sum_{1 \leq i \leq m} r_i$, where each r_i is the enumerative sequence of the leaves of a k-ary rational tree.

Proofs and algorithms used to establish the results are based on automata theory and symbolic dynamics. In particular, we use the state splitting algorithm which has been introduced by R. Adler, D. Coppersmith and M. Hassner in [1] to solve coding problems for constrained channels by constructing finite-state codes with sliding block decoders. This was partly based on earlier work of B. Marcus in [7].

A variant of the problem considered here consists in replacing the enumerative sequence of leaves by the enumerative sequence of all nodes. Soittola ([11]) has characterized the series which are the enumerative sequence of nodes in a rational tree. The problem of a similar characterization for rational k-ary trees remains open in the general case.

In [9], a particular case is treated. It allows to solve the problem for the enumerative sequence of leaves in the equality case under the additional assumption of a unique pole of minimal modulus.

The paper is organized as follows. We first give basic definitions and properties of rational objects, sequences and trees. We then give some definitions coming from the theory of symbolic dynamics. We define the notions of state splitting, approximate eigenvector and recall the algorithm of [1]. In section 3, we establish the announced results and give examples for the constructions.

2 Definitions and background

2.1 Rational sequences of nonnegative numbers

We denote by G a directed graph with E as its set of edges. We actually use multigraphs instead of ordinary graphs in order to be able to have several distinct edges with the same origin and end. Formally a multigraph is given by two sets E (the edges) and V (the vertices) and two functions from E to V which define the origin and the end of an edge. An edge in a multigraph going from p to q will be noted (p, x, q) where $x \in \mathbb{N}$. This is equivalent to number the edges going from p to q in order to distinguish them. We shall always say "graph" instead of "multigraph".

In this paper, we consider sequences of nonnegative numbers. Such a sequence $s = (s_n)_{n \geq 0}$ will be said to be ℕ-*rational* if s_n is the number of paths of length n going from a state in I to a state in F in a finite directed graph G, where I and F are two special subsets of states, the initial and final states respectively. We say that the triple (G, I, F) is a *representation* of the sequence s.

This definition is usually given for the series $\sum_{n\geq 0} s_n z^n$ instead of the sequence s. Any IN-rational sequence s satisfies a recurrence relation with integer coefficients. However, it is not true that a sequence of nonnegative integers satisfying a linear recurrence relation is IN-rational. An example can be found in [5] p. 93.

A well known result in automata theory allows us to use a particular representation of an IN-rational sequence s. One can choose a representation (G, i, F) of s with a unique initial state i and such that :

- no edge is coming in state i
- no edge is going out of any state of F.

Such a representation is called a *normalized representation*. Moreover, it is possible to reduce to one state the set of final states (see for example [10] p. 14).

We now give some basic definitions about trees. A *tree* T on a set of nodes N with a root r is a function $T : N - \{r\} \longrightarrow N$ which associates to each node distinct from the root its father $T(n)$ in such a way that, for each node n, there is a nonnegative integer h such that $T^h(n) = r$. The integer h is the height of the node n. A tree is k-ary if each node has at most k sons. A leaf is a node without son. We denote by $l(T)$ the enumerative sequence of its leaves by height, that is the sequence of numbers s_n, where s_n is the number of leaves of T at height n. A tree is said to be *rational* if it admits only a finite number of non isomorphic subtrees. If T is a rational tree, the sequence $l(T)$ is an IN-rational sequence.

The sequence $s = l(T)$ of a k-ary tree is the length distribution of a prefix code over a k-letter alphabet. The associate series $s(z) = \sum_{n\geq 1} s_n z^n$ satisfies then Kraft's inequality : $s(1/k) \leq 1$. We shall say that Kraft's strict inequality is satisfied when $s(1/k) < 1$. The equality is reached when each node of the tree has exactly zero or k sons. Conversely, the McMillan construction establishes that for any series s satisfying Kraft's inequality, there is a k-ary tree such that $s = l(T)$. Moreover, if the series satisfies Kraft's equality, then the internal nodes will have exactly k sons. But the tree obtained is not rational in general.

It is also easy to see that an IN-rational sequence is the enumerative sequence of the leaves of a rational tree. A normalized representation can be used to do that by "developing" the tree. The root will correspond to the initial state of the graph. If a node of the tree at height n corresponds to a state i in the graph which has r outgoing edges ending at states $j_1, j_2, \ldots, j_r$, it will admit r sons at height $n+1$, each of them corresponding respectively to the states $j_1, j_2, \ldots, j_r$ of the graph. The leaves of the tree will correspond to the final states of the normalized representation. The maximal number of sons of a node we get is then equal to the maximal number of edges going out of any state of the graph of this representation.

If s satisfies Kraft's inequality, the above construction does not lead in general to a k-ary rational tree. The aim of this paper is to get a k-ary rational tree T such that $s = l(T)$. This result was conjectured in [9]. We solve it for all IN-rational sequences satisfying Kraft's strict inequality and give a weaker result for the equality case.

2.2 Approximate eigenvector and state splitting

Let s be an $\mathbb{N}$-rational sequence and let (G, i, F) be a normalized representation of s. If we identify the initial state i and all final states of F in a single state still denoted i, we get a new graph denoted $\overline{G}$, which is strongly connected. The sequence s is then the length distribution of the paths of first returns to state i, that is of finite paths going from i to i without going through state i. Using the terminology of symbolic dynamics, the graph $\overline{G}$ can be seen as an irreducible shift of finite type (see, for example, [3], [4] or [6]).

We denote by M the adjacency matrix associated to the graph $\overline{G}$, that is the matrix $M = (m_{ij})_{1 \leq i,j \leq n}$, where n is the number of nodes of $\overline{G}$ and where m_{ij} is the number of edges going from state i to state j. By the Perron-Frobenius theorem (see [6]), the positive matrix M associated to the strongly connected graph $\overline{G}$ has a positive eigenvalue of maximal modulus denoted by λ, also called the spectral radius of the matrix. Actually, λ only depends on the series s, $1/\lambda$ is the minimal modulus of the poles of $\frac{1}{1-s}$. The dimension of the eigenspace of λ is equal to one. There is a positive eigenvector (componentwise) associated to λ. Moreover, if there is a positive eigenvector associated to an eigenvalue ρ, then $\rho = \lambda$.

When λ is an integer, the matrix admits a *positive integral* eigenvector. When $\lambda < k$, where k is an integer, the matrix admits a *k-approximate* eigenvector, that is, by definition, a positive integral vector $\mathbf{v}$ with $M\mathbf{v} \leq k\mathbf{v}$.

For example the left side of the figure below gives a representation (G, i, F) of the serie $s(z) = \frac{3z^2}{1-z^2}$, and the right side gives the associated graph $\overline{G}$. The adjacency matrix of $\overline{G}$ is

$$M = \begin{pmatrix} 0 & 3 & 0 \\ 1 & 0 & 1 \\ 0 & 1 & 0 \end{pmatrix}$$

Its maximal eigenvalue is $\lambda = 2$. The components of a positive integral eigenvector are written on the nodes.

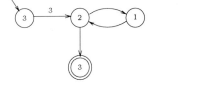
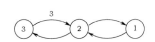

Fig. 2. Representation (G, i, F) **Fig. 3.** Graph $\overline{G}$

Proposition 1. *If s satisfies Kraft's inequality $s(1/k) \leq 1$, then $\lambda \leq k$. In the equality case where $s(1/k) = 1$ we have $\lambda = k$.*

For a proof, we refer the reader to [3], [4] or [6].

We now define the operation of *output state splitting* in a graph $G = (V, E)$. Let q be a vertex of G and let I (resp. O) be the set of edges coming in q (resp. going out of q). Let $O = O' + O''$ be a partition of O. The operation of *(output) state splitting* relative to (O', O'') transforms G into the graph $G' = (V', E')$ where $V' = (V \setminus \{q\}) \cup q' \cup q''$ is obtained from V by splitting state q into two states q' and q'', and where E' is defined as follows:

1. all edges of E that are not incident to q are left unchanged.
2. the both states q' and q'' have the same input edges as q.
3. the output edges of q are distributed between q' and q'' according to the partition of O into O' and O''. We denote U' and U'' the sets of output edges of q' and q'' respectively :
$U' = \{(q', x, p) \mid (q, x, p) \in O'\}$ and $U'' = \{(q'', x, p) \mid (q, x, p) \in O''\}$.

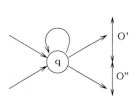 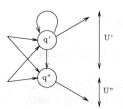

Fig. 4. Graph G **Fig. 5.** Graph G'

Let us now assume that $\mathbf{v}$ is a k-approximate eigenvector for the graph G. We denote by v_p the component of index p of $\mathbf{v}$. All components v_p are positive integers. A state splitting of a state q is said to be *admissible* according to k, if the partition in O' and O'' is such that O' and O'' are not empty and:

$$k \text{ divides} \sum_{(q,x,r) \in O'} v_r$$

If the state splitting is admissible (according to k), the vector $\mathbf{v}'$ defined as follows will be a k-approximate eigenvector for the new graph G'. If p is a state distinct from q' and q'' then $v'_p = v_p$. For states q' and q'' we have:

$$v'_{q'} = \frac{1}{k} \sum_{(q,x,r) \in O'} v_r \quad \text{and} \quad v'_{q''} = v_q - v'_{q'}.$$

By the state splitting construction, one can check that $M'\mathbf{v}' \leq k\mathbf{v}'$, where M' is the adjacency matrix of G'.

The state splitting algorithm of [1] ensures that there is a finite number of state splittings leading to a k-ary graph, that is a graph such that at most k edges are going out of any state. For the sake of completeness, we briefly recall the proof. If there is a state q which admits more than k edges going out of it, we choose k of them and denote by $r_1, r_2, \ldots, r_k$ the sequence of end states of these edges. We then choose a subset O' of these k edges such that k divides $\sum_{(q,x,r) \in O'} v_r$. This is always possible. Indeed, by considering the $k+1$ numbers $v_{r_1}, v_{r_1} + v_{r_2}, \ldots, v_{r_1} + v_{r_2} + \cdots v_{r_k}$, we can see that at least two of them are equal modulo k, and then their difference is equal to zero modulo k. The partition of the output edges of q in O' and O'' leads to an admissible state splitting and v'_q is strictly less than v_q. This point ensures that the process stops after a finite number of splits, the final number of states being bounded by the sum of the components of the initial approximate eigenvector. The final graph obtained is k-ary.

We shall compute approximate eigenvectors for the strongly connected graphs $\overline{G}$ associated to normalized representations (G, i, F) of sequences. We shall then perform admissible state splittings that can be seen either on the graph G or on the graph $\overline{G}$. To do that, we shall associate to each node of G a *value* equal to the corresponding component of the approximate eigenvector of the graph $\overline{G}$. The initial and the final states will have same value since they correspond to the same state of $\overline{G}$.

3 The results

We now state the result in the case of Kraft strict inequality.

Theorem 2. *Let $s = (s_n)_{n \geq 1}$ be an $\mathbb{N}$-rational sequence of nonnegative integers et let k be an integer such that $\sum_{n \geq 1} s_n k^{-n} < 1$. Then there is a k-ary rational tree such that s is the enumerative sequence of its leaves.*

In order to prove this result, we first prove one lemma that remains true in the equality case. We therefore consider an $\mathbb{N}$-rational sequence s and an integer k such that $\sum_{n \geq 1} s_n k^{-n} \leq 1$. We begin with a normalized representation (G, i, F) of the $\mathbb{N}$-rational sequence s. We denote by M the adjacency matrix of G and by λ its spectral radius. Then $\lambda \leq k$. We then compute a k-approximate eigenvector $\mathbf{v} = (v_1, v_2, \ldots, v_n)^t$ of the graph $\overline{G}$. By definition, we have $M\mathbf{v} \leq k\mathbf{v}$. Without loss of generality, we can assume that state 1 is the initial state in all normalized representations.

Lemma 3. *If k divides v_1, then there is another normalized representation for s and a new corresponding approximate eigenvector v' with $v'_1 = v_1$ div k.*

Proof. We denote by P the set of states q such that there is in G an edge denoted (q, x, t) going from q to a final state t of F. Remark that, as state t is equal to state 1 in $\overline{G}$, the value of state t is equal to the value of state 1.

Let us first suppose that the initial state 1 does not belong to the set P. If there is in P a state q which admits more than one (say n) outgoing edges, we split q in q' and q'' according to partition (O', O'') where $O' = \{(q, x, t)\}$. Since k divides v_1, this state splitting is admissible and $v'_{q'} = v_1$ div k. Moreover, in the new graph G', q' admits only one outgoing edge (going to t) and q'' is either not in P or admits less than n outgoing edges. By successive state splittings of all states in P having more than one outgoing edges, we will get, in a finite number of steps, a representation such that all states with one outgoing edge ending in F have no other outgoing edges. Under the hypothesis that state 1 does not belong to P, the initial state has not been split during this processand so each new computed graph is still a normalized representation of the sequence. We denote again by $(G, 1, F)$ the final representation obtained for s and by P_{last} the set of states having one outgoing edge ending in F in this graph. Remark that the values of states of P_{last} are greater than or equal to v_1 div k. We turn all values of states of P_{last} greater than v_1 div k into v_1 div k; the vector **v** remains a k-approximate eigenvector.

We then transform the representation $(G, 1, F)$ in a new one, (H, i, P_{last}), where H is the graph obtained from G by adding a state i, an edge from i to 1 and by removing all edges of G going out of a state of P_{last}. If we look at paths in G going from 1 to F, we have just cut the last edge and added one at the beginning. We assign to state i the value v_1 div k, and the values of all states correspond now to a new k-approximate eigenvector for $\overline{H}$. We call this tranformation the "shift" transformation.

Let us now suppose that the initial state 1 belongs to P. We first split, as explained above, all states of P having more that one outgoing edge. In this case, state 1 may have been split. We denote by $1_{(1)}, 1_{(2)}, 1_{(3)}, \ldots 1_{(r)}$ the copies of state 1 obtained by successive state splittings of the initial state 1. We still denote by G the graph obtained by this transformation and by P_{last} the set of states having one outgoing edge ending in F in this graph. We then transform the representation $(G, 1, F)$ into a new one, (H, i, P_{last}), where H is the graph obtained from G by adding a state i, an edge from i to each $1_{(j)}, 1 \leq j \leq r$ and by removing all edges of G going out of a state of P_{last}. Remark that $(r-1)$ states among $1_{(1)}, 1_{(2)}, 1_{(3)}, \ldots 1_{(r)}$ belong to P_{last}. We again assign to the state i the value v_1 div k, and the values of all states correspond now to a new k-approximate eigenvector for $\overline{H}$.

Corollary 4. *If v_1 is a power of k, then there is another normalized representation and a new corresponding approximate eigenvector v' with $v'_1 = 1$.*

Proof. If $v_1 = k^m$, we iterate the construction given in previous lemma and get $v'_1 = 1$ in m steps.

Example

Let s be the following series:

$$s(z) = 2z^3 + 2z^2 \left(z^2(z^2)^*\right)^*$$

Here, $k = 2$ and $s(1/2) = 1$.
In the following pictures, the nodes are labeled with their value.

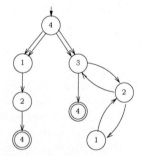

Fig. 6. Initial normalized representation

First step

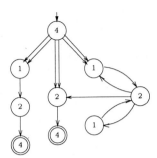

Fig. 7. First state splitting

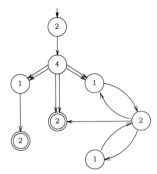

Fig. 8. First "shift"

Second step

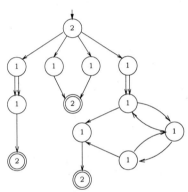

Fig. 9. Other state splittings

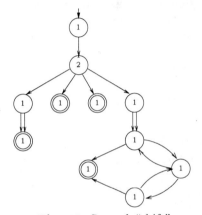

Fig. 10. Second "shift"

The *last step* is described in the proof of Theorem 2.
It corresponds here to a state splitting of all states of the graph of value different from 1.

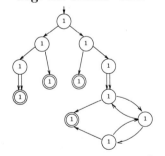

Fig. 11. Last representation

We now prove another lemma which is true only in the case of Kraft strict inequality.

Lemma 5. *Let M be a nonnegative integral matrix. If its spectral radius is strictly less than k, then there is a k-approximate eigenvector $\mathbf{w}$ of M such that w_1 is a power of k.*

Proof. Let λ ($\lambda < k$) be the positive real eigenvalue of maximal modulus of M and let $\mathbf{v}$ be an eigenvector associated to λ. We denote by P the set of positive vectors $\mathbf{w}$ such that $M\mathbf{w} < k\mathbf{w}$. The set P is an open set and $\mathbf{v}$ belongs to P. By dividing all components of $\mathbf{v}$ par v_1, we can assume that v_1 is equal to 1. As P is open, there is a positive real ϵ such that $B(v, \epsilon) \subset P$, where $B(v, \epsilon) = \{\mathbf{w} \mid v_i - \epsilon \leq w_i \leq v_i + \epsilon\}$. Let us now choose an integer m such that $1/k^m < \epsilon$. As $B(v, 1/k^m) \subset P$, we have $\{k^m \mathbf{w} \mid \mathbf{w} \in B(v, 1/k^m)\} \subset P$. This set is $\{\mathbf{w} \mid k^m v_i - 1 \leq w_i \leq k^m v_i + 1\}$ and contains $\mathbf{w}$ where $w_i = \lceil k^m v_i \rceil$. This vector is a positive integer vector $\mathbf{w}$ with $M\mathbf{w} < k\mathbf{w}$: it is a k-approximate eigenvector. Moreover $w_1 = k^m$.

Proof. (Theorem 2) We begin with a normalized representation of s and compute, by Lemma 5, a k-approximate eigenvector whose component for the initial state is a power of k. We then compute, by Corollary 4, a normalized representation $(G, 1, F)$ of s which admits a k-approximate eigenvector of component 1 for the initial state. Finally, we apply to G the state splitting algorithm described in the previous section to obtain a k-ary graph. As the component of the approximate eigenvector on the initial state is 1 and as the state splittings have to be admissible, this state will never be split during the process. A state splitting of a state of G different from state 1 leads by construction to a graph G' still representing the same sequence. The result follows then from the fact that the final normalized representation has a k-ary graph.

We can apply the construction given above to the case of Kraft equality when it is possible to find a representation of s which admits a k-eigenvector with a power of k as component on the initial state. This may perhaps not always be the case. We do not know, for example, if the series $s(z) = \frac{z^2}{(1-z^2)} + \frac{z^2}{(1-5z^3)}$ (communicated to us by Christophe Reutenauer) has such a representation for $k = 2$:

As a consequence of the previous result, we get the following proposition in the equality case, where an ultimately k-ary tree is a tree where all nodes but a finite number have at most k sons.

Proposition 6. *Let $s = (s_n)_{n \geq 1}$ be an $\mathbb{N}$-rational sequence of nonnegative integers and let k be an integer such that $\sum_{n \geq 1} s_n k^{-n} = 1$. Then there is an ultimately k-ary rational tree such that s is the enumerative sequence of its leaves.*

Proof. If we remove one term of the sequence, the remainder satisfies Kraft's strict inequality and is still $\mathbb{N}$-rational. This proves that one can construct a rational tree T for s which will be k-ary for all nodes except the root which will have $k + 1$ sons.

We now state another result for the equality case which is weaker than the previous theorem. We show that if s is an $\mathbb{N}$-rational sequence of nonnegative integers satisfying Kraft's equality for an integer k, then there is an integer m such that ms is the sum of m enumerative sequences of leaves of m k-ary rational trees.

Theorem 7. Let s be a an $\mathbb{N}$-rational sequence satisfying $\sum_{n\geq 1} s_n k^{-n} = 1$. There is a positive integer m and m k-ary rational trees $T_1, \ldots, T_m$ such that $ms = l(T_1) + \cdots + l(T_m)$.

Proof. We begin with a normalized represention of $s : (G, 1, F)$. Let **v** be a positive integral eigenvector associated to the spectral radius k of the adjacency matrix of G. The component v_1 on the initial state 1 is denoted by m. If $m = k^r m'$, where m' and k are relatively prime, we compute, by Corollary 4, a normalized represention of s such that $v_1 = m'$ in order to get a smaller integer m. After this step, m and k are relatively prime. If $m = 1$, we finish by the same proof as the proof of Theorem 2.

Otherwise, we denote by r the positive integer such that $k^{r-1} < m < k^r$ and $x = k^r - m$. We define a new graph H by adding to G $(r+1)$ new states $i_1, i_2, \ldots, i_r$ and j, and with the following new edges :

$$i_1 \longrightarrow i_2 \longrightarrow \cdots i_r \longrightarrow 1$$

and x edges (in the multigraph) going from i_r to j :

$$i_r \longrightarrow j \ (x \text{ edges})$$

We assign to state i_l the value k^{l-1} and to state j the value 1 (state 1 has value m). For each state t in F we make the following transformation. We replace t by m copies of this state and we duplicate each edge entering t in m edges entering the m copies of t. We give to all copies of t the value 1. We denote by g the new sequence which admits as normalized representation $(H, i_1, F \cup \{j\})$. Note that the values of states of H correspond now to a k-integer eigenvector of $\overline{H}$ which admits 1 as component on the initial state i_1. Using the series notations, one can verify that

$$g(z) = \sum_{n\geq 1} g_n z^n = xz^r + mz^r s(z),$$

and by construction: $g(1/k) = 1$.

Since the representation of g has an eigenvector of component 1 at the initial state, g is the height distribution of leaves of a k-ary rational tree (by applying the construction of theorem 1 in the case where the value of the initial state is 1). The series g is either equal to 1 or to $zg_1 + zg_2 + \cdots + zg_k$, where g_l are again series of this type (g_l is the height distribution of the leaves of the subtree rooted by a son of the root of the tree representing g). By iterating this decomposition for each g_l, we can write

$$g(z) = z^r (f_1(z) + f_2(z) + \cdots + f_{k^r}(z)) = xz^r + mz^r s(z),$$

where f_i are height distributions of leaves of k-ary rational trees, which we simplify into:

$$f_1 + f_2 + \cdots + f_{k^r} = x + ms$$

As all f_i have nonnegative integer coefficients and satisfy $f_i(1/k) = 1$, this implies that x series among $f_1, f_2, \ldots, f_{k^r}$ are equal to 1. The m remainding series that we renumber $f_1, f_2, \ldots, f_m$ verify the equality : $f_1 + f_2 + \cdots + f_m = ms$, which is the announced result.

Example Let s be the following series:

$$s(z) = \frac{z^2}{1-z^2} + \frac{2z^2}{1-2z^3}.$$

We get that $3s = f_A + f_B + f_C$, where f_X is the height distribution of the leaves of the tree rooted by the node X in the last picture.

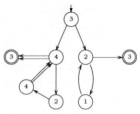

Fig. 12. The sequence s

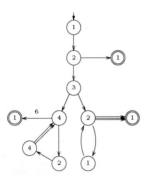

Fig. 13. The sequence g

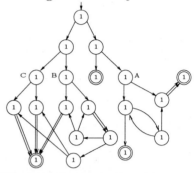

Fig. 14. The sequence g after state splittings

References

1. R. L. Adler, D. Coppersmith, and M. Hassner. Algorithms for sliding block codes. *I.E.E.E. Trans. Inform. Theory*, IT-29:5–22, 1983.
2. F. Bassino. *Séries rationnelles et distributions de longueurs*. Thèse, Université de Marne-La-Vallée, 1996.
3. M.-P. Béal. *Codage Symbolique*. Masson, 1993.
4. M.-P. Béal and D. Perrin. Symbolic dynamics and finite automata. In G. Rozenberg and A. Salomaa, editors, *Handbook of Formal Languages*. Springer-Verlag, 1997.
5. J. Berstel and C. Reutenauer. *Rational Series and their Languages*. Springer-Verlag, 1988.
6. D. Lind and B. Marcus. *An Introduction to Symbolic Dynamics and Coding*. Cambridge, 1995.
7. B. Marcus. Factors and extensions of full shifts. *Monats.Math*, 88:239–247, 1979.
8. D. Perrin. Arbres et séries rationnelles. *C.R.A.S. Paris, Série I*, 309:713–716, 1989.
9. D. Perrin. A conjecture on rational sequences. In R. Capocelli, editor, *Sequences*, pages 267–274. Springer-Verlag, 1990.
10. D. Perrin. Finite automata. In J. V. Leeuven, editor, *Handbook of Theoretical Computer Science*, volume B, chapter 1. Elsevier, 1990.
11. A. Salomaa and M. Soittola. *Automata-theoretic Aspect of Formal Power Series*. Springer-Verlag, Berlin, 1978.

A Completion Algorithm for Codes with Bounded Synchronization Delay

Véronique Bruyère

University of Mons-Hainaut, 15 Avenue Maistriau,
B-7000 Mons, Belgium.
Veronique.Bruyere@umh.ac.be

Abstract. We show that any rational code with bounded synchronization delay is included in a rational maximal code with bounded synchronization delay.

1 Introduction

The theory of codes is originated from Shannon's works on information theory. It is now a well-developed branch of theoretical computer science. We refer to [7, 31] for a systematic exposition of the topic, to [6, 21] for application of codes in symbolic dynamics and coding for constrained channels, and to [17] for a survey on codes used in the context of information transmission systems.

A lot of beautiful properties provide a good understanding of the structure of codes. Nevertheless, several problems on codes remain unsolved despite the effort of researchers [8, 9]. In this paper, we are interested in the following problem : given a code X with some property $\mathcal{P}$, find (if it exists) an effective procedure to embed X into a maximal code Y with the same property $\mathcal{P}$. Effective embedding procedures exist for rational codes [14], rational codes with bounded deciphering delay [10, 3], rational biprefix codes [23, 33]. The case of finite codes is particular : there exist finite codes included in no finite maximal codes [27, 19]. One of the main open problems on codes is whether the inclusion of a finite code in a finite maximal code is decidable.

We here show that any rational code with bounded synchronization delay is effectively embeddable into a rational maximal code again with bounded synchronization delay. Codes with bounded synchronization delay [16] are part of the family of circular codes, i.e., codes defining a unique factorization of words written on a circle [20] or of biinfinite words [12]. Circular codes and codes with bounded synchronization delay have numerous interesting properties. For instance, sequences of integers which are the length distribution of a circular code are completely characterized [30, 28, 4]; codes appearing in factorizations of free monoids are necessarily circular [29] (see also [32, 18, 13] for the description of circular codes used in finite factorizations); codes with bounded synchronization delay satisfy the commutative equivalence conjecture [24]; encoding digital data for transmission through constrained channels involve circular codes [15, 1, 5]; recently a set of codons constituting a circular code has been identified in the study of the repartition of trinucleotides in the protein of coding genes [2].

2 Codes with bounded synchronization delay

For the notions given in this section and the next one, we refer to [7] and [6].

Given a finite alphabet A, a *code* $X \subseteq A^*$ is a set of words such that for all $x_1, \ldots, x_n, y_1, \ldots, y_m \in X$,

$$x_1 \cdots x_n = y_1 \cdots y_m \quad \Rightarrow \quad n = m, \; x_i = y_i \; \forall i.$$

This definition means that any coded message $x_1 \cdots x_n$ is uniquely decoded into the code-words $x_1, \ldots, x_n$.

We are here interested in codes with bounded synchronization delay. Such codes allow to easily localize a position into a coded message through which the decoding must pass, and thus to decode the two parts separately. Formally, a code has a *bounded synchronization delay* if there exists $\sigma \geq 0$ such that (see Figure 1)

$$uxyv \in X^*, \text{ with } x, y \in X^\sigma \quad \Rightarrow \quad ux, yv \in X^*. \tag{1}$$

The smallest integer σ satisfying (1) is called the *synchronization delay* of the code X.

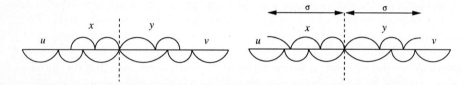

Fig. 1. Synchronization delay σ.

Fig. 2. Synchronization delay σ when counting with letters.

Example 1. The code $X = a^*b$ has synchronization delay 1, since the letter b only occurs at the end of the code-words. On the opposite, the code $X = ab^*c \cup b$ has no bounded synchronization delay, because $b^{2\sigma}$ is factor of the word $ab^{2\sigma}c$ of X, for any σ.

There is another way to define the synchronization delay of codes. Instead of counting with words as done in Definition (1), one can count with letters as follows. We denote by $P(X^*)$ the set of the prefixes of the words of X^*, and by $P_\sigma(X^*)$ the set $P(X^*) \cap A^\sigma$. For suffixes we use the notations $S(X^*)$ and $S_\sigma(X^*)$. A code $X \subseteq A^*$ has a *bounded letter-synchronization delay* if there exists $\sigma \geq 0$ such that (see Figure 2)

$$uxyv \in X^*, \text{ with } x \in S_\sigma(X^*), y \in P_\sigma(X^*) \quad \Rightarrow \quad ux, yv \in X^*. \tag{2}$$

The smallest integer σ satisfying (2) is called the *letter-synchronization delay* of the code X.

For finite codes, both synchronization delays (on words or on letters) are bounded simultaneously. This is no longer true for infinite codes.

Example 2. Let $A = \{a, b, c\}$ and $X \subseteq A^*$ equal to $a^*b \cup ca^*ba^*b$. The code X has a synchronization delay 2 if counting with words, but no bounded synchronization delay if counting with letters.

The family of codes of bounded synchronization delay or bounded letter-synchronization delay is included in the one of circular codes. We recall that a code is *circular* if

$$ux_2 \cdots x_n v = y_1 \cdots y_m, \quad x_1 = vu \quad \Rightarrow \quad n = m, \ v = 1, \ x_i = y_i \ \forall i.$$

In the case of finite codes $X \subseteq A^*$, the concepts of circular code, code with bounded synchronization delay, code with bounded letter-synchronization delay coincide. Moreover, these code properties are equivalent to :

$$X^* = P \cup (UA^* \cap A^*V \setminus A^*WA^*)$$

with P, U, V, W finite subsets of A^*.

However for *rational* codes (that is, codes recognized by a finite automaton), there exist circular codes with an infinite synchronization delay, as the code $X = ab^*c \cup b$ mentioned in Example 1. As a matter of fact, a rational circular code has a bounded synchronization delay if and only if $\exists p, X \cap A^* X^p A^* = \emptyset$.

Remark. Any code $X \subseteq A^*$ with synchronization delay 0 (on words or on letters) is necessarily included in the alphabet A. From now on, we suppose that $\sigma \geq 1$ in a way to discard such trivial codes.

3 Completion's problem

In this paper, we solve Problem 8 of [9] about the completion of codes with bounded synchronization delay.

Recall that *complete* codes $X \subseteq A^*$ are codes such that any word over A is factor of a coded message :

$$\forall w \in A^*, \quad A^*wA^* \cap X^* \neq \emptyset.$$

It is well-known [7] that for rational codes, this combinatorial property is equivalent to the extremal property of being a *maximal* code (with respect to the inclusion).

We here prove that codes with bounded synchronization delay (on words or on letters) can be embedded into a complete one. The case of bounded letter-synchronisation delay is solved separately, since the two notions of delay differ for infinite codes (see Example 2).

Theorem 1. *Let $X \subseteq A^*$ be a code with synchronization delay σ. Then X can be embedded into a complete code $Y \subseteq A^*$ with synchronization delay $\sigma' \leq 2\sigma$. Moreover if X is rational, then Y is also rational.*

Theorem 2. *Let $X \subseteq A^*$ be a code with letter-synchronization delay σ. Then X can be embedded into a complete code $Y \subseteq A^*$ with letter-synchronization delay $\sigma' \leq 3\sigma - 2$. Moreover if X is rational, then Y is also rational.*

The method given in [14] for embedding a rational code X into a rational maximal code Y, also works for rational circular codes [11, 4]. Hence any rational circular code is included in a maximal one. However this method is not able to keep the bounded synchronization delay from X to Y.

The proofs of Theorems 1 and 2 are given below in the next two sections. They are based on the following propositions which state a simple combinatorial property of complete codes with a bounded synchronization (letter-synchronization resp.) delay.

Proposition 3. *Let $X \subseteq A^*$ be a code. Then X is a complete code with synchronization delay $\leq \sigma$ if and only if $X^\sigma A^* X^\sigma \subseteq X^*$.* □

Proposition 4. *Let $X \subseteq A^*$ be a code. Then X is a complete code with letter-synchronization delay $\leq \sigma$ if and only if $P_\sigma(X^*)A^*S_\sigma(X^*) \subseteq X^*$.* □

The next example shows that the bound 2σ of Theorem 1 is tight. The bound $3\sigma - 2$ of Theorem 2 is also tight, but the example, more elaborated, is omitted.

Example 3. Consider the alphabet $A = \{a, b, c, d\}$ and an integer $\sigma \geq 1$. The set $X = \{a, ca^{2\sigma-1}b, ba^{2\sigma-1}d, cb^{4\sigma-2}d\}$ is a code over A with synchronization delay σ. Assume that X can be included in a complete code $Y \subseteq A^*$ with a synchronization delay $\sigma' \leq 2\sigma - 1$. By Proposition 3, one has $Y^{\sigma'}A^*Y^{\sigma'} \subseteq Y^*$. Then
$$a^{\sigma'}ca^{2\sigma-1} \in Y^*, \quad a^{2\sigma-1}da^{\sigma'} \in Y^*.$$
The word $a^{\sigma'}ca^{2\sigma-1}ba^{2\sigma-1}da^{\sigma'}$ decomposes into words of Y^* as indicated in Figure 3. As Y is a code, b must belong to Y. But $b^{2\sigma'}$ is factor of the word $cb^{4\sigma-2}d$ of Y, in contradiction with the synchronization delay σ' of Y.

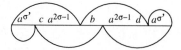
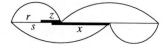

Fig. 3. The bound 2σ is tight.   **Fig. 4.** z is not factor of x.

4 Embedding when counting with words

In this section, X is a given code over the alphabet A with a bounded synchronization delay σ. The way to embed X into the code Y mentioned in Theorem 1 is done in two steps : construct
$$M = (X^{2\sigma}A^* \cap A^*X^{2\sigma}) \cup X^*,$$
$$Y = \text{Base}(M) = (M \setminus 1) \setminus (M \setminus 1)^2. \qquad (3)$$

In addition of X^*, the monoid M contains all the words beginning and ending with *markers* $z \in X^{2\sigma}$. The use of markers already appears in the method of [14] : one marker only is used, given by an unbordered word which is factor of no word of X^*.

The following simple lemmas, together with Proposition 3 lead to a proof of Theorem 1 : Lemmas 5 and 6 show that Y is a code containing X. This code is proved to be complete with synchronization delay $\sigma' \leq 2\sigma$ thanks to Lemma 7 and Proposition 3. Clearly if X is rational, so is Y.

Lemma 5. *Y is a code.*

Proof. To prove that Y is a code, we show that the monoid M is *stable* (see [7]) : if $u, wv, uw, v \in M$, then $w \in M$. Assume that this is not the case, and consider a word uwv of minimal length such that

$$u, wv, uw, v \in M \quad \text{but} \quad w \notin M. \tag{4}$$

We begin with three claims concerning u and wv. Symmetrically they hold for v and uw.

Claim 1. If $u = rz$ with $z \in X^{2\sigma}$ and $uw = sx$ with $x \in X^*$, then z is not factor of x (see Figure 4).
If z is factor of x, then due to the synchronization delay σ of X, we get $z = z_1 z_2$, $x = x_1 x_2$ with $z_1, z_2 \in X^\sigma$, $x_1, x_2 \in X^*$ and $rz_1 = sx_1$, $z_2 wv = x_2 v$. The second equality links shorter words satisfying a relation similar to (4). This is impossible.

Claim 2. If $wv = zr$ with $z \in X^{2\sigma}$ and $uw = sx$, $v = x's'$ with $x, x' \in X^*$, then z is not factor of xx' (see Figure 5).

Fig. 5. z is not factor of xx'.   **Fig. 6.** w is a proper prefix of z.

Assume that z is factor of xx' and let $z = z_1 z_2$ with $z_1, z_2 \in X^\sigma$. If w is prefix of z_1, we get the same contradiction as done in Claim 1. So z_1 is prefix of w. By the synchronization delay σ of X, we have $x = x_1 x_2$ with $x_1, x_2 \in X^*$ and $uz_1 = sx_1$. It follows that $w = z_1 x_2$ belongs to X^*, a contradiction with (4).

Claim 3. If $wv = zr$ with $z \in X^{2\sigma}$, then w is a proper prefix of z (see Figure 6).
Assume the contrary, i.e., z is prefix of w. By Claim 2, uw belongs to $M \setminus X^*$. Moreover any suffix $z' \in X^{2\sigma}$ of uw is a proper suffix of w, again by Claim 2. It follows that $w \in X^{2\sigma} A^* \cap A^* X^{2\sigma} \subseteq M$, a contradiction with (4).

We now end the proof. In (4), at least one of the words u, wv, uw and v is in $M \setminus X^*$ since X is a code.

Assume that $u \in M \setminus X^*$ and let $u = rz$ with $z \in X^{2\sigma}$. It follows by Claim 1 that $uw \in M \setminus X^*$. Let $uw = r'z'$ with $z' \in X^{2\sigma}$. Again by Claim 1, we get

$|r| < |r'|$. Consider now the word wv. It must belong to $M \setminus X^*$ by Claim 2 applied to uw. Let $wv = z''r''$ with $z'' \in X^{2\sigma}$. Then $|z''| > |w|$ by Claim 3. But this is in contradiction with Claim 2 applied to uw.

Assume that $u \in X^*$ and $wv \in M \setminus X^*$. Then $wv = zr$ with $z \in X^{2\sigma}$ and $|z| > |w|$ by Claim 3. Claim 2 applied to uw shows that $uw \in X^*$. It follows that $v \in M \setminus X^*$ again by Claim 2 applied to wv. Let $v = z'r'$ with $z' \in X^{2\sigma}$. Claim 1 applied to v leads to z being factor of $uwz' \in X^*$. This is in contradiction with Claim 2.

The other cases are symmetrical. Therefore, assumption (4) is false, $w \in M$ showing that Y is a code. □

Lemma 6. $X \subseteq Y$.

Proof. Assume the contrary, that is, some word $x \in X$ factorizes as $y_1 \cdots y_n$ with $n \geq 2$ and $y_1, \ldots, y_n \in Y$. At least one of these words, say y_i, belongs to $Y \setminus X$ since X is a code. As $y_i \in X^{2\sigma} A^* \cap A^* X^{2\sigma}$ and y_i is factor of x, this leads to a contradiction with the synchronization delay σ of X. □

Lemma 7. $Y^{2\sigma} A^* Y^{2\sigma} \subseteq Y^*$.

Proof. By (3), we have $Y^{2\sigma} A^* Y^{2\sigma} \subseteq X^{2\sigma} A^* X^{2\sigma} \subseteq M = Y^*$. □

5 Embedding when counting with letters

In this section, $X \subseteq A^*$ is a code with letter-synchronization delay σ. We show how to construct the complete code Y of Theorem 2 and we prove the correctness of the construction. We denote by τ the constant $3\sigma - 2$.

The algorithm uses a particular operation $\mathcal{Z}(M)$ defined by (see Figure 7)

$$\begin{aligned}\mathcal{Z}(M) = \ &\{w \in A^* \setminus X^* \mid w = zu = u'z', \text{with} \\ &\qquad z \in P_\tau(M), z' \in S_\tau(M)\} \\ &\cup \{w \in A^* \setminus X^* \mid \text{there exist } u \in S(M), u' \in P(M) \text{ with} \\ &\qquad z = wu' \in P_\tau(M), z' = uw \in S_\tau(M)\}.\end{aligned}$$

Notice that $\mathcal{Z}(M) \cap X^* = \emptyset$ and that $\mathcal{Z}(M)$ is the union of two sets, one with words of length greater than or equal to τ, the other with words of length less than or equal to τ. As done above in Section 4, the operation $\mathcal{Z}$ uses markers z in $P_\tau(M)$ or $S_\tau(M)$ (instead of $X^{2\sigma}$).

Fig. 7. Operation $\mathcal{Z}$.

The algorithm works as follows :

$$M = X^*$$
Repeat
$$M' = M$$
$$M = (\mathcal{Z}(M) \cup M)^*$$
until $M = M'$
$$Y = \text{Base}(M)$$

The proof of Theorem 2 is done in a similar way as for Theorem 1. We begin with a technical lemma.

Lemma 8. $M \setminus X^* \subseteq P_\sigma(X^*)A^* \cap A^*S_\sigma(X^*)$.

Proof. We are going to prove the next four statements. Lemma 8 is a corollary of (4) since $2\sigma - 1 \geq \sigma$.

1. Let $w \in \mathcal{Z}(M)$ with length $|w| \leq \tau$ and $u \in S(M)$, $u' \in P(M)$ such that $z = wu' \in P_\tau(M)$ and $z' = uw \in S_\tau(M)$. Then uwu' has no factor in $S_\sigma(X^*)P_\sigma(X^*)$.
2. Any $w \in \mathcal{Z}(M)$ has length at least equal to $2\sigma - 1$.
3. For any $w \in \mathcal{Z}(M)$, let $z = w$ if $|w| \leq \tau$, let z be the prefix of length τ of w otherwise. Then either $z \in P(X^*)$ or z has a proper prefix in $X^*(\mathcal{Z}(M) \cap P(X^*))$.
   Symmetrically, let $z' = w$ if $|w| \leq \tau$, let z' be the suffix of length τ of w otherwise. Then either $z' \in S(X^*)$ or z' has a proper suffix in $(\mathcal{Z}(M) \cap S(X^*))X^*$.
4. For any $w \in M \setminus X^*$, w has a prefix (resp. suffix) with length $2\sigma - 1$ in $P(X^*)$ (resp. $S(X^*)$).

The four statements are proved by induction on the passes through the repeat instruction of the previous algorithm. We denote by M_i the value of M at the beginning of the repeat instruction at pass i. Initially $M_1 = X^*$. Notice that $M_i \subseteq M_{i+1} \ \forall i$.

• *Pass 1.* At this stage, consider $\mathcal{Z}(M_1) = \mathcal{Z}(X^*)$.

(1) As X has letter-synchronization delay σ and $\tau \geq \sigma$, we have $|u|, |u'| < \sigma$ otherwise $w \in X^*$. Assume that uwu' has a factor in $S_\sigma(X^*)P_\sigma(X^*)$, i.e.,

$$uwu' = rx_1x_2r' \quad \text{with } x_1 \in S_\sigma(X^*), x_2 \in P_\sigma(X^*).$$

Then $|u| < |rx_1|$, $|u'| < |x_2r'|$. Let $w = w_1w_2$ such that $uw_1 = rx_1$, $w_2u' = x_2r'$. By the letter-synchronization delay σ of X, we get $w_1, w_2 \in X^*$, a contradiction with $w \notin X^*$.

(2) The statement holds for words w of $\mathcal{Z}(X^*)$ with length $|w| \geq \tau$. For the other words we use the notations of (1). We already know that $|u|, |u'| < \sigma$. As $|uw| = |wu'| = \tau$, it follows that $|w| \geq 2\sigma - 1$.

(3) Clearly $z \in P(X^*)$ (resp. $z' \in S(X^*)$).

(4) As a consequence of (2) and (3), any word of $(\mathcal{Z}(M_1) \cup M_1)^* \setminus X^* = M_2 \setminus X^*$ has a prefix in $P_{2\sigma-1}(X^*)$ and a suffix in $S_{2\sigma-1}(X^*)$.

- *Pass i, with $i > 1$.* We suppose that $\mathcal{Z}(M_{i-1})$ satisfies (1)–(3) and $(\mathcal{Z}(M_{i-1}) \cup M_{i-1})^* = M_i$ satisfies (4). Let us consider $\mathcal{Z}(M_i)$.

(1) Let $u \in S(M_i)$, $u' \in P(M_i)$ such that $z = wu' \in P_\tau(M_i)$ and $z' = uw \in S_\tau(M_i)$.

Assume that uwu' has a factor in $S_\sigma(X^*)P_\sigma(X^*)$, i.e.,

$$uwu' = rx_1x_2r' \quad \text{with } x_1 \in S_\sigma(X^*), x_2 \in P_\sigma(X^*).$$

To get a contradiction, the idea is the following. We first suppose that $|u| \leq |rx_1|$ and $|u'| \leq |x_2r'|$. Let $w = w_1w_2$ such that $uw_1 = rx_1$ and $w_2u' = x_2r'$. We will prove that

$$w_1, w_2 \in X^*$$

showing that $w \in X^*$, which is impossible. If $|u| > |rx_1|$ or $|u'| > |x_2r'|$, the contradiction is obtained in the same way. Indeed, suppose that $|u| > |rx_1| \geq \sigma$. Let x'_1 (resp. x'_2) be the suffix of u (resp. prefix of z) with length σ. By induction hypothesis (4), $x'_1 \in S_\sigma(X^*)$, $x'_2 \in P_\sigma(X^*)$. We then replace x_1x_2 by $x'_1x'_2$ and we repeat the situation just described, showing that $w \in X^*$.

So, consider that $|u| \leq |rx_1|$ and $|u'| \leq |x_2r'|$. Let us show that $w_1 \in X^*$ (a symmetrical argument shows that $w_2 \in X^*$). Since $|x_2| = \sigma$ and $|z| = \tau$, we have

$$|w_1| < 2\sigma - 1.$$

Let $u'' = u$ if $|u| \leq \sigma$, let u'' be the suffix of u with length σ otherwise. Then

$$u'' \in S(X^*)$$

by induction hypothesis (4). This situation is summarized in Figure 8.

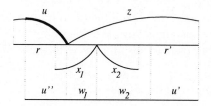

Fig. 8. Case $|u| \leq |rx_1|$, $|u'| \leq |x_2r'|$.

If $z \in P(X^*) = P(M_1)$, then x_1x_2 is factor of $u''z \in S(X^*)P(X^*)$. By the letter-synchronization delay σ of X, it follows that $w_1 \in X^*$.

If $z \notin P(X^*)$, let $z = xw's$ such that $x \in X^*$, $w' \in \mathcal{Z}(M_{i-1})$ and $s \in P(M_i)$. By induction hypothesis (1), w' has no factor in $S_\sigma(X^*)P_\sigma(X^*)$. Hence either $|r| < |ux|$ or $|uxw'| < |rx_1x_2|$.

Consider the first case. We know that w' has a prefix p of length $2\sigma - 1$ in $P(X^*)$ by induction hypothesis (4). Therefore, we have done as just before because x_1x_2 (of length 2σ) is factor of the word $u''xp \in S(X^*)P(X^*)$.

Suppose now that

$$|ux| \leq |r| \text{ and } |uxw'| < |rx_1x_2|$$

and let us show that this case cannot occur (see Figure 9). By induction hypothesis (2), we have $|w'| \geq 2\sigma - 1$ and then $|s| < \sigma$. Thus $|rx_1| < |uxw'|$ because $|w_1| < 2\sigma - 1$.

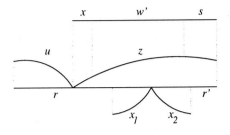

Fig. 9. Case $|ux| \leq |r|$, $|uxw'| < |rx_1x_2|$.

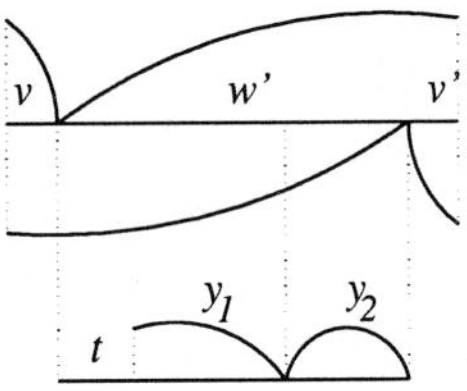

Fig. 10. Construction of w'.

By induction hypothesis (4), $w' = ty$ with $y \in S_{2\sigma-1}(X^*)$. Similarly, $s \in P(X^*)$ since $|s| < \sigma$. Therefore x_1x_2 is factor of $ys \in S(X^*)P(X^*)$. By the letter-synchronization delay σ of X, it follows that $y = y_1y_2$ with $uxty_1 = rx_1$, $y_2s = x_2r'$ and $y_2 \in X^*$.

As $w' \in \mathcal{Z}(M_{i-1})$ and $|w'| \leq \tau$,

$$vw' \in S_\tau(M_{i-1}) \quad \text{and} \quad w'v' \in P_\tau(M_{i-1})$$

for some $v \in S(M_{i-1})$, $v' \in P(M_{i-1})$ (see Figure 10).

We have $\sigma \leq |ty_1| < 2\sigma - 1$ because x_1 is factor of y_1 and $|w_1| < 2\sigma - 1$. Hence $\sigma \leq |y_2v'| < 2\sigma - 1$ since $|y_2v'| = |w'v'| - |ty_1|$, and $v' \in P(X^*)$ by induction hypothesis (4). It follows that ty_1 has a suffix $x_1 \in S_\sigma(X^*)$ and y_2v' has a prefix in $P_\sigma(X^*)$. This is impossible with respect to induction hypothesis (1) applied to $vw'v'$.

This concludes the proof.

(2)   We only have to give the proof for words w of $\mathcal{Z}(M_i)$ with length $|w| \leq \tau$. Let $u \in S(M_i)$, $u' \in P(M_i)$ such that $z = wu' \in P_\tau(M_i)$ and $z' = uw \in S_\tau(M_i)$. Assume that $|w| < 2\sigma - 1$, then $|u|, |u'| \geq \sigma$. By induction hypothesis (4), z' has a suffix in $S_\sigma(X^*)$ and u' has a prefix in $P_\sigma(X^*)$. This is impossible by (1).

(3)   Either $z \in P(M_1) = P(X^*)$ or $z = xw's$ with $x \in X^*$, $w' \in \mathcal{Z}(M_{i-1})$ and $s \in P(M_i)$. By induction hypothesis (3), w' is either in $P(X^*)$ or has a proper prefix in $X^*(\mathcal{Z}(M_{i-2}) \cap P(X^*))$.

(4)   Consequence of (2) and (3). □

Lemma 9. *Y is a code.*

Proof. We show that the monoid M constructed by the algorithm is stable. Let $u, wv, uw, v \in M$. If $|wv|, |uw| \geq \tau$, then $w \in \mathcal{Z}(M) \subseteq M$. Otherwise, we obtain the same conclusion with the word $w'uwvw'$ such that $w' \in M$ has length $|w'| \geq \tau$. □

Lemma 10. $X \subseteq Y$.

Proof. By construction, $X \subseteq M$. Assume that some $x \in X$ belongs to Y^+, i.e.,

$$x = y_1 \cdots y_n \quad \text{with} \quad y_1, \ldots, y_n \in Y \text{ and } n \geq 2.$$

At least one of these words, say y_i, is in $Y \setminus X$ since X is a code.

Suppose that $i \neq 1$. By Lemma 8, y_i has a prefix in $P_\sigma(X^*)$. Take $y \in X^*$ such that $|yy_1 \cdots y_{i-1}| \geq \sigma$. Either $yy_1 \cdots y_{i-1} \in X^*$ or $yy_1 \cdots y_{i-1} \in A^*(Y \setminus X)X^*$. In both cases, this word has a suffix in $S_\sigma(X^*)$ (Lemma 8). Then the word yx of X^* has a factor in $S_\sigma(X^*)P_\sigma(X^*)$. Due to the letter-synchronization delay σ of X, it follows $x \in X^+$. This is impossible.

The case $i = 1$ is solved in a similar way, by working with $y_{i+1} \cdots y_n$ instead of $y_1 \cdots y_{i-1}$. □

Lemma 11. $P_\tau(Y^*)A^*S_\tau(Y^*) \subseteq Y^*$.

Proof. Immediate since $P_\tau(Y^*)A^*S_\tau(Y^*) \subseteq \mathcal{Z}(M) \subseteq M = Y^*$. □

Lemmas 9, 10 and 11 together with Proposition 4 show that Y is a complete code with letter-synchronization delay $\leq \tau$. The property that Y is rational if X is rational is proved below. Consequently, Theorem 2 is proved.

Lemma 12. *If X is rational, then Y is rational.*

Proof. It is enough to show that $\mathcal{Z}(M)$ is rational, and the execution of the algorithm needs a finite number of passes trough the repeat instruction.

The set $\mathcal{Z}(M)$ is composed of two subsets. The first one equals $P_\tau(M)A^* \cap A^*S_\tau(M) \setminus X^*$ which is rational since $P_\tau(M)$ and $S_\tau(M)$ are finite. The second one is composed of some words with length less or equal to τ. It is therefore rational.

Inside the repeat instruction, we have $M' \neq M$ if the operation $\mathcal{Z}$ gives new words of length less than τ. Such words are in finite number, showing that the repeat instruction is executed finitely many times. □

References

1. R.L. Adler, D. Coppersmith, and M. Hassner. Algorithms for sliding block codes. *IEEE Trans. Inform. Theory*, IT-29:5–22, 1983.
2. D. Arquès and C. Michel. A possible code in the genetic code. In *STACS'95 Proceedings*, volume 900 of *Lecture Notes in Comput. Sci.*, pages 640–651, 1995.
3. J. Ashley, B. Marcus, D. Perrin, and S. Tuncel. Surjective extensions of sliding block codes. *SIAM J. Discrete Math.*, 6:582–611, 1993.
4. F. Bassino. *Séries rationnelles et distributions de longueurs*. PhD thesis, University of Marne-La-Vallée, 1996.
5. M.-P. Béal. The method of poles : a coding method for constrained channels. *IEEE Trans. Inf. Theory*, 36:763–772, 1990.
6. M.-P. Béal. *Codage symbolique*. Masson, 1993.
7. J. Berstel and D. Perrin. *Theory of codes*. Academic Press, 1985.

8. J. Berstel and D. Perrin. Trends in the theory of codes. *Bull. EATCS*, 29:84–95, 1986.
9. V. Bruyère and M. Latteux. Variable-length maximal codes. In *ICALP'96 Proceedings*, volume 1099 of *Lecture Notes in Comput. Sci.*, pages 24–47, 1996.
10. V. Bruyère, L. Wang, and L. Zhang. On completion of codes with finite deciphering delay. *European J. Combin.*, 11:513–521, 1990.
11. J. Devolder. *Codes, mots infinis et bi-infinis*. PhD thesis, University of Lille I, 1993.
12. J. Devolder. Precircular codes and periodic biinfinite words. *Inf. Comput.*, 107:185–201, 1993.
13. G. Duchamp and J.-Y. Thibon. Bisections reconnaissables. *RAIRO Inform. Théor. Appl.*, 22:113–128, 1988.
14. A. Ehrenfeucht and G. Rozenberg. Each regular code is included in a regular maximal code. *RAIRO Inform. Théor. Appl.*, 20:89–96, 1985.
15. P.A. Franaszek. A general method for channel coding. *IBM J. Res. Dev.*, 24:638–641, 1980.
16. S.W. Gordon and B. Gordon. Codes with bounded synchronization delay. *Inf. Control*, 8:355–372, 1965.
17. H. Jürgensen and S. Konstantinidis. *Handbook of Formal Languages*, chapter Codes. Springer-Verlag, 1997. to appear.
18. D. Krob. Codes limités et factorisations finies du monoïde libre. *RAIRO Inform. Théor. Appl.*, 21:437–467, 1987.
19. N.H. Lam. On codes having no finite completions. preprint, 1996.
20. V.I. Levenshtein. Some properties of coding and self-adjusting automata for decoding messages. *Problemy Kibernet.*, 11:63–121, 1964.
21. D. Lind and B. Marcus. *Symbolic Dynamics and Coding*. Cambridge University Press, 1996.
22. R. Montalbano. Local automata and completion. In *STACS'93 Proceedings*, volume 665 of *Lecture Notes in Comput. Sci.*, pages 333–342, 1993.
23. D. Perrin. Completing biprefix codes. *Theoret. Comput. Sci.*, 28:329–336, 1984.
24. D. Perrin and M.-P. Schützenberger. Un problème élémentaire de la théorie de l'information. In *Théorie de l'Information, Colloques Internat. CNRS*, volume 276, pages 249–260, Cachan, 1977.
25. A. Restivo. On a question of McNaughton and Papert. *Inf. Control*, 1:1, 1974.
26. A. Restivo. A combinatorial property of codes having finite synchronization delay. *Theoret. Comput. Sci.*, 1:95–101, 1975.
27. A. Restivo. On codes having no finite completions. *Discrete Math.*, 17:309–316, 1977.
28. R.A. Scholtz. Codes with synchronization capability. *IEEE Trans. Inform. Theory*, IT-2:135–142, 1966.
29. M.-P. Schützenberger. On a factorization of free monoids. *Proc. Amer. Math. Soc.*, 16:21–24, 1965.
30. M.-P. Schützenberger. Sur une question concernant certains sous-monoïdes libres. *C. R. Acad. Sci. Paris*, 261:2419–2420, 1965.
31. H.J. Shyr. *Free Monoids and Languages*. Hon Min Book Company, Taichung, second edition, 1991.
32. G. Viennot. *Algèbres de Lie libres et monoïdes libres*. PhD thesis, University Paris 7, 1974.
33. L. Zhang and Z. Shen. Completion of recognizable bifix codes. *Theoret. Comput. Sci.*, 145:345–355, 1995.

The Expressibility of Languages and Relations by Word Equations

Juhani Karhumäki [*,1] and Wojciech Plandowski [**,1]
and Filippo Mignosi [***,2]

[1] Turku Centre for Computer Science and
Department of Mathematics, Turku University, 20 014, Turku, Finland.
[2] Dipartimento di Matematica ed Applicazioni,
Universitá di Palermo via Archirafi, 90 123 Palermo, Italy

Abstract. Classically, several properties and relations of words, such as "being a power of a same word", can be expressed by using word equations. This paper is devoted to study in general the expressive power of word equations. As main results we prove theorems which allow us to show that certain properties of words are not expressible as components of solutions of word equations. In particular, "the primitiveness" and "the equal length" are such properties, as well as being "any word over a proper subalphabet".

1 Introduction

Several authors in the existing literature, cf. [16], used word equations in order to describe properties and relations of words, but, to our knowledge no attempt to synthesis or of a systematization of this topic has been done. This was emphasized also in a recent survey [6] where some results of the field were collected.

Classical relations on words that are characterized as solutions sets of word equations are for instance, "two words X and Y are powers of a same word" if and only if they constitute a solution of the equation $XY = YX$, and "two words X and Y are conjugates" if and only if they constitute a solution of the equation $XZ = ZY$. In the first case we need no extra variables, while in the second case an additional variable seems to be needed. As above we identify names of variables and particular solutions of an equation.

Motivated by above, we say that a property of words - either a language $\mathcal{L} \subseteq \Sigma^*$ or a k-ary relation $\mathcal{R} \subseteq (\Sigma^*)^k$ - is *expressible* by a word equation, if there exists an equation e with $t \geq k$ variables over Σ such that

---

[*] Supported by Academy of Finland under grant 14047. Email:`karhumak@cs.utu.fi`.
[**] Supported partially by the grant KBN 8T11C01208. On leave from Instytut Informatyki, Uniwersytet Warszawski, Banacha 2, 02–097 Warszawa, Poland. Email:`wojtekpl@mimuw.edu.pl`.
[***] Supported by Academy of Finland under grant 14047. Email: `mignosi@altair.math.unipa.it`.

- $\mathcal{L}$ coincides with the values of a fixed component of all solutions of e,

or

- $\mathcal{R}$ coincides with the values of k fixed components of all solutions of e.

Obviously, languages are k-ary relations with $k = 1$, but, due to the importance of this particular case, we have chosen to define those separately. We allow e to contain constants from Σ. An important feature here is also that t can be larger than k, i.e. additional variables are allowed. This increases essentially the expressive power of equations, and in particular makes it much easier to express certain properties by equations.

As an illustration we recall the following. The union of solution sets of two equations can be expressed as a solution set of one equation, as was shown in [4] using 4 additional variables, and later improved to require only 2 additional ones by [8], cf. also [6]. Similarly, the inequality, that is the set of t-tuples of words which does not satisfy a given equation e with t variables, can be expressed as a union of the solution sets of a finitely many equations each of those using 3 extra variables, cf. e.g. [6], and consequently the inequality is expressible by one equation if additional variables are allowed.

This way of expressing relations on words using word equations is very natural and resembles the way of expressing enumerable relations on integers by diophantine equations. However, the expressive power of our method is weaker. Namely, while diophantine equations can express all recursively enumerable sets (of integers), cf. [18], the word equations can express only recursive relations on words due to Makanin's result, cf. [17]. And actually our results show that not even all of those can be expressed.

A central problem in the study of the expressive power of word equations is to show that some relations are not expressible. A similar situation - a need to show that certain languages are not generated by a certain type of devices - was encountered at the early stages of the formal language theory. By now there are a lot of tools for the latter problem, while there seems to be none for the former.

As the main contribution of this paper we introduce such tools for word equations. More precisely, we prove theorems resembling pumping lemmas of formal languages, which allow to prove the nonexpressibility. Very intuitively, we show that if a given equation defines a certain language, or, in fact, just a certain word of it via a variable X, then X actually contains some "unfixed parts" which can be filled arbitrarily, and thus leads outside the considered language.

The contents of this paper are summarized as follows. In the next section we state several properties of words which are expressible by equations, including some closure properties of expressible languages, such as the closure under catenation, union and Kleene star of a word. Most of the material in this section can be considered as a folklore, although we have at least one new proof.

Then in Section 4 we prove our main results, namely tools for showing the nonexpressibility. In Section 5 we use our theorems to show that particular languages or relations, such as "the set of primitive words", "the language $(a \cup b)^*$

over $\{a, b, c\}$" or "the relation equal length", are not expressible. As a consequence we conclude that expressible languages are not closed under operations of Kleene star, complementation or shuffle. In Section 6 we compare the family of expressible languages to a few much studied families, and finally, in Section 7 we state several open problems.

Due to a limited space all proofs are omitted; a complete version of the paper can be found in http://www.tucs.abo.fi/publications/techreports.

2 On the power of expressibility

In this section we give — without trying to be exhaustive — several examples of properties of words which are expressible as solutions of word equations and some closure properties of languages and relations. All results presented here are either very simple or presented before, however, some of those seem to be not very generally known, and moreover we seem to have a simplified proof.

Let Σ be an alphabet of constants and Θ be an alphabet of variables. We assume that these alphabets are disjoint. We use the convention that lower case letters represent constants and capital letters represent variables.

A word equation is a pair of words $(u, v) \in (\Sigma \cup \Theta)^* \times (\Sigma \cup \Theta)^*$ usually denoted by $u = v$. A *size* of an equation is the sum of lengths of u and v. A *solution* of a word equation $u = v$ is a morphism $h : (\Sigma \cup \Theta)^* \to \Sigma^*$ such that $h(a) = a$, for $a \in \Sigma$, and $h(u) = h(v)$. We say that a language L is *expressible*, if there is an equation e and a variable X such that

$$L = \{h(X) : h \text{ is a solution of } e\}.$$

Similarly, we say that a property $\mathcal{R} \in (\Sigma^*)^k$ is *expressible* by an equation e if there are variables $X_1, \ldots, X_k$ such that

$$\mathcal{R} = \{(h(X_1), \ldots, h(X_k)) : h \text{ is a solution of } e\}.$$

The property of the expressibility depends on the sizes of the alphabets Σ and Θ. In this paper we concentrate to the case when the alphabet Σ is finite. We also assume that $|\Sigma| \geq 2$. In the case of a unary alphabet all expressible languages are trivially regular. Denote by $\mathcal{L}(\Sigma)$ the family of expressible languages over the alphabet Σ.

Example 1. The properties:

- W is not square-free, and
- those words W in $\{a, b, c\}^*$ which contain a letter c

are expressible. Indeed, the former is obtained from the equation $W = XUUY$ under the extra condition $U \neq \varepsilon$, so that, by Theorem 2, the whole property can be encoded into one equation. The latter one is expressed by the equation $W = XcY$.

Example 2. Every finite and co-finite language over a finite alphabet Σ is expressible. Indeed, for $L = \{w_1, \ldots, w_t\} \subseteq \Sigma^*$, L and $\Sigma^* - L$ are expressed by the formulae

$$\bigvee_{i=1}^{t} X = w_i$$

and

$$(\bigvee_{w \neq w_i, |w| \leq N} X = w) \text{ or } (\bigvee_{|w|=N+1} X = wY),$$

where $N = \max\{|w_i| : i = 1, \ldots, t\}$. As above Theorem 2 makes it possible to express these formulae using only one equation.

Example 3. The properties

- W is imprimitive, and
- W is not minimal in its conjugacy class with respect to the lexicographic ordering $\prec$

are expressible, too. This follows, as above, from the formula

$$WZ = ZW \text{ and } Z = WT \text{ and } T \neq 1$$

and

$$W = UV \text{ and } W' = VU \text{ and } W' \prec W$$

after the observation that the relation $W' \prec W$ is expressible by the formula

$$\bigvee_{a \prec b} (W' = RaT \text{ and } W = RbT').$$

After these examples we formulate several closure properties of expressible languages and relations. Our first result is very easy.

Theorem 1. *The family of expressible languages is closed under the following operations: catenation, cyclic closure, and Klenee star of a single word.*

Our second result, which we have already used several times, deals with the closure properties under Boolean operations.

Theorem 2. *Let $e : u = v$ and $e' : u' = v'$ be two equations. Then*

1. *A property expressible by e and e' is expressible by a single equation without any additional variables.*
2. *A property expressible by e or e' is expressible by a single equation using two additional variables.*
3. *The relation satisfying $u \neq v$ is expressible by a single equation using a finite number of additional variables.*

Theorem 2 deserves a few comments. First, we have a new proof of case 2 which is a simplification of the proof presented in [6] which, in turn, was based on ideas of S. Grigorieff [8]. Second, it clearly gives more closure properties of expressible languages and relations, such as

Corollary 3. *Any language or relation of words expressible by a formula built on word equations using operations of conjunction, disjunction and negation is expressible by a single equation.*

Third, with the case 3 one has to be carefull. It says that the complement of the relation defined by an equation $u = v$ using **all variables of the equation** is expressible by a single equation (using additional variables). This, however, does not mean that expressible languages are closed under the complementation. In fact, they are not, as we shall show in Section 5. Of course, in some special cases, such a closure might hold.

We conclude this section by stating two more closure properties of the family of expressible languages.

Theorem 4. *The expressible languages are closed under*

1. *finite intersections, and*
2. *finite unions.*

3 Expressibility of languages by equations with two variables

In this section we introduce technical tools and apply those to languages expressible using only two variables. First, given a vector $\mathbf{z}$ of natural numbers, we define an equivalence relation $\mathcal{R}_\mathbf{z}$ on positions in words determined by solutions specified by a vector $\mathbf{z}$ of lengths of words constituting a solution of an equation. The intuition behind the definition of $\mathcal{R}_\mathbf{z}$ is as follows. Consider a fixed equation $u = v$, and fix the lengths of the components of a solution by the vector $\mathbf{z}$. This fixes the lengths of both sides of $h(u) = h(v)$. But this is an identity in Σ^+ so that corresponding positions on both sides must be filled with the same letter. This induces via $\mathcal{R}_\mathbf{z}$ the equivalence classes $\mathcal{X}$ above. These classes may contain constants, i.e. pairs of the form $(1, a)$ with $a \in \Sigma$, or unfixed parts of the variables, i.e. pairs of the form (i, X) corresponding to the i-th letter of X. Of course, in a concrete solution the second components of an equivalence class must coincide.

Assume that an equation e contains t variables $X_1, X_2, \ldots X_t$ and $\mathbf{z} = (z_1, \ldots, z_t)$ is a vector of t natural numbers. We say that h is a $\mathbf{z}$-*solution* of e if h is a solution of e and $|h(X_j)| = z_j$, for $1 \leq j \leq t$. For a vector $\mathbf{z} = (z_1, \ldots, z_t)$ we define a function $|\cdot|_\mathbf{z} : (\Theta \cup \Sigma)^* \to N$ by

$$|u|_\mathbf{z} = \begin{cases} z_m & \text{if } u = X_m \in \Theta, \\ 1 & \text{if } u \in \Sigma, \\ \sum_{k=1}^s |u_k|_\mathbf{z} & \text{if } u = a_1 a_2 \ldots a_s \text{ with } a_j \in \Theta \cup \Sigma. \end{cases}$$

In other words $|w|_\mathbf{z}$ is the length of the word $h(w)$ if h is a z-solution of some equation.

Now, assume that we are given an equation $u_1 \ldots u_k = v_1 \ldots v_s$ over t variables and a vector $\mathbf{z} \in N^t$ such that $|u|_\mathbf{z} = |v|_\mathbf{z}$. We define a function $left_\mathbf{z}$: $\{1, \ldots, |u|_\mathbf{z}\} \to N \times (\Theta \cup \Sigma)$ in the following way:

$$left_\mathbf{z}(j) = (r, x) \text{ iff}$$
$$|u_1 \ldots u_p|_\mathbf{z} < j \leq |u_1 \ldots u_{p+1}|_\mathbf{z} \text{ and } r = j - |u_1 \ldots u_p|_\mathbf{z} \text{ and } u_{p+1} = x$$

Similarly, we define the function $right_\mathbf{z}$:

$$right_\mathbf{z}(j) = (r, x) \text{ iff}$$
$$|v_1 \ldots v_p|_\mathbf{z} < j \leq |v_1 \ldots v_{p+1}|_\mathbf{z} \text{ and } r = j - |v_1 \ldots v_p|_\mathbf{z} \text{ and } v_{p+1} = x$$

An equivalence relation $\mathcal{R}_\mathbf{z}$ on positions $\{1 \ldots |u|_\mathbf{z}\}$ is the transitive closure of the relation $\mathcal{R}'_\mathbf{z}$ defined by

$i\mathcal{R}'_\mathbf{z} j$ iff $left_\mathbf{z}(i) = right_\mathbf{z}(j)$ or $left_\mathbf{z}(i) = left_\mathbf{z}(j)$ or $right_\mathbf{z}(i) = right_\mathbf{z}(j)$.

We say that a position i *belongs* to a variable X if either $left_\mathbf{z}(i) = (j, X)$ or $right_\mathbf{z}(i) = (j, X)$, for some j. Let $\mathcal{X}$ be an equivalence class of the relation $\mathcal{R}_\mathbf{z}$. We say that $\mathcal{X}$ *corresponds* to a constant a if there is a position i in $\mathcal{X}$ such that either $left_\mathbf{z}(i) = (1, a)$ or $right_\mathbf{z}(i) = (1, a)$.

Example 4. Consider an equation $e : aX_1X_2bX_1 = X_3X_4X_3$. Let $\mathbf{z} = (2, 4, 5, 0)$. Then the values of the functions $left_\mathbf{z}$ and $right_\mathbf{z}$ are listed below.

	1	2	3	4	5
$left_\mathbf{z}$	$(1, a)$	$(1, X_1)$	$(2, X_1)$	$(1, X_2)$	$(2, X_2)$
$right_\mathbf{z}$	$(1, X_3)$	$(2, X_3)$	$(3, X_3)$	$(4, X_3)$	$(5, X_3)$

	6	7	8	9	10
$left_\mathbf{z}$	$(3, X_2)$	$(4, X_2)$	$(1, b)$	$(1, X_1)$	$(2, X_1)$
$right_\mathbf{z}$	$(1, X_3)$	$(2, X_3)$	$(3, X_3)$	$(4, X_3)$	$(5, X_3)$

Then the equivalence classes of $\mathcal{R}_\mathbf{z}$ are $\mathcal{X} = \{1, 6\}$, $\mathcal{Y} = \{3, 5, 8, 10\}$ and $\mathcal{Z} = \{2, 4, 7, 9\}$. The equivalence classes $\mathcal{X}$ and $\mathcal{Y}$ correspond to the constants a and b since $left_\mathbf{z}(1) = (1, a)$ and $left_\mathbf{z}(8) = (1, b)$, respectively. The equivalence class $\mathcal{Z}$ does not correspond to any constant. Hence, the positions in $\mathcal{Z}$ can be filled with any letter and, by case 4 of Lemma 5, they can be replaced by any word as well. This gives the following family of solutions of the equation e:

$$X_1 = \beta a, X_2 = \beta b a \beta, X_3 = a \beta b \beta b, X_4 = \varepsilon,$$

where β can be replaced by any word.

The above procedure, illustrated in Example 4, can be seen as a method of filling the positions of the variables in an equation. This simple method, which was first used in [15], can be used, for example, to give a very illustrative proof for the periodicity theorem of Fine and Wilf, cf. e.g. [6].

Now the following lemma is obvious. Denote by $w[i]$ the i-th letter of the word w.

Lemma 5. *Let $\mathcal{C}$ be an equivalence class of the relation $\mathcal{R}_\mathbf{z}$ connected to an equation $e : u = v$. Then the following conditions are satisfied:*

1. *For any two positions $i, j \in \mathcal{C}$ and a $\mathbf{z}$-solution h of e, $h(u)[i] = h(u)[j]$.*
2. *If $\mathcal{C}$ corresponds to a constant a, then for each $\mathbf{z}$-solution h of e, $h(u)[i] = a$.*
3. *If $\mathcal{C}$ corresponds to two different constants a and b, then the equation e has no $\mathbf{z}$-solution.*
4. *If $\mathcal{C}$ does not correspond to any constant and e has a $\mathbf{z}$-solution, then replacing the positions in $\mathcal{C}$ by any word produces a new solution of e.*

Note, that in case 4 the new solutions obtained need not be $\mathbf{z}$-solutions anymore.

In a formulation of our results we need a notion of a pattern language from [2], cf. also [11]. A *pattern* is a word over the alphabet $\Theta \cup \Sigma$. A *pattern language* generated by a pattern w is the set of all words which are morphic images of w under all morphisms $h : (\Theta \cup \Sigma)^* \to \Sigma^*$ satisfing $h(a) = a$, for a in Σ. In particular, it is natural to denote by $p((\Sigma^*)^k)$ the pattern language generated by a pattern $p(X_1, X_2, \ldots, X_k)$ containing k variables $X_1, X_2, \ldots, X_k$. We have an obvious connection:

Example 5. Each pattern language is expressible. Let u be a pattern and Z be a variable which does not occur in u. A variable Z in equation $Z = u$ expresses the pattern language generated by u.

We also need an auxilary lemma which follows rather straightforwardly from Lemma 5 and which holds for any number of variables.

Lemma 6. *Let L be an expressible language via a variable X in an equation e. Suppose that there is no one variable pattern $p(Y)$ such that $p(\Sigma^*) \subseteq L$. Then for each vector $\mathbf{z}$ there is a word $w \in L$ such that for each $\mathbf{z}$-solution h of e $h(X) = w$.*

Now denoting by $\#_L(n)$ the number of words of length n in the langauge L, we are ready to prove the main result of this section.

Theorem 7. *Let L be an expressible language by an equation on two variables. Then either $\#_L(n) = O(n)$ or there is a pattern $p(Y)$ with one variable such that $p(\Sigma^*) \subseteq L$.*

As a straightforward consequence of Theorem 7 we obtain a gap theorem for possible complexities of the function $\#_L(n)$. Note here that for each language L we have $\#_L(n) = 2^{O(n)}$.

Corollary 8. *Let L be expressible by an equation with two variables. Then either $\#_L(n) = O(n)$ or $\#_L(an + b) = 2^{\Omega(n)}$, for some constants a, b.*

4 Main results

This section is devoted to prove some pumping-like properties of expressible languages. These are achieved by using the tools of the previous section, and, more importantly, by considernig special types of factorizations of words to generalize a technique in [5], cf. also [14], which was used to prove an upper bound for an index of the periodicity of a minimal solution of a word equation.

We recall that an F-factorization of a word w is any sequence $w_1, \ldots, w_k$ of words from a language F such that $w = w_1 \ldots w_k$. We generalize it as follows. Let $\mathcal{F}$ be a property of sequences of words. We say that a sequence $w_1, \ldots, w_k$ is an $\mathcal{F}$-factorization of w if $w = w_1 \ldots w_k$ and the sequence $w_1, \ldots, w_k$ satisfies $\mathcal{F}$. The factors w_1 and w_k are called *outer factors* of w and the other factors are called *inner factors* of w. Further we say that a property $\mathcal{F}$ defines *synchronizing factorizations*, or briefly that $\mathcal{F}$ is *synchronizing*, if the following holds:

1. Each word admits a unique $\mathcal{F}$-factorization.
2. If a word w admits an $\mathcal{F}$-factorization $v_1, \ldots, v_k$ then, for each symbol a in Σ the word aw admits either an $\mathcal{F}$-factorization $u, v, v_2, \ldots, v_k$, where $uv = av_1$, or an $\mathcal{F}$-factorization $av_1, \ldots, v_k$, and the word wa admits either an $\mathcal{F}$-factorization $v_1, \ldots, v_{k-1}, u, v$, where $uv = v_k a$ or an $\mathcal{F}$-factorization $v_1, \ldots, v_k a$.

Note that our notion of an $\mathcal{F}$-factorization is connected to but not the same as that of a factorization of a free monoid, cf. [3, 16]. These factorizations are used to decompose free monoids, while in our considerations a focus is on factorizations of a single word. Note also that the above conditions (1) and (2) could be named separately: factorizations satisfying (1) could be called uniquely deciphering and those satisfying (2) synchronizing. We prefered the chosen terminology since all factorizations considered here satisfy (1). Finally note that conditions (1) and (2) could be defined with respect to a language L: each word of L should satisfy these conditions.

With the above notions we have the following obvious lemma.

Lemma 9. *Assume that a property $\mathcal{F}$ defines a synchronizing factorization and that $x_1, x_2, \ldots, x_k$ and $y_1, y_2, \ldots, y_l$ are $\mathcal{F}$-factorizations of words x and y, respectively. Then, if y is a subword of x and the factor y_1 of y ends inside factor x_i of x and factor y_l starts inside a factor x_j, then $j - i = l - 1$ and $y_2 = x_{i+1}, y_3 = x_{i+2}, \ldots, y_{l-1} = x_{j-1}$.*

We say that an $\mathcal{F}$-factorization is synchronizing with a *finite delay*, if there are numbers q, r such that for each word x with an $\mathcal{F}$-factorization $x_1, \ldots, x_k$ and each subword y of x with an $\mathcal{F}$-factorization $y_1, \ldots, y_l$ if the factor y_1 of y ends in factor x_i of x and the factor y_l starts in x_j, then y_1 is a suffix of $x_{\max\{i-q,1\}} \ldots x_i$ and y_l is a prefix of $x_j \ldots x_{\min\{j+r,k\}}$.

Let $w_1, \ldots, w_k$ be a factorization of w. We say that this factorization of w *synchronizes* with a pattern p iff $p = u_1 u_2 \ldots u_k$ where, for all i, either u_i is a variable or $u_i = w_i$. Let $n_{\mathcal{F}}(w)$ be the number of different words in the factorization of w. For a language L, denote $n_{\mathcal{F}}(L) = \max\{n_{\mathcal{F}}(w) : w \in L\}$. Now we formulate the first tool to show the nonexpressibility.

Theorem 10. *Let L be an expressible language and $\mathcal{F}$ a property definining finite delay synchronizing factorizations. Then there exists a number k such that for each $w \in L$ satisfying $n_{\mathcal{F}}(w) > k$ there is a pattern $p(X_1, \ldots, X_s)$ with $s = n_{\mathcal{F}}(w) - k$ variables synchronizing with a $\mathcal{F}$-factorization of w and satisfying $p((\Sigma^*)^s) \subseteq L$.*

Next we define, for each primitive word P, particular factorizations which turns out to be synchronizing. Let P be a primitive word. Then, as is well-known, each word w can be uniquely written in the form $w = w_1 P^{x_1} w_2 \ldots P^{x_{k-1}} w_k$, where

- w_i does not contain P^2 as a subword,
- P is a proper prefix of w_i, for $1 < i \leq k$,
- P is a proper suffix of w_i, for $1 \leq i < k$,
- $x_i \geq 0$, for $1 \leq i \leq k-1$.

These conditions clearly defines an instance of an $\mathcal{F}$-factorization, we call it $\mathcal{F}_P$-factorization. Moreover, as is straightforward to see it is synchronizing and with a finite delay. Next we set

$$T(w) = \{x_i : P^{x_i} \text{ is a factor in a } P\text{-factorization of } w\}$$

and define *the index* of w with respect to P, by the formula

$$exp_P(w) = \max\{x_i : x_i \in T(w)\}.$$

Now we formulate our second tool to show the nonexpressibility.

Theorem 11. *Let L be an expressible language and P be a primitive word. Then there exists a natural number k such that for each word u in L satisfying $exp_P(u) > k$ there is a word w in L with $exp_P(w) \leq k$ and which is obtained from u by removing some occurrences of P.*

Theorem 11 can be used to prove the following characterization of expressible relations concerning lengths of variables.

Let f be a function $f : N^r \to N$. We say that a property $f(|X_1|, \ldots, |X_r|) = 0$ is expressible, if the relation

$$\{(w_1, \ldots, w_r) : f(|w_1|, \ldots, |w_r|) = 0\}$$

is expressible.

Corollary 12. *Let X_1, X_2, $\ldots$, X_r be r different variables. If a property*

$$f(|X_1|, |X_2|, \ldots, |X_r|) = 0$$

is expressible, then there is a constant k such that if $f(i_1, i_2, \ldots, i_r) = 0$, for some $i_1, \ldots, i_r$, and $i_s > k$ then also

$$f(i_1, i_2, \ldots, i_{s-1}, p, i_{s+1}, \ldots, i_r) = 0, \text{ for some } p < k.$$

5 Applications of main results

In this section we apply our results of the last section to achieve our original goal: to prove that several very natural properties of words are not expressible. We recall that to our knowledge no such result is known in litterature except for the property "X being a prefix of Y" that cannot be expressed without using additional unknowns, cf. [19].

Example 6. The language $L_1 = \{a^n b^n : n \geq 1\}$ is not expressible. We prove it by a contradiction applying Theorem 11, for $P = a$. Let k be a constant from Theorem 11. Take a word $w = a^{k+1} b^{k+1}$. Since $w \in L_2$ and $exp_a(w) > k$ there is a word u in L_1, which is obtained from w by removing some occurrences of the word a. A contradiction.

Example 7. The property "w is primitive" is not expressible. Now we can apply Theorem 10. Let $\mathcal{F}$ be a factorization defined by dividing word into blocks of the same letters. Clearly, $\mathcal{F}$ has the synchronizing property. Assume the property "w is primitive" is expressible and let k be a constant from Theorem 10. Consider a word $w = a^{k+1} b a^k b \ldots a b$ which admits the factorization $a^{k+1}, b, a^k, b, \ldots, b, a, b$. Since $n_{\mathcal{F}}(w) = k+2$, by Theorem 10, there is a pattern with two variables and one of them corresponds to a factor of w of the form a^i. Since each factor of this form occurs in w exactly once, the variable occurs exactly once in the pattern. The results now follows from the fact that the word $w_1 X w_2$ is a square if $X = w_2 w_1$.

Example 8. The language $L_2 = (a \cup b)^*$ over three-letter alphabet $\Sigma = \{a, b, c\}$ is not expressible. In the same way as in the previous example we prove that if L_2 is expressible, then there is a pattern $p(X)$ such that $p(\Sigma^*) \subseteq L_2$. Substituting $X = c$ we obtain a contradiction.

Example 9. The relation "x and y are of equal length", i.e.

$$\mathcal{T} = \{(x,y) \in \Sigma^* \times \Sigma^* : |x| - |y| = 0\}$$

is not expressible. This is due to Corollary 12. Observe here, that the relation $\mathcal{T}$ is expressible if $|\Sigma| = 1$.

As a consequence of the above examples we easily obtain.

Theorem 13. *The family of expressible languages is not closed under the operations of complementation, morphic image, inverse morphic image and shuffle.*

We conclude this section by emphasizing that the combination of closure and nonclosure properties of expressible languages, especially closure under intersection and union and nonclosure under complementation and morphic image, makes the family quite different from usually considered families of formal languages.

6 Comparisons with other families of languages

We already pointed out that the nonclosure and closure properties of $\mathcal{L}(\Sigma)$ makes this family different from most of the usually studied families of languages. We further emphasize this fact by the following theorem which is proved by considering particular languages.

Theorem 14. *1. $\mathcal{L}(\Sigma)$ is a proper subset of the family of recursive languages over Σ.*
2. $\mathcal{L}(\Sigma)$ is incomparable with the families of D0L, regular and context-free languages.

7 Concluding remarks

As a major contribution of this paper we introduced - according to our knowledge - first tools to show that certain properties of words are not expressible as solutions of word equations, or more precisely as values of some components of solutions of word equations. Our tools were based on special factorizations of words, which we called synchronizing.

As applications of our results several concrete properties of words were shown to be nonexpressible by word equations, as well as several nonclosure properties of expressible languages were obtained.

On the other hand, we also stated many known closure properties of expressible languages, and in particular gave a shorter proof for the fact that expressible properties are closed under disjunction.

Finally, it is worth mentioning that there remains a lot of research to be done on this interesting and fundamental field. We point out here just a few open problems:

- **Problem 1.** Is the relation "u is a sparse subword (subsequence) of v" expressible?
- **Problem 2.** Are the properties "w is square-free" and "w is a Fibonacci word" expressible? Recall, that Fibonacci words are defined by recurrence formulae $w_0 = a$, $w_1 = b$, $w_{n+2} = w_{n+1}w_n$, for $n \geq 2$.
- **Problem 3.** When is the complement of an expressible language expressible?
- **Problem 4.** Is our gap theorem true for languages expressible by word equations with more than two variables?

References

1. Albert, M.H., and Lawrence, J., A proof of Ehrenfeucht's Conjecture, *Theoret. Comput. Sci.* **41**, 121-123, 1985.
2. Angluin D., Finding pattern common to a set of strings, *in* Proceedings of STOC'79, 130-141, 1979.
3. Berstel, J., and Perrin D., *Theory of Codes*, Academic Press, 1985.

4. Büchi, R. and Senger, S., Coding in the existential theory of concatenation, *Arch. Math. Logik*, **26**, 101-106, 1986/87.
5. Bulitko, V.K., Equations and inequalities in a free group and a free semigroup, *Tul. Gos. Ped. Inst. Ucen. Zap. Mat. Kafedr. Geometr. i Algebra*, **2**, 242-252, 1970 (in Russian).
6. Choffrut, C., and Karhumäki, J., Combinatorics of words, *in* G.Rozenberg and A.Salomaa (eds), *Handbook of Formal Languages*, Springer, 1997.
7. Culik II, K., and Karhumäki, J., Systems of equations and Ehrenfeucht's conjecture, *Discr. Math.*, **43**, 139-153, 1983.
8. Grigorieff, S., Personal comunication.
9. Guba, V., The equivalence of infinite systems of equations in free groups and free semigroups to their finite subsystems, *Matem.Zametki*, 40 (3), September 1986 (in Russian).
10. Harrison, M.A., *Introduction to Formal Language Theory*, Addison-Wesley Publishing Company, 1978.
11. Jiang T., Salomaa A., Salomaa K., Yu S., Decision problems for patterns, *J. Comput. Sys. Sciences* **50**, 53-63, 1995.
12. Khmelevski, Yu. I., Solution of word equations in three variables, *Dokl.Akad.Nauk. SSSR*, 177, 1023-1025, 1967 (in Russian).
13. Khmelevski, Yu. I., Equations in free semigroups, *Trudy Mat. Inst. Steklov*, 107, 1971 (English translation: *Proc. Steklov Inst. of Mathematics 107 (1971)*, American Mathematical Society, 1976.)
14. Koscielski, A., and Pacholski, L., Complexity of Makanin's algorithm, *J. ACM* **43**(4), 670-684, 1996.
15. Lentin, A., *Equations dans des Monoides Libres*, Gouthiers-Villars, 1972.
16. Lothaire, M., *Combinatorics on Words*, Addison-Wesley, 1993.
17. Makanin, G.S., The problem of solvability of equations in a free semigroup, *Mat. Sb.*, Vol. 103,(145), 147-233, 1977. English transl. in *Math. U.S.S.R.* Sb. Vol 32, 1977.
18. Matijasevich, Y., Enumerable sets are diophantine, *Soviet. Math. Doklady* 11, 354-357, 1970. English transl. in *Dokl. Akad. Nauk SSSR* 191, 279-282, 1971.
19. Seibert, S., Quantifier hierarchies and word relations, Springer LNCS 626, 329-338 (1992).

Finite Loops Recognize Exactly the Regular Open Languages*

Martin Beaudry[†]　　François Lemieux[‡]　　Denis Thérien[§]

Abstract

In this paper, we characterize exactly the class of languages that are recognizable by finite loops, i.e. by cancellative binary algebras with an identity. This turns out to be the well-studied class of regular open languages. Our proof technique is interesting in itself: we generalize the operation of block product of monoids, which is so useful in the associative case, to the situation where the left factor in the product is non-associative.

1 Introduction

The algebraic approach in the study of regular languages, based on considering finite monoids as language recognizers, certainly is the most powerful tool available for understanding computations realized by finite-state automata. It has developed into a rich and coherent framework to relate combinatorial descriptions of regular languages and algebraic properties of their recognizers [10, 16]. An early example of such relationship is the famous theorem of Schützenberger [24]: a subset of A^* is star-free (i.e. can be obtained from finite sets using Boolean operations and concatenation) iff it can be recognized by a group-free monoid (i.e. in which no subset forms a non-trivial group).

In much the same way that monoids can be used to recognize languages, one may also consider other types of algebras and study their computational power. For example, non-associative binary algebras, usually called groupoids, are exactly the recognizers needed for context-free languages: this relationship has been well-known in theory of tree automata (see [12]) and can be traced back to the paper of Mezei and Wright [15]. It has also been used in complexity theoretic work (e.g. see [26, 5]). In view of this connection, it is natural to try to characterize the languages recognized by various specific subclasses of finite groupoids. One such class, that has been extensively studied in the past

---

*Work supported by FCAR (Québec) and CRSNG (Canada)

[†]Département de mathématiques et d'informatique, Université de Sherbrooke, Sherbrooke (Qc) Canada, J1K 2R1, beaudry@dmi.usherb.ca

[‡]School of Computer Science, McGill University, 3480 rue University, Montréal (Qc) Canada, H3A 2A7, lemieux@cs.mcgill.ca

[§]**Corresponding author**: School of Computer Science, McGill University, 3480 rue University, Montréal (Qc) Canada, H3A 2A7, denis@cs.mcgill.ca

[7, 8, 1, 6], consists of *loops*, i.e. groupoids with an identity and for which every row and every column of the multiplication table contains every element. In [9] it was proved that any language recognized by a finite loop must be regular. The main result of our paper gives an exact characterization of which languages can be recognized by loops.

The answer is surprizing and elegant: a language $L \subseteq A^*$ can be recognized by a finite loop iff L is regular and open in the group topology on A^*. This topology, introduced by [21, 22], is the smallest one such that every morphism from A^* onto a finite group is continuous; investigations of its properties were motivated early on by several deep connections with important questions about finite monoids [17, 18]. Our result thus adds on a new perspective to a class of languages which already has a significant history.

The paper is organized as follows: in section 2, we introduce most of the relevant definitions that will be needed. In section 3, we present some tools that are useful in constructing loops to recognize languages. In section 4, we show that loops recognize only regular open languages. In section 5, we prove that every regular open language is recognizable by a finite loop. We derive some consequences of this theorem in the last section where we also present some ideas for further applications of our techniques.

2 Preliminaries

In this section, we introduce our notation and review some elementary facts about monoids and groupoids.

Let A be a finite set: we write A^* for the *free monoid* generated by A, i.e. for the set of all words of finite length over the alphabet A, concatenation being the associative operation. The *length* of a word x is denoted by $|x|$ and ϵ stands for the unique word of length 0, which is the identity element of the free monoid. A *congruence* on A^* is an equivalence relation α that is compatible with concatenation, i.e. $x_1 \,\alpha\, y_1$ and $x_2 \,\alpha\, y_2$ imply $x_1 x_2 \,\alpha\, y_1 y_2$. The quotient A^*/α is then an A-generated monoid, and every A-generated monoid is of this form. A language $L \subseteq A^*$ is *recognized* by the monoid M iff there exist a morphism $\phi : A^* \to M$ and a subset F of M such that $L = \{x \in A^* : \phi(x) \in F\}$; equivalently we can view the morphism as going from A^* to M^*, transforming a word x into a string of monoid elements which is then evaluated in M; in this point of view, only alphabetical morphisms need to be considered, that is morphisms mapping letters to letters. We observe that when L is recognized by M, $\phi(A^*)$ is a submonoid of M isomorphic to A^*/α for some congruence α and L is a union of α-classes. It is well-known that a language is regular iff it can be recognized by a finite monoid, i.e. iff it is a union of α-classes for some congruence α of finite index. We will say that $L \subseteq A^*$ is a group language iff L can be recognized by a finite group.

The notions above have natural counterparts in the non-associative world. A *groupoid* is given by a set and a binary operation: we will assume here that every groupoid contains a 2-sided identity element. The *free groupoid* generated by A will be denoted by $A^{(*)}$. It will be convenient to think of an element in $A^{(*)}$ as a pair (t, x) where x is a word in the free monoid and t is a rooted binary tree with $|x|$ leaves; in particular, the identity of the free groupoid is then $(\emptyset, \epsilon)$. The product of (t_1, x_1) with (t_2, x_2) is then the pair $(t_1 t_2, x_1 x_2)$ where $t_1 t_2 = t_1$ if t_2 is empty, $t_1 t_2 = t_2$ if t_1 is empty, and otherwise $t_1 t_2 = t$ is the tree obtained by joining the root of t_1 and the root of t_2 to a new node, which becomes the root of t. If $g \in A^{(*)}$ is identified in this way with a pair (t, x) we define Tree$(g) = t$ and Yield$(g) = x$. We say that $g_1, g_2 \in A^{(*)}$ are *yield-equivalent* if Yield$(g_1) =$ Yield(g_2). We can view each row and each column in the multiplication table of the groupoid G as defining a mapping from G to G. The closure of these mappings under the operation of composition is called the *multiplicative monoid* of G, denoted by $\mathcal{M}(G)$.

Congruences on $A^{(*)}$ are defined in the same way as in the associative case. If α is a congruence on $A^{(*)}$, the quotient $A^{(*)}/\alpha$ is an A-generated groupoid, and every A-generated groupoid arises in this way. A *loop* G is a groupoid whose multiplication table contains every element in each row and in each column; clearly, this will be the case iff $\mathcal{M}(G)$ is a group. Equivalently, a loop is a groupoid that is left and right cancellative, i.e. $ab = ac$ implies $b = c$ and $ba = ca$ implies $b = c$, for any $a, b, c \in G$.

We now wish to use groupoids to recognize subsets of A^*; note that if G is not associative, the notion of a morphism from A^* to G is not well-defined. We say that the language $L \subseteq A^*$ is recognized by the groupoid G if there exist an alphabetic morphism $\phi : A^* \to G^*$ and a subset F of G such that $L = \{x \in A^* : G(\phi(x)) \cap F \neq \emptyset\}$, where $G(\phi(x))$ is the set of elements of G obtained by evaluating the string $\phi(x)$ of G^* in all possible ways. Note that if G is associative, there is only one way of evaluating $\phi(x)$ and we are back to the definition given for monoids. We will say that $L \subseteq A^*$ is a *loop language* iff L can be recognized by a loop. In terms of congruences, the groupoid G recognizes the language L iff there exist an A-generated subgroupoid of G isomorphic to $A^{(*)}/\alpha$ and a subset F of this subgroupoid such that $x \in L$ iff there is some tree t such that $[(t, x)]_\alpha$ is in F. One pleasant feature of this notion of language recognition is

Lemma 2.1 *[5] L is recognizable by a finite groupoid iff L is context-free.*

The *finite group topology* on A^* is the smallest topology such that every morphism from A^* onto a finite group is continuous. It is equivalent to say that the group languages form a basis for this topology. It was first introduced by Hall [13] for the free group, and by Reutenauer for the free monoid [21, 22]. Connections were soon discovered between some classical problems about finite monoids and computing the closure of a given regular language for this topology [17, 18]; it thus became an important question to characterize which regular languages are open or closed. A sequence of deep results [2, 3] finally led to the following combinatorial characterization for the regular open sets [19].

Lemma 2.2 *A regular language is open iff it is a finite union of languages of the form $L_0 a_1 L_1 \ldots a_k L_k$ where the a_i's are letters and the L_i's are group languages.*

3 Recognizing languages with loops

The aim of this section is to prove that any language of the form $B^*a_1B^* \cdots B^*a_kB^*$, where B is a finite alphabet and $a_i \in B$, can be recognized by a finite loop. This result will be of great help in proving Theorem 5.3.

In general, it is not an easy task to construct directly a loop B that recognizes a given language L. What can be done instead, is to construct a partially defined loop G that recognizes L, and then, embed G into a loop B. This motivates the following definition.

A groupoid G with an absorbing element, denoted 0, is *weakly cancellative* if for any $a, x, y \in G$, the two properties $(ax = ay \neq 0) \Rightarrow (x = y)$ and $(xa = ya \neq 0) \Rightarrow (x = y)$ are satisfied.

The Cayley table of a weakly cancellative groupoid is such that in each row and each column no nonzero element appears twice. Hence, the nonzero elements of such a groupoid form a partially defined groupoid which we call an *incomplete loop*. This terminology is justified by the following lemma.

Lemma 3.1 ([11]) *An incomplete loop containing n elements can be embedded in a loop containing t elements, for any $t \geq 2n$.* □

Recall that if Q is a loop and $w \in Q^+$ then $Q(w)$ is the set of elements that result from evaluating w using all possible bracketings.

Lemma 3.2 *Let Q be a loop and let $u, v, w \in Q^+$. Then, the cardinality of $Q(uwv)$ is at least as large as that of $Q(w)$.*

Weakly cancellative groupoids will be useful to prove that a language can be recognized by a loop. This is a consequence of the following lemma.

Lemma 3.3 *Any language recognized by a weakly cancellative groupoid, with 0 in the accepting set, is also recognized by a loop.*

Proof. Let G be a weakly cancellative groupoid, and let $L \subseteq G^*$ be a language recognized by G. Assume that 0 belongs to the accepting set. Let $B = G - \{0\}$, let $B^{(*)}$ be the free groupoid over the set B, and let β be the cardinality of B. We also denote by B the incomplete loop induced by the elements of B in G.

We will define a sequence of incomplete loops B_i, for $i \geq 0$. Let $B_0 = B$ and define B_{i+1} from B_i as follows. All products defined in B_i are defined identically in B_{i+1}. Moreover, for any undefined product $a \cdot b$ in B_i, we define $a \cdot b = (ab)$ in B_{i+1}

Remark. *Observe that for any $a, b \in B$, if the product ab is not defined in B_k, then $c = ab$ is a new element in B_{k+1}. Moreover, for any $d \in B_{k+1}$, the products cd and dc are not defined in B_{k+1}. Those products generate two new elements in B_{k+2}, and so on. This and Lemma 3.2 imply that for any $u, v \in B^*$ such that $k = |u| + |v|$, $B_{k+1}(uabv)$ contains at least k elements.*

Let $k = \beta + 2$ and let B_k be embedded in a finite loop H. We will argue that L is recognized by H with the accepting set containing all nonzero elements of the accepting set of G plus all elements not in B.

If $w \in B^*$ can be evaluated to a nonzero element in G, then w can be evaluated to the same element in H using the same parenthesization. This shows that if $w \in B^*$ cannot be evaluated to 0 in G, then w is accepted by G if and only if it is accepted by H.

Suppose that w can be evaluated to 0 in G. Then, there exists a segment u of w of minimal length that can be evaluated to 0, i.e. $w = sut$, $0 \in G(u)$ and for any strict segment v of u, $0 \notin G(v)$. So, there exist $u_1, u_2 \in B^+$ and $a, b \in B$ such that $u = u_1 u_2$, $a \in G(u_1)$, $b \in G(u_2)$ and $ab = 0$ in G, but $a \neq 0$ and $b \neq 0$. This implies that w can be partially evaluated to $sabt$ both in G and in H. Now, there are two possibilities. First, if $|s| + |t| < k$, then $s(ab)t$ can only be evaluated, in H, to an element in $B^{(*)} - B$: in this case H accepts w. Otherwise, by the above remark, $H(w)$ contains at least $\beta + 1$ different elements, and so, at least one of them is not in B. Thus, H accepts w if and only if G accepts w. □

As an example of application of Lemma 3.3, we can show that any cofinite language is recognized by a finite loop. Since it is easy to see that no finite language can be recognized by a finite loop, the class of loop languages is not closed under complement.

Loops can also recognize languages that are not cofinite and are not recognized by a group. A simple example is the set OR $\subseteq \{0,1\}^*$, composed of all words that contain at least one 1. This language is recognized by U_1, the monoid defined by $00 = 0$ and $01 = 10 = 11 = 1$. Here, 0 is an identity and 1 is absorbing. Since U_1 is a weakly cancellative groupoid, the language OR can be recognized by a finite loop. It is easy to verify that the complement of OR cannot be recognized by any finite loop.

We close this section with a lemma that will play an important role in the proof of Theorem 5.3.

Lemma 3.4 *Let A be a finite alphabet and let $a_1, \ldots, a_k$ be elements of A ($k > 0$). Then $L_k = A^* a_1 A^* \cdots A^* a_k A^*$ is recognized by a finite loop.*

Proof. The proof is by induction on k. Observe first that $L_1 = A^* a_1 A^*$ can be recognized by the weakly cancellative groupoid U_1 discussed above. By Lemma 3.3, L_1 can also be recognized by a finite loop.

Let $k = i + j$, where $i, j > 0$. Then, there exists a finite loop Q_i that recognizes $L_i = A^* a_1 A^* \cdots A^* a_i A^*$ with the accepting set $F_i \subseteq Q_i$ and there exists a finite loop Q_j that recognizes $L_j = A^* a_{i+1} A^* \cdots A^* a_k A^*$ with the accepting set $F_j \subseteq Q_j$.

Consider the weakly cancellative groupoid Q defined as the loop $Q_i \times Q_j$ except that $(a,b)(c,d) = 0$ whenever $a \in F_i$ and $d \in F_j$. Then, Q recognizes the language $L_k = A^* a_1 A^* \cdots A^* a_k A^*$ with 0 as the accepting element. By Lemma 3.3, L is also recognized by a finite loop. □

4 Finite loops recognize only open regular languages

In [9], it is shown that finite loops only recognize regular languages. In this section we refine this result by showing that only open regular languages can be recognized by such algebras. The following can be observed.

Lemma 4.1 *Any language $L \subseteq A^*$ of the form $L_0 \cdots L_k$, where L_i is recognized by a finite group, is open.*

To prove the next theorem, we will use the following definition. Let A be an alphabet and S a set of variables. A special tree t over A with variables in S is a binary tree where each element of S appears exactly once as a label of a leaf. We will use special trees in two particular situations: when S contains a single variable X; and when each leaf of t is labeled with a variable in S. In this last case we say that $t = t(x_1, \ldots, x_n)$ is a special tree with n leaves.

Let t be a special tree over A with the variable X and let t' be any tree. We denote by $t \cdot t'$ the tree obtained when the leaf in t labeled with X is replaced by t'. Observe that when t' is also a special tree with variable X, the result is still a special tree with variable X.

Observe also that $\cdot$ is an associative operation. Hence, for any special trees $t_1, \ldots, t_k$ over Q with variable X, the expression $t_1 \cdot t_2 \cdot \cdots \cdot t_k$ defines the same special tree no matter which parenthesization is used. This will be denoted by $\prod_{i=1}^{k} t_i$.

Similarly, if $s(x_1, \ldots, x_n)$ is a special tree with n leaves and $t_1, \ldots t_n$ are arbitrary trees, then $s(t_1, \ldots, t_n)$ is the tree obtained by substituting the tree t_i for the leaf labeled with x_i, for all i.

Theorem 4.2 *Finite loops recognize only open regular languages.*

Proof. We will use the technique of [9].

Let Q be a finite loop. We define a *comb* over Q recursively as follows. Any $a \in Q \cup \{\epsilon\}$ is a comb. If $a \in Q$ and $u \in Q^{(*)}$ is a comb then $w = (au)$ is also a comb. No other element of $Q^{(*)}$ is a comb. Hence, a comb $c \in Q^{(*)}$ corresponds to the left-to-right bracketing of Yield(c).

Any $t \in Q^{(*)}$ can be decomposed into $t = s(t_1, \ldots, t_n)$, where $n \geq 1$, $s(x_1, \ldots, x_n)$ is a special tree with n leaves, and t_i is a comb over Q. Let comb(t) be the smallest n for which such a decomposition exists.

We will show that, for any tree $t \in Q^{(*)}$, there exists a yield-equivalent tree $s \in Q^{(*)}$ evaluating to the same element and such that comb(s) is bounded by a constant. By Lemma 4.1, this will prove the theorem because the set of words in Q^* that left-to-right evaluate to a given element forms a language recognized by the multiplication group of Q.

More precisely, we will show that for any tree $t \in Q^{(*)}$ such that comb$(t) > 8^q$, where q is the order of Q, we can find a yield-equivalent tree $t' \in Q^{(*)}$ evaluating to the same element as t and such that comb$(t') <$ comb(t).

Suppose that $t \in Q^{(*)}$ is such that $\mathrm{comb}(t) = n > 8^q$, and let $t = s(t_1, \ldots, t_n)$ be decomposed as explained above. Since s has more than 8^q leaves, it must possess a path of length $k > 3q$. Let the nodes on this path be $d_0, d_1, \ldots, d_k$, where d_0 is the root of s and d_{i+1} is a child of d_i.

For $0 \leq i \leq q$, let s_i be the tree rooted at d_{3i}. Moreover, for $0 \leq i < q$, let v_i be the special tree constructed from s_i by substituting the variable X for s_{i+1}. Hence, for each v_i there exist four indices $1 \leq \alpha_i \leq \beta_i < \gamma_i \leq \delta_i \leq n$ such that the leaves at the left of X in v_i are labeled with $x_{\alpha_i}, \ldots, x_{\beta_i}$ and those on its right are labeled with $x_{\gamma_i}, \ldots, x_{\delta_i}$. Moreover, the leaves of s_q are labeled with $x_{\alpha_q}, \ldots, x_{\beta_q}$ for some $1 \leq \alpha_q \leq \beta_q \leq n$. We have $s = (\prod_{i=0}^{q-1} v_i) \cdot s_q$, where $v_i = v_i P(x_{\alpha_i}, \ldots, x_{\beta_i}, X, x_{\gamma_i}, \ldots, x_{\delta_i})$, and $s_q = s_q(x_{\alpha_q}, \ldots, x_{\beta_q})$.

We can thus write: $t = (\prod_{i=0}^{q-1} z_i) \cdot z_q$, where $z_i = v_i(t_{\alpha_i}, \ldots, t_{\beta_i}, X, t_{\gamma_i}, \ldots, t_{\delta_i})$ and $[z_q = s_q(t_{\alpha_q}, \ldots, t_{\beta_q})$.

Let $w_i = \mathrm{Yield}(t_i)$ and define l_i to be the comb whose yield is $w_{\alpha_i} \cdots w_{\beta_i}$, and r_i the comb whose yield is $w_{\gamma_i} \cdots w_{\delta_i}$. Then $\bar{z}_i = ((l_i X) r_i)$ is yield equivalent to z_i. Using the fact that our loop is both cancellative and finite, it is easily verified (Lemma 7 of [9]) that there exist two integers a and b such that t and t' evaluate to the same element, where t' is defined as $t' = (\prod_{i=0}^{a-1} z_i) \cdot (\prod_{i=a}^{b} \bar{z}_i) \cdot (\prod_{i=b+1}^{q-1} z_i) \cdot z_q$. We observe that $\mathrm{comb}(z_i) \geq 3$ while $\mathrm{comb}(\bar{z}_i) \leq 2$. This implies that $\mathrm{comb}(t') < \mathrm{comb}(t)$, proving the theorem. $\square$

5 Every regular open language is recognized by a finite loop

In this section, we will conclude the proof of our main result by establishing the converse of Theorem 4.2. In order to do so, we will introduce the *block product* of a monoid with a groupoid. If the second operand is also a monoid, our construction reduces to the known notion of block product applied to associative structures [23, 25], which have proved itself to be extremely useful as a decomposition tool for finite monoids. Note that the block product is a two-sided version of the classical notion of wreath product: our construction below can be trivially modified to define the wreath product of a monoid with a groupoid. Actually, the wreath product is sufficient to prove our main result. We choose to give the more general construction as the potential for future applications seems more important.

Let α be a finite-index congruence on A^*, let B be the finite alphabet $A^*/\alpha \times A \times A^*/\alpha$ and let β be a finite-index congruence on the free groupoid $B^{(*)}$. For any u, v in A^*, we define the mapping $\theta_{u,v} : A^* \to B^*$ by $\theta_{u,v}(\epsilon) = \epsilon$ and $\theta_{u,v}(a_1 \ldots a_n) = b_1 \ldots b_n$, where $b_i = ([ua_1 \ldots a_{i-1}]_\alpha, a_i, [a_{i+1} \ldots a_n v]_\alpha)$.

We now define a binary relation on $A^{(*)}$;

$(t,x)\,\beta\,\square\,\alpha\,(s,y)$ iff 1) $x\,\alpha\,y$ and

2) $(t,\theta_{u,v}(x))\,\beta\,(s,\theta_{u,v}(y))$ for all $u,v \in A^*$

Lemma 5.1 $\beta\,\square\,\alpha$ *is a congruence of finite index on* $A^{(*)}$.

Proof. That it is an equivalence of finite index is easily checked. Suppose now that $(t_1,x_1)\,\beta\,\square\,\alpha\,(s_1,y_1)$ and $(t_2,x_2)\,\beta\,\square\,\alpha\,(s_2,y_2)$; we want to show that $(t_1t_2,x_1x_2)\,\beta\,\square\,\alpha\,(s_1s_2,y_1y_2)$.

Since α is a congruence, we have that $x_1x_2\,\alpha\,y_1y_2$. Fix now u and v arbitrarily in A^*; we have

$$\begin{aligned}(t_1t_2,\theta_{u,v}(x_1x_2)) &= (t_1,\theta_{u,x_2v}(x_1))(t_2,\theta_{ux_1,v}(x_2)) \\ &= (t_1,\theta_{u,y_2v}(x_1))(t_2,\theta_{uy_1,v}(x_2)) \\ &\beta\ (s_1,\theta_{u,y_2v}(y_1))(s_2,\theta_{uy_1,v}(y_2)) \\ &= (s_1s_2,\theta_{u,v}(y_1y_2)).\end{aligned}$$

$\square$

The next lemma says that the cancellation properties are preserved by the block product.

Lemma 5.2 *If* A^*/α *is a group and* $B^{(*)}/\beta$ *is a loop, then* $A^{(*)}/\beta\,\square\,\alpha$ *is a loop.*

Proof. It is clear that the $\beta\,\square\,\alpha$-class containing the identity of $A^{(*)}$ is an identity for the groupoid $A^{(*)}/\beta\,\square\,\alpha$. We next show that $\beta\,\square\,\alpha$ is a left-cancellative congruence, i.e. $(t,x)(s,y)\,\beta\,\square\,\alpha\,(t,x)(q,z)$ implies $(s,y)\,\beta\,\square\,\alpha\,(q,z)$.

The hypothesis says that $(ts,xy)\,\beta\,\square\,\alpha\,(tq,xz)$; hence $xy\,\alpha\,xz$ and because α is a group congruence, hence left-cancellative, we deduce $y\,\alpha\,z$.

Consider some arbitrary u and v in A^*: we now need to show that $(s,\theta_{u,v}(y))\,\beta\,(q,\theta_{u,v}(z))$. Choose $w \in A^*$ such that $wx\,\alpha\,u$ (such w exists since α is a group congruence). The hypothesis implies that $(ts,\theta_{w,v}(xy))\,\beta\,(tq,\theta_{w,v}(xz))$, hence $(t,\theta_{w,yv}(x))(s,\theta_{wx,v}(y))\,\beta\,(t,\theta_{w,zv}(x))(q,\theta_{wx,v}(z))$. Since y and z are α-congruent, we have $(t,\theta_{w,yv}(x)) = (t,\theta_{w,zv}(x))$. Since β is a left cancellative congruence, we infer $(s,\theta_{wx,v}(y))\,\beta\,(q,\theta_{wx,v}(z))$, i.e. $(s,\theta_{u,v}(y))\,\beta\,(q,\theta_{u,v}(z))$. Hence $(s,y)\,\beta\,\square\,\alpha\,(q,z)$.

By symmetry, we get that $\beta\,\square\,\alpha$ is right cancellative as well, so that $A^{(*)}/\beta\,\square\,\alpha$ is a loop.

$\square$

We are now ready to complete the proof of our main result.

Theorem 5.3 *If* $L \subseteq A^*$ *is a regular open language, then* L *can be recognized by a finite loop.*

Proof. Suppose that L is an open regular language; by Theorem 2.2, L is a finite union of languages of the form $L_0 a_1 L_1 \ldots a_k L_k$, where each L_i is recognized by a group. Using the classical construction for the associative case, it is readily verified that the class of loop languages is closed under union, so it suffices to prove that any language of the above form is recognized by a loop. If $k = 0$, the claim is clearly true.

Let now $k \geq 1$; without loss of generality we can assume that all L_i's are recognized by the same group G, e.g. by taking the direct product of the syntactic monoid of each language L_i. Let $G = A^*/\alpha$, so that each L_i is a union of α-classes; since concatenation distributes over union, and using once more closure under union, it suffices to consider the case where each L_i is a single class of the congruence, i.e. $L_i = [u_i]_\alpha$ for some u_i in A^*. Let $B = A^*/\alpha \times A \times A^*/\alpha$ and $H = B^{(*)}/\beta$ be the loop recognizing the language $B^* b_1 B^* \ldots B^* b_k B^*$, as given by lemma 3.4, where $b_i = ([u_0 a_1 u_1 \ldots u_{i-1}]_\alpha, a_i, [u_i a_{i+1} u_{i+1} \ldots u_k]_\alpha)$. We claim that L is recognized by the loop $A^{(*)}/\beta \square \alpha$; in fact we will show that $x \in L$ iff there is some tree t with $|x|$ leaves such that $[(t, \theta_{\epsilon,\epsilon}(x))]_\beta$ is an accepting element of H.

Let first x be in L; thus $x = x_0 a_1 x_1 \ldots a_k x_k$, with $x_i \alpha u_i$ for each i. Thus $\theta_{\epsilon,\epsilon}(x)$ is in $B^* b_1 B^* \ldots b_k B^*$, hence for some tree t, $[(t, \theta_{\epsilon,\epsilon}(x))]_\beta$ is an accepting element of H.

Conversely, suppose $x \in A^*$ is such that, for some tree t, $[(t, \theta_{\epsilon,\epsilon}(x))]_\beta$ is an accepting element of H. Thus $(t, \theta_{\epsilon,\epsilon}(x)) = y_0 b_1 y_1 \ldots b_k y_k$, where $b_i = ([u_0 a_1 u_1 \ldots u_{i-1}]_\alpha, a_i, [u_i a_{i+1} u_{i+1} \ldots u_k]_\alpha)$. Therefore $x = x_0 a_1 x_1 \ldots a_k x_k$, where $x_0 a_1 x_1 \ldots x_{i-1} \alpha u_0 a_1 u_1 \ldots u_{k-1}$ for $i = 1, \ldots, k$, and also $x_k \alpha u_k$. Using the fact that α is a group congruence, we deduce $x_i \alpha u_i$ for each i, so that x is in L. $\square$

6 Conclusion

Our characterization has a number of consequences, from the point of view of algebra, language theory and computational complexity.

First, we get a new combinatorial description of the regular open languages.

Corollary 6.1 *Any regular open language is a finite union of languages of the form $L_1 \ldots L_k$, where each L_i is a group language.*

Proof. By the proof of theorem 4.2, every loop language is of this form, hence by Theorem 5.3, this is also true for regular open languages. $\square$

It is also appropriate to note the following structural representation that we get for loops. By the proof of theorem 4.2, we see that a loop G recognizes only regular open languages where the group languages that are needed are recognizable by the multiplication group of G. By the proof of theorem 5.3, any language of the form $L_0 a_1 L_1 \ldots a_k L_k$, where each L_i is recognized by the group

$\mathcal{M}(G)$, can be recognized by a loop of the form $A^{(*)}/\beta\square\alpha$, where $A^*/\alpha \simeq \mathcal{M}(G)$ and β is the loop congruence induced by the construction of Lemma 3.4. Thus, in some sense, computing over the loop G is similar to computing over the group $\mathcal{M}(G)$, the non-associativity being taken care of by the very simple loops given in 3.4. It would be very interesting to see to what extent this phenomenon holds for groupoids in general.

Another consequence of this work is that the computational complexity of testing membership in a language recognized by a loop G can be infered from the algebraic structure of its multiplication group $\mathcal{M}(G)$. Any language L recognized by G is a finite union of languages of the form $L_1 L_2 \cdots L_k$ where the L_i's are recognized by $\mathcal{M}(G)$. When $\mathcal{M}(G)$ is solvable, each of these languages is in ACC^0 ([4]), where ACC^0 is the class of languages that are recognized by a family polynomial-size constant-depth Boolean circuits using NOT, AND, OR, and modular gates. In such a case, it is easy to see that L is also in ACC^0. This shows the following corollary.

Corollary 6.2 *Any language recognized by a loop whose multiplication group is solvable belongs to* ACC^0.

Note that when G is a group, it can be shown that $\mathcal{M}(G)$ is solvable precisely when G is solvable. Hence, the above result naturally fits in the structural complexity framework of [4].

It is remarkable that non-associative algebras such as loops could be related to such natural class of languages as the regular open languages. Our generalization of the block product yields a loop decomposition that shows that absence of associativity does not necessarily imply absence of structure. This is also confirmed by other recent works, such as [6]. We strongly believe that a better understanding of non-associative algebras, in particular finite groupoids, could have important consequences in language theory and computational complexity.

References

[1] A.A. Albert, *Quasigroups I*, Trans. Amer. Math. Soc., **54** (1943) 507–519. *Quasigroups II*, Trans. Amer. Math. Soc., **55** (1944) 401–419.

[2] C.J. Ash, *Inevitable sequences and a proof of the type II conjecture*, Proc. of the Monash Conf. on Semigroup Theory, World Scientific, Singapore (1991) 31–42.

[3] C.J. Ash, *Inevitable graphs: a proof of the type II conjecture and some related decision procedures*, Int. J. Alg. and Comp. **1** (1991) 127–146.

[4] D. Barrington and D. Thérien, *"Finite Monoids and the Fine Structure of NC^1"*, JACM **35**4(1988)941–952

[5] F. Bédard, F. Lemieux and P.McKenzie, *Extensions to Barrington's M-program model*, TCS **107** (1993), pp. 31-61.

[6] J. Berman, A. Drisko, F. Lemieux, C. Moore, and D. Thérien, *Circuits and Expressions with Non-Associative Gates*, Submitted to 12th Annual Conference on Computational Complexity (CCC'97)

[7] R.H. Bruck, *Contributions to the Theory of Loops*, Trans. AMS **60** (1946) 245–354.

[8] R.H. Bruck, *A Survey of Binary Systems*, Springer-Verlag, 1966.

[9] H. Caussinus and F. Lemieux, *The complexity of computing over quasigroups*, Proc. 14th annual FST&TCS, 1994, pp.36-47.

[10] S. Eilenberg, *Automata, Languages and Machines, vol. B*, Academic Press, New York, 1976.

[11] T. Evans, *Embedding Incomplete Latin Squares*, Amer. Math. Monthly, 67 pp.958-961, 1960.

[12] F. Gécseg and M. Steinby, *Tree Automata*, Akadémiai Kiadó, Budapest, 1984.

[13] M. Hall Jr., *A topology for free groups and related groups*, Ann. of Maths **52** (1950) 127–139.

[14] F. Lemieux, *Finite groupoids and their applications to computational complexity*, Ph.D. Thesis, McGill University, May 1996.

[15] J. Mezei and J.B. Wright, *Algebraic automata and context-free sets*, Inform. and Contr. **11** (1967) 3–29.

[16] J.-E. Pin, *Varieties of Formal Languages*, Plenum Press, New York, 1986.

[17] J.-E. Pin, *A topological approach to a conjecture of Rhodes*, Bulletin of the Australian Mathematical Society **38** (1988) 421–431.

[18] J.-E. Pin, *Topologies for the free monoid*, Journal of Algebra **137** (1991) 297–337.

[19] J.-E. Pin, *Polynomial closure of group languages and open sets of the Hall topology*, 21th ICALP, Springer-Verlag, LNCS 820, 1994, 424–435.

[20] J.-E. Pin, *BG = PG: A Success Story*, Proc. of Intern. Conf. on Groups, Semigroups, and Formal Languages, York 1993, Kluwer Publisher.

[21] C. Reutenauer, *Une topologie du monoïde libre*, Semigroup Forum **18** (1979), 33–49.

[22] C. Reutenauer, *Sur mon article "Une topologie du monoïde libre"*, Semigroup Forum **22** (1981), 93–95.

[23] J. Rhodes and B. Tilson, *The kernel of monoid morphisms*, J. Pure and Applied Algebra **62** (1989) 227–268.

[24] M.-P. Schützenberger *On finite monoids having only trivial subgroups*, Information and Control **8** (1965) 190–194.

[25] D. Thérien, *Two-sided wreath product of categories*, J. Pure and Applied Algebra **74** (1991) 307–315.

[26] L.G Valiant, *General context-free recognition in less than cubic time*, J. Comput. System Sci. **10** (2) (1975) 308–315.

An Abstract Data Type for Real Numbers. *

Pietro Di Gianantonio

Dipartimento di Matematica e Informatica, Università di Udine
via delle Scienze 206 I-33100 Udine Italy
E-mail: digianantonio@dimi.uniud.it

Abstract. We present a PCF-like calculus having real numbers as a basic data type. The calculus is defined by its denotational semantics. We prove the universality of the calculus (i.e. every computable element is definable). We address the general problem of providing an operational semantics to calculi for the real numbers. We present a possible solution based on a new representation for the real numbers.

keywords: real number computability, domain theory, denotational and operational semantics, abstract data types.

1 Introduction

The aim of this work is to relate two different approaches to computability on real numbers: a practical approach based on programming languages, and a more theoretical one based on domain theory. Several implementations of exact computations on real numbers have been proposed so far ([BC90], [MM], [Vui88]). In these works, real numbers are represented by programs generating sequences of discrete elements, e.g. digits. On the other hand, different theoretical works on computability on real numbers are based on domain theory: [Lac59,ML70], [EE96], [DG96]. In all these works domains of approximations for the real numbers are considered. A point in these domains represents either a real number or the approximation of a real number. Approximated reals are normally described by intervals of the real line.

The relation existing between the two approaches is described in several steps. First we present a domain of approximations which is directly derived from a representation for the real number used in some implementations of the exact real number computation ([BC90,MM]). From this domain of approximations we derive a calculus for the real numbers. The calculus we present is an extension of PCF having the real numbers as ground type. We call it $\mathcal{L}_r$. We define $\mathcal{L}_r$ giving its denotational semantics.

The next natural step consists in giving an operational semantics to the calculus, possibly using the representation for the real numbers we start with. If this would be possible, we will have established a close connection between the

---
* Work partially supported by an EPSRC grant: "Techniques of Real Number Computation" at Imperial College of Science, Technology and Medicine, London and by EEC/HCM Network "Lambda Calcul Typé".

domain of approximations for the real numbers and the implementations. We will have a calculus that is for many aspect similar to the calculi used in the implementations and whose terms can be directly interpreted in the approximations domain. Unfortunately we prove that it is impossible to define the operational semantics in this way. We prove this negative result in a general manner, the impossibility holds not only if we consider the particular representation for the real numbers we chose, the domain of approximations obtained from it and the calculus $\mathcal{L}_r$. The negative result holds for a large class of representations, domains, and calculi.

Finally we define an operational semantics for $\mathcal{L}_r$. In order to do this however we need to introduce a new representation for the real numbers. This new representation is quite different from the classical ones, in it real numbers can be represented also by sequences of digits undefined on some elements. In order to compute with this new representation is absolutely necessary to use parallel operators. The use of parallel operators is the price we need to pay to have a faithful calculus for the real numbers.

Acknowledgements: I would like to thank Abbas Edalat, Martin Escardo, Peter Potts and Michael Smith for several discussions on the subject.

2 Real Number Computation in PCF

We consider the following representation for real numbers:

Definition 1. A real number x is represented by a computable sequence of integers $\langle s_0, \ldots, s_i, \ldots \rangle$ such that:

$$(i) \ \forall n . \ 2s_n - 1 \leq s_{n+1} \leq 2s_n + 1$$
$$(ii) \ x = \bigcap_{n \in \mathbb{N}} \left[\frac{s_n - 1}{2^n}, \frac{s_n + 1}{2^n} \right]$$

In this representation a sequence of integers is used to describe a sequence of rational intervals. The intervals in the sequence are contained one into the other. For practical purposes this representation is quite convenient. It allows to reduce exact real number computation to computation on integers. In this way it is possible to exploit the implementation of integer arithmetic already available on computers. In [BCRO86] and [MM] a similar representation has been used to develop quite efficient algorithms for the arithmetic operations.

We refer to [Plo77] for a definition of PCF. In order to represent real numbers in PCF it is sufficient to translate in PCF the representation of Definition 1. In the following, given a type σ, $\mathcal{L}_{PA+\exists}^{\sigma}$ indicates the set of closed terms in $\mathcal{L}_{PA+\exists}$ having type σ.

Definition 2. A partial representation function $\text{Eval}_\mathbb{R} : \mathcal{L}_{PA+\exists}^{\iota \to \iota} \rightharpoonup \mathbb{R}$ is defined by: $\text{Eval}_\mathbb{R}(M_{\iota \to \iota}) = x$ if there exists a sequence of integers s such that:
(i) $\forall n \in \mathbb{N} . \text{Eval}(M_{\iota \to \iota} n)) = s_n$;
(ii) $\forall n . 2s_n - 1 \leq s_{n+1} \leq 2s_n + 1$
(iii) $x = \bigcap_{n \in \mathbb{N}} [\frac{s_n - 1}{2^n}, \frac{s_n + 1}{2^n}]$.

A real number x is said $\mathcal{L}$-computable, if belongs to the image of the $\text{Eval}_\mathbb{R}$.

We indicate with $\mathbb{R}_l$ the set of the $\mathcal{L}$-computable real numbers. The definition of computability can be extended to functions on real numbers.

Definition 3. The function $\mathrm{Eval}_{\mathbb{R}}^1 : \mathcal{L}_{PA+\exists}^{(\iota \to \iota) \to (\iota \to \iota)} \rightharpoonup (\mathbb{R}_l \to \mathbb{R}_l)$ is defined by:
$\mathrm{Eval}_{\mathbb{R}}(M) = f$ iff
$\forall x \in \mathbb{R}_l. \forall N \in \mathcal{L}_{PA+\exists}^{\iota \to \iota}. \mathrm{Eval}_{\mathbb{R}}(N) = x \Rightarrow \mathrm{Eval}_{\mathbb{R}}(MN) = f(x)$.
A function $f : \mathbb{R}_l \to \mathbb{R}_l$ is said $\mathcal{L}$-*computable* if belongs to the image of $\mathrm{Eval}_{\mathbb{R}}^1$.

It is worthwhile to observe that the sequential operators are sufficient to define every computable function. That is every $\mathcal{L}$-computable function on reals can be defined by a term not containing the parallel test or the existential quantifier. The form of computation presented in this section, is very similar to the one used in implementations of exact real number computation and described in [BC90] and in [MM].

3 A Domain of Approximations for Real Numbers

In the literature there are several approaches to computability on real numbers which use of domain theory. Early works in this ambit are [Lac59], [ML70], and [Sco70]. In all these approaches the real line is embedded in a space of approximations where a notion of computability can be defined in a natural way. Many results concerning the computability theory on real numbers are given in these contexts. Here we are going to present a space of approximations that is similar in many respects to the ones mentioned above but has two important differences. First, we base our construction on the representation of Definition 1. As result our space has less approximation points and is more closely related to the computation describe in [BC90] and [MM]. A second important difference is the following: our space of approximations turns out to be a Scott-domain. The other approaches use spaces of approximations that are continuous but not algebraic cpos. The space of approximations presented here has been extensively studied in [DG96]. Here we resume the main results without giving the proofs.

The domain of approximations defined next is called Reals Domain (RD). We present a construction of RD starting with the integer sequence representation for real numbers. Let $\langle s_i \rangle_{i \in \mathbb{N}}$ be a sequence of integers defining a real number x according to Definition 1 and let $\langle s_i \rangle_{i < n}$ be an initial subsequence. $\langle s_i \rangle_{i < n}$ gives partial information about the value x. Examining $\langle s_i \rangle_{i < n}$ we can deduce that the value x is contained in an interval of real numbers.

Definition 4. Let S be the subset of sequences of integers defined by:
$$S = \{\langle s_i \rangle_{i < n} \mid \forall i < n-1 . 2s_i - 1 \leq s_{i+1} \leq 2s_i + 1\}.$$

The function ϕ from S to the set of rational intervals is defined by:
$$\phi(\langle s_0, s_1, \ldots, s_n \rangle) = [\frac{s_n - 1}{2^n}, \frac{s_n + 1}{2^n}])$$

The set S contains the "valid" sequences of integers. The function ϕ associates to any finite sequence $\langle s_i \rangle_{i<n}$ the interval $[a, b]$ containing the real numbers that can be represented by sequences having as initial subsequence $\langle s_i \rangle_{i<n}$. The interval $[a, b]$ represents the information contained in the sequence $\langle s_i \rangle_{i<n}$.

Let $(DI, \sqsubseteq)$ denote the partial order formed by the set of rational intervals in the image of the function ϕ. The order relation $\sqsubseteq$ on DI is the superset relation, that is $[a, b] \sqsubseteq [a', b']$ if $[a', b'] \subseteq [a, b]$ (if $[a'b']$ is a more precise approximation of a real number that $[a, b]$). The set DI forms the base of the domain RD.

Definition 5. Let RD be the cpo obtained by the ideal completion of $(DI, \sqsubseteq)$.

Proposition 6. *RD is a consistently complete ω-algebraic cpo (Scott-domain). RD is an effective Scott-domain when we consider the following enumeration of finite elements:*
$$e^r(0) = \bot \quad e^r(\langle\langle n_1, n_2\rangle, n_3\rangle + 1) = \downarrow [(n_1 - n_2 - 1)/2^{n_3}, (n_1 - n_2 + 1)/2^{n_3}].$$
Where $\langle \rangle$ is an effective coding function for pairs of natural numbers.

The elements of RD can be thought as equivalence classes of (partial) sequences of integers. Each equivalence class is composed by sequences containing identical information about the real value they approximate. The relationship existing between the real line and the infinite elements of RD can be clarified by means of following functions:

Definition 7. A function $q_\mathcal{P} : RD \to \mathcal{P}(\mathbb{R})$ is defined by:
$$q_\mathcal{P}(d) = \bigcap_{[a,b] \in d} [a, b]$$

Conversely, three functions $e, e^-, e^+ : \mathbb{R} \to RD$ are defined by:
$e(x) = \{[a, b] \in DI \mid x \in (a, b)\}$
$e^-(x) = \{[a, b] \in DI \mid x \in (a, b]\} \quad e^+(x) = \{[a, b] \in DI \mid x \in [a, b)\}$
where (a, b) indicates the open interval from a to b and $(a, b]$ and $[a, b)$ indicate the obvious part open, part closed intervals.

Proposition 8. *The following statements hold:*
i) for every infinite element $d \in RD$ there exists a real number x such that $q_\mathcal{P}(d) = \{x\}$
ii) for every real number x, $\{x\} = q_\mathcal{P} \circ e(x) = q_\mathcal{P} \circ e^-(x) = q_\mathcal{P} \circ e^-(x)$,
iii) for every non-dyadic number $x \in \mathbb{R}/D$, $e(x) = e^-(x) = e^+(x)$,
iv) for every dyadic number $x \in D$, $e(x) \sqsubset e^-(x)$, $e(x) \sqsubset e^+(x)$ and $e^-(x)$ is not consistent with $e^+(x)$,
v) $e(\mathbb{R}) \cup e^-(\mathbb{R}) \cup e^+(\mathbb{R})$ is equal to the set of infinite elements of RD.

We can say that the infinite elements of RD are a close representation of the real line, the set of infinite elements in RD looks like the real line except that each dyadic number is triplicated.

In [DG96] it is shown how to solve the problem of multiple representations by means of a retract construction.

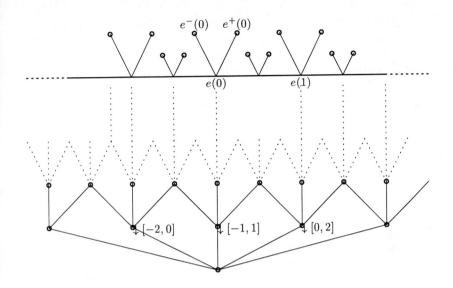

Fig. 1. The diagram representing RD.

4 PCF Extended with Real Numbers

In this section we use the domain RD introduced above, to define an extension of the language PCF having a ground type for the real numbers. We call $\mathcal{L}_r$ this extension. We will prove that any computable function on RD is definable by a suitable expression in $\mathcal{L}_r$. A programming language similar to $\mathcal{L}_r$ has been introduced in [DG93]. An extension of PCF based on a different domain of approximation for the real numbers has been presented in [Esc96].

Compared with the real computation described in Section 2, the real computation in $\mathcal{L}_r$ has several advantages. Given a closed term $M \in \mathcal{L}^{(\iota \to \iota) \to (\iota \to \iota)}$ the value $\text{Eval}_{\mathbb{R}}(M)$[1] can be undefined for several reasons. For example:
(i) there can be a term N representing a real number such that the sequence of $((MN)0), \ldots, ((MN)n), \ldots$ does not define a real number.
(ii) there can be two terms N_1 and N_2 defining the same real number and such that (MN_1) and (MN_2) define different real numbers.

The language $\mathcal{L}_r$ is free from these inadequacies. Terms of type r in $\mathcal{L}_r$ can always be interpreted as an (approximated) real and more importantly terms of type $r \to r$ preserve the equivalence between different representations of the same real number. We can say that $\mathcal{L}_r$ defines an abstract data type for real numbers. It defines a collection of primitive functions on reals which generate any other computable function.

The types of $\mathcal{L}_r$ are the PCF types extended with a new ground r. The set T of type expressions is defined by the grammar:

$$\sigma := \iota \mid o \mid r \mid \sigma \to \tau$$

The terms of $\mathcal{L}_r$ are the terms of $\mathcal{L}_{PA+\exists}$ extended with the new constants:

$$(-1), (+1), (\times 2), (\div 2), \quad \mathrm{PR} : r \to r,$$
$$(\leq 0) : r \to o \qquad \mathrm{pif}_r : o \to r \to r \to r,$$

We define $\mathcal{L}_r$ giving its denotational semantics. To this end we use the set of Scott-domains, $UD = \{D_\sigma \mid \sigma \in T\}$, where $D_\iota = \mathbb{Z}_\bot$, $D_o = \{\mathrm{tt}, \mathrm{ff}\}_\bot$, $D_r = RD$ and $D_{\sigma \to \tau} = [D_\sigma \to D_\tau]$.

The denotation of the new constants is the following:
the constants $(+1), (-1), (\times 2), (\div 2)$ realize the corresponding functions on reals.

$$[\![(+1)]\!]_\rho(d) = \{[a+1, b+1] \mid [a,b] \in d\}$$
$$[\![(-1)]\!]_\rho(d) = \{[a-1, b-1] \mid [a,b] \in d\}$$
$$[\![(\times 2)]\!]_\rho(d) = \{[a \times 2, b \times 2] \mid [a,b] \in d \wedge [a \times 2, b \times 2] \in RI\}$$
$$[\![(\div 2)]\!]_\rho(d) = \bigcup_{[a,b] \in d} \downarrow [a \div 2, b \div 2]$$

The constant (≤ 0) tests if a number is smaller or larger than 0.

$$[\![(\leq 0)]\!]_\rho(d) = \begin{cases} \mathrm{tt} & \text{if it exists } [a,b] \in d, \ b \leq 0 \\ \mathrm{ff} & \text{if it exists } [a,b] \in d, \ 0 \leq a \\ \bot & \text{otherwise} \end{cases}$$

The constant PR defines a kind of projection on the interval $[-1, 1]$.

$$[\![\mathrm{PR}]\!]_\rho(d) = \begin{cases} d \sqcup \downarrow [-1, 1] & \text{if } d \text{ is consistent with } \downarrow [-1, 1] \\ e^+(-1) & \text{if } \exists [a,b] \in d. b \leq -1 \\ e^-(1) & \text{if } \exists [a,b] \in d. a \geq 1 \end{cases}$$

The constant pif_r defines a parallel test.

$$[\![\mathrm{pif}_r]\!]_\rho(e)(d)(d') := \begin{cases} d & \text{if } e = \mathrm{tt} \\ d' & \text{if } e = \mathrm{ff} \\ d \sqcap d' & \text{if } e = \bot \end{cases}$$

If the boolean argument is undefined the function $[\![\mathrm{pif}_r]\!]_\rho$ gives as output the most precise approximation of the second and third argument.

It is not difficult to prove that for every closed expression M^σ and environment ρ, $[\![M^\sigma]\!]_\rho$ is a computable element of D_σ. Next we prove the universality of $\mathcal{L}_r$, that is, we prove that every computable functions on RD is definable by a suitable term in $\mathcal{L}_r$. In order to do this we present a generalisation of the universality theorem for PCF [Plo77, Theorem 5.1]. The generalisation applies to any extension of PCF where ground types are denoted by coherent domains. The proof in [Plo77] works only for flats domains. An equivalent generalisation has already been given in [Str94]. In that work the proof is based on categorical arguments and uses as a lemma the original result in [Plo77]. Our proof follows the line of the original proof and it is more direct. Some definitions and lemmata are necessary here.

Definition 9. A subset A of a partial order P is *coherent* if any pair of elements has an upper bound. A *coherent domain* is a Scott-domain for which any coherent subset has an upper bound.

Coherent domains are closed for many semantics functors. In particular if D_1 and D_2 are coherent domains then $[D_1 \to D_2]$ is a coherent domain. Moreover the domain RD is coherent.

A fundamental step in the proof of universality consists in showing that for every type σ it is possible to define three functions, namely, c_σ, p_σ and $\#_\sigma$. Where c_σ and p_σ are respectively a test and a projection function for the types σ, while $\#_\sigma(n)(d)$ checks if the element d is inconsistent with the finite element $e^\sigma(n)$ (where e^σ is the effective enumeration of the finite elements of the domain D_σ ([Plo77, page 249])). Formally:

Definition 10. A partial function $f : D_{\sigma_1} \to \ldots D_{\sigma_n} \rightharpoonup D_\sigma$ is *definable* in $\mathcal{L}_r$ if there exists a closed term M such that for all $d_1 \in D_{\sigma_1} \ldots d_n \in D_{\sigma_n}$ if $f(d_1)\ldots(d_n)$ is defined then $[\![M]\!]_\rho(d_1)\ldots(d_n) = f(d_1)\ldots(d_n)$.

Definition 11. Given a coherent-domain D_σ the function $c_\sigma : B_\perp \to D_\sigma \to D_\sigma \to D_\sigma$, and the partial functions $\#_\sigma : \mathbb{Z}_\perp \to D_\sigma \rightharpoonup B_\perp$, $p_\sigma : \mathbb{Z}_\perp \to D_\sigma \rightharpoonup D_\sigma$ are defined by:

$$c_\sigma(b)(d_1)(d_2) = \begin{cases} d_1 & if\ b = tt \\ d_2 & if\ b = ff \\ d_1 \sqcap d_2 & if\ b = \perp \end{cases}$$

$$\#_\sigma(n)(d) = \begin{cases} ff & \text{if } n \in \mathbb{N}, e^\sigma(n) \sqsubseteq d \\ tt & \text{if } n \in \mathbb{N}, e^\sigma(n) \text{ and } d \text{ are inconsistent} \\ \text{undefined} & \text{if } n \text{ is a negative number} \\ \perp & \text{otherwise} \end{cases}$$

$$p_\sigma(n)(d) = \begin{cases} d \sqcup e^\sigma(n) & \text{if } n \in \mathbb{N}, d, e^\sigma(n) \text{ are consistent} \\ \text{undefined} & \text{otherwise} \end{cases}$$

Lemma 12. *If, in a language extending $\mathcal{L}_{PA+\exists}$ with new ground types, for every ground type τ the function $c_\tau, p_\tau, \#_\tau$ are definable by some terms $\text{pif}_\tau, P_\tau, T_\tau$ then for any other type σ the functions $c_\sigma, p_\sigma, t_\sigma$ are definable by some suitable terms $\text{pif}_\sigma, P_\sigma, T_\sigma$.*

Lemma 13. *If in an extension of the language $\mathcal{L}$ for a type σ the function p_σ is definable then every computable element in D_σ is definable.*

Theorem 14. *For every computable element d in D_σ there exists a closed expression M in $\mathcal{L}_r$ such that: $[\![M]\!]_\rho = d$.*

5 Operational Semantics, a First Attempt

In this section we discuss the problem of defining an operational semantics for $\mathcal{L}_r$. In Section 3 the elements of RD are constructed as equivalence classes of partial sequences of integers. One can use functions in $[\mathbb{Z}_\perp \to \mathbb{Z}_\perp]$ to represent sequences of integers and hence elements in RD. Following this approach one can use higher order function of $[\mathbb{Z}_\perp \to \mathbb{Z}_\perp]$ to represent functions on RD. The construction is the following. Let S' be the subset of $[\mathbb{Z}_\perp \to \mathbb{Z}_\perp]$ defined by,
$S' = \{s \mid \forall i \in \mathbb{N}. \ (\ s(i+1) \neq \perp \Rightarrow (\ s(i) \neq \perp \wedge 2s(i) - 1 \leq s(i+1) \leq 2(i)+1\ ))\}$
the elements of S' define the partial sequences of digits representing elements in RD. Let $\phi' : S' \to RD$ be the function,
$\phi'(s) = \bigcap \{[\frac{s(i)-1}{2^i}, \frac{s(i)+1}{2^i}] \mid i \in \mathbb{N}, s(i) \neq \perp\}$.
Given a function g on RD, for example, $g : RD \to RD \to RD$, we say that g is represented by a function $f : [\mathbb{Z}_\perp \to \mathbb{Z}_\perp] \to [\mathbb{Z}_\perp \to \mathbb{Z}_\perp] \to [\mathbb{Z}_\perp \to \mathbb{Z}_\perp]$ if for all $s_1, s_2 \in S'$, $g(\phi'(s_1))(\phi'(s_2)) = \phi'(f(s_1)(s_2))$.

The above representation for functions on RD suggests the following approach to operational semantics: for each new constant c in $\mathcal{L}_r$ one try to find a computable function f_c on $[\mathbb{Z}_\perp \to \mathbb{Z}_\perp]$ representing the function $[\![c]\!]$. If the functions f_c would exist then a set of closed $\mathcal{L}_{PA+\exists}$-terms M_c such that $\mathcal{E}[\![M_c]\!]_\rho = f_c$, would define an operational semantics for $\mathcal{L}_r$. The operational semantics would be given by the reductions rules $c \to M_c$. In fact the operational behaviour of M_c is in accordance with the denotational semantics of c. Unfortunately this natural approach is doomed to failure. In fact the function $[\![\text{pif}_r]\!]_\rho$ cannot be represented by any functional on integers. We state this negative result in a more general setting, considering not only the real number representation of Definition 1 and the corresponding domain RD but a large class of real number representations and domains of approximations.

In almost all the representations considered in the literature a real number is represented by a sequence of elements of a countable set C. For example C can be a set of digits, the set of integers, the set of p-adic rational numbers, the set of rational numbers, the set of rational intervals.

Definition 15. A *sequence representation* for the real numbers is given by a countable set C, a subset S of $\mathbb{N} \to C$ and a representation function $v : S \to \mathbb{R}$. The set S is the subset of sequences defining real numbers.

Repeating the construction of Section 3 we map finite sequences to subsets of reals.

Definition 16. Given a sequence representation $v : S \to \mathbb{R}$, its extension to partial sequences $\overline{v} : [\mathbb{N} \to C_\perp] \to \mathcal{P}(\mathbb{R})$, is defined by,

$$\overline{v}(s) = \{v(t) \mid t \in S, s \sqsubseteq t\}.$$

Given a sequence s and a natural number n we indicate with $s|_n$ the partial sequence containing the first n elements of s: $s|_n(m) = s(m)$ if $m \leq n$, $s|_n(m) = \perp$ otherwise. In [Wei87, pages 479–482] it has been introduced the notion of admissible representation for real numbers. That definition can be reformulated as follows.

Definition 17. A sequence representation $\langle S, v \rangle$ is *admissible* if it satisfies the following conditions,
(i) $\forall s \in S . \forall \epsilon \in \mathbb{R} . \exists n \in \mathbb{N} . \overline{v}(s|_n)$ is contained in an interval having width ϵ,
(ii) For each real number x there exists a sequence s such that for each n, x is contained in the interior of $\overline{v}(s|_n)$.

Condition (i) states that the function $v : S \to \mathbb{R}$ is continuous, w.r.t. the Cantor topology on S and the Euclidean topology on $\mathbb{R}$. Almost all the representation functions used in computable analysis are admissible.

Any sequence representation induces an information order on partial sequences: s is below t in the information order if $\overline{v}(s) \supseteq \overline{v}(t)$. We have the following negative result.

Theorem 18. *For any admissible representation v, and there is no continuous functional $g : [\mathbb{N} \to C_\perp] \to [\mathbb{N} \to C_\perp] \to [\mathbb{N} \to C_\perp]$ such that:*
(i) g implements addition, that is: for all s, t in S, $v(g(s)(t)) = v(s) + v(t)$
(ii) g respects the induced order relation on partial functions that is: for all s, s', t, t' in $[\mathbb{N} \to C_\perp]$, $\overline{v}(s) \supseteq \overline{v}(s')$ and $\overline{v}(t) \supseteq \overline{v}(t')$ implies $\overline{v}(g(s)(t)) \supseteq \overline{v}(g(s)(t))$.

The previous theorem implies that, if we use an admissible then the operational semantics of $\mathcal{L}_r$ cannot be given in terms of computations on sequences. This result generalises to any domain derived from an admissible representation and to any calculus define on the derived domain. There are two possible solutions to this problem. The first one consists in introducing non deterministic or intensional operators in the language. The second one consists in using representations that are not admissible, but that are suitable for real number computations. The first approach has been followed in [Esc96], there the operational semantics of a language similar to $\mathcal{L}_r$ is given using a non deterministic operator. Here we will follow the second approach.

6 An Operational Semantics

The notations considered so far in the literature represent real numbers using sequences that are completely defined. It is possible to represent real numbers using sequences that are undefined on some elements. An example is the following.

Definition 19. A real number x in the interval $[-1, 1]$ is represented by a sequence s of digits $-1, 1$ such that: $x = \sum_{i \in N} \prod_{0 \leq j \leq i} s_j / 2$

This notation is similar to the binary digit notation. The main differences consist in the use of the digit -1 instead of the digit 0 and in the fact that in this notation the value of a digit affects the weights of all the consecutive digits. In this notation the real number 0 has two representations: the sequence

$\langle -1, -1, 1, 1, 1 \ldots \rangle$ and the sequence $\langle 1, -1, 1, 1, 1 \ldots \rangle$. The two representations differ just for the first digit. Hence 0 can also be represented by the sequence $\langle \bot, -1, 1, 1, 1 \ldots \rangle$ undefined on the first element. Moreover examining the finite initial parts of the incomplete sequence it is possible to determine the number represented by it with an arbitrary precision. Similar considerations hold for any other dyadic rational number. Every real number that is not rational dyadic has exactly one representation. If we allow as possible representations for the dyadic rational numbers also the sequences undefined on one element we obtain a representation suitable for the real number computation.

In order to represent the whole real line we consider the following notation.

Definition 20. A representation function $v : (\mathbb{N} \to \{-1, 1\}) \to \mathbb{R}$ is defined by:

$$v(s) = s(0) \times (k + \sum_{i \geq k} \prod_{k \leq j \leq i} s(j)/2)$$

where $k = min\{i \mid i > 0, s(i) = -1\}$

This is a sort "sign, integer part, mantissa" notation for the real numbers. The first digit gives the sign, the next consecutive positive digits determine the integer part, the remaining part of the sequence is the mantissa. Also in this case every dyadic rational number is represented by two functions that differ just for one element and every real number that is not rational dyadic has exactly one representation.

Definition 21. The extension of v to partial functions is the function $\overline{v} : (\mathbb{N} \to \{-1, 1\}_\bot) \to \mathcal{P}(\mathbb{R})$ defined by:

$$\overline{v}(s) = \{v(t) \mid t : \mathbb{N} \to \{-1, 1\}, s \sqsubseteq t\}.$$

The set $\overline{v}(s)$ is an interval if and only if

$$\forall n . (s(n)\uparrow \wedge s(n+1)\downarrow)$$
$$\Rightarrow \forall m < n . s(m)\downarrow \wedge s(n+1) = -1 \wedge \forall m > n+1 . (s(m)\uparrow \vee s(m) = 1).$$

Let S^∞ denote the set of partial functions s such that $\overline{v}(s)$ is an interval. S^∞ is a complete partial order. If we repeat the construction of Section 3, with the representation v and the set S^∞ of partial elements we obtained a new domain for real numbers. We call the new domain RD'. In this case no pair of elements in S^∞ contain the same information. It follows that S^∞ and RD' are isomorphic. The structures of RD and RD' are quite similar. The main difference consists in the fact that RD' contains for each natural number n the intervals $[-\infty, -n]$ and $[n, +\infty]$ and, as a consequence, the infinite points $-\infty$ and $+\infty$.

Proposition 22. *There exists an effective embedding-projection pair $\langle e, p \rangle$ from S^∞ to $[N \to \{-1, 1\}_\bot]$, $p : [N \to \{-1, 1\}_\bot] \to S^\infty$ is defined by:*

$$p(s) = \bigsqcup \{s' \in S^\infty \mid s' \sqsubseteq s\}$$

$e : S^\infty \to [N \to \{-1, 1\}_\bot]$ *is the identity functions.*

It follows that there exists an effective embedding-projection pair $\langle e_r, p_r \rangle$ from RD' to $[Z_\perp \to \{tt, ff\}_\perp]$. The embedding-projection can be extended to the functions spaces.

$$e_{\sigma \to \tau}(f) = e_\tau \circ f \circ q_\sigma$$
$$q_{\sigma \to \tau}(f) = q_\tau \circ f \circ e_\sigma$$

Repeating the considerations presented in Section 5, it is possible to represent elements in RD' by theirs embeddings in $[Z_\perp \to \{tt, ff\}_\perp]$ and functions on RD' (S^∞) by the corresponding embeddings on functions spaces of $[Z_\perp \to \{tt, ff\}_\perp]$. Let C be the set of the new constants in $\mathcal{L}_r$, for each $c^\sigma \in C$ let M_{c^σ} be a term in $\mathcal{L}_{PA+\exists}$ defining the function $e_\sigma(\llbracket c^\sigma \rrbracket_\rho)$. By the universality of $\mathcal{L}_{PA+\exists}$ the terms M_{c^σ} exists. An operational semantics for $\mathcal{L}_r$ can be given adding to the set single-step reduction rules for $\mathcal{L}_{PA+\exists}$ the new set of rules $\{c \to M_c \mid c \in C\}$. For lack of space we do not present the actual set of rules.

References

[BC90] H.-J. Boehm and R. Cartwright. Exact real arithmetic: formulating real numbers as functions. In David Turner, editor, *Research topics in functional programming*, pages 43–64. Addison-Wesley, 1990.

[BCRO86] H.-J. Boehm, R. Cartwright, M. Riggle, and M.J. O'Donell. Exact real arithmetic: a case study in higher order programming. In *ACM Symposium on lisp and functional programming*, 1986.

[DG93] P. Di Gianantonio. *A functional approach to real number computation*. PhD thesis, University of Pisa, 1993.

[DG96] P. Di Gianantonio. Real number computability and domain theory. *Information and Computation*, 127(1):11–25, May 1996.

[EE96] A. Edalat and M. Escardo. Integration in real pcf. In *IEEE Symposium on Logic in Computer Science*, 1996.

[Esc96] M. Escardo. Pcf extended with real numbers. *Theoret. Comput. Sci*, July 1996.

[Lac59] D. Lacombe. Quelques procédés de définitions en topologie recursif. In *Constructivity in mathematics*, pages 129–158. North-Holland, 1959.

[ML70] P. Martin-Löf. *Note on Constructive Mathematics*. Almqvist and Wiksell, Stockholm, 1970.

[MM] V. Ménissier-Morain. Arbitrary precission real arithmetic: design and algorithms. Submitted to the Journal of Symbolic Computation. Available at http://pauillac.inria.fr/ menissier.

[Plo77] G.D. Plotkin. Lcf considered as a programing language. *Theoret. Comput. Sci.*, 5:223–255, 1977.

[Sco70] Dana Scott. Outline of the mathematical theory of computation. In *Proc. 4th Princeton Conference on Information Science*, 1970.

[Str94] T. Streicher. A universality theorem for pcf with recursive types, parallel-or and $\exists$. *Mathematical Structures for Computing Science*, 4(1):111–115, 1994.

[Vui88] J. Vuillemin. Exact real computer arithmetic with continued fraction. In *Proc. A.C.M. conference on Lisp and functional Programming*, pages 14–27, 1988.

[Wei87] K. Weihrauch. *Computability*. Springer-Verlag, Berlin, Heidelberg, 1987.

Recursive Computational Depth *

James I. Lathrop and Jack H. Lutz

Department of Computer Science
Iowa State University
Ames, Iowa 50011
U.S.A.

Abstract. In the 1980's, Bennett introduced computational depth as a formal measure of the amount of computational history that is evident in an object's structure. In particular, Bennett identified the classes of *weakly deep* and *strongly deep* sequences, and showed that the halting problem is strongly deep. Juedes, Lathrop, and Lutz subsequently extended this result by defining the class of *weakly useful* sequences, and proving that every weakly useful sequence is strongly deep.
The present paper investigates refinements of Bennett's notions of weak and strong depth, called *recursively weak depth* (introduced by Fenner, Lutz and Mayordomo) and *recursively strong depth* (introduced here). It is argued that these refinements naturally capture Bennett's idea that deep objects are those which "contain internal evidence of a nontrivial causal history." The fundamental properties of recursive computational depth are developed, and it is shown that the recursively weakly (respectively, strongly) deep sequences form a proper subclass of the class of weakly (respectively, strongly) deep sequences. The above-mentioned theorem of Juedes, Lathrop, and Lutz is then strengthened by proving that every weakly useful sequence is recursively strongly deep. It follows from these results that not every strongly deep sequence is weakly useful, thereby answering a question posed by Juedes.

1 Introduction

Computational depth was introduced by Bennett [2,3] as a formal measure of the amount of computational history that is evident in the structure of a computational, physical, or biological object. Roughly speaking, if x is an object (such as a computer program, a point in a phase space, or a DNA sequence) that can be encoded in binary in a natural way — in which case we identify x with its encoding — then the computational depth of x is the amount of time required for a computation to derive x from its shortest binary description. Like Solomonoff [13], Bennett regards a description of x as a formal analog of a scientific explanation of x. By Occam's razor, then, the shortest description of x

---

* This research was supported in part by National Science Foundation Grant CCR-9157382, with matching funds from Rockwell, Microware Systems Corporation, and Amoco Foundation.

is the most plausible explanation of x, and the computational depth of x is the amount of time required for an effective process to generate x from its most plausible explanation. Bennett thus says that a deep object is "one whose most plausible origin, via an effective process, entails a lengthy computation," and, more succinctly, that a deep object is one that contains "internal evidence of a nontrivial causal history" [3].

In order to avoid undue sensitivity to the underlying computational model, Bennett's definition of depth refers not only to an object's shortest description, but to all descriptions of the object that have nearly minimal length. This is achieved by adding a significance parameter to the definition. Specifically, for $c \in \mathbb{N}$, the computational depth of an object x at significance level c is the time required for a computation to derive x from a binary description π that is itself compressible by no more than c bits. (That is, every description of π consists of at least $|\pi| - c$ bits.)

For (infinite, binary) sequences, Bennett [2,3] introduced two interesting depth conditions, strong depth and weak depth. A sequence S is *strongly deep* if, for every computable time bound $t : \mathbb{N} \to \mathbb{N}$ and every constant $c \in \mathbb{N}$, for all but finitely many $n \in \mathbb{N}$, the n-bit prefix $S[0..n-1]$ of S has depth greater than $t(n)$ at significance level c. If we regard a description π from which $S[0..n-1]$ can be derived in at most $t(n)$ computation steps as a $t(n)$-*compression* of $S[0..n-1]$, then this says that, for all computable time bounds t and constants c, for all but finitely many n, every $t(n)$-compression of $S[0..n-1]$ is itself compressible by more than c bits. Thus a sequence is strongly deep if no computable time bound suffices to compress infinitely many of its prefixes to within a constant number of bits of the optimal compression.

To put the matter more fancifully, no matter how (computably) much time is spent looking for inner structure (i.e., basis for compression) in a strongly deep sequence, an unbounded quantity of such inner structure remains undiscovered. A strongly deep sequence is thus analogous to a great work of literature for which no number of readings suffices to exhaust its value.

It was shown by Bennett [3] (and also in [7]) that no sequence that is either decidable or random (i.e., algorithmically random in the sense of Martin-Löf [10]) can be strongly deep. However, strongly deep sequences do exist. For example, Bennett [3] noted that K, the diagonal halting problem, is strongly deep. This is because K, unlike a decidable or random sequence, can be used (as an oracle) to decide any decidable sequence within a computable (in fact, polynomial) time bound that does not depend on the sequence.

This relationship between depth and usefulness (as an oracle) was investigated more explicitly and generally by Juedes, Lathrop, and Lutz [7], who defined strong and weak usefulness conditions for sequences. A sequence S is *strongly useful* if there is a fixed computable time bound $t : \mathbb{N} \to \mathbb{N}$ such that the set $\text{DTIME}^S(t)$, consisting of all sequences that can be decided in $t(n)$ time using the oracle S, contains every decidable sequence, i.e., $\text{REC} \subseteq \text{DTIME}^S(t)$, where REC is the set of all decidable sequences. A sequence S is *weakly useful* if there is a fixed computable time bound $t : \mathbb{N} \to \mathbb{N}$ such that the set $\text{DTIME}^S(t)$ does not have measure 0 in REC, i.e., $\text{DTIME}^S(t) \cap \text{REC}$ is a nonnegligible subset of REC in the sense of the recursive case of the resource-bounded measure theory developed by Lutz [9]. That is, S is weakly useful if a nonnegligible set of decidable sequences can be decided within a computable time bound that may depend on S but does not depend on the sequence being decided. By the above remark, K is strongly useful. It is evident that every strongly useful sequence is weakly useful, and Fenner, Lutz, and Mayordomo [4] have shown that the converse does not hold, so the set of strongly useful sequences is properly contained in the set of weakly useful sequences.

Juedes, Lathrop, and Lutz [7] proved that every weakly useful sequence is strongly deep. This generalized Bennett's observation that K is strongly deep and gave formal support to Bennett's informal arguments relating depth and usefulness. Strong depth is a necessary condition for weak usefulness. Juedes [6] subsequently asked whether the converse is true, i.e., whether strong depth actually characterizes weak usefulness.

In this paper, we show that weakly useful sequences have a strictly stronger depth property than strong depth, thereby answering Juedes's question negatively. In fact, this stronger depth property, a constructive refinement of strong depth called *recursively strong depth*, is the main topic of this paper.

In the terminology used above to describe strong depth, a sequence S is *recursively strongly deep* (briefly, rec-*strongly* deep) if, for every computable time bound t and constant c, there exists a computable time bound l such that, for all but finitely many n, every $t(n)$-compression of $S[0..n-1]$ is itself $l(n)$-compressible by more than c bits. It is the existence of this computable time bound l that distinguishes rec-strong depth from strong depth. Returning to the more fanciful language used earlier, no matter how (computably) much time is spent looking for inner structure in a rec-strongly deep sequence, and no matter now much additional structure (any constant number of bits) one wishes to find, there is always a greater (computable) amount of time that suffices to find that much more structure. A rec-strongly deep sequence is thus analogous to a great work of literature with the property that, no matter how many times it has been read, there is a greater number of readings from which one can derive significantly more value.

In this paper, we establish the existence of sequences that are strongly deep but not rec-strongly deep. Such a sequence S must have the following two properties.

(i) There exist a *fixed* computable time bound $t_0 : \mathbb{N} \to \mathbb{N}$ and a *fixed* constant $c_0 \in \mathbb{N}$ such that, for *every* computable time bound $l : \mathbb{N} \to \mathbb{N}$, there are infinitely many prefixes $S[0..n-1]$ of S that have $t_0(n)$-compressions that are not $l(n)$-compressible by c_0 or more bits.

(ii) For *every* constant $c \in \mathbb{N}$ (no matter now much larger than c_0), for all but finitely many prefixes $S[0..n-1]$ of S, every $t_0(n)$-compression of $S[0..n-1]$ is itself compressible by more than c bits.

By (i), none of the additional compression (beyond c_0 bits) promised in (ii) can be realized within any computable time bound. Once again comparing a sequence to a work of literature and taking a number of readings as an analogy for a computable time bound, a sequence that is strongly deep but not rec-strongly deep is analogous to a work of literature for which no number of readings exhausts its value, but some number of readings does exhaust all the value that can be exhausted by any number of readings.

Using Bennett's terminology, a rec-strongly deep sequence S shows evidence of a nontrivial causal (computational) history in the constructive, incremental sense that every explanation of S that can be realized by an effective process of computable duration is significantly less plausible than some other explanation of S that can also be realized by an effective process of some greater computable duration. In contrast, a sequence that is strongly deep but not rec-strongly deep has an explanation that (i) can be realized by an effective process of computable duration, and (ii) is as plausible as any other explanation that can be realized by an effective process of computable duration. Although such a sequence does have a more plausible explanation, there is no constructive evidence of this fact.

None of the above should be taken to imply that rec-strong depth is a better (or worse) notion than strong depth. Both notions merit further investigation. In the case of rec-strong depth, there are several reasons for this. First, as noted above, rec-strongly deep sequences show evidence of a "nontrivial causal history" in a natural, constructive, incremental sense. Second, as we show in this paper, rec-strong depth enjoys the same useful slow-growth property (and consequent upward closure under truth-table reductions) that Bennett [3] proved for strong depth. Third, as we show in this paper, rec-strong depth can be used to separate weak usefulness from strong depth, thereby answering Juedes's question. Fourth, as developed below, rec-strong depth is based on a *recursive* depth function (with an additional latency parameter), and therefore provide a useful model for the design and analysis of *implementable* depth measures such as the compression depth introduced by Lathrop [8]. Fifth, and perhaps most compelling, we show that the relationships among rec-strong depth, the notion of rec-weak depth introduced by Fenner, Lutz and Mayordomo [4], and the notion of rec-randomness that has been investigated by Schnorr [11,12], van Lambalgen [14], Lutz [9], Wang [15], and others correspond closely to the relationships among strong depth, weak depth and algorithmic randomness.

This paper is largely self-contained. It can be read independently of [3,7], but we assume that [7] is at hand for reference. At the end of this section, we introduce a small amount of terminology and notation. Section 2, the main section of this paper, presents rec-strong depth, rec-weak depth, and our results on these notions. Section 2 is divided into a preamble and four (sub-)sections. In the preamble, we develop the above-mentioned recursive depth function, $depth_c^l(w)$. In section 2.1 we use this function to introduce rec-strong depth. In section 2.2 we prove the deterministic slow growth law for recursive computational depth and establish the basic inclusion relations among the weak, strong, rec-weak,

and rec-strong depth classes, namely,

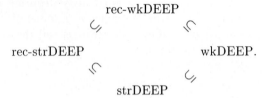

In section 2.3 we prove that all these inclusions are proper by proving that the classes rec-wkDEEP and strDEEP are incomparable. Both directions of the incomparability proof are nontrivial. One direction yields the stronger fact that rec-random sequences can be strongly deep, while the other direction uses the recursive version of the first Borel-Cantelli lemma [9] in a Baire category argument. In section 2.4 we prove that every weakly useful sequence is rec-strongly deep, thereby answering Juedes's question. Proofs of our results appear in the full version of this paper.

We work in the *Cantor space* $\mathbf{C}$, consisting of all (infinite, binary) *sequences*. A string $w \in \{0,1\}^*$ is a *prefix* of a sequence $S \in \mathbf{C}$, and we write $w \sqsubseteq S$ if there is a sequence $A \in \mathbf{C}$ such that $S = wA$. For $S \in \mathbf{C}$ and $n \in \mathbb{N}$, we write $S[n]$ for the nth bit of S and $S[0..n-1]$ for the n-bit prefix of S. The *complement* of a set $X \subseteq \mathbf{C}$ is the set $X^c = \mathbf{C} - X$.

We write REC for the set of all decidable sequences in $\mathbf{C}$ and rec for the set of all computable (total) functions from $\{0,1\}^*$ to $\{0,1\}^*$. Identifying strings s_n with their indices n in the standard enumeration of $\{0,1\}^*$, we also write rec for the set of all computable functions from $\mathbb{N}$ to $\mathbb{N}$.

2 Recursive Computational Depth

As noted by Bennett [3], the value $\text{depth}_c(w)$ – the computational depth of a string w at significance level c – is not computable from w and c. The following definition remedies this at the expense of introducing an additional variable.

Definition. For $w \in \{0,1\}^*$ and $c, l \in \mathbb{N}$, the *recursive computational depth of w at significance level c with latency l* is

$$\text{depth}_c^l(w) = \min\left\{ t \in \mathbb{N} \;\middle|\; (\exists \pi \in \text{PROG}^t(w)) \, |\pi| < K^l(\pi) + c \right\}.$$

That is, $\text{depth}_c^l(w)$ is the minimum amount of time required to obtain w from a program π that cannot itself be obtained in time l from a program that is c or more bits shorter than π. It is clear that $\text{depth}_c^l(w)$ is computable from w, c, and l; this is why it is called the *recursive* computational depth. Two other properties of $\text{depth}_c^l(w)$ are immediately evident. For each $w \in \{0,1\}^*$ and $c \in \mathbb{N}$, $\text{depth}_c^l(w)$ is nondecreasing in l, and $\lim_{l \to \infty} \text{depth}_c^l(w) = \text{depth}_c(w)$. For each $w \in \{0,1\}^*$ and $l \in \mathbb{N}$, the value $\text{depth}_c^l(w)$ is, like $\text{depth}_c(w)$, nonincreasing in c.

2.1 Recursive Depth Classes

We begin by defining the recursive analogs of the depth classes $D_g^t(n)$ and D_g^t introduced in [7].

Definition. For $t, g, l : \mathbb{N} \to \mathbb{N}$ and $n \in \mathbb{N}$, define the sets

$$D_g^{t,\, l}(n) = \left\{ S \in \mathbf{C} \,\Big|\, \text{depth}_{g(n)}^{l(n)}(S[0..n-1]) > t(n) \right\}$$

and

$$D_g^{t,\, l} = \bigcup_{m=0}^{\infty} \bigcap_{n=m}^{\infty} D_g^{t,\, l}(n) = \left\{ S \in \mathbf{C} \,\Big|\, (\forall^\infty n)\, S \in D_g^{t,\, l}(n) \right\}.$$

Note that

$$D_g^{t,\, l}(n) = \left\{ S \in \mathbf{C} \,\Big|\, (\forall \pi \in \text{PROG}^t(S[0..n-1]))\, K^{l(n)}(\pi) \leq |\pi| - g(n) \right\}.$$

(It is crucial here that the left-hand side of the inequality is $K^{l(n)}(\pi)$, not $K^l(\pi)$, i.e., that the time bound is $l(n)$, not $l(|\pi|)$.)

Definition. Let $t, g : \mathbb{N} \to \mathbb{N}$. A sequence $S \in \mathbf{C}$ is *recursively t-deep at significance level g*, and we write $S \in D_g^{t,\, \text{rec}}$, if there is a computable function $l : \mathbb{N} \to \mathbb{N}$ such that $S \in D_g^{t,\, l}$.

It is clear that, for all $t, g, l : \mathbb{N} \to \mathbb{N}$ with l computable, $D_g^{t,\, l} \subseteq D_g^{t,\, \text{rec}} \subseteq D_g^t$. To define recursively strong depth, we substitute $D_g^{t,\, \text{rec}}$ for D_g^t in the definition of strong depth.

Definition. A sequence $S \in \mathbf{C}$ is *recursively strongly deep* (or, briefly, rec-strongly deep), and we write $S \in$ rec-strDEEP, if for every computable time bound $t : \mathbb{N} \to \mathbb{N}$ and every constant $c \in \mathbb{N}$, $S \in D_c^{t,\, \text{rec}}$.

We note that every rec-strongly deep sequence is strongly deep. Since REC $\cap$ strDEEP $= \emptyset$ [3] (see also [7]), it follows immediately that no recursive sequence can be rec-strongly deep.

Recall that a sequence S is strongly deep if, for every computable time bound t and constant c, all but finitely many prefixes of S can be described at least c bits more succinctly without a time bound than with the time bound t. In contrast, a sequence S is rec-strongly deep if, for every computable time bound t and constant c, there exists a *computable* time bound l such that all but finitely many prefixes of S can be described at least c bits more succinctly *with the time bound l* than with the time bound t. Very informally, a sequence is strongly deep if it has more regularity than can be explained by a causal (computational) history of any computable duration. For a sequence to be rec-strongly deep, it must also be the case that, for every computable duration t there is a larger computable duration l such that more of the sequence's regularity can be explained by a causal history of duration l than can be explained by a causal history of duration t.

Our next result states that rec-strongly deep sequences cannot be rec-random.

Theorem 1. $\mathrm{RAND}(\mathrm{rec}) \cap \textit{rec-strDEEP} = \emptyset$. *In fact, there exist a computable function* $t(n) = O(n \log n)$ *and a constant* $c \in \mathbb{N}$ *such that* $\mathrm{RAND}(\mathrm{rec}) \cap \mathrm{D}_c^{t,\,\mathrm{rec}} = \emptyset$.

Recursively weak depth was introduced by Fenner, Lutz, and Mayordomo [4]. We write rec-wkDEEP for the class of all rec-weakly deep sequences.

2.2 Class Inclusions

In this section, we establish the basic inclusion relations that hold among the weak and strong depth classes defined in [7] and section 2.1. For this and later purposes, we need a technical lemma. This result, called the *deterministic slow-growth law for recursive computational depth*, places a quantitative upper bound on the ability of a time-bounded oracle Turing machine to amplify the depth of its oracle. Details appear in the full version of this paper.

An easy consequence of the Slow Growth Lemma is the fact that the class of rec-strongly deep sequences is (like the class of strongly deep sequences [7]) closed upwards under tt-reductions.

Theorem 2. *Let* $A, B \in \mathbf{C}$. *If* $B \leq_{\mathrm{tt}} A$ *and* B *is rec-strongly deep, then* A *is rec-strongly deep.*

We now come to the main result of section 2.2. The following theorem gives the inclusion relations that hold among the weak, strong, rec-weak, and rec-strong depth classes.

Theorem 3. *The following diagram of inclusions holds.*

$$\begin{array}{ccc}
 & \textit{rec-wkDEEP} & \\
 \subseteq \nearrow & & \nwarrow \supseteq \\
\textit{rec-strDEEP} & & \textit{wkDEEP} \\
 \nwarrow \supseteq & & \subseteq \nearrow \\
 & \textit{strDEEP} &
\end{array}$$

2.3 Class Separations

We now show that all four inclusions in Theorem 3 are proper. It is most efficient (and most informative) to prove this by proving the two non-inclusions

$$\mathrm{strDEEP} \not\subseteq \textit{rec-wkDEEP}$$

and

$$\text{rec-wkDEEP} \not\subseteq \text{strDEEP}.$$

We prove these in succession.

We prove that strDEEP $\subsetneq$ rec-wkDEEP by proving the much stronger fact that strongly deep sequences can be recursively random. We do this by examining the Kolmogorov and the time-bounded Kolmogorov complexities of recursively random sequences.

We first prove that rec-random sequences have very high time-bounded Kolmogorov complexities.

Theorem 4. *Assume that S is rec-random and that $t, g : \mathbb{N} \to \mathbb{N}$ are computable functions with g nondecreasing and unbounded. Then, for all but finitely many $n \in \mathbb{N}$,*

$$K^t(S[0..n-1]) > n - g(n).$$

The function g above may be very slowly growing, e.g., an inverse Ackermann function. Theorem 4 thus says that, for every rec-random sequence S and computable time bound t, all but finitely many of the prefixes of S have K^t-complexities that are nearly as large as their lengths.

We next show that the situation is very different in the absence of the time bound t.

Definition. A sequence $S \in \mathbf{C}$ is *ultracompressible* if, for every computable, nondecreasing, unbounded function $g : \mathbb{N} \to \mathbb{N}$, there exists $n_g \in \mathbb{N}$ such that, for all $n \geq n_g$,

$$K(S[0..n-1]) < K(n) + g(n). \tag{1}$$

It is clear that every n-bit string w must satisfy $K(w) \geq K(n) - O(1)$. A sequence S is thus ultracompressible if, for every computable, nondecreasing, unbounded (but perhaps very slowly growing) function g, for all but finitely many n, the n-bit prefix of S has K-complexity that is within $g(n)$ bits of the minimum possible K-complexity for an n-bit string.

We now show that a rec-random sequence can be ultracompressible. Similar results have been proven by Wang [15] and Ambos-Spies and Wang [1] for the monotone Kolmogorov complexities of rec-random sequences. The present result is slightly stronger than these results in that it gives a single rec-random sequence S that has property (1) for every computable, nondecreasing, unbounded function g. The proof is based in part on a simpler, unpublished construction by Gasarch and Lutz [5] of a rec-random sequence that is not algorithmically random.

Theorem 5. *There is a rec-random sequence that is ultracompressible.*

We now note that rec-random sequences can be strongly deep.

Theorem 6. *There is a rec-random sequence that is strongly deep.*

Theorem 6 contrasts sharply with Theorem 1 and the fact that RAND ∩ strDEEP = ∅. There is of course nothing paradoxical in this contrast. It is merely a consequence of the strong, quantitative separation of RAND(rec) from RAND given by Theorem 5.

We now have the first of the desired noninclusions.

Corollary 7. strDEEP ⊄ rec-wkDEEP.

The following known theorem says that the set of strongly deep sequences is small in the sense of Baire category.

Theorem 8 (Juedes, Lathrop, and Lutz [7]). *The class* strDEEP *is meager.*

We show that rec-wkDEEP ⊄ strDEEP by showing that rec-wkDEEP is comeager. Our proof of this fact is somewhat more involved than the proof by Juedes, Lathrop, and Lutz [7] that wkDEEP is comeager.

Theorem 9. *For each uniform reducibility F, the class rec-F-deep is rec-comeager, hence comeager in* REC.

Theorem 10. *The class rec-wkDEEP is comeager.*

Corollary 11. *rec-wkDEEP* ⊄ strDEEP.

We now have the main result of section 2.3.

Theorem 12. *The following diagram of proper inclusions holds.*

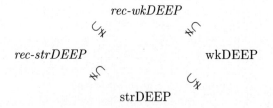

By Theorem 12, there exist sequences that are strongly deep, but not rec-strongly deep. Let S be such a sequence. Since S is not rec-strongly deep, there exist a *fixed* computable time bound $t_0 : \mathbb{N} \to \mathbb{N}$ and a *fixed* constant $c_0 \in \mathbb{N}$ such that, for *every* computable time bound $l : \mathbb{N} \to \mathbb{N}$, there are infinitely many prefixes of S that *cannot* be described c_0 bits more succinctly with the time bound l than with the time bound t_0. Nevertheless, since S is strongly deep, it must be the case that, for *every* constant $c \in \mathbb{N}$ (even when c is much greater than c_0), all but finitely many prefixes of S can be described at least c bits more succinctly without a time bound than with the time bound t_0. *None* of this additional succinctness (beyond c_0 bits) can be realized within any computable time bound; *all* of it requires greater-than-computable running time. The depth of such a sequence S appears not to come from so much from a nontrivial causal (computational) history as from something utterly noncomputational.

If F is a uniform reducibility that is (like all standard reducibilities) reflexive, then the measure and category of the class rec-F-DEEP are of some interest. First, rec-F-DEEP must be disjoint from RAND(rec), so rec-F-DEEP must be a measure 0 subset of **C**. Also, by Theorem 9, rec-F-DEEP must be comeager. Thus, the class rec-F-DEEP is small in the sense of measure, but large in the sense of Baire category. This state of affairs is not unusual and would not be worth mention, were it not for the fact that the situation changes when we look at the measure and category of rec-F-DEEP in REC. By [4] and Theorem 9, rec-F-DEEP is large in REC in the senses of both measure and category. The class rec-F-DEEP is thus one concerning which measure and category agree in REC, but disagree in **C**.

2.4 Weakly Useful Sequences

Juedes, Lathrop, and Lutz [7] defined the class of *weakly useful* sequences and proved that every weakly useful sequence is strongly deep. Fenner, Lutz, and Mayordomo [4] subsequently proved that every weakly useful sequence is rec-weakly deep. In this section, we strengthen both these results by proving that every weakly useful sequence is rec-strongly deep. Our argument closely follows that of [7].

Definition (Juedes, Lathrop, and Lutz [7]). A sequence $A \in \mathbf{C}$ is *strongly useful*, and we write $A \in \text{strUSEFUL}$, if there is a computable time bound $s : \mathbb{N} \to \mathbb{N}$ such that $\text{REC} \subseteq \text{DTIME}^A(s)$. A sequence $A \in \mathbf{C}$ is *weakly useful*, and we write $A \in \text{wkUSEFUL}$, if there is a computable time bound $s : \mathbb{N} \to \mathbb{N}$ such that $\text{DTIME}^A(s)$ does not have measure 0 in REC.

Thus a sequence is strongly useful if it enables one to solve all decidable sequences in some fixed, computable amount of time. A sequence is weakly useful if it enables one to solve all elements of a nonnegligible set of decidable sequences in some fixed, computable amount of time.

Recall that the diagonal halting problem is the sequence K whose n^{th} bit is

$$K[n] = [\![M_n(n) \text{ halts}]\!],$$

where $M_0, M_1, \ldots$ is a standard enumeration of all deterministic Turing machines. It is well-known that K is polynomial-time many-one complete for the set of all recursively enumerable subsets of $\mathbb{N}$, so K is strongly useful.

It is clear that every strongly useful sequence is weakly useful. Fenner, Lutz, and Mayordomo [4] used martingale diagonalization to construct a sequence that is weakly useful but not strongly useful, so $\text{strUSEFUL} \subsetneq \text{wkUSEFUL}$.

We now establish the rec-strong depth of weakly useful sequences.

Theorem 13. *Every weakly useful sequence is rec-strongly deep.*

Juedes [6] asked whether every strongly deep sequence is weakly useful. We now answer this question negatively.

Corollary 14. wkUSEFUL $\subsetneq$ strDEEP

Acknowledgments

We thank Bas Terwijn, David Juedes, Bill Gasarch, and Giora Slutzki for useful discussions. We also thank David Juedes for helpful remarks on a preliminary draft of this paper.

References

1. K. Ambos-Spies and Y. Wang. Algorithmic randomness concepts: a comparison. Talk by K. Ambos-Spies at the Workshop on Information and Randomness in Complexity Classes, Schloss Dagstuhl, Germany, July 17, 1996.
2. C. H. Bennett. Dissipation, information, computational complexity and the definition of organization. In D. Pines, editor, *Emerging Syntheses in Science, Proceedings of the Founding Workshops of the Santa Fe Institute*, pages 297–313, 1985.
3. C. H. Bennett. Logical depth and physical complexity. In R. Herken, editor, *The Universal Turing Machine: A Half-Century Survey*, pages 227–257. Oxford University Press, 1988.
4. S. A. Fenner, J. H. Lutz, and E. Mayordomo. Weakly useful sequences. In *Proceedings of the 22nd International Colloquium on Automata, Languages, and Programming*, pages 393–404. Springer–Verlag, 1995.
5. W. I. Gasarch and J. H. Lutz. Unpublished manuscript, 1991.
6. D. W. Juedes. *The Complexity and Distribution of Computationally Useful Problems*. PhD thesis, Department of Computer Science, Iowa State University, 1994.
7. D. W. Juedes, J. I. Lathrop, and J. H. Lutz. Computational depth and reducibility. *Theoretical Computer Science*, 132:37–70, 1994.
8. J. I. Lathrop. Compression depth and the behavior of cellular automata. *Complex Systems*, 1997. To appear.
9. J. H. Lutz. Almost everywhere high nonuniform complexity. *Journal of Computer and System Sciences*, 44:220–258, 1992.
10. P. Martin-Löf. On the definition of random sequences. *Information and Control*, 9:602–619, 1966.
11. C. P. Schnorr. A unified approach to the definition of random sequences. *Mathematical Systems Theory*, 5:246–258, 1971.
12. C. P. Schnorr. Zufälligkeit und Wahrscheinlichkeit. *Lecture Notes in Mathematics*, 218, 1971.
13. R. J. Solomonoff. A formal theory of inductive inference. *Information and Control*, 7:1–22, 224–254, 1964.
14. M. van Lambalgen. *Random Sequences*. PhD thesis, Department of Mathematics, University of Amsterdam, 1987.
15. Y. Wang. *Randomness and Complexity*. PhD thesis, Department of Mathematics, University of Heidelberg, 1996.

Some Bounds on the Computational Power of Piecewise Constant Derivative Systems (Extended Abstract)

Olivier Bournez

Laboratoire de l'Informatique du Parallélisme
Ecole Normale Supérieure de Lyon
46, Allée d'Italie F-69364 Lyon Cedex 07, France
obournez@lip.ens-lyon.fr

Abstract. We study the computational power of Piecewise Constant Derivative (PCD) systems. PCD systems are dynamical systems defined by a piecewise constant differential equation and can be considered as computational machines working on a continuous space with a continuous time. We show that the computation time of these machines can be measured either as a discrete value, called discrete time, or as a continuous value, called continuous time. We prove that the languages recognized by PCD systems in dimension d in finite continuous time are precisely the languages of the $d - 2^{th}$ level of the arithmetical hierarchy. Hence we provide a precise characterization of the computational power of purely rational PCD systems in continuous time according to their dimension and we solve a problem left open by [2].

1 Introduction

There has been recently an increasing interest in the community of control and verification theory about hybrid systems. A hybrid system is a system that combines discrete and continuous dynamics. Hybrid systems can be also be considered as computational machines: they can be seen either as machines working on a continuous space with a discrete time or as machines working on a continuous space with a continuous time.

The first point of view has been investigated in [1, 2, 4, 5]. In particular, in [1, 2, 3] the attention is focused on a very simple type of hybrid systems: Piecewise Constant Derivative Systems (PCD systems) are dynamical systems defined by a piecewise constant differential equation. It is shown that the reachability problem for PCD systems is decidable in dimension $d = 2$ and undecidable in dimension $d \geq 3$ [1, 3]. In [4], the computational power of Piecewise Constant Derivative systems is characterized as *P/poly* in polynomial discrete time, and as unbounded in exponential discrete time.

This paper deals with the second point of view that considers hybrid systems as machines that work on a continuous space with a continuous time. The study of computational machines that work in a continuous time is only beginning: in [6], Moore proposed a recursion theory for computations on the reals in continuous time. Recently, Asarin and Maler [2] showed, using Zeno's paradox, that every set of the arithmetical hierarchy can be recognized in finite continuous time and in finite dimension by a PCD system: every set of the arithmetical hierarchy in $\Sigma_k \cup \Pi_k$ can be recognized by a rational PCD system in dimension $5k+1$. Unfortunately, no precise characterization of the PCD recognizable sets was given in [2]. In this paper, we improve the results of Asarin and Maler and we provide a full characterization of the sets recognized by purely rational PCD systems: we show that the sets that are recognized by purely rational PCD systems in dimension d are precisely the sets of the $d-2^{th}$ level of the arithmetical hierarchy.

Section 2 is devoted to some general definitions: PCD systems, computations on PCD systems, discrete and continuous time. In section 3, we improve 5 times the result of Asarin and Maler: any arithmetical set in Σ_k can be recognized in dimension $2+k$. In section 4 we prove that this bound is optimal for purely rational PCD systems: no other set can be recognized in that dimension.

2 Definitions

A convex polyhedron of $\mathbb{R}^d$ is any finite intersection of open or closed half spaces of $\mathbb{R}^d$. A polyhedron of $\mathbb{R}^d$ is a finite union of convex polyhedral of $\mathbb{R}^d$. In particular, a polyhedron may be unbounded or flat. For $V \subset \mathbb{R}^d$, we denote by $\overline{V}$ the topological closure of V. We denote by d the Euclidean distance of $\mathbb{R}^d$. A rational point of $\mathbb{R}^d$ is a point of $\mathbb{R}^d$ with rational coordinates.

Definition 1 PCD System [1, 2]. A *Piecewise Constant Derivative (PCD) system of dimension d is a couple* $\mathcal{H} = (X, f)$ *with* $X = \mathbb{R}^d$, $f : X \to X$, *where the range of f is a finite set* $C \subset X$, *such that for any* $c \in C$ *(c is called a slope)* $f^{-1}(c)$ *is a finite union of convex polyhedral sets (called regions). A trajectory of $\mathcal{H}$ starting from x_0 is a continuous solution to the differential equation* $\dot{x}_d = f(x)$, *with initial condition x_0, where $\dot{x}_d$ denotes the right derivative: that is* $\Phi : D \subset \mathbb{R}^+ \to X$ *where D is an interval of $\mathbb{R}^+$ containing 0, $\Phi(0) = x_0$, and* $\forall t \in D, \dot{\Phi}_d(t) = f(\Phi(t))$. *Trajectory Φ is said to continue for ever if $D = \mathbb{R}^+$.*

In other words a PCD system consists of partionning the space into convex polyhedral regions, and assigning a constant derivative c, called *slope*, to all the points sharing the same region. The trajectories of such systems are broken lines with the breakpoints occuring on the boundaries of the regions [2]. See figure 1. The *signature* of a trajectory is the sequence of the regions that are crossed by the trajectory.

Definition 2 Rational, purely rational PCD systems. – A PCD system is called *rational* if all the slopes as well as all the polyhedral regions can be described using only rational coefficients.
  – A PCD system is called *purely rational*, if in addition, for all trajectory Φ starting from a rational point, each time Φ enters a region in a point x, necessarily x has rational coordinates.

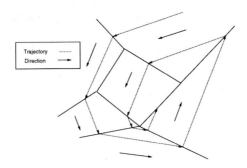

Fig. 1. A PCD system in dimension 2.

Some comments are in order: one must understand that a trajectory Φ can enter a region either by a discrete transition or by converging to a point of the region: see figure 2. Thus, in other words, in a purely rational PCD system any converging process converges towards a point with rational coordinates. Note that one can construct a rational PCD system of dimension 5 that is not purely rational.

We can say some words on the existence of trajectories in a PCD system: let $x_0 \in X$. We say that x_0 is *trajectory well-defined* if there exists a $\epsilon > 0$ such that $f(x) = f(x_0)$ for all $x \in [x_0, x_0 + \epsilon * f(x_0)]$. It is clear that, for any $x_0 \in X$, there exists a trajectory starting from x_0 iff x_0 is trajectory well-defined. Given a rational PCD system $\mathcal{H}$, one can effectively compute the set $NoEvolution(\mathcal{H})$ of the points of X that are not trajectory well-defined. See that a trajectory can continue for ever iff it does not reach $NoEvolution(\mathcal{H})$.

Definition 3 Computation [2]. – Let $\mathcal{H} = (X, f)$ be a PCD system of dimension d. Let $I = [0,1]$ and let $r : \mathbb{N} \to I$ be an injective coding function, let x^1, x^0 be two distinct points of $\mathbb{R}^d$. A computation of system $\hat{H} = (\mathbb{R}^d, f, r, I, x^1, x^0)$ on entry $n \in \mathbb{N}$ is a trajectory that can continue forever (defined on all $\mathbb{R}^+$) of $\mathcal{H} = (X, f)$ starting from $(r(n), 0, \ldots, 0)$. The computation is accepting if the trajectory eventually reaches x^1, and refusing if it reaches x^0. It is assumed that the derivatives at x^1 and x^0 are zero.
– Language $L \subset \mathbb{N}$ is semi-recognized by $\hat{H}$ if, for every $n \in \mathbb{N}$, there is a computation on entry n and the computation is accepting iff $n \in L$. L is said to be (fully-)recognized by $\hat{H}$ when, in addition, this trajectory reaches x^0 iff $n \notin L$.

Definition 4 Continuous and Discrete time. Let $\Phi_n : \mathbb{R}^+ \to X$ be an accepting computation on entry $n \in \mathbb{N}$.

– The continuous time $T_c(n)$ of the computation is $T = \min\{t \in \mathbb{R}^+ / \Phi_n(t) = x^1\}$.
– Let $T_n = \{t / \Phi_n(t)$ crosses a boundary of a region at time $t\}$. It is easy to see that T_n is a well ordered set. The discrete time $T_d(n)$ of the computation is

defined as the order type of well ordered set T_n (= the ordinal corresponding to T_n).

Note that Zeno's paradox appears: to a continuous finite time can correspond a transfinite discrete time: see figure 2.

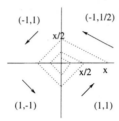

Fig. 2. Zeno's paradox: at finite continuous time $5x = 2.5(x + x/2 + x/4 + \ldots)$ the trajectory is in $(0,0)$, but it takes a transfinite discrete time ω to reach this point.

We recall the following definition:

Definition 5 Arithmetical hierarchy [8, 7]. The classes $\Sigma_k, \Pi_k, \Delta_k$, for $k \in \mathbb{N}$, are defined inductively by:

- Σ_0 is the class of the languages that are recursive.
- For $k \geq 1$, Σ_k is the class of the languages that are recursively enumerable in a set in Σ_{k-1} (that is semi-recognized by a Turing machine with an oracle in Σ_{k-1})
- For $k \in \mathbb{N}$, Π_k is defined as the class of languages whose complement are in Σ_k, and Δ_k is defined as $\Delta_k = \Pi_k \cap \Sigma_k$.

Several characterizations of the sets of the arithmetical hierarchy are known: see [7, 8]. In particular we will assume the reader familiar with Tarski-Kuratowski computations: assume a first order formula F, over some recursive predicates, characterizing the elements of a set $S \subset \mathbb{N}$, is given. Then S is in the arithmetical hierarchy and the Tarski-Kuratowski algorithm on formula F returns a level of the arithmetical hierarchy containing S: see [7, 8] for the full details.

3 PCD Systems can Recognize Arithmetical Sets

It was shown in [2] that every set of the arithmetical hierarchy can be recognized in finite continuous time: more precisely, it is shown that $L \in \Sigma_k \cup \Pi_k$ can be recognized by a PCD system of dimension $5k + 1$. Therefore, five dimensions are used in [2] to climb each level of the arithmetical hierarchy: one for a timer, one used for the divisions by 2, one used to do the homogenization, and two dimensions used to go from quantifier elimination to semi-recognition. We show here that only one dimension is needed (the one used to do the homogenization), and that the construction only requires purely rational PCD systems.

Theorem 6.
- Any language L of Σ_k is semi-recognized by a purely rational PCD system in dimension $2 + k$.
- Any language L of Δ_k is fully-recognized by a purely rational PCD system in dimension $2 + k$.

The proof is rather technical: timers are suppressed by using machines that cross a given hyper-plane at regular time, divisions by two are done by reusing the variables defining the machines, and the two variables used in [2] to go from quantifier elimination to semi-recognition are suppressed by storing some information in the variable used to do the homogenization.

4 PCD Systems Cannot Recognize Any Other Set

4.1 Local dimension

We define:

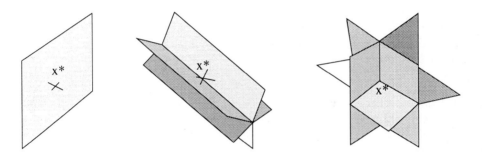

Fig. 3. From left to right: x^* is of local dimension $1^+, 2^+, 3$ in a PCD system of dimension 3.

Definition 7 Local dimension. Let $\mathcal{H} = (X, f)$ be a PCD system in dimension d. Let x^* be a point of X. Let Δ be a polyhedral subset $\Delta \subset X$ of maximal dimension $d - d'$ ($1 \leq d' \leq d$) such that there exists an open convex polyhedron $V \subset X$, with $x^* \in \Delta \cap V$, and such that, for any region F of $\mathcal{H}$, $F \cap V \neq \emptyset$ implies $\Delta \subset \overline{F}$ ($\overline{F}$ is the topological closure of F).

If $d' < d$ then x^* is said to be of *local dimension* d'^+. If $d' = d$ then x^* is said to be of *local dimension* d' and we can always choose V small enough such that x^* is the only point of local dimension d' in $\overline{V}$: see figure 3.

Note that given a rational PCD system $\mathcal{H} = (X, f)$ and $k = d'$ or $k = d'^+$ one can effectively compute $LocDim(\mathcal{H}, k)$ defined as the set of the points $x \in X$ that have a local dimension equals to k.

The main idea behind definition 7 is given by the following lemma: see figure 4.

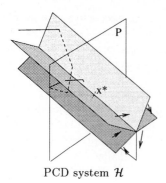
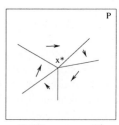

PCD system $\mathcal{H}$ PCD system $\mathcal{H}'$

Fig. 4. Proposition 8: if x^* is of local dimension 2^+ in a PCD system $\mathcal{H}$ of dimension 3, the projections on P of the trajectories of $\mathcal{H}$ in neighborhood V of x^* are precisely the trajectories of some PCD system $\mathcal{H}'$ of dimension 2.

Proposition 8. *Let $\mathcal{H} = (X, f)$ be a PCD system in dimension d. Let x^* be a point of local dimension $(d')^+$ with $d' < d$. Call P the affine variety of dimension d' which is the orthogonal of Δ in x^*. It is possible to construct a PCD system $\mathcal{H}' = (X' = \mathbb{R}^{d'}, f')$ in dimension d' such that the trajectories of $\mathcal{H}'$ are the orthogonal projections on P of the trajectories of $\mathcal{H}$ in V.*

For any point x^*, the corresponding V is denoted by V_{x^*}. $\mathcal{H}'$, Δ are respectively denoted by $\mathcal{H}_{x^*}$ and Δ_{x^*}. If $d' < d$ we denote by p_{x^*} and q_{x^*} the functions that map all point $x \in X$ onto its orthogonal projection on P and onto its orthogonal projection on Δ respectively. If $d' = d$, we define p_{x^*} and q_{x^*} as respectively the identity function and the null function. We assume the natural order $1 < 1^+ < 2 < 2^+ < \ldots$.

Lemma 9. *Let $\mathcal{H} = (X, f)$ be a PCD system of dimension d. Let Φ be a trajectory of $\mathcal{H}$ that reaches x^* at finite continuous time T_c. Assume that x^* is of local dimension $k = d'$ or $k = (d')^+$. For any l, denote by S_l the set of the points $x \in X$ that are reached by Φ at some time $0 \leq t < T_c$ and that have local dimension l. Assume $S_l = \emptyset$, for all $l > k$.*

- *S_k is a finite set.*
- *Assume $S_k = \emptyset$. Fix the origin in x^*. Then either $S_{(d'-1)^+}$ is a finite set or there exist $y_1, y_2 \in X$ that are reached by Φ, there exists $0 < \lambda < 1$ such that $p_{x^*}(y_2) = \lambda p_{x^*}(y_1)$ and such that, for all $n \geq 1$, Φ reaches at a time $t_n \leq T_c$ the point y_n defined by $p_{x^*}(y_n) = \lambda^n p_{x^*}(y_1)$ and $q_{x^*}(y_n) = q_{x^*}(y_1) + \sum_{i=1}^{n} \lambda^i (q_{x^*}(y_2) - q_{x^*}(y_1))$.*

Proof. Let $m \leq k$. We prove first that if S_m is not a finite set, then Φ reaches a point of local dimension $> m$ at some time $\leq T_c$: assume that S_m is not a finite set. $T_m = \{t | \Phi(t) \in S_m\}$ is a well ordered set. Denote its elements by $t_1^m, t_2^m, \ldots, t_\omega^m, \ldots$. Take $t_\infty^m = \sup_{i \in \mathbb{N}} t_i^m$. We have $t_\infty^m \leq T_c$. Consider $x_\infty^m = \Phi(t_\infty^m)$. By continuity of Φ, there exists $t^m < t_\infty^m$ such that $t \in [t^m, t_\infty^m] \Rightarrow \Phi(t) \in$

$V_{x_{\infty}^m}$. Take $t \in [t^m, t_{\infty}^m] \cap S_m$. From considerations of dimensions about point $\Phi(t)$ of local dimension m in $V_{x_{\infty}^m}$, we get that the local dimension d'' of x_{∞}^m is $\geq m$. From the definition of t_{∞}^m, we get $d'' \neq m$. Hence $d'' \geq m$ and our claim is proved: if S_m is not a finite set then Φ reaches some x_{∞}^m of local dimension $> m$.

The first assertion of the lemma is an easy consequence of this claim with $m = k$.

For the second assertion, take $m = (d'-1)^+$, and assume that $S_{(d'-1)^+}$ is not a finite set. From $S_k = \emptyset$, we must have $x_{\infty}^m = x^*$ and $t_{\infty}^m = T_c$. If $k < d$ denote $\mathcal{H}' = \mathcal{H}_{x^*}$ else take $\mathcal{H}' = \mathcal{H}$. Define Φ' as $p_{x^*}(\Phi)$. From time t^m up to time T_c, Φ' is a trajectory of $\mathcal{H}' = (X', f')$ (apply proposition 8 for $k < d$), reaching $p_{x^*}(x^*)$ at time T_c. Let $\mathcal{L}$ be the set of the one-dimensional regions of $\mathcal{H}'$ that intersect $V'_{x^*} = p_{x^*}(V_{x^*})$. We claim that each time Φ' reaches a point of $S_{(d'-1)^+}$, Φ' reaches an element of $\mathcal{L}$: if Φ' reaches some point $x^{*\prime} \in X'$ of local dimension $(d-1)^+$ at some time $t \in [t^m, T_c]$, then $p_{x^*}(\Delta_{x^{*\prime}})$ is an element of $\mathcal{L}$ and contains $x^{*\prime}$. See figure 5.

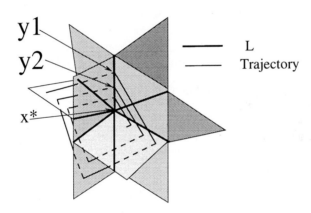

Fig. 5. Proof of lemma 9: here $d = d' = 3$. $\mathcal{L}$ defined as the set of the one dimensional regions that intersect $p_{x^*}(V_{x^*})$. $\mathcal{L}$ is made of a finite number of segments. Each time the trajectory reaches a point of local dimension 2^+, it reaches $\mathcal{L}$. If the trajectory reaches two times $\mathcal{L}$ in a same segment then the trajectory is ultimately cycling.

Since Φ' converges to $p_{x^*}(x^*)$, since $\mathcal{L}$ is a finite set, since $S_{(d'-1)^+}$ is infinite, $p_{x^*}(\Phi)$ reaches two times the same element of $\mathcal{L}$ in $p_{x^*}(y_1)$ and $p_{x^*}(y_2)$ with $p_{x^*}(y_2) = \lambda p_{x^*}(y_1)$ for some $0 < \lambda < 1$, at some times t_{y_1}, t_{y_2} with $t^m \leq t_{y_1} < t_{y_2} < T_c$. Now see that by definition of V'_{x^*} all the regions of $\mathcal{H}'$ intersecting V'_{x^*} contain $p_{x^*}(x^*)$ in their topological closure. Hence we have $f'(x) = f'(\mu x)$, for all $x \in V'_{x^*}, \mu \in (0, 1]$. If $\Phi'(t)$ is solution to differential equation $\dot{x}_d = f'(x)$, $\Psi'(t) = \lambda \Phi'(t/\lambda)$ is also solution. As a consequence trajectory Φ' must reach $\lambda^n p_{x^*}(y_1)$ for all n. From the definition of $\mathcal{H}'$ this implies that Φ reaches the y_n of the lemma for all n : see figure 5.

4.2 Problems *Reach* and *Conv*

Define the following problems:

Definition 10 Problems $Reach_{d'}$, $Reach_{d'^+}$. Let k be either of type $k = d'$ or of type $k = d'^+$, where d' is an integer.

- *Instance*: A purely rational PCD system $\mathcal{H} = (X, f)$ of dimension d, a polyhedral convex subset $V \subset X$, a rational polygon $x^1 \subset X$, a rational number $t_{sup} \in \mathbb{Q}$, a rational number $t_{inf} \in \mathbb{Q}$, a rational point $x_0 \in X$.
  Question "$Reach_k(\mathcal{H}, V, x_0, x^1, t_{inf}, t_{sup})$": "Do all the following conditions hold simultaneously:
  - trajectory Φ starting from x^0 reaches x^1 at some finite continuous time T_c
  - $t_{inf} < T_c \leq t_{sup}$
  - for any $0 \leq t \leq T_c$, $x = \Phi(t)$ is in V and is of local dimension $\leq k$."
- *Instance*: A purely rational PCD system $\mathcal{H} = (X, f)$ of dimension d, a polyhedral convex subset $V \subset X$, a rational point $x^* \in X$, a rational number $t_{sup} \in \mathbb{Q}$, a rational number $t_{inf} \in \mathbb{Q}$, a rational point $x_0 \in X$.
  Question "$Conv_k(\mathcal{H}, V, x_0, x^*, t_{inf}, t_{sup})$": "Do all the following conditions hold simultaneously:
  - the trajectory Φ starting from x_0 reaches point x^* at some finite continuous time T_c
  - x^* is of local dimension k and is in V
  - $t_{inf} < T_c \leq t_{sup}$
  - for any $0 \leq t < T_c$, $x = \Phi(t)$ is in V and is of local dimension $< k$."

4.3 Case $d = 3$

Using topological considerations (the sphere of $\mathbb{R}^3$ verifies Jordan Theorem and the arguments of [3]) we prove:

Lemma 11. *Let $\mathcal{H} = (X, f)$ be a PCD system of dimension d. Let Φ be a trajectory of $\mathcal{H}$ of finite continuous time T_c and discrete time $T_d \geq \omega$ converging towards $x^* = \Phi(T_c)$. Assume that x^* is of local dimension $\leq 3^+$. Then necessarily the signature of Φ is ultimately cyclic.*

Lemma 12. *The following problem is decidable:*
Instance: a rational PCD system $\mathcal{H} = (X, f)$ of dimension d, a finite sequence of distinct regions $(F_0, F_1, \ldots, F_j)$ of $\mathcal{H}$, a rational point $x_0 \in X$.
Question: "Does the trajectory Φ starting from x_0 have a periodic signature of type $(F_0, F_1, \ldots, F_j)^\omega$ and then reach some point $x^* \in X$ of local dimension $\leq 3^+$ at some finite continuous time t^*"
Moreover, given a positive instance, one can effectively compute t^* and x^* as a function of the coordinates of x_0.

With these lemmas, we prove:

Theorem 13. *The problems $Reach_3$ and $Reach_{3+}$ are in Σ_1.*

Proof (sketch). We prove the assertion by providing a Turing machine algorithm that (semi-)computes the predicates: to reply to $Reach_{3+}(\mathcal{H}, V, x_0, x^1, t_{inf}, t_{sup})$, the general idea is the following: we simulate step by step the evolution of the trajectory Φ starting from x_0. Simultaneously, if we detect that Φ crosses for the second time a given region, we use lemma 12 to see if the signature of Φ is entering or not an infinite cycle. If it is so, still by lemma 12, we compute directly the limit of the cycle x^* and the corresponding time t^* and the simulation goes on directly from new position x^* and time t^*. We stop if we reach x^1 or the complement of V, or if the time reaches a value greater than t_{sup}. From lemma 9, we know that every point of local dimension $k = 3$ or $k = 3^+$ can only be reached using a finite number of points of local dimension k. From lemma 11 each such point x of local dimension k is reached by a cyclic signature and is dropped by the algorithm.

4.4 Case $d \geq 4$

We generalize theorem 13 to higher dimensions. We prove first:

Lemma 14. *Let $d' \geq 4$. Assume that $Reach_{(d'-1)+} \in \Sigma_p$ and that $Reach_{(d'-2)+} \in \Sigma_q$ for some integers p, q. Then*

- *$Conv_{d'} \in \Sigma_{max(p,q+2)}$.*
- *$Conv_{d'+} \in \Sigma_{max(p,q+2)}$.*

Proof. Denote by $B(x^*, 1/n_1)$ the ball of radius $1/n_1$ centered in x^* for the norm of the maximum. For a subset $U \subset X$, denote its complement by U^c. Let $k = d'$ or $k = d'^+$. We claim:

$Conv_k(\mathcal{H}, V, x_0, x^*, t_{inf}, t_{sup})$
$\Leftrightarrow \quad x^* \in LocDim(\mathcal{H}, k) \wedge x^* \in V \wedge t_{inf} < t_{sup}$
$\wedge \exists y_1 \in \mathbb{Q}^d \; \exists t_1, t_2 \in \mathbb{Q} \; y_1 \in V_{x^*} \wedge Reach_{(d'-1)+}(\mathcal{H}, V, x_0, y_1, t_1, t_2)$

$$\wedge \left\{ \begin{array}{l} \left\{ \begin{array}{l} \exists y_2 \in \mathbb{Q}^d \; \exists t_3, t_4 \in Q \; \exists \lambda \in \mathbb{R}^+ \\ Reach_{(d'-1)+}(\mathcal{H}, V \cap V_{x^*}, y_1, y_2, t_3, t_4) \\ p_{x^*}(y_2) = \lambda p_{x^*}(y_1) \\ \lambda < 1 \\ t_1 + \sum_{i=1}^{\infty} \lambda^i t_3 > t_{inf} \\ t_2 + \sum_{i=1}^{\infty} \lambda^i t_4 \leq t_{sup} \\ q_{x^*}(y_1) + \sum_{i=1}^{\infty} \lambda^i (q_{x^*}(y_2) - q_{x^*}(y_1)) = q_{x^*}(x^*) \end{array} \right. \\ \vee \; \forall n_1 \in \mathbb{N} \; Reach_{(d'-2)+}(\mathcal{H}, V, y_1, B(x^*, 1/n_1), t_{inf} - t_1, t_{sup} - t_2) \end{array} \right.$$

Assume that we have a positive instance to formula $Conv_k$: use the notations of definition 10. Denote by S the set of the points that are reached by Φ before time T_c and that have local dimension $(d' - 1)^+$. Since Φ converges to x^*, there must exist an $y_1 = \Phi(t_{y_1}) \in V_{x^*}$, $t_{y_1} < T_c$ that is reached by Φ, and such that

Φ stays in V_{x^*} between time t_{y_1} and time T_c. y_1 is reached using points of local dimension $\leq (d'-1)^+$. If S is not a finite set, by lemma 9 the first clause of the disjunction is true. Assume now that S is a finite set: we can assume that t_{y_1} is chosen big enough such that Φ does not reach any point of S between time t_{y_1} and time T_c. For all $n_1 \in \mathbb{N}$ we get that the trajectory starting from y_1 reaches $B(x_*, 1/n_1)$ using only points of local dimension $\leq (d'-2)^+$. Hence the second clause of the disjunction is true.

Conversely, assume that the right hand side of the formula is true. If the first clause of the disjunction is true, the trajectory is cycling and the formula $Conv_k$ should be true. Assume now that the second clause is true. For all $n_1 \in \mathbb{N}$, we get that there exists t_{n_1} such that $\Phi(t_{n_1}) \in B(x^*, 1/n_1)$. Denote $T_c = sup_{n_1 \in \mathbb{N}} t_{n_1}$. From the continuity of Φ we get that[1] $\Phi(T_c) = x^*$. Hence Φ reaches x^* of local dimension k and formula $Conv_k$ must be true.

The result is now immediate by applying the Tarski-Kuratowski algorithm on the formula [8].

We also prove in a similar way:

Lemma 15. *Let $d' \geq 4$. Assume $Reach_{(d'-1)^+} \in \Sigma_p$ for some integer p. Then $Conv_{d'} \in \Sigma_{p+1}$.*

Proof (sketch). For a point $x^* \in X$ of local dimension d, define Out_{x^*} as the set of the points $x \in X$ such that the trajectory starting from x intersects the complement of V_{x^*} at a discrete time less or equal to one. We prove that, now, the following formula holds:

$Conv_{d'}(\mathcal{H}, V, x_0, x^*, t_{inf}, t_{sup})$
$\Leftrightarrow \quad x^* \in LocDim(\mathcal{H}, k) \wedge x^* \in V \wedge t_{inf} < t_{sup} \wedge dimension(\mathcal{H}) = d'$
$\wedge \exists y_1 \in \mathbb{Q}^d \ \exists t_1, t_2 \in \mathbb{Q} \ y_1 \in V_{x^*} \wedge Reach_{(d'-1)^+}(\mathcal{H}, V, x_0, y_1, t_1, t_2)$
$\begin{cases} Reach_{(d'-1)^+}(\mathcal{H}, X, y_1, X, t_{inf} - t_1, t_{inf} - t_1 + 1) \\ \wedge \neg Reach_{(d'-1)^+}(\mathcal{H}, X, y_1, V^c \cup NoEvolution(\mathcal{H}) \cup Out_{x^*}, 0, t_2 - t_{sup}) \\ \wedge \neg Reach_{(d'-1)^+}(\mathcal{H}, X, y_1, X, t_{sup} - t_2, t_{sup} - t_2 + 1) \end{cases}$

We get:

Theorem 16. *Let $d' \geq 3$.*

- *$Reach_{d'}$ is in $\Sigma_{d'-2}$.*
- *$Reach_{d'^+}$ is in $\Sigma_{d'-1}$ if d' is even.*
- *$Reach_{d'^+}$ is in $\Sigma_{d'-2}$ if d' is odd.*

Proof. The assertion is proved by recurrence over d' using theorem 13, lemmas 15 and 14, by Tarski-Kuratowski and the fact that we have for $k = d'$ or $k = d'^+$:

---

[1] Note that if function Φ is not defined on value T_c, since Φ is continuous with a bounded right derivative, Φ can always be extended to a continuous functions defined on value T_c.

$Reach_k(\mathcal{H}, V, x_0, x^1, t_{inf}, t_{sup})$
$\Leftrightarrow Reach_{(d'-1)+}(\mathcal{H}, V, x_0, x^1, t_{inf}, t_{sup})$
$\vee \exists n \in \mathbb{N} \; \exists <x_0^*, x_1^*, x_2^*, \ldots, x_n^*> \in \mathbb{Q}^d \; \exists <t_0, \ldots, t_n>$
$\exists <t'_0, \ldots, t'_n>$
$\begin{cases} x_0^* = x_0 \\ \forall 0 \le i < n \; Conv_k(\mathcal{H}, V, x_i^*, x_{i+1}^*, t_i, t'_i) \\ Reach_{(d'-1)+}(\mathcal{H}, V, x_n^*, x^1, t_n, t'_n) \\ t_0 + t_1 + \ldots + t_n > t_{inf} \\ t'_0 + t'_2 + \ldots + t'_n \le t_{sup} \end{cases}$

By Tarski-Kuratowski on formula $n \in L \Leftrightarrow \exists t_1 \in \mathbb{N} \; Reach_d(\mathcal{H}, X, r(n), x^1, 0, t_1)$, we get the main result of this section:

Corollary 17. – *If L is semi-recognized by a purely rational PCD system of dimension d, then $L \in \Sigma_{d-2}$.*
– *If L is recognized by a purely rational PCD system of dimension d, then $L \in \Delta_{d-2}$.*

And by using theorem 6:

Corollary 18. – *The languages that are semi-recognized by purely rational PCD systems of dimension d in finite continuous time are precisely the languages of Σ_{d-2}*
– *The languages that are recognized by purely rational PCD systems of dimension d in finite continuous time are precisely the languages of Δ_{d-2}*

References

1. Eugene Asarin and Oded Maler. On some Relations between Dynamical Systems and Transition Systems. In *Proceedings of ICALP*, pages 59–72, 1994. Lecture Notes in Computer Science, 820.
2. Eugene Asarin and Oded Maler. Achilles and the Tortoise Climbing Up the Arithmetical Hierarchy. In *Proceedings of FSTTCS*, pages 471–483, 1995. Lecture Notes in Computer Science, 1026.
3. Eugene Asarin, Oded Maler, and Amir Pnueli. Reachability analysis of dynamical systems having piecewise-constant derivatives. *Theoretical Computer Science*, 138:33–65, 1995.
4. Olivier Bournez and Michel Cosnard. On the computational power of hybrid and dynamical systems. *Theoretical Computer Science*, 168(2):417–459, 1996.
5. Michael S. Branicky. Universal computation and other capabilities of hybrid and continuous dynamical systems. *Theoretical Computer Science*, 138:67–100, 1995.
6. Cristopher Moore. Recursion theory on the reals and continuous-time computation. *Theoretical Computer Science*, 162:23–44, 1996.
7. P. Odifreddi. *Classical Recursion Theory*, volume 125 of *Studies in Logic and the foundations of mathematics*. Elsevier, 1992.
8. H. Rogers. *Theory of Recursive Functions and Effective Computability*. McGraw-Hill, 1967.

Monadic Simultaneous Rigid E-Unification and Related Problems

Yuri Gurevich[1]* and Andrei Voronkov[2]**

[1] EECS Department
University of Michigan
Ann Arbor, MI, 48109-2122, USA
[2] Computing Science Department, Uppsala University
Box 311, S-751 05 Uppsala, Sweden

Abstract. We study the monadic case of a decision problem know as simultaneous rigid E-unification. We show its equivalence to an extension of word equations. We prove decidability and complexity results for special cases of this problem.

1 Introduction

Simultaneous rigid E-unification is a combinatorial problem in equational logic which is closely connected with some formulations of the Herbrand theorem and with automated theorem proving by the tableau method and the connection (or mating) method. In this section we define simultaneous rigid E-unification, discuss its connection with several decision problems in logic and survey some known results.

We shall consider *equational logic*, i.e. logic whose only predicate is the equality predicate $\simeq$. Let $s_1, t_1, \ldots, s_n, t_n, s, t$ be terms. All atomic formulas in equational logic are *equations*, i.e. expressions of the form $s \simeq t$. We do not distinguish an equation $s \simeq t$ from the equation $t \simeq s$. We write $s_1 \simeq t_1, \ldots, s_n \simeq t_n \vdash s \simeq t$ to denote that the formula $\forall (s_1 \simeq t_1 \wedge \ldots \wedge s_n \simeq t_n \supset s \simeq t)$ is true, i.e. it is provable in first-order (classical or intuitionistic) logic. Equivalently, we can say that s and t lie in the same class of the congruence induced by $\{s_1 \simeq t_1, \ldots, s_n \simeq t_n\}$.

A *rigid equation* is an expression $\mathcal{E} \vdash_\forall s \simeq t$, where $\mathcal{E}$ is a finite set of equations. The set $\mathcal{E}$ is called the *left-hand side* of this rigid equation, and the equation $s \simeq t$ — its *right-hand side*. A *solution to a rigid equation* $\{s_1 \simeq t_1, \ldots, s_n \simeq t_n\} \vdash_\forall s \simeq t$ is any substitution θ such that $s_1\theta \simeq t_1\theta, \ldots, s_n\theta \simeq t_n\theta \vdash s\theta \simeq t\theta$. A *system of rigid equations* is a finite set of rigid equations. A *solution to a system of rigid equations* $\mathcal{R}$ is any substitution that is a solution to every rigid equation in $\mathcal{R}$. The problem of solvability of rigid equations is known as *rigid E-unification*. The problem of solvability of systems of rigid equations is known as *simultaneous rigid E-unification*, in the sequel abbreviated as SREU.

---
* Partially supported by grants from NSF, ONR and the Faculty of Science and Technology of Uppsala University.
** Supported by a TFR grant.

We shall denote sets of equations by $\mathcal{E}$, systems of rigid equations by $\mathcal{R}$ and rigid equations by R. We shall sometimes write the left-hand side of a rigid equation as a *sequence* of equations, for example $x \simeq a \vdash_\forall g(x) \simeq x$ instead of $\{x \simeq a\} \vdash_\forall g(x) \simeq x$.

In [2] it is shown that the decidability of SREU is equivalent to the decidability of some other fundamental problems, for example the decidability of the prenex fragment of intuitionistic logic with equality. We refer to [2, 6] for the discussion of these problems.

Best known (un)decidability results on SREU are the following: (i) SREU with ground left-hand sides, two variables and three rigid equation is undecidable (Veanes [16]); (ii) SREU with one variable is DEXPTIME-complete (Degtyarev, Gurevich, Narendran, Veanes and Voronkov [3]). The last two results imply a complete classification of decidable prenex fragments of intuitionistic predicate calculus with equality: the $\exists\exists$ fragment is undecidable and the $\forall^*\exists\forall^*$ fragment is decidable. All the above mentioned undecidability results require that the signature contain a function symbol of arity ≥ 2.

The special case of SREU when all function symbols have arity ≤ 1, is called *monadic SREU*. The decidability of monadic SREU is an open problem. The following facts are known about monadic SREU (Degtyarev, Matiyasevich and Voronkov [4]).

- The word equation problem is effectively reducible to monadic SREU. (This fact shows that if this problem is decidable, its decidability should be uneasy to prove.)
- Monadic SREU with one function symbol is decidable (this fact has a nonelementary proof).
- Monadic SREU is decidable if and only if it is decidable in the signature with two function symbols.

This paper studies monadic SREU. Although the general case remains an open problem, we prove its equivalence to a combinatorial problem of words defined in Section 5. This problem is defined in terms of *ideals* on the set of pairs of words and called *the ideal equation problem*. We prove

Theorem 4 Monadic SREU is decidable if and only if the ideal equation problem is decidable.

We also prove the decidability of some special cases of monadic SREU. In Section 4 we prove a result similar to the main result of [3]:

Theorem 3 Monadic SREU with one variable is PSPACE-complete.

Plaisted [13] proved that SREU with ground left-hand sides is undecidable. The corresponding monadic case is shown to be decidable in Section 3:

Theorem 2 Monadic SREU with ground left-hand sides is decidable.

The complexity of monadic SREU with ground left-hand sides is not known. We prove

Theorem 1 Monadic SREU with one variable and ground left-hand sides is PSPACE-hard.

2 Preliminaries

In this section we introduce basic definitions concerning terms, equations, words, word equations, automata and rewrite rules. We have to define so many concepts since it is unreasonable to expect the reader to know everything. We also assert some statements proved elsewhere and prove some properties of the introduced notions which will be used in subsequent sections.

The symbol $\rightleftharpoons$ means "equal by definition".

Terms and equations. The set of all variables of a term t is denoted $var(t)$. A term is *ground* iff it has no variables, i.e. $var(t) = \emptyset$. The symbol $\vdash$ denotes provability in first-order logic. When we write $\varphi_1, \ldots, \varphi_n \vdash \varphi$, where $\varphi_1, \ldots, \varphi_n, \varphi$ are formulas, it means provability of the formula $\varphi_1 \wedge \ldots \wedge \varphi_n \supset \varphi$. *Substitutions* of terms $t_1, \ldots, t_n$ for variables $x_1, \ldots, x_n$ are denoted $\{t_1/x_1, \ldots, t_n/x_n\}$. The *application of such a substitution θ to a term t*, is the operation of simultaneous replacement of all occurrences of x_i by t_i. The result of the application is the term denoted $t\theta$. We shall also apply substitutions to equations and sets of equations and use the same notation for the result of the application.

For any expression E (for example, term, or a set of equations), we denote by E_c^t the expressions obtained from E by the replacement of all occurrences of the constant c by a term t. We write $s[t]$ to denote a particular occurrence of a subterm t of a term s.

In this paper, we shall only consider *monadic signatures* consisting of a finite set $\mathcal{F}$ of unary function symbols and a finite set $\mathcal{C}$ of constants. Such signatures are denoted $(\mathcal{F}, \mathcal{C})$. The set of ground terms of this signature is denoted by $T_{(\mathcal{F},\mathcal{C})}$. We always assume $\mathcal{C} \neq \emptyset$ and hence $T_{(\mathcal{F},\mathcal{C})} \neq \emptyset$. For any set of equations $\mathcal{E}$ we denote by $T(\mathcal{E})$ the set of all terms occurring in $\mathcal{E}$ and their subterms. For example, if $\mathcal{E} = \{f(x) \simeq g(c), c \simeq g(f(x))\}$, then $T(\mathcal{E}) = \{x, f(x), c, g(c), g(f(x))\}$.

We shall denote variables by x, y, z, constants by a, b, c, d, function symbols by f, g, h, terms by r, s, t and substitutions by θ.

We shall use the following statement proved in Kozen [9] or Shostak [15].

Lemma 1 (Derivability of equations is in PTIME) *There is a polynomial-time algorithm checking, by a given finite set of equations $\mathcal{E}$ and terms s, t, whether $\mathcal{E} \vdash s \simeq t$.*

We write $\mathcal{E}' \vdash \mathcal{E}$ iff for any equation $(s \simeq t) \in \mathcal{E}$ we have $\mathcal{E}' \vdash s \simeq t$. In the sequel we shall use the following lemma whose proof is standard.

Lemma 2 (Lemma on constants) *Let $\mathcal{E}$ and $\mathcal{E}'$ be sets of equations. For any constant c and term t, if $\mathcal{E} \vdash \mathcal{E}'$, then $\mathcal{E}_c^t \vdash \mathcal{E}'{}_c^t$.*

Words and finite automata. This section defines *words* and *finite automata*. We shall also introduce a notation for monadic terms which allows us to easily come from terms to words and back.

Let $\mathcal{F}$ be a finite non-empty set, called the *alphabet*. Its elements are called *letters*. *Words* are finite sequences of letters. We denote words by a juxtaposition of its letters, as $W = a_1 a_2 \ldots a_n$. The natural number n is called the *length* of the word W and denoted $|W|$. We denote by ε the *empty word*, which is the unique word of length zero. The set of all words with letters in $\mathcal{F}$ is denoted by $\mathcal{F}^*$.

It will be convenient for us to use the alphabet $\mathcal{F}$ also as the set of unary function symbols of a monadic signature $(\mathcal{F}, \mathcal{C})$. Every term s in such a signature has the form $f_1(f_2(\ldots f_n(t)\ldots))$ where $n \geq 0$, $f_1, \ldots, f_n$ are unary function symbols and t is a constant or a variable. We shall denote such a term s in the reversed Polish notation, i.e. as $t f_n \ldots f_2 f_1$. Thus, every term can be represented in the form tW, where t is a constant or a variable and W is a word. Similarly, any term of the form $f_1(f_2(\ldots f_n(t)\ldots))$, where t is an arbitrary term, will be written as $t f_n \ldots f_2 f_1$.

A *finite automaton* $\mathcal{A}$ on the alphabet $\mathcal{F}$ is a quadruple (Q, I, T, E), where Q is a finite set, called *the set of states*, I and T are distinguished subsets of Q, called the sets of *initial* and *terminal* states, respectively. The set $E \subseteq Q \times \mathcal{F} \times Q$ is *the set of edges of* $\mathcal{A}$. An edge (p, f, q) is also denoted $p \xrightarrow{f} q$. The automaton is *deterministic* iff whenever $(p, f, q_1) \in E$ and $(p, f, q_2) \in E$, then $q_1 = q_2$.

A word $f_1 \ldots f_n$ is *recognized by an automaton* (Q, I, T, E) iff there is a sequence of states $q_0 \ldots q_n$ such that $q_0 \in I$, $q_n \in T$ and $q_{i-1} \xrightarrow{f_i} q_i$ for all $i \in \{1, \ldots, n\}$. A set of words is *regular* iff it is the set of words recognized by some automaton.

The *intersection nonemptyness of deterministic finite automata problem* is the following decision problem. Given any finite set $\{\mathcal{A}_1, \ldots, \mathcal{A}_n\}$ of deterministic finite automata, is there a word recognized by each automaton in this set. The following statement is proved in Kozen [10]:

Lemma 3 *The intersection nonemptyness of deterministic finite automata problem is PSPACE-complete.*

Word equations. In addition to the alphabet $\mathcal{F}$, we shall also consider a countable set $\mathcal{V}$ of *word variables*, denoted u, v, w. A *word equation* is any expression of the form $V \simeq W$, where $V, W \in (\mathcal{F} \cup \mathcal{V})^*$. A *word substitution* is any expression $\sigma = \{V_1/v_1, \ldots, V_n/v_n\}$, where v_i are word variables and V_i are words in $\mathcal{F}^*$. Its *domain*, denoted $dom(\sigma)$ is the set $\{v_1, \ldots, v_n\}$. The *application of such a word substitution* θ *to a word* $W \in (\mathcal{F} \cup \mathcal{V})^*$, is the operation of simultaneous replacement of all occurrences of v_i by V_i. The result of the application is the word denoted $W\sigma$. A word substitution σ is a *solution to a word equation* $U \simeq V$ iff all variables in U, V belong to $dom(\sigma)$ and we have $U\sigma = V\sigma$. A *system of word equations* is any finite set of word equations, its *solution* is any substitution solving all equations in the system. Words will be denoted by U, V, W, word variables by u, v, w and word substitutions by ρ, σ, τ.

Makanin [11] proved that word equations are decidable. Analyzing Makanin's algorithm, Schultz [14] proves the following result.

Lemma 4 (Decidability of word equations with regular constraints)
The problem of solvability of word equations where every word variable u_i ranges over a regular set S_i, is decidable.

It is known that the problem of solvability of word equations is NP-hard. No good upper bound for the complexity of this problem has been obtained so far, it is only known that the problem is in 3-NEXP (Kościelski and Pacholski [7, 8]).

Equational logic and rigid equations. Let $\mathcal{R}$ be a system of rigid equations. The *signature of* $\mathcal{R}$ is defined as the signature consisting of all constants and function symbols occurring in $\mathcal{R}$; and in addition a fixed constant if $\mathcal{R}$ contains no constants. A solution θ to $\mathcal{R}$ is called *grounding for* $\mathcal{R}$ iff for every variable x occurring in $\mathcal{R}$ the term $x\theta$ is ground. A substitution θ is called *relevant* for $\mathcal{R}$ iff for every variable x the term $x\theta$ is in the signature of $\mathcal{R}$.

In the sequel, we shall need the following technical property of systems of rigid equations.

Lemma 5 (Existence of relevant grounding solutions) *Let $\mathcal{R}$ be a solvable system of rigid equations. Then there exists a solution θ to $\mathcal{R}$ that is grounding and relevant for $\mathcal{R}$.*

We shall introduce one particular kind of rigid equations that will be used as a technical tool for proofs in this paper. For any monadic signature $(\mathcal{F}, \mathcal{C})$, any variable x and any constant $c \in C$ introduce the following rigid equation:

$$Gr_{(\mathcal{F},\mathcal{C})}(x) \rightleftharpoons \{d \simeq c \mid d \in \mathcal{C}\} \cup \{cf \simeq c \mid f \in \mathcal{F}\} \vdash_\forall x \simeq c$$

We shall use the following obvious lemma:

Lemma 6 *A substitution θ is a solution to $Gr_{(\mathcal{F},\mathcal{C})}(x)$ iff $x\theta \in T_{(\mathcal{F},\mathcal{C})}$.*

As a consequence, we have

Lemma 7 *For any system $\mathcal{R}$ of rigid equations there is a system $\mathcal{R}'$ of rigid equations such that for any substitution θ, θ is a solution to $\mathcal{R}'$ if and only if θ is a grounding relevant solution to $\mathcal{R}$. In addition, $\mathcal{R}'$ can be found by $\mathcal{R}$ using a polynomial-time algorithm; and $\mathcal{R}'$ has ground left-hand sides if $\mathcal{R}$ has ground left-hand sides.*

Proof. Let $x_1, \ldots, x_n$ be all variables in $\mathcal{R}$ and $(\mathcal{F}, \mathcal{C})$ be the signature of $\mathcal{R}$. Define $\mathcal{R}' \rightleftharpoons \mathcal{R} \cup \{Gr_{(\mathcal{F},\mathcal{C})}(x_i) \mid i \in \{1, \ldots, n\}\}$. Then apply Lemma 6.

Rewrite rules. This section introduces a technique standard in the theory of ground systems of rewrite rules. However, we shall use ordinary equations instead of rewrite rules.

Introduce an ordering $\succ$ on terms in $T_{(\mathcal{F},\mathcal{C})}$ in the following way. Let $>$ be any total ordering on $\mathcal{F} \cup \mathcal{C}$ and $s = cf_1 \ldots f_m$, $t = dg_1 \ldots g_n$. Then $s \succ t$ iff one of the following conditions is true:

1. $m > n$;
2. $m = n$ and the string $cf_1 \ldots f_m$ is greater than $dg_1 \ldots g_n$ in the lexicographic ordering induced by $>$.

The ordering $\succ$ is total, noetherian and can be extended to a simplification ordering [1]. Some properties of the ordering formulated below are simple consequence of standard statements in the theory of rewrite systems. Their proofs may be found in e.g. [1]. Note that the ordering $\succ$ depends on the ordering of $>$. In the definitions below we assume that we have chosen a fixed ordering $>$ on $\mathcal{F} \cup \mathcal{C}$, and hence $\succ$ is also fixed.

Let $\mathcal{E}, \mathcal{E}'$ be finite sets of ground equations and $\mathcal{E}$ contains distinct equations $s \simeq t$ and $r[s] \simeq u$. We say that $\mathcal{E}'$ *is obtained from $\mathcal{E}$ by simplification from $s \simeq t$ into $r[s] \simeq u$*, denoted $\mathcal{E} \to \mathcal{E}'$ iff

$$\mathcal{E}' = (\mathcal{E} \setminus \{r[s] \simeq u\}) \cup \{r[t] \simeq u\}$$

The reflexive and transitive closure of the relation $\to$ on sets of ground equations is denoted by $\to^*$. A set of equations $\mathcal{E}$ is called *irreducible* iff there exists no $\mathcal{E}'$ such that $\mathcal{E} \to \mathcal{E}'$.

Let $\mathcal{E}$ be an irreducible set of ground equations. We write $t \to_\mathcal{E} t'$ if there exists an equation $(r \simeq s) \in \mathcal{E}$ such that $r \succ s$, and t' is obtained from t by the replacement of one occurrence of the subterm r by s. The relation $\to^*_\mathcal{E}$ is the reflexive and transitive closure of $\to_\mathcal{E}$. A term t is called *irreducible with respect to $\mathcal{E}$* iff there is no term s such that $t \to_\mathcal{E} s$. The *normal form of a term t w.r.t. $\mathcal{E}$*, denoted $t \downarrow_\mathcal{E}$, is the term s such that $t \to^*_\mathcal{E} s$ and s is irreducible w.r.t $\mathcal{E}$. The normal form of any term exists and is unique. We shall use the following statements which are easy to prove.

Lemma 8 *Let $\mathcal{E}$ be an irreducible set of ground equations and s, t be terms. Then $\mathcal{E} \vdash s \simeq t$ if and only if $s \downarrow_\mathcal{E} = t \downarrow_\mathcal{E}$.*

Mixing words and rigid equations. We call a *word term*, or simply *w-term*, in the signature $(\mathcal{F}, \mathcal{C})$ any expression of the form cW such that $c \in \mathcal{C}$ and $W \in (\mathcal{F} \cup \mathcal{V})^*$. A *w-equation* is any expression $cV \simeq dW$, where cV and dW are w-terms. A *rigid w-equation* is any expression of the form $\mathcal{W} \vdash_\forall cV \simeq dW$, where $\mathcal{W}$ is a finite set of w-equations, cV and dW are w-terms. A *system of rigid w-equations* is any finite set of rigid w-equations. *The signature of a system of rigid w-equations* is defined similar to that of a system of rigid equations. Sets of w-equations will be denoted by $\mathcal{W}$, and sets of rigid w-equations by $\mathcal{S}$.

A *solution to a rigid w-equation* $\mathcal{W} \vdash_\forall cV \simeq dW$ is any word substitution σ whose domain contains all word variables in $\mathcal{W}, V, W$ such that $\mathcal{W}\sigma \vdash cV\sigma \simeq dW\sigma$. A *solution to a system* $\mathcal{S}$ *of rigid w-equations* is any word substitution that is a solution to every rigid w-equation in $\mathcal{S}$.

Note that a ground w-equation is also an ordinary equation.

In Lemma 9 below we show that one can consider systems of rigid w-equations instead of systems of rigid equations. The following technical lemma is proved in [6]:

Lemma 9 *The problem of solvability of systems of rigid w-equations is polynomial-time reducible to monadic SREU. Monadic SREU is effectively reducible to the problem of solvability of systems of rigid w-equations.*

3 Ground left-hand sides

In this section we prove that monadic SREU with ground left-hand sides is decidable and PSPACE-hard.

SREU with ground left-hand sides is PSPACE-hard.

Lemma 10 *Let $\mathcal{A} = (Q, I, T, E)$ be a deterministic finite automaton over $\mathcal{F}$. There exists a system $\mathcal{R}$ of two monadic rigid equations of one variable x with the following properties:*

1. *$\mathcal{R}$ has ground left-hand sides;*
2. *for every solution θ to $\mathcal{R}$ we have $x\theta = cW$, where $W \in \mathcal{F}^*$ and c is a fixed constant;*
3. *for any word $W \in \mathcal{F}$, the substitution $\{cW/x\}$ is a solution to $\mathcal{R}$ if and only if W is recognized by $\mathcal{A}$.*

In addition, $\mathcal{R}$ can be effectively constructed from $\mathcal{A}$ using a polynomial-time algorithm.

Proof. Without loss of generality we can assume that I consists of one state (see e.g. [12]). By renaming states, we can assume that $I = \{c\}$. Let F be a unary function symbol fresh for $\mathcal{F}$ and d be a constant fresh for Q. Define $\mathcal{R}$ as $\{R_1, R_2\}$, where

$$R_1 = \{pf \simeq q \mid (p \xrightarrow{f} q) \in E\} \cup \{rF \simeq d \mid r \in T\} \vdash_\forall xF \simeq d$$
$$R_2 = Gr_{(\mathcal{F}, \{c\})}(x)$$

Consider any substitution $\theta = \{t/x\}$. By Lemma 6, θ is a solution to R_2 if and only if t has the form cW such that $W \in \mathcal{F}^*$. Consider when such substitution $\{cW/x\}$ is also a solution to R_1. By definition, this means

$$\{pf \simeq q \mid (p \xrightarrow{f} q) \in E\} \cup \{rF \simeq d \mid r \in T\} \vdash cWF \simeq d \qquad (1)$$

Since the automaton is deterministic, the left-hand side of (1) is irreducible. Using Lemma 8, one can see that (1) holds if and only if W is recognizable by $\mathcal{A}$. Evidently, $\mathcal{R}$ is constructed by $\mathcal{A}$ in polynomial time.

Lemma 11 *The intersection nonemptyness of deterministic finite automata problem is polynomial-time reducible to monadic SREU with one variable and ground left-hand sides.*

Proof. Let $\mathcal{A}_1, \ldots, \mathcal{A}_n$ be deterministic finite automata. Let $\mathcal{R}_i$, where $i \in \{1, \ldots, n\}$ be the system of rigid equations constructed by $\mathcal{A}_i$ as in Lemma 10. Define $\mathcal{R} = \bigcup_{i=1}^{n} \mathcal{R}_i$. By Lemma 10, every solution to $\mathcal{R}$ has the form $\{cW/x\}$ and any substitution $\{cW/x\}$ is a solution to $\mathcal{R}$ if and only if W is recognized by each $\mathcal{A}_i$. Hence, $\mathcal{R}$ is solvable if and only if there is a word recognizable by all $\mathcal{A}_i$. Evidently, $\mathcal{R}$ is constructed by $\mathcal{A}_1, \ldots, \mathcal{A}_n$ in polynomial time.

Combining Lemmas 3 and 11 we obtain

Theorem 1 *Monadic SREU with one variable and ground left-hand sides is PSPACE-hard.*

Monadic SREU with ground left-hand sides is decidable. A finite set $\mathcal{E}$ of equations is *in the automaton form* iff

1. every equation in $\mathcal{E}$ has the form $cf \simeq d$;
2. for every two w-equations $cf \simeq d_1$ and $cf \simeq d_2$ in $\mathcal{E}$ we have $d_1 = d_2$;

Note that any set of equations in the automaton form is irreducible. The following statement is proved in [6]:

Lemma 12 *Given any rigid w-equation S with ground left-hand side, one can effectively find in polynomial time a rigid w-equation S' with ground left-hand side such that*

1. *S and S' have the same solutions;*
2. *the left-hand side of S' is in the automaton form.*

Let $\mathcal{E}$ be a set of equations in the automaton form and c, d be any constants. Denote by $\mathcal{A}(\mathcal{E}, c, d)$ the following automaton (Q, I, T, E). Its alphabet is the set of function symbols occurring in $\mathcal{E}$. The set of states Q is the set of all constants occurring in $\mathcal{E}, c, d$. The sets of initial states and terminal states are defined by $I \rightleftharpoons \{c\}$ and $T \rightleftharpoons \{d\}$. Finally, the set of edges is defined by

$$E \rightleftharpoons \{a \xrightarrow{f} b \mid (af \simeq b) \in \mathcal{E}\}.$$

Lemma 13 *A word W is recognized by $\mathcal{A}(\mathcal{E}, c, d)$ if and only if $\mathcal{E} \vdash cW \simeq d$.*

Proof. Immediate by Lemma 8.

Lemma 14 *Let $\mathcal{E}$ be a set of equations in the automaton form, $W, W' \in \mathcal{F}^*$ and c, c' be constants. Then $\mathcal{E} \vdash cW \simeq c'W'$ if and only if there is a constant d and words U, U', V such that $W = UV$, $W' = U'V$, U is recognized by $\mathcal{A}(\mathcal{E}, c, d)$ and U' is recognized by $\mathcal{A}(\mathcal{E}, c', d)$.*

Proof.
($\Rightarrow$) We have $\mathcal{E} \vdash cW \simeq c'W'$. By Lemma 8 we have $cW \downarrow_\mathcal{E} = c'W' \downarrow_\mathcal{E}$. Choose d and V such that $cW \downarrow_\mathcal{E} = dV$. Define U and U' such that $W = UV$ and $W' = U'V$. We have $\mathcal{E} \vdash cU \simeq d$ and $\mathcal{E} \vdash c'U' \simeq d$. By Lemma 13 words U and U' are recognized by $\mathcal{A}(\mathcal{E}, c, d)$ and $\mathcal{A}(\mathcal{E}, c', d)$, respectively.
($\Leftarrow$) We have $W = UV$, $W' = U'V$, U is recognized by $\mathcal{A}(\mathcal{E}, c, d)$ and U' is recognized by $\mathcal{A}(\mathcal{E}, c', d)$. By Lemma 13 we have $\mathcal{E} \vdash cU \simeq d$ and $\mathcal{E} \vdash c'U' \simeq d$. Hence, $\mathcal{E} \vdash cUV \simeq dV$ and $\mathcal{E} \vdash c'U'V \simeq dV$. Then $\mathcal{E} \vdash cUV \simeq c'U'V$, i.e. $\mathcal{E} \vdash cW \simeq c'W'$.
□

Lemma 15 *The problem of solvability of systems of rigid w-equations with ground left-hand sides effectively reduces to word equations with regular constraints.*

Proof. Let $\mathcal{S} = \{S_1, \ldots, S_n\}$ be such a system of rigid w-equations. By Lemma 12 we can assume that the left-hand sides of all S_i are in the automaton form. Let $S_i = (\mathcal{E}_i \vdash_\forall c_i W_i \simeq c'_i W'_i)$, for all $i \in \{1, \ldots, n\}$. Let $u_1, \ldots, u_n, v_1, \ldots, v_n$ and $u'_1, \ldots, u'_n$ be word variables fresh for $\mathcal{S}$. By Lemma 14, the system $\mathcal{S}$ is solvable if and only if there are constants d_i occurring in S_i, for all $i \in \{1, \ldots, n\}$ such that the following system of word equations and regular constraints is solvable:

$$W_1 \simeq u_1 v_1 \quad u_1 \text{ is recognized by } \mathcal{A}(\mathcal{E}_1, c_1, d_1)$$
$$\ldots \quad \ldots$$
$$W_n \simeq u_n v_n \quad u_n \text{ is recognized by } \mathcal{A}(\mathcal{E}_n, c_n, d_n)$$
$$W'_1 \simeq u'_1 v_1 \quad u'_1 \text{ is recognized by } \mathcal{A}(\mathcal{E}_1, c'_1, d_1)$$
$$\ldots \quad \ldots$$
$$W'_n \simeq u'_n v_n \quad u'_n \text{ is recognized by } \mathcal{A}(\mathcal{E}_n, c'_n, d_n)$$

To conclude the proof we note that there is only a finite number of choices for d_i.

Theorem 2 *Monadic SREU with ground left-hand sides is decidable.*

Proof. By Lemma 9 monadic SREU with ground left-hand sides is effectively reducible to the problem of solvability of systems of rigid w-equations. By Lemma 15 the latter problem is effectively reducible to word equations with regular constraints. Then apply Lemma 4.

4 One-variable case

In this section we consider rigid equations with one variable x. We shall write $\mathcal{E}(x)$ to denote all occurrences of a variable x in $\mathcal{E}$, and write $\mathcal{E}(t)$ to denote the set of equations obtained from $\mathcal{E}$ by replacement of all occurrences of x by t. We shall use similar notation for terms, for example $s(x)$. Using this notation, we can write any rigid equation of one variable x as $\mathcal{E}(x) \vdash_\forall s(x) \simeq t(x)$. The following statement is proved in [6]:

Lemma 16 *Let $\mathcal{E}(x)$ be a finite set of equations of one variable x and $s(x), t(x)$ be terms of one variable x such that $\mathcal{E}(x) \not\vdash s(x) \simeq t(x)$. Let c be a constant fresh for $\mathcal{E}(x), s(x), t(x)$ and r be a ground term such that c does not occur in r. If $\mathcal{E}(r) \vdash s(r) \simeq t(r)$, then there exists a ground term $r' \in T(\mathcal{E}(c) \cup \{s(c) \simeq t(c)\})$ such that $\mathcal{E}(c) \vdash r \simeq r'$.*

Lemma 17 *Let $\mathcal{E}(x) \vdash_\forall s(x) \simeq t(x)$ be a rigid equation of one variable x, c be a constant fresh for this rigid equation, r be a ground term in which c does not occur and $\mathcal{E}(x) \not\vdash s(x) \simeq t(x)$. Then the substitution $\theta = \{r/x\}$ is a solution to this rigid equation if and only if there is a ground term $r' \in T(\mathcal{E}(c) \cup \{s(c) \simeq t(c)\})$ such that $\mathcal{E}(c), \mathcal{E}(r') \vdash s(r') \simeq t(r')$ and θ is a solution to $\mathcal{E}(c) \vdash_\forall r' \simeq x$.*

Proof.
$\Rightarrow$ We have that θ is a solution to $\mathcal{E}(x) \vdash_\forall s(x) \simeq t(x)$. Then $\mathcal{E}(r) \vdash s(r) \simeq t(r)$. By Lemma 16 there is a term $r' \in T(\mathcal{E}(c) \cup \{s(c) \simeq t(c)\})$ such that $\mathcal{E}(c) \vdash r \simeq r'$. Then $\mathcal{E}(r), \mathcal{E}(c) \vdash s(r') \simeq t(r')$.
$\Leftarrow$ We have $\mathcal{E}(c), \mathcal{E}(r') \vdash s(r') \simeq t(r')$ and $\mathcal{E}(c) \vdash_\forall r' \simeq r$. Then $\mathcal{E}(c), \mathcal{E}(r) \vdash s(r) \simeq t(r)$. By Lemma 2 we can substitute r for c obtaining $\mathcal{E}(r) \vdash s(r) \simeq t(r)$. □

Lemmas 16 and 17 also hold for non-monadic signatures [3].

Lemma 18 *Monadic SREU with one variable is in PSPACE.*

Proof. We shall give a non-deterministic algorithm reducing monadic SREU with one variable to the intersection nonemptyness of deterministic finite automata problem.

Let $\mathcal{R}$ be a system of rigid equations of one variable x whose signature is $(\mathcal{F}, \mathcal{C})$. It has the form

$$\mathcal{E}_1 \vdash_\forall s_1(x) \simeq t_1(x) \quad \cdots \quad \mathcal{E}_n \vdash_\forall s_n(x) \simeq t_n(x)$$

By Lemma 5 we can restrict ourselves to relevant grounding solutions $\theta = \{r/x\}$ only. Let c be a variable fresh for $(\mathcal{F}, \mathcal{C})$. By Lemma 17 θ is a solution to $\mathcal{R}$ if and only if there are ground terms $r_i' \in T(\mathcal{E}_i(c) \cup \{s_i(c) \simeq t_i(c)\})$, where $i \in \{1, \ldots, n\}$ such that $\mathcal{E}(c), \mathcal{E}(r') \vdash s(r') \simeq t(r')$ and θ is a solution to the system

$$\mathcal{E}_1(c) \vdash_\forall r_1' \simeq x \quad \cdots \quad \mathcal{E}_n(c) \vdash_\forall r_n' \simeq x$$

Nondeterministically select such $r_1', \ldots, r_n'$ and verify the condition $\mathcal{E}(c), \mathcal{E}(r') \vdash s(r') \simeq t(r')$ (it can be checked in polynomial time using Lemma 1).

Such θ is a solution to this system of rigid equations if and only if there is a constant $d \in \mathcal{C}$ such that the following system of rigid w-equations is solvable:

$$\mathcal{E}_1(c) \vdash_\forall r_1' \simeq dx \quad \cdots \quad \mathcal{E}_n(c) \vdash_\forall r_n' \simeq dx$$

Nondeterministically select such d. By Lemma 12 we can equivalently replace this system with a system

$$\mathcal{E}_1' \vdash_\forall c_1 \simeq d_1 x \quad \cdots \quad \mathcal{E}_n' \vdash_\forall c_n \simeq d_n x$$

where $\mathcal{E}_i'$ are in the automaton form. By Lemma 13, this system is solvable if and only if the intersection of automata $A(\mathcal{E}_1', d_1, c_1), \ldots, A(\mathcal{E}_n', d_n, c_n)$ is non-empty.

We have given a non-deterministic algorithm reducing monadic SREU with one variable to the intersection nonemptyness of deterministic finite automata problem. On each branch, the algorithm makes polynomially many steps. Applying Lemma 3 on the complexity of the intersection nonemptyness of deterministic finite automata problem we get that monadic SREU with one variable is in NPSPACE, and hence in PSPACE.

Combining Theorem 1 and Lemma 18, we obtain

Theorem 3 *Monadic SREU with one variable is PSPACE-complete.*

5 General case

Denote by **W** the set of pairs of words on $\mathcal{F}$. Introduce on **W** a binary function $*$, a unary function ^{r} and a binary relation $\leq$ in the following way:

$$(U_1, U_2) * (V_1, V_2) \rightleftharpoons \begin{cases} (U_2W, V_2) & \text{if } V_1 \text{ has the form } U_1W \\ (V_1, V_2) & \text{otherwise} \end{cases}$$

$$(U_1, U_2)^r \rightleftharpoons (U_2, U_1)$$

$(U_1, U_2) \leq (V_1, V_2) \rightleftharpoons$ there is a word W such that $(V_1, V_2) = (U_1W, U_2W)$

An *ideal* on **W** is any set of pairs containing $(\varepsilon, \varepsilon)$ and closed under $*$, ^{r} and upward closed under $\leq$. The *ideal generated by a set of pairs S*, denoted $ideal(S)$ is defined as the least ideal containing S.

An ideal equation is an expression

$$(U, V) \in ideal(\{(U_1, V_1), \ldots, (U_n, V_n)\}),$$

where $n \geq 0$ and $U, V, U_1, \ldots, U_n, V_1, \ldots, V_n \in (\mathcal{F} \cup \mathcal{V})^*$. A *solution to such ideal equation* is any word substitution σ such that

1. words $U\sigma, V\sigma, U_1\sigma, \ldots, U_n\sigma, V_1\sigma, \ldots, V_n\sigma$ are words over $\mathcal{F}$;
2. the word $(U\sigma, V\sigma)$ belongs to the ideal generated by

$$\{(U_1\sigma, V_1\sigma), \ldots, (U_n\sigma, V_n\sigma)\}.$$

A *system of ideal equations* is any finite set of ideal equations. *Solutions to a system of ideal equations* are substitutions that solve each equation in the system. The *ideal equations problem* is the decision problem of solvability of systems of ideal equations. The aim of this section is to show that monadic SREU is equivalent to the ideal equations problem.

The following lemma proved in [6] is the main reason for introducing the notion of an ideal.

Lemma 19 *Let $U_1, \ldots, U_n, V_1, \ldots, V_n, U, V$ be words on $\mathcal{F}$ and a be any constant. Then $aU_1 \simeq aV_1, \ldots, aU_n \simeq aV_n \vdash aU \simeq aV$ if and only if $(U, V) \in ideal(\{(U_1, V_1), \ldots, (U_n, V_n)\})$.*

Theorem 4 *Monadic SREU is decidable if and only if the ideal equation problem is decidable.*

Proof. See [6].

Technical report [6] discusses ideal equations in more detail. In particular, it is shown that ideal equations are decidable if and only if word equations extended by a family of predicates behaving like a greatest common divisor on word are decidable. In addition, the following statement is proved:

Lemma 20 *Ideal equations are decidable if and only if ideal equations with regular constraints and the inequality constraints $U \not\simeq V$ are decidable.*

Acknowledgments. We thank Anatoli Degtyarev and Gennadi Makanin.

References

1. N. Dershowitz and J.-P. Jouannaud. Rewrite systems. In J. Van Leeuwen, editor, *Handbook of Theoretical Computer Science*, volume B: Formal Methods and Semantics, chapter 6, pages 243–309. North Holland, Amsterdam, 1990.
2. A. Degtyarev, Yu. Gurevich, and A. Voronkov. Herbrand's theorem and equational reasoning: Problems and solutions. In *Bulletin of the European Association for Theoretical Computer Science*, volume 60, page ???. October 1996. The "Logic in Computer Science" column.
3. A. Degtyarev, Yu. Gurevich, P. Narendran, M. Veanes, and A. Voronkov. The decidability of simultaneous rigid E-unification with one variable. UPMAIL Technical Report 139, Uppsala University, Computing Science Department, March 1997.
4. A. Degtyarev, Yu. Matiyasevich, and A. Voronkov. Simultaneous rigid E-unification and related algorithmic problems. In *Eleventh Annual IEEE Symposium on Logic in Computer Science (LICS'96)*, pages 494–502, New Brunswick, NJ, July 1996. IEEE Computer Society Press.
5. A. Degtyarev and A. Voronkov. The undecidability of simultaneous rigid E-unification. *Theoretical Computer Science*, 166(1–2):291–300, 1996.
6. Yu. Gurevich and A. Voronkov. Monadic simultaneous rigid E-unification and related problems. UPMAIL Technical Report 137, Uppsala University, Computing Science Department, February 1997.
7. A. Kościelski and L. Pacholski. Complexity of unification in free groups and free semigroups. In *Proc. 31st Annual IEEE Symposium on Foundations of Computer Science*, pages 824–829, Los Alamitos, 1990.
8. A. Kościelski and L. Pacholski. Complexity of Makanin's algorithm. *Journal of the Association for Computing Machinery*, 43(4):670–684, 1996.
9. D. Kozen. Complexity of finitely presented algebras. In *Proc. of the 9th Annual Symposium on Theory of Computing*, pages 164–177, New York, 1977. ACM.
10. D. Kozen. Lower bounds for natural proof systems. In *Proc. 18th IEEE Symposium on Foundations of Computer Science (FOCS)*, pages 254–266, 1977.
11. G.S. Makanin. The problem of solvability of equations in free semigroups. *Mat. Sbornik (in Russian)*, 103(2):147–236, 1977. English Translation in American Mathematical Soc. Translations (2), vol. 117, 1981.
12. D. Perrin. Finite automata. In J. Van Leeuwen, editor, *Handbook of Theoretical Computer Science*, volume B: Formal Methods and Semantics, chapter 1, pages 1–57. Elsevier Science, Amsterdam, 1990.
13. D.A. Plaisted. Special cases and substitutes for rigid E-unification. Technical Report MPI-I-95-2-010, Max-Planck-Institut für Informatik, November 1995.
14. K.U. Schulz. Makanin's algorithm: Two improvements and a generalization. In K.U. Schulz, editor, *Word Equations and Related Topics*, volume 572 of *Lecture Notes in Computer Science*, Tübingen, Germany, October 1990.
15. R. Shostak. An algorithm for reasoning about equality. *Communications of the ACM*, 21:583–585, July 1978.
16. M. Veanes. Uniform representation of recursively enumerable sets with simultaneous rigid E-unification. UPMAIL Technical Report 126, Uppsala University, Computing Science Department, 1996.

Computability on the Probability Measures on the Borel Sets of the Unit Interval

Klaus Weihrauch

FernUniversität, D-58084 Hagen, klaus.weihrauch@fernuni-hagen.de

Abstract. While computability theory on many countable sets is well established and for computability on the real numbers several (mutually non-equivalent) definitions are applied, for most other uncountable sets, in particular for measures, no generally accepted computability concepts at all have been available until now. In this contribution we introduce computability on the set **M** of probability measures on the Borel subsets of the unit interval [0; 1]. Its main purpose is to demonstrate that this concept of computability is not merely an ad hoc definition but has very natural properties. Although the definitions and many results can of course be transferred to more general spaces of measures, we restrict our attention to **M** in order to keep the technical details simple and concentrate on the central ideas. In particular, we show that simple obvious reqirements exclude a number of similar definitions, that the definition leads to the expected computability results, that there are other natural definitions inducing the same computability theory and that the theory is embedded smoothly into classical measure theory. As background we consider TTE, Type 2 Theory of Effectivity [KW84, KW85], which provides a frame for very realistic computability definitions. In this approach, computability is defined on finite and infinite sequences of symbols explicitly by Turing machines and on other sets by means of notations and representations. Canonical representations are derived from information structures [Wei97] . We introduce a standard representation $\delta_m :\subseteq \Sigma^\omega \longrightarrow$ **M** via some natural information structure defined by a subbase σ (the atomic properties) of some topology τ on **M** and a standard notation of σ. While several modifications of δ_m suggesting themselves at first glance, violate simple and obvious requirements, δ_m has several very natural properties and hence should induce an important computability theory. Many interesting functions on measures turn out to be computable, in particular linear combination, integration of continuous functions and any transformation defined by a computable iterated function system with probabilities. Some other natural representations of **M** are introduced, among them a Cauchy representation associated with the Hutchinson metric, and proved to be equivalent to δ_m. As a corollary, the final topology τ of δ_m is the well known weak topology on **M**.

1 Introduction

Measure and integration is a central branch of mathematics pervading almost all parts of abstract analysis. Several authors have already considered questions of effectivity, constructivity, computability or computational complexity in measure or integration theory. Kushner [Kus85] studies computability and Ko [Ko91] computational complexity of integration. Bishop and Bridges [BB85] present constructive measure theory extensively. Although they do not consider computability, certainly many of their concepts and results have computational counterparts. Edalat gives a domain theoretic approach to effective integration [Eda95, Eda96]. He also does not consider computability, but it should be possible to extend his topological approach by computability concepts. Traub et al. [TWW88] investigate the computational complexity of numerical algorithms for integration in the real number model of computation. However, this model is unrealistic in many situations and therefore not generally accepted. A systematic study of computability in integration and measure theory does not yet exist. In this paper we introduce a very natural and realistic computability theory on probability measures. We achieve this by extending TTE, Type 2 Theory of Effectivity, to measure theory. TTE has been introduced by Kreitz and Weihrauch [KW84, KW85] as a general framework for studying effectivity, i.e. continuity, computability and computational complexity, in Analysis. For details the reader is referred to the introduction [Wei95] and a recent short survey [Wei97] containing most of the notations we shall use in this paper. More details can be found in [KW85, Wei87]. Since this paper is a first attempt, we consider only the space of probability measures on the Borel subsets of the real unit interval.

By $f :\subseteq A \longrightarrow B$ we denote a partial function, i.e. a function from a subset of A to B. Throughout this paper let Σ be a sufficiently large finite alphabet. Let Σ^* be the set of finite and $\Sigma^\omega = \{p \mid p : \omega \longrightarrow \Sigma\}$ the set of "ega-words over Σ. On Σ^* we consider the discrete topology and on Σ^ω the cantor topology defined by the basis $\{w\Sigma^\omega \mid w \in \Sigma^*\}$. For $Y_0, Y_1, \ldots, Y_k \in \{\Sigma^*, \Sigma^\omega\}$, a function $f :\subseteq Y_1 \times \ldots \times Y_k \longrightarrow Y_0$ is called computable, iff it is computed by a Turing machine with a one-way output tape. Every computable function is continuous. The basic idea of TTE is to use finite or infinite sequences as names of "abstract" objects. As naming systems we consider notations, i.e. surjections $\nu :\subseteq \Sigma^* \longrightarrow S$, and representations, i.e. surjections $\delta :\subseteq \Sigma^\omega \longrightarrow M$. Continuity and computability concepts are transferred from Σ^* and Σ^ω via notations and representations, respectively, to the named sets straightforwardly, see [KW85, Wei87, Wei95, Wei97]. Mainly notations or representations which are compatible with some relevant structure on the set under consideration are of practical interest. We do not discuss this for notations (see [RW80, Wei87] and Appendix C in [Wei95]), but we will introduce "effective" notations explicitly whenever necessary. In particular, for the rational numbers let $\nu_Q :\subseteq \Sigma^* \longrightarrow \mathbb{Q}$ be the standard representation via fractions of integers in binary notation. We shall abbreviate $\nu_Q(w)$ by $\bar{w}$. Standard notations of the natural numbers, pairs of rational numbers etc. will be used without further definitions. For uncountable sets M we shall consider mainly representations derived from "information structures" (M, σ, ν), where σ is a countable subset of 2^M of "atomic properties" which identifies points, and ν is a notation of σ [Wei97]. It is assumed that a computer (Turing machine) manipulates ν–names of atomic properties. As a name of an object $x \in M$ we consider any infinite list of all properties $A \in \sigma$ which hold for x. Concretely, the standard representation $\delta_\nu :\subseteq \Sigma^\omega \longrightarrow M$ is defined by

$$\delta_\nu(p) = x \iff p = w_0 \sharp w_1 \sharp \ldots \text{ and } \{w_i \mid i \in \omega\} = \{w \mid x \in \nu(w)\}.$$

Every finite prefix of a δ_ν–name p of x contains finitely many atomic properties of x which "approximate" x. Mathematically, this kind of approximation is described by the topology τ_σ on M, which has σ as a subbase. Computability on σ and via δ_ν on M are fixed by the notation ν which expresses how atomic properties can be handled concretely. Thus, for any information structure (M, σ, ν), σ characterizes approximation and ν computability on M. The topology τ_σ and the standard representation δ_ν are closely related: $X \in \tau_\sigma \iff \delta_\nu^{-1} X$ is open in $dom(\delta_\nu)$ (for all $X \subseteq M$), i.e. τ_σ is the final topology of δ_ν. Let $\delta :\subseteq \Sigma^* \longrightarrow M$ and $\delta' :\subseteq \Sigma^* \longrightarrow M'$ be representations and let $f :\subseteq M \longrightarrow M'$ be a function. An element $x \in M$ is called δ–computable, iff $\delta(p) = x$ for some computable sequence $p \in \Sigma^*$. By definition, $\delta \leq_t \delta'$ ($\delta \leq \delta'$), iff $\delta = \delta' g$ for some continuous (computable) function $g :\subseteq \Sigma^* \longrightarrow \Sigma^*$, and f is (δ, δ')–continuous (–computable), iff $f\delta = \delta' g$ for some continuous (computable) function $g :\subseteq \Sigma^* \longrightarrow \Sigma^*$. (Accordingly for functions with two or more arguments.) By the "main theorem for admissible representations" [KW85] a function is continuous relative to standard representations, iff it is continuous w.r.t. the associated final topologies in the usual sense. For more details see [KW85, Wei87, Wei95, Wei97]. For the real numbers, we need three representations $\rho_<, \rho_>, \rho :\subseteq \Sigma^\omega \longrightarrow \mathbb{R}$, derived from information structures. They can be defined explicitly as follows [Wei87, Wei97]:

$$\rho_<(p) = x :\iff p = w_0 \sharp w_1 \sharp \ldots \text{ with } \{w_i \mid i \in \omega\} = \{w \mid \bar{w} < x\},$$
$$\rho_>(p) = x :\iff p = w_0 \sharp w_1 \sharp \ldots \text{ with } \{w_i \mid i \in \omega\} = \{w \mid \bar{w} > x\},$$
$$\rho(p) = x \;\;:\iff p = v_0 \sharp w_0 \sharp v_1 \sharp w_1 \ldots \text{ with } \{(v_i, w_i) \mid i \in \omega\} = \{(v, w) \mid \bar{v} < x < \bar{w}\}.$$

The final topologies are $\tau_< = \{(y; \infty) \mid y \in \mathbb{R}\} \cup \{\mathbb{R}\}$, $\tau_> = \{(-\infty; y) \mid y \in \mathbb{R}\} \cup \{\mathbb{R}\}$ and the set $\tau_\mathbb{R}$ of ordinary open subsets of $\mathbb{R}$, respectively. Notice that ρ induces the standard computability theory on the real line. The translatability or reducibility properties [Wei87, Wei97] $\rho \leq \rho_<$, $\rho \leq \rho_>$, $\rho_< \not\leq_t \rho$, $\rho_> \not\leq_t \rho$, $\rho_< \not\leq_t \rho_>$, $\rho_> \not\leq_t \rho_<$ can be proved easily.

In Section 2 we introduce a standard representation δ_m of the set $\mathbf{M}$ of probability measures on the Borel sets of the interval $[0;1]$ by a very natural information structure. We prove a stability theorem for this definition. We discuss some further modifications of the definition and show that that they have undesirable properties. The results indicate that the computability theory on $\mathbf{M}$ induced by the representation δ_m is indeed very natural. In Section 3 we prove computability of several interesting functions on measures, in particular linear combination and integration of continuous functions. Also the measure transformation induced by a computable iterated function system with probabilities

[Hut81, Bar93] is computable. Finally in Section 4, we introduce representations based on other natural information structures and a Cauchy representation for the Hutchinson metric [Hut81, Bar93] . We prove that all these representations are equivalent and that their final topology is the well known weak topology [Bau74].

2 The standard representation of measures

In this section we introduce the standard representation δ_m of the probability measures and show that it induces a very natural computability theory. Let $Int := \{(a;b), [0;a), (b;1], [0;1] \mid a, b \in \mathbb{Q}, 0 < a < b < 1\}$ be the set of open subintervals of $[0;1]$ with rational boundaries, and let $I :\subseteq \Sigma^* \longrightarrow Int$ be some standard notation of Int with $dom(I) \subseteq (\Sigma \backslash \{\natural, \sharp\})^*$. We write I_w for $I(w)$. By $\mathbf{B}$ we denote the set of Borel subsets of $[0;1]$, i.e. the smallest σ-algebra containing Int. By $\mathbf{M}$ we denote the set of probability measures $\mu : \mathbf{B} \longrightarrow \mathbb{R}$ on the space $([0;1], \mathbf{B})$. By a basic theorem of measure theory [Bau74], every measure $\mu \in \mathbf{M}$ is defined uniquely by its values on the generating set Int. We introduce a standard representation of $\mathbf{M}$ via an information structure. The informations available from some standard name of a measure μ shall be all (r, J) with $r \in \mathbb{Q}$ and $J \in Int$ such that $r < \mu(J)$.

Definition 1. Define an information structure $(\mathbf{M}, \sigma, \nu)$ by $\sigma := range(\nu)$, where $\mu \in \nu(u \natural v) :\iff \bar{u} < \mu(I_v)$ for all $u \in dom(\nu_Q)$, $v \in dom(I)$ and $\mu \in \mathbf{M}$. Let τ_m be the topology on $\mathbf{M}$ with subbase σ and let δ_m be the standard representation of $\mathbf{M}$ derived form ν.

It remains to show that σ identifies the points of $\mathbf{M}$. Consider measures $\mu, \mu' \in \mathbf{M}$ such that $r < \mu(J) \iff r < \mu'(J)$ for all $r \in \mathbb{Q}$ and $J \in Int$. Then obviously, $\mu(J) = \mu'(J)$ for all $J \in Int$, i.e. $\mu = \mu'$. The definition of the representation δ_m looks somewhat arbitrary. By the next stability lemma, we obtain an equivalent representation, if we replace ν_Q and I by adequate other notations. For any $X \subseteq \mathbb{R}$ let $cls(X)$ be the closure of X.

Lemma 2. *(stability of δ_m) Let $\nu_S :\subseteq \Sigma^* \longrightarrow S$ be a notation of a set S which is dense in $\mathbb{R}$ such that $\{(u,v) \mid \nu_S(u) < \nu_Q(v)\}$ and $\{(u,v) \mid \nu_Q(u) < \nu_S(v)\}$ are r.e. . Let D be a countable dense subset of $[0, 1]$ and let I' be a notation of $Int' := \{(a;b), [0;a)(a;1], [0;1] \mid a,b \in D, 0 < a < b < 1\}$ such that $\{(u,v) \mid cls(I'_u) \subseteq I_v\}$ and $\{(u,v) \mid cls(I_u) \subseteq I'_v\}$ are r.e. Define τ'_m and δ'_m by substituting ν_S for ν_Q and I' for I in Definition 1.*
Then $\tau'_m = \tau_m$ and $\delta'_m \equiv \delta_m$.

If we replace, for example, rational numbers by finite binary fractions or by finite decimal fractions in the definition of the set Int and in Definition 1, we obtain an equivalent representation with the same final topology.
If we replace the relation "<" in Definition 1 by "≤", ">" or "≥", we obtain

rerpesentations which violate Lemma 2. Remember that by definition, the topology τ_m has the subbase $\sigma = \{U_{r,J} \mid r \in \mathbb{Q} \text{ and } J \in Int\}$ where $U_{r,J} = \{\mu \in \mathbf{M} \mid r < \mu(J)\}$. We prepare the proof of the theorem by two lemmas. First, we consider the cases "$r \leq \mu(J)$", "$r \geq \mu(J)$".

Lemma 3. *For $Q \subseteq \mathbb{R}$ let $\tau(Q)$ be the topology on $\mathbf{M}$ generated by the subbase $\sigma(Q) := \{U_{r,J} \mid r \in Q, J \in Int\}$, where $U_{r,J} = \{\mu \in \mathbf{M} \mid r \leq \mu(J)\}$. Then $\tau(P) \not\subseteq \tau(Q)$, if $t \in P \setminus Q$ for some $t \in (0;1)$ (for all $P, Q \subseteq \mathbb{R}$). The statement holds accordingly, if "$\leq$" is replaced by "$\geq$".*

The next lemma considers the case "$r > \nu(J)$".

Lemma 4. *For $D \subseteq (0;1)$ let $Int(D) := \{(a;b), [0;a), (a;1], [0;1] \mid a,b \in D, 0 < a < b < 1\}$. Let $\tau(D)$ be the topology on $\mathbf{M}$ generated by the subbase $\sigma(D) := \{U_{r,J} \mid r \in \mathbb{Q}, J \in Int(D)\}$ where $U_{r,J} = \{\mu \in \mathbf{M} \mid r > \mu(J)\}$. Then*

$$D \subseteq E \iff \tau(D) \subseteq \tau(E) \text{ (for all } D, E \subseteq (0;1)\text{)}.$$

Theorem 5. *If in Definition 1 the relation "$\bar{u} < \mu(I_v)$" is replaced by "$\bar{u} > \mu(I_v)$", "$\bar{u} \geq \mu(I_v)$" or "$\bar{u} < \mu(I_v)$", the resulting representations δ_m violate the stability lemma 2.*

By Definition 1 and Lemmata 3 and 4, many different more or less natural representations and hence computability theories for the set $\mathbf{M}$ of probability measures on $([0;1], \mathbf{B})$ can be introduced. The "user" has to decide, which of them is adequate for his application. The stable representation δ_m from Definition 1 is certainly the most important one, since its computability theory will occur most frequently. We shall study it in the following exclusively.

As a simple consequence of Definition 1, all rational lower bounds of $\mu(J)$ can be obtained from any δ_m–name of μ and any I–name of J. This property characterizes the representation δ_m except for equivalence: The representation δ_m is $\leq$–complete in the set of all representations δ of $\mathbf{M}$, for which $(\mu, J) \mapsto \mu(J)$ is $(\delta, I, \rho_<)$–computable.

Theorem 6. *For any representation δ of $\mathbf{M}$: $\delta \leq \delta_m \iff (\mu, J) \mapsto \mu(J)$ is $(\delta, I, \rho_<)$–computable.*

Notice, that in particular $(\mu, J) \mapsto \mu(J)$ is $(\delta_m, I, \rho_<)$–computable. Computing only lower rational bounds does not seem to be satisfactory. We would like to compute also arbitrarily close upper bounds of $\mu(I_v)$. We prove a negative and a positive answer. For any $x \in [0;1]$ define $\mu_x \in \mathbf{M}$ by $\mu_x(A) := (1$ if $x \in A, 0$ otherwise). For any good and useful representation δ of $\mathbf{M}$ it should be possible to determine a δ–name of the measure μ_x effectively from a name of x. Let $M' := \{\mu_x \mid x \in [0;1]\}$.

Theorem 7. *For any representation δ of $\mathbf{M}$, for which $x \mapsto \mu_x$ is (ρ, δ)–continuous on $(0;1)$, $\mu \mapsto \mu[0;1/2)$ is not $(\delta, \rho_>)$–continuous on M'. δ_m is such a representation.*

Therefore, for reasonable representations δ of $\mathbf{M}$, in particular for our standard representation δ_m, arbitrarily close rational upper bounds of measures of open intervals cannot be computed. Although this contradicts intuition at first glance, it has to be accepted as a matter of fact. Notice, that for proving Lemma 3, Lemma 4 and Theorem 7 we have used measures $\mu \in \mathbf{M}$ with $\mu\{x\} > 0$ for some $x \in \mathbb{R}$. Since the arguments have been purely topological without reference to computability, we have also shown that the final topology τ_m of the representation δ_m, which formalizes a concept of "approximation" on the set $\mathbf{M}$ of measures, is quite natural. If we exclude measures μ with $\mu\{x\} > 0$ for some $x \in [0;1]$, $(\mu, J) \mapsto \mu(J)$ becomes (δ_m, I, ρ)-computable. Let $\mathbf{M}^0 := \{\mu \in \mathbf{M} \mid \forall x \in [0;1].\mu\{x\} = 0\}$

Theorem 8. *The function* $(\mu, J) \mapsto \mu(J)$ *is* (δ_m, I, ρ)-*computable for* $J \in Int$ *and* $\mu \in \mathbf{M}^0$.

3 Computable Functions on Measures

In this section we prove computability of some interesting functions on probability measures. By the next theorem, the linear combination of measures is computable in all variables.

Theorem 9. *The function* $(a, \mu, \mu') \mapsto a\mu + (1-a)\mu'$ *is* $(\rho, \delta_m, \delta_m, \delta_m)$-*computable for* $0 \leq a \leq 1$.

By Theorem 6, $(\mu, J) \mapsto \mu(J)$ is $(\delta_m, I, \rho_<)$-computable on $\mathbf{M} \times Int$. We extend this result to $\tau'_{\mathbb{R}} = \{U \cap [0;1] \mid U \in \tau_{\mathbb{R}}\}$, the set of all open subsets of $[0;1]$. First we need a representation of this topology. For the set $\tau_{\mathbb{R}}$ of open subsets of $\mathbb{R}$, the following information structure $(\tau_{\mathbb{R}}, \sigma, \nu)$ and its derived representation δ_o and topology τ_o are natural (see [Wei97]): For any $U \in \tau_{\mathbb{R}}$ and $u, v \in \Sigma^*$ let $U \in \nu(u\natural v)$ iff $[\bar{u}; \bar{v}] \subseteq U$. Consequently, $\delta_o(p) = U$ iff p is a list of all closed intervals with rational boundaries contained in U. We define our standard representation of $\tau'_{\mathbb{R}}$ accordingly: $\delta'_o(p) = U :\iff p$ is a list of all $w \in \Sigma^*$ with $cls(I_w) \subseteq U$ ($p \in \Sigma^\omega$, $U \in \tau'_{\mathbb{R}}$). Let $\mu_L \in \mathbf{M}$ be the Lebesgue measure on $([0;1], \mathbf{B})$.

Theorem 10. *(1)* $(\mu, U) \mapsto \mu(U)$ *for* $\mu \in \mathbf{M}$ *and* $U \in \tau'_{\mathbb{R}}$ *is* $(\delta_m, \delta'_o, \rho_<)$-*computable.* *(2)* $(\mu, U) \mapsto \mu(U)$ *for* $\mu = \mu_L$ *and* $U \in \tau'_{\mathbb{R}}$ *is not* $(\delta_m, \delta'_o, \rho_>)$-*continuous.*

For uniform formulations in the next theorems we need a standard representation $\delta_\to$ of the set $C[0;1]$ of continuous functions $f : [0;1] \longrightarrow \mathbb{R}$. We define $\delta_\to$ and the corresponding final topology $\tau_\to$ by the following information structure $(C[0;1], \sigma, \nu)$: $f \in \nu(u\natural v\natural w) :\iff \bar{u} < f(cls\, I_v) < \bar{w}$ for all $f \in C[0;1]$ and $u, v, w \in \Sigma^*$. Properties of $\delta_\to$ are discussed in [Wei87, Wei95, Wei97]. In particular, $\tau_\to$ is the compact–open topology on $C[0;1]$, which is also generated by the metric $d(f, g) := \max\{|f(x) - g(x)| \mid 0 \leq x \leq 1\}$ on $C[0;1]$. For any measure $\mu \in \mathbf{M}$ and any continuous function $f : [0;1] \longrightarrow [0;1]$ define the measure $T_f(\mu)$ by $T_f(\mu)(A) := \mu f^{-1}(A)$ for every Borel set $A \subseteq [0;1]$ (see [Bau74], page 42).

Theorem 11. *The function* $(f, \mu) \mapsto T_f(\mu)$ *for continuous* $f : [0; 1] \to [0; 1]$ *and* $\mu \in \mathbf{M}$ *is* $(\delta_\to, \delta_m, \delta_m)$-*computable.*

We apply this theorem to iterated function systems with probabilities [Hut81, Bar93]. An interated function system (IFS) on $[0; 1]$ with probabilities is a tuple $\mathbf{S} = ([0; 1], f_1, \ldots, f_k, p_1, \ldots, p_k)$ where $f_1, \ldots, f_k : [0; 1] \longrightarrow [0; 1]$ are continuous functions and $p_1, \ldots, p_k$ are positive real numbers with $p_1 + \ldots + p_k = 1$. With $\mathbf{S}$ one associates the function $T_\mathbf{S} : \mathbf{M} \longrightarrow \mathbf{M}$ defined by $T_\mathbf{S}(\mu) := \sum_{i=1}^{k} p_i T_{f_i}(\mu)$

Corollary 12. *Let* $\mathbf{S} = ([0; 1], f_1, \ldots, f_k, p_1, \ldots, p_k)$ *be an IFS with probabilities such that* $f_1, \ldots, f_k$ *are* $\delta_\to$-*computable and* $p_1, \ldots, p_k$ *are* ρ-*computable. Then* $T_\mathbf{S} : \mathbf{M} \longrightarrow \mathbf{M}$ *is* (δ_m, δ_m)-*computable.*

Therefore, for any computable iterated function system $\mathbf{S}$ with probabilities, the associated measure transformation $T_\mathbf{S} : \mathbf{M} \longrightarrow \mathbf{M}$ is a (δ_m, δ_m)-computable function. We shall show below (Theorem 23) that its unique fixed point $\mu_\mathbf{S} \in \mathbf{M}$ is δ_m-computable, if the system $\mathbf{S}$ is hyperbolic [Hut81]. We shall show that integration of continuous functions is computable in both arguments. The integral of a continuous function can be defined via summations over finite partitions. Consider $\mu \in \mathbf{M}$ and $f \in C[0; 1]$. Let $Part$ be the set of all finite partitions Z of $[0; 1]$ into intervals with rational boundaries (remember: $\bigcup Z = [0; 1]$ and $I \cap J =$ for $I, J \in Z$). For $Z \in Part$ define $s_+(Z) := \sum_{J \in Z} \mu(J) \cdot \sup_{x \in J} f(x)$ and $s_-(Z) := \sum_{J \in Z} \mu(J) \cdot \inf_{x \in J} f(x)$. Since f is continuous, we have $\sup_{Z \in Part} s_-(Z) = \inf_{Z \in Part} s_+(Z) =: \int f d\mu$. The following lemma is the key to the next proof.

Lemma 13. *For any* $\beta, \gamma > 0$ *there are a finite set* $T \subseteq Int$ *of (pairwise disjoint) open intervals and a finite set* L *of closed intervals such that* $T \cup L \in Part$, $length(J) < \gamma$ *for every* $J \in T$ *and* $\mu(\bigcup L) < \beta$. *(L can be chosen, such that each $J \in L$ has length 0.)*

Theorem 14. *The function* $(f, \mu) \mapsto \int f d\mu$ *for* $f \in C[0; 1]$ *and* $\mu \in \mathbf{M}$ *is* $(\delta_\to, \delta_m, \rho)$-*computable.*

Proof: For any $T \subseteq Int$ let $s_-(T) := \sum\{\mu(J) \cdot \inf f(J) \mid J \in T\}$. Consider $f \in C[0; 1]$ and $\varepsilon > 0$. By uniform continuity of f there is some $\gamma > 0$ such that $|x - y| < \gamma \implies |fx - fy| < \varepsilon/4$. Let $M := \max\{|f(x)| \mid 0 \leq x \leq 1\}$, choose $\beta := \varepsilon/(4(1 + M))$. By Lemma 13 there is some set $T \subseteq Int$ of pairwise disjoint intervals such that $1 - \beta < \mu \bigcup T \leq 1$ and $\forall J \in T.length(J) < \gamma$. Furthermore, there are $z_J \in \mathbb{Q}$ such that $z_J < \mu(J)$ for $J \in T$ and $1 - \beta < \sum\{z_J \mid J \in T\} < 1$. We describe a procedure for determining from (p, q, n) a number $r \in \mathbb{Q}$ with $|r - \int f d\mu| < 2^{-n}$ where $\delta_\to(p) = f$ and $\delta_m(q) = \mu$.

- From p and n determine some $k \in \omega$ such that $|x - y| < 2^{-k} \implies |fx - fy| < 2^{-n-2}$ [Wei95, Wei97].
- From p determine some integer upper bound m of M.

- Let $\beta := 2^{-n-2}/(1+m)$.
- By systematic search find a finite set $T \subseteq Int$ of pairwise disjoint intervals and rational numbers z_J ($J \in T$) with $length(J) < 2^{-k}$ and $z_J < \mu(J)$ for $J \in T$ and $1 - \beta < \sum\{z_J \mid J \in T\}$.
- Determine some $r \in \mathbb{Q}$ such that $|\sum\{z_J \cdot \inf f(J) \mid J \in T\} - r| < 2^{-n-2}$.

The existence of T and the numbers z_J has already been shown. We prove $|r - \int f d\mu| < 2^{-n}$. Let L be the set from Lemma 13 and let $T' := T \cup L$. We have:

$$|\int f d\mu - s_-(T')| \leq |s_+(T') - s_-(T')|$$
$$\leq |\sum\{\mu(J)(\sup f(J) - \inf f(J))\| J \in T'\}$$
$$< \sum\{\mu(J) \cdot 2^{-n-2} | J \in T'\}$$
$$< 2^{-n-2}$$

$$|s_-(T') - s_-(T)| \leq \sum\{\mu(L) \cdot \inf f(J) | J \in L\}$$
$$\leq \mu \bigcup L \cdot m$$
$$< \beta \cdot m$$
$$< 2^{-n-2}$$

$$|s_-(T) - \sum\{z_J \cdot \inf f(J) | J \in T\}| \leq \sum\{(\mu(J) - z_J) \inf f(J) \mid J \in T\}$$
$$< \beta \cdot m$$
$$< 2^{-n-2}$$

By the triangle inequality we obtain $|\int f d\mu - r| < 2^{-n}$.

There is a computable procedure for determing r, i.e. there is some computable function $g : \subseteq \Sigma^\omega \times \Sigma^\omega \times \Sigma^* \longrightarrow \Sigma^*$ such that for $f = \delta_\rightarrow(p)$, $\mu = \delta_m(q)$ and $n = \bar{u}$ we have $|\bar{v} - \int f d\mu| < 2^{-n}$ where $v = g(p, q, u)$. Using a machine for g one can define easily a machine for a function $h : \subseteq \Sigma^\omega \times \Sigma^\omega \longrightarrow \Sigma^\omega$ such that $\int \delta_\rightarrow(p) d\delta_m(q) = \rho h(p, q)$ for all $p \in dom(\delta_\rightarrow)$ and $q \in dom(\delta_m)$.
□

As a corollary of Theorem 7, Theorem 14 cannot be extended from $C[0;1]$ to the measurable functions, not even to step functions.

Corollary 15. *Let $f : [0;1] \longrightarrow \mathbb{R}$ be the characteristic function of $[0;1/2)$. Then $\mu \mapsto \int f d\mu$ is not $(\delta_m, \rho_>)$-continuous on $\mathbf{M}$.*

4 Further Representations of Measures

In Definition 1 we have used atomic properties $r < \mu(J)$ with $r \in \mathbb{Q}$ and $J \in Int$ for identifying measures. By Theorem 14, $(f, \mu) \mapsto \int f d\mu$ is $(\delta_\rightarrow, \delta_m, \rho)$-computable for continuous functions. In the following we indentify measures μ by atomic properties $r < \int t d\mu$ or $r < \int t d\mu < s$, where $r, s \in \mathbb{Q}$ and t is from a set of simple continuous "test functions".

Definition 16. *For $n \in \omega$ and $0 \leq m \leq 2^n$ define the triangle function $t_{nm} \in C[0;1]$ by*

$$t_{nm}(x) := \begin{cases} x - (m-1)2^{-n} & \text{if } (m-1)2^{-n} \leq x \leq m \cdot 2^{-n} \\ (m+1)2^{-n} - x & \text{if } m \cdot 2^{-n} < x \leq (m+1) \cdot 2^{-n} \\ 0 & \text{otherwise.} \end{cases}$$

Let δ'_m and δ''_m be the standard representartion of $\mathbf{M}$ induced by the information structures $(\mathbf{M}, \sigma', \nu')$ and $(\mathbf{M}, \sigma'', \nu'')$, respectively, defined as follows: $\mu \in \nu'(0^n \natural 0^m \natural u) :\iff \bar{u} < \int t_{nm} d\mu$, $\mu \in \nu''(0^n \natural 0^m \natural u \natural v) :\iff \bar{u} < \int t_{nm} d\mu < \bar{v}$ for all $\mu \in \mathbf{M}$, $n \in \omega$, $0 \le m < 2^n$ and $u, v \in dom(\nu_Q)$.

We have not yet shown, that the systems σ' and σ'' from Definition 16 identify points, i.e. δ'_m and δ''_m may still be representations of partitions of $\mathbf{M}$ which are coarser than $\{\{\mu\} \mid \mu \in M\}$.

Theorem 17. δ'_m and δ''_m are representations of $\mathbf{M}$ such that $\delta_m \equiv \delta'_m \equiv \delta''_m$.

By definition, the weak topology τ_w on the set $\mathbf{M}$ of probability measures on $([0;1], \mathbf{B})$ is the coarsest, i.e. smallest, topology τ, such that $\mu \mapsto \int f d\mu$ is $(\tau, \tau_\mathbb{R})$–continuous for every $f \in C[0;1]$ [Bau74]. As a corollary of Theorem 17 we obtain:

Corollary 18. *The weak topology τ_w is the final topology τ_m of the representation δ_m.*

The weak topology τ_w on $([0;1], \mathbf{B})$ can be generated by a metric [Bau74].

Definition 19. (*Hutchinson metric*) Let $Lip := \{f \in C[0;1] \mid f(x) = 0$ and $\forall x, y. |f(x) - f(y)| \le |x - y|\}$. Define $d^H : \mathbf{M} \times \mathbf{M} \longrightarrow \mathbb{R}$ by $d^H(\mu, \mu') := \sup\{|\int f d\mu - \int f d\mu'| \mid f \in Lip\}$.

The metric d^H is called the Hutchinson metric [Hut81, Bar93].

Lemma 20. d^H *is a metric on* $\mathbf{M}$.

Theorem 21. $d^H : \mathbf{M} \times \mathbf{M} \longrightarrow \mathbb{R}$ *is* $(\delta_m, \delta_m, \rho)$-*computable*.

By Lemma 2.1 from [Wei93], the metric space $(\mathbf{M}, d^H)$ has a countable dense subset. By Corollary 45.4 from [Bau74], the discrete measures are dense. We shall use the discrete measures determined by rational numbers as a dense subset. Let $\mathbf{M}_\alpha$ be the set of all probability measures $\mu \in \mathbf{M}$ such that there are a finite set K and rational numbers $r_k, s_k \in [0;1]$ for all $k \in K$ such that $\sum \{s_k \mid k \in K\} = 1$ and $\mu = \sum s_k \mu_{r_k}$, where $\mu_x(A) = (1 \text{ if } x \in A, 0 \text{ otherwise})$. Let ν_d be a standard notation of $\mathbf{M}_\alpha$. A computable metric space is a quadruple (M, d, A, ν) such that (M, d) is a metric space, A is a dense countable subset and ν is a notation $\nu :\subseteq \Sigma^* \longrightarrow A$ of A such that the set $\{(u, v, w, x) \mid \bar{u} < d(\nu(v), \nu(w)) < \bar{x}\}$ is r.e. [Wei93]. This definition is somewhat stronger than that in [Wei87]. For a computable metric space (M, d, A, ν), the Cauchy rerpresentation δ_C [Wei97] is defined as follows (we assume w.l.o.g. $dom(\nu) \subseteq (\Sigma \setminus \{\sharp\})^*$) : $\delta_C(p) = x :\iff p = u_0 \sharp u_1 \sharp \dots$ such that $\forall i > k \ \ d(\nu(u_i), \nu(u_k)) < 2^{-k}$ and $x = \lim_{i \to \infty} \nu(u_i)$.

Theorem 22. *(1)* $\nu_d \le \delta_m$ *(2)* $(\mathbf{M}, d^H, \mathbf{M}_d, \nu_d)$ *is a computable metric space. (3) The Cauchy representation δ_m^C for this space is equivalent to δ_m.*

Since $\delta_m \equiv \delta'_m \equiv \delta''_m \equiv \delta^C_m$, these four representations of the probability measures **M** on the space $([0;1], \mathbf{B})$ induce the same computability theory and in particular have the same final topology, which is the topology τ generated by the Hutchinson-metric. As a consequence, for a hyperbolic [Hut81] computable IFS with probabilities as in Corollary 12 the unique invariant measure is computable w.r.t. any of these representations. For a domain-theoretic approach see [Eda96].

Theorem 23. *Let* $\mathbf{S} = ([0;1], f_1, \ldots, f_k, p_1, \ldots, p_k)$ *be a hyperbolic IFS with probabilities such that* $f_1, \ldots, f_k$ *are* $\delta_{\rightarrow}$*-computable and* $p_1, \ldots, p_k$ *are* ρ*-computable. Then the unique fixed point* $\mu_\mathbf{S}$ *of the operator* $T_\mathbf{S} : \mathbf{M} \longrightarrow \mathbf{M}$ *defined by*
$$T_\mathbf{S}(\mu)(A) := \sum_{i=1}^{k} p_i \mu(f_i^{-1}(A)) \text{ is } \delta_m\text{-computable.}$$

In measure theory not only probability measures but arbitrary measures $\mu : \mathbf{B} \longrightarrow \mathbb{R} \cup \{\infty\}$ are studied. Let $\mathbf{M}^b$ be the set of all measures $\mu : \mathbf{B} \longrightarrow \mathbb{R}$, i.e. all bounded measures on $([0;1], \mathbf{B})$. Let $\delta^<$ be the representation of $\mathbf{M}^b$ obtained from Definition 1, where **M** is replaced by $\mathbf{M}^b$. While $\delta_m(p)[0;1] = 1$, $\delta^<(p)[0;1]$ may be any non–negative real number. An easy proof shows that $\mu \mapsto \mu[0;1]$ is only $(\delta^<, \rho_<)$-computable and not $(\delta^<, \rho)$-continuous. This means, that informations about upper bounds of $\delta^<(p)[0;1]$ are not available from prefixes of p. As a consequence, Theorem 14 on integration fails for $\delta^<$. Only the following weak version can be proved: $(f, \mu) \mapsto \int f d\mu$ for non–negative $f \in C[0;1]$ and $\mu \in \mathbf{M}^b$ is $(\delta_{\rightarrow}, \delta^<, \rho_<)$-computable. We can, however, include informations about upper bounds of $\mu[0;1]$ in the names. Let δ^b be the representation of $\mathbf{M}^b$ defined by the following notation ν of atomic pieces of information: $\mu \in \nu(u\mathfrak{k}v\mathfrak{k}w) \iff \bar{u} < \mu(I_v)$ and $\mu[0;1] < \bar{w}$. Then the theorems we have proved for δ_m hold accordingly for δ^b, in particular Theorem 14 on integration. The connection to δ_m is given by the following lemma.

Lemma 24. *The function* $\mu \mapsto \mu[0;1]$ *on* $\mathbf{M}^b$ *is* (δ^b, ρ)*-computable, and the function* $\mu \mapsto \mu/\mu[0;1]$ *is* (δ^b, δ_m)*-computable for* $\mu \in \mathbf{M}^b, \mu[0;1] \neq 0$.

5 Conclusion

In this paper we have introduced and discussed a very natural and canonical computability theory on the set **M** of probability measures on the Borel subsets of the unit interval $[0;1]$. In particular, we have shown that simple obvious requirements exclude a number of similar definitions, that the definition leads to the expected computability results, that there are other natural definitions inducing the same computability theory and that the theory is embedded smoothly into classical measure theory. Although we have only stated the existence of computable functions throughout the paper, all the proofs provide algorithms, which can be realized by programs from some common programming language like PASCAL or C. Of course the basic definitions and many results can be transferred from the space **M** to more general spaces of measures.

References

[Bar93] M.F. Barnsley. *Fractals everywhere*. Academic Press, Boston, 1993.
[Bau74] Heinz Bauer. *Wahrscheinlichkeitstheorie und Grundzüge der Maßtheorie*. de Gruyter, Berlin, 1974.
[BB85]   Errett Bishop und Douglas S. Bridges. *Constructive Analysis*, Band 279 der Reihe *Grundlehren der mathematischen Wissenschaft*. Springer, Berlin, 1985.
[Eda95] Abbas Edalat. Domain theory and integration. *Theoretical Computer Science*, 151:163–193, 1995.
[Eda96] Abbas Edalat. Power domains and iterated function systems. *Information and Computation*, 124(2):182–197, 1996.
[Hut81] J. Hutchinson. Fractals and self-similarity. *Indiana University Journal of Mathematics*, 30:713–747, 1981.
[Ko91]   Ker-I Ko. *Complexity Theory of Real Functions*. Progress in Theoretical Computer Science. Birkhäuser, Boston, 1991.
[Kus85] Boris Abramovich Kushner. *Lectures on Constructive Mathematical Analysis*, Band 60 der Reihe *Translation of Mathematical Monographs*. American Mathematical Society, Providence, 1985.
[KW84]   Christoph Kreitz und Klaus Weihrauch. A unified approach to constructive and recursive analysis. In M.M. Richter, E. Börger, W. Oberschelp, B. Schinzel und W. Thomas, Hrsg., *Computation and Proof Theory*, Band 1104 der Reihe *Lecture Notes in Mathematics*, Seiten 259–278, Berlin, 1984. Springer. Proceedings of the Logic Colloquium, Aachen, July 18-23, 1983, Part II.
[KW85]   Cristoph Kreitz und Klaus Weihrauch. Theory of representations. *Theoretical Computer Science*, 38:35–53, 1985.
[RW80]   Angelika Reiser und Klaus Weihrauch. Natural numberings and generalized computability. *Elektronische Informationsverarbeitung und Kybernetik*, 16:11–20, 1980.
[TWW88] Joseph F. Traub, G.W. Wasilkowski und H. Woźniakowski. *Information-Based Complexity*. Computer Science and Scientific Computing. Academic Press, New York, 1988.
[Wei87] Klaus Weihrauch. *Computability*, Band 9 der Reihe *EATCS Monographs on Theoretical Computer Science*. Springer, Berlin, 1987.
[Wei93] Klaus Weihrauch. Computability on computable metric spaces. *Theoretical Computer Science*, 113:191–210, 1993. Fundamental Study.
[Wei95] Klaus Weihrauch. A foundation of computable analysis. *Bulletin of the European Association for Theoretical Computer Science*, 57:167–182, Oktober 1995. The Structural Complexity Column by Juris Hartmanis.
[Wei97] Klaus Weihrauch. A Foundation for Computable Analysis. In Douglas S. Bridges, Cristian S. Calude, Jeremy Gibbons, Steve Reeves und Ian H. Witten, Hrsg., *Combinatorics, Complexity, and Logic*, Discrete Mathematics and Theoretical Computer Science, Seiten 66–89, Singapore, 1997. Springer. Proceedings of DMTCS'96.

Worst-Case Hardness Suffices for Derandomization: A New Method for Hardness-Randomness Trade-Offs

Alexander E. Andreev[1], Andrea E. F. Clementi[2], José D. P. Rolim[3]

[1] Dept. of Mathematics, University of Moscow,
andreev@mntn.msk.su
[2] Dip. di Scienze dell'Informazione, University "La Sapienza" of Rome
clementi@dsi.uniroma1.it
[3] Centre Universitaire d'Informatique, University of Geneva, CH,
rolim@cui.unige.ch

Abstract. Up to know, the known derandomization methods have been derived assuming average-case hardness conditions. In this paper we instead present the first worst-case hardness conditions sufficient to obtain $P = BPP$.
Our conditions refer to the worst-case circuit complexity of Boolean operators computable in time exponential in the input size. Such results are achieved by a new method that departs significantly from the usual known methods based on pseudo-random generators.
Our method also gives a worst-case hardness condition for the circuit complexity of Boolean operators computable in NC (with respect to their output size) to obtain $NC = BPNC$.

1 Introduction

1.1 Motivations and previous results. A major goal in complexity theory is the study of the real power of randomized algorithms, that is algorithms that make decisions based on the output of a random source of bits. To this aim, several recent works have been focused on the design of general methods that decrease (or remove) the amount of random bits used by these algorithms. A central question in this area is the relationship between the existence of computationally-hard functions and the existence of efficient derandomization methods. Yao [12], and Blum and Micali [5] introduced the concept of *Pseudo-Random Generator* (PSRG), any Boolean operator $G = \{G_n : \{0,1\}^{k(n)} \to \{0,1\}^n, n > 0\}$, (denoted by $G : k(n) \to n$) that, for a.e. n and for any Boolean function $f : \{0,1\}^n \to \{0,1\}$ whose circuit complexity $L(f)$ is at most n, satisfies: $|\mathbf{Pr}(f(\mathbf{y}) = 1) - \mathbf{Pr}(f(G_n(\mathbf{x})) = 1)| \leq 1/n$ (where $\mathbf{y}$ is chosen uniformly at random from $\{0,1\}^n$, and $\mathbf{x}$ from $\{0,1\}^{k(n)}$). The output sets of PSRG are also called *discrepancy* sets for circuits of linear size.
According to the definition used in [10], a Boolean operator $Op : k(n) \to n$ is *quick* if it can be computed in time polynomial in n (note in passing that if $k(n) = O(\log n)$ then the "quick" condition is equivalent to assume that Op

belongs to EXP). It is not hard to show [10] that the existence of a quick PSRG $G : k(n) \to n$ with $k(n) = O(\log n)$ implies $P = BPP$. Nisan and Wigderson [10] showed a method to construct *quick* PSRG based on the existence of Boolean functions in EXP that have exponential *hardness* [10]. The hardness condition used by Nisan and Wigderson requires the existence of a function in EXP that not only has a hard *worst-case* circuit complexity[4] but also a hard *average-case* circuit complexity. More formally, a function $f : \{0,1\}^n \to \{0,1\}$ is (ϵ, L)-*hard* if, for any circuit C of size at most L, $|\mathbf{Pr}\,(C(\mathbf{x}) = f(\mathbf{x})) - 1/2| \le \epsilon/2$. Given a Boolean function $F = \{F_n : \{0,1\}^n \to \{0,1\}, n > 0\}$, the *hardness* at n of F (denoted as $H_F(n)$) is defined as the maximum integer h_n such that F_n is $(1/h_n, h_n)$-hard. Then, F has exponential hardness if $H_F(n) \ge 2^{\Omega(n)}$. Nisan and Wigderson showed a fundamental "Hardness vs Randomness" result.

Theorem 1. *[10] If a Boolean function F exists such that* i) $F \in EXP$, *and* ii) F *has exponential hardness, then there exists a quick PSRG $G : k(n) \to n$ where $k(n) = O(\log n)$, and consequently $P = BPP$.*

The hardness required by Nisan and Wigderson's construction of quick PSRG thus refers to average-case complexity. Then a consequent and natural question is the following: Does any "worst-case" hardness assumption on the circuit complexity of Boolean functions computable in time exponential in the input size exist which allows to derive an efficient derandomization method (in particular, to obtain $P = BPP$)?

We give two answers to this question. Both answers make use of a new method (informally described in Section 1.3) that relies on a particular class of Boolean operators (different from PSRG), denoted as *Hitting Set Generators*, which have been recently introduced in [3]. Let $L(f)$ denote the circuit complexity of a finite function $f : \{0,1\}^n \to \{0,1\}$ and, given any positive number dp, the term $L_{dp}(f)$ denotes the minimum size of circuits of depth dp which are able to compute f.

Definition 2. *Let $\epsilon(n)$, $\beta(n)$, and $\gamma(n)$ be polynomial-time computable functions such that for any $n \ge 1$: $0 < \epsilon(n) < 1$, $n \le \beta(n) \le 2^n$, and $\gamma(n) \ge \log n$. Then, a Boolean operator $H : k(n) \to n$ is an $(\epsilon(n), \beta(n), \gamma(n))$-Hitting Set Generator (in short, $(\epsilon(n), \beta(n), \gamma(n))$-HSG) if, for any Boolean function f such that $L_{\gamma(n)}(f) \le \beta(n)$ and $\mathbf{Pr}\,(f = 1) \ge \epsilon(n)$, H is required to provide one "example" $\mathbf{y}$ for which $f(\mathbf{y}) = 1$, i.e., there exists $\mathbf{a} \in \{0,1\}^{k(n)}$ such that $f(H_n(\mathbf{a})) = 1$. When no depth constraint $\gamma(n)$ is imposed, we will use notation $(\epsilon(n), \beta(n))$-HSG.*

By making a simple comparison between the definition of discrepancy sets and that of hitting sets it should be clear that HSG satisfy a property significantly weaker than that of PSRG. Nevertheless, Andreev *et al* [3] proved that, given any BPP-algorithm A, the output of any quick HSG can be transformed into an *ad hoc* discrepancy set for A by means of a deterministic polynomial-time algorithm.

---

[4] As circuit complexity of a finite Boolean function f, we will always mean the size of the smallest circuit that computes f.

Theorem 3. *[3] Let $k(n) = O(\log n)$ and let ϵ be any constant such that $0 < \epsilon < 1$. If there exists a quick (ϵ, n)-HSG $H : k(n) \to n$ then $P = BPP$.*

As we will describe in Section 1.3, the polynomial-time algorithm in [3] is of independent interest and it is used in this paper to obtain Theorem 5. On the other hand, more recently (after the submission of our paper), a different algorithmic proof of Theorem 3 has been given in [4]. This algorithm is simpler and runs in NC^1.

1.2 Our results. We give two worst-case hardness conditions which are sufficient to construct quick HSG that satisfy Theorem 3 thus obtaining $P = BPP$. The circuit complexity of a Boolean operator H will be denoted as $L^{op}(H)$. Observe that if $L^{op}(k,n)$ denotes the worst-case circuit complexity of Boolean operators $H : k(n) \to n$, then it is known [9, 11] that, for any $\log n \le k \le n$, $L^{op}(k,n) = (1 + o(1))(2^k n)/(k + \log n)$. Furthermore, for a.e. Boolean operator $H : k \to n$, we have $L^{op}(H) = \Theta((2^k n)/(k + \log n))$. The first condition deals with the worst-case circuit-complexity of characteristic functions of sets generated by Boolean operators.

Theorem 4. *Let δ be such that $0 < \delta \le 1/2$, and let $k(n) = (1 + \Theta(1))\log n$. If there exists a quick operator $H : k(n) \to n$ such that the characteristic function of its output sets $F^H = \{F_n^H : \{0,1\}^n \to \{0,1\}$, where $F_n^H(\mathbf{x}) = 1$ if $\exists\ \mathbf{y} \in \{0,1\}^{k(n)}$ s.t. $H_n(\mathbf{y}) = \mathbf{x},\ n > 0\}$ satisfies*

$$L(F_n^H) \ge (1/2 + \delta)(2^{k(n)} n)/(k(n) + \log n),$$

then it is possible to construct a quick operator $H' : k'(n) \to n$ where $k'(n) = \Theta(\log n)$ such that H' is an (ϵ, n)-HSG for some constant $0 < \epsilon < 1$, thus $P = BPP$.

Another way to state the above theorem is the following. Assume that there exists a sparse language $S = \{S_n \subseteq \{0,1\}^n, n > 0\}$ that can be generated by an uniform algorithm which runs in time polynomial in n, and such that the worst-case circuit complexity of deciding S is not smaller (up to some constant factor) than the worst-case circuit complexity of generating languages S' having the same sparsity factor of S. Then $P = BPP$.

The second sufficient condition to obtain a quick HSG refers directly to the worst-case circuit complexity of Boolean operators instead of the characteristic functions of their output sets.

Theorem 5. *Let $k(n) = \Theta(\log n)$. Let $H : k(n) \to n$ be a quick operator such that for a.e. n,*

$$L^{op}(H_n) \ge L^{op}(k,n) - (2^{k(n)})/(k(n)^2).$$

Then, for any constant $0 < \epsilon < 1$, and for any positive integer q, it is possible to construct a quick $(1 - \epsilon, n^q)$-HSG $H' : k'(n) \to n$, where $k'(n) = \Theta(\log n)$, thus obtaining $P = BPP$.

Furthermore, using the new "parallel" proof of Theorem 3, we provide here a worst-case hardness condition for Boolean operators sufficient to derandomize any BPNC algorithm (i.e. to obtain $BPNC = NC$).

Theorem 6. *A constant $0 < c_0 < 1$ exists such that if an operator $H : k(n) \to n$ with $k(n) = O(\log n)$ exists such that 1) H is an NC operator[5], and 2) for any $d \geq 1$ there exists a constant c with $0 < c_0 \leq c < 1$ such that the characteristic function F^H of its output sets satisfies $L_{\log^d n}(F_n^H) \geq c(2^{k(n)}n)/(k(n) + \log n)$, then $NC = BPNC$.*

1.3 Our method and further connections with other works. All of our proofs share a common method based on the following fact. There is a precise trade-off between the worst-case circuit complexity of partial Boolean functions and the number of 1's in their outputs. In particular, we formalize the intuitive fact that a partial Boolean function having a hard worst-case circuit complexity cannot return 0 for a "large" number of inputs. This property is used to construct the preliminary versions of our HSG which are then combined with a convenient use of the properties of *expanders* graphs [2] (to obtain Theorem 4) and with a new analysis of the performances of the already mentioned Andreev *et al*'s algorithm [3] (to obtain Theorem 5).

Finally, we remark that hardness vs randomness results similar to those obtained in our paper have been obtained, independently from our work, by Impagliazzo and Wigderson in [6]. Their method (based on the derandomization of the XOR-lemma) achieves a trade-off which is stronger than ours in the case of sequential algorithms (i.e. BPP algorithms). However it is not clear, to our present knowledge, whether their method can be applied to obtain trade-offs for parallel computation (like ours) since they use, in a rather envolved way, *expander* walks which seem to be hard to parallelize.

Due to the lack of space, proofs will be given in the full version of this paper.

2 Preliminary results on the circuit complexity of partial Boolean functions

Let $\mathcal{F}(n, N, m)$ be the set of all partial Boolean functions $f(x_1, \ldots, x_n)$ defined on $N \leq 2^n$ inputs and assuming 1 on $m \leq N$ inputs. Furthermore, $L(n, N, m)$ denotes the worst-case circuit complexity of functions from $\mathcal{F}(n, N, m)$, and $L_{depth}(n, N, m)$ denotes the maximum value $L_{depth}(f)$ among all functions f from $\mathcal{F}(n, N, m)$. Lupanov [9] obtained the asymptotical bounds result for the case of total Boolean functions.

However, in order to construct quick HSG we need that Lupanov's results hold also for partial Boolean functions. In particular, the generalization of the upper bounds cannot be derived directly from the proofs in [9]. Then we give a

---

[5] With *"NC operator"*, we will always mean an operator which is computable in NC with respect to the size of its output

reduction from general Boolean functions to the restricted case of total Boolean functions which is based on a probabilistic construction of suitable linear operators.

Theorem 7.
$$L(n, N, m) = (1 + o(1)) \left(\log \binom{N}{m} \right) / \left(\log \log \binom{N}{m} \right) + O(n) .$$

Furthermore a constant $c > 0$ exists such that
$$L_{c \log n}(n, N, m) = (1 + o(1)) \left(\log \binom{N}{m} \right) / \left(\log \log \binom{N}{m} \right) + O(n) .$$

3 Hard characteristic functions and HSG

The following theorem provides a first trade-offs between the hardness of characteristic functions of Boolean subsets and their hitting properties[6].

Theorem 8. *Let $0 < c_2 < 1$ be a constant [and $d \geq 1$], and let $S_n \subseteq \{0,1\}^n$ be any subset such that $|S_n| \leq b_n$, where $b_n = n^{\Theta(1)}$. Suppose that for the characteristic function F_n of S_n we have*

$$i) \ L(F_n) \geq c_2 \frac{b_n n}{\log b_n + \log n} \quad [\ i') \ L_{\log^{d+1} n}(F_n) \geq c_2 \frac{b_n n}{\log b_n + \log n} \ ] .$$

Then, for any constant c_1, such that $0 < c_1 < c_2$, for any Boolean function $f(x_1, \ldots, x_n)$ such that

$$ii) \ \mathbf{Pr}\,(f = 1) \geq 1 - 2^{(c_1 - 1)n}, \ \text{and} \ iii) \ L(f) \leq b_n \ [\ iii') \ L_{\log^d n}(f) \leq b_n \ ],$$

there exists $\mathbf{a} \in S_n$ for which $f(\mathbf{a}) = 1$.

Sketch of the proof. Suppose, by contradiction, that f satisfies conditions ii) and iii) but for any $\mathbf{a} \in S_n$ we have $f(\mathbf{a}) = 0$. Let $Z \subseteq \{0,1\}^n$ be the subset of all inputs on which $f = 0$. Clearly, we have $S_n \subseteq Z \subseteq \{0,1\}^n$. Then consider the partial Boolean function $g(x_1, \ldots, x_n)$ defined as follows: $g(\mathbf{a}) = 1$ if $\mathbf{a} \in S_n$, $g(\mathbf{a}) = 0$ if $\mathbf{a} \in Z \setminus S_n$, and $g(\mathbf{a})$ is not defined if $\mathbf{a} \in \{0,1\}^n \setminus Z$. Since $|Z| \leq 2^{c_1 n}$ and $|S_n| \leq b_n$, from Theorem 7, we have

$$L(g) \leq (1 + o(1)) \left(\log \binom{2^{c_1 n}}{b_n} \right) / \left(\log \log \binom{2^{c_1 n}}{b_n} \right) + O(n)$$

$$\leq (1 + o(1)) c_1 (b_n n) / (\log b_n + \log n) .$$

From $S_n \subseteq Z$, it is easy to prove that, given any $\mathbf{a}$, $F_n(\mathbf{a})$ can be computed as $g(\mathbf{a}) \wedge \neg f(\mathbf{a})$. Hence

---

[6] Each result will be given in both sequential and "parallel" version. The latter will be included in square brackets.

$$L(F_n) \leq L(g) + L(f) + O(1) \leq (1+o(1))c_1\frac{b_n n}{\log b_n + \log n} + b_n + O(1) \leq$$

$$\leq (1+o(1))c_1\frac{b_n n}{\log b_n + \log n}.$$

For sufficiently large n, this last upper bound is in contradiction with hypothesis (i) of our theorem. The "parallel" version of the theorem can be easily derived using the same contradiction argument. □

In which follows, we will consider HSG which always have a monotone function prize $k(n)$ such that, for any $n > 0$, $k(n+1)-k(n) \leq 1$ and $n^\alpha \geq k(n) \geq \log n$ where $0 < \alpha < 1$. Let $H : k(n) \to n$ be a Boolean operator with $k(n) = \Theta(\log n)$, and let $F^H = \{F_n^H : \{0,1\}^n \to \{0,1\}, n > 0\}$ be the corresponding family of the characteristic functions.

Corollary 9. *Suppose that a quick [NC] operator $H : k(n) \to n$ exists such that $k(n) = (1+\Theta(1))\log n$ and a constant $0 < c_2 < 1$ exists such that, for a.e. n, $L(F_n^H) \geq c_2(2^{k(n)}n)/(k(n)+\log n)$ [$L_{\log^{d+1} n}(F_n^H) \geq c_2(2^{k(n)}n)/(k(n)+\log n)$ for some $d \geq 1$]. Then, for any positive constant q and for any constant c_1 such that $0 < c_1 < c_2$, it is possible to construct a quick [NC] operator $H' : k'(n) \to n$ with $k'(n) = \Theta(\log n)$ and such that H' is an $(1-2^{(c_1-1)n}, n^q)$-HSG [H' is an $(1-2^{(c_1-1)n}, n^q, \log^d n)$-HSG].*

3.1 Improved HSG using expanders

Corollary 9 gives a quick HSG for the class of polynomial size circuits (functions) C that have a very large fraction of 1's, i.e. $\Pr(C = 1) \geq 1 - 2^{-cn}$ for some positive constant smaller than 1. However, this hitting property does not suffice to derandomize BPP-algorithms (see Theorem 3). It is in fact required to hit all linear-size circuits having "only" a constant fraction of $1's$. To this aim, we will combine the HSG in Corollary 9 with a *random walk* on *expanders*, a tool that has been often used in decreasing randomness in probabilistic algorithms.

An undirected graph $G(V, E)$ is a (d,c)-*expander* if the maximum degree of a vertex is d, and for every set $W \subseteq V$ of cardinality $|W| \leq |V|/2$, the inequality $|N(W) - W| \geq c|W|$ holds, where $N(W)$ denotes the set of all vertices adjacent to some vertex in W. The expanding properties of a graph can be established by determining the value of its second largest eigenvalue. Indeed, if λ is an upper bound on the second largest eigenvalue of any d-regular graph $G(V, E)$, then G is a (d,c)-expander for $c = (d - \lambda)/2d$. Expander graphs have the following important "hitting" property proved by *Ajtai et al* [1].

Theorem 10. *Let $G(V, E)$ be a d-regular graph, and assume that its second largest eigenvalue is at most $\lambda > 0$. Given any subset $W \subseteq V$ such that $|W| = \alpha n$ $(\alpha < 1)$. Then, for every $t > 0$, the number of walks of length t in G that avoid W is at most $n(1-\alpha)^{1/2}((1-\alpha)d^2 + \lambda^2)^{t/2}$.*

In [7], a polynomial-time algorithm is presented that, given $n > 0$, and $d \leq n$, constructs a d'-regular expanders G such that $d' = O(d)$, $|V| = O(n)$, and its second largest eigenvalues $\lambda > 0$ is such that $\lambda \leq 2\sqrt{d-1}$ (such graphs are called *Ramanujan* graphs).

For any $n > 0$, consider a d-regular *Ramanujan expander* $EP_n = (V_n, X_n)$ where $2^n < |V_n| \leq 2^{n+1}$ [7]. Observe that the Boolean strings with last component equal 0 correspond to the input set of the function we want to hit. This assumption is required when EP_n cannot be constructed on vertex sets whose size is exactly a power of 2. Let $l = \lceil \log d \rceil$. We suppose that d is a large but constant value. Then, we consider the operator $EPR_{n,t} : \{0,1\}^{n+l\cdot(2^t-1)+t} \to \{0,1\}^n$, such that

$$EPR_{n,t}(\mathbf{a}, \mathbf{u}_1, \ldots, \mathbf{u}_{2^t-1}, \mathbf{s}) , \qquad \mathbf{a} \in \{0,1\}^n , \ \mathbf{u}_i \in \{0,1\}^l , \ \mathbf{s} \in \{0,1\}^t ,$$

are the first n components of the $\phi(\mathbf{s})$-th vertex of the EP_n-walk of length 2^t which starts from vertex $(\mathbf{a}, 0)$ and is uniquely determined by the sequence of edge choices in the neighborhood of each vertex: $\phi(\mathbf{u}_1), \ldots, \phi(\mathbf{u}_{2^t-1})$. Observe that if $t = \Theta(\log n)$, the operator $EPR_{n,t}$ can be computed in time polynomial in n. Consider now a Boolean function $g(x_1, \ldots, x_n)$, and the operator $EPR^g_{n,t} : \{0,1\}^{n+l\cdot 2^t} \to \{0,1\}$ that performs the OR among the values of g computed on the input points visited by a fixed EP_n-walk of length 2^t, i.e.,

$$EPR^g_{n,t}(\mathbf{a}, \mathbf{u}_1, \ldots, \mathbf{u}_{2^t}) = \bigvee_{\mathbf{s} \in \{0,1\}^t} g\left(EPR_{n,t}(\mathbf{a}, \mathbf{u}_1, \ldots, \mathbf{u}_{2^t-1}, \mathbf{s})\right) . \quad (1)$$

As consequence of Theorem 10, we can prove the following bound.

Lemma 11. *If* $\mathbf{Pr}\,(g=0) \leq c < \frac{1}{2}$, *then* $\mathbf{Pr}\left(EPR^g_{n,t} = 0\right) \leq \left(c + \frac{\lambda}{d}\right)^{2^t-2}$.

Theorem 12. *Assume that there exists a quick operator* $H : k(n) \to n$, *such that* $k(n) = (1 + \Theta(1))\log n$ *and the characteristic functions of its output sets satisfies*

$$L(F^H_n) \geq ((\log(4\lambda)/(\log d) + \delta)(2^{k(n)}n)/(k(n) + \log n)$$

for some constant $\delta > 0$. *Then it is possible to construct a quick operator* $H'' : k''(n) \to n$ *with* $k''(n) = \Theta(\log n)$ *and such that* H'' *is an* $(1-\epsilon, n)$-*HSG for some constant* $0 < \epsilon < 1$, *thus* $P = BPP$.

4 Hitting Set Generators for BPNC

Ramanujan's graphs cannot be used to derive NC Hitting Set Generators since no efficient parallel method to perform random walks on such graphs is presently available. However, Zuckermann [13] recently introduced an NC construction of *samplers* [13] which can replace the role of expanders in our construction. In particular, we can use the following result.

Theorem 13. *[13] Any BPNC algorithm that uses n random bits and has error probability bounded by 1/3 can be simulated by a BPNC algorithm that uses $r(n) = O(n)$ random bits and has error probability bounded by $(1/2)^n$.*

Informally speaking, this result allows us to consider only "parallel" circuits having a fraction of 1's not smaller than $1 - 2^{-cn}$ for some fixed constant $0 < c < 1$. By using the same method of Section 3.1, we can combine Corollary 9 and Theorem 13 to obtain the following result

Theorem 14. *A constant $0 \leq c_z < 1$ exists such that the following holds. Assume that there exists an NC operator $H : k(n) \to n$ with $k(n) = (1+\Theta(1))\log n$ and such that, for any constant $d \geq 1$, the characteristic functions of its output sets satisfy $L_{\log^{d+1} n}(F_n^H) \geq \delta(2^{k(n)}n/(k(n) + \log n)$, for some constant $\delta \geq c_z$. Then it is possible to construct an NC operator $H' : k'(n) \to n$ with $k'(n) = \Theta(\log n)$ and such that H' is an $(1 - \epsilon, n, \log^d n)$-HSG for any constant $0 < \epsilon < 1$ and $d \geq 1$.*

In the next corollary, the above HSG is combined with the new "parallel" proof of Theorem 3 given in [4].

Corollary 15. *A constant $0 < c_z < 1$ exists such that if an NC operator $H : k(n) \to n$ exists that satisfies the same conditions of Theorem 14 then $NC = BPNC$.*

Note. In the previous version of this paper (when the new proof of Theorem 3 was still unknown) we were able to provide only sufficient hardness conditons to obtain $ZNC = BPNC$. The proof of this weaker result is of independent interest and has been used in [4] to obtain some results in the context of *weak random sources*. A new version of this proof can be found in [4].

5 Hitting sets from hard Boolean operators

The construction of an efficient HSG from a Boolean operator which has hard circuit-complexity is based on the following "contradiction" argument. Suppose that a Boolean operator $T : \{0,1\}^m \longrightarrow \{0,1\}^n$ is not a HSG for a certain class of circuits defined by the parameters $\epsilon(n)$ and $\beta(n)$ (see Def. 2). Roughly speaking, this negative fact implies that the output sequence of T can be represented by a new binary sequence which contains a "large" number of 0's (this number depends on $\epsilon(n)$ and $\beta(n)$). Then, using Andreev et al's technique shown in [3], it is possible to compress this new binary sequence in order to prove an upper bound on the circuit complexity of T. This bound is obtained by a new analysis of the compression rate achieved by this technique and by applying the upper bound for the Shannon function $L(n, N, m)$ in Theorem 7. If T is supposed to have a hard circuit complexity, we get a contradiction.

5.1 Compressing Boolean operators

Let $T : \{0,1\}^m \to \{0,1\}^n$ and $C(x_1,\ldots,x_n)$ be a circuit with n inputs. Given $\alpha \in \{0,1\}^n$, consider the function $Med(f, T, \alpha) = 2^{-m}\sum_{u\in\{0,1\}^m} C(T(u) \oplus \alpha)$ (as in the proof of Corollary 15). It is easy to prove that $\mathbf{E}\left(Med(C,T,\alpha)\right) = \mathbf{Pr}\left(C(x_1,\ldots,x_n) = 1\right)$ where the expected value is computed with respect to α. We briefly describe here the *Andreev et al*'s technique introduced in [3]. Let α_1 and α_2 be two different elements in $\{0,1\}^n$. Define $d_1 = Med(C,T,\alpha_1)$ and $d_2 = Med(C,T,\alpha_2)$ and assume that $D = d_2 - d_1 > 0$. The j-th component of $\mathbf{a}$ will be denoted as $[\mathbf{a}]^j$. Since we are considering the case in which $D > 0$, we can assume that there exists an index s for which $[\alpha_1]^s \neq [\alpha_2]^s$. Consider the operator $T^\# : \{0,1\}^m \to \{0,1\}^n$ defined as follows $T^\#(\mathbf{u}) = T(\mathbf{u}) \oplus ([T(\mathbf{u})]^s \cdot (\alpha_1 \oplus \alpha_2))$ where the operation "$\cdot$" is the standard scalar product. The s-th component of $T^\#(\mathbf{u})$ satisfies the following equations:

$$[T^\#(\mathbf{u})]^s = [T(\mathbf{u})]^s \oplus ([T(\mathbf{u})]^s \cdot ([\alpha_1]^s \oplus [\alpha_2]^s)) = [T(\mathbf{u})]^s \oplus [T(\mathbf{u})]^s \cdot 1 = 0 . \quad (2)$$

Observe also that the set $\{T^\#(\mathbf{u}) \oplus \alpha_1, T^\#(\mathbf{u}) \oplus \alpha_2\}$ is equal to the set $\{T(\mathbf{u}) \oplus \alpha_1 , T(\mathbf{u}) \oplus \alpha_2\}$. Let

$$N(\sigma, \phi_1, \phi_2) = |\{\mathbf{u} : [T(\mathbf{u})]^s = \sigma,\ C(T(\mathbf{u})\oplus\alpha_1) = \phi_1 \text{ and } C(T(\mathbf{u})\oplus\alpha_2) = \phi_2\}| . \quad (3)$$

We can now introduce the function which approximates the s-th component of $T(\mathbf{u})$. Consider the function Q defined as follows:

$$Q_{N(\sigma,\phi_1,\phi_2)}(x,y) = \begin{cases} x & \text{if } x \neq y \\ 1 & \text{if } x = y = 0 \text{ and } N(1,0,0) \geq N(0,0,0) \\ 0 & \text{if } x = y = 0 \text{ and } N(1,0,0) < N(0,0,0) \\ 1 & \text{if } x = y = 1 \text{ and } N(1,1,1) \geq N(0,1,1) \\ 0 & \text{if } x = y = 1 \text{ and } N(1,1,1) \geq N(0,1,1) \end{cases}$$

In which follows we will consider the function N as a fixed parameter, and thus we will omit the index $N(\sigma, \phi_1, \phi_2)$ in the definition of Q. Then the approximation function for the s-th bit of $T(\mathbf{u})$ is $Z(\mathbf{u}) = Q(C(T^\#(\mathbf{u})\oplus\alpha_1), C(T^\#(\mathbf{u})\oplus\alpha_2))$, $i = 1,\ldots,m$. Our next goal is to estimate the number of errors generated by $Z(\mathbf{u})$. Let $ND(\sigma, \phi_1, \phi_2)$ be the number of inputs $\mathbf{u}$ such that the following conditions are satisfied: $i)$ $[T(\mathbf{u})]^s \oplus Z(\mathbf{u}) = 1$ (i.e. there is an error); $ii)$ $[T(\mathbf{u})]^s = \sigma$; $iii)$ $C(T(\mathbf{u}) \oplus \alpha_1) = \phi_1$; $iv)$ $C(T(\mathbf{u}) \oplus \alpha_2) = \phi_2$.

The following Lemma gives an upper bound on the number of errors in approximating the s-th bit of $T(\mathbf{u})$.

Lemma 16. *[3]* $\sum_{(\sigma,\phi_1,\phi_2)\in\{0,1\}^3} ND(\sigma,\phi_1,\phi_2) \leq m\left(\frac{1}{2} - \frac{d_2-d_1}{2}\right)$.

Some new hardness-compression trade-offs Using Lemma 16, we are now able to give an useful bound on the circuit complexity of T. Observe that function $U(\mathbf{u}) = [T(\mathbf{u})]^s \oplus Z(\mathbf{u})$ with $\mathbf{u} \in \{0,1\}^m$, singles out the positions in T for which an error occurs.

Lemma 17. $L(T) \leq L^{op}(m, n-1) + L(U) + O(L(C)) + O(n)$.

Lemma 18. *If for some constant c_1 we have that $D \geq c_1$, then there exists a constant $c_2 < 1$ such that $L(U) \leq c_2(2^m/m)$.*

5.2 The Hitting Set Generator

In order to derive our HSG, we will make use of the following result given by Lupanov (see also [11]). Let $L^{op}(k, n)$ denote the worst-case circuit complexity of Boolean operators having k variables and n outputs. Then $L^{op}(k,n) = (1 + o(1))(2^k n)/(k + \log n)$.

Theorem 19. *Assume that a quick operator $H : k(n) \to n$ exists such that $k(n) = (1 + \Theta(1))\log n$, and for a.e. n $L^{op}(H_n) \geq L(k(n), n) - (2^{k(n)})/(k(n)^2)$. Then, it is possible to construct a $(1/2, n)$-HSG $H' : k'(n) \to n$ such that $k'(n) = \Theta(\log n)$. Hence, $P = BPP$.*

Acknowledgements. We are grateful to Luca Trevisan for several interesting discussions.

References

1. Ajtai M, Komlos J, and Szemeredi E. (1987), Deterministic simulation in LOGSPACE, Proc. of *19th ACM STOC*, 132-140.
2. Alon N. (1986), "Eigenvalues and Expanders", *Combinatorica*, 6, 83-96.
3. Andreev A., Clementi A., and Rolim J. (1996), "Hitting Sets Derandomize BPP", in *XXIII International Colloquium on Algorithms, Logic and Programming (ICALP'96)*, LNCS. Also available via ftp/WWW in the electronic journal *ECCC* (TR95-061)
4. Andreev A., Clementi A., Rolim J and Trevisan L. (1997), Weak Random Sources, Hitting Sets, and BPP Simulations", Manuscript, February 1997.
5. Blum M., and Micali S. (1984), "How to generate cryptographically strong sequences of pseudorandom bits", *SIAM J. of Computing*, 13(4), 850-864.
6. Impagliazzo R., Wigderson A. (1997), "P=BPP unless E has subexponential circuits: Derandomizing the XOR Lemma", to appear in *29th ACM STOC*.
7. A. Lubotzky, R. Phillips, and P. Sarnak. (1988), "Ramanujan graphs", *Combinatorica*, 8(3):261–277, 1988.
8. Lupanov, O.B. (1956) "About gating and contact-gating circuits", *Dokl. Akad. Nauk SSSR* 111, 1171-11744.
9. Lupanov, O.B. (1965), "About a method circuits design – local coding principle", *Problemy Kibernet.* 10, 31-110 (in Russian).
10. Nisan N., and Wigderson A. (1994), "Hardness vs Randomness", *J. Comput. System Sci.* 49, 149-167 (also presented at the *29th IEEE FOCS*, 1988).

11. Wegener, I. (1987), *The complexity of finite Boolean functions*, Wiley-Teubner Series in Computer Science.
12. Yao A. (1982), "Theory and applications of trapdoor functions", in *23th IEEE FOCS*, 80-91.
13. Zuckermann D. (1996), "Randomness-Optimal Sampling, Extractors, and Constructive leader Election", in *28th ACM STOC*, 286-295.

Results on Resource-Bounded Measure

Harry Buhrman[*1], and Stephen Fenner[**2], and Lance Fortnow[***3]

^{1} Centrum voor Wiskunde en Informatica
^{2} University of Southern Maine
^{3} CWI & The University of Chicago

Abstract. We construct an oracle relative to which NP has p-measure 0 but D^p has measure 1 in EXP. This gives a strong relativized negative answer to a question posed by Lutz [Lut96]. Secondly, we give strong evidence that BPP is small. We show that BPP has p-measure 0 unless EXP = MA and thus the polynomial-time hierarchy collapses. This contrasts with the work of Regan et. al. [RSC95], where it is shown that P/*poly* does *not* have p-measure 0 if exponentially strong pseudorandom generators exist.

1 Introduction

Since the introduction of resource-bounded measure by Lutz [Lut92], many researchers investigated the size (measure) of complexity classes in exponential time (EXP). A particular point of interest is the *hypothesis* that NP does not have p-measure 0. Recent results have shown that many reasonable conjectures in computational complexity theory follow from the hypothesis that NP is not small (i.e., $\mu_p(\text{NP}) \neq 0$), and hence it seems to be a plausible scientific hypothesis [LM96, Lut96].

In [Lut96], Lutz shows that if $\mu_p(\text{NP}) \neq 0$ then BPP is low for Δ_2^P. He shows that this even follows from the seemingly weaker hypothesis that $\mu_p(\Delta_2^P) \neq 0$. He asks whether the latter assumption is weaker or equivalent to $\mu_p(\text{NP}) \neq 0$. In this paper we show that, relative to some oracle, the two assumptions are *not* equivalent.

We show a relativized world where D^p = EXP whereas NP has no P-bi-immune sets. This immediately implies, via a result of Mayordomo [May94a], that in this relativized world, NP has p-measure 0 and D^p, and hence Δ_2^P, has measure 1 in EXP, and thus does not have p-measure 0, or even p_2-measure 0.

---

[*] URL: http://www.cwi.nl/cwi/people/Harry.Buhrman.html. E-mail: buhrman@cwi.nl. Partially supported by the Dutch foundation for scientific research (NWO) by SION project 612-34-002, and by the European Union through Neuro-COLT ESPRIT Working Group Nr. 8556, and HC&M grant nr. ERB4050PL93-0516.
[**] URL: http://www.cs.usm.maine.edu/~fenner/. Email: fenner@cs.usm.maine.edu. Partially supported by NSF grant CCR 92-09833.
[***] URL: http://www.cs.uchicago.edu/~fortnow. Email: fortnow@cs.uchicago.edu. Supported in part by NSF grant CCR 92-53582, the Dutch Foundation for Scientific Research (NWO) and a Fulbright Scholar award.

This shows in a very strong way that relativized measure for NP and P^{NP} differ: $\mu_p(\text{NP}) = 0$ whereas $\mu_p(P^{NP[2]}) \neq 0$. Here $P^{NP[2]}$ is the class of sets recognized by polynomial time Turing machines that are allowed two queries to an NP oracle. We show that our results cannot be improved to $P^{NP[1]}$.

Secondly, we investigate the possibility that BPP does not have p-measure 0. Intuitively BPP is a feasible complexity class close to P and therefore it should be the case that BPP is small. We give very strong evidence supporting this intuition. We show that $\mu_p(\text{BPP}) = 0$ unless EXP = MA and thus the polynomial-time hierarchy collapses.

Since BPP $\subseteq$ P/$poly$ our result contrasts with the one by Regan, Sivakumar and Cai [RSC95], where it is shown that $\mu_p(\text{P}/poly) \neq 0$, unless exponentially strong pseudorandom generators do not exist.

2   Preliminaries

We let $\Sigma = \{0,1\}$ and identify strings in Σ^* with natural numbers via the usual binary representation. We fix $N_1, N_2, \ldots$ to be a standard enumeration of all nondeterministic polynomial-time oracle Turing machines (NOTMs), where for each i and input of length n, N_i runs in time n^i for all oracles. All our machines run using symbols 0, 1 and blanks. Fix a deterministic oracle TM M which accepts some standard $\leq_m^p$-complete language for EXP^A for all $A \subseteq \Sigma^*$. We may assume that M runs in time 2^n. We let $\langle \cdot, \cdot \rangle$ be the standard pairing function, and we note that $x, y \leq \langle x, y \rangle$ for all $x, y \in \Sigma^*$. A set is in D^p if it can be expressed as the difference of two sets in NP.

The notations $\mathcal{R}$, $\mathcal{Q}$, $\mathcal{R}^+$ and $\mathcal{Q}^+$ denote the real numbers, the rational numbers, the positive real numbers and the positive rational numbers respectively.

2.1   Resource Bounded Measure

Classical Lebesque measure is an unusable tool in complexity classes. As these classes are all countable, everything we define in such a class has measure 0. Yet, we might wish to have a notion of "abundance" and "randomness" in complexity classes. Lutz [Lut87, Lut90] introduced the notion of *resource bounded measure*, and gave a tool to talk about these notions inside complexity classes.

Definition 1. A *martingale* d is a function from Σ^* to $\mathcal{R}^+$ with the property that $d(w0) + d(w1) = 2d(w)$ for every $w \in \Sigma^*$.

Definition 2. A p-*martingale* is a martingale $d : \Sigma^* \mapsto \mathcal{Q}^+$ that is polynomial time computable.

Definition 3. A martingale d *succeeds* on a language A if
$$\limsup_{n \mapsto \infty} d(\chi_A[0\ldots n-1]) = +\infty$$

We write $S^\infty[d] = \{A \mid d \text{ succeeds on } A\}$

Definition 4. Let $\mathcal{X}$ be a class of languages.

- $\mathcal{X}$ has p-measure 0 ($\mu_p(\mathcal{X}) = 0$) iff there exists a p-martingale d such that $\mathcal{X} \subseteq S^\infty[d]$.
- $\mathcal{X}$ has p-measure 1 ($\mu_p(\mathcal{X}) = 1$) iff $\mu_p(\overline{\mathcal{X}}) = 0$
- $\mathcal{X}$ has p-measure 0 in EXP ($\mu_p(\mathcal{X}|\text{EXP}) = 0$) iff $\mu_p(\mathcal{X} \cap \text{EXP}) = 0$
- $\mathcal{X}$ has p-measure 1 in EXP ($\mu_p(\mathcal{X}|\text{EXP}) = 1$) iff $\mu_p(\overline{\mathcal{X}} \cap \text{EXP}) = 0$

One often defines measure in EXP using p_2-measure where the martingale can use $2^{\log^{O(1)} n}$ time. All of our results also hold in this weaker model.

3 Measure of NP versus Measure of P^{NP}

In this section we concentrate on the question posed by Lutz [Lut96]. We show that relative to some oracle $\mu_p(\text{NP}) = 0$ does not imply that $\mu_p(\text{P}^{\text{NP}}) = 0$. We do this in a very strong way by constructing an oracle such that NP does not contain P-bi-immune sets and $\text{D}^p = \text{EXP}$.

Theorem 5. *There exists an oracle A such that, relative to A, NP has no P-bi-immune sets and $\text{D}^p = \text{EXP}$.*

Proof. We will code EXP into D^p on one "side" of the oracle and prevent P-bi-immunity on the other, i.e., strings in $\Sigma^*0 = \{x0 \mid x \in \Sigma^*\}$ will be used to code EXP into D^p, while strings in $\Sigma^*1 = \{x1 \mid x \in \Sigma^*\}$ will code the information to find an infinite subset of each NP set or its complement. Some diagonalization will also be necessary to force certain NP computations.

To mix coding with diagonalization, we employ a simplified version of the trick used to construct an oracle for $\text{P}^{\text{NP}} = \text{NEXP}$ [BT94, FF95]. For each x, we reserve two potential regions—*left* and *right*—in which to code $M^A(x)$, only one of which will actually be used. To code correctly in a region we must let exactly one string in the region enter A. We will code in the left region unless we have to diagonalize against some NP machine, which may necessitate adding several strings of the left region to A. If this happens, we scrap the left region and code in the right region, but we can do this only if our diagonalization hasn't already put strings of the *right* region into A.

We now proceed with the formal treatment. For every $x \in \Sigma^*$ with $|x| = n$ and $b \in \Sigma$, we call s an (x, b, left)-*coding string* (respectively, an (x, b, right)-*coding string*) if $s = xyb00$ (respectively, $s = xyb10$) for some $y \in \Sigma^*$ of length $3n$. We identify *left* and *right* with 0 and 1, respectively. We build the oracle A in stages, each successive stage extending a finite portion of A's characteristic function. If $\alpha: \Sigma^* \to \Sigma$ is some partial characteristic function, N an oracle machine, and $x \in \Sigma^*$, then the computation $N^\alpha(x)$ is defined as usual, except that when N makes any query outside domain(α), it is answered negatively. As is customary, we regard α as a set of ordered pairs. If β is another characteristic function, we write $\beta \succeq \alpha$ to mean that β extends α. Finally, define the "tower of 2's" function $t(n)$ for $n \geq 0$ by

$$t(0) = 1$$
$$t(n+1) = 2^{t(n)}.$$

Stage -1.
$\alpha_{-1} := \emptyset$.
End Stage.

Stage $n \geq 0$.
We are given α_{n-1}. Set $\alpha := \alpha_{n-1}$.

1. (*Forcing an NP computation*) If $n \neq t(k)$ for any k, then set

$$d_n := \begin{cases} \text{right} & \text{if } \alpha(s) = 1 \text{ for some } (x, b, \text{left})\text{-coding string } s \text{ with } |x| = n, \\ \text{left} & \text{otherwise,} \end{cases}$$

    and go to step 2. Otherwise, let $n = t(k)$ for some $k = \langle i, j \rangle$. If there exists a minimal $\beta \succeq \alpha$ such that both
    (a) $N_i^\beta(0^n)$ has an accepting path in which all queries are in domain(β), and
    (b) for no x with $|x| \geq n$ and no (x, b, right)-coding string s does $\beta(s) = 1$,
    then set $\alpha := \beta \cup \{(0^{n^i} 1, 1)\}$ and set $d_n := \text{right}$ (note that β is only defined on strings no longer than n^i). Otherwise, set $\alpha := \alpha \cup \{(0^{n^i} 1, 0)\}$ and set $d_n := \text{left}$.

2. (*Preserving computations of M*) For all x of length n, run $M^\alpha(x)$, and extend α with just enough 0's to "cover" all queries made by $M^\alpha(x)$ not in domain(α).

3. (*Coding computations of M*) For all $x \in \Sigma^*$ of length n, let $y \in \Sigma^*$ be the lexicographically least string (if one exists) such that $|y| = 3n$ and neither the $(x, 0, d_n)$-coding string nor the $(x, 1, d_n)$-coding string corresponding to y is in domain(α). If M^α accepts, set $\alpha := \alpha \cup \{(xy1d_n0, 1)\}$; otherwise, set $\alpha := \alpha \cup \{(xy0d_n0, 1)\}$.

4. Set α_n to be α extended with just enough 0's to cover all remaining (x, b, d)-coding strings for all $b \in \Sigma$, $d \in \{\text{left}, \text{right}\}$, and x of length n.

End Stage.

Let A be such that χ_A extends α_n for all n ($\chi_A(x) = 0$ for any $x \notin \bigcup_n \alpha_n$). For any $B \subseteq \Sigma^*$, define the language L^B by

$$L^B(x) = \begin{cases} 1 & \text{if either } B \text{ contains an } (x, 1, \text{right})\text{-coding string, or} \\ & B \text{ contains no } (x, 0, d)\text{-coding strings for any } d \in \{\text{left}, \text{right}\}, \\ 0 & \text{otherwise.} \end{cases}$$

Clearly, $L^B \in \text{coD}^{p,B}$. We now show that $L^A(x) = M^A(x)$ for all $x \in \Sigma^*$, and hence $\text{coD}^{p,A} = \text{EXP}^A = \text{D}^{p,A}$.

Pick an n large enough, and fix an input x of length n. In Step 3 of Stage n, such a y must exist: there are at most $2^n \cdot (2^{n+1} - 1)$ (x, b, d)-coding strings queried by M on inputs of length $\leq n$, because of the running time of M, and less than $n \cdot n^{\log^* n} < 2^{(\log n)^2}$ total strings queried by the N_i in Step 1 of Stages 0 through n. Thus there are less than 2^{3n} (x,b,d)-coding strings in domain(α) at Step 3 of Stage n.

The fact that
$$M^A(x) = L^A(x) \tag{1}$$
is now easily seen: first we observe that no (x, b, right)-coding string (for any $b \in \Sigma$) gets into A in Steps 1 or 2 of any stage. Thus we have two cases:

$d_n = $ left: For any $b \in \Sigma$ and $d \in \{\text{left}, \text{right}\}$, the only (x, b, d)-coding string that ever enters A does so in Step 3 of Stage n. This unique string is an $(x, 1, \text{left})$-coding string if $M^A(x)$ accepts, and is otherwise an $(x, 0, \text{left})$-coding string; thus, (1) is satisfied.

$d_n = $ right: Exactly one (x, b, right)-coding string enters A. It is an $(x, 1, \text{right})$-coding string iff $M^A(x)$ accepts. Again, (1) is satisfied.

It remains to show that NP^A has no P^A-bi-immune sets. This will be done if we can show that for any $L \in \text{NP}^A$, there exist P^A sets Q and R with Q infinite, such that $L \cap Q = R$ (or at least the symmetric difference of $L \cap Q$ and R is finite). Let $L = L(N_i^A)$ for some fixed i. Let

$$Q = \{0^n \mid (\exists j) n = t(\langle i, j \rangle)\},$$
$$R = Q \cap \{0^n \mid 0^{n^i} 1 \in A\}.$$

The sets Q and R are clearly in P^A. Pick $n = t(\langle i, j \rangle)$ for j large enough so that $t(\langle i, j \rangle + 1) = 2^n > n^i$, and consider Step 1 of Stage n. If β exists, then $N_i^A(0^n)$ accepts and $0^{n^i} 1 \in A$, so $0^n \in R$. If no such β exists, then $0^n \notin R$. To see that $N_i^A(0^n)$ rejects, we simply observe that $d_n = d_{n+1} = \cdots = d_{n^i-1} = d_{n^i} = $ left, so no (x, b, right)-coding strings enter A in any of the stages n through n^i. Therefore, A preserves our conditions on the nonexistence of β, and so $N_i^A(0^n)$ rejects.

Corollary 6. *There exists an oracle relative to which* NP *has p-measure 0 and* $\text{D}^p = \text{EXP}$ *(and thus has p-measure 1 in* E *and in* EXP*).*

We actually get something more from the construction above: relative to A, we have $\text{EXP} \subseteq (\text{NP} \cap \text{coNP})/1$. That is, EXP can be computed in $\text{NP} \cap \text{coNP}$ with one bit of advice for strings of length n, namely d_n. On input x of length n, an NP^A machine accepting $L(M^A)$ (respectively $\overline{L(M^A)}$) simply checks if there is some $(x, 1, d_n)$-coding string (respectively, some $(x, 0, d_n)$-coding string) in A.

A natural question is whether Theorem 5 and Corollary 6 are tight. It could still happen that $\mu_p(\text{NP}) = 0$ and $\mu_p(\text{P}^{\text{NP}[1]}) \neq 0$. The next theorem discards this possibility.

Theorem 7. *If* $\mu_p(\text{P}^{\text{NP}[1]}) \neq 0$ *then* $\mu_p(\text{NP}) \neq 0$.

Proof. $\mu_p(P^{NP[1]}) \neq 0$ implies that SAT is weakly $\leq_{1tt}^p$-complete for EXP. Ambos-Spies, Mayordomo, and Zheng [ASMZ96] have shown that the weakly $\leq_{1tt}^p$-completeness notion coincides with weakly $\leq_m^p$-completeness for EXP. Hence SAT is weakly $\leq_m^p$-complete for EXP and thus $\mu_p(NP) \neq 0$.

Corollary 8. *Relative to the oracle constructed in Theorem 5 it holds that* $D^p = coD^p \neq P^{NP[1]}$.

4 BPP likely has measure 0

In this section we investigate the consequences of BPP not having p-measure 0. We will see that this is unlikely since it would collapse the polynomial-time hierarchy. Hence we provide strong evidence that $\mu_p(BPP) = 0$.

Theorem 9. *If* $\mu_p(BPP) \neq 0$ *then* $EXP = MA$.

Since $MA \in \Sigma_2^p \cap \Pi_2^p$ [BM89], $EXP = MA$ implies that $PH = \Sigma_2^p$.

We use the following Theorem from Babai, Fortnow, Nisan and Wigderson [BFNW93] stating that if $EXP \neq MA$ then BPP can be simulated in subexponential time for infinitely many input lengths.

Theorem 10 [BFNW93]. *If* $EXP \neq MA$ *then for all* $L \in BPP$, *and for all* ϵ *there exists a set* $L' \in DTIME(2^{n^\epsilon})$ *such that for infinitely many* n, $L \cap \Sigma^n = L' \cap \Sigma^n$.

We will see that if BPP can be simulated in subexponential time for infinitely many input lengths, then it has p-measure 0. Taking this together with Theorem 10 yields that $EXP \neq MA$ implies that $\mu_p(BPP) = 0$, which proves Theorem 9.

Theorem 11. *If for all languages* $L \in BPP$ *there exists an* $\epsilon < 1$ *and a set* $L' \in DTIME(2^{n^\epsilon})$ *such that for infinitely many* n, $L \cap \Sigma^n = L' \cap \Sigma^n$, *then* $\mu_p(BPP) = 0$.

Proof. (Sketch) We will construct a martingale that succeeds on all sets in BPP that runs in time n^k for some fixed k. Let $L \in BPP$ and let $M_{L'}$ be the machine that runs in subexponential time and accepts L'. If we are betting on strings of length n such that $L \cap \Sigma^n = L' \cap \Sigma^n$ then we can use $M_{L'}$ to predict exactly the next bit, and hence we win 2^n times. The problem however is that we do not know for which n, $M_{L'}$ is going to be correct. We overcome this problem by the following strategy.

Assume that our initial capital is 1. We reserve 2^{-n} to bet against the strings of length n, using $M_{L'}$ to predict the next bit (i.e. whether the next string of length n is in L'). We bet everything won so far on the strings of length n to the outcome of $M_{L'}$. At the last string of length n we set aside what (if any) we have won betting on the strings of length n.

Observe that if n is a length such that $L \cap \Sigma^n = L' \cap \Sigma^n$ then we win $2^{2^n} * 2^{-n}$ and this is greater than n. So for infinitely many n we add n to our capital and hence the lim-inf of this martingale goes to infinity.

To make the construction work uniformly for all $L \in$ BPP we simulate all the DTIME(2^n) machines with a single DTIME(2^{2n}) machine allocating 2^{-i} of our initial capital to machine i (see [Lut92, May94b]).

Acknowledgment

We thank Leen Torenvliet for comments on an earlier version and Dieter van Melkebeek for helpful discussions on the writeup of the proof of Theorem 11.

References

[ASMZ96] K. Ambos-Spies, E. Mayordomo, and Xizhong Zheng. A comparison of weak completeness notions. In *Proeceedings of Eleventh Annual Conference on Computational Complexity*, pages 171 – 178, 1996.

[BFNW93] L. Babai, L. Fortnow, N. Nisan, and A. Wigderson. BPP has subexponential simulations unless EXPTIME has publishable proofs. *Computational Complexity*, 3:307–318, 1993.

[BM89] László Babai and Shlomo Moran. Proving properties of interactive proofs by a generalized counting technique. *Information and Computation*, 82(2):185–197, August 1989.

[BT94] Buhrman and Torenvliet. On the cutting edge of relativization: The resource bounded injury method. In *Annual International Colloquium on Automata, Languages and Programming*, pages 263–273, 1994.

[FF95] S. Fenner and L. Fortnow. Beyond $P^{NP} = NEXP$. In *STACS 95*, volume 900 of *Lecture Notes in Computer Science*, pages 619–627. Springer, 1995.

[LM96] J. Lutz and E. Mayordomo. Cook versus Karp-Levin: Separating completeness notions if NP is not small. *Theoretical Computer Science*, 164(1-2):141–163, 1996.

[Lut87] J. Lutz. *Resource-Bounded Category and Measure in Exponential Complexity Classes*. PhD thesis, Department of Mathematics, California Institute of Technology, 1987.

[Lut90] J. Lutz. Category and measure in complexity classes. *SIAM J. Comput.*, 19(6):1100–1131, December 1990.

[Lut92] J. Lutz. Almost everywhere high nonuniform complexity. *J. Computer and System Sciences*, 44:220–258, 1992.

[Lut96] J. Lutz. Observations on measure and lowness for Δ_2^P. In *STACS 96*, volume 1046 of *Lecture Notes in Computer Science*, pages 87 – 98. Springer, 1996.

[May94a] E. Mayordomo. Almost every set in exponential time is p-bi-immune. *Theoretical Computer Science*, 136(2):487–506, 1994.

[May94b] E. Mayordomo. *Contributions to the study of resource-bounded measure*. PhD thesis, Universitat Politècnica de Catalunya, 1994.

[RSC95] K. Regan, D. Sivakumar, and J. Cai. Pseudorandom generators, measure theory, and natural proofs. In *36th Annual Symposium on Foundations of Computer Science*, pages 26 – 35, 1995.

Randomization and Nondeterminism Are Comparable for Ordered Read-Once Branching Programs

Farid Ablayev[1]*

Dept. of theoretical cybernetics Kazan University Kazan 420008, Russia

Abstract. In [3] we exhibited a simple boolean functions f_n in n variables such that:
1) f_n can be computed by polynomial size randomized ordered read-once branching program with one sided small error;
2) any nondeterministic ordered read-once branching program that computes f_n has exponential size.
In this paper we present a simple boolean function g_n in n variables such that:
1) g_n can be computed by polynomial size nondeterministic ordered read-once branching program;
2) any two-sided error randomized ordered read-once branching program that computes f_n has exponential size.
These mean that BPP and NP are incomparable in the context of ordered read-once branching program.

1 Preliminaries

Branching programs is well known model of computation for discrete functions [14]. Many types of restricted branching programs have been investigated as important theoretical model of computations [9]. Ordered read-once branching program or ordered binary decision diagrams (OBDD) [4, 15] also important for practical computer science. They are used in circuits verifications. But many important functions cannot be computed by determinsitc read-once branching programs of polynomial size [4, 13, 8].

In [2] we introduced the model of randomized branching programs and showed that randomized ordered read-once branching programs can be more effective than determinstic ones. In [3] we defined exclusive boolean function f_n in n variables which can be computed by polynomial size randomized ordered read-once branching program, but any nondeterminstic ordered read-once branching program needs exponetial size to compute f_n. Martin Sauerhoff [10] considered function from theorem 3 [6]. He proved that this function needs (also as in the deterministic case) exponential size randomized read-once branching programs for

---

* Work done in part while visiting Steklov Mathematical Institute in Moscow. The research supported by Russia Fund for Basic Research 96-01-01962. ablayev@ksu.ru http://www.ksu.ru/~ablayev

one-sided error. In this paper we presented exclusive function g_n which is "simple" for nondeterminstic ordered read-once branching programs, but is "hard" for randomized read-once branching programs with two-sided error of computation.

Together with the result from [3] this proves that complexity classes BPP and NP are incomparable in the context of ordered read-once branching programs.

Note that the results of the paper for ordered read-once branching programs are true for a more common model — weak-ordered branching program that we define in the paper. Informaly speaking weak-ordered property for branching program P means existence of partition of its set $\{x_1, x_2, \ldots, x_n\}$ of variables into two parts X_1 and X_2, $X_1 \cap X_2 \neq \emptyset$, such that for any computation path of P the following is true. If a variable from X_2 is tested then no variable from X_1 can be tested in the rest part of this path.

A *deterministic* branching program P for computing a function $g : \{0,1\}^n \to \{0,1\}$ is a directed acyclic multi-graph with a distinguished source node s and a distinguished sink node t. The out degree of of each non-sink node is exactly 2 and the two outgoing edges are labeled by $x_i = 0$ and $x_i = 1$ for variable x_i associated with the node. Call such node an x_i-node. The label "$x_i = \delta$" indicates that only inputs satisfying $x_i = \delta$ may follow this edge in the computation. The branching program P computes function g in the obvious way: for each $\sigma \in \{0,1\}^n$ we let $f(\sigma) = 1$ iff there is a directed $s - t$ path starting in the source s and leading to to the accepting node t such that all labels $x_i = \sigma_i$ along this path are consistent with $\sigma = \sigma_1, \sigma_2, \ldots, \sigma_n$.

The branching program becomes *nondeterministic* [5] if we allow "guessing nodes" that is nodes with two outgoing edges being unlabeled. Unlabeled edges allow all inputs to produced. A nondeterministic branching program P computes a function g, in the obvious way; that is, $g(\sigma) = 1$ iff there exists (at least one) computation on σ starting in the source node s and leading to the accepting node t.

Define a *randomized* branching program [2] as a one which has in addition to its standard inputs specially designated inputs called "random inputs". When values of these "random inputs" are chosen from the uniform distribution, the output of the branching program is a random variable.

Say that a randomized branching program (a, b)-computes a boolean function f if it outputs 1 with probability at most a for input σ such that $f(\sigma) = 0$ and outputs 1 with probability at least b for inputs σ such that $f(\sigma) = 1$.

As usual for a branching program P (deterministic or random), we define size(P) (complexity of the branching program P) as the number of internal nodes in P. Define, following [5], the size(P) of the nondeterminstic branching program P as the number of internal nodes in P minus the number of guessing nodes.

Read-once branching programs is branching program in which for each path each variable is tested no more than once. An ordered read-once branching program is a read-once branching program which respects a fixed ordering π of the variables, i.e. if an edge leads from an x_i-node to an x_j-node, the condition $\pi(i) < \pi(j)$ has to be fulfilled.

2 Results

We specify a boolean function f_n of $n = 4l$ variables as follows. For a sequence $\sigma \in \{0,1\}^{4l}$ call odd bits a "type" bits and even bits a "value" bits. Say that even bit $\sigma_i \in \sigma$, $i \in \{2, 4, \ldots, 4l\}$, has type 0 (1) if corresponding odd bit σ_{i-1} is 0 (1). For a sequence $\sigma \in \{0,1\}^{4l}$ denote σ^0 (σ^1) subsequence of σ that consists of all even bits of type 0 (1).

For every $\sigma \in \{0,1\}^n$ boolean function $f_n : \{0,1\}^n \to \{0,1\}$ is defined as $f_n(\sigma) = 1$ iff $\sigma^0 = \sigma^1$.

Definition 1. Call branching program a π-weak-ordered branching program if its respects a partition π of variables $\{x_1, x_2, \ldots, x_n\}$ into two parts X_1 and X_2 such that if an edge leads from an x_i-node to an x_j-node, where $x_i \in X_t$ and $x_j \in X_m$, then the condition $t \leq m$ has to be fulfilled.

Call branching program P an weak-ordered if it is π-weak-ordered for some partition π of the set of variables of P into two sets.

Clearly that ordered read-once branching program is also weak-ordered. We proved the following result in [3] (we use here a restrictive variant of this result).

Theorem 2. *For the function f_n the following is true:*

1. f_n can be $(\varepsilon(n), 1)$-computed by randomized ordered read-once branching program of the size

$$O\left(\frac{n^6}{\varepsilon^3(n)} \log^2 \frac{n}{\varepsilon(n)}\right).$$

2. Any nondeterministic ordered read-once branching program that computes function f_n has the size no less than $2^{n/4-1}$.

Now define function g_n which is "hard" for randomized computation but is "simple" for nondeterminstic computation for our model of branching program. This boolean function presented in [11]. Let n be an integer and let $p[n]$ be the smallest prime greater or equal to n. Then, for every integer s, let $\omega_n(s)$ be defined as follows. Let j be the unique integer satisfying $j = s \bmod p[n]$ and $1 \leq j \leq p[n]$. Then, $\omega_n(s) = j$, if $1 \leq j \leq n$, and $\omega_n(s) = 1$ otherwise.

For every n, the boolean function $g_n : \{0,1\}^n \to \{0,1\}$ is defined as $g_n(\sigma) = \sigma_j$, where $j = \omega_n(\sum_{i=1}^n i\sigma_i)$.

We will use the following notations in the rest part of the paper. Let $h : \{0,1\}^n \to \{0,1\}$ be a boolean function. Consider a partition π of variables $\{x_1, x_2, \ldots, x_n\}$ into two parts $X_1 = \{x_i : i \in I\}$ and $X_2 = \{x_j : j \in J\}$, where $I \subset \{1, 2, \ldots, n\}$, $|I| = l$ and $J = \{1, 2, \ldots, n\} \setminus I$, $|J| = t$.

Denote L, R sets of binary sequences of length l and t with indexes from I and J respectively. For $u \in L$ and $w \in R$ let (u, w) mean the sequence σ from $\{0,1\}^n$ in wich bits with indexes from I respectively J have the same values as in u respectively w. We will also use the notation $h(u,w)$ instead of $h(\sigma)$ where it will be convenient.

Consider one-way randomized communication computation. We use the following standard model of one-way randomized communication computation for function h. Two players A and B receive respectively $u \in L$ and $w \in R$. In the randomized one-way model, A sends the messages $\beta_1, \beta_2, ..., \beta_d$ with probabilities $p_1, p_2, ..., p_d$ respectively ($\sum_{i=1}^{d} p_i = 1$). B, on the receipt of β_i, outputs 1 with probability q_i and 0 with probability $1 - q_i$. The probability distribution on the set of messages sent by A is entirely determined by the input at A alone, and is not influenced by the input at B. Similarly, the probabilities q_i at B depend only on its input and the message β_i received.

In the computation $T_\phi(u,w)$, the probability of outputting the bit $b = 1$ is $\sum_{i=1}^{d} p_i(u) q_i(w)$ and the bit $b = 0$ is $1 - \sum_{i=1}^{d} p_i(u) q_i(w)$.

Let $p = \frac{1}{2} + \varepsilon$ for $0 \le \varepsilon \le 1/2$. Say that the probabilistic protocol ϕ p-computes a function h if for every input $\sigma = (u,w)$ it holds that $h(\sigma) = b$ iff the probability of outputting the bit b in the computation $T_\phi(u,w)$ is no less than p.

Let a set $U \subseteq \{0,1\}^n$ be such that $U = L \times R$. The randomized communication complexity $C(\phi)$ of the probabilistic protocol ϕ on the inputs from U is $\lceil \log |M(\phi)| \rceil$, where $M(\phi)$ is the set of messages used by ϕ during computations on inputs from U. For $p \in [1/2, 1]$ the randomized communication complexity $PC_{p,\pi}^{U}(h)$ of a boolean function h is

$$\min\{C(\phi) : \text{protocol } \phi \text{ } p\text{-computes } h \text{ for the partion } \pi \text{ of inputs from } U\}.$$

The proof of following lemma is based on simulation technique of weak-ordered branching program by communication protocol and is similar to simulation technique from [1] (lemma 6.1).

Lemma 3. *Let $\varepsilon \in [0, 1/2]$, $p = 1/2 + \varepsilon$. Let randomized π-weak-ordered branching program P $(1-p, p)$-computes function $h : \{0,1\}^n \to \{0,1\}$. Let $U \subseteq \{0,1\}^n$ be such that $U = L \times R$, where L and R are defined in according to partition π of inputs. Then*

$$size(P) \ge 2^{PC_{p,\pi}^{U}(h)-1}.$$

Proof. Describe the following communication protocol Φ, which p-computes function h for the partion π of inputs.

Let $\sigma \in U$ be a valuation of x, $\sigma = (u,w)$, $u \in L$, $w \in R$. Players A and B receive respectively u and w in according to partition π of inputs. Let $v_1, \ldots, v_d$ be all internal nodes of P that are reachable during paths of computation on the part u of input σ with non zero probabilities $p_1(u), \ldots, p_d(u)$.

During the computation on the input u, player A sends node v_i with probability $p_i(u)$ to player B. Player B on obtaining message v_i from A starts its computation (simulation of the branching program P) from the node v_i on the part w of the nput σ.

From the definition of the protocol Φ results the statement of the lemma. ∎

We use the lower bound for probabilistic one-way complexity from [1] in the proof of the theorem 6 below. Recall notations and the statement we need from [1] in the convinient for us form.

For $U = L \times R$ with a boolean function h we associate a $|L| \times |R|$ communication matrix CM whose (u,w)-th entry, $CM[u,w]$ is $h(u,w)$ for all $(u,w) \in L \times R$. As it is mentioned in [16] the one-way deterministic communication complexity $DC_\pi^U(h)$ for partition π of inputs from U of a boolean function h is easily seen to be $\lceil \log(nrow(CM)) \rceil$, where $nrow(CM)$ is the number of distinct rows of communication matrix CM of the function h.

Consider w.l.g. the case when all rows of CM are different, $nrow(CM) = |L|$.

Choose a $Y \subseteq R$ such that for an arbitrary two words $u, u' \in L$ there exists a word $y \in Y$ such that $h(u,y) \neq h(u',y)$. The set Y is called the control set for the matrix CM.

Denote
$$ts(CM) = \min\{|Y| : Y \text{ is a control set for } CM\}.$$

It is evident that $\lceil \log nrow(CM) \rceil \leq ts(CM) \leq nrow(CM)$.

For number $p \in [1/2, 1]$, define $pcc_p^U(h) = \frac{ts(CM)}{\log nrow(CM)} H(p)$, where $H(p) = -p \log p - (1-p) \log(1-p)$ is the Shannon entropy. Call $pcc_p^U(h)$ the p-probabilistic communication characteristic of the function h.

Theorem 4. [1] Let $\varepsilon \in [0, 1/2]$, $p = 1/2 + \varepsilon$. Let $U \subseteq \{0,1\}^n$ be such that $U = L \times R$, where L and R are defined in according to partition π of inputs of function $h : \{0,1\}^n \to \{0,1\}$. Then

$$PC_{p,\pi}^U(h) \geq DC_\pi^U(h)(1 - pcc_p^U(h)) - 1.$$

In the proof of the theorem 6 below we use the following result from number theory (see [7] and [12] for additional citation).

For every natural number n let $p(n)$ be the smallest prime greater or equal than n. Consider $Z_{p(n)}$ the field of the residue classes modulo p.

Lemma 5. For every n large enough, the following is true. If $A \subseteq Z_{p(n)}$ and $|A| \geq 3\sqrt{n}$, then, for every $t \in Z_{p(n)}$, there is a subset $B \subseteq A$ such that the sum of the elements of B is equal to t.

Theorem 6. Let $\varepsilon \in [0, 1/2]$, $p = 1/2 + \varepsilon$. Then for arbitrary $\delta > 0$ for every n large enough it holds that any randomized ordered read-once branching program that $(1-p, p)$-computes function g_n has the size no less than

$$1/4 \left(\frac{2^{n - \lceil 3\sqrt{n} \rceil}}{n} \right)^{1 - (1+\delta)H(p)}.$$

Proof. Let P be a randomized ordered read-once branching program with an ordering τ of variables which computes function g_n. For ordering $\tau = \{i_1, i_2, \ldots, i_n\}$ consider the partition π of variables x of g_n into two parts $X_1 = \{x_{i_1}, \ldots, x_{i_l}\}$ and $X_2 = \{x_{i_{l+1}}, \ldots, x_{i_n}\}$, where $l = n - \lceil 3\sqrt{n} \rceil$. Denote $t = \lceil 3\sqrt{n} \rceil$.

Describe below a subset $U \subset \{0,1\}^n$ in the form $U = L \times R$ where $|L| = l$, $|R| = t$.

Denote by I and J sets of indexes of variables from sets X_1 and X_2 respectively. For $s \in \{1, \ldots, n\}$ denote L_s a subset of binary sequences of length l with indexes from I such that $L_s = \{u : \omega_n(\sum_{i \in I} i u_i) = s\}$. Denote L a maximum among sets $L_1, \ldots, L_n$.

$$|L| = \max_{s \in \{1, \ldots, n\}} \{|L_s|\}.$$

Clearly that

$$|L| \geq \frac{2^{n - \lceil 3\sqrt{n} \rceil}}{n}.$$

Let $L = L_s$. Then denote $R = \{w : \omega_n(\sum_{j \in J} j w_j + s) = k, k \in I\}$. From the definition of R we have the following properties:
1) $|R| = l$;
2) for arbitrary u and u' from L there exists $w \in R$ such that $g_n(u, w) \neq g_n(u', w)$.

We will prove the second property (the first one is evident). Let $i \in I$ be an index such that i-th bits in sequences u and u' are different, $u_i \neq u'_i$. From the lemma 5 it follows that for every n large enough, for our number s and the number i there exists a sequence $w \in R$ such that $s + \sum_{j \in J} j w_j = i \mod p(n)$. Then from the definition of g_n it follows that $g_n(u, w) \neq g_n(u', w)$.

Now define set U as $U = L \times R$. From the above it follows that for the set U $|L| \times |R|$ communication matrix CM of g_n has the following properties:
1) $nrow(CM) = |L|$;
2) the set R is the control set for CM.

This means that $DC^U(g_n) = \log |L|$ and that for p-probabilistic communication characteristic of $pcc_p^U(g_n)$ of function g_n it is true that

$$pcc_p^U(g_n) = (l/\log |L|) H(p) \leq ((n - \lceil 3\sqrt{n} \rceil)/(n - \lceil 3\sqrt{n} \rceil - \log n)) H(p).$$

From this it follows that for arbitrary $\delta > 0$ for every n large enough it holds that

$$pcc_p^U(g_n) \leq (1 + \delta) H(n).$$

From the above property and the theorem 4 it follows that for every n large enough the following is true

$$PC_p^U(g_n) \geq (n - \lceil 3\sqrt{n} \rceil - \log n)(1 - (1 + \delta) H(p)) - 1.$$

From this and the lemma 3 the lower bound for $size(P)$ results. ∎

Note that in the proof of the theorem 6 from the property of P that it is ordered read-once we use only the following fact. Set x of variables of P can be partition into two parts X_1 and X_2 such that $|X_1| = n - \lceil 3\sqrt{n} \rceil$ and $|X_2| = \lceil 3\sqrt{n} \rceil$. The cardinality of X_2 is essential for application of lemma 5. This means that the following statement is true.

Theorem 7. *Let $\varepsilon \in [0, 1/2]$, $p = 1/2 + \varepsilon$. Let P be a randomized π-weak-ordered branching program that $(1 - p, p)$-computes function g_n. Let π be a partition of x in two two parts X_1, X_2 such that $|X_2| = t \geq \lceil 3\sqrt{n} \rceil$ and $|X_1| = l = n - t$. Then for arbitrary $\delta > 0$ for every n large enough it holds that*

$$size(P) \geq 1/4 \left(\frac{2^l}{n} \right)^{1-(1+\delta)H(p)}.$$

Theorem 8. *There is polynomial size nondeterministic ordered read-once branching program that computes function g_n.*

Proof. The proof is simple. For arbitrary input σ nondeterministic ordered read-once branching program P that computes function g_n works as follows. On the first (nondeterminstic) phase P nondeterministicaly selects number $s \in \{1, \ldots, n\}$. Then on the second (deterministic) phase P reads inputs in the order $x_1, \ldots, x_n$. During computation path on input σ P 1) counts number $a = \omega_n(\sum_{i=1}^{n} i\sigma_i)$ and 2) store s-ths bit σ_s. If $a = s$ then P ouputs bit σ_s of the input σ else P outputs 0. Clearly, that P has polynomial size. ∎

Acknowledgments. I would like to thank Sasha Razborov for his invitation to me to spend my sabbatical semester in Steklov Mathematical Institute and for a number of valuable discussions on the subject of the paper, to participants of Complexity seminar in Steklov Mathematical Institute and Moscow University for listening to results presented in this paper.

References

1. F.Ablayev, Lower bounds for one-way probabilistic communication complexity and their application to space complexity, *Theoretical Computer Science*, 157, (1996), 139-159.
2. F. Ablayev and M. Karpinski, On the power of randomized branching programs, *in Proceedings of the ICALP'96, Lecture Notes in Computer Science, Springer-Verlag*, 1099, (1996), 348-356.
3. F. Ablayev and M. Karpinski, On the power of randomized branching programs, *manuscript* (generalization of ICALP'96 paper results for the case of pure boolean function), available at http://www.ksu.ru/~ablayev
4. R. Bryant, Symbolic boolean manipulation with ordered binary decision diagrams, *ACM Computing Surveys*, 24, No. 3, (1992), 293-318.

5. A. Borodin, A. Razborov, and R. Smolensky, On lower bounds for read-k-times branching programs, *Computational Complexity*, 3, (1993), 1-18.
6. Y. Breitbart, H.Hunt III, and D. Rosenkratz, On the size of binary decision diagrams representing Boolean functions, *Theoretical Computer Science*, 145, (1995), 45-69.
7. J. Dias da Silva and Y. Hamidoune, Cyclic spaces for Grassmann derivatives and additive theory, *Bull. London Math. Soc.*, 26, (1994), 140-146.
8. S. Ponsio, A lower bound for integer multiplication with read-once branching programs, *Proceedings of the 27-th STOC*, (1995), 130-139.
9. A. Razborov, Lower bounds for deterministic and nondeterministic branching programs, *in Proceedings of the FCT'91, Lecture Notes in Computer Science, Springer-Verlag*, 529, (1991), 47–60.
10. M. Sauerhoff, Lower bounds for the RP-OBDD-Size, *manuscript*, personal communication.
11. P. Savicky, S. Zak, A large lower bound for 1-branching programs, *Electronic Colloquium on Computational Complexity*, Revision 01 of TR96-036, (1996), available at http://www.eccc.uni-trier.de/eccc/
12. P. Savicky, S. Zak, A hierarchy for $(1,+k)$-branching programs with respect to k, *Electronic Colloquium on Computational Complexity*, TR96-050, (1996), available at http://www.eccc.uni-trier.de/eccc/
13. J. Simon and M. Szegedy, A new lower bound theorem for read-only-once branching programs and its applications, *Advances in Computational Complexity Theory*, ed. Jin-Yi Cai, DIMACS Series, 13, AMS (1993), 183-193.
14. I. Wegener, *The complexity of Boolean functions*. Wiley-Teubner Series in Comp. Sci., New York – Stuttgart, 1987.
15. I. Wegener, Efficient data structures for boolean functions, *Discrete Mathematics*, 136, (1994), 347-372.
16. A. C. Yao, Some Complexity Questions Related to Distributive Computing, *in Proc. of the 11th Annual ACM Symposium on the Theory of Computing*, (1979), 209-213.

Checking Properties of Polynomials *

(Extended Abstract)

Bruno Codenotti,[1] Funda Ergün,[2] Peter Gemmell,[3] and S Ravi Kumar[2]

[1] IMC-CNR, Via S. Maria 46, 56126-Pisa, Italy. (codenotti@imc.pi.cnr.it)
[2] Cornell University, Ithaca, NY 14853. ({ergun, ravi}@cs.cornell.edu)
[3] Sandia National Labs, Albuquerque, NM 87185. (psgemme@cs.sandia.gov)

Abstract. In this paper we show how to construct efficient checkers for programs that supposedly compute *properties* of polynomials. The properties we consider are roots, norms, and other analytic/algebraic functions of polynomials. In our model, both the program Π and the polynomial p are available to the checker each as a black box. We show how to check programs that compute a specific root (e.g., the largest) or a subset of roots of the given polynomial.

The checkers, in addition to never computing the root(s) themselves, strive to minimize both the running time (preferably $o(\deg^2 p)$) and the number of black box evaluations of p (preferably $o(\deg p)$). We obtain deterministic checkers when a separation bound between the roots is known and probabilistic checkers when the roots can be arbitrarily close. We then extend the checkers to handle the situations when the program Π returns an approximation to the root and when the evaluation of the polynomial p is approximate. Our results translate into efficient checkers for matrix spectra computations both in the exact and approximate settings, operating in the library model of [BLR93]. Next we show that the usual characterization of norms using the triangle inequality is not suited for self-testing in the exact case, but surprisingly, could be used in the approximate case.

Our results are complementary to most of the existing results on testing polynomials. The testers in the latter have the goal of determining whether a program computes a polynomial of given degree, whereas we are interested in checking the properties of a *given* polynomial.

1 Introduction

The paradigm of program checking and its extensions, self-testing, and self-correcting, have received considerable attention (e.g., [Blu88, BK89, BLR93, Lip91, GLR+91, RS96, ABC+93, GGR96, EKR96].) The results in this field

---

* This work was done while the first, second, and fourth authors were visiting Sandia National Labs. The second and fourth authors are also supported by the NSF Career Award CCR-9624552, the Alfred P. Sloan Research Award, and the NSF grant DMI-91157199.

have practical value as tools for efficient verification of the correctness of programs. Furthermore, they have been applied to develop efficient probabilistically checkable proofs [ALM+92].

In this paper we investigate the problem of checking and testing (both in the exact and approximate cases) programs that compute *properties* (i.e., functions or relations) of polynomials. The properties we consider include the set of all roots, the largest root, the smallest root, norms, multiplication, differentiation, resultants, etc. Our checkers for root-finding problems only assume an oracle access to the polynomial p. Note that this is a weaker requirement than the availability of an explicit representation of p. This model lets us view the checkers for matrix spectra computations in the library setting of [BLR93]. In this framework, checkers call already tested programs in the library, counting each call as a unit time call. Such calls naturally correspond to the evaluation of the polynomial in our model. Consequently, it is imperative that the number of evaluations of p be minimized.

Our approach is complementary to previous work on checking and testing polynomials. The main difference is the following. Most of the existing results are concerned with checking/testing programs purportedly *evaluating* polynomials. In this paper we are interested in checking programs that take a polynomial as an input and compute its properties.

Our Results. We describe efficient checkers for programs that compute one, few, all, or specific roots (e.g., the largest) of a polynomial p. We address at length the checking of programs computing the largest root. For this problem, we construct some checkers that run in time $o(\deg^2 p)$ and make only $o(\deg p)$ calls to p (thus ruling out an explicit interpolation) using powerful tools from analysis. This translates into more efficient checkers than ones offered by several other methods that use explicit interpolation. We obtain deterministic checkers when a separation bound between the roots is known and probabilistic checkers when the roots can be arbitrarily close (Section 3). We also consider the situations where (i) the program $\Pi(p)$ is computing an approximation to the root(s) of polynomial p and (ii) the oracle returns an approximate evaluation of the polynomial p (Section 4). In these cases, we provide checkers for some of the problems.

Next we consider programs that claim to compute some (unspecified) norm on the domain (Section 5). There are several norms for polynomials (see [Z93]); the goal is to test whether there exists a norm that agrees with the program on most inputs. We show that the standard characterization of norms (using triangle inequality) cannot be used to construct exact testers. I.e., there are extremely "bad" programs (those that do not agree with any one norm for any non-trivial fraction of the inputs) that still pass the test. The same test, however, can be used to verify that the program *approximates* some norm for a non-trivial fraction of the domain. Our result, which applies to norms defined on any domain, is intriguing because most of the current techniques for testing use an exact characterization to build an exact tester and an approximate characterization (where the equalities are relaxed to approximations, see [ABC+93, EKR96] for further exposition) to build an approximate tester. The exact characterization for norms is too lenient to lead to an exact tester, however, surprisingly, is strong enough for an approximate one (even without resorting to an approximate characterization). Additionally, this is the first instance where an unbounded inequality (i.e., an inequality of the form $|h(\cdot)| > 0$, where h is an expression) has been addressed in testing.

The nature of these properties entails the use of techniques from several disciplines (like numerical and complex analysis, geometry, and in particular, geometry of polynomials) that are new to checking.

Applications. We show how to check programs that perform matrix spectra computations, which are fundamental in scientific computing (Section 6). We exploit the fact that the eigenvalues of a matrix are the roots of its characteristic polynomial. The characteristic polynomial is evaluated using a library program for the determinant that has been tested, for instance using the exact checker of [Kan90] or the approximate checker of [ABC+93]. Several vital parameters in control theory (e.g., stability of a system) are related to the location of the roots of certain polynomials. Programs that compute these parameters are very common in practice [BCL82]; our checkers could be used to check such programs. Another application of property testing of polynomials is in verifying parts of computational algebra systems. We have taken an initial step in this direction but many interesting questions remain.

Previous and Related Work. The problem of testing root-finding programs is considered as early as 1975 in [JT75]. Here, the authors lay down some concrete requirements for an efficient testing of such programs. The setting proposed, however, is very different from ours and is mostly heuristic and informal.

A number of papers deal with testing whether a program is computing a low-degree polynomial in the exact [GLR+91, AS92, GLR+91, RS96] and approximate [EKR96] settings. Testing certain polynomial functions like polynomial multiplication and FFT is investigated in [BLR93, Erg95]. Checkers for several linear algebra computations like matrix rank, determinant, matrix multiplication are given in [Fre79, BK89, Kan90, BLR93]. Approximate testers for several linear algebra computations can be found in [ABC+93]. Testing graph properties is considered in [GGR96].

2 Preliminaries

Our Model. We consider properties f of polynomials p. In this context, we assume that properties are relations such as those binding p to one or more of its roots. For shorthand, we sometimes use "$f(p)$" to denote *one* of the values to which f binds p.

Although checkers are defined for properties and are otherwise independent of the programs that they check, we sometimes refer to a checker for a program Π. Implicit in these references is that the checker is for the property f that Π purports to compute, i.e. that the checker verifies that $\Pi(x) \in f(x)$ for the input x in question.

Definition 1. Let Π be a program that purports to compute a property f. Let $\beta > 0$ be a security parameter. Then, a $(q(n), t(n); \epsilon_1, \epsilon_2)$-*checker* for f is a (probabilistic) oracle program $T^{\Pi,p}$ that has oracle access to both Π and p such that it

1. makes $O(q(n))$ oracle accesses to p (i.e., it evaluates p at $O(q(n))$ points)
2. runs in time $O(t(n))$, counting oracle calls as one unit of time
3. if $\exists y \in f(p) : |\Pi(p) - y| < \epsilon_1$, outputs "PASS" with probability $\geq 1 - \beta$
4. if $\forall y \in f(p) : |\Pi(p) - y| > \epsilon_2$, outputs "FAIL" with probability $\geq 1 - \beta$.

To simplify notation, we adopt the following conventions: (i) if $q(n) = t(n)$, we omit one of them, (ii) if $\epsilon_1 = \epsilon_2$, we omit one of them, and (iii) if $\epsilon_1 = \epsilon_2 = 0$, we omit both from the checker's parameters.

Note that the above model is more general than the standard checking model in that p is available as an oracle rather than in an explicit form. (It is often unrealistic or less efficient to assume that an explicit representation of p is available to T.) We will see that this model (i) captures the library setting of [BLR93] and helps us build efficient checkers, (ii) is useful in our applications to checking matrix spectra computations, and (iii) elegantly extends our checkers to the approximate setting.

Variations of the Model. Our model permits the following variations and their combinations: (i) The program purports to return an approximation to f. In this case, the program is denoted by $\tilde{\Pi}$. (ii) Each oracle call to evaluate p returns an approximation. In this case, the oracle is denoted by $\tilde{p}$. and (iii) p is "close" to a polynomial (as in the PCP setting). We will address the first and second variations. They make the problem more appealing since in practice we are seldom guaranteed an exact answer to any numerical question. In this paper, we will call a checker for the second scenario an *approximate checker*.

Self-Testing, Self-Correcting, Checking, and Libraries. Self-testing ensures that Π equals the target function f (from a function family F) on most inputs. A self-tester usually has two stages [BLR93]: (i) testing if Π is a member of F (the *property test*) and (ii) testing if Π is *the* specific member, i.e., f (the *equality test*). Self-correction involves taking a Π that is correct on most inputs and converting it into a program that is correct on all inputs. A self-tester together with a self-corrector gives a result-checker. In the library setting, a collection of previously checked programs is used to build checkers for new functions. For details see [BLR93].

Mathematical Notation. We consider polynomials over a field $\mathcal{F}$. Let $\mathbb{R}$ denote the real numbers and $\mathbb{C}$ denote the complex numbers.

Let $\mathcal{F}_n[x]$ denote the ring of polynomials of degree $\leq n$ with coefficients from $\mathcal{F}$. Let $p(\cdot)$ be a degree n polynomial (i.e., $p \in \mathcal{F}_n[x]$). Assuming p factors completely in $\mathcal{F}$, let the roots of p be $|\lambda_1| \geq \cdots \geq |\lambda_n|$. When $\mathcal{F} = \mathbb{R}$, we call p a *real polynomial* and if all the roots of p are real, we call p a *real-root polynomial*.

For any $a \in \mathbb{C}$, let $\bar{a} \in \mathbb{C}$ denote its complex conjugate. For any curve (line segment, interval) $\mathcal{C}$, let $|\mathcal{C}|$ denote its length and int $\mathcal{C}$ its interior. For $x, y \in \mathbb{R}^2$, let $\overline{xy}$ denote the line segment between x and y. A *convex* curve in $\mathbb{R}^2$ is called a *contour* if it encloses the origin. A curve $\mathcal{C} : [0, 2\pi) \to \mathbb{R}^2$ is called *star-shaped* if it is an injective closed curve.

Let $p(x) = \prod_{i=1}^{n}(x - \lambda_i) = \sum_{i=0}^{n} a_i x^i, a_n = 1$. Then, it easily follows that $p'(x) = \sum_{i=1}^{n} \prod_{j \neq i}(x - \lambda_j)$.

$\mathbf{g(x)}$: We will use $\mathbf{g(x)}$ to denote $p'(x)/p(x) = \sum_{i=1}^{n} 1/(x-\lambda_i) = d \ln|p(x)|/dx$.

$\lambda_{\inf}, \lambda_{\max}$: Cauchy's inequality [BCL82] gives bounds on the roots of p as $\lambda_{\inf} = |a_n|/(|a_n| + \max_{i=0}^{n-1}\{|a_i|\}) < |\lambda_{\min}| \leq |\lambda_{\max}| < 1 + \max_{i=1}^{n}\{|a_i|\}/|a_0| = \lambda_{\sup}$.

δ: A separation bound between the roots of p is given by [BCL82] as $\delta = \min_{\lambda_i \neq \lambda_j} |\lambda_i - \lambda_j| > \sqrt{3} n^{-(n+1)/2} \|p\|^{1-n} \sqrt{\operatorname{disc}(p)}$, where the discriminant $\operatorname{disc}(p) = |\prod_{i \neq j}(\lambda_i - \lambda_j)| = |\operatorname{res}(p, p')|$ and $\|p\|^2 = \sum_{i=0}^{n} |a_i|^2$ [Z93]. Here, resultant $\operatorname{res}(p, q) = \prod_{i=1}^{\deg p} q(\lambda_i)$ where λ_i is a root of p. Some of our checkers assume that a lower bound on δ is known.

Problem Definitions. Let Π be a program that purports to compute one or more roots of p and let $\{\mu_i\}$ be the value(s) computed by Π. Let $\{\lambda_i\}$ be the actual root(s) that Π should have output. (Thus for instance, $\Pi_{\max}$, which purports to compute $\lambda_{\max}$, outputs $\mu_{\max}$ to be the largest root.) Given a polynomial p of degree n, let:

- $\mathcal{R}_1(p)$ be a relation mapping p to any one of its roots. We refer to programs that purport to compute a value $\mathcal{R}_1(p)$ as Π_1.
- $\mathcal{R}_r(p)$ be a relation mapping p to any r of its roots. We refer to programs that purport to compute a set $\mathcal{R}_r(p)$ as Π_r.
- $\mathcal{R}_{\langle k \rangle}(p)$ be the kth largest root of p in absolute value (i.e., λ_k). $\Pi_{\langle k \rangle}$ refers to a program that purports to compute $\mathcal{R}_{\langle k \rangle}$.
- $\mathcal{R}_{\max} = \mathcal{R}_{\langle 1 \rangle}, \mathcal{R}_{\min} = \mathcal{R}_{\langle n \rangle}$ and $\Pi_{\max}, \Pi_{\min}$ refer to programs that supposedly compute $\mathcal{R}_{\max}, \mathcal{R}_{\min}$.

In general, we use a tilde to denote programs that purport to return approximations to the corresponding exact relation (e.g., $\widetilde{\Pi}_{\max}$).

3 Checking Roots: Exact Setting

Checkers for $\mathcal{R}_1, \mathcal{R}_r, \mathcal{R}_n, \mathcal{R}_{\max}$.

Theorem 2. *Let* $|\mathcal{F}| \geq n + \Omega(n)$. *There is:*

1. *a (1)-checker for $\mathcal{R}_1(p)$,*
2. *a $(1,n)$-checker for $\mathcal{R}_n(p)$, and*
3. *a $(\min\{r, n-r\}, \max\{r, n-r\})$-checker for $\mathcal{R}_r(p)$.*

In the exact setting, given $\mu_{\max}$, it is trivial to verify that it is *a root of p*. It is non-trivial to verify the maximality claim. Theorem 3 below states a checker for $\Pi_{\max}(p)$. In the next section, we will show more efficient checkers (that avoid explicit interpolation) for $\Pi_{\max}(p)$.

Theorem 3. $\forall \epsilon > 0$, *there is an $(n, n^2; \epsilon)$-checker for $\mathcal{R}_{\min}(p)$. If p is a real-root polynomial, $\forall \epsilon > 0$, there is an $(n, n^2; \epsilon)$-checker for $\mathcal{R}_{\max}(p)$.*

A checker for $\Pi_{\max}$ is constructed from a checker for $\Pi_{\min}$ in an obvious manner by observing that $1/\lambda_i$ are the roots of $x^n p(1/x)$. Note that the checkers given by Theorem 3 can also be used to check $\Pi_{\langle k \rangle}$.

Improved Checkers for $\mathcal{R}_{\max}$: δ known. For the rest of this section, we will take either $\mathcal{F} = \mathbb{C}$ or $\mathcal{F} = \mathbb{R}$. We use the following theorems from complex analysis (see [Con78]). Let $n(\mathcal{C}; z)$ be the number of times $\mathcal{C}$ "winds" around the point $z \in \mathbb{C}$.

Theorem 4 Cauchy's Residue Theorem. *Let G be an open subset of the plane and $f : G \to \mathbb{C}$ an analytic function. If $\mathcal{C}$ is a closed rectifiable (finite length) curve in G such that $n(\mathcal{C}; z) = 0 \; \forall z \in \mathbb{C} \backslash G$, then for $\lambda \in G \backslash \mathcal{C}$, $2\pi i f(\lambda) n(\mathcal{C}; \lambda) = \int_\mathcal{C} f(z)/(z - \lambda) dz$.*

Corollary 5. *Let G be an open subset of the plane and f be an analytic function on G with zeros $\lambda_1, \ldots, \lambda_n$ (repeated according to multiplicity). If $\mathcal{C}$ is a closed rectifiable curve in G which does not pass through any point λ_k, and $n(\mathcal{C}; z) = 0, \forall z \in \mathbb{C} \backslash G$, then $\int_\mathcal{C} f'(z)/f(z) \, dz = 2\pi i \sum_{i=1}^n n(\mathcal{C}; \lambda_i)$ counts (with multiplicities) the number of roots of $f(z)$ within $\mathcal{C}$.*

Theorem 6. *There is a $((|\mu_{\max}|/\delta)^{3/2}; \delta/2)$-checker for $\mathcal{R}_{\max}(p)$.*

Proof. If $\mathcal{C}$ is a circle, then by Corollary 5, $\int_\mathcal{C} p'(z)/p(z) dz$ computes $2\pi i$ times the number of zeros of p that are within $\mathcal{C}$ (noting that $\mathcal{C}$ winds once around each root). So, our goal is to check that $\int_\mathcal{C} p'(z)/p(z) dz = 2\pi n$, where $\mathcal{C}(t) = (|\mu_{\max}| + \delta/2)e^{it}$. Recall that $g(z) = p'(z)/p(z)$. We compute an approximation S to $\int_\mathcal{C} g(z) dz$, which must satisfy $|S - 2\pi n| < \pi$. If we use trapezoidal rule, we have $\left|\int_\mathcal{C} g(z) dz - \sum_{i=1}^N \alpha_i g(z_i)\right| \leq (|\mathcal{C}|^3/N^2) \max_{z \in \mathcal{C}} |g''(z)|$, where α_i's are constants and N is the number of points of evaluation.

Since we can only approximate $p'(z)$, we actually end up computing $\sum_{i=1}^N \alpha_i (g(z_i) + \epsilon_i)$. Therefore, we can evaluate the overall error as

$$\left|\int_\mathcal{C} g(z) dz - \sum_{i=1}^N \alpha_i (g(z_i) + \epsilon_i)\right| \leq \frac{|\mathcal{C}|^3}{N^2} \max_z |g''(z)| + c_1 \epsilon |\mu_{\max}|$$

where c_1 is a small constant and $\epsilon = \max_i |\epsilon_i|$. Our goal is to find conditions under which $(|\mathcal{C}|^3/N^2) \max_z |g''(z)| + c_1 \epsilon |\mu_{\max}| < \pi$, such that rounding always gives the correct value. We first find a bound on ϵ for which $c_1 \epsilon |\mu_{\max}| < \pi/2$. If p' is approximated by finite differences, then $\epsilon = \max_\zeta |(p(\zeta + \Delta) - p(\zeta))/(\Delta p(\zeta)) - g(\zeta)| \leq \Delta p''_{\max}(\zeta')/p(\zeta)$, for some $\zeta' \in (\zeta, \zeta + \Delta)$. An upper bound on Δ is dictated by these conditions. Now, we have $p''_{\max}(\zeta')/p(\zeta) = \sum_{i \neq j} (\zeta - \lambda_i)^{-1} (\zeta - \lambda_j)^{-1} \prod_{k \neq i,j} (\zeta' - \lambda_k)/(\zeta - \lambda_k) \leq e^{n\Delta/\delta}(n^2/\delta^2)$. Thus, Δ must satisfy $\Delta \leq (c_2 \delta^2)/(n^2 |\mu_{\max}|)$, where c_2 is a small constant.

The other error term $(|\mathcal{C}|^3/N^2) \max_z |g''(z)|$, can now be upper bounded as $(|\mathcal{C}|^3/N^2) \max_z |g''(z)| \leq (c_3 |\mu_{\max}|^3)/(N^2 \delta^3)$, from which the number of evaluation points $N = O(|\mu_{\max}|/\delta)^{3/2}$.

For real-root polynomials, the number of oracle calls to p can be reduced by stronger bounds on $\max_z |g''(z)|$. The proof is omitted.

Corollary 7. *If p is a real-root polynomial, then there exists a $(\sqrt{n}|\mu_{\max}| + (1/\delta)^{1/2+o(1)}; \delta/2)$-checker for $\mathcal{R}_{\max}(p)$.*

If $\delta = O(1/n)$, then the above corollary yields asymptotically better checkers than those given by Theorems 3 and 6. We also give a different checker (proof omitted) that can be extended to work in the approximate setting.

Theorem 8. *There is a $((\lambda_{\sup}^2 - |\mu_{\max}|^2)/(\delta^{5/2}\epsilon); \delta/2)$-checker for $\mathcal{R}_{\max}(p)$ where $\epsilon \leq 1/|p''(x)|, \forall x$.*

Improved Checkers for $\mathcal{R}_{\max}$: δ unknown. We obtain the following checkers for the case when δ is not known. The proofs of the theorems are omitted.

Theorem 9. *Let p be a real-root polynomial. $\forall \epsilon, \beta > 0$, there is a $(\sqrt{n}\log^{3/2}(n/\epsilon) \log(1/\beta); \epsilon/2, \epsilon)$-checker for $\mathcal{R}_{\max}(p)$ that is correct with probability $\geq 1 - \beta$.*

The above theorem can be extended to the case when the roots are complex. The checker is still attractive in terms of its running time, but has more evaluations of p.

Corollary 10. *$\forall \epsilon, \beta > 0$, there is an $(n^{3/2}/\epsilon \log(1/\beta); \epsilon/2, \epsilon)$-checker for $\mathcal{R}_{\max}(p)$ that is correct with probability $\geq 1 - \beta$.*

The checkers in this section can be extended to check $\Pi_{\langle k \rangle}$.

4 Checking Roots: Approximate Setting

So far, we have been using the assumption that the programs being checked should return the exact root(s) and the oracle returns the exact value. As we stated in the description of our model, we have two variants – $\tilde{\Pi}$ and $\tilde{p}$. The former turns out to be easier than the latter.

Case I: $\tilde{\Pi}$. Suppose $\tilde{\Pi}_1(p)$ returns an ϵ-approximation (i.e., it claims $|\mu_i - \lambda_i| \leq \epsilon < \delta$). When p is real-root, with two oracle calls to p we can check if there is a sign-change in $[\mu_i - \epsilon, \mu_i + \epsilon]$. For $\tilde{\Pi}_n(p)$ (resp. $\tilde{\Pi}_r(p)$), we can extend the above checker with $2n$ (resp. $2r$) calls to p. Since we do not have a nice analog of Rolle's theorem in complex analysis, the problem becomes harder when p is not real-root.

All our checkers for $\mathcal{R}_{\max}$ in Section 3 can be extended to this approximate setting. This can be done as follows: (i) first we check if $\mu_{\max}$ is an approximate root and (ii) then we check if it is indeed the maximum root. The former is accomplished by checking if there is a root inside a small circle around $\mu_{\max}$ (see previous paragraph) and the latter is accomplished by selecting two curves separated by ϵ and then performing the numerical integration twice. Thus, we obtain the following theorem:

Theorem 11. $\forall \epsilon, \beta > 0$, there is (i) a $((|\mu_{\max}|/\delta)^{3/2}; \delta/2)$-checker for $\tilde{\mathcal{R}}_{\max}(p)$ and (ii) an $(n^{3/2}/\epsilon \log(1/\beta); \epsilon/2, \epsilon)$-checker for $\tilde{\mathcal{R}}_{\max}(p)$ that is correct with probability $\geq 1-\beta$. If p is a real-root polynomial, $\forall \epsilon, \beta > 0$, there is a (i) $(\sqrt{n}|\mu_{\max}| + (1/\delta)^{1/2+\epsilon}; \delta/2)$-checker for $\tilde{\mathcal{R}}_{\max}(p)$, (ii) $(\sqrt{n} \log^{3/2}(n/\epsilon) \log(1/\beta); \epsilon/2, \epsilon)$-checker for $\mathcal{R}_{\max}(p)$ that is correct with probability $\geq 1-\beta$.

Case II: $\tilde{p}$. Theorem 8 can be extended to the case when we have only $\tilde{p}$ (i.e., the evaluation of p is approximate). The proof is omitted.

Corollary 12. There is a $((\lambda_{\sup}^2 - |\mu_{\max}|^2)/(\delta^{5/2}\epsilon); \delta/2)$-approximate checker for $\tilde{\mathcal{R}}_1(\tilde{p})$.

The checkers in Section 3 are not directly usable in this case because of the instability of numerical derivative computation in the presence of errors.

5 Testing Norms

A function $f : \mathcal{V}(\mathcal{F}) \to \mathbb{R}^+$, where $\mathcal{V}$ is a vector space over $\mathcal{F}$, is called a *norm* if it satisfies: (i) $f(\mathbf{x}) = 0 \iff \mathbf{x} = \mathbf{0}$, (ii) $\forall \mathbf{x} \in \mathcal{V}, k \in \mathcal{F}, f(k\mathbf{x}) = kf(\mathbf{x})$ (*scalability*), and (iii) $\forall \mathbf{x}, \mathbf{y} \in \mathcal{V}, f(\mathbf{x}+\mathbf{y}) \leq f(\mathbf{x}) + f(\mathbf{y})$ (*triangle inequality*).

In this section we investigate the problem of checking whether the function computed by a program Π_{norm} is close to a norm (i.e., there is a norm that agrees with Π_{norm} on most inputs). In the specific case of vector p-norms on $\mathbb{R}^n$, which are of the form $|\mathbf{x}|_p = (\sum_{i=1}^n x_i^p)^{1/p}$, the problem reduces to the well-studied problem of multivariate degree-testing [AS92, RS96]. In fact, matrix spectral norms can be checked using our techniques in Section 3 and Section 4. In the more general case of checking whether the function is close to *any* norm, we show that the properties characterizing a norm are *not usable* for exact self-testing. This result is already interesting in that our tests are almost exactly the same as the standard linearity test except for an inequality in the second test. This, however, makes a big difference in the validity of the test, which leads us to believe that inequalities in general do not lead to (exact) self-testers. In a striking contrast, we show that these properties characterizing norms can lead to approximate self-testers. The following discussion is for $\mathbb{R}^2$ and can be extended to $\mathbb{R}^n$.

Exact Testing. To check scalability of Π_{norm}, note that along a vector $\mathbf{x}$, scalability defines the same set of functions as linearity. Checking $\Pi_{\text{norm}}(a\mathbf{x}) + \Pi_{\text{norm}}(b\mathbf{x}) = \Pi_{\text{norm}}((a+b)\mathbf{x})$ for $\mathbf{x}, |\mathbf{x}| = 1$ will determine if Π_{norm} is scalable along $\mathbf{x}$ (this is the linearity test of [BLR93]). By performing this test at many $\mathbf{x}$, we can ensure that Π_{norm} is scalable for many $\mathbf{x}$. Therefore, for the rest of this discussion, we can assume that Π_{norm} is scalable. $\forall i \in \mathbb{R}$, define the "concentric" contours $\mathcal{C}_i = \{\mathbf{x} \mid \Pi_{\text{norm}}(\mathbf{x}) = i\}$. We first show that checking the triangle inequality is equivalent to checking the convexity of $\mathcal{C}_i$ in $\mathbb{R}^2$ for any $i \in \mathbb{R}$.

Lemma 13. *Let f be a scalable function, i.e., $f(k\mathbf{x}) = kf(\mathbf{x})$. Then, $\exists \mathbf{a}, \mathbf{b} \in \mathcal{V}$, such that $f(\mathbf{a}+\mathbf{b}) > f(\mathbf{a}) + f(\mathbf{b}) \iff \forall i$, the i-th contour C_i is not convex (the non-convexity occurs along $\mathbf{a} + \mathbf{b}$.)*

We show in Theorem 14 that random sampling of condition 3 does not work: there are extremely "bad" programs that pass it.

Theorem 14. *$\forall 0 \leq \delta \leq 1$, there exists a scalable Π_{norm} that is at least δ away from the nearest convex function g, i.e., $\Pr_x[\Pi_{\text{norm}}(x) \neq g(x)] \geq \delta$, but Π_{norm} passes the test for condition 3 with arbitrarily high probability.*

Approximate Testing. In contrast to exact testing, we show that the properties characterizing norms can be used to test if a program approximately computes a norm at a non-trivial fraction of the inputs.

For a given star-shaped C, let the diameter be $\operatorname{diam} C = \sup_{t_2,t_2}\{|C(t_1) - C(t_2)|\}$. Given two curves C_1, C_2, let the distance between them be $|C_1 - C_2| = \sup_t\{|C_1(t) - C_2(t)|\}$. For two contours C_1, C_2 and for any other star-shaped C, let the deviation measure be $\operatorname{dev}_{C_1,C_2}(C) = \Pr_t[C(t) \geq C_1(t), C_2(t) \text{ or } C(t) \leq C_1(t), C_2(t)]$. This measures the fraction of C not lying between C_1 and C_2. For a star-shaped C, let $\Delta = \Delta(C) = \Pr_{s,t,z \in \overline{C(s)C(t)}}[z \notin C \cup \operatorname{int} C]$. In other words, Δ is the probability that, if we pick random $s, t \in C$ and a random point z on the line joining them, then z lies outside C. Testing condition 3 on random $\mathbf{x}, \mathbf{y}$, we can estimate Δ corresponding to the contour defined by Π_{norm} (assuming it is star-shaped, which is easy to check).

Theorem 15. *Given $\rho > 0$, $\exists \epsilon = \epsilon(\rho) < 1, \gamma = \gamma(\epsilon) > 0$ such that for any star-shaped C with $\operatorname{diam} C \leq 1$, if $\Delta(C) \leq \gamma$ then there is a contour $\overline{C}$ such that $\Pr_t[|\overline{C}(t) - C(t)| > \rho] \leq \epsilon$.*

6 Some Applications: Matrix Computations

In this section, we show applications of our checkers for polynomial roots to matrix spectra computations. Let the eigenvalues of $\mathbf{A} \in \mathcal{F}^{n \times n}$ be $\Lambda(\mathbf{a}) = \{\lambda_i \mid 1 \leq i \leq n\}$ with $|\lambda_{\max} = \lambda_1| \geq \cdots \geq |\lambda_n|$. It is easy to find an upper bound $\lambda_{\sup}$ on $\lambda_{\max}$ (e.g., set $\lambda_{\sup} = \|\mathbf{A}\|_\infty$). We denote by δ a separation bound between the eigenvalues of $\mathbf{A}$. Let DET be a correct program available in the library for computing the determinant of a matrix. DET corresponds to the oracle p in our model.

Eigenvalues in the Exact Setting. All the checkers in Section 3 and Section 4 translate to checkers for eigenvalues. We now illustrate more efficient checkers for some special cases which are of interest in practice:

Lemma 16. *Let $\mathbf{A} \in \mathbb{R}^{n \times n}$ with $\mathbf{A} = \mathbf{A}^T$. There is an (n)-checker for program computing the largest eigenvalue of $\mathbf{A}$. If $\mathbf{A}$ is tridiagonal, there is an (n)-checker for a program computing the k-th largest eigenvalue of $\mathbf{A}$.*

These checkers can be used to check programs designed for computing the second largest eigenvalue of a regular graph (or the largest eigenvalue of its Laplacian), which is related to the expansion of the graph. Another natural application of these checkers is to check programs that decide whether a matrix is positive definite.

Eigenvalues in the Approximate Setting. All of our approximate checkers for roots can be used in this case. We consider an interesting special case of this problem. Let $\langle \lambda, \mathbf{x} \rangle$ be an exact eigenvalue–eigenvector pair of $\mathbf{A}$ with $\|x\| = 1$. Let $\tilde{\mathcal{R}}_1$ be a relation that binds matrix $\mathbf{A}$ to pairs $\langle \mu, \tilde{\mathbf{x}} \rangle$ with $|\mu - \lambda| < \epsilon_1$ and $\|\tilde{\mathbf{x}} - \mathbf{x}\| < \epsilon_2$. Let $\tilde{\Pi}_1$ be a program that, on input $\mathbf{A}$, purports to compute a $\langle \mu, \tilde{\mathbf{x}} \rangle \in \tilde{\mathcal{R}}_1(\mathbf{A})$. The most natural way of checking $\tilde{\mathcal{R}}_1$ would be by checking that $\|\mathbf{A}\tilde{\mathbf{x}} - \mu\tilde{\mathbf{x}}\| < \epsilon$, for a certain threshold ϵ, and passing $\tilde{\Pi}_1$ if the above inequality is satisfied. Unfortunately, from perturbation theory [GV89] we have that the value of ϵ above can be small, but $|\mu - \lambda|$ be as large as $\epsilon/|\mathbf{y}^H \mathbf{x}|$, where $\mathbf{y}^H$ is a unit length left eigenvector of $\mathbf{A}$ ($\mathbf{y}^H \mathbf{A} = \lambda \mathbf{y}^H$), assuming λ to be a simple eigenvalue. Thus, we might need to set ϵ to a very small value if we want to make sure that $|\mu - \lambda| < \epsilon_1$ for a reasonably small value ϵ_1. Note however that for normal matrices, $|\mathbf{y}^H \mathbf{x}| = 1$ so that $\|\mathbf{A}\mu - \tilde{\mathbf{z}}\mu\| < \epsilon \Rightarrow |\lambda - \mu| < \epsilon$. Thus, we have the following lemma:

Lemma 17. $\tilde{\mathcal{R}}_1(\mathbf{A})$ *can be approximately checked when* $\mathbf{A}$ *is normal.*

In general, if we do not make assumptions on the problem condition, the approximate checker may yield very poor bounds. This is because the determinant of a matrix can be very close to zero (e.g., $1/2^n$) despite all eigenvalues being well-separated from zero (e.g., $\lambda_i = 1/2$).

Singular Values. Suppose MULT is a correct library program for matrix multiplication. If Π_{sing} is a program that purports to compute the singular values $\sigma_1, \ldots, \sigma_n$ of $\mathbf{A}$, construct a checker for Π_{sing} as follows: (i) check if $\sigma_i \geq 0, 1 \leq i \leq n$, (ii) compute the matrix $\mathbf{A}^T \mathbf{A} \in \mathcal{F}^{n \times n}$ using MULT, and (iii) use the checkers for eigenvalues to verify if $\{\sigma_1^2, \ldots, \sigma_n^2\} = \Lambda(\mathbf{A}^T \mathbf{A})$. The correctness of this construction is from the definition of singular values (see [GV89]).

7 Further Work

All of our checkers are assumed to perform exact arithmetic. This assumption is not always true in practice. It will be interesting to design checkers when the checker's numerical errors are critical. Many issues are still unresolved in the case of $\tilde{p}$. Are there efficient checkers for programs that compute Gröbner bases, programs that solve Diophantine problems and lattice problems? Such checkers would find numerous applications in computational algebra systems. Can we get efficient checkers for sparse-matrix computations?

Acknowledgements

We thank Richard Allen (Sandia), Bruce Hendrickson (Sandia), Mahan Mitra (Berkeley), Ronitt Rubinfeld (Cornell), Nick Trefethen (Cornell), Divakar Vishwanath (Cornell), and Richard Zippel (Cornell) for interesting discussions and suggestions.

References

[ABC+93] S. Ar, M. Blum, B. Codenotti, and P. Gemmell. Checking approximate computations over the reals. *Proc. 25th STOC*, pp. 786–795, 1993.

[ALM+92] S. Arora, C. Lund, R. Motwani, M. Sudan, and M. Szegedy. Proof verification and hardness of approximation problems. *Proc. 33rd FOCS*, pp. 14–23, 1992.

[AS92] S. Arora and S. Safra. Probabilistic checking of proofs: A new characterization of NP. *Proc. 33rd FOCS*, pp. 2–13, 1992.

[Blu88] M. Blum. Designing programs to check their work. TR 88-009, ICSI, 1988.

[BK89] M. Blum and S. Kannan. Program correctness checking ... and the design of programs that check their work. *Proc. 21st STOC*, pp. 86–97, 1989.

[BLR93] M. Blum, M. Luby, and R. Rubinfeld. Self-testing/correcting with applications to numerical problems. *JCSS*, 47(3):549–595, 1993.

[BCL82] B. Buchberger, G.E. Collins, and R. Loos. *Computer Algebra – Symbolic and Algebraic Computation*. Springer-Verlag, 1982.

[Con78] J.B. Conway. *Functions of One Complex Variable*. Springer-Verlag, 1978.

[Erg95] F. Ergün. Testing multivariate linear functions: Overcoming the generator bottleneck. *Proc. 27th STOC*, pp. 407–416, 1995.

[Fre79] R. Freivalds. Fast probabilistic algorithms. *Proc. 8th MFCS*, LNCS 74, pp. 57–69, 1979.

[EKR96] F. Ergün, S. Ravi Kumar, and R. Rubinfeld. Approximate checking of polynomials and functional equations. *Proc. 37th FOCS*, To appear.

[GLR+91] P. Gemmell, R. Lipton, R. Rubinfeld, M. Sudan, and A. Wigderson. Self-testing/correcting for polynomials and for approximate functions. *Proc. 23rd STOC*, pp. 32–42, 1991.

[GGR96] O. Goldreich, S. Goldwasser, and D. Ron. Property testing and its connection to learning and approximation. *Proc. 37th FOCS*, pp. 339–348, 1996.

[GV89] G.H. Golub and C. Van Loan. *Matrix Computations*. Johns Hopkins U. Press, 1989.

[JT75] M.A. Jenkins and J.F. Traub. Principles for testing polynomial zero-finding programs. *ACM Trans. on Mathematical Software*, 1:26–34, 1975.

[Kan90] S. Kannan. *Program Result Checking with Applications*. PhD thesis, U. of California at Berkeley, 1990.

[Lip91] R. Lipton. New directions in testing. *Proc. DIMACS Workshop on Distributed Computing and Cryptography*, pp. 191–202, 1991.

[Mar66] M. Marden. *Geometry of Polynomials*. AMS, 1966.

[PTV92] W. H. Press, S. A. Teukolsky, W. T. Vetterling, and B. P. Flannery. *Numerical Recipes in C*. Cambridge U. Press, 1992.

[RS96] R. Rubinfeld and M. Sudan. Robust characterizations of polynomials and their applications to program testing. *SICOMP*, 25(2):252–271, 1996.

[Wil65] J. H. Wilkinson. *The Algebraic Eigenvalue Problem*. Clarendon Press, 1966.

[Z93] R.E. Zippel. *Effective Polynomial Computation*. Kluwer Academic Press, 1993.

Exact Analysis of Dodgson Elections: Lewis Carroll's 1876 Voting System is Complete for Parallel Access to NP[*]

Edith Hemaspaandra,[1][**] Lane A. Hemaspaandra,[2][***] and Jörg Rothe[3][†]

[1] Department of Mathematics, Le Moyne College, Syracuse, NY 13214, USA
[2] Department of Computer Science, University of Rochester, Rochester, NY 14627, USA
[3] Institut für Informatik, Friedrich-Schiller-Universität Jena, 07743 Jena, Germany

Abstract. In 1876, Lewis Carroll proposed a voting system in which the winner is the candidate who with the fewest changes in voters' preferences becomes a Condorcet winner—a candidate who beats all other candidates in pairwise majority-rule elections. Bartholdi, Tovey, and Trick provided a lower bound—NP-hardness—on the computational complexity of determining the election winner in Carroll's system. We provide a stronger lower bound and an upper bound that matches our lower bound. In particular, determining the winner in Carroll's system is complete for parallel access to NP, i.e., it is complete for Θ_2^p, for which it becomes the most natural complete problem known. It follows that determining the winner in Carroll's elections is not NP-complete unless the polynomial hierarchy collapses.

1 Introduction

The Condorcet criterion is that an election is won by any candidate who defeats all others in pairwise majority-rule elections ([Con85], see [Bla58]). The Condorcet Paradox, dating from 1785 [Con85], notes that not only is it not always the case that Condorcet winners exist but, far worse, when there are more than two candidates, pairwise majority-rule elections may yield strict cycles in the aggregate preference even if each voter has non-cyclic preferences.[4] This is a widely discussed and troubling feature of majority rule (see, e.g., the discussion in [Mue89]).

In 1876, Charles Lutwidge Dodgson—more commonly referred to today by his pen name, Lewis Carroll—proposed an election system that is inspired by the Condorcet

---

[*] A full version of this paper, including all proofs, can be found at http://www.cs.rochester.edu/trs as UR-CS-TR-96-640. Supported in part by grants NSF-CCR-9322513 and NSF-INT-9513368/DAAD-315-PRO-fo-ab, and a University of Rochester Bridging Fellowship.
[**] edith@bamboo.lemoyne.edu. Work done in part while visiting Friedrich-Schiller-Universität Jena and the University of Amsterdam.
[***] lane@cs.rochester.edu. Work done in part while visiting Friedrich-Schiller-Universität Jena and the University of Amsterdam.
[†] rothe@informatik.uni-jena.de. Work done in part while visiting Le Moyne College.
[4] The standard example is an election over candidates a, b, and c in which 1/3 of the voters have preference $\langle a < b < c \rangle$, 1/3 of the voters have preference $\langle b < c < a \rangle$, and 1/3 of the voters have preference $\langle c < a < b \rangle$. In this case, though each voter individually has well-ordered preferences, the aggregate preference of the electorate is that b trounces a, c trounces b, and a trounces c. In short, individually well-ordered preferences do not necessarily aggregate to a well-ordered societal preference.

criterion,[5] yet that sidesteps the abovementioned problem [Dod76]. In particular, a Condorcet winner is a candidate who defeats each other candidate in pairwise majority-rule elections. In Carroll's system, an election is won by the candidate who is "closest" to being a Condorcet winner. In particular, each candidate is given a score that is the smallest number of exchanges of adjacent preferences in the voters' preference orders needed to make the candidate a Condorcet winner with respect to the resulting preference orders. Whatever candidate (or candidates, in the case of a tie) has the lowest score is the winner. This system admits ties but, as each candidate is assigned an integer score, no strict-preference cycles are possible.

Bartholdi, Tovey, and Trick, in their paper "Voting Schemes for which It Can Be Difficult to Tell Who Won the Election" [BTT89], raise a difficulty regarding Carroll's election system. Though the notion of winner(s) in Carroll's election system is mathematically well-defined, Bartholdi et al. raise the issue of what the *computational complexity* is of determining who is the winner. Though most natural election schemes admit obvious polynomial-time algorithms for determining who won, in sharp contrast Bartholdi et al. prove that Carroll's election scheme has the disturbing property that it is NP-hard to determine whether a given candidate has won a given election (a problem they dub CarrollWinner—they use the name "Dodgson" throughout, but we treat this as if they had written the equivalent "Carroll"), and that it is NP-hard even to determine whether a given candidate has tied-or-defeated another given candidate (a problem they dub CarrollRanking).

Bartholdi, Tovey, and Trick's NP-hardness results establish lower bounds for the complexity of CarrollRanking and CarrollWinner. *We optimally improve their two complexity lower bounds by proving that both problems are hard for Θ_2^p, the class of problems that can be solved via parallel access to NP, and we provide matching upper bounds. Thus, we establish that both problems are Θ_2^p-complete.* Bartholdi et al. explicitly leave open the issue of whether CarrollRanking is NP-complete: "...Thus CarrollRanking is as hard as an NP-complete problem, but since we do not know whether CarrollRanking is in NP, we can say only that it is NP-hard" [BTT89, p. 161]. From our optimal lower bounds, it follows that neither CarrollWinner nor CarrollRanking is NP-complete unless the polynomial hierarchy collapses.

As to our proof method, in order to raise the known lower bound on the complexity of Carroll elections, we first study the ways in which feasible algorithms can control Carroll elections. In particular, we establish a series of lemmas showing how polynomial-time algorithms can control oddness and evenness of election scores, "sum" over election scores, and merge elections. These lemmas then lead to our hardness results.

We remark that it is somewhat curious finding "parallel access to NP"-complete (i.e., Θ_2^p-complete) problems that were introduced almost one hundred years before complexity theory itself existed. In addition, CarrollWinner, which we prove complete for this class, is extremely natural when compared with previously known complete problems for this class, essentially all of which have quite convoluted forms, e.g., asking whether a given list of boolean formulas has the property that the number of formulas in the list that are satisfiable is itself an odd number (see the discussion

---

[5] Carroll did not use this term. Indeed, Black has shown that Carroll "almost beyond a doubt" was unfamiliar with Condorcet's work [Bla58, p. 193–194].

in [Wag87]). In contrast, the class NP, which is contained in Θ_2^p, has countless natural complete problems. Also, we mention that Papadimitriou [Pap84] has shown that UniqueOptimalTravelingSalesperson is complete for P^{NP}, which contains Θ_2^p.

2 Preliminaries

In this section, we introduce some standard concepts and notations from computational complexity theory [Pap94,BC93]. NP is the class of languages solvable in nondeterministic polynomial time. The polynomial hierarchy, PH, is defined as $\text{PH} = \text{P} \cup \text{NP} \cup \text{NP}^{\text{NP}} \cup \text{NP}^{\text{NP}^{\text{NP}}} \cup \cdots$ where, for any class $\mathcal{C}$, $\text{NP}^{\mathcal{C}} = \bigcup_{C \in \mathcal{C}} \text{NP}^C$, and NP^C is the class of all languages that can be accepted by some NP machine that is given a black box that in unit time answers membership queries to C. The polynomial hierarchy is said to collapse if for some k the kth term in the preceding infinite union equals the entire infinite union. Computer scientists strongly suspect that the polynomial hierarchy does not collapse, though proving (or disproving) this remains a major open research issue.

The polynomial hierarchy has a number of intermediate levels. Of particular interest to us will be the level Θ_2^p. Θ_2^p is the class of all languages that can be solved via $\mathcal{O}(\log n)$ queries to some NP set (see [Wag90]). Equivalently, and more to the point for the purposes of this paper, Θ_2^p equals the class of problems that can be solved via parallel access to NP, as explained formally below. Θ_2^p falls between the first and second levels of the polynomial hierarchy: $\text{NP} \subseteq \Theta_2^p \subseteq \text{P}^{\text{NP}} \subseteq \text{NP}^{\text{NP}}$. Kadin [Kad89] has proven that if NP has a sparse Turing-complete set then the polynomial hierarchy collapses to Θ_2^p, Wagner [Wag90] has shown that the definition of Θ_2^p is extremely robust, and Jenner and Torán [JT95] have shown that the robustness of the class Θ_2^p seems to fail for its function analogs.

Problems are encoded as languages of strings over some fixed alphabet Σ having at least two letters. Σ^* denotes the set of all strings over Σ. For any string $x \in \Sigma^*$, let $|x|$ denote the length of x. For any set $A \subseteq \Sigma^*$, let $\overline{A}$ denote $\Sigma^* \setminus A$. For any set $A \subseteq \Sigma^*$, let $||A||$ denote the cardinality of A. For any multiset A, $||A||$ will denote the cardinality of A. For example, if A is the multiset containing one occurrence of the preference order $\langle w < x < y \rangle$ and seventeen occurrences of the preference order $\langle w < y < x \rangle$, then $||A|| = 18$. As is standard, for each language $A \subseteq \Sigma^*$ we use χ_A to denote the characteristic function of A, i.e., $\chi_A(x) = 1$ if $x \in A$ and $\chi_A(x) = 0$ if $x \notin A$. Let $\langle \cdots \rangle$ be any standard, multi-arity, easily computable, easily invertible pairing function. We will also use the notation $\langle \cdots \rangle$ to denote preference orders, e.g., $\langle w < x < y \rangle$. Which use is intended will be clear from context.

In computational complexity theory, reductions are used to relate the complexity of problems. Very informally, if A reduces to B that means that, given B, one can solve A. For any a and b such that $\leq_a^b$ is a defined reduction type, and any complexity class $\mathcal{C}$, let $\text{R}_a^b(\mathcal{C})$ denote $\{L \mid (\exists C \in \mathcal{C}) [L \leq_a^b C]\}$. We refer readers to the standard source, Ladner, Lynch, and Selman [LLS75], for definitions and discussion of the standard reductions. However, we briefly and informally present to the reader the definitions of the reductions to be used in this paper. $A \leq_m^p B$ ("A polynomial-time many-one reduces to B") if there is a polynomial-time computable function f such that $(\forall x \in \Sigma^*)[x \in A \iff f(x) \in B]$. $A \leq^p_{tt} B$ ("A polynomial-time truth-table reduces to B") if there is a polynomial-time Turing machine that, on input x, computes a query that itself consists of a list of strings and, given that the machine after writing the query is then given as its answer a list telling which of the listed strings are in B, the machine then correctly determines whether x is in A (this is not the original Ladner-Lynch-Selman definition, as we have merged their querying machine and their evaluation machine, however this formulation is common and equivalent). Since a $\leq^p_{tt}$-reducing machine, on a given input, asks all its questions in a *parallel* (also called *non-adaptive*) manner, the informal statement above that Θ^p_2 captures the complexity of "parallel access to NP" can now be expressed formally as the claim $\Theta^p_2 = \mathrm{R}^p_{tt}(\mathrm{NP})$, which is known to hold [KSW87,Hem89].

As has become the norm, we always use hardness to denote hardness with respect to $\leq^p_m$ reductions. That is, for any class $\mathcal{C}$ and any problem A, we say that A is $\mathcal{C}$-*hard* if $(\forall C \in \mathcal{C})[C \leq^p_m A]$. For any class $\mathcal{C}$ and any problem A, we say that A is $\mathcal{C}$-*complete* if A is $\mathcal{C}$-hard and $A \in \mathcal{C}$. Completeness results are the standard method in computational complexity theory of categorizing the complexity of a problem, as a $\mathcal{C}$-complete problem A is both in $\mathcal{C}$, and is the hardest problem in $\mathcal{C}$ (in the sense that every problem in $\mathcal{C}$ can be easily solved using A).

3 The Complexity of Carroll Elections

Lewis Carroll's voting system ([Dod76], see also [NR76,BTT89]) works as follows. Each voter has strict preferences over the candidates. Each candidate is assigned a score, namely, the smallest number of sequential *exchanges of two adjacent candidates in the voters' preference orders* (henceforward called "switches") needed to make the given candidate a Condorcet winner. We say that a candidate c *ties-or-defeats* a candidate d if the score of d is not less than that of c. (Bartholdi et al. [BTT89] use the term "defeats" to denote what we, for clarity, denote by ties-or-defeats; though the notations are different, the sets being defined by Bartholdi et al. and in this paper are identical.) A candidate c is said to win the Carroll-type election if c ties-or-defeats all other candidates. Of course, due to ties it is possible for two candidates to tie-or-defeat each other, and so it is possible for more than one candidate to be a winner of the election.

Recall that all preferences are assumed to be strict. A candidate c is a *Condorcet winner* (with respect to a given collection of voter preferences) if c defeats (i.e., is preferred by strictly more than half of the voters) each other candidate in pairwise majority-rule elections. Of course, Condorcet winners do not necessarily exist for a given set of preferences, but if a Condorcet winner does exist, it is unique.

We now return to Carroll's scoring notion to clarify what is meant by the sequential nature of the switches, and to clarify by example that one switch changes only one voter's preferences. The *(Carroll) score* of any Condorcet winner is 0. If a candidate is not a Condorcet winner, but one switch (recall that a switch is an exchange of two *adjacent* preferences in the preference order of *one* voter) would make the candidate a Condorcet winner, then the candidate has a score of 1. If a candidate does not have a score of 0 or 1, but two switches would make the candidate a Condorcet winner, then the candidate has a score of 2. Note that the two switches could both be in the same voter's preferences, or could be one in one voter's preferences and one in another voter's preferences. Note also that switches are sequential. For example, with two switches, one could change a single voter's preferences from $\langle a < b < c < d \rangle$ to $\langle c < a < b < d \rangle$, where $e < f$ will denote the preference: "f is strictly preferred to e." With two switches, one could also change a single voter's preferences from $\langle a < b < c < d \rangle$ to $\langle b < a < d < c \rangle$. With two switches (not one), one could also change two voters with initial preferences of $\langle a < b < c < d \rangle$ and $\langle a < b < c < d \rangle$ to the new preferences $\langle b < a < c < d \rangle$ and $\langle b < a < c < d \rangle$. As noted earlier in this section, Carroll scores of 3, 4, etc., are defined analogously, i.e., the Carroll score of a candidate is the smallest number of sequential switches needed to make the given candidate a Condorcet winner. (We note in passing that Carroll was before his time in more ways than one. His definition is closely related to an important concept that is now known in computer science as "edit-distance"—the minimum number of operations (from some specified set of operations) required to transform one string into another. Though Carroll's single "switch" operation is not the richer set of operations most commonly used today when doing string-to-string editing (see, e.g., [SK83]), it does form a valid basis operation for transforming between permutations, which after all are what preferences are.)

Bartholdi et al. [BTT89] define a number of decision problems related to Carroll's system. They prove that given preference lists, and a candidate, and a number k, it is NP-complete to determine whether the candidate's score is at most k in the election specified by the preference lists (they call this problem CarrollScore). They define the problem CarrollRanking to be the problem of determining, given preference lists and the names of two voters, c and d, whether c ties-or-defeats d. They prove that this problem is NP-hard. They also prove that, given a candidate and preference lists, it is NP-hard to determine whether the candidate is a winner of the election.

For the formal definitions of these three decision problems, a preference order is strict (i.e., irreflexive and antisymmetric), transitive, and complete. Since we will freely identify voters with their preference orders, and two different voters can have the same preference order, we define a set of voters as a multiset of preference orders.

We will say that $\langle C, c, V \rangle$ is a *Carroll triple* if C is a set of candidates, c is a member of C, and V is a multiset of preference orders on C. Throughout this paper, we assume that, as inputs, multisets are coded as lists, i.e., if there are m voters in the voter set then $V = \langle P_1, P_2, \ldots, P_m \rangle$, where P_i is the preference order of the ith voter. $Score(\langle C, c, V \rangle)$ will denote the Carroll score of c in the vote specified by C and V.

Decision Problem: CarrollScore

Instance: A Carroll triple $\langle C, c, V \rangle$; a positive integer k.

Question: Is $Score(\langle C, c, V \rangle)$, the Carroll score of candidate c in the election specified by $\langle C, V \rangle$, less than or equal to k?

Decision Problem: CarrollRanking

Instance: A set of candidates C; two distinguished members of C, c and d; a multiset V of preference orders on C (encoded as a list, as discussed above).

Question: Does c tie-or-defeat d in the election? That is, is $Score(\langle C, c, V \rangle) \leq Score(\langle C, d, V \rangle)$?

Decision Problem: CarrollWinner

Instance: A Carroll triple $\langle C, c, V \rangle$.

Question: Is c a winner of the election? That is, does c tie-or-defeat all other candidates in the election?

We now state the complexity of CarrollRanking.

Theorem 1. *CarrollRanking is Θ_2^p-complete.*

It follows immediately—since (a) $\Theta_2^p = \mathrm{NP} \Rightarrow \mathrm{PH} = \mathrm{NP}$, and (b) $\mathrm{R}_m^p(\mathrm{NP}) = \mathrm{NP}$—that CarrollRanking, though known to be NP-hard [BTT89], cannot be NP-complete unless the polynomial hierarchy collapses quite dramatically.

Corollary 2. *If CarrollRanking is NP-complete, then $\mathrm{PH} = \mathrm{NP}$.*

Wagner has provided a useful tool for proving Θ_2^p-hardness, and we state his result below as Lemma 3. However, to be able to exploit this tool we must explore the structure of Carroll elections. In particular, we have to learn how to control oddness and evenness of election scores, how to add election scores, and how to merge elections. We do so as Lemmas 4, 5, and 7, respectively. On our way towards establishing Theorem 1, using Lemmas 3, 4, and 5 we will first establish Θ_2^p-hardness of a special problem that is closely related to CarrollRanking. This result is stated as Lemma 6 below. It is not hard to prove Theorem 1 using Lemma 6 and Lemma 7. Note that Lemma 7 gives more than is needed merely to establish Theorem 1. In fact, the way this lemma is stated even suffices to provide—jointly with Lemma 6—a direct proof of the Θ_2^p-hardness of CarrollWinner.

Lemma 3. [Wag87] *Let A be some NP-complete set, and let B be any set. If there exists a polynomial-time computable function g such that, for all $k \geq 1$ and all strings $x_1, \ldots, x_{2k} \in \Sigma^*$ satisfying $\chi_A(x_1) \geq \chi_A(x_2) \geq \cdots \geq \chi_A(x_{2k})$, it holds that*

$$\|\{i \mid x_i \in A\}\| \text{ is odd} \iff g(x_1, \ldots, x_{2k}) \in B,$$

then B is Θ_2^p-hard.

Lemma 4. *There exists an NP-complete set A and a polynomial-time computable function f that reduces A to CarrollScore in such a way that, for every $x \in \Sigma^*$, $f(x) = \langle \langle C, c, V \rangle, k \rangle$ is an instance of CarrollScore with an odd number of voters and (1) if $x \in A$ then $Score(\langle C, c, V \rangle) = k$, and (2) if $x \notin A$ then $Score(\langle C, c, V \rangle) = k + 1$.*

Proof of Lemma 4. Bartholdi et al. [BTT89] prove the NP-hardness of CarrollScore by reducing ExactCoverByThreeSets to it. However, their reduction doesn't have the additional properties that we need in this lemma. We will construct a reduction from the standard NP-complete problem

ThreeDimensionalMatching (3DM) to CarrollScore that *does* have the additional properties we need. Let us first give the definition of 3DM:

Decision Problem: ThreeDimensionalMatching (3DM)

Instance: Sets M, W, X, and Y, where $M \subseteq W \times X \times Y$ and W, X, and Y are disjoint, nonempty sets having the same number of elements.

Question: Does M contain a *matching*, i.e., a subset $M' \subseteq M$ such that $||M'|| = ||W||$ and no two elements of M' agree in any coordinate?

We now describe a polynomial-time reduction f (from 3DM to CarrollScore) having the desired properties. Our reduction is defined by $f(x) = f'(f''(x))$, where f' and f'' are as described below. Informally, f'' turns all inputs into a standard format (instances of 3DM having $||M|| > 1$), and f' assumes its input has this format and implements the actual reduction.

Let f'' be a polynomial-time function that has the following properties.

1. If x is not an instance of 3DM or is an instance of 3DM having $||M|| \leq 1$, then $f''(x)$ will output an instance y of 3DM for which $||M|| > 1$ and, furthermore, it will hold that $y \in$ 3DM $\iff x \in$ 3DM.

2. If x is an instance of 3DM having $||M|| > 1$, then $f''(x) = x$.

It is clear that such functions exist. In particular, for concreteness, let $f''(x)$ be $\langle \{(d,e,p),(d,e,p')\}, \{d,d'\}, \{e,e'\}, \{p,p'\}\rangle$ if x is not an instance of 3DM or both $x \notin$ 3DM and x is an instance of 3DM having $||M|| \leq 1$; let $f''(x)$ be $\langle\{(d,e,p),(d',e',p')\}, \{d,d'\}, \{e,e'\}, \{p,p'\}\rangle$ if x is an instance of 3DM having $||M|| \leq 1$ and such that $x \in$ 3DM; let $f''(x)$ be x otherwise.

We now describe f'. Let x be our input. If x is not an instance of 3DM for which $||M|| > 1$ then $f'(x) = 0$; this is just for definiteness, as due to f'', the only actions of f' that matter are when the input is an instance of 3DM for which $||M|| > 1$. So, suppose $x = \langle M, W, X, Y\rangle$ is an instance of 3DM for which $||M|| > 1$. Let $q = ||W||$. Define $f'(\langle M, W, X, Y\rangle) = \langle\langle C, c, V\rangle, 3q\rangle$ as follows: Let c, s, and t be elements not in $W \cup X \cup Y$. Let $C = W \cup X \cup Y \cup \{c, s, t\}$ and let V consist of the following two subparts:

1. Voters simulating elements of M. Suppose the elements of M are enumerated as $\{(w_i, x_i, y_i) \mid 1 \leq i \leq ||M||\}$. (The w_i are not intended to be an enumeration of W. Rather, they take on values from W as specified by M. In particular, w_j may equal w_k even if $j \neq k$. The analogous comments apply to the x_i and y_i variables.) For every triple (w_i, x_i, y_i) in M, we will create a voter. If i is odd, we create the voter $\langle s < c < w_i < x_i < y_i < t < \cdots\rangle$, where the elements after t are the elements of $C \setminus \{s, c, w_i, x_i, y_i, t\}$ in arbitrary order. If i is even, we do the same, except that we exchange s and t. That is, we create the voter $\langle t < c < w_i < x_i < y_i < s < \cdots\rangle$, where the elements after s are the elements of $C \setminus \{s, c, w_i, x_i, y_i, t\}$ in arbitrary order.

2. $||M|| - 1$ voters who prefer c to all other candidates.

We will now show that f has the desired properties. It is immediately clear that f'' and f', and thus f, are polynomial-time computable. It is also clear from our construction that, for each x, $f(x)$ is an instance of CarrollScore having an odd number of voters since, for every instance $\langle M, W, X, Y \rangle$ of 3DM with $||M|| > 1$, $f'(\langle M, W, X, Y \rangle)$ is an instance of CarrollScore with $||M|| + (||M|| - 1)$ voters, and since f'' always outputs instances of this form. It remains to show that, for every instance $\langle M, W, X, Y \rangle$ of 3DM with $||M|| > 1$:

(a) if M contains a matching, then $Score(\langle C, c, V \rangle) = 3q$, and

(b) if M does not contain a matching, then $Score(\langle C, c, V \rangle) = 3q + 1$.

Note that if we prove this, it is clear that f has the properties (1) and (2) of Lemma 4, in light of the properties of f''. Note that, recalling that we may now assume that $||M|| > 1$, by construction c is preferred to s and t by more than half of the voters, and is preferred to all other candidates by $||M|| - 1$ of the $2||M|| - 1$ voters.

Now suppose that M contains a matching M'. Then $||M'|| = q$, and every element in $W \cup X \cup Y$ occurs in M'. $3q$ switches turn c into a Condorcet winner as follows. For every element $(w_i, x_i, y_i) \in M'$, switch c upwards 3 times in the voter corresponding to (w_i, x_i, y_i). For example, if i is odd, this voter changes from $\langle s < c < w_i < x_i < y_i < t < \cdots \rangle$ to $\langle s < w_i < x_i < y_i < c < t < \cdots \rangle$. Let z be an arbitrary element of $W \cup X \cup Y$. Since z occurs in M', c has gained one vote over z. Thus, c is preferred to z by $||M||$ of the $2||M|| - 1$ voters. Since z was arbitrary, c is a Condorcet winner.

On the other hand, c's Carroll score can never be less than $3q$, because to turn c into a Condorcet winner, c needs to gain one vote over z for every $z \in W \cup X \cup Y$. Since c can gain only *one* vote over *one* candidate for each switch, we need at least $3q$ switches to turn c into a Condorcet winner. This proves condition (a).

To prove condition (b), first note that there is a "trivial" way to turn c into a Condorcet winner with $3q + 1$ switches: Just switch c to the top of the preference order of the first voter. The first voter was of the form $\langle s < c < w_1 < x_1 < y_1 < t < \cdots \rangle$, where the elements after t are exactly all elements in $W \cup X \cup Y \setminus \{w_1, x_1, y_1\}$, in arbitrary order. Switching c upwards $3q + 1$ times moves c to the top of the preference order for this voter, and gains one vote for c over all candidates in $W \cup X \cup Y$, which turns c into a Condorcet winner. This shows that $Score(C, c, V) \leq 3q + 1$, regardless of whether M has a matching or not.

Finally, note that a Carroll score of $3q$ implies that M has a matching. As before, every switch has to involve c and an element of $W \cup X \cup Y$. (This is because c must gain a vote over $3q$ other candidates—$W \cup X \cup Y$—and so any switch involving s or t would ensure that at most $3q - 1$ switches were available for gaining against the $3q$ members of $W \cup X \cup Y$, thus ensuring failure.) Thus, for every voter, c switches at most three times to become a Condorcet winner. Since c has to gain one vote in particular over each element in Y, and to "reach" an element in Y it must hold that c first switches over the elements of W and X that due to our construction fall between it and the nearest y element (among the $||M||$ voters simulating elements of M—it is clear that if any switch involves at least one of the $||M|| - 1$ dummy voters this could never lead to a Carroll score of $3q$ for c), it must be the case that c switches upwards exactly three times for exactly q voters corresponding to elements of M. This implies that the q elements of M that correspond to these q voters form a matching, thus proving condition (b). ∎

Lemma 5. *There exists a polynomial-time computable function CarrollSum such that, for all k and for all $\langle C_1, c_1, V_1\rangle, \langle C_2, c_2, V_2\rangle, \ldots, \langle C_k, c_k, V_k\rangle$ satisfying $(\forall j)[\|V_j\|$ is odd], it holds that CarrollSum($\langle \langle C_1, c_1, V_1\rangle, \langle C_2, c_2, V_2\rangle, \ldots, \langle C_k, c_k, V_k\rangle\rangle$) is a Carroll triple having an odd number of voters and such that $\sum_j Score(\langle C_j, c_j, V_j\rangle) = Score(CarrollSum(\langle \langle C_1, c_1, V_1\rangle, \langle C_2, c_2, V_2\rangle, \ldots, \langle C_k, c_k, V_k\rangle\rangle))$.*

Lemma 3, Lemma 4, and Lemma 5 together establish the Θ_2^p-hardness of a special problem that is closely related to the problems that we are interested in, CarrollRanking and CarrollWinner. Let us define the decision problem TwoElectionRanking (2ER).

Decision Problem: TwoElectionRanking (2ER)

Instance: A pair of Carroll triples $\langle\langle C, c, V\rangle, \langle D, d, W\rangle\rangle$ both having an odd number of voters and such that $c \neq d$.

Question: Is $Score(\langle C, c, V\rangle) \leq Score(\langle D, d, W\rangle)$?

Lemma 6. TwoElectionRanking *is Θ_2^p-hard.*

We note in passing that 2ER clearly is in $R_{tt}^p(\text{NP})$, and so from the fact that $\Theta_2^p = R_{tt}^p(\text{NP})$, it is clear that 2ER is in Θ_2^p. Thus, in light of Lemma 6, 2ER is Θ_2^p-complete. We also note in passing that, since one can trivially rename candidates, 2ER remains Θ_2^p-complete in the variant in which "and such that $c \neq d$" is removed from the problem's definition.

In order to make the results obtained so far applicable to CarrollRanking and CarrollWinner, we need the following lemma that tells us how to merge two elections into a single election in a controlled manner.

Lemma 7. *There exist polynomial-time computable functions Merge and Merge$'$ such that, for all Carroll triples $\langle C, c, V\rangle$ and $\langle D, d, W\rangle$ for which $c \neq d$ and both V and W represent odd numbers of voters, there exist $\widehat{C}$ and $\widehat{V}$ such that*

(i) *Merge($\langle C, c, V\rangle, \langle D, d, W\rangle$) is an instance of* CarrollRanking *and Merge$'$($\langle C, c, V\rangle, \langle D, d, W\rangle$) is an instance of* CarrollWinner,

(ii) *Merge($\langle C, c, V\rangle, \langle D, d, W\rangle$) = $\langle \widehat{C}, c, d, \widehat{V}\rangle$ and Merge$'$($\langle C, c, V\rangle, \langle D, d, W\rangle$) = $\langle \widehat{C}, c, \widehat{V}\rangle$,*

(iii) $Score(\langle \widehat{C}, c, \widehat{V}\rangle) = Score(\langle C, c, V\rangle) + 1$,

(iv) $Score(\langle \widehat{C}, d, \widehat{V}\rangle) = Score(\langle D, d, W\rangle) + 1$, *and*

(v) *for each $e \in \widehat{C} \setminus \{c, d\}$, $Score(\langle \widehat{C}, c, \widehat{V}\rangle) < Score(\langle \widehat{C}, e, \widehat{V}\rangle)$.*

The results we now have established suffice to prove both Theorem 1 above and Theorem 8 below—which states that CarrollWinner is Θ_2^p-complete, the main result of this paper. Full proofs of the results in this paper can be found in the full version [HHR96].[6]

Theorem 8. CarrollWinner *is Θ_2^p-complete.*

Corollary 9. *If* CarrollWinner *is NP-complete, then* PH $=$ NP.

Acknowledgments: We are indebted to J. Banks and R. Calvert for recommending Carroll elections to us as an interesting open topic worthy of study, and for providing us with the literature on this topic. We thank D. Austen-Smith, J. Banks, R. Calvert, A. Rutten, M. Scott, and J. Seiferas for helpful conversations and suggestions. L. Hemaspaandra thanks J. Banks and R. Calvert for arranging, and J. Banks for supervising, his Bridging Fellowship at the University of Rochester's Department of Political Science, during which this project was started.

References

[BC93] D. Bovet and P. Crescenzi. *Introduction to the Theory of Complexity.* Prentice Hall, 1993.

[Bla58] D. Black. *Theory of Committees and Elections.* Cambridge University Press, 1958.

[BTT89] J. Bartholdi III, C. Tovey, and M. Trick. Voting schemes for which it can be difficult to tell who won the election. *Social Choice and Welfare,* 6:157–165, 1989.

[Con85] M. J. A. N. de Caritat, Marquis de Condorcet. *Essai sur l'Application de L'Analyse à la Probabilité des Décisions Rendues à la Pluraliste des Voix.* 1785. Facsimile reprint of original published in Paris, 1972, by the Imprimerie Royale.

[Dod76] C. Dodgson. A method of taking votes on more than two issues, 1876. Pamphlet printed by the Clarendon Press, Oxford, and headed "not yet published" (see the discussions in [MU95,Bla58], both of which reprint this paper).

[Hem89] L. Hemachandra. The strong exponential hierarchy collapses. *Journal of Computer and System Sciences,* 39(3):299–322, 1989.

[HHR96] E. Hemaspaandra, L. Hemaspaandra, and J. Rothe. Exact analysis of Dodgson elections: Lewis Carroll's 1876 voting system is complete for parallel access to NP. Technical Report TR-640, University of Rochester, Department of Computer Science, Rochester, NY, October 1996.

---

[6] Bartholdi et al. [BTT89] have stated without proof that CarrollRanking $\leq_m^p$ CarrollWinner. Theorem 1 plus this assertion would yield Theorem 8. However, as we wish our proof to be complete, we have proven Theorem 8 without relying on their assertion (see [HHR96]). We note in passing that our full paper implicitly provides an indirect proof of their assertion. In particular, given that one has proven Theorem 1 and Theorem 8, the assertion follows, since it follows from the definition of Θ_2^p-completeness that all Θ_2^p-complete problems are $\leq_m^p$-interreducible.

[JT95]   B. Jenner and J. Torán. Computing functions with parallel queries to NP. *Theoretical Computer Science*, 141(1–2):175–193, 1995.

[Kad89] J. Kadin. $P^{NP[\log n]}$ and sparse Turing-complete sets for NP. *Journal of Computer and System Sciences*, 39(3):282–298, 1989.

[KSW87] J. Köbler, U. Schöning, and K. Wagner. The difference and truth-table hierarchies for NP. *RAIRO Theoretical Informatics and Applications*, 21:419–435, 1987.

[LLS75] R. Ladner, N. Lynch, and A. Selman. A comparison of polynomial time reducibilities. *Theoretical Computer Science*, 1(2):103–124, 1975.

[MU95]   I. McLean and A. Urken. *Classics of Social Choice*. University of Michigan Press, 1995.

[Mue89] D. Mueller. *Public Choice II*. Cambridge University Press, 1989.

[NR76]   R. Niemi and W. Riker. The choice of voting systems. *Scientific American*, 234:21–27, 1976.

[Pap84] C. Papadimitriou. On the complexity of unique solutions. *Journal of the ACM*, 31(2):392–400, 1984.

[Pap94] C. Papadimitriou. *Computational Complexity*. Addison-Wesley, 1994.

[SK83]   D. Sankoff and J. Kruskal, editors. *Time Warps, String Edits, and Macromolecules: The Theory and Practice of Sequence Comparison*. Addison-Wesley, 1983.

[Wag87] K. Wagner. More complicated questions about maxima and minima, and some closures of NP. *Theoretical Computer Science*, 51(1–2):53–80, 1987.

[Wag90] K. Wagner. Bounded query classes. *SIAM Journal on Computing*, 19(5):833–846, 1990.

Game Theoretic Analysis
of Call-by-Value Computation

KOHEI HONDA   NOBUKO YOSHIDA

ABSTRACT. We present a general semantic universe of call-by-value computation based on elements of game semantics, and validate its appropriateness as a semantic universe by the full abstraction result for call-by-value PCF, a generic typed programming language with call-by-value evaluation. The key idea is to consider the distinction between call-by-name and call-by-value as that of the structure of information flow, which determines the basic form of games. In this way call-by-name computation and call-by-value computation arise as two independent instances of sequential functional computation with distinct algebraic structures. We elucidate the type structures of the universe following the standard categorical framework developed in the context of domain theory. Mutual relationship between the presented category of games and the corresponding call-by-name universe is also clarified.

1. INTRODUCTION

The *call-by-value* is a mode of calling procedures widely used in imperative and functional programming languages, e.g. [1, 30], in which one evaluates arguments before applying them to a concerned procedure. The semantics of higher-order computation based on call-by-value evaluation has been widely studied by many researchers in the context of domain theory, cf. [35, 23, 32, 12, 40, 11], through which it has become clear that the semantic framework for the call-by-value computation has a basic difference from the one for call-by-name computation (see [15, 42] for introduction to the topic). The difference between the semantics of call-by-value and that of call-by-name in this context may roughly be captured as the difference in the classes of involved functions: in call-by-name, we take any continuous functions between pointed cpos, while, in call-by-value, one takes *strict* continuous functions. The latter is also equivalently presentable as partial continuous functions between (possibly bottomless) cpos. This distinction leads to a basic algebraic difference of the induced categorical universes, cf.[11, 12].

The present paper offers a semantic analysis of call-by-value computation from a different angle, based on elements of game semantics. In game semantics, computation is modelled as specific classes of interacting processes (called *strategies*), which, together with a suitable notion of composition, form a categorical universe with appropriate type structures. One may compare this approach to Böhm trees or to sequential algorithms [6, 22], in both of which computation is modelled not by set-theoretic functions of a certain kind but by objects with internal structures which reflect computational behaviour of the concerned class of computation. Game semantics has its origin in Logics [7, 10] and has been used for the semantic analysis of programming languages, especially for characterising the notion of sequentiality [8, 34]. By concentrating on specific forms of interaction which obey a few basic constraints, the approach makes it possible to extract desired classes of interacting processes at a high-level of abstraction, offering suitable semantic universes for varied calculi and programming languages, cf. [2, 3, 4, 19, 20, 24]. The forms of interaction in these universes are however inherently call-by-name: it has

---

LFCS, Department of Computer Science, University of Edinburgh. e-mail: kohei@dcs.ed.ac.uk, ny@dcs.ed.ac.uk. Supported in part by EPSRC Fellowships and JSPS Research Fellowships.

not been clear how the call-by-value computation can be captured in the setting of game semantics, in spite of its equally significant status as a mode of computation.

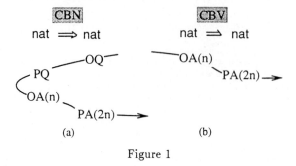

Figure 1

In the present work it will be shown that a general semantic universe of the call-by-value higher-order computation can indeed be simply constructed, employing basic elements of the foregoing game semantics, but with a key difference in the structures of interaction. More specifically, we find that the distinction between call-by-name and call-by-value in game semantics arises as the one in the form of the flow of information. Let us illustrate this point by simple examples. Figure 1 (a) depicts how a function which doubles a given natural number is modelled in the foregoing game semantics ("O" for Opponent, "P" for Player, "A" for Answer, and "Q" for Question). Computation starts when Opponent asks a question on the right, requesting an answer: then Player (the function) asks what the argument is on the left, from which the number is received, and finally it returns to the right to answer the initial question by the double of the received number. In Figure 1 (b), the same function is modelled in the call-by-value game. This time the flow starts at the *left* component, which already carries a value: then the function just returns the answer on the right. One may notice that this means the interaction should start from an *answer*, which might be regarded as an anomaly in the preceding convention in game semantics. However, it turns out that this parameter of games — whether one initiates a game by answers or by questions — is orthogonal to other basic elements of the game semantics, leading to a simple construction of a categorical universe in which representative functional calculi based on call-by-value evaluation can be faithfully interpreted. The independence of the parameter suggests we may obtain a suitable universe to model, say, imperative call-by-value computation by simply altering other parameters, cf. [4, 21]. We also note that the possibility to model "data-driven computation" in contrast to "demand-driven computation" as games is discussed in an early paper on game semantics by Abramsky and Jagadeesan [2].

The main technical contribution of the present work is the validation of the semantic exactness with which the induced universe captures the call-by-value sequential higher-order computation through the full abstraction result for the call-by-value version of PCF [35, 40], a paradigmatic functional calculus. The result seems the first one in this context[1] and is easily extendable to other languages as we shall indicate in Section 6. We also clarify the relationship between the present universe of games and the corresponding call-by-name universe by showing they are faithfully embeddable to each other. These results indicate, together with the preceding results on call-by-name PCF [3, 19], that the two basic notions of calling procedures in higher-order computation are representable in the game-based semantic framework in an exact way, and that they

---

[1] Independently and concurrently Riecke and Sandholm [38] obtained a similar result, see Section 6.

arise as two independent, though mutually related, semantic universes with equal status (which parallels the findings in domain theory, cf. [11]). It is also notable that, as we clarify later, the universe of call-by-value games assumes basic type structures which have arisen through the categorical analysis of domain-theoretic universes for call-by-value, or partial, computation, cf.[11, 12, 23, 31, 32, 36, 39], though with a strong intensional flavour. This suggests an abstract notion of "call-by-value computation" may be delineated apart from the standard domain theoretic constructions, cf. [11, 12].

The structure of games we shall use is a conservative extension of the construction by Hyland and Ong [19]. The relationship is detailed in [18].

This is an extended abstract of [18]. The reader may refer to [18] for proofs and detailed technical discussions. In the remainder, Section 2 introduces the basic notion of games and strategies. Sections 3 and 4 outline the algebraic structures of the category of games and its extensional quotient. Section 5 establishes the main result of the paper, the inequational full abstraction for call-by-value PCF. Section 6 discusses further results and remaining topics. Appendix briefly reviews call-by-value PCF.

2. Games and Strategies

This section introduces the basic construction of games and strategies which are to become objects and morphisms in the categorical universe. We start from sorting (the terminology is from [29]), from which call-by-value types arise as its specific subclass.

2.1. Sorting and Type.

(i) (sorting) A *sorting* $\mathbb{S}$ is a triple of: (1) $\underline{\mathbb{S}}$, which is a collection of mutually disjoint non-empty sets ranged over by $S, S', \ldots$ each called a *sort*, (2) $\lambda : \underline{\mathbb{S}} \to \{ [, (,],) \}$, a labelling function and (3) $Obs : \underline{\mathbb{S}} \to 2^{\underline{\mathbb{S}}}$, the justification relation (if $S' \in Obs(S)$ we say S justifies S'), where $S' \in Obs(S)$ implies:
- $\lambda(S) = [$ then $\lambda(S') \in \{ (,] \}$. Dually $\lambda(S) = ($ then $\lambda(S') \in \{ [,) \}$.
- $\lambda(S) =]$ then $\lambda(S') = [$ always. Dually $\lambda(S) =)$ then $\lambda(S') = ($ always.

Elements of a sort are called *actions*, denoted $x, y, \ldots$, writing e.g. x^S when $x \in S$. The set of *initial sorts*, denoted $\text{init}(\mathbb{S})$, is given as $\{ S \mid \text{for no } S' \in \underline{\mathbb{S}}. S \in Obs(S') \}$.

(ii) (type) A *cbv-type*, or simply a *type*, is a sorting such that all initial sorts are labelled by "]" and any of its sorts is reachable from some initial sort, where reachability is understood regarding sortings as graphs (nodes are sorts, directed edges are given by Obs). Types are denoted by $A, B, C, \ldots$.

An action of a sort labelled by each of "$(, [,],)$" is called, respectively, *Player Question*, *Opponent Question*, *Player Answer*, and *Opponent Answer*, the first two collectively *Question*, the last two *Answer*, the first and third *P-action*, and the second and fourth *O-action*. Answers of initial sorts are often called *signals*. On labels we define a self-inverse function $\overline{(\cdot)}$, giving the *dual* of a label, satisfying: $\overline{[} = ($ and $\overline{]} =)$.

2.2. Examples. (sorting)

(i) **0** is the empty sorting, which is a type. **1** is a sorting whose unique]-labelled sort is a singleton, which is again a type. **nat** is made as **1** replacing a singleton with ω (the set of natural numbers), similarly **bool** with $\{\text{true}, \text{false}\}$.

(ii) Given $\mathbb{S}$, write $\overline{\mathbb{S}}$ for the sorting which is the result of changing labels by $\overline{(\cdot)}$. So $\overline{\text{nat}}$ is the sorting with the same sort as **nat** which is however labelled by ")". Next, given $\mathbb{S}_1$ and $\mathbb{S}_2$, let $\mathbb{S}_1 \uplus \mathbb{S}_2$ denote their disjoint union, i.e. the sorts are the disjoint union of $\mathbb{S}_1$ and $\mathbb{S}_2$, inheriting labelling and justification. Then $\overline{\text{nat}} \uplus \text{nat}$ is the sorting with two copies of ω labelled by ")" and "]".

(iii) We define nat$\rightleftharpoons$nat as a type with three sorts, one is a singleton written $]^{\mathsf{nat}\rightleftharpoons\mathsf{nat}}$, another a copy of ω written $[^{\mathsf{nat}}$, and the third again a copy of ω written $]^{\mathsf{nat}}$, for which labels are given as these notations indicate. The justification is given so that $]^{\mathsf{nat}\rightleftharpoons\mathsf{nat}}$ only justifies $[^{\mathsf{nat}}$, which in turn only justifies $]^{\mathsf{nat}}$.

By a *sequence from a set* X we mean a partial function from ω to X defined for a finite initial segment of ω (called *indices*) and undefined for the rest. As an example, abc has $\{0,1,2\}$ as its indices. We often confuse elements and their occurrences in a sequence. ε denotes the empty sequence. We are interested in sequences of actions representing a certain kind of interaction between an agent (Player) and the outside (Opponent).

2.3. Action Sequence. Given a sorting $\mathbb{S}$, an *action sequence in* $\mathbb{S}$ or often simply a *sequence in* $\mathbb{S}$ is a sequence from actions in $\mathbb{S}$ (let is be $x_0 x_1 ... x_{n-1}$), together with the relation on its indices denoted $\mapsto$ (writing $x_i \mapsto x_j$ for $i \mapsto j$), satisfying:

(consistency) (1) $x_i \mapsto x_j \Rightarrow i \lneq j$, (2) $(x_i \mapsto x_k \wedge x_j \mapsto x_k) \Rightarrow i = j$, (3) $x_i^S \mapsto x_k^{S'} \Rightarrow S' \in Obs(S)$, (4) $\neg \exists x_i . \, x_i \mapsto x_j^S \Rightarrow S$ initial (then x_j *occurs free*),

(linearity in answers) (5) $(x_i \mapsto x_j \wedge x_i \mapsto x_k \wedge x_k$ an answer$) \Rightarrow j = k$, (6) A free O-answer (resp. a free P-answer) occurs at most once, and:

(strict alternation) (7) If x_i is a P-action (resp. O-action), then x_{i+1} is an O-action (resp. P-action) for $0 \leq i \leq n-2$.

$s, s', \ldots$ range over action sequences, often leaving the associated $\mapsto$ implicit. We say x_i *justifies* x_j when $x_i \mapsto x_j$. On action sequences we define two functions, $\ulcorner s \urcorner$, the *P-view* of s, and $\llcorner s \lrcorner$, the *O-view* of s, as, inheriting $\mapsto$ whenever possible: **(pv0)** $\ulcorner \varepsilon \urcorner = \varepsilon$, **(pv1)** $\ulcorner sx_i \urcorner = x_i$ when x_i is a free O-action, **(pv2)** $\ulcorner s_0 x_i s_1 x_j \urcorner = \ulcorner s_0 \urcorner x_i x_j$ when $x_i \mapsto x_j$ and x_j is an O-action, and **(pv3)** $\ulcorner s_0 x_i \urcorner = \ulcorner s_0 \urcorner x_i$ if x_i is a P-action; $\llcorner s \lrcorner$ is defined dually, i.e. by exchanging "O-action" and "P-action" throughout. We then say:

(i) s is *well-bracketed* when: if $s_0 x_i s_1 x_j$ is a prefix of s such that (1) x_i is a question (2) x_j is an answer and (3) either x_j occurs free or x_j is justified by a question in s_0, then x_i justifies an answer in s_1.

(ii) s *satisfies the visibility condition* when, in any of its prefix $s_0 x_i$ where x_i is a P-action (resp. O-action) which is y_j s.t. $y_j \mapsto x_i$ always occurs in $\ulcorner s_0 \urcorner$ (resp. $\llcorner s_0 \lrcorner$).

An action sequence is *legal* when it is well-bracketed and satisfies the visibility condition. Legal action sequences are sometimes called *legal positions*. We can verify the set of legal sequences of any sorting is closed under prefix and view constructions.

We are now ready to give the main definition of this section, which determines the class of interacting processes we are concerned with in the present study.

2.4. Definition. (strategy) An *innocent strategy from A to B*, or simply a *strategy from A to B*, is a prefix-closed set σ of legal positions in $\overline{A} \uplus B$, such that:

(O-initial) $s \in \sigma$ implies the initial action of s (if any) is an O-action.

(contingency completeness) $s \in \sigma$ and sx_i is legal for an O-action x_i imply $sx_i \in \sigma$.

(innocence) If $s_1 x, s_2 \in \sigma$, x is a P-action and $\ulcorner s_1 \urcorner = \ulcorner s_2 \urcorner$, then $s_2 y \in \sigma$ such that (1) $\ulcorner s_1 x \urcorner = \ulcorner s_2 y \urcorner$ and (2) $s_2 z \in \sigma \Rightarrow s_2 z = s_2 y$.

We write $\sigma : A \to B$ when σ is a strategy from A to B. f_σ denotes the partial function determined by σ, mapping even-length P-views to next actions (if any) with justification. Given $\sigma, \tau : A \to B$, we set $\sigma \leq \tau$ when $\sigma \subset \tau$, equivalently when $f_\sigma \subset f_\tau$.

Using the function representation, it is easy to see the set of strategies from A to B forms a dI-domain under $\leq$, where compact elements are those with finite graphs. Further, given $sx_ix_{i+1} \in \sigma : A \to B$, if x_i and x_{i+1} come from different types then x_{i+1} is necessarily a P-action (switching condition). Also the projection of $s \in \sigma : A_1 \to A_2$ onto A_i ($i = 1, 2$), written $s \upharpoonright A_i$, is always legal in A_i.

2.5. Examples. (strategies)

(i) (undefined) For each A and B, there is a strategy from A to B which is totally undefined, so that it is least w.r.t. the ordering $\leq$. We write this strategy $\bot_{A \to B}$.

(ii) (first-order function) The set of strategies from nat to nat precisely correspond to the set of partial functions from ω to ω.

(iii) (higher-order function) We describe a strategy $\sigma : \mathsf{nat} \Rightarrow \mathsf{nat} \to \mathsf{nat}$ which corresponds to the behaviour of an open call-by-value PCF-term, $x : \iota \to \iota \rhd \mathsf{succ}(x3) : \iota$. After receiving a signal on the left, which is a function, σ asks the result of applying 3 to that function, and, on receiving the answer, returns its successor to the right. Except the last free answer, each action is justified by the preceding one.

Strategies denote a certain kind of deterministic processes, and are, as such, precisely representable as (name passing) synchronisation trees, see [18]. The presentation is often useful for describing, and reasoning about, strategies: indeed the full abstraction result was originally obtained in this setting [17]. The following inductive definition of composition of strategies is suggested by such representation.

2.6. Definition. (composition) Given $\sigma : A \to B$ and $\tau : B \to C$, we set:

$$\{s_1; s_2 \mid s_1 \in \sigma,\ s_2 \in \tau,\ s_1 \upharpoonright B = s_2 \upharpoonright \overline{B}\}$$

where $s_1; s_2$ with s_1 and s_2 as above is given: (1) $\varepsilon; \varepsilon = \varepsilon$, (2) $s_1 x^B; s_2 \overline{x^B} = s_1; s_2$ ($\overline{x^B}$ is the corresponding dual action of x^B), and (3) $s_1 x^A; s_2 = (s_1; s_2) x^A$, $s_1; s_2 x^C = (s_1; s_2) x^C$, in each case inheriting the justification relation from the original pair.

(3) above is well-defined since two cases are always disjoint due to the switching condition. We can also verify: (i) $\sigma; \tau$ is a strategy from A to C, (ii) ; is associative with identity given by the copy-cat strategy, i.e. that which exactly copies actions between $\overline{A}$ and A, and (iii) ; is bi-continuous with respect to $\leq$. Thus we define:

2.7. Definition. $\mathcal{CBV}$ denotes the category of cbv-types and innocent strategies.

By the preceding discussions, $\mathcal{CBV}$ is enriched over CPO, the category of possibly bottomless cpos and continuous functions. Each homset has a least element $\bot$ for which the composition is left strict, that is $\bot; \sigma = \bot$ always.

3. Intensional Universe

Type structures of a semantic universe offer the basic articulation of its algebraic structures needed, for example, for interpreting various programming languages in it. This section clarifies the basic type structure of $\mathcal{CBV}$ in the light of the distinction between total and partial maps. We first introduce the notion of totality, cf. [13].

3.1. Definition. σ is *total* when $\tau; \sigma = \bot$ implies $\tau = \bot$. We write $\sigma \Downarrow$ when σ is total.

The totality of $\sigma : A \to B$ is equivalent to any one of: (1) $\forall \tau : \mathbf{1} \to A.\ \tau \Downarrow \Rightarrow \tau; \sigma \Downarrow$, (2) the square $\langle \mathbf{0} \to A \xrightarrow{\sigma} B,\ \mathbf{0} \to \mathbf{0} \to B \rangle$ is a weak pullback (notice $\mathbf{0}$ is initial and weakly terminal), and (3) σ immediately emits the P-signal for each initial O-signal. (1) relates to a familiar idea of totality, (2) is a categorically basic one, and (3) gives the behavioural characterisation, clarifying the dynamic aspect of totality.

3.2. **Examples.** (total maps)

(i) The unique arrow $\bot$ from $\mathbf{0}$ to any type is total, by definition. All isomorphisms are total. Also, there is no total map to $\mathbf{0}$, except from itself.

(ii) There is a unique total map $!_A : A \to \mathbf{1}$ for each A. It reacts to the initial signal (if any) by the unique P-signal at $\mathbf{1}$, and no more action is possible.

(iii) $\sigma : \mathsf{nat} \to \mathsf{nat}$ is total iff the underlying number-theoretic function is total.

Let us denote $\mathcal{CBV}_t$ for the category of types and total strategies. Since totality is closed upwards w.r.t. $\leq$, $\mathcal{CBV}_t$ again CPO-enriches. It has finite products: 3.2 (ii) above shows $\mathbf{1}$ is terminal, while the product of A and B is given by a type $A \otimes B$ whose sorts are the disjoint union of non-initial sorts of A and B together with, for each pair of $S \in \mathsf{init}(A)$ and $S' \in \mathsf{init}(B)$, a sort $]^{S,S'} = S \times S'$ (the set theoretic product), which justifies what S and S' justify in A and B, the rest as in A and B ($A \otimes \mathbf{0}$ and $\mathbf{0} \otimes A$ are set as $\mathbf{0}$). Projection maps are evidently given. $\otimes$ is often denoted $\times$ in $\mathcal{CBV}_t$. We also note that $\mathcal{CBV}_t$ has arbitrary (small) products and co-products, but we do not need them here.

The relationship between total maps and usual (often called *partial*) maps is clarified by the notion of *lifting*. Write $A_\bot$ for the type given by adding two singleton sorts to A, one initial which justifies the other one, the latter justifying all $S \in \mathsf{init}(A)$, the rest as in A. Then we can see the set of *total* arrows from $\mathbf{1}$ to $A_\bot$ is order-isomorphic to the set of *partial* arrows from $\mathbf{1}$ to A. These two are mediated by two copy-cat like strategies, $\mathsf{up} : A \to A_\bot$ and $\mathsf{dn} : A_\bot \to A$, with obvious behaviours (up reacts to an initial action at A by going though two added actions at $A_\bot$ then does the copy-cat; dn just does the dual). In a familiar way this induces the adjoint situation as described below.

3.3. **Proposition.**

(i) Let F be the inclusion functor from $\mathcal{CBV}_t$ to $\mathcal{CBV}$. Then F has the right adjoint T, with $T(A) = A_\bot$, the unit $\eta_A = \mathsf{up}$, and the co-unit $\epsilon = \mathsf{dn}$, which CPO-enriches. The monad $\langle T, \eta_A, T(\mathsf{dn}) \rangle$ is denoted $\mathbf{T}$, which has a tensorial strength $\mathsf{st}_{A,B}$ and a co-strength (in the sense of [37]) $\mathsf{st}'_{A,B}$.

(ii) The Kleisli category of $\mathbf{T}$ on $\mathcal{CBV}_t$ is isomorphic to $\mathcal{CBV}$. We write $\sigma^\dagger$ for $\mathsf{up}; T(\sigma) : A \to B_\bot$ where $\sigma : A \to B$ is partial, and $\sigma_\dagger$ for $\sigma; \mathsf{dn} : A \to B$ where $\sigma : A \to B_\bot$ is total.

Using the monad $\mathbf{T}$, we can now present the basic type structures of $\mathcal{CBV}$. In (iii) below $A \rightrightarrows B$ is a type whose sorts are the disjoint union of those of A and B together with new $]^{A \rightrightarrows B}$ which is a singleton, with the label of each $S \in \mathsf{init}(A)$ changed into [and those of A's non-initial sorts dualised. Justification is as in A and B, with the addition of $]^{A \rightrightarrows B}$ justifying what were in $\mathsf{init}(A)$, each of which in turn justifying what were in $\mathsf{init}(B)$ ($\mathbf{0} \rightrightarrows B$ is set as $\mathbf{1}$). Notice the similarity with the construction of $\overline{A} \uplus B$.

3.4. **Definition and Proposition.**

(i) (partial pairing [32]) Given $\sigma_1 : C \to A$ and $\sigma_2 : C \to B$, their *left pairing*, $\langle\!\langle \sigma_1, \sigma_2 \rangle\!\rangle_l : C \to A \otimes B$, and the *right pairing*, $\langle\!\langle \sigma_1, \sigma_2 \rangle\!\rangle_r : C \to A \mathbin{\mathcal{S}} B$ are given as: $\langle\!\langle \sigma_1, \sigma_2 \rangle\!\rangle_l \stackrel{\text{def}}{=} (\langle \sigma_1^\dagger, \sigma_2^\dagger \rangle; \psi_{A,B})_\dagger$ and $\langle\!\langle \sigma_1, \sigma_2 \rangle\!\rangle_r \stackrel{\text{def}}{=} (\langle \sigma_1^\dagger, \sigma_2^\dagger \rangle; \tilde{\psi}_{A,B})_\dagger$ where $\psi_{A,B} = \mathsf{st}'_{A,TB}; T(\mathsf{st}_{A,B}; \mathsf{dn})$ and $\tilde{\psi}_{A,B} = \mathsf{st}_{A,TB}; T(\mathsf{st}'_{A,B}; \mathsf{dn})$.

(ii) (premonoidal tensor [37]) Given A, we define $A \otimes$ and $\otimes A$ by: (i) $A \mathbin{\mathcal{S}} \cdot B = A \otimes B$ and $B \cdot \otimes A = B \otimes A$, and (ii) $A \otimes \sigma \stackrel{\text{def}}{=} \langle\!\langle \pi_1, \pi_2; \sigma \rangle\!\rangle$ and $\sigma \otimes A \stackrel{\text{def}}{=} \langle\!\langle \pi_1; \sigma, \pi_2 \rangle\!\rangle$ where π_i denote projections. Then $A \otimes$ and $\otimes A$ both define functors on $\mathcal{CBV}$

which CPO-enrich. We then define, for $\sigma : A \to B$ and $\tau : C \to D$: (i) $\sigma \otimes_l \tau = (\sigma \otimes C); (B \otimes \tau)$ and (ii) $\sigma \otimes_r \tau = (A \otimes \sigma); (\tau \otimes C)$.

(iii) (partial exponential [23]) The functor $\_ \otimes A : CBV_t \to CBV$ has the right adjoint $A \Rightarrow \_ : CBV \to CBV_t$, which CPO-enriches. Equivalently, there exists an arrow $\text{ev} : (A \Rightarrow B) \otimes A \to B$ such that, for any $\sigma : C \otimes A \to B$, there is a unique total arrow $p\lambda(\sigma) : C \to B$ satisfying $(p\lambda(\sigma) \otimes \text{id}); \text{ev} = \sigma$, and $p\lambda$ is a continuous operator.

An outstanding fact on partial pairing is that the right and left pairings of the same tuple do *not* coincide in general. This exhibits a strongly intensional character of CBV, substantiating Moggi's remark that $\langle\!\langle \sigma_1, \sigma_2 \rangle\!\rangle_l$ and $\langle\!\langle \sigma_1, \sigma_2 \rangle\!\rangle_r$ reflect the "order of evaluation" [32]. This also implies the tensor in CBV does not give a bifunctor, cf. Corollary 4.3 of [37]. We write $\langle\!\langle \sigma_1, \sigma_2 \rangle\!\rangle$ when two versions coincide (as when either is total).

The final structure we need is *recursion*, here presented as an operator on each homset. [18] gives an alternative presentation as constants. Below we say A is *pointed* when it has a unique initial sort which is a singleton, equivalently when $\hom(1, A)$ in CBV_t is a pointed cpo. For such A, $\text{dn}'_A : TA \to A$ denotes the unique total map such that $\text{up}_A; \text{dn}'_A = \text{id}_A$. Pointed types are precisely objects in the category of Eilenberg-Moore algebra of the monad **T**. Also, any type of form $A \Rightarrow B$ is pointed.

3.5. Proposition. Let A be pointed and $\sigma : C \times A \to A$. Then there is a strategy $\text{rec}(\sigma) : C \to A$ which satisfies: (i) $\tau; \text{rec}(\sigma) = \tau; \langle\!\langle \text{id}_C, \text{rec}(\sigma) \rangle\!\rangle; \sigma$ for $\tau : 1 \to C$ (if σ is total we can take off τ from the equation). (ii) $\text{rec}(\tau \otimes \text{id}_A; \sigma) = \tau; \text{rec}(\sigma)$ for each $\tau : B \to C$, and: (iii) Given $\tau : 1 \to C$, if $\{\rho_i : 1 \to A\}_{i \in \omega}$ is defined as: (1) $\rho_0 = \bot$, (2) $\rho_{i+1} = \langle\!\langle \tau, \rho_i^\dagger; \text{dn}' \rangle\!\rangle; \sigma$, then $\{\rho_i\}$ is an increasing ω-chain such that $\sqcup \rho_i = \tau; \text{rec}(\sigma)$.

4. Extensional Universe

CBV represents an abstract notion of execution of call-by-value computation. For the interpretation of programming languages at the same abstraction level as in the standard semantic universe like the category of domains, we may need a more abstract universe, which we construct from CBV by a simple quotient construction. The universe is also useful for understanding the behaviour of arrows in CBV in an abstract way. Below we briefly outline the basic structure of this universe, leaving details to [18]. We start from the following ordering (cf. [36, 11]):

$$\sigma_1 \precsim \sigma_2 \stackrel{\text{def}}{\Leftrightarrow} \forall C, C', \tau : C \to A, \tau' : B \to C'. \; \tau; \sigma_1; \tau' \Downarrow \;\Rightarrow\; \tau; \sigma_2; \tau' \Downarrow.$$

Immediately $\precsim$ is a preorder for which the composition is monotone (thus the quotient is well-defined), and $\leq \subseteq \precsim$. We now define $\widehat{CBV}$ as the category of types and $\precsim$-equivalence classes of strategies. $f, g, \ldots$ range over arrows in $\widehat{CBV}$. The induced partial order is still denoted $\precsim$. $\widehat{CBV}$ is enriched over Poset, the category of posets with monotone maps, since monotonicity carries over from CBV. Observing **0** is the zero object in $\widehat{CBV}$ (i.e. both terminal and initial), we define $\bot : A \to B$ as the unique map that factors through **0**, cf. [13]. Then $\bot$ is indeed the least element in each homset, and the composition is strict at both sides. We can then define total maps as before: $f \Downarrow$ when $g; f = \bot$ implies $g = \bot$ for each g, equivalently when the square $\langle 0 \to A \stackrel{f}{\to} B, 0 \to 0 \to B \rangle$ is a pullback, from which all properties of total maps as in CBV follow. Notice also $f \Downarrow \Leftrightarrow \forall \sigma \in f. \sigma \Downarrow \Leftrightarrow \exists \sigma \in f. \sigma \Downarrow$. We write $\widehat{CBV}_t$ for the subcategory of total maps.

We can then show $\widehat{CBV}_t$ is well-pointed, with finite products (indeed all small products and co-products) inducing Poset-enriched bi-functors, all inheriting from CBV_t. Again as in CBV, the inclusion functor from $\widehat{CBV}_t$ to $\widehat{CBV}$ has the right adjoint inheriting constructions from T, which we write again T, which Poset-enriches. The corresponding monad, again denoted $\mathbf{T}$, has strengths and is now commutative, i.e. $\psi_{A,B} = \tilde{\psi}_{A,B}$ in 3.4 (i). Again the Kleisli category of $\mathbf{T}$ on $\widehat{CBV}_t$ is isomorphic to $\widehat{CBV}$. Using the monad, we can now clarify the basic type structures of $\widehat{CBV}$. Thus, again from the general result by Power and Robinson [37], we know $\widehat{CBV}$ is a Poset-enriched symmetric monoidal category, which has all type structures as given in Proposition 3.4 (i)(ii)(iii) inheriting the constructions from CBV, where left and right pairings are identified. Finally the recursion in CBV carries over to $\widehat{CBV}$, though all $\tau : 1 \to C$ in 3.5 can be replaced with id_C. We also note that $\widehat{CBV}$ allows the treatment of recursive types for a large class of functors, but we do not use them in the present paper.

5. Interpretation of $\mathrm{PCF_V}$

$\mathrm{PCF_V}$[35, 36] is a typed programming language based on call-by-value evaluation. The syntax and evaluation rules can be found in the standard literature, cf.[15, 42, 40], which are briefly reviewed in Appendix (following [15] except the recursion is only defined for function types, cf.[42, 40]). CBV and its extensional quotient are conceived to represent call-by-value, or partial, higher-order functional computation. Moreover it has a type structure which does include that of $\mathrm{PCF_V}$. Thus we may seek to represent $\mathrm{PCF_V}$-terms and its computation in these universes. We primarily consider the interpretation in CBV, and only move to $\widehat{CBV}$ at the last step. The interpretation follows.

5.1. Definition. First we define the mapping from the set of types and environments of $\mathrm{PCF_V}$ to objects in CBV as: $[\![\iota]\!] \stackrel{\mathrm{def}}{=} \mathbf{nat}$, $[\![o]\!] \stackrel{\mathrm{def}}{=} \mathbf{bool}$, $[\![\alpha \Rightarrow \beta]\!] \stackrel{\mathrm{def}}{=} [\![\alpha]\!] \Rightarrow [\![\beta]\!]$, $[\![\varepsilon]\!] \stackrel{\mathrm{def}}{=} 1$ and $[\![\Gamma, x : \alpha]\!] \stackrel{\mathrm{def}}{=} [\![\Gamma]\!] \otimes [\![\alpha]\!]$. Then the mapping from $\mathrm{PCF_V}$-terms to arrows in CBV is given inductively as follows, assuming either of the left/right pairings is selected uniformly.

(i) $[\![\Gamma, x : \alpha, \Delta \triangleright x : \alpha]\!] \stackrel{\mathrm{def}}{=} \pi : [\![\Gamma]\!] \otimes [\![\alpha]\!] \otimes [\![\Delta]\!]$, where π is an appropriate projection.

(ii) $[\![\Gamma \triangleright \lambda x^\alpha . M : \alpha \Rightarrow \beta]\!] \stackrel{\mathrm{def}}{=} p\lambda(\sigma) : [\![\Gamma]\!] \to [\![\beta]\!]$, where $[\![\Gamma, x : \alpha \triangleright M : \beta]\!] = \sigma$.

(iii) $[\![\Gamma \triangleright MN : \beta]\!] \stackrel{\mathrm{def}}{=} \langle\!\langle \sigma_1, \sigma_2 \rangle\!\rangle; \mathbf{ev} : [\![\Gamma]\!] \to [\![\beta]\!]$, where $[\![\Gamma \triangleright M : \alpha \Rightarrow \beta]\!] = \sigma_1$ and $[\![\Gamma \triangleright N : \alpha]\!] = \sigma_2$.

(iv) $[\![\Gamma \triangleright \mu x^\alpha . M : \alpha]\!] \stackrel{\mathrm{def}}{=} \mathbf{rec}(\sigma) : [\![\Gamma]\!] \to [\![\alpha]\!]$, where $[\![\Gamma, x : \alpha \triangleright M : \alpha]\!] = \sigma$

(v) $[\![\Gamma \triangleright \mathbf{cond}\ L\ M_1\ M_2 : \alpha]\!] \stackrel{\mathrm{def}}{=} (\langle\!\langle \tau, \langle\!\langle \sigma_1^\dagger, \sigma_2^\dagger \rangle\!\rangle \rangle\!\rangle; \gamma_{T([\![\alpha]\!])})_\dagger : [\![\Gamma]\!] \to [\![\alpha]\!]$ where $[\![\Gamma \triangleright L : o]\!] = \tau$, $[\![\Gamma \triangleright M_1 : \alpha]\!] = \sigma_1$, $[\![\Gamma \triangleright M_2 : \alpha]\!] = \sigma_2$ and $\gamma_A : \mathbf{bool} \otimes A \otimes A \to A$ is a strategy with an appropriate behaviour.

(vi) For a constant c of type α, we set: $[\![\Gamma \triangleright c : \alpha]\!] \stackrel{\mathrm{def}}{=} !_{[\![\Gamma]\!]}; \bar{c} : [\![\Gamma]\!] \to 1 \to [\![\alpha]\!]$ where $\bar{c} : 1 \to [\![\alpha]\!]$ is given as a strategy with obvious behaviour for each c.

The descriptions of γ and $\bar{c}$ for each c are given in [18]. As basic properties of the mapping, we know $[\![\Gamma \triangleright V : \alpha]\!]$ is always total, where V denotes a value, i.e. an abstraction or a non-Ω constant; $[\![\Gamma \triangleright M\{V/x\} : \beta]\!] = \langle\!\langle \mathrm{id}_{[\![\Gamma]\!]}, \tau \rangle\!\rangle; \sigma : [\![\Gamma]\!] \to [\![\beta]\!]$ for any $\tau = [\![\Gamma \triangleright V : \alpha]\!]$ and $\sigma = [\![\Gamma, x : \alpha \triangleright M : \beta]\!]$; and that $\Gamma \triangleright M \Downarrow V$ implies $[\![M]\!] = [\![V]\!]$. We can then verify the following key properties of the interpretation.

5.2. Proposition.

(i) (computational adequacy) $[\![M]\!] \neq \bot$ iff $\exists V.\ M \Downarrow V$ for a closed M.

(ii) (adequacy) $[\![M]\!] \precsim [\![N]\!]$ implies $M \precsim_{obs} N$ for closed M, N of the same type.

Given the adequacy result, if we show its converse, i.e. $\precsim_{obs}$ implies $\precsim$ via the interpretation, then we obtain the full abstraction. For the purpose it suffices to prove all compact elements of appropriate types are PCF$_V$-definable, cf.[25, 35]. The definability argument is carried out using a subset of PCF$_V$-terms defined as follows.

5.3. Definition. *Finite canonical forms* (FCFs for short) are inductively given as:
 (i) $\Gamma \triangleright \Omega : \alpha$ and $\Gamma \triangleright n : \iota$ are FCF's.
 (ii) $\Gamma \triangleright \lambda y^\alpha . M : \alpha \to \beta$ is a FCF if $\Gamma, y : \alpha \triangleright M : \beta$ is.
 (iii) $\Gamma \triangleright \text{let } y^\alpha = zV \text{ in } N : \beta$ is a FCF if (1) $\Gamma, y : \alpha \triangleright N : \beta$ is a FCF, (2) z has a type $\beta \Rightarrow \alpha$ in Γ, and (3) $\Gamma \triangleright V : \beta$ is a FCF (which is also a value).
 (iv) $\Gamma \triangleright (\text{case } x \text{ of } n_1 : M_1 \,[\!]\, n_2 : M_2 \,[\!].\!.\,[\!]\, n_k : M_k\,) : \alpha$ is a FCF if $x : \iota \in \Gamma$ and, for each i, $\Gamma \triangleright M_i : \alpha$ is a FCF.
where, in (iii), **let** $y^\alpha = zM$ **in** N stands for $(\lambda y^\alpha.N)(zM)$, and, in (iv), **case** y **of** n_1: $M_1\,[\!].\!.\,[\!]\,n_k : M_k$ stands for cond $(y = n_1)\, M_1(...(\text{cond } (y = n_k)\, M_k\, \Omega)..)$, the latter assuming the equality check is suitably encoded in PCF$_V$.

FCFs faithfully capture the behaviour of compact strategies of PCF-types:
 (i) Ω denotes $\bot$. $m : \iota$ immediately returns m after an initial O-signal.
 (ii) $\lambda x^\alpha . M : \alpha \Rightarrow \beta$ represents a strategy which, after an initial O-signal, does a sequence of actions $]^{\alpha \Rightarrow \beta}\, [^\alpha$ (here an annotated label denotes an action of that kind) where $]^{\alpha \Rightarrow \beta} \mapsto [^\alpha$, then behaves as M.
 (iii) $\Gamma, x_i : \gamma_1 \Rightarrow \gamma_2, \Delta \triangleright \text{let } y^\alpha = x_i M \text{ in } N : \beta$ first interacts at x_i by $(^{\gamma_i}$, then Opponent may ask at M (when γ_1 is a higher-order type) which, after some interactions, will be answered by Player, followed by an Opponent Answer $)^{\gamma_2}$. Then the actions move to N. Here the "let" construct is used to make the order of evaluation explicit (see [32] for a similar use of the construct in a different context).
 (iv) The case statement corresponds to the situation when a strategy acts according to the received ground values (here natural numbers). A vector of values can be handled by nesting the construct.

Using FCFs we can prove:

5.4. Theorem. (definability) For each compact element $\sigma : 1 \to [\![\alpha]\!]$ for any PCF$_V$-type α in $\mathcal{CBV}$, there is a FCF $F : \alpha$ such that $[\![F : \alpha]\!] = \sigma$. Conversely, the interpretation of any FCF is a compact element in the respective type.

The proof is by induction on the cardinality of compact elements, translating the behaviour of strategies into the corresponding FCFs based on the correspondence between actions and strategies we illustrated above. We note that, like FCFs themselves, the argument is much simpler than the corresponding one in call-by-name PCF, cf.[19]. See [18] for details. Write $[\![\Gamma \triangleright M : \alpha]\!]_e$ for $[\![\Gamma \triangleright M : \alpha]\!]_{\precsim}$. From the definability result we can now conclude:

5.5. Theorem. (full abstraction) For closed PCF$_V$-terms $M : \alpha$ and $N : \alpha$, we have $M : \alpha \precsim_{obs} N : \alpha$ iff $[\![M : \alpha]\!]_e \precsim [\![N : \alpha]\!]_e$.

6. DISCUSSIONS

6.1. Further Results. First we briefly outline how call-by-name universe and the call-by-value universe are mutually embeddable, as in the context of domains. Let *cbn-types* be sortings in which (1) initial sorts are all opponent questions and (2) each sort is reachable from some initial sort. The strategies are then as in Definition 2.4 with an added condition which ensures the switching condition. The composition of strategies is just as in Section 2, based on which we obtain the category of cbn-types and innocent strategies which is cartesian-closed and is enriched over CPO, which we denote $\mathcal{CBN}$. There is a full embedding of $\mathbb{CA}$ of [19] in $\mathcal{CBN}$ and its extensional quotient allows interpretation of call-by-name FPC as in the category in [24]. Now we say a $\mathcal{CBN}$ type is *pointed* when it has a unique initial sort which is a singleton, just as in $\mathcal{CBV}$. Let us also say a strategy in $\mathcal{CBN}$ is *linear* when, after the initial question at the codomain, it immediately asks the question at the domain, and never asks an initial question at the domain again. Writing $\mathcal{CBN}_1$ for the subcategory of $\mathcal{CBN}$ of pointed types and linear strategies, the embedding result says (i) $\mathcal{CBN}$ is isomorphic to the full subcategory of $\mathcal{CBV}_t$ of pointed types, and (ii) $\mathcal{CBV}$ is isomorphic to the full subcategory of $\mathcal{CBN}_1$ of pointed types whose initial questions justify no questions. The proof is by the translation of information flow. See [18] for details.

Next we discuss how we would extend the full abstraction result in Section 5 to other call-by-value programming languages. Firstly it is straightforward to extend the argument in Section 5 to PCF$_V$ with sums and products or to the untyped call-by-value λ-calculus. Recursively typed languages such as FPC [15] can also be handled (though the premonoidal tensor in $\mathcal{CBV}$ poses a problem), as observed by Fiore and as will be reported elsewhere. For the interpretation of imperative constructs, we would consider, as noted in Introduction, variants of the present universe by changing parameters of games following [4, 21], which does lead to coherent semantic universes. One interesting topic in this context would be whether one needs refined type structures as in [4] for the interpretation of the impure constructs: indeed a much simpler, and more direct, approach seems possible in the present setting. Some results on these topics will be reported elsewhere.

6.2. Related works. After completing the full version of this paper [18], the authors were informed of an independent (and essentially concurrent) work by Riecke and Sandholm [38] in which they obtained a full abstraction for call-by-value FPC (which easily implies that of PCF$_V$). The construction is based on Kripke logical relations on pCPO, and is thus quite different from the present one. No quotienting is necessary to reach the semantic universe, while the construction of the universe itself is substantially more complicated. In a brief comparison, one may say that their approach would give better insights for understanding why some (continuous) function is *not* sequential; while their construction does not directly model the dynamic aspects of sequential call-by-value computation, thus may not lead to the insights in that context. Thus two methods would play different roles in semantic analysis.

In game semantics, Abramsky and McCusker are working on game semantics on call-by-value languages, based on McCusker's early idea and also suggested by the present work, which tries to extract call-by-value strategies from the universes of call-by-name games in [24, 4] (personal communication).[2] In another vein, Harmer and Malacaria are working on game semantics for call-by-value computation based on games originally introduced in [3]. [16] gives a preliminary study in this direction.

6.3. Intensionality and relationship with process theories. The strongly intensional character of $\mathcal{CBV}$ is not at the same level of abstraction as, say, pCPO. The same can be said about its call-by-name counterpart and other categories of games, in the sense

---

[2]At the final stage of preparation of this camera-ready version, we obtained their typescript [5], which exploits the type structures of the original universe in [4] to interpret a functional language with a certain imperative feature. Detailed discussions, especially the comparison with an approach we mentioned in 6.1, should be left for a future occasion.

that they reflect some notion of execution, albeit abstractly, cf. [9, 19]. From the viewpoint that the primary purpose of semantic representation of programming languages lies in giving (in)equations over programs as general as possible, this feature may be considered as a drawback. However we can take a different perspective, and ask whether this novel way of representing programs can be put to a significant use, especially once given the full abstraction result as the semantic justification of the representation. As a first such step, one may exploit the representation for the development of abstract theory of execution, including the formal optimisation techniques. Type structures as we studied in Section 4 may be put to an effective use in this context. One interest in this regard is that our interpretation of PCF_V in CBV already gives a concise abstract implementation of the language in the form name passing processes. The representation is comparable to Milner's direct encoding in [27], performing the β_v-reduction by three name passing interactions. Such a "physical" character of the abstract universe suggests we may study the execution of, say, call-by-value programming languages from a new level of mathematical abstraction (this is in line with Girard's studies on the semantics of cut elimination [14]). Relatedly the induced encodings also suggest the possibility of relating game semantics and process theories at the fundamental level. The study of behavioural types by Milner [28] may suggest possible directions (from which the present study actually started).

Acknowledgments. Special thanks go to Marcelo Fiore for his suggestions concerning pertinent categorical structures. We thank Samson Abramsky, Paul Mellies, Pasquale Malacaria, Guy McCusker, Jon Riecke and anonymous referees for comments and/or discussions, and N. Raja for his hospitality in Bombay.

REFERENCES

[1] Abelson, H., Sussman, G.J., *Structure and Interpretation of Computer Program*, MIT Press, 1985.
[2] Abramsky, S. and Jagadeesan, R., Games and Full Completeness for Multiplicative Linear Logic, Journal of Symbolic Logic, 59(2), pp. 543–574, 1994.
[3] Abramsky, S., Jagadeesan, R. and Malacaria, P., Full Abstraction for PCF, 1994. To appear.
[4] Abramsky, S. and McCusker, G., Linearity, Sharing and State: a fully abstract game semantics for Idealized Algol with active expressions, ENTCS, Vol.3, North Holland, 1996.
[5] Abramsky, S. and McCusker, G., Call-by-value games, a typescript, 12p, Apr. 1997.
[6] Berry, G. and Curien, P. L., Sequential algorithms on concrete data structures. *TCS* Vol.20, pp.265–321, North-Holland, 1982.
[7] Blass, A., A game semantics for linear logic, *Annuals of Pure and Applied Logic*, 56:183–220, 1992.
[8] Curien, P. L., Sequentiality and full abstraction. In Proc. of *Application of Categories in Computer Science*, LNM 177, pp.86–94, Cambridge Press, 1995.
[9] Danos, V. and Regnier, L., Games and abstract machines. *LICS'96*, IEEE, 1994.
[10] Felsheer, W., Dialogue games as a foundation for intuitionistic logic, *Handbook of Philosophical logic*, Vol.3, pp.341–372, D. Reidel Publishing Company, 1986.
[11] Fiore, M., *Axiomatic Domain Theory in Category of Partial Maps*, PhD thesis, ECS-LFCS-94-307, Univ. of Edinburgh, 1994.
[12] Fiore, M. and Plotkin, G., An Axiomatisation of Computationally Adequate Domain Theoretic Models of FPC, *LICS'94*, pp.92–102, IEEE, 1994.
[13] Freyd, P., Algebraically Complete Categories, In Proc. of *Como. Category Theory Conference*, LNM 1488, pp.95–104, Springer Verlag, 1991.
[14] Girard, J.-Y., Linear Logic, *TCS*, Vol.50, pp.1–102, North-Holland, 1987.
[15] Gunter, C., *Semantics of Programming Languages: Structures and Techniques*, MIT Press, 1992.
[16] Harmer, R., Malacaria, P., Linear foundations of game semantics, a typescript, Sep. 1996.
[17] Honda, K., Yoshida, N., Name-Passing Games: a functional universe, a typescript, 35p, Nov. 1996.
[18] Honda, K. and Yoshida, N., Game-theoretic Analysis of Call-by-value Computation (full version of this paper), ftp-able at ftp.dcs.ed.ac.uk/export/kohei/cbvfull.ps.gz, Feb, 1997.
[19] Hyland, M. and Ong, L., On Full Abstraction for PCF: I, II and III, 130 pages, ftp-able at theory.doc.ic.ac.uk/papers/Ong, 1994.
[20] Hyland, M. and Ong, L., Pi-calculus, dialogue games and PCF, *FPCA'93*, ACM, 1995.

[21] Laird, J., Full abstraction for functional languages with control, *LICS'97*, IEEE, 1997.
[22] Kahn, G. and Plotkin, D., Domaines Concrets. INRIA Report 336, 1978.
[23] Longo, G. and Moggi, E., Cartesian closed categories of enumarations for effective type-structures, LNCS 173, Springer-Varlag, 1984.
[24] McCusker, G., Games and Full Abstraction for FPC. *LICS'96*, IEEE, 1996.
[25] Milner. R., Fully abstract models of typed lambda calculi. *TCS*, Vol.4, 1–22, North-Holland, 1977.
[26] Milner, R., *A Calculus of Communicating Systems*, LNCS 76, Springer-Verlag, 1980.
[27] Milner, R., Functions as Processes. *MSCS*, 2(2), pp.119–146, 1992.
[28] Milner, R., Sorts and Types of π-Calculus, a manuscript, 43pp, 1990.
[29] Milner, R., Polyadic π-Calculus: a tutorial. *Proceedings of the International Summer School on Logic Algebra of Specification*, Marktoberdorf, 1992.
[30] Milner, R., Tofte, M. and Harper, R., *The Definition of Standard ML*, MIT Press, 1990.
[31] Moggi, E., Partial morphisms in categories of effective objects, *Info.&Comp.*, 76:250–277, 1988.
[32] Moggi, E., Notions of Computations and Monads. *Info.&Comp.*, 93(1):55–92, 1991.
[33] Nickau, M., Hereditarily Sequential Functionals, LNCS 813, pp.253–264, Springer-Verlag, 1994.
[34] Ong, L., Correspondence between Operational Semantics and Denotational Semantics, *Handbook of Logic in Computer Science*, Vol.4, pp.269–356, Oxford University Press, 1995.
[35] Plotkin, G., LCF considered as a programming language, *TCS*, 5:223–255, North-Holland, 1975.
[36] Plotkin, G., Lecture on Predomains and Partial Functions. Notes for a course given at the Center for the Study of Language and Information, Stanford, 1985.
[37] Power, J., Robinson, E., Premonoidal Categories and Notions of Computation, To appear in *MSCS*.
[38] Riecke, J., and Sandholm,A. Relational Account of Call-by-value Sequentiality, *LICS'97*, 1997.
[39] Robinson, E. and Rosolini, P., Categories of Partial Maps, *Info.&Comp.*, 79:95–130, 1988.
[40] Sieber, K., Relating Full Abstraction Results for Different Programming Languages, *FST/TCS'10*, pp. 373-387, LNCS 472, Springer-Verlag, 1990.
[41] Winskel, G., Synchronization Trees, *TCS*, Vol.34, pp. 33–82, North-Holland, 1985.
[42] Winskel, G., *The Formal Semantics of Programming Languages*, MIT Press, 1993.

Appendix: PCF_V

We give a brief review of syntax and operational semantics of the call-by-value PCF [15, 42, 40]: our treatment is nearest to [15]. Given an infinite set of *variables*, ranged over by $x, y, z, \ldots$, the syntax of the language is given as follows.

$$\alpha ::= \iota \mid o \mid \alpha \Rightarrow \beta \qquad M ::= x \mid \lambda x^\alpha.M \mid MM \mid \text{cond } L\, M_1\, M_2 \mid \mu x^{\alpha \Rightarrow \beta}.M \mid c$$

where c is a constant. An *environment* is a list of pairs of a variable and a type, where all variables are distinct, ranged over by $\Gamma, \Delta, \ldots$ The typing rules of PCF_V is given as:

$$\Gamma, x:\alpha, \Gamma' \triangleright x:\alpha \qquad \frac{c \text{ is a constant of type } \alpha}{\Gamma \triangleright c:\alpha} \qquad \frac{\Gamma \triangleright M:\alpha \Rightarrow \beta \quad \Gamma \triangleright N:\alpha}{\Gamma \triangleright MN:\beta}$$

$$\frac{\Gamma, x:\alpha \triangleright M:\beta}{\Gamma \triangleright \lambda x^\alpha.M:\alpha \Rightarrow \beta} \qquad \frac{\Gamma \triangleright L:o \quad \Gamma \triangleright M:\alpha \quad \Gamma \triangleright N:\alpha}{\Gamma \triangleright \text{cond } L\, M\, N:\alpha} \qquad \frac{\Gamma, x:\alpha \Rightarrow \beta \triangleright M:\alpha \Rightarrow \beta}{\Gamma \triangleright \mu x.M:\alpha \Rightarrow \beta}$$

As a set of constants, we assume: $n : \iota$ for each numeral n, $\Omega : \alpha$ for each α, succ $: \iota \Rightarrow \iota$, and zero? $: \iota \Rightarrow o$. Terms of form $\triangleright M : \alpha$ (often written $M : \alpha$) are called *closed terms*. Abstractions and constants except Ω are called *values*.

On the set of terms we define an evaluation relation $\Downarrow$ in the style of natural semantics.

$$V \Downarrow V \qquad \frac{M \Downarrow \lambda x.M_0 \quad N \Downarrow V \quad M_0\{V/x\} \Downarrow U}{MN \Downarrow U} \qquad \frac{M\{\mu x.M/x\} \Downarrow V}{\mu x.M \Downarrow V} \qquad \frac{M \Downarrow n}{\text{succ } M \Downarrow n+1}$$

$$\frac{M \Downarrow 0}{\text{zero?}M \Downarrow \text{true}} \qquad \frac{M \Downarrow n+1}{\text{zero?}M \Downarrow \text{false}} \qquad \frac{L \Downarrow \text{true} \quad M_1 \Downarrow V}{\text{cond } L\, M_1\, M_2 \Downarrow V} \qquad \frac{L \Downarrow \text{false} \quad M_2 \Downarrow U}{\text{cond } L\, M_1\, M_2 \Downarrow U}$$

Finally an *observational preorder* on closed terms is defined as follows: $M \preceq_{obs} N$ iff, for any well-typed context of a program type $C[\cdot]$, we have $C[M] \Downarrow n$ iff $C[N] \Downarrow n$. We note that this is the same thing as considering convergence at *all* types, a situation quite different from the case of call-by-name evaluation.

On Modular Properties of Higher Order Extensional Lambda Calculi

Roberto Di Cosmo   Neil Ghani

DMI-LIENS (CNRS URA 1327)
Ecole Normale Supérieure - 45, Rue d'Ulm - 75230 Paris, France
e-mail:dicosmo@dmi.ens.fr, nxg@cs.bham.ac.uk

Abstract. We prove that confluence and strong normalisation are both modular properties for the addition of algebraic term rewriting systems to Girard's F^ω equipped with either β-equality or $\beta\eta$-equality. The key innovation is the use of η-expansions over the more traditional η-contractions.

We then discuss the difficulties encountered in generalising these results to type theories with dependent types. Here confluence remains modular, but results concerning strong normalisation await further basic research into the use of η-expansions in dependent type theory.

1 Introduction

A property P is *modular* for the combination of rewrite systems $\mathcal{T}_1$ and $\mathcal{T}_2$ iff whenever both $\mathcal{T}_1$ and $\mathcal{T}_2$ satisfy P, then so does the combined rewrite system $\mathcal{T}_1 \cup \mathcal{T}_2$. This paper studies the modularity of confluence and strong normalization for combinations of higher order lambda calculi and algebraic term rewriting systems. That is, does the addition of a confluent algebraic TRS to a higher order lambda calculus (with or without rewrite rules for η-conversion) produce a system which is still confluent? Similarly, is the combination of a strongly normalising algebraic TRS and a higher order lambda calculus (again, with or without rewrite rules for η-conversion) still SN? And do these results generalise to dependent type theories such as the Calculus of Constructions? These questions are important from both a theoretical point of view, where one looks for general results on combination of rewriting systems, and from a practical point of view, when one develops higher order semi-unification algorithms, or establishes the formal properties of algebraic-functional languages.

Tannen [9] showed that strong normalization and confluence are both moldular properties for the combination of algebraic TRS's with the simply typed lambda calculus equipped with β-reduction. Gallier and Tannen [10, 11] extended these results to System F. Although strong normalisation remains modular in these type theories if we work with both β- and η-reductions, confluence is no longer a modular property. For example, if s is a base type with constants $f : s \to s$ and $* : s$ and with a rewrite rule $fx \Rightarrow *$, then $\Rightarrow$ is confluent. However, the combination of $\Rightarrow$ with the contractive η-rewrite rule fails to be confluent: $\lambda x.* \Leftarrow \lambda x.fx \Rightarrow f$. Because of these problems with η-contractions, later research was restricted to adding more expressive TRSs to systems equipped only with β-reduction. In particular, translations into intersection type-assignment systems [3, 29, 26, 6, 5, 7, 4] were used to prove the modularity of strong normalisation and *completeness*, i.e. the property of strong normalisation and confluence together, with confluence following from strong normalisation by Newman's lemma. As far as the authors are aware, modularity of confluence *alone* was not pursued any further and no attempts were made to study modularity results for calculi equipped with $\beta\eta$-equality.

This paper extends the works of Tannen and Gallier in several ways. Firstly, we shall consider more expressive calculi such as Girard's F^ω and Coquand and Huet's Calculus of Constructions, henceforth denoted CoC. We show that confluence is modular for the combination of algebraic TRS's with these calculi (without η-conversion). As mentioned earlier, these results are surprisingly missing in the literature. Our second contribution is to extend these modularity results to calculi equipped with $\beta\eta$-equality. This is done by replacing the problematic interpretation of η-conversion as a contractive rewrite relation with its more recent interpretation as an expansionary rewrite rule. Eta-expansions in the simply typed λ-calculus were first studied in the 70's but only recently they made the object of accurate study in a number of papers [1, 16, 13, 19, 27, 17] (for an up-to-date survey, the interested reader can refer to [15]). This paper relies on Ghani's recent results on η-expansions in F^ω [23] and CoC [22].

2 Extensional and Non-extensional F^ω

We use the standard notions of substitutions, reduction, normal form, confluence, normalization, etc., from the theory of λ-calculus and rewriting systems [8, 14]. The *free variables* of a term M are denoted $FV(M)$ and we write $M\theta$ for the result of applying a substitution θ to the term M. The *domain* of a substitution θ is denoted $\text{dom}(\theta)$. If $\mathcal{R}$ is a rewrite relation with unique normal forms, then reduction to $\mathcal{R}$-normal form is denoted $\mathcal{R} \downarrow$ and the unique $\mathcal{R}$-normal form of t is denoted $\mathcal{R}(t)$. Finally, a relation R commutes with S iff $(R^*)^{-1}; S^* \subseteq S^*; (R^*)^{-1}$ where ; is the usual composition of relations. If two confluent relations commute, then their union is also confluent.

In this section, two versions of F^ω will be defined. *Extensional* F^ω uses $\beta\eta$-equality for type conversion while *non-extensional* F^ω has only β-equality for type conversion — our presentation is based on Gallier's [21]. Formally, let $*$ be a distinguished symbol and let TVar and Var be disjoint sets of type variables and term variables. These variables are used to define the *kinds*, *types* (also called *type constructors*) and *terms* of F^ω as follows:

$(Kinds)\ K := *|K \to K$

$(Types)\ T := t|T{\to}T|\forall t : K.T|\lambda t : K.T|T\,T$

$(Terms)\ M := x|\lambda x : T.M|M\,M|\Lambda t : K.M|M[T]$

where $t \in$ TVar is a type variable and $x \in$ Var is a term variable. A term is called an *abstraction* iff it is of the form $\lambda x : T.M$ or $\Lambda t : K.M$. In order to ensure that types inhabit unique kinds, we assign to each type variable t a unique kind and denote the set of type variables having kind K as TVar(K). This kinding information is used to define the kinding judgements of F^ω as follows

$$\frac{t \in \text{TVar}(K)}{t : K} \qquad \frac{s : K_2 \quad t \in \text{TVar}(K_1)}{(\lambda t : K_1.s) : K_1 \to K_2} \qquad \frac{t : K_1 \to K_2 \quad s : K_1}{ts : K_2}$$

$$\frac{t \in \text{TVar}(K) \quad s : *}{\forall t : K.s : *} \qquad \frac{t : * \quad s : *}{t \to s : *}$$

In order to give the typing judgements of extensional F^ω we define the usual $\beta\eta$-equality relation on well-kinded types; if two types t and s are $\beta\eta$-equal, we denote this by writing $t =_{\beta\eta} s$. The following lemma is proved in [23]

Lemma 1. *$\beta\eta$-equality over types can be generated by a confluent, strongly normalizing reduction relation containing β reduction and restricted η-expansions. The unique normal form of a type A is its long $\beta\eta$-normal form and is denoted* NF(A).

The typing judgements of extensional F^ω are defined by the following rules, while the typing judgements of non-extensional F^ω use only β-equality for type conversion.

$$\frac{x : T \in \text{dom}(\Gamma)}{\Gamma \vdash x : T} \qquad \frac{\Gamma \vdash M : t \quad t =_{\beta\eta} s \quad s : K}{\Gamma \vdash M : s}$$

$$\frac{\Gamma, x : t_1 \vdash M : t_2}{\Gamma \vdash (\lambda x : t_1.M) : t_1 \to t_2} \qquad \frac{\Gamma \vdash M : t_1 \to t_2 \quad \Gamma \vdash N : t_1}{\Gamma \vdash MN : t_2}$$

$$\frac{\cdot\, \Gamma, t_1 : K \vdash M : t_2}{\Gamma \vdash \Lambda t_1 : K.M : \forall t_1 : K.t_2} \qquad \frac{\Gamma \vdash M : \forall t_1 : K.t_2 \quad \Gamma \vdash s : K}{\Gamma \vdash M[s] : t_2[s/t_1]}$$

In the rest of this paper, we confine our attention to only those types that kind check and those terms that type check. In addition, we increase legibility by dropping all reference to the context in which a typing judgement occurs whenever there is no danger of confusion arising.

2.1 Eta-expansions in F^ω

As argued in the introduction, any robust result concerning the modularity of confluence in the presence of η-conversion requires its interpretation as an expansion. In the simply typed λ-calculus, one permits an expansion $t \Rightarrow \lambda x : A.tx$ providing that t is neither a λ-abstraction nor applied to another term. This restricted expansion relation is SN, confluent and its reflexive, symmetric and transitive closure is $\beta\eta$-equality. Thus $\beta\eta$-equality can be decided by reduction to normal form in this restricted fragment.

However, defining η-expansion in F^ω requires further care so as to avoid pitfalls caused by the presence of multiple typings for terms. For instance, if an expansion $M \xrightarrow{\eta} \lambda x : A.Mx$ is permitted providing $M : A \to B$, then η-expansion alone is not even confluent as there are rewrites

$$\lambda x : A'.Mx \xleftarrow{\eta} M \xrightarrow{\eta} \lambda x : A.Mx$$

where we only know that $A =_{\beta\eta} A'$ in the type-conversion relation. Worse, η-expansion defined this way does not have unique normal forms and hence the usual strategy for computing long normal forms (first contract β redexes and then perform all remaining expansions) would no longer be valid. For these reasons we define a *type normalised* form of η-expansion as follows

$$M \xrightarrow{\bar{\eta}} \lambda x : A.Mx, \text{ if } \begin{cases} x \ fresh \\ M : A \to C, \text{with } A \to C \ in \ type \ normal \ form \\ M \text{ is not a } \lambda\text{-abstraction} \\ M \text{ is not applied} \end{cases} \qquad (1)$$

Note that the existence of type normal forms is assured by lemma 1. There is no need for a type-normalised form of the higher order η-rewrite rule because if a term inhabits the types $\forall t : K.A$ and $\forall t : K'.A'$, then we must have $K = K'$. Hence our higher order η-expansion is:

$$M \xrightarrow{\eta} (\Lambda t : K.M[t]) \text{ if } \begin{cases} t \ fresh \\ M : (\forall t : K.A) \\ M \text{ is not a polymorphic } \lambda\text{-abstraction} \\ M \text{ is not applied} \end{cases} \qquad (2)$$

Definition 2. Let β be the rewrite relation consisting of all β-reductions on types and term. Also, let $\bar{\eta}$ be the rewrite relation consisting of all restricted expansions on types and those expansions given in rules 1 and 2. The relation η is defined by ommiting the restriction to type normal forms in rule 1. Finally define $\beta\bar{\eta} = \beta \cup \bar{\eta}$ and $\beta\eta = \beta \cup \eta$.

Results such as the modularity of confluence and strong normalisation are proven first for $\beta\bar{\eta}$ and then lifted to the more general $\beta\eta$ via the following lemma.

Lemma 3. *The reflexive, symmetric and transitive closure of* $\xrightarrow{\beta\eta}$ *and* $\xrightarrow{\beta\bar{\eta}}$ *are both the usual $\beta\eta$-equality over terms of F^ω.*

Proof. Firstly, all η equalities $M = \lambda x : A.Mx$ that seem to be forbidden by the restrictions of $\xrightarrow{\beta\bar{\eta}}$ can be obtained by β-reduction of $\lambda x : A.Mx$. Thus the reflexive, symmetric, transitive closure of $\xrightarrow{\beta\bar{\eta}}$ is $\beta\eta$-equality. For the second part of the lemma, notice that $\xrightarrow{\beta\bar{\eta}}$-expansions are examples of $\xrightarrow{\beta\eta}$-expansions. In addition, if $M \xrightarrow{\beta\eta} \lambda x : A.Mx$, but A is not a type normal form, then both of these terms $\xrightarrow{\beta\bar{\eta}}$-reduce to $\lambda x : \text{NF}(A).Mx$.

The major theorems concerning $\beta\eta$ and $\beta\overline{\eta}$ are

Theorem 4. *The rewrite relations $\beta\overline{\eta}$ and $\beta\eta$ are confluent and strongly normalizing to the long $\beta\eta$-normal forms. The long $\beta\eta$-normal form of a term may be calculated by first contracting all β-redexes and then performing any remaining type-normalised η-expansions.*

3 Modularity Results for F^ω

In this section we define algebraic TRSs and show the modularity of confluence and strong normalisation for the unions of algebraic TRSs with F^ω. First some definitions.

Definition 5. A signature Σ consists of disjoint sets $\mathcal{T}$ of *base types* and $\mathcal{F}$ of *function symbols* together with a function which assigns to every function symbol $f \in \mathcal{F}$, a *typing* of the form $f : \alpha_1 \to \ldots \to \alpha_n \to \alpha$, where $\alpha_1, \ldots, \alpha_n, \alpha \in \mathcal{T}$ and $n \geq 0$. We say the *arity* of f is n.

Definition 6. An *algebraic rewrite rule* is an ordered pair (T, U) of algebraic terms such that T is not a variable, and every variable of U also appears in T. An *algebraic term rewriting system* $\mathcal{T}$ is a finite set $\{(T_i, U_i)\}_{i=1}^n$ of algebraic rewrite rules.

Definition 7. Given an algebraic TRS $\mathcal{T}$, the associated *algebraic rewrite relation* is the least binary relation $\xrightarrow{\mathcal{T}}$ on terms such that if $(T, U) \in \mathcal{T}$, θ is a substitution and C is a context, then $C[T\theta] \xrightarrow{\mathcal{T}} C[U\theta]$

Given an algebraic TRS, its union with calculi such as F^ω is defined as expected. A term of the union of an algebraic TRS and F^ω is *algebraic* if it is either a variable of base type or has the form $f\,t_1 \ldots t_n$, where $f \in \mathcal{F}$ has arity n, and every t_i is an algebraic term. Note that an algebraic term is always of base type. The key concept in modular term rewriting is the *layer structure*, i.e. the ability to decompose a term constructed from symbols in the union of two disjoint signatures into a term constructed from symbols in only one signature and strictly smaller subterms whose head symbol comes from the other signature. We follow [10] in using the following defintions relating to layer structure.

Definition 8. A typing judgement $\Gamma \vdash M : s$ is called *trunk* iff M is of the form $fM_1, \ldots, M_k$ where f is a constant of arity k, otherwise it is called *non-trunk*.

Definition 9. An *algebraic trunk decomposition* of a typing judgement $\Gamma \vdash M : s$ consists of a typing judgement $\Delta \vdash A : s$, where A is an algewbraic term, and a term-valued substitution ϕ such that $M = A\phi$, $\text{dom}(\phi) = FV(A)$ and

- Each free variable in A occurs only once

- For each $x \in FV(A)$, the typing judgement $\Gamma \vdash \phi(x) : s$ is non-trunk.

Note that all judgements $\Gamma \vdash M : s$ are either trunk or non-trunk because M is of basesort. Induction shows that all typing judgements $\Gamma \vdash M : s$ have algebraic trunk decompositions which are unique upto the renaming of the free variables of A. We therefore write $M = A[\phi]$ for an algebraic trunk decomposition of M and refer to A as a trunk of the term M.

Example 1. If f is a binary function symbol and a is a non-trunk term, then a trunk decomposition for the term $f\,aa$ is $f\,xy[a/x, a/y]$. If g is a unary function symbol and a is a constant, then a trunk decomposition of $g((\lambda x : s.x)(a))$ is $gy[(\lambda x : s.x)(a)/y]$

Definition 10. A reduction $M = A[\phi] \xrightarrow{\mathcal{T}} N$ is a *trunk* reduction iff the redex contracted is not a subterm of one of the $\phi(x)$'s, otherwise it is a *non-trunk* reduction.

Example 2. Using the terms of example 1, and given a rewrite rule $fxx \xrightarrow{T} x$, there is a trunk reduction $faa \xrightarrow{T} a$. There is a non-trunk reduction $g((\lambda x : s.x)(a)) \xrightarrow{\beta} ga$.

Example 2 show two undesirable properties of reduction. Firstly, the presence of non-left linear rewrite rules means that trunk reductions do not induce reductions of the trunk of the redex. For instance $faa \xrightarrow{T} a$ but there is no reduction $fxy \xrightarrow{T} x$. Also β-reduction may collapse the *layer structure* of a term and hence a non-trunk reduction need not preserve the trunk of the redex, eg the trunk of $g((\lambda x : s.x)(a))$ is gy but the trunk of ga is ga. We solve the first problem by introducing a special term variable j^s for each sort and then defining a special substitution j which maps every term variable of type s to j^s. There is also solution for the second problem.

Lemma 11. *Let $A\phi$ be a trunk decomposition for M*

- *If $M \xrightarrow{T} N$ is not a trunk reduction, then there is an algebraic trunk decomposition $N = A\phi'$ such that for some $x \in FV(A)$, $\phi(x) \xrightarrow{T} \phi'(x)$, while for all other $y \in FV(A)$, $\phi(y) = \phi'(y)$.*

- *If $M \xrightarrow{T} N$ is a trunk reduction, then there is an algebraic trunk decomposition $N = A'\phi'$ such that $Aj \xrightarrow{T} A'j$ and for every $y \in FV(A')$, there exists an $x \in FV(A)$ such that $\phi'(y) = \phi(x)$.*

- *If $M \xrightarrow{\beta\bar{\eta}} N$, then there is an algebraic trunk decomposition $N = A'\phi'$ and for every $y \in FV(A')$, there exists an $x \in FV(A)$ such that either $\phi(x) \xrightarrow{\beta\bar{\eta}} N_x$ and $\phi'(y)$ is a subterm of N_x, or $\phi(x) = \phi'(y)$.*

Proof. The lemma is proved by induction on the term M.

3.1 Modularity of Confluence

The proof strategy of [11] is used to show the modularity of confluence for the combination of algebraic TRSs with both extensional and non-extensional F^ω. In particular, reduction to long $\beta\eta$-normal form in F^ω commutes with algebraic reductions.

Lemma 12. *If $\mathcal{T}$ is a confluent algebraic rewriting system (over algebraic terms), then it is confluent over the terms of $F^\omega \cup \mathcal{T}$ (mixed terms).*

Proof. This proof of [11] generalises to F^ω and CoC because the *only* property required of mixed terms is that the trunk of a term is preserved by non-trunk, algebraic reductions, as proven in lemma 11.

Lemma 13. *Reduction to β normal form commutes w.r.t. algebraic reduction, i.e.*

$$\begin{array}{ccc} & \xrightarrow{\mathcal{T}} & \\ \beta\downarrow & & \downarrow\beta \\ & \dashrightarrow{\mathcal{T}} & \end{array}$$

Proof. See lemma 31 in the appendix for the proof.

These lemmas allow us to derive our first modularity result, namely that of confluence for the addition of algebraic TRSs to non-extensional F^ω. This is a new result as it shows modularity of confluence alone, and not of confluence and strong normalization together as in [7]:

Corollary 14. *The union of non-extensional F^ω with a confluent algebraic TRS is confluent.*

Proof. By lemma 13, if $t =_{\beta \cup \mathcal{T}} t'$, then $\beta(t) =_{\mathcal{T}} \beta(t')$. By lemma 12, $\mathcal{T}$ is confluent *over mixed terms*. Hence $\beta(t)$ and $\beta(t')$ have a common $\mathcal{T}$-reduct and hence t and t' have a common reduct.

Proving that confluence is modular for the addition of algebraic TRSs to extensional F^ω requires us to relate algebraic rewriting to expansive normal forms, extending [17]:

Lemma 15. *Reduction to $\overline{\eta}$ normal form commutes w.r.t. algebraic reduction, i.e.*

$$\begin{array}{ccc} \cdot & \xrightarrow{\mathcal{T}} & \cdot \\ {\scriptstyle \overline{\eta}}\downarrow & & \downarrow{\scriptstyle \overline{\eta}} \\ \cdot & \dashrightarrow{\mathcal{T}} & \cdot \end{array}$$

Proof. The proof is by induction on the structure of terms. The fact that the η normal form of a term is unique is necessary for the lemma to hold with arbitrary TRSs and not only left-linear ones.

As a consequence of the previous lemmas, we have the following

Corollary 16. *Reduction to $\beta\overline{\eta}$ normal form commutes with algebraic reduction, i.e.*

$$\begin{array}{ccc} \cdot & \xrightarrow{\mathcal{T}} & \cdot \\ {\scriptstyle \beta\overline{\eta}}\downarrow & & \downarrow{\scriptstyle \beta\overline{\eta}} \\ \cdot & \dashrightarrow{\mathcal{T}} & \cdot \end{array}$$

Proof. By theorem 4, the long $\beta\eta$-normal form of a term can be computed by first contracting all β-redexes and then performing any remaining (restricted) η-expansions. Thus the corollary follows from lemma 13 and lemma 15.

Theorem 17. *The union of $\beta\overline{\eta}$ with a confluent algebraic TRS is confluent.*

Proof. As in corollary 14 using corollary 16.

Corollary 18. *The union of $\beta\eta$ (where η is not restricted to type normal forms) with a confluent algebraic TRS $\mathcal{T}$ is confluent.*

Proof. If two terms are $\mathcal{T} \cup \beta\eta$ equivalent, they are $\mathcal{T} \cup \beta\overline{\eta}$ equivalent and hence by theorem 17 there is a $\mathcal{T} \cup \beta\overline{\eta}$ completion for these terms. But this is also a $\mathcal{T} \cup \beta\eta$ completion.

3.2 Modularity of Strong Normalization

The relations $\beta\overline{\eta}$ and $\beta\eta$ were proved confluent and SN in [23] by a modified reducibility argument, adapted from traditional reducibility proofs to cope with the presence of expansionary η-rewrite rules. Reducibility arguments are designed to cope with the higher order features at the level of kinds and type constructors, while the effect of adding algebraic TRSs is only felt at the level of base types. Thus these reducibility arguments generalise to prove the modularity of strong normalisation for the combination of algebraic TRSs with extensional F^ω.

Lemma 19. *If $\mathcal{T}$ is a SN algebraic TRS, then its extension to F^ω is also SN.*

Proof. The lemma is proved by induction on the structure of terms with the only interesting case being a trunk term $M = A\phi$. By lemma 11, any infinite reduction sequence of M induces either an infinite reduction sequence of a $\phi(x)$, or an infinite reduction sequence of $A\jmath$. The first possiblity is impossible by the induction hypothesis, while the second possibility is also impossible as $\mathcal{T}$ is SN on algebraic terms and $A\jmath$ is algebraic.

We now prove the main result of this section, namely that the union of a SN algebraic TRS and $\beta\bar{\eta}$-reduction in F^ω is SN. The proof follows the modified reduciblity argument of [23] and thus we only sketch the general reducibility argument and concentrate instead on the particular novelties which arise via the addition of algebraic TRSs. One defines a notion of *reducibility candidate* and *reducibility parameter* exactly as in [23] and proves that if T is a type and θ is a reducibility parameter, then $T\theta$ is a reducibility candidate. The only new case is when T is a sort s and here the reducibility candidate $s\theta$ is defined to be the SN terms of type s. The following pair of lemmas are the key to completeing the proof.

Lemma 20. *If the terms $t_1, \ldots, t_n$ are SN, then so is $ft_1 \ldots t_n$.*

Proof. That there are no infinite $\beta\bar{\eta}$ reduction sequences is proved in [23]. By corollary 16, a rewrite $ft_1 \ldots t_n = M \xrightarrow{\mathcal{T}} N$ induces a sequence of rewrites $M_0 \xrightarrow{\mathcal{T}} N_0$ where M_0 and N_0 are the long $\beta\eta$-normal forms of M and N. Close inspection of the proof shows that if the initial rewrite is of the trunk, then this induced rewrite sequence is of length at least one. Hence there can be no infinite reduction sequences containing an infinite number of trunk rewrites. By lemma 11, all other infinite reduction sequences of $ft_1 \ldots t_n$ induce infinite reduction sequences of one of the terms t_i which is prohibited by assumption.

Lemma 21. *If t_i is a SN term of sort s_i for $i = 1, \cdots, m$, and f has type $s_1 \to \ldots \to s_n$ where $m < n$, then $ft_1 \ldots t_m$ is reducible.*

Proof. The proof is by induction on the type of the term $ft_1 \ldots t_m$. If this type is a sort, then we must show that $ft_1 \ldots t_m$ is SN under the assumption that each of the t_i are SN. But this is precisely lemma 20. If however the type of $ft_1 \ldots t_m$ is of the form $s \to T$, then we must show that if t is a reducible term of type s, then $ft_1 \ldots t_m t$ is reducible. Since the reducible terms of type s are exactly the SN ones, this follows from the induction hypothesis.

Lemma 22. *If $\mathcal{T}$ is a SN algebraic TRS, then so are $\beta\bar{\eta} \cup \mathcal{T}$ and $\beta \cup \mathcal{T}$.*

Proof. Having defined reducibility candidates as in [23], the proof concludes by showing that if t is an arbitrary term, θ is a reduciblity parameter, the free term variables of t are among $x_j : T_j$ and u_j are members of the reducibility candidate $T_j\theta$, then $t[|\theta|][u_j/x_j]$ is a member of the reduciblity candidate $T\theta$ (note $|\theta|$ is the type-valued substitution underlying the reduciblity parameter θ).

The only new case is when t is of the form $ft_1 \ldots t_n$ and one must show $(ft_1 \ldots t_n)[|\theta|][u_j/x_j]$ is reducible when each of the terms $t_i[|\theta|][u_j/x_j]$ is reducible. But this follows from lemma 21. Strong normalisation of $\beta\bar{\eta} \cup \mathcal{T}$ follows by taking the identity substitution and identity reducibility parameter, while strong normalisation of $\beta \cup \mathcal{T}$ follows as this is a subrelation of $\beta\bar{\eta} \cup \mathcal{T}$.

There is a simple trick to extend strong normalisation of $\beta\bar{\eta} \cup \mathcal{T}$ to $\beta\eta \cup \mathcal{T}$. If t is a term, let $\text{TNF}(t)$ be the *type normal form* of t, ie the term that is obtained by normalising all the types occuring as subterms and in λ-abstractions in t. A reduction $t \xrightarrow{\beta\eta} t'$ is called *type induced* iff the redex contracted occurs inside a subterm of t which is actually a type.

Lemma 23. *If there is a rewrite $t \xrightarrow{\beta\eta\cup\mathcal{T}} t'$, then there is a rewrite $\text{TNF}(t) \xrightarrow{\beta\bar{\eta}\cup\mathcal{T}} \text{TNF}(t')$. If the original rewrite is not type induced then the final rewrite sequence is not of zero length.*

Proof. The lemma is proved exactly as in [23]

Corollary 24. *If $\mathcal{T}$ is a SN algebraic TRS, then $\beta\eta \cup \mathcal{T}$ is also SN.*

Proof. There are no infinite sequences of type induced reductions because reduction on types is SN. In addition, if $t \xrightarrow{\beta\eta} t'$ is type induced, then $\text{TNF}(t) = \text{TNF}(t')$. Thus any infinite $\beta\eta \cup \mathcal{T}$ reduction sequence is mapped by type normalisation to an infinite $\beta\bar{\eta} \cup \mathcal{T}$ reduction sequence.

4 Modularity for Algebraic TRS and CoC

We have proven a series of modularity results concerning the addition of algebraic TRSs to F^ω. The next logical step is to apply the same ideas to the much more powerful Calculus of Constructions [12]. Due to lack of space, we cannot introduce it here in detail, but we recall that the most important feature is that the distinction between types and terms is blurred and types can contain terms embedded within them; let β and η refer to the Calclulus of Constructions rules in this section. Type dependency introduces infinite reduction sequences which are not present in non-dependent type theories. For example, if we define expansions by

$$\frac{\Gamma \vdash t : \Pi x : A.B}{\Gamma \vdash t \Rightarrow \lambda x : A.tx}$$

and define the term $B(x) = (\lambda z : X \to X.X)(x)$, then there is a typing judgement $X : *, x : X \to X \vdash x : \Pi z : B(x).X$ and hence an infinite reduction sequence

$$X : *, x : X \to X \vdash x \Rightarrow \lambda z : B(x).xz \Rightarrow \lambda z : B(\lambda z : B(x).xz).xz \Rightarrow \ldots$$

Notice that this example does not use any higher order types and so can be formulated in simpler dependent type theories such as LF. The existence of infinite reduction sequences such as the one above forces us to restrict our attention to a type normalised form of restricted η-expansion which we again denote by $\overline{\eta}$. Further, let $\beta\overline{\eta}$ be the rewrite relation containing all β-reductions and type normalised restricted expansions and $\beta\eta$ be defined as in $\beta\overline{\eta}$ but without the type normal form requirement.

In F^ω the existence of type normal forms is easy to prove as reduction at the level of types is defined independently to reduction at the level of terms. However in a dependent type theory such as CoC the existence of long $\beta\eta$-normal forms is much harder to prove. One can either use the standard theory of η-contractions as in [20] or prove their existence while simultaneously developing the theory of expansions as in [22]. The following lemma is proved in [22] – we conjecture that $\beta\overline{\eta}$ is actually SN but a proof awaits further research.

Theorem 25. *$\beta\overline{\eta}$ and $\beta\eta$ are confluent and weakly normalising to the long $\beta\eta$-normal forms.*

4.1 Modularity of Confluence

As we have described above, the theory of strong normalization for η-expansions in Coc is not settled. Nevertheless, we can use confluence and weak normalization of $\beta\overline{\eta}$ to good avail and get the modularity of confluence for the union of algebraic TRSs with CoC.

Lemma 26. *Algebraic reduction commutes with β-normalization in CoC.*

Proof. As in [11]. Again, see lemma 31.

Corollary 27. *If $\mathcal{T}$ is a confluent algebraic TRS, then $\beta \cup \mathcal{T}$ is also confluent*

Proof. As in corollary 14 and using lemma 26

Proving that confluence is modular for the union of algebraic TRSs with extensional CoC requires another commutation lemma.

Lemma 28. *Algebraic reduction commutes with $\overline{\eta}$-normalisation.*

Proof. Similar to lemma 15.

Corollary 29. *If $\mathcal{T}$ is a confluent algebraic TRS, then $\beta\overline{\eta} \cup \mathcal{T}$ and $\beta\eta \cup \mathcal{T}$ are also confluent.*

Proof. $\beta\overline{\eta} \cup \mathcal{T}$ is proven confluent by a similar argument to theorem 17 using the commutation lemmas 26 and 28. The confluence of $\beta\eta \cup \mathcal{T}$ is proved as in corollary 18.

5 Conclusions

We have proved a variety of modularity results for the combination of algebraic TRSs with higher order typed λ-calculi. In generalising the previous results in the literature, our key innovation is the use of η-expansions instead of the more problematic η-contractions.

There are several directions in which we wish to persue this research. Most importantly we want a modularity result for strong normalisation for the addition of algebraic TRSs to CoC. As we remarked in the paper, this research awaits further basic research into the use of η-expansions in CoC. In particular we conjecture that $\beta\overline{\eta}$ is SN and we further conjecture that the combination of a SN algebraic TRS with $\beta\overline{\eta}$ remains SN.

Acknowledgements

The first author would like to thank Delia Kesner and Adolfo Piperno for many enlightening discussions on all these matters, without which this paper would not have seen the light.

References

1. Y. Akama. On Mints' reductions for ccc-Calculus. In *Typed Lambda Calculus and Applications*, number 664 in LNCS, pages 1–12. Springer Verlag, 1993.
2. F. Barbanera. Combining term-rewriting and type-assignment systems. In *Third Italian Conference on Theoretical Computer Science*, Mantova, 1989. World Scientific Publishing Company.
3. F. Barbanera. Combining term rewriting and type assignment systems. *Int. Journal of Found. of Comp. Science*, 1:165–184, 1990.
4. F. Barbanera and M. Fernandez. Intersection type assignment systems with higher-order algebraic rewriting. *Theoretical Computer Science*. To appear.
5. F. Barbanera and M. Fernandez. Modularity of termination and confluence in combinations of rewrite systems with λ_ω. In A.Lingas, R.Karlsson, and S.Carlsson, editors, *Intern. Conf. on Automata, Languages and Programming (ICALP)*, number 700 in Lecture Notes in Computer Science, Lund, 1993.
6. F. Barbanera and M. Fernandez. Modularity of termination and confluence in combinations of rewrite systems with the typed lambda-calculus of order omega. Technical report, Universit Paris Sud, 1994.
7. F. Barbanera, M. Fernandez, and H. Geuvers. Modularity of strong normalization and confluence in the algebraic-λ-cube. In *Proceedings of the Symposium on Logic in Computer Science (LICS)*, Paris, 1994. IEEE Computer Society Press.
8. H. Barendregt. *The Lambda Calculus; Its syntax and Semantics (revised edition)*. North Holland, 1984.
9. V. Breazu-Tannen. Combining algebra and higher order types. In IEEE, editor, *Proceedings of the Symposium on Logic in Computer Science (LICS)*, pages 82–90, July 1988.
10. V. Breazu-Tannen and J. Gallier. Polymorphic rewriting preserves algebraic strong normalization. *Theoretical Computer Science*, 83:3–28, 1991.
11. V. Breazu-Tannen and J. Gallier. Polymorphic rewiting preserves algebraic confluence. *Information and Computation*, 114:1–29, 1994.
12. T. Coquand and G. Huet. Constructions: a higher-order proof system for mechanizing mathematics. *EUROCAL85* in LNCS 203, 1985.
13. D. Cubric. On free CCC. Distributed on the `types` mailing list, 1992.
14. N. Dershowitz and J.-P. Jouannaud. Rewrite systems. In J. Van Leeuwen, editor, *Handbook of theoretical computer science*, volume Vol. B : Formal Models and Semantics, chapter 6, pages 243–320. The MIT Press, 1990.
15. R. Di Cosmo. A brief history of rewriting with extensionality. In Kluwer, editor, *Proceedings of the 1996 Glasgow Summer School*, 1996. To appear. A set of slides is availables from http://www.dmi.ens.fr/~dicsmo.
16. R. Di Cosmo and D. Kesner. Simulating expansions without expansions. *Mathematical Structures in Computer Science*, 4:1–48, 1994. A preliminary version is available as Technical Report LIENS-93-11/INRIA 1911.
17. R. Di Cosmo and D. Kesner. Combining algebraic rewriting, extensional lambda calculi and fixpoints. *Theoretical Computer Science*, 1995. To appear.
18. D. J. Dougherty. Adding algebraic rewriting to the untyped lambda calculus. *Information and Computation*, 101(2):251–267, Dec. 1992.
19. D. J. Dougherty. Some lambda calculi with categorical sums and products. In *Proc. of the Fifth International Conference on Rewriting Techniques and Applications (RTA)*, 1993.
20. G. Dowek, G. Huet, and B. Werner. On the definition of the eta-long normal form in the type systems of the cube. In *Informal Proceedings of the Workshop "Types"*, Nijmegen, 1993.

21. J. Gallier. *On Girard's "Candidats de Reductibilité"*, pages 123–203. Logic and Computer Science. Academic Press, 1990. Odifreddi, editor.
22. N. Ghani. Eta-expansions in dependent type theory - the calculus of constructions. In Proceedings, TLCA 97 LNCS 1210, Nancy, France 1997. Eds de Groote and JR Hindley
23. N. Ghani. Eta-expansions in F^ω. Presented at CSL'96 Utrecht Holland. To appear in CSL'96 proceedings.
24. N. Ghani. $\beta\eta$-equality for coproducts. In M. Dezani-Ciancaglini and G. Plotkin, editors, *Typed Lambda Calculus and Applications*, volume 902 of *Lecture Notes in Computer Science*, Apr. 1995.
25. N. Ghani. Extensionality and polymorphism. University of Edinburgh, Submitted, 1995.
26. B. Howard and J. Mitchell. Operational and axiomatic semantics of pcf. In *Proceedings of the LISP and Functional Programming Conference*, pages 298–306. ACM, 1990.
27. C. B. Jay and N. Ghani. The Virtues of Eta-expansion. Technical Report ECS-LFCS-92-243, LFCS, 1992. University of Edinburgh, preliminary version of [28].
28. C. B. Jay and N. Ghani. The Virtues of Eta-expansion. *Journal of Functional Programming*, 5(2):135–154, Apr. 1995.
29. J.-P. Jouannaud and M. Okada. A computation model for executable higher-order algebraic specification languages. In *Proceedings, Sixth Annual IEEE Symposium on Logic in Computer Science*, pages 350–361, Amsterdam, The Netherlands, 15–18 July 1991. IEEE Computer Society Press.
30. G. Mints. Teorija categorii i teoria dokazatelstv.I. *Aktualnye problemy logiki i metodologii nauky*, pages 252–278, 1979.
31. V. van Oostrom. Developing developments. Submitted to Theoretical Computer Science should appear in volume 145, 1994.

A Commutation of algebraic reduction with reduction to β or Coc normal form.

In this section we simply reformulate lemma 4.1 of [11] in the framework of non extensional F^ω and Coc. It is to be noticed that there is really nothing new in the proof, as the clever argument used in that lemma is tight enough to only involve the first order fragment of the caculi, so that extensions to other calculi is straightforward.

In the following, let $A \xrightarrow{r} B$ be an algebraic rewrite rule, with s being the sort of the algebraic term A (and B) and $\vec{x} = x_1 : s_1, \ldots, x_n : s_n = FV(A) \cup FV(B)$ with the s_i's being the sorts of the variables used in the algebraic rule. Let also z be a chosen variable of type $s_1 \to \ldots \to s_n \to s$. We also suppose a given typing and kinding context that we omit for readability.

We say that a term has the *z-algebraic* property if all occurrences of the variable z in it are fully applied, i.e. at the head of a subterm $zP_1 \ldots P_n$ that possesses the type s with all the P_i's possessing the type s_i. This property is clearly inherited by subterms.

The central property which is needed is the following (where by $\beta - n.f.$ we mean reduction to n.f. only w.r.t. the first order rule β while F^ω (resp. Coc)-n.f. is w.r.t the full non extensional reduction system, which we will also call *full normal form*):

Proposition 30. *If Z is an F^ω (resp. Coc) normal form having the z-algebraic property, then*

$$X \equiv \beta - n.f.(Z[\lambda \vec{x} : \vec{s}.A/z]) \quad and \quad Y \equiv \beta - n.f.(Z[\lambda \vec{x} : \vec{s}.B/z])$$

are F^ω (resp. Coc) normal forms and moreover $X \xrightarrow{r} Y$.

Proof. This is by induction on the size of Z. Since Z is a normal form, it must be of the shape $\lambda v_1 \ldots v_k.hT_1 \ldots T_m$ with v_i being either a term variable $x_i : S_i$ with S_i a normal form, or a type variable $t_i : K$.
We have now two cases:

$h \not\equiv z$ then $X \equiv \beta - n.f.(Z[\lambda \vec{x} : \vec{s}.A/z]) = \lambda \vec{v}.hT_1^A \ldots T_m^A$ and $Y \equiv \beta - n.f.(Z[\lambda \vec{x} : \vec{s}.B/z]) = \lambda \vec{v}.hT_1^B \ldots T_m^B$ with $T_i^A = \beta - n.f.(T_i[\lambda \vec{x} : \vec{s}.A/z])$ and $T_i^B = \beta - n.f.(T_i[\lambda \vec{x} : \vec{s}.B/z])$. But T_i is still a full normal form, of size strictly smaller than Z (as at least h is removed), and it still possesses the z-algebraic property as it is a subterm of Z. So, by induction hypothesis, T_i^A is a full normal form and $T_i^A \xrightarrow{r} T_i^B$, hence X is a full normal form and $X \xrightarrow{r} Y$.

$h \equiv z$ In this case, $k = m$ and we have that

$$Z[\lambda \vec{x} : \vec{s}.A/z]) = \lambda \vec{v}.(\lambda \vec{x} : \vec{s}.A)T_1 \ldots T_m \xrightarrow{\beta} \lambda \vec{v}.A[T_1/x_1 \ldots T_m/x_n]$$

and

$$Z[\lambda \vec{x} : \vec{s}.B/z]) = \lambda \vec{v}.(\lambda \vec{x} : \vec{s}.B)T_1 \ldots T_m \xrightarrow{\beta} \lambda \vec{v}.B[T_1/x_1 \ldots T_m/x_n]$$

Then, since no β-reduction can take place at the junction points of the T_i with A, as they have as type a base sort, $X \equiv \beta - n.f.(Z[\lambda \vec{x} : \vec{s}.A/z]) = \lambda \vec{v}.A[T_1^A/x_1 \ldots T_m^A/x_n]$ and $Y \equiv \beta - n.f.(Z[\lambda \vec{x} : \vec{s}.B/z]) = \lambda \vec{v}.B[T_1^B/x_1 \ldots T_m^B/x_n]$. As above, the T_i^A (resp. T_i^B) are smaller normal forms than X (resp. Y), so by induction hypothesis we have that the T_i^A and T_i^B are full normal forms and that $T_i^A \xrightarrow{r} T_i^B$. Then, both X and Y are full normal forms and moreover $X \equiv \lambda \vec{v}.A[T_1^A/x_1 \ldots T_m^A/x_n] \xrightarrow{r} \lambda \vec{v}.A[T_1^B/x_1 \ldots T_m^B/x_n] \xrightarrow{r} \lambda \vec{v}.B[T_1^B/x_1 \ldots T_m^B/x_n]$. We are done.

Using this crucial result it is then quite easy to show the equivalent of Lemma 4.1 of [11]:

Lemma 31. *Let $A \xrightarrow{r} B$ be an algebraic rewrite rule. If $M \xrightarrow{r} N$, then $fnf(M) \xrightarrow{r} fnf(N)$, where $fnf(M)$ is the full non-extensional normal form w.r.t. F^ω or Coc.*

Proof. If $M \xrightarrow{r} N$, then $M \equiv C[A\phi]$ and $N \equiv C[B\phi]$ with ϕ a substitution $[P_1/x_1, \ldots, P_n/x_n]$. Then, for a suitable variable z of type $s_1 \to \ldots \to s_n \to s$, we can write terms

$$M' \equiv C[zP_1 \ldots P_n][\lambda \vec{x} : \vec{s}.A/z] \quad and \quad N' \equiv C[zP_1 \ldots P_n][\lambda \vec{x} : \vec{s}.B/z]$$

s.t. $M' \xrightarrow{\beta} M$ and $N' \xrightarrow{\beta} N$. Now, $C[zP_1 \ldots P_n]$ has the z-algebraic property, and since this property is preserved by the non-extensional F^ω and Coc reductions, also $fnf(C[zP_1 \ldots P_n])$ has it.

Now, we can apply the previous theorem to such a full normal form and obtain that $M'' = \beta - n.f.(fnf(C[zP_1 \ldots P_n])[\lambda \vec{x} : \vec{s}.A/z])$ and $N'' = \beta - n.f.(fnf(C[zP_1 \ldots P_n])[\lambda \vec{x} : \vec{s}.B/z])$ are full normal forms and that $M'' \xrightarrow{r} N''$. Since $M' \Longrightarrow M''$ (resp. $N' \Longrightarrow N''$) and $M' \xrightarrow{\beta} M$ (resp. $N' \xrightarrow{\beta} N$), we have, due to confluence of F^ω and Coc, that $M'' = fnf(M)$ and $N'' = fnf(N)$, and we are done.

On Explicit Substitutions and Names (Extended Abstract)

Eike Ritter and Valeria de Paiva *

School of Computer Science, University of Birmingham

Abstract. Calculi with explicit substitutions have found widespread acceptance as a basis for abstract machines for functional languages. In this paper we investigate the relations between variants with de Bruijn-numbers, with variable names, with reduction based on raw expressions and calculi with equational judgements. We show the equivalence between these variants, which is crucial in establishing the correspondence between the semantics of the calculus and its implementations.

1 Introduction

Explicit substitution calculi (or $\lambda\sigma$-calculi for short) first appeared in a seminal paper by Abadi et al. [1]. The basic idea is that instead of having substitutions as a meta-level operation, as in traditional λ-calculus, we should make them part of the object-level calculus. The advantages of this approach are twofold. Firstly, it makes it possible to design much more efficient abstract machines as we are allowed to delay substitutions, and secondly it makes it much easier to prove them correct since the calculus and its implementation are closer.

There are several variants of calculi with explicit substitutions. Some of these variants are geared towards semantics [15], [3], others are derived with implementations in mind [9], [8], [2]. Rather than listing all variants, we explain in this paper what we take to be the principal differences between them. This way we describe what appears at first sight as various "design choices" for lambda-calculi. But we then justify why we have to develop calculi for each possible choice if we want to prove semantics and syntax equivalent. Moreover, by using the context handling of type theory as a guide, we are able to define a confluent calculus with explicit substitutions and names—something that Abadi et al. were not able to do.

1.1 Equations first versus Reductions first

There are two main approaches when defining typed λ-calculi with or without explicit substitutions. The first one, in the spirit of Martin Löf's type theory [10], defines the calculus with equations-in-context. Reduction is then a derived

---

*Research supported under the EPSRC project no. GR/L28296, **x-SLAM: The Explicit Substitutions Linear Abstract Machine**.

notion, obtained by orienting the equations. The second approach considers the set of typed terms as a subset of the set of raw terms, and hence reduction is defined on raw terms, which are not necessarily well-formed. Equality is now the derived notion, namely it is the symmetric and transitive closure of the relation generated by the reduction rules.

The first approach is required when giving semantics to λ-calculi because only well-formed objects have a meaning. The second approach avoids the need to check for well-formedness during reduction, which is incorporated in the first approach. As a consequence, this approach is well-suited for implementations, but a semantics for terms can only be given by showing the equivalence of this presentation to the Martin Löf-style presentation. Whereas this equivalence is easy to prove in the case of the simply-typed λ-calculus (and hence it is not really necessary to differentiate between the two approaches in this case), the difference becomes crucial as soon as we add, for example, dependent types [14][1]. This difference becomes crucial again when we consider calculi with explicit substitutions.

This paper presents calculi for both approaches and shows their equivalence (see section 3). This is because we want to connect the implementation, which is based on the second approach, with the semantics, which is based on the first approach.

1.2 Typed versus untyped calculi

There are typed and untyped calculi with explicit substitutions, both of which are presented already in [1]. The typing rules enforce two different restrictions: firstly, they eliminate expressions with misuse of variables, *e.g.*, ones where we try to substitute two different terms for the same variable simultaneously. Secondly, they ensure that the only well-typed λ-terms are the ones of the simply-typed λ-calculus.

1.3 Names versus de Bruijn numbers

Another important kind of choice the designer of a explicit substitution λ-calculus can make concerns the difference between variable names and de Bruijn numbers. De Bruijn numbers were initially considered, as an implementational trick for Automath: instead of using variables like x, y, z de Bruijn proposed to use natural numbers (that correspond to the binding level of the variable), in such a way that a class of α-congruent terms correspond to a single syntactic object. Hence two expressions with variable names are α-equivalent if and only if the corresponding terms with de Bruijn numbers are syntactically equal. More than simply an implementational trick, de Bruijn numbers are helpful when defining the semantics of the calculus in question. The point is that a de Bruijn-number n corresponds exactly to the n-th projection $A_n \times \cdots \times A_1 \to A_n$.

---

[1] The equivalence proofs can still be done [6], but some of the required properties of the type theories, like confluence and subject reduction, are very hard to establish.

There is a trade-off between a version of the calculus with de Bruijn numbers and a version with names. Expressions with variable names are much easier to read. The difference becomes apparent even for relatively small terms (*e.g.*, compare the expressions $\lambda x.(\lambda yz.x)(\lambda z.x)$ and $\lambda.(\lambda.\lambda.3)(\lambda.2)$). The main drawback of the version with names is the need to identify terms which only differ in the name of bound variables: the semantics of terms can only be defined modulo α-equivalence. This complicates the definition of the syntax significantly, as the definition of α-equivalence is rather involved (see section 3). On the other hand, α-conversion is not needed for the version with de Bruijn numbers, and the absence of α-equivalence makes this better suited for implementations.

So a judicious use of both versions seems the best option: for the presentation of results in the meta-theory, the version with names is used, and for implementations one uses de Bruijn-terms to handle variable access. Of course a good implementation keeps the variable names as extra information during reduction so that terms can be printed with names rather than with de Bruijn numbers.

1.4 Iterated Substitutions

The fourth choice concerns the need (or not) for composition of substitutions.

The precursor of the $\lambda\sigma$-calculus, Curien's $\lambda\rho$-calculus [5], was designed to capture environment machines and had no notion of iterated substitutions. This is rather restrictive, as nested substitutions arise in several situations: during reduction to normal form rather than weak head normal form, when modelling sharing in environment machines, when modelling instantiation in theorem provers, and as the counterpart of composition in the categorical semantics of λ-calculi. The $\lambda\sigma$-calculus was developed by Abadi et al. [1] with these applications in mind. Iterated substitutions seem to us an essential part of any $\lambda\sigma$-calculus.

Summing up

Summarising, it seems to us that the first "design choices" are not choices at all. We must have both the equations-in-context and the reductions-first versions, both the typed and untyped versions and both the de Bruijn and the names versions, as our goal is the implementation of abstract machines. It also seems essential to have composition of substitution for the reasons outlined above. Explicit weakening or not is, as far as this paper is concerned, a matter of taste.

The paper is structured as follows. We define our calculus of explicit substitutions and equations in context in the next section. Next we discuss issues relating binding operations and α-equivalences in explicit substitutions calculi. We prove the necessary syntactical properties (confluence and normalisation) of our calculus and then we examine the equivalence between the versions of the $\lambda\sigma$-calculus with typed and untyped reduction rules. We conclude by briefly discussing implementations and applications, which are mostly future work.

2 A calculus with equational judgements

In this section we present (with minor modifications) Martin-Löf's λ-calculus with explicit substitutions. This calculus is the $\lambda\sigma$-calculus by Abadi et al. but with names and equations-in-context. Tasistro [15] describes this calculus and gives ample motivation about the form of the judgements and their interpretation. [2]

2.1 Well-formed expressions

We start by presenting raw expressions and defining the judgements for well-formed expressions and then give a few intuitions about the calculus.

Definition 1 Raw Expressions. *The types of the $\lambda\sigma$-calculus with names are base types and function types $A \Rightarrow B$. The raw expressions of the calculus are given by the following grammar:*

$$t ::= x \mid \lambda x\colon A.t \mid tt \mid f*t \qquad f ::= \langle\rangle \mid \langle f, t/x\rangle \mid f;f$$

We call expressions of the first kind terms and expressions of the second kind substitutions[3]. Moreover, we write $\langle t_n/x_n, \ldots, t_1/x_1\rangle$ for $\langle \cdots \langle\langle\rangle, t_n/x_n\rangle, t_{n-1}/x_{n-1}\rangle, \ldots, t_1/x_1\rangle$.

We identify terms which are identical up to change of bound variables. Because not only the λ-abstraction but also the explicit substitution $f*t$ binds variables, the definition of bound variable is significantly more complex than in the λ-calculus; for a precise definition of the notion of bound variable and of α-equivalence see Section 3.

Judgements for well-formed expressions require an additional kind of raw expressions, namely contexts. Such a context is a list $x_1\colon A_1, \ldots, x_n\colon A_n$ of assignments of a type to a variable. (Contexts are called environments in [1].) We call a context well-formed if no variable occurs twice in it. From now on we tacitly assume contexts to be well-formed. We denote the empty context, which is the special case of $n = 0$, by [ ]. Note that contexts are lists rather than multisets; in other words the order is relevant. This approach generalises to dependent type theory and is compatible with categorical semantics. Because contexts like $x\colon A, y\colon B$ and $y\colon B, x\colon A$ are not identified, there is an explicit representation of the exchange rule. This avoids problems with the existence of normal forms of substitutions; for details see Section 4.

We have two judgements for the well-formedness of raw expressions, namely $\Gamma \vdash t\colon A$, the usual "t is a term of type A in context Γ", and $\Gamma \vdash f\colon \Delta$. The last judgement should be interpreted as "f is an (explicit) substitution for variables in Δ where the free variables of the terms to be substituted are contained in Γ". Such a substitution roughly corresponds to a list of substitutions in the λ-calculus. We call any context Γ' arising from Γ by deleting some assignments $x_i\colon A_i$ a *subcontext*; in that case we write $\Gamma' \subseteq \Gamma$ and call Γ an extension of Γ'.

---

[2] We use the term $\lambda\sigma$-calculus as a generic term for any variant of the calculi presented in [1].
[3] Note in particular the existence of an *explicit substitution operator*, denoted by $*$, which takes a substitution f and a term t and returns a term $f*t$.

Definition 2 Typing Judgements. *The inference rules for the judgements* $\Gamma \vdash t\colon A$ *and* $\Gamma \vdash f\colon \Delta$ *are as follows:*

(i) *On terms:*

$$\frac{}{\Gamma, x\colon A, \Gamma' \vdash x\colon A} \qquad \frac{\Gamma, x\colon A \vdash t\colon B}{\Gamma \vdash \lambda x\colon A.t\colon A \Rightarrow B} \qquad \frac{\Gamma \vdash t\colon A \Rightarrow B \quad \Gamma \vdash s\colon A}{\Gamma \vdash ts\colon B} \qquad \frac{\Gamma \vdash f\colon \Delta \quad \Delta \vdash t\colon A}{\Gamma \vdash f*t\colon A}$$

(ii) *On substitutions:*

$$\frac{}{\Gamma \vdash \langle\rangle\colon \Gamma'}\ (\Gamma' \subseteq \Gamma) \qquad \frac{\Gamma \vdash f\colon \Delta \quad \Gamma \vdash t\colon A}{\Gamma \vdash \langle f, t/x\rangle\colon \Delta, x\colon A} \qquad \frac{\Gamma \vdash f\colon \Gamma' \quad \Gamma' \vdash g\colon \Gamma''}{\Gamma \vdash f; g\colon \Gamma''}$$

The new syntax is best explained by relating the terms with explicit substitutions to terms with the usual implicit substitution of the simply-typed λ-calculus. The basic idea is that a substitution $\Gamma \vdash f\colon \mathbf{y}\colon \mathbf{B}$ [4] in the $\lambda\sigma$-calculus corresponds to a list of terms $\mathbf{t} = (t_1, \ldots, t_n)$ such that $\Gamma \vdash t_i\colon B_i$ in the λ-calculus. Moreover, the operation $*$ models explicit substitution: a term $f*t$ in the $\lambda\sigma$-calculus corresponds to a term $t[t_i/x_i]$ (with the simultaneous substitution of all terms t_i for x_i in t) in the λ-calculus.

The operations ";" and "$\langle \_, \_\rangle$" model sequential and parallel composition of substitutions respectively. If $\Gamma \vdash f\colon (\mathbf{x}\colon \mathbf{A})$ and $\mathbf{x}\colon \mathbf{A} \vdash g\colon \Delta$ and f and g correspond to the lists $\mathbf{t}$ and $\mathbf{s}$ respectively, then the substitution $f; g$ corresponds to the list $(s_1[\mathbf{t}/\mathbf{x}], \ldots, s_m[\mathbf{t}/\mathbf{x}])$ and hence models sequential composition of the substitutions f and g. The substitution $\langle\rangle$ acts not only as the identity substitution in the sense that the term $\langle\rangle * t$ corresponds to t but also as weakening: If $\Gamma \vdash t\colon A$ and Γ' is an extension of Γ then the term $\Gamma' \vdash \langle\rangle * t\colon A$ corresponds to the λ-term $\Gamma' \vdash t\colon A$ in the extended context Γ'.

2.2 Equations and Reductions

Now we turn to the equations-in-context, which are judgements $\Gamma \vdash f = g\colon \Delta$ and $\Gamma \vdash t = s\colon A$. This notion of equality is sometimes called *judgemental equality*. If a judgement $\Gamma \vdash f = g\colon \Delta$ can be stated for any contexts Γ and Δ such that $\Gamma \vdash f\colon \Delta$ implies $\Gamma \vdash g\colon \Delta$, we will write $f = g$ for $\Gamma \vdash f = g\colon \Delta$. Similarly, if a judgement $\Gamma \vdash t = s\colon A$ can be stated for any context Γ and type A such that $\Gamma \vdash t\colon A$ implies $\Gamma \vdash s\colon A$, we will write $t = s$ for this judgement. In section 5 we will relate this version of the calculus to a version with equations derived from reduction defined on raw terms.

Definition 3. *The equations of the $\lambda\sigma$-calculus with names are as follows:*

(i) *Equations modelling (traditional) λ-calculus-reductions:*

$$(\lambda x\colon A.t)s = \langle\langle\rangle, s/x\rangle * t \qquad \lambda x\colon A.tx = t \quad \text{if } x \text{ not free in } t$$

---

[4] We abbreviate a context $x_1\colon A_1, \ldots, x_n \cdots A_n$ to $\mathbf{x}\colon \mathbf{A}$. Similarly we write $t[\mathbf{s}/\mathbf{x}]$ for $t[s_1/x_1, \ldots, s_n/x_n]$.

(ii) Equations for substitutions (In the third rule, $y = x$ if y is neither a free variable nor a substitution variable in f, or y is a variable which is neither a free variable of t and f nor a substitution variable in f) [5]:

$$\langle f, t/x \rangle * x = t \quad (1) \qquad \langle f, t/y \rangle * x = f * x \text{ if } x \neq y \quad (2)$$
$$f * \lambda x{:}A.t = \lambda y{:}A.\langle f, y/x \rangle * t \quad (3) \qquad f * (ts) = (f * t)(f * s) \quad (4)$$
$$\langle \rangle; f = f \quad (5) \qquad \langle \rangle * t = t \quad (6)$$
$$f; \langle g, t/x \rangle = \langle f; g, f * t/x \rangle \quad (7) \qquad f; (g; h) = (f; g); h \quad (8)$$
$$f * (g * t) = (f; g) * t \quad (9)$$

$$\frac{\Gamma \vdash f : \Delta = x_1 : A_1, \ldots, x_n : A_n}{\Gamma \vdash f = \langle f * x_1/x_1, \ldots, f * x_n/x_n \rangle : \Delta}$$

The first two equations are the equations corresponding to β-and η-reduction in the λ-calculus respectively. The equation for the β-rule has a term with an explicit substitution on the right hand side rather than an implicit substitution as in the λ-calculus. This is the place where explicit substitutions are introduced during the reduction of λ-terms to normal form in order to make the delay of substitution possible. The equations (1)–(4) push substitutions over the constructors of λ-terms. The equation $\langle f, t/x \rangle * x = x$ is the one where the replacement of the term t for x actually takes place. The equations $f; (g; h) = (f; g); h$ and $f * (g * t) = (f; g) * t$ express associativity of substitution. The last equation for substitution expresses the fact that substitution is determined by its effect on variables. In particular, this equation causes the substitutions $(\mathbf{x} : \mathbf{A}) \vdash \langle \rangle : (\mathbf{x} : \mathbf{A})$ and $(\mathbf{x} : \mathbf{A}) \vdash \langle x_i/x_i \rangle : (\mathbf{x} : \mathbf{A})$ to be equal. This equation can be thought of as an η-rule for the explicit substitutions. It is necessary for the definition of an extensional semantics, *e.g.*, a categorical semantics.

Definition 4 Reduction Relations. *The (typed) reduction relations $\Gamma \vdash t \leadsto t' : A$ (over terms), and $\Gamma \vdash f \leadsto f' : \Delta$ (over substitutions) are defined by orienting the above equations from left to right.*

Again, if a reduction rule can be stated for any contexts Γ, Δ and types A such that $\Gamma \vdash f : \Delta$ implies $\Gamma \vdash f' : \Delta$ and $\Gamma \vdash t : A$ implies $\Gamma' \vdash t' : A$, we will write $f \leadsto f'$ and $t \leadsto t'$ respectively.

Before we investigate the meta-theoretical properties of this calculus, we examine α-equivalence in detail in the next section.

3 α-equivalence

In this section we examine α-equivalence in a $\lambda\sigma$-calculus with names, which is more complex than in the λ-calculus.

We aim to retain the results for the λ-calculus, in particular we want two expressions to be α-equivalent iff their corresponding de Bruijn-terms are equal,

---

[5] the substitution variables in a substitution f are all variables x occurring in an expression $\langle g, t/x \rangle$; for a precise definition see Section 3.

and reduction should preserve α-equivalence. The latter causes problems which are not apparent in the λ-calculus. If we define α-equivalence to be the smallest congruence such that $\lambda x\colon A.t = \lambda y\colon A.t[y/x]$, then β-reduction does not preserve α-equivalence: the two terms $(\lambda x\colon A.t)s$ and $(\lambda y\colon A.t[y/x])s$ are α-equivalent, but the contracta $\langle s/x \rangle * t$ and $\langle s/y \rangle * t[y/x]$ are not.

Hence we have to define α-equivalence in such a way that terms like $\langle s/x \rangle * t$ and $\langle s/y \rangle * t[y/x]$ are α-equivalent. This means that the substitution operator $*$ acts as another binding operator. However, this is a different kind of binding from the one λ-abstraction provides: the substitution operator binds in any expression $f * t$ those variables in t where there is a term contained in f which is to be substituted in t. In the example $\langle s/x \rangle * t$, the variable x is bound by $*$. Note that the substitution operator $*$ does not indicate the scope nor the name of the variables that it binds.

We define the sets of free variables and *substitution* variables (which are all those variables in a substitution f which are bound in a term $f * t$ or in a substitution $f; g$) by a mutual induction. The interesting cases for the free variables are $\mathrm{FV}(\lambda x\colon A.t) = \mathrm{FV}(t) \setminus \{x\}$ and $\mathrm{FV}(f * t) = \mathrm{FV}(f) \cup (\mathrm{FV}(t) \setminus \mathrm{SV}(f))$. The substitution variables are defined by $\mathrm{SV}(\langle \rangle) = \emptyset$, $\mathrm{SV}(\langle f, t/x \rangle) = \mathrm{SV}(f) \cup \{x\}$ and $\mathrm{SV}(f; g) = (\mathrm{SV}(f) \setminus \mathrm{FV}(g)) \cup \mathrm{SV}(g)$. A variable occurring in t is called *bound* in the term t if it is not a free variable in t. A variable occurring in f is called *bound* in f if it is neither a free variable nor a substitution variable in f.

In the λ-calculus Curry defines substitution before he defines α-equivalence. As the substitution has been made explicit, we only need to define *renaming* (*i.e.*, the replacement of one variable by another) as an operation in the meta-theory to state α-equivalence. This definition of renaming requires an auxiliary notion to change the name of the substitution variable x in $\langle f, t/x \rangle$ to y, *i.e.*, we define an operation $f\{y/x\}$, which satisfies $\langle f, t/x \rangle \{y/x\} = \langle f, t/y \rangle$. This name-changing substitution is given by $\langle \rangle \{y/x\} = \langle \rangle$; $(f; g) \{y/x\} = f; (g \{y/x\})$ if $x \in \mathrm{SV}(g)$ and $f; g \{y/x\} = f \{y/x\}; g$ if $x \notin \mathrm{SV}(g)$; $\langle f, t/x \rangle \{y/x\} = \langle f, t/y \rangle$ and $\langle f, t/z \rangle \{y/x\} = \langle f \{y/x\}, t/z \rangle$ if $x \neq y$.

Definition 5. *We define the renaming of the variable x by the variable y in t or f by induction over the structure of raw expressions.*

$$x[y/x] = y \qquad z[y/x] = z \text{ if } z \neq x$$
$$(\lambda x\colon A.t)[y/x] = \lambda x\colon A.t \qquad (\lambda z\colon A.t)[y/x] = \lambda w\colon A.t[w/z][y/x] \,(z \neq x)$$
$$(tu)[y/x] = (t[y/x])(u[y/x]) \qquad (f * t)[y/x] = f\{z_i/y_i\}[y/x] * t[z_i/y_i][y/x]$$
$$\langle \rangle [y/x] = \langle \rangle \qquad \langle f, t/z \rangle [y/x] = \langle f[y/x], t[y/x]/z \rangle$$
$$(f; g)[y/x] = (f\{z_i/y_i\})[y/x]; (g[z_i/y_i])[y/x]$$

In the second rule for λ-abstraction, w is equal to y if $x \notin \mathrm{FV}(t)$ or $y \notin \mathrm{FV}(s)$, otherwise w occurs neither free nor bound in t or s. In the rule for $f * t$, the variable z_i is equal to y_i if $y_i \notin \mathrm{FV}(s)$, otherwise it is a fresh variable. The same condition applies for the case $f; g$.

The definition of α-equivalence can now be stated.

Definition 6. *We define α-equivalence in the $\lambda\sigma$-calculus to be the smallest congruence relation on raw expressions including*

$$\lambda x\colon A.t \equiv_\alpha \lambda y\colon A.t[y/x] \quad f * t \equiv_\alpha f\{y/x\} * t[y/x] \quad f; g \equiv_\alpha f\{y/x\}; g[y/x]$$

The variable y is either x or it is not free in t, f and g nor is it contained in $\text{SV}(f)$. In the last two rules x is bound by $*$ and $;$ respectively.

Next we examine the interaction between α-equivalence which is defined on raw expressions, and the judgements. For the typing judgements, $\Gamma \vdash t \colon A$ means there exists an α-equivalent term t' such that $\Gamma \vdash t' \colon A$ according to the rules presented in Section 2. A similar convention is adopted for all other judgements and for reduction on raw expressions. The next theorem justifies this convention.

Theorem 7. *Assume that t_1 and t_2 are two α-equivalent $\lambda\sigma$-terms, and assume that f_1 and f_2 are two α-equivalent substitutions. If $\Gamma \vdash t_1 \colon A$, then also $\Gamma \vdash t_2 \colon A$, and similarly if $\Gamma \vdash f_1 \colon \Delta$, then also $\Gamma \vdash f_2 \colon \Delta$. If $\Gamma \vdash t_1 = s \colon A$, then also $\Gamma \vdash t_2 = s \colon A$, and if $\Gamma \vdash f_1 = g \colon \Delta$, then $\Gamma \vdash f_2 = g \colon \Delta$. If t and s are α-equivalent terms in the λ-calculus, then they are α-equivalent in the $\lambda\sigma$-calculus, too.*

The $\lambda\sigma$-calculus with de Bruijn numbers has no variable names and hence also no α-equivalence. The intuition is that α-equivalence is in fact only a consequence of the existence of names and does not affect the $\lambda\sigma$-calculus in any other way. More precisely, equality modulo α-equivalence in the calculus with names and equality in the $\lambda\sigma$-calculus with de Bruijn numbers coincide. The translation from the $\lambda\sigma$-calculus with names into the $\lambda\sigma$-calculus with de Bruijn-numbers is defined by an induction over the derivation and replaces each variable x in a context $\Gamma, x \colon A, \Gamma'$ by the length $|\Gamma'|$ of the context Γ'. For details, see [12].

The results of this section imply that Barendregt's variable convention can be adopted in the rest of this paper when we prove meta-theoretic properties. To be precise, we consider α-equivalent terms to be syntactically equal, and in the sequel we assume that all bound variables occur nowhere else in a given mathematical context (e.g., neither as free variables as in $x(\lambda x \colon A.x)$ nor as substitution variables as in $\langle f, t/x \rangle * \lambda x \colon A.s$).

4 Confluence and Normalisation

This section investigates confluence and normalisation for the (equational) $\lambda\sigma$-calculus. We deduce confluence from the confluence of the simply-typed λ-calculus, using a modularity argument, first described in [7] and familiar under the name "interpretation method". The argument is well-known, here we just make an effort to present it in its generic form.

Definition 8 Modularity Properties. *Assume that there is a translation $[\![-]\!]$ of the extended calculus into the confluent one satisfying the following modularity properties: Firstly, if $t \leadsto s$ in the extended system, then also $[\![t]\!] \leadsto^* [\![s]\!]$. Secondly, for each term t in the extended system we have $t \leadsto^* [\![t]\!]$. Thirdly, for each reduction $t \leadsto s$ in the confluent system there exists a reduction sequence $t \leadsto^* s$ in the extended system.*

In our case, this general argument works as follows. The translation $[\![-]\!]$ works by "carrying out the substitutions", i.e., $[\![\langle t_i/x_i \rangle * t]\!] = [\![t]\!][[\![t_i]\!]/x_i]$.

All reduction rules except the β-rule $(\lambda x\colon A.t)s \leadsto \langle s/x \rangle * t$ and the η-rule $\lambda x\colon A.tx \leadsto t$ model explicit substitution. We call these rules σ-rules, and we denote a σ-reduction by $t \stackrel{\sigma}{\leadsto} s$. We expect the translation from the $\lambda\sigma$-calculus to the λ-calculus to map the redex and the contractum of a σ-reduction to the same λ-term. We obtain the modularity properties as a consequence; for details see the technical report [12].

Note that the modularity properties do not hold for the original version of the $\lambda\sigma$-calculus with names. In particular, the reduction $\langle t/x, s/y \rangle \leadsto \langle s/y, t/x \rangle$ violates the first modularity property.

Formalising the argument given before to establish confluence of the $\lambda\sigma$-calculus we obtain the desired confluence.

Theorem 9 Typed confluence. *Let $\Gamma \vdash t\colon A$ be any well-formed $\lambda\sigma$-term. If $t \leadsto^* t_1$ and $t \leadsto^* t_2$, then there exists a well-formed $\lambda\sigma$-term $\Gamma \vdash u\colon A$ such that $t_1 \leadsto^* u$ and $t_2 \leadsto^* u$. Similarly, let $\Gamma \vdash f\colon \Delta$ be any well-formed substitution. If $f \leadsto^* f_1$ and $f \leadsto^* f_2$, then there exists a well-formed substitution $\Gamma \vdash g\colon \Delta$ such that $f_1 \leadsto^* g$ and $f_2 \leadsto^* g$.*

Normalisation also arises as a consequence of the modularity properties of the translation. Because the proof consists of giving an effective normalisation strategy, we obtain decidability of equality in the $\lambda\sigma$-calculus as a corollary.

Theorem 10 Normalisation. *Every well-formed term t and substitution f of the $\lambda\sigma$-calculus has a normal form, which can be effectively computed. The normal form for a term is a normal λ-term, and the normal form for a substitution is a lists of normal λ-terms.*

Mellies [11] shows that strong normalisation does not hold. As a counterexample, he gives a λ-term which reduces to the identity but which admits a reduction sequence where a term t reduces to a term t' which contains t as a subterm. But it is possible to show that all reduction strategies that reduce an expression first to one in weak head-normal form (*i.e.*, substitution is pushed under λ-abstraction only if the λ-abstraction is the outermost constructor) lead only to finite sequences of reductions [13].

5 Reduction on Raw Terms

The main part of this section examines a typed calculus with reduction defined on raw terms. At the end we mention briefly untyped calculi.

Apart from the extensionality rule for substitution $\Gamma \vdash f \leadsto \langle f*x_i/x_i \rangle\colon \mathbf{x}\colon \mathbf{A}$, all reduction rules do not use typing information. Hence we omit this rule, and write $\leadsto_r$ for the notion of reduction on raw terms given by turning all reduction rules $\Gamma \vdash f \leadsto g\colon \Delta$ except $\Gamma \vdash f \leadsto \langle f * x_i/x_i \rangle\colon \mathbf{x}\colon \mathbf{A}$ into rules $f \leadsto_r u$, and all reduction rules $\Gamma \vdash t \leadsto s\colon A$ into rules $t \leadsto_r s$. For this restricted fragment, which suffices for the design of abstract machines, we show in this section that reduction based on raw terms and the reduction derived from equational judgements (see Section 2) coincide. The important properties for this proof are uniqueness of types and subject reduction, which says that well-typed expressions reduce to well-typed expressions. The same proofs that work for the simply-typed λ-calculus with reduction defined on raw expressions work also for the system with explicit substititution.

Now we turn to the confluence proof for the calculus based on reduction on raw terms. The proof follows the general outline established in the previous section but it does not work directly because the previous proof uses the fact that every substitution reduces to a list of terms. This is no longer true if we use reduction on raw expressions: substitutions can no longer be reduced to lists of terms in general, but only to so-called *canonical forms*, *i.e.*, lists of terms with an additional weakening at the end. In particular, the substitution $\langle t/x \rangle; \langle \rangle$ is a normal form if t is a normal form.

The details and the adaptation of the confluence proof are given in the technical report [12]. We only cite the final theorem.

Theorem 11. *Let $\Gamma \vdash t: A$ be any well-formed $\lambda\sigma$-term. If $t \leadsto_r^* t_1$ and $t \leadsto_r^* t_2$, then there exists a well-formed $\lambda\sigma$-term $\Gamma \vdash u: A$ such that $t_1 \leadsto_r^* u$ and $t_2 \leadsto_r^* u$. Similarly, let $\Gamma \vdash f: \Delta$ be any well-formed substitution. If $f \leadsto_r^* f_1$ and $f \leadsto_r^* f_2$, then there exists a well-formed substitution $\Gamma \vdash g: \Delta$ such that $f_1 \leadsto_r^* g$ and $f_2 \leadsto_r^* g$.*

Remark Curien et al. [4] showed that confluence on open terms fails for the untyped $\lambda\sigma$-calculus. To obtain confluence they introduce a special syntactic construction, which describes the effect of pushing a substitution under a λ-abstraction. (They consider a version with de Bruijn-numbers, but the idea should work as well with a calculus with variables.)

The result of good design now follows: the judgemental equality presentation of our $\lambda\sigma$-calculus with names is equivalent to its presentation based on reduction on raw terms.

Theorem 12. *The $\lambda\sigma$-calculus with judgemental equality is equivalent to the $\lambda\sigma$-calculus based on reduction on raw terms. Thus $\Gamma \vdash t = s: A$ if and only if $\Gamma \vdash t: A$, $\Gamma \vdash s: A$ and $t \leftrightarrow^* s$, where $\leftrightarrow^*$ is the equivalence relation generated by $\leadsto$, and similarly for substitutions.*

This confluence proof can also be applied to the untyped $\lambda\sigma$-calculus. The reason is that the translation of explicit substitutions into list of λ-terms still can be done. In this way the confluence of the untyped λ-calculus can be lifted. Obviously, normalisation fails as any counterexample to normalisation in the untyped λ-calculus can be reproduced in the calculus with explicit substitutions.

6 Conclusions

We examined choices for designing calculi with explicit substitutions. We presented our own version of a calculus of explicit substitutions, for the simply typed λ-calculus, for which we showed the equivalence between its version arising from semantical (equations-in-context) considerations and syntactic (reduction on raw term) ones. (This equivalence is crucial in establishing the correspondence between the semantics of the calculus and its implementations.) We discussed its typed and untyped variants and the names and de Bruijn flavours of the calculus. Also we proved all the necessary, standard, properties of our calculus. The proofs are also standard.

This calculus contains what we take to be the essential points of our approach of using categorical type theory to inform the implementation of abstract machines. Ritter's PhD thesis is perhaps a more impressive example of the same approach, dealing with the Calculus of Constructions. But the point of the paper is to show how "inevitable" this calculus is, given our original goals. This is to be contrasted with the multitude of other explicit substitution calculi. Also it was necessary to clarify the case of the simply typed-lambda-calculus, to modify it appropriately, to deal with the linear lambda-calculus. Linearity introduces several new challenges that we are tackling at the moment.

References

1. M. Abadi, L. Cardelli, P.-L. Curien, and J.-J. Lévy. Explicit substitutions. *Journal of Functional Programming*, 1(4):375–416, 1991.
2. R. Bloo and K.H. Rose. Preesrvation of strong normalisation in named lambda calculi with explicit substitution and garbage collection. In *Proc. CSN'95—Computer Science in the Netherlands*, pages 62–72, 1995.
3. C. Coquand. From semantics to rules: A machine assisted analysis. In *CSL'93*, volume 832 of *LNCS*, 1994.
4. P.-L. Curien, Th. Hardin, and J.-J.Lévy. Confluence properties of weak and strong calculi of explicit substitutions. *Journal of the ACM*, 43:362–397, March 1996.
5. Pierre-Louis Curien. An abstract framework for environment machines (Note). *Theoretical Computer Science*, 82(2):389–402, 1991.
6. Herman Geuvers. *Logics and Type Systems*. PhD thesis, Univ. of Nijmegen, 1993.
7. Thérèse Hardin. Confluence results for the pure strong categorical logic CCL. λ-calculi as subsystems of CCL. *Theoretical Computer Science*, 65:291–342, 1989.
8. F. Kamareddine and A. Rios. A lambda-calculus a la de bruijn with explicit substitutions. In *PLILP'95*, volume 982 of *LNCS*, 1995.
9. P. Lescanne. From $\lambda\sigma$ to $\lambda\upsilon$: a journey through calculi of explicit substitutions. *POPL'94*, pages 60–69, Portland, Oregon, 1994.
10. Per Martin-Löf. *Intuitionistic Type Theory*. Bibliopolis, Napoli, 1984.
11. P.-A. Mellies. Typed λ-calculi with explicit substitution may not terminate. *TLCA'95*, pages 328–334. LNCS No. 902, 1995.
12. E. Ritter and V. de Paiva. On explicit substitution and names. Technical report, Univ. of Birmingham, School of Computer Science, 1997.
13. Eike Ritter. Normalization for typed lambda calculi with explicit substitution. *CSL'93*, pages 295–304. LNCS No. 832, 1994.
14. Thomas Streicher. *Correctness and Completeness of a Categorical Semantics of the Calculus of Constructions*. PhD thesis, Universität Passau, June 1989.
15. A. Tasistro. Formulation of Martin-Löf's theory types with explicit substitutions. Licenciate Thesis, Chalmers University, Dept. of Computer Science, May 1993.

On the Dynamics of Sharing Graphs *

Andrea Asperti and Cosimo Laneve

Dipartimento di Scienze dell'Informazione,
Mura Anteo Zamboni 7, 40127, Bologna, Italy

Abstract. We provide a characterization of *fan annihilation rules* of Lamping's optimal algorithm through suitable paths on the initial graphs of the evaluation. This allows to recast the computational complexity issues of the algorithm in terms of statics. The fruitfulness of the path characterization is pointed out by proving the relationship between the computational complexity of the Krivine machine and Lamping's algorithm.

1 Introduction

At the end of 80'ies, Lamping discovered a complex graph reduction technique [6] of λ-terms that was optimal in the sense that no redex is ever duplicated by the algorithm (*cf.* [8]). This goal was achieved by an ingenious management of shared contexts, using suitable sharing (fan-in) and unsharing (fan-out) nodes in the graphs.

Recently Asperti [1] and, independently, Lawall and Mairson [7] have shown that Lamping's management of shared expressions may have an exponential cost with respect to the number of β-reductions. They also conjectured that the total number of fan-annihilations in the reduction of a term could provide a reasonable lower bound to its "intrinsic complexity". Unfortunately, very little is known about the dynamic aspects of Lamping's algorithm, such as the growth of Lamping's graphs (called *sharing graphs* in the following), the ratio between application-abstraction nodes and the other nodes, the exact cost of the sharing management (which is our utmost goal).

So far, the only dynamic results concern beta-reductions. In particular, in [3, 2] we provided a bijective correspondence between families of β-reductions fired along the evaluation of a term t and suitable *paths* in the *initial* graph of t. This result has been used for proving the correctness and coincidence of several optimal algorithms, proving the fruitfulness of our approach. In this paper, we apply the same technique to cover other dynamic aspects of Lamping's algorithm, giving a precise and simple description of fan-annihilations as suitable *paths* in the initial term.

It turns out that the computational complexity of Lamping's abstract algorithm for λI-terms is a function of fan annihilations and β-reductions. In other words, the computational complexity issues may be recast in terms of statics,

---
* This work is partly supported by the ESPRIT CONFER-2 WG-21836

hopefully a more easily comprehensible view. Indeed we exploit the static view for proving that the complexity of the Krivine machine cannot be better than the number of fan annihilations in Lamping's algorithm. This is not striking, but just aims at emphasizing the relevance of path characterizations for reasoning about (even different) machines.

Technical developments and proofs are missing in this extended abstract. They may be found at ftp://ftp.cs.unibo.it/pub/laneve/fullicalp.ps.gz.

1.1 Lamping's abstract algorithm

We said that Lamping's algorithm implements optimality through a suitable sharing of subexpressions, performed by explicit nodes called *fan*. Fan nodes, together with application and abstraction nodes are the core set of nodes of Lamping's algorithm. The rules governing their interaction are illustrated in Figure 1 below.

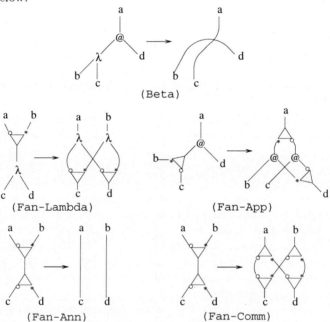

Fig. 1. Interaction rules of Lamping's abstract algorithm

There is no space here for introducing Lamping's algorithm. The reader can find a smooth introduction in [1]. Remark only that there are two rules for evaluating fan-interactions: one annihilating the two fans and the other performing duplication. In the *abstract* algorithm described above we have assumed the presence of an oracle solving the problem of which rule to apply (at each time exactly one rule may be used). Lamping's implementation of the oracle is described in Section 2.

1.2 The producer/consumer analogy

The set of rules in Figure 1 may be split in two groups: annihilations and duplications. Annihilations and duplications are strongly related: in a sense the first ones "consume" nodes, the latters "produce" (by duplicating) nodes. This relation is evident in those terms of the λI-calculus whose normal form is an atomic value (the final graph is a single edge). Let d, f and a be respectively the number of duplications, β-reductions and annihilations along the reduction. Let moreover $|M|$ be the number of applications, abstractions and fan nodes in M. Since each duplication adds two new nodes in the graph, each β-redex or fan-annihilation removes two nodes from it, and we have no nodes at the end of the computation, the following equation holds:

$$|M| + 2d - 2f - 2a = 0$$

So, $f + a = d + |M|/2$. This immediately gives the following property:

Property 1. *The length of the abstract Lamping-evaluation of λI-expressions yielding constant values only depends on the families of β-redexes and of fan-annihilations.*

1.3 Dynamics vs. Statics

By Property 1, the computational complexity of λ-terms only depends on β-redexes and fan annihilations. In [3], β-reductions have been successfully recast in terms of suitable paths on syntax trees of λ-terms. In this paper we are going to apply the same methodology to fan-annihilation rules, thus covering every interesting dynamic aspect of Lamping's abstract algorithm. For instance, the reader may observe that, in the evaluation of (2 Δ), the rule (FAN-ANN) is used twice. Consider one of them and, going backward along the reduction, follow the path traversed by the two interacting fans. When you get back to the initial graph, you will discover that each annihilation rule corresponds to a path in Figure 2. Both paths have a very precise and similar structure: they start and terminate at the same fan, and can be uniquely decomposed as $\xi \, \lambda \, \psi \, @ \, \phi \, @ \, \psi^r \, \lambda \, \xi^r$, where ξ is a discriminant (the path from the fan to the variable port of the λ), ψ is a virtual redex, followed by a @-cycle ϕ (see Definition 9) and $(\ )^r$ is the "reversing" operation. In the present paper we prove that this decomposition is general:

Property 2. *Fan-annihilations are in bijective correspondence with legal paths in the initial graph consisting of a discriminant, a virtual redex, an @-cycle, the virtual redex reverted and the discriminant reverted.*

1.4 The comparison with Krivine machine

Paths offer a fine grain description of the evaluation of λ-terms. For this reason other reduction mechanisms may be reduced to path computations. As a consequence the characterization of fan annihilations in terms of paths becomes

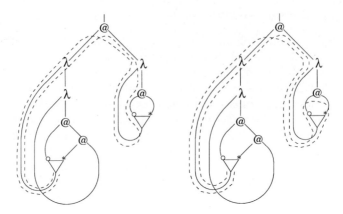

Fig. 2. Virtual fan-annihilations in $(2\ \Delta)$

an important step towards the comparison of Lamping's optimal algorithm with other reduction techniques.

For instance, Danos and Regnier have recently proved that each move of Krivine machine, a well known environment machine for functional languages, actually corresponds to a path computation. A close inspection of this correspondence, together with Property 2, allows us to draw the following consequence:

Property 3. *Let M be a λI-term reducing to a constant. The length of computation of M in Lamping's abstract algorithm is at most $O(n)$, where n is the length of the Krivine machine computation.*

We observe that this property also gives more evidence to the thesis that the total number of fan-annihilations in the reduction of a term provides a reasonable lower bound to its "intrinsic" computational complexity [1, 7]. We finally recall that the Krivine machine may have an exponential slow-down with respect to Lamping abstract algorithm (for instance the evaluation of **n2Ic**, where **n** and **2** are Church numbers, I is the identity and c is a constant, is $O(2^n)$ in Krivine machines and $O(n)$ in Lamping's algorithm). This is not very surprising, since the Krivine machine implements a call-by-name strategy, which is very inefficient for evaluating terms.

2 Pairing fans: Lamping's full algorithm

In order to solve the problem of correct fan pairing, Lamping added a local level structure to the bidimensional graphs presented in the Introduction. Each node is decorated with an integer tag which specifies the level at which it lives: two fans match if they meet at the same level; they mismatch otherwise. Furthermore there are two new control nodes which operate on the level structure: the *croissant*, which opens or closes a level, and the *bracket*, which temporarily closes a level or restores a temporarily closed one.

More precisely, sharing graphs are unoriented graphs built from the indexed nodes in Figure 3.

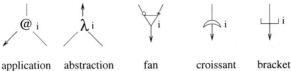

application   abstraction   fan   croissant   bracket

Fig. 3. Sharing nodes

The port of a node depicted with an arrow is called its *principal port*. This is the only port where a node can possibly interact with other nodes in a graph reduction rule. The other ports of each node are called *auxiliary*. It is convenient to introduce particular names for the auxiliary ports of application and abstraction nodes. In particular, the port of the application leading to the context (usual depicted at its top) will be called *context* port, while the other auxiliary port will be the *argument* port. In the case of an abstraction node, the port leading to the body of the function (usually depicted at the right of the other auxiliary port) will be called *body* port, while the other auxiliary port is the *bound* port (since it leads to the variable bound by the abstraction).

Two nodes (nodes of the graph) annihilates if they meet along their principal ports at the same level. In Section 1.1 we have already introduced two annihilation rules: (BETA) and (FAN-ANN). The other two annihilation rules are described in Figure 4.

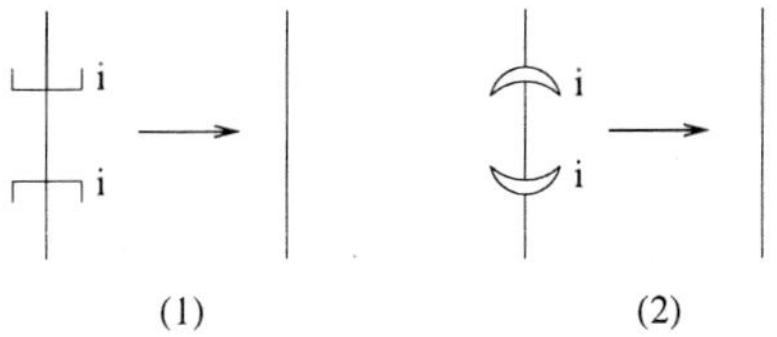

Fig. 4. (1) The rule (BRACKET-ANN); (2) the rule (CROISSANT-ANN)

A node at a given level can also act upon any other node f at a higher level (reached at its principal port), according to the rules in Figure 5 (f represents a generic node). In these rules, the nodes are simply propagated through each

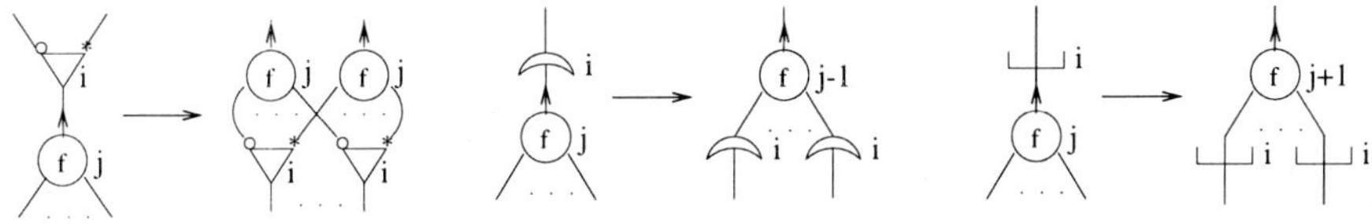

Fig. 5. Commutation rules

other in such a way that their effect on the level structure is left unchanged.

Observe that rules (FAN-COMM), (FAN-APP) and (FAN-LAMBDA) are instances of the leftmost rule in Figure 5.

2.1 The initial encoding of λ-terms

A λ-term N with n free variables will be represented by a graph with $n+1$ entries (free edges): n for the free variables (the inputs), and one for the "root" of the term (the output). The translation is inductively defined by the rules in Figure 6. The translation function is indexed by an integer which can be thought of as being the level at which we want the root to be; the translation starts at level 0, *i.e.* $[M] = [M]_0$.

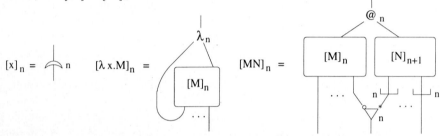

Fig. 6. Initial translation

2.2 Consistent paths and the correctness

The correctness of Lamping's algorithm was proved by means of suitable paths called *consistent paths* [6]. Let us recall the notions in [5].

The (finite) *contexts* are the terms generated by the following grammar:

$$a ::= \Box \mid \circ \cdot a \mid * \cdot a \mid \sharp \cdot a \mid \natural \cdot a \mid \langle a, a \rangle$$

We denote by $A_n[a]$ a context of the form $\langle \cdots \langle a, a_n \rangle, \cdots a_1 \rangle$.

Definition 4. (Consistent path) A *consistent path* in a graph $[M]$ is a path such that

1. every edge of the path is labeled with a context;
2. consecutive pairs of edges satisfy one of the following constraints:

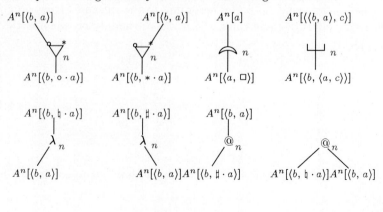

Consistent paths are taken equivalent up to contexts. That is, two consistent paths having pairwise equal edges are considered equal, even if the contexts differ.

The above definition describes how the nodes of the graphs modify the contexts when traversed. Notice that, the traversal of a node n can be forbidden if the external context does not allow the transformation performed by n. As a consequence, there are illegal (better, not consistent) paths.

In order to formalize the statement of correctness we need the preliminary notions of residual and ancestor path. To this aim, let $[M] \to^* G \xrightarrow{u} G'$ and ρ be a path in G which starts and terminates at two principal ports [2]. The notion of residual path is not defined when u is an annihilation rule and u involves the endpoints of ρ.

By definition of sharing rules, u is "local", namely it involves exactly two nodes n and n' in G and the edges starting at these nodes. Therefore, let $G_\flat$ be the subgraph of G where n, n' and the edges starting at n and n' are missing. Then $G_\flat$ is also a subgraph of G'. If ρ is internal to $G_\flat$ of G then the *residual* of ρ is the corresponding path in $G_\flat$ of G'. Otherwise $\rho = \rho_1 e_1 m_1 u p_1 e_1' \cdots e_k m_k u p_k e_k' \rho_{k+1}$ such that ρ_i are internal to the subgraph $G_\flat$ of G, $\rho_1 e_1$ or $e_k' \rho_{k+1}$ may miss and $\{m_i, p_i\} = \{n, n'\}$. There are two cases:

(u is a commutation rule) Let us define the cases when $\rho_1 e_1$ or $\rho_1 e_1 m_1 u$ are missing: the other cases may be defined in a similar way. In this case the residual of ρ is $m_1' c_1' \rho_2' \cdots c_k p_k' v_k m_k' c_k' \rho_{k+1}'$, where ρ_i' are the residuals of ρ_i, $c_i p_i' v_i m_i' c_i'$ are the unique paths traversing the part of G' which is not in $G_\flat$ such that they connect the ports which correspond to the initial port of e_i and the final port of e_i'. The node m_1' is consecutive to the initial node of ρ_2' through the port corresponding to the final port of e_1'.

(u is an annihilation rule) Remark that $\rho_1 e_1$ and $e_k' \rho_{k+1}$ cannot be missing in this case. The residual of ρ is $\rho_1' c_1 \cdots c_k \rho_{k+1}'$, where ρ_i' are the residuals of ρ_i and c_i are the edges connecting the ports which correspond to the initial port of e_i and the final port of e_i'.

There is an obvious consequence of the above definition:

Proposition 5. *The residual of a path (if any) is unique.*

The unicity of residuals allows to define the "inverse" notion, called *ancestor*.

A close inspection of Lamping's graph rewriting rules reveals that they preserve the consistency of paths:

Property 6. (The context semantics [5]) *Let $[M] \to^* G \xrightarrow{u} G'$, φ be a consistent path in G and u is not an annihilation rule involving the endpoints of φ. Then the residual of φ does exist and it is consistent. Similarly for ancestors.*

---

[2] This constraint guarantees the unicity of the ancestor. Indeed, assume ρ starting at the auxiliary port of a node m, m is involved in u and u is not the first edge of ρ. Then the residuals of ρ and $u\rho$ should be the same.

The context semantics has been remarkably used for proving the correctness of Lamping's algorithm. In particular, in [5], the authors noticed that consistent paths starting and terminating at root nodes (the root and the free variables) are invariant with respect to reduction rules. Since these paths suffice for defining the Böhm tree of a λ-term, it follows that the implementation is correct.

3 Legality, fan annihilations and cycles

An alternative definition of consistency, called *legality*, has been provided in [3] (the coincidence of consistent paths and legal paths is in [2]). The notion of legality has the advantage (with respect to the others) of clarifying the *symmetries* inside paths, which, as we will see, are crucial for defining the path characterization of fan annihilations. Therefore, let us recall briefly the main definitions and properties of legal paths.

A path is *straight* if it traverses nodes form auxiliary to principal ports (in particular, the path cannot "bouncing back", exiting from the same port it entered through). A straight path is *elementary* if one end is connected at a port of a @-node or a λ-node, it traverses control nodes only and the other end is connected at a λ-node or a @-node or at a (free or bound) variable node. A *discriminant* is an elementary path starting at a bound variable (a discriminant represents an occurrence of the bound variable).

Definition 7. Let φ be a straight path connecting the principal port of an application @ and an abstraction λ. These two nodes are *paired* (along φ) if and only if either φ is a redex, or every other application and abstraction internal to φ is paired (along a subpath of φ).

Definition 8. A straight path φ is a *well-balanced path* (shortly wbp) if and only if, for each application @ and abstraction λ paired along a subpath of φ, the following conditions are satisfied:

1. φ traverses @ through the context port if and only if it traverses λ through the body port;
2. φ traverses @ through the argument port if and only if it traverses λ through the bound port.

Next we define by crossed induction two other types of paths: @-*cycles* and v-*cycles*.

Definition 9.

(@-**cycle**) Let @ be an application node in $[M]$, u be the argument edge of @ and N be the second argument of @. An @-*cycle* of @ is a path:
  1. $@u\psi u^r@$, where ψ is internal to N;
  2. or $@u\psi_1\xi_1\psi_2\cdots\psi_n\xi_n\psi_{n+1}u^r@$, where ψ_i are internal to N and ξ are v-cycles over some free variable in N;

(v-cycle) Let γ be a discriminant of λ and starting at **v**. A *v-cycle* over **v** is a path $\mathbf{v}\gamma\lambda\psi^r@\phi@\psi\lambda\gamma^r\mathbf{v}$ where ϕ is a wbp starting at @ and terminating at λ.

Definition 10. (Legal paths) A wbp φ is a *legal path* if and only if, for every @-cycle ϕ contained in φ, φ can be decomposed in one one of the following possible ways:

(1) $\varphi = \zeta@\phi@(\zeta)^r\zeta_1$,
(2) $\varphi = \zeta_1\zeta@\phi@(\zeta)^r$,
(3) $\varphi = \zeta_1\gamma\lambda\psi@\phi@(\psi)^r\lambda(\gamma)^r\zeta_2$, where @ and λ are paired along ψ and γ is a discriminant.

In case (3), we shall say that ψ and ψ^r are respectively the *call* and *return* paths of the @-cycle.

The above definition essentially says that whenever we have a @-cycle in φ the call and return paths, together with the associated discriminants, must be the same. Cases (1) and (2) are used to cover the cases in which the call or return paths are not complete.

Definition 11. A legal path $@\phi\lambda$ where @ and λ are paired along ϕ is called a *virtual redex*.

This definition is justified by the following:

Theorem 12. *[3] Given a λ-term M, there is a one-to-one correspondence between virtual redexes in $[M]$ and all the possible redex families obtained by evaluating M.*

As proved in [2] consistent paths and legal paths are strongly related:

Theorem 13. *Every wbp is legal if and only if it is consistent.*

The main result of the paper, namely the path characterization of (FAN-ANN) moves is stated in Theorem 14 below.

Theorem 14. *Let $[M] \to^* G$ and u be an edge in G connecting the principal ports of two fans. The interaction u annihilates the two fans if and only if the ancestor of u is a path $\xi\lambda\psi@\phi@\psi^r\lambda\xi^r$, where ξ is a discriminant, ψ is a virtual redex and ϕ is a @-cycle.*

It is evident that Property 2 is a smooth statement for Theorem 14.

4 The comparison with Krivine machine

Theorem 14 looks particularly appealing since it provides a new insight for reasoning about dynamics of Lamping abstract algorithm. This insight relies on path computations, which indeed is an alternative evaluation of λ-tems.

A meaningful application of Theorem 14, which we are going to show, allows to clarify the *computational* correspondence between two algorithms for evaluating λ-terms: Krivine machine and Lamping's abstract algorithm. With this we mean that it is possible to fix the relationship between the lengths of computations in the two algorithms. To this aim we use a result recently put forward by Danos and Regnier [4]: each step of the Krivine machine is actually a suitable path in the sharing graph. Let us give the key intuition of Danos and Regnier, omitting the details of [4]. The intuition follows by the symmetries inside paths and two properties about @-cycles and well balanced paths (which we omit to recall since they are not relevant in the following discussion). Take for instance the consistent path

$$\varphi = \xi \lambda \psi @ \phi @ \psi^r \lambda \xi^r$$

where ξ is a discriminant starting at a croissant at depth n, ψ is a virtual redex between a λ at depth p $(p \leq n)$ and an application @ at depth q. Let also $A = \langle \cdots \langle S, a_n \rangle, \cdots, a_1 \rangle$ be the initial context of φ. A may be rewritten as the following pair

$$a_n :: \cdots :: a_p :: \mathcal{E},\ S$$

where $a_n :: \cdots :: a_p :: \mathcal{E}$ is called the *environment* and S is called the *stack*. Now we observe that:

1. By definition of ξ, its final context will have the shape $\mathcal{E}, \sigma :: S$, where σ is a suitable context depending on $a_n, \cdots, a_p$.
2. So we start ψ with the context $\mathcal{E}, \sigma :: S$. Since ψ is a wbp, by the Rendez-vous property in [2], the final context of ψ will have the shape $\mathcal{E}', \sigma :: S$, for some $\mathcal{E}'$.
3. Now, by the @-cycle property in [2], if we start the @-cycle ϕ with a context $\mathcal{E}', \sigma :: S$, we shall terminate with a context $\mathcal{E}', \sigma :: S'$, for some S'.
4. The reverse path of ψ performs the reverse transformation on contexts. So at the end of ψ^r we have the context $\mathcal{E}, \sigma :: S'$.
5. For a similar reason, the context at the end of ξ^r has to be $a_n :: \cdots :: a_p :: \mathcal{E}, \sigma :: S'$.

Observe that, the purpose of steps 4 and 5 is to restore the initial environment by using the informations on the top of the stack and the environment $\mathcal{E}'$. This steps may be skipped if we were more careful in steps 1 and 2. That is, let us save the address d of the croissant at the beginning of ξ and the environment $a_n :: \cdots :: a_p :: \mathcal{E}$ on top of the stack S. Namely, the stack at the end of steps 1 and 2 is $(\mathbf{d}, a_n :: \cdots :: a_p) :: S$. Then, at the end of the @-cycle, we have a context $\mathcal{E}', (\mathbf{d}, a_n :: \cdots :: a_p :: \mathcal{E}) :: S'$ and we may safely skip steps 3 and 4, just by restoring what is on top of the stack and jumping to d.

It turns out that the above optimization corresponds to the step of the Krivine machine performed when a bound name is met, while the step where the pair $(\mathbf{d}, a_n :: \cdots :: a_p)$ is saved on the stack corresponds to the "stacking" move in the Krivine machine. The third reduction of the Krivine machine, the β-move, corresponds obviously to subpaths which are virtual redexes. Therefore the following result:

Theorem 15. *[4] The optimized path computation is isomorphic to the Krivine machine.*

Since Theorem 15 gives a path characterization of steps of Krivine machine, we may establish the computational correspondence between Krivine machine and Lamping's abstract algorithm.

Theorem 16. *Let M be a λI-term reducing to a constant. The length of computation of M in Lamping's abstract algorithm is at most $O(n)$, where n is the length of the Krivine machine computation.*

5 Conclusions

The path characterization of fan annihilations has to be meant as the first step towards the goal of determining the total amount of work required by Lamping's (abstract) algorithm. We observe that a direct evaluation of this parameter looks very problematic, especially since not all sharing-graphs can be obtained by the reduction of a λ-term, and *nothing* is known about the structure of these "legal graphs". As a consequence, no reasoning by induction on the size or the structure of these graphs seem possible. Vice-versa, computing paths in a λ-term looks as a more realistic and promising research direction, since in this case we can profit of all the theoretical machinery of the geometry of interaction and its dynamic algebra [2].

References

1. A. Asperti. On the complexity of beta-reduction. In *Proceedings 23rd ACM Symposium on Principles of Programmining Languages*, 1996.
2. A. Asperti, V. Danos, C. Laneve, and L. Regnier. Paths in the λ-calculus. In *Proceedings 9th Annual Symposium on Logic in Computer Science*, Paris, pages 426 – 436, 1994.
3. A. Asperti and C. Laneve. Paths, computations and labels in the λ-calculus. *Theoretical Computer Science*, 142, 1995.
4. V. Danos and L. Regnier. Reversible and irreversible computations. Autumn 1994.
5. G. Gonthier, M. Abadi, and J.J. Lévy. The geometry of optimal lambda reduction. In *Proceedings 19th ACM Symposium on Principles of Programmining Languages*, pages 15 – 26, 1992.
6. J. Lamping. An algorithm for optimal lambda calculus reductions. In *Proceedings 17th ACM Symposium on Principles of Programmining Languages*, pages 16 – 30, 1990.
7. J.L. Lawall and H.G. Mairson. Optimality and inefficiency: what isn't a cost model of the λ-calculus. In *International Conference on Functional Programming*, 1996. Philadelphia, Pennsylvania, USA.
8. J.J. Lévy. *Réductions correctes et optimales dans le lambda calcul*. PhD thesis, Université Paris VII, 1978.

Minimizing Diameters of Dynamic Trees

Stephen Alstrup[1] Jacob Holm[1]
Kristian de Lichtenberg[1] Mikkel Thorup[1]

Department of Computer Science, University of Copenhagen, Universitetsparken 1,
DK-2100 Copenhagen, Denmark (e-mail : stephen,samson,morat,mthorup@diku.dk,
www : http://www.diku.dk/~stephen,~mthorup)

Abstract. In this paper we consider an on–line problem related to minimizing the diameter of a dynamic tree T. A new edge f is added, and our task is to delete the edge e of the induced cycle so as to minimize the diameter of the resulting tree $T \cup \{f\} \setminus \{e\}$. Starting with a tree with n nodes, we show how each such *best swap* can be found in worst–case $O(\log^2 n)$ time. The problem was raised by Italiano and Ramaswami at ICALP'94 together with a related problem for edge deletions. Italiano and Ramaswami solved both problems in $O(n)$ time per operation.

1 Introduction

The diameter of a tree is the length of a longest simple path in the tree and such a path is called a diameter path. The unique midpoint on all diameter paths is called the center, hence the center is the point whose maximal distance to any node is as small as possible. In 1973 Handler [4] showed how one in linear time can compute the diameter (and center) of a tree. However, as pointed out by Rauch [8], too little work has been done to dynamically maintain information about the diameter. To the best of our knowledge, the only dynamic algorithms concerning diameters are those given by Italiano and Ramaswami in ICALP'94 [5], motivated by problems in high-speed wide-area networks (see [6, 7] for details). They consider how to minimize the diameter of a dynamic tree T with n nodes and non–negative edge cost. Let f be a new edge which introduce a cycle C in the dynamic tree. Then removing an edge e from the cycle C is called a *swap(e, f)*. The *best swap* is the swap which minimizes the diameter of the resulting tree $T_{f/e} = T \cup \{f\} \setminus \{e\}$. In this paper we present an on-line algorithm for maintaining a dynamic tree, such that given a new edge, the tree computed is the tree resulting from the best *swap*. Italiano and Ramaswami [5] presented an $O(n)$ time algorithm for finding a best swap. In this paper, we show how to improve the complexity to $O(\log^2 n)$ worst-case time.

Italiano and Ramaswami [5] considered the above incremental best swap problem as part of a fully dynamic type heuristic for maintaining a small diameter spanning tree T in a dynamic connected graph G. If an edge e is added to G, the above incremental algorithm is called to find a best swap for T with e. If an edge e is deleted and it belongs to T, they have a complementing decremental algorithm that finds a "best swap" edge from G reconnecting T minimizing the resulting diameter of T. They supported both insertions and deletions in time $O(n)$. Note that the above scheme does not maintain the spanning tree of the smallest possible diameter. As mentioned, this paper does not consider the decremental problem.

As an intermediate step to our algorithm we show how to maintain a dynamic forest of trees under *link* and *cut* where given a node from a tree it returns the diameter of the tree the node belongs to. The time-complexity is $O(\log n)$ for each operation, where n is the number of nodes in the tree(s) involved. We show this, since to the best of our knowledge, no such algorithm has been presented before.

All our results are based on topology trees [3, 2] (the terminology of topology trees is recalled in Section 2). Our algorithm for maintaining the diameter is straightforward, based on a simple observation. Our algorithm for finding a best swap is much more involved. One complication is that when we want to merge two clusters, we need to consider not only the information associated with the clusters being merged, but the information associated with $O(\log n)$ sub-clusters of each of the two clusters. This implies that a merge takes $O(\log n)$ time, and each best swap gives rise to $O(\log n)$ merges. Thus our $O(\log^2 n)$ time algorithm for best swap is derived.

The paper is organized as follows: In section 2 preliminaries are given. In section 3 we present an algorithm for maintaining the diameters of trees in a dynamic forest. Finally in section 4 we give an algorithm which compute a best *swap* in $O(\log^2 n)$ time.

2 Preliminaries

In this section we give a short presentation of the topology trees by Frederickson [3, 2]. Our presentation differ slighty from the original topology trees. We provide a more simple interface in order to simplify the use of the topology trees.

Let T be a tree with n nodes. For a connected subtree of T, we call a node which has edges out of the subtree a *boundary node*. A *cluster* is a connected subtree of T with at most two boundary nodes. The set of boundary nodes of a cluster C is denoted ∂C. We say that $\partial C = \{a, b\}$ if C has boundary nodes a and b even if a and b are identical. Two clusters are said to be *neighbours* if they intersect in exactly one node. A *topology tree* τ of T is a binary tree such that:[1]

1. The nodes of τ represents clusters of T.
2. The leaves of τ represents the edges of T.
3. If C is represented by an internal node of τ with children representing A and B, then $C = A \cup B$ and A and B are neighbours.
4. The root of τ represents T.
5. The height of τ is $O(\log n)$.

---

[1] In this description all leaf clusters contains only one edge, however the simplification presented in this paper holds for any size of the leaf clusters.

A tree with a single node has an empty topology tree.

In order to maintain topology trees for a forest of dynamic trees we make use of the following operations: *Merge* takes two topology tree root nodes a and b and creates a new topology tree root with children a and b. By the definition above we have that only nodes representing neighbouring clusters may be *Merge*d. *DeleteRoot* is the reverse operation, deleting the root of a topology tree.

This presentation of the topology trees differ in the interface from those in [2, 3]. For the topology trees presented we have made no restriction on the degree of the tree for which the topology tree is used and we have reduced the number of different ways clusters can be related to each other. Because of the lack of space, we defer the description of the modification to the full journal version in which we will show that it does not change the complexity of the topology tree operations. From Frederickson [2, lemma 1,theorem 2] and Frederickson [3, lemma 2.3] we have the following proposition for topology trees.

Proposition 1. *A topology tree τ of a tree T with n nodes, can be computed using a linear number of* Merge *operations. Topology trees for a forest of trees can be maintained under* link *and* cut*, using $O(\log n)$* Merges *and* DeleteRoots *per* link *and* cut *operation.* □

Consequently

Theorem 2. *Let* info *be some information of clusters in a dynamic forest with n nodes so that*

1. *For any edge e,* info($\{e\}$) *can be computed in time t_1.*
2. *For any neighbouring clusters C_1 and C_2,* info($C_1 \cup C_2$) *can be computed in time t_2, given* info(C_1) *and* info(C_2).

Then we can maintain info *for all trees in a dynamic forest in $O(t_1 + t_2 \log n)$ time per* link *and* cut*, given the ability to use $O(n * (t_1 + t_2))$ time and $O(n)$ space for preprocessing.* □

3 Dynamic Diameters

In this section we will present a simple algorithm for maintaining information about the size of diameters of trees in a dynamic forest under *link* and *cut*. The algorithm will be used in the following section. It builds on a generalization of former exploitations of properties of diameters and spanning trees (see e.g. [1, 4, 5]). This generalization, given in the following lemma, makes it possible to construct efficient divide and conquer algorithms.

Let $T = (V, E)$ be a tree with n nodes. With each edge e in E is associated a nonnegative number $cost(e)$. For two nodes $a, b \in V$ we then define the distance, $dist(a, b)$, to be the sum of $costs$ for all edges on the simple path from a to b in the tree. For a subset of nodes $W \subseteq V$ we define $diam_T(W) = \max_{a,b \in W} dist(a, b)$, hence the diameter in the tree is $diam_T(V)$. By the path from a to b, denoted $a \cdots b$, we mean both the set of edges and the set of nodes on that path.

Lemma 3. *Let $T = (V, E)$ be a tree, $\{a, b\} \subseteq V' \subseteq V, \{c, d\} \subseteq V'' \subseteq V$, where $\text{dist}(a, b) = \text{diam}_T(V')$ and $\text{dist}(c, d) = \text{diam}_T(V'')$ then $\text{diam}_T(V' \cup V'') = \text{diam}_T(\{a, b, c, d\})$.*

Proof. Assume for contradiction that $\text{diam}_T(\{a,b,c,d\}) < \text{diam}_T(V' \cup V'')$. Then there exists $e \in V' \setminus V'', f \in V'' \setminus V'$, so $\text{dist}(e, f) = \text{diam}_T(V' \cup V'') > \text{diam}_T(\{a,b,c,d\})$. Now either $e \notin \{a, b\}$ or $f \notin \{c, d\}$. Say $e \notin \{a, b\}$. Let P denote the path $e \cdots f$. Let $x, y \in P$ be the nodes, such that $a \cdots x \cap P = \{x\}$ and $b \cdots y \cap P = \{y\}$. Now assume w.l.o.g. that a, x, b, y is arranged such that $\text{dist}(e, x) \leq \text{dist}(e, y)$. We now have $\text{dist}(a, x) + \text{dist}(x, b) \geq \text{dist}(a, b) \geq \text{dist}(e, b) = \text{dist}(e, x) + \text{dist}(x, b)$, hence $\text{dist}(a, x) \geq \text{dist}(e, x)$ which yields $\text{dist}(a, f) \geq \text{dist}(e, f)$ contradicting our assumption. We therefore conclude that $e \in \{a, b\}$ and symmetrically $f \in \{c, d\}$, which concludes the proof. □

We now show how to use lemma 3 with theorem 2. Given two neighbouring clusters C_1 and C_2 of a tree T we can compute $\text{diam}_T(C_1 \cup C_2)$ given the following information, $\text{info}(C)$, for each of the clusters C_1 and C_2.

1. The boundary nodes ∂C.
2. Two nodes $a, b \in C$ with $\text{dist}(a, b) = \text{diam}_T(C)$.
3. The distances between the nodes above.

As we will show in the journal version, it is now straightforward to prove:

Theorem 4. *There exists an algorithm that maintains the diameters of trees in a dynamic forest in time $O(\log n)$ under link and cut, given the ability to use $O(n)$ time and space for preprocessing, where n is the number of nodes in the tree(s) involved in the operation.* □

4 Best swap

Given a tree T with n nodes and an edge $f = (b_1, b_2)$ not in T, we wish to find an edge e on the cycle $C = b_1 \cdots b_2 \cup \{f\}$ that yields the smallest diameter of $T_{f/e} = T \cup \{f\} \setminus \{e\}$.

Using theorem 4 we can maintain the diameter of a tree dynamically under *link* and *cut* using $O(\log n)$ time per operation. So when an edge f is presented we can solve the best *swap* problem in $O(k \log n)$ where k is the number of edges on the cycle C, by simply trying them one by one. But in general this is worse than the $O(n)$ algorithm given by Italiano and Ramaswami [5], however in this section we will provide an $O(\log^2 n)$ time solution to the problem.

4.1 Outline of the algorithm

If we dynamically maintain the diameter of the tree, using theorem 4, we already know the diameter of the tree $T = T_{f/f}$. Therefore we only need to concentrate on finding the edge e on the path $b_1 \cdots b_2$ which minimize the diameter of $T_{f/e}$.

If we remove an edge $e \in b_1 \cdots b_2$ from T, we divide it into two subtrees dependent on e: $T_{b_1}(e), T_{b_2}(e)$ where $b_1 \in T_{b_1}(e)$ and $b_2 \in T_{b_2}(e)$. We know from lemma 3 that when linking $T_{b_1}(e)$ and $T_{b_2}(e)$ with f, the diameter of the combined tree $T_{f/e}$ is the maximum of $diam(T_{b_1}(e)), diam(T_{b_2}(e))$ and the longest path in $T_{f/e}$ which includes f, denoted $maxpath_f(e)$. From now on we assume that $T_{b_1}(e)$ is rooted in b_1 and $T_{b_2}(e)$ is rooted in b_2. Then the length of the longest path containing the edge f in $T_{f/e}$, $maxpath_f(e)$, becomes $height(T_{b_1}(e)) + cost(f) + height(T_{b_2}(e))$. Since $cost(f)$ is constant, minimizing $maxpath_f(e)$ means minimizing $height(T_{b_1}(e)) + height(T_{b_2}(e))$.

To ease the following discussion, we will introduce notation regarding the order of edges and nodes on a path. Let $a \cdots b$ be a path, and let $e, e' \in a \cdots b$. We then have the order relation $\prec$ with respect to $a \cdots b$: $e \prec e'$ iff $dist(a, e) < dist(a, e')$, similar $e \preceq e'$ iff $dist(a, e) \leq dist(a, e')$.

The following theorem, proven in section 4.2, is the basis of our algorithm.

Theorem 5. *Let b_1, b_2 be nodes in a tree T and let $f = (b_1, b_2)$ be an edge not in T. Then there exists two nodes $v_1, v_2 \in b_1 \cdots b_2$ such that $v_1 \preceq v_2$ and for any edge $e \in b_1 \cdots b_2$:*

$$diam(T_{f/e}) = \begin{cases} diam(T_{b_2}(e)), & \text{if } e \in b_1 \cdots v_1 \\ maxpath_f(e), & \text{if } e \in v_1 \cdots v_2 \\ diam(T_{b_1}(e)), & \text{if } e \in v_2 \cdots b_2 \end{cases}$$

The algorithm consists of the following steps.

Algorithm 1
1. Find v_1 and v_2.
2. Minimize $diam(T_{b_2}(e))$ on $b_1 \cdots v_1$.
3. Minimize $maxpath_f(e)$ on $v_1 \cdots v_2$.
4. Minimize $diam(T_{b_1}(e))$ on $v_2 \cdots b_2$.
5. Compare with $diam(T)$ and select the best swap.

In section 4.2 we prove theorem 5 and we show how to find v_1, v_2 and how to minimize the diameters of the subtrees. In section 4.3 we show how to minimize $maxpath_f(e)$, which is the difficult part of the algorithm.

4.2 What and how to minimize

In order to prove theorem 5 we now proceed to investigate the behavior of $diam(T_{b_1}(e))$, $diam(T_{b_2}(e))$ and $maxpath_f(e)$ when $e \in b_1 \cdots b_2$. We know that when linking two trees, the diameter of the resulting tree is greater or equal to the diameters of both the original trees. Whereas $maxpath_f(e)$, is not a simple monotone function, it still bears some relationship with $diam(T_{b_1}(e))$ and $diam(T_{b_2}(e))$ as we will show in the next two lemmas.

Lemma 6. *Let T_1 be a tree with root x. Let T_2 be another tree and let T be the tree rooted in x obtained by linking T_1 and T_2 with some arbitrary edge e. Then $height(T) - height(T_1) \leq diam(T) - diam(T_1)$.* □

Lemma 7. *There exists a node* $w \in b_1 \cdots b_2$ *such that* $diam(T_{b_1}(e))$ $< maxpath_f(e)$ *when* $e \in b_1 \cdots w$ *and* $diam(T_{b_1}(e)) \geq maxpath_f(e)$ *when* $e \in w \cdots b_2$.

Proof. We prove it by showing that as we move the edge e from b_1 to b_2, $diam(T_{b_1}(e))$ grows at least as much as $maxpath_f(e)$.

Formally let $e', e'' \in b_1 \cdots b_2$ be edges such that $e' \preceq e''$. Then
$maxpath_f(e'') - maxpath_f(e')$
$= height(T_{b_1}(e'')) + height(T_{b_2}(e'')) - height(T_{b_1}(e')) - height(T_{b_2}(e'))$
$\leq height(T_{b_1}(e'')) - height(T_{b_1}(e'))$, since $height(T_{b_2}(e'')) - height(T_{b_2}(e')) \leq 0$
$\leq diam(T_{b_1}(e'')) - diam(T_{b_1}(e'))$, by lemma 6.

Thus, if there exists an edge $e = (x_1, x_2)$ such that $diam(T_{b_1}(e)) \geq maxpath_f(e)$ then $diam(T_{b_1}(e)) \geq maxpath_f(e)$ for all edges $e \in x_1 \cdots b_2$. By the same argument if there exists an edge $e = (y_1, y_2)$ such that $diam(T_{b_1}(e)) \leq maxpath_f(e)$ then $diam(T_{b_1}(e)) \leq maxpath_f(e)$ for all $e \in b_1 \cdots y_2$. $w \in b_1 \cdots b_2$ is then the node with greatest distance to b_1 such that $diam(T_{b_1}(e)) < maxpath_f(e)$ for $e \in b_1 \cdots w$. □

Proof of theorem 5. By lemma 7 we know that there exists a node w_2 such that $diam(T_{b_1}(e)) \geq maxpath_f(e)$ when $e \in w_2 \cdots b_2$ and $diam(T_{b_1}(e)) < maxpath_f(e)$ when $e \in b_1 \cdots w_2$. By symmetry there exists a node w_1 so $diam(T_{b_2}(e)) \geq maxpath_f(e)$ when $e \in b_1 \cdots w_1$ and $diam(T_{b_2}(e)) < maxpath_f(e)$ when $e \in w_2 \cdots b_2$. From this we see that if $w_1 \preceq w_2$ on $b_1 \cdots b_2$ then we can choose $v_1 = w_1, v_2 = w_2$ and $maxpath_f(e)$ is greater or equal to both $diam(T_{b_1}(e))$ and $diam(T_{b_2}(e))$ when $e \in v_1 \cdots v_2$. Otherwise, if $w_1 \succeq w_2$ then for all $e \in b_1 \cdots b_2$ either $diam(T_{b_1}(e)) \geq maxpath_f(e)$ or $diam(T_{b_2}(e)) \geq maxpath_f(e)$ since the diameter of both the subtrees are as least as great as $maxpath_f(e)$ when $e \in w_2 \cdots w_1$. If this is the case then there exists a node $v \in b_1 \cdots b_2$ such that $diam(T_{b_2}(e)) \geq diam(T_{b_1}(e))$ when $e \in b_1 \cdots v$ and $diam(T_{b_1}(e)) \geq diam(T_{b_2}(e))$ when $e \in v \cdots b_2$. In this case we choose $v_1 = v_2 = v$ which concludes the proof. □

Proposition 8. *The nodes v_1 and v_2 can be computed in $O(\log^2 n)$ time.*

Proof. We have $diam(T_{b_2}(e)) = diam(T_{f/e})$ for $e \in b_1 \cdots v_1$ and $diam(T_{b_2}(e)) < diam(T_{f/e})$ for $e \in v_1 \cdots b_2$. Thus, using the topology tree structure of section 2, v_1 is found by a simple binary search where each query is based on linking and cutting trees in $O(\log n)$ time, as described in theorem 4. The node v_2 is found symmetrically. □

Proposition 9. *We can minimize $diam(T_{b_2}(e))$ on $b_1 \cdots v_1$ and $diam(T_{b_1}(e))$ on $v_2 \cdots b_2$ in $O(\log n)$ time.*

Proof. The edge $e \in b_1 \cdots v_1$ which minimizes $diam(T_{b_2}(e))$ is simply the edge with the greatest distance to b_1 since $diam(T_{b_2}(e))$ is monotonically decreasing as e moves from b_1 to v_1. Similarly the edge minimizing $diam(T_{b_1}(e))$ is the edge with greatest distance to b_2 on $v_2 \cdots b_2$. These edges are easily found in $O(\log n)$ time, using the topology tree structure described in section 2. □

In this section we have shown that v_1 and v_2 can be found in $O(\log^2 n)$ time. In fact these two nodes can be found in $O(\log n)$ time, which we will show in the journal version, however because it is rather technical and since it would not change the overall complexity we have only given the simple argument above.

4.3 Minimizing the sum of the heights

Recall from section 4.1 that if the new edge f is involved in the diameter of $T_{f/e}$, $e \in b_1 \cdots b_2$, then $diam(T_{f/e}) = maxpath_f(e) = height(T_{b_1}(e)) + cost(f) + height(T_{b_2}(e))$ and minimizing $maxpath_f(e)$ means minimizing $height(T_{b_1}(e)) + height(T_{b_2}(e))$. By theorem 5 we know that we only need to minimize $maxpath_f(e)$ on the path $v_1 \cdots v_2$.

For any node v and any set of edges E', let $maxdist_{E'}(v)$ denote the maximum distance from v reachable in E'. For any path $P = p_1 \cdots p_2$ with $p_1 \neq p_2$ let $First(P)$ and $Last(P)$ denote the edges on P incident to p_1 and p_2 respectively.

Let U be the subtree of T, which consists of all the nodes reachable from v_1 (and v_2) without using any edges from $b_1 \cdots v_1 \cup v_2 \cdots b_2$. Then U is a cluster of T with $\partial U \subseteq \{v_1, v_2\}$.

For any edge $e \in v_1 \cdots v_2$ we have

$height(T_{b_1}(e)) = \max\{dist(b_1, v_1) + maxdist_{U\setminus\{e\}}(v_1), maxdist_{T\setminus U}(b_1)\}$
$= \max\{maxdist_{U\setminus\{e\}}(v_1), maxdist_{T\setminus U}(b_1) - dist(b_1, v_1)\} + dist(b_1, v_1)$
$= \max\{maxdist_{U\setminus\{e\}}(v_1), h_{v_1}\} + dist(b_1, v_1)$

where $h_{v_1} = maxdist_{T\setminus U}(b_1) - dist(b_1, v_1)$

$height(T_{b_2}(e)) = \max\{maxdist_{U\setminus\{e\}}(v_2), h_{v_2}\} + dist(b_2, v_2)$

where $h_{v_2} = maxdist_{T\setminus U}(b_2) - dist(b_2, v_2)$

Thus in order to solve the problem, all we need to know about the tree outside U, is the constant values h_{v_1} and h_{v_2}.

Definition 10. Let C be a cluster with $\partial C = \{a, b\}$, let $e \in a \cdots b$ be an edge and let h_a and h_b be any nonnegative numbers. Then define
$hsum_C(e, h_a, h_b) = \max\{maxdist_{C\setminus\{e\}}(a), h_a\} + \max\{maxdist_{C\setminus\{e\}}(b), h_b\}$.

With this definition, we have $height(T_{b_1}(e)) + height(T_{b_2}(e)) = hsum_U(e, h_{v_1}, h_{v_2})$ for $e \in v_1 \cdots v_2$.

Lemma 11. Let A, B and $C = A \cup B$ be clusters with $\partial A = \{a, c\}, \partial B = \{b, c\}$ and $\partial C = \{a, b\}, a \neq b$ and let h_a and h_b be any nonnegative numbers. For any edge $e_1 \in a \cdots c$ and $e_2 \in c \cdots b$ we have:

$hsum_C(e_1, h_a, h_b) = hsum_A(e_1, h_a, \max\{maxdist_B(b), h_b\} - dist(b, c)) + dist(b, c)$
$hsum_C(e_2, h_a, h_b) = hsum_B(e_2, \max\{maxdist_A(a), h_a\} - dist(a, c), h_b) + dist(a, c)$

Proof.

$$hsum_C(e_1, h_a, h_b) = \max\{maxdist_{C\setminus\{e_1\}}(a), h_a\} + \max\{maxdist_{C\setminus\{e_1\}}(b), h_b\}$$
$$= \max\{maxdist_{A\setminus\{e_1\}}(a), h_a\} +$$
$$\quad \max\{\max\{maxdist_{A\setminus\{e_1\}}(c) + dist(b,c), maxdist_B(b)\}, h_b\}$$
$$= \max\{maxdist_{A\setminus\{e_1\}}(a), h_a\} +$$
$$\quad \max\{maxdist_{A\setminus\{e_1\}}(c) + dist(b,c), \max\{maxdist_B(b), h_b\}\}$$
$$= \max\{maxdist_{A\setminus\{e_1\}}(a), h_a\} +$$
$$\quad \max\{maxdist_{A\setminus\{e_1\}}(c), \max\{maxdist_B(b), h_b\} - dist(b,c)\} + dist(b,c)$$
$$= hsum_A(e_1, h_a, \max\{maxdist_B(b), h_b\} - dist(b,c)) + dist(b,c)$$

The second equation follows by symmetry. □

Definition 12. Let C, a, b, h_a and h_b be defined as in lemma 11, then define $BestCuts_C(h_a, h_b)$ to be the set of edges $e \in a \cdots b$ minimizing $hsum_C(e, h_a, h_b)$.

This definition of *BestCuts* satisfies the following two lemmas.

Lemma 13. *Let A, B, C, a, b, c, h_a and h_b be as in lemma 11, then*

$$BestCuts_C(h_a, h_b) \cap A = BestCuts_A(h_a, \max\{maxdist_B(b), h_b\} - dist(b,c)) \bigvee$$
$$BestCuts_C(h_a, h_b) \cap B = BestCuts_B(\max\{maxdist_A(a), h_a\} - dist(a,c), h_b).$$

Proof. If there exists an edge e_A in $BestCuts_C(h_a, h_b) \cap A$ then e_A must minimize $hsum_C(e_A, h_a, h_b)$ on the path $a \cdots c$ and by lemma 11 it must also minimize $hsum_A(e_A, h_a, \max\{maxdist_B(b), h_b\} - dist(b,c))$ on that path. But then $BestCuts_C(h_a, h_b) \cap A = BestCuts_A(h_a, \max\{maxdist_B(b), h_b\} - dist(b,c))$ as desired. By symmetry, if there exists an edge e_B in $BestCuts_C(h_a, h_b) \cap B$ then $BestCuts_C(h_a, h_b) \cap B = BestCuts_B(\max\{maxdist_A(a), h_a\} - dist(a,c), h_b)$. And since $a \neq b$ then at least one of e_A and e_B must exist, yielding the desired result. □

Lemma 14. *Let C, a, b, h_a and h_b be as in lemma 11, then*

$$h_a \geq maxdist_C(a) \Rightarrow Last(a \cdots b) \in BestCuts_C(h_a, h_b)$$
$$h_b \geq maxdist_C(b) \Rightarrow First(a \cdots b) \in BestCuts_C(h_a, h_b).$$

Proof. Assume $h_a > maxdist_C(a)$. Then $hsum_C(e, h_a, h_b) = h_a + \max\{maxdist_{C\setminus\{e\}}(b), h_b\}$ for all $e \in a \cdots b$. But then any edge minimizing $maxdist_{C\setminus\{e\}}(b)$ will also minimize $hsum_C(e, h_a, h_b)$ and since $Last(a \cdots b)$ is such an edge we have $Last(a \cdots b) \in BestCuts_C(h_a, h_b)$ which proves the first part. The second part follows by symmetry. □

With this in hand we may now proceed to provide a procedure *bestcutedge* that finds an edge from $BestCuts_C(h_a, h_b)$. For any cluster with only one edge, it should just return that edge. For all other clusters we have the following proposition.

Proposition 15. *Let A,B,C,a,b,c,h_a and h_b be as in lemma 11. If we define*

$$bestcutedge_C(h_a, h_b) = \begin{cases} e_A, & \text{if } b = c, (|\partial B| = 1) \\ e_B, & \text{if } a = c, (|\partial A| = 1) \\ e_A, & \text{if } hsum_C(e_A, h_a, h_b) \leq hsum_C(e_B, h_a, h_b) \\ e_B, & \text{otherwise} \end{cases}$$

$$\text{where } e_A = \begin{cases} Last(a \cdots c), & \text{if } h_a \geq maxdist_A(a) \\ bestcutedge_A(h_a, \max\{maxdist_B(b), h_b\} - dist(b, c)) \\ & \text{otherwise} \end{cases}$$

$$\text{and } e_B = \begin{cases} First(c \cdots b), & \text{if } h_b \geq maxdist_B(b) \\ bestcutedge_B(\max\{maxdist_A(a), h_a\} - dist(a, c), h_b) \\ & \text{otherwise} \end{cases}$$

Then $bestcutedge_C(h_a, h_b) \in \text{BestCuts}_C(h_a, h_b)$.

Proof. By lemma 13 and lemma 14 we have that either e_A or e_B belongs to $BestCuts_C(h_a, h_b)$ and since $bestcutedge_C(h_a, h_b)$ picks the one minimizing $hsum_C(e, h_a, h_b)$ we have $bestcutedge_C(h_a, h_b) \in BestCuts_C(h_a, h_b)$ as desired. □

Proposition 15 gives us a recursive way of finding a *best cut edge* for a cluster in a topology tree. The idea is now, for each cluster C, $\partial C = \{a, b\}$, to save the latest value found by a call $bestcutedge_C(h_a, h_b)$ together with h_a, h_b and $hsum_C(h_a, h_b)$. Then if the next call $bestcutedge_C(h'_a, h'_b)$ has $h'_a = h_a$ and $h'_b = h_b$, we can immediately return the desired values in constant time. Otherwise, if $h'_a \geq h_a$ and $h'_b \geq h_b$ the memorization means that we only need to do a logarithmic number of recalculations, as stated in the following lemma:

Lemma 16. *Let C be a node in a topology tree, with $\partial C = \{a, b\}$, let h_a,h_b and $h'_a \geq h_a$ be any nonnegative numbers and suppose the last call to $bestcutedge_C$ was $bestcutedge_C(h_a, h_b)$. Then the number of recalculations needed to compute $bestcutedge_C(h'_a, h_b)$ is $O(\log n)$.*

Proof. From the definition of *bestcutedge* it is clear that whenever $bestcutedge_C(h'_a, h_b)$ makes two recursive calls, so does $bestcutedge_C(h_a, h_b)$, and at least one of these calls is identical to one made by $bestcutedge_C(h'_a, h_b)$. Furthermore the one new call made by $bestcutedge_C(h'_a, h_b)$ only differ in one parameter, so the same argument can be applied recursively. Thus by induction at most one recalculation can occur for each level in the topology tree, yielding a total of $O(\log n)$ recalculations. □

Formally, for every cluster C in the topology tree, with $\partial C = \{a, b\}$, $info(C)$ should include the following information in order for each of the recalculations to take constant time:
- $dist(a, b)$
- $maxdist_C(a), maxdist_C(b)$

And if C has more than one boundary node:

- $e_1 = First(a \cdots b), e_2 = Last(a \cdots b)$
- $maxdist_{C \setminus \{e_1\}}(a), maxdist_{C \setminus \{e_1\}}(b), maxdist_{C \setminus \{e_2\}}(a), maxdist_{C \setminus \{e_2\}}(b)$
- e, h_a, h_b and $hsum_C(e, h_a, h_b)$, where $e = bestcutedge_C(h_a, h_b)$ was the last call to $bestcutedge_C$.

Whenever a cluster C with two boundary nodes becomes the root on a topology tree, either by a *Merge* or a *DeleteRoot*, it should be initialized with a call $bestcutedge_C(0, 0)$.

Lemma 17. *The time needed to update* info *during a* Merge *or a* DeleteRoot *is* $O(\log n)$.

Proof. Let A and B be the clusters we want to merge, let C denote the cluster $A \cup B$ and let $\partial A = \{a, c\}, \partial B = \{b, c\}$ and $\partial C = \{a, b\}$. In order to do the *Merge*, we need to compute $bestcutedge_C(0, 0)$. By proposition 15 this can be done by computing $bestcutedge_A(0, maxdist_B(b) - dist(b, c))$ and $bestcutedge_B(maxdist_A(a) - dist(a, c), 0)$. In the structure we already have $bestcutedge_A(0, 0)$ and $bestcutedge_B(0, 0)$, and so by lemma 16 we only need to recalculate $bestcutedge$ for $O(\log n)$ clusters. Using lemma 11 and proposition 15 each recalculation can be done in constant time given the information available, yielding a total of $O(\log n)$ time for the update.

When deleting C again, we have to recalculate $bestcutedge_A(0, 0)$ and $bestcutedge_B(0, 0)$. The update made by the *Merge* operation that created C changed at most $O(\log n)$ clusters, and using lemma 11 and proposition 15 each value can be recalculated in constant time given the information available. Thus updating the structure under a *DeleteRoot* can be done in $O(\log n)$ time. □

By theorem 2 we now have that we can maintain *info* for a topology tree τ in time $O(\log^2 n)$ per operation, such that if C is an internal node in τ with $\partial C = \{a, b\}$, and the last call of $bestcutedge_C$ was $bestcutedge_C(h_a, h_b)$ then for any $h'_a \geq h_a$ and $h'_b \geq h_b$ we can find an edge $e \in a \cdots b$ minimizing $hsum_C(e, h'_a, h'_b)$ in time $O(\log n)$ according to lemma 16.

Given a topology tree τ and an arbitrary path $P = p_1 \cdots p_2$, there may not be a cluster in τ where p_1 and p_2 are boundary nodes. Thus in order for the search described above to work for the path P, we will have to change the topology tree to create such a cluster. To do this we will introduce the concept of *external* boundary nodes.

Let τ be a topology tree, let C be an internal node of τ with $\partial C = \{a, b\}$, and let τ' be the subtree of τ with root C. If we restrict ourselves to looking at τ' then C has no boundary nodes in the normal sense. But the structure of τ' is still exactly as if $\{a, b\}$ were boundary nodes. Formally we say that:
- $\{a, b\}$ are *external* boundary nodes of C in τ'.
- $\{a, b\}$ are *internal* boundary nodes of C in τ

And we say that τ' is a *topology tree with external boundary nodes* $\{a, b\}$. Formally: For any tree T, and any nodes $a, b \in T$ there obviously exists a tree T' with topology tree τ', such that T is represented by a node in τ' and $\partial T = \{a, b\}$. If we let τ be the subtree of τ' with root T, then τ is said to be a *topology tree with external boundary nodes a and b*. In the journal version we will prove the following lemma

Lemma 18. *Given a tree T with topology tree τ and two nodes p_1 and p_2 from T. Then we can change τ into a topology tree τ' with external boundary nodes p_1 and p_2, and back, using $O(\log n)$ Merge and DeleteRoot operations.* □

Theorem 19. *There exists an algorithm for maintaining a dynamic forest supporting* link, cut *and* best swap *operations in $O(\log^2 n)$ time, given the ability to use $O(n \log n)$ time and $O(n)$ space for preprocessing, where n is the number of nodes in the tree(s) involved in the operation.*

Proof. By theorem 5 algorithm 1 solves the *best swap* problem. By proposition 8 we can perform step 1 in $O(\log^2 n)$ time. By proposition 9 we can compute step 2 and 4 in $O(\log n)$ time. To solve step 3 we do the following. We cut at most two edges to obtain the subtree U containing the path $v_1 \cdots v_2$. By lemma 18 we can make v_1 and v_2 external boundary nodes in a topology tree structure for U using $O(\log n)$ *Merges* and *DeleteRoots*. Then we can apply lemma 16 to find the edge minimizing $maxpath_f$ in $O(\log n)$ time. Then we relink the topology tree back to its normal form without external boundary nodes. This is done to rebuild the structure that we may use it again. This can be done using $O(\log n)$ *Merges* and *DeleteRoots* by lemma 18. Since step 5 amounts to comparing four numbers and picking the smallest, this step clearly runs in constant time. Thus all steps in the algorithm can be done using $O(\log n)$ *Merges* and *DeleteRoots*. By lemma 17 both a *Merge* and a *DeleteRoot* takes $O(\log n)$ time and using proposition 1 we can update the structure in $O(\log^2 n)$ time under *link* and *cut*. By proposition 1 and lemma 17 the preprocessing takes $O(n \log n)$ time and $O(n)$ space. □

References

1. Z-Z. Chen. A simple parallel algorithm for computing the diameters of all vertices in a tree and its application. *Information Processing Letters*, 42:243–248, 1992.
2. G. N. Frederickson. Data structures for on-line updating of minimum spanning trees, with applications. *SIAM J. Computing*, 14(4):781–798, 1985.
3. G. N. Frederickson. Ambivalent data structures for dynamic 2–edge–connectivity and k smallest spanning trees. In *IEEE Symposium on Foundations of Computer Science (FOCS)*, pages 632–641, 1991.
4. G.Y. Handler. Minimax location of a facility in an undirected tree network. *Transportation. Sci.*, 7:287–293, 1973.
5. G. F. Italiano and R. Ramaswami. Mantaining spanning trees of small diameter. In *Proc. 21st Int. Coll. on Automata, Languages and Programming*. Lecture Notes in Computer Science, Springer-Verlag, Berlin, 1994.
6. G. F. Italiano and R. Ramaswami. Mantaining spanning trees of small diameter. Unpublished revised version of the ICALP paper, 1996.
7. R. Ramaswami. Multi-wavelength lightwave networks for computer communication. *IEEE Communications Magazine*, 31:78–88, 1993.
8. M. Rauch. *Fully dynamic graph algorithms and their data structures*. PhD thesis, Department of computer science, Princeton University, December 1992.

Improving Spanning Trees by Upgrading Nodes

S. O. Krumke,[1] M. V. Marathe,[2] H. Noltemeier,[1]
R. Ravi,[3] S. S. Ravi,[4] R. Sundaram,[5] H. C. Wirth[1]

[1] Dept. of Computer Science, University of Würzburg, Am Hubland, 97074 Würzburg, Germany.
Email: {krumke,noltemei,wirth}@informatik.uni-wuerzburg.de
[2] Los Alamos National Laboratory, P.O. Box 1663, MS K990, Los Alamos, NM 87545, USA. Email: madhav@c3.lanl.gov[§]
[3] GSIA, Carnegie Mellon University, Pittsburgh, PA 15213. Email: ravi+@cmu.edu[¶]
[4] Dept. of Computer Science, University at Albany – SUNY, Albany, NY 12222, USA. Email: ravi@cs.albany.edu
[5] Delta Trading Co. Work done while at MIT, Cambridge MA 02139. Email: koods@theory.lcs.mit.edu.[‖]

Abstract. We study *budget constrained optimal network upgrading problems*. We are given an edge weighted graph $G = (V, E)$ where node $v \in V$ can be upgraded at a cost of $c(v)$. This upgrade reduces the delay of each link emanating from v. The goal is to find a minimum cost set of nodes to be upgraded so that the resulting network has a good performance. We consider two performance measures, namely, the weight of a minimum spanning tree and the bottleneck weight of a minimum bottleneck spanning tree, and present approximation algorithms.

1 Introduction, Motivation and Summary of Results

Several problems arising in areas such as communication networks and VLSI design can be expressed in the following general form: Enhance the performance of a given network by upgrading a suitable subset of nodes. In communication networks, upgrading a node corresponds to installing faster communication equipment at that node. Such an upgrade reduces the communication delay along each edge emanating from the node. In signal flow networks used in VLSI design, upgrading a node corresponds to replacing a circuit module at the node by a functionally equivalent module containing suitable drivers. Such an upgrade decreases the signal transmission delay along the wires connected to the module. There is a cost associated with upgrading a node, and there is often a budget on the total upgrading cost. Therefore, it is of interest to study the problem of upgrading a network so that the total upgrading cost obeys the budget constraint and the resulting network has the best possible performance among all upgrades that satisfy the budget constraint.

[§] Supported by the Department of Energy under Contract W-7405-ENG-36.
[¶] Supported by NSF CAREER grant CCR-9625297.
[‖] Supported by DARPA contract N0014-92-J-1799 and NSF CCR 92-12184.

The performance of the upgraded network can be quantified in a number of ways. In this paper, we consider two such measures, namely, the weight of a minimum spanning tree in the upgraded network and the bottleneck cost (i.e., the maximum weight of an edge) in a spanning tree of the upgraded network. Under either measure, the upgrading problem can be shown to be NP-hard. So, the focus of the paper is on the design of efficient approximation algorithms.

1.1 Background: Bicriteria Problems and Approximation

The problems considered in this paper involve two optimization objectives, namely, the upgrading cost and the performance of the upgraded network. A framework for such bicriteria problems has been developed in [7]. A generic bicriteria problem can be specified as a triple (A, B, Γ) where A and B are two objectives and Γ specifies a class of subgraphs. An instance specifies a budget on the objective A and the goal is to find a subgraph in the class Γ that minimizes the objective B for the upgraded network. As an example, the problem of upgrading a network so that the modified network has a spanning tree of weight at most D while minimizing the node upgrading cost can be expressed as (TOTAL WEIGHT, NODE UPGRADING COST, SPANNING TREE).

Definition 1. A polynomial time algorithm for a bicriteria problem (A, B, Γ) is said to have *performance* (α, β), if it has the following property: For any instance of (A, B, Γ) the algorithm

1. either produces a solution from the subgraph class Γ for which the value of objective A is at most α times the specified budget and the value of objective B is at most β times the minimum value of a solution from Γ that satisfies the budget constraint, or
2. correctly provides the information that there is no subgraph from Γ which satisfies the budget constraint on A.

1.2 Problem Definitions

The *node based upgrading model* discussed in this paper can be formally described as follows. Let $G = (V, E)$ be a connected undirected graph. For each edge $e \in E$, we are given three integers $d_0(e) \geq d_1(e) \geq d_2(e) \geq 0$. The value $d_i(e)$ represents the *length* or *delay* of the edge e if exactly i of its endpoints are upgraded.

Thus, the upgrade of a node v reduces the delay of each edge incident with v. The (integral) value $c(v)$ specifies how expensive it is to upgrade the node v. The cost of upgrading all vertices in $W \subseteq V$, denoted by $c(W)$, is equal to $\sum_{v \in W} c(v)$.

For a set $W \subseteq V$ of vertices, denote by d_W the edge weight function resulting from the upgrade of the vertices in W; that is, for an edge $(u, v) \in E$

$$d_W(u, v) := d_i(u, v) \qquad \text{where } i = |W \cap \{u, v\}|.$$

We denote the total length of a minimum spanning tree (MST) in G with respect to the weight function d_W by $\text{MST}(G, d_W)$.

Definition 2. Given an edge and node weighted graph $G = (V, E)$ as above and a bound D, the *upgrading minimum spanning tree problem*, denoted by (TOTAL WEIGHT, NODE UPGRADING COST, SPANNING TREE), is to upgrade a set $W \subseteq V$ of nodes such that $\text{MST}(G, d_W) \leq D$ and $c(W)$ is minimized.

We also consider the node based upgrading problem to obtain a spanning tree with the bottleneck cost at most a given value. We denote the bottleneck weight (i.e., the maximum weight of an edge) of a minimum bottleneck spanning tree of G with respect to the weight function d_W by $\text{MBOT}(G, d_W)$.

Definition 3. Given an edge and node weighted graph $G = (V, E)$ as above and a bound D, the *upgrading minimum bottleneck spanning tree problem*, denoted by (BOTTLENECK WEIGHT, NODE UPGRADING COST, SPANNING TREE), is to upgrade a set $W \subseteq V$ of nodes such that $\text{MBOT}(G, d_W) \leq D$ and $c(W)$ is minimized.

Dual Problems The problem (TOTAL WEIGHT, NODE UPGRADING COST, SPANNING TREE) is formulated by specifying a budget on the weight of a tree while the upgrading cost is to be minimized. It is also meaningful to consider the corresponding *dual problem*, denoted by (NODE UPGRADING COST, TOTAL WEIGHT, SPANNING TREE), where we are given a budget on the upgrading cost and the goal is to minimize the weight of a spanning tree in the resulting graph.

Lemma 4. *If there exists an approximation algorithm for* (TOTAL WEIGHT, NODE UPGRADING COST, SPANNING TREE) *with a performance of* (α, β), *then there is an approximation algorithm for* (NODE UPGRADING COST, TOTAL WEIGHT, SPANNING TREE) *with performance of* (β, α).

Proof. Let A be an (α, β)-approximation algorithm for (TOTAL WEIGHT, NODE UPGRADING COST, SPANNING TREE). We will show how to use A to construct a (β, α)-approximation algorithm for the dual problem.

An instance of (NODE UPGRADING COST, TOTAL WEIGHT, SPANNING TREE) is specified by a graph $G = (V, E)$, the node cost function c, the weight functions d_i, $i = 0, 1, 2$, on the edges and the bound B on the node upgrading cost. We denote by OPT the optimum weight of an MST after upgrading a vertex set of cost at most B. Observe that OPT is an integer such that $(n-1)D_2 \leq \text{OPT} \leq (n-1)D_0$ where $D_2 := \min_{e \in E} d_2(e)$ and $D_0 := \max_{e \in E} d_0(e)$.

We use binary search to find the minimum integer D such that $(n-1)D_2 \leq D \leq (n-1)D_0$ and algorithm A applied to the instance of (NODE UPGRADING COST, BOTTLENECK WEIGHT, SPANNING TREE) given by the weighted graph G as above and the bound D on the weight of an MST after the upgrade outputs an upgrading set of cost at most αB. It is easy to see that this binary search indeed works and terminates with a value $D \leq \text{OPT}$. The corresponding upgrading set W then satisfies $\text{MST}(G, d_W) \leq \beta D \leq \beta \text{OPT}$ and $c(W) \leq \alpha B$. □

A result similar to Lemma 4 can be shown for the bottleneck case. In view of these results, we express our results for the problems (TOTAL WEIGHT, NODE UPGRADING COST, SPANNING TREE) and (BOTTLENECK WEIGHT, NODE UPGRADING COST, SPANNING TREE).

1.3 Summary of Results

For the total weight MST upgrading problem, we derive our approximation results under the following assumption:

Assumption 5. *There is a polynomial p such that $D_0 - D_2 \leq p(n)$, where $D_0 := \max_{e \in E} d_0(e)$ and $D_2 := \min_{e \in E} d_2(e)$ are the maximum and minimum edge weight, respectively, and n denotes the number of nodes in the graph.*

Theorem 6. *For any fixed $\varepsilon > 0$, there is a polynomial time algorithm which, for any instance of (TOTAL WEIGHT, NODE UPGRADING COST, SPANNING TREE) satisfying Assumption 5, provides a performance of $(1, (1+\varepsilon)^2 \mathcal{O}(\log n))$.*

For the bottleneck case, we do not need any assumption about the edge weights.

Theorem 7. *There is an approximation algorithm for the (BOTTLENECK WEIGHT, NODE UPGRADING COST, SPANNING TREE) problem with performance $(1, 2 \ln n)$.*

Our approximation results are complemented by the following hardness results:

Theorem 8. *Unless $\mathsf{NP} \subseteq \mathsf{DTIME}(n^{\mathcal{O}(\log \log n)})$, there can be no polynomial time approximation algorithm for either (TOTAL WEIGHT, NODE UPGRADING COST, SPANNING TREE) or (BOTTLENECK WEIGHT, NODE UPGRADING COST, SPANNING TREE) with a performance of $(f(n), \alpha)$ for any $\alpha < \ln n$ and any polynomial time computable function f. This result continues to hold with $f(n) = n^k$ being any polynomial, even if Assumption 5 holds.*

Due to space limitations, the remainder of this paper discusses mainly the algorithm mentioned in Theorem 6 above. Proofs of other results will appear in a complete version of this paper.

1.4 Related Work

Some node upgrading problems have been investigated under a simpler model by Paik and Sahni [9]. In their model, the delay of an edge is decreased by constant factors of δ or δ^2, when one or two of its endpoints are upgraded, respectively. Clearly, this model is a special case of the model treated in our paper.

Under their model, Paik and Sahni studied the upgrading problem for several performance measures including the maximum delay on an edge and the diameter of the network. They presented NP-hardness results for several problems. Their focus was on the development of polynomial time algorithms for special classes of networks (e.g. trees, series-parallel graphs) rather than on the development of approximation algorithms. Our constructions can be modified to show that all the problems considered here remain NP-hard even under the Paik-Sahni model.

Edge-based network upgrading problems have also been considered in the literature [1, 4, 5]. There, each edge has a current weight and a minimum weight (below which the edge weight cannot be decreased). Upgrading an edge corresponds to decreasing the weight of that particular edge and there is a cost associated with such an upgrade. The goal is to obtain an upgraded network with the best performance. In [4] the authors consider the problem of edge-based upgrading to obtain the best possible MST subject to a budget constraint on the upgrading cost and present a $(1+\varepsilon, 1+1/\varepsilon)$-approximation algorithm. Generalized versions where there are other constraints (e.g. bound on maximum node degree) and the goal is to obtain a good Steiner tree, are considered in [5]. Other references that address problems that can be interpreted as edge-based improvement problems include [3, 8, 10].

2 Upgrading Under Total Weight Constraint

In this section we develop our approximation algorithm for the (TOTAL WEIGHT, NODE UPGRADING COST, SPANNING TREE) problem. Without loss of generality we assume that for a given instance of (TOTAL WEIGHT, NODE UPGRADING COST, SPANNING TREE) the bound D on the weight of the minimum spanning tree after the upgrade satisfies $D \geq \text{MST}(G, d_2)$, i.e., the weight of an MST with respect to d_2, since node upgrading cannot reduce the weight of the minimum spanning tree below this value. Thus, there always exists a subset of the nodes which, when upgraded, leads to an MST of weight at most D. We remind the reader that our algorithm also uses Assumption 5 (stated in Section 1.3) regarding the edge weights in the given instance.

2.1 Overview of the Algorithm

Our approximation algorithm can be thought of as a *local improvement* type algorithm. To begin with, we compute an MST in the given graph with edge weights given by $d_0(e)$. Now, during each iteration, we select a node and a subset of its neighbors and upgrade them. The policy used in the selection process is that of finding a set which gives us the best ratio improvement, which is defined as the ratio of the improvement in the total weight of the spanning tree to the total cost spent on upgrading the nodes. Having selected such a set, we recompute the MST and repeat our procedure. The procedure is halted when the weight of the MST is at most the required threshold D. To find a subset of node with the best ratio improvement in each iteration, we use an approximate solution to the *Two Cost Spanning Tree Problem* defined below.

Definition 9 *Two Cost Spanning Tree Problem.* Given a connected undirected graph $G = (V, E)$, two edge weight functions, c and l, and a bound B, find a spanning tree T of G such that the total cost $c(T)$ is at most B and the total cost $l(T)$ is a minimum among all spanning trees that obey the budget constraint.

The above problem can be expressed as the bicriteria problem (c-TOTAL WEIGHT, l-TOTAL WEIGHT, SPANNING TREE). This problem has been addressed by Ravi and Goemans [11] who obtained the following result.

Theorem 10. *For all $\varepsilon > 0$, there is a polynomial time approximation algorithm for the Two Cost Spanning Tree problem with a performance of $(1+\varepsilon, 1)$.*

2.2 Algorithm and Performance Guarantee

The steps of our algorithm are shown in Figure 1. This algorithm uses Procedure COMPUTE QC whose description appears in Figure 2.

ALGORITHM UPGRADE MST(Ω)
- *Input:* A graph $G = (V, E)$, three edge weight functions $d_0 \geq d_1 \geq d_2$, a node weight function c, and a number D, which is a bound on the weight of an MST in the upgraded graph; a "guess value" Ω for the optimal upgrading cost.
  1. Initialize the set of upgraded nodes: $W_0 := \emptyset$.
  2. Let $T_0 := \text{MST}(G, d_{W_0})$.
  3. Initialize the iteration count: $i := 1$.
  4. Repeat the following steps until for the current tree T_{i-1} and the weight function $d_{W_{i-1}}$ we have: $d_{W_{i-1}}(T_{i-1}) \leq D$:
     (a) Let $T_{i-1} := \text{MST}(G, d_{W_{i-1}})$ be an MST w.r.t. the weight function $d_{W_{i-1}}$.
     (b) Call Procedure COMPUTE QC to find a marked claw C with "good" quotient cost $q(C)$. Procedure COMPUTE QC is called with the graph G, the current MST T_{i-1}, the current weight function $d_{W_{i-1}}$ and the bound Ω.
     (c) If Procedure COMPUTE QC reports failure, then report failure and stop.
     (d) Upgrade the marked vertices $M(C)$ in C: $W_i := W_{i-1} \cup M(C)$.
     (e) Increment the iteration count: $i := i + 1$.
- *Output:* A spanning tree with weight at most D, such that total cost of upgrading the nodes is no more than $(1+\varepsilon)\Omega \cdot \mathcal{O}(\log n)$, provided $\Omega \geq$ OPT. Here, OPT denotes the optimal upgrading cost to reduce the weight of an MST to be at most D.

Fig. 1. Approximation algorithm for node upgrading under total weight constraint.

Before we embark on a proof of Theorem 6, we give the overall idea behind the proof. Recall that each basic step of the algorithm consists of finding a node and a subset of neighbors to upgrade.

Definition 11. A graph $C = (V, E)$ is called a *claw*, if E is of the form $E = \{(v, w) : w \in V \setminus \{v\}\}$ for some node $v \in V$. The node v is said to be the *center* of the claw. A claw with at least two nodes is called a *nontrivial claw*.

Let W be a subset of the nodes upgraded so far and let T be an MST with respect to d_W; that is, $T = \text{MST}(G, d_W)$. For a claw C with nodes $M(C) \subseteq C$ marked, we define its *quotient cost* $q(C)$ to be

$$q(C) := \frac{c(M(C))}{d_W(T) - \text{MST}(T \cup C, d_{W \cup M(C)})} \quad , \text{ if } M(C) \neq \emptyset,$$

and $+\infty$ otherwise. In other words, $q(C)$ is the cost of the vertices in $M(C)$ divided by the decrease in the weight of the MST when the vertices in $M(C)$ are also upgraded and edges in the current tree T can be exchanged for edges in the claw C. Notice that this way the real profit of upgrading the vertices $M(C)$ is underestimated, since the weight of edges outside of C might also decrease.

Our analysis essentially shows that in each iteration there exists a claw of quotient cost at most $\frac{2\,\text{OPT}}{d_W(T)-D}$, where T is the weight of an MST at the beginning of the iteration and W are the nodes upgraded so far. We can then use a potential function argument to show that this yields a logarithmic performance guarantee.

PROCEDURE COMPUTE QC(Ω)
- *Input:* A graph $G = (V, E)$, a spanning tree T and a weight function d on E; $W \subseteq V$ is the set of upgraded nodes; a "guess" Ω for the optimal upgrading cost.
1. Let $s := \lceil \log_{1+\varepsilon} \Omega \rceil$.
2. For each node $v \notin W$ and all $K \in \{1, (1+\varepsilon), (1+\varepsilon)^2, \ldots, (1+\varepsilon)^s\}$ do
   (a) Set up an instance $I_{v,K}$ of the *Two Cost Spanning Tree Problem* as follows:
       - The vertex set of the graph G_v contains all the vertices in G and an additional "dummy node" x.
       - There is an edge (v, x) joining v to the dummy node x of length $l(v, x) = 0$ and cost $c(v, x) = c(v)$ thus modeling the upgrading cost of v.
       - For each edge $(v, w) \in E$, G_v contains two parallel edges h and h_{up}. The edge h models the situation where w is not upgraded:

       $$c(h) := 0 \qquad\qquad l(h) := \begin{cases} d_2(v, w) & \text{if } w \in W \\ d_1(v, w) & \text{if } w \notin W \end{cases}$$

       Similarly, h_{up} models an upgrade of w:

       $$c(h_{\text{up}}) := \begin{cases} 0 & \text{if } w \in W \\ c(w) & \text{if } w \notin W \end{cases} \qquad l(h_{\text{up}}) := d_2(v, w).$$

       - For each edge $(u, w) \in T$, there is one edge $(u, w) \in E$ which has length $l(u, w) = d(u, w)$ and cost $c(u, w) = 0$.
       - The bound B on the c-cost of the tree is set to K.
   (b) Using the algorithm mentioned in Theorem 10, find a tree of c-cost at most $(1+\varepsilon)K$ and l-cost no more than that of a minimum budget K bounded spanning tree (if one exists). Let $T_{v,K}$ be the tree produced by the algorithm.
3. If the algorithm fails for *all* instances $I_{v,K}$ then report failure and stop.
4. Among all the trees $T_{v,K}$ find a tree T_{v^*,K^*} which minimizes the ratio $c(T_{v^*,K^*})/(d(T) - l(T_{v^*,K^*}))$.
5. Construct a marked claw C from T_{v^*,K^*} as follows:
   - The center of C is v^* and v^* is marked.
   - The edge (v^*, w) is in the claw C if T_{v^*,K^*} contains an edge between v^* and w. The node w is marked if and only if the edge in T_{v^*,K^*} between v^* and w has c-cost greater than zero.
- *Output:* A marked claw C (with its center also marked) with quotient cost $q(C)$ satisfying $q(C) \leq 2(1+\varepsilon)^2 \frac{\text{OPT}}{d(T)-D}$ and cost $c(M(C)) \leq (1+\varepsilon)\Omega$.

Fig. 2. Algorithm for computing a good claw.

2.3 Bounded Claw Decompositions

Definition 12. Let $G = (V, E)$ be a graph and $W \subseteq V$ a subset of marked vertices. Let $\kappa \geq 1$ be an integer constant. A *κ-bounded claw decomposition* of G with respect to W is a collection $C_1, \ldots, C_r$ of nontrivial claws, which are all subgraphs of G, with the following properties:

1. $\bigcup_{i=1}^r V(C_i) = V$ and $\bigcup_{i=1}^r E(C_i) = E$.
2. No node from W appears in more than κ claws.
3. The claws are edge-disjoint.
4. If a claw C_i contains nodes from W, then its center belongs also to W.

Lemma 13. *Let F be a forest in $G = (V, E)$ and let $W \subseteq V$ be a set of marked nodes. Then there is a 2-bounded claw decomposition of F with respect to W.* □

Lemma 14. *Let $T := T_{i-1}$ be an MST at the beginning of iteration i with $W := W_{i-1}$ being the nodes upgraded so far. Let $U \subseteq V$ be a set of nodes. Let $T' = \mathrm{MST}(G, d_{W \cup U})$ be a minimum spanning tree after the additional upgrade of the vertices in U. Then, there is a bijection $\varphi : T \to T'$ with the following properties:*
1. For all edges $e \in T \cap T'$ we have $\varphi(e) = e$, 2. $d_{W \cup U}(\varphi(e)) \leq d_W(e)$ for all $e \in T$, 3. the "swaps" $e \mapsto \varphi(e)$ transform T into T', and 4. $\sum_{e \in T}(d_W(e) - d_{W \cup U}(\varphi(e))) = d_W(T) - d_{W \cup U}(T')$. □

Lemma 15. *Let $T := T_{i-1}$ be an MST at the beginning of iteration i, i.e., $T = \mathrm{MST}(G, d_W)$, where $W := W_{i-1}$ is the upgrading set constructed so far. Then there is a marked claw C (where its center v is also marked and $v \notin W$) with quotient cost $q(C)$ satisfying*

$$q(C) \leq \frac{2\,\mathrm{OPT}}{d_W(T) - D} \qquad \text{and} \qquad c(M(C)) \leq \mathrm{OPT}.$$

Proof. Let $T' = \mathrm{MST}(G, d_{W \cup \mathrm{OPT}})$ be an MST after the additional upgrade of the vertices in OPT. Clearly, $d_{W \cup \mathrm{OPT}}(T') \leq D$. Apply Lemma 13 to T' with the vertices in $Z := \mathrm{OPT} \setminus W$ marked. The lemma shows that there is a 2-bounded claw decomposition of T' with respect to Z. Let the claws be $C_1, \ldots, C_r$. In each claw C_j, the corresponding nodes $M(C_j) := C_j \cap Z$ from Z are marked. Since the decomposition is 2-bounded with respect to Z, it follows that

$$\sum_{j=1}^r c(M(C_j)) \leq 2 \cdot \mathrm{OPT}. \qquad (1)$$

Moreover, the cost $c(M(C_j))$ of the marked nodes in each single claw C_j does not exceed OPT, since we have marked only nodes from Z. By Lemma 14, there exists a bijection $\varphi : T \to T'$ such that

$$\sum_{e \in T}\left(d_W(e) - d_{W \cup \mathrm{OPT}}(\varphi(e))\right) = d_W(T) - d_{W \cup \mathrm{OPT}}(T') \geq d_W(T) - D. \qquad (2)$$

For each of the claws C_j with $M(C_j) \neq \emptyset$ in the 2-bounded decomposition of T' its quotient cost $q(C_j)$ satisfies

$$q(C_j) \leq \frac{c(M(C_j))}{\sum_{\varphi(e) \in C_j}(d_W(e) - d_{W \cup \text{OPT}}(\varphi(e)))}, \qquad (3)$$

since we can exchange the edges $\varphi(e)$ ($e \in C_j$) for the corresponding edges e in the current tree T after the upgrade and thus decrease the weight of the tree by at least $\sum_{\varphi(e) \in C_j}(d_W(e) - d_{W \cup \text{OPT}}(\varphi(e)))$.

Let C be a claw among all the claws C_j with minimum $q(C)$. Then,

$$q(C) \cdot \sum_{e \in C_j} \Big(d_W(e) - d_{W \cup \text{OPT}}(\varphi(e))\Big) \leq c(M(C_j)) \quad \text{for } j = 1, \ldots, r. \qquad (4)$$

Notice that the above equation holds, regardless of whether $M(C_j)$ is empty or not. Summing the inequalities in (4) over $j = 1, \ldots, r$, and using Equations (1) and (2), it can be seen that C is a claw with the desired properties. □

2.4 Finding a good claw in each iteration

Lemma 15 implies the existence of a marked claw with the required properties. We will now deal with the problem of finding such a claw.

Lemma 16. *Suppose that the bound Ω given to Algorithm* UPGRADE MST *satisfies $\Omega \geq$ OPT. Then, for each stage i of the algorithm, it chooses a marked claw C' such that*

$$q(C') \leq 2(1+\varepsilon)^2 \frac{\text{OPT}}{d_W(T) - D} \qquad \text{and} \qquad c(M(C')) \leq (1+\varepsilon)\Omega,$$

where $T := T_{i-1}$ is an MST at the beginning of iteration i and $W := W_{i-1}$ is the set of nodes upgraded so far.

Proof. By Lemma 15, there is a marked claw C with quotient cost $q(C)$ at most $2\frac{\text{OPT}}{d_W(T)-D}$. Let v be the center of this claw. By Lemma 15, v is marked. Let $c(C) := c(M(C))$ be the cost of the marked nodes in C and $L := \text{MST}(T \cup C, d_{W \cup M(C)})$ be the weight of the MST in $T \cup C$ resulting from the upgrade of the marked vertices in C. Then, by definition of the quotient cost $q(C)$ we have

$$q(C) = \frac{c(C)}{d_W(T) - L} \leq 2\frac{\text{OPT}}{d_W(T) - D}. \qquad (5)$$

Consider the iteration of PROCEDURE COMPUTE QC when it processes the instance $I_{v,K}$ of *Two Cost Spanning Tree Problem* with graph G_v and $c(C) \leq K < (1+\varepsilon) \cdot c(C)$. The tree $\text{MST}(T \cup C, d_{W \cup M(C)})$ induces a spanning tree in G_v of total c-cost at most $c(C)$ (which is at most K) and of total l-length no more than L. Thus, the algorithm from Theorem 10 will find a tree $T_{v,K}$ such that its total c-cost $c(T_{v,K})$ is bounded from above by $(1+\varepsilon)K \leq (1+\varepsilon)^2 c(C)$ and of total l-length $l(T_{v,K})$ no more than L.

By construction, the marked claw C' computed by PROCEDURE COMPUTE QC from $T_{v,K}$ has quotient cost at most $c(T_{v,K})/(d_W(T) - l(T_{v,K}))$, which is at most $(1+\varepsilon)^2 c(C)/(d_W(T) - L)$. The lemma now follows from (5). □

2.5 Guessing an Upper Bound on the Improvement Cost

We run our Algorithm UPGRADE MST depicted in Figure 1 for all values of

$$\Omega \in \{1, (1+\varepsilon), (1+\varepsilon)^2, \ldots, (1+\varepsilon)^t\}, \qquad \text{where } t := \lceil \log_{1+\varepsilon} c(V) \rceil.$$

We then choose the best solution among all solutions produced. Our analysis shows that when $\text{OPT} \leq \Omega < (1+\varepsilon) \cdot \text{OPT}$, the algorithm will indeed produce a solution. In the sequel, we estimate the quality of this solution. Assume that the algorithm uses $f+1$ iterations and denote by $C_1, \ldots, C_f, C_{f+1}$ the claws chosen in Step 4b of the algorithm. Let $c_i := c(M(C_i))$ denote the cost of the vertices upgraded in iteration i. Then, by construction

$$c_i \leq (1+\varepsilon)\Omega \leq (1+\varepsilon)^2 \text{OPT} \quad \text{for } i = 1, \ldots, f+1. \tag{6}$$

2.6 Potential Function Argument

We are now ready to complete the proof of the performance stated in Theorem 6. Let MST_i denote the weight of the MST at the end of iteration i, i.e., $\text{MST}_i := d_{W_i}(T_i)$. Define $\phi_i := \text{MST}_i - D$. Since we have assumed that the algorithm uses $f+1$ iterations, we have $\phi_i \geq 1$ for $i = 0, \ldots, f$ and $\phi_{f+1} \leq 0$. As before, let $c_i := c(M(C_i))$ denote the cost of the vertices upgraded in iteration i. Then

$$\phi_{i+1} = \phi_i - (\text{MST}_i - \text{MST}_{i+1}) \overset{\text{Lemma 16}}{\leq} \left(1 - \frac{c_{i+1}}{\alpha \cdot \text{OPT}}\right) \phi_i, \tag{7}$$

where $\alpha := 2(1+\varepsilon)^2$. We now use an analysis technique due to Leighton and Rao [6]. The recurrence (7) and the estimate $\ln(1-\tau) \leq -\tau$ give us

$$\sum_{i=1}^{f} c_i \leq \alpha \cdot \text{OPT} \cdot \ln \frac{\phi_0}{\phi_f}. \tag{8}$$

Notice that the total cost of the nodes chosen by the algorithm is exactly the sum $\sum_{i=1}^{f+1} c_i$. By (8) and (6) we have

$$\sum_{i=1}^{f+1} c_i = c_{f+1} + \sum_{i=1}^{f} c_i \leq (1+\varepsilon)^2 \text{OPT} + 2(1+\varepsilon)^2 \text{OPT} \cdot \ln \frac{\phi_0}{\phi_f}. \tag{9}$$

We will now show how to bound $\ln \frac{\phi_0}{\phi_f}$. Notice that $\phi_f = \text{MST}_f - D \geq 1$, since the algorithm uses $f+1$ iterations and does not stop after the fth iteration. We have $\phi_0 = \text{MST}_0 - D \leq (n-1)(D_0 - D_2)$, where D_0 and D_2 denote the maximum and the minimum edge weight in the graph. It now follows from Assumption 5 that $\ln \phi_0 \in \mathcal{O}(\log(np(n))) \subseteq \mathcal{O}(\log n)$. Using this result in (9) yields

$$\sum_{i=1}^{f+1} c_i \leq (1+\varepsilon)^2 \cdot \text{OPT} + 2(1+\varepsilon)^2 \mathcal{O}(\log n) \cdot \text{OPT} \in (1+\varepsilon)^2 \mathcal{O}(\log n) \cdot \text{OPT}. \quad \square$$

3 Concluding Remarks

Our algorithms produced solutions in which the budget constraints were strictly satisfied. This is unlike many bicriteria network design problems where it is necessary to violate the budget constraint to obtain a solution that is near-optimal with respect to the objective function [7].

An open problem that arises immediately from our work is whether there is a good approximation algorithm for the (TOTAL WEIGHT, NODE UPGRADING COST, SPANNING TREE) problem even when Assumption 5 is not satisfied. It is also of interest to investigate whether our results for spanning trees can be extended to Steiner trees. Other open problems under the node-based upgrading model can be formulated using different performance measures for the upgraded network. Some measures which are of interest in this context include bottleneck weight, diameter and lengths of paths between specified pairs of vertices.

References

1. O. Berman, "Improving The Location of Minisum Facilities Through Network Modification," *Annals of Operations Research*, Vol. 40, 1992, pp. 1–16.
2. U. Feige, "A threshold of $\ln n$ for approximating set cover," *Proc. 28th Annual ACM Symposium on the Theory of Computing*, Philadelphia, PA, May 1996, pp. 314–318.
3. G. N. Frederickson and R. Solis-Oba, "Increasing the Weight of Minimum Spanning Trees", *Proc. 6th Annual ACM-SIAM Symposium on Discrete Algorithms*, January 1996, pp. 539–546.
4. S. O. Krumke, H. Noltemeier, M. V. Marathe, S. S. Ravi and K. U. Drangmeister, "Modifying Networks to Obtain Low Cost Trees," *Proc. Workshop on Graph Theoretic Concepts in Computer Science*, Cadenabbia, Italy, June 1996, pp. 293–307.
5. S. O. Krumke, H. Noltemeier, M. V. Marathe, R. Ravi and S. S. Ravi, "Improving Steiner Trees of a Network Under Multiple Constraints", Technical Report, LA-UR 96-1374, Los Alamos National Laboratory, Los Alamos, NM, 1996.
6. F. T. Leighton and S. Rao, "An Approximate Max-Flow Min-Cut Theorem for Uniform Multicommodity Flow Problems with Application to Approximation Algorithms", *Proc. 29th Annual IEEE Conference on Foundations of Computer Science*, Oct. 1988, pp. 422–431.
7. M. V. Marathe, R. Ravi, R. Sundaram, S. S. Ravi, D. J. Rosenkrantz and H. B. Hunt III, "Bicriteria Network Design Problems", In *Proc. 22nd International Colloquium on Automata, Languages and Programming*, July 1995, Vol. 944 of *Lecture Notes in Computer Science*, pp. 487–498.
8. J. Plesnik, "The Complexity of Designing a Network with Minimum Diameter", *Networks*, Vol. 11, 1981, pp. 77–85.
9. D. Paik and S. Sahni, "Network Upgrading Problems," *Networks*, Vol. 26, 1995, pp. 45–58.
10. C. Phillips, "The Network Inhibition Problem," *Proc. 25th Annual ACM Symposium on Theory of Computing*, San Diego,CA, May 1993, pp. 288–293.
11. R. Ravi and M. X. Goemans, "The Constrained Minimum Spanning Tree Problem", *Proc. Scandinavian Workshop on Algorithmic Theory*, Reykjavik, July 1996.

Dynamic Algorithms for graphs of Bounded Treewidth

Torben Hagerup

Max-Planck-Institut für Informatik, D–66123 Saarbrücken, Germany
`torben@mpi-sb.mpg.de`

Abstract. The formalism of monadic second-order (MS) logic has been very successful in unifying a large number of algorithms for graphs of bounded treewidth. We extend the elegant framework of MS logic from static problems to dynamic problems, in which queries about MS properties of a graph of bounded treewidth are interspersed with updates of vertex and edge labels. This allows us to unify and occasionally strengthen a number of scattered previous results obtained in an ad-hoc manner and to enable solutions to a wide range of additional problems to be derived automatically.

As an auxiliary result of independent interest, we dynamize a data structure of Chazelle and Alon and Schieber for answering queries about sums of labels along paths in a tree with edges labeled by elements of a semigroup.

1 Introduction

Many graph properties can be expressed via formulas in a suitable logic. E.g., for given vertices s and t in a directed graph, the fact that the subgraph spanned by a set A of edges contains a path from s to t can be expressed by saying that every vertex set U containing s, but not t, can be left via an edge in A, i.e., by the formula

$$Joins(A, s, t) \equiv \forall U((s \in U \wedge t \notin U) \Rightarrow \\ \exists e \exists u \exists v (tail(u, e) \wedge head(v, e) \wedge e \in A \wedge u \in U \wedge v \notin U)),$$

where e ranges over all edges, u and v range over all vertices, U ranges over all sets of vertices, and $tail(u, e)$ and $head(v, e)$ express that u and v are the tail and the head of e, respectively. If we want the graph spanned by A to be just a single (simple) path from s to t, we can additionally require A to be minimal, i.e., $Path(A, s, t) \equiv Joins(A, s, t) \wedge \forall B((B \subseteq A \wedge Joins(B, s, t)) \Rightarrow B = A)$, where B ranges over all sets of edges.

Expressing computational problems such as "Is there a path from s to t?" in a formal framework holds out the prospect of deriving algorithms to solve such problems in an automatic way. Indeed, every graph property expressible in first-order logic can be decided in polynomial time. The catch is that first-order logic is too weak to express most graph properties of interest (see, e.g., (Courcelle, 1990a)). It allows variables ranging over vertices and edges, existential and universal quantification over such variables, the usual logic connectives $\wedge$, $\vee$, and $\neg$, and predicates such as *tail* and *head* for accessing the basic connectivity structure of the graph under consideration. Very frequently, however, one is led, as in the examples above, to introduce variables ranging not over individual vertices or edges, but over sets of vertices or edges. Extending first-order logic with this possibility, we arrive at *monadic second-order (MS) logic*. As noted by many researchers, MS logic is a powerful language that allows the expression of a wide range of graph properties. Indeed, the collection of decision problems on graphs that can defined by MS formulas is so large that it includes many NP-complete problems, leaving little hope of obtaining efficient algorithms for the general case. Rather than reverting to a less expressive logic, one can try to evade this problem by restricting the class of input graphs. Arnborg et al. (1991) argue that a particularly felicitous combination is to consider problems definable by an MS formula on graphs of *bounded treewidth*, i.e., on graphs drawn from a class with a uniform upper bound on the *treewidth* of all graphs in the class. Loosely speaking, the treewidth of a graph is a measure of how far the graph deviates from being a tree. The details of the definition will be provided in the next section.

Consider a single MS formula Φ with l free set variables (such as "A" in the formula "$Path(A, s, t)$") and without free simple variables. Φ gives rise to several computational graph problems: First, there is the *decision problem* of determining whether there are sets $A_1, \ldots, A_l$ of vertices or edges that satisfy Φ if substituted for its free variables (e.g., "Is there a path from s to t?"). For this first type of problem it is not necessary to allow Φ to have free variables—we might as well quantify them existentially; still, we keep the present formulation for the sake of uniformity. Second, the *counting problem* of detecting the number of such tuples (e.g., "How many (simple) paths are there from s to t?"). Third, if the input additionally associates each vertex or edge a with an l-tuple $(f_1(a), \ldots, f_l(a))$ of real numbers, whose ith element is interpreted as the cost of including a in A_i, for $i = 1, \ldots, l$, the *optimization problem* of computing the minimal cost of a tuple $(A_1, \ldots, A_l)$ that satisfies Φ (e.g., "What is the distance from s to t?"). Fourth, in the same setting, the *construction problem* (this is not a standard term) of actually computing a tuple $(A_1, \ldots, A_l)$ satisfying Φ and of minimal cost (e.g., "Which path from s to t is shortest?"). And fifth, if $f_i(a)$ is reinterpreted as the probability of a stepping into A_i, for $i = 1, \ldots, l$, with each vertex or edge entering each set independently of all other such random decisions, the *reliability problem* of computing the probability of obtaining a tuple $(A_1, \ldots, A_l)$ that satisfies Φ (e.g., "What is the probability of having an operational path from s to t?").

Results by Courcelle (1990b) and Bodlaender (1996a) imply that every decision problem defined by an MS property can be solved in linear time on graphs of bounded treewidth. Generalizations of these and related earlier results to counting, optimization, construction, and reliability problems were investigated by a number of authors (Arnborg et al., 1991; Bern et al., 1987; Bodlaender, 1993a; Borie et al., 1992; Courcelle and Mosbah, 1993; Stearns and Hunt, 1996). One of the simplest and most general extensions was suggested by Courcelle and Mosbah (1993), and we will essentially use their framework. In our formulation, a generic algorithm is instantiated by choosing a particular *commutative semiring* $\mathcal{R} = (R, \oplus, \otimes, \bar{0}, \bar{1})$, i.e., an algebraic structure consisting of a set R, equipped with two associative and commutative operations $\oplus$ and $\otimes$ with neutral elements $\bar{0}$ and $\bar{1}$, respectively, such that $\otimes$ distributes over $\oplus$ (i.e., $a \otimes (b \oplus c) = (a \otimes b) \oplus (a \otimes c)$ for all $a, b, c \in R$) and $a \otimes \bar{0} = \bar{0}$ for all $a \in R$. Given an input graph G with associated functions $f_1, \ldots, f_l$ (which will be called *cost functions*, independently of their interpretation), the generic algorithm computes the *value* of G under Φ and $\mathcal{R}$, defined as the quantity

$$|G|_{\Phi,\mathcal{R}} = \bigoplus_{G \models \Phi[A_1,\ldots,A_l]} \bigotimes_{i=1}^{l} \bigotimes_{a \in A_i} f_i(a),$$

i.e., the "sum", over all tuples $(A_1, \ldots, A_l)$ that satisfy Φ, of the "products", over the sets A_i, of the appropriate costs. With suitably chosen commutative semirings, this can be shown to solve the problems mentioned above as well as a number of additional problems. For example, with $\mathcal{R} = (\{0, 1, 2, \ldots\}, +, \cdot, 0, 1)$, we obtain a solution to the counting problem.

The problem of computing $|G|_{\Phi,\mathcal{R}}$ for fixed Φ and $\mathcal{R}$ will be called *static*, meaning that the entire input as well as the question to be answered are known from the outset. The focus of this paper is to extend the elegant framework of MS logic to a dynamic setting in which, following a certain *initialization* or *preprocessing* based on the input graph, a sequence of *attribute updates* and *queries* must be executed online, i.e., each query must be answered before the next operation to be executed is revealed. An (attribute) update changes a single attribute of a vertex or edge without affecting the structure of the graph. One might also consider *structural updates* that insert or delete vertices or edges. The data structures and algorithms described here can easily be extended to allow deletions, but supporting insertions of vertices and edges appears to be considerably more difficult; see (Bodlaender, 1993b) for results in this direction in the case of graphs of treewidth 2. We allow *boolean attributes*, which take values in $\{false, true\}$, indicate (non)membership in "user-defined" sets, and may be tested in Φ through corresponding predicates, and *ring attributes*, which take values in R, together define the cost functions, and cannot be referred to in Φ. A query temporarily (for the duration of the query) carries out a constant number of updates of boolean and/or ring attributes, thereby changing G into G', and then computes and returns $|G'|_{\Phi,\mathcal{R}}$, after which all attributes revert to their values before the query. This view of a query operation may be unfamiliar, but it is general and permits a convenient statement of our results.

Our running example centered around the the MS formula $Path(A, s, t)$ will be used to clarify some of the concepts introduced above. We have already seen that $Path(A, s, t)$ expresses that the edges in A span a (simple) path from s to t, and if we give each edge e a ring attribute $f(e)$ equal to its length, the length of the path spanned by A is $\sum_{e \in A} f(e)$, which can be minimized by choosing $\mathcal{R} = (\mathbb{R} \cup \{\infty\}, \min, +, \infty, 0)$. What is lacking is that we would like to support queries asking for the distance from s to t (call this an (s, t)-query), where s

and t are variable. We can achieve this effect within the general framework by introducing two "user-defined" sets, S and T, both initialized to $\emptyset$, letting an (s,t)-query temporarily change two boolean vertex attributes to make $S = \{s\}$ and $T = \{t\}$, and using instead of the original formula the formula

$$\exists s \exists t (Origin(s) \wedge Destination(t) \wedge Path(A, s, t)),$$

where *Origin* and *Destination* are predicate symbols corresponding to the sets S and T. It should be clear that other traditional types of queries can be formulated in a similar way. We show that for all $\tau \geq 1$, the dynamic version of every problem defined by an MS formula Φ and a commutative semiring $\mathcal{R}$ whose operations can be carried out in constant time (call such a semiring *efficient*) can be solved on n-vertex graphs of bounded treewidth with initialization time $O(n)$, (attribute-)update time $O(\tau n^{1/\tau})$, and query time $O(\tau + \alpha(n))$, where α is a slowly-growing "inverse Ackermann" function. Alternatively, for arbitrary integer $k \geq 1$, with the same update time, but initialization time $O(nI_k(n))$ and query time $O(\tau + k)$, where I_k, for every integer $k \geq 1$, is another slowly-growing function. Both α and the functions I_k are defined in the next section. In the special case of the dynamic *distance* and *shortest-path* problems considered above, this result was obtained previously by Chaudhuri and Zaroliagis (1995) for $\tau = O(1)$ as well as with a worse tradeoff between initialization time, update time, and query time. In more detail, Chaudhuri and Zaroliagis indicate the following bounds, for all integers $r \geq 1$: Initialization time $O(c^r n)$, update time $O(c^{2r} n^{1-r})$, and query time $O(c^{2r} \alpha(n))$, where $c = \Theta(3^r)$. In order to compare these bounds with ours, observe, e.g., that in order to achieve an update time of $O(2^{\sqrt{\log n}})$ with the bounds of Chaudhuri and Zaroliagis, it is necessary to choose r larger than $\frac{1}{2} \log \log n$, which yields a query time of $(\log n)^{\Omega(\log \log n)}$, whereas our bounds associate an update time of $O(2^{\sqrt{\log n}})$ with a query time of $O(\sqrt{\log n})$. Our bounds are never worse than those of Chaudhuri and Zaroliagis, and strictly better for all nonconstant τ and r. One end of the tradeoff, with update and query times both $O(\log n)$, was demonstrated previously by Bodlaender (1993b).

If only queries but no updates are to be supported, we achieve initialization time $O(n)$ and query time $O(\alpha(n))$ or, for every integer $k \geq 1$, initialization time $O(nI_k(n))$ and query time $O(k)$. This result was found previously by Chaudhuri and Zaroliagis (1995) for the distance and shortest-path problems and by Arikati et al. (1995) for the problem of computing (the value of) a minimum cut separating two given vertices. (The value of a minimum cut separating s and t can be found by minimizing $\sum_{e \in A} f(e)$, where $f(e)$ denotes the capacity of the edge e, subject to $\forall B(Path(B, s, t) \Rightarrow (A \cap B \neq \emptyset))$, where A and B range over all sets of edges.)

In some cases, queries may become cheaper if they can be batched. We consider queries that (temporarily) change at most d boolean attributes and no ring attributes and use the term *exhaustive d-dimensional query* to denote a set of all possible queries of this type (e.g., the well-known all-pairs shortest-paths problem is to answer an exhaustive 2-dimensional query). We can show that for all $d \geq 1$, exhaustive d-dimensional queries defined by an MS formula and an efficient commutative semiring can be answered in $O(n^d)$ time for n-vertex input graphs of bounded treewidth. This was proved previously for $d = 1$ and $d = 2$ for the distance problem by Radhakrishnan et al. (1992) and for $d = 2$ for the problem of computing (the value of) a minimum cut by Arikati et al. (1995).

All of the algorithms described above translate into parallel algorithms for the EREW PRAM. Due to space limitations, we omit further discussion of exhaustive queries and parallel algorithms.

2 Definitions

As introduced by Robertson and Seymour (1986), a *tree decomposition* of a graph $G = (V, E)$ is a pair $(T_D, \mathcal{U})$, where $T_D = (V_D, E_D)$ is a tree and $\mathcal{U} = \{U_x \mid x \in V_D\}$ is a family of subsets of V called *bags*, one for each node in T_D, such that

(1) $\bigcup_{x \in V_D} U_x = V$ (every vertex in G occurs in some bag);
(2) for all $u, v \in V$, if u and v are the endpoints of some edge in E, then there exists an $x \in V_D$ with $\{u, v\} \subseteq U_x$ (every edge in G is "internal" to some bag);
(3) for all $x, y, z \in V_D$, if y is on the (simple) path from x to z in T_D, then $U_x \cap U_z \subseteq U_y$ (every vertex in G occurs in the bags in a connected part of T_D, i.e., in a subtree).

The *width* of a tree decomposition $(T_D = (V_D, E_D), \{U_x \mid x \in V_D\})$ is $\max_{x \in V_D} |U_x| - 1$. The *treewidth* of a graph G is the smallest treewidth of any tree decomposition of G. Many important graph classes are of bounded treewidth, including those of outerplanar and series-parallel graphs; for surveys of results of this kind, see (van Leeuwen, 1990) and (Bodlaender, 1996b).

Define $I_0 : \mathbb{N} = \{1, 2, \ldots\} \to \mathbb{N}$ by $I_0(n) = \lceil n/2 \rceil$, for all $n \in \mathbb{N}$. Inductively, for $k = 1, 2, \ldots$, define $I_k : \mathbb{N} \to \mathbb{N}$ by $I_k(n) = \min\{i \in \mathbb{N} \mid I_{k-1}^{(i)}(n) = 1\}$, for all $n \in \mathbb{N}$, where superscript (i) denotes i-fold repeated application. Finally, for all $n \in \mathbb{N}$, take $\alpha(n) = \min\{k \in \mathbb{N} \mid I_k(n) \leq 3\}$.

3 Static Algorithms

Given an MS formula Φ with l free set variables and without free simple variables and a commutative semiring $\mathcal{R} = (R, \oplus, \otimes, \bar{0}, \bar{1})$, we say that a graph $G = (V, E)$ is *appropriate* for the pair $(\Phi, \mathcal{R})$ if each $a \in V \cup E$ has a boolean attribute for each unary predicate symbol occurring in Φ and l ring attributes $f_1(a), \ldots, f_l(a) \in R$. The (static) RMS problem defined by Φ and $\mathcal{R}$ is, given a graph G appropriate for $(\Phi, \mathcal{R})$, to compute the value $|G|_{\Phi, \mathcal{R}}$ of G under Φ and $\mathcal{R}$. Courcelle and Mosbah (1993) show that every RMS problem defined by an MS formula Φ without free simple variables and an efficient commutative semiring $\mathcal{R}$ can be solved in linear time on graphs of bounded treewidth. Our dynamic algorithms are based on a different proof of their result, which uses techniques of Arnborg et al. (1991), and which we now sketch.

Theorem 1. (Courcelle and Mosbah, 1993) *For all constants $t \geq 1$ and all integers $n \geq 1$, every RMS problem defined by an MS formula Φ without free simple variables and an efficient commutative semiring $\mathcal{R}$ can be solved in $O(n)$ time on n-vertex input graphs appropriate for $(\Phi, \mathcal{R})$ and of treewidth at most t.*

Proof. Let $G = (V_G, E_G)$ be a n-vertex input graph appropriate for $(\Phi, \mathcal{R})$ and of treewidth at most t and take $\Omega = V_G \cup E_G$. Arnborg et al. (1991) show that $O(n)$ time suffices to construct an MS formula Ψ with the same free variables as Φ, a rooted binary tree $T^* = (V^*, E^*)$ appropriate for $(\Psi, \mathcal{R})$, and an injective function $\pi : \Omega \to V^*$ so that the following holds: Suppose that Φ has l free set variables. Then, for all $A_1, \ldots, A_l \subseteq \Omega$, $G \models \Phi[A_1, \ldots, A_l]$ if and only if $T^* \models \Psi[\pi(A_1), \ldots, \pi(A_l)]$; moreover, $T^* \not\models \Psi[B_1, \ldots, B_l]$ whenever $\bigcup_{i=1}^{l} B_i \not\subseteq \pi(\Omega)$. Intuitively, if we identify a and $\pi(a)$, for all $a \in \Omega$, then T^* satisfies Ψ under a particular assignment (association of free set variables with sets of vertices and/or edges) if and only if G satisfies Φ under the same assignment.

Informally, a finite tree automaton is the natural generalization of a usual finite automaton from inputs that are strings to inputs that are binary trees. Formally, we can take a finite tree automaton to be a 5-tuple $(S, \Sigma, \delta, s_0, F)$, where S is a finite set of *states*, Σ is a finite *alphabet*, δ is a *transition function* from $S \times S \times \Sigma$ to S, $s_0 \in S$ is a distinguished *initial state*, and $F \subseteq S$ is a distinguished set of *accepting states*. Given a binary tree, each of whose vertices is labeled with an element of Σ, the tree automaton assigns a state to each vertex in the tree, working from the leaves to the root (i.e., processing each vertex after all of its children). If the left and right children of a vertex v are assigned states s and t, respectively, and v is labeled σ, the state $\delta(s, t, \sigma)$ is assigned to v; if one or both children are missing, the initial state s_0 is used in place of their states. The tree automaton *accepts* the input tree exactly if the state assigned to the root belongs to F. Arnborg et al. (1991) show how to construct a tree automaton $M = (S, \Sigma, \delta, s_0, F)$ with the following property: Suppose that the unary predicates appearing in Ψ are $P_1, \ldots, P_k$. Then $\Sigma = \{false, true\}^{k+l}$, and for arbitrary subsets $A_1, \ldots, A_l$ of V^*, if each vertex $v \in V^*$ is labeled with the bit vector $(P_1(v), \ldots, P_k(v), b_1, \ldots, b_l) \in \Sigma$, where $b_i = true$ iff $v \in A_i$, for $i = 1, \ldots, l$, then M accepts T^* exactly if $T^* \models \Psi[A_1, \ldots, A_l]$.

We show how to derive from M another tree automaton $M' = (S', \Sigma', \delta', s'_0)$ to solve the RMS problem at hand. M' is not a finite automaton, since both its alphabet and its state set may be infinite, and it will compute a value (namely, $|T^*|_{\Psi, \mathcal{R}}$) rather than just accepting or rejecting, for which reason it has no set of accepting states; in other respects, M' behaves exactly as a finite tree automaton.

Write $\mathcal{R} = (R, \oplus, \otimes, \bar{0}, \bar{1})$, let $m = |S|$, and identify the states of M with the integers $1, \ldots, m$, with 1 being the initial state. We take the state set S' of M' to be R^m, the set of vectors of length m with components in R, and define the initial state s'_0 as $(\bar{1}, \bar{0}, \ldots, \bar{0})$. The alphabet of M' is $\Sigma' = \{false, true\}^k \times R^l$, and the label of a vertex $v \in V^*$ is $(P_1(v), \ldots, P_k(v), f_1(v), \ldots, f_l(v))$, where $f_1, \ldots, f_l$ are the cost functions copied to T^* from the input graph G according to π, i.e., for $i = 1, \ldots, l$, $f_i(\pi(a)) = f_i(a)$ for all $a \in \Omega$, and $f_i(a) = \bar{0}$ for all $a \in V^* \setminus \pi(\Omega)$. We next define the transition function δ'. Assume that the states of the (possibly fictitious) left and right children of a vertex $u \in V^*$ are

$(s_1, \ldots, s_m)$ and $(t_1, \ldots, t_m)$, respectively. Then the state of u is $(r_1, \ldots, r_m)$, where

$$r_j = \bigoplus_{p=1}^{m} \bigoplus_{q=1}^{m} \bigoplus_{\substack{(b_1,\ldots,b_l) \in \{false,true\}^l \\ \delta(p,q,(P_1(u),\ldots,P_k(u),b_1,\ldots,b_l))=j}} \left(s_p \otimes t_q \otimes \bigotimes_{\substack{1 \leq i \leq l \\ b_i = true}} f_i(u) \right),$$

for $j = 1, \ldots, m$. It can be seen that the sum for the jth component (corresponding to the jth state of M) is over those pairs of states of M and those choices of (non)membership of u in $A_1, \ldots, A_l$ that would lead the original automaton to give u the state j, for $j = 1, \ldots, m$. It can then be proved by induction that for each vertex $u \in V^*$ and for $j = 1, \ldots, m$, the jth component of the state assigned to u is

$$\bigoplus_{\substack{A_1,\ldots,A_l \subseteq U \\ M(u,A_1,\ldots,A_l)=j}} \bigotimes_{i=1}^{l} \bigotimes_{a \in A_i} f_i(a),$$

where U is the set of descendants of u in T^*, and $M(u, A_1, \ldots, A_l)$ denotes the state assigned by M to u if the vertex labels are set according to $A_1, \ldots, A_l$. If the state of the root of T^* computed by M' is $(s_1, \ldots, s_m)$, this observation shows that $|G|_{\Phi,\mathcal{R}} = |T^*|_{\Psi,\mathcal{R}} = \bigoplus_{j \in F} s_j$, which the automaton therefore computes and returns. Since m is a constant (for fixed Φ), each application of δ' takes constant time, so that the entire processing of T^* by M' can be carried out in $O(n)$ time.

4 Data Structures for Queries

In this section we describe data structures that support queries efficiently, but not updates. Given an MS formula Φ without free simple variables, a commutative semiring $\mathcal{R}$, and a constant $d \in \mathbb{N}$, the *d-dimensional RMS query problem* defined by Φ and $\mathcal{R}$ is, given a graph G appropriate for $(\Phi, \mathcal{R})$, to preprocess G for subsequent queries for quantities of the form $|G'|_{\Phi,\mathcal{R}}$, where G' is obtained from G by (temporarily) changing at most d boolean and/or ring attributes.

Let the tree T^* and the machines M and M' be as in the proof of Theorem 1 and consider a vertex $u \in V^*$ with left and right children v and w, respectively. Let $(r_1, \ldots, r_m)$, $(s_1, \ldots, s_m)$, and $(t_1, \ldots, t_m)$ be the states assigned by M' to u, v, and w, respectively. Then, by definition of the transition function δ', we have $r_j = \bigoplus_{p=1}^{m} c_{jp} s_p$ for $j = 1, \ldots, m$, where

$$c_{jp} = \bigoplus_{q=1}^{m} \bigoplus_{\substack{(b_1,\ldots,b_l) \in \{false,true\}^l \\ \delta(p,q,(P_1(u),\ldots,P_k(u),b_1,\ldots,b_l))=j}} \left(t_q \otimes \bigotimes_{\substack{1 \leq i \leq l \\ b_i = true}} f_i(u) \right),$$

for $p = 1, \ldots, m$. In other words, provided that the state of w remains constant, the function that maps the state of v to the state of u is premultiplication with an $m \times m$ matrix (over the semiring $\mathcal{R}$). We call this function the *relay function* of the edge $\{u, v\}$ (for the input graph G). The relay functions of edges between vertices and their right children are defined in complete analogy and have the same form.

Suppose that G is changed into G' by modifying a single boolean or ring attribute of some vertex or edge $a \in \Omega$. This translates into a change of a single boolean or ring attribute of $v = \pi(a)$ in T^* or, as seen from the point of view of M', into a change of the label of v. We can compute $|G'|_{\Phi,\mathcal{R}}$ by simulating the execution of M' on the new label settings. One way to do this is to compose the relay functions of all edges on the path from v to the root r^* of T^*, and then to apply the resulting function to the new state of v; this yields the new state of the root, from which $|G'|_{\Phi,\mathcal{R}}$ can be computed in constant time. Similarly, a query that changes the labels of two vertices v and w can be handled by composing relay functions along the two paths from v and w to the children of the lowest common ancestor (LCA) u of v and w, using the result to compute the new state of u, and then propagating the change to r^* by composing the relay functions on the path from u to r^*. Answering queries therefore essentially reduces to composing functions along paths in T^*, a problem that has been studied in a more general setting.

Let us call a semigroup $\mathcal{S} = (S, \odot)$ *efficient* if $a \odot b$ can be computed from a and b in constant time for all $a, b \in S$. We can assume without loss of generality that $\mathcal{S}$ contains a neutral element. In the context of a tree T, each of whose edges is labeled by an element of a semigroup $(S, \odot)$ called its *weight*, we define the weight of a (simple) path in T of length k as the quantity $\lambda_1 \odot \cdots \odot \lambda_k$, where λ_i is the weight of the ith edge on the path, for $i = 1, \ldots, k$, and we define a *path-weight query* as a query that specifies two vertices u and v and asks for the weight of the (unique) path in T from u to v. The following lemma was proved by Chazelle (1987) and Alon and Schieber (1987).

Lemma 2. *For all $n, k \in \mathbb{N}$, an n-vertex tree with edge weights drawn from a efficient semigroup $(S, \odot)$ can be preprocessed for path-weight queries with preprocessing time $O(nI_k(n))$ and query time $O(k)$.*

A particularly interesting special case of preprocessing time $O(n)$ and query time $O(\alpha(n))$ is obtained by choosing $k = \alpha(n)$. Similar remarks apply below.

Theorem 3. *For all constants $t \geq 1$ and all integers $n, k \geq 1$, every t-dimensional RMS query problem defined by an MS formula Φ without free simple variables and an efficient commutative semiring $\mathcal{R}$ can be solved on n-vertex input graphs appropriate for $(\Phi, \mathcal{R})$ and of treewidth bounded by t with preprocessing time $O(nI_k(n))$ and query time $O(k)$.*

Proof. We preprocess the tree T^* of the proof of Theorem 1 according to Lemma 2, the weight of each edge being its relay function and $\odot$ being function composition (i.e., matrix multiplication over $\mathcal{R}$). We also preprocess T^* so that subsequent queries for the LCA of two arbitrary vertices can be answered in constant time; it is known how to do this in $O(n)$ time (Harel and Tarjan, 1984; Schieber and Vishkin, 1988).

Suppose that a query changes the labels of the vertices in some set $U \subseteq V^*$ (thus $|U| \leq t$). Let $Q = U \cup W \cup \{r^*\}$, where W is the set of all lowest common ancestors of two vertices in U and r^* is the root of T^*; Q is still of bounded size. Let $T = (V, E)$ be the tree obtained from T^* by contracting each vertex in $V^* \setminus Q$ into its closest ancestor whose parent belongs to Q. With the aid of still more LCA queries, to determine for all $u, v \in Q$ whether u is an ancestor of v in T^*, T can be constructed in constant time. We now process T from the leaves to the root, for each vertex v in T computing the new state assigned to v by M' after the label changes caused by the update. For a vertex v in Q, this is trivial, since the new states of its children in T^*, if any, will be known when v is processed. For a vertex v in $V \setminus Q$, on the other hand, all descendants of v in T^* that belong to U, if any, are descendants of a single vertex in Q whose new state is known when v is processed. Thus the new value of v can be computed in $O(k)$ time by composing relay functions according to Lemma 2. Once the new state of r^* is known, the query can be answered in constant time.

5 Dynamic Data Structures

In this section we dynamize the path-weight-query data structure of Lemma 2 to allow updates of edge weights as well as path-weight queries and state the implications for dynamic RMS problems.

Theorem 4. *For all integers $n, k, \tau \geq 1$, every n-edge tree with edge weights drawn from an efficient semigroup $(S, \odot)$ can be preprocessed for path-weight queries with preprocessing time $O(nI_k(n))$, query time $O(\tau + k)$, and update time $O(\tau n^{1/\tau})$.*

Proof. We reuse part of a scheme developed by Chazelle (1987) in order to prove Lemma 2. For a parameter m with $1 \leq m \leq n$ to be chosen below, we partition the edge set E of the input tree $T = (V, E)$ into at most $3n/m$ sets, each of which spans a subtree of T, called a *piece*, with at most m edges. Chazelle shows how to do this in $O(n)$ time (Lemma 3). Call a vertex of T a *fringe vertex* if it is shared between two or more pieces. In order to make what follows clearer, let us assume that we separate the pieces by replacing each fringe vertex v, shared between d pieces, by a star consisting of a central vertex, which we identify with v, connected to d new vertices; each of the d new vertices is associated with a different piece containing v, is called the *representative* of v in that piece, and replaces v as an endpoint of each edge belonging to the piece and incident on v. Provided that each star edge is given a weight of $\bar{0}$, the neutral element of $(S, \odot)$, this transformation does not change the weight of the path between any two vertices in T. It at most triples the number of edges and is easily carried out in $O(n)$ time.

The number of fringe vertices is bounded by $3n/m$, and Chazelle shows how to construct an edge-weighted tree T^* with at most $6n/m$ edges that contains all fringe vertices and assigns the same weight as T to the path between any two fringe vertices; T^* is obtained in $O(n)$ time from T by removing all nonfringe vertices that have fewer than three incident edges lying on paths between fringe vertices and replacing paths of such vertices by single edges with the same weight.

Each piece is preprocessed independently for path-weight queries, and the global tree T^* is processed recursively as just described. For all $u, v \in V$, denote by $\Lambda(u, v)$ the weight of the path in T from u to v. Consider two vertices u and v in T and let x and y be the first and last fringe vertices on the path in T from u to v, respectively, if any. If x and y do not exist, u and v belong to the same piece, and the weight λ of the path from u to v can be obtained from the data structure maintained for that piece. Otherwise $\lambda = \Lambda(u, x) \odot \Lambda(x, y) \odot \Lambda(y, v)$. If $x = u$ (u is a fringe vertex), $\Lambda(u, x) = \bar{0}$; otherwise $\Lambda(u, x) = \Lambda(u, r_x)$, where r_x is the representative of x in the piece of u, and the latter quantity can be obtained from the data structure maintained for the piece of u. $\Lambda(y, v)$ is computed similarly, and $\Lambda(x, y)$ is obtained recursively from the data structures maintained for T^*. One small issue, how to determine x and y and possibly r_x and r_y, is resolved with the help of yet another tree T^+, obtained from T by replacing all edges within each piece by edges from each (nonfringe) vertex in the piece to a new vertex representing the piece. The vertices of interest occur among the first four and the last four vertices on the path in T^+ from u to v and can be identified by two applications of the algorithm of Lemma 2: The weight of each edge is its identity, considered as a string of length 1, and $\odot$ is "truncated concatenation", which concatenates its two arguments but, if the resulting string is of length ≥ 4, keeps only its suffix of length 3.

Without loss of generality assume that $k \geq 2$. On the first recursive level we choose $m = m_0 = \lceil \sqrt{I_1(n)} \rceil$ and preprocess the pieces for path-weight queries according to Lemma 2. This needs a total of $O(nI_k(n))$ time and provides a query time of $O(k)$. On all subsequent recursive levels we choose $m = m_1 = \max\{\lceil n^{1/(2\tau)} \rceil, 12\}$ and preprocess the pieces for path-weight queries according to Lemma 2 with $k = 2$, ending the recursion when the number of edges drops below 12. This provides a query time of $O(1)$ per recursive level, and since $m_0 = \Omega(I_2(n))$, the preprocessing effort sums to $O(n)$ over all levels. Because the recursive depth is $O(\log n / \log m_1) = O(\tau)$, the overall query time is $O(\tau + k)$. An update of an edge weight requires recomputation of data structures maintained for a single piece on each recursive level, and thus needs $O(m_0 I_k(n))$ time on the first level and $O(m_1 I_2(m_1))$ time on all subsequent levels, resulting in an overall update time of $O(\tau n^{1/\tau})$.

Given an MS formula Φ without free simple variables, a commutative semiring $\mathcal{R}$, and a constant $d \in \mathbb{N}$, the d-dimensional *dynamic RMS problem* defined by Φ and $\mathcal{R}$ is, given a graph G appropriate for $(\Phi, \mathcal{R})$, to preprocess G for subsequent updates of single boolean or ring attributes and queries for quantities of the form $|G'|_{\Phi, \mathcal{R}}$, where G' is obtained from (the current) G by (temporarily) changing at most d boolean and/or ring attributes. As an immediate consequence of Theorem 4 and the methods introduced in Section 4, we obtain:

Theorem 5. *For all constants $t \geq 1$ and all integers $n, k, \tau \geq 1$, every t-dimensional dynamic RMS query problem defined by an MS formula Φ without free simple variables and an efficient commutative semiring $\mathcal{R}$ can be solved on n-vertex input graphs appropriate for $(\Phi, \mathcal{R})$ and of treewidth bounded by t with preprocessing time $O(nI_k(n))$, query time $O(\tau + k)$, and update time $O(\tau n^{1/\tau})$.*

References

Alon, N., and Schieber, B. (1987), Optimal preprocessing for answering on-line product queries, Tech. Rep. No. 71/87, Tel Aviv University.

Arikati, S. R., Chaudhuri, S., and Zaroliagis, C. D. (1995), All-pairs min-cut in sparse networks, in Proc. 15th Conference on Foundations of Software Technology and Theoretical Computer Science (FST&TCS), Springer Lecture Notes in Computer Science, Vol. 1026, pp. 363–376.

Arnborg, S., Lagergren, J., and Seese, D. (1991), Easy problems for tree-decomposable graphs, *J. Algorithms* **12**, pp. 308–340.

Bern, M. W., Lawler, E. L., and Wong, A. L. (1987), Linear-time computation of optimal subgraphs of decomposable graphs, *J. Algorithms* **8**, pp. 216–235.

Bodlaender, H. L. (1993a), On reduction algorithms for graphs with small treewidth, in Proc. 19th International Workshop on Graph-Theoretic Concepts in Computer Science (WG), Springer Lecture Notes in Computer Science, Vol. 790, pp. 45–56.

Bodlaender, H. L. (1993b), Dynamic algorithms for graphs with treewidth 2, in Proc. 19th International Workshop on Graph-Theoretic Concepts in Computer Science (WG), Springer Lecture Notes in Computer Science, Vol. 790, pp. 112–124.

Bodlaender, H. L. (1996a), A linear-time algorithm for finding tree-decompositions of small treewidth, *SIAM J. Comput.* **25**, pp. 1305–1317.

Bodlaender, H. L. (1996b), A partial k-arboretum of graphs with bounded treewidth, Tech. Rep. No. UU–CS–1996–02, Dept. of Computer Science, Utrecht University.

Borie, R. B., Parker, R. G., and Tovey, C. A. (1992), Automatic generation of linear-time algorithms from predicate calculus descriptions of problems on recursively constructed graph families, *Algorithmica* **7**, pp. 555–581.

Chaudhuri, S., and Zaroliagis, C. D. (1995), Shortest path queries in digraphs of small treewidth, Proc. 22nd International Colloquium on Automata, Languages and Programming (ICALP), Springer LNCS, Vol. 944, pp. 244–255.

Chazelle, B. (1987), Computing on a free tree via complexity-preserving mappings, *Algorithmica* **2**, pp. 337–361.

Courcelle, B. (1990a), Graph rewriting: An algebraic and logic approach, in *Handbook of Theoretical Computer Science, Vol. B: Formal Models and Semantics* (J. van Leeuwen, ed.), Chap. 5, pp. 193–242, Elsevier, Amsterdam.

Courcelle, B. (1990b), The monadic second-order logic of graphs. I. Recognizable sets of finite graphs, *Inform. and Comput.* **85**, pp. 12–75.

Courcelle, B., and Mosbah, M. (1993), Monadic second-order evaluations on tree-decomposable graphs, *Theor. Comput. Sci.* **109**, pp. 49–82.

Harel, D., and Tarjan, R. E. (1984), Fast algorithms for finding nearest common ancestors, *SIAM J. Comput.* **13**, pp. 338–355.

Radhakrishnan, V., Hunt, H B., III, and Stearns, R. E. (1992), Efficient algorithms for solving systems of linear equations and path problems, Proc. 9th Annual Symposium on Theoretical Aspects of Computer Science (STACS), Springer Lecture Notes in Computer Science, Vol. 577, pp. 109–119.

Robertson, N., and Seymour, P. D. (1986), Graph Minors. II. Algorithmic aspects of tree-width, *J. Algorithms* **7**, pp. 309–322.

Schieber, B., and Vishkin, U. (1988), On finding lowest common ancestors: Simplification and parallelization, *SIAM J. Comput.* **17**, pp. 1253–1262.

Stearns, R. E., and Hunt, H. B., III (1996), An algebraic model for combinatorial problems, *SIAM J. Comput.* **25**, pp. 448–476.

van Leeuwen, J. (1990), Graph algorithms, in *Handbook of Theoretical Computer Science, Vol. A: Algorithms and Complexity* (J. van Leeuwen, ed.), Chap. 10, pp. 525–631, Elsevier, Amsterdam.

The Name Discipline of Uniform Receptiveness
(Extended Abstract)

Davide Sangiorgi

INRIA - Sophia Antipolis, France.

1 Introduction

The π-calculus [9] is a paradigmatical process calculus for message-passing concurrency. Two processes with acquaintance of a given name can use it to interact with each other. Names themselves may be exchanged in communications, which can model modifications of the linkage structure among processes. These are the basic process constructs (using lower case for names and upper case for processes): $\overline{a}\langle b\rangle. P$, the output of b at a with P as continuation; $a(b). P$, an input at a with b placeholder for the name received in the input; $P_1 \mid P_2$, the parallel composition of the two processes; $\nu a\, P$, which makes name a local to P; and $!P$, which denotes a potentially-infinite number of copies of P in parallel.

In this paper, we study the situation in which certain names are *uniformly receptive*. A name x is receptive in a process P if at any time P is able of offering an input at x (at least as long as there are processes that could send messages at x). The receptiveness of x is uniform if all inputs at x have the same continuation. Receptiveness ensures that any message sent at x can be immediately processed; unformity ensures that there is a unique way in which a message at x may be processed (that is, the input end of x is "functional").

These are *semantic* conditions, and are undecidable. To obtain decidable conditions we impose some restrictions. Roughly, we guarantee receptiveness by demanding that the name is available in input-replicated form as soon as created. For instance, x is receptive in

$$P_1 \stackrel{\text{def}}{=} \nu x\, (!x(p). P \mid Q) \qquad P_2 \stackrel{\text{def}}{=} \nu x\, (\overline{r}\langle x\rangle. !x(p). P) \qquad (1)$$

(On the right, name x is created when the output $\overline{r}\langle x\rangle$ is consumed since, before this, x is frozen.[1]) We guarantee uniformity by demanding that there is only one input occurrence of the name; hence in (1) name x should not occur free in input position in P and Q. To preserve the uniformity property in a network of processes, we then also demand that only the output capability of the name may be transmitted; that is, as all π-calculus names, so uniform receptive names can be transmitted but, in contrast with the other names, they can be used by a recipient only in output (retransmitting the name, or sending a message at it).

In the processes P_1 and P_2 above, the receptiveness at x is persistent, which is necessary if unboundedly many messages could be sent at x. It is useful to

---
[1] Indeed P_2 is behaviourally the same as $\nu x\, (!x(p). P \mid \overline{r}\langle x\rangle)$, which is of the same form as P_1.

consider separately the case in which at most one message can be sent. Then the replication in front of the input at x is unnecessary. We call the first form ω-*receptiveness*, the second *linear receptiveness*.

Uniform receptiveness corresponds to a precise discipline in the usage of names; it could by formulated by syntactic means, but it is easier and more elegant to do so using a type system along the lines of type systems for the π-calculus. The impact of receptiveness on behavioural equivalences and process reasoning is the main focus of this paper. We shall develop some theory and proof techniques for processes with receptive names, and then illustrate their usefulness by means of some non-trivial examples, like the proof of some transformations that introduce parallelism in a resource, and the proof of the correctness of an optimisation of the translation of higher-order process calculi into the π-calculus [18, 13], which is adopted in the compiler of Pict [12].

The challenge in these examples is that the equalities implied by the transformations fail in the ordinary π-calculus (even w.r.t. the very coarse notion of trace equivalence). That is, there are contexts of the ordinary π-calculus that are able to detect the difference between the processes of the equalities. By imposing the type system for receptiveness, these contexts are ruled out as ill-typed.

Uniform receptiveness often occurs in the π-calculus. Our first example is the coding of functions. A process Q with a local function $\lambda r.\, M$, accessible via a name z, is normally written $\nu z\,(!z(r,y).\, P \mid Q)$ where P is the coding of M and y is (a placeholder for) the name where the result of a function call will be delivered. Within Q, a call of the function with argument n is written $\nu x\,(\overline{z}\langle n,x\rangle.\, x(p).\, Q')$ where p is (a placeholder for) the result of the call. In the function declaration, z is ω-receptive; in the function call, x is linear receptive. Similar combinations of linear and ω-receptiveness occur in the coding of higher-order communications and of Object-Oriented languages. Typically, ω-receptiveness occurs in the modelling of resources which are private to one or more client processes (above, the resource is a function). A discipline similar to ω-receptiveness is presently used in the compiler of Pict [12], to allow optimisations of the code implementing communications. An important example of linear receptiveness (indeed, perhaps the most important) is found in process interactions based on the *Remote Procedure Call* (RPC) paradigm. An RPC interaction involves two synchronisations between a caller and a callee where, after the first synchronisation, the caller waits the time necessary for the callee to elaborate a response. When we are modeling RPC's in the π-calculus, the return name at which the callee delivers its response is used as linear receptive. (The function call above too is an example of an RPC interaction.)

As behavioural equivalence on processes, we use *barbed equivalence*. This equates processes which, very roughly, in all contexts give rise to the same patterns of interactions. The main inconvenience of barbed equivalence is that it uses quantification over contexts in the definition, and this can make proofs of processes equality heavy. Against this, one looks for *direct* characterisations, without context quantification. For instance, in CCS and in the ordinary π-calculus barbed equivalence coincides with the well-known *labeled* bisimilarities

[13]. (In a labeled bisimilarity the bisimulation game is played not only on silent actions, as for barbed bisimilarity, but also on input and output actions.)

We sketch the essential points of our theory for processes with receptive names. The schema is the same for linear and for ω receptiveness. We first introduce a type system which forces the receptiveness discipline, and prove some basic properties for it. Secondly, we isolate a subclass of the well-typed processes, called *discreet processes*, roughly characterised by the property that all receptive names which are emitted are private to the sender. Discreet processes are defined by means of syntactic restrictions on the output prefix similar to those in the language πI [15]. Thirdly, we introduce a simple but powerful algebraic law, with which any well-typed process can be transformed into a discreet process. Remarkably, this law equates a process whose first action is the output of a *global* name with a process whose first action is the output of a *private* name. The law is not valid in the untyped π-calculus, but it is valid under the receptiveness type system. Finally, we prove a direct characterisation of barbed equivalence on discreet processes, as a labelled bisimilarity called *receptive bisimilarity*. The latter differs from the ordinary bisimilarity in the requirement for input actions, but otherwise it can be used with the standard co-inductive techniques of labelled bisimilarities, including proof techniques such as "bisimulation up to expansion".

For lack of space, some definitions and most of the proofs are omitted. More examples can be found in [17].

2 Some background on the π-calculus

We use lower case letters $p, q, r, \ldots$ to range over names, and upper case letters P, Q, R to range over the set $\mathcal{P}$ of processes. This is the π-calculus grammar (for simplicity, we develop our theory on the monadic calculus):

$$P := 0 \mid p(q).P \mid \overline{p}\langle q\rangle.P \mid \overline{p}(q).P \mid [p=q]P$$
$$\mid P_1 \mid P_2 \mid \nu p\, P \mid P_1 + P_2 \mid !p(q).P$$

We allow the bound-output prefix $\overline{p}(q).P$ in the syntax; often in the π-calculus literature, $\overline{p}(q).P$ is given as an abbreviation for $\nu q\,\overline{p}\langle q\rangle.P$. We use σ to range over substitutions; for any expression E, we write $E\sigma$ for the result of applying σ to E, with the usual renaming convention to avoid captures. We assign sum and parallel composition the lowest precedence among the operators. We write $\overline{p}.P$ and $p.P$ when the name transmitted at p is not important, and we often abbreviate $\alpha.0$ as α. The labeled transition system is the usual one, in the early style. *Actions*, ranged over by μ, can be of four forms: τ (interaction), $p\langle q\rangle$ (an input at p in which q is received), $\overline{p}\langle q\rangle$ (free output) and $\overline{p}(q)$ (bound output). In these actions, p is the *subject*. Free and bound names of actions and processes are defined as usual. In a statement, we say that a name is *fresh* to mean that it is different from the names of other processes or actions in the statement. Relation $\Longrightarrow$ is the reflexive and transitive closure of $\xrightarrow{\tau}$, and $\stackrel{\mu}{\Longrightarrow}$ stands for $\Longrightarrow \xrightarrow{\mu} \Longrightarrow$. $P \Downarrow_p$ holds if there is P' and an action μ with subject p s.t. $P \stackrel{\mu}{\Longrightarrow} P'$. A context C is *static* if it has the form $\nu \widetilde{p}\,(P \mid [\cdot])$, for some P and $\widetilde{p}$.

Definition 1 barbed bisimulation, equivalence and congruence. *Barbed bisimulation* is the largest symmetric relation $\dot{\approx}$ on processes s.t. $P \dot{\approx} Q$ implies:

1. whenever $P \Longrightarrow P'$ then there exists Q' such that $Q \Longrightarrow Q'$ and $P' \dot{\approx} Q'$;
2. for each name p, $P \Downarrow_p$ iff $Q \Downarrow_p$.

Two processes P and Q are *barbed equivalent*, written $P \approx Q$, if for each static context C it holds that $C[P] \dot{\approx} C[Q]$; they are *barbed congruent*, written $P \simeq Q$, if $C[P] \dot{\approx} C[Q]$ for all contexts.

Barbed equivalence and congruence usually coincide with the ordinary labeled (early) bisimilarity and congruence of the π-calculus [13]. The proof of this fact is simple on the class of the image finite processes (to which most of the processes one would like to write belongs) by exploiting the n-approximants of the labeled equivalences. We recall that the class of *image-finite processes* is the largest subset $\mathcal{I}$ of $\mathcal{P}$ which is derivation closed and s.t. $P \in \mathcal{I}$ implies that, for all μ, the set $\{P' \ : \ P \stackrel{\mu}{\Longrightarrow} P'\}$, quotiented by α conversion, is finite.

3 Linear receptiveness

The discipline of uniform receptiveness (briefly receptiveness) can be added to any of the main existing type systems for the π-calculus. In this paper, our base type system will be Milner's *sorting*, that we now briefly recall. Names are partitioned into a collection of *sorts*. Then a sorting function is defined which maps sorts onto sorts (in the polyadic calculus it maps sorts onto sequences of sorts). If a sort s is mapped onto a sort t this means that names in s may only carry names in t; moreover, t is the *object sort* of s. In the remainder, we shall assume that there is a sorting system under which all processes are well-typed. We separate the base type system (Milner's sorting) from the typing rules for receptiveness so as to show the essence of the latter rules.

We begin our analysis of receptiveness from the case of *linear receptiveness*. We call the non-linear-receptive names *plain names*. There are no constraints on plain names except those imposed by the underlying sorting. We shall omit the adjective "linear" when there is no ambiguity. For simplicity, we assume that: There is a single sort `L-recep` of linear receptive names; linear receptive names carry plain names. These two assumptions can be relaxed without difficulties. We also assume the existence of a sort `trig` of names, different from `L-recep` but with the same object sort as `L-recep` (note that names in `trig` are plain names). The sort `trig` will be used to derive simpler characterisations of our bisimilarities. *In the remainder, $x, y, z \ldots$ range over linear receptive names, $a, b, \ldots$ over plain names, and v over names in* `trig`*. We recall that p, q, r range over the set of all names. Δ, Γ range over finite sets of linear receptive names.* We sometimes write $\Delta - x$ as abbreviation for $\Delta - \{x\}$ and Δ, x for $\Delta \cup \{x\}$, and also x for $\{x\}$. The type system for linear receptiveness is in Table 3. A rule with double conclusion is an abbreviation for more rules with same premises but separate conclusions. Judgements have the form $\Delta; \Gamma \vdash P$. As sets, the order in

$$\text{(T-inp-mat)} \; \frac{\emptyset; \Gamma \vdash P}{\emptyset; \Gamma \vdash a(b).\, P,\, [a=b]P} \qquad \text{(T-inp-2)} \; \frac{x \notin \Gamma \quad \emptyset; \Gamma, x \vdash P}{\emptyset; \Gamma \vdash a(x).\, P}$$

$$\text{(T-inp-3)} \; \frac{x \notin \Gamma \quad \emptyset; \Gamma \vdash P}{x; \Gamma \vdash x(b).\, P} \qquad \text{(T-rep)} \; \frac{\emptyset; \emptyset \vdash a(b)P}{\emptyset; \emptyset \vdash !a(b).\, P}$$

$$\text{(T-res-1)} \; \frac{x \notin \Gamma \cup \Delta \quad \Delta, x; \Gamma, x \vdash P}{\Delta; \Gamma \vdash \nu x\, P} \qquad \text{(T-res-2)} \; \frac{p \notin \Gamma \cup \Delta \quad \Delta; \Gamma \vdash P}{\Delta; \Gamma \vdash \nu p\, P}$$

$$\text{(T-out-1)} \; \frac{\emptyset; \Gamma \vdash P}{\emptyset; \Gamma \vdash \overline{a}\langle b\rangle.\, P,\; \overline{a}(b).\, P} \qquad \text{(T-out-2)} \; \frac{x \notin \Gamma \quad \emptyset; \Gamma \vdash P}{\emptyset; \Gamma, x \vdash \overline{a}\langle x\rangle.\, P}$$

$$\text{(T-out-3)} \; \frac{x \notin \Gamma \quad \emptyset; \Gamma \vdash P}{\emptyset; \Gamma, x \vdash \overline{x}\langle b\rangle.\, P,\; \overline{x}(b).\, P} \qquad \text{(T-bout)} \; \frac{x \notin \Gamma \quad x; \Gamma \vdash P}{\emptyset; \Gamma \vdash \overline{a}(x).\, P}$$

$$\text{(T-par)} \; \frac{\Delta_1; \Gamma_1 \vdash P_1 \quad \Delta_2; \Gamma_2 \vdash P_2 \quad \Delta_1 \cap \Delta_2 = \emptyset \quad \Gamma_1 \cap \Gamma_2 = \emptyset}{\Delta_1, \Delta_2; \Gamma_1, \Gamma_2 \vdash P_1 \mid P_2}$$

$$\text{(T-nil)} \; \frac{}{\emptyset; \emptyset \vdash 0} \qquad \text{(T-sum)} \; \frac{\emptyset; \Gamma \vdash P_1 \quad \emptyset; \Gamma \vdash P_2}{\emptyset; \Gamma \vdash P_1 + P_2}$$

Table 1. Typing rules for linear receptiveness.

which names appear in Δ and Γ does not matter. Intuitively, if $\Delta; \Gamma \vdash P$ then $\Delta \cup \Gamma$ are the only receptive names which appear free in P; process P must use any name in Γ *exactly once* in output position (that is, either performing an output at that name or transmitting this capability to another process), and names in Δ *immediately* and *only once* in input. This intuition is formalised in Theorem 2, which relates types and operational semantics of processes. We say that P is *well typed* if there are Δ, Γ s.t. $\Delta; \Gamma \vdash P$ holds.

Theorem 2 soundness theorem. *Suppose* $\Delta; \Gamma \vdash P$.

1. *if* $x \in \Delta$ *then for all* a *there is a unique* P' *s.t.* $P \xrightarrow{x\langle a\rangle} P'$;
2. *If* $P \xrightarrow{a\langle x\rangle} P'$ *and* $x \notin \Gamma$ *then* $\Delta; \Gamma, x \vdash P'$;
3. *if* $P \xrightarrow{x\langle a\rangle} P'$ *then* $x \in \Delta$ *and* $\Delta - x; \Gamma \vdash P'$;
4. *if* $P \xrightarrow{a\langle b\rangle} P'$ *or* $P \xrightarrow{\overline{a}\langle b\rangle} P'$ *or* $P \xrightarrow{\overline{a}(b)} P'$, *then* $\Delta; \Gamma \vdash P'$;
5. *if* $P \xrightarrow{\tau} P'$ *then either* $\Delta; \Gamma \vdash P'$ *or there is* $x \in \Delta \cap \Gamma$ *and* $\Delta - x; \Gamma - x \vdash P'$;
6. *if* $P \xrightarrow{\overline{x}\langle a\rangle} P'$ *or* $P \xrightarrow{\overline{x}(a)} P'$ *or* $P \xrightarrow{\overline{a}\langle x\rangle} P'$, *then* $x \in \Gamma$ *and* $\Delta; \Gamma - x \vdash P'$;
7. *if* $P \xrightarrow{\overline{a}(x)} P'$ *and* $x \notin \Delta \cup \Gamma$ *then* $\Delta, x; \Gamma \vdash P'$.

Behavioural equivalences under linear receptiveness As usual in typed calculi, the definitions of the barbed relations take typing into account, so that the composition of a context and a process be well-typed. With receptiveness, an additional ingredient has to be taken into account, namely the input availability of receptive names. If a process has the possibility of using certain receptive names in output, then a context in which the process is tested should guarantee the input-availability at these names, otherwise the essence of receptiveness — outputs at receptive names can be immediately consumed — is lost.

Definition 3 complete processes and contexts. A process P is *complete* if $\Delta; \emptyset \vdash P$, for some Δ. We say that context C is *complete on* $(\Delta; \Gamma)$ if $C[P]$ is complete, for all P s.t. $\Delta; \Gamma \vdash P$.

Definition 4 barbed equivalences under linear receptiveness.
Suppose $\Delta; \Gamma \vdash P, Q$. Then we say that P and Q are *barbed equivalent under linear receptiveness at* $(\Delta; \Gamma)$, briefly $P \approx_L^{\Delta;\Gamma} Q$, if for each static context C which is complete on $(\Delta; \Gamma)$ it holds that $C[P] \mathrel{\dot\approx} C[Q]$ (where $\mathrel{\dot\approx}$ is barbed bisimulation, Definition 1). *Barbed congruence under linear receptiveness at* $(\Delta; \Gamma)$, briefly $\simeq_L^{\Delta;\Gamma}$, is defined similarly — just remove the constraint on C being static.

We write $\Delta; \Gamma \vdash_D P$ if $\Delta; \Gamma \vdash P$ can be proved without using rule T-out-2; in this case we say that P is *discreet*. In a discreet process, all receptive names which are exported must be private: Syntactically, this means that outputs of global receptive names are disallowed (that is, using the terminology in [15], only *internal mobility* — the sending of fresh names — is allowed on receptive names). We write $p \triangleright q$ as abbreviation for a process $p(r).\overline{q}\langle r\rangle.0$ (a 1-place ephemeral buffer from p to q). We can transform well-typed processes into discreet processes using the law

$$\overline{b}\langle x\rangle. P = \overline{b}\langle y\rangle.(y \triangleright x \mid P) \quad \text{for } y \text{ fresh} \tag{2}$$

This law makes the output of a *global* name into the output of a *local* (i.e., private) name. The law is not valid in the ordinary π-calculus, but it is valid under receptiveness:

Lemma 5. *If $\Delta; \Gamma \vdash \overline{b}\langle x\rangle.P$ and y is fresh, then $\overline{b}\langle x\rangle. P \simeq_L^{\Delta;\Gamma} \overline{b}\langle y\rangle.(y \triangleright x \mid P)$.*

We now derive a characterisation of the receptive barbed equivalence as a labeled bisimulation on discreet processes. We begin by defining the labeled bisimilarity on *complete* discreet processes. We say that an action μ is a *plain input* if μ is the input of a plain name, i.e., $\mu = p\langle a\rangle$ for some plain name a.

Definition 6 linear-receptive bisimilarity, $\asymp_L$. Linear-receptive bisimilarity is the largest relation $\asymp_L$ on complete discreet processes s.t. $P \asymp_L Q$ implies:

1. if $P \xrightarrow{\mu} P'$ with bound name of μ (if it exists) fresh for P and Q, and μ is an output or a plain input then there is Q' s.t. $Q \xRightarrow{\mu} Q'$ and $P' \asymp_L Q'$;
2. if $P \xrightarrow{\tau} P'$ then there is Q' s.t. $Q \Longrightarrow Q'$ and $P' \asymp_L Q'$;
3. if $P \xrightarrow{p\langle x\rangle} P'$ and x is fresh for P and Q then, for some fresh name v, there are Q' and Q'' s.t.: (a) $Q \xRightarrow{p\langle x\rangle} Q'$; (b) $\nu x\,(x \triangleright v \mid Q') \Longrightarrow Q''$; (c) $\nu x\,(x \triangleright v \mid P') \asymp_L Q''$.

The main novelty of receptive bisimulation is the use of a process $x \triangleright v$ in the input clause (3). To understand this addition, recall that x represents a private receptive name that the observer exports; if the observer behaves as a well-typed process, then it must make x immediately available in input, as a process of the form $x(p).R$. It is perhaps surprising that we do not test the behaviour of the derivatives P' and Q' for all infinite choices of the process $x(p).R$, but only on a single, simple, process, namely a link $x \triangleright v$.

Definition 7 linear-receptive bisimilarity on all discreet processes.
Suppose $\Delta; \Gamma \vdash_D P, Q$. Let $\tilde{x} = \Delta \cap \Gamma$ and $\tilde{y} = \Gamma - \Delta$ (therefore $\Gamma = \tilde{x} \cup \tilde{y}$); and let $\tilde{v}$ be fresh and pairwise distinct names with $|\tilde{y}| = |\tilde{v}|$.
We then set $P \asymp_L^{\Delta;\Gamma} Q$ if $(\boldsymbol{\nu}\tilde{x},\tilde{y})(\tilde{y} \triangleright \tilde{v} \mid P) \asymp_L (\boldsymbol{\nu}\tilde{x},\tilde{y})(\tilde{y} \triangleright \tilde{v} \mid Q)$.

The definition makes sense because processes $(\boldsymbol{\nu}\tilde{x},\tilde{y})(\tilde{y} \triangleright \tilde{v} \mid P)$ and $(\boldsymbol{\nu}\tilde{x},\tilde{y})(\tilde{y} \triangleright \tilde{v} \mid Q)$ are complete and discreet, and we have already defined $\asymp_L$ on this class. Moreover, since on complete processes $\asymp_L$ is preserved by structural equality and injective renaming, the above definition does not depend on the order of names in $\tilde{x}, \tilde{y}$ and $\tilde{v}$, and on the choice of names $\tilde{v}$.

The closure of barbed bisimulation w.r.t. the static contexts gives the ordinary (early) labeled bisimulation [13]; the closure w.r.t. the *complete* static contexts gives receptive bisimulation. The proofs for the ordinary bisimulation can be adapted to receptive bisimulation. Here are further useful laws for receptive barbed equivalence that are easy to prove using the labeled bisimilarity $\asymp_L^{\Delta;\Gamma}$, and that are not valid in the ordinary π-calculus:

$$\text{If } \Delta; \Gamma \vdash \overline{x}\langle p \rangle. P, \text{ then } \overline{x}\langle p \rangle. P \approx_L^{\Delta;\Gamma} \overline{x}\langle p \rangle \mid P. \tag{3}$$

Suppose that $\Delta; \Gamma \vdash P, Q$, for some Δ and Γ with $x \in \Delta - \Gamma$, and let v be a fresh name; then

$$P \approx_L^{\Delta;\Gamma} Q \quad \text{iff} \quad \boldsymbol{\nu}x\,(v \triangleright x \mid P) \approx_L^{\Delta-x;\Gamma} \boldsymbol{\nu}x\,(v \triangleright x \mid Q) \tag{4}$$

Suppose that $\Delta; \Gamma \vdash P, Q$, for some Δ and Γ with $y \in \Gamma - \Delta$, and let v be a fresh name; then

$$P \approx_L^{\Delta;\Gamma} Q \quad \text{iff} \quad \boldsymbol{\nu}y\,(y \triangleright v \mid P) \approx_L^{\Delta;\Gamma-y} \boldsymbol{\nu}y\,(y \triangleright v \mid Q). \tag{5}$$

Law (3) transforms a "synchronous" output into an "asynchronous" one; (4) transforms a global input into a local input; (5) does the same for outputs.

4 ω-receptiveness

The other interesting example of uniform receptiveness is ω-*receptiveness*, where: The input of a name is always available, and always with the same continuation; there are no limitations on the utilisation of the name in output. A simple way of ensuring the uniformity condition on inputs is to require that the only input occurrence be replicated, i.e., of the form $!x(p).P$.

When adapting the theory of linear receptiveness to ω-receptiveness, there are several, but not surprising, modifications to make. In the typing system, the interpretation of a judgement $\Delta; \Gamma \vdash P$ is now that P must make names in Δ immediately available, in input-replicated form; whereas it may use names in Γ arbitrarily many times in output. We only show the new version of rules T-par and T-out-2, and one of the rules for replication:

$$\frac{\Delta_1; \Gamma \vdash P_1 \quad \Delta_2; \Gamma \vdash P_2 \quad \Delta_1 \cap \Delta_2 = \emptyset}{\Delta_1, \Delta_2; \Gamma \vdash P_1 \mid P_2}$$

$$\frac{x \in \Gamma \quad \emptyset; \Gamma \vdash P}{\emptyset; \Gamma \vdash \overline{a}\langle x \rangle . P} \qquad \frac{\emptyset; \Gamma \vdash P}{x; \Gamma \vdash !x(b) . P}$$

In the definitions of the typed barbed relations, typed labeled bisimilarities and the algebraic laws for the ω-case, the main modification w.r.t. the linear case is that the links $p \triangleright q$ have to become persistent. Using $\simeq_\omega^{\Delta;\Gamma}$ for barbed congruence under ω-receptiveness at $\Delta; \Gamma$, law (2) becomes

$$\overline{b}\langle x \rangle . P \simeq_\omega^{\Delta;\Gamma} \overline{b}\langle y \rangle . (!y \triangleright x \mid P) \qquad \text{for } y \text{ fresh} \tag{6}$$

5 Examples

Parallelisation of resources We can use linear receptiveness to validate transformations that increase the parallelism in processes. In the processes below, we use recursion, polyadicity and communication of integers, which are straightforward to accommodate in the theory of bisimulation previously developed (recursion can be coded up). Thus m, n range over integers and variables over integers. Consider the process:

$$A_1 \langle b \rangle \stackrel{\text{def}}{=} a(x) . b(n, c) . \nu d\, \overline{c}\langle d \rangle . \overline{x}\langle n \rangle . A_1 \langle d \rangle$$

A client can interrogate $A_1 \langle b \rangle$ at a, and it will receive at the return channel x an integer n that $A_1 \langle b \rangle$ has received at another channel b (this channel is renewed at each cycle using c). Interactions between A_1 and the clients are Remote Procedure Calls (RPC), therefore the return channels are used according to the discipline of linear receptiveness (see the discussion on RPC in the introductory section). The behaviour of A_1 is strictly sequential. Let us introduce some parallelism:

$$A_2 \langle b \rangle \stackrel{\text{def}}{=} a(x) . b(n, c) . \nu d \left(\overline{x}\langle n \rangle . \overline{c}\langle d \rangle \mid A_2 \langle d \rangle \right)$$

$$A_3 \langle b \rangle \stackrel{\text{def}}{=} a(x) . \nu d \left(b(n, c) . \overline{c}\langle d \rangle . \overline{x}\langle n \rangle \mid A_3 \langle d \rangle \right)$$

Process $A_2 \langle b \rangle$ can accept a second request at a before the answer to the fist request has been delivered; however answers cannot overtake one another — they are delivered in the same order in which the requests were made. Process $A_3 \langle b \rangle$ can even accept a request before receiving an integer at b; answers can overtake. Let now $I \langle n \rangle$ be a counter $I \langle n, b \rangle \stackrel{\text{def}}{=} \nu c\, \overline{b}\langle n, c \rangle . c(d) . I \langle n+1, d \rangle$ and consider the systems (n is any integer) $S_i \langle n \rangle \stackrel{\text{def}}{=} \nu b\, (A_i \langle b \rangle \mid I \langle n, b \rangle)$, for $i \in \{1, 2, 3\}$.

All these systems are distinguished in the ordinary π-calculus — the different degrees of parallelism that they exhibit are observable. We can prove that they *are* equivalent exploiting the linear receptiveness of the return channels x, y. [2] For $\text{S1}\langle n \rangle \approx_{\text{L}}^{\emptyset;\emptyset} \text{S2}\langle n \rangle$ one proves that the relation composed by all pairs of the form
$$\left(\text{S1}\langle n' \rangle \mid \prod_{i=1}^{m} \overline{v_i}\langle n_i \rangle, \text{S2}\langle n' \rangle \mid \prod_{i=1}^{m} \overline{v_i}\langle n_i \rangle\right)$$
for some channel v_i and integers m, n_i, n' is a $\asymp_{\text{L}}$-bisimulation up to expansion. The other equalities can be proved in a similar way.

The above processes are simple. A more interesting example of parallelisation of resources is Cliff Jones's parallelisation transformation problem [5]. We analyse this in [16], where we prove Jones's transformation using a combination of the techniques for linear and ω receptiveness.

Encoding of higher-order process calculi We now present an example with ω-receptiveness. Below x, y are supposed to be *ω-receptive names*. We prove the correctness of an optimisation of the translation of higher-order process calculi into the π-calculus [13, 18]. In a higher-order calculus, terms of the languages may be transmitted. For simplicity of presentation, we consider the simpler case of a calculus where *only* processes may be communicated. The operators are those for sending a process ($\overline{p}\langle P_1 \rangle . P_2$), receiving a process ($p(X) . P$), process variable ($X$), plus the usual operators of restriction, parallel composition, summation, replication. This calculus, which we call HOPC, is the core of Plain CHOCS [18], and is a second-order fragment of the Higher-Order π-calculus [13]. Upper case letter X ranges over process variables. A process is *closed* if it does not contain free variables. The compilation $\mathcal{C}$ of this calculus into the π-calculus in [13, 18] acts as a homomorphism on all process constructs except input, output prefixes and process variables where it is so defined:

$$\mathcal{C}[\![\overline{p}\langle P \rangle . Q]\!] \stackrel{\text{def}}{=} \nu x\, \overline{p}\langle x \rangle . (!x . \mathcal{C}[\![P]\!] \mid \mathcal{C}[\![Q]\!]) \quad \text{for } x \text{ fresh}$$
$$\mathcal{C}[\![p(X) . Q]\!] \stackrel{\text{def}}{=} p(x) . \mathcal{C}[\![Q]\!] \qquad\qquad \mathcal{C}[\![X]\!] \stackrel{\text{def}}{=} \overline{x} . 0$$

In the compilation, the communication of a process P is translated as the communication of a private name which acts as a pointer to (the translation of) P and which the recipient can use to trigger a copy of (the translation of) P. These pointers, introduced in the compilation, are used as ω-receptive names.

In [13], the correctness of compilation $\mathcal{C}$ is established, by proving that it is fully abstract w.r.t. barbed congruence (that is, for all closed HOPC processes P and Q, $P \simeq Q$ iff $\mathcal{C}[\![P]\!] \simeq \mathcal{C}[\![Q]\!]$). The optimisation that we consider acts on outputs of process variables. Let us call $\mathcal{O}$ the optimised compilation. It is defined as $\mathcal{C}$ except for the case of an output of a variable, for which we have:

$$\mathcal{O}[\![\overline{p}\langle X \rangle . Q]\!] \stackrel{\text{def}}{=} \overline{p}\langle x \rangle . \mathcal{O}[\![Q]\!]$$

---

[2] In these definitions, also name c is linear receptive. We do not need this fact for the proofs (and it is reasonable not to use it, because the linear receptiveness of c is accidental — one can modify the definitions so that c is not linear receptive.)

For instance, when translating $p(X).\overline{q}\langle X\rangle.0$, the result of $\mathcal{O}$ is $p(x).\overline{q}\langle x\rangle.0$ while that of $\mathcal{C}$ is $p(x).\nu y\,\overline{q}\langle y\rangle.!y.\overline{x}.0$. The optimisation avoids us one level of indirection through pointers. This optimisation is analysed in [13] and is shown to be unsound for untyped barbed equivalence. However, we can show that the optimisation is sound if we take into account the receptiveness of names. The proof is an immediate consequence of law (6), since, for all P, $\mathcal{O}[\![P]\!]$ can be transformed into $\mathcal{C}[\![P]\!]$ by repeatedly applying the law:

Theorem 8. *Let P be a HOPC process with free variables in $\{X_1,\ldots,X_n\}$, and let $\Gamma \stackrel{\mathrm{def}}{=} \{x_1,\ldots,x_n\}$. It holds that $\mathcal{C}[\![P]\!] \simeq_\omega^{\emptyset;\Gamma} \mathcal{O}[\![P]\!]$.*

Combining this with the theorem on $\mathcal{C}$ in [13], we can prove: for all closed HOPC processes P and Q, $P \simeq Q$ iff $\mathcal{O}[\![P]\!] \simeq_\omega^{\emptyset;\emptyset} \mathcal{O}[\![Q]\!]$.

In an expanded paper [17], other examples of application of ω-receptiveness are reported: The proof of the equivalence between the target processes of Milner's two encodings of call-by-values λ-calculus into π-calculus [8] (this is a novel result); the proofs of some stronger versions of π-calculus replication theorems [10] (these results were already proved in [10]; exploiting receptiveness we get simpler proofs).

6 Final remarks

Several type systems have been proposed for process calculi. The most relevant for this work are [10], where the type system has input/output modalities to distinguish between the capabilities of reading and writing on names, and the type systems expressing linearity information [3, 7, 4]. The type system for receptiveness represents a refinement of [10] and, in the case of linear receptiveness, also of [7]. Also [10] and [7] contain studies of the effect of types on process behaviours, using barbed congruence. The proof techniques developed in this paper are easier to apply, mainly because based on labeled bisimilarities.

Other papers with results on behavioural consequences of π-calculus types include the following. [6] defines a type system for the asynchronous π-calculus that guarantees deadlock freedom in certain cases; a subsystem of this system is similar to ours for ω-receptiveness. [19] uses a type system where types have a graph structure to prove the full abstraction of an encoding of the polyadic π-calculus into the monadic calculus. Graphs allow expressing sophisticated communication protocols but introduce some complications in the typing. [14] uses a type system with input/output modalities and variant types to guarantee the adequacy of a translation of a typed object-oriented calculus into the π-calculus. [11] studies the constraints imposed by parametric polymorphism.

Some of the ideas in this paper should be useful to develop reasoning techniques for other type systems, in particular those with input/output modalities and with linearity. They might also be useful in cases where either the receptiveness or the uniformity condition fails; for instance the calculus in [2], where all names are uniform but not necessarily receptive, or that in [1], where all names are receptive but not necessarily uniform.

Acknowledgements. The author would like to thank G. Boudol, N. Kobayashi, C. Jones, B. Pierce, D. Walker, N. Yoshida and two of the anonymous referees for useful discussions or suggestions. This research has been supported by FRANCE TÉLÉCOM, CTI-CNET 95-1B-182, "Modélisation de Systèmes Mobiles".

References

1. R. Amadio. Locality and failures II. To appear as a Technical Report, INRIA-Sophia Antipolis, 1997.
2. Fournet C. and Gonthier G. The Reflexive Chemical Abstract Machine and the Join calculus. In *Proc. 23th POPL*. ACM Press, 1996.
3. K. Honda. Types for dydadic interaction. In E. Best, editor, *Proc. CONCUR '93*, LNCS 715, pages 509–523. Springer Verlag, 1993.
4. K. Honda. Composing processes. In *Proc. 23th POPL*. ACM Press, 1996.
5. C.B. Jones. Constraining interference in an object-based design method. *Proc. TAPSOFT'93*, LNCS 668, pages 136–150. Springer Verlag, 1993.
6. N. Kobayashi. A partially deadlock-free typed process calculus. To appear as a technical report, Department of Information Science, University of Tokyo, 1997. A summary is to appear in *Proc. 12th LICS Conf.*, 1997.
7. N. Kobayashi, B.C. Pierce, and D.N. Turner. Linearity and the pi-calculus. In *Proc. 23th POPL*. ACM Press, 1996.
8. R. Milner. Functions as processes. Research Report 1154, INRIA, Sophia Antipolis. Final version in *J. Mathem. Struct. in Computer Science* 2(2):119–141, 1992.
9. R. Milner, J. Parrow, and D. Walker. A calculus of mobile processes, (Parts I and II). *Information and Computation*, 100:1–77, 1992.
10. B. Pierce and D. Sangiorgi. Typing and subtyping for mobile processes. *Journal of Mathematical Structures in Computer Science*, 6(5):409–454, 1996.
11. B. Pierce and D. Sangiorgi. Behavioral equivalence in the polymorphic pi-calculus. In *24th POPL*. ACM Press, 1997.
12. B. C. Pierce and D. N. Turner. Pict: A programming language based on the pi-calculus. In preparation, 1997.
13. D. Sangiorgi. *Expressing Mobility in Process Algebras: First-Order and Higher-Order Paradigms*. PhD thesis CST-99-93, University of Edinburgh, 1992.
14. D. Sangiorgi. An interpretation of typed objects into typed π-calculus. Technical Report RR-3000, INRIA-Sophia Antipolis, 1996.
15. D. Sangiorgi. π-calculus, internal mobility and agent-passing calculi. *Theoretical Computer Science*, 167(2):235–274, 1996.
16. D. Sangiorgi. Typed π-calculus at work: a proof of Jones's parallelisation transformation on concurrent objects. Presented at Fourth Workshop on Foundations of Object-Oriented Languages, 1997. Available as ftp://zenon.inria.fr/meije/theorie-par/davides/JonesTransform.ps.gz.
17. D. Sangiorgi. The name discipline of uniform receptiveness. (This is a slightly longer version of the present paper.) Available as ftp://zenon.inria.fr/meije/theorie-par/davides/RecePi.ps.Z.
18. B. Thomsen. Plain CHOCS, a second generation calculus for higher-order processes. *Acta Informatica*, 30:1–59, 1993.
19. Nobuko Yoshida. Graph types for monadic mobile processes. In *Proc. FST & TCS*, LNCS 1180, pages 371–386. Springer Verlag, 1996.

On Confluence in the π-Calculus

Anna Philippou and David Walker

Department of Computer Science, University of Warwick
Coventry CV4 7AL, U.K.

Abstract. An account of the basic theory of confluence in the π-calculus is presented, techniques for showing confluence of mobile systems are given, and the utility of some of the theory presented is illustrated via an analysis of a distributed algorithm.

1 Introduction

Confluence arises in a variety of forms in computation theory. It was first studied in the context of concurrent systems by Milner in [6]. Its essence, to quote [7], is that "of any two possible actions, the occurrence of one will never preclude the other". As shown in the works cited, for pure CCS agents confluence implies determinacy and semantic-invariance under silent actions, and is preserved by several important system-building operations. These facts make it possible to guarantee by construction that certain systems are confluent and to exploit this fact fruitfully when analysing their behaviours. A more general study was made in [1] which in particular clarified the relationships among various notions of confluence and semantic-invariance under silent actions, and illustrated the utility of the ideas for state-space reduction and protocol analysis; see also [1] for further references. Confluence of value-passing CCS agents was studied first in [18] and later in [22] where consideration was given to conditions under which confluent systems result from combinations of 'semi-confluent' agents and the ideas were utilized to show determinacy of programs in a fragment of a concurrent imperative programming language.

The elaboration of techniques for reasoning about *mobile* systems expressed in the π-calculus [9] and variants of it has involved extension of established methods and development of new concepts specific to the richer setting. Stemming from [8] there have been several works on disciplines of name-use respected by agents, sometimes expressed via type systems; see for instance [2, 15, 23, 25, 20, 16]. Such disciplines contribute much to the effectiveness of π-calculi as descriptive formalisms and analytical tools. One promising strand of development concerns varieties of confluence. These have been used in showing determinacy of systems prescribed by concurrent object-oriented programs [13], in justifying optimizations in the Pict compiler [3, 17], and in proving the soundness of transformation rules for concurrent object-oriented programs [4, 14]. The aims of this paper are to give an account of the basic theory of confluence in the π-calculus, to develop techniques for showing that mobile systems are confluent, and to illustrate the utility of some of the theory presented via an analysis of a distributed algorithm. The extension of the theory from pure and simple value-passing agents to mobile agents is at some places fairly straightforward: we then proceed quickly, drawing attention only to significant points. Due to the richness of name-passing, however, techniques for showing mobile systems to be confluent are more involved. This paper contains a sample of results obtained on this topic in the first author's thesis [12]. Independently, Uwe Nestmann in his thesis [11] has developed a static type system concerned with sharing of ports (polarized names) by mobile agents and shown that well-typed agents are confluent.

A summary of the paper follows. Preliminary material is collected in the next section, while in section 3 the basic definitions and results on confluence in the π-calculus are given. Section 4 is concerned with techniques for showing that complex systems are confluent. The final section is devoted to an illustration of the utility of some of the theory presented: an analysis of a distributed algorithm. Due to lack of space all proofs are omitted; see [12] for a detailed technical account.

We are grateful to an anonymous referee for helpful comments.

2 Preliminaries

In this section we recall briefly background material on the (polyadic) π-calculus [9, 8]. For undefined terms and explanation we refer to these papers.

We assume an infinite set N of *names*, ranged over by lower-case letters, a partition S of N into a set of infinite *(subject) sorts*, and a *sorting* $\lambda : \mathsf{S} \longrightarrow \mathsf{S}^*$. For $S \in \mathsf{S}$, $\lambda(S)$ is the *object sort* associated with S. The *agents* are the expressions given as follows which *respect* the sorting λ:

$$P ::= \mathbf{0} \mid \pi.P \mid P+Q \mid P \mid Q \mid (\nu y)P \mid A\langle \widetilde{y} \rangle .$$

Here π ranges over the *prefixes* τ, $x(\widetilde{y})$ and $\overline{x}\langle \widetilde{y} \rangle$, in the latter two of which x is the *subject* and the tuple $\widetilde{y}$ is the *object*. In a prefix $x(\widetilde{y})$ the occurrences of the pairwise-distinct names $\widetilde{y}$ are binding; the occurrence of y in (νy) is also binding. We write $\mathsf{fn}(P)$ (resp. $\mathsf{bn}(P)$) for the set of free (resp. bound) names of P, and $\mathsf{n}(P)$ for the set of all names occurring in P. We write also $\mathsf{fn}_S(P)$ for the free names of P of sort S. Each agent constant A has a defining equation $A(\widetilde{x}) \stackrel{\text{def}}{=} P$ where $\mathsf{fn}(P) \subseteq \widetilde{x}$ and $\widetilde{x}$ are pairwise distinct. We regard as identical agents which differ only by change of bound names. We write $\equiv$ for *structural congruence* of agents. A *substitution* is a sort-respecting mapping from N to N. We write $P\sigma$ for the agent obtained from P by applying the substitution σ. We write $\{\widetilde{y}/\widetilde{x}\}$ for the substitution which maps each component of $\widetilde{x}$ to the corresponding component of $\widetilde{y}$ and is otherwise the identity.

Here we give the behaviour of agents by the early transition rules [10, 19]. In this system there are three kinds of action: input actions of the form $x\langle \widetilde{y} \rangle$; output actions of the form $(\nu \widetilde{z})\overline{x}\langle \widetilde{y} \rangle$, where the set $\widetilde{z}$ of bound names of the action (which is omitted when it is empty) satisfies $\widetilde{z} \subseteq \widetilde{y}$; and the silent action τ representing communication between agents. We write $\mathsf{bn}(\alpha)$ for the set of bound names of the output α and set $\mathsf{bn}(\alpha) = \emptyset$ if α is an input or τ. We write Act for the set of actions. The subject/object terminology carries over from prefixes to visible actions. The transition rules are as follows where $\mathsf{n}(\alpha)$ is the set of names occurring in the action α. The third, fourth and fifth have symmetric forms.

1. $x(\widetilde{y}).P \xrightarrow{x(\widetilde{z})} P\{\widetilde{z}/\widetilde{y}\}$ if the sorts of the components of $\widetilde{y}$ and $\widetilde{z}$ agree.
2. $\pi.P \xrightarrow{\pi} P$ if π is τ or $\overline{x}\langle \widetilde{y} \rangle$.
3. If $P \xrightarrow{\alpha} P'$ then $P + Q \xrightarrow{\alpha} P'$.
4. If $P \xrightarrow{\alpha} P'$ then $P \mid Q \xrightarrow{\alpha} P' \mid Q$ if $\mathsf{bn}(\alpha) \cap \mathsf{fn}(Q) = \emptyset$.
5. If $P \xrightarrow{(\nu \widetilde{z})\overline{x}\langle \widetilde{y} \rangle} P'$ and $Q \xrightarrow{x\langle \widetilde{y} \rangle} Q'$ then $P \mid Q \xrightarrow{\tau} (\nu \widetilde{z})(P' \mid Q')$ if $\widetilde{z} \cap \mathsf{fn}(Q) = \emptyset$.
6. If $P \xrightarrow{\alpha} P'$ and $y \notin \mathsf{n}(\alpha)$ then $(\nu y)P \xrightarrow{\alpha} (\nu y)P'$.
7. If $P \xrightarrow{(\nu \widetilde{z})\overline{x}\langle \widetilde{y} \rangle} P'$ and $w \in \widetilde{y} - (\widetilde{z} \cup \{x\})$ then $(\nu w)P \xrightarrow{(\nu w \widetilde{z})\overline{x}\langle \widetilde{y} \rangle} P'$.
8. If $P\{\widetilde{y}/\widetilde{x}\} \xrightarrow{\alpha} P'$ and $A(\widetilde{x}) \stackrel{\text{def}}{=} P$ then $A\langle \widetilde{y} \rangle \xrightarrow{\alpha} P'$.

We write $\Longrightarrow$ for the reflexive and transitive closure of $\xrightarrow{\tau}$, $\stackrel{\alpha}{\Longrightarrow}$ for the composition $\Longrightarrow \xrightarrow{\alpha} \Longrightarrow$, and $\stackrel{\widehat{\alpha}}{\Longrightarrow}$ for $\Longrightarrow$ if $\alpha = \tau$ and $\stackrel{\alpha}{\Longrightarrow}$ otherwise. We further write $P \xrightarrow{\widehat{\alpha}} Q$ if $P \xrightarrow{\alpha} Q$, or $\alpha = \tau$ and $P \equiv Q$.

We often tacitly assume that bound names of actions are fresh. *(Early) bisimilarity* is the largest symmetric relation $\approx$ such that if $P \approx Q$ and $P \xrightarrow{\alpha} P'$, for some Q', $Q \stackrel{\widehat{\alpha}}{\Longrightarrow} Q'$ and $P' \approx Q'$. *Branching bisimilarity* is the largest symmetric relation $\simeq$ such that if $P \simeq Q$ and $P \xrightarrow{\alpha} P'$, then either $\alpha = \tau$ and $P' \simeq Q$, or for some Q', Q'', $Q \Longrightarrow Q'' \xrightarrow{\alpha} Q'$, $P \simeq Q''$ and $P' \simeq Q'$. The standard notations for these relations have a dot to differentiate them from the congruences defined as bisimilarity under all substitutions. Since we do not consider the latter here we use the less cumbersome symbols. Finally, an agent P *diverges*, written $P \uparrow$, if P can perform an infinite sequence of τ actions; otherwise P *converges*, $P \downarrow$; and P is *fully convergent* if for each derivative P' of P, $P' \downarrow$.

3 Confluence

In [7] confluence for pure CCS agents was defined using bisimilarity, and it was shown that a wide range of behavioural equivalences coincide on confluent agents. In developing a theory of confluence for the π-calculus we choose here to base it on early bisimilarity. The connections between this treatment and the various other possibilities are straightforward. In our view, in applications of the theory there is likely to be little substantial difference between the variants. With this choice 'determinacy' can be defined as it can for pure CCS agents.

Definition 1. *P is determinate if for each derivative Q of P and action α, if $Q \xrightarrow{\alpha} Q'$ and $Q \stackrel{\widehat{\alpha}}{\Longrightarrow} Q''$ then $Q' \approx Q''$.* □

Note for instance that $P \equiv a(x).(x(y).\bar{a}\langle y\rangle.\mathbf{0} + b(y).\mathbf{0})$ is not determinate if x, b have the same sort as $P \xrightarrow{a\langle b\rangle} Q \equiv b(y).\bar{a}\langle y\rangle.\mathbf{0} + b(y).\mathbf{0}$ and Q has non-bisimilar $b\langle y\rangle$-derivatives. As in pure CCS, an agent bisimilar to a determinate agent is determinate, and determinate agents are bisimilar if they may perform the same sequences of visible actions. The following lemma summarizes conditions under which determinacy is preserved by operators. In the last part, sort(M) is the set of sorts of the names in M.

Lemma 2.

1. If P is determinate so are $\tau.P$, $\bar{x}\langle\tilde{y}\rangle.P$ and $(\nu y)P$.
2. If P is determinate and for each $y \in \tilde{y}$, if y is of sort S then $\mathsf{fn}_S(P) \subseteq \{y\}$, then $a(\tilde{y}).P$ is determinate.
3. If each $\pi_i.P_i$ is determinate, no π_i is τ and no two of the π_i are inputs or outputs with the same subject, then $\sum_i \pi_i.P_i$ is determinate.
4. If P_1, P_2 are determinate, $\mathsf{fn}(P_1) \cap \mathsf{fn}(P_2) = \emptyset$, $\mathsf{sort}(\mathsf{bn}(P_1)) \cap \mathsf{sort}(\mathsf{n}(P_2)) = \emptyset$ and $\mathsf{sort}(\mathsf{bn}(P_2)) \cap \mathsf{sort}(\mathsf{n}(P_1)) = \emptyset$, then $P_1 \mid P_2$ is determinate. □

The condition in (2) cannot be dropped: consider $R \equiv x(y).\bar{a}\langle y\rangle.\mathbf{0} + b(y).\mathbf{0}$ where x, b have the same sort. Clearly R is determinate but $P \equiv a(x).R$ above is not as $R\{b/x\}$ is not. Note, however, that if x, b were of different sorts, P would be determinate. Using sorts to make distinctions among names in this way is often helpful in applications of the calculus. Similarly, the condition in (4) cannot be dropped: as in CCS, P_1, P_2 cannot share free names (consider $a.\mathbf{0} \mid \bar{a}.\mathbf{0}$), but in addition in the mobile setting more must be said as that property need not be preserved under transition; for instance if $P \equiv w(z).z(x).\bar{b}.\mathbf{0}$, $Q \equiv a(y).\bar{c}.\mathbf{0}$ and a, z are of the same sort, then $P \mid Q \xrightarrow{w\langle a\rangle} R \equiv a(x).\bar{b}.\mathbf{0} \mid a(y).\bar{c}.\mathbf{0}$ and R is not determinate. The condition in (4) ensures that a bound name of one component cannot be instantiated with a name free in the other.

A pure CCS agent P is confluent if for each derivative Q of it and distinct α, β, (i) if $Q \xrightarrow{\alpha} Q_1$ and $Q \xRightarrow{\alpha} Q_2$, then $Q_1 \Longrightarrow Q_1'$ and $Q_2 \Longrightarrow Q_2' \approx Q_1'$, and (ii) if $Q \xrightarrow{\alpha} Q_1$ and $Q \xRightarrow{\beta} Q_2$, then $Q_1 \xRightarrow{\hat{\beta}} Q_1'$ and $Q_2 \xRightarrow{\hat{\alpha}} Q_2' \approx Q_1'$. For value-passing CCS agents the definition must be refined to account for different inputs with the same subject [18, 22]. This holds also for mobile agents with the additional point that data received are names which may be used for interaction: consider $P \stackrel{\text{def}}{=} a(x).\bar{x}\langle y\rangle.\mathbf{0}$ which one would expect to be determinate and the transitions $P \xrightarrow{a\langle b\rangle} \bar{b}\langle y\rangle.\mathbf{0}$ and $P \xrightarrow{a\langle c\rangle} \bar{c}\langle y\rangle.\mathbf{0}$. In the π-calculus a further consideration arises: consider $P \equiv (\nu z)(\bar{a}\langle z\rangle.\mathbf{0} \mid \bar{b}\langle z\rangle.\mathbf{0})$ and its transitions $P \xrightarrow{(\nu z)\bar{a}\langle z\rangle} P_1 \equiv \bar{b}\langle z\rangle.\mathbf{0}$ and $P \xrightarrow{(\nu z)\bar{b}\langle z\rangle} P_2 \equiv \bar{a}\langle z\rangle.\mathbf{0}$. Note that P_1 has no $(\nu z)\bar{b}\langle z\rangle$-transition, and dually for P_2. In our view P should none the less be regarded as confluent. To give the definition we introduce two pieces of notation.

Notation 3 We write $\alpha \bowtie \beta$ if α and β are different actions and are not both inputs with the same subject. The *weight* $\alpha \lfloor \beta$ of action α over action β is α except if $\alpha = (\nu \widetilde{z})\overline{a}\langle \widetilde{y} \rangle$ when it is $(\nu \widetilde{z} - \mathsf{bn}(\beta))\overline{a}\langle \widetilde{y} \rangle$. □

Thus for instance, $(\nu yz)\overline{a}\langle y, z\rangle \lfloor (\nu z)\overline{b}\langle x, z\rangle$ is $(\nu y)\overline{a}\langle y, z\rangle$. We then have:

Definition 4. An agent P is *confluent* if for each derivative Q of P and α, β with $\alpha \bowtie \beta$, (i) if $Q \xrightarrow{\alpha} Q_1$ and $Q \xrightarrow{\alpha} Q_2$, then $Q_1 \Longrightarrow Q_1'$ and $Q_2 \Longrightarrow Q_2' \approx Q_1'$, and (ii) if $Q \xrightarrow{\alpha} Q_1$ and $Q \xrightarrow{\beta} Q_2$, then $Q_1 \xrightarrow{\widehat{\beta \lfloor \alpha}} Q_1'$ and $Q_2 \xrightarrow{\widehat{\alpha \lfloor \beta}} Q_2' \approx Q_1'$. □

Thus for instance $P \equiv (\nu z)(\overline{a}\langle z\rangle.\mathbf{0} \mid \overline{b}\langle z\rangle.\mathbf{0})$ above is confluent as after $P \xrightarrow{(\nu z)\overline{a}\langle z\rangle} P_1 \equiv \overline{b}\langle z\rangle.\mathbf{0}$ and $P \xrightarrow{(\nu z)\overline{b}\langle z\rangle} P_2 \equiv \overline{a}\langle z\rangle.\mathbf{0}$ we have $P_1 \xrightarrow{\overline{b}\langle z\rangle} \mathbf{0}$ and $P_2 \xrightarrow{\overline{a}\langle z\rangle} \mathbf{0}$.

It is easy to see that an agent bisimilar to a confluent agent is itself confluent. An agent P is τ-*inert* if for each derivative Q of P, if $Q \xrightarrow{\tau} Q'$ then $Q' \approx Q$. By a generalization of the argument from the CCS case we have:

Lemma 5. If P is confluent then P is τ-inert. □

The following result is a useful characterization of confluence in which only single transitions need be considered. It holds only for fully convergent agents. In [1] it was observed that for fully convergent ('τ-well founded') agents, τ-inertness implies confluence. A similar observation is included here.

Lemma 6. Suppose P is fully convergent. Then P is confluent iff P is τ-inert and for each derivative Q of P and α, β with $\alpha \bowtie \beta$, (i) if $Q \xrightarrow{\alpha} Q_1$ and $Q \xrightarrow{\alpha} Q_2$ then $Q_1 \approx Q_2$, and (ii) if $Q \xrightarrow{\alpha} Q_1$ and $Q \xrightarrow{\beta} Q_2$, then $Q_1 \xrightarrow{\widehat{\beta \lfloor \alpha}} Q_1'$ and $Q_2 \xrightarrow{\widehat{\alpha \lfloor \beta}} Q_2' \approx Q_1'$. □

The proof shows that if P is fully convergent and τ-inert and satisfies (i), then P is determinate. The assumption that P is fully convergent cannot be dropped: consider $P \stackrel{\text{def}}{=} a.b.\mathbf{0} + \tau.(a.\mathbf{0} + \tau.P)$. It is easy to see that P is τ-inert and that all of its derivatives satisfy (i) and (ii). However, P is not determinate.

We record the analogues for confluence of the earlier results on preservation of determinacy by operators.

Lemma 7.

1. If P is confluent so are $\tau.P$, $\overline{x}\langle \widetilde{y}\rangle.P$ and $(\nu y)P$.
2. If P is confluent and for each $y \in \widetilde{y}$, if y is of sort S then $\mathsf{fn}_S(P) \subseteq \{y\}$, then $a(\widetilde{y}).P$ is confluent.
3. If P_1, P_2 are confluent, $\mathsf{fn}(P_1) \cap \mathsf{fn}(P_2) = \emptyset$, $\mathsf{sort}(\mathsf{bn}(P_1)) \cap \mathsf{sort}(\mathsf{n}(P_2)) = \emptyset$ and $\mathsf{sort}(\mathsf{bn}(P_2)) \cap \mathsf{sort}(\mathsf{n}(P_1)) = \emptyset$, then $P_1 \mid P_2$ is confluent. □

Of course here the guarded summation clause is missing.

In the following section we will consider further techniques for showing systems to be confluent. Before doing so we consider a variant of confluence based on branching bisimilarity.

Definition 8. P is $\simeq$-confluent if for each derivative Q of P and α, β with $\alpha \bowtie \beta$, (i) if $Q \xrightarrow{\alpha} Q_1$ and $Q \Longrightarrow\xrightarrow{\alpha} Q_2$, then $Q_1 \Longrightarrow Q_1'$ and $Q_2 \Longrightarrow Q_2' \approx Q_1'$, and (ii) if $Q \xrightarrow{\alpha} Q_1$ and $Q \Longrightarrow\xrightarrow{\beta} Q_2$, then $Q_1 \Longrightarrow\xrightarrow{\widehat{\beta\lfloor\alpha}} Q_1'$ and $Q_2 \Longrightarrow\xrightarrow{\widehat{\alpha\lfloor\beta}} Q_2' \simeq Q_1'$. □

The following observations were made in [4]. Confluence (for non-mobile labelled transition systems) based on branching bisimilarity was also considered in [1] and observations similar to some of these made. An agent P is $\tau_{\simeq}$-inert if for each derivative Q of P, if $Q \xrightarrow{\tau} Q'$ then $Q' \simeq Q$.

Lemma 9.

1. If P is $\simeq$-confluent then P is $\tau_{\simeq}$-inert.
2. If P, Q are τ-inert and $P \approx Q$ then $P \simeq Q$.
3. P is $\tau_{\simeq}$-inert iff P is τ-inert.
4. P is confluent iff P is $\simeq$-confluent. □

In contrast to these coincidences, to obtain a satisfactory notion of 'partial' confluence which is not τ-inert it is essential to base the theory on branching bisimilarity rather than bisimilarity; see [4].

4 Confluence by construction

A main motivation in [7] for studying confluence was to find an interesting property implying determinacy which can be guaranteed to hold simply by confining the use of combinators in building systems. Work elaborating this view and showing its fruitfulness has been described in the Introduction. Here the emphasis is on sample results of this kind in the richer setting of name-passing. The approach is complementary to development of static type systems as in [11, 20]. A useful definition: an agent P is *o-determinate* if for each derivative Q of P, there are not two distinct output actions α, β with the same subject such that $Q \xrightarrow{\alpha}$ and $Q \xRightarrow{\beta}$. The first result gives conditions under which a combination of confluent agents is confluent.

Theorem 10. Suppose $P \equiv (\nu \widetilde{z})(P_1 \mid \ldots \mid P_n)$ where each P_i is confluent and o-determinate. Suppose that for each derivative $P' \equiv (\nu \widetilde{z}')(P_1' \mid \ldots \mid P_n')$ of P, no name occurs free in more than two components of P', and a free name of P' occurs in exactly one component of P'. Then P is confluent. □

Note that in this theorem it is not possible to replace 'confluent' by 'determinate': consider $(\nu a)(\overline{a}.\mathbf{0} \mid (a.\mathbf{0} + b.\mathbf{0}))$.

It is often the case that although the components of a system are not themselves confluent, the constraints they place upon one another's behaviour ensure that the system itself is confluent. The second theorem is an instance of this idea. To state it we need some definitions. We refer to a set of agents closed under derivation as a *system*. For $S \in \mathsf{S}$ we say a system is *S-closed* if none of its agents may perform an input or an output via an S-name.

Definition 11. Suppose S and $\tilde{S} = S_1 \ldots S_n$ are distinct sorts and the sorting λ is such that $\lambda(S) = (\tilde{S})$ and no S_i occurs in any other $\lambda(S')$. A system $\mathcal{P}$ is $S, \tilde{S}$-*sensitive* if there is a partition $\{\mathcal{P}^{\tilde{p}} \mid \tilde{p} \text{ a finite subset of } S_1 \times \ldots \times S_n\}$ of $\mathcal{P}$ such that:

1. if $P \in \mathcal{P}^{\tilde{p}}$ and $P \xrightarrow{\alpha} P'$ where α is not an input or output via an S-name or an input via an S_i-name, then $P' \in \mathcal{P}^{\tilde{p}}$;
2. if $P \in \mathcal{P}^{\tilde{p}}$ and $P \xrightarrow{\alpha} P'$ where α is an output via an S-name, then $\alpha = (\nu \tilde{z}) \overline{x} \langle \tilde{z} \rangle$ $P' \in \mathcal{P}^{\tilde{p} \cup \{\tilde{z}\}}$;
3. if $P \in \mathcal{P}^{\tilde{p}}$ and $P \xrightarrow{\alpha} P'$ where $\alpha = x \langle z_1, \ldots, z_n \rangle$ with $x : S$, then at most one of the z_i occurs free in P';
4. if $P \in \mathcal{P}^{\tilde{p}}$ and $P \xrightarrow{\alpha} P'$ where α is an input via an S_i-name, then there is $\tilde{z} = (z_1, \ldots, z_n) \in \tilde{p}$ such that the subject of α is z_i and $P' \in \mathcal{P}^{\tilde{p} - \{\tilde{z}\}}$.

Further, $\mathcal{P}$ is $S, \tilde{S}$-*confluent* if it is $S, \tilde{S}$-sensitive and whenever $P \in \mathcal{P}^{\tilde{p}}$, $P \xrightarrow{\alpha} P_1$ and $P \xrightarrow{\beta} P_2$, then unless for some $(z_1, \ldots, z_n) \in \tilde{p}$, α and β are inputs via distinct z_i and z_j, $P_1 \xrightarrow{\widehat{\beta \mid \alpha}} P'_1$ and $P_2 \xrightarrow{\widehat{\alpha \mid \beta}} P'_2 \approx P'_1$. □

We then have:

Theorem 12. Suppose $P \equiv (\nu \tilde{z})(P_1 \mid \ldots \mid P_n)$ and $\mathcal{P} = \{Q \mid Q \text{ is a derivative of a } P_i\}$ is S-closed and $S, \tilde{S}$-confluent with partition $\{\mathcal{P}^{\tilde{p}}\}_{\tilde{p}}$. Suppose each $P_i \in \mathcal{P}^{\emptyset}$ and is o-determinate. Suppose that for each derivative $P' \equiv (\nu \tilde{z}')(P'_1 \mid \ldots \mid P'_n)$ of P, no name occurs free in more than two components of P', and a free name of P' occurs in exactly one component of P'. Then P is confluent. □

In closing this section we mention that related results of a synthetic nature can also be obtained for useful varieties of 'partial confluence' as described in the Introduction, and that static type systems as in for instance the papers cited earlier complement them effectively.

5 An application

The aim of this section is to illustrate the utility of some of the theory presented via an analysis of a distributed algorithm. It is a variant of the Propagation of Information with Feedback protocol of [21] studied in [24]. Consider a network of m processes connected by communication links, where the graph having the processes as nodes and the links as edges is connected. Each process stores an integer, its *value*. A distinguished process, the *root*, conducts the interaction between the network and its environment. The intended behaviour of the algorithm is that on receiving a request from the environment, the root should emit to it the *value* of the network, i.e. the sum of the values of the m processes. We proceed to give and explain the process-calculus description of the algorithm.

We use the following sorts: E, T, D, I, O. The sorting λ is as follows: $\lambda(\mathsf{E}) = (\mathsf{T}, \mathsf{D})$, $\lambda(\mathsf{T}) = (\text{int})$, $\lambda(\mathsf{D}) = ()$, $\lambda(\mathsf{I}) = (\mathsf{O})$, $\lambda(\mathsf{O}) = (\text{int})$. Here int is the type of integers; we allow simple arithmetic expressions in the descriptions – the foregoing theory extends easily to accommodate this. It is intended that each process passes from its initial quiescent state through some active states to a final inactive state. The behaviour of a non-root process is described as follows, where Q represents the quiescent state, A the active states, I the inactive state, and ε is the empty tuple.

$$Q(\widetilde{e}, v) \stackrel{\text{def}}{=} \Sigma_{e \in \widetilde{e}}\, e(t, d).\, A\langle t, \widetilde{e} - e, \widetilde{e} - e, \varepsilon, \varepsilon, v\rangle$$
$$A(t, \varepsilon, \varepsilon, \varepsilon, \varepsilon, v) \stackrel{\text{def}}{=} \overline{t}\langle v\rangle.\, I$$
$$I \stackrel{\text{def}}{=} 0$$
$$A\langle t, \widetilde{s}, \widetilde{r}, \widetilde{d}, \widetilde{p}, v\rangle \stackrel{\text{def}}{=} \Sigma_{e \in \widetilde{s}}\, (\nu t' d)\, \overline{e}\langle t', d\rangle.\, A\langle t, \widetilde{s} - e, \widetilde{r}, \widetilde{d}, \widetilde{p}\langle t', d\rangle, v\rangle$$
$$+ \Sigma_{e \in \widetilde{r}}\, e(t', d).\, A\langle t, \widetilde{s}, \widetilde{r} - e, \widetilde{dd}, \widetilde{p}, v\rangle$$
$$+ \Sigma_{d \in \widetilde{d}}\, \overline{d}.\, A\langle t, \widetilde{s}, \widetilde{r}, \widetilde{d} - d, \widetilde{p}, v\rangle$$
$$+ \Sigma_{\langle t', d\rangle \in \widetilde{p}}\, (t'(v').\, A\langle t, \widetilde{s}, \widetilde{r}, \widetilde{d}, \widetilde{p} - \langle t', d\rangle, v + v'\rangle$$
$$+ d.\, A\langle t, \widetilde{s}, \widetilde{r}, \widetilde{d}, \widetilde{p} - \langle t', d\rangle, v\rangle).$$

In $Q\langle \widetilde{e}, v\rangle$, v is the value of the process and the names $\widetilde{e}$ of sort E represent the edges incident on it in the network. In the quiescent state the agent may receive via any such name a pair of names, t of sort T and d of sort D. It discards d and undertakes to send an integer along t which it does when it has all but completed its activity (second and third clauses). That activity is described in the fourth clause: $A\langle t, \widetilde{s}, \widetilde{r}, \widetilde{d}, \widetilde{p}, v\rangle$ represents the state in which the process is storing v, has yet to send data along each E-name in $\widetilde{s}$, has yet to receive data along each E-name in $\widetilde{r}$, has yet to send a signal along each D-name in $\widetilde{d}$, and for each T-name, D-name pair in $\widetilde{p}$, has yet to receive either an integer along the T-name or a signal along the D-name.

The behaviour of the root is given as follows:

$$Q_0(\text{in}, \widetilde{e}, v) \stackrel{\text{def}}{=} \text{in}(\text{out}).\, A_0\langle \text{out}, \widetilde{e} - e, \widetilde{e} - e, \varepsilon, \varepsilon, v\rangle$$
$$A_0(\text{out}, \varepsilon, \varepsilon, \varepsilon, \varepsilon, v) \stackrel{\text{def}}{=} \overline{\text{out}}\langle v\rangle.\, I_0$$
$$I_0 \stackrel{\text{def}}{=} 0$$
$$A_0\langle \text{out}, \widetilde{s}, \widetilde{r}, \widetilde{d}, \widetilde{p}, v\rangle \stackrel{\text{def}}{=} \ldots$$

where the fourth clause is as for A but with 'A_0' in place of 'A' and 'out' in place of 't'. Thus the root behaves similarly to the other nodes except that it is activated by receiving along the name in of sort I a name of sort O via which it undertakes to send the network's value. The network is represented by

$$P_0 \stackrel{\text{def}}{=} (\nu \widetilde{e})(Q_0\langle \text{in}, \widetilde{e}_0, v_0\rangle \mid \Pi_{1 \leq i < m}\, Q\langle \widetilde{e}_i, v_i\rangle)$$

where $\widetilde{e}$ are the E-names representing all the edges and for each i, $\widetilde{e}_i$ those incident on the ith process. We will prove the following correctness result:

Theorem 13. $P_0 \approx \text{in(out)}.\overline{\text{out}}\langle v \rangle.\mathbf{0}$, where $v = \Sigma_{i=0}^{m-1} v_i$.

The algorithm may be thought of as consisting of two phases. In the first a spanning tree for the network is established, and in the second each non-root process passes to its parent the sum of the values stored in its descendants, and the root then emits to the environment the network's value. The sending by A_0 or A along a name e of a pair t', d of fresh names is an invitation to the receiver either to become a child of the sender and to undertake to send it an integer along t', or, if the receiver is already active (and so has a parent), to decline to do so by sending a signal via d. A process sends an integer to its parent only when it has determined the sum of the values of its descendants.

First we give a characterization of derivatives of P_0. For $S \in \mathsf{S}$, in an agent of the form $(\nu \widetilde{z}) \Pi_i Z_i$, we say *there is an S-path between* components Z' and Z'' if there are S-names $x_1, \ldots, x_p$ such that $x_i \in \mathsf{fn}(Z_i, Z_{i+1})$ for each i, $Z' \equiv Z_1$ and $Z'' \equiv Z_{p+1}$.

Lemma 14. If $P_0 \xrightarrow{w} P$ where $w \in Act^*$ then $P \equiv (\nu \widetilde{etd})(R \mid \Pi_{1 \leq i < m} N_i)$ where: (a) $\mathsf{fn}(P)$ is $\{\mathsf{in}\}$, $\{\mathsf{out}\}$ or $\emptyset$, and in and out may occur only in R (the derivative of the root Q_0); (b) no name occurs free in more than two components of P; (c) if a T-name occurs free in a component of P, there is a unique T-path between that component and R; (d) the sum of the integers stored in the components which are quiescent or active is the network's value. □

Some useful notation: $P_1 \equiv (\nu \widetilde{e})(A_0 \langle \mathsf{out}, \widetilde{e}_0, \widetilde{e}_0, \varepsilon, \varepsilon, v_0 \rangle \mid \Pi_{1 \leq i < m} Q \langle \widetilde{e}_i, v_i \rangle)$, $P_\psi \equiv (\nu \widetilde{e})(A_0 \langle \mathsf{out}, \varepsilon, \varepsilon, \varepsilon, \varepsilon, v \rangle \mid \Pi_{1 \leq i < m} I)$, and $P_\omega \equiv (\nu \widetilde{e})(I_0 \mid \Pi_{1 \leq i < m} I)$. We will later show that P_ψ and P_ω are derivatives of P_0. We use P to range over derivatives of P_0. Key in proving the theorem will be the agents of the form

$$Q'\langle e, \widetilde{e}, v \rangle \stackrel{\text{def}}{=} e(t,d).A\langle t, \widetilde{e} - e, \widetilde{e} - e, \varepsilon, \varepsilon, v \rangle$$

where $e \in \widetilde{e}$. Q' is similar to Q except that it may be activated only by an interaction along the specific name e. Note that, where $\sim$ is strong bisimilarity,

$$Q\langle \widetilde{e}, v \rangle \sim \Sigma_{e \in \widetilde{e}} Q'\langle e, \widetilde{e}, v \rangle. \tag{1}$$

Let $\mathcal{T}$ be the set of agents of the form

$$T_0 \stackrel{\text{def}}{=} (\nu \widetilde{e})(Q_0 \langle \mathsf{in}, \widetilde{e}_0, v_0 \rangle \mid \Pi_{1 \leq i < m} Q'\langle e_i, \widetilde{e}_i, v_i \rangle)$$

where $e_1, \ldots, e_{m-1}$ represent a spanning tree of the graph, with $e_i \in \widetilde{e}_i$ for each i. Note that such a T_0 differs from P_0 just in having Q' where P_0 has Q: the edge via which each non-root node will receive its first communication is determined; intuitively, T_0 represents the fragment of the behaviour of P_0 in which the spanning tree is given by those edges. Let $T_0 \in \mathcal{T}$. Directly from (1) and Lemma 14 we have:

Corollary 15. If $T_0 \xrightarrow{w} T$ where $w \in Act^*$ then $T \equiv (\nu \widetilde{etd})(R \mid \Pi_{1 \leq i < m} N_i')$ where (a)–(d) as in Lemma 14 (with 'T' for 'P') hold. □

Some useful notation: $T_1 \equiv (\nu\widetilde{e})(A_0\langle\text{out}, \widetilde{e}_0, \widetilde{e}_0, \varepsilon, \varepsilon, v_0\rangle \mid \Pi_{1 \leq i < m} Q'\langle e_i, \widetilde{e}_i, v_i\rangle)$, $T_\psi \equiv (\nu\widetilde{e})(A_0\langle\text{out}, \varepsilon, \varepsilon, \varepsilon, \varepsilon, v\rangle \mid \Pi_{1 \leq i < m} I)$ and $T_\omega \equiv (\nu\widetilde{e})(I_0 \mid \Pi_{1 \leq i < m} I)$. We use T to range over derivatives T_0. We analyse T_0, noting first that it has a specific behaviour:

Lemma 16. $T_0 \xrightarrow{\text{in}\langle\text{out}\rangle} T_1 \Longrightarrow T_\psi \xrightarrow{\overline{\text{out}}\langle v\rangle} T_\omega$. □

We now have the key observation whose proof appeals Theorem 12.

Lemma 17. T_0 is confluent. □

¿From these two results we have:

Corollary 18. $T_0 \approx \text{in}(\text{out}).\overline{\text{out}}\langle v\rangle.\mathbf{0}$. □

Having used confluence to analyse the behaviour of T_0 we now relate it to that of P_0. We say P and T are *similar* if they differ only in that where P has a quiescent component Q, T has a quiescent component Q'.

Lemma 19. $\{\langle T, P\rangle \mid P \text{ and } T \text{ are similar}\}$ is a strong simulation. □

By Lemma 16 and 19 we have that $P_0 \xrightarrow{\text{in}\langle\text{out}\rangle} P_1 \Longrightarrow P_\psi \xrightarrow{\overline{\text{out}}\langle v\rangle} P_\omega$. We say that T_0 is *compatible* with a computation $P_0 \xrightarrow{\alpha_1} P_1 \xrightarrow{\alpha_2} \ldots \xrightarrow{\alpha_r} P_r$ if for each i, if α_i is τ and arises from complementary actions $(\nu t', d)\overline{e}\langle t', d\rangle$, $e\langle t', d\rangle$ where the second is performed by a quiescent component $Q\langle\widetilde{e}_j, v_j\rangle$, then in T_0 that component is $Q'\langle e, \widetilde{e}_j, v_j\rangle$; i.e. the E-names used to activate components in the computation are those via which the Q'-components of T_0 may be activated.

Lemma 20. If $P_0 \xrightarrow{w} P$ then for any T_0 compatible with the computation, $T_0 \xrightarrow{w} T$ with P and T similar. □

We can now prove the theorem. Since $P_0 \sim \text{in}(\text{out}).P_1$ it suffices to show that $P_1 \approx \overline{\text{out}}\langle v\rangle.\mathbf{0}$. We have seen that $P_1 \xrightarrow{\overline{\text{out}}\langle v\rangle} P_\omega \sim \mathbf{0}$. Choose one such computation and, by Lemma 14, choose T_0 compatible with it. Then not $(P_1 \xrightarrow{\alpha})$ with $\alpha \neq \overline{\text{out}}\langle v\rangle$ as otherwise by Lemma 20, $(T_1 \xrightarrow{\alpha})$, contradicting Lemma 18. Finally, and for the same reason, not $(P_1 \Longrightarrow P_1' \not\rightarrow)$. □

We conclude by briefly comparing this analysis with that in [24]. The latter uses a static I/O-automaton model [5] of the algorithm and establishes that the fair traces of the automaton representing it are included in those of an automaton akin to the agent $\text{in}(\text{out}).\overline{\text{out}}\langle v\rangle.\mathbf{0}$. In our view name-passing and careful use of sorts allow a very direct and perspicuous description of the algorithm's behaviour: the construction and use of the spanning tree are manifest in the description. Moreover the use of reasoning techniques involving name-passing aids the analysis, and the proof illustrates the idea that when studying the behaviour of a confluent system it may suffice to examine in detail only a (small) part of it. Finally, here the correctness criterion is bisimilarity, rather than inclusion of fair traces.

References

1. J. F. Groote and M. Sellink. Confluence for process verification. In *Proceedings of CONCUR'95*, pages 204–218. Springer, 1995.
2. K. Honda. Types for dyadic interaction. In *CONCUR'93*, pages 509–523. Springer, 1993.
3. N. Kobayashi, B. Pierce, and D. Turner. Linearity and the pi-calculus. *Principles of Programming Languages*, 1996.
4. X. Liu and D. Walker. Confluence of processes and systems of objects. In *Proceedings of TAPSOFT'95*, pages 217–231. Springer, 1995.
5. N. Lynch, M. Merritt, W. Weihl, and A. Fekete. *Atomic Transactions*. Morgan Kaufmann, 1994.
6. R. Milner. *A Calculus of Communicating Systems*. Springer, 1980.
7. R. Milner. *Communication and Concurrency*. Prentice-Hall, 1989.
8. R. Milner. The polyadic π-calculus: a tutorial. In *Logic and Algebra of Specification*. Springer, 1992.
9. R. Milner, J. Parrow, and D. Walker. A calculus of mobile processes, parts 1 and 2. *Information and Computation*, 100:1–77, 1992.
10. R. Milner, J. Parrow, and D. Walker. Modal logics for mobile processes. *Theoretical Computer Science*, 114:149–171, 1993.
11. U. Nestmann. *On determinacy and nondeterminacy in concurrent programming*. PhD thesis, University of Erlangen, 1996.
12. A. Philippou. *Reasoning about systems with evolving structure*. PhD thesis, University of Warwick, 1996.
13. A. Philippou and D. Walker. On sharing and determinacy in concurrent systems. In *Proceedings of CONCUR'95*, pages 456–470. Springer, 1995.
14. A. Philippou and D. Walker. On transformations of concurrent object programs. In *Proceedings of CONCUR'96*, pages 131–146. Springer, 1996.
15. B. Pierce and D. Sangiorgi. Typing and subtyping for mobile processes. In *Proceedings of LICS'93*, pages 376–385. Computer Society Press, 1993.
16. B. Pierce and D. Sangiorgi. Behavioral equivalence in the polymorphic pi-calculus. In *Proceedings of POPL'97*, to appear.
17. B. Pierce and D. Turner. *Pict language definition*, 1996.
18. M. Sanderson. *Proof techniques for CCS*. PhD thesis, University of Edinburgh, 1982.
19. D. Sangiorgi. *Expressing mobility in process algebras: first-order and higher-order paradigms*. PhD thesis, University of Edinburgh, 1992.
20. D. Sangiorgi. The name discipline of receptiveness. Technical report, INRIA, to appear.
21. A. Segall. Distributed network protocols. *IEEE Transactions on Information Theory*, IT-29(2):319–340, 1983.
22. C. Tofts. *Proof methods and pragmatics for parallel programming*. PhD thesis, University of Edinburgh, 1990.
23. D. Turner. *The polymorphic pi-calculus: theory and implementation*. PhD thesis, University of Edinburgh, 1996.
24. F. Vaandrager. Verification of a distributed summation algorithm. In *Proceedings of CONCUR'95*, pages 190–203. Springer, 1995.
25. N. Yoshida. Graph types for monadic mobile processes. In *Proceedings of FST/TCS'96*, pages 371–386. Springer, 1996.

A Proof Theoretical Approach to Communication*

Yuxi Fu

Department of Computer Science
Shanghai Jiao Tong University
1954 Hua Shan Road, Shanghai 200030, China

Abstract. The paper investigates a concurrent computation model, chi calculus, in which communications resemble cut eliminations for classical proofs. The algebraic properties of the model are studied. Its relationship to sequential computation is illustrated by showing that it incorporates the operational semantics of the call-by-name lambda calculus. Practically the model has pi calculus as a submodel.

1 Communication as Cut Elimination

Concurrent computation is currently an open-ended issue. The situation is in contrast with sequential computation whose operational semantics is formalized by, among others, the λ-calculus ([2]). In retrospect, the λ-calculus can be seen as a fallout of proof theory. Curry-Howard's proposition-as-type principle allows one to code up constructive proofs as typed terms. At the core of the constructive logic is the minimal logic, whose type theoretical formulation gives rise to, roughly, the simply typed λ-calculus. Now the untyped λ-calculus is obtained from the simply typed λ-calculus by removing all the typing information.

In recent years, classical proofs have been investigated in a computational setting. Girard proposed proof nets ([4]) as term representations of classical linear proofs. These classical terms are typed. The conclusion of a proof derivation is the type of the proof net corresponding to that proof derivation. The computations of these terms are cut eliminations modeled by rewritings of graphs. As the terms are typed, cuts happen between nodes of correlated types. Abramsky's proof-as-process interpretation ([1, 3]) relates proof nets to processes. At operational level, this interpretation is supported by a cut-elimination-as-communication paradigm. It looks like a type-erasing interpretation similar to the one found in a constructive world.

This paper investigates a concurrent computation model obtained by reversing the roles of proofs and processes in Abramsky's paradigm. That is to say that we regard communications as cut eliminations. The way to arrive at such a model of communication echoes that in the sequential world. First we take the multiplicative linear logic as the 'minimal logic' in a classical framework. There is nothing canonical about this choice. As the typed classical terms we take the

---

* Supported by NNSF of China, grant number 69503006.

proof nets. The following left diagram is a proof net:

The first step towards the model is to abstract away the logical aspect of proof nets but keep its proof theoretical content. The above proof net becomes the right diagram in the above. There are two kinds of edge in the net. So the second step is to transform the net into a graph with only directed arrows:

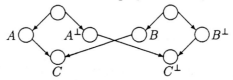

We then forget about the typing information while recording positive and negative information by labels on arrows, arriving at an untyped graph (left below).

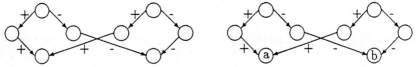

This is the untyped version of the original classical typed term. Notice that there are two kinds of node in the proof net: the internal nodes and the conclusion nodes. In order to distinguish them in the untyped graph, we label the conclusion nodes with small letters (above right). We call graphs of this kind reaction graphs. In a reaction graph, a node without (with) a label is called *local* (*global*). Reaction graphs can be seen as the underlying graphs of proof derivations in a generalized and distilled form. Computations with reaction graphs are cut eliminations. Here is an example of two consecutive cut-eliminations:

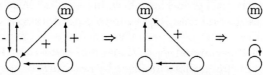

In the left graph, the two upper nodes show up opposite polarities to the left bottom node. This cut is eliminated in the first reduction. The two arrows are removed and the two upper nodes are coerced with the resulting node labeled by m. In the middle graph, the two bottom nodes with the arrows pointing to the node labeled m form a cut. The second reduction eliminates the cut. The idea of this paper is to think of these cut-eliminations as communications. To develop the idea, we need a process-like notation for reaction graphs. Let us define graph terms by abstract syntax as follows: $G := \mathbf{0} \mid m[x] \mid \overline{m}[x] \mid (x)G \mid G|G'$. Here $\mathbf{0}$ is the empty reaction graph; $m[x]$ and $\overline{m}[x]$ are respectively the following graphs:

$$(x) \xrightarrow{+} (m) \qquad (x) \xrightarrow{-} (m)$$

$(x)G$ is obtained from G by removing the label x from G; $G|G'$ is the amalgamation of G and G', coercing nodes with same labels. The two consecutive cut-eliminations in the above can now be described by the following reductions:

$$(x)(y)(z)(m[x]|\overline{y}[x]|y[m]|\overline{y}[z]|\overline{z}[y]) \to (x)(y)(m[x]|\overline{y}[x]|\overline{m}[y]) \to (x)(\overline{x}[x]).$$

This term representation gives rise to a calculus of reaction graphs.

The calculus of graphs only deals with finite computations. To achieve Turing computability, we extend the language with standard process combinators. The resulting language will be referred to as χ-calculus, where χ stands for ex*ch*ange of *in*formation. The paper initiates a study of this computation model.

2 A Model for Concurrent Computation

Let $\mathcal{N}$ be a set of names ranged over by lower case letters and $\overline{\mathcal{N}} \stackrel{\text{def}}{=} \{\overline{a} \mid a \in \mathcal{N}\}$ be the set of conames. The union $\mathcal{N} \cup \overline{\mathcal{N}}$ will be ranged over by α. Define $\overline{\overline{\alpha}}$ to be m ($\overline{m}$) whenever α is $\overline{m}$ (m). Let $\mathcal{T}$ be the set of χ-terms defined as follows:

$$P := 0 \mid \alpha[x].P \mid P|P' \mid (x)P \mid \alpha(x)*P.$$

Here $m[x].P$ and $\overline{m}[x].P$ are terms that must first perform a communication through name m and then enacts $P[y/x]$, where y is the name received in the communication. In $(x)P$, the (x)-part is a localization combinator. In both $(x)P$ and $\alpha(x)*P$, x is local. The set of local names appeared in P is denoted by $ln(P)$, whereas the set of global names, or non local names, in P is designated by $gn(P)$. Set $n(P)$ is the union of $ln(P)$ and $gn(P)$. We adopt the α-convention saying that a local name in a term can be replaced by a fresh name without changing its syntax.

The effect of a substitution $[y_1/x_1]\ldots[y_n/x_n]$ on a term is defined as follows: $P[y_1/x_1]\ldots[y_n/x_n] \stackrel{\text{def}}{=} (\ldots P[y_1/x_1]\ldots)[y_n/x_n]$. Substitutions will be ranged over by σ.

For simplicity, a structural congruence is imposed on the members of $\mathcal{T}$.

Definition 1. The relation $=$ is the least congruence on χ-terms that contains:
(i) $P|0 = P$, $P_1|P_2 = P_2|P_1$, and $P_1|(P_2|P_3) = (P_1|P_2)|P_3$;
(ii) $(x)0 = 0$, $(x)(y)P = (y)(x)P$, and $(x)(P|Q) = P|(x)Q$ if $x \notin gn(P)$;
(iii) $P = Q$ if P and Q are α-convertible.

We regard $=$ as a grammatic equality. So $P = Q$ means that P and Q are syntactically the same. The operational semantics of the language can be defined in terms of a labeled transition system. We prefer however a reductional semantics for χ-calculus in the style of [5]:

$$(x)(R|\alpha[x].P|\overline{\alpha}[y].Q) \to (x)(R[y/x]|P[y/x]|Q[y/x])$$

$$\alpha(x)*P|\overline{\alpha}[y].Q \to \alpha(x)*P|P[y/x]|Q$$

$$\frac{P \to P'}{P|Q \to P'|Q} \qquad \frac{P \to P'}{(x)P \to (x)P'}.$$

To help understand the communication rules, we now give some examples, assuming x and y are distinct:

$$(x)(R|\overline{m}[y].P|m[x].Q) \to R[y/x]|P[y/x]|Q[y/x]$$
$$\overline{m}[y].P|(x)(R|m[x].Q) \to P|R[y/x]|Q[y/x]$$
$$(y)(\overline{m}[y].P|(x)(R|m[x].Q)) \to (y)(P|R[y/x]|Q[y/x])$$
$$(x)\overline{m}[x].P|(y)m[y].Q \to (z)(P[z/x]|Q[z/y]), \text{ where } z \text{ is fresh}$$
$$(x)(\overline{m}[x].P|m[x].Q) \to (x)(P|Q).$$

It is clear from these examples that the localization operator in χ-calculus acts as an effect delimiter. A communication either instantiates a local name by a global name or identifies two local names.

Let $\to^+$ ($\to^*$) be the (reflexive and) transitive closure of $\to$. We will denote by $\mathbf{x}$ a sequence $x_1, \ldots, x_n$ of names. We will also abbreviate $(x_1)\ldots(x_n)P$ to $(\mathbf{x})P$. When the length of the sequence $\mathbf{x}$ is zero, $(\mathbf{x})P$ is just P.

3   Algebraic Properties

To study the algebraic semantics of χ-terms, a labeled transition system is defined as follows, where δ ranges over $\{\stackrel{\alpha x}{\to}, \stackrel{\alpha[x]}{\to}, \stackrel{\alpha(x)}{\to} | \alpha \in \mathcal{N} \cup \overline{\mathcal{N}}, x \in \mathcal{N}\}$:

$$\overline{(y)(R|\alpha[y].P) \stackrel{\alpha x}{\to} (R|P)[x/y]} \quad \overline{\alpha(y)*P \stackrel{\alpha x}{\to} \alpha(y)*P|P[x/y]} \quad \overline{\alpha[x].P \stackrel{\alpha[x]}{\to} P}$$

$$\frac{P \stackrel{\alpha[x]}{\to} P'}{(x)P \stackrel{\alpha(x)}{\to} P'} \quad \frac{P \stackrel{\delta}{\to} P' \;\; ln(\delta) \cap gn(Q) = \emptyset}{P|Q \stackrel{\delta}{\to} P'|Q} \quad \frac{P \stackrel{\delta}{\to} P' \;\; x \notin n(\delta)}{(x)P \stackrel{\delta}{\to} (x)P'}.$$

In the rules, $ln(\delta)$ is $\{x\}$ when δ is $\alpha(x)$; it is the empty set otherwise. $n(\delta)$ is the set of names in δ. Let $\stackrel{\delta}{\Rightarrow}$ denote relation $\to^* \stackrel{\delta}{\to} \to^*$.

A bisimulation equivalence for χ-terms should take into account the distinguished feature of the localization operators of the language. The equivalence we introduce in this section is based upon the old idea that two terms are considered observationally equivalent if and only if placing them in a same context results in two observationally equivalent terms. Working explicitly with contexts is unnecessary in our setting due to the presence of the structural equality $=$.

Definition 2. Suppose $\mathcal{R} \subseteq \mathcal{T} \times \mathcal{T}$. The relation $\mathcal{R}$ is a local simulation if whenever $P\mathcal{R}Q$ then for any term R and any sequence $\mathbf{x}$ of names it holds that
(i) if $(\mathbf{x})(P|R) \to P'$ then Q' exists such that $(\mathbf{x})(Q|R) \to^* Q'$ and $P'\mathcal{R}Q'$;
(ii) if $(\mathbf{x})(P|R) \stackrel{\delta}{\to} P'$ then Q' exists such that $(\mathbf{x})(Q|R) \stackrel{\delta}{\Rightarrow} Q'$ and $P'\mathcal{R}Q'$.
The relation $\mathcal{R}$ is a local bisimulation if both $\mathcal{R}$ and its inverse are local simulations. The local bisimilarity $\approx$ is the largest local bisimulation.

As usual, local bisimulation up to $\approx$ is a useful tool for proving two χ-terms being locally bisimilar. We omit the standard definition.

In the rest of this section, we prove that $\approx$ is a congruence relation. The fact that $\approx$ is closed under parasition and localization combinators can be proved already at this point.

Proposition 3. *If $P \approx Q$ then (i) $P|O \approx Q|O$ and (ii) $(x)P \approx (x)Q$.*

The next lemma is crucial in showing that $\approx$ is a congruence relation. It is the first indication that local bisimilarity is algebraically appropriate. The property is not enjoyed by local bisimilarity for π-processes.

Lemma 4. *If $P \approx Q$ then $P\sigma \approx Q\sigma$ for an arbitrary substitution σ.*

Proof. Let $\mathcal{R}$ be the union of $\approx$ and the following

$$\left\{ ((z)(P\sigma|R), (z)(Q\sigma|R)) \;\middle|\; \begin{array}{l} P \approx Q,\; R \in \mathcal{T},\; \mathbf{z} \text{ a sequence of names,} \\ \sigma \text{ a substitution } [y_1/x_1]\ldots[y_n/x_n] \text{ such} \\ \text{that } x_1,\ldots,x_n \text{ are pairwise distinct} \end{array} \right\}.$$

Suppose $(z)(P\sigma|R)\mathcal{R}(z)(Q\sigma|R)$ and $(z)(P\sigma|R) \xrightarrow{\delta} P'$, where σ is the substitution $[y_1/x_1]\ldots[y_n/x_n]$ with $x_1,\ldots,x_n$ being pairwise distinct. Let a and b be fresh names. Then for the sequence $\mathbf{z}$ of names

$$(z)((x)(a)(b)(b[b].P|a[x_1].\ldots.a[x_n]|\overline{a}[y_1].\ldots.\overline{a}[y_n].\overline{b}[b])|R) \to^* (z)(P\sigma|R)$$
$$\xrightarrow{\delta} P'.$$

As $b \notin gn(P,Q)$, $b[b].P \approx b[b].Q$ follows easily. By Proposition 3,

$$(x)(a)(b)(b[b].P|a[x_1].\ldots.a[x_n]|\overline{a}[y_1].\ldots.\overline{a}[y_n].\overline{b}[b])$$
$$\approx (x)(a)(b)(b[b].Q|a[x_1].\ldots.a[x_n]|\overline{a}[y_1].\ldots.\overline{a}[y_n].\overline{b}[b]).$$

So by definition, there exists some Q' such that $P' \approx Q'$ and

$$(z)((x)(a)(b)(b[b].Q|a[x_1].\ldots.a[x_n]|\overline{a}[y_1].\ldots.\overline{a}[y_n].\overline{b}[b])|R) \xrightarrow{\delta} Q'.$$

During the above reduction every $a[x_i]$ must have reacted upon $\overline{a}[y_i]$, for $1 \leq i \leq n$, and $b[b]$ upon $\overline{b}[b]$. It can be easily proved that all the communications through a and that through b can happen in the very beginning. That is

$$(z)((x)(a)(b[b].Q|a[x_1].\ldots.a[x_n]|\overline{a}[y_1].\ldots.\overline{a}[y_n].\overline{b}[b])|R) \to^* (z)(Q\sigma|R)$$
$$\xrightarrow{\delta} Q'.$$

So $(z)(P\sigma|R) \xrightarrow{\delta} P'$ is matched by $(z)(Q\sigma|R) \xrightarrow{\delta} Q'$. The case when $(z)(P\sigma|R) \to P'$ is similar. So $\mathcal{R}$ is a local bisimulation. It follows that $P \approx Q$ implies $P[y/x] \approx Q[y/x]$. Therefore $P \approx Q$ implies $P\sigma \approx Q\sigma$ for a substitution σ. $\square$

We now come to the main result of the section.

Theorem 5. $\approx$ *is a congruence equivalence: if $P \approx Q$ and $O \in \mathcal{T}$ then*
(i) $\alpha[x].P \approx \alpha[x].Q$;   (ii) $P|O \approx Q|O$;
(iii) $(x)P \approx (x)Q$;   (iv) $\alpha(x)*P \approx \alpha(x)*Q$.

Proof. We sketch the proof of (iv). The proof of (i) is simpler. Let $\mathcal{R}$ be

$$\{((\mathbf{x})(m(y)*P|R), (\mathbf{x})(m(y)*Q|R)) \mid P \approx Q,\ R \in \mathcal{T},\ m, \mathbf{x}\ \text{names}\}.$$

Suppose $(\mathbf{x})(m(y)*P|R) \rightarrow P'$ and that $(\mathbf{x})(m(y)*P|R) \rightarrow P'$ is caused by a communication between $m(y)*P$ and R. Then P' is $(\mathbf{x})(m(y)*P|P[a/y]|R')$. Similarly $(\mathbf{x})(m(y)*Q|R) \rightarrow (\mathbf{x})(m(y)*Q|Q[a/y]|R')$. By Lemma 4, $P[a/y] \approx Q[a/y]$. By Proposition 3, $(\mathbf{x})(m(y)*Q|P[a/y]|R') \approx (\mathbf{x})(m(y)*Q|Q[a/y]|R')$. It is then easy to see that $\mathcal{R}$ is a local bisimulation up to $\approx$. □

4 π-Processes as χ-Terms

A question naturally arises as to the relationship between π-calculus and χ-calculus. We give a first answer in this section. Let $\mathcal{P}$ be the set of π-processes defined as follows: $P := \mathbf{0} \mid m(x).P \mid \overline{m}x.P \mid P|P' \mid (x)P \mid m(x)*P$. We refer the reader to [6] for background material on π-calculus.

There are many bisimulation equivalences on π-processes. What is most relevant in this section is the open bisimilarity defined in [8]. Actually we will use a version of open bisimilarity stronger than Sangiorgi's.

Definition 6. Let $\mathcal{R}$ be a binary relation on the set of π-processes. The relation $\mathcal{R}$ is an open bisimulation if whenever $P\mathcal{R}Q$ then for any π-process R, any sequence $\mathbf{x}$ of names and any substitution σ it holds that
(i) if $(\mathbf{x})(P\sigma|R) \overset{\beta}{\Rightarrow} P'$ then Q' exists such that $(\mathbf{x})(Q\sigma|R) \overset{\beta}{\Rightarrow} Q'$ and $P'\mathcal{R}Q'$;
(ii) if $(\mathbf{x})(Q\sigma|R) \overset{\beta}{\Rightarrow} Q'$ then P' exists such that $(\mathbf{x})(P\sigma|R) \overset{\beta}{\Rightarrow} P'$ and $P'\mathcal{R}Q'$.
The open bisimilarity $\approx^o$ is the largest open bisimulation.

$\approx^o$ is a congruence equivalence and is closed under substitution.

A structural translation from π to χ has as nontrivial clauses the following:

$$(m(x).P)^o \overset{\text{def}}{=} (x)m[x].P^o,$$
$$(\overline{m}x.P)^o \overset{\text{def}}{=} \overline{m}[x].P^o.$$

Imposing on $\mathcal{P}$ a same structural congruence as given in Definition 1, one has

Theorem 7. *For $P, Q \in \mathcal{P}$, it holds that*
(i) $P \rightarrow Q$ iff $P^o \rightarrow Q^o$;   (ii) $P \overset{mx}{\rightarrow} Q$ iff $P^o \overset{mx}{\rightarrow} Q^o$;
(iii) $P \overset{\overline{m}x}{\rightarrow} Q$ iff $P^o \overset{\overline{m}[x]}{\rightarrow} Q^o$;   (iv) $P \overset{\overline{m}(x)}{\rightarrow} Q$ iff $P^o \overset{\overline{m}(x)}{\rightarrow} Q^o$.

Theorem 8. *For $P, Q \in \mathcal{P}$, $P \approx^o Q$ iff $P^o \approx Q^o$.*

5 Call-by-Name in χ-Calculus

A concurrent computation model has to answer the question of whether it captures sequential computation successfully. The issue is often addressed by relating variants of λ-calculus to the model. Our focus in this section is on the call-by-name λ-calculus ([7]), whose semantics is defined by the following rules:

$$\frac{}{(\lambda x.M)N \to M[N/x]} \quad \frac{M \to M'}{MN \to M'N} \quad \frac{M \to M'}{\lambda x.M \to \lambda x.M'}.$$

The following translation, which is Milner's encoding of the lazy λ-calculus with modification, serves as an encoding of the call-by-name λ-calculus in χ-calculus:

$$[\![x]\!]u \stackrel{def}{=} \overline{x}[u]$$
$$[\![\lambda x.M]\!]u \stackrel{def}{=} (v)(x)(u[x].u[v]|[\![M]\!]v)$$
$$[\![MN]\!]u \stackrel{def}{=} (v)(x)([\![M]\!]v|\overline{v}[x].\overline{v}[u].x(w)*[\![N]\!]w).$$

The parasition of $\overline{u}[x].\overline{u}[v]$ and $[\![M]\!]v$ in $[\![\lambda x.M]\!]u$ allows $[\![M]\!]v$ to evolve independently, thus modeling reduction under λ-abstraction. The encoding preserves the operational semantics of the call-by-name λ-calculus in the sense the operational semantics of the lazy λ-calculus is preserved by Milner's encoding ([5]). A formal treatment is omitted in this extended abstract.

The call-by-name λ-calculus is one example which can not be treated successfully in π-calculus.

6 Towards an Integration of χ and λ

There are two problems one encounters when trying to simulate the operational semantics of the full λ-calculus. The first is how to model reduction under λ-abstraction. The second is how to model reduction $MN \to MN'$ caused by $N \to N'$. The former is to do with parallel computation. There is no reason why it should pose any problem for concurrent computation. This view is supported by the result in Sect. 5. The latter is to do with recursion because the λ-term N may be duplicated in future reduction. In any *structural* interpretation, this N must be translated into the body of a replicator or guarded recursion. So if the N induces an infinite reduction, the interpretation of MN would have no terminating reduction sequences. It is our view that the second problem is orthogonal to concurrent computation. It is caused essentially by the incompatibility of the two recursion mechanisms.

In this section we take a look at a higher order calculus combining the communication mechanism of the χ-calculus and the recursion mechanism of the λ-calculus. The purpose of this investigation is to see if the two mechanisms fit coherently and if local bisimulation suffices as a tool for studying the algebraic properties of the language.

6.1 χ with Call-by-Name λ

Let the set $\mathcal{H}$ of higher order χ-terms be defined by the following abstract syntax:

$$E := X \mid \alpha[x].E \mid E|E' \mid (x)E \mid \alpha(X)E \mid \alpha[E],$$

where X is a term variable. Let $\mathbf{0}$ abbreviate $(a)a(X)X$. The semantics of the higher order χ-calculus is defined by the relevant rules of the first order χ-calculus together with the following rules incorporating a call-by-name mechanism:

$$\frac{}{\alpha(X)E|\overline{\alpha}[F] \to E[F/X]} \qquad \frac{E \to F}{\alpha(X)E \to \alpha(X)F}$$

A structural equality is imposed on the members of $\mathcal{H}$, whose definition is the same as Definition 1. Usually a bisimulation equivalence for a higher order process calculus is defined for closed processes. This is a tractable approach. But in the presence of the second reduction rule given above, the method breaks down. A bisimulation equivalence for higher order χ-calculus has to be defined on all terms. For that purpose, let's say that a binary relation $\mathcal{R}$ on $\mathcal{H}$ is substitution closed if whenever $E\mathcal{R}F$ then $E[E_1/X_1, \ldots, E_i/X_i]\mathcal{R}F[E_1/X_1, \ldots, E_i/X_i]$ for $E_1, \ldots, E_i \in \mathcal{H}$ and $X_1, \ldots, X_i$ that are among the free variables of $E|F$.

Definition 9. A substitution closed binary relation $\mathcal{R}$ on $\mathcal{H}$ is a local bisimulation if whenever $E\mathcal{R}F$ then for any $H \in \mathcal{H}$ and $\{\mathbf{x}\} \subseteq \mathcal{N}$ it holds that
(i) if $(\mathbf{x})(E|H) \stackrel{\delta}{\Rightarrow} E'$ then F' exists such that $(\mathbf{x})(F|H) \stackrel{\delta}{\Rightarrow} F'$ and $E'\mathcal{R}F'$;
(ii) if $(\mathbf{x})(F|H) \stackrel{\delta}{\Rightarrow} F'$ then E' exists such that $(\mathbf{x})(E|H) \stackrel{\delta}{\Rightarrow} E'$ and $E'\mathcal{R}F'$.
The local bisimilarity $\approx^\omega$ is the largest local bisimulation on higher order terms.

The above definition is given in terms of a labeled transition system on $\mathcal{H}$ that is defined by the relevant rules in Sect. 3. It should be remarked that $\approx^\omega$ is by definition substitution closed.

Theorem 10. $\approx^\omega$ *is a congruence equivalence: if* $E \approx^\omega F$ *and* $G \in \mathcal{H}$ *then*
(i) $\alpha[x].E \approx^\omega \alpha[x].F;$ (ii) $E|G \approx^\omega F|G;$ (iii) $(x)E \approx^\omega (x)F;$
(iv) $\alpha(X)E \approx^\omega \alpha(X)F;$ (v) $\alpha[E] \approx^\omega \alpha[F].$

Proof. We only prove (v). For the sake of this proof, let's define $\mathcal{H}_o[X]$ to be the set of all higher order terms E such that each occurrence of X is within $\alpha[G]$ for some $\alpha \in \mathcal{N} \cup \overline{\mathcal{N}}$ and some $G \in \mathcal{H}$. Let $\mathcal{R}$ be

$$\{(E[A/X], E[B/X]) \mid A \approx^\omega B, \ E \in \mathcal{H}_o[X], \ X \text{ a variable}\}.$$

Suppose $E[A/X] \to G$. Then $G \equiv F[A]$ for some $F \in \mathcal{H}_o[X]$. It can be easily shown that some $H \in \mathcal{H}$ exists such that $E[B/X] \to H$ and $F[B/X] \approx^\omega H$. It follows that $\mathcal{R}$ is a local bisimulation up to $\approx^\omega$. Thus $\alpha[E] \approx^\omega \alpha[F]$ since $\alpha[X] \in \mathcal{H}_o[X]$. □

In the remaining of the section, we justify our claim that the higher order calculus is a combination of χ and λ.

6.2 Recursion

As a test for local bisimilarity, we examine Thomsen's recursion ([9]) in this section. Suppose that E contains free variable X and a does not occur in E. The following abbreviations will be used:

$$W_X^a(E) \stackrel{\text{def}}{=} a[a] | a(X)(\overline{a}[a].E|\overline{a}[X]),$$
$$\text{rec} X.E \stackrel{\text{def}}{=} (a)(W_X^a(E)|\overline{a}[W_X^a(E)]).$$

We remark that $\text{rec} X.E$ defined here is slightly different from Thomsen's. The idea is to make $W_X^a(E)$ inert. Before proving the main property concerning $\text{rec} X.E$, we first establish the following result.

Lemma 11. $(a)(F[W_X^a(E)/X]|\overline{a}[W_X^a(E)]) \approx^\omega (a)(b)(F'|\overline{a}[W_X^a(E)]|\overline{b}[W_X^b(E)])$, where E and F have free variable X and F' is obtained from $F[W_X^a(E)/X]$ by replacing some occurrences of $W_X^a(E)$ by $W_X^b(E)$. Here a and b are fresh.

Theorem 12. Suppose E contains free X. Then $\text{rec} X.E \approx^\omega E[\text{rec} X.E/X]$.

Proof. Suppose E and F contain free variable X, $a \notin n(E,F)$ and $gn(E) \cap ln(F) = \emptyset$. Using Lemma 11, one proves that $(a)(F[W_X^a(E)/X]|\overline{a}[W_X^a(E)]) \approx F[\text{rec} X.E/X]$. So $\text{rec} X.E \approx^\omega (a)(E[W_X^a(E)/X]|\overline{a}[W_X^a(E)]) \approx^\omega E[\text{rec} X.E/X]$, which is what we are after. □

6.3 Projecting Out Guarded Recursion

In this section we show that the higher order χ can be seen as an extension of the first order χ. A fallout of the result is a justification of the claim that the guarded recursion is completely unnecessary in the higher order χ-calculus. Let χ^+ be the higher order χ-calculus enriched with the guarded recursion. The language χ^+ can be investigated along the same line as the higher order χ has been. $\mathcal{H}_+$ and $\approx^+$ are defined accordingly. It can also be shown that $\approx^+$ is a congruence relation. The definition of a structural translation $\widehat{}$ from χ^+-terms to χ^ω-terms is nontrivial only on guarded recursion:

$$\widehat{\alpha(x)*E} \stackrel{\text{def}}{=} (a)((x)\alpha[x].(\widehat{E}|a(X)(X|\overline{a}[X]))|\overline{a}[(x)\alpha[x].(\widehat{E}|a(X)(X|\overline{a}[X]))]).$$

The translation $\widehat{}$ projects the guarded recursion out, as it were.

Theorem 13. For $P \in \mathcal{H}_+$, $P \approx^+ \widehat{P}$.

Theorem 14. (i) Suppose P and Q are in $\mathcal{H}$. Then $P \approx^+ Q$ iff $P \approx^\omega Q$.
(ii) Suppose P and Q are in $\mathcal{H}_+$. Then $P \approx^+ Q$ iff $\widehat{P} \approx^\omega \widehat{Q}$.
(iii) (a) if $P \stackrel{\delta}{\to} P'$ ($P \to P'$) then $\widehat{P} \stackrel{\delta}{\to} P''$ ($\widehat{P} \to P''$) such that $P'' \approx^\omega \widehat{P'}$;
(b) if $\widehat{P} \stackrel{\delta}{\to} P''$ ($\widehat{P} \to P''$) then $P \stackrel{\delta}{\to} P'$ ($P \to P'$) such that $P'' \approx^\omega \widehat{P'}$.

Proof. (i) Suppose P, Q are in $\mathcal{H}$. $P \approx^+ Q$ clearly implies $P \approx^\omega Q$. Suppose $P \approx^\omega Q$. Then $(\mathbf{x})\widehat{(P|R)} \equiv (\mathbf{x})(P|\widehat{R})$ and $(\mathbf{x})\widehat{(Q|R)} \equiv (\mathbf{x})(Q|\widehat{R})$, where $R \in \mathcal{H}_+$. By theorem 13, $(\mathbf{x})(P|R) \approx^+ (\mathbf{x})(P|\widehat{R})$ and $(\mathbf{x})(Q|R) \approx^+ (\mathbf{x})(Q|\widehat{R})$. It is now easy to see that $\approx^\omega$ is a local bisimulation up to $\approx^+$.
(ii) By theorem 13, $P \approx^+ Q$ iff $\widehat{P} \approx^+ \widehat{Q}$. By (i) $\widehat{P} \approx^+ \widehat{Q}$ iff $\widehat{P} \approx^\omega \widehat{Q}$. □

As χ^+ extends the first order χ, so does the higher order χ-calculus in view of Theorem 13 and Theorem 14.

6.4 Full Integration

An integration of χ with the full λ is the higher order calculus extended with

$$\frac{E \to F}{\alpha[E] \to \alpha[F]}.$$

The operational semantics of the full λ-calculus can be simulated in the fully integrated calculus. The encoding is the following:

$$[\![x]\!]_u \stackrel{\text{def}}{=} \overline{x}[u]|X$$

$$[\![\lambda x.M]\!]_u \stackrel{\text{def}}{=} (x)(v)(u[v].u[x]|x(X)[\![M]\!]_v)$$

$$[\![MN]\!]_u \stackrel{\text{def}}{=} (x)(v)([\![M]\!]_v|\overline{v}[u].\overline{v}[x]|\overline{x}[(w)(x[w]|[\![N]\!]_w)]).$$

Theorem 15. *Suppose M is a λ-term. If $M \to N$ then $[\![M]\!]_u \to^+ [\![N]\!]_u$.*

Definition 9 now gives rise to an equivalence relation on the set of all terms of the fully integrated calculus. The results in Sect. 6.2 and Sect. 6.3 also hold for this language. The (i) through (iv) of Theorem 10 also hold. But so far we haven't been able to prove the (v) of Theorem 10 for the fully integrated calculus.

7 Remark on Pragmatics

In the formulation of χ-calculus, we use the same set of names for both global and local names. But conceptually the identification is not always helpful. The standard bisimilarity ([6]) for the π-processes is not closed under input prefixing operation. This is because the variable names and the free names are regarded as semantically different in this approach. Sangiorgi's open bisimilarity is congruent. But in that approach local names are treated differently. In χ-calculus, both local and global names are variable names, which is what local bisimilarity assumes. The situation is similar to that in λ-calculus, where both free and closed variables are, well, variables that can be instantiated by any λ-terms.

But variable names alone do not suffice in practice. This is clear from the mobile process interpretation of object oriented languages ([10]). The usual practice is to postulate that $\mathcal{N}$ consists of two parts: a set $\mathcal{N}_v$ of variable names and a set $\mathcal{N}_c$ of constant names. We can now define a χ-process to be a χ-term in which all variable names are localized. So in χ-processes there are two kinds of local names: local variable names and local constant names. A communication either identifies two local variable names or replaces a local variable name by a local or global constant name. A communication between two constant names is prohibited. Let β range over $\{\rightarrow, \xrightarrow{aa}, \xrightarrow{a[a]}, \xrightarrow{\alpha(a)} | a \in \mathcal{N}_c, \alpha \in \mathcal{N}_c \cup \overline{\mathcal{N}_c}\}$.

Definition 16. Let $\mathcal{R}$ be a binary relation on the set of χ-processes. $\mathcal{R}$ is a simulation if $P\mathcal{R}Q$ implies that if $P \xrightarrow{\beta} P'$ then there exists some Q' such that $Q \xRightarrow{\hat{\beta}} Q'$ and $P'\mathcal{R}Q'$. The relation $\mathcal{R}$ is a bisimulation if both $\mathcal{R}$ and its reverse are simulations. The bisimilarity $\approx^\chi$ is the largest bisimulation.

The π-calculus can be reexamined in this new setting. The input prefix operation restricts variable names whereas the localization operation always restricts constant names. π-processes are now defined to be those processes in which all variable names are restricted by input prefixes. Let γ range over $\{\rightarrow, \xrightarrow{ca}, \xrightarrow{\bar{c}a}, \xrightarrow{\bar{c}(a)} | a, c \in \mathcal{N}_c\}$.

Definition 17. Let $\mathcal{R}$ be a binary relation on the set of π-processes. $\mathcal{R}$ is a simulation if $P\mathcal{R}Q$ implies that if $P \xrightarrow{\gamma} P'$ then there exists some Q' such that $Q \xRightarrow{\hat{\gamma}} Q'$ and $P'\mathcal{R}Q'$. The relation $\mathcal{R}$ is a bisimulation if both $\mathcal{R}$ and its inverse are simulations. The bisimilarity $\approx^\pi$ is the largest bisimulation.

The translation given in Sect. 4 works in this practical setting. It establishes an operational correspondence in the sense of Theorem 7. In addition one has

Theorem 18. *For π-processes P and Q, $P \approx^\pi Q$ if and only if $P^\circ \approx^\chi Q^\circ$.*

So practically speaking, π is a subcalculus of χ.

References

1. Abramsky, S.: Proofs as Processes. Theoretical Computer Science **135** (1994) 5–9
2. Barendregt, H.: The Lambda Calculus: Its Syntax and Semantics. 1984
3. Bellin, G., Scott, P.: Remarks on the π-Calculus and Linear Logic. Theoretical Computer Science **135** (1994) 11–65
4. Girard, J.: Linear Logic. Theoretical Computer Science **50** (1987) 1–102
5. Milner, R.: Functions as Processes. Journal of Mathematical Structures in Computer Science **2** (1992) 119–141
6. Milner, R., Parrow, J., Walker, D.: A Calculus of Mobile Processes. Information and Computation **100** (1992) 1–40 (Part I), 41–77 (Part II)
7. Plotkin, G.: Call-by-Name, Call-by-Value and the λ-Calculus. Theoretical Computer Science **1** (1975) 125-159
8. Sangiorgi, D.: A Theory of Bisimulation for π-Calculus. Proc. CONCUR 93. LNCS **715** (1993) 127–142
9. Thomsen, B.: Plain CHOCS—A Second Generation Calculus for Higher Order Processes. Acta Informatica **30** (1993) 1–59
10. Walker, D.: Objects in the π-Calculus. Information and Computation **116** (1995) 253–271

Solving Trace Equations Using Lexicographical Normal Forms

Volker Diekert[1], Yuri Matiyasevich[2]*, and Anca Muscholl[1]

[1] Institut für Informatik, Universität Stuttgart,
Breitwiesenstr. 20-22, 70565 Stuttgart, Germany
[2] Steklov Institute of Mathematics at St.Petersburg
Fontanka 27, St. Petersburg, 191011 Russia

Abstract. Very recently, the second author showed that the question whether an equation over a trace monoid has a solution or not is decidable [11,12]. In the original proof this question is reduced to the solvability of word equations with constraints, by induction on the size of the commutation relation. In the present paper we give another proof of this result using lexicographical normal forms. Our method is a direct reduction of a trace equation system to a word equation system with regular constraints, using a new result on lexicographical normal forms.

1 Introduction

Solving equations is a central topic in various fields of computer science, especially concerning unification, as required by automated theorem proving or logic programming. A celebrated result of Makanin [10] states that the question whether an equation over words has a solution or not is decidable: There exists an algorithm deciding for a given equation $L = R$, where $L, R \in (\Omega \cup \Sigma)^*$ contain both unknowns from Ω and constants from Σ, whether an assignment $\sigma: \Omega \to \Sigma^*$ exists, satisfying $\sigma(L) = \sigma(R)$. Slightly more general, the existential theory of equations over free monoids is decidable, i.e., given an existentially quantified, closed first-order formula S over atomic predicates of the form $L = R$ and $L \neq R$, it is decidable whether S is valid over a given free monoid. Moreover, adding regular constraints, i.e., atomic predicates of the form $x \in C$, where C is a regular language, preserves decidability [14].

In this paper we prove the generalization of Makanin's result to trace monoids, which were originally studied in combinatorics [4]. They became meaningful for computer science in concurrency theory, where they were introduced by Mazurkiewicz [13] in connection with the semantics of labelled Petri nets. For an overview of trace theory and related topics see "The Book of Traces" [7].

Most results obtained so far in the area of equations on traces were restricted to equations without constants, see [8,5]. The decidability of the solvability of equations with constants was stated as an important open question.

---
* This work was done during a stay at the University of Stuttgart.

2 Notations, Preliminaries and Lexicographical Normal Forms

An *independence alphabet* is a pair (Σ, I), where Σ is a finite alphabet and $I \subseteq \Sigma \times \Sigma$ is an irreflexive and symmetric relation, called *independence relation*. With a given independence alphabet (Σ, I) we associate the *trace monoid* $\mathbb{M}(\Sigma, I)$. This is the quotient monoid $\Sigma^*/\equiv_I$, where $\equiv_I$ denotes the congruence being the equivalence relation generated by the set $\{uabv = ubav \mid (a,b) \in I, u, v \in \Sigma^*\}$; an element $t \in \mathbb{M}(\Sigma, I)$ is called a *trace*, the length $|t|$ of a trace t is given by the length of any representing word. By alph(t) we denote the alphabet of a trace t, being the set of letters occurring in t.

By 1 we denote both the empty word and the empty trace. Words $v, w \in \Sigma^*$ are called *independent* (w.r.t. I), if alph$(v) \times$ alph$(w) \subseteq I$. In this case we simply write $(v, w) \in I$ or $v \in I(w)$ where $I(w)$ for $w \in \Sigma^*$ is a shorthand for $\{a \in \Sigma \mid \{a\} \times \text{alph}(w) \subseteq I\}$.

The *initial alphabet* of $w \in \Sigma^*$ is the set init$(w) = \{a \in \Sigma \mid \exists w', w'' \in \Sigma^*$ with $w \equiv_I w'$ and $w' = aw''\}$.

A word language $L \subseteq \Sigma^*$ is called *I-closed* if whenever $v \in L$ and $w \equiv_I v$ then we have $w \in L$.

Throughout the paper we will suppose that (Σ, I) denotes an independence alphabet, where Σ has the cardinality $n \geq 1$. We suppose that Σ is totally ordered by $<$ and we identify Σ with the set $\{1, \ldots, n\}$. The order on Σ is extended to the lexicographical order on Σ^*.

A word $v \in \Sigma^*$ is in *lexicographical normal form* (w.r.t. I and $<$) if $v \leq w$ holds for all w such that $v \equiv_I w$. Let LNF denote the set of lexicographical normal forms, i.e., LNF $\subseteq \Sigma^*$ is the set of minimal representatives for $\mathbb{M}(\Sigma, I)$. For $v \in \Sigma^*$ we denote by lex(v) the unique word $w \in$ LNF such that $w \equiv_I v$. We view lex as a mapping lex $: \Sigma^* \to$ LNF.

There is a simple characterization of lexicographical normal forms due to Anisimov and Knuth:

Proposition 1 ([3]). *Let Σ be totally ordered by $<$. Then a word $v \in \Sigma^*$ is in lexicographical normal form (w.r.t. I, $<$) if and only for every factor aub of v with $a, b \in \Sigma$, $u \in \Sigma^*$ and $(au, b) \in I$ we have $a < b$.*

Definition 2. *Let Σ be totally ordered by $<$. For $\emptyset \neq A \subseteq \Sigma$ let the height $h(A)$ be $h(A) = \max\{a \mid a \in A\}$. Let also $h(\emptyset) = 0$. (Thus, $h(A) \in \{0, \ldots, n\}$.) The height $h(v)$ of a word $v \in \Sigma^*$ is defined as $h(v) = h(\text{alph}(v))$.*

Remark 3. Let $m \geq 1$ and $s, t, v, s_1, \ldots, s_m, t_1, \ldots, t_m \in \Sigma^*$ be words satisfying the following conditions:

$$s = s_1 \cdots s_m,$$
$$t \equiv_I t_1 \cdots t_m,$$
$$v = s_1 t_1 \cdots s_m t_m,$$
$$t_j \in I(s_{j+1} \cdots s_m) \text{ for all } 1 \leq j < m.$$

Then we have $st \equiv_I v$.

The previous remark is clear and its converse will be stated for lexicographical normal forms in the Main Lemma below. It is the crucial correctness argument for our reduction from trace equations to word equations. The important point is that the value of m (given below) can be bounded as a function in the size of the alphabet, and that the height decreases.

Lemma 4 (Main Lemma). *Let $s, t, v \in \text{LNF}$ be words in lexicographical normal form such that $st \equiv_I v$.*
Let $h = h(s)$ denote the height of s and suppose $h > 0$.
Then there exist an integer m, $1 \leq m \leq \frac{(n-1)(h-1)}{2} + 1$, and words $s_1, \ldots, s_m$, $t_1, \ldots, t_m \in \text{LNF}$ in lexicographical normal form such that the following conditions hold:
$$s = s_1 \cdots s_m,$$
$$t \equiv_I t_1 \cdots t_m,$$
$$v = s_1 t_1 \cdots s_m t_m,$$
$$s_i \neq 1, \text{ for all } 1 < i \leq m,$$
$$t_j \neq 1 \text{ for all } 1 \leq j < m,$$
$$t_j \in I(s_{j+1} \cdots s_m) \text{ for all } 1 \leq j < m,$$
$$h(t_j) < h \text{ for all } 1 \leq j < m.$$

Remark 5. Before giving the proof of the Main Lemma, let us note that the trace equality $st \equiv_I v$ above cannot be replaced by word equalities of type $s = s_1 \cdots s_m$, $t = t_1 \cdots t_m$, $v = s_1 t_1 \cdots s_m t_m$. For example, consider $\text{M}(\Sigma, I) = \{a, b, c\}^* / \{ab = ba, bc = cb\}$ and $s = c$, $t = ab$. Then the lexicographical normal form of st is $v = bca$.

Proof of the Main Lemma. We have $st \equiv_I v$ with $s, t, v \in \text{LNF}$ and $h = h(s) > 0$. Consider the decomposition of v, $v = s_1 t_1 \cdots s_m t_m$, where $m \geq 1$ is minimal such that $s \equiv_I s_1 \cdots s_m$, $t \equiv_I t_1 \cdots t_m$, and $t_j \in I(s_{j+1} \cdots s_m)$ for all j, $1 \leq j < m$. Clearly, since m is minimal, we have $s_i \neq 1$ and $t_j \neq 1$ for all $1 < i \leq m$, $1 \leq j < m$. Moreover, the words s_i, t_j are in lexicographical normal form.
Let us first show that $s = s_1 \cdots s_m$. Assume aub is a factor of $s_1 \cdots s_m$ with $a, b \in \Sigma$, $u \in \Sigma^*$ and $b \in I(au)$. If aub is a factor of some s_i, then $a < b$ follows by Prop. 1 and we are done. Otherwise let $i < j$ be such that $s_i \in \Sigma^* au'$, $s_j \in u''b\Sigma^*$ and $u = u's_{i+1} \cdots s_{j-1}u''$. Since $t_k \in I(s_j)$ for $k < j$ we obtain $b \in I(au's_{i+1}t_{i+1} \cdots s_{j-1}t_{j-1}u'')$, hence $a < b$ due to v being in lexicographical normal form. Thus $s_1 \cdots s_m$ is in lexicographical normal form, again by Prop. 1, and it follows that $s = s_1 \cdots s_m$.
Suppose that $1 \leq j < m$ and let b denote the first letter of s_{j+1}. Let $a \in \text{alph}(t_j)$, i.e. $t_j = uau'$ for some words u, u'. Then $au'b$ is a factor of $v \in \text{LNF}$ satisfying $b \in I(au')$, thus we have $a < b$. Therefore $h(t_j) < h(b) \leq h$ for every $1 \leq j < m$.
Finally, assume by contradiction that $m > (n - 1)(h - 1)/2 + 1$. Let b_i, a_j denote the first letter of s_i, t_j respectively, $1 < i \leq m$, $1 \leq j < m$. Consider the chain of alphabets $I(s_2 \cdots s_m) \subseteq I(s_3 \cdots s_m) \subseteq \cdots \subseteq I(s_m)$. Note that we have $I(s_2 \cdots s_m) \neq \emptyset$ due to $t_1 \neq 1$, and also $I(s_m) \neq \Sigma$ due to $s_m \neq 1$. Therefore by the pigeon-hole principle there exist some indices $1 \leq i, j < m$ with

$j - i \geq (h-1)/2$ satisfying $I(s_{i+1} \cdots s_m) = I(s_{j+1} \cdots s_m)$. Consider the factor $t_i s_{i+1} t_{i+1} \cdots t_j s_{j+1}$ of v. Note that $(t_k, s_l) \in I$ holds for every k, l such that $i \leq k, l-1 \leq j$, since $t_k \in I(s_{k+1} \cdots s_m) = I(s_{i+1} \cdots s_m)$. Therefore, $v \in \text{LNF}$ implies $a_i < b_{i+1} < a_{i+1} < \cdots < a_j < b_{j+1}$ and we obtain $h(s) \geq h(b_{j+1}) \geq 2(j-i+1) > h$, a contradiction.

3 Trace Equation Systems

Definition 6. Let Ω denote a finite set of unknowns with $\Sigma \cap \Omega = \emptyset$.

i) A *word equation* over Σ and Ω has the form $L = R$, with $L, R \in (\Sigma \cup \Omega)^*$.

ii) An *assignment for* an equation over Σ and Ω is a mapping $\sigma: \Omega \to \Sigma^*$ being extended in a natural way to a homomorphism $\sigma: (\Sigma \cup \Omega)^* \to \Sigma^*$, by $\sigma|_\Sigma = \text{id}_\Sigma$.
A *solution for* the equation $L = R$ is an assignment σ satisfying the equality $\sigma(L) = \sigma(R)$ in Σ^*.

Makanin [10] showed in 1977 that the question whether a word equation has a solution or not is decidable. Moreover, the solvability of a system of word equations can be reduced by well-known techniques to the solvability of a single equation. The problem can also be generalized by introducing regular constraints for the unknowns, i.e. regular sets $C_x \subseteq \Sigma^*$ for $x \in \Omega$. Here, a solution σ for an equation is required to satisfy $\sigma(x) \in C_x$ for all x. It has been shown by Schulz [14] that the solvability of word equations with regular constraints remains decidable. We are going to show that this more general result generalizes to traces.

Definition 7. Let (Σ, I) denote an independence alphabet and Ω a finite set of unknowns, $\Sigma \cap \Omega = \emptyset$.

i) A *trace equation* over (Σ, I) and Ω has the form $L \equiv R$, with $L, R \in (\Sigma \cup \Omega)^*$.
A *solution for* the equation $L \equiv R$ is an assignment $\sigma: \Omega \to \Sigma^*$ satisfying $\sigma(L) \equiv_I \sigma(R)$.

ii) A *system of trace equations* is a formula built with the connectives **and** (&), **or** ($\vee$), **not** ($\neg$) over atomic predicates of the form $L \equiv R$ (trace equation) and $x \in C$ (constraint), where $C \subseteq \Sigma^*$ denotes an I-closed regular language.
A *solution for* a system S over $(\Sigma, I), \Omega$ is an assignment $\sigma: \Omega \to \Sigma^*$ such that S evaluates to **true** when the atomic predicates $L \equiv R$, $x \in C$ are replaced by the truth value of $\sigma(L) \equiv_I \sigma(R)$, $\sigma(x) \in C$, respectively.

Remark 8. Later we will deal simultaneously with trace and word equations, so we distinguish notationally between $L = R$ for a word equation, whereas $L \equiv R$ denotes a trace equation. The difference is that equality under an assignment is interpreted in the free monoid Σ^*, resp. in the trace monoid $\mathbb{M}(\Sigma, I)$.

Remark 9. A system of word equations (with regular constraints) is just a special case of Def. 7 where one takes $I = \emptyset$. Since negations can be eliminated (see also 3.1), we note that the question whether a system of word equations has a solution or not is decidable.

Remark 10. Adding arbitrary (i.e., not I-closed) regular constraints to a system of trace equations makes the question of solvability undecidable. This is due to the fact that the solvability of the equation $x \equiv y$ with $x \in C, y \in C'$ is equivalent to the non-emptiness of the intersection $\{w \in \Sigma^* \mid w \equiv_I v \text{ for some } v \in C\} \cap \{w \in \Sigma^* \mid w \equiv_I v \text{ for some } v \in C'\}$. For regular languages C, C' this last question is known to be undecidable, see [1].

Remark 11. Similar to the word case, the solvability of a trace equations system could be reduced to the solvability of a single trace equation (with additional constraints). However, this would be of no use here.

The aim of this section is to reduce the solvability problem for trace equations to word equations with regular constraints. We will give a direct proof using lexicographical normal forms to show the following

Theorem 12 ([11,12]). *Let S be a trace equation system over (Σ, I) and Ω. Then a set $\Omega' \supseteq \Omega$ of unknowns and a system of word equations S' over Σ, Ω' can be effectively constructed, such that S is solvable if and only if S' is solvable.*

Corollary 13. *It is decidable whether a system of trace equations has a solution.*

3.1 Basic Reductions

For a given trace equation system S we first eliminate constants by introducing new unknowns x_a and constraints $x_a \in \{a\}$, for $a \in \Sigma$. Then we replace a by x_a in each equation $L \equiv R$ of S. Hence, without loss of generality atomic predicates are of the form $L \equiv R$, where $L, R \in \Omega^*$.

Furthermore, we may assume that the given system is written in disjunctive normal form. Then we replace every negation **not**$(L \equiv R)$ by the disjunction of formulas of the type

$$L \equiv xy \ \& \ R \equiv xz \ \& \ \text{init}(y) = A \ \& \ \text{init}(z) = A' \tag{1}$$

where x, y, z denote new unknowns and the disjunction is taken over all alphabets $A, A' \subseteq \Sigma$ such that $A \cap A' = \emptyset$ and $A \cup A' \neq \emptyset$. Clearly, constraints of the form $\text{init}(x) = A$ or $\text{alph}(x) = A$, $A \subseteq \Sigma$, can be expressed by I-closed regular languages.

Since the set of I-closed regular languages forms an effective boolean algebra (as the family of recognizable subsets of a monoid [9]) we may also suppose that the formula contains no negated constraints, i.e. no formula of type **not**$(x \in C)$. Moreover, it suffices to consider trace equations of the form $x_1 \cdots x_k \equiv y_1 \cdots y_l$ with $k \geq l > 0$, $x_i, y_j \in \Omega$. (The equation $x_1 \cdots x_k \equiv 1$ and the occurrences of each x_i can be deleted from all equations, adding the constraints $\text{alph}(x_i) = \emptyset$.)

3.2 From Traces to Words

The main idea for reducing trace equations to word equations will consist in replacing a trace equation $L \equiv R$ by some word equations $L_1 = R_1, \ldots, L_k = R_k$ with additional constraints and unknowns. Moreover, for every solution σ for $L \equiv R$ the mapping lex $\circ\, \sigma\colon \Omega \to \Sigma^* \to$ LNF can be extended to a solution for the equations $L_1 = R_1, \ldots, L_k = R_k$. Vice versa, each solution for the new equations will also be a solution for $L \equiv R$ when restricted to its unknowns.

This reduction actually goes by a chain of intermediate trace equations. By choosing an appropriate ordering we will show that the reduction process terminates yielding a system of word equations (with constraints).

We will consider in the following formulas $S(T, W, C)$ in disjunctive normal form with atomic predicates from some finite sets T, W, C, containing no negations. T will denote a set of trace equations, W a set of word equations and $C = \{x \in C_x \mid x \in \Omega\}$ a set of constraints, where each C_x is an I-closed regular language. Moreover, every $L \equiv R$ in T has the form $x_1 \cdots x_k \equiv y_1 \cdots y_l$ with $k \geq l \geq 1$, $x_i, y_j \in \Omega$. A solution for $S(T, W, C)$ is an assignment $\sigma\colon \Omega \to \Sigma^*$ which makes the formula evaluate to **true** when $(L \equiv R)$ from T, $(L = R)$ from W and $x \in C_x$ from C are replaced by the truth value of $\sigma(L) \equiv_I \sigma(R)$, $\sigma(L) = \sigma(R)$, and $\sigma(x) \in C_x$, respectively.

Definition 14. A formula $S(T, W, C)$ as above is called *normalized* if for every solution σ for S the mapping lex $\circ\, \sigma$ is a solution for S, too.

Remark 15. Note that a formula $S(T, \emptyset, C)$ with I-closed constraints C is always normalized.

Remark 16. Suppose $S = S(T, W, C)$ is normalized and let $x \equiv y$ belong to T, where $x, y \in \Omega$. Consider the new formula $S' = S'(T', W', C)$ obtained from S by replacing every occurrence of $x \equiv y$ by $x = y$ and letting $T' = T \setminus \{x \equiv y\}$, $W' = W \cup \{x = y\}$. Then S is solvable if and only if S' is solvable. Note that a solution for S' is a solution for S, too. However, the converse is true only because S is a normalized system. Without this assumption about S it cannot be guaranteed that every solution for S also solves S', see the example below. Moreover, S' is a normalized system, too.

Example 17. Consider the trace equation system $S = (\{x \equiv y\}, \{x = ab, y = ba\}, \emptyset)$ given as the conjunction $(x \equiv y)\ \&\ (x = ab)\ \&\ (y = ba)$, where $(a, b) \in I$. Then S is not normalized, but of course it has a solution. However, replacing $x \equiv y$ by the word equation $x = y$ yields a system with no solution.

Proof of Thm. 12. Recall that an equation system with I-closed constraints $S = S(T, \emptyset, \{x \in C_x\}_{x \in \Omega})$ over (Σ, I), Ω is a normalized system. As previously noted it suffices to consider a formula S with trace equations of the form

$$x_1 \cdots x_k \equiv y_1 \cdots y_l, \quad k \geq l \geq 1, \quad (k, l) \neq (1, 1). \tag{2}$$

We suppose without loss of generality that for all unknowns $x \in \Omega$ some $A_x \subseteq \Sigma$ exists such that $h(A_x) > 0$, and $x \in C_x$ implies alph$(x) \subseteq A_x$, for all x. Moreover, let S be a conjunction of trace equations as in (2), of word equations and of I-closed regular constraints $x \in C_x$.

We define the *weight of* a trace equation $x_1 \cdots x_k \equiv y_1 \cdots y_l$ as in (2) as the triple of natural numbers $(l, h(\cup_{i=1}^{k-1} A_{x_i}), k)$ and we consider the lexicographical ordering on $\mathbb{N} \times \mathbb{N} \times \mathbb{N}$. We will show in the following that every such trace equation can be replaced by a formula over word equations and trace equations of lower weight, together with some additional constraints. Concretely, we apply the following rules.

Rule 1: Suppose $l > 1$ and let z denote a new unknown. Then we replace the equation $x_1 \cdots x_k \equiv y_1 \cdots y_l$ by

$$x_1 \cdots x_k \equiv z \quad \& \quad y_1 \cdots y_l \equiv z \quad \& \quad \text{alph}(z) \subseteq \cup_{i=1}^{k} A_{x_i} .$$

Rule 2: Suppose $l = 1$ and $k > 2$, and let z denote a new unknown. Then we replace the equation $x_1 \cdots x_k \equiv y_1$ by

$$x_1 z \equiv y_1 \quad \& \quad x_2 \cdots x_k \equiv z \quad \& \quad \text{alph}(z) \subseteq \cup_{i=2}^{k} A_{x_i} .$$

Rule 3: Suppose $l = 1$ and $k = 2$ and, in order to simplify notation, consider the equation $xy \equiv z$ (rather than uniformly $x_1 x_2 \equiv y_1$). Moreover, let $h = h(A_x)$ denote the height of A_x (where alph$(x) \subseteq A_x$ follows from the constraint $x \in C_x$). We replace $xy \equiv z$ by the disjunction of the word equation

$$xy = z \tag{3}$$

and of formulas of the type

$$\begin{aligned} x = x_1 \cdots x_m \quad \& \quad y \equiv y_1 \cdots y_m \quad \& \quad z = x_1 y_1 \cdots x_m y_m \quad \& \\ \text{alph}(x_1) \subseteq A_1 \quad \& \quad \cdots \quad \& \quad \text{alph}(x_m) \subseteq A_m \quad \& \\ \text{alph}(y_1) \subseteq B_1 \quad \& \quad \cdots \quad \& \quad \text{alph}(y_m) \subseteq B_m , \end{aligned} \tag{4}$$

where x_i, y_j are new unknowns and the disjunction is taken over all values of m such that $1 < m \leq (n-1)(h-1)/2 + 1$ and over all alphabets $A_1, \ldots, A_m$, $B_1, \ldots, B_m \subseteq \Sigma$ such that[1]

$$\begin{aligned} &A_i \neq \emptyset \text{ for all } 1 < i \leq m, \text{ and} \\ &1 \leq h(B_j) < h \text{ for all } 1 \leq j < m, \text{ and} \\ &B_j \times A_i \subseteq I \text{ for all } 1 \leq j < i \leq m, \text{ and} \\ &A_1 \cup \cdots \cup A_m \subseteq A_x, \text{ and } B_1 \cup \cdots \cup B_m \subseteq A_y . \end{aligned} \tag{5}$$

The word equation $xy = z$ in (3) corresponds to the case $m = 1$ in (4) (this is in particular the case when $h = 1$ in (5)). It is actually the main case where the number of trace equations in S decreases.

Let S' denote the formula obtained from S by applying one of the three rules described above. Note that none of the rules adds negations.

---

[1] Obviously some equations become redundant and they can be actually omitted in the disjunction.

Lemma 18. *Let S be a normalized equation system. Then the new system S' is normalized, too. Moreover, S' is solvable if and only if S is solvable.*

Proof. The claim is easily seen for the first two rules above, since there is a natural bijection between the set of solutions of S and of S', respectively.
Clearly, if S' has been obtained from S by the third rule, then every solution for S' is a solution for S, too, see Rem. 3. Therefore, let us consider an equation $xy \equiv z$ in S and a solution $\sigma: \Omega \to \Sigma^*$ for S. Then $\sigma' = \text{lex} \circ \sigma$ is also solution for S, since S is normalized. We show that σ' can be extended to a solution for S'. Let $s = \sigma'(x)$, $t = \sigma'(y)$ and $v = \sigma'(z)$. Hence, $st \equiv_I v$ with $s, t, v \in \text{LNF}$. If $h(s) = 1$, then in the Main Lemma we have $m = 1$, hence $v = st$. Therefore σ' is a solution of the new system S'.
Suppose that $st \equiv_I v$ with $s, t, v \in \text{LNF}$, $h(s) = h > 1$. Then some m, $1 \leq m \leq (n-1)(h-1)/2 + 1$, and words $s_1, \ldots, s_m, t_1, \ldots, t_m$ exist, satisfying the conditions of the Main Lemma. With $\sigma'(x_i) = s_i$, $\sigma'(y_j) = t_j$ it is easily verified that σ' is a solution for S'.
The relation between the solution set of S and the solution set of S', together with the fact that S is normalized, imply that S' is normalized, too. This shows the lemma.

Finally, note that the new trace equation $y_1 \cdots y_m \equiv y$ in (4) has lower weight than $xy \equiv z$ due to $h(\bigcup_{j=1}^{m-1} B_j) < h = h(A_x)$. Hence the reduction rules establish a noetherian rewriting system on trace equation systems. Applying the rules as long as possible we end with a system of word equations $S' = (\emptyset, W', C')$. This concludes our proof.

4 Computing Lexicographical Normal Forms

The aim of this section is to give a formula for computing the product of lexicographical normal forms. This yields an alternative proof of Thm. 12 and the so far best known upper bound on the number of new unknowns needed for the reduction. We conclude the section with two remarks concerning the parallel complexity of computing lexicographical normal forms.

Definition 19. *Let $\sim_I$ be a relation on $(\Sigma^*)^*$ defined as*

$$(x_1, \ldots, x_m) \sim_I (x'_1, \ldots, x'_{m'})$$

if $m = m'$ and there exists some i, $1 \leq i < m$ such that

$$x_j = x'_j \quad \text{for all } 1 \leq j \leq m, \ j \notin \{i, i+1\}, \text{ and}$$
$$(x_i, x_{i+1}) = (x'_{i+1}, x'_i) \text{ and } (x_i, x_{i+1}) \in I.$$

By $\approx_I$ we denote the equivalence relation generated on $(\Sigma^*)^*$ by $\sim_I$.

Let $x \in \Sigma^*$, by abuse of language we write $(x_1, \ldots, x_m) \approx_I x$ if some words $x'_1, \ldots, x'_m$ exist such that

$$(x_1, \ldots, x_m) \approx_I (x'_1, \ldots, x'_m) \quad \text{and} \quad x = x'_1 \cdots x'_m.$$

Theorem 20. *Let $s, t, v \in \text{LNF}$ be words in lexicographical normal form such that $st \equiv_I v$.*
Then there exist positive integers m, p with $m \leq \frac{(n-1)^2}{2} + 1$, $p \leq n^n n!$ such that

$$s = s_1 \cdots s_m,$$
$$t = t_1 \cdots t_p,$$
$$(s_1, \ldots, s_m, t_1, \ldots, t_p) \approx_I v,$$

for some words $s_1, \ldots, s_m, t_1, \ldots, t_p \in \Sigma^*$.

Proof. Let $h = h(s)$ denote the height of s. Let $m(h), p(h)$ denote the minimal integers such that

$$s = s_1 \cdots s_{m(h)},$$
$$t = t_1 \cdots t_{p(h)},$$
$$(s_1, \ldots, s_{m(h)}, t_1, \ldots, t_{p(h)}) \approx_I v,$$

for some words s_i, t_j. Note that $m(h), p(h) \leq |v|$. For $h = 0$ we have $s = 1$, thus $m(0) = p(0) = 1$, which satisfies the theorem.
For $h \geq 1$ we will show by induction on h that $m(h) \leq (n-1)(h-1)/2 + 1$ and $p(h) \leq n^h h!$, thereby proving the theorem.
Let $h \geq 1$. By the Main Lemma there exist an integer $m \leq (n-1)(h-1)/2 + 1$ and words $s_1, \ldots, s_m, t_1, \ldots, t_m$ in lexicographical normal form satisfying

$$s = s_1 \cdots s_m,$$
$$t \equiv_I t_1 \cdots t_m,$$
$$v = s_1 t_1 \cdots s_m t_m,$$
$$s_i \neq 1, \ t_j \neq 1 \text{ for } 1 < i \leq m, \ 1 \leq j < m,$$
$$t_j \in I(s_{j+1} \cdots s_m) \text{ and } h(t_j) < h \text{ for } 1 \leq j < m. \tag{6}$$

If $h = 1$, then $m = 1$ in (6), so we can take $m(h) = p(h) = 1$, since $t = t_1 \in \text{LNF}$, which satisfies the claim. Hence let $h, m \geq 2$.
Let $\bar{t}_1 = t_1$ and $\bar{t}_i = \text{lex}(\bar{t}_{i-1} t_i)$ for $i = 2, \ldots, m$. Clearly, $\bar{t}_m = t$, $h(\bar{t}_i) < h$ for $1 \leq i < m$ and

$$\bar{t}_{i-1} t_i \equiv_I \bar{t}_i, \ \text{for } 1 < i \leq m. \tag{7}$$

Now we can apply the induction hypothesis to each of the $(m-1)$ equivalences (7) obtaining

$$t \approx_I (t'_1, \ldots, t'_p), \tag{8}$$

for some $p \leq (m-1)[m(h-1) + p(h-1)]$, some words $t'_1, \ldots, t'_p$ and some integers $1 = l_0 < l_1 < \cdots < l_m = p+1$ such that

$$t_i = t'_{l_{i-1}} \cdots t'_{l_i - 1} \text{ for every } 1 \leq i \leq m. \tag{9}$$

The above claim can be verified by noting that

$$t \approx_I (t'_1, \ldots, t'_i, \ldots, t'_j, \ldots, t'_q) \text{ and } t'_i \cdots t'_j \approx_I (v_1, \ldots, v_k)$$

implies that
$$t \approx_I (t'_1, \ldots, t'_{i-1}, v'_1, \ldots, v'_l, t'_{j+1}, \ldots, t'_q),$$
for some $l \leq j - i + k$ and $v'_1, \ldots, v'_l \in \Sigma^*$, such that $v'_1 \cdots v'_l = v_1 \cdots v_k$ and each v'_q is a factor of some v_r. Hence, we obtain from (8), (9) for suitable words $t''_1, \ldots, t''_p$:
$$t = t''_1 \cdots t''_p,$$
$$v \approx_I (s_1, \ldots, s_m, t_1, \ldots, t_m) \approx_I (s_1, \ldots, s_m, t'_1, \ldots, t'_p)$$
$$\approx_I (s_1, \ldots, s_m, t''_1, \ldots, t''_p).$$

Hence by the induction hypothesis we get
$$p(h) \leq (m-1)[m(h-1) + p(h-1)]$$
$$\leq (n-1)(h-1)/2 \left[(n-1)(h-2)/2 + 1 + n^{h-1}(h-1)!\right] \leq n^h h!,$$
which concludes the proof.

Remark 21. We can also use Thm. 20 in order to prove the main result, Thm. 12. Recall that the main difficulty consists in replacing a trace equation of the form $xy \equiv z$, where $x, y, z \in \Omega$. By Thm. 20 we simply replace such an equation $xy \equiv z$ by a disjunction over clauses of the form
$$x = x_1 \cdots x_m \quad \& \quad y = y_1 \cdots y_p \quad \&$$
$$z = z_{\pi(1)} \cdots z_{\pi(m+p)} \quad \& \quad \text{alph}(z_i) \subseteq A_i,$$
for all $1 \leq m \leq \frac{(n-1)^2}{2} + 1$, $1 \leq p \leq n^n n!$, $\pi \in S^I_{m+p}$ and $A_i \subseteq \Sigma$. Here x_i, y_j denote new variables and $z_i = x_i$ for $1 \leq i \leq m$, resp. $z_{m+j} = y_j$ for $1 \leq j \leq p$. S^I_{m+p} denotes the set of permutations over $\{1, \ldots, m+p\}$ such that for $i < j$ the inequality $\pi(i) > \pi(j)$ implies $A_i \times A_j \subseteq I$. This reduction of a single trace equation to word equations roughly yields an increase in the number of word equations by $(N+2)! 2^{n(N+1)}$, where $N = n^n n! + (n-1)^2/2 + 1$. Hereby we need N additional unknowns.

We conclude this section with two remarks concerning the parallel complexity of computing lexicographical normal forms. We consider uniform circuit complexity classes like AC^0 and TC^0. Let $f: \Sigma^* \to \Sigma^*$ be a function such that $|f(w)| = p(|w|)$ for some polynomial p and every $w \in \Sigma^*$. Let $k \geq 0$. Then f is AC^k-computable if there is a family $(C_n)_{n \geq 0}$ of polynomial-size circuits of depth $O(\log^k(n))$ with AND and OR gates of unbounded fan-in/out and unary NOT gates, such that $C_{|w|}$ computes $f(w)$ for all $w \in \Sigma^*$. A function f is TC^k-computable if there is a family of circuits as above which in addition to AND, OR and NOT gates contain MAJORITY gates of unbounded fan-in/out. A MAJORITY gate yields 1 if and only if more than half of its inputs are 1. In order to be able to deal with arbitrary alphabets Σ one usually assumes that the circuits have special input/output gates testing $x = a$ for each input position x and letter $a \in \Sigma$ (analogously for the outputs). Uniformity means that given $n \geq 0$

(a fixed coding of) the circuit C_n can be easily computed (e.g. in logarithmic space). It is not very hard to verify that $\text{AC}^k \subseteq \text{TC}^k \subseteq \text{AC}^{k+1}$, $k \geq 0$. For more details about circuit complexity see e.g. [15]. We state the results below without proofs (being sketched in [6]). With Thm. 20 we obtain

Corollary 22. *Let (Σ, I) denote an independence alphabet. Then we can compute $\text{lex}(st)$ on input $s, t \in \text{LNF}$ in uniform AC^0.*

Remark 23. We could apply Cor. 22 in order to compute the function lex in AC^1. However, we can do better: the mapping lex: $\Sigma^* \to \text{LNF}$ is computable in uniform TC^0. This result can be compared with the fact that the equivalence $s \equiv_I t$ can be verified in uniform TC^0, too (see [2]).

References

1. IJ. J. Aalbersberg and H. J. Hoogeboom. Characterizations of the decidability of some problems for regular trace languages. *Mathematical Systems Theory*, 22:1–19, 1989.
2. C. Àlvarez and J. Gabarró. The parallel complexity of two problems on concurrency. *Information Processing Letters*, 38:61–70, 1991.
3. A. V. Anisimov and D. E. Knuth. Inhomogeneous sorting. *International Journal of Computer and Information Sciences*, 8:255–260, 1979.
4. P. Cartier and D. Foata. *Problèmes combinatoires de commutation et réarrangements.* Number 85 in Lecture Notes in Mathematics. Springer, 1969.
5. C. Choffrut. Combinatorics in trace monoids I. In [7].
6. V. Diekert, Yu. Matiyasevich, and A. Muscholl. Solving trace equations using lexicographical normal forms. Technical report, Universität Stuttgart, Fakultät Informatik, Bericht 1997/01, 1997.
7. V. Diekert and G. Rozenberg, editors. *The Book of Traces.* World Scientific, Singapore, 1995.
8. C. Duboc. On some equations in free partially commutative monoids. *Theoretical Computer Science*, 46:159–174, 1986.
9. S. Eilenberg. *Automata, Languages, and Machines*, volume A. Academic Press, New York and London, 1974.
10. G. S. Makanin. The problem of solvability of equations in a free semigroup. *Math. Sbornik*, 103:147–236, 1977. English transl. in Math. USSR Sbornik 32 (1977).
11. Yu. Matiyasevich. Reduction of trace equations to word equations, 1996. Talk given at the "Colloquium on Computability, Complexity, and Logic", 5-6 Dec. 1996, Institut für Informatik, Universität Stuttgart, Germany.
12. Yu. Matiyasevich. Some decision problems for traces. In S. Adian and A. Nerode, editors, *Proc. of the 4th International Symposium on Logical Foundations of Computer Science (LFCS'97)*, number 1234 in Lecture Notes in Computer Science, pages 248–257, 1997. Springer. Invited lecture. To appear.
13. A. Mazurkiewicz. Concurrent program schemes and their interpretations. DAIMI Rep. PB 78, Aarhus University, Aarhus, 1977.
14. K. U. Schulz. Makanin's algorithm for word equations — Two improvements and a generalization. In K. U. Schulz, ed., *Word Equations and Related Topics*, number 572 in Lecture Notes in Computer Science, pp. 85–150, Springer, 1991.
15. H. Straubing. *Finite automata, formal logic, and circuit complexity.* Birkhäuser, 1994.

Star-Free Picture Expressions Are Strictly Weaker than First-Order Logic

Thomas Wilke

Christian-Albrechts-Universität zu Kiel, Institut für Informatik und Praktische Mathematik, 24098 Kiel, Germany

Abstract. We exhibit a first-order definable picture language which we prove is *not* expressible by any star-free picture expression, i. e., it is not star-free. Thus first-order logic over pictures is strictly more powerful than star-free picture expressions are. This is in sharp contrast with the situation with words: the well-known McNaughton-Papert theorem states that a word language is expressible by a first-order formula if and only if it is expressible by a star-free (word) expression.
The main ingredients of the non-expressibility result are a Fraïssé-style algebraic characterization of star freeness for picture languages and combinatorics on words.

1 Introduction

There are two fundamental results connecting logical definability with concepts in the theory of regular languages: 1) Büchi's theorem (see [1]) which states that a word language is recognized by a finite automaton if and only if it is definable in (existential) monadic second-order logic, and 2) McNaughton and Papert's theorem (see [9]) which says that a word language is star-free if and only if it is definable in first-order logic. In [6], it was shown that the first result essentially carries over to picture (or "two-dimensional") languages in the following sense: a picture language is recognized by a tiling system if and only if it is definable in existential monadic second-order logic (while in [5], see also [6], full monadic second-order logic had been proven to be strictly more powerful).

In this paper, we show that the second result does *not* carry over to picture languages. More precisely, we exhibit a simple, first-order definable picture language, denoted L_+ (see page 3), and show that L_+ is not expressed by any star-free picture expression. On the other hand, it is straightforward to see that every star-free picture language is definable in first-order logic. We thus conclude that the class of star-free picture languages is strictly contained in the class of first-order definable picture languages. This clarifies an interesting question about the fine structure of the class of all recognizable picture languages, which was brought up in [6]. It should also be noted that by a result from [5], the class of first-order definable picture languages is strictly contained in the class of all recognizable picture languages.

As with star-free word expressions, star-free picture expressions are built from singleton sets using boolean combinations and concatenation. Of course, due to the two-dimensional structure of pictures there are two kinds of concatenation: "horizontal" and "vertical", sometimes also called "row" and "column" concatenation. Similarly, in first-order formulas over pictures one can use a "horizontal" and a "vertical" order relation to specify spatial relations between positions. It is the unrestricted use of these two order relations that makes first-order logic over pictures more powerful than star-free expressions.

The proof that L_+ is not star-free is based on a characterization of star-free picture languages in the style of Fraïssé's algebraic characterization of first-order definability ([3], see also [2]). The other ingredient of the proof is an encoding of certain pictures by words, which allows us to apply "one-dimensional" combinatorial arguments. Although Fraïssé's idea is quite old, this is the first time that it has been applied to a problem on picture languages.

The paper is organized as follows. In Section 2, basic terminology and notation is introduced and the main result is stated. Section 3 then describes the algebraic characterization of star freeness, Section 4 focuses on the encoding of "diagonal" pictures in words, in Section 5 a combinatorial lemma about words is established, and in Section 6, the proof of the main theorem is completed.

For a survey on picture languages, see the forthcoming handbook chapter [4].

Thanks to Kousha Etessami, Oliver Matz, and Sebastian Seibert for fruitful discussions and comments on drafts of this paper.

2 Basic Terminology and Main Result

A *picture*[1] over an alphabet A is a matrix with entries from A. We say $(m \times n)$-picture for a picture with m rows and n columns. An *atomic* picture is a (1×1)-picture. Words can and should be thought of as $(1 \times n)$-pictures.

There are two concatenations defined for pictures: juxtaposition and supraposition.[2] The *juxtaposition* of an $(m \times n)$-picture with an $(m' \times n')$-picture is defined when $m = m'$ and is the $(m \times (n+n'))$-picture denoted $P \boxempty Q$ where

$$(P \boxempty Q)_{ij} = \begin{cases} P_{ij} & \text{if } j \leq n, \\ Q_{i,j-n} & \text{if } j > n. \end{cases} \quad (1)$$

The *supraposition* of P and Q is defined when $n = n'$ and is the $((m+m') \times n)$-picture denoted $P \boxminus Q$ where

$$(P \boxminus Q)_{ij} = \begin{cases} P_{ij} & \text{if } i \leq m, \\ Q_{i-m,j} & \text{if } i > m. \end{cases} \quad (2)$$

Juxtaposition and supraposition are extended to sets of pictures just as concatenation of words is extended to sets of words.

A *star-free picture expression* over an alphabet A is built from the letters of A (each letter a standing for the singleton set with the atomic picture a) using the additional symbols $\emptyset$ (for the empty set), $+$ (for set-theoretic union), $\sim$ (for set-theoretic complementation with respect to the set of all pictures over A), and $\boxempty$ and $\boxminus$. Each star-free picture expression over A defines a picture language over

---
[1] Not to be confused with the notion of picture defined in [8].
[2] "Juxtaposition" and "supraposition" are also known as "horizontal" and "vertical" as well as "row" and "column" concatenation.

A in a canonical way. For instance, given an alphabet A with two letters a and b, the expression $a + (a \boxminus {\sim}\emptyset) + (a \boxminus {\sim}\emptyset) + (a \boxminus {\sim}\emptyset) \boxminus {\sim}\emptyset$ defines the set of all pictures whose upper left entry is a. Notice that we don't consider the empty picture. A picture language is said to be *star-free* if it can be expressed by a star-free picture expression.

The first-order vocabulary we use consists of built-in predicates $<_v$ and $<_h$ for horizontal and vertical order relation and a unary predicate P_a for each letter a. First-order formulas in this language are interpreted in pictures, where the first-order variables range over the positions of the picture in question. For example, consider the formula

$$\exists x \exists y_1 \ldots \exists y_4 (y_1 <_h x <_h y_2 \wedge y_3 <_v x <_v y_4 \wedge P_1 x \wedge P_1 y_1 \wedge \ldots \wedge P_1 y_4) \ . \quad (3)$$

This formula defines the set of all pictures satisfying the following condition: there is a position labeled 1 to the left and right of which there is an occurrence of 1 and over and under which there is an occurrence of 1. We write L_+ for the picture language containing all pictures over $\{0,1\}$ satisfying (3).

The main result of this paper is:

Theorem 1. *The language L_+ is not star-free.*

Every star-free picture expression can be converted into an equivalent first-order sentence in a straightforward way, in fact, by reusing variables one can even show that five first-order variables are always sufficient. (The interested reader may want to notice that in order to define L_+ two variables are actually enough.)

As a consequence, we have:

Corollary 2. *The class of star-free picture languages is strictly contained in the class of first-order definable picture languages.*

In the notation of [4], we thus have: $\mathcal{L}(\text{SFRE}) \subsetneq \mathcal{L}(\text{FO}) \subsetneq \mathcal{L}(\text{EMSO}) \subsetneq \mathcal{L}(\text{MSO})$.

3 Algebraic Characterization of Star Freeness

Fix an alphabet A and let $k \geq 0$. There are only a finite number of picture languages over A that can be defined by star-free picture expressions over A of nesting depth at most k in the concatenation operations. The set of all these picture languages is not only finite but also a boolean algebra, i.e., there is a finite partition of all pictures over A such that an arbitrary picture language over A is definable by a star-free expression of concatenation depth at most k if and only if it is a union of the blocks of this partition. In this section, we describe this partition in terms of the corresponding equivalence relation.

Concatenation depth, denoted cd, is defined by:

$$\text{cd}(\emptyset) = \text{cd}(a) = 0 \ ,$$
$$\text{cd}({\sim}E) = \text{cd}(E) \ ,$$

$$\operatorname{cd}(E+F) = \max(\operatorname{cd}(E), \operatorname{cd}(F)) \; ,$$
$$\operatorname{cd}(E \boxdot F) = \operatorname{cd}(E \boxminus F) = \max(\operatorname{cd}(E), \operatorname{cd}(F)) + 1 \; ,$$

where a stands for an arbitrary letter and E and F stand for arbitrary picture expressions.

We define k-*equivalence*, $\equiv_k$ in symbols, as a relation over pictures inductively as follows.

1. Pictures are 0-equivalent if they are identical or both not atomic.
2. Pictures P and Q are $(k+1)$-equivalent if the following conditions hold:
   (K) P and Q are k-equivalent.
   (J) For all pictures P_1, P_2 such that $P = P_1 \boxdot P_2$ there exist pictures Q_1, Q_2 such that $Q = Q_1 \boxdot Q_2$, $P_1 \equiv_k Q_1$, and $P_2 \equiv_k Q_2$.
   (S) For all pictures P_1, P_2 such that $P = P_1 \boxminus P_2$ there exist pictures Q_1, Q_2 such that $Q = Q_1 \boxminus Q_2$, $P_1 \equiv_k Q_1$, and $P_2 \equiv_k Q_2$.
   (J') & (S') Conditions (J) and (S) hold when the roles of P and Q are exchanged.

This means P and Q are $(k+1)$-equivalent if and only if they are k-equivalent and for any decomposition of P into two pictures, one can find a decomposition of Q into two pictures such that corresponding "factors" are k-equivalent, and vice versa.

The key fact about this equivalence relation is:

Theorem 3 (correctness and completeness). *A picture language L is definable by a star-free expression of concatenation depth at most k if and only if L is a union of $\equiv_k$-classes.*

We leave out the proof, which follows proofs of similar claims in the literature, see, e. g., [10], where the fine structure of the class of all star-free word languages is characterized.

Thus, in order to prove that a picture language L is not star-free we only have to show that for every k there are two pictures P and Q such that $P \in L$, $Q \notin L$, but $P \equiv_k Q$. That is what we will do in Section 6 (for $L = L_+$).

We need some facts about k-equivalence on words, all of which can be proven by a straightforward induction on k.

Lemma 4 (projections). *Let A and B be alphabets and $\pi \colon A^+ \to B^+$ a homomorphism. If u and v are strings over A such that $u \equiv_k v$, then $\pi(u) \equiv_k \pi(v)$.*

Lemma 5 (congruence property for words). *Let $k \geq 0$. The relation $\equiv_k$ restricted to words is a congruence relation, i. e., $uu' \equiv_k vv'$ whenever $u \equiv_k v$ and $u' \equiv_k v'$ for words u, u', v, and v'.*

Lemma 6 (aperiodicity). *For each $k \geq 0$, there exists $l_k > 0$ such that*

$$u^{l_k + m} \equiv_k u^{l_k} \qquad \text{for every word } u \text{ and every } m \geq 0. \tag{4}$$

Corollary 7. Let $k \geq 0$. Assume u is a word of the form

$$0^{i_0} 1 0^{i_1} 1 0^{i_2} \ldots 0^{n_2} 1 0^{i_{n-1}} 1 0^{i_n} , \qquad (5)$$

where $n > l_k + 1$ and $i_j \geq l_k$ for every $j \in \{1, \ldots, n-1\}$.

(a) If $i_0, i_n \geq l_k$ and v is obtained from u by changing one occurrence of 1 to 0, then $u \equiv_k v$.
(b) If v is obtained by changing one inner (i.e., neither the first nor the last) occurrence of 1 to 0, then $u \equiv_k v$.

We also need the following very simple (and weak) congruence-like property of k-equivalence over pictures, which can also be proven by induction on k.

Lemma 8. Let $k \geq 0, l > 0$. Assume A is an alphabet, $a \in A$, P is an $(m \times n)$-picture over A, and Q is an $(m' \times n')$-picture over A. Define P' and Q' to be the unique $(m \times l)$- and $(m' \times l)$-picture over the alphabet $\{a\}$. If $P \equiv_k Q$, then $P \square P' \equiv_k Q \square Q'$. The dual claim holds for supraposition instead of juxtaposition.

4 Diagonal Pictures

As explained above, in order to prove that L_+ is not star-free we have to find pictures P^k and Q^k such that $P^k \in L_+$, $Q^k \notin L_+$, and $P^k \equiv_k Q^k$ for every k. We will choose the pictures P^k and Q^k from a class of specifically designed pictures, so-called "diagonal pictures".

We will introduce diagonal pictures as certain pictures over $\{0, 1\}$ determined by words over an alphabet denoted by D. This alphabet is defined to be the set of subsets of the five-element set $C = \{1, \mathtt{n}, \mathtt{s}, \mathtt{w}, \mathtt{e}\}$, where $\mathtt{n}$ stands for "north", $\mathtt{s}$ for "south", etc. Given an element $a \in C$ and a string u over D, the a-projection of u, denoted $u \downarrow a$, is the unique string v over $\{0, 1\}$ of length $|u|$ satisfying $v_i = 1$ if and only if $a \in u_i$. Given a word u of length l over D, the corresponding diagonal picture, $P(u)$ in symbols, is given by:

$$P(u) = \begin{bmatrix} u_1 \downarrow 1 & u_2 \downarrow \mathtt{n} & u_3 \downarrow \mathtt{n} & \cdots & u_i \downarrow \mathtt{n} & \cdots & u_{l-1} \downarrow \mathtt{n} & u_l \downarrow \mathtt{n} \\ u_2 \downarrow \mathtt{w} & u_2 \downarrow 1 & 0 & \cdots & 0 & \cdots & 0 & u_2 \downarrow \mathtt{e} \\ u_3 \downarrow \mathtt{w} & 0 & u_3 \downarrow 1 & \cdots & 0 & \cdots & 0 & u_3 \downarrow \mathtt{e} \\ \vdots & \vdots & \vdots & \ddots & \vdots & & \vdots & \vdots \\ u_i \downarrow \mathtt{w} & 0 & 0 & \cdots & u_i \downarrow 1 & \cdots & 0 & u_i \downarrow \mathtt{e} \\ \vdots & \vdots & \vdots & & \vdots & \ddots & \vdots & \vdots \\ u_{l-1} \downarrow \mathtt{w} & 0 & 0 & \cdots & 0 & \cdots & u_{l-1} \downarrow 1 & u_{l-1} \downarrow \mathtt{e} \\ u_l \downarrow \mathtt{w} & u_2 \downarrow \mathtt{s} & u_3 \downarrow \mathtt{s} & \cdots & u_i \downarrow \mathtt{s} & \cdots & u_{l-1} \downarrow \mathtt{s} & u_l \downarrow 1 \end{bmatrix}. \qquad (6)$$

Given an arbitrary $(m \times n)$-picture P and r_0, r_1, r_2, and r_3 with $1 \leq r_0 \leq r_1 \leq m$ and $1 \leq r_2 \leq r_3 \leq n$, we define the *subpicture* of P determined by r_0,

..., r_3, denoted $P[r_0, r_1, r_2, r_3]$, or, $P[\bar{r}]$ for short, to be

$$\begin{bmatrix} P_{r_0,r_2} & P_{r_0,r_2+1} & \cdots & P_{r_0,r_3} \\ P_{r_0+1,r_2} & P_{r_0+1,r_2+1} & \cdots & P_{r_0+1,r_3} \\ P_{r_0+2,r_2} & P_{r_0+2,r_2+1} & \cdots & P_{r_0+2,r_3} \\ \vdots & \vdots & \ddots & \vdots \\ P_{r_1,r_2} & P_{r_1,r_2+1} & \cdots & P_{r_1,r_3} \end{bmatrix} . \tag{7}$$

As with diagonal pictures, we describe subpictures of diagonal pictures by words. For this, we use two additional symbols: $\mathbf{h}$, the *horizontal clipping mark*, and $\mathbf{v}$, the *vertical clipping mark*. We write E for the sets of subsets of $C \cup \{\mathbf{h}, \mathbf{v}\}$.

With $P(u)[\bar{r}]$ we associate the word $u[\bar{r}]$ over E defined by:

- $|u[\bar{r}]| = |u|$,
- $u[\bar{r}]_i \cap C = u_i$ for $i \in \{1, \ldots, |u|\}$,
- $\mathbf{h} \in u[\bar{r}]_i$ iff $r_0 = i$ or $r_1 = i$, and
- $\mathbf{v} \in u[\bar{r}]_i$ iff $r_2 = i$ or $r_3 = i$.

The important lemma about diagonal pictures is the following.

Lemma 9. *Let $k \geq 0$. Assume u and v are words over D and $\bar{r}$ and $\bar{s}$ are such that $u[\bar{r}] \equiv_{7k+4} v[\bar{s}]$. Then $P(u)[\bar{r}] \equiv_k P(v)[\bar{s}]$.*

Proof. The proof goes by induction on k.

Induction base, $k = 0$. By symmetry, it is sufficient to show that under the assumption $u[\bar{r}] \equiv_4 v[\bar{s}]$, if $P(u)[\bar{r}]$ is atomic, then so is $P(v)[\bar{s}]$ and both pictures are identical.

In general, a picture $P(u)[\bar{r}]$ is atomic if and only if $r_0 = r_1$ and $r_2 = r_3$. This is true if and only if $u[\bar{r}]$ contains exactly one position i such that $\mathbf{h} \in u[\bar{r}]_i$ and exactly one position j such that $\mathbf{v} \in u[\bar{r}]_j$. Whether or not this is true is easily seen to be determined by the 4-equivalence class of $u[r]$. Hence, if $u[\bar{r}] \equiv_4 v[\bar{s}]$ and $P(u)[\bar{r}]$ is atomic, then $P(v)[\bar{s}]$ is atomic as well.

Furthermore, if a picture $P(u)[\bar{r}]$ is atomic, then:

$$P(u)[\bar{r}] = \begin{cases} u[\bar{r}]_{r_0} \downarrow 1 & \text{if } r_0 = r_2, \\ u[\bar{r}]_{r_2} \downarrow \mathbf{n} & \text{if } 1 = r_0 < r_2, \\ u[\bar{r}]_{r_0} \downarrow \mathbf{w} & \text{if } 1 = r_2 < r_0, \\ u[\bar{r}]_{r_2} \downarrow \mathbf{s} & \text{if } 1 < r_2 < r_0 = |u|, \\ u[\bar{r}]_{r_0} \downarrow \mathbf{e} & \text{if } 1 < r_0 < r_2 = |u|, \\ 0 & \text{otherwise.} \end{cases}$$

It is now easily seen that the order relation between r_0 and r_2 as well as which of these two values is 1 or $|u|$ is determined by the 4-equivalence class of $u[\bar{r}]$, i.e., if $u[\bar{r}] \equiv_4 v[\bar{s}]$ and $P(u)[\bar{r}]$ is atomic, then we are in the same of the above cases for both pictures $P(u)[\bar{r}]$ and $P(v)[\bar{s}]$.

Also, if $u[\bar{r}] \equiv_4 v[\bar{s}]$, $r_0 = r_1$, $r_2 = r_3$, $s_0 = s_1$, and $s_2 = s_3$, then $u[\bar{r}]_{r_0} = v[\bar{s}]_{s_0}$ and $u[\bar{r}]_{r_1} = u[\bar{s}]_{s_1}$. Thus, if $u[\bar{r}] \equiv_4 v[\bar{s}]$ and $P(u)[\bar{r}]$ is atomic, then $P(u)[\bar{r}]$ and $P(v)[\bar{s}]$ are identical.

Induction step. Assume that for all $k' \leq k$ and all $\bar{r}'$ and $\bar{s}'$, if $u[\bar{r}'] \equiv_{7k'+4} v[\bar{s}']$, then $P(u)[\bar{r}'] \equiv_{k'} P(v)[\bar{s}']$. Assume also that $u[\bar{r}] \equiv_{7k+11} v[\bar{s}]$. Write P and Q for $P(u)[\bar{r}]$ and $P(v)[\bar{s}]$, respectively. We want to show $P \equiv_{k+1} Q$.

First, notice that we have $P \equiv_k Q$ by induction hypothesis, as $u[\bar{r}] \equiv_{7k+11} v[\bar{s}]$ implies $u[\bar{r}] \equiv_{7k+4} v[\bar{s}]$. So (K) holds. What we need to show in addition is that (J), (S), (J'), and (S') hold. By symmetry, it is enough to consider only (J).

Let P_1 and P_2 be such that $P = P_1 \square P_2$. There exists r_4 such that $r_2 \leq r_4$, $r_4 + 1 \leq r_3$, and
$$P_1 = P(u)[r_0, r_1, r_2, r_4] ,$$
$$P_2 = P(u)[r_0, r_1, r_4 + 1, r_3] .$$

In the rest of this proof we will only analyze the situation where $r_2 < r_4$ and $r_4 + 1 < r_3$; the other three cases (where $r_2 = r_4$ and $r_4 + 1 = r_3$, or $r_2 < r_4$ and $r_4 + 1 = r_3$, or $r_2 = r_4$ and $r_4 + 1 < r_3$) are simpler and can be dealt with in a similar way.

There exist unique (possibly empty) words u_1, u_2, u_3, and u_4 and letters a_1, a_2, a_3, and a_4 such that
$$u[\bar{r}] = u_1(a_1 \cup \{\mathbf{v}\})u_2 a_2 a_3 u_3 (a_4 \cup \{\mathbf{v}\})u_4 ,$$
$$u[r_0, r_1, r_2, r_4] = u_1(a_1 \cup \{\mathbf{v}\})u_2(a_2 \cup \{\mathbf{v}\})a_3 u_3 a_4 u_4 ,$$
$$u[r_0, r_1, r_4+1, r_3] = u_1 a_1 u_2 a_2 (a_3 \cup \{\mathbf{v}\})u_3 (a_4 \cup \{\mathbf{v}\})u_4 .$$

Since we assume $u[\bar{r}] \equiv_{7k+11} v[\bar{s}]$, we can conclude there are v_1, v_2, v_3, and v_4 such that
- $v[\bar{s}] = v_1(a_1 \cup \{\mathbf{v}\})v_2 a_2 a_3 v_3 (a_4 \cup \{\mathbf{v}\})v_4$, and
- $u_1 \equiv_{7k+4} v_1$, $u_2 \equiv_{7k+4} v_2$, $u_3 \equiv_{7k+4} v_3$, and $u_4 \equiv_{7k+4} v_4$.

Let $s_4 = |v_1 a_1 v_2 a_2|$ and define
$$Q_1 = P(v)[s_0, s_1, s_2, s_4] ,$$
$$Q_2 = P(v)[s_0, s_1, s_4 + 1, s_3] .$$

Then $Q = Q_1 \square Q_2$, and to finish the proof we need only show $P_1 \equiv_k Q_1$ and $P_2 \equiv_k Q_2$.

From the definition of s_4, we know:
$$v[s_0, s_1, s_2, s_4] = v_1(a_1 \cup \{\mathbf{v}\})v_2(a_2 \cup \{\mathbf{v}\})a_3 v_3 a_4 v_4 ,$$
$$v[s_0, s_1, s_4+1, s_3] = v_1 a_1 v_2 a_2 (a_3 \cup \{\mathbf{v}\})v_3 (a_4 \cup \{\mathbf{v}\})v_4 .$$

Since $(7k + 4)$-equivalence is a congruence relation on words, we obtain:
$$u_1(a_1 \cup \{\mathbf{v}\})u_2(a_2 \cup \{\mathbf{v}\})a_3 u_3 a_4 u_4 \equiv_{7k+4} v_1(a_1 \cup \{\mathbf{v}\})v_2(a_2 \cup \{\mathbf{v}\})a_3 v_3 a_4 v_4 ,$$
$$u_1 a_1 u_2 a_2 (a_3 \cup \{\mathbf{v}\})u_3 (a_4 \cup \{\mathbf{v}\})u_4 \equiv_{7k+4} v_1 a_1 v_2 a_2 (a_3 \cup \{\mathbf{v}\})v_3 (a_4 \cup \{\mathbf{v}\})v_4 .$$

This implies, by the induction hypothesis,
$$P(u)[r_0, r_1, r_2, r_4] \equiv_k P(v)[s_0, s_1, s_2, s_4] ,$$
$$P(u)[r_0, r_1, r_4+1, r_3] \equiv_k P(v)[r_0, r_1, r_4+1, r_3] ,$$

hence $P_1 \equiv_k Q_1$ and $P_2 \equiv_k Q_2$.

5 A Combinatorial Lemma about Words

As said above, we will define the pictures P^k and Q^k we are looking for as certain diagonal pictures: we will set $P^k = P(s_k)$ and $Q^k = P(t_k)$ for appropriate words s_k and t_k with specific combinatorial properties. The building blocks in the construction of these words are the words described in the following lemma.

Lemma 10. Let A be an arbitrary finite alphabet, $a \in A$, and $k, m \geq 0$. there exists a word $u_{k,m}$ such that

- all words in $(u_{k,m}^+ A)^* u_{k,m}^+$ are k-equivalent,
- $u_{k,m} \in (a^{\geq m} A)^* a^{\geq m}$, and
- $u_{k,m} \in A^*(bA^*)^{m+1}$ for every $b \in A$.

Proof. By induction on k. Let $A = \{a_1, \ldots, a_n\}$ and assume $a = a_1$.

Induction base. For $k = 0$, the following choice is obviously correct:

$$u_{0,m} = a_1 a_1 (a_1^m a_2 a_1^m a_3 a_1^m \cdots a_1^m a_n a_1^m)^{m+1} \, .$$

Induction step. Suppose $u_{k,m}$ is a word such that the above three conditions hold. Set

$$u_{k+1,m} = u_{k,m} u_{k,m} a_1 u_{k,m} u_{k,m} a_2 u_{k,m} u_{k,m} \cdots u_{k,m} u_{k,m} a_n u_{k,m} u_{k,m} \, . \quad (8)$$

We claim that this choice is correct. The second and third condition are obviously satisfied.

By the induction hypothesis, all words in $(u_{k+1,m}^+ A)^* u_{k+1,m}^+$ are k-equivalent. Furthermore, if $u = u'u''$ is a decomposition of a word from the set denoted by $(u_{k+1,m}^+ A)^* u_{k+1,m}^+$, then

- there exist w, w', w'', and w''' such that $u' = ww'$, $u'' = w''w'''$, $w'w'' \in u_{k,m} A u_{k,m}$, and $w \equiv_k u_{k,m} \equiv_k w'''$ (by the induction hypothesis); or
- $|u'| < |u_{k,m}|$ and there exists w and w' such that $u'w = u_{k,m}$, $ww' = u''$, and $w' \equiv_k u_{k,m}$ (by the induction hypothesis); or, symmetrically,
- $|u''| < |u_{k,m}|$ and there exists w and w' such that $w'u'' = u_{k,m}$, $ww' = u'$, and $w \equiv_k u_{k,m}$ (by the induction hypothesis).

On the other hand, every word from $(u_{k+1,m}^+ A)^* u_{k+1,m}^+$ allows all the decompositions described above. Therefore, all words in $(u_{k+1,m}^+ A)^* u_{k+1,m}^+$ are $(k+1)$-equivalent.

6 Tying Things Together

As pointed out in Section 3, all we need to do in order to prove that L_+ is not star-free is to define pictures $P^k \in L_+$ and $Q^k \notin L_+$ and show $P^k \equiv_k Q^k$, for $k \geq 0$.

For notational convenience, write $+$ for $\{1, \mathtt{n}, \mathtt{s}, \mathtt{w}, \mathtt{e}\}$ and $\top$ for $\{1, \mathtt{s}, \mathtt{w}, \mathtt{e}\}$.

Let w_k be the word $u_{8k, 2l_{7k}}$ from Lemma 10 with $A = D \setminus \{+\}$ and $a = \emptyset$. Set $s_k = w_k + w_k$ and $t_k = w_k \emptyset w_k$ and define $P^k = P(s_k)$ and $Q^k = P(t_k)$. These are the pictures we are looking for:

Proposition 11. For $k \geq 0$,

1. $P^k \in L_+$ and $Q^k \notin L_+$, and
2. $P^{k'} \equiv_k Q^{k'}$ for all $k' \geq k$.

Proof. The first claim is obvious (cf. (6)).

The proof of the second claim goes by induction on k. The induction base, $k = 0$, is trivial. In the induction step, we assume $P^{k'} \equiv_k Q^{k'}$ for all $k' \geq k$ and need to show $P^{k'} \equiv_{k+1} Q^{k'}$ for all $k' > k$.

Let $k' > k$. Write P and Q for $P^{k'}$ and $Q^{k'}$. We have to verify (K), (J), (S), (J'), and (S'). By induction hypothesis, we know $P \equiv_k Q$, hence (K) holds. Of the other four requirements, we will only consider (S) in the rest of this proof. That (J), (J'), and (S') hold can be proven in a similar fashion.

Let $P = P_1 \boxminus P_2$. Without loss of generality, assume P_1 has less rows than P_2. (Notice that P_1 and P_2 cannot have the same number of rows.)

Also, assume that P_1 and P_2 have at least 2 rows. If this is not the case, the situation is simpler but changes would have to be made to the notation in the following.

We have to find Q_1 and Q_2 such that $Q = Q_1 \boxminus Q_2$, $P_1 \equiv_k Q_1$, and $P_2 \equiv_k Q_2$.

Let p be the number of rows of P (which is also the number of columns of P and the number of rows and columns of Q) and p_1 the number of rows of P_1. Then $P_1 = P(w_{k'} \dotplus w_{k'})[1, p_1, 1, p]$ and $P_2 = P(w_{k'} \dotplus w_{k'})[p_1 + 1, p, 1, p]$.

Write $w_{k'} \dotplus w_{k'}$ as $a_1 s' a_2 a_3 s'' \dotplus s''' a_4$ such that

$$(w_{k'} \dotplus w_{k'})[1, p_1, 1, p] = (a_1 \cup \{\mathbf{h}, \mathbf{v}\}) s' (a_2 \cup \{\mathbf{h}\}) a_3 s'' \dotplus s''' (a_4 \cup \{\mathbf{v}\}) ,$$
$$(w_{k'} \dotplus w_{k'})[p_1 + 1, p, 1, p] = (a_1 \cup \{\mathbf{v}\}) s' a_2 (a_3 \cup \{\mathbf{h}\}) s'' \dotplus s''' (a_4 \cup \{\mathbf{h}, \mathbf{v}\}) .$$

By definition of $w_{k'}$, we know $w_{k'} \top w_{k'} \equiv_{8k'} w_{k'}$. Therefore, there exist t' and t'' such that $w_{k'} = a_1 t' a_2 a_3 t''$ and

$$s' \equiv_{8k'-4} t' , \tag{9}$$
$$s'' \top w_{k'} \equiv_{8k'-4} t'' . \tag{10}$$

Let $q_1 = |a_1 t' a_2|$, and define

$$Q_1 = P(w_{k'} \emptyset w_{k'})[1, q_1, 1, p] , \tag{11}$$
$$Q_2 = P(w_{k'} \emptyset w_{k'})[q_1 + 1, p, 1, p] . \tag{12}$$

Clearly, $Q = Q_1 \boxminus Q_2$. To conclude the proof, we show $P_1 \equiv_k Q_1$ and $P_2 \equiv_k Q_2$.

Proof of $P_1 \equiv_k Q_1$. First note the following. Since $\equiv_{8k'-4}$ is a congruence relation, (10) implies $s'' \top w_{k'} \emptyset w_{k'} \equiv_{8k'-4} t'' \emptyset w_{k'}$, which, in turn, by assumption about $w_{k'}$, implies

$$s'' \top w_{k'} \equiv_{8k'-4} t'' \emptyset w_{k'} , \tag{13}$$

hence

$$s'' \top s''' \equiv_{8k'-5} t'' \emptyset s''' . \tag{14}$$

We now proceed by a case distinction on $|s''|$.

First case, $|s''| \geq l_{7k'}$. First of all, observe

$$P_1 = P(a_1 s' a_2 a_3 (s'' \widetilde{+} s''' \downarrow \mathbf{n}) a_4)[1, p_1, 1, p] ,\tag{15}$$

$$Q_1 = P(a_1 t' a_2 a_3 (t'' \emptyset s''' \downarrow \mathbf{n}) a_4)[1, q_1, 1, p] .\tag{16}$$

Since $|s''| \geq l_{7k'}$, we can use Corollary 7 (part (a) or (b)) in combination with the definition of $w_{k'}$ to conclude:

$$s'' \widetilde{+} s''' \downarrow \mathbf{n} \equiv_{7k'} s'' \top s''' \downarrow \mathbf{n} .\tag{17}$$

This, together with (14) and Lemma 4, yields

$$s'' \widetilde{+} s''' \downarrow \mathbf{n} \equiv_{7k'} t'' \emptyset s''' \downarrow \mathbf{n} .\tag{18}$$

Using the congruence property again and combining (9) and (18), we obtain:

$$(a_1 \cup \{\mathbf{h}, \mathbf{v}\}) s' (a_2 \cup \{\mathbf{h}\}) a_3 (s'' \widetilde{+} s''' \downarrow \mathbf{n}) (a_4 \cup \{\mathbf{v}\})$$
$$\equiv_{7k'-3} (a_1 \cup \{\mathbf{h}, \mathbf{v}\}) t' (a_2 \cup \{\mathbf{h}\}) a_3 (t'' \emptyset s''' \downarrow \mathbf{n}) (a_4 \cup \{\mathbf{v}\}) ,$$

which means, as $k' > k$,

$$(a_1 s' a_2 a_3 (s'' \top s''' \downarrow \mathbf{n}) a_4)[1, p_1, 1, p] \equiv_{7k+4} P(a_1 t' a_2 a_3 (t'' \emptyset s''' \downarrow \mathbf{n}) a_4)[1, q_1, 1, p] .$$

From this, (15), and (16), together with Lemma 9, now follows $P_1 \equiv_k Q_1$.

Second case, $|s''| < l_{7k'}$. Write l for $l_{7k'}$. Then $|s'| > l$, and, by construction of $w_{k'}$, we can write s' as $s_0 \emptyset^l$ for an appropriate s_0. Define pictures R and R' as follows:

$$R = P(a_1 s_0 (\emptyset^l a_2 a_3 s'' \widetilde{+} w_{k'} \downarrow \mathbf{n}))[1, |a_1 s_0|, 1, p] ,$$
$$R' = P(a_1 s_0 (\emptyset^l a_2 a_3 s'' \emptyset w_{k'} \downarrow \mathbf{n}))[1, |a_1 s_0|, 1, p] .$$

We have $\emptyset^l a_2 a_3 s'' a_4 \widetilde{+} w_{k'} \downarrow \mathbf{n} \equiv_{7k'} \emptyset^l a_2 a_3 s'' a_4 \emptyset w_{k'} \downarrow \mathbf{n}$ by Corollary 7(a), hence $R \equiv_k R'$ by Lemma 9.

Let Z be the unique $((l+1) \times p)$-picture over $\{0\}$. Then, by Lemma 8, $R \boxminus Z \equiv_k R' \boxminus Z$. On the other hand, $R \boxminus Z = P_1$ (recall that $a_2 = \emptyset$ by definition of $w_{k'}$). So for the rest it is enough to show $R' \boxminus Z \equiv_k Q_1$.

By construction of R' and Z, we know

$$R' \boxminus Z = P(a_1 s' a_2 a_3 s'' \emptyset w_{k'})[1, p_1, 1, p] .\tag{19}$$

Combining (9) and (13), we obtain

$$(a_1 s' a_2 a_3 s'' \top w_{k'})[1, p_1, 1, p] \equiv_{8k'-4} (a_1 t' a_2 a_3 t'' \emptyset w_{k'})[1, q_1, 1, p] .\tag{20}$$

Thus, by Lemma 9, $R' \boxminus Z \equiv_k Q_1$.

Proof of $P_2 \equiv_k Q_2$. Combining (9) and (14), we obtain

$$(a_1 \cup \{\mathbf{v}\}) s' a_2 (a_3 \cup \{\mathbf{h}\}) s'' \top s''' (a_4 \cup \{\mathbf{h}, \mathbf{v}\})$$
$$\equiv_{8k'-5} (a_1 \cup \{\mathbf{v}\}) t' a_2 (a_3 \cup \{\mathbf{h}\}) t'' \emptyset s''' (a_4 \cup \{\mathbf{h}, \mathbf{v}\}) ,$$

which means
$$(a_1 s' a_2 a_3 s'' \top w_{k'})[p_1+1,p,1,p] \equiv_{7k+4} (a_1 t' a_2 a_3 t'' \emptyset w_{k'})[q_1+1,p,1,p] \ .$$
Using Lemma 9, we conclude
$$P(w_{k'} \top w_{k'})[p_1+1,p,1,p] \equiv_k P(w_{k'} \emptyset w_{k'})[q_1+1,p,1,p] \ ,$$
which means $P_2 \equiv_k Q_2$.

7 Concluding Remarks

We have seen that the class of star-free picture languages is strictly included in the class of first-order definable picture languages, which clarifies an important aspect of the fine structure of the class of all recognizable picture languages.

One obvious question is: what happens when the power of star-free picture expression is enhanced, for instance, by introducing a concatenation with four arguments? The proof methods presented here yield the following result: star-free picture expressions are strictly less expressive than star-free picture expressions augmented by the four-place concatenation, and these expressions are strictly less expressive than first-order logic.

The second question that is interesting here is whether there is a constant k such that each first-order sentence over pictures is equivalent to a first-order sentence using k variables. This is true for words and $k = 3$, see [7].

References

1. J. R. Büchi. Weak second-order arithmetic and finite automata. *Zeitschrift für mathematische Logik und Grundlagen der Mathematik*, 6:66–92, 1960.
2. H.-D. Ebbinghaus, J. Flum, and W. Thomas. *Mathematical Logic*. Springer-Verlag, New York, 1984.
3. R. Fraïssé. Sur quelques classifications des relations, basés sur des isomorphismes restreints, Publ. Sci. de l'Univ. Alger, Sér. A 1, pp. 35–182, 1954.
4. D. Giammarresi and A. Restivo. Two-dimensional languages, in G. Rozenberg and A. Salomaa, ed., *Handbook of Formal Languages*. Springer-Verlag, Berlin. To appear. Preprint available as: http://www.dsi.unive.it/%7Edora/Papers/chap96.ps.Z.
5. D. Giammarresi and A. Restivo. Two-dimensional finite state recognizability. *Fundamenta Informaticae*. To appear.
6. D. Giammarresi, A. Restivo, S. Seibert, and W. Thomas. Monadic second-order logic over rectangular pictures and recognizability by tiling systems. *Inform. and Comput.*, 125(1):32–45, 1996.
7. N. Immerman and D. Kozen. Definability with bounded number of bound variables. *Inform. and Comput.*, 83(2):121–139, 1989.
8. H. A. Maurer, G. Rozenberg, and E. Welzl. Using string languages to describe picture languages. *Inform. and Control*, 54(3):155–185, 1982.
9. R. McNaughton and S. Papert. *Counter-Free Automata*, vol. 69 of Research Monograph. MIT Press, Cambridge, Mass., 1971.
10. W. Thomas. A concatenation game and the dot-depth hierarchy. In E. Börger, editor, *Computation Theory and Logic*, volume 270 of Lecture Notes in Comput. Science, pages 415–426. Springer-Verlag, 1987.

An Algebra-Based Method to Associate Rewards with EMPA Terms

Marco Bernardo

Università di Bologna, Dipartimento di Scienze dell'Informazione
Mura Anteo Zamboni 7, 40127 Bologna, Italy
E-mail: bernardo@cs.unibo.it

Abstract. We present a simple method to associate rewards with terms of the stochastic process algebra EMPA in order to make the specification and the computation of performance measures easier. The basic idea behind this method is to specify rewards within actions of EMPA terms, so it substantially differs from methods based on modal logic. The main motivations of this method are its ease of use as well as the possibility of defining a notion of equivalence which relates terms having the same reward, thus allowing for simplification without altering the performance index. We prove that such an equivalence is a congruence finer than the strong extended Markovian bisimulation equivalence, and we present its axiomatization.

1 Introduction

A commonly used method to specify steady-state performance measures for Markovian models is based on *rewards* [6]. The basic idea is that a number describing a reward (or weight) is attached to every state of the Markovian model, and the performance index is defined as the weighted sum of the steady-state probabilities of the states of the Markovian model.

So far the specification of performance measures in the field of stochastic process algebras has received a scarce attention. The main negative consequence is that the whole Markovian model underlying a given term has to be manually scanned by the designer in order to assign rewards to states.

Recently, in [3] a technique to formally specify rewards for the stochastic process algebra PEPA [5] has been proposed. The idea is to express rewards by means of the Hennessy-Milner logic [4]: a logical formula is specified together with an arithmetical expression, and every state satisfying the formula is assigned the reward specified by means of the arithmetical expression. We shall call such a method *logic-based*.

The idea of describing rewards through a modal logic seems to be quite adequate because modal logic formulae make assertions about changing state, hence they constitute an adequate link between algebraic terms, which describe the behavior of concurrent systems, and rewards, which are associated with states.

In this paper we propose a different way to associate rewards with terms of stochastic process algebras. The idea is not to use a separate formalism in order to specify rewards: they are directly described within the actions forming the algebraic terms. This method, which we shall call *algebra-based*, closely resembles the manual method consisting of associating rewards while scanning the state space of the Markovian model: the difference is that in the algebra-based method the algebraic term, which is much more compact than its underlying state space, is scanned and the appropriate actions are assigned a reward. The algebra-based method could be convenient due to its ease of use, since the designer is not forced to know the modal logic formalism, its low computational cost, as rewards are associated with states during the construction of the semantic models without the need to check for a modal logic formula, and the possibility of defining a congruence which equates terms having the same reward, thereby allowing for simplification without altering the performance measure.

The purpose of this paper is to extend the theory developed for the stochastic process algebra EMPA [1, 2] in order to deal with rewards according to the algebra-based method. In Sect. 2 we show that several performance measures can be derived using the algebra-based method. In Sect. 3 we introduce the syntax and the semantics for EMPA augmented with rewards. In Sect. 4 we define an equivalence which relates two terms if they have the same reward, we prove that such an equivalence is a congruence strictly contained in the strong extended Markovian bisimulation equivalence, and we present its axiomatization. In Sect. 5 we report some concluding remarks.

2 Deriving Performance Measures

In this section we show by means of an example that the algebra-based method we are going to introduce, though less powerful in general than the logic-based method proposed in [3], allows the designer to easily specify several steady-state performance measures frequently occurring in practice such as those identified in [3]: rate type (e.g. throughput of a service center), counting type (e.g. mean number of customers waiting in a service center), delay type (e.g. mean response time experienced by customers in a service center), and percentage type (e.g. the fraction of time during which a server is busy).

The example we consider is taken from queueing theory, and concerns a queueing system $M/M/n/n$ with arrival rate λ and service rate μ [7]. Such a queueing system represents a service center composed of n independent servers, such that the customer interarrival time is exponentially distributed with rate λ and the service time of each server is exponentially distributed with rate μ. The queueing system at hand can be given two different descriptions with EMPA: a state-oriented description where the focus is on the state of the set of servers (intended as the number of servers that are currently busy), and a resource-oriented description where the servers are modeled separately [9]. Recalling that "$<a,\tilde{\lambda}>.\_$" is the prefix operator where a is the action type and $\tilde{\lambda}$ is the action rate (a positive real number in the case of exponentially timed actions, $\infty_{l,w}$ in the case of prioritized weighted immediate actions, and $*$ in the case of passive actions), "$\_ + \_$" is the alternative composition operator, and "$\_\|_S\_$" is the parallel composition operator with synchronization set S, the state-oriented description is given by

$System_{M/M/n/n}^{so} \triangleq Arrivals \|_{\{a\}} Servers_0$
$Arrivals \triangleq <a, \lambda>.Arrivals$
$Servers_0 \triangleq <a, *>.Servers_1$
$Servers_h \triangleq <a, *>.Servers_{h+1} + <s, h \cdot \mu>.Servers_{h-1}, \quad 1 \leq h \leq n-1$
$Servers_n \triangleq <s, n \cdot \mu>.Servers_{n-1}$

whereas the resource-oriented description is given by

$System_{M/M/n/n}^{ro} \triangleq Arrivals \|_{\{a\}} Servers$
$Arrivals \triangleq <a, \lambda>.Arrivals$
$Servers \triangleq \underbrace{S \|_\emptyset S \|_\emptyset \ldots \|_\emptyset S}_{n}$
$S \triangleq <a, *>.<s, \mu>.S$

In order to highlight the difference between the logic-based method and the algebra-based method for assigning rewards to stochastic process algebra terms, we compute for the queueing system above the mean number of customers in the system. Since every state must be given a reward equal to the number of customers in that state, we proceed as follows:

- In the case of $System_{M/M/n/n}^{so}$, the reward specification used in the logic-based method is $\langle s \rangle tt \implies rate(s)/\mu$, i.e. every state having an outgoing transition with type s is given a reward equal to the rate of that transition divided by μ. Using the algebra-based method, every action of the form $<s, h \cdot \mu>$ must be replaced by $<s, h \cdot \mu, h>$ (and any other action must be replaced by a triple with zero reward). Thus, in such a case the two methods are equally simple.
- In the case of $System_{M/M/n/n}^{ro}$, the logic-based method turns out to be more complex because the modal logic formula must somehow count the number of possible consecutive actions with type s that can be executed: as a consequence, the rewards can be specified through the set composed of $\langle s \rangle \neg \langle s \rangle tt \implies 1, \langle s \rangle \langle s \rangle \neg \langle s \rangle tt \implies 2, \ldots, \langle s \rangle \langle s \rangle \ldots \langle s \rangle \neg \langle s \rangle tt \implies n$. If we use instead the algebra-based method, all we have to do is to replace every action of the form $<s, \mu>$ with $<s, \mu, 1>$ as we assume that rewards are additive (by analogy with rates of exponentially timed actions and weights of immediate actions), i.e. the reward gained by a state is the sum of the rewards labeling its outgoing transitions. Therefore, in such a case the ease of use of the algebra-based method becomes evident, and it would be even more evident if we considered e.g. a queueing system similar to the previous one where a FIFO queue with a given capacity is introduced in front of the set of servers: since the delivery of a customer from the queue to the server has to be modeled by means of an action, and since actions of type s are interleaved with actions of this kind, the formalization of modal logic formulae that capture the number of customers in the system is really difficult.

To conclude this section, we show that other performance measures for the queueing system above can be easily specified with the algebra-based method, and that this capability depends on the style used to represent the system:

- If we want to compute the throughput of the system, defined as the mean number of customers served per time unit, we have to take into account the rate of actions having type s. As a consequence, in the case of $System^{so}_{M/M/n/n}$ we must replace every action of the form $<s, h \cdot \mu>$ with $<s, h \cdot \mu, h \cdot \mu>$, while in the case of $System^{ro}_{M/M/n/n}$ we must replace every action of the form $<s, \mu>$ with $<s, \mu, \mu>$.
- If we want to compute the mean response time of the system, defined as the mean time spent by the customers in the system, we can exploit Little's law [7] which states that the mean response time of the system is equal to the mean number of customers in the system divided by the customer arrival rate. Therefore, in the case of $System^{so}_{M/M/n/n}$ we must replace every action of the form $<s, h \cdot \mu>$ with $<s, h \cdot \mu, h/\lambda>$, while in the case of $System^{ro}_{M/M/n/n}$ we must replace every action of the form $<s, \mu>$ with $<s, \mu, 1/\lambda>$.
- If we want to compute the utilization of the system, defined as the fraction of time during which servers are busy, we have to single out states having an outgoing transition labeled with s. Thus, in the case of $System^{so}_{M/M/n/n}$ we must replace every action of the form $<s, h \cdot \mu>$ with $<s, h \cdot \mu, 1>$. We observe that, unlike the logic-based method, in the case of $System^{ro}_{M/M/n/n}$ the algebra-based method cannot be used to determine the utilization of the system due to the additivity assumption: the rate to associate with actions of the form $<s, \mu>$ would be the reciprocal of the number of transitions labeled with s exiting from the same state. Since the main objective of the algebra-based method is its ease of use, we prefer to keep the specification of rewards as simple as possible, i.e. just by means of numbers: thus we avoid the introduction of arithmetical expressions as well as particular functions such as the one determining the number of transitions of a given type exiting from the same state. Incidentally, the inability to compute the utilization in the case of the resource-oriented description should not come as a surprise, since this description is more suited to the determination of performance indices concerning a single server instead of the whole set of servers. As it turns out, it is quite easy to measure the utilization of a given server specified in $System^{ro}_{M/M/n/n}$, whereas this is not possible for $System^{so}_{M/M/n/n}$. This means that the style [9] used to describe a given system through an algebraic term is strongly related to the possibility of deriving certain performance measures through the algebra-based method.

3   Syntax and Semantics for EMPA$_r$

In this section we extend the syntax and the semantics for EMPA [1, 2] in order to cope with the presence of rewards treated according to the algebra-based method outlined in the previous section: the resulting stochastic process algebra is called *EMPA$_r$*.

As usual, the building blocks of $EMPA_r$ are actions. Each action is a triple $<a, \tilde{\lambda}, r>$ consisting of the type of the action, the rate of the action and the reward of the action: the third component is new with respect to the structure of EMPA actions. Like in EMPA, actions are divided into external and internal (τ) according to types, while they are classified as exponentially timed, immediate or passive according to rates. Since exponentially timed actions model activities that are relevant from the performance standpoint, nonzero rewards can be assigned only to them. We denote by $AType$ the set of types, by $ARate = \mathbf{R}_+ \cup Inf \cup \{*\}$, with $Inf = \{\infty_{l,w} \mid l \in \mathbf{N}_+ \wedge w \in \mathbf{R}_+\}$, the set of rates, by $AReward = \mathbf{R}$ the set of rewards, and by $Act_r = \{<a, \tilde{\lambda}, r> \in AType \times ARate \times AReward \mid \tilde{\lambda} \in Inf \cup \{*\} \Longrightarrow r = 0\}$ the set of actions. We use $a, b, c, \ldots$ as metavariables for $AType$, $\tilde{\lambda}, \tilde{\mu}, \tilde{\gamma}, \ldots$ for $ARate$, $\lambda, \mu, \gamma, \ldots$ for $\mathbf{R}_+$, and $r, r', r'', \ldots$ for $AReward$. Finally, we denote by $PLevel = \{-1\} \cup \mathbf{N}$ the set of priority levels, and we assume that $* < \lambda < \infty_{l,w}$ for all $\lambda \in \mathbf{R}_+$ and $\infty_{l,w} \in Inf$.

Let $Const$ be a set of constants, ranged over by $A, B, C, \ldots$, and let $RFun = \{\varphi : AType \longrightarrow AType \mid \varphi(\tau) = \tau \wedge \varphi(AType - \{\tau\}) \subseteq AType - \{\tau\}\}$ be a set of relabeling functions.

Definition 1. The set $\mathcal{L}_r$ of *process terms* of $EMPA_r$ is generated by the following syntax
$$E ::= \underline{0} \mid <a, \tilde{\lambda}, r>.E \mid E/L \mid E[\varphi] \mid E + E \mid E \|_S E \mid A$$
where $L, S \subseteq AType - \{\tau\}$. The set $\mathcal{L}_r$ will be ranged over by $E, F, G, \ldots$. We denote by $\mathcal{G}_r$ the set of guarded and closed terms of $\mathcal{L}_r$. ∎

We recall from [1, 2] that the alternative composition operator is parametric in the nature of the choice: the choice is solved according to durations in the case of exponentially timed actions (race policy) and according to priorities and weights in the case of immediate actions (preselection policy), while it is purely nondeterministic in the case of passive actions. We also remind that, concerning the parallel composition operator, a synchronization can occur if and only if the involved actions have the same type belonging to the synchronization set, and at most one of the involved actions is not passive.

The integrated semantics of $EMPA_r$ terms can be defined by exploiting again the idea of potential move: the multiset [1] of the potential moves of a given term is inductively computed, then those potential moves having the highest priority level are selected and appropriately merged. The formal definition is based on the transition relation $\longrightarrow$, which is the least subset of $\mathcal{G}_r \times Act_r \times \mathcal{G}_r$ satisfying the inference rule reported in the first part of Table 1. This rule selects the potential moves having the highest priority level, and then merges together those having the same action type, the same priority level and the same

---

[1] We use "$\{\!|$" and "$|\!\}$" as brackets for multisets, "$\_ \oplus \_$" to denote multiset union, $\mathcal{M}u_{fin}(S)$ ($\mathcal{P}_{fin}(S)$) to denote the collection of finite multisets (sets) over set S, $M(s)$ to denote the multiplicity of element s in multiset M, and $\pi_i(M)$ to denote the multiset obtained by projecting the tuples in multiset M on their i-th component. Thus, e.g., $(\pi_1(PM_2))(<a, *, 0>)$ in the fifth part of Table 1 denotes the multiplicity of tuples of PM_2 whose first component is $<a, *, 0>$.

$$\frac{(<a,\tilde{\lambda},r>,E') \in Melt_r(Select_r(PM_r(E)))}{E \xrightarrow{a,\tilde{\lambda},r} E'}$$

$PM_r(\underline{0}) = \emptyset$

$PM_r(<a,\tilde{\lambda},r>.E) = \{|(<a,\tilde{\lambda},r>,E)|\}$

$PM_r(E/L) = \{|(<a,\tilde{\lambda},r>,E'/L) \mid (<a,\tilde{\lambda},r>,E') \in PM_r(E) \wedge a \notin L|\} \oplus$
$\qquad \{|(<\tau,\tilde{\lambda},r>,E'/L) \mid (<a,\tilde{\lambda},r>,E') \in PM_r(E) \wedge a \in L|\}$

$PM_r(E[\varphi]) = \{|(<\varphi(a),\tilde{\lambda},r>,E'[\varphi]) \mid (<a,\tilde{\lambda},r>,E') \in PM_r(E)|\}$

$PM_r(E_1 + E_2) = PM_r(E_1) \oplus PM_r(E_2)$

$PM_r(E_1 \|_S E_2) = \{|(<a,\tilde{\lambda},r>,E_1' \|_S E_2) \mid a \notin S \wedge (<a,\tilde{\lambda},r>,E_1') \in PM_r(E_1)|\} \oplus$
$\qquad \{|(<a,\tilde{\lambda},r>,E_1 \|_S E_2') \mid a \notin S \wedge (<a,\tilde{\lambda},r>,E_2') \in PM_r(E_2)|\} \oplus$
$\qquad \{|(<a,\tilde{\gamma},r>,E_1' \|_S E_2') \mid a \in S \wedge$
$\qquad\qquad (<a,\tilde{\lambda}_1,r_1>,E_1') \in PM_r(E_1) \wedge$
$\qquad\qquad (<a,\tilde{\lambda}_2,r_2>,E_2') \in PM_r(E_2) \wedge$
$\qquad\qquad \tilde{\gamma} = Norm_{r,rate}(a,\tilde{\lambda}_1,\tilde{\lambda}_2,PM_r(E_1),PM_r(E_2)) \wedge$
$\qquad\qquad r = Norm_{r,reward}(a,r_1,r_2,PM_r(E_1),PM_r(E_2))|\}$

$PM_r(A) = PM_r(E) \quad \text{if } A \stackrel{\Delta}{=} E$

$Select_r(PM) = \{|(<a,\tilde{\lambda},r>,E) \in PM \mid PL_r(<a,\tilde{\lambda},r>) = -1 \vee$
$\qquad \forall (<b,\tilde{\mu},r'>,E') \in PM. \; PL_r(<a,\tilde{\lambda},r>) \geq PL_r(<b,\tilde{\mu},r'>)|\}$

$PL_r(<a,*,0>) = -1 \qquad PL_r(<a,\lambda,r>) = 0 \qquad PL_r(<a,\infty_{l,w},0>) = l$

$Melt_r(PM) = \{(<a,\tilde{\lambda},r>,E) \mid (<a,\tilde{\mu},r'>,E) \in PM \wedge$
$\qquad \tilde{\lambda} = Min\{|\tilde{\gamma} \mid (<a,\tilde{\gamma},r''>,E) \in PM \wedge PL_r(<a,\tilde{\gamma},r''>) = PL_r(<a,\tilde{\mu},r'>)|\} \wedge$
$\qquad r = \sum\{|r'' \mid (<a,\tilde{\gamma},r''>,E) \in PM \wedge PL_r(<a,\tilde{\gamma},r''>) = PL_r(<a,\tilde{\mu},r'>)|\}\}$

$* \, Min \, * = * \qquad \lambda_1 \, Min \, \lambda_2 = \lambda_1 + \lambda_2 \qquad \infty_{l,w_1} \, Min \, \infty_{l,w_2} = \infty_{l,w_1+w_2}$

$Norm_{r,rate}(a,\tilde{\lambda}_1,\tilde{\lambda}_2,PM_1,PM_2) = \begin{cases} Split(\tilde{\lambda}_1, 1/(\pi_1(PM_2))(<a,*,0>)) \text{ if } \tilde{\lambda}_2 = * \\ Split(\tilde{\lambda}_2, 1/(\pi_1(PM_1))(<a,*,0>)) \text{ if } \tilde{\lambda}_1 = * \end{cases}$

$Norm_{r,reward}(a,r_1,r_2,PM_1,PM_2) = \begin{cases} r_1/(\pi_1(PM_2))(<a,*,0>) \text{ if } r_2 = 0 \\ r_2/(\pi_1(PM_1))(<a,*,0>) \text{ if } r_1 = 0 \end{cases}$

$Split(*,\alpha) = * \qquad Split(\lambda,\alpha) = \lambda \cdot \alpha \qquad Split(\infty_{l,w},\alpha) = \infty_{l,w \cdot \alpha}$

Table 1. Inductive rules for EMPA$_r$ integrated interleaving semantics

derivative term. The first operation is carried out through functions $Select_r$: $\mathcal{M}u_{fin}(Act_r \times \mathcal{G}_r) \longrightarrow \mathcal{M}u_{fin}(Act_r \times \mathcal{G}_r)$ and $PL_r : Act_r \longrightarrow PLevel$, which are defined in the third part of Table 1. The second operation is carried out through function $Melt_r : \mathcal{M}u_{fin}(Act_r \times \mathcal{G}_r) \longrightarrow \mathcal{P}_{fin}(Act_r \times \mathcal{G}_r)$ and partial function $Min : (ARate \times ARate) \longrightarrow ARate$, which are defined in the fourth part of Table 1. Observe that function $Melt_r$ sums the rewards of the potential moves to merge: this is consistent with the additivity assumption about rewards.

The multiset $PM_r(E) \in \mathcal{M}u_{fin}(Act_r \times \mathcal{G}_r)$ of potential moves of $E \in \mathcal{G}_r$ is defined by structural induction in the second part of Table 1. The normalization of rates and rewards of potential moves resulting from the synchronization of an action with several independent or alternative passive actions is carried out through partial functions $Norm_{r,rate} : (AType \times ARate \times ARate \times \mathcal{M}u_{fin}(Act_r \times \mathcal{G}_r) \times \mathcal{M}u_{fin}(Act_r \times \mathcal{G}_r)) \longrightarrow ARate$ and $Norm_{r,reward} : (AType \times AReward \times AReward \times \mathcal{M}u_{fin}(Act_r \times \mathcal{G}_r) \times \mathcal{M}u_{fin}(Act_r \times \mathcal{G}_r)) \longrightarrow AReward$, and function $Split : (ARate \times \mathbf{R}_{]0,1]}) \longrightarrow ARate$, which are defined in the fifth part of Table 1. Observe that the normalization of rewards is consistent with the additivity assumption about rewards.

Definition 2. The *integrated interleaving semantics* of $E \in \mathcal{G}_r$ is the labeled transition system $\mathcal{I}_r[\![E]\!] = (\uparrow E, Act_r, \longrightarrow_E, E)$ where $\uparrow E$ is the set of states reachable from E, and $\longrightarrow_E$ is $\longrightarrow$ restricted to $\uparrow E \times Act_r \times \uparrow E$. ∎

As in [1, 2], from the integrated semantic model it is possible to obtain a functional semantic model (by dropping action rates and rewards) as well as a performance semantic model (basically by dropping action types and by lifting rewards from transitions to states according to the additivity assumption). Due to lack of space, we do not show the related definitions here.

4 A Notion of Equivalence for EMPA$_r$

In [1, 2] we developed a notion of equivalence for EMPA called strong extended Markovian bisimulation equivalence and denoted $\sim_{EMB}$. Such an equivalence was defined according to the idea of probabilistic bisimulation [8] on the integrated semantic model, and we proved that it is necessary to define it on the integrated semantic model in order for the congruence property to hold. For the sake of convenience, we can extend $\sim_{EMB}$ to EMPA$_r$ since it disregards rewards, provided that like in [1, 2] we introduce a priority operator "$\Theta(\_)$" and we consider the language $\mathcal{L}_{r,\Theta}$ generated by the following syntax

$$E ::= \underline{0} \mid <a, \tilde{\lambda}, r>.E \mid E/L \mid E[\varphi] \mid \Theta(E) \mid E + E \mid E\|_S E \mid A$$

whose semantic rules are those in Table 1 except that the rule in the first part is replaced by

$$\frac{(<a, \tilde{\lambda}, r>, E') \in Melt_r(PM_r(E))}{E \xrightarrow{a, \tilde{\lambda}, r} E'}$$

and the following rule for the priority operator is introduced in the second part

$$PM_r(\Theta(E)) = Select_r(PM_r(E))$$

It is easily seen that $EMPA_r$ coincides with the set of terms $\{\Theta(E) \mid E \in \mathcal{L}_r\}$. We denote by $\mathcal{G}_{r,\Theta}$ the set of guarded and closed terms of $\mathcal{L}_{r,\Theta}$.

One of the advantages of the algebra-based method, besides its ease of use, is the possibility of defining a notion of equivalence for $EMPA_r$ which relates terms having the same reward, thus allowing for simplification without altering the value of the performance index we are interested in. Exploiting the lesson learnt with $\sim_{EMB}$, we define this new equivalence on the integrated semantic model. For simplicity, one may be tempted to relate strongly extended-Markovian bisimilar terms having the same total reward, intended as the sum of the rewards attached to the actions it can execute. However, in this way one would fail both to capture an equivalence preserving the performance measure at hand and to obtain a congruence.

Example 1. Consider terms
$$A \stackrel{\Delta}{=} <a,\lambda,r>.<b,\mu,r_1>.A$$
$$B \stackrel{\Delta}{=} <a,\lambda,r>.<b,\mu,r_2>.B$$
where $r_1 \neq r_2$. Then $A \sim_{EMB} B$ and A and B have the same total reward r, but if we solve the two underlying performance models we obtain two different values of the performance measure we are interested in: $r \cdot \mu/(\lambda+\mu) + r_1 \cdot \lambda/(\lambda+\mu)$ and $r \cdot \mu/(\lambda+\mu) + r_2 \cdot \lambda/(\lambda+\mu)$. ∎

Example 2. Consider terms
$$E_1 \equiv <a,\lambda,r_1>.\underline{0} + <b,\mu,r_2>.\underline{0}$$
$$E_2 \equiv <a,\lambda,r_2>.\underline{0} + <b,\mu,r_1>.\underline{0}$$
where $r_1 \neq r_2$. Then $E_1 \sim_{EMB} E_2$ and E_1 and E_2 have the same total reward $r_1 + r_2$. but e.g. $E_1 \|_{\{b\}} \underline{0}$ has total reward r_1 while $E_2 \|_{\{b\}} \underline{0}$ has total reward r_2. ∎

The examples above show that if we want to preserve the performance measure and to obtain a congruence, we cannot treat rewards separately from the rest of the actions: rewards must be checked in the bisimilarity clause in order to guarantee that, given two equivalent terms, they have the same total reward and any pair of equivalent terms reachable from them have the same total reward.

Below we show that it is really easy to extend the definition of $\sim_{EMB}$ in such a way that both objectives are achieved. Proofs of results are omitted whenever they are smooth adaptations of the corresponding proofs in [2].

Definition 3. We define partial functions *Rate*, *Reward*, *RR* with domain $\mathcal{G}_{r,\Theta} \times AType \times PLevel \times \mathcal{P}(\mathcal{G}_{r,\Theta})$ and ranges *ARate*, *AReward*, *ARate* × *AReward* respectively, by

$$Rate(E,a,l,C) = Min\{\!|\tilde{\lambda} \mid E \xrightarrow{a,\tilde{\lambda},r} E' \wedge PL_r(<a,\tilde{\lambda},r>) = l \wedge E' \in C |\!\}$$
$$Reward(E,a,l,C) = \sum \{\!| r \mid E \xrightarrow{a,\tilde{\lambda},r} E' \wedge PL_r(<a,\tilde{\lambda},r>) = l \wedge E' \in C |\!\}$$
$$RR(E,a,l,C) = (Rate(E,a,l,C), Reward(E,a,l,C))$$
∎

Definition 4. An equivalence relation $\mathcal{B} \subseteq \mathcal{G}_{r,\Theta} \times \mathcal{G}_{r,\Theta}$ is a *strong extended Markovian reward bisimulation (strong EMRB)* iff, whenever $(E_1, E_2) \in \mathcal{B}$, then for all $a \in AType$, $l \in PLevel$ and $C \in \mathcal{G}_{r,\Theta}/\mathcal{B}$
$$RR(E_1, a, l, C) = RR(E_2, a, l, C)$$
In this case we say that E_1 and E_2 are *strongly extended-Markovian reward bisimilar (strongly EMRB)*. ∎

Proposition 5. *Let $\sim_{EMRB}$ be the union of all the strong EMRBs. Then $\sim_{EMRB}$ is the largest strong EMRB.* ∎

Definition 6. We call $\sim_{EMRB}$ the *strong extended Markovian reward bisimulation equivalence (strong EMRBE)*, and we say that $E_1, E_2 \in \mathcal{G}_{r,\Theta}$ are *strongly extended-Markovian reward bisimulation equivalent (strongly EMRBE)* if and only if $E_1 \sim_{EMRB} E_2$. ∎

Proposition 7. $\sim_{EMRB} \subseteq \sim_{EMB}$.

Proof. It follows immediately from the fact that every strong EMRB is a strong EMB too. ∎

The following example shows that the inclusion is strict. We would like to point out that this is not inconsistent with $\sim_{EMB}$. The purpose of $\sim_{EMB}$ is to relate terms describing concurrent systems having the same functional and performance properties: if $E_1 \sim_{EMB} E_2$ but $E_1 \not\sim_{EMRB} E_2$, this simply means that we are measuring two different performance indices for E_1 and E_2.

Example 3. Consider terms
$$A \triangleq <a, \lambda, 1>.<b, \mu, 0>.A$$
$$B \triangleq <a, \lambda, 0>.<b, \mu, 1>.B$$
Then $A \sim_{EMB} B$ but $A \not\sim_{EMRB} B$. If we regard a and b as the transmission over two different channels, then by means of A we can compute the utilization of the former channel, whereas by means of B we can compute the utilization of the latter channel. ∎

Theorem 8. *Let $E_1, E_2 \in \mathcal{G}_{r,\Theta}$. If $E_1 \sim_{EMRB} E_2$ then:*

1. *For every $<a, \tilde{\lambda}, r> \in Act_r$, $<a, \tilde{\lambda}, r>.E_1 \sim_{EMRB} <a, \tilde{\lambda}, r>.E_2$.*
2. *For every $L \subseteq AType - \{\tau\}$, $E_1/L \sim_{EMRB} E_2/L$.*
3. *For every $\varphi \in RFun$, $E_1[\varphi] \sim_{EMRB} E_2[\varphi]$.*
4. *$\Theta(E_1) \sim_{EMRB} \Theta(E_2)$.*
5. *For every $F \in \mathcal{G}_{r,\Theta}$, $E_1 + F \sim_{EMRB} E_2 + F$ and $F + E_1 \sim_{EMRB} F + E_2$.*
6. *For every $F \in \mathcal{G}_{r,\Theta}$ and $S \subseteq AType - \{\tau\}$, $E_1 \|_S F \sim_{EMRB} E_2 \|_S F$ and $F \|_S E_1 \sim_{EMRB} F \|_S E_2$.* ∎

Theorem 9. *$\sim_{EMRB}$ is preserved by recursive definitions.* ∎

Theorem 10. *Let $\mathcal{A}_r$ be the set of axioms in Table 2. The deductive system $Ded(\mathcal{A}_r)$ is sound and complete with respect to $\sim_{EMRB}$ for the set of nonrecursive terms of $\mathcal{G}_{r,\Theta}$.* ∎

$(\mathcal{A}_{r,1})$ $(E_1 + E_2) + E_3 = E_1 + (E_2 + E_3)$
$(\mathcal{A}_{r,2})$ $E_1 + E_2 = E_2 + E_1$
$(\mathcal{A}_{r,3})$ $E + \underline{0} = E$
$(\mathcal{A}_{r,4})$ $<a, \tilde{\lambda}_1, r_1>.E + <a, \tilde{\lambda}_2, r_2>.E = <a, \tilde{\lambda}_1\ Min\ \tilde{\lambda}_2, r_1 + r_2>.E$
    if $PL_r(<a, \tilde{\lambda}_1, r_1>) = PL_r(<a, \tilde{\lambda}_2, r_2>)$

$(\mathcal{A}_{r,5})$ $\underline{0}/L = \underline{0}$
$(\mathcal{A}_{r,6})$ $(<a, \tilde{\lambda}, r>.E)/L = \begin{cases} <a, \tilde{\lambda}, r>.(E/L) \text{ if } a \notin L \\ <\tau, \tilde{\lambda}, r>.(E/L) \text{ if } a \in L \end{cases}$
$(\mathcal{A}_{r,7})$ $(E_1 + E_2)/L = E_1/L + E_2/L$

$(\mathcal{A}_{r,8})$ $\underline{0}[\varphi] = \underline{0}$
$(\mathcal{A}_{r,9})$ $(<a, \tilde{\lambda}, r>.E)[\varphi] = <\varphi(a), \tilde{\lambda}, r>.(E[\varphi])$
$(\mathcal{A}_{r,10})$ $(E_1 + E_2)[\varphi] = E_1[\varphi] + E_2[\varphi]$

$(\mathcal{A}_{r,11})$ $\Theta(\underline{0}) = \underline{0}$
$(\mathcal{A}_{r,12})$ $\Theta(\sum_{i \in I} <a_i, \tilde{\lambda}_i, r_i>.E_i) = \sum_{j \in J} <a_j, \tilde{\lambda}_j, r_j>.\Theta(E_j)$
    where $J = \{i \in I \mid \tilde{\lambda}_i = * \vee \forall h \in I.\ PL_r(<a_i, \tilde{\lambda}_i, r_i>) \geq PL_r(<a_h, \tilde{\lambda}_h, r_h>)\}$

$(\mathcal{A}_{r,13})$ $\underline{0} \|_S \underline{0} = \underline{0}$
$(\mathcal{A}_{r,14})$ $(\sum_{i \in I} <a_i, \tilde{\lambda}_i, r_i>.E_i) \|_S \underline{0} = \sum_{j \in J} <a_j, \tilde{\lambda}_j, r_j>.(E_j \|_S \underline{0})$ where $J = \{i \in I \mid a_i \notin S\}$
$(\mathcal{A}_{r,15})$ $\underline{0} \|_S (\sum_{i \in I} <a_i, \tilde{\lambda}_i, r_i>.E_i) = \sum_{j \in J} <a_j, \tilde{\lambda}_j, r_j>.(\underline{0} \|_S E_j)$ where $J = \{i \in I \mid a_i \notin S\}$
$(\mathcal{A}_{r,16})$ $(\sum_{i \in I_1} <a_i, \tilde{\lambda}_i, r_i>.E_i) \|_S (\sum_{i \in I_2} <a_i, \tilde{\lambda}_i, r_i>.E_i) =$
    $\sum_{j \in J_1} <a_j, \tilde{\lambda}_j, r_j>.(E_j \|_S \sum_{i \in I_2} <a_i, \tilde{\lambda}_i, r_i>.E_i) +$
    $\sum_{j \in J_2} <a_j, \tilde{\lambda}_j, r_j>.(\sum_{i \in I_1} <a_i, \tilde{\lambda}_i, r_i>.E_i \|_S E_j) +$
    $\sum_{k \in K_1 \wedge h \in H_k} <a_k, Split(\tilde{\lambda}_k, 1/n_k).r_k/n_k>.(E_k \|_S E_h) +$
    $\sum_{k \in K_2 \wedge h \in H_k} <a_k, Split(\tilde{\lambda}_k, 1/n_k).r_k/n_k>.(E_h \|_S E_k)$
    where $J_1 = \{i \in I_1 \mid a_i \notin S\}$
    $J_2 = \{i \in I_2 \mid a_i \notin S\}$
    $K_1 = \{i_1 \in I_1 \mid \exists i_2 \in I_2.\ a_{i_1} = a_{i_2} \in S \wedge \tilde{\lambda}_{i_2} = *\}$
    $K_2 = \{i_2 \in I_2 \mid \exists i_1 \in I_1.\ a_{i_1} = a_{i_2} \in S \wedge \tilde{\lambda}_{i_1} = *\}$
    $H_k = \{h \in I_2 \mid a_k = a_h \wedge \tilde{\lambda}_h = *\}$ with $k \in K_1$
    $H_k = \{h \in I_1 \mid a_k = a_h \wedge \tilde{\lambda}_h = *\}$ with $k \in K_2$
    $n_k = |H_k|$

Table 2. Axioms for $\sim_{EMRB}$

5 Conclusion

In this paper we have introduced an algebra-based method to attach rewards with EMPA terms in order to derive performance measures. As observed in Sect. 2, though less powerful in general than the logic-based method proposed in [3], the algebra-based method may be convenient due to its ease of use, its low computational cost and the possibility of defining a notion of equivalence accounting for rewards. Furthermore, it has been a really easy task to extend the theory developed for EMPA in order to take into account rewards according to the algebra-based method.

Concerning future work, we could allow the designer to associate rewards with immediate actions as well, because in this way we could derive performance measures also when we restrict ourselves to the probabilistic kernel [2] of EMPA. Finally, the algebra-based method will be implemented in a software tool (we are currently developing) based on EMPA for the modeling and analysis of functional and performance properties of concurrent systems.

Acknowledgements

This research has been partially funded by MURST and CNR.

References

1. M. Bernardo, R. Gorrieri, *"Extended Markovian Process Algebra"*, in Proc. of the *7th Int. Conf. on Concurrency Theory (CONCUR '96)*, LNCS 1119:315-330, Pisa (Italy), 1996
2. M. Bernardo, R. Gorrieri, *"A Tutorial on EMPA: A Theory of Concurrent Processes with Nondeterminism, Priorities, Probabilities and Time"*, Technical Report UBLCS-96-17, University of Bologna (Italy), 1996
3. G. Clark. *"Formalising the Specification of Rewards with PEPA"*, in Proc. of the *4th Workshop on Process Algebras and Performance Modelling (PAPM '96)*, CLUT, pp. 139-160, Torino (Italy), 1996
4. M. Hennessy, R. Milner, *"Algebraic Laws for Nondeterminism and Concurrency"*, in Journal of the ACM 32:137-161, 1985
5. J. Hillston, *"A Compositional Approach to Performance Modelling"*, Cambridge University Press, 1996
6. R.A. Howard, *"Dynamic Probabilistic Systems"*, John Wiley & Sons, 1971
7. L. Kleinrock, *"Queueing Systems"*, John Wiley & Sons, 1975
8. K.G. Larsen, A. Skou, *"Bisimulation through Probabilistic Testing"*, in Information and Computation 94(1):1-28, 1991
9. C.A. Vissers, G. Scollo, M. van Sinderen, E. Brinksma, *"Specification Styles in Distributed Systems Design and Verification"*, in Theoretical Computer Science 89:179-206, 1991

A Semantics Preserving Actor Translation

Ian A. Mason
University of New England
iam@turing.une.edu.au

Carolyn L. Talcott
Stanford University
clt@sail.stanford.edu

Abstract

In this paper we present two actor languages and a semantics preserving translation between them. The source of the translation is a high-level language that provides object-based programming abstractions. The target is a simple functional language extended with basic primitives for actor computation. The semantics preserved is the interaction semantics of actor systems—sets of possible interactions of a system with its environment. The proof itself is of interest since it demonstrates a methodology based on the actor theory framework for reasoning about correctness of transformations and translations of actor programs and languages and more generally of concurrent object languages.

1 Introduction

In this paper we continue our investigation of the actor model of computation [Hew77, Agh86, Agh90, AMST97, Tal96b, Tal96a]. Actors are independent computational agents that interact solely via *asynchronous* message passing. An actor can create other actors; send messages; and modify its own local state. An actor can only effect the local state of other actors by sending them messages, and it can only send messages to its acquaintances – addresses of actors it was given upon creation, it received in a message, or that it created. Actor semantics requires computations to be fair.

We take two views of actors: as individuals and as elements of components. Individual actors provide units of encapsulation and integrity. Components are collections of actors (and messages) provided with an interface specifying the receptionists (actors accessible from outside the component) and external actors (accessible from but not existing inside the component). Collecting actors into components provides for composability and coordination. Individual actors are described in terms of local transitions. Components are described in terms of interactions with their environment.

The actor model provides a natural framework for inter-operation of multiple languages since the details of the code describing an individual actors behavior are not visible outside that actor. All that needs to be common is the messages communicated among the different actors. In [Tal96b], this intuition is formalized using the notion of an *abstract actor structure*. Here we generalize the notion of an abstract actor structure to an *actor theory*. Actor theories provide a general semantic framework for specifying and reasoning about actor systems as well as for reasoning about relations between different actor languages. An actor theory plays the role of a theory that axiomatizes the behavior of individual actors. The models of an actor theory account directly for the interaction (exchange of messages) of a actor component with its environment. Each model of an actor theory gives rise to a corresponding semantics of actor components. Two important models are: computation paths—analogous to labelled transition system semantics; and interaction paths—obtained from computation paths by omitting details of internal computation. These give rise to computation path and interaction semantics, respectively. Both semantics are composable and as we will see, interaction semantics is largely insensitive to the particular choice of actor language.

In this paper we illustrate the ideas and techniques based on actor theories by showing how they can be used to establish the correctness of a translation from a high-level actor language to low level actor language such as might be found in compiler preprocessor. The low-level kernel language, $^k\mathcal{L}$, is an extension of a simple functional language based on the call-by-value λ-calculus with primitives for actor computation. The high-level user language, $^u\mathcal{L}$, provides object-based programming abstractions. Each of the languages is given a semantics by defining a corresponding actor theory. We give a separate semantics for the user language in order to be able to reason directly about user programs. The correctness theorem shows that we can also reason about user programs by translating to the kernel language and reasoning in terms of the kernel semantics. The translation, $u2k$, from the user language to the kernel language eliminates the object-based programming abstractions in favor of the simple actor primitives. The main result presented there is that the translation, $u2k$, preserves the interaction semantics.
Theorem (user-to-kernel): $Isem(^uP) = Isem(u2k(^uP))\lceil^u\mathrm{M}$ where ^uP is a user language program, $Isem$ maps programs to their interaction semantics, and $\lceil^u\mathrm{M}$ restricts the kernel interactions to user language messages.

The proof that the translation preserves interaction semantics itself is of interest since it demonstrates a methodology for proving correctness of transformations and translations of actor languages and more generally of concurrent object languages. For the proof we lift the translation to semantic configurations that correspond to the possible actor system states and show that the following diagram commutes

$$\begin{array}{ccc} ^uP & \stackrel{u2k}{\longrightarrow} & ^kP \\ {\scriptstyle [\_]}\downarrow & & \downarrow{\scriptstyle [\_]} \\ ^uK & \stackrel{u2k}{\longrightarrow} & ^kK \end{array}$$

where P is a top-level program, K, is a configuration, and $[\_]$ gives the semantics of a program in terms of the initial configuration that it describes. (We use the following convention: if X is some entity, then we use the super-prescript ^uX to indicate that X belongs to the user language and ^kX to indicate that X belongs to the kernel language. So for example ^uK is an user language configuration.) The proof is completed by showing that interaction semantics is preserved by translation at the semantic level $Isem(^uK) = Isem(u2k(^uK))\lceil^uM$. This proof involves establishing a correspondence between the (possibly infinite) computations of two systems. The actor theories defined for each of the languages correspond to standard transition system semantics with transitions that are small and easy to understand, but expose much irrelevant detail. We make use of a general interaction semantics preserving actor theory transformation that can be thought of as moving from a small step a big step operational semantics. Changing the level of abstraction of the operational semantics of a fixed language is a general technique useful for reasoning about systems at the desired level of detail. Reasoning about the level changing transformation on actor theories and the language changing translation is simplified by using ideas from the rewriting logic model of concurrent computation [Mes92, Tal96a] to define notions of computation path equivalence.

Notation: We use the usual notation for sets, functions, finite sequences, etc. Let Y be a set. We specify meta-variable conventions in the form: let y range over Y, which should be read as: the meta-variable y and decorated variants such as y', y_0, ..., range over the set Y. $\mathcal{M}_\omega[Y]$ is the set of (finite) multi-sets with elements in Y. $\emptyset$ is the empty multiset and if X_1 and X_2 are multisets, then X_0, X_1 is the multiset union of the two.

2 A Semantic Framework for Actors

In this section we introduce *actor theories* as a general semantic framework for actor computation. The notion of actor theory provides an axiomatic characterization of actor languages: the basic features, capabilities, and constraints. Actor theories can be considered as an operational alternative to the domain theoretic behaviors used by Clinger [Cli81]. Actor theories are a simplification and generalization of the notion of abstract actor structures presented in [Tal96b, Tal96a].

An actor theory describes individual actor behaviors and their local interactions in a representation independent manner. An actor theory specifies sets of *actor names*, *actor states*, *message contents*, and *labelled reaction rules*. Actor names are the means of uniquely identifying individual actors. Actor states are intended to carry information traditionally contained in the script (methods) and acquaintances (values of instance variables), as well as the local message queue and the current processing state. Message contents represent the information that can be communicated between actors, both locally and as interactions with the environment. Reaction rules determine what an actor in a given state can do next and how it will respond to messages with given contents. More generally reaction rules describe synchronous interactions of groups of actors and messages. Reaction rules are labelled. These labels are used in deriving a labelled transition system semantics. In this way the labels provide information concerning the basic observations that can be made as an actor system evolves. An actor theory must obey the fundamental acquaintance (locality) laws of actors [BH77, Cli81] in addition to renaming laws that express the fact that computation is uniformly parameterized in the choice of actor names—renaming commutes with everything. To state these laws an actor theory also provides a primitive operation to determine the acquaintances of (actor names occurring in) the various entities and a primitive operation to rename them.

The operational semantics of an actor theory is given by the transition relation on configurations derived from the reaction rules. A configuration can be thought of as representing a global snapshot of an actor system with respect to some idealized observer [Agh86]. It contains a set of receptionist names, a set of external actor names, and a collection of actors and messages. The sets of receptionist names and external actor names are the interface of an actor configuration to its environment. They specify what internal actors are visible from the environment, and what actor connections must be provided for the configuration to function. Both the set of receptionist names and the set of external actor names may grow as the configuration evolves. The collection of actors and messages is the *interior* of the configuration. It specifies the internal actors and their current states, and the state of the internal message system. Configurations evolve either by internal computation or by interaction with the environment. The transition relation expresses the ways a configuration might evolve and interact with its environment. The *computation path semantics* of a configuration is the set of fair computations possible starting with that configuration. *Interaction semantics* gives a more abstract view of an actor system, specifying only the possible interactions (patterns of message passing) a system can have with its environment. Interaction semantics is the result of hiding all information concerning the internal computations and what actors may be present beyond the receptionists.

The term *reaction rule* is used here in the same spirit as in the Chemical Abstract Machine [BB92] to indicate local interactions of reactive entities. Actors and messages can be thought of as special kinds of molecules and interiors are like solutions. Actor theories are in fact a special case of rewrite theories and we the mechanisms we use to derive the computations of a actor system are based on those of rewriting logic [Mes92].

An actor theory is a structure AT of the following form: $AT = \langle\,\langle\mathbf{A},\mathbf{S},\mathbf{M},\mathbf{L}\rangle,\,\langle acq,\widehat{\ }\rangle,\,RR\,\rangle$. $\mathbf{A}$, $\mathbf{S}$, $\mathbf{M}$, $\mathbf{L}$ are the primitive sorts of AT. $\mathbf{A}$ is a countable set of actor names, $\mathbf{S}$ is a set of actor states, $\mathbf{M}$ is a set of message contents, and $\mathbf{L}$ is a set of labels. From the primitive sorts we form actor entities (briefly actors), $\mathbf{AE}$, messages, $\mathbf{Msg}$, and configuration interiors, $\mathbf{I}$. We let a range over $\mathbf{A}$, M range over $\mathbf{M}$, s range over $\mathbf{S}$, l range over $\mathbf{L}$, and I range over $\mathbf{I}$. $[s]_a$ is an *actor* with *name*, a, in *state*, s and $a \triangleleft M$ is a *message* with *addressee*, a, and *contents*, M. A configuration interior, I, is a multiset of actors and messages in which no two actor entities have the same name.

RR is a set of reaction rules that specify the behavior of individual actors and their synchronization with other internal actors and messages. Elements of RR are triples of the form $l : I \Rightarrow I'$ where l is the rule label, I is rule source, and I' is the rule target.

The primitive operations of AT are: acq and $\widehat{\ }$. The acquaintance function, $acq : \mathbf{S} \cup \mathbf{M} \cup \mathbf{L} \to \mathcal{P}_\omega[\mathbf{A}]$, gives the (finite) set of actor names occurring in a state, message contents, or label. acq extends homomorphically to structures built from the primitive sorts. Actor addresses cannot be explicitly created by actors, and the semantics cannot depend on the particular choice of addresses of a group of actors. A renaming mechanism is used to formulate this requirement. We let $\mathbf{Bij}(\mathbf{A})$ be the set of bijections on $\mathbf{A}$ (renamings) and let α range over $\mathbf{Bij}(\mathbf{A})$. For any such α, $\widehat{\alpha}$ is the associated renaming function on states, message contents, and labels. Renaming is extended naturally to structures built from addresses, states, and values. For example $\widehat{\alpha}([s]_a) = [\widehat{\alpha}(s)]_{\alpha(a)}$. Renaming, $\widehat{\alpha}$, commutes with the acquaintance function and is determined by the restriction of α to the acquaintances of an object. It is a bijection on $\mathbf{A} \cup \mathbf{M} \cup \mathbf{L}$. To state the axioms for reaction rules, we define two auxiliary functions: $InAct, ExtAct : \mathbf{I} \to \mathcal{P}_\omega[\mathbf{A}]$. $InAct(I)$ is the set of names of actors that occur in I, and $ExtAct(I)$ is the set of names of external actors referred to in I: $InAct(I) = \{a \in \mathbf{A} \mid (\exists s \in \mathbf{S})([s]_a \in I)\}$ $ExtAct(I) = acq(I) - InAct(I)$.

Axioms for Reaction rules (RR)   If $l : I \Rightarrow I' \in RR$, then
(i) $InAct(I) \neq \emptyset$
(ii) $l : I_0 \Rightarrow I_0' \in RR$ implies $InAct(I) = InAct(I_0)$ and $InAct(I') = InAct(I_0')$
(iii) $InAct(I) \subseteq InAct(I') \subseteq acq(l)$
(iv) $ExtAct(I') \subseteq ExtAct(I)$
(v) $\widehat{\alpha}(l) : \widehat{\alpha}(I) \Rightarrow \widehat{\alpha}(I') \in RR$ for any renaming α in $\mathbf{Bij}(\mathbf{A})$

(i) states that reactions must involve at least one existing actor; (ii) states that a label uniquely determines the actors involve in a reaction; (iii) states that actors cannot disappear and that the actors involved in a reaction must be made explicit as acquaintances of the reaction label; (iv) states that no references to external actors are acquired in an internal transition, although some may be forgotten; and (v) states that the set of rules is closed under renaming.
If $l : I \Rightarrow I' \in RR$, we call $InAct(I)$ the *old* actors of l and $InAct(I') - InAct(I)$ the *new* actors of l.

An actor configuration is a configuration interior, I, together with two sets of actor names: the receptionists ρ, which are a subset of the internal actors of the interior; and the externals χ which include all actors mentioned in the interior that are not internal actors.

Definition (Configurations, K): $\mathbf{K} = \{ \langle\!\langle I \rangle\!\rangle_\chi^\rho \mid \rho \subseteq InAct(I) \wedge ExtAct(I) \subseteq \chi \}$. We let K range over $\mathbf{K}$.

The computations of a configuration are given by the labelled transition relation $K \xrightarrow{l} K'$. K is the *source* of the transition and K' is the *target* and l is the *label*. Transition labels are either rule labels, input/output labels, or a special idle label, $\mathtt{idle}$. An input label has the form $\mathtt{in}(a \triangleleft M)$, indicating a message coming in from the environment. An output label has the form $\mathtt{out}(a \triangleleft M)$ indicating a message transmitted to the environment. We now let the range of l include these additional transition labels.

Definition (Transition rules):

(internal) $\langle\!\langle I_0, I \rangle\!\rangle_\chi^\rho \xrightarrow{l} \langle\!\langle I_1, I \rangle\!\rangle_\chi^\rho$ if $l : I_0 \Rightarrow I_1 \in RR$

(in) $\langle\!\langle I \rangle\!\rangle_\chi^\rho \xrightarrow{\mathtt{in}(a.M)} \langle\!\langle I, a \triangleleft M \rangle\!\rangle_{\chi \cup (acq(M) - \rho)}^\rho$ if $a \in \rho \wedge acq(M) \cap InAct(I) \subseteq \rho$

(out) $\langle\!\langle I, a \triangleleft M \rangle\!\rangle_\chi^\rho \xrightarrow{\mathtt{out}(a,M)} \langle\!\langle I \rangle\!\rangle_\chi^{\rho \cup (acq(M) - \chi)}$ if $a \notin InAct(I)$

(idle) $\langle\!\langle I \rangle\!\rangle_\chi^\rho \xrightarrow{\mathtt{idle}} \langle\!\langle I \rangle\!\rangle_\chi^\rho$

In (**internal**) we assume that the configurations are well-formed – $InAct(I_1) \cap InAct(I) = \emptyset$, $\rho \subseteq InAct(I_0) \cup InAct(I)$, and $ExtAct(I_0, I) \subseteq \chi$.

The computation paths of a configuration, $\mathcal{P}(K)$ are the computation paths whose initial configuration is K.

Definition (Computation Paths, $\mathcal{P}, \mathcal{P}(K)$): $\mathcal{P}$ is the set of sequences, π, of the form

$$\pi = [K_i \xrightarrow{l_i} K_{i+1} \mid i \in \mathbf{N}] \qquad \mathcal{P}(K) = \{\pi \in \mathcal{P} \mid K \text{ is the source of } \pi(0)\}$$

A finite computation is a path in which all but a finite number of the transition labels are $\mathtt{idle}$. Recall that actor computations are required to be fair. Thus we do not want to consider arbitrary paths, only the fair ones. A computation is fair if whenever a transition is enabled, either it eventually fires or it becomes permanently disabled. We only consider enabledness for transitions whose label is a reaction rule label or an output label. We can not force the environment to do an input and the idle transitions are simply ignored for the purpose of fairness. $\mathcal{F}(K)$ is the fair paths for K.

In analogy to thinking of a sequential procedure as a black box characterized by its input/output relation, we would like to think of an actor system as a black box characterized by the set of possible interactions with its environment. Thus we define the interaction semantics of an actor system in such a way as to hide the details of internal transitions. The interaction semantics of a configuration is its set of possible interaction paths. An interaction path of a configuration is an infinite sequence of interaction labels together with an initial interface consisting of a pair of finite sets of actor names (the receptionists and externals). An interaction label is either an input/output label or the special sign, τ^*, standing for possible internal activity. The infinite sequence of interaction labels in an interaction path is obtained from a computation path by mapping internal transitions to silent transitions.

The function *isem* maps transition labels to interaction labels and computation paths to interaction paths. The receptionists and externals of $isem(\pi)$ are those of the initial configuration of π. The interaction sequence of $isem(\pi)$ is the sequence of labels obtained by replacing internal and idle transition labels to τ^*.

Definition ($isem(\pi)$ $Isem(K)$):

$isem(l) = \begin{cases} \tau^* & \text{if } l \in \mathbf{L} \cup \{\mathtt{idle}\} \\ l & \text{if } l \in \mathtt{in}(\mathbf{Msg}) \cup \mathtt{out}(\mathbf{Msg}) \end{cases}$

$isem(\pi) = \vartheta_{\chi_0}^{\rho_0}$ where $\pi(i) = \langle\!\langle I_i \rangle\!\rangle_{\chi_i}^{\rho_i} \xrightarrow{l_i} \langle\!\langle I_{i+1} \rangle\!\rangle_{\chi_{i+1}}^{\rho_{i+1}}$ and $\vartheta(i) = isem(l_i)$ for $i \in \mathbf{N}$

$Isem(K) = \{isem(\pi) \mid \pi \in \mathcal{F}(K)\}$

So far we have been working in the context of a fixed, but arbitrary actor theory. In the case that we consider interaction semantics in more than one actor theory, we index $Isem$ by the name of the actor theory, writing $Isem_{AT}(K)$. It is sometimes convenient to restrict the interactions of a configuration with its environment by restricting the possible set of input messages. For $V \subset \mathbf{M}$, we define $Isem(K)\lceil V$ to be set of interaction paths $\zeta \in Isem(K)$ whose input labels are messages with contents in V.
Definition ($Isem(K)\lceil V$):

$$Isem(K)\lceil V = \{isem(\pi) \mid \pi \in \mathcal{F}(K) \land (\forall i \in \mathbf{N}, a \in \mathbf{A}, M \in \mathbf{M})(\pi(i) = \text{in}(a \triangleleft M) \Rightarrow M \in V)\}$$

For a given actor language, we usually define the reaction rules for an actor theory by giving the semantics in terms of basic reduction steps for expressions of the language. We call this a small step actor theory. It is simple to define, but gives rise to computations with many small and mostly uninteresting steps. In the following we show how to transform such an actor theory in to big step actor theory which preserves the interaction semantics of the language. In the big step theory internal computation steps are those that create actors, send messages, or involve some synchronization of actors and messages, thus suppressing further details of internal computation of an actor.

The key ideas motivating the transformation are the notions of silent step and that of a path being in *big-step* form. A silent step is one involving a single actor that creates no new messages or actors. A path in big step form consists of input/output transistions and non-silent steps each preceded by the necessary silent steps to prepare the reacting actors. For $AT = \langle \langle \mathbf{A}, \mathbf{S}, \mathbf{M}, \mathbf{L} \rangle, \langle acq, \hat{} \rangle, RR \rangle$ we define its big-step variant $AT^\dagger$ by $AT^\dagger = \langle \langle \mathbf{A}, \mathbf{S}, \mathbf{M}, \mathbf{L}^\dagger \rangle, \langle acq^\dagger, \hat{}^\dagger \rangle, RR^\dagger \rangle$ where

$$RR^\dagger = \{l^\dagger : I \Rightarrow I' \mid (\exists I'')(I \xrightarrow{*} I'' \land l : I'' \Rightarrow I' \in RR \quad \text{a non-silent rule}\}$$

and $I \xrightarrow{*} I''$ is sequence of silent steps. The crucial property of the big step operation is that it preserves interaction semantics. Let AT be an actor theory and let K be a configuration of AT. Then
Theorem (small2big): $Isem_{AT}(K) = Isem_{AT^\dagger}(K)$
The proof relies on the ability to put paths into big-step form.

3 The Kernel Language

We assume given an infinite set of variables, $\mathbf{X}$. We also assume as given a collection of basic or atomic data, $\mathbf{At}$, that includes the booleans $\mathtt{t, f} \in \mathbf{Bool}$, Scheme style symbols, $\mathbf{Sym}$, ($\mathbf{Sym}$ includes $\mathtt{nil}$, the empty or null list), (constants denoting the elements of) the integers, $\mathbf{Z}$, and actor names, $\mathbf{A}$. Expressions are built from atoms and variables by the following operations: λ-abstraction, application of primitive operations to sequences of expressions, conditional branching, and an actor creation construct. The primitive operations include operations on basic data and pairs, and kernel primitives manipulating actors, procedures, and local continuations. The data operations $\mathbf{dOp}$ contains the recognizers: $\mathtt{boolean?}$ for booleans, $\mathtt{symbol?}$ for symbols, $\mathtt{integer?}$ for integers, $\mathtt{cons?}$ for pairs, and $\mathtt{actor?}$ for actors (all of arity 1); pairing $\mathtt{cons}$, $\mathtt{car}$, $\mathtt{cdr}$ (arities 2, 1, 1); the equality predicate, $\mathtt{equal?}$, on atomic data; and the usual arithmetic operations, $\mathbf{aOp}$. We consider actor addresses to be atomic data and consequently can tell one address from another. The functional specific primitives are $\mathtt{procedure?}$, the recognizer for procedures (arity 1), $\mathtt{app}$, lambda application (arity 2), and $\mathtt{clc}$, control abstraction (arity 1). We include $\mathtt{app}$ in the list of primitive operations as a technical convenience, to make the syntax more concise. The actor primitives consists of an actor creation construct plus the operations: $\mathtt{self}$ (of arity 0), the name of the executing actor; $\mathtt{send}$, asynchronous send (arity 2); $\mathtt{ready}$, establishing behavior for receiving (arity 1). Actor creation expressions are of the form $\mathtt{letactor}\{x_0 := e_0, \ldots x_k := e_k\}\, e$ where the x_i are pairwise distinct variables. Executing a $\mathtt{letactor}$ expression creates a new actor entity a_i for each x_i executing expressions e_i with x_i bound to a_i. The original executing actor then proceeds by executing e (with x_i bound to a_i).

The top level syntactic construct is a *kernel program* which describes a configuration. For convenience, kernel programs may include a library of mutually recursive definitions. For this purpose we reserve a subset $^k\mathbf{FunId}$ of $\mathbf{X}$ to be used as function names.
Definition (Kernel Programs):

$^k\mathbf{M} = \mathbf{At} \cup \mathtt{cons}(^k\mathbf{M}, ^k\mathbf{M})$ $^k\mathbf{Program} = \mathtt{program}(\mathtt{receptionists} : \mathcal{P}_\omega[\mathbf{A}],\ \mathtt{externals} : \mathcal{P}_\omega[\mathbf{A}]$
$\qquad\qquad\qquad\qquad\qquad\qquad\qquad\quad \mathtt{library} : \mathcal{P}_\omega[\mathbf{FunId} := \lambda \mathbf{X}.^k\mathbf{E}]$
$\qquad\qquad\qquad\qquad\qquad\qquad\qquad\quad \mathtt{actors} : \mathcal{P}_\omega[\mathbf{A} := {^k}\mathbf{E}]$
$\qquad\qquad\qquad\qquad\qquad\qquad\qquad\quad \mathtt{messages} : \mathcal{M}_\omega[\mathbf{A} \triangleleft {^k}\mathbf{M}])$

where the function identifiers in the $\mathtt{library}$ part and actor names in the $\mathtt{actors}$ part must be distinct, and all actor names occurring in an actor state or message contents must either be one of the actor names defined in the $\mathtt{actors}$ part or one of the names occurring in the $\mathtt{externals}$ part. Message contents are simply values built up from the atomic data via the pairing operation $\mathtt{cons}$. Lambda abstractions and structures containing lambda abstractions are not allowed to be communicated in messages.
Definition (Kernel Expressions):

$^k\mathbf{O} = \mathbf{dOp} \cup \{\mathtt{procedure?}, \mathtt{app}, \mathtt{clc}\} \cup \{\mathtt{self}, \mathtt{ready}, \mathtt{send}\}$

$\mathbf{At} = \mathbf{A} \cup \mathbf{Bool} \cup \mathbf{Z} \cup \mathbf{Sym}$

$^k\mathbf{E} = \mathbf{X} \cup \mathbf{At} \cup \lambda \mathbf{X}.^k\mathbf{E} \cup \mathbf{O}_n(\mathbf{E}_n) \cup \mathtt{if}(\mathbf{E}, \mathbf{E}, \mathbf{E}) \cup \mathtt{letactor}\{(\mathbf{X} := {^k}\mathbf{E})^+\}^k\mathbf{E}$

We let x, y, z range over $\mathbf{X}$, a ranges over $\mathbf{A}$, $^k e$ ranges over $^k\mathbf{E}$, $^k M$ ranges over $^k\mathbf{M}$. The binding constructs are letactor and λ. $\lambda x.e$ binds the variable x in the expression e. letactor$\{\ldots x_i := {}^k e_i \ldots\}^k e$ binds the x_i in each of the $^k e_j$, and also in $^k e$. Two expressions are considered equal if they are the same up to the renaming of bound variables. For any expression e, we write $\mathrm{FV}(e)$ for the set of free variables of e. We write $e'[\bar{x} := \bar{e}]$ to denote the expression obtained from e' by simultaneously replacing all free occurrences of $\bar{x}$ by $\bar{e}$, avoiding the capture of free variables in $\bar{e}$. We use standard abbreviations: let, for lambda application; boolean functions not, and, boolean functions; and letrec$\{^k\!fid_j = \lambda x.{}^k e\}_{1 \leq j \leq k}$ $^k e_j$, for mutual recursive definition. We also use the following definitions for structuring message contents.

$$\mathrm{list}_n = \lambda x_1.\lambda x_2.\ldots.\lambda x_n \mathrm{cons}(x_1, \mathrm{cons}(x_2, \ldots \mathrm{cons}(x_n, \mathrm{nil})))$$

$$\mathrm{msgMk} = \lambda x_{\mathrm{mid}}.\lambda x_{\mathrm{args}}.\lambda x_{\mathrm{cust}}.\mathrm{list}_3(x_{\mathrm{mid}}, x_{\mathrm{args}}, x_{\mathrm{cust}})$$

As indicated earlier, the semantics is given by defining an actor theory, $^k AT$. The only primitive sort of $^k AT$ that remains to be defined is the set of kernel actor theory states, $^k\mathbf{S}$.

Definition ($^k\mathbf{S}$): $^k\mathbf{S} = \{^k e \in {}^k\mathbf{E} \mid \mathrm{FV}(^k e) = \emptyset\}$

The acquaintances of a state (or message contents for that matter) is simply just the actor names occurring therein and renaming is simply substitution. The meaning of a kernel program is defined to be a configuration of $^k AT$ as follows.

Definition ($[\![^k P]\!]$): Let $^k P$ be given by

$$\texttt{program}(\texttt{receptionists}: \rho, \ \texttt{externals}: \chi, \ \texttt{library}: \{^k\!fid_j = \lambda x.{}^k e\}_{1 \leq j \leq l}$$
$$\texttt{actors}: \{ [{}^k e_j]_{a_j} \}_{1 \leq j \leq m}, \ \texttt{messages}: \{ a'_j \triangleleft {}^k M_j \}_{1 \leq j \leq n})$$

then $[\![^k P]\!] = \langle\!\langle {}^k I \rangle\!\rangle_\chi^\rho$ where $^k I = \{ [\texttt{letrec}\{^k\!fid_j := \lambda x.{}^k e\}_{1 \leq j \leq l} \ {}^k e_j]_{a_j} \}_{1 \leq j \leq m}, \ \{ [{}^k M_j]_{a'_j} \}_{1 \leq j \leq n}$

To complete the semantics all that remains is to define the reaction rules. To do this we decompose each non-value expression as a reduction context filled with a redex. Reduction contexts identify the subexpression of an expression that is to be evaluated next using the standard call-by-value reduction strategy of [Plo75] and were first An expression e is either a value or it can be decomposed uniquely into a reduction context filled with a redex. Thus, local actor computation is deterministic.

Definition ($^k\mathbf{V}$ $^k\mathbf{E}_{\mathrm{rdx}}$ $^k\mathbf{R}$): The set of *values*, $^k\mathbf{V}$, the set of *redexes*, $^k\mathbf{E}_{\mathrm{rdx}}$, and the set of *reduction contexts*, $^k\mathbf{R}$, are defined by

$$^k\mathbf{V} = \mathbf{At} \cup \mathrm{cons}(^k\mathbf{V}, {}^k\mathbf{V}) \cup \lambda \mathbf{X}.{}^k\mathbf{E}$$

$$^k\mathbf{E}_{\mathrm{rdx}} = (^k\mathbf{O}_n(^k\mathbf{V}^n) - \mathrm{cons}(^k\mathbf{V}, {}^k\mathbf{V})) \cup \texttt{if}(^k\mathbf{V}, {}^k\mathbf{E}, {}^k\mathbf{E}) \cup \texttt{letactor}\{(\mathbf{X} := {}^k\mathbf{E})^+\}^k\mathbf{E}$$

$$^k\mathbf{R} = \{\bullet\} \cup {}^k\mathbf{O}_{n+m+1}(^k\mathbf{V}^n, {}^k\mathbf{R}, {}^k\mathbf{E}^m) \cup \texttt{if}(^k\mathbf{R}, {}^k\mathbf{E}, {}^k\mathbf{E})$$

We let $^k R$ range over $^k\mathbf{R}$. With the exception of the actor primitives letactor send, and ready, reduction steps are silent – they only depend on information local to the executing actor and only effect the state of the executing actor. Thus we define a sequential reduction relation, $e \xrightarrow{s}_\kappa e'$, on expressions that lifts uniformly to define the silent reaction rules. The decoration $^k\zeta$ is an abstract context introduced to make the dependence on local context explicit. We use a function $\mathit{self}(^k\zeta)$ that extracts the name of the executing actor from $^k\zeta$. To define the sequential relation, we first define the purely functional reduction relation $r \xrightarrow{f}_\kappa e$ which gives the rules for redexes that do not manipulate the reduction context. The rules are standard and are omitted. The sequential reduction relation is then defined by lifting functional reduction and adding the rule for clc.

Definition (Sequential steps ($\xrightarrow{s}_\kappa$)):

(rdx) $\quad ^k R[e] \xrightarrow{s}_\kappa {}^k R[e']$ if $^k e \xrightarrow{f}_\kappa {}^k e'$

(clc) $\quad ^k R[\mathrm{clc}(^k v)] \xrightarrow{s}_\kappa \mathrm{app}(^k v, \lambda x.{}^k R[x]) \quad x \notin \mathrm{FV}(^k R[\mathrm{nil}])$

clc captures the actors local continuation, R, as a function, $\lambda x.R[x]$, and applies its argument $^k v$ to this function, in the empty reduction context (the local top level). We let $\xrightarrow{s}\!\!\!\!\twoheadrightarrow_\kappa$ be the reflexive, transitive closure of $\xrightarrow{s}_\kappa$. Now we are ready to define the reaction rules of $^k AT$.

Definition ($^k RR$):

$\mathrm{seq}(a): [{}^k e]_a \Rightarrow [{}^k e']_a \quad$ if $^k e \xrightarrow{s}_\kappa {}^k e' \quad$ where $\quad \mathit{self}(^k\zeta) = a$

$\mathrm{send}(a): [{}^k R[\mathrm{send}(^k v_0, {}^k v_1)]]_a \Rightarrow [{}^k R[\mathrm{nil}]]_a, \ {}^k Emit(^k v_0 \triangleleft {}^k v_1)$

$\mathrm{ready}(a, {}^k M): [{}^k R[\mathrm{ready}(^k v)]]_a, \ a \triangleleft {}^k M \Rightarrow [\mathrm{app}(^k v, {}^k M)]_a$

$\mathrm{leta}(a, \vec{a}): [{}^k R[\texttt{letactor}\{\bar{x} := {}^k \bar{e}\} \ {}^k e]]_a \Rightarrow [{}^k R[{}^k e[\bar{x} := \vec{a}]]]_a, \ \{[{}^k e_i[\bar{x} := \vec{a}]]_{a_i}\}_{1 \leq i \leq m}$

$\quad$ if $\mathrm{Len}(\vec{a}) = \mathrm{Len}(\bar{x}) = m$ and $\vec{a} \cap \mathrm{acq}(^k R[{}^k e, {}^k \bar{e}]) = \emptyset$

Where ${}^k\textit{Emit}({}^k v_0 \triangleleft {}^k v_1) = {}^k v_0 \triangleleft {}^k v_1$ if ${}^k v_0 \in \mathbf{A}$ and ${}^k v_1 \in {}^k\mathbf{M}$, otherwise it is $\emptyset$. The meta-function ${}^k\textit{Emit}$ prevents ill-formed messages from getting into the system. The labels of ${}^k AT$ are

$$\mathrm{seq}(\mathbf{A}) \cup \mathrm{send}(\mathbf{A}) \cup \mathrm{ready}(\mathbf{A}, {}^k\mathbf{M}) \cup \mathrm{leta}(\mathbf{A}, \mathbf{A}^*)$$

where in the case of $\mathrm{leta}(a, \vec{a})$ we require $a \notin \vec{a}$. $acq(l)$ is just the union of old and new except for the delivery label where $acq(\mathrm{ready}(a, {}^k M)) = \{a\} \cup acq({}^k M)$. Again, renaming is just substitution.

4 The User Language

The user language has the same variables, basic data, actor names, and data operations as the kernel language. In addition we assume given two disjoint, countably infinite sets of identifiers: **FunId** for functions; and **BehId** for behaviors. Expressions are built from atoms and variables by the following operations: application of primitive operations to sequences of expressions, let binding, conditional branching, the letactor actor creation construct, and asynchronous and synchronous method invocation. The primitive operations include **dOp** and following user primitives: self, as in the kernel; customer, the customer of the current message (arity 0); $\textit{fid}_i$, user defined operations (arity i) for $i \in \mathbf{N}$, $\textit{fid} \in \mathbf{FunId}$; and $\mathrm{ready}_{bid,i}$, specifying the behavior for the next message (arity i) for $i \in \mathbf{N}$, $bid \in \mathbf{BehId}$. An *asynchronous invocation* is of the form ${}^u e_a \triangleleft mid[{}^u\vec{e}]@{}^u e_c$. The target of the request is the value of ${}^u e_a$ and the message contents has method name mid, arguments ${}^u\vec{e}$ and customer, ${}^u e_c$. Once the target, arguments, and customer are evaluated, nil is returned as the value and the requesting actor proceeds with its computation without waiting for a reply. A *synchronous invocation* (also referred to as a *request* or *remote procedure call*) is of the form ${}^u e_a . mid[{}^u\vec{e}]$. The target of the request is the value of ${}^u e_a$ and the message contents has method name mid, arguments ${}^u\vec{e}$. The requesting actor suspends execution until a reply is received. A *ready* expression is of the form $\mathrm{ready}_{bid,n}({}^u e_1, \ldots, {}^u e_n)$ (also written $\mathrm{ready}(bid({}^u e_1, \ldots, {}^u e_n)))$. Execution of a ready expression terminates processing of the current message and looks for the next message enabled for the behavior bid with parameters given by the values of the ${}^u e_i$. If there is no enabled message in the local message queue the actor waits for one to be delivered. In the user language there is no lambda abstraction and thus no functions as values. Instead, each program contains a library of (mutually recursive) function and behavior definitions. A *behavior definition* has the form

behavior $bid(p)(methodDefs)$.

where bid a the behavior identifier, p is a parameter list (a list of distinct variables), and $methodDefs$ is a set of method definitions. A *method definition* has the form

method $mid(p)[\mathrm{disable} - \mathrm{when}^u e^d]^u e^m$

where mid is a method name (a symbol from **Sym**), p is a parameter list, ${}^u e^d c$ is the [optional] disabling condition (assumed false when not present) that specifies when a method can be invoked, and ${}^u e^m$ is the method body. ${}^u e_c$ is required to be functional, i.e. its evaluation involves no actor primitives other than self or customer. For consistency we require that a method (i.e a method identifier) should have a unique definition within a given behavior. The free variables of constraints and method bodies must be among the method parameters or the behavior parameters. A *function definition* has the form

function $\textit{fid}(p)^u e$

where $\textit{fid}$ is a function identifier, p is a parameter list, and ${}^u e$ is an expression, the function body. The free variables of the function body must be among the function parameters.

Definition (User Programs and Libraries):

$${}^u\mathbf{M} = \mathbf{MethId}[{}^u\mathbf{V}^*]@(\mathbf{A} \cup \{\mathrm{nil}\}) \quad {}^u\mathrm{Program} = \mathrm{program}(\mathrm{receptionists} : \mathcal{P}_\omega[\mathbf{A}], \; \mathrm{externals} : \mathcal{P}_\omega[\mathbf{A}]$$
$$\mathrm{library} : \mathcal{P}_\omega[(\mathbf{BehDef} \cup \mathbf{FunDef})]$$
$$\mathrm{actors} : \mathcal{P}_\omega[\mathbf{A} := {}^u\mathbf{E}], \; \mathrm{messages} : \mathcal{M}_\omega[\mathbf{A} \triangleleft {}^u\mathbf{M}])$$

where the actor names in the actors part must be distinct, and all actor names occurring in an actor state or message contents must either be one of the actor names defined in the actors part or one of the names occurring in the externals part. We let ${}^u M$ range over ${}^u\mathbf{M}$. We let c range over $\mathbf{A} \cup \{\mathrm{nil}\}$ and we may omit the customer part if it is nil. Message contents consist of a method identifier (symbol), an argument list (a list of values), and a customer (an actor name or nil signifying no customer). We identify **MethId** with **Sym** and let mid range over **MethId** and use mid to stand for a symbol used as a method identifier. $mid[\vec{v}]@c$ abbreviates the list construction $\mathrm{msgMk}(mid, \vec{v}, c)$. We use the the following meta functions: $\mathrm{msgMeth}(M)$ selects the method component; $\mathrm{msgArgs}(M)$ selects the arguments component; and $\mathrm{msgCust}(M)$ selects the customer component. A library is well-formed if it contains at most one definition of each $\textit{fid} \in \mathbf{FunId}$, and $bid \in \mathbf{BehId}$, and these definitions themselves are well-formed. We let $\bar{x}, \bar{y}$, and p range over lists of distinct variables. User expressions, ${}^u\mathbf{E}$, are defined in a manner similar to kernel expressions and we omit the details. We let ${}^u e$ range over ${}^u\mathbf{E}$.

As for the kernel language, the semantics is given by defining an actor theory ${}^u AT$. Since libraries do not evolve we parameterize the actor theory by the library of definitions in force, letting library be just part of the auxiliary axioms describing the actor theory. Thus to give the semantics we need only define user states, ${}^u\mathbf{S}$ (since the message contents, ${}^u\mathbf{M}$, have been explained above) and give ${}^u RR$, relative to the given library.
There are five kinds of actor states:

- $({}^u e, c, Q)$ – processing message with customer c, with current state ${}^u e$;

- $(bid(^u\bar{v}), Q_l, Q_r)$ – traversing the queue of delivered but unprocessed messages, $Q = Q_l * Q_r$, looking for an message that is not disabled. The current behavior is $bid(^u\bar{v})$, The message (contents) in Q_r have been checked and rejected (i.e. they are disabled). The messages in Q_l are yet to be checked;

- $(\varphi, ^uM, bid(^u\bar{v}), Q_l, Q_r)$ checking disabling constraints of behavior bid for message ^uM.

- $(a', ^uR, c, Q)$ – waiting for a reply to a request. ^uR is a reduction context – the continuation of the computation upon receipt of an answer. a' is an actor created to serve as a reply address, to distinguish the request-reply from other arriving messages.

- (a) – the state of an actor serving as a reply address for a request sent by a;

where a, a' are actor names, ^ue is an expression (of the user language), c is a customer – and actor name or nil, φ is functional expression, ^uM is the contents of a user message, and Q is a mail queue – a sequence of messages (or more precisely their contents).

A state of the form $(^ue, c, Q)$ is an execution state. It attempts to step by decomposing ^ue into a reduction context and redex and reducing the redex. It is hung if the redex fails to reduce. A state of the form $(bid(^u\bar{v}), Q_l, Q_r)$ is an execution state if Q_l is not empty. If $^u\bar{v}$ does not match the parameter list of the definition of bid then the state is hung. Otherwise it steps by starting evaluation of the constraints associated, in the behavior definition for bid, with the method name of the first message in Q_l. If Q_l empty then the state is waiting for delivery of a message (having already walked through its queue and found no enabled messages). A state of the form $(\varphi, ^uM, bid(^u\bar{v}), Q_r, Q_1)$ steps by evaluating φ one step if it is not a value expression. If φ is the value f, then it starts evaluation of the method body associated with the method of ^uM in the behavior definition for bid. If φ is a value other than f then ^uM is considered disabled and put on the end of rejects queue, Q_r. States of the last two forms occur in pairs $[a', ^uR, c, Q]_a$, $[a]_{a'}$ that are waiting for a reply to a request by a that will arrive as a message to a', serving as a unique request identifier.

The meaning of a user program is defined to be a configuration of ^uAT as follows.

Definition ($[^uP]$): Let ^uP be given by

$$^uP = \text{program}(\text{receptionists}: \rho, \text{ externals}: \chi, \text{ library}: Lib$$
$$\text{actors}: \{a_j := {^ue_j}\}_{1 \le j \le m}, \text{ messages}: \{a'_j \triangleleft {^uM_j}\}_{1 \le j \le n})$$

then $[^uP] = \left\langle\!\!\left\langle {^uI} \right\rangle\!\!\right\rangle^\rho_\chi$ where $^uI = \{[{^ue_j}, \text{nil}, \text{nil}]_{a_j}\}_{1 \le j \le m}, \{a'_j \triangleleft {^uM_j}\}_{1 \le j \le n}$

To complete the definition of ^uAT we must give the reaction rules. We first define some auxiliary meta functions and predicates to ease definition of rules concerning behaviors and methods: $behMatch(Lib, bid, ^u\bar{v})$ tests whether the parameters of ready expressions match those of the behavior definition; $cstrExp(Lib, bid, ^u\bar{v}, mid(^u\bar{v}')@c)$ extracts the constraint associated with a method; and $methExp(Lib, bid, ^u\bar{v}, mid(^u\bar{v}')@c)$ extracts a method body from a library given a behavior identifier, a parameter list, and a message. We write $^ue[p := \bar{v}]$ for the simultaneous substitution of the ith value in $\bar{v}$ for the ith variable in p. This is defined only when p and $\bar{v}$ have the same length. The definition of $cstrExp$ reflects the fact that a message is considered disabled if there is no matching method definition and that messages with matching method definitions are by default enabled if there is no explicit disabling constraint.

As in the kernel language, to give the reaction rules, we first define the sequential reduction relation $^ue \xrightarrow{s}_\chi {^ue'}$ parameterized by and abstract context $^u\zeta$. We also use a function $customer(^u\zeta)$ to extract the customer of the current message. We define the values $^u\mathbf{V}$, reduction contexts $^u\mathbf{R}$ and redexes $^u\mathbf{E}_{\text{rdx}}$ of the user language, anaogous to the kernel language definitions, again giving the unique decomposition property for non-value expressions. We let ^uR range over $^u\mathbf{R}$. The relations $\xrightarrow{f}_\chi$ and $\xrightarrow{s}_\chi$ are defined similarly to the kernel case and again we omit details. We let $\xrightarrow{s}\!\!\!\!\twoheadrightarrow_\chi$ be the reflexive, transitive closure of $\xrightarrow{s}_\chi$. Notice that the sequential rules are sufficient to evaluate functional expressions, in particular we only need the sequential rules to check constraints.

The labelled reaction rules for the user language are given by the following.

Definition (^uRR):

$\text{seq}(a): [{^ue}, c, Q]_a \Rightarrow [{^ue'}, c, Q]_a \quad \text{if } {^ue} \xrightarrow{s}_{a,c} {^ue'}$

$\text{send}(a): [{^uR}[{^uv} \triangleleft mid({^u\bar{v}})@{^uv'}], c, Q]_a \Rightarrow [{^uR}[\text{nil}], c, Q]_a, \quad {^u\text{Emit}}({^uv} \triangleleft mid({^u\bar{v}})@{^uv'})$

$\text{rpc}(a, a_0): [{^uR}[{^uv} \cdot mid({^u\bar{v}})], c, Q]_a \Rightarrow [a_0, {^uR}, c, Q]_a, \quad [a]_{a_0}, \quad {^u\text{Emit}}({^uv} \triangleleft mid({^u\bar{v}})@a_0)$

$\quad \text{if } a_0 \notin acq([{^uR}[a' \cdot mid({^u\bar{v}})], c, Q]_a)$

$\text{rcv}(a, a_0, {^uv}): [a_0, {^uR}, c, Q]_a, \quad [a]_{a_0}, \quad a_0 \triangleleft mid([{^uv}] * {^u\bar{v}})@c' \Rightarrow [{^uR}[v], c, Q]_a, \quad [\text{nil}, \text{nil}, []]_{a_0}$

$\text{deliver}(a, {^uM}): [bid({^u\bar{v}}), [\,], Q]_a, \quad a \triangleleft {^uM} \Rightarrow [bid({^u\bar{v}}), [{^uM}], Q]_a$

$\text{walk}(a): [{^uR}[\text{ready}(bid({^u\bar{v}}))], {^uM}, Q]_a \Rightarrow [bid({^u\bar{v}}), Q, [\,]]_a \quad \text{if } behMatch(Lib, bid, {^u\bar{v}})$

$\text{leta}(a, \bar{a}): [{^uR}[\text{letactor}\{\bar{x} := {^u\bar{e}}\}{^ue}], c, Q]_a \Rightarrow [{^uR}[{^ue}[\bar{x} := \bar{a}]], c, Q]_a, \quad \{[{^ue_i}[\bar{x} := \bar{a}], c, [\,]]_{a_i}\}_{1 \le i \le k}$

$\quad \text{if } \text{Len}(\bar{a}) = \text{Len}(\bar{x}) = k \text{ and } \bar{a} \cap acq({^uR}[\text{letactor}\{\bar{x} := {^u\bar{e}}\}{^ue}], c, Q) = \emptyset$

$\text{cstr}(a): [bid({^u\bar{v}}), [{^uM}] * Q_l, Q_r]_a \Rightarrow [cstrExp(Lib, bid, {^u\bar{v}}, {^uM}), {^uM}, bid({^u\bar{v}}), Q_l, Q_r]_a$

$\text{enable}(a): [\text{f}, {^uM}, bid({^u\bar{v}}), Q_l, Q_r]_a \Rightarrow [methExp(Lib, bid, {^u\bar{v}}, {^uM}), msgCust({^uM}), Q_l * Q_r]_a$

$\text{disable}(a): [{^uv}, {^uM}, bid({^u\bar{v}}), Q_l, Q_r]_a \Rightarrow [bid({^u\bar{v}}), Q_l, Q_r * [{^uM}]]_a \quad \text{if } {^uv} \ne \text{f}$

$\text{check}(a): [{^ue_0}, {^uM}, bid({^u\bar{v}}), Q_l, Q_r]_a \Rightarrow [{^ue_1}, {^uM}, bid({^u\bar{v}}), Q_l, Q_r]_a \quad \text{if } {^ue_0} \xrightarrow{s}_{a, msgCust({^uM})} {^ue_1}$

where $^uEmit(^uv \triangleleft {}^uM) = {}^uv \triangleleft {}^uM$ if $^uv \in \mathbf{A}$ and $msgCust^uM \in \mathbf{A} \cup \{\text{nil}\}$, otherwise it is $\emptyset$. As in the kernel language, the meta-function ^uEmit prevents ill-formed messages from getting into the system.

5 A Semantics Preserving User to Kernel Translation

In this section we define a translation, $u2k : {}^u\mathcal{L} \to {}^k\mathcal{L}$ and show that this translation preserves interaction semantics. $u2k$ is a family of maps, one for each syntactic category. The members of the family are distinguished by context of application rather than by name. Programs are translated by translating the library, actors, and messages parts. A library is translated by translating the function and behavior definitions, producing a kernel language library. An actor description is translated by translating the expression part assuming it executes in a local context in which the current message has no customer and message queue is empty. A message description is translated by simply eliminating the syntactic sugar. The core of the translation is its behavior on expressions. Expressions are translated in the context of a user library. In order to leave this dependence implicit, we adopt a standard convention about converting user function and behavior identifiers into variables and assume sufficient renaming has been done to avoid conflicts. The translation $u2k(^ue)$ of a user expression is a lambda term of the form $\lambda c.\lambda q.^ke$ which when applied to a *customer*, c, and a *message queue*, q, (represented as a list) reduces to a kernel expression that corresponds to the user expression executing in a local context where the current message has customer c and message queue elements are the elements of q. We use the following abbreviation $u2k^*(^ue, c, q) \triangleq \text{app}(u2k(^ue), c, q)$ in defining the translation.

The translation of the expression forms that are common to the two languages as well as customer and asynchronous send are straightforward. It amounts to passing the customer and message queue parameters to the translated subexpressions. The translation of synchronous invocations (requests) and ready_{bid} are where care is needed. In the user language, the transition that delivers the reply to a request involves two actors, the actor requesting the reply and the actor created to serve as the reply address, as well as the reply message. Kernel actor transitions involve at most one actor. The three-body interaction is replaced by a delivery to the reply address followed by a forwarding and delivery to the requesting actor. To avoid forgery, we introduce a third actor which has null behavior and simply serves as a secret key known only by the requestor and the actor serving as the reply address. The forwarded message is tagged with this key.

The translation of a ready_{bid} expression must produce code to walk the message queue, checking the disabling constraints for the method of each message. If an enabled message is found, then the translated method body is executed. If the end of the queue is reached, then the actor executes ready with a behavior that treats the next message delivered as the next element of the message queue to check.

We begin by defining the mapping on programs, and work our way down to expressions. Programs are translated as follows:

$u2k(\text{program}(\text{receptionists} : \rho \quad \text{externals} : \chi \quad \triangleq \quad \text{program}(\text{receptionists} : \rho \quad \text{externals} : \chi$
$\qquad \text{library} : {}^uLib \qquad\qquad\qquad\qquad\qquad \text{library} : u2k(^uLib)$
$\qquad \text{actors} : \{a_i := {}^ue_i\}_{1 \leq i \leq k} \qquad\qquad \text{actors} : \{a_i := u2k^*(^ue_i, \text{nil}, \text{nil})\}_{1 \leq i \leq k}$
$\qquad \text{messages} : \{a'_j \triangleleft M_j\}_{1 \leq i \leq n})) \qquad \text{messages} : \{a'_j \triangleleft M_j\}_{1 \leq i \leq n})$

To translate function definitions we associate a kernel function symbol ^kfid to each user function identifier, fid. The translation of a definition of fid yields a definition of ^kfid. The translation of behavior definitions is a little more complex. For each defined behavior identifier, bid, the translation consists of definitions of three operations: ^kbid, the top level behavior function; $\text{Qwalk}[bid]$, controls the queue walking for bid; and $\text{Mcheck}[bid]$, checks constraints for a particular message.

User functions, behaviors and methods have parameter lists. The translated operations will be applied to a list of arguments and must check if the number of arguments is correct and then bind these to the individual parameters. For this purpose, we define a family of abbreviations $\text{parBind}[p, v, e]$ that binds elements of p to corresponding elements of v in e. The translation of a function definition is given by:

$u2k(\text{function } fid(p)^ue) \triangleq {}^kfid := \lambda c.\lambda q.\lambda y.\text{if}(\text{not}(\text{equal?}(\text{length}(y), n)),$
$\qquad\qquad\qquad\qquad\qquad\qquad\qquad\qquad \text{hang},$
$\qquad\qquad\qquad\qquad\qquad\qquad\qquad\qquad \text{parBind}(p, y, u2k^*(^ue, c, q)))$

where $\text{Len}(p) = n$ and hang is some functional redex that fails to reduce, for example $\text{car}(\text{nil})$, thus hanging the computation if the arguments do not match the parameters.

Let $methodDefs$ be $\{\text{method } mid_i(p_i)[\text{disable} - \text{when } {}^ue_i^d]^ue_i^m \mid i \leq m_{bid}\}$ ($^ue_i^d$ is taken to be to f if no disabling constraint is present). The translation $u2k(\text{behavior } bid(p)(methodDefs))$ of a behavior definition with $\text{Len}(p) = n$ is $^kbid := \lambda q.\lambda y.\text{if}(\text{not}(\text{equal?}(\text{length}(y), n)), \text{nil}, \text{Qwalk}[bid](q, \text{nil}, y))$ $\text{Qwalk}[bid](q_l, q_r, y)$ waits for a message to be delivered, if $q_l = \text{nil}$ and otherwise calls $\text{Mcheck}[bid](\text{cdr}(q_l), q_r, y)(\text{car}(q_l)$ to check the first element of q_l. $\text{Mcheck}[bid](q_l, q_r, y)(^uM)$ looks for a method definition matching ^uM. If none exists, or if the matching method method is disabled relative to the behavior parameters, y and the message arguments, then it calls $\text{Qwalk}[bid](q_l, \text{append}(q_r, \text{list}(^uM)), y)$. Otherwise it reduces to the appropriately instantiated method body.

We give the clauses for the most interesting cases:

$u2k(\text{customer}()) \triangleq \lambda c.\lambda q.c$

$u2k(fid(^ue_1, \ldots, {}^ue_n)) \triangleq \lambda c.\lambda q.\text{app}(^kfid, c, q, \text{list}_n(u2k^*(^ue_1, c, q), \ldots u2k^*(^ue_n, c, q)))$

$u2k(\texttt{letactor}\{a_i := {}^ue_i\}_{1\leq i\leq n} {}^ue) \triangleq \lambda c.\lambda q.\texttt{letactor}\{a_i := u2k^*({}^ue_i,c,\texttt{nil})\}_{1\leq i\leq n} u2k^*({}^ue,c,q)$

$u2k(\texttt{ready}(bid({}^ue_1,\ldots,{}^ue_n))) \triangleq \lambda c.\lambda q.\texttt{let}\{x_i := u2k^*({}^ue_i,c,q)\}_{1\leq i\leq n}$
$\qquad\qquad\qquad\qquad\qquad\qquad \texttt{let}\{y := \texttt{list}_n(x_1,\ldots,x_n)\}$
$\qquad\qquad\qquad\qquad\qquad\qquad \texttt{clc}(\lambda f.\texttt{app}({}^kbid,q,y))$

$u2k({}^ue_0 \,.\, mid[{}^ue_1,\ldots,{}^ue_n]) \triangleq$
$\quad \lambda c.\lambda q.\texttt{let}\{x_i := u2k^*({}^ue_i,c,q)\}_{0\leq i\leq n}$
$\qquad\quad \texttt{let}\{m_{\text{args}} := \texttt{list}_n(x_1,\ldots,x_n)\}$
$\qquad\qquad \texttt{letactor}\{a_{\text{key}} := \texttt{nil}\}$
$\qquad\qquad\quad \texttt{let}\{b := \texttt{RpcAux}(\texttt{self}(),a_{\text{key}})\}$
$\qquad\qquad\qquad \texttt{letactor}\{w := \texttt{ready}(b)\}$
$\qquad\qquad\qquad\quad \texttt{seq}(\texttt{send}(x_0,\texttt{msgMk}(mid,m_{\text{args}},w)),$
$\qquad\qquad\qquad\qquad \texttt{clc}(\lambda k.\texttt{ready}(\texttt{RpcWait}(k,a_{\text{key}}))))$

where the following definitions are also added to the generated kernel library of any program translation

$\texttt{RpcAux} = \lambda x_a.\lambda x_{\text{key}}.\lambda m.\texttt{send}(x_a,\texttt{msgMk}(\texttt{nil},\texttt{list}_1(\texttt{car}(\texttt{msgArgs}(m))),x_{\text{key}})$

$\texttt{RpcWait} = \lambda k.\lambda x_{\text{key}}.\lambda m.\texttt{if}(\texttt{equal?}(x_{\text{key}},\texttt{msgCust}(m)),$
$\qquad\qquad\qquad \texttt{app}(k,\texttt{car}(\texttt{msgArgs}(m))),$
$\qquad\qquad\qquad \texttt{seq}(\texttt{send}(\texttt{self}(),m),\texttt{ready}(\texttt{RpcWait}(k,x_{\text{key}}))))$

To establish correctness of the user–kernel translation, we extend it to actor theory configurations and show that this mapping preserves interaction semantics. The following lemma says that the user-to-kernel translation commutes with the meaning function on programs. This formalizes the commuting diagram of §1.
Lemma (user-to-kernel.1): For any user program, ${}^uP$, $[\![u2k({}^uP)]\!] = u2k([\![{}^uP]\!])$
Proof : By calculation using the definitions of $[\![\,]\!]$, and $u2k$.
The main work of the proof is in the following lemma.
Lemma (user-to-kernel.2): For any user configuration ${}^uK$ we have

$$Isem({}^uK) = Isem(u2k({}^uK))\lceil {}^u\mathbb{M}$$

The main theorem is an easy consequence of the above lemmas.
Theorem (user-to-kernel):

$$Isem({}^uP) = Isem(u2k({}^uP))\lceil {}^u\mathbb{M}$$

6 Conclusions

The main technical contribution of this paper is to present a method for establishing equivalence of actor systems, or more generally for distributed object-based systems. The main result of this paper is a proof of correctness of what is essentially a stage of compilation of a high-level actor language. In [PT94] high-level object programming constructs are explained by expansion in the Pict language. In [Wal95] a semantics for a variant of POOL is given via translation to a sorted Pi calculus. This is shown to be a simulation (up to bisimulation) of a direct transition system operational semantics of POOL. Core Facile is a synthesis of the typed lambda calculus and pi-calculus style concurrency primitives. In [Ama94] a translation from Core Facile to a variant based on asynchronous communication is given. The translation of a process is shown to preserve barbed bisimilarity and barbed congruence of the translation of two expressions implies congruence of the expressions. The converse is left open. The translation goes by an intermediate language obtained by adding a control operator to the asynchronous Facile much as we have done in the kernel language. In [AP94] an extension of the Pi-calculus to model locality and failure is translated in to a simply sorted Pi-calculus and similar properties are proved for the translation. Our approach differs in giving both languages an abstract, composable semantics in the same semantic domain and showing that the translation preserves the abstract semantics. The notion of barbed bisimulation seems to share with abstract actor structures and interaction semantics the objective of hiding details of internal computation. More detailed investigation of the relation between these approaches is an interesting topic for future work.

Acknowledgements

The authors would like to thank Gul Agha, Scott Smith and the three anonymous ICALP referees for many helpful comments and corrections. This work was done while the first author was partially supported by ARC grant IA131.84. The second author was partially supported by ONR grant N00014-94-1-0857, NSF grant CCR-9312580, and ARPA/SRI subcontract C-Q0483, ARPA/AirForce grant F30602-96-1-0300, NSF grant CRR-9633419.

References

[Agh86] G. Agha. *Actors: A Model of Concurrent Computation in Distributed Systems*. MIT Press, Cambridge, Mass., 1986.

[Agh90] G. Agha. Concurrent object-oriented programming. *Communications of the ACM*, 33(9):125–141, September 1990.

[Ama94] R. M. Amadio. Translating core facile. Technical Report ECRC-1994-3, European Computer-Industry Research Centre, 1994.

[AMST97] G. Agha, I. A. Mason, S. F. Smith, and C. L. Talcott. A foundation for actor computation. *Journal of Functional Programming*, 7:1–72, 1997.

[AP94] R. M. Amadio and S. Prasad. Localities and failures. Technical Report ECRC-1994-18, European Computer-Industry Research Centre, 1994.

[BB92] G Berry and G. Boudol. The Chemical Abstract Machine. *Theoretical Computer Science*, 96:217–248, 1992.

[BH77] Henry G. Baker and Carl Hewitt. Laws for communicating parallel processes. In *IFIP Congress*, pages 987–992. IFIP, August 1977.

[Cli81] W. D. Clinger. Foundations of actor semantics. AI-TR- 633, MIT Artificial Intelligence Laboratory, May 1981.

[FF86] M. Felleisen and D.P. Friedman. Control operators, the SECD-machine, and the λ-calculus. In M. Wirsing, editor, *Formal Description of Programming Concepts III*, pages 193–217. North-Holland, 1986.

[Hew77] C. Hewitt. Viewing control structures as patterns of passing messages. *Journal of Artificial Intelligence*, 8(3):323–364, 1977.

[Mes92] J. Meseguer. Conditional rewriting logic as a unified model of concurrency. *Theoretical Computer Science*, 96(1):73–155, 1992.

[Plo75] G. Plotkin. Call-by-name, call-by-value and the lambda calculus. *Theoretical Computer Science*, 1:125–159, 1975.

[PT94] Benjamin C. Pierce and David N. Turner. Concurrent objects in a process calculus. In *Theory and Practice of Parallel Programming (TPPP), Sendai, Japan*, Lecture Notes in Computer Science. Springer-Verlag, November 1994. To appear, 1995.

[Tal96a] C. L. Talcott. An actor rewriting theory. In *Workshop on Rewriting Logic*, number 4 in Electronic Notes in Theoretical Computer Science, 1996.

[Tal96b] C. L. Talcott. Interaction semantics for components of distributed systems. In E. Najm and J-B. Stefani, editors, *1st IFIP Workshop on Formal Methods for Open Object-based Distributed Systems, FMOODS'96*, 1996. proceedings published in 1997 by Chapman & Hall.

[Wal95] D. Walker. Objects in the π-calculus. *Information and Computation*, 116:253–271, 1995.

Periodic and Non-periodic Min-Max Equations

Uwe Schwiegelshohn[1] and Lothar Thiele[2]

[1] Computer Engineering Institute, University Dortmund, D-44221 Dortmund, Germany, uwe@carla.e-technik.uni-dortmund.de
[2] Computer Engineering and Communications Laboratory, Swiss Federal Institute of Technology (ETH), CH-8092 Zürich, Switzerland, thiele@tik.ee.ethz.ch

Abstract. In this paper we address min-max equations for periodic and non-periodic problems. In the non-periodic case a simple algorithm is presented to determine whether a graph has a potential satisfying the min-max equations. This method can also be used to solve a more general quasi periodic min-max problem on periodic graphs. Also some results regarding the uniqueness of solutions in the latter case are given.

1 Introduction

Min-max problems can be considered to be a generalization of a variety of graph problems involving potentials. There is a close relationship with network flow problems (non-periodic case), see e.g. [1], and the well known maximum cycle mean problem (periodic case), see e.g. [10], [8]. In particular, previous results in the *non-periodic case* can be related to a feasible potential function p observing lower linear constraints

$$p(v_j) \leq p(v_i) + w(v_i, v_j) \quad \forall v_j \in V^-, (v_i, v_j) \in E \tag{1}$$

and an optimal potential function p using min constraints

$$p(v_j) = \min_{(v_i, v_j) \in E} \{p(v_i) + w(v_i, v_j)\} \quad \forall v_j \in V^- \tag{2}$$

associated with some network $\mathcal{G}^-(V^-, E, w)$. Our paper addresses a generalization where these sets of inequalities are mixed for a given network $\mathcal{G}(V, E, w)$ with their dual forms, that is upper linear constraints

$$p(v_j) \geq p(v_i) + w(v_i, v_j) \quad \forall v_j \in V^+, (v_i, v_j) \in E \tag{3}$$

and max constraints

$$p(v_j) = \max_{(v_i, v_j) \in E} \{p(v_i) + w(v_i, v_j)\} \quad \forall v_j \in V^+ \tag{4}$$

with $V^+ \cup V^- = V$. If a distance function d is given additionally, the corresponding *quasi periodic problem* deals with edge weights $w(v_i, v_j) - \lambda(v_j) d(v_i, v_j)$ where a specific period $\lambda(v_j)$ is associated with each node v_j.

In the area of interface timing verification, see [11, 17, 19], problems related to the existence of min and/or max constraints frequently occur. There, the difference between the potentials of two nodes must be maximized under various constraints. In particular, it is possible to transform one of the problems addressed in [11], [17] and [19] to a problem with mixed constraints (1), (2) and (3). Different pseudo-polynomial algorithms are derived for the solution of this problem based on iterative tightening [11], removing negative cycles [17] and maximum separations [19]. However so far, neither a polynomial algorithm nor a proof of intractability is known.

In comparison to these results, we are mainly dealing with constraints (2) and (4). Note that constraints of the form (1) or (3) can easily be converted into constraints (2) and (4) by a simply adding one additional node and two edges for each node v_j with constraints of type (1) or (3). Regarding the non-periodic case our paper presents efficient pseudo-polynomial algorithms for finding optimal potentials satisfying constraints (2) and (4).

The consideration of constraints (4) in connection with periodic graphs has raised significant interest in the past, as it is the root for many problems from different application areas, see [9, 14, 10, 8]. This includes e.g. control theory and manufacturing [5], timing properties of discrete event systems [15], parallel algorithms [16], and other areas of computer science. A comprehensive treatment of the theory and its applications can be found in [2]. Especially the use of linear equations over a new max-plus algebra [5, 2] has produced many results. Some of these results have even been generalized to problems which are periodic in multiple dimensions, see [3].

Driven by application areas like asynchronous circuit design, timing and protocol verification, and timing behavior of general Petri nets, some recent approaches addressed the generalization of these results to dynamic graphs with constraints of the form (4) *and* (2). These dynamic min-max systems have been investigated in [12, 13, 2]. Further results in this direction are described in [6, 7]. However, the models used in these two groups of papers are quite different. Olsder [12, 13] describes a periodic min-max problem in terms of an eigenvalue problem, whereas Gunawardena [6, 7] defines a certain class of min-max functions. Both models are special cases of those used in our paper.

Also with respect to numerical procedures and the uniqueness of the period, the results in [12, 13] are restricted to a subclass of min-max problems. On the other hand, [6, 7] contain "complete" results in the case that only two distances have the value 1 while all others are zero. For all other considered cases $(d(v_i, v_j) \in \{0, 1\})$, there is no procedure which decides whether a min-max system has a period or not. Moreover, the given algorithm for the computation of the period is exponential in the size of the graph. In this area our paper contains the following new results:

- A relation between potential functions of dynamic and weight transformed static graphs is derived. This is similar to a known result for max-plus problems [4].
- Results on the uniqueness of the periods in the quasi-periodic case are given as well as algorithms to determine these periods.

2 The Static Min-Max Problem

2.1 Definitions and Properties

We start this section by defining various forms of graph potentials.

Definition 1 Min-Max Potential. Assume a weighted digraph $\mathcal{G}(V = V^+ \cup V^-, E, w)$ with $V^+ \cap V^- = \emptyset$, $E \subseteq V \times V$ and $w : E \to \mathcal{Q}$, also called min-max graph subsequently. Then, a potential $p : V \to \mathcal{Q}$ is called feasible if

$$p(v_i) \begin{cases} \geq p(v_j) + w(v_j, v_i) & \forall (v_j, v_i) \in E, v_i \in V^+ \\ \leq p(v_j) + w(v_j, v_i) & \forall (v_j, v_i) \in E, v_i \in V^-. \end{cases}$$

Further, a feasible potential $p : V \to \mathcal{Q}$ is a min potential if

$$p(v_i) = \min_{(v_j, v_i) \in E} \{p(v_j) + w(v_j, v_i)\} \quad \forall v_i \in V^-.$$

Similarly, a feasible potential $p : V \to \mathcal{Q}$ is a max potential, if

$$p(v_i) = \max_{(v_j, v_i) \in E} \{p(v_j) + w(v_j, v_i)\} \quad \forall v_i \in V^+.$$

Finally, a potential $p : V \to \mathcal{Q}$ is a min-max potential if it is a min potential and a max potential at the same time. ∎

The definition of a min-max potential directly leads to our first key problem:

Problem 2. Is there a min-max potential for a given min-max graph $\mathcal{G}$?

The problem can be simplified by using the following few observations:

1. If $\mathcal{G}$ consists of two independent graphs, it is sufficient to consider each graph separately.
2. If $\mathcal{G}^+ = (V, E \cap (V^+ \times V^+), w)$ contains a positive weight cycle, then there is no min-max potential for $\mathcal{G}$ (positive cycle in a longest path problem).
3. If $\mathcal{G}^- = (V, E \cap (V^- \times V^-), w)$ contains a negative weight cycle, then there is no min-max potential for $\mathcal{G}$ (negative cycle in a shortest path problem).
4. It suffices to only consider bipartite min-max graphs where $E \subseteq (V^+ \times V^-) \cup (V^- \times V^+)$ as additional nodes can be inserted without changing the problem substantially. A proof of this claim is given in [18].

Therefore, we assume for the remainder of this section that $\mathcal{G}$ is a connected bipartite graph and that for each node $v_j \in V$ there is at least one edge $(v_i, v_j) \in E$.

In the next corollary we show that knowledge about a min potential for a graph can provide some information about min potentials for related graphs.

Corollary 3. *If a bipartite graph $\mathcal{G}(V, E, w)$ has a min potential, then there is also a min potential for any graph $\mathcal{G}'(V, E, w')$ with $w'(v_i, v_j) \leq w(v_i, v_j)$ for all $(v_i, v_j) \in E$. On the other hand, if a bipartite graph $\mathcal{G}(V, E, w)$ has no min potential, then no min potential exists for any graph $\mathcal{G}'(V, E, w')$ with $w'(v_i, v_j) \geq w(v_i, v_j)$ for all $(v_i, v_j) \in E$.*

Proof. Let p be a min potential of $\mathcal{G}$ and $w'(v_i, v_j) \leq w(v_i, v_j)$ for all $(v_i, v_j) \in E$. Then p' with

$$p'(v_i) = \begin{cases} p(v_i) & \text{for } v_i \in V^+ \\ \min_{(v_j, v_i) \in E} \{p(v_j) + w'(v_j, v_i)\} & \text{for } v_i \in V^- \end{cases}$$

is a min potential for $\mathcal{G}'$, as $p'(v_i) \geq p(v_j) + w(v_j, v_i) \geq p'(v_j) + w'(v_j, v_i)$ for all $v_i \in V^+$ and $(v_j, v_i) \in E$. The second claim of the corollary is a direct consequence of the first one. ∎

Of course, a similar corollary holds for max potentials as well. It is easy to see that 'tight' edges (v_i, v_j) of a min-max graph with $p(v_j) = p(v_i) + w(v_i, v_j)$ are especially important. For any min-max potential p, there must be a tight input edge for each node $v_j \in V$. Also, a min-max potential for a graph $\mathcal{G}$ implies the existence of a cycle C consisting of tight edges. In $\mathcal{G}$ this cycle C must be a *zero weight cycle*, i.e. $\sum_{(v_i, v_j) \in C} w(v_i, v_j) = 0$.

Moreover, we can restrict ourselves to those min-max potentials where $\mathcal{G}_p$ is connected. Then, the difference between the min-max potential values of any two vertices $|p(v_i) - p(v_j)|$ is bounded by the maximum length of any simple (undirected) path in $\mathcal{G}$. For such a path we can use the following upper bound s:

$$s = \sum_{v_j \in V} (\max_{(v_i, v_j) \in E} \{|w(v_i, v_j)|\}). \tag{5}$$

2.2 Algorithms

Now, we describe a method to determine whether a bipartite weighted digraph has a min-max potential. This method is based on Function *increase* in Table 1.

The following corollary describes the possible outcome of Function *increase*, see [18] for a detailed proof.

Corollary 4. *If and only if the bipartite min-max graph $\mathcal{G}$ has a min potential, then Function increase returns 'true' and the generated potential p is a min potential.* ∎

Any change of the potential of a node $v_i \in V^+$ requires that at some time during the execution of the function there was a node $v_j \in V^-$ with $p(v_i) = p(v_j) + w(v_j, v_i)$. On the other hand, if $p(v_i) = p(v_j) + w(v_j, v_i)$ at any time during the execution of the loop for a node $v_i \in V^+$, then there is a tight edge (v_k, v_i) for some node $v_k \in V^-$ provided the function returns 'true'. Hence, if

```
Boolean Function increase($\mathcal{G}, p, \mathcal{G}_t$) {
    in $\mathcal{G}$; inout $p$; out $\mathcal{G}_t$;
    $a = \max\{p(v) \mid v \in V^+\}$;
loop: $p(v_j) = \min\{p(v_i) + w(v_i, v_j) \mid (v_i, v_j) \in E\}$ for all $v_j \in V^-$;
    if ($\exists (v_j, v_i) \in E$ with $v_i \in V^+$ and $p(v_i) < p(v_j) + w(v_j, v_i)$) {
        $p(v_i) = p(v_j) + w(v_j, v_i)$; }
    else { $\mathcal{G}_t = \mathcal{G}$; return 'true'; }
    if (there is no change in the potential of any node $v_i$ with $p(v_i) \leq a + s$) {
        $\mathcal{G}_t$ = subgraph of $\mathcal{G}$ induced by all nodes with $p(v_i) \leq a + s$;
        return 'false'; }
    goto loop;
}
```

Table 1. Function *increase*

Function *increase* starts with a max potential and returns 'true', the generated potential p will be a min-max potential. This leads directly to the following theorem:

Theorem 5. *$\mathcal{G}$ has a min-max potential, if and only if it has a min potential and a max potential.* ∎

Therefore, a min-max potential of $\mathcal{G}$ can be detected by first applying Function *increase* to an arbitrary initial potential and then applying its dual counterpart Function *decrease* to the resulting potential. The existence of a min-max potential for $\mathcal{G}$ requires that Function *increase* and Function *decrease* both return 'true'.

This procedure constitutes a pseudo polynomial way to solve the min-max problem. However, cycles with a small weight sum, like e.g. $w(v_i, v_j) + w(v_j, v_i) = \epsilon \to 0$, in connection with large edge weights may lead to a large number of iterations. This problem is addressed in [18], where we introduce improved alternatives to functions *increase* and *decrease*.

3 The Dynamic Min-Max Problem

In this section we address the quasi-periodic min-max problem on dynamic graphs. To this end, we first define dynamic graphs as usual via static graphs, see e.g. [9]. Then, Problem 2 is extended to dynamic graphs.

Definition 6 Static Graph. *A (bipartite) static graph $\mathcal{G}_s(V, E, w, d)$ with $V = V^+ \cup V^-$ is a bipartite weighted digraph with a weight function $w : E \to \mathcal{Z}$ and a distance function $d : E \to \mathcal{Z}_{\geq 0}$.* ∎

Definition 7 Dynamic Graph. *The dynamic graph corresponding to a given static graph $\mathcal{G}_s(V, E, w, d)$ is an infinite weighted bipartite graph $\mathcal{G}_d(V_d, E_d, w_d)$ where*

$$\begin{aligned}
V_d &= \{v_i(k) \mid v_i \in V, k \in \mathcal{Z}_{\geq 0}\}, \\
E_d &= \{(v_i(k - d(v_i, v_j)), v_j(k)) \mid \\
&\quad (v_i, v_j) \in E, k \in \mathcal{Z}, k \geq d(v_i, v_j)\}, \\
w_d(v_i(k - d(v_i, v_j)), v_j(k)) &= w(v_i, v_j) \text{ for all } (v_i(k - d(v_i, v_j)), v_j(k)) \in E_d.
\end{aligned}$$

∎

Definition 8 Quasi-Periodic Min-Max Potential. The quasi-periodic min-max potential $p_d : V_d \to \mathcal{Q}$ of a dynamic graph $\mathcal{G}_d(V_d, E_d, w_d)$ is a min-max potential p_d for all $k \geq K = \max_{(v_i, v_j) \in E}\{d(v_i, v_j)\}$. Moreover, there is a period function $\lambda : V \to \mathcal{Q}$ such that

$$p_d(v_i(k+1)) = p_d(v_i(k)) + \lambda(v_i) \text{ for all } v_i(k) \in V_d.$$

∎

Problem 9. Is there a quasi-periodic min-max potential for a dynamic graph $\mathcal{G}_d(V_d, E_d, w_d)$?

In order to avoid dealing with infinite dynamic graphs, we use the regularity of those graphs to describe them with a cycle graph, see also [1].

Definition 10 Quasi-Periodic Cycle Graph. For a static graph $\mathcal{G}_s(V, E, w, d)$ and a period function $\lambda : V \to \mathcal{Q}$ with $\lambda(v_i) \geq \lambda(v_j)$ for all $v_i \in V^+$ and v_i, v_j adjacent in $\mathcal{G}_s$, the quasi-periodic cycle graph $\mathcal{G}_c(V, E_c, w_c)$ is defined by

$$E_c = \{(v_i, v_j) \in E \mid \lambda(v_i) = \lambda(v_j)\},$$

$$w_c(v_i, v_j) = w(v_i, v_j) - \lambda(v_j)d(v_i, v_j) \text{ for all } (v_i, v_j) \in E_c.$$

∎

Now, the following corollary establishes a close relation between the quasi-periodic min-max problem of a dynamic graph and the min-max problem of the corresponding quasi-periodic cycle graph.

Corollary 11. *Assume a static graph $\mathcal{G}_s(V, E, w, d)$. Then, the following two statements are equivalent:*

- *The dynamic graph $\mathcal{G}_d$ corresponding to $\mathcal{G}_s$ has a quasi-periodic min-max potential p_d with the period function λ.*
- *The quasi-periodic cycle graph $\mathcal{G}_c$ corresponding to $\mathcal{G}_s$ and λ has a min-max potential.*

Proof. A max potential of $\mathcal{G}_d$ requires for all $v_i(k) \in V_d^+$ and $k \geq K$ the correctness of the equation

$$p_d(v_i(k)) = \max_{(v_j(k-d(v_j,v_i)),v_i(k)) \in E_d} \{p_d(v_j(k-d(v_j,v_i))) + w_d(v_j(k-d(v_j,v_i)), v_i(k))\}.$$

Using the definition of a dynamic graph and the periodicity of p_d we obtain the equivalent conditions

$$p_d(v_i(0)) = \max_{(v_j,v_i) \in E} \{p_d(v_j(0)) + w_c(v_j,v_i) + (k - d(v_j,v_i))(\lambda(v_j) - \lambda(v_i))\} \quad \forall k > K \tag{6}$$

1. $\mathcal{G}_d$ has a quasi-periodic min-max potential $\to$ $\mathcal{G}_c$ has a min-max potential. First assume that $\lambda(v_i) < \lambda(v_j)$. Then, $\mathcal{G}_c$ does not exist. Also, the validity of Equation (6) *for all $k > K$* prevents $p_d(v_i(0))$ from being finite.
   On the other hand if $\lambda(v_i) > \lambda(v_j)$, then edge (v_j, v_i) cannot affect Equation (6) for $k \to \infty$. Therefore, it need not be considered in Equation (6). This results in

$$p_d(v_i(0)) = \max_{(v_j,v_i) \in E_c} \{p_d(v_j(0)) + w_c(v_j,v_i)\} \quad \forall k > K,$$

   which leads to a max potential $p_c(v_i) = p_d(v_i(0))$ for all $v_i \in V^+$.

2. $\mathcal{G}_c$ has a min-max potential $\to$ $\mathcal{G}_d$ has a quasi-periodic min-max potential. Suppose that a cycle graph is given with $p_c(v_i) = \max_{(v_j,v_i) \in E_c} \{p_c(v_j) + w_c(v_j,v_i)\}$ for all $v_i \in V_c$. If we set $p_d(v_i(0)) = p_c(v_i)$ for all $v_i \in E_c$, then Equation (6) holds as $\lambda(v_i) \geq \lambda(v_j)$ and the third term in the max-expression becomes sufficiently negative for all edges in $E \setminus E_c$ and $k \to \infty$.

Similar arguments are used for all $v_i(k) \in V_d^-$. ∎

Due to Corollary 11, the solution in the quasi-periodic case divides $\mathcal{G}_s$ into subgraphs with different periods. In other words, the quasi-periodic cycle graph $\mathcal{G}_c$ of $\mathcal{G}_s$ consists of unconnected subgraphs, where each subgraph has a min-max potential and a period common to all nodes. This suggests an algorithm to determine the periods and subgraphs by iteratively peeling off subgraphs with decreasing periods from the static graph. Therefore, at first the case of a single period λ for all nodes $v_i \in \mathcal{G}_s$ will be considered. The functions *lower-period* and *upper-period* are introduced to determine the *single* period λ for all nodes $v_i \in \mathcal{G}_s$.

Note that if a dynamic graph $\mathcal{G}_d$ has a periodic min-max potential with period λ, then the corresponding cycle graph $\mathcal{G}_c$ has at least one directed cycle C with $\sum_{(v_i,v_j) \in C} w_c(v_i,v_j) = 0$. Assuming $\sum_{(v_i,v_j) \in C} d(v_i,v_j) > 0$ this results in

$$\lambda = \frac{\sum_{(v_i,v_j) \in C} w(v_i,v_j)}{\sum_{(v_i,v_j) \in C} d(v_i,v_j)}. \tag{7}$$

```
Boolean Function lower-period(𝒢_s, λ_l, p) {
    in 𝒢_s; out λ_l; inout p;
    determine s and t;
    λ_l = -s; λ_u = s;
    generate the periodic cycle graph 𝒢_c of 𝒢_s and λ_l;
    if (increase(𝒢_c, p, 𝒢_t)) { λ_l = -∞; return 'true'; }
    generate the periodic cycle graph 𝒢_c of 𝒢_s and λ_u;
    if (!increase(𝒢_c, p, 𝒢_t)) { return 'false'; }
    loop  λ = (λ_u + λ_l)/2;
    generate the periodic cycle graph 𝒢_c of 𝒢_s and λ;
    if (!increase(𝒢_c, p, G_t)) { λ_l = λ; } else { λ_u = λ; }
    if (λ_u - λ_l ≤ 1/t²) { return 'true'; }
    goto loop;
}
```

Table 2. Function *lower-period*

Now, we can introduce Function *lower-period* in Table 2 to determine the minimal period λ_l for which a periodic min-max potential may exist. This function is based on binary search, see also [10], [1] and uses the following upper bound t for the sum of distances in any simple path in $\mathcal{G}_s$:

$$t = \sum_{v_j \in V} (\max_{(v_i, v_j) \in E} \{|d(v_i, v_j)|\}). \tag{8}$$

The correctness of Function *lower-period* is addressed in Corollary 12. In the remaining part of this section all corollaries and theorems are given without proofs due to space restrictions. Regarding the proofs the interested reader is referred to [18].

Corollary 12. *If Function* lower-period *returns 'false', then the dynamic graph $\mathcal{G}_d$ corresponding to $\mathcal{G}_s$ has no periodic min-max potential. Otherwise, $\mathcal{G}_d$ has min potentials for all $k \geq K$ and for all periods $\lambda \geq \lambda_l$, while there is no min-potential for all periods $\lambda < \lambda_l$.* ∎

Similarly, a Function *upper-period* based on Function *decrease* is used to determine the maximal period λ_l, for which a periodic min-max potential may exist. The combination of both functions yields an algorithm to determine whether there are periodic min-max potentials for a dynamic graph $\mathcal{G}_d$. The proof is a direct consequence of Corollary 12, its counterpart for Function *upper-period* and Theorem 5.

Theorem 13. *If either Function* lower-period *or Function* upper-period *return 'false' or if $\lambda_l > \lambda_u$ is produced, then there is no periodic min-max potential for the dynamic graph $\mathcal{G}_d$ corresponding to $\mathcal{G}_s$. Otherwise, there are periodic min-max potentials for all periods $\lambda_l \leq \lambda \leq \lambda_u$.* ∎

In the following corollary we address the computational complexity for the presented method.

Corollary 14. *There is an algorithm which computes a periodic min-max potential in pseudo-polynomial time.* ∎

Finally, a result on the uniqueness of a period λ is derived.

Theorem 15. *If the static graph $\mathcal{G}_s$ corresponding to a dynamic graph $\mathcal{G}_d$ contains only edges with distance > 0, then $\mathcal{G}_d$ either has no periodic min-max potential or a min-max potential with a unique period.* ∎

Now, we can return to the main task of addressing the general case of different periods associated with the nodes of the dynamic graph. In order to simplify the following discussions, we suppose that there is no directed cycle with a zero sum of distances in the given static graph $\mathcal{G}_s$, i.e.

$$\sum_{(v_i,v_j)\in C} d(v_i,v_j) > 0 \text{ for all directed cycles } C \text{ of } \mathcal{G}_s.$$

Remember that the period function defines a partition of the dynamic graphs into subgraphs whose nodes have equal periods. At first, these subgraphs are defined formally. This is done using the weighted bipartite graph $\mathcal{G}$ in a similar fashion as in the static min-max problem, see Section 2 and Definition 1.

Definition 16 Dominating Subgraph. A dominating subgraph $\mathcal{G}_t$ of a digraph $\mathcal{G}$ (as defined in Definition 1) is a subgraph of $\mathcal{G}$ with the following properties:

1. There exists a min potential p of $\mathcal{G}$ which is a min-max potential of $\mathcal{G}_t$.
2. There are no edges (v_i, v_j) or (v_j, v_i) with $v_i \in V_t^-$ and $v_j \in (V^+ \setminus V_t^+)$.

∎

Next, the following theorem provides results on one step of a procedure which determines the quasi-periodic min-max potential of a given static graph. It is shown that the concatenation of Functions *lower-period* and *decrease*

- peals off a subgraph of a given static graph,
- produces a period λ_{max} and a corresponding min-max potential for this subgraph and
- that the remaining static graph has a min potential for a period less than λ_{max}.

Corollary 17. *Given a static graph $\mathcal{G}_s$. After execution of Functions lower-period $(\mathcal{G}_s, \lambda_{max}, p)$ with initial potentials $p(v_i) = 0$ and decrease$(\mathcal{G}_c, p, \mathcal{G}_t)$ with the periodic cycle graph $\mathcal{G}_c$ corresponding to $\mathcal{G}_s$ and λ_{max}, the following properties hold:*

1. $\mathcal{G}_t$ is a dominating subgraph of $\mathcal{G}_c$.
2. The application of Function lower-period$((\mathcal{G}_s\backslash\mathcal{G}_t), \lambda, p)$ returns 'true' with $\lambda < \lambda_{max}$.

∎

Now, we are ready to present the complete algorithm for the calculation of the quasi-periodic min-max potential, see Table 3. Input to the Function period$(\mathcal{G}_s, \lambda())$ is the given static graph $\mathcal{G}_s$, while its output is the resulting period function λ. The corresponding min-max potentials can be either extracted during execution of Function period or by using the proof of Corollary 11.

```
Boolean Function period(𝒢ₛ, λ()) {
    in 𝒢ₛ; out λ();
loop  p(vᵢ) = 0 for all vᵢ ∈ Vₛ;
      lower-period(𝒢ₛ, λₘₐₓ, p);
      generate the periodic cycle graph 𝒢_c of 𝒢ₛ and λₘₐₓ;
      if (decrease(𝒢_c, p, 𝒢_t)) {
          λ(vᵢ) = λₘₐₓ for all vᵢ ∈ V_t;
          return 'true'; }
      else {
          λ(vᵢ) = λₘₐₓ for all vᵢ ∈ V_t;
          𝒢ₛ = 𝒢ₛ\𝒢_t;
          goto loop;
      }
}
```

Table 3. Function *period*

Finally, the following theorem states one of the main results of this paper.

Theorem 18. *Any dynamic graph $\mathcal{G}_d$ has a quasi-periodic min-max potential. The potential is unique.* ∎

4 Conclusion

In this paper, we demonstrate a close relationship between static and dynamic min-max problems. Also, pseudo polynomial algorithms for the solution of min-max equations systems in the quasi-periodic and non-periodic case are presented. Further, we show that any dynamic graph has a unique period function.

References

1. AHUJA, R. K., MAGNANTI, T. L., AND ORLIN, J. *Network Flows*. Prentice Hall, 1993.

2. BACCELLI, F., COHEN, G., OLSDER, G., AND QUADRAT, J.-P. *Synchronization and Linearity*. John Wiley, Sons, New York, 1992.
3. BACKES, W., SCHWIEGELSHOHN, U., AND THIELE, L. Analysis of free schedule in periodic graphs. In *4th Annual ACM Symposium on Parallel Algorithms and Architectures*, San Diego, CA, USA, June 1992, pp. 333-342.
4. COHEN, G., DUBOIS, D., QUADRAT, J. P., AND VIOT, M. A linear-system-theoretic view of discrete-event processes and its use for performance evaluation in manufacturing. *IEEE Transactions on Automatic Control AC-30, No. 3*, March 1985, 210-220.
5. CUNNINGHAME-GREEN, R. Describing industrial processes and approximating their steady-state behaviour. *Opt. Res. Quart. 13*, 1962, 95-100.
6. GUNAWARDENA, J. Timing analysis of digital circuits and the theory of min-max functions. In *TAU'93, ACM International Workshop on Timing Issues in the Specification and Synthesis of Digital Systems*, September 1993.
7. GUNAWARDENA, J. Min-max functions. Tech. Rep. to be published in Discrete Event Dynamic Systems, Department of Computer Science, Stanford University, Stanford, CA 94305, USA, March 1994.
8. KARP, R. A characterization of the minimum cycle mean in a digraph. *Discrete Mathematics 23*, 1978, 309-311.
9. KOSARAJU, S. R., AND SULLIVAN, G. F. Detecting cycles in dynamic graphs in polynomial time. In *20th Annual ACM Symposium on Theory of Computing*, 1988, pp. 398-406.
10. LAWLER, E. Optimal cycles in doubly weighted directed linear graphs. In *Thèorie des Graphes*, 1966, P. Rosenstiehl, Ed., pp. 209-213.
11. MCMILLAN, K., AND DILL, D. Algorithms for interface timing verification. In *IEEE Int. Conference on Computer Design*, 1992, pp. 48-51
12. OLSDER, G. Eigenvalues of dynamic max-min systems. *Discrete Event Dynamic Systems: Theory and Applications*, 1991, 1:177-207.
13. OLSDER, G. Analyse de systèmes min-max. Tech. Rep. 1904, Institute National de Recherche en Informatique et en Automatique, May 1993.
14. ORLIN, J. Some problems in dynamic and periodic graphs. In *Progress in Combinatorial Optimization*, Academic Press, Orlando, Florida, 1984, W.R. Pulleyblank, Ed., pp. 215-225.
15. RAMAMOORTHY, C. Performance evaluation of asynchronous concurrent systems using Petri nets. *IEEE Transactions on Software Engineering*, 1980, 440-449.
16. REITER, R. Scheduling parallel computations. *Journal of the Association for Computing Machinery 15, No. 4*, October 1968, 590-599.
17. WALKUP, E., AND BORRIELLO, G. Interface timing verification with application to synthesis. In *IEEE/ACM Design Automation Conference*, 1994, pp. 106-112.
18. SCHWIEGELSHOHN, U., AND THIELE, L. Dynamic min max problems. *Technical Report ETH Zurich, Computer Engineering and Networks Laboratory*, January 1997.
19. YEN, T., ISHII, A., CASAVANT, A., AND WOLF, W. Efficient algorithms for interface timing verification. *Technical Report Princeton University*, June 1995.

Efficient Parallel Graph Algorithms for Coarse Grained Multicomputers and BSP*

E. Cáceres[1], F. Dehne[2], A. Ferreira[3], P. Flocchini[4],
I. Rieping[5], A. Roncato[6], N. Santoro[7], and S. W. Song[8]

[1] Univ. Federal de Mato Grosso do Sul, Campo Grande, Brasil, *edson@dct.ufms.br*
[2] Carleton Univ., Ottawa, Canada, *dehne@scs.carleton.ca*
[3] ENS Lyon, Lyon, France, *ferreira@lip.ens-lyon.fr*
[4] Univ. de Montreal, Montreal, Canada, *flocchin@iro.umontreal.ca*
[5] Univ. Paderborn, Paderborn, Germany, *inri@uni-paderborn.de*
[6] Facolta di Scienze Mat. Fis. e Nat., Mestre, Italia, *roncato@dsi.unive.it*
[7] Carleton Univ., Ottawa, Canada, *santoro@scs.carleton.ca*
[8] Univ. of São Paulo, São Paulo, Brazil, *song@ime.usp.br*

Abstract. In this paper, we present *deterministic* parallel algorithms for the coarse grained multicomputer (CGM) and bulk-synchronous parallel computer (BSP) models which solve the following well known graph problems: (1) list ranking, (2) Euler tour construction, (3) computing the connected components and spanning forest, (4) lowest common ancestor preprocessing, (5) tree contraction and expression tree evaluation, (6) computing an ear decomposition or open ear decomposition, (7) 2-edge connectivity and biconnectivity (testing and component computation), and (8) cordal graph recognition (finding a perfect elimination ordering). The algorithms for Problems 1-7 require $O(\log p)$ communication rounds and linear sequential work per round. Our results for Problems 1 and 2 hold for arbitrary ratios $\frac{n}{p}$, i.e. they are *fully scalable*, and for Problems 3-8 it is assumed that $\frac{n}{p} \geq p^\epsilon$, $\epsilon > 0$, which is true for all commercially available multiprocessors. We view the algorithms presented as an important step towards the final goal of $O(1)$ communication rounds. Note that, the number of communication rounds obtained in this paper is independent of n and grows only very slowly with respect to p. Hence, for most practical purposes, the number of communication rounds can be considered as constant. The result for Problem 1 is a considerable improvement over those previously reported. The algorithms for Problems 2-7 are the first practically relevant deterministic parallel algorithms for these problems to be used for commercially available coarse grained parallel machines.

---

* Research partially supported by the Natural Sciences and Engineering Research Council of Canada, FAPESP (Brasil), CNPq (Brasil), PROTEM-2-TCPAC (Brasil), the Commission of the European Communities (ESPRIT Long Term Research Project 20244, ALCOM-IT), DFG-SFB 376 "Massive Parallelität" (Germany), and the Région Rhône-Alpes (France).

1 Introduction

The Models: Speedup results for theoretical PRAM algorithms do not necessarily match the speedups observed on real machines [2] [31]. Given sufficient slackness in the number of processors, Valiant's BSP approach [34] simulates PRAM algorithms optimally on distributed memory parallel systems. Valiant points out, however, that one may want to design algorithms that utilize local computations and minimize global operations [33] [34]. The BSP approach requires that g (= local computation speed / router bandwidth) is low, or fixed, even for increasing number of processors. Gerbessiotis and Valiant [17] describe circumstances where PRAM simulations can not be performed efficiently, among others, if the factor g is high. Unfortunately, this is true for most currently available multiprocessors. The parallel algorithms presented in this paper consider this case for graph problems.

As pointed out in [34], the cost of a message also contains a constant overhead cost s. The value of s can be fairly large and the total message overhead cost can have a considerable impact on the speedup observed (see e.g. [8]). We are therefore also using a more practical version of the BSP model, referred to as the *coarse grained multicomputer* model (CGM) [8], [9], [10]. It is comprised of a set of p processors $P_1, \ldots, P_p$ with $O(n/p)$ local memory per processor and an arbitrary communication network (or shared memory). All algorithms consist of alternating local computation and global communication rounds. Each communication round consists of routing a single h-relation with $h = O(n/p)$, i.e. each processor sends $O(n/p)$ data and receives $O(n/p)$ data. We require that all information sent from a given processor to another processor in one communication round is packed into one long message, thereby minimizing the message overhead. In the BSP model, a computation/communication round is equivalent to a superstep with $L = \frac{n}{p}g$ (plus the above "packing requirement").

Finding an optimal algorithm in the coarse grained multicomputer model (CGM) is equivalent to minimizing the number of communication rounds as well as the total local computation time. This considers all parameters discussed above that are affecting the final observed speedup and it requires no assumption on g. Furthermore, it has been shown that minimizing the number of supersteps also leads to improved portability across different parallel architectures ([33] [34] [13]). The above model has been used (explicitly or implicitly) in parallel algorithm design for various problems ([4], [8], [9], [14], [12], [22], [10]) and shown very good practical timing results.

The Results: In this paper, we study deterministic parallel graph algorithms for the CGM and BSP models. We consider the following well known graph problems:

1. list ranking
2. Euler tour construction
3. computing the connected components and spanning forest
4. lowest common ancestor preprocessing

5. tree contraction and expression tree evaluation
6. computing an ear decomposition or open ear decomposition
7. 2-edge connectivity and biconnectivity (testing and component computation)
8. cordal graph recognition, finding a perfect elimination ordering

These problems have been extensively studied for the PRAM (see e.g. [28]) and for fine-grained parallel network models of computation (see e.g. [1]). However, for the practically much more relevant CGM/BSP model there exist, to the best of our knowledge, only a few results on parallel graph algorithms.

Reid-Miller's [27] presented an empirical study of parallel list ranking for the Cray C-90. The paper followed essentially the CGM/BSP model and claimed that this was the fastest list ranking implementation so far. The algorithm in [27] required $O(\log n)$ communication rounds. In [11], an improved algorithm was presented which required, with high probability, only $O(k \log p)$ rounds, where $k \leq \log^* n$. In [13], $O(\log p)$ communication rounds are achieved by a randomized algorithm. Bäumker and Dittrich [3] presented a randomized connected components algorithm for planar graphs using $O(\log p)$ communication rounds. They suggest an extension of this algorithm for general graphs with the same number of communication rounds.

We improve these results by giving the first *deterministic* algorithms for list ranking and computing connected components using $O(\log p)$ rounds. This improvement is an important step towards the ultimate goal, a deterministic algorithm with only $O(1)$ communication rounds. In fact, it is an open problem whether this is possible for these graph problems. Algorithms with $O(1)$ rounds have been presented for various Computational Geometry problems [8, 9, 10, 11, 16], but the graph problems studied in this paper have considerably less "internal structure" which could be exploited to obtain such solutions. Note that, in practice, the number of processors is usually fixed. In contrast to the previous deterministic results, the improved number of communication rounds obtained in this paper, $O(\log p)$, is *independent* of n and grows only very slowly with respect to p. Hence, for most practical purposes, the number of communication rounds can be considered as constant. We expect, that this will be of considerable practical relevance.

As in [27] we will, in general, assume that $n \gg p$ (coarse grained), because this is usually the case in practice. Note, however, that our results for Problems 1 and 2 hold for arbitrary ratios $\frac{n}{p}$. Goodrich [18] calls such algorithms *fully scalable*. For Problems 3-8 we will assume that $\frac{n}{p} \geq p^\epsilon$, $\epsilon > 0$, which is true for all commercially available multiprocessors.

2 List Ranking

Let L be a list represented by a vector s s.t. $s[i]$ is the node following i in the list L. The last element l of the list L is the one with $s[l] = l$. The distance between i and j, $d_L(i,j)$, is the number of nodes between i and j plus 1 (i.e. the distance is 0 iff $i = j$, and it is one if and only if one node follows the other). The list ranking problem consists of computing for each $i \in L$ the distance between i and l, referred to as $rank_L(i) = d_L(i, l)$.

For our algorithm, we need the following definitions. A *r-ruling set* is defined as a subset of selected list elements that has the following properties: (1) No two neighboring elements are selected. (2) The distance of any unselected element to the next selected element is at most r.

An overview of our CGM list ranking algorithm is as follows. First, we compute a $O(p^2)$-ruling set R with $|R| = O(n/p)$ and broadcast R to all processors. More precisely, the $O(p^2)$-ruling set R is represented as a linked list where each element i is assigned a pointer to the next element j of R with respect to the order implied by L as well as the distance between i and j in L. Then, every processor sequentially performs a list ranking of R, computing for each $i \in R$ its distance to the last element of L. All other list elements have at most distance $O(p^2)$ from the next element of R in the list. Their distance is determined by simulating standard PRAM pointer jumping until the next element of R is reached.

All steps, except for the computation of the $O(p^2)$-ruling set R, can be easily implemented in $O(\log p)$ communication rounds.

In the remainder of this section we introduce a new technique, called *deterministic list compression*, which will allows us to compute a $O(p^2)$-ruling set in $O(\log p)$ communication rounds.

The basic idea behind *deterministic list compression* is to have an alternating sequence of *compress* and *concatenate* phases. In a compress phase, we select a subset of list elements, and in a *concatenate* phase we use pointer jumping to work our way towards building a linked list of selected elements.

For the compress phase, we apply the *deterministic coin tossing* technique of [7] but with a different set of labels. Instead of the memory address used in [7], we use the number of the processor storing list item i as its label $l(i)$. During the computation, we select sequentially the elements of R in the sublists of subsequent nodes in L which are stored at the same processor. The term "subsequent" refers to successor with respect to the *current* value of s.

Note that, there are at most p different labels, and subsequent nodes in those parts of L that are not processed sequentially have different labels. We call list element $s[i]$ a *local maximum* if $l(i) < l(s[i]) > l(s[s[i]])$. We apply deterministic coin tossing to those parts of L that are not processed sequentially.

The naive approach of applying this procedure $O(\log p)$ times would yield a $O(p^2)$-ruling set, but unfortunately it would require more than $O(\log p)$ communication rounds. Note that, when we want to apply it for a second, third, etc. time, the elements selected previously need to be linked by pointers. Since two subsequent elements selected by deterministic coin tossing can have distance $O(p)$, this may require $O(\log p)$ communication rounds, each. Hence, this straight forward approach requires a total of $O(\log^2 p)$ communication rounds.

Notice, however, that if two selected elements are at distance $\Theta(p)$ at a given moment, then it is unnecessary to further apply deterministic coin tossing in order to reduce the number of selected elements. The basic approach of our algorithm is therefore to interleave pointer jumping and deterministic coin tossing operations with respect to our new labeling scheme. More precisely, we will have only one pointer jumping step between subsequent deterministic coin tossing steps, and such pointer jumping operations will not be applied to those list elements that are pointing to selected elements.

This concludes the high level overview of our *deterministic list compression* techniques. The following describes the algorithm in detail.

Algorithm 1 CGM Algorithm for computing a p^2-ruling set.
Input: A linked list L and a vector s where $s[i]$ is the node following i in the list L. L and s are stored on a p processor CGM with total $O(n)$ memory.
Output: A set of selected nodes of L (which is a p^2-ruling set).

(1) Mark all list elements as *not selected*.

(2) FOR EVERY list element i IN PARALLEL:
    IF $l(i) < l(s[i]) > l(s[s[i]])$ THEN mark $s[i]$ as *selected*.

(3) Sequentially, at each processor, process the sublists of subsequent list elements which are stored at the same processor. For each such sublist, mark every second element as *selected*. If a sublist has only two elements, and not both neighbors have a smaller label, then mark both elements of the sublist as *not selected*.

(4) FOR $k = 1 \ldots \log p$ DO
    (4.1) FOR EVERY list element i IN PARALLEL:
        IF $s[i]$ is not selected THEN set $s[i] := s[s[i]]$.
    (4.2) FOR EVERY list element i IN PARALLEL:
        IF $(i, s[i]$ and $s[s[i]]$ are selected) AND NOT $(l(i) < l(s[i]) > l(s[s[i]]))$ AND $(l(i) \neq l(s[i]))$ AND $(l(s[i]) \neq l(s[s[i]]))$ THEN mark $s[i]$ as *not selected*.
    (4.3) Sequentially, at each processor, process the sublists of subsequent selected list elements which are stored at the same processor. For each such sublist, mark every second selected element as *not selected*. If a sublist has only two elements, and not both neighbors have a smaller label, then mark both elements of the sublist as *not selected*.

(5) Select the last element of L.

— End of Algorithm —

We first prove that the set of elements selected at the end of Algorithm 1 is of size at most $O(n/p)$.

Lemma 1. *After the kth iteration in Step 4, there are no more than two selected elements among any 2^k subsequent elements of the original list L.*

Proof. Due to space limitations, the proof is omitted. It can be found in the full version of this paper [5].

In order to show that subsequent elements selected at the end of Algorithm 1 have distance at most $O(p^2)$, we need the following lemmas.

Lemma 2. *After every execution of Step 4.3, the distance of two subsequent selected elements with respect to the current pointers (represented by vector s) is at most $O(p)$.*

Proof. Due to space limitations, the proof is omitted. It can be found in the full version of this paper [5].

Lemma 3. *After the k-th execution of Step 4.3, two subsequent elements with respect to the current pointers (represented by vector s) have distance $O(2^k)$ with respect to the original list L.*

Proof. Obvious consequence of the fact that only k pointer jumping operations were so far executed in Step 4.1.

Lemma 4. *No two subsequent selected elements have a distance of more than $O(p^2)$ with respect to the original list L.*

Proof. Follows from Lemma 2 and Lemma 3.

In summary, we obtain

Theorem 5. *The list ranking problem for a linked list with n vertices can be solved on a CGM with p processors and $O(\frac{n}{p})$ local memory per processor using $O(\log p)$ communication rounds and $O(\frac{n}{p})$ local computation per round.*

3 Euler Tour in a Tree

Let $T = (V, E)$ be an undirected tree and $T^* = (V, E^*)$ be a directed graph with $E^* = \{(v, w), (w, v) | \{v, w\} \in E\}$. Thus, T^* is Eulerian because $indegree(v) = outdegree(v)$ for each vertex v. The Euler Tour problem for T consists of computing for T^* a path that traverses each edge exactly once and returns to its starting point, as well as for each vertex its rank in this path.

Theorem 6. *The Euler Tour of a tree T with n vertices can be computed on a CGM with p processors and $O(\frac{n}{p})$ local memory per processor using $O(\log p)$ communication rounds and $O(\frac{n}{p})$ local computation per round.*

Proof. Due to space limitations, the algorithm and proof are omitted. They can be found in the full version of this paper [5].

4 Connected Components and Spanning Forest

Consider an undirected graph $G = (V, E)$ with n vertices and m edges. Each vertex $v \in V$ has a unique label between 1 and n. Two vertices u and v are connected if there is an undirected path of edges from u to v. A connected subset of vertices is a subset of vertices where each pair of vertices is connected. A *connected component* of G is defined as a maximal connected subset.

In this section, we study the problem of computing the connected components of G on a CGM with p processors and $O(\frac{n+m}{p})$ local memory per processor. We introduce a new technique, called *clipping*, which refers to the idea of taking a PRAM algorithm for the same problem but running it for only $O(\log p)$ rounds and then finishing the computation with some other $O(\log p)$ rounds CGM algorithm. (See also JaJa's *accelerated cascading* technique for the PRAM [19].)

Steps 1 and 2 of Algorithm 2 simulate Shiloch and Vishkin's PRAM algorithm [30], but for $\log p$ phases only. Each vertex v has a pointer to a vertex $parent(v)$ such that the $parent(v)$ pointers always form trees. The trees are also referred to as *supervertices*. A tree of height one is called a star. An edge (u, v) is *live* if $parent(u) \neq parent(v)$. Shiloch and Vishkin's PRAM algorithm merges supervertices along live edges until they equal the connected components. When simulated on a CGM or BSP computer, Shiloch and Vishkin's PRAM algorithm results in $\log n$ communication rounds or supersteps, respectively.

Our CGM algorithm requires $O(\log p)$ rounds only. It simulates only the first $\log p$ iterations of the main loop in the PRAM algorithm by Shiloch and Vishkin and then completes the computation in another $\log p$ communication rounds (Steps 3 - 7).

Algorithm 2 CGM Algorithm for Connected Component Computation
Input: An undirected graph $G = (V, E)$ with n vertices and m edges stored on a p processor CGM with total $O(n + m)$ memory. **Output:** The connected components of G represented by the the values $parent(v)$ for all vertices $v \in V$.

(1) FOR all $v \in V$ IN PARALLEL DO $parent(v) := v$.

(2) FOR $k := 1$ to $\log p$ DO
   (2.1) FOR all $v \in V$ IN PARALLEL DO $parent(v) := parent(parent(v))$.
   (2.2) FOR every live edge (u, v) IN PARALLEL DO (simulating concurrent write)
     (a) IF $parent(parent(v)) = parent(v)$ AND $parent(parent(u)) = parent(u)$ THEN { IF $parent(u) > parent(v)$ THEN $parent(parent(u)) := parent(v)$ ELSE $parent(parent(v)) := parent(u)$ }
     (b) IF $parent(u) = parent(parent(u))$ AND $parent(u)$ did not get new links in steps 2.1 and 2.2(a) THEN $parent(parent(u)) := parent(v)$
     (c) IF $parent(v) = parent(parent(v))$ AND $parent(v)$ did not get new links in steps 2.1 and 2.2.1 THEN $parent(parent(v)) := parent(u)$
   (2.3) FOR all $v \in V$ IN PARALLEL DO $parent(v) := parent(parent(v))$.

(3) Use the Euler Tour algorithm in Section 3 to convert all trees into stars. For each $v \in V$, set $parent(v)$ to be the root of the star containing v. Let $G' = (V', E')$ be the graph consisting of the supervertices and live edges obtained. Distribute G' such that each processor stores the entire set V' and a subset of $\frac{m}{p}$ edges of E'. Let E_i be the edges stored at processor i, $0 \le i \le p-1$.

(4) Mark all processors as *active*.

(5) FOR $k := 1$ to $\log p$ DO
   (5.1) Partition the active processors into groups of size two.
   (5.2) FOR each group P_i, P_j of active processors, $i < j$ IN PARALLEL DO
      (a) processor P_j sends it's edge set E_j to processor P_i.
      (b) processor P_j is marked as *passive*.
      (c) processor P_i computes the spanning forest (V', E_s) of the graph $SF = (V', E_i \cup E_j)$ and sets $E_i := E_s$.

(6) Mark all processors as *active* and broadcast E_0.

(7) Each processor i computes sequentially the connected components of the graph $G''' = (V', E_0)$. For each vertex v of V' let $parent'(v)$ be the smallest label $parent(w)$ of a vertex $w \in V'$ which is in the same connected component with respect to $G''' = (V', E_0)$. For each vertex $u \in V$ stored at processor P_i set $parent(u) := parent'(parent(u))$. (Note that $parent(u) \in V'$.)

— End of Algorithm —

Lemma 7. [30] *The number of different trees after iteration k of Step 2 is bounded by $(\frac{2}{3})^k n$.*

We obtain

Theorem 8. *Algorithm 2 computes the connected components and spanning forest of a graph $G = (V, E)$ with n vertices and m edges on a CGM with p processors and $O(\frac{n+m}{p})$ local memory per processor, $\frac{n+m}{p} \ge p^\epsilon$ ($\epsilon > 0$), using $O(\log p)$ communication rounds and $O(\frac{n+m}{p})$ local computation per round.*

Proof. Due to space limitations, the proof is omitted. It can be found in the full version of this paper [5].

5 Other Graph Problems

In the remainder, we summarize our solutions for Problems 4-8. Due to space limitations, the algorithms and proofs are omitted. They can be found in the full version of this paper [5].

Lowest Common Ancestor: The *lowest common ancestor*, $LCA(u,v)$, of two vertices u and v of a rooted tree $T = (V, E)$ is the vertex w that is an ancestor to both u and v, and is farthest from the root. The problem of preprocessing T in order to answer a query $LCA(u,v)$ quickly for any pair (u,v) is called the *lowest-common-ancestor (LCA)* problem.

Theorem 9. *Consider a rooted tree $T = (V, E)$ with n vertices. The LCA problem can be solved on a CGM with p processors and $O(\frac{n}{p})$ local memory per processor using $O(\log p)$ communication rounds and $O(\frac{n}{p})$ local computation per round.*

Tree Contraction and Expression Tree Evaluation: We observe that the classical tree contraction and expression tree evaluation algorithm of [24] can be easily implemented on a CGM to run in $O(\log p)$ communication rounds.

Observation 1 *Tree contraction and expression tree evaluation on a tree T with n nodes can be performed on a CGM with p processors and $O(\frac{n}{p})$ local memory per processor, $\frac{n}{p} \geq p^\epsilon$ ($\epsilon > 0$), using $O(\log p)$ communication rounds and $O(\frac{n}{p})$ local computation per round.*

Open Ear Decomposition and Biconnected Components: Consider an undirected graph $G = (V, E)$ with n vertices and m edges. For the remainder, we assume that G is connected. An *ear decomposition* of G is an ordered partition of E into r simple paths $P_1, \ldots, P_r$ such that P_1 is a cycle, and, for each $2 \leq i \leq r$, P_i is a simple path with endpoints belonging to $P_1 \cup \ldots \cup P_{i-1}$ but with none of its internal vertices belonging to P_j, $j < i$. The paths P_i are called *ears*. If none of the $P_i, i > 1$, is a cycle, then the decomposition is called an *open ear decomposition*. For an edge e in P_i, let i be the *ear number* of e. An edge $e \in E$ is a *cut-edge* if e does not lie on a cycle in G. A connected undirected G is *2-edge connected* if it contains no cut-edge. G has an ear decomposition if and only if G is 2-edge connected. A *cut-vertex* is a vertex whose removal leaves G disconnected. G is *biconnected* if it contains at least three vertices and has no cut-vertex.

Theorem 10. *For a graph $G = (V, E)$ with n vertices and m edges, the ear decomposition, open ear decomposition, as well as its 2-edge connected and biconnected components can be computed on a CGM with p processors and $O(\frac{n+m}{p})$ local memory per processor using $O(\log p)$ communication rounds and $O(\frac{n}{p})$ local computation per round.*

Chordal Graph Recognition: A graph $G = (V, E)$ is *chordal*, if every cycle of length greater than three has a *chord*, i.e., an edge connecting two non-consecutive nodes of the cycle. A *simplicial* node is a node whose neighbors form a clique. Dirac [15] showed that every chordal graph has a simplicial node. It is easy to see that removing an arbitrary node from a chordal graph yields another chordal graph. Therefore, after removing the simplicial node of a chordal graph, the new graph has another simplicial node. Successively removing all simplicial nodes gives an ordering of the nodes of G. This ordering is called *perfect elimination ordering (PEO)*.

Theorem 11. *Finding the PEO of a given graph $G = (V, E)$ with n vertices and m edges can be solved on a CGM with p processors and $O(\frac{n+m}{p})$ local memory per processor, $\frac{n+m}{p} \geq p^\epsilon$ ($\epsilon > 0$), using $O(\log n \log p)$ communication rounds and $O(\frac{n+m}{p})$ local computation per round.*

References

1. S.G. Akl, *Parallel Computation*, Prentice Hall, 1997.
2. R.J. Anderson, and L. Snyder, "A Comparison of Shared and Nonshared Memory Models of Computation," in Proc. of the IEEE, 79(4), pp. 480-487.
3. A. Bäumker and W. Dittrich, "Parallel Algorithms for Image Processing: Practical Algorithms with Experiments," *International Parallel Processing Symposium*, IEEE Computer Society Press, 1996, pp. 429 - 433.
4. G.E. Blelloch, C.E. Leiserson, B.M. Maggs, C.G. Plaxton, "A Comparison of Sorting Algorithms for the Connection Machine CM-2.," in Proc. ACM Symp. on Parallel Algorithms and Architectures, 1991, pp. 3-16.
5. E. Cáceres, F. Dehne, A. Ferreira, P. Flocchini, I. Rieping, A. Roncato, N. Santoro, and S.W. Song, "Efficient Parallel Graph Algorithms For Coarse Grained Multicomputers and BSP,", on-line Postscript at http://www.scs.carleton.ca/scs/faculty/dehne.html.
6. R. Cole, "Parallel merge sort," SIAM J. Comput., 17(4), 1988, pp. 770-785.
7. R. Cole and U. Vishkin, "Approximate parallel scheduling. Part I: The basic technique with applications to optimal parallel list ranking in logarithmic time", *SIAM Journal of Computing*, Vol. 17, No. 1, 1988.
8. F. Dehne, A. Fabri, and A. Rau-Chaplin, "Scalable Parallel Geometric Algorithms for Coarse Grained Multicomputers," in Proc. *ACM 9th Annual Computational Geometry*, pages 298–307, 1993.
9. F. Dehne, A. Fabri, and C. Kenyon, "Scalable and Architecture Independent Parallel Geometric Algorithms with High Probability Optimal Time," in Proc. *6th IEEE Symposium on Parallel and Distributed Processing*, pages 586–593, 1994.
10. F. Dehne, X. Deng, P. Dymond, A. Fabri, and A. A. Kokhar, "A randomized parallel 3D convex hull algorithm for coarse grained multicomputers," in Proc. *ACM Symposium on Parallel Algorithms and Architectures* (SPAA'95), pp. 27–33, 1995.
11. F.Dehne, S.W. Song, "Randomized parallel list ranking for distributed memory multiprocessors," in Proc. *Second Asian Computing Science Conference*, ASIAN'96, Singapore, Dec. 1996, Springer Lecture Notes in Computer Science 1179, pp. 1–10.
12. X. Deng, "A Convex Hull Algorithm for Coarse Grained Multiprocessors," in Proc. 5th International Symposium on Algorithms and Computation, 1994.
13. X. Deng and P. Dymond, "Efficient Routing and Message Bounds for Optimal Parallel Algorithms," in Proc. Int. Parallel Proc. Symp., 1995.

14. X. Deng and N. Gu, "Good Programming Style on Multiprocessors," in Proc. IEEE Symposium on Parallel and Distributed Processing, 1994, pp. 538-543.
15. G.A. Dirac. "On rigid circuit graphs". *Abh. Math. Sem. Univ. Hamburg* 25, 1961, pp. 71–76.
16. A. Ferreira, A. Rau-Chaplin, and S. Ubeda, "Scalable 2d convex hull and triangulation algorithms for coarse-grained multicomputers," in *Proceedings of the 7th IEEE Symposium on Parallel and Distributed Processing – SPDP'95*, pages 561–569, San Antonio (USA), October 1995. IEEE Press.
17. A.V. Gerbessiotis and L.G. Valiant, "Direct Bulk-Synchronous Parallel Algorithms," in Proc. 3rd Scandinavian Workshop on Algorithm Theory, Lecture Notes in Computer Science, Vol. 621, 1992, pp. 1-18.
18. M.T. Goodrich, "Communication efficient parallel sorting," ACM Symposium on Theory of Computing (STOC), 1996.
19. Ja'Ja', *An Introduction to Parallel Algorithms*, Addison Wesley, 1992.
20. P. Klein. "Efficient Parallel Algorithms for Chordal Graphs". *Proc. 29th Symp. Found. of Comp. Sci., FOCS* 1989, pp. 150–161.
21. P. Klein. "Parallel Algorithms for Chordal Graphs". In *Synthesis of parallel algorithms*, J. H. Reif (editor). Morgan Kaufmann Publishers, 1993, pp. 341–407.
22. Hui Li, and K. C. Sevcik, "Parallel Sorting by Overpartitioning," in Proc. ACM Symp. on Parallel Algorithms and Architectures, 1994, pp. 46-56.
23. Y. Maon, B. Schieber, U. Vishkin. "Parallel ear decomposition search (EDS) and st-numbering in graphs". *Theoretical Computer Science*, vol. 47, 1986, pp. 277 - 298.
24. G.L. Miller, J.H. Reif, "Parallel tree contraction and its application," IEEE Symp. on Foundations of Computer Science, 1985, pp. 478–489.
25. G. L. Miller, V. Ramachandran. "Efficient parallel ear decomposition with applications", manuscript, MSRI, Berkeley, January 1986.
26. V. Ramachandran. "Parallel open ear decomposition with applications to graph biconnectivity and triconnectivity", in [28], pp. 276 - 340.
27. M. Reid-Miller, "List ranking and list scan on the Cray C-90," in Proc. ACM Symp. on Parallel Algorithms and Architectures, 1994, pp. 104-113.
28. J. H. Reif (editor), *Synthesis of parallel algorithms*, Morgan Kaufmann Publishers, 1993.
29. D.J. Rose, R.E. Tarjan, and G.S. Lueker. "Algorithmic Aspects of Vertex Elimination on Graphs". *SIAM J. Comp.* 5, 1976, pp. 266–283.
30. Y. Shiloch, U. Vishkin, "An $O(\log n)$ parallel connectivity algorithm," *Journal of Algorithms*, 3(1), pp. 57–67, 1983.
31. L. Snyder, "Type architectures, shared memory and the corollary of modest potential," *Annu. Rev. Comput. Sci. 1*, 1986, pp. 289-317.
32. R.E. Tarjan, U. Vishkin, "An efficient parallel biconnectivity algorithm," SIAM J. Comput., 14(4), 1985, pp. 862–874.
33. L. Valiant, "A bridging model for parallel computation," Communications of the ACM, Vol. 33, No. 8, August 1990.
34. L.G. Valiant et al., "General Purpose Parallel Architectures," *Handbook of Theoretical Computer Science*, Edited by J. van Leeuwen, MIT Press/Elsevier, 1990, pp.943-972.
35. H. Whitney. "Non-separable and planar graphs". *Trans. Amer. Math. Soc.* 34, 1932, pp. 339 - 362.

Upper Bound on the Communication Complexity of Private Information Retrieval*

Andris Ambainis

Institute of Mathematics and Computer Science, University of Latvia, Raina bulv. 29, Riga, Latvia, e-mail: ambainis@cclu.lv. From August 1997 at Computer Science Division, University of California at Berkeley.

Abstract. We construct a scheme for private information retrieval with k databases and communication complexity $O(n^{1/(2k-1)})$.

1 Introduction

Much attention has been given to the problem of protecting a database from a user that tries to retrieve an information that he is not allowed to access[2, 8, 12].

In some scenarios, an opposite problem can appear: a user wishes to retrieve some infomation from a database without revealing to the database what information he needs. For example[7], an investor wishes to receive an information about a certain stock but he does not wish others (even the database) to know in which particular stock he is interested.

However, there is only one way to reach a complete privacy: the user should ask for the copy of the entire database. Otherwise, the database will get some information what the user wishes to know. This is not a good solution because it requires much time and much communication from the database to the user.

If there are several identical copies of the database, an another scenario is possible[7]:

The user asks a query to each database and combines the results of the queries, obtaining the desired information. Each query alone gives no information what the user is interested in.

Chor, Coldreich, Kushilevitz, Sudan[7] introduced this model and constructed several schemes for a private retrieval of one bit from a database:

1. A scheme for 2 databases with $O(n^{1/3})$ communication. (n is the size of the database)
2. A scheme for k databases with $O(n^{1/k})$ communication.
3. A scheme for $O(\log n)$ databases with $O(\log^2 n \log \log n)$ communication.

In this paper, we improve their result, constructing a protocol for k databases with $O(n^{1/(2k-1)})$ communication.

---
* The author was supported by Latvia Science Council Grant 96.0282 and scholarship "SWH Izglītībai, Zinātnei un Kultūrai" from Latvia Education Foundation

Related work. Protocols for private information retrieval in [7] and this paper have used ideas from several related problems (instance hiding and multiparty communication complexity).

Instance hiding[1, 5, 6] is the problem of obtaining the ith bit from the oracle so that i remains secret. There are some similarities and some substantial differences between instance hiding and private information retrieval (see [7] for more detailed discussion).

Techniques from instance hiding were relevant to protocols for private information retrieval in [7]. However, they are not used in this paper.

Multiparty communication complexity is also related to private information retrieval. Pudlak, Rödl, Sgall[11] and Ambainis[3] have considered the problem of computing $x_{(i+j) \bmod n}$ where x is a string of n bits and i, j are integers in the following model:

Player 1 knows x, i, Player 2 knows x, j. Each of them sends one message to Player 3. Player 3 computes the result, using only the messages received from Players 1 and 2.

Any protocol for the above problem can be easily transformed into protocol for private information retrieval. Thus, we can obtain nontrivial protocols for private information retrieval with $o(n)$ communication.

Another communication complexity problem was studied by Babai, Kimmel and Lokam[4]. It also can be applied to private information retrieval.

However, all these protocols are less efficient than the protocols for private information retrieval designed in [7]. Still, the ideas from [3, 4, 11] (not explicit protocols) can be useful in the study of private information retrieval. In particular, this paper is based on the idea of combining two protocols which appeared in the setting of multiparty communication complexity[3, 11].

2 Model

Formally, we view the database as a string x consisting of n bits. k denotes the number of identical databases. We assume that the user wishes to retrieve a single bit x_i from the database.

We require that, for every database, indices i, j and any message from the user, the probability of the database receiving this message is equal when the user retrieves the ith bit and when the user retrieves the jth bit. This means that database does not get any information about i.

There are several extensions of this model. [7] considered schemes which allow to retrieve blocks of information and give a higher degree of privacy (knowing $k - 1$ of k queries gives no information about the bit that the user retrieves). Ostrovsky and Shoup[9] have extended the results of [7] and designed schemes for private information storage. Using their schemes, the user can both read and write to the database without revealing which bit is accessed. They have shown that any protocol for private information retrieval can be transformed to the protocol for private information storage with a slight increase in the number of databases and communication.

However, in this paper, we consider only the basic one-bit model of [7]. $S \oplus i$ denotes $S \cup \{i\}$, if $i \notin S$ and $S - \{i\}$ if $i \in S$.

3 Result

Consider some protocol for private information retrieval. Does the user use all bits in the messages from the databases? In some protocols, only a few bits are really neccessary. If the user knows in advance which bits are necessary, two protocols can be combined, obtaining the third with more databases and less communication.

Below, we show how to combine a protocol for 2 databases and a protocol for $k - 1$ databases, obtaining a protocol for k databases with less communication.

1. The user in the k database protocol simulates the user in the protocol for 2 databases. Let x_1 denote the message sent to the 1st database and x_2 the message sent to the 2nd database in the 2 database protocol.
   The user sends x_1 to the 1st database and x_2 to the 2nd, ..., the kth database.
2. Then, the user computes the length of the reply from the 2nd database in the 2 database protocol and the positions of necessary bits in this reply. Further, m denotes the length and $n_1, \ldots, n_i$ denote the positions of the necessary bits.
   The user simulates the user in the protocols for $k - 1$ databases where n_1^{th}, ..., n_i^{th} bits from an m bit database are retrieved, sending to the $(i + 1)^{st}$ database the messages which are sent to the ith database in the $(k - 1)$ database protocol.
3. The 1st database simulates the 1st database in 2 database protocol and sends the user the same message.
4. The 2nd, ..., the kth database simulate the 2nd database in the 2 database protocol. Instead of sending the message to the user, they consider it as a new m-bit database.
   Further, they simulate databases in the $(k - 1)$ database protocol for the retrieval of the n_1^{th}, ..., the n_i^{th} bit and send the messages from these protocols to the user.
5. The user simulates the user in the $(k - 1)$ database protocol for the retrieval of the n_1^{th}, ..., the n_i^{th} bits. Then, knowing the message from the 1st database and all the necessary bits from the second message, the user simulates the user in the 2 database protocol. The result of this simulation is the bit that the user wishes to retrieve.

If we wish to apply this idea, 2 database protocol should satisfy certain constraints:

1. The most of communication goes from the databases to the user. (The amount of communication from the user to databases increases when two protocols are combined. Hence, if it is already large, the combination is useless.)

2. Only few bits from the messages received by the user are necessary.
3. The user knows in advance which bits are necessary, i.e. the positions of these bits do not depend on the databases' contents.

Below, we use the idea of combining two protocols to prove

Theorem 1. *Let $k \geq 2$. There exists a protocol for private information retrieval with k databases and $O(n^{1/(2k-1)})$ bits of communication.*

Proof. By induction.

The protocol for 2 databases was constructed by Chor, Goldreich, Kushilevitz and Sudan[7]. The protocol for k databases is obtained as the combination of the protocols for 2 databases and $(k-1)$ databases.

First, we describe the 2 database protocol that we use to obtain a k database protocol from a $(k-1)$ database protocol.

1. Let $l = \lceil \sqrt[2k-1]{n} \rceil$. The database can be considered as a $2k-1$ dimensional cube $\{0, \ldots, l-1\}^{2k-1}$. Each position $i \in \{0, \ldots, n-1\}$ in the database coresponds to some position $(i_1, \ldots, i_{2k-1})$ in the cube.

    The user chooses independently $(2k-1)$ random subsets of $\{0, \ldots, l-1\}$: S_1^1, $\ldots$, S_{2k-1}^1. Let $S_1^2 = S_1^1 \oplus i_1$, $\ldots$, $S_{2k-1}^2 = S_{2k-1}^1 \oplus i_{2k-1}$ where $(i_1, \ldots, i_{2k-1})$ is the position of the required bit in the $(2k-1)$ dimensional cube.

    He sends $S_1^1, \ldots, S_{2k-1}^1$ to the 1st database and $S_1^2, \ldots, S_{2k-1}^2$ to the 2nd database.

2. The 1st database computes the exclusive-or of the bits in positions $(j_1, \ldots, j_{2k-1})$ such that $j_1 \in S_1^1, \ldots, j_{2k-1} \in S_{2k-1}^1$ and sends it to the user.

    The database also computes the exclusive-or of the bits in positions $(j_1, \ldots, j_{2k-1})$ such that $j_1 \in S_1', \ldots, j_{2k-1} \in S_{2k-1}'$ for each possible $S_1', \ldots, S_{2k-1}'$ such that
    (a) $S_j' = S_j^1 \oplus t$ for some $j \in \{1, \ldots, 2k-1\}$ and $t \in \{0, \ldots, l-1\}$;
    (b) $S_i' = S_i^1$ for all $i \neq j$.
    The exclusive-xor for each possible $S_1', \ldots, S_{2k-1}'$ is sent to the user, too.

3. The 2nd database computes the exclusive-or of the bits in positions $(j_1, \ldots, j_{2k-1})$ such that $j_1 \in S_1^2, \ldots, j_{2k-1} \in S_{2k-1}^2$ and sends it to the user.

    Further, the 2nd database computes the exclusive-or of the bits in positions $(j_1, \ldots, j_{2k-1})$ such that $j_1 \in S_1', \ldots, j_{2k-1} \in S_{2k-1}'$ for each possible $S_1', \ldots, S_{2k-1}'$ such that
    (a) For each $i \in \{1, \ldots, 2k-1\}$ S_i' is equal to S_i^2 or $S_i^2 \oplus t_i$ for some $t_i \in \{0, \ldots, l-1\}$;
    (b) There exist at least two $i \in \{1, \ldots, 2k-1\}$ such that $S_i' = S_i^2$.
    The exclusive-xor for each possible $S_1', \ldots, S_{2k-1}'$ is sent to the user, too.

4. For each possible $S_1', \ldots, S_{2k-1}'$ such that S_i' is either S_i^1 or S_i^2, the user finds the exclusive-or of bits in positions $(j_1, \ldots, j_{2k-1})$ satisfying $j_1 \in S_1'$, $\ldots, j_{2k-1} \in S_{2k-1}'$:
    (a) If $S_i' = S_i^2$ for at most one i, then the exclusive-or is one of the bits sent by the 1st database.

(b) If $S'_i = S^2_i$ for at least two i, then the exclusive-or is one of the bits sent by the 2nd database.

The user computes the exclusive-or of all these values. It is the necessary bit from the database.

($S^2_j = S^1_j \oplus i_j$. Hence, i_j belongs to exactly one of S^1_j and S^2_j and $i_1 \in S'_1$, ..., $i_{2k-1} \in S'_{2k-1}$ for exactly one choice of $S'_1, \ldots, S'_{2k-1}$.

For each other position $(i'_1, \ldots, i'_{2k-1})$ we have $i'_1 \in S'_1, \ldots, i'_{2k-1} \in S'_{2k-1}$ for an even number (possibly zero) of combinations $S'_1, \ldots, S'_{2k-1}$.

Hence, the exclusive-or computed by the user contains the bit in the position $(i_1, \ldots, i_{2k-1})$ exactly once and any other bit an even number of times. It follows that this exclusive-or is equal to the bit in the position $(i_1, \ldots, i_{2k-1})$, i.e. the bit that the user wishes to retrieve.)

The amount of transmitted bits.

1. Communication from the user to the databases.
   To transmit a set S^i_j, the user needs $l = \sqrt[2k-1]{n}$ bits. (For each $x \in \{0, \ldots, l-1\}$, the user must say whether $x \in S^i_j$.) The user transmits $2k - 1$ sets ($S^1_1, \ldots, S^1_{2k-1}$) to the 1st database and $2k - 1$ sets to the 2nd database. So, the total amount of communication in this direction is $2(2k-1)\sqrt[2k-1]{n} = O(\sqrt[2k-1]{n})$.

2. Communication from the 1st database to the user.
   The 1st database computes the exclusive-or of the bits for several combinations of $S'_1, \ldots, S'_{2k-1}$ and sends it to the user. The amount of bits transmitted by the 1st database is equal to the number of the combinations of $S'_1, \ldots, S'_{2k-1}$, i.e. $(2k-1)l + 1$.
   k is a constant and $l = \lceil \sqrt[2k-1]{n} \rceil$. Hence, the amount of communication in this direction is $O(\sqrt[2k-1]{n})$, too.

3. Communication from the 2nd database to the user.
   Similarly to the previous case, the amount of bits transmitted by the 2nd database is equal to the number of combinations $S'_1, \ldots, S'_{2k-1}$.
   For the 2nd database, the amount of such combinations is at most $(2^{2k-1} - 2k)l^{2k-3} = O(n^{(2k-3)/(2k-1)})$ because:

   (a) Those i for which $S'_i \neq S^2_i$ form a subset of $\{1, \ldots, 2k-1\}$ with at most $2k - 3$ elements. (For at least two $i \in \{1, \ldots, 2k-1\}$, $S^2_i = S'_i$.)
   The amount of such subsets is $2^{2k-1} - 2k$.

   (b) If we have chosen i for which $S^2_i \neq S'_i$, it remains to choose t_i. There are l possible values of t_i for each i.
   t_i is chosen for at most $2k - 3$ values of i. Hence, there are at most l^{2k-3} possible combinations of t_i.

So, the user transmits $O(\sqrt[2k-1]{n})$ bits, the 1st database $O(\sqrt[2k-1]{n})$ bits and the 2nd database $O(n^{(2k-3)/(2k-1)})$ bits.

From the 2nd database's answer the user needs a constant amount $(2^{2k-1} - 2k)$ of bits. The positions of these bits in the message from the 2nd database do not depend on the contents of the database.

Hence, we can combine the described protocol with a $(k-1)$ database protocol, using the method described at the beginning of this section.

Communication in the k database protocol.

1. Communication from the user to the databases.
   The user sends to the databases:
   (a) The information from the 2 database protocol: $O(\sqrt[2k-1]{n})$ bits to each database.
   (b) The information for the simulations of the $(k-1)$ database protocol: $O(\sqrt[2k-3]{m})$ bits where m is the length of the message from the 2nd database in the 2 database protocol. We have
   $$m = O(n^{(2k-3)/(2k-1)}).$$
   Hence, $O(\sqrt[2k-1]{n})$ bits are transmitted for this purpose.
2. Communication from the 1st database to the user. It is the same as in the 2 database protocol, i.e. $O(\sqrt[2k-1]{n})$ bits.
3. Communication from the 2nd, ..., the kth database to the user.
   In each simulation of $(k-1)$ database protocol, these databases communicate
   $$O(\sqrt[2k-3]{m}) = O(\sqrt[2k-3]{n^{(2k-3)/(2k-1)}}) = O(\sqrt[2k-1]{n})$$
   bits. The amount of simulations performed by the databases is equal to the amount of bits needed by the user from the 2nd database's message, i.e. constant. Hence, the communication by these databases is $O(\sqrt[2k-1]{n})$, too.

We have constructed a protocol with k databases and $O(\sqrt[2k-1]{n})$ communication from a protocol with $(k-1)$ databases and $O(\sqrt[2k-3]{n})$ communication.

Using the construction of Ostrovsky and Shoup[9] and the protocol described above, we can obtain a scheme in which both reading and writing are private. This scheme has $(k+1)$ databases and $O(n^{1/(2k-1)} \log n)$ communication complexity for any $k \geq 2$.

References

1. M. Abadi, J. Feigenbaum and J. Kilian, *On hiding information from an oracle*, Journal of Computer and System Sciences, 39(1989), pp. 21-50
2. N. Adam and J. Wortmann, *Security control methods for statistical databases: a comparative study*, ACM Somputing Surveys, 21(1989), pp. 515-555
3. A. Ambainis, *Upper bounds on multiparty communication complexity of shifts*, Proceedings of STACS'96, Lecture Notes in Computer Science, vol. 1047(1996), pp. 631-642
4. L. Babai, P. Kimmel, S.V. Lokam, *Simultaneous messages versus communication,,* Proceedings of STACS'95, Lecture Notes in Computer Science, vol. 900(1995), pp. 361-372
5. R. Beaver, J. Feigenbaum, *Hiding instances in multioracle queries*, Proceedings of STACS'90, Lecture Notes in Computer Science, vol. 415(1990), pp. 37-48

6. R. Beaver, J. Feigenbaum, J. Kilian, P. Rogaway, *Security with low communication overhead*, Crypto'90
7. B. Chor, O. Goldreich, E. Kushilevitz, M. Sudan, *Private information retrieval*, Proceedings of FOCS'95, pp. 41-50. To appear in the Journal of the ACM.
8. D. Denning, *Cryptography and Data Security*, Addison-Wesley, 1982
9. R. Ostrovsky, V. Shoup, *Private information storage*, to appear in STOC'97.
10. P. Pudlak, V. Rödl, *Modified ranks of tensors and the size of circuits*, Proceedings of 25-th ACM STOC, 1993, pp. 523-531. Preliminary version of [11].
11. P. Pudlak, V. Rödl, J. Sgall, *Boolean circuits, ranks of tensors and communication complexity*, to appear in SIAM J. Computing.
12. J. D. Ullman, *Principles of Database Systems*. 1982.

Computation Paths Logic:
An Expressive, yet Elementary, Process Logic
(abridged version)

David Harel and Eli Singerman
The Weizmann Institute of Science, Rehovot, ISRAEL

Abstract: A new process logic, is defined, called computation paths logic (CPL), which treats formulas and programs essentially alike. CPL is a pathwise extension of PDL in the spirit of the logic R of Harel and Peleg. It enjoys most of the advantages of previous process logics, yet is decidable in elementary time. We also offer extensions for modeling asynchronous/synchronous concurrency and infinite computations.

1 Introduction

Two major approaches to modal logics of programs are dynamic logic [Pr] and temporal logic [Pn]. Propositional dynamic logic, PDL [FL] is a natural 'dynamic' extension of the propositional calculus, in which programs are intermixed with propositions in a modal-like fashion. Formulas of PDL can express many input/output properties of programs in a natural way. Moreover, validity/satisfiability in PDL is decidable in exponential time, and the logic has a simple complete axiomatization [KP]. PDL is thus a suitable system for reasoning about the input/output behavior of sequential programs on the propositional level. However, PDL is unsuited for dealing with the continuous, or progressive behavior of programs, i.e., the situations occuring *during* computations. The need for reasoning about continuous behavior arises naturally in the study of reactive and concurrent programs.

The main approach proposed in response to this need is temporal logic, TL [Pn], in which assertions can be made naturally about the progressive behavior of programs. In particular, TL can easily express freedom from deadlock, liveness, and mutual exclusion. The basic versions of TL, however, are not compositional, in the sense that their treatment of a well-structured program does not derive directly from their treatment of its components. Indeed, TL usually does not name programs at all, but refers to instructions and labels in a fixed program. Although TL can discuss the synthesis of complex programs from simpler ones to some extent using *at* predicates, this method is rather cumbersome.

This dichotomy between the dynamic and temporal logic approaches has prompted researchers to try to combine the best of the two in what is generally called *process logic*. Accordingly, a system called PL was proposed in [HKP]. It borrows the program constructs and modal operators [ ] and ⟨ ⟩ from DL, and the temporal connectives **suf** (similar to **until**) and **f** (standing for **first**) from TL, and combines them into a single system. The expressive power of PL is greater than that of PDL and of TL, and its validity/satisfiability problem was shown in [HKP] to be decidable, though it is not known to be elementary.[1]

There are some inconvenient features of PL, including the asymmetry of its central path operator, **suf**, and the fact that its formula connectives are somewhat weaker than its program operators. A proposal that overcomes these problems is the regular process logic, RPL, of [HP]. In RPL, the operators **suf** and **f** are replaced by **chop** and **slice**, corresponding essentially to Kleene's regular operations of concatenation and star. In this way, the regular operations on programs, $\alpha \cup \beta$, $\alpha\beta$, α^*, have natural counterparts on formulas: $X \vee Y$, X **chop** Y and **slice** X. It is shown in [HP] that RPL is even more expressive than PL, and that its validity problem is also decidable but nonelementary.

Using the fact that in RPL both program and path operators are those of regular expressions, and that programs and formulas are interpreted over paths, a *uniform* process logic R was defined in [HP]. In R, formulas are constructed inductively from atomic propositions and binary atomic programs, using a single set of regular operators. It was shown in [HP] that R is more expressive than RPL with binary atomic programs, and is decidable (though, again, nonelementary).

In the interest of obtaining a useful process logic decidable in elementary time, an automata-oriented logic, YAPL, was defined in [VW]. In YAPL, formulas are constructed using finite automata for both temporal (path) connectives and for constructing compound programs from basic (atomic) ones. There is a clear distinction between state and path formulas in YAPL, atomic programs are binary and atomic formulas are restricted to being state formulas. YAPL is indeed shown in [VW] to be decidable in elementary time (even over infinite paths). YAPL formulas, however, can be somewhat less intuitive and not that easy to comprehend.

In the present paper, we try to combine some of the advantages of previous methods by introducing a new process logic that is compositional, uniform in its treatment of programs and formulas, expressive enough to capture the interesting path properties mentioned in the literature in a natural way, explicit in its treatment of concurrency, and elementary decidable.

We term our basic formalism *computation paths logic* (CPL). A single set of regular operators acts on both transition formulas (programs) and state formulas. For example, $a^* \cdot P \cdot b$ is a CPL formula. (Here a and b are atomic programs and P is an atomic state formula.) Intuitively this formula means: "perform action a some nondeterministic number of times, check for property P and then do action b". An important operator in CPL is '$\cap$' — pathwise intersection. Thus, $f \cap g$ is true on paths that satisfy both f and g. Using this operator, it is possible to express a large variety of properties of computation paths. For example, $\alpha \cap (skip^* \cdot P \cdot skip^*)$, where α is a program and P is a proposition, is true on α-paths that contain some P-state. Note that $a \cap b$, for atomic programs a and b is true only for paths which are both a-paths and b-paths, and is not expressible by PDL programs or formulas.

---

[1] Some versions of PL have been shown to be nonelementary [Ha], but it is still not known whether PL itself is elementary.

Unlike PL and its descendants, RPL and R, we have decided not to include the modal operators [ ] and ⟨ ⟩ in CPL. The reason is as follows. Consider a PL/RPL/R formula of the form $[\alpha]\varphi$, where α is a program and φ is a path property. While one might expect this formula to be true on all α-paths that satisfy φ, in PL it is defined to be true on all paths p which, when extended by an α-path r, result in a path $p \cdot r$ satisfying φ. This, however, corresponds to the above intuition only when p is a path of length 0, i.e., a state. This broader (and somewhat complicated) definition in PL is an unavoidable outcome of the wish of the authors of [HKP] to have only path formulas, but at the same time use ⟨ ⟩ and [ ] as in PDL. (For example, they wanted $\langle\alpha\beta\rangle\varphi$ to be equivalent to $\langle\alpha\rangle\langle\beta\rangle\varphi$.)

To make our logic elementary decidable, we use a special form of negation. Specifically, negation in CPL is not taken relative to the set of all paths (as is done, e.g., in PL/RPL/R). In fact, a negated formula is a *state* property, made true in any state that is not the initial state of a path that satisfies the argument formula. For example, $\neg(a \cdot P)$ asserts "it is not possible to carry out a computation of $a \cdot P$ from the present state". While this form of negation is weaker than negation relative to all paths, most interesting path properties are still expressible.

In Section 3, we show that CPL is elementary decidable, by reducing its satisfiability problem to that of APDL, the version of PDL in which programs are represented by finite automata rather than regular expressions [HS2]. The reduction is rather involved, and combines ideas from both [Pe] and [SPH].

In Section 4 we propose an extension of CPL for modeling concurrent processes, called ICPL. It uses '||' to denote interleaving. This might be termed *asynchronous* concurrency. Even though the interleaving operator itself is very intuitive, combining it with other operators (especially '∩') turns out to be rather technically involved. Nevertheless, ICPL is also decidable in elementary time.

To model *synchronous* concurrency, we introduce a further extension in Section 5, called SICPL (ICPL with synchronization). In SICPL, which is shown to be elementary, interleaving can be synchronized with respect to subsets of atomic programs. For such a subset syn, and formulas f and g, the interleaving of f and g synchronized on syn is expressed by $f \mid syn \mid g$ (the notation is apt, since '||' denotes the special case where $syn = \emptyset$). For example, the formula $(a \cup b) \cdot c \mid a, b \mid (a \cup c) \cdot P \cdot (b \cup c)$ is true only in paths of the form:

$$\xrightarrow{a}\xrightarrow{c}\xrightarrow{P}\xrightarrow{c} \qquad \xrightarrow{a}\xrightarrow{P}\xrightarrow{c}\xrightarrow{c} \qquad \xrightarrow{c}\xrightarrow{P}\xrightarrow{b}\xrightarrow{c}$$

A further elementary extension of CPL for expressing properties of infinite computations, ωCPL, is defined in Section 6.

2 Definitions and Basic Observations

Definition 1. A *path* over a set S is simply a non-empty finite sequence of elements of S. The notions *first*, *last* and *fusion*, denoted $p \cdot q$, are defined in the usual way.

We now define *computation paths logic*, CPL for short. It has two sorts, a set ASF of atomic state formulas (propositions), and a set ATF of atomic transition formulas (programs). The set of formulas is defined as the least set containing ASF and ATF, and such that if f and g are formulas, then so are $(\neg f)$, (f^*), $(f \cdot g)$, $(f \cup g)$ and $(f \cap g)$. (We often omit the parentheses where there is no confusion.)

CPL formulas are interpreted over models $M = (S_M, \rho_M)$, where S_M is the set of *states*, $\rho_M(P) \subseteq (S_M)$ for each $P \in$ ASF, and $\rho_M(a) \subseteq S_M \times S_M$ for each $a \in$ ATF. In addition, ρ_M is is extended to all formulas as follows:

$$\rho_M(f \cdot g) = \rho_M(f) \cdot \rho_M(g) \qquad \rho_M(f \cup g) = \rho_M(f) \cup \rho_M(g)$$
$$\rho_M(f \cap g) = \rho_M(f) \cap \rho_M(g) \qquad \rho_M(f^*) = \rho_M(f)^*$$
$$\rho_M(\neg f) = (S_M) \setminus first(\rho_M(f))$$

(We often leave out the M subscript of S and ρ.) A path p in a model M *satisfies* a CPL formula f, written $M, p \models f$, when $p \in \rho_M(f)$. A formula f is *satisfiable* iff $M, p \models f$ for some path p in some model M. A state s in a model M *satisfies* a CPL formula f, written $M, s \models f$ iff there exist a path satisfying f whose first state is s.

Example: Consider the CPL formula $\varphi : (P \cdot a)^* \cdot Q \cap (b \cup c)^* \cdot \neg(b \cdot P) \cdot a$, where $P, Q \in$ ASF and $a, b \in$ ATF. In the model illustrated in the figure below, paths that satisfy φ are (among others): $(1, 2, 3, 4, 5)$, $(1, 2, 3, 1, 2, 3)$ and $(1, 2, 3, 1, 2, 3)$. On the other hand, a path that does *not* satisfy φ is $(1, 2, 7, 8, 9)$ (this is because $(8) \not\models \neg(b \cdot P)$).

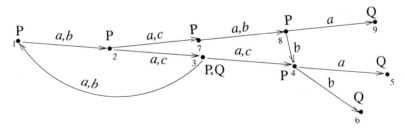

For CPL formulas f and g, it is sometimes convenient to use the following abbreviations: $f?$ instead of $\neg\neg f$, $f \vee g$ instead of $f? \cup g?$ and $f \wedge g$ instead of $f? \cap g?$. Regarding transitions, it useful to use the following abbreviations: $skip$ instead of $\cup_{a \in ATF} a$, $path$ instead of $skip^*$ and $true$ instead of $path?$. Note that $path$ holds in every path in which consecutive states are connected by some atomic transition. Moreover, it follows from the semantics of CPL that for every $f \in$ CPL and every path p in any model M, if $p \in \rho_M(f)$ then $p \models path$. So that $path$ plays the role of 'true' for paths that correspond to formulas. The formula $true$ is a 'state version' of $path$ and is true in every path of length 0, i.e., in every state in every model.

Let us demonstrate how to express some useful path properties in CPL.

- The existence of some segment of the path satisfying f is expressed by $someseg(f) = path \cdot f \cdot path$.

- The existence of some prefix of the path satisfying f is expressed by $somepre\ (f) = f \cdot path$.
- The existence of some suffix of the path satisfying f is expressed by $somesuf\ (f) = path \cdot f$.
- The existence of some state in the path satisfying f is expressed by $somestate\ (f) = someseg\ (f?)$.
- An operator similar to $\bigcirc$ (*nexttime*) of TL is $next\ (f) = skip \cdot f$.
- An operator similar to $\mathcal{U}$ of TL is $f\ until\ g = (f \cdot skip)^* \cdot g$.

CPL can clearly be viewed as a pathwise extension of PDL. It is not too difficult to show that PDL$<_{\neq}$CPL in expressive power, where we only consider state formulas of CPL. Considering other process logics, CPL can be thought of as a restricted version of the logic R of [HP], so that: PDL$<_{\neq}$CPL$\leq$R. From this, and the fact that R is decidable [HP], we can conclude that CPL is decidable. Since R is nonelementary [Pe], this yields a nonelementary decision procedure for CPL. We will show in the next section, however, that CPL is in fact elementary.

3   CPL is Elementary Decidable

In this section we show that satisfiability of CPL formulas is decidable in elementary time. This will be done in two steps. In the first, we carry out a reduction from the satisfiability problem of CPL to the satisfiability problem of CPL over one-action-per-transition models. These *one-action-per-transition* (*oapt*, for short) models are defined below. (These models were used in [Pe] for the logic R.) In the second step we carry out a reduction from the satisfiability problem of CPL over oapt-models to the satisfiability problem of APDL.

Definition 2. A model M is called an *oapt-model relative to the set* ATF $a_1, ..., a_n$ if for every $1 \leq i \neq j \leq n$, $\rho(a_i) \cap \rho(a_j) = \emptyset$. A CPL formula f is oapt-satisfiable iff there exist some oapt-model which satisfies f.

Lemma 3. *For every* CPL *formula f over* $\{a_1 \ldots a_n\}$, *there exists a* CPL *formula* f' *(over a new* ATF*) such that f is satisfiable iff* f' *is oapt-satisfiable.*

Proof: Let f be a formula f over $\{a_1, ..., a_n\}$. We define a set ATF′ of $2^n - 1$ new symbols (to be used as the atomic transition formulas of f'), each of the form $a_{c_1 \ldots c_n}$ where $c_k \in \{k, \bar{k}\}$,

$$\text{ATF}' = \{a_{c_1 \ldots c_n} \mid \forall\ 1 \leq k \leq n\ ,\ c_k \in \{k, \bar{k}\}\} \setminus \{a_{\bar{1}\bar{2}\ldots\bar{n}}\}.$$

Let f' be the formula obtained from f by replacing every appearance of a_k, for $1 \leq k \leq n$, with $\beta_k = \bigcup_{\{c_1 \ldots c_n \mid c_k = k\}} a_{c_1 \ldots c_n}$ The following claim (the proof of which we omit here) completes the proof of the lemma.
Claim: f *is satisfiable* $\iff$ f' *is oapt-satisfiable.* ∎

As preparation for the reduction to APDL, let us start the discussion in the framework of PDL. Recall that a PDL model is also a CPL model; note, however, that while CPL formulas are interpreted over *paths*, PDL formulas are interpreted over *states*. To overcome this dichotomy we shall relate paths to PDL programs in the following way:

Definition 4. For a PDL program α and a path $p = (p_1,\ldots,p_k)$ in a model (S, τ, R), $p \in \alpha$ is defined by induction on the structure of α: If $\alpha \in \text{ATF}$ then $p \in \alpha$ iff $k = 1$ and $(p_0, p_1) \in R(\alpha)$; $p \in \alpha \cup \beta$ iff $p \in \alpha$ or $p \in \beta$; $p \in \alpha; \beta$ iff there are paths $q \in \alpha$ and $r \in \beta$ with $p = q \cdot r$; $p \in \alpha^*$ iff $p \in \alpha^i$ for some $i \geq 1$ or $p = (p_0)$; $p \in \varphi$? iff $p = (p_0)$ and $(p_0, p_0) \in R(\varphi?)$.

Via this association we can view PDL programs as being carried out along paths rather than as binary relations. For the reduction, however, it is more convenient to use the automata version of PDL, namely APDL [HS2]. The reason for this is that '∩' can be handled more economically by automata than by regular expressions. (This also applies to other operators used in the extensions of CPL we define later on.) Even though APDL formulas are, in general, more succinct than their equivalent PDL formulas, satisfiability for APDL can be decided in EXPTIME [HS2]. This is also the case for deciding satisfiability over oapt-models. For if M, $s \models \varphi$, then M can be transformed into an oapt-model of φ (by duplicating states).

We shall use this to get an elementary decision procedure for CPL by carrying out a reduction from CPL into APDL. Relating paths in a model to APDL programs is done as in Def.4, i.e., if α is an automaton (APDL program) then $p \in \alpha$ iff $p \in r(\alpha)$, where $r(\alpha)$ is a regular expression denoting the language of α.

Lemma 5. *For every CPL formula f there exists an APDL program (NFA) A_f, such that for every path p in every oapt-model, $p \in \rho(f)$ iff $p \in A_f$.*

Proof: The APDL automaton (program) A_f corresponding to the CPL formula f is built by induction on the structure of f. Here we briefly describe the following two (non-routine) cases. For $\neg g$ we let $A_{\neg g}$ be a two state NFA accepting the (one word) language $\{([A_g]false)?\}$. For $g \cap h$ we have to be careful since the '∩' in CPL is intersection in the *path* sense rather than in the *language* sense. We use the fact that we are dealing with oapt-models and build $A_{g \cap h}$ that simulates both A_g and A_h synchronizing on ATF-letters. ∎

Theorem 6. *If we fix ATF to be a subset of $\{a_1,\ldots,a_n\}$, then satisfiability of CPL formulas can be decided in 2EXPTIME .*

Proof: Let f be a CPL formula over $\text{ATF} \subseteq \{a_1,\ldots,a_n\}$. Use Lemma 3 to construct f' with new atomic transition formulas ATF', such that f is satisfiable iff f' is oapt-satisfiable. Note that since the set $\{a_1,\ldots,a_n\}$ is fixed, $|f'| = c_1 \cdot |f|$, for some constant c_1. By Lemma 5, there exist an APDL program $A_{f'}$ (in the form of an NFA over the alphabet $\text{ATF}' \cup Prop_A$) such that for every path p in every oapt-model M: $p \in \rho(f)$ iff $p \in A_{f'}$. In other words: $p \in \rho(f)$ iff $first(p) \models <A_{f'}>true$. It is known [HS2], that satisfiability of APDL formulas can be decided in deterministic exponential time. One can easily prove by induction on the structure of f' that $|A_{f'}| \leq 2^{c_2 \cdot |f'|}$, for some constant c_2 (actually, the exponent is needed only for the '∩' case). So that the overall time complexity of deciding satisfiability of the original CPL formula f is bounded by $2^{2^{c_3 \cdot |f|}}$, for some constant c_3. ∎

4 CPL with Interleaving

The motivation for adding the interleaving operator to CPL is twofold. Our primary motivation is that the interleaving operator can be interpreted as the simplest case of composition used in algebraic approaches to modeling concurrent computation (see, e.g., [M]). Interleaving represents the case where processes run concurrently in such a fashion that their atomic steps can be arbitrarily interleaved but where no communication between them takes place. This form of concurrency, modeled by interleaving, might also be described as *asynchronous*. Second, as discussed in the sequel, using interleaving we gain succinctness.

Let us now define ICPL (CPL with interleaving). The syntax of ICPL extends that of CPL as follows: if f and g are formulas, then so is $(f \parallel g)$. Turning to the semantics, the basic difficulty is that our ρ, which associates paths with formulas, is not informative enough to capture interleaving. For example, we would like the formula $(a \cdot P) \parallel b$ to be satisfied by the paths: $a \cdot P \cdot b$ (i.e., an a-transition followed by a b-transition, with P true in the intermediate state), $a \cdot b \cdot P$ and $b \cdot a \cdot P$. However, paths of the second form would not appear if we used $\rho(a \cdot P)$ and $\rho(b)$, since $\rho(a \cdot P)$ contains only 'a'-paths with P at the last state. To solve this problem we shall use a more detailed version of ρ. The idea is that now $\rho_M(f)$ will contain, in addition to paths in M that are associated with f, some 'evidence' of this association. We will associate with each formula (via this extended ρ) a set of *computation paths* (defined below) rather than a set of (ordinary) paths. A computation path in a model M consists of two objects: a *computation*, which is a sequence of transitions accompanied by a sequence of properties (state formulas); and an ordinary path over M, i.e., a sequence of states of M. To get a feeling for this, the figure below illustrates a computation path:

$$\begin{array}{ccccc} P & a & \neg(b \cdot Q) & a, c & R \\ \xrightarrow{} & & \xrightarrow{} & & \xrightarrow{} \\ s & & t & & r \end{array}$$

Here, the *path* is (s, t, r), i.e., the sequence of states, and the *computation* is $\langle (a, \{a, c\}), (P, \neg(b \cdot Q), R) \rangle$.

Definition 7. The set of *state formulas* SF is the minimal set of ICPL formulas that contains ASF, contains all formulas of the form $\neg f$, and is closed under $\cdot$ and $\cap$. For state formulas f and g of the form $f = f^1 \cdot f^2 \cdot \ldots \cdot f^k$, $g = g^1 \cdot g^2 \cdot \ldots \cdot g^l$, where $k, l \geq 1$, let

$$f \dot\cap g = \begin{cases} (f^1 \cap g^1) \cdot \ldots \cdot (f^k \cap g^l), & k = l \\ (f^1 \cap g^1) \cdot \ldots \cdot (f^l \cap g^l) \cdot f^{l+1} \cdot \ldots \cdot f^k, & k > l \\ (f^1 \cap g^1) \cdot \ldots \cdot (f^k \cap g^k) \cdot g^{k+1} \cdot \ldots \cdot g^l, & k < l \end{cases}$$

Definition 8. A *computation* is a pair $c = \langle Tran_c, Val_c \rangle$, where $Tran_c$ is a path over the set $2^{\text{ATF}} - \emptyset$ and Val_c is a path of length $|Tran_c| + 1$ over the set SF. The length of c, denoted $|c|$, is $|Val_c|$.

We now define several operations on computations. For this we use the two computations:

$$c = \langle \overbrace{(t_1, \ldots, t_k)}^{Tran_c}, \overbrace{(f_0, \ldots, f_k)}^{Val_c} \rangle \text{ and } d = \langle \overbrace{(r_1, \ldots, r_l)}^{Tran_d}, \overbrace{(g_0, \ldots, g_l)}^{Val_d} \rangle$$

- $c \cdot d \stackrel{\text{def}}{=} \langle (Tran_c); (Tran_d), (Val_c)\hat{\,}(Val_d) \rangle$, where $(t_1, \ldots, t_k); (r_1, \ldots, r_l) = (t_1, \ldots, t_k, r_1, \ldots, r_l)$ and $(f_0, \ldots, f_k)\hat{\,}(g_0, \ldots, g_l) = (f_0, \ldots, f_k \cdot g_0, \ldots, g_l)$.
- If c and d are of the same length (i.e., $k = l$), then $c \cap d \stackrel{\text{def}}{=} \langle (t_1 \cup r_1, \ldots, t_k \cup r_k), (f_0 \dot\cap g_0, \ldots, f_k \dot\cap g_k) \rangle$.

The next operation we want to define is $c \parallel d$. In general, $c \parallel d$ is a set of computations. A computation in $c \parallel d$ is obtained by sequentially executing portions from c or from d. Let us make this notion more precise. First, denote by $I_c \subseteq \{0, \ldots, k\}$ the set of indices s.t. $i \in I_c$ iff f_i is of the form $f_i^1 \cdot f_i^2 \cdots f_i^{last(f_i)}$ (and $last(f_i) \geq 2$). Next, define a *formula portion* of c to be any element of the set

$$\left(\bigcup_{i \in \{0,\ldots,k\} - I_c} f_i \right) \cup \left(\bigcup_{i \in I_c} \bigcup_{m=1}^{j_f^{(i)}} f_i^m \right).$$

Finally, a *portion* of c is a formula portion or a *transition portion* of c, where a transition portion of c is an element of $\{r_i\}_{i=1}^l$. Portions of d are defined in a similar way.

Constructing a computation $e \in c \parallel d$ is carried out as follows. Initialize $Tran_e$ and Val_e with ( ), and set pointers to the leftmost formula portions of c and d. While there remain portions of c and d, that have not been dealt with, non-deterministically add to e the next portion of c or that of d, and advance the corresponding pointer to the next portion, where the successor of a transition is a formula and the successor of a formula portion is either the next portion of the same formula or the next transition, if the current portion is last in the formula. When one of c or d has been consumed, simply add to e the remaining portions of the other.

Definition 9. A *computation path* in a model M is a pair $p = (Stat_p, c_p)$, where $Stat_p$ is a nonempty path over S_M (i.e., an ordinary path in the model M) and c_p is a computation with $|Val_{c_p}| = |Stat_p|$.

For a computation path $p = (Stat_p, c_p)$, we denote $Tran_{c_p}$ and Val_{c_p} by $Tran_p$ and Val_p, respectively. We intend to use a computation path p as follows: $Stat_p$ will be the states along p, $Tran_p$ will be the sequence of transitions along p, and Val_p will be the sequence of state formulas satisfied in states along p. For example, a computation path p with $Stat_p = (s, t, r)$, $Tran_p = (a, \{a, c\})$ and $Val_p = (P, \neg(b \cdot Q), R)$ is illustrated in the figure prior to Def.7.

We have defined · both on computations and on paths, and we now use these together to define $p \cdot q$, for computation paths p and q (and then, extend it to sets of computation paths in the usual way): $p \cdot q \stackrel{\text{def}}{=} (Stat_p \cdot Stat_q, c_p \cdot c_q)$.

Definition 10. Let CP be a set of computation paths in a model M. A path $p = (s_0, \ldots, s_k)$ in M is CP *consistent with a computation* $c = \langle (t_1, \ldots, t_l), (f_0, \ldots, f_l) \rangle$, if the following conditions are satisfied: (i) $|p| = |c|$ (i.e., $k = l$), (ii) For every $0 \le i \le k-1$, there exist $q \in CP$ s.t. $Stat_q = (s_i, \Sigma_{i+1})$ and $Tran_q = (t_i)$, and (iii) For every $0 \le i \le k$, there exist $q \in CP$ s.t. $Stat_q = (s_i)$ and $Val_q = (f_i)$.

We can now define the semantics of ICPL. Formulas are interpreted over the same models as in CPL, that is, models of the form $M = (S_M, \rho_M^0)$, where S_M is the set of *states*, $\rho_M^0(P) \subseteq (S)$, for every element $P \in$ ASF, and $\rho_M^0(a) \subseteq S \times S$, for every element $a \in$ ATF.

Next, ρ_M^0 is extended by induction to a function ρ_M, which assigns a set $\rho_M(f)$ of computation paths to every ICPL formula f. The set of all computation paths assigned to formulas in this way (i.e., those that are in $\rho_M(f)$ for some f) is denoted $CP(M)$. All the inductive cases in the definition of ρ_M are straightforward, except for the following two:

$\rho_M(f \cap g) = \{r \mid \exists p \in \rho_M(f), q \in \rho_M(g) \text{ s.t. } Stat_r = Stat_p = Stat_q \text{ and } c_r = c_p \cap c_q\}$

$\rho_M(f \parallel g) = \{r \mid Stat_r \text{ is } CP(M) \text{ consistent with } c_r \text{ and } c_r \in (c_p \parallel c_q), \text{ for some } p \in \rho_M(f), q \in \rho_M(g)\}$.

Definition 11. An ICPL formula f is *satisfied in a path p* of a model M, written $M, p \models f$, iff $p = Stat_q$ for some computation path $q \in \rho_M(f)$. f is satisfiable iff $M, p \models f$ for some path p of some model M.

How does ICPL relate to CPL? Recall that ICPL is intended to be CPL extended with the '$\parallel$'-operator. While syntactically it is clear that CPL¡ICPL, semantically this may seem less obvious due to the differences in the definitions. We therefore proceed by showing that under the canonical correspondence between CPL models and ICPL models, that is, $\rho_{CPL} = \rho_{ICPL}^0$, this is indeed the case.

Proposition 12 *For every CPL formula f and every (ordinary) path p in any model M, $M, p \models_{CPL} f$ iff $M, p \models_{ICPL} f$.*

Proof: Omitted. ■

In what sense is ICPL 'better' than CPL? Well, using the well known fact that regular sets are closed under interleaving it is not difficult to prove that ICPL and CPL have the same expressive power. Nevertheless, ICPL has two important advantages over CPL. The first is *clarity* in modeling asynchronous concurrent computations. For example, consider the following two computations: (i) Execute a, observe P and then perform b. (ii) Observe Q and then execute b followed by a. In ICPL, we can use the formula $a \cdot P \cdot b \parallel Q \cdot b \cdot a$ to model computations that arise from running these two in parallel, while in CPL it appears that one must use a much more cumbersome formula that explicity lists many of the possible interleavings. The second (and related) advantage of ICPL over CPL is *succinctness*. It is known that the use of the interleaving operator can shorten a regular expression by an exponential amount [F, MS]. It is true that interleaving in ICPL is (in general) not interleaving in the language sense. However, ICPL formulas that use only ATF and the operators '·', '*', '∪' and '∥' correspond essentially to regular expressions (extended with interleaving operator) over the alphabet ATF. As to decidability, we have:

Theorem 13. *Satisfiability of* ICPL *formulas with* ATF$\subseteq \{a_1, \ldots, a_n\}$ *can be decided in* 2EXPTIME.

5 ICPL With Synchronization

ICPL is suited for modeling asynchronous concurrency. To model synchronous concurrency as well, we introduce ICPL with synchronization (SICPL). All ICPL formulas are SICPL formulas. In addition, if f and g are SICPL formulas and syn is a subset of ATF, then $f \mid syn \mid g$ is a SICPL formula. (The set syn has to be written out in full, for example as in $(a \cdot b)^* \cdot P \mid a, b \mid (a \cup b)$.) Intuitively, $f \mid syn \mid g$ represents the interleaving of f and g synchronized w.r.t. syn. See the example in Section 1.

To present the formal semantics of SICPL (which will not be given here), one has to modify each step in the definition of $\rho_M(f \parallel g)$. Here we have:

Theorem 14. *Satisfiability of* SICPL *formulas with* ATF$\subseteq \{a_1, \ldots, a_n\}$ *can be decided in* 2EXPTIME.

6 Infinite Computations

CPL (and its extensions ICPL, SICPL) are input/output oriented and are therefore appropriate for stating properties concerning programs with finite computations. We wish, however, to make it possible to reason about processes with possible *infinite* computations. For example, we would like to say that the infinite model $P \cdot a \cdot P \cdot a \ldots$ (the a's are transitions, and the P's signify truth in the intermediate states), admits in addition to the finite computations described by $(P \cdot a)^*$ also the infinite computation $(P \cdot a)^\omega$. With this idea in mind, we introduce the extension ωCPL.

Basically, one would like ωCPL to extend CPL by employing the new operator 'ω' and to use formulas of the form f^ω, where f is a CPL formula. The most intuitive interpretation of f^ω is simply to associate with it infinite paths that result by fusing infinitely many (finite) paths of f (that is, take $\rho(f^\omega)$ as $\rho(f)^\omega$). Choosing this interpretation, however, forces one to make a distinction between 'ω-formulas' (those with possibly infinite paths corresponding to the ω) and 'finite formulas'. This is necessary in order to interpret (or to forbid) formulas of the form $f^\omega \cdot g$, $f^\omega \cdot g^\omega$, $(f^\omega)^*$ etc.

To enable a uniform representation, we have decided to adopt a more modest interpretation of f^ω, as follows. We shall consider f^ω rather as a test, true in states (i.e., paths of length 0) from which it possible to repeatedly carry out computations of f infinitely often. The advantage of using this interpretation is that even though paths associated with formulas are finite, and hence all CPL operators are applicable and retain their usual meaning, it is still possible to make assertions concerning infinite computations.

Definition 15. An ω-*path* over a set S is an infinite sequence of elements of S. For a set P of finite paths, let $\mathcal{P}^\omega = \{p_1 \cdot p_2 \cdot p_3 \cdots \mid \forall i \geq 1,\ p_i \in \mathcal{P}\}$. That is, P^ω is the set of finite and infinite paths obtained by repeatedly fusing (finite) paths from P infinitely often.

The syntax is such that ωCPL contains all CPL formulas, and in addition if f and g are ωCPL formulas, then so are $(\neg f)$, (f^*), (f^ω), $(f \cdot g)$, $(f \cup g)$ and $(f \cap g)$. As for semantics, ωCPL is interpreted over the same models as CPL. Given a model M and an ωCPL formula f, $\rho_M(f)$ is defined exactly as in CPL with the addition of the clause: $\rho_M(f^\omega) = \mathit{first}\,((\rho_M(f))^\omega)$.

ωCPL can be considered to be a 'path version' of RPDL [HS1]. Indeed, we can extend the embedding of PDL in CPL to an embedding of RPDL in ωCPL by: $(\mathit{repeat}(\beta))' = (\beta')\omega$. Thus, ωCPL's expressive power is at least as that of RPDL, which is known to be high (for example it exceeds that of CTL* [E].)

Proving that ωCPL is elementary decidable is done by reducing its satisfiability problem to that of ARPDL (the automata version of PDL+*repeat*). Here, we omit the details, and only mention that this reduction costs at most an exponential in added size. Thus, using the fact that ARPDL is decidable in 3EXPTIME [VW], we have:

Theorem 16. *Satisfiability of* ωCPL *formulas with* ATF$\subseteq \{a_1, \ldots, a_n\}$ *can be decided in* 4EXPTIME .

References

[E]   E. A. Emerson, *Handbook of Theoretical Computer Science*, (J. Van Leeuwen, ed.), Vol. B, Elsevier Science Publishers B.V., Amsterdam (1990), 996–1072.
[F]   M. Fürer, *Proc. 7th Intl. Colloq. on Automata, Languages, and Programming*, LNCS, Springer-Verlag, Vol. 85 (1980), 234–245.
[FL] M. J. Fischer and R. E. Ladner, *J. Comput. Sys. Sci.* **18** (1979), 194–211.
[HKP] D. Harel, D. Kozen & R. Parikh, *J. Comput. Sys. Sci.* **25** (1982), 144–170.
[HP] D. Harel and D. Peleg, *Theor. Comput. Sci.* **38** (1985), 307–322.
[HS1] D. Harel and R. Sherman, *Inf. and Control* **55** (1982), 175–192.
[HS2] D. Harel and R. Sherman, *Inf. and Control* **64** (1985), 119–135.
[Ha] J. Y. Halpern, *Proc. 23rd IEEE Found. Comput. Sci.*, pp. 204-216, 1982.
[KP] D. Kozen and R. Parikh, *Theor. Comput. Sci.* **14**(1) (1981), 113–118.
[M]   R. Milner, LNCS, Springer-Verlag, Vol. 92 (1980).
[MS] A. J. Mayer and L. J. Stockmeyer, *Inf. and Comp.* **115**(2) (1994), 293–311.
[Pe] D. Peleg, M.Sc. Thesis, Bar-Ilan Univ., Ramat Gan, Israel , 1982 (in Hebrew).
[Pn] A. Pnueli, *Proc. 18th IEEE Symp. Found. Comput. Sci.* (1977), 46-57.
[Pr] V. R. Pratt, *Proc. 17th IEEE Symp. Found. Comput. Sci.*, (1976), 109–121.
[SPH] R. Sherman, A. Pnueli & D. Harel, *SIAM J. Comput.* **13** (1984), 825–839.
[VW] M. Vardi and P. Wolper, *Proc. Symp. on Logics of Programs* , LNCS, Springer-Verlag, Vol. 164 (1983), 501–512.

Model Checking the Full Modal Mu-Calculus for Infinite Sequential Processes

Olaf Burkart*[1], and Bernhard Steffen[2]

[1] LFCS, University of Edinburgh, JCMB, King's Buildings, Edinburgh EH9 3JZ, UK
<olaf@dcs.ed.ac.uk>
[2] FMI, Universität Passau, Innstraße 33, 94032 Passau, Germany
<steffen@fmi.uni-passau.de>

Abstract. In this paper we develop a new exponential algorithm for model-checking infinite sequential processes, including *context-free processes, pushdown processes*, and *regular graphs*, that decides the full modal mu-calculus. Whereas the actual model checking algorithm results from considering conditional semantics together with backtracking caused by alternation, the corresponding correctness proof requires a stronger framework, which uses *dynamic environments* modelled by finite-state automata.

1 Introduction

Over the past decade model-checking has emerged as a powerful tool for the automatic analysis of concurrent systems. Whereas model-checking for finite-state systems is nowadays well-established, the theory for infinite systems is a current research topic (cf. [BE97]). Since even weak branching time logics are undecidable for infinite-state systems incorporating parallel operators, much work has focused on the verification of *sequential processes*. The strongest results obtained so far show the decidability of monadic second order logic (MSOL) for the infinite binary tree [Rab69], pushdown transition graphs [MS85], regular graphs [Cou90], and rational restricted recognizable graphs [Cau96]. However, all decision procedures are non-elementary and thus not applicable to practical problems. Moreover, MSOL is usually too expressive, since it allows to distinguish even bisimilar models. For these reasons, the modal mu-calculus is seen as an attractive alternative for specifying behavioural properties.

The model-checking problem for sequential processes and the modal mu-calculus was first considered in [BS92]. The authors developed an iterative model-checking algorithm that decides the *alternation-free* part of the modal mu-calculus for *context-free* processes based on a conditional formulation of the semantics of μ-formulas. Moreover, in [HS94] it is shown how this can be done using tableaux-based techniques, allowing local model checking. Finally,

---

* This work was supported during my stay at IRISA by the European Community under HCM grant ERBCHBGCT 920017, and during my stay at the LFCS by the DAAD under grant D/95/14834 of the NATO science committee.

the approach was also extended to the strictly larger classes of *pushdown processes* [BS95] and *regular graphs* [BQ97]. Since alternation of fixpoints gives rise to a strict hierarchy [Bra96] the problem of model-checking the full modal mu-calculus has still been open. Only recently, Walukiewicz presented a first exponential model-checking algorithm for pushdown processes based on games [Wal96].

In this paper we develop an alternative algorithm which, essentially, arises as a combination of extending the standard iterative model-checking techniques with *conditional reasoning*, in order to capture sequential model structures in an *alternation-free* setting [BS92, BS95, BQ97], and the observation that alternating fixpoints require some kind of *backtracking*, as it is known from regular model checking (cf. e.g. [CKS92]). Whereas the actual model checker results from this combination, the corresponding correctness proof requires a stronger framework, which uses *dynamic environments*. In contrast to the 'standard' assertions, which suffice algorithmically, dynamic environments also explicitly model valuations of variables that occur *free* in the actual fixpoint computation. This explicit treatment is necessary in order to establish the link between the result of the fixpoint iteration and the semantics of the full modal mu-calculus.

Fortunately, all this additional complexity is only required for the proof and need not be considered for an implementation. Taking $|\mathcal{C}|$ as the number of transitions, and $|Q|$ as the branching degree in the finite sequential process representation, as well as $|\Phi|$ as the size of the formula, and "ad" as the alternation depth of the formula under consideration, the overall complexity[1] is

$$O(\ |\Phi| * (|Q| * |\mathcal{C}|)^{\mathrm{ad}(\Phi)+1} * 2^{|\Phi|*(\mathrm{ad}(\Phi)+|Q|)}\ ).$$

Note that this does not only cover context-free and pushdown processes, but also regular graphs, which are not covered by the algorithm proposed by Walukiewics. It is not at all clear, whether a similar extension is also possible for Walukiewics' algorithms.

The plan of the paper is now as follows. The next section describes the class of processes we will consider, and presents the modal mu-calculus. Subsequently, we develop our model-checking algorithm which is proved to be correct in Section 4. The final section contains our conclusions and directions for future research. Proofs and further details can be found in the full version [BS97].

2 Processes and Specifications

Infinite sequential processes comprise context-free processes, pushdown processes, and regular graphs. In this paper we will mainly concentrate on the model-checking problem for context-free processes, as the extension to pushdown processes, respectively regular graphs, can be obtained following the lines of [BS95], respectively [BQ97].

---

[1] In this paper we neglect the optimization of [LBC+94] which exploits monotonicity arguments and would reduce $\mathrm{ad}(\Phi)$ to $\mathrm{ad}(\Phi)/2$.

2.1 Context-Free Processes

As usual, we consider *labelled transition graphs* as models for the behaviour of concurrent systems, since they allow to represent the underlying semantics of many process calculi. In particular, we are interested in classes of infinite transition graphs which can be finitely represented by labelled rewrite systems.

Definition 2.1. A *labelled transition graph* is a triple $\mathcal{T} = (\mathcal{S}, Act, \to)$ where $\mathcal{S}$ is the set of *states*, Act is the set of *transition labels* (or *actions*), and $\to \subseteq \mathcal{S} \times Act \times \mathcal{S}$ is the *transition relation*.

Definition 2.2. A *labelled rewrite system* is a triple $\mathcal{R} = (V, Act, R)$ where V is an *alphabet*, Act is a set of *labels*, and $R \subseteq V^* \times Act \times V^*$ is a finite set of *rewrite rules*. If the rewrite rules are of the form $R \subseteq V \times Act \times V^*$ the rewrite system is called *alphabetic*.

In the remainder of the paper, a rewrite rule $(u, a, v) \in R$ is also written as $u \xrightarrow{a} v$. In general, rewrite systems are used to define a rewrite relation on words of V^* where a rewrite rule may be applied at any position. The technical development of this paper concentrates on rewritings of the following restricted form.

Definition 2.3. Let $\mathcal{R} = (V, Act, R)$ be a rewrite system. Then the *prefix rewriting relation* of R is defined by $\longmapsto_R =_{df} \{ (uw, a, vw) \mid (u \xrightarrow{a} v) \in R, w \in V^* \}$, and the labelled transition graph $\mathcal{T}_\mathcal{R} =_{df} (V^*, Act, \longmapsto_R)$ is called the *prefix transition graph* of $\mathcal{R}$. By abuse of notation, we will henceforth write $uw \xrightarrow{a} vw$ instead of $uw \longmapsto_R^a vw$.

An alphabetic rewrite system which is interpreted wrt. prefix rewriting is called a *context-free system*, and a *context-free process* is then the rooted prefix transition graph of a context-free system. Note that the states of a context-free process are words over V, and we will henceforth use lower greek letters $\alpha, \beta, \ldots$ to denote them. One standard example for a context-free process is the prefix transition graph of $\mathcal{C}_{ex} = \{ A \xrightarrow{a} AB, A \xrightarrow{b} \epsilon, B \xrightarrow{b} \epsilon \}$ rooted at A.

2.2 The Modal Mu-Calculus

Nowadays it is widely accepted that system properties can conveniently be expressed by temporal logic formulas. Particularly, the modal mu-calculus as introduced by Kozen [Koz83] is a powerful branching time logic. It combines standard modal logic with least and greatest fixpoint operators which allows to express very complex temporal properties within this formalism. Due to its expressiveness and its conciseness the mu-calculus can be regarded as the "assembly language" of temporal logics. Formulas of the mu-calculus, given in positive form, are defined by the following grammar

$$\Phi ::= \mathtt{tt} \mid \mathtt{ff} \mid X \mid \Phi \vee \Phi \mid \Phi \wedge \Phi \mid [a]\Phi \mid \langle a \rangle \Phi \mid \mu X.\Phi \mid \nu X.\Phi$$

where X ranges over a (countable) set of variables Var, and a over a set of actions Act. We will use $L\mu$ to denote the set of all mu-calculus formulas.

Standard Semantics Given $\mathcal{T}_\mathcal{R} = (V^*, Act, \rightarrow)$, and a valuation $\mathcal{V} : Var \rightarrow 2^{V^*}$, the inductive definition below stipulates when a context-free process $\alpha \in V^*$ has the property Φ, written as $\alpha \models_\mathcal{V} \Phi$. If α fails to satisfy Φ, we will write $\alpha \not\models_\mathcal{V} \Phi$.

$\alpha \models_\mathcal{V} \text{tt}$
$\alpha \not\models_\mathcal{V} \text{ff}$
$\alpha \models_\mathcal{V} X$ iff $\alpha \in \mathcal{V}(X)$
$\alpha \models_\mathcal{V} \Phi_1 \vee \Phi_2$ iff $\alpha \models_\mathcal{V} \Phi_1 \vee \alpha \models_\mathcal{V} \Phi_2$
$\alpha \models_\mathcal{V} \Phi_1 \wedge \Phi_2$ iff $\alpha \models_\mathcal{V} \Phi_1 \wedge \alpha \models_\mathcal{V} \Phi_2$
$\alpha \models_\mathcal{V} \langle a \rangle \Phi$ iff $\exists \alpha'. \alpha \xrightarrow{a} \alpha' \wedge \alpha' \models_\mathcal{V} \Phi$
$\alpha \models_\mathcal{V} [a]\Phi$ iff $\forall \alpha'. \alpha \xrightarrow{a} \alpha' \Rightarrow \alpha' \models_\mathcal{V} \Phi$
$\alpha \models_\mathcal{V} \mu X.\Phi$ iff $\forall S \subseteq V^*. (\forall \beta \in V^*. \beta \models_{\mathcal{V}[X \mapsto S]} \Phi \Rightarrow \beta \in S) \Rightarrow \alpha \in S$
$\alpha \models_\mathcal{V} \nu X.\Phi$ iff $\exists S \subseteq V^*. (\forall \beta \in V^*. \beta \in S \Rightarrow \beta \models_{\mathcal{V}[X \mapsto S]} \Phi) \wedge \alpha \in S$

where $\mathcal{V}[X \mapsto S]$ is the valuation resulting from $\mathcal{V}$ by updating the binding of X to S. The clauses for the fixpoints are a reformulation of the Tarski-Knaster theorem which states that the least fixpoint is the intersection of all pre-fixpoints and the greatest fixpoint is the union of all post-fixpoints. As a consequence, states satisfy a fixpoint formula iff they satisfy the *unfolding* of the formula, i.e. $\alpha \models_\mathcal{V} \sigma X.\Phi$ iff $\alpha \models_\mathcal{V} \Phi[\sigma X.\Phi/X]$ where $\sigma \in \{\mu, \nu\}$ and $\Phi[\Psi/X]$ denotes the simultaneous replacement of all free occurrences of X in Φ by Ψ.

The satisfaction relation defined above is independent of the valuation if the considered formula has no free variables in which case we will drop the index $\mathcal{V}$. We extend our satisfaction relation, moreover, to sets of formulas by writing $\alpha \models \Gamma$ if $\alpha \models \Phi$, for all $\Phi \in \Gamma$. Finally, we observe that the usual denotation of formulas as the set of states where the formula holds is obtained in our presentation by $[\![\Phi]\!]_\mathcal{V} = \{\alpha \mid \alpha \models_\mathcal{V} \Phi\}$. Next we define some standard notions which will allow us to deal with occurrences of subformulas in a given formula, as well as to measure the complexity of a formula.

Definition 2.4 (Binding). A formula Φ is called *well named* if every fixpoint operator in Φ binds a distinct variable, and free variables are distinct from bound variables. With each well named formula Φ we then associate its *binding function* $\mathcal{D}_\Phi$ which assigns to every bound variable X of Φ the unique subformula $\sigma X.\Psi(X)$ of Φ, called the *binding definition* of X in Φ.

From now on we assume that every formula is well named.

Definition 2.5 (Dependency order, Expansion). Given a formula Φ, we define the *dependency order* over the bound variables of Φ, denoted by $\leq_\Phi$, as the least partial order such that if X occurs free in $\mathcal{D}_\Phi(Y)$ then $X \leq_\Phi Y$. Moreover, for every subformula Ψ of Φ, we define the *expansion* of Ψ with respect to $\mathcal{D}_\Phi$ as: $(\!(\Psi)\!)_{\mathcal{D}_\Phi} =_{df} \Psi\,[\mathcal{D}_\Phi(X_n)/X_n]\ldots[\mathcal{D}_\Phi(X_1)/X_1]$ where the sequence $(X_1, \ldots, X_n)$ is a linear ordering of all bound variables of Φ compatible with the dependency order, i.e. if $X_i \leq_\Phi X_j$ then $i \leq j$.

Definition 2.6 (Subformulas, Closure). The *subformula relation* on $L\mu$, denoted by $\preceq$, is the least partial order on $L\mu$ such that $\Psi_i \preceq \Psi_1 \vee \Psi_2$, $\Psi_i \preceq \Psi_1 \wedge \Psi_2$, $\Psi \preceq \langle a \rangle \Psi$, $\Psi \preceq [a]\Psi$, $\Psi \preceq \mu X.\Psi$, and $\Psi \preceq \nu X.\Psi$, for $i = 1, 2$ and $a \in Act$. Given a formula Φ, we define the *closure* of Φ as $CL(\Phi) = \{\Psi \mid \Psi \preceq \Phi\}$. Furthermore, if $CL(\Phi) = \{\Psi_1, \ldots, \Psi_n\}$ we will henceforth assume that the subformulas Ψ_i are linearly ordered compatible with $\preceq$, i.e. if $\Psi_i \preceq \Psi_j$ then $i \geq j$.

Definition 2.7 (Alternation Depth).
A formula Φ is said to be in the classes Σ_0 and Π_0 iff it contains no fixpoint operators. To form the class Σ_{n+1}, take $\Sigma_n \cup \Pi_n$, and close under (i) boolean and modal combinators, (ii) $\mu X.\Phi$, for $\Phi \in \Sigma_{n+1}$, and (iii) substitution of $\Phi' \in \Sigma_{n+1}$ for a free variable of $\Phi \in \Sigma_{n+1}$ provided that no free variable of Φ' is captured by Φ; and dually for Π_{n+1}. The (Niwinski) *alternation depth* of a formula Φ, denoted by $ad(\Phi)$, is then the least n such that $\Phi \in \Sigma_{n+1} \cap \Pi_{n+1}$.

Assertion-Based Semantics As pointed out in [BS92], context-free processes can be verified by considering Hoare-logic style pre-condition/post-condition pairs of sets of formulas for each of the nonterminals occurring in the context-free system. A triple $\{\Gamma\} \, \alpha \, \{\Delta\}$ is then interpreted as α satisfies all formulas of Γ if we assert that after termination of α exactly the set of formulas Δ holds. This intuition is formally captured by the following definition of *assertion-based semantics* which generalises standard semantics by taking into account the set of formulas which hold after termination of a process.

Given $\mathcal{T_C} = (V^*, Act, \longmapsto_\mathcal{C})$, and a valuation $\mathcal{V} : Var \to 2^{V^*}$, the inductive definition below stipulates when a context-free process $\alpha \in V^*$ has the property Φ under the hypothesis that after termination of α the formulas Δ hold, written as $\alpha \models_\mathcal{V} (\Phi, \Delta)$. If α fails to satisfy Φ under the hypothesis Δ, we will write $\alpha \not\models_\mathcal{V} (\Phi, \Delta)$. First we have $\epsilon \models_\mathcal{V} (\Phi, \Delta)$ iff $\Phi \in \Delta$ and then, for $\alpha \neq \epsilon$,

$\alpha \models_\mathcal{V} (\mathtt{tt}, \Delta)$
$\alpha \not\models_\mathcal{V} (\mathtt{ff}, \Delta)$
$\alpha \models_\mathcal{V} (X, \Delta)$ iff $\alpha \in \mathcal{V}(X)$
$\alpha \models_\mathcal{V} (\Phi_1 \vee \Phi_2, \Delta)$ iff $\alpha \models_\mathcal{V} (\Phi_1, \Delta) \vee \alpha \models_\mathcal{V} (\Phi_2, \Delta)$
$\alpha \models_\mathcal{V} (\Phi_1 \wedge \Phi_2, \Delta)$ iff $\alpha \models_\mathcal{V} (\Phi_1, \Delta) \wedge \alpha \models_\mathcal{V} (\Phi_2, \Delta)$
$\alpha \models_\mathcal{V} (\langle a \rangle \Phi, \Delta)$ iff $\exists \alpha'.\ \alpha \xrightarrow{a} \alpha' \wedge \alpha' \models_\mathcal{V} (\Phi, \Delta)$
$\alpha \models_\mathcal{V} ([a]\Phi, \Delta)$ iff $\forall \alpha'.\ \alpha \xrightarrow{a} \alpha' \Rightarrow \alpha' \models_\mathcal{V} (\Phi, \Delta)$
$\alpha \models_\mathcal{V} (\mu X.\Phi, \Delta)$ iff $\forall S \subseteq V^*.\ (\forall \beta \in V^*.\ \beta \models_{\mathcal{V}[X \mapsto S]} (\Phi, \Delta) \Rightarrow \beta \in S)$
$\Rightarrow \alpha \in S$
$\alpha \models_\mathcal{V} (\nu X.\Phi, \Delta)$ iff $\exists S \subseteq V^*.\ (\forall \beta \in V^*.\ \beta \in S \Rightarrow \beta \models_{\mathcal{V}[X \mapsto S]} (\Phi, \Delta))$
$\wedge \alpha \in S$

As in the case of the standard semantics, we will use $\alpha \models_\mathcal{V} (\Gamma, \Delta)$ to denote $\alpha \models_\mathcal{V} (\Phi, \Delta)$, for all $\Phi \in \Gamma$.

The usefulness of the assertion-based semantics is underpinned by the following proposition [BS92] which states that, firstly, the assertion-based semantics extend the standard semantics, and secondly, that they allow to reason compositionally about context-free processes.

Proposition 2.8. *The assertion-based semantics is*
1. *an* extension *of standard semantics, i.e. given a closed formula Φ, we have,*
$$\alpha \models \Phi \quad \textit{iff} \quad \alpha \models (\Phi, \Delta_\epsilon) \qquad \textit{for } \Delta_\epsilon = \{\Psi \in CL(\Phi) \mid \epsilon \models \langle\!\langle \Psi \rangle\!\rangle_{\mathcal{D}_\Phi}\}.$$
2. compositional *wrt. context-free processes, i.e. for all $\Delta, \Gamma \subseteq L\mu$,*
$$\alpha\beta \models (\Gamma, \Delta) \quad \textit{iff} \quad \exists\, \Sigma \subseteq L\mu.\ \alpha \models (\Gamma, \Sigma) \textit{ and } \beta \models (\Sigma, \Delta)$$

The effectiveness of our algorithm, which is presented in the next section, relies, in particular, on Proposition 2.8.1, as it shows that Φ can be verified by taking into account merely the semantics of all subformulas of Φ.

3 The Model-Checking Algorithm

In this section we develop our model-checking algorithm which checks closed μ-formulas with arbitrary alternation depth for context-free processes in exponential time. In fact, the algorithm coincides with a backtracking extension of the model-checker of [BS92] which deals only with the alternation-free fragment of the modal mu-calculus.

Given a context-free system $\mathcal{C}$ and a closed formula Φ, each nonterminal $A \in V = \{A_1, \ldots, A_n\}$ defines a mapping $[\![A]\!] : 2^{CL(\Phi)} \to 2^{CL(\Phi)}$ from post- to pre-conditions. As we are, however, in particular interested in the question whether a given subformula $\Psi \in CL(\Phi)$ belongs to the pre-condition set or not, we refine this notion by defining the following functions, called *characteristic property transformers* (CPT).

$$[\![A]\!]^\Psi(\Delta) =_{\text{df}} \begin{cases} 1 & \text{if } A \models (\Psi, \Delta) \\ 0 & \text{otherwise} \end{cases}$$

Writing $\mathbb{B}$ for the usual lattice of boolean values, characteristic property transformers are elements of the boolean lattice consisting of all functions from $2^{CL(\Phi)}$ to $\mathbb{B}$, where the ordering, and the meet and join operations respectively, are defined argument-wise. More importantly, they can be obtained as a fixpoint solution of an appropriate function scheme, called the *property transformer scheme* (PTS). This scheme is defined by the rules given in Figure 1, and consists of two parts. The first part copes with the structure of the context-free system, as well as with the semantics of the formula, and defines an equation for each pair $(A, \Psi) \in V \times CL(\Phi)$. The second part deals with the empty process according to the first clause of the assertion-based semantics, as well as with composed processes according to Proposition 2.8.2. Whereas the rules for the basic cases mimic directly the semantics of the subformula, the fixpoint related equations are slightly more complicated and require a simultaneous computation of all their corresponding transformers. sel_A then simply selects the A component of the resulting tuple. The other auxiliary function, mem_Ψ, tests the membership of Ψ in the given set of formulas. It returns 1 if $\Psi \in \Delta$ and 0 otherwise.

The overall structure of the model-checking algorithm consists now of the following three steps.

$$
\begin{aligned}
[\![A]\!]_{\mathcal{V}}^{\text{tt}} &= 1 & [\![A]\!]_{\mathcal{V}}^{\Psi_1 \vee \Psi_2} &= [\![A]\!]_{\mathcal{V}}^{\Psi_1} \sqcup [\![A]\!]_{\mathcal{V}}^{\Psi_2} \\
[\![A]\!]_{\mathcal{V}}^{\text{ff}} &= 0 & [\![A]\!]_{\mathcal{V}}^{\Psi_1 \wedge \Psi_2} &= [\![A]\!]_{\mathcal{V}}^{\Psi_1} \sqcap [\![A]\!]_{\mathcal{V}}^{\Psi_2} \\
[\![A]\!]_{\mathcal{V}}^{X} &= \mathcal{V}(X, A) & [\![A]\!]_{\mathcal{V}}^{\langle a \rangle \Psi} &= \bigsqcup_{A \xrightarrow{a} \alpha} [\![\alpha]\!]_{\mathcal{V}}^{\Psi} \\
& & [\![A]\!]_{\mathcal{V}}^{[a]\Psi} &= \bigsqcap_{A \xrightarrow{a} \alpha} [\![\alpha]\!]_{\mathcal{V}}^{\Psi}
\end{aligned}
$$

$$
[\![A]\!]_{\mathcal{V}}^{\mu X.\Psi} = \text{sel}_A(\sqcap\{\,(h_{A_1}^X, \ldots, h_{A_n}^X) \mid \forall_{i \in [1,n]}\ [\![A_i]\!]_{\mathcal{V}[(X,A_j) \mapsto h_{A_j}^X,\, j \in [1,n]]}^{\Psi} \sqsubseteq h_{A_i}^X \,\})
$$
$$
[\![A]\!]_{\mathcal{V}}^{\nu X.\Psi} = \text{sel}_A(\sqcup\{\,(h_{A_1}^X, \ldots, h_{A_n}^X) \mid \forall_{i \in [1,n]}\ h_{A_i}^X \sqsubseteq [\![A_i]\!]_{\mathcal{V}[(X,A_j) \mapsto h_{A_j}^X,\, j \in [1,n]]}^{\Psi} \,\})
$$

$$
[\![\epsilon]\!]_{\mathcal{V}}^{\Psi}(\Delta) = \text{mem}_{\Psi}(\Delta)
$$
$$
[\![A\alpha]\!]_{\mathcal{V}}^{\Psi}(\Delta) = [\![A]\!]_{\mathcal{V}}^{\Psi}(\{\,\Upsilon \in CL(\Phi) \mid [\![\alpha]\!]_{\mathcal{V}}^{\Upsilon}(\Delta) = 1\,\})
$$

Figure 1. The property transformer scheme.

1. Given a context-free system $\mathcal{C}$ and a closed μ-formula Φ construct the property transformer scheme according to the rules given in Figure 1.
2. Solve the (finite) fixpoint problem for the property transformer scheme.
3. Check whether $[\![A_1]\!]^{\Phi}(\Delta_\epsilon) = 1$ where A_1 is the root of the context-free system, and $\Delta_\epsilon = \{\,\Psi \in CL(\Phi) \mid \epsilon \models \langle\!\langle \Psi \rangle\!\rangle_{\mathcal{D}_\Phi}\,\}$.

In Section 4 we prove that the second step of the algorithm computes transformers which reflect the assertion-based semantics, while Proposition 2.8.1 now ensures that the third step solves the model-checking problem, as we have

$$[\![A_1]\!]^{\Phi}(\Delta_\epsilon) = 1 \quad \text{iff} \quad A_1 \models (\Phi, \Delta_\epsilon) \quad \text{iff} \quad A_1 \models \Phi.$$

Moreover, the ordinary semantics of Φ can be obtained from the set of CPT's by means of $[\![\Phi]\!] = \{\,\alpha \in V^* \mid [\![\alpha]\!]^{\Phi}(\Delta_\epsilon) = 1\,\}$. This set can always be shown to be a regular set of states.

As expected, the required backtracking for alternating μ-formulas yields a worst-case time complexity for the algorithm, which is exponentially worse (in the alternation depth) than the estimation given for the alternation-free case [BS92, BS95].

Theorem 3.1 (Complexity).
Let $\mathcal{C}$ be a context-free system, and Φ be a closed μ-formula. Then the worst-case time complexity of solving the property transformer scheme is

$$O(\,|\Phi| * (|\mathcal{C}| * 2^{|\Phi|})^{ad(\Phi)+1}\,)$$

4 Dynamic Environments

In the presence of formulas containing free variables the simple composition property of Proposition 2.8.2 no longer captures correctly the behaviour of context-free processes wrt. the specification at hand. This defect is eliminated by the slight modification given below.

$$\{\Gamma, \mathcal{V}'\} \, \alpha\beta \, \{\Delta, \mathcal{V}\} \quad \text{iff} \quad \exists \, \Sigma, \mathcal{V}'' \, \{\Gamma, \mathcal{V}'\} \, \alpha \, \{\Sigma, \mathcal{V}''\} \text{ and } \{\Sigma, \mathcal{V}''\} \, \beta \, \{\Delta, \mathcal{V}\}$$

Intuitively, the modified composition rule expresses that in addition to assertions also *environments* must be adapted when considered at intermediate states. In general, the valuation $\mathcal{V}''$ is obtained from $\mathcal{V}$ by right cancellation of β, i.e. for all $X \in dom(\mathcal{V})$, $\mathcal{V}''(X) = (\mathcal{V}(X) \cap V^*\beta)\beta^{-1}$. As an example, $\alpha\beta \in \mathcal{V}(X)$ would imply $\alpha \in \mathcal{V}''(X)$.

In the remainder of this section we fix now a context-free system $\mathcal{C}$, and a formula Φ with closure $\{\Psi_1, \ldots, \Psi_n\}$. Our aim is to develop a formalism, the *dynamic environments*, which faithfully models the adaptations of valuations needed for composition. Dynamic environments will be partitioned into levels $k \in [1, n]$ where a dynamic environment of level k defines the valuations for $\{\Psi_1, \ldots, \Psi_k\}$. This change from valuations for variables to valuations for subformulas is reflected in the semanics by adding the rule "if $\Psi \in dom(\mathcal{V})$ then $(\alpha \models_\mathcal{V} (\Psi, \Delta)$ if $\alpha \in \mathcal{V}(\Psi))$". The original model-checking problem is then reduced to a corresponding fixpoint problem on the finite domain of dynamic environments, such that the semantics of the original formula is captured by the final environment of level n.

Definition 4.1 (Dynamic Environment).

A *dynamic environment* $\bar{\mathcal{A}}_k$ of level $k \in [1, n]$ is a sequence of deterministic finite-state automata $\mathcal{A}_i = (Q_{\mathcal{A}_i}, V, \delta_{\mathcal{A}_i}, F_{\mathcal{A}_i})$, $i \in [1, k]$, where $Q_{\mathcal{A}_i} = (2^{CL(\Phi)})^i$ are the state sets of the automata, V is the input alphabet, $\delta_{\mathcal{A}_i} : Q_{\mathcal{A}_i} \times V \to Q_{\mathcal{A}_i}$ are the transition functions obeying the constraints $\delta_{\mathcal{A}_i}(\bar{\Delta}_i, A) = \bar{\Gamma}_i$ implies $\delta_{\mathcal{A}_{i-1}}(\bar{\Delta}_{i-1}, A) = \bar{\Gamma}_{i-1}$ where $\bar{\Delta}_i$ denotes $(\Delta_1, \ldots, \Delta_i)$, and $F_{\mathcal{A}_i} = \{\bar{\Delta}_i \in Q_{\mathcal{A}_i} \mid \Psi_i \in \Delta_i\}$ is the set of accepting states. Denoting the transitive closure of $\delta_{\mathcal{A}_i}$, as usual, also by $\delta_{\mathcal{A}_i}$ the language accepted by $\mathcal{A}_i$ starting in the state $\bar{\Delta}_i$ is $\mathcal{L}_{\mathcal{A}_i}(\bar{\Delta}_i) = \{\alpha \in V^* \mid \delta_{\mathcal{A}_i}(\bar{\Delta}_i, \tilde{\alpha}) \in F_{\mathcal{A}_i}\}$ where $\tilde{\alpha}$ is the reverse of α^2.

A dynamic environment $\bar{\mathcal{A}}_k$ together with a state $\bar{\Delta}_k$ is then interpreted as an environment which defines valuations for $\Psi_1, \ldots, \Psi_k$ by means of

$$\bar{\mathcal{A}}_1\langle\bar{\Delta}_1\rangle =_{df} [\Psi_1 \mapsto \mathcal{L}_{\mathcal{A}_1}(\bar{\Delta}_1)]$$
$$\bar{\mathcal{A}}_k\langle\bar{\Delta}_k\rangle =_{df} \bar{\mathcal{A}}_{k-1}\langle\bar{\Delta}_{k-1}\rangle [\Psi_k \mapsto \mathcal{L}_{\mathcal{A}_k}(\bar{\Delta}_k)] \text{ for } 2 \leq k \leq n$$

Dynamic environments are a convenient formalism to describe the semantics of μ-formulas on context-free processes since they model compositionality simply by transitions in the finite automaton.

Lemma 4.2. *Let* $\{\Gamma, \mathcal{V}'\} \, A \, \{\Delta, \bar{\mathcal{A}}_k\langle\bar{\Delta}_k\rangle\}$. *Then*

1. *For all* $i \leq k$, $\Psi_i \in \Gamma$ *iff* $A \in \bar{\mathcal{A}}_k\langle\bar{\Delta}_k\rangle(\Psi_i)$, *and* 2. $\mathcal{V}' = \bar{\mathcal{A}}_k\langle\delta_{\mathcal{A}_k}(\bar{\Delta}_k, A)\rangle$.

---

[2] Here we have to use $\tilde{\alpha}$ as the automaton has to model the above mentioned right cancellation

The first property expresses that a dynamic environment of level k captures the semantics of all subformulas up to level k, while the second property states that the environment to be considered in the pre-condition of A coincides with the interpretation of the A-successor of $\bar{\Delta}_k$ in $\mathcal{A}_k$.

The granularity of the transition functions of dynamic environments is not sufficient to obtain a match between the semantic and the iterative intuition behind the model checking problem. We therefore split these transition functions into *characteristic transition functions* as follows.

$$\delta^{i,j}(\bar{\Delta}_i, A) = \begin{cases} 1 & \text{if } \delta_{\mathcal{A}_i}(\bar{\Delta}_i, A) = \bar{\Gamma}_i \text{ and } \Psi_j \in \Gamma_i \\ 0 & \text{otherwise} \end{cases}$$

The split into characteristic transition functions allows us to view a dynamic environment $\bar{\mathcal{A}}_k$ as a matrix of CTF's as depicted below.

$$\begin{matrix} \delta^{1,1} & \delta^{1,2} & \ldots & \delta^{1,k} & \ldots & \delta^{1,n} \\ \vdots & & \ddots & & & \vdots \\ \delta^{k,1} & \delta^{k,2} & \ldots & \delta^{k,k} & \ldots & \delta^{k,n} \end{matrix}$$

This matrix can be systematically extended to a matrix for $\bar{\mathcal{A}}_{k+1}$ with new row $(\delta^{k+1,1}, \ldots, \delta^{k+1,n})$ by means of a fixpoint computation such that the final result will capture the semantics of the formula Φ on the given process[3].

As will be elaborated on in the next subsection, these matrices are adequate for proving our main result, Theorem 4.6, i.e. the equivalence of the semantic and the iterative algorithm presented in Section 4.1, because it is possible to "synchronize" their corresponding computations on the diagonal.

4.1 Semantic and Iterative Solutions

Given the semantics of the formulas $\Psi_1, \ldots, \Psi_{k-1}$ in terms of a dynamic environment $\bar{\mathcal{A}}_{k-1}$ we will now consider the semantics of the remaining formulas $\Psi_k, \ldots, \Psi_n$.

Definition 4.3 (Semantic Solutions).
We call $\bar{\mathcal{A}}_k$, for $k \in [1, n]$, the *semantic solution* of $\bar{\mathcal{A}}_{k-1}$, written as $\mathcal{S}(\bar{\mathcal{A}}_{k-1})$, if the transition function of $\mathcal{A}_k$ satisfies

$$\delta_{\mathcal{A}_k}(\bar{\Delta}_k, A) = \bar{\Gamma}_k \quad \text{iff} \quad (\Gamma_k, \bar{\mathcal{A}}_{k-1}\langle\bar{\Gamma}_{k-1}\rangle) \ A \ (\Delta_k, \bar{\mathcal{A}}_{k-1}\langle\bar{\Delta}_{k-1}\rangle).$$

Moreover, we call $(\bar{\mathcal{A}}_k, \ldots, \bar{\mathcal{A}}_n)$ the *semantic solutions* of $\bar{\mathcal{A}}_{k-1}$, denoted by $\bar{\mathcal{S}}(\bar{\mathcal{A}}_{k-1})$, if $\bar{\mathcal{A}}_i = \mathcal{S}(\bar{\mathcal{A}}_{i-1})$, for $i \in [k, n]$.

It turns out that the semantic solution respects the standard substitution lemma.

---

[3] More precisely, since the arity of characteristic transition functions depends on the row, they have to be adapted as described in [BS97] during this computation.

Lemma 4.4. *Let* $(\Gamma_k, \bar{\mathcal{A}}_{k-1}\langle \bar{\Gamma}_{k-1}\rangle)$ *A* $(\Delta_k, \bar{\mathcal{A}}_{k-1}\langle \bar{\Delta}_{k-1}\rangle)$ *and let* $\bar{\mathcal{A}}_k$ *be the semantic solution of* $\bar{\mathcal{A}}_{k-1}$. *Then*

$$(\Gamma_k, \bar{\mathcal{A}}_{k-1}\langle \bar{\Gamma}_{k-1}\rangle[\Psi_k \mapsto \mathcal{L}_{\mathcal{A}_k}(\bar{\Gamma}_k)])\ A\ (\Delta_k, \bar{\mathcal{A}}_{k-1}\langle \bar{\Delta}_{k-1}\rangle[\Psi_k \mapsto \mathcal{L}_{\mathcal{A}_k}(\bar{\Delta}_k)])$$

Corollary 4.5 (Diagonal Consistency). *If* $\bar{\mathcal{A}}_k, \ldots, \bar{\mathcal{A}}_n$ *are the semantic solutions of* $\bar{\mathcal{A}}_{k-1}$ *then* $\delta^{i,j} = \delta^{j,j}$, *for* $i \in [k,n], j \in [1,n]$.

Due to this corollary we may simply identify the semantic solutions $\bar{\mathcal{A}}_k, \ldots, \bar{\mathcal{A}}_n$ with the characteristic transition functions $\delta^{k,k}, \ldots, \delta^{n,n}$.

Let us finally sketch the resulting (conceptual) algorithm which iteratively computes the semantic solutions for $\bar{\mathcal{A}}_{k-1}$. Given $\bar{\mathcal{A}}_{k-1}$, we would like to compute $\delta^{i,j}$ for $i \in [k,n], j \in [1,n]$. By Corollary 4.5 we already know that $\delta^{i,j} = \delta^{j,j}$ for $i \in [k,n], j \in [1, k-1]$. The remaining characteristic transition functions are then computed level-wise by a two-level fixpoint computation. During the inner-level computation we have fixed some approximant $\delta^{k,k}$ and vary the values of $\delta^{k,k+1}, \ldots, \delta^{k,n}$. The idea is that $(\delta^{k,1}, \ldots, \delta^{k,n})$ together with $\bar{\mathcal{A}}_{k-1}$ defines a dynamic environment $\bar{\mathcal{A}}_k$ for which we can compute the semantic solutions $\theta^{k+1,k+1}, \ldots, \theta^{n,n}$ by induction. We may therefore update $\delta^{k,k+1}, \ldots, \delta^{k,n}$ by $\theta^{k+1,k+1}, \ldots, \theta^{n,n}$, and repeat this iteration until we reach consistency. In the outer-level fixpoint computation we may now update the fixed $\delta^{k,k}$ by evaluating the characteristic transition function for the "unfolding" of Ψ_k in the current setting, and start the inner fixpoint computation again. Our main theorem then states that if we have reached consistency also at the outer-level then the iterative and the semantic solutions for $\bar{\mathcal{A}}_{k-1}$ coincide.

Theorem 4.6. *For any given dynamic environment* $\bar{\mathcal{A}}_k$, *the semantic and the iterative solutions coincides.*

The observation that only the characteristic transition functions on the diagonal have to be taken into account when updating $\delta^{k,k}$ wrt. the current dynamic environment $\bar{\mathcal{A}}_n$, allows us to replace the "conceptual" algorithm used in the correctness proof to the "actual" model-checking algorithm presented in Section 3. This optimization is the key for proving the claimed complexity result.

5 Conclusions and Further Research

In this paper we have presented an iterative, exponential model-checking algorithm for context-free processes which deals with the full modal mu-calculus. This basic algorithm can also be extended to the class of pushdown processes following the lines of [BS95], as well as to the class of regular graphs following the lines of [BQ97], respectively. Essentially, both extensions are obtained by taking into account the arity Q of pushdown processes (i.e. the number of states in the finite control), respectively regular graphs (i.e. the maximal arity of an hyperedge), which yields characteristic property transformers with multiple arguments. For these extensions our algorithm has the worst-time complexity $O(\,|\Phi| * (|Q| * |\mathcal{C}|)^{\mathrm{ad}(\Phi)+1} * 2^{|\Phi|*(\mathrm{ad}(\Phi)+|Q|)}\,)$.

Recently, Walukiewicz presented another model-checker for pushdown processes which uses games [Wal96]. His algorithm has the different complexity estimation $O(\,|\mathcal{C}| * (2^{|Q|*|\Phi|*\mathrm{ad}(\Phi)})^{\mathrm{ad}(\Phi)}\,)$ and behaves hence worse for increasing degrees of alternation depths.

Since our algorithm directly mimics the behavioural intuition behind sequential processes and, in particular, keeps process and formula structure transparent, it gives a direct handle to extending the underlying process structure. Intended future work includes plans to extend model-checking to the class of rational restricted recognizable graphs as introduced in [Cau96], and second, to develop a local variant. Both extensions will exploit the structural transparency of our approach and, in particular, use the framework of dynamic environments.

References

BE97. O. Burkart and J. Esparza. More Infinite Results. In *INFINITY '96*, volume 6 of *ENTCS*, 23 pages. Elsevier Science B.V., 1997.

BQ97. O. Burkart and Y.-M. Quemener. Model-Checking of Infinite Graphs Defined by Graph Grammars. In *INFINITY '96*, volume 6 of *ENTCS*, 15 pages. Elsevier Science B.V., 1997.

Bra96. J.C. Bradfield. The Modal mu-Calculus Alternation Hierarchy is Strict. In *CONCUR '96*, LNCS 1119, pages 233–246. Springer, 1996.

BS92. O. Burkart and B. Steffen. Model Checking for Context-Free Processes. In *CONCUR '92*, LNCS 630, pages 123–137. Springer, 1992.

BS95. O. Burkart and B. Steffen. Composition, Decomposition and Model-Checking of Pushdown Processes. *Nordic Journal of Computing*, 2:89–125, 1995.

BS97. O. Burkart and B. Steffen. Model Checking the Full-Modal Mu-Calculus for Infinite Sequential Processes. Technical Report LFCS-97-355, University of Edinburgh, April 1997.

Cau96. D. Caucal. On Infinite Transition Graphs Having a Decidable Monadic Theory. In *ICALP '96*, LNCS 1099, pages 194–205. Springer, 1996.

CKS92. R. Cleaveland, M. Klein, and B. Steffen. Faster Model Checking for the Modal Mu-Calculus. In *CAV '92*, LNCS 663, pages 410–422, 1992.

Cou90. B. Courcelle. Graph Rewriting: An Algebraic and Logic Approach. In J. van Leeuwen, editor, *Handbook of Theoretical Computer Science*, chapter 5, pages 193–242. Elsevier Science Publisher B.V., 1990.

HS94. H. Hungar and B. Steffen. Local Model-Checking for Context-Free Processes. *Nordic Journal of Computing*, 1(3):364–385, 1994.

Koz83. D. Kozen. Results on the Propositional μ-Calculus. *Theoretical Computer Science*, 27:333–354, 1983.

LBC+94. D.E. Long, A. Browne, E.M. Clarke, S. Jha, and W.R. Marrero. An Improved Algorithm for the Evaluation of Fixpoint Expressions. In *CAV '94*, LNCS 818, pages 338–350. Springer, 1994.

MS85. D.E. Muller and P.E. Schupp. The Theory of Ends, Pushdown Automata, and Second-Order Logic. *Theoretical Computer Science*, 37:51–75, 1985.

Rab69. R.O. Rabin. Decidability of Second-Order Theories and Automata on Infinite Trees. *Transactions of the AMS*, 141:1–35, 1969.

Wal96. I. Walukiewicz. Pushdown Processes: Games and Model-Checking. In *CAV '96*, LNCS 1102. Springer, 1996.

Symbolic Model Checking for Probabilistic Processes

Christel Baier[1], Edmund M. Clarke[2]*, Vasiliki Hartonas-Garmhausen[2],
Marta Kwiatkowska[3] and Mark Ryan[3]**

[1] Fakultät für Mathematik & Informatik
Universität Mannheim
68131 Mannheim, Germany
baier@pi1.informatik.uni-mannheim.de

[2] Department of Computer Science
Carnegie Mellon University
Pittsburgh, PA 15213, USA
{emc,hartonas}@cs.cmu.edu

[3] School of Computer Science
University of Birmingham
Birmingham B15 2TT, UK
{mzk,mdr}@cs.bham.ac.uk

Abstract. We introduce a symbolic model checking procedure for Probabilistic Computation Tree Logic PCTL over labelled Markov chains as models. Model checking for probabilistic logics typically involves solving linear equation systems in order to ascertain the probability of a given formula holding in a state. Our algorithm is based on the idea of representing the matrices used in the linear equation systems by Multi-Terminal Binary Decision Diagrams (MTBDDs) introduced in Clarke *et al* [14]. Our procedure, based on the algorithm used by Hansson and Jonsson [24], uses BDDs to represent formulas and MTBDDs to represent Markov chains, and is efficient because it avoids explicit state space construction. A PCTL model checker is being implemented in Verus [9].

1 Introduction

Probabilistic techniques, and in particular probabilistic logics, have proved successful in the specification and verification of systems that exhibit uncertainty, such as fault-tolerant systems, randomized distributed systems and communication protocols. Models for such systems are variants of probabilistic automata (such as labelled Markov chains used in e.g. [24, 34, 35, 17]), in which the usual (boolean) transition relation is replaced with its probabilistic version given in the form of a Markov probability transition matrix. The probabilistic logics are typically obtained by "lifting" a non-probabilistic logic to the probabilistic case by constructing for each formula ϕ and a real number p in the $[0, 1]$-interval the formula $[\phi]_{\geq p}$ in which p acts as a *threshold for truth* in the sense that for the formula $[\phi]_{\geq p}$ to be satisfied (in the state s) the probability that ϕ holds in s must be *at least* p (see [26, 32, 25] for a different approach). With such logics one can express *quantitative* properties such as "the probability of the message being delivered within t time steps is at least 0.75" (see e.g. the timing or average-case analysis of real-time or randomized distributed systems [24, 23, 5, 6, 2]) or (the more prevalent) *qualitative* properties, for which ϕ is required to be satisfied by almost all executions (which amounts to showing that ϕ is satisfied with probability 1, see e.g. [1, 17, 23, 24, 21, 22, 29, 30, 34]).

---

* This research was sponsored in part by the National Science Foundation under grant no. CCR-8722633, by the Semiconductor Research Corporation under contract 92-DJ-294, and by the Wright Laboratory, Aeronautical Systems Center, Air Force Materiel Command, USAF, the Advanced Research Projects Agency (ARPA) under grant F33615-93-1-1330.

** This research was sponsored in part by the European Union ESPRIT projects ASPIRE and FIREworks, British Telecom, and the Nuffield Foundation.

Much has been published concerning the verification methods for probabilistic logics. Probabilistic extensions of dynamic logic [26] and temporal and modal logics, e.g. [2, 6, 17, 24, 21, 27, 30, 31, 34], and automatic procedures for checking satisfaction for such logics have been proposed. The latter are based on reducing the calculation of the probability of formulas being satisfied to a linear algebra problem: for example, in [24], the calculation of the probability of 'until' formulas is based on solving the linear equation system given by an $n \times n$ matrix where n is the size of the state space. Optimal methods are known (for sequential Markov chains, the lower bound is single exponential in the size of the formula and polynomial in the size of the Markov chain [18]), but these algorithms are not of much practical use when verifying realistic systems. As a result, efficiency of probabilistic analysis lags behind efficient model checking techniques for conventional logics, such as symbolic model checking [11, 12, 10, 8, 15, 28], for which tools capable of tackling industrial scale applications are available (cf. smv). This is undesirable as probabilistic approaches allow one to establish that certain properties hold (in some meaningful probabilistic sense) where conventional model checkers fail, either because the property simply is not true in the state (but holds in that state with some acceptable probability), or because exhaustive search of only a portion of the system is feasible.

The main difficulty with current probabilistic model checking is the need to integrate a linear algebra package with a conventional model checker. Despite the power of existing linear algebra packages, this can lead to inefficient and time consuming computation through the implicit requirement for the construction of the state space. This paper proposes an alternative, which is based on expressing the probability calculations in terms of Multi-Terminal Binary Decision Diagrams (MTBDDs) [16]. MTBDDs are a generalization of (ordered) BDDs in the sense that they allow arbitrary real numbers in the terminal nodes instead of just 0 and 1, and so can provide a compact representation for matrices. As a matter of fact, in [13] MTBDDs have been shown to perform *no worse* than sparse matrices. Thus, converting to MTBDDs ensures smooth integration with a symbolic model checker such as smv and has the potential to outperform sparse matrices due to the compactness of the representation, in the same way as BDDs have outperformed other methods. As with BDDs, the precise time complexity estimates of model checking for MTBDDs are difficult to obtain, but the success of BDDs in practice [8, 28] serves as sufficient encouragement to develop the foundations of MTBDD-based probabilistic model checkers.

In this paper we consider a probabilistic extension of CTL called Probabilistic Computation Tree Logic (PCTL), and give a *symbolic* model checking procedure which avoids the explicit construction of the state space. We use finite-state labelled Markov chains as models. The model checking procedure is based on that of [24, 18], but we use BDDs to represent the boolean formulas, and a suitable combination of BDDs and MTBDDs for probabilistic formulas. Currently, we are implementing the PCTL symbolic model checking in Verus [9]. For reasons of space we omit much detail from this paper, which will be reported in [4]. We assume some familiarity with BDDs, automata on infinite sequences, probability and measure theory [8, 33, 20].

2 Labelled Markov chains

We use discrete time Markov chains as models (we do not consider nondeterminism). Let AP denote a finite set of atomic propositions. A *labelled Markov chain* over a set of atomic propositions AP is a tuple $M = (S, \mathbf{P}, L)$ where S is a finite set of *states*, $\mathbf{P} : S \times S \to [0,1]$ a *transition matrix*, i.e. $\sum_{t \in S} \mathbf{P}(s,t) = 1$ for all $s \in S$, and $L : S \to 2^{AP}$ a *labelling function* which assigns to each state $s \in S$ a set of atomic propositions. We assume that there are 2^n states for some n, and that there are sufficiently many atomic propositions to distinguish them (i.e. $L(s) \neq L(s')$ for all states s, s' with $s \neq s'$). Any labelled Markov chain may be transformed into one satisfying these conditions by adding dummy states and new propositions.

Execution sequences arise by resolving the probabilistic choices. Formally, an *execution sequence* in M is a nonempty (finite or infinite) sequence $\pi = s_0 s_1 s_2, \ldots$ where s_i are states and $\mathbf{P}(s_{i-1}, s_i) > 0$, $i = 1, 2, \ldots$. The first state of π is denoted by $first(\pi)$. $\pi(k)$ denotes the $k+1$-th state of π. An execution sequence π is also called a *path*, and a *full path* iff it is infinite. $Path_\omega(s)$ is the set of full paths π with $first(\pi) = s$. For $s \in S$, let $\Sigma(s)$ be the smallest σ-algebra on $Path_\omega(s)$ which contains the basic cylinders $\{\pi \in Path_\omega(s) : \rho \text{ is a prefix of } \pi\}$ where ρ ranges over all finite execution sequences starting in s. The probability measure $Prob$ on $\Sigma(s)$ is the unique measure with $Prob\{\pi \in Path_\omega(s) : \rho \text{ is a prefix of } \pi\} = \mathbf{P}(\rho)$ where $\mathbf{P}(s_0 s_1 \ldots s_k) = \mathbf{P}(s_0, s_1) \cdot \mathbf{P}(s_1, s_2) \cdot \ldots \cdot \mathbf{P}(s_{k-1}, s_k)$.

Example 1. We consider a simple communication protocol similar to that in [24]. The system consists of three entities: a sender, a medium and a receiver. The sender sends a message to the medium, which in turn tries to deliver the message to the receiver. With probability $\frac{1}{100}$, the messages get lost, in which case the medium tries again to deliver the message. With probability $\frac{1}{100}$, the message is corrupted (but delivered); with probability $\frac{98}{100}$, the correct message is delivered. When the (correct or faulty) message is delivered the receiver acknowledges the receipt of the message. For simplicity, we assume that the acknowledgement cannot be corrupted or lost. We describe the system in a simplified way where we omit all irrelevant states (e.g. the state where the receiver acknowledges the receipt of the correct message).

We use the following four states:

s_{init} the state in which the sender passes the message to the medium

s_{del} the state in which the medium tries to deliver the message

s_{lost} the state reached when the message is lost

s_{error} the state reached when the message is corrupted

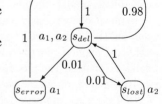

The transition $s_{del} \to s_{init}$ stands for the acknowledgement of the receipt of the correct message, $s_{error} \to s_{init}$ for the acknowledgement of the receipt of the corrupted message. We use two atomic propositions a_1, a_2 and the labelling function $L(s_{init}) = \emptyset$, $L(s_{del}) = \{a_1, a_2\}$, $L(s_{lost}) = \{a_2\}$, $L(s_{error}) = \{a_1\}$. ■

3 Probabilistic branching time temporal logic

In this section we present the syntax and semantics of the logic PCTL (Probabilistic Computation Tree Logic) introduced by Hansson & Jonsson [24][4]. PCTL is a probabilistic extension of CTL which allows one to express quantitative properties of probabilistic processes such as "the system terminates with probability at least 0.75". PCTL contains atomic propositions and the operators: next-step X and until U. The operators X and U are used in connection with an interval of probabilities. The syntax of PCTL is as follows:

$$\Phi ::= tt \mid a \mid \Phi_1 \wedge \Phi_2 \mid \neg \Phi \mid [X\Phi]_{\sqsupseteq p} \mid [\Phi_1 U \Phi_2]_{\sqsupseteq p}$$

where a is an atomic proposition, $p \in [0,1]$, $\sqsupseteq$ is either $\geq$ or $>$. Formulas of the form $X\Phi$ or $\Phi_1 U \Phi_2$, where Φ, Φ_1, Φ_2 are PCTL formulas, are called *path formulas*. PCTL formulas are interpreted over the states of a labelled Markov chain, whereas path formulas are interpreted over paths. The subscript $\sqsupseteq p$ denotes that the probability of paths starting in the current state fulfilling the path formula is $\sqsupseteq p$. Thus, PCTL is like CTL, except that the path operators A and E in CTL have been replaced by the operator $[\cdot]_{\sqsupseteq p}$. The usual derived constants and operators are: $f\!f = \neg tt$, $\Phi_1 \vee \Phi_2 = \neg(\neg \Phi_1 \wedge \neg \Phi_2)$, $\Phi_1 \to \Phi_2 = \neg \Phi_1 \vee \Phi_2$. Operators for modelling "eventually" or "always" can be derived by: $[\Diamond \Phi]_{\geq p} = [tt U \Phi]_{\geq p}$, $[\Box \Phi]_{\geq p} = \neg[\Diamond \neg \Phi]_{>1-p}$, and similarly for $[\cdot]_{>p}$.

Let $M = (S, \mathbf{P}, L)$ be a labelled Markov chain. The satisfaction relation $\models \; \subseteq S \times PCTL$ is given by

$s \models tt$ for all $s \in S$ $\quad s \models \Phi_1 \wedge \Phi_2$ iff $s \models \Phi_1$ and $s \models \Phi_2$

$s \models a$ iff $a \in L(s)$ $\quad s \models \neg \Phi$ iff $s \not\models \Phi$

$s \models [X\Phi]_{\sqsupseteq p}$ iff $Prob\{\pi \in Path_\omega(s) : \pi \models X\Phi\} \sqsupseteq p$

$s \models [\Phi_1 U \Phi_2]_{\sqsupseteq p}$ iff $Prob\{\pi \in Path_\omega(s) : \pi \models \Phi_1 U \Phi_2\} \sqsupseteq p$

$\pi \models X\Phi$ iff $\pi(1) \models \Phi$

$\pi \models \Phi_1 U \Phi_2$ iff there exists $k \geq 0$ with $\pi(i) \models \Phi_1, i = 0, 1, \ldots, k-1$ and $\pi(k) \models \Phi_2$.

For a path formula f the set $\{\pi \in Path_\omega(s) : \pi \models f\}$ is measurable [34, 18]. If $s \models \Phi$ then we say s satisfies Φ (or Φ holds in s). The truth value of formulas involving the linear time quantifiers $\Diamond$ and $\Box$ can be derived:

$s \models [\Diamond \Phi]_{\sqsupseteq p}$ iff $Prob\{\pi \in Path_\omega(s) : \pi(k) \models \Phi$ for some $k \geq 0\} \sqsupseteq p$

$s \models [\Box \Phi]_{\sqsupseteq p}$ iff $Prob\{\pi \in Path_\omega(s) : \pi(k) \models \Phi$ for all $k \geq 0\} \sqsupseteq p$.

Given a probabilistic process $\mathcal{P}$, described by a labelled Markov chain $M = (S, \mathbf{P}, L)$ with an initial state s, we say $\mathcal{P}$ satisfies a PCTL formula Φ iff $s \models \Phi$. For instance, if a is an atomic proposition which stands for termination and $\mathcal{P}$ satisfies $[\Diamond a]_{\geq p}$ then $\mathcal{P}$ terminates with probability at least p.

4 Multi-terminal binary decision diagrams

Ordered Binary Decision Diagrams (BDDs) [7, 8, 15, 28] are a compact representation of boolean functions $f : \{0,1\}^n \to \{0,1\}$. They are based on the canonical representation of the binary tree of the function as a directed graph obtained through folding

---

[4] For simplicity we omit the bounded 'until' operator of [24].

internal nodes representing identical subfunctions (subject to an ordering of the variables to guarantee uniqueness of the representation) and using 0 and 1 as leaves. In [16] it is shown how one can generalize BDDs to cogently and efficiently represent matrices in terms of so-called *multi-terminal* binary decision diagrams (MTBDDs).

Formally, MTBDDs can be defined as follows. Let $x_1, \ldots, x_n$ be distinct variables, which we order by $x_i < x_j$ iff $i < j$. A *multi-terminal binary decision diagram* (MTBDD) over $(x_1, \ldots, x_n)$ is a rooted, directed graph with vertex set V containing two types of vertices, *nonterminal* and *terminal*. Each nonterminal vertex v is labelled by a variable $var(v) \in \{x_1, \ldots, x_n\}$ and two children $left(v), right(v) \in V$. Each terminal vertex v is labelled by a real number $value(v)$. For each nonterminal node v, we require $var(v) < var(left(v))$ if $left(v)$ is nonterminal, and similarly, $var(v) < var(right(v))$ if $right(v)$ is nonterminal. A suitable adaptation of the operator $REDUCE(\cdot)$ [7] yields an operator which accepts an MTBDD as its input and returns the corresponding reduced MTBDD.

Each MTBDD Q over $\{x_1, \ldots, x_n\}$ represents a function $F_Q : \{0,1\}^n \to I\!R$, and, vice versa, each function $F : \{0,1\}^n \to I\!R$ can be described by a unique reduced MTBDD over $(x_1, \ldots, x_n)$. In the sequel, by the MTBDD for a function $F : \{0,1\}^n \to I\!R$ we mean the unique reduced MTBDD Q with $F_Q = F$. If all terminal vertices are labelled by 0 or 1, i.e. if the associated function F_Q is a boolean function, the MTBDD specializes to a BDD over $(x_1, \ldots, x_n)$.

MTBDDs are used to represent D–valued matrices as follows. Consider a $2^m \times 2^m$–matrix A. Its elements a_{ij} can be viewed as the values of a function $f_A : \{1, \ldots 2^m\} \times \{1, \ldots 2^m\} \to D$, where $f_A(i,j) = a_{ij}$. Using the standard encoding $c : \{0,1\}^m \to \{1, \ldots 2^m\}$ of boolean sequences of length m into the integers, this function may be interpreted as a D–valued boolean function $f : \{0,1\}^m \to D$ where $f(x,y) = f_A(c(x), c(y))$ for $x = (x_1 \ldots x_m)$ and $y = (y_1 \ldots y_m)$. This transformation now allows matrices to be represented as MTBDDs. In order to obtain an efficient MTBDD–representation, the variables of f are permuted. Instead of the MTBDD for $f(x_1 \ldots x_m, y_1 \ldots y_m)$, we use the MTBDD obtained from $f(x_1, y_1, x_2, y_2, \ldots x_m, y_m)$. This convention imposes a recursive structure on the matrix from which efficient recursive algorithms for all standard matrix operations are derived [16].

4.1 Representing labelled Markov chains by MTBDDs

To represent the transition matrix of a labelled Markov chain by a MTBDD we abstract from the names of states and instead, similarly to [8, 15], use binary tuples of atomic propositions that are true in the state. Let $M = (S, \mathbf{P}, L)$ be a labelled Markov chain. We fix an enumeration $a_1, \ldots, a_n$ of the atomic propositions and identify each state s with the boolean n-tuple $e(s) = (b_1, \ldots, b_n)$ where $b_i = 1$ iff $a_i \in L(s)$. In what follows, we identify $\mathbf{P}$ with the function $F : \{0,1\}^{2n} \to [0,1]$, $F(x_1, y_1, \ldots, x_n, y_n) = \mathbf{P}((x_1, \ldots, x_n), (y_1, \ldots, y_n))$, and represent M by the MTBDD for $\mathbf{P}$ over $(x_1, y_1, \ldots, x_n, y_n)$. The associated MTBDD is denoted by P.

Example 2. For the system in Example 1 we use the encoding $e(s_{init}) = 00$, $e(s_{del}) = 11$, $e(s_{lost}) = 01$ $e(s_{error}) = 10$. The values of the matrix $\mathbf{P}$, the function F and the MTBDD P for F are are given by:

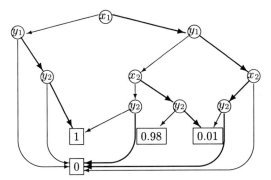

$$F(x_1, y_1, x_2, y_2) = \begin{cases} 1 & : \text{if } x_1y_1x_2y_2 \in \{0101, 0111, 1000\} \\ \frac{1}{100} & : \text{if } x_1y_1x_2y_2 \in \{1011, 1110\} \\ \frac{98}{100} & : \text{if } x_1y_1x_2y_2 = 1010 \\ 0 & : \text{otherwise.} \end{cases}$$

(The thick lines stand for the "right" edges, the thin lines for the "left" edges.) ■

4.2 Operators on MTBDDs

Our model checking algorithm makes use of several operators on MTBDDs proposed in Bryant [7] and Clarke *et al* [14]. We briefly describe them below.

Operator $BDD(\cdot)$: takes an MTBDD Q and an interval I, and returns the BDD representing the function $F(\bar{x}) = 1$ if $F_Q(\bar{x}) \in I$, else $F(\bar{x}) = 0$. We obtain $B = BDD(Q, I)$ from Q by changing the values of the terminal vertices (into 1 or 0 depending on whether or not $value(v) \in I$) and applying Bryant's reduction procedure $REDUCE(\cdot)$. We write $BDD(Q, > p)$ rather than $BDD(Q,]p, \infty[)$ and $BDD(Q, \geq p)$ rather than $BDD(Q, [p, \infty[)$.

Operator $APPLY(\cdot)$: allows elementwise application of the binary operator op to two MTBDDs. If op is a binary operator on reals (e.g. multiplication $*$ or minus $-$) and Q_1, Q_2 are MTBDDs over $\bar{x}$ then $APPLY(Q_1, Q_2, op)$ yields a MTBDD over $\bar{x}$ which represents the function $f(\bar{x}) = f_{Q_1}(\bar{x}) \, op \, f_{Q_2}(\bar{x})$.

Operator $COMPOSE^k(\cdot)$: This operator allows the composition of a real function $F : \{0,1\}^{n+k} \to \mathbb{R}$ and boolean functions $G_i : \{0,1\}^n \to \{0,1\}, i = 1, \ldots, k$ giving $H(\bar{x}) = F(\bar{x}, G_1(\bar{x}), \ldots, G_k(\bar{x}))$.

Matrix and vector operators: The standard operations on matrices and vectors have corresponding operations on the MTBDDs that represent them [13]. If MTBDDs A and Q over $2n$ and n variables represent the matrix $\mathbf{A}$ and vector $\mathbf{q}$ respectively, then $MV\_MULTI(A, Q)$ denotes the MTBDD over n variables that represents the vector $\mathbf{A} \cdot \mathbf{q}$.

Operator $SOLVE(\cdot)$: [8] presents a method to decompose a regular matrix $\mathbf{A}$ into a lower and upper triangular matrices and a permutation matrix. Using this LU-decomposition we can obtain an operator $SOLVE(A, Q)$ that takes as its input a MTBDD A over $2n$ variables where the corresponding matrix $\mathbf{A}$ is regular and a MTBDD Q over n variables which represents a vector $\mathbf{q}$, and returns a MTBDD Q' over n variables which represents the unique solution of the linear equation system $\mathbf{A} \cdot \mathbf{x} = \mathbf{q}$. Alternatively, we can use iterative techniques to solve the equations; our experiments indicate that this performs better.

4.3 Description of (MT)BDDs by relational terms of the μ-calculus

We will use the μ-calculus as a notation for describing (MT)BDDs. In the algorithm in the next section, all our (MT)BDDs are either over $2n$ variables (in which case they represent $2^n \times 2^n$ matrices), or over n variables (in which case they represent vectors of length 2^n). For example, if B, C are BDDs over n variables and $\overline{u} = (u_1, \ldots, u_n)$, $\overline{v} = (v_1, \ldots, v_n)$, then $D = \lambda \overline{u}\,\overline{v}\,[B(\overline{u}) \wedge C(\overline{v})]$ is a BDD over $2n$ variables; if B, C represent the vectors $(b_i)_{1 \leq i \leq n}$ and $(c_i)_{1 \leq i \leq n}$ respectively, then D represents the matrix whose element in the ith row and jth column is $b_i \wedge c_j$. The BDD $E = \lambda \overline{u}\,[B(\overline{u}) \wedge C(\overline{u})]$ is a BDD over n variables, representing the vector $(b_i \wedge c_i)_{1 \leq i \leq n}$.

We write $TRUE$ for the BDD over n variables which returns 1 in all cases of its arguments. We write $\neg B$ instead of $\lambda \overline{x}[\neg B(\overline{x})]$, and $B_1 \wedge B_2$ for the BDD $\lambda \overline{x}[B_1(\overline{x}) \wedge B_2(\overline{x})]$. If $\overline{x} = (x_1, \ldots, x_n)$, $\overline{y} = (y_1, \ldots, y_n)$ then $\overline{x} = \overline{y}$ abbreviates the formula $\bigwedge_{1 \leq i \leq n}(x_i \leftrightarrow y_i)$.

We require one further operator. If the labelled Markov chain $M = (S, \mathbf{P}, L)$ is represented by a MTBDD P as described in Section 4.1, and B_1, B_2 are BDDs that represent the characteristic functions of subsets S_1, S_2 of S, then $REACH(B_1, B_2, BDD(P, > 0))$ represents the set of states $s \in S$ from which there exists an execution sequence $s = s_0, s_1, \ldots, s_k$ with $k \geq 0$ and $s_0, \ldots, s_{k-1} \in S_1$, $s_k \in S_2$, and which is used in the operator $UNTIL(\cdot)$ defined in Section 5.

Operator $REACH(\cdot)$ Let B_1, B_2 be BDDs with n variables and T a BDD with $2n$ variables. We define $REACH(B_1, B_2, T)$ to be the BDD over n variables which is given by the μ-calculus formula $\mu Z\,\lambda \overline{x}\,[B_2(\overline{x}) \vee (B_1(\overline{x}) \wedge \exists \overline{y}[Z(\overline{y}) \wedge T(\overline{x}, \overline{y})])]$. This operator uses the method of [8] to obtain the BDD for a term involving the least fixed point operator μ.

5 Model checking for PCTL

Our model checking algorithm for PCTL is based on established BDD techniques (i.e. converting boolean formulas to their BDD representation), which it combines with a new method, namely expressing the probability calculation for the probabilistic formulas in terms of MTBDDs. In the case of $[X\Phi]_{\sqsupseteq p}$ the probability is calculated by multiplying the transition matrix by the boolean vector set to 1 iff the state satisfies Φ, whereas for $[\Phi_1 U \Phi_2]_{\sqsupseteq p}$ we derive an operator called $UNTIL(\cdot)$, based on [24], which we express in terms of MTBDDs.

Let $M = (S, \mathbf{P}, L)$ be a labelled Markov chain which is represented by a MTBDD P over $2n$ variables as described in Section 4.1. For each PCTL formula Φ, we define a BDD $B[\Phi]$ over $\overline{x} = (x_1, \ldots, x_n)$ that represents $Sat(\Phi) = \{s \in S : s \models \Phi\}$. We compute the BDD representation $B[\Phi]$ of a PCTL formula Φ by structural induction:
$B[tt] = TRUE \qquad B[a_i] = \lambda \overline{x}\,[x_i]$
$B[\neg \Phi] = \neg B[\Phi] \qquad B[\Phi_1 \wedge \Phi_2] = B[\Phi_1] \wedge B[\Phi_2]$

$B[\,[X\Phi]_{\sqsupseteq p}\,] \;=\; BDD\,(\;MV\_MULTI(P,B[\Phi]),\sqsupseteq p\,)$
$B[\,[\Phi_1 U \Phi_2]_{\sqsupseteq p}\,] \;=\; BDD\,(\,UNTIL(B[\Phi_1],B[\Phi_2],P),\sqsupseteq p\,)\,)$

The operator $UNTIL(B[\Phi_1],B[\Phi_2],P)$ assigns to each state $s \in S$ the *probability* of the set of full paths from s satisfying $\Phi_1 U \Phi_2$; formally, it represents the function $S \to [0,1]$, $s \mapsto p_s$, where $p_s = Prob\,\{\pi \in Path_\omega(s) : \pi \models \Phi_1 U \Phi_2\}$. Our method for computing p_s is based on the partition of S introduced in [24, 18], but we must compute with BDDs. We first compute the set $V = \{s \in S : p_s > 0\}$ and then set $V' = V \setminus Sat(\Phi_2)$. We then have: $p_s = 1$ if $s \models \Phi_2$; $p_s = 0$ if $s \notin V$; and for the remaining cases (i.e. those such that $s \in V'$)

$$p_s = \sum_{t \in V'} \mathbf{P}(s,t) \cdot p_t + \sum_{t \in Sat(\Phi_2)} \mathbf{P}(s,t) \cdot p_t + \sum_{t \in S \setminus V} \mathbf{P}(s,t) \cdot p_t.$$

In the second term, each $p_t = 1$ and in the third term, each $p_t = 0$. Therefore p_s ($s \in V'$) satisfies a $|V'|$-dimensional equation system of the form $\mathbf{x} = \mathbf{Ax} + \mathbf{b}$, or equivalently $(\mathbf{I} - \mathbf{A})\mathbf{x} = \mathbf{b}$ where $\mathbf{I}$ is the $|V'| \times |V'|$ identity matrix. One can show this system has a unique solution using the method in [24, 18].

We now demonstrate how $UNTIL(\cdot)$ can be expressed in terms of MTBDDs. Let $B_i = B[\Phi_i]$, $i = 1,2$. The set V is given by the BDD $B = REACH(B_1, B_2, BDD(P, > 0))$, V' by $B' = \lambda \bar{x}\,[B(\bar{x}) \wedge \neg B_2(\bar{x})]$. In order to avoid the BDD for the "new" transition matrix $\mathbf{A}$ with $\lceil \log_2 |V'| \rceil$ variables, we instead reformulate the equation in terms of the matrix $\mathbf{P}' = (p'_{s,t})_{s,t \in S}$ which is given by: $p'_{s,t} = \mathbf{P}(s,t)$ if $s, t \in V'$ and $p'_{s,t} = 0$ in all other cases. The MTBDD P' for $\mathbf{P}'$ can be obtained from the MTBDD P representing the Markov transition matrix. The following lemma shows that $\mathbf{I} - \mathbf{P}'$ is regular (we omit the proof).

Lemma 1. *Let V', $\mathbf{P}'$, $\mathbf{I}$ be as as above. Then, $\mathbf{I} - \mathbf{P}'$ is regular. The unique solution $\mathbf{x} = (x_s)_{s \in S}$ of the linear equation system $(\mathbf{I} - \mathbf{P}') \cdot \mathbf{x} = \mathbf{q}$ where $\mathbf{q} = (q_s)$, $q_s = \sum_{t \in Sat(\Phi_2)} \mathbf{P}(s,t)$ satisfies: $x_s = p_s$ if $s \in V'$.*

The algorithm for the operator $UNTIL(\cdot)$ is shown in Figure 1. It first calculates the MTBDDs B and B', for V and V'. B^2 is used as a mask to obtain P' from P; it sets to 0 the entries not corresponding to states in V'. We next calculate the MTBDD Q for the vector $\mathbf{q}$, and use the operator $SOLVE(\cdot)$ to obtain the MTBDD Q' satisfying $F_{Q'}(s) = p_s$ for all $s \in V'$. The result, the MTBDD Q'' for the vector $\mathbf{p} = (p_s)_{s \in S}$, is obtained from the MTBDD for the function $F(\bar{x}) = \max\{\,F_{B_2}(\bar{x}),\,F_{Q'}(\bar{x}) \cdot F_{B'}(\bar{x})\,\}$ which uses Q' for all $s \in V'$ and ensures that 1 is returned as the probability of the states already satisfying Φ_2.

Example 3. Let $\Phi \;=\; [\,try\_to\_deliver\;U\;correctly\_delivered\,]_{\geq 0.9}$ where $try\_to\_deliver = a_2$ and $correctly\_delivered = \neg a_1 \wedge \neg a_2$. We consider the system in Example 1. Our algorithm first computes the BDDs B_1 for $Sat(try\_to\_deliver) = \{s_{del}, s_{lost}\}$, B_2 for $Sat(correctly\_delivered) = \{s_{init}\}$, and then applies Algorithm $UNTIL(B_1, B_2, P)$. $V = \{s_{init}, s_{del}, s_{lost}\}$ is represented by the BDD B, $V' = \{s_{del}, s_{lost}\}$ by the BDD B'. Thus, B^2, P' and A stand for the matrices

$$\mathbf{B}^2 = \begin{pmatrix} 0 & 0 & 0 & 0 \\ 0 & 1 & 0 & 1 \\ 0 & 0 & 0 & 0 \\ 0 & 1 & 0 & 1 \end{pmatrix} \quad \mathbf{P}' = \begin{pmatrix} 0 & 0 & 0 & 0 \\ 0 & 0 & 0 & 1 \\ 0 & 0 & 0 & 0 \\ 0 & \frac{1}{100} & 0 & 0 \end{pmatrix} \quad \mathbf{A} = \begin{pmatrix} 1 & 0 & 0 & 0 \\ 0 & 1 & 0 & -1 \\ 0 & 0 & 1 & 0 \\ 0 & -\frac{1}{100} & 0 & 1 \end{pmatrix}$$

```
Algorithm: UNTIL(B₁, B₂, P)
```

Input: A labelled Markov chain represented by a MTBDD P over $2n$ variables,
          BDDs B_1, B_2 over n variables
Output: MTBDD X over n variables which represents the function that assigns to each
          state the probability of a path from the state reaching a B_2-state via an execution
          sequence through B_1-states
Method: $B := REACH(B_1, B_2, BDD(P, > 0)); B' := \lambda \bar{x} \, [\, B(\bar{x}) \land \neg B_2(\bar{x}) \,];$
          $B^2 := \lambda x_1 y_1 \ldots x_n y_n \, [B'(x_1, \ldots, x_n) \land B'(y_1, \ldots, y_n)];$
          $P' := APPLY(P, B^2, *); I := \lambda x_1 y_1 \ldots x_n y_n \, [\bar{x} = \bar{y}];$
          $A := APPLY(I, P', -); Q := MV\_MULTI(P, B_2);$
          $Q' := SOLVE(A, Q); Q'' := APPLY(B_2, APPLY(Q', B', *), \max);$
          Return($REDUCE(Q'')$).

Fig. 1. Algorithm $UNTIL(B_1, B_2, P)$

B_2 (viewed as a vector) is $\mathbf{q}_2 = (1, 0, 0, 0)$. Thus, Q is the MTBDD for the vector $\mathbf{P} \cdot \mathbf{q}_2 = (0, 0, 1, 0.98)$. We solve the linear equation system

$$\begin{pmatrix} 1 & 0 & 0 & 0 \\ 0 & 1 & 0 & -1 \\ 0 & 0 & 1 & 0 \\ 0 & -\frac{1}{100} & 0 & 1 \end{pmatrix} \cdot \mathbf{x} = \begin{pmatrix} 0 \\ 0 \\ 1 \\ \frac{98}{100} \end{pmatrix}$$

which yields the solution $\mathbf{x} = (0, \frac{98}{99}, 1, \frac{98}{99})$ (represented by the MTBDD Q'). Moreover, the MTBDD $APPLY(Q', B', *)$ can be identified with the vector $(0, \frac{98}{99}, 0, \frac{98}{99})$. $UNTIL(B_1, B_2, P)$ and the BDD $B[\Phi]$ are of the following form.

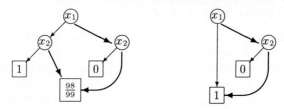

Thus, $B[\Phi]$ represents the characteristic function for $Sat(\Phi) = \{s_{init}, s_{del}, s_{lost}\}$. ∎

6 Implementing PCTL model checking

We are integrating PCTL symbolic model checking within Verus [9], which is a tool specifically designed for the verification of finite-state real-time systems. Verus has been used already to verify several interesting real-time systems: an aircraft controller, a medical monitor, the PCI local bus, and a robotics controller. These examples have not been originally modeled using probabilities. However, these systems exhibit behaviors which can best be described probabilistically. The integration of PCTL model checking with Verus allows us to verify stochastic properties of these and other interesting applications.

The Verus language is an imperative language with a syntax resembling that of the C language with additional special primitives to express timing aspects such as deadlines, priorities, and delays. An important feature of Verus is the use of the `wait` statement to control the passage of time. In Verus time only passes when a wait statement is executed: non-wait statements execute in zero time. This feature allows a more accurate control of time and leads to models with less states, since consecutive statements not separated by a `wait` statement are compiled into a single state. To describe probabilistic transitions we extend the Verus language with the probabilistic `select` statement.

From the Verus description of the application, the tool generates automatically a labeled state-transition graph and the corresponding transition probability matrix using BDDs and MTBDDs respectively.

The first experimental results of our PCTL symbolic model checking implementation are promising: Parrow's Protocol (which is of a similar size to Example 1) can be verified in less than a second. We have modeled a fault tolerant system [23, p. 168–171] with three processors that has about 35000 reachable states (out of 10^8 states). A safety property of this system took only a few seconds to check. Next we plan to evaluate how well PCTL symbolic model checking performs as a formal verification tool in real applications by modeling industrial size systems.

7 Concluding remarks and further directions

We have proposed a symbolic model checking procedure for the logic PCTL which we are implementing using MTBDDs in Verus, thus forming the basis of an efficient tool for verifying probabilistic systems. Our algorithm can be extended to cater for "bounded until" of [24] which is useful in timing analysis of systems. We expect that MTBDDs can be used to derive PCTL* model checking by applying the methods of [18]. Likewise, testing of probabilistic bisimulation and simulation [3, 19] can be implemented using MTBDDs. An extension to the case of infinite state systems, perhaps by appropriate combination with induction, as well as a generalization to allow non-determinism, would be desirable.

References

1. R. Alur, C. Courcoubetis, D. Dill. Verifying Automata Specifications of Probabilistic Real-Time Systems. In *Proc. Real-Time: Theory and Practice*, LNCS 600, pp 27-44, Springer, 1991.
2. L. de Alfaro. Formal Verification of Performance and Reliability of Real-Time Systems. Techn. Report, Stanford University, 1996.
3. C. Baier. Polynomial Time Algorithms for Testing Probabilistic Bisimulation and Simulation. In *Proc. CAV'96*, LNCS 1102, pp 38-49, Springer, 1996.
4. C. Baier, S. Campos, E. Clarke, V. Hartonas-Garmhausen, M. Kwiatkowska, M. Minea, and M. Ryan. Probabilistic model checking using multi terminal binary decision diagrams. In preparation.
5. C. Baier, M. Kwiatkowska. Model Checking for a Probabilistic Branching Time Logic with Fairness. Techn. Report CSR-96-12, University of Birmingham, 1996.
6. A. Bianco, L. de Alfaro. Model Checking of Probabilistic and Nondeterministic Systems. In *Proc. Foundations of Software Technology and Theoretical Computer Science*, LNCS 1026, pp 499-513, Springer, 1995.
7. R. Bryant. Graph-Based Algorithms for Boolean Function Manipulation. *IEEE Transactions on Computers*, C-35(8), pp 677-691, 1986.

8. J. Burch, E. Clarke, K. McMillan, D. Dill, L. Hwang. Symbolic Model Checking: 10^{20} States and Beyond. *Information and Computation*, 98(2), pp 142-170, 1992.
9. S. V. Campos, E. M. Clarke, W. Marrero, and M. Minea. Verus: a tool for quantitative analysis of finite-state real-time systems. In *Proc. Workshop on Languages, Compilers and Tools for Real-Time Systems*, 1995.
10. E. M. Clarke and E. A. Emerson. Synthesis of synchronization skeletons for branching time temporal logic. In D. Kozen, eds, *Proc. Logic of Programs*, LNCS 131, Springer, 1981.
11. E. M. Clarke, E. A. Emerson, and A. P. Sistla. Automatic verification of finite-state concurrent systems using temporal logic specifications: A practical approach. In *Proc. 10th Annual Symp. of Programming Languages*, 1983.
12. E. M. Clarke, E. A. Emerson, and A. P. Sistla. Automatic verification of finite-state concurrent systems using temporal logic specifications. *ACM Trans. Programming Lnaguages and Systems*, 1(2), 1986.
13. E. Clarke, M. Fujita, P. McGeer, J. Yang, and X. Zhao. Multi-terminal binary decision diagrams: An efficient data structure for matrix representation. In *IWLS '93: International Workshop on Logic Synthesis, Tahoe City*, May 1993.
14. E. Clarke, M. Fujita, X. Zhao. Multi-Terminal Binary Decision Diagrams and Hybrid Decision Diagrams. In T. Sasao and M. Fujita, eds, *Representations of Discrete Functions*, pp 93-108, Kluwer Academic Publishers, 1996.
15. E. Clarke, O. Grumberg, D. Long. Verfication Tools for Finite-State Concurrent Programs. In *Proc. A Decade of Concurrency*, LNCS 803, pp 124-175, Springer, 1993.
16. E. M. Clarke, K. L. McMillan, X. Zhao, M. Fujita, and J. Yang. Spectral transforms for large boolean functions with applications to technology mapping. In *Proc. 30th ACM/IEEE Design Automation Conference*, pp 54–60, IEEE, 1993.
17. C. Courcoubetis, M. Yannakakis. Verifying Temporal Properties of Finite-State Probabilistic Programs. In *Proc. FOCS'88*, pp 338–345, IEEE, 1988.
18. C. Courcoubetis, M. Yannakakis. The Complexity of Probabilistic Verification. *J. ACM*, 42(4), pp 857-907, 1995.
19. R. Enders, T. Filkorn, D. Taubner. Generating BDDs for Symbolic Model Checking in CCS. *Distributed Computing*, 6, 1993.
20. P. Halmos. *Measure Theory*, Springer, 1950.
21. S. Hart, M. Sharir. Probabilistic Temporal Logic for Finite and Bounded Models. In *Proc. 16th ACM Symposium on Theory of Computing*, pp 1-13, 1984.
22. S. Hart, M. Sharir, A. Pnueli. Termination of Probabilistic Concurrent Programs. *ACM Trans. Programming Languages and Systems*, 5, pp 356-380, 1983.
23. H. Hansson. *Time and Probability in Formal Design of Distributed Systems*, Elsevier, 1994.
24. H. Hansson, B. Jonsson. A Logic for Reasoning about Time and Probability. *Formal Aspects of Computing*, 6, pp 512-535, 1994.
25. M. Huth, M. Kwiatkowska. Quantitative Analysis and Model Checking, In *Proc. LICS'97*, IEEE Computer Society Press, 1997.
26. D. Kozen. A Probabilistic PDL, *JCSS*, 30(2), pp 162-178, 1985.
27. K. Larsen, A. Skou. Bisimulation through Probabilistic Testing. *Information and Computation*, 94, pp 1-28, 1991.
28. K. McMillan. *Symbolic Model Checking: An Approach to the State Explosion Problem*, Kluwer Academic Publishers, 1993.
29. A. Pnueli, L. Zuck. Verification of Multiprocess Probabilistic Protocols. *Distributed Computing*, 1(1), pp 53-72, 1986.
30. A. Pnueli, L. Zuck. Probabilistic Verification. *Information and Computation*, 103, pp 1-29, 1993.
31. R. Segala, N. Lynch. Probabilistic Simulations for Probabilistic Processes. In *Proc. CONCUR*, LNCS 836, pp 481-496, Springer, 1994.
32. K. Seidel C. Morgan, A. McIver and J.W. Sanders. Probabilistic Predicate Transformers. Techn. Report PRG-TR-4-95, Oxford University Computing Laboratory, 1995.
33. W. Thomas. Automata on Infinite Objects. In *Handbook of Theoretical Computer Science*, Vol. B, pp 135-191, North-Holland, 1990.
34. M. Vardi. Automatic Verification of Probabilistic Concurrent Finite-State Programs. In *Proc. FOCS'85*, pp 327-338, IEEE, 1985.
35. M. Vardi, P. Wolper. An Automata-Theoretic Approach to Automatic Program Verification. In *Proc. LICS'86*, pp 332-344, IEEE Computer Society Press, 1986.

On the Concentration of the Height of Binary Search Trees

J. M. Robson[1]

LaBRI
Université Bordeaux 1
robson@labri.u-bordeaux.fr

Abstract. The expectation of the absolute value of the difference between the heights of two random binary search trees of n nodes is less than 6.25 for infinitely many n. Given a plausible assumption, this expectation is less than 4.96 for all but a finite number of values of n.

1 Introduction

A binary search tree (BST) of n nodes is constructed from n distinct keys in random order by inserting each key in turn into an initially empty tree by the familiar algorithm which inserts a key into an empty tree by constructing a new root node with this key and otherwise inserts the key into the left or right subtree depending on whether it is smaller or larger than the key at the root. Two other equivalent definitions are often useful in considering the shape or in particular the height of such a tree:

- A random tree of n nodes is empty if n is zero and otherwise consists of a root node and a left subtree of l nodes and a right subtree of $n - 1 - l$ nodes where l is an integer chosen uniformly on $0 \ldots n-1$; the subtrees are constructed in the same way, all the random choices being independent.
- The i-th node is inserted into the tree by choosing one of the i external nodes of the tree, each with the same probability $1/i$, and replacing it by a new internal node. Hence we have the important result that the probability of this insertion increasing the height of the tree is $1/i$ times the number of external nodes at the deepest level containing any external nodes, or alternatively $2/i$ times the number of internal nodes at the deepest level containing any internal nodes; we call these internal nodes at the deepest level *critical nodes*.

We are interested in the distribution of the random variable $h(n)$ which is the height of a tree constructed in any of these ways. $h(n)$ is also the stack depth used by a straightforward version of Quicksort to sort n randomly ordered distinct values.

The mean value of $h(n)$ is known to be close to $c \log n$ where $c \approx 4.3011$ is the larger root of $c = 2e^{1-1/c}$. An upper bound of the form $(c + o(1)) \log n$ was shown in [4]; a lower bound of the form $(c - o(1)) \log n$ was shown in [1];

and finally the height was shown with high probability to lie within bounds $c \log n \pm O(\log \log n)$ in [2].

Direct calculation for small to moderate values of n and random construction of larger trees [3] have shown that the variance of the height remains small for quite large n and shows no sign of diverging. To date the only explanation of these results has been an upper bound of $O(\log^2 \log n)$ on the variance in [2].

In this paper we will show that there are indefinitely large values of n for which $E[|h(n)-E[h(n)]|]$ is less than 6.25. Although this does not prove anything about the variance and does necessarily not apply to all n, it suggests very strongly that the distribution remains tightly concentrated around its mean. If we make a simple and plausible assumption about the convergence of the number of critical nodes, we can both strengthen the bound and show that the conclusion applies to all sufficiently large n.

In section 2 we will prove a very weak version of the main theorem which we hope will illustrate the essential (and very simple) ideas in the clearest possible way. In section 3 we will give the strongest form that we yet know of the theorem. In section 4 we show briefly how the theorem can be strengthened further if we assume that the expected number of critical nodes converges. Finally in section 5 we sketch some directions for further work.

2 A weak upper bound

We take $h_1(n)$ and $h_2(n)$ as two independent random variables each distributed as the height of an n node BST. Let c be the limit of $E[h(n)]/\log n$. Let ϵ be an arbitrary positive number.

Theorem 1. $E[|h_1(n) - h_2(n)|] < 6c \log 3 - 6 + \epsilon$ infinitely often.

(It follows immediately that $E[|h(n)-E[h(n)]|] < 6c \log 3 - 6 + \epsilon$ ($\approx 22.417 + \epsilon$) infinitely often.)

Proof by contradiction: Suppose the contrary. Then for large enough N, $E[|h_1(n) - h_2(n)|] \geq 6c \log 3 - 6 + \epsilon$ for all n greater than N. Now choose a ν greater than N such that $E[h(3\nu)] < E[h(\nu)] + c \log 3 + \epsilon/6$. (Infinitely many such ν exist since $E[h(n)]/\log n \to c$.)

Consider a random tree of size 3ν and the following algorithm to choose one of its immediate subtrees (L is its larger immediate subtree and S its smaller):

```
if |L| ≥ 2ν
then
        choose L
else
        if the height of S is greater than that of L
        then
                choose S
        else
                choose L
        fi
fi
```

Note that the probability of taking the first case is 2/3.

Consider the height of the subtree chosen:

In case 1 we choose a random subtree of size greater than ν: $height > E[h(\nu)]$;

For the case of $|L| < 2\nu$, it is clear that the mean height is greater than it would be for two subtrees of size ν, namely $E[max(h_1(\nu), h_2(\nu))] = E[(h_1(\nu) + h_2(\nu))/2 + |h_1(\nu) - h_2(\nu)|/2]$ which is greater than $E[h(\nu)] + 3c\log 3 - 3 + \epsilon/2$ by the choice of ν.

Hence the expected height of the subtree chosen is at least $E[h(\nu)] + c\log 3 - 1 + \epsilon/6$ making it greater than $E[h(3\nu)] - 1$ which is clearly impossible since the maximum possible height for a subtree is one less than the height of the tree. □

3 Three ways to improve the bound

The argument giving the upper bound of 22.417 can be strengthened in (at least) three ways:

- Choose a value other than 3ν for the size of the tree. Any size greater than 2ν will give some non-trivial upper bound.
- Consider not only the immediate subtrees of the tree but possibly deeper subtrees. As long as a subtree has size greater than 2ν we can consider its split into two subtrees, certain that one of them will have at least ν nodes.
- Where a subtree has size $\alpha\nu$ ($\alpha > 1$), we have lower bounded its height by that of a tree of size ν. In fact, since every tree has at least one critical node, the mean height of an $\alpha\nu$ node tree must exceed that of an ν node tree by at least $2\sum_{i=\nu+1}^{\alpha\nu} 1/i$ which we can approximate by $2\log\alpha$ for large ν.

Hence we have a general scheme for a method of choosing a subtree and an associated upper bound:

Starting with a tree of size $k\nu$, with probability $1 - p$ find a subtree of size $\alpha\nu$ ($\alpha \geq 1$), at a depth A from the root. Otherwise (with probability p) find two disjoint subtrees with sizes $\beta\nu$ and $\gamma\nu$ ($\beta, \gamma \geq 1$), at depths B and C respectively from the root; choose one of these two according to which was higher at the moments when each contained exactly ν nodes.

We consider the expected value of the depth of the deepest node in the subtree thus chosen. If only one subtree is found (probability $1-p$) this is at least $E[A + 2\log\alpha + h(\nu)]$. Otherwise (probability p) it is the expected depth of the root of the subtree chosen plus the expected height of that subtree when its size was ν plus the amount by which the height has increased since then, giving at least $(E[(B+C+2\log\beta+2\log\gamma)/2+h(\nu)+E[|h_1(\nu)-h_2(\nu)|/2])$. Putting these two together we find that the expected depth is at least $(1-p)(E[A+2\log\alpha+h(\nu)]) + p(E[(B+C+2\log\beta+2\log\gamma)/2+h(\nu)+E[|h_1(\nu)-h_2(\nu)|/2])$. Since this must be no greater than $E[h(k\nu)]$ and we can choose infinitely many ν with $E[h(k\nu)] < E[h(\nu)] + c\log k + \epsilon$, we can deduce an upper bound, valid infinitely often of

$$E[|h_1(\nu) - h_2(\nu)|/2] < c\log k/p - E[(B+C+2\log\beta+2\log\gamma)/2] - \frac{1-p}{p}E[A+2\log\alpha] + O(\epsilon).$$

If we define a random variable $f(n)$ as the value of $A + 2\log\alpha$ if only one subtree is found and $(B+C+2\log\beta+2\log\gamma)/2$ if two subtrees are found, when the scheme is applied to an n node tree, we can rewrite this inequality as

$$E[|h_1(\nu) - h_2(\nu)|] \leq 2(c\log k - E[f(k\nu)])/p \qquad (1)$$

Theorem 2. $E[|h_1(n) - h_2(n)|] < 6.247$ infinitely often.

Proof: We consider the particular instance of this scheme in which we choose a cut off depth d and apply the algorithm $choose2(T, d)$ where $choose2$ is defined as

```
choose2 (tree,depth);
        if depth = 0 then return tree fi;
        if size(tree) < 2ν
            return tree
        else
            let L and S be the larger and smaller immediate subtrees of tree
            if size(S) ≥ ν
                then return {choose1(L,depth-1), choose1(S,depth-1)}
                else return {choose2(L,depth-1)}
            fi
        fi
end;
```

where *choose*1 chooses some subtree of size at least ν:

*choose*1(*tree, depth*);
    if depth = 0 *then return tree fi*;
    if size(*tree*) < 2ν *then*
        return tree
    else
        *return choose*1(*larger subtree of tree, depth* − 1)
    fi
end;

To apply inequality (1) we need to know the value of p and that of $E[f(k\nu)]$. Defining $p(n, depth)$ as the probability of finding two subtrees of size at least ν when starting with a tree of size n ($> \nu$) at *depth* levels above the cut off level, we obtain

$p(n, depth) =$
    if $n < 2\nu$ *or depth* = 0
    then
        0
    else
        $1 - 2\nu/n + 2\nu/n \times E[p(n', depth - 1)]$
    fi

where the expectation is taken over subtrees of sizes n' from $n - \nu$ to $n - 1$

Further defining $pp(x, depth)$ as the limit of $p(x\nu, depth)$ as ν tends to infinity, we find that

$pp(x, depth) =$
if $x < 2$ *or depth* = 0
then
    0
else
    $1 - 2/x + 2/x \int_{x-1}^{x} pp(y, depth - 1) \, dy$
fi

Similarly defining $ff(x, depth)$ and $gg(x, depth)$ as the limits as ν tends to infinity of the expected values of (root depth + 2 log size) averaged over the nodes returned by *choose*2($T, depth$) and *choose*1($T, depth$) applied to $x\nu$ node trees T, we obtain

$$ff(x, depth) =$$
$$if\ x < 2\ or\ depth = 0$$
$$then$$
$$\qquad 2 * \log(x)$$
$$else$$
$$\qquad 1 + 1/x \int_1^{x-1} gg(y, depth - 1)\ dy\ + 2/x \int_{x-1}^{x} ff(y, depth - 1)\ dy$$
$$fi$$

where

$$gg(x, depth) =$$
$$if\ x < 2\ or\ depth = 0$$
$$then$$
$$\qquad 2 * \log(x)$$
$$else$$
$$\qquad 1 + 2/x \int_{x/2}^{x} gg(y, depth - 1)\ dy$$
$$fi$$

and $\lim_{\nu \to \infty} E[f(k\nu)] = ff(k, d)$
giving an upper bound on $E[|h_1(n) - h_2(n)|]$ as close as required to $2(c \log k - ff(k, d))/pp(k, d)$.

The strongest bound yet found was obtained by taking $k = 3.9$, $d = 5$, giving a bound slightly less than 6.247. □

(The computation of $ff(3.9, 5)$ and $pp(3.9, 5)$ was done by Maple after definition of multiple functions such as $ff[d, i](x)$ defined only for $\lfloor x \rfloor = i$ and there equal to $ff(x, d)$; this enables the integrals to be written as sums of integrals so that no $if\ then\ else\ fi$ constructs remain in the definitions.)

4 The Convergence Hypothesis

The result of the previous section can be strengthened both by replacing the 'infinitely often' by 'almost always' and by reducing the bound, if we accept a very plausible and empirically justified hypothesis.

Definition: $ec(n)$ is the expected number of critical nodes of an n-node tree.

Note: the probability that the addition of the $(n + 1)$-st node increases the height of the tree is $2ec(n)/(n + 1)$. Calculation of $ec(n)$ for n up to $100,000$ and approximation by constructing random trees for larger n both suggest that $ec(n)$ is monotonically increasing after initial fluctuations while n is less than 8. (See [3] for methods of rapid construction of very large random trees.)

The Convergence Hypothesis: $ec(n)$ tends to a limit as n increases.
Note: if this hypothesis is correct the limit must be $c/2$.

Theorem 3. If the Convergence Hypothesis holds then $E[|h_1(n)-h_2(n)|] < 4.96$ except for (possibly) finitely many n.

(Again, this implies immediately that the same bound applies to $E[|h(n) - E[h(n)]|]$).

Proof.

- **extending the result to almost all n:**
  In the proof of the main theorem, we relied on the fact that since $E[h(n)/\log n]$ tends to c, there exist infinitely many ν for which $E[h(k\nu)] - E[h(\nu)] < c\log k + \epsilon$. Now given the convergence hypothesis, for any k and ϵ, by choosing N large enough we can guarantee that for all ν larger than N, $ec(\nu) < c/2 + \epsilon/2\log k$ so that $E[h(k\nu) - h(\nu)] < c\log k + \epsilon$ and the argument of section 3 goes through unchanged.
- **reducing the bound:** The proof in section 3 used the fact that a random tree of size $\alpha\nu$ had height at least $E[h(\nu)]+2\log\alpha$ (and a similar result for the higher of two trees of sizes $\beta\nu$ and $\gamma\nu$). Given the Convergence Hypothesis, provided we choose ν large enough, the first of these results remains true with any constant less than c instead of the "2". (The second does not since the higher of two random trees is not a random tree.) Hence we can replace the "$2\log x$" by "$(c-\epsilon)\log x$" in the definition of ff (but not gg) and recompute the bound obtained by the modified version of inequality (1). This time the best result obtained has $d = 1$ and $k = 2.67$, giving a bound of just under 4.953. □

5 Further work

As has been shown, a significant improvement in the main theorem would be obtained if the expected value of the number of critical nodes was shown to converge. Even showing that this expectation is bounded for all n would prove that $E[|h_1(n) - h_2(n)|]$ is bounded though it would not directly give an explicit bound. It seems extremely implausible that this expectation should oscillate unboundedly but a proof that it does not do so has not been easy to find.

Alternatively, improving the algorithm for choosing a subtree could further decrease the numerical value of the bound. Two ways of doing this seem worth exploring: firstly, when the $k\nu$ node tree turns out to have three or more ν node subtrees, a careful choice between these should give a deeper leaf than the current choice between the first two found; secondly, when two subtrees are found with depths B and C and sizes $\beta\nu$ and $\gamma\nu$, biasing the choice to the one with larger depth and size must give a deeper leaf on average. Also there may be other parameters for which the existing algorithm gives a better bound. Unfortunately the computations are very slow with $d > 3$ so not very many have been done (the computation of the bound in theorem 2 took Maple a weekend).

The methods and results developed here tell us nothing about the variance of $h(n)$. We continue to conjecture that this variance is bounded as n goes to infinity.

References

1. L. DEVROYE. A note on the height of binary search trees. *Journal of the ACM*, 33:489–498, 1986.
2. L. DEVROYE and B. REED. On the variance of the height of random binary search trees. *SIAM J. Computing*, 24:1157–1162, december 1995.
3. L. DEVROYE and J. M. ROBSON. On the generation of random binary search trees. *SIAM J. Computing*, 24:1141–1156, december 1995.
4. J. M. ROBSON. The height of binary search trees. *Australian Computer Journal*, 11:151–153, 1979.

An Improved Master Theorem for Divide-and-Conquer Recurrences *

Salvador Roura

Departament de Llenguatges i Sistemes Informàtics
Universitat Politècnica de Catalunya
E-08028 Barcelona, Catalonia, Spain
E-mail:roura@lsi.upc.es

Abstract. We present a new master theorem for the study of divide-and-conquer recursive definitions, which improves the old one in several aspects. In particular, it provides more information, frees us completely from technicalities like floors and ceilings, and covers a wider set of toll functions and weight distributions, stochastic recurrences included.

1 Introduction

Let F_n denote a variable related to some divide-and-conquer (d.a.c., for short) algorithm or data structure, such as the number of comparisons made in quicksort or the number visited nodes while a search in a BST, while dealing with an instance of size n. By the recursive structure of the algorithm or data structure it is always possible to get a recurrence that defines F_n from the values of the variable for instances of smaller size. From this recurrence it is necessary to deduce explicit or asymptotic expressions for F_n (that is, we need to "solve" the recurrence).

To this end, we can make use of the (classical) *master theorem* (see [1, 5]). See also [7, 8] for several improved versions. It is a set of simple rules that provide quick (albeit partial) information on the value of F_n. Assume that we have the recurrence

$$F_n = t_n + W \cdot F_{S_n}, \qquad (1)$$

where t_n is the *toll function* or cost of the divide and combine steps needed to solve a problem of size n, W is the (fixed) number of recursive calls at each step, and $S_n = Z \cdot n + \mathcal{O}(1)$ is the size of the subproblems to be recursively solved, for some $0 < Z < 1$. Notice that expresions with floors and ceilings in the argument of the recursive call, like $\lfloor n/2 \rfloor$, are covered by the term $\mathcal{O}(1)$ above. Let $\alpha = -\log_Z W$. Then, the classical master theorem states that the solution to this recurrence is

$$F_n = \begin{cases} \Theta(n^\alpha), & \text{if } t_n = \mathcal{O}(n^a) \text{ for } a < \alpha; \\ \Theta(t_n \log n), & \text{if } t_n = \Theta(n^\alpha \log^c n) \text{ for } c \geq 0; \\ \Theta(t_n), & \text{if } t_n = \Omega(n^a) \text{ for } a > \alpha. \end{cases} \qquad (2)$$

---
* This research was supported by the ESPRIT LTR Project ALCOM-IT, contract # 20244 and by a grant from CIRIT (Comissió Interdepartamental de Recerca i Innovació Tecnològica).

Notice that there are two gaps for the values of t_n where we cannot use the master theorem.

Although this theorem is sometimes enough for simple purposes, it presents some drawbacks. For instance, consider the recurrence

$$B_n = 1 + \frac{\lfloor (n-1)/2 \rfloor}{n} \cdot B_{\lfloor (n-1)/2 \rfloor} + \frac{\lceil (n-1)/2 \rceil}{n} \cdot B_{\lceil (n-1)/2 \rceil}, \quad \text{if } n \geq 2, \quad (3)$$

with $B_0 = 0$ and $B_1 = 1$, defining the expected number of comparisons during a binary search in an array of size n, when we search for some key in the array chosen at random. It does not follow the master theorem pattern utterly, since we do not have exactly one expected recursive call at each step but $1 - 1/n$, since the central item in the current search range could be, by chance, the sought item. In other words, the number of recursive calls is not constant but tends to a constant. Despite this, we can *assume* that the solution to the recurrence $F_n = 1 + F_{n/2}$ must be close to B_n (which is true) and therefore deduce that $B_n = \Theta(\log n)$. Posterior reasoning can rigorously prove that this approximation does not lead to a wrong answer.

Much more difficulties presents the analysis of stochastic recurrences like

$$S_0 = 0, \quad S_n = n - 1 + \frac{2}{n^2} \sum_{0 \leq k < n} k S_k, \quad \text{if } n \geq 1, \quad (4)$$

defining the expected number of comparisons to select the i-th of the n keys of an array (where i is chosen at random) when using Hoare's FIND [3]. Here we would need to make further approximations, which could easily lead to wrong conclusions.

The theorems presented in this paper improve previous theorems in several aspects. On the one hand, we will show how technicalities like floors and ceilings will not need to be treated any more, not previously to the analysis of the recurrence nor afterwards. On the other, recurrences where the asymptotic sizes of the subproblems to be recursively solved consist in a set of several fixed fractions of the original problem (this improvement was already considered in [8]) and the number of recursive calls to each one tends to a constant (but is not constant) like

$$F_n = t_n + (2 - 1/\sqrt{n})F_{\lfloor n/3 \rfloor} + 4F_{\lfloor n/2 - \sqrt[3]{n} \rfloor} + (1 + 1/n)F_{\lceil 4n/5 + \ln^2 n \rceil} \quad (5)$$

for n large enough, can be easily analysed through our theorems as well. Depending on t_n, we can also deduce the constant of the main term of the solution (for the basic recurrence (1) this constant can also be found; see [5], for instance). Furthermore, we will be able to deal with stochastic recurrences like (4), and a simple application of our theorems will sometimes yield several of the main terms of the explicit solution of a recurrence with their corresponding multiplicative factors, and not only the growing order of the dominating term. Finally, we will see how the new theorems cover a wider set of toll functions. In terms of the classical master theorem, the new ranges for t_n are $t_n = \mathcal{O}(n^\alpha \log^c n)$ for $c < -1$, $t_n = \Theta(n^\alpha \log^c n)$ for $c > -1$ and $t_n = \Omega(n^a)$ for $a > \alpha$, thus closing almost completely the gap between the first and the second case. The results for the first range were first given in [7].

Next sections are organized as follows. In Sections 2 and 3 we present the two types of recurrences that are more likely to appear in practical situations, and give a master theorem for each one. Section 4 includes the main results from which both theorems can be derived. Section 5 ends the paper with some final remarks.

2 The Discrete Master Theorem

To begin with, let us introduce the concept of divide-and-conquer recursive definition formally.

Definition 1. Let $F_n \geq 0$ be a function defined for all $n \geq 0$. We say that

$$\mathcal{F} = \left[N, \{b_n\}_{0 \leq n < N}, \{t_n\}_{n \geq N}, \{w_{n,k}\}_{n \geq N}^{0 \leq k < n}\right]$$

is a *d.a.c. recursive definition* of F_n iff $N \geq 1$, $F_n = b_n$ for all $0 \leq n < N$ and

$$F_n = t_n + \sum_{0 \leq k < n} w_{n,k} F_k \tag{6}$$

for every $n \geq N$, where $t_n \geq 0$ and $w_{n,k} \geq 0$.

The weight $w_{n,k}$ is the (expected) number of recursive calls to the algorithm to deal with a subproblem of size k when the original problem has size n, while t_n includes the cost to divide a problem of size n into smaller subproblems that will be recursively solved, and to combine the solutions of the recursive calls to find the answer to the whole problem.

Definition 2. Let $\mathcal{F}$ be a d.a.c. recursive definition of a function F_n. We say that $\mathcal{F}$ is a *discrete* recursive definition if it follows the pattern

$$F_n = t_n + \sum_{1 \leq d \leq D} R_{d,n} F_{S_{d,n}} \tag{7}$$

for every $n \geq N$, where $D \geq 1$ is the (finite) number of subproblems to be recursively solved; $R_{d,n} = w_d + r_{d,n} \geq 0$ is the number of recursive calls to deal with the d-th subproblem, where $w_d > 0$ is the *asymptotic* number of calls to it and

$$\sum_{1 \leq d \leq D} |r_{d,n}| = \mathcal{O}(n^{-\rho}) \tag{8}$$

for some $\rho > 0$; $S_{d,n} = z_d \cdot n + s_{d,n}$ is the (integer) size of the d-th subproblem to be recursively solved, where $0 < z_d < 1$ and

$$\sum_{1 \leq d \leq D} \frac{|s_{d,n}|}{n} = \mathcal{O}(n^{-\sigma}) \tag{9}$$

for some $\sigma > 0$.

For example, (3) is a discrete recursive definition. Here we have two subproblems to recursively deal with ($D = 2$) that are both asymptotically 1/2 the size of the whole problem ($z_1 = z_2 = 1/2$, $-3/2 \leq s_{1,n} \leq -1/2$, $-1/2 \leq s_{2,n} \leq 1/2$) and 1/2 expected calls to each one ($w_1 = w_2 = 1/2$, $-3/2n \leq r_{1,n} \leq -1/2n$, $-1/2n \leq r_{2,n} \leq 1/2n$), where for the bounds of $s_{n,k}$ and $r_{n,k}$ we have used the fact that $r - 1 \leq \lfloor r \rfloor \leq r$ and $r \leq \lceil r \rceil \leq r + 1$ for every real r. Notice than $\rho = \sigma = 1$ is a possible choice here.

Theorem 3 (Discrete Master Theorem). *Let $\mathcal{F}$ be a discrete recursive definition of a function F_n, and let $Bn^a \ln^c n$ be the main term of t_n, for some constants B, a and c. Let us define*

$$\Phi(x) = \sum_{1 \leq d \leq D} w_d \cdot z_d^x,$$

and let $\mathcal{H} = 1 - \Phi(a)$. Then,
1) if $\mathcal{H} > 0$ then $F_n \sim t_n/\mathcal{H}$;
2) if $\mathcal{H} = 0$ then
   2.1) if $c > -1$ then $F_n \sim t_n \ln n / \mathcal{H}'$, where

$$\mathcal{H}' = -(c+1) \sum_{1 \leq d \leq D} w_d \cdot z_d^a \ln z_d;$$

   2.2) if $c = -1$ then $F_n = \mathcal{O}(n^a \log^\epsilon n)$ for any $\epsilon > 0$;
   2.3) if $c < -1$ then $F_n = \Theta(n^a)$;
3) if $\mathcal{H} < 0$ then $F_n = \Theta(n^\alpha)$, where α is the unique solution of $\Phi(\alpha) = 1$.

Some Examples of the Use of the Discrete Master Theorem

Let us solve (3). To begin with, the main term in its toll function is $n^0 \log^0 n$. Now we can use the master theorem as follows.

1. First, we identify the set of values $\{w_d\}_{1 \leq d \leq D}$ and $\{z_d\}_{1 \leq d \leq D}$. This yields $w_1 = w_2 = 1/2$ and $z_1 = z_2 = 1/2$. We should make sure that properties (8) and (9) hold, but this is trivial here (floors and ceilings are never a problem).
2. We define $\Phi(x) = (1/2)^x$ and hence $\mathcal{H} = 1 - \Phi(0) = 0$.
3. Since $\mathcal{H} = 0$ and $c > -1$, we define $\mathcal{H}' = -(0+1)((1/2)^0 \ln(1/2)) = -\ln(1/2) = \ln 2$, and finally $B_n \sim \ln n / \ln 2 = \log_2 n$.

Let us consider now (5). Assume that $t_n = 6n^2/\ln^5 n$.

1. $w_1 = 2$, $w_2 = 4$ and $w_3 = 1$; $z_1 = 1/3$, $z_2 = 1/2$ and $z_3 = 4/5$. (It is a simple matter to check that this recurrence is indeed a discrete recursive definition).
2. $\Phi(x) = 2(1/3)^x + 4(1/2)^x + (4/5)^x$ and hence $\mathcal{H} = 1 - \Phi(2) = -194/225$.
3. Since $\mathcal{H} < 0$, $F_n = \Theta(n^\alpha)$, where α is the unique solution of $\Phi(\alpha) = 1$, which numerically is $\alpha \simeq 3.16756$.

Finally, let us set $t_n = n^2$ for the recurrence

$$F_n = t_n + F_{\lfloor n/4 \rfloor}. \tag{10}$$

Notice that we do not need to explicitly state the values of F_n at small indices, since they are irrelevant to the master theorem. Solving it is very easy.

1. $w_1 = 1$ and $z_1 = 1/4$.
2. $\Phi(x) = (1/4)^x$ and hence $\mathcal{H} = 1 - \Phi(2) = 15/16$.
3. Since $\mathcal{H} > 0$, $F_n \sim n^2/(15/16) = 16n^2/15$.

Note that the last example above follows the pattern of Equation 1, and hence we can analyse it through the old master theorem (2), which becomes a particular case of the new one.

3 The Continuous Master Theorem

This section covers the analysis of recursive definitions like (4). To begin with, let us define the concept of shape function (the reason for this name will be clear after Definition 5).

Definition 4. Let $w(z) \geq 0$ be a function over $[0,1]$ such that $w'(z)$ exists and is bounded for every $0 \leq z \leq 1$. Furthermore, let $\int_0^1 w(z)\,dz$ be greater or equal than 1. Then we say that $w(z)$ is a *shape function*.

Definition 5. Let $\mathcal{F}$ be a d.a.c. recursive definition of a function F_n. We say that $\mathcal{F}$ is a *continuous* recursive definition if it follows the pattern

$$F_n = t_n + \sum_{0 \leq k < n} w_{n,k} F_k \tag{11}$$

for every $n \geq N$, and if there exists some shape function $w(z)$ such that

$$\sum_{0 \leq k < n} \left| w_{n,k} - \int_{\frac{k}{n}}^{\frac{k+1}{n}} w(z)\,dz \right| = \mathcal{O}(n^{-\rho}) \tag{12}$$

for some $\rho > 0$.

Loosely speaking, the last definition allows us to use the integral in the right of the expression above to find a good approximation to $w_{n,k}$. For instance, the shape function for (4) is $w(z) = 2z$ (notice that this function follows the conditions required for a shape function), since

$$\int_{\frac{k}{n}}^{\frac{k+1}{n}} w(z)\,dz = \int_{\frac{k}{n}}^{\frac{k+1}{n}} 2z\,dz = z^2 \Big|_{\frac{k}{n}}^{\frac{k+1}{n}} = \frac{2k}{n^2} + \frac{1}{n^2} = w_{n,k} + \frac{1}{n^2},$$

and hence the sum of errors is $1/n = \mathcal{O}(n^{-\rho})$ for $\rho = 1$.

Therefore, $\omega(z)$ is nothing except the asymptotic shape of the distribution of weights, which now does not consist in a finite number of fixed fractions of the original size of the problem (as it was in previous section), but is very similar to a continuous probability distribution, where the area beneath the function is the asymptotic number of recursive calls. Recall that, by definition, $\int_0^1 \omega(z)dz \geq 1$, and therefore we are assuming that there is at least one asymptotic recursive call. This condition (very likely to hold in practice) simplifies the study of these recurrences.

Theorem 6 (Continuous Master Theorem). *Let $\mathcal{F}$ be a continuous recursive definition of a function F_n, and let $Bn^a \ln^c n$ be the main term of t_n, where B, a and c are constants. Let us define*

$$\varphi(x) = \int_0^1 \omega(z) z^x \, dz,$$

and let $\mathcal{H} = 1 - \varphi(a)$. Then,
  1) if $\mathcal{H} > 0$ then $F_n \sim t_n/\mathcal{H}$;
  2) if $\mathcal{H} = 0$ then
    2.1) if $c > -1$ then $F_n \sim t_n \ln n/\mathcal{H}'$, where

$$\mathcal{H}' = -(c+1) \int_0^1 \omega(z) z^a \ln z \, dz;$$

  2.2) if $c = -1$ then $F_n = \mathcal{O}(n^a \log^\epsilon n)$ for any $\epsilon > 0$;
  2.3) if $c < -1$ then $F_n = \Theta(n^a)$;
  3) if $\mathcal{H} < 0$ (including the case $\mathcal{H} = -\infty$) then $F_n = \Theta(n^\alpha)$, where α is the unique solution of $\varphi(\alpha) = 1$.

Some Examples of the Use of the Continuous Master Theorem

Let us solve the recursive definition

$$Q_0 = 0, \qquad Q_n = 1 + \frac{4}{n(n+1)} \sum_{0 \leq k < n} (n-k) Q_k, \quad \text{if } n \geq 1, \qquad (13)$$

related to the number of comparisons in a half-defined search in a quad-tree [2]. Notice that the main term in the toll function is n^0. Hence,

1. First, we identify the shape function of the weights. As first chance, we can try the following.
   (a) From $\omega_{n,k}$ we compute a set of new weights $\sigma_{n,k}$, by replacing terms like $n+1$ or $(n-1-k)$ by n or $(n-k)$, respectively. This yields $\sigma_{n,k} = 4(n-k)/n^2$.
   (b) Now we have to check that $|\omega_{n,k} - \sigma_{n,k}| = \mathcal{O}(n^{-2})$. This is true in our example, since $|\omega_{n,k} - \sigma_{n,k}| = 4(n-k)/(n^2(n+1)) = \mathcal{O}(n^{-2})$.
   (c) We compute $\omega(z) = n \cdot \sigma_{n,zn}$. This step produces an expression without n's, $\omega(z) = n \cdot 4(n-zn)/n^2 = 4(1-z)$.

(d) Finally, we should prove that $w'(z)$ exists and is bounded, and also that $\int_0^1 w(z)dz \geq 1$, which is trivial here.

2. We define
$$\varphi(x) = 4\int_0^1 (1-z)z^x \, dz = 4\left[\frac{z^{x+1}}{x+1} - \frac{z^{x+2}}{x+2}\right]_0^1 = 4\left(\frac{1}{x+1} - \frac{1}{x+2}\right)$$

if $x > -1$ (and $\varphi(x) = +\infty$ otherwise). Hence, $\mathcal{H} = 1 - \varphi(0) = -1$.

3. Since $\mathcal{H} < 0$ we have that $F_n = \Theta(n^\alpha)$, where α is defined as the unique solution of $\varphi(\alpha) = 1$, which yields $\alpha = (\sqrt{17}-3)/2 \simeq 0.56155$ (this is the only solution to $\alpha^2 + 3\alpha - 2 = 0$ that is greater than -1).

Let us analyse (4).

1. We already know that $w(z) = 2z$.
2. We define
$$\varphi(x) = \int_0^1 w(z)z^x \, dz = 2\int_0^1 z^{x+1} \, dz = 2\left[\frac{z^{x+2}}{x+2}\right]_0^1 = \frac{2}{x+2}$$

if $x > -2$ (and $\varphi(x) = +\infty$ otherwise). Hence, $\mathcal{H} = 1 - \varphi(1) = 1/3$.

3. Since $\mathcal{H} > 0$, $S_n \sim n/(1/3) = 3n$.

In this example we can get even more information, by means of a simple trick. Define $G_n = S_n - 3n$. Then
$$G_n = n - 1 + \frac{2}{n^2}\sum_{0 \leq k < n} kS_k - 3n = -2n - 1 + \frac{6}{n^2}\sum_{0 \leq k < n} k^2 + \frac{2}{n^2}\sum_{0 \leq k < n} kG_k.$$

It is well known that $\sum_{0 \leq k < n} k^2 = n^3/3 - n^2/2 + n/6$. Therefore,
$$G_n = -4 + \frac{1}{n} + \frac{2}{n^2}\sum_{0 \leq k < n} kG_k.$$

We can now solve this recurrence using Theorem 6 again. The first step is already done, since the distribution of weights remains the same, as is the case for $\varphi(x)$. Computing $\mathcal{H}$ produces $\mathcal{H} = 1 - \varphi(0) = 0$. Since $\mathcal{H} = 0$ and $c > -1$, we define
$$\mathcal{H}' = -(0+1)\int_0^1 2zz^0 \ln z \, dz = -2\int_0^1 z \ln z \, dz = -2\left[\frac{z^2}{2}\ln z - \frac{z^2}{4}\right]_0^1 = \frac{1}{2},$$

and get $G_n \sim -4\ln n/(1/2) = -8\ln n$. We can make one more step, defining $I_n = G_n + 8\ln n$ for $n > 0$, which produces
$$I_n = \frac{8\ln n}{n} + \frac{1}{n} - \frac{4\ln n}{3n^2} + \Theta\left(\frac{1}{n^2}\right) + \frac{2}{n^2}\sum_{0 \leq k < n} kI_k,$$

where we have used the equality $\sum_{0<k<n} k \ln k = \frac{n^2 \ln n}{2} - \frac{n^2}{4} - \frac{n \ln n}{2} + \frac{\ln n}{12} + \Theta(1)$. Now we get $\mathcal{H} = 1 - \varphi(-1) = -1 < 0$, and hence $I_n = \mathcal{O}(n^\alpha) = \mathcal{O}(1)$. Notice that we cannot deduce that $I_n = \Theta(1)$, since the toll function in the definition of I_n includes positive and negative terms together, and their contributions could cancel each other. As a final conclusion we have that $S_n = 3n - 8 \ln n + \mathcal{O}(1)$.

We end this section analysing the number of comparisons while sorting an array of n keys through the variant of quicksort that uses as the pivot of the partition stage the median of a random sample of $2k+1$ keys, for some fixed $k \geq 0$ ($k = 0$ reduces to basic quicksort). This method was suggested by Hoare himself in [4], and later Van Emden [6] analysed it by means of information-theoretic arguments and sensible approximations. We can now prove the same results as a simple consequence of the continuous master theorem.

Let $Q_n^{(k)}$ be the number of expected comparisons while using quicksort to sort an array with n items when the sample has $2k+1$ keys at each stage (except for small n). The recurrence for any k is

$$Q_n^{(k)} = n - 1 + S_k + 2\binom{n}{2k+1}^{-1} \sum_{0 \leq i < n} \binom{i}{k}\binom{n-1-i}{k} Q_i^{(k)}, \quad (14)$$

where S_k is the (linear in k, but constant in n) number of comparisons to find the median of the sample. Therefore,

1. (a) We compute
$$\sigma_{n,i}^{(k)} = \frac{2(2k+1)!}{n^{2k+1}} \cdot \frac{i^k}{k!} \cdot \frac{(n-i)^k}{k!}$$
as a good approximation for $w_{n,i}^{(k)}$.
   (b) It is routine work to check that $\left| w_{n,i}^{(k)} - \sigma_{n,i}^{(k)} \right| = \mathcal{O}(n^{-2})$.
   (c) We compute $w_k(z) = n \cdot \sigma_{n,zn}^{(k)} = 2(2k+1)!/k!^2 \cdot z^k(1-z)^k$.
   (d) Since $w_k(z)$ is a polynomial on z, we have that $w_k'(z)$ exists and is bounded. Furthermore, we know that the asymptotic number of recursive calls is 2.
2. We define
$$\varphi_k(x) = \int_0^1 w_k(z) z^x \, dz = \frac{2(2k+1)!}{k!^2} \int_0^1 z^{k+x}(1-z)^k \, dz,$$
and evaluate $\mathcal{H}(k) = 1 - \varphi_k(1)$. For this step, we can use the equality
$$\int_0^1 z^\alpha (1-z)^\beta \, dz = \frac{\alpha!\beta!}{(\alpha+\beta+1)!}$$
(see [2], page 479, for instance) to find that, as expected, $\mathcal{H}(k) = 0$.
3. Therefore, we define
$$\mathcal{H}'(k) = -\int_0^1 w_k(z) z^1 \ln z \, dz = -\frac{2(2k+1)!}{k!^2} \int_0^1 z^{k+1}(1-z)^k \ln z \, dz.$$
This step yields $\mathcal{H}'(k) = 1/(k+2) + 1/(k+3) + \ldots + 1/(2k+2)$ (see [6], page 565). Finally, $Q_n^{(k)} \sim n \ln n / \mathcal{H}'(k)$.

4 The Theorems

In this section we present, without proof, the main technical results from which both Theorem 3 and Theorem 6 can be derived. Most of them refer to canonical recursive definitions, which are defined as follows.

Definition 7. Let $\mathcal{F}$ be a d.a.c. recursive definition of a function F_n. Let $W_n = \sum_{0 \leq k < n} w_{n,k}$. We say that $\mathcal{F}$ is a *canonical* recursive definition if and only if both these properties hold: 1) It exists some $\rho > 0$ such that $|W_n - 1| = \mathcal{O}(n^{-\rho})$. 2) It exists some upper bound $U < 1$ such that

$$\sum_{0 \leq k < n} \frac{w_{n,k}}{W_n} \cdot \frac{k}{n} \leq U$$

for n large enough.

Intuitively, the first condition requires that the total number of recursive calls to solve a problem of size n tends to 1 (with a minimum convergence speed). Notice that, opposed to the old master theorem, the number of recursive calls depends on n. The sum in the second condition above is the average fraction of the original problem that is solved by a recursive call. Therefore, this condition implies that the problem is broken into pieces that are (on average) a fraction of the original one.

Let $\mathcal{F}$ be a recursive definition of F_n. As an immediate consequence of (6) we have that $F_n = \Omega(t_n)$. The natural question that arises is: Can F_n grow faster than t_n and, if so, under which conditions? For the recursive definitions we deal with and roughly speaking, we could say that there is a *growing order associated to every distribution of weights, irrespective of how small t_n is.* Let us call it $\Theta(n^\alpha)$. Then, the growing order of F_n should be $\text{Max}\{\Theta(t_n), \Theta(n^\alpha)\}$. And this is almost true.

For instance, let us consider (10). We will see in a moment that $\alpha = 0$ for any canonical recurrence, such as this one. Therefore, for "big" values of t_n, such as n, n^3 or 2^n we should get $F_n = \text{Max}\{\Theta(t_n), \Theta(1)\} = \Theta(t_n)$, which is true. For "small" values of t_n like $1/n$, $1/n^3$ or 2^{-n}, F_n should be $\Theta(1)$, which is also true. However, things are not so easy for values of t_n close to $\Theta(1)$. For example, for $t_n = 1$, F_n turns out to be $\Theta(\log n)$ instead of $\Theta(1)$. We will see in this section how to cope with this additional factor.

There is another remark about the results in this section that we should make, namely that recursive definitions with "small" toll function (and thus inside the zone dominated by the term $\Theta(n^\alpha)$ associated to the distribution of weights) are the most difficult to analyse. Indeed, there is no way to find the lower order terms in the asymptotic expression of F_n nor even the multiplicative factor of the main term n^α, but to consider *all the values of F_n*, the values at small indices included. In terms of a recursion tree (see [1], for example) this situation corresponds to the case in which the solution to the recurrence is dominated by the values at the leaves. Therefore, a method for the study of recursive definitions based only in the asymptotic properties of the toll function and the distribution of weights (like our master theorems) cannot be used to get the multiplicative factor of the main term n^α in F_n. Moreover, for some recursive definitions that factor is not asymptotically constant.

Next theorems formalize one of the claims stated above, namely that any canonical recursive definition is, on the one hand $\Omega(1)$ (under some minimum additional conditions), and on the other, $\mathcal{O}(1)$ for "small" t_n.

Theorem 8. *Let $\mathcal{F}$ be a canonical recursive definition of a function F_n, such that $t_n > 0$ and $\sum_{0 \le k < N} w_{n,k} = \mathcal{O}(n^{-\rho})$ for some $\rho > 0$. Then $F_n = \Omega(1)$.*

Notice that, apart from being canonical, $\mathcal{F}$ must follow two additional conditions. They are mainly technical properties that hold in most cases. Roughly speaking, $t_n > 0$ avoids the case "everything is zero", whilst the second condition makes the values of F_n at $n < N$ completely irrelevant.

Theorem 9. *Let $\mathcal{F}$ be a canonical recursive definition of a function F_n, such that $t_n = \mathcal{O}(\log^c n)$ for some $c < -1$. Then $F_n = \mathcal{O}(1)$.*

Now we present the main results related to the canonical recursive definitions that are dominated by the toll function. In contrast to the last recurrences, where the toll function lay in the influence zone of the distribution of weights, recursive definitions whose toll function is big enough to dominate the recurrence are typically easier to analyse, and in most cases we can get the multiplicative factor of the main term of the asymptotic expression of F_n Moreover, as we have already seen in Section 3, sometimes it is possible to get several of the main terms of the solution, their multiplicative factors included.

Theorem 10. *Let $\mathcal{F}$ be a canonical recursive definition of a function F_n, and let $t_n = n^a \delta_n$, where $a > 0$ and δ_n is a strictly positive increasing (eventually constant) function for n large enough. Then $F_n = \Theta(t_n)$.*

Theorem 11. *Let $\mathcal{F}$ be a canonical recursive definition of a function F_n, and let $t_n = \ln^c n \cdot \delta_n$, where $c > -1$ and δ_n is a strictly positive increasing (eventually constant) function for n large enough. Then $F_n = \mathcal{O}(t_n \log n)$.*

Notice that, according to the old master theorem, $\Theta(1)$ seemed to be the threshold value for t_n above which F_n became $\omega(1)$. Combining last theorem with Theorem 9 allows us to state that in fact, this threshold lies close to $t_n = 1/\log n$.

The remaining theorems in this section are crucial to both master theorems.

Theorem 12. *Let $\mathcal{F}$ be a canonical recursive definition of a function F_n, and let $t_n = n^a \delta_n$, where $a > 0$ and δ_n is a strictly positive increasing function for n large enough. Furthermore, let*

$$\mathcal{H} = \lim_{n \to \infty} \left(1 - \sum_{M \le k < n} w_{n,k} \cdot \frac{t_k}{t_n} \right)$$

exist for some M. Then $F_n = t_n / \mathcal{H} + o(t_n)$.

Theorem 13. *Let $\mathcal{F}$ be a canonical recursive definition of a function F_n, and let $t_n = \ln^c n \cdot \delta_n$, where $c > -1$ and δ_n is a strictly positive increasing function for n large enough. Furthermore, let*

$$\mathcal{H} = \lim_{n \to \infty} \left(\ln n - \sum_{M \leq k < n} w_{n,k} \cdot \frac{t_k}{t_n} \cdot \ln k \right)$$

exist for some M. Then $F_n = t_n \ln n / \mathcal{H} + o(t_n \log n)$.

Theorem 14. *Let $\mathcal{F}$ be a discrete (continuous) recursive definition of a function F_n, and let α be the unique solution of the equation $\Phi(\alpha) = 1$ ($\varphi(\alpha) = 1$). Let $\mathcal{B}$ be the recursive definition that we get after the substitution $B_n = F_n/n^\alpha$. Then $\mathcal{B}$ is a canonical discrete (continuous) recursive definition.*

5 Final Remarks

We have shown how to extract useful information from the most common recursive definitions, exclusively through the analysis of the asymptotic behaviour of the toll function and distribution of weights. We have only given restricted versions of the master theorems, which can be further generalized. For instance, they could be adapted to deal with toll functions that include sublogarithmical factors (like $\log \log n$). On the other hand, the definition of shape function is also a bit restrictive, but it is enough for general purposes.

Acknowledgements

Several referees made useful suggestions on the previous version of this paper. Many thanks are due to Conrado Martínez for his wise comments, constant support and $\Theta(2^n)$ patience. This work was partially written while the author was visiting Princeton University.

References

1. T.H. Cormen, C.E. Leiserson, and R.L. Rivest. *Introduction to Algorithms*. The MIT Press, Cambridge, MA, 1990.
2. Philippe Flajolet, Gaston Gonnet, Claude Puech, and J.M. Robson. Analytic variations on quadtrees. *Algorithmica*, 10:473–500, 1993.
3. C. A. R. Hoare. Find (Algorithm 65). *Communications of the ACM*, 4:321–322, 1961.
4. C. A. R. Hoare. Quicksort. *Computer Journal*, 5:10–15, 1962.
5. Robert Sedgewick and Philippe Flajolet. *An Introduction to the Analysis of Algorithms*. Addison-Wesley, 1996.
6. M.H. van Emden. Increasing the efficiency of quicksort. *Communications of the ACM*, 13:563–567, 1970.
7. Rakesh M. Verma. A general method and a master theorem for divide-and-conquer recurrences with applications. *Journal of Algorithms*, 16:67–79, 1994.
8. Rakesh M. Verma. General techniques for analyzing recursive algorithms with applications. *SIAM Journal on Computing*, 26(2):568–581, 1997.

Bisimulation for Probabilistic Transition Systems: A Coalgebraic Approach

E.P. de Vink[1] and J.J.M.M. Rutten[2]

[1] Faculty of Mathematics and Computer Science, Vrije Universiteit, De Boelelaan 1081a, 1081 HV Amsterdam, The Netherlands, e-mail: vink@cs.vu.nl
[2] Department of Software Technology, CWI, P.O.Box 94079, 1090 GB Amsterdam, The Netherlands, e-mail: janr@cwi.nl

Abstract. The notion of bisimulation as proposed by Larsen and Skou for discrete probabilistic transition systems is shown to coincide with a coalgebraic definition in the sense of Aczel and Mendler in terms of a set functor. This coalgebraic formulation makes it possible to generalize the concepts to a continuous setting involving Borel probability measures. Under reasonable conditions, generalized probabilistic bisimilarity can be characterized categorically. Application of the final coalgebra paradigm then yields an internally fully abstract semantical domain with respect to probabilistic bisimulation.

Keywords. Bisimulation, probabilistic transition system, coalgebra, ultrametric space, Borel measure, final coalgebra.

1 Introduction

For discrete probabilistic transition systems the notion of probabilistic bisimilarity of Larsen and Skou [LS91] is regarded as the basic process equivalence. The definition was given for reactive systems. However, Van Glabbeek, Smolka and Steffen showed in joint work with Tofts [GSS95], that for a concrete process language the usual notion of strong bisimilarity and the probabilistic concepts of reactive, generative and so-called stratified bisimulation constitute a hierarchy of observational congruences. Several other probabilistic equivalences are proposed as well in the literature. However, in all papers, *discrete* probability distributions are used, and hence the transition systems that are treated are in essence of a finitely branching or image-finite nature. The recent work of Blute et al. [BDEP97] is the single execption that we know of.

For the exploration of probabilistic transition systems and stochastic equivalences in the setting of modeling continuous systems, such as real-time or hybrid systems, one usually wants to allow more general probability measures than the more limited discrete probability distributions. [BDEP97] use stochastic kernels and spans of zigzags to underpin their notion of process equivalence. They prove that their notion of bisimulation agrees in the discrete case with the Larsen-Skou definition, but do not provide a characterization of bisimilarity in terms of transition steps, i.e., they do not give a continuous analogue for the Larsen-Skou bisimulation.

Here we attack the problem of continuous probabilistic transition systems and bisimulation by exploiting the transition-systems-as-coalgebras paradigm. Using a minimal amount of category theory, it can be summarized as follows: Let $\mathcal{F}: \mathcal{C} \to \mathcal{C}$ be any functor on a category $\mathcal{C}$. A *coalgebra* of $\mathcal{F}$ is an object S in $\mathcal{C}$ together with an arrow $\alpha: S \to \mathcal{F}(S)$. For many categories and functors, such a pair (S, α) represents a transition system, the type of which is determined by the functor $\mathcal{F}$. Vice versa, many types of transition systems can be captured by a functor this way. For instance, consider the familiar *labeled transition systems* $(S, A, \to)$, consisting of a set S of states, a set A of actions, and a transition relation $\to \subseteq S \times A \times S$. Put $\mathcal{L}(X) = \mathcal{P}(A \times X)$, the collection of all subsets of $A \times X$, for any set X, and, for $f: X \to Y$, $\mathcal{L}(f): \mathcal{L}(X) \to \mathcal{L}(Y)$, by $\mathcal{L}(f)(\{(a_i, x_i) \mid i \in I\}) = \{(a_i, f(x_i)) \mid i \in I\}$. It can be easily shown that $\mathcal{L}$ is a functor on the category of sets and functions. A labeled transition system $(S, A, \to)$ can now be represented as an $\mathcal{L}$-coalgebra by defining

$$\alpha: S \to \mathcal{L}(S), \quad s \mapsto \{(a, s') \mid (s, a, s') \in \to\}.$$

Conversely, any $\mathcal{L}$-coalgebra corresponds to a transition system: If (S, α) is a coalgebra for $\mathcal{L}$, then $(S, A, \to)$, with $\to \subseteq S \times A \times S$ given by $(s, a, s') \in \to$ iff $(a, s') \in \alpha(s)$, is clearly a transition system. (See [Rut96] for more details.)

One of the advantages of the coalgebraic view on transition systems is the existence of a general definition of $\mathcal{F}$-bisimulation, for any functor $\mathcal{F}$ (cf. [AM89]). For instance, applying that definition to the functor $\mathcal{L}$ above yields the standard notion of strong bisimulation. In general, the coalgebraic theory gives a generic approach to the definition and description of bisimulation: First define or characterize the transition systems one is interested in as coalgebras of a suitably chosen functor $\mathcal{F}$. Then obtain a definition of bisimulation for those systems by applying the categorical definition of $\mathcal{F}$-bisimulation.

The coalgebraic approach is applicable to many kinds of transition systems—see [Rut96] for many examples. In the present paper, this scheme is used to describe discrete and continuous probabilistic transition systems and bisimulations. The functor $\mathcal{M}_1$ assigns to a metric space its collection of Borel probability measures. It is shown that the corresponding notion of $\mathcal{M}_1$-bisimulation coincides, under mild conditions, with the continuous analogue of Larsen-Skou bisimulation. This extends a similar result for the discrete case, which is in fact given first: the functor $\mathcal{D}$, which assigns to a set the collection of its simple probability distributions, is shown to yield a categorical characterization of Larsen-Skou bisimulation. Hence, in agreement with general opinion, also from the coalgebraic point of view the latter equivalence is suggested as the canonical one.

Another appealing aspect of the coalgebraic approach is a canonical way of finding internally fully abstract domains of bisimulation, where two elements are equal if and only if they are bisimilar. It follows from a simple but very general argument that *final* coalgebras are fully abstract (see Aczel's final coalgebra model for nonwellfounded sets [Acz88], and also [RT93]). We shall show that it follows from general coalgebraic considerations [AR89,Bar93,RT93] that both our functors $\mathcal{D}$ and $\mathcal{M}_1$ have a final coalgebra, which consequently are internally fully abstract with respect to (discrete and continuous) probabilistic bisimulation. Therefore these final coalgebras can be exploited as semantic domains for probabilistic bisimulation (an important direction for future research).

As mentioned above, the functor $\mathcal{M}_1$ is defined on ultrametric spaces, and the Borel σ-algebras and associated measures are taken with respect to the metric topology. Our reasons for considering metric spaces rather than the, in semantical contexts, more standard use of ordered structures, as studied, e.g., by Jones and Plotkin [JP89] and by Edalat [Eda94] are twofold. Firstly, one can resort to the rich literature for standard measure theory on metric spaces. Secondly, we can apply the recently developed theory on coalgebraic bisimulation and final coalgebras in the metric setting [AM89,RT94]. Notably, we shall see that $\mathcal{M}_1$ is *locally contractive*, from which it follows that it has a final coalgebra. Because of the coalgebraic definition of bisimulation, we thus obtain an internally fully abstract domain. Such a full abstractness result has been lacking so far in the literature.

In conclusion, $\mathcal{D}$-bisimilarity and Larsen-Skou bisimilarity coincide for discrete probabilistic transition systems. For the continuous case, the functor $\mathcal{M}_1$ captures the generalization of probabilistic transition systems, and, under conditions, characterizes the associated notion of probabilistic bisimulation. For both functors a final coalgebra and hence, internally fully abstract domain exists, which can be exploited in the construction of domains for probabilistic bisimulation semantics.

Acknowledgments We are grateful to Henno Brandsma, Prakash Panangaden, Jaco de Bakker, and, as always, the members of the Amsterdam Concurrency Group for discussions on various aspects of this paper.

Note A technical report version of this paper is available by anonymous ftp from `ftp.cs.vu.nl` as `/pub/papers/theory/IR-423.ps.Z`.

2 Mathematical Preliminaries

Basic measure theoretic definitions (See, e.g., the standard textbook [Rud66].) A σ-algebra Σ on a set X is a collection of subsets which contains X and is closed under complement and countable union. Elements E of Σ are called measurable subsets of X. Trivially, the powerset $\mathcal{P}(X)$ is a σ-algebra for X. If X is a topological space, the Borel σ-algebra $\mathcal{B}(X)$ is defined as the least σ-algebra containing all open sets.

A function $\mu\colon \Sigma \to [0,1]$, where Σ is a σ-algebra on a set X, is called a Σ-probability measure if $\mu(X) = 1$ and μ is σ-additive, i.e., $\mu(\bigcup_{i \in I} E_k) = \sum_{i \in I} \mu(E_i)$ for any countable disjoint collection of measurable sets $\{E_i \mid i \in I\}$. For X a topological space, a Borel probability measure is a probability measure on X taken with respect to the Borel σ-algebra $\mathcal{B}(X)$. For $x \in X$, the Dirac-measure δ_x is given by $\delta_x(E) = 1$ if $x \in E$, and $\delta_x(E) = 0$ otherwise. A function $\mu\colon X \to [0,1]$ is called a simple probability distribution if there exist n distinct points $x_1, \ldots, x_n$, $n > 0$, such that $\mu(x_1) + \cdots + \mu(x_n) = 1$ and $\mu(x) = 0$ for $x \notin \{x_1, \ldots, x_n\}$. $\mathcal{D}(X)$ denotes the collection of all simple probability distributions on X. For $E \subseteq X$, $\mu[E]$ is short for $\sum_{x \in E} \mu(x)$. This way, a simple probability distribution corresponds to a convex linear combination of Dirac-measures.

Metric spaces (See, e.g., the monograph [BV96].) A pair (M, d) with M a nonempty set and $d: M^2 \to [0, 1]$ is called an ultrametric space if, for all $x, y, z \in M$: $d(x, y) = d(y, x)$, $d(x, y) = 0 \Leftrightarrow x = y$, and $d(x, z) \leq \max\{d(x, y), d(y, z)\}$. The last expression is referred to as the strong triangle inequality. For metric spaces M_1, M_2, a function $f: M_1 \to M_2$ is called nonexpansive if $d_2(f(x), f(y)) \leq d_1(x, y)$, for all $x, y \in M$. In case $d_2(f(x), f(y)) \leq \kappa \cdot d_1(x, y)$, for all $x, y \in M$, the function f is called κ-contractive, where κ is a constant with $0 \leq \kappa < 1$. The collection of all nonexpansive mappings from M_1 to M_2 is denoted by $M_1 \to_1 M_2$. We use the notation $\mathcal{O}$, or more explicit $\mathcal{O}(M)$, for the collection of all open subsets of M. For $\varepsilon > 0$ we put $\mathcal{O}_\varepsilon = \{\, O \in \mathcal{O} \mid \forall x \in O : B_\varepsilon(x) \subseteq O \,\}$.

Binary relations For a binary relation $R \subseteq S \times T$ we use π_1 and π_2 for the projections of R on S and T, respectively. R is called total if the two projections π_1 and π_2 are surjective. We say that R is z-closed if, for all $s, s' \in S$, $t, t' \in T$, $R(s, t) \wedge R(s', t) \wedge R(s', t') \Rightarrow R(s, t')$. If we put, for $n \in \mathbb{N}$, $R_0 = R$, $R_{n+1} = \{(s, t') \in S \times T \mid \exists s' \in S, t' \in T : R(s, t) \wedge R_n(s', t) \wedge R(s', t')\}$, and $R^* = \bigcup_{n \in \mathbb{N}} R_n$, we have that R^* is the least z-closed binary relation on $S \times T$ containing R. Below we will employ, for $s \in S$, the notation $F(s) = \{t \in T \mid R(s, t)\}$ and, for $U \subseteq S$, $F[U] = \bigcup_{s \in U} F(s)$, and, likewise, for $t \in T$, $E(t) = \{s \in S \mid R(s, t)\}$, and, for $V \subseteq T$, $E[V] = \bigcup_{t \in V} E(t)$.

Coalgebras (See, e.g., [Rut96].) Let $\mathcal{C}$ be either the category of sets and functions, or the category of ultrametric spaces and nonexpansive mappings. (These are the only categories playing a role in this paper.) Let $\mathcal{F}: \mathcal{C} \to \mathcal{C}$ be a functor. An $\mathcal{F}$-coalgebra is a pair (S, α) consisting of an object S in $\mathcal{C}$ together with an arrow $\alpha: S \to \mathcal{F}(S)$ in $\mathcal{C}$ called a coalgebra structure on S. A homomorphism between two $\mathcal{F}$-coalgebras (S, α) and (T, β) is an arrow $f: S \to T$ in $\mathcal{C}$ such that $\mathcal{F}(f) \circ \alpha = \beta \circ f$.

An *$\mathcal{F}$-bisimulation* between two $\mathcal{F}$-coalgebras (S, α) and (T, β) is a relation $R \subseteq S \times T$ for which there exists a coalgebra structure $\gamma: R \to \mathcal{F}(R)$ such that the projections $\pi_1: R \to S$ and $\pi_2: R \to T$ are homomorphisms: $\mathcal{F}(\pi_1) \circ \alpha = \gamma \circ \pi_1$ and $\mathcal{F}(\pi_2) \circ \beta = \gamma \circ \pi_2$. We then say that R is an $\mathcal{F}$-bisimulation for α and β. The arrow γ is called mediating for α and β. We write $x \sim y$ ('x and y are $\mathcal{F}$-bisimilar') whenever there exists an $\mathcal{F}$-bisimulation R with $(x, y) \in R$. An $\mathcal{F}$-coalgebra (D, δ) is called *final* if there exists for any $\mathcal{F}$-coalgebra (S, α) a unique homomorphism from (S, α) to (D, δ). We have the following result.

Theorem 1. *(Internal full abstractness) For a final $\mathcal{F}$-coalgebra (D, δ) and $x, y \in D$, $x = y$ if and only if $x \sim y$.*

The proof is easy, see, e.g., [Rut96], Theorem 9.2. The main difficulty in obtaining full abstractness lies in the *construction* of a final coalgebra, which in general is nontrivial.

3 A coalgebraic interpretation of Larsen-Skou bisimulation

Starting from the definitions of a discrete probabilistic transition system and probabilistic bisimulation as proposed in the literature, we will consider generalizations of (discrete) probabilistic transition systems as coalgebras of a functor $\mathcal{D}$ on **Set**. We argue that $\mathcal{D}$-bisimilarity implies probabilistic bisimilarity, and, using the notion of z-closure, that probabilistic bisimulation and totality imply $\mathcal{D}$-bisimilarity. Then it is shown how this leads to the existence of a fully abstract domain.

Definition 2. [LS91,GSS95] A discrete probabilistic transition system is a tuple (Pr, Act, μ) where Pr is a given set of processes, Act is a given set actions, and $\mu: Pr \times Act \times Pr \to [0,1]$ is a so-called transition probability function, i.e., for all $P \in Pr$, $a \in Act$, $\mu(P, a, \cdot)$ is either the zero-map or a simple probability distribution.

A probabilistic bisimulation for a discrete probabilistic transition system is an equivalence '$\equiv$' on Pr such that

$$P \equiv Q \Rightarrow \sum_{P' \in E} \mu(P, a, P') = \sum_{P' \in E} \mu(Q, a, P')$$

for all $P, Q \in Pr$, $a \in Act$, and equivalence classes $E \in Pr/\equiv$. (Using the conventions of Section 2, the implication can also be written as $P \equiv Q \Rightarrow \mu[P, a, E] = \mu[Q, a, E]$.) Two processes P and Q are said to be probabilistic bisimilar if some probabilistic bisimulation contains the pair (P, Q).

Above we introduced the notation $\mathcal{D}(S)$ for the collection of all simple probability distributions over a set S. In fact, $\mathcal{D}$ can be extended to a **Set**-functor by defining for a mapping $f: S \to T$ a function $\mathcal{D}(f): \mathcal{D}(S) \to \mathcal{D}(T)$ which maps a simple distribution μ on S to a simple distribution $\mathcal{D}(f)(\mu)$ on T such that $\mathcal{D}(f)(\mu)(t) = \mu[f^{-1}(\{t\})]$.

Let **0** represent termination. Note that a probabilistic transition system is just a mapping $\mu: Pr \times Act \to \mathcal{D}(Pr) + \{\mathbf{0}\}$ or, equivalently, a function $\mu: Pr \to (Act \to (\mathcal{D}(Pr) + \{\mathbf{0}\}))$. In other words, a probabilistic transition system is precisely a coalgebra of the functor $Act \to (\mathcal{D}(\cdot) + \{\mathbf{0}\})$. Applying the category theoretical machinery as described in Section 2 now gives us the coalgebraic notion of bisimulation. We will show that it corresponds to (actually generalizes) the notion of probabilistic bisimulation of Definition 2, thus providing categorical evidence for the Larsen-Skou bisimulation as the canonical process equivalence for discrete probabilistic transition systems.

For clarity of presentation we suppress, for the moment, the action component of a probabilistic transition system, and also do not bother about termination. Thus we consider coalgebras of the functor $\mathcal{D}$ itself. As it turns out, the presence of labels and termination does not make any essential difference for the technical content of what follows. Before we relate probabilistic bisimulation with $\mathcal{D}$-bisimulation, we first give a generalization of Definition 2, by allowing bisimulations between different transition systems, which are not necessarily equivalence relations.

Definition 3. Let $\alpha\colon S \to \mathcal{D}(S)$, $\beta\colon T \to \mathcal{D}(T)$ be two (stripped) discrete probabilistic transition systems. A binary relation $R \subseteq S \times T$ is called a probabilistic bisimulation for α, β iff $R(s,t) \Rightarrow \alpha(s)[U] = \beta(t)[V]$, for all $s \in S, t \in T$ and $U \subseteq S, V \subseteq T$ such that $\pi_1^{-1}(U) = \pi_2^{-1}(V)$. Two elements $s \in S$, $t \in T$ are said to be probabilistic bisimilar if some probabilistic bisimulation contains the pair (s,t).

Note that if R is an equivalence relation, then $\pi_1^{-1}(U) = \pi_2^{-1}(V)$ if and only if $U = \bigcup_{i \in I} E_i = V$, for some collection of equivalence classes $\{E_i | i \in I\}$ of R. Thus in this case, the condition on U and V in Definition 3 amounts to the assumption of E being an equivalence class in Definition 2, or, following the terminology of [Hen95], U and V are the same '$\equiv$'-block. This shows that Definition 2 is a special instance of Definition 3 ('modulo' the presence of labels and termination).

By exploitation of the various definitions one straightforwardly verifies that $\mathcal{D}$-bisimulation implies probabilistic bisimulation.

Lemma 4. Let $\alpha\colon S \to \mathcal{D}(S)$ and $\beta\colon T \to \mathcal{D}(T)$ be two discrete probabilistic transition systems. Let R be a $\mathcal{D}$-bisimulation for α, β. Then R is a probabilistic bisimulation for α, β.

The reverse of the above lemma is more intricate. We will first use the concept of z-closure and associated properties as developed in Section 2.

Lemma 5. If $R \subseteq S \times T$ is a probabilistic bisimulation for $\alpha\colon S \to \mathcal{D}(S)$, $\beta\colon T \to \mathcal{D}(T)$, then so is R^*, the z-closure of R.

So, if $s \in S$ and $t \in T$ are probabilistic bisimilar, we can assume —without loss of generality— that there exists a z-closed probabilistic bisimulation containing (s,t). We will need, for technical reasons, that R is total. This is equivalent with the common assumption of transition systems to have a distinguished initial state and considering reachable states only.

Theorem 6. Let $R \subseteq S \times T$ be a probabilistic bisimulation for $\alpha\colon S \to \mathcal{D}(S)$ and $\beta\colon T \to \mathcal{D}(T)$. Moreover, assume R to be z-closed and total. Then R is a $\mathcal{D}$-bisimulation.

Proof. The mapping $\gamma\colon R \to \mathcal{D}(R)$ given by

$$\gamma(s,t)(s',t') = \begin{cases} 0 & \text{if } \beta(t)[F(s')] = 0 \\ \dfrac{\alpha(s)(s') \cdot \beta(t)(t')}{\beta(t)[F(s')]} & \text{otherwise,} \end{cases}$$

for $(s,t) \in R$, is mediating for α and β. $\square$

The format of the definition of $\gamma(s,t)$ is reminiscent of the discrete probability distributions of [JL91]. It is however not clear how their notion of probabilistic specification extends to the continuous setting of Section 4.

It is straightforward to adapt the above line of reasoning to a functor $\mathcal{D}'$ given by $\mathcal{D}' = Act \to (\mathcal{D}(\cdot) + \{\mathbf{0}\})$. The discrete probabilistic transition systems of Definition 2 are in 1–1 correspondence with the coalgebras of this functor, and the notion of $\mathcal{D}'$-bisimulation coincides with that of probabilistic bisimulation of Definition 2 (for total relations R).

We can now benefit from some general insights in the theory of coalgebras, by applying (a minor variation on) a result from [Bar93] involving boundedness of a set functor.

Theorem 7. *The functor $\mathcal{D}'$ (and also $\mathcal{D}$) has a final coalgebra.*

The final coalgebra for $\mathcal{D}'$ is nontrivial. The final coalgebra for $\mathcal{D}$, though, is degenerate: it equals the one element set. This is equivalent to the fact that, due to the absence of labels and a concept of termination as present for $\mathcal{D}'$, all elements in any two $\mathcal{D}$-coalgebras are probablisitically bisimilar.

Let $\mathbb{P}$ be the final $\mathcal{D}'$-coalgebra, so $\mathbb{P} \cong Act \to (\mathcal{D}(\mathbb{P}) + \{\mathbf{0}\})$. (Note that final coalgebras are always fixed points. See, e.g., [Rut96], Theorem 9.1.) The following is immediate by Theorem 1.

Corollary 8. *The system $\mathbb{P}$ is internally fully abstract with respect to the original notion of probabilistic bisimulation of Definition 2.*

4 $\mathcal{M}_1$-Bisimilarity for Probabilistic Transition Systems

The previous section illustrates that in a discrete probabilistic setting, a coalgebraic interpretation of probabilistic transition systems and bisimulation can be given, which is equivalent with the usual 'direct' approach. One of the advantages of the abstract coalgebraic approach is that it can fairly easily be generalized to the continuous setting of stochastic systems. We will now, in fact, allow probability measures to play the role of the simple distributions in the definition of a probabilistic transition system.

Probability measures only make sense in the context of a σ-algebra. When the collection of processes comes equipped with a topology —as is the case if the set of processes is endowed with an order or a metric structure— the obvious choice for this σ-algebra is the Borel σ-algebra, i.e. the least σ-algebra containing all the open sets. As mentioned in the introduction, we prefer the use of ultrametric (cf. [BV96]) above order, because of a combination of the following two reasons: (1) the technical advantage of a close relationship between standard measure theory and metric topology, and (2) the availability of a final coalgebra theorem in the metric setting, leading to a fully abstract domain for general probabilistic bisimulation.

The generalization of the notion of a discrete probabilistic transition system and the associated concept of bisimulation as proposed by Larsen and Skou is as follows.

Definition 9. A (general) probabilistic transition system is a tuple (Pr, Act, μ) where Pr is a given ultrametric space of processes, Act is a given set of actions, and $\mu\colon Pr \times Act \times \mathcal{B}(Pr) \to [0,1]$ is a so-called (general) transition probability function, i.e., $\mu(P, a, \cdot)$ is either the zero-map, or a Borel probability measure, for all $P \in Pr$, $a \in Act$. (Here $\mathcal{B}(Pr)$ denotes the collection of Borel measurable subsets of Pr.)

A probabilistic bisimulation for a probabilistic transition system (Pr, Act, μ) is an equivalence '$\equiv$' on Pr such that every equivalence class $E \subseteq Pr$ of '$\equiv$' is measurable, and
$$P \equiv Q \Rightarrow \mu(P, a, E) = \mu(Q, a, E)$$
for all $P, Q \in Pr$, $a \in Act$, and $E \in Pr/\equiv$. Two processes P and Q in Pr are said to be probabilistic bisimilar if there exists a probabilistic bisimulation containing the pair (P, Q).

Note that the equivalence classes E of '$\equiv$' must be measurable, since only then the values $\mu(P, a, E), \mu(Q, a, E)$ are well-defined.

For reasons of presentation, we dispense with the actions and with the treatment of termination. They can be added again later. In this way, a probabilistic transition system becomes a function $\alpha\colon S \to \mathcal{M}_1(S)$ where $\mathcal{M}_1(S)$ denotes the collection of all Borel probability measures. In the reformulation of the related notion of probabilistic bisimulation we give, as before, first a slightly more general definition of bisimilarity of systems with different carriers.

Definition 10. Let $\alpha\colon S \to \mathcal{M}_1(S)$ and $\beta\colon T \to \mathcal{M}_1(T)$ be two probabilistic transition systems. A relation $R \subseteq S \times T$ is called a probabilistic bisimulation for α, β iff $R(s,t) \Rightarrow \alpha(s)(U) = \beta(t)(V)$ for all $s \in S$, $t \in T$ and $U \in \mathcal{B}(S)$, $V \in \mathcal{B}(T)$ such that $\pi_1^{-1}(U) = \pi_2^{-1}(V)$. Two elements $s \in S$, $t \in T$ are said to be probabilistic bisimilar iff some probabilistic bisimulation contains the pair (s,t).

As for $\mathcal{D}$ in the previous section, $\mathcal{M}_1$ can be regarded as a functor, viz. a functor on the category **UMS** of ultrametric spaces and nonexpansive mappings.

Definition 11. The functor $\mathcal{M}_1\colon \textbf{UMS} \to \textbf{UMS}$ is given as follows: $\mathcal{M}_1(M)$ is the collection of all Borel probability measures endowed with the metric d such that $d(\mu, \nu) \leq \varepsilon \iff \forall O \in \mathcal{O}_\varepsilon\colon \mu(O) = \nu(O)$, for all $\mu, \nu \in \mathcal{M}_1(M)$, $\varepsilon > 0$. For nonexpansive $f\colon M \to N$ the mapping $\mathcal{M}_1(f)\colon \mathcal{M}_1(M) \to \mathcal{M}_1(N)$ is defined by $\mathcal{M}_1(f)(\mu)(V) = \mu(f^{-1}(V))$, for all $V \in \mathcal{B}(N)$.

Elementary considerations concerning Borel-σ-algebras and nonexpansive maps show that $\mathcal{M}_1$ is a well-defined functor on **UMS**. Following the coalgebraic paradigm, $\mathcal{M}_1$ induces a notion of $\mathcal{M}_1$-bisimulation. One half of the relationship of $\mathcal{M}_1$-bisimulation and probabilistic bisimulation can be shown directly.

Lemma 12. Let $\alpha\colon S \to \mathcal{M}_1(S)$, $\beta\colon T \to \mathcal{M}_1(T)$ be two probabilistic transition systems. Any $\mathcal{M}_1$-bisimulation R for α and β is also a probabilistic bisimulation for α, β.

Below we show that the reverse also holds under reasonable conditions. The technicality to be dealt with concerns the proper generalization of the measurability condition of the equivalence classes E.

For a probabilistic bisimulation '$\equiv$' in the sense of Definition 9 we have, by an elementary set-theoretic argument, a partitioning into squares of subsets. Moreover, these subsets are measurable by assumption. So, we have $\equiv = \bigcup_{i \in I} E_i \times E_i$. Similarly, for the general set-up, we want a decomposition $R = \bigcup_{k \in K} E_k \times F_k$ where the E_k and F_k are Borel sets in S and T, respectively. Additionally, for measure theoretical considerations, we will assume the number of rectangles $E_k \times F_k$ that constitute R to be countable.

Definition 13. A binary relation $R \subseteq S \times T$ on two ultrametric spaces S and T is said to have a Borel decomposition iff $R = \bigcup_{k \in K} E_k \times F_k$ where $\{E_k \mid k \in K\}$, $\{F_k \mid k \in K\}$ are countable partitions of Borel sets of S and T, respectively.

In the construction of a mediating probabilistic transition system $\gamma: R \to \mathcal{M}_1(R)$, for a given probabilistic bisimulation R, we can again assume that R is z-closed. Since no measure theoretical considerations are involved, the proof of this is literally as for Lemma 5. The property is used in the next result.

Theorem 14. *Let $\alpha: S \to \mathcal{M}_1(S)$, $\beta: T \to \mathcal{M}_1(T)$ be two probabilistic transition systems. Let R be a probabilistic bisimulation for α, β in the sense of Definition 10. Assume that R is z-closed. If R has a Borel decomposition, then R is an $\mathcal{M}_1$-bisimulation for α, β.*

Proof. Let $\{E_k \times F_k \mid k \in K\}$ be a Borel decomposition of R. Suppose $R(s, t)$ holds. The mapping $\gamma(s, t): \mathcal{B}(R) \to [0, 1]$ is then given by

$$\gamma(s,t)((U \times V) \cap R) = \sum_{k \in K} \frac{\alpha(s)(U \cap E_k) \cdot \beta(t)(V \cap F_k)}{\beta(t)(F_k)} \quad (4.1)$$

for $U \in \mathcal{B}(S)$, $V \in \mathcal{B}(T)$. The verification that $\gamma(s, t)$ is well-defined and mediating for $\alpha(s)$, $\beta(t)$ is nontrivial but omitted for reasons of space. □

In the remainder of this section, we shall again use some general insights from the theory of coalgebras, this time by applying a result from [AR89,RT93].

In turns out, that we are only able to show the existence of a final coalgebra when we consider an adaptation of $\mathcal{M}_1$, say $\mathcal{M}'_1$, which delivers Borel probability measures with so-called compact support, i.e., measures that vanish outside a compact set. More precisely, for a metric space M, $\mu: \mathcal{B}(M) \to [0, 1]$ is said to have a compact support if, for some compact subset $K \subseteq M$, we have that $U \cap K = \emptyset \Rightarrow \mu(U) = 0$, for all $U \in \mathcal{B}(M)$. Let $\mathcal{M}'_1(M)$ denote the collection of all Borel probability measures of an ultrametric space M. Similarly as for $\mathcal{M}_1$, the new $\mathcal{M}'_1$ extends to a functor on **UMS**.

Additionally, to ensure the property of local contractivity (see, e.g., [RT93]), we put in a scaling functor $\cdot/2$. This operation is harmless from a semantical point of view. The usage of $\mathcal{M}'_1$, though, does narrow the type of transition systems falling within the framework. However, we stress that the established relationship of coalgebraic and probabilistic bisimulation, still carry through for the modified setting. Additionally, for the class of transition systems, now captured by the functor $Act \to (\mathcal{M}'_1(\cdot)/2 + \{\mathbf{0}\})$, the existence of a final coalgebra is guaranteed.

Theorem 15. *Let the functor $\mathcal{F}$: **UMS** $\to$ **UMS** be given by $\mathcal{F} = Act \to (\mathcal{M}'_1(\cdot)/2 + \{\mathbf{0}\})$. Then the following holds:*

(a) $\mathcal{F}$ is locally contractive, i.e., for some κ, $0 \leq \kappa < 1$, and all ultrametric spaces M and N, the function $\mathcal{F}_{M,N}: (M \to_1 N) \to (\mathcal{F}(M) \to_1 \mathcal{F}(N))$ given by $\mathcal{F}_{M,N}(f) = \mathcal{F}(f)$ is κ-contractive.
(b) If M is complete, then $\mathcal{F}(M)$ is complete.
(c) The functor $\mathcal{F}$ has a final coalgebra.

The presence of '$\cdot/2$' in the definition of $\mathcal{F}$ results in (a). (The other constituent functors are locally nonexpansive.) Only for part (b) the assumption of measures having a compact support is necessary. Its proof is non-trivial. Finally, part (c) follows from (a), (b), and (a minor variation of) [RT93], Theorem 4.8.

Let $\mathbb{Q}$ be the final $\mathcal{F}$-coalgebra: $\mathbb{Q} \cong Act \to (\mathcal{M}'_1(\mathbb{Q})/2 + \{\mathbf{0}\})$. From Theorem 1 and 15 we then immediately obtain the following result.

Corollary 16. *The system $\mathbb{Q}$ is internally fully abstract with respect to probabilistic bisimulation.*

5 Conclusion and future research

In this paper, a framework is proposed for probabilistic transition systems, involving general probability measures, and an associated notion of probabilistic bisimulation. Most research reported in the literature so far deals with *discrete* probabilistic transition systems, employing simple probability distributions only. The use of Borel measures allows for an extension of this to a continuous setting, which is necessary for the further development of models for dynamical, real-time, and in particular hybrid systems, for which discreteness and image-finiteness are often too restrictive.

Following the transition-systems-as-coalgebras paradigm, the categorical set-up provides a characterization of the Larsen-Skou bisimulation in terms of a set functor. For the continuous case, a similar result is shown for a functor on the category of ultrametric spaces. Moreover, exploiting parts of the theory of coalgebras, both for the discrete case and for the continuous case, internally fully abstract domains are constructed.

Further investigations of the proposed notion of Borel decomposition should clarify how the latter relates to the use of Polish spaces as in [BDEP97]. We expect that the technical result obtained there, on the existence of weak pullbacks, applies also to our setting. Also, once a suitable continuous process language is identified (such as PCCS [GJS90] for the discrete case), the process equivalences and fully abstract domains presented in this paper may be fruitfully applied in the semantical study of dynamical and hybrid systems.

References

[Acz88] P. Aczel. *Non-Well-Founded Sets*. CSLI Lecture Notes 14. Center for the Study of Languages and Information, Stanford, 1988.

[AM89] P. Aczel and N. Mendler. A final coalgebra theorem. In D.H. Pitt et al., editors, *Proc. Category Theory and Computer Science*, pages 357–365. LNCS 389, 1989.

[AR89] P. America and J.J.M.M. Rutten. Solving reflexive domain equations in a category of complete metric spaces. *Journal of Computer Systems and Sciences*, 39:343–375, 1989.

[Bar93] M. Barr. Terminal coalgebras in well-founded set theory. *Theoretical Computer Science*, 114:299–315, 1993. See also the addendum in Theoretical Computer Science, 124:189–192, 1994.

[BDEP97] R. Blute, J. Desharnais, A. Edalat, and P. Panangaden. Bisimulation for labelled Markov processes. In *Proc. LICS'97*. Warzaw, 1997.

[BV96] J.W. de Bakker and E.P. de Vink. *Control Flow Semantics*. The MIT Press, 1996.

[Eda94] A. Edalat. Domain theory and integration. In *Proc. LICS'94*, pages 115–124. Paris, 1994.

[GJS90] A. Giacalone, C. Jou, and S.A. Smolka. Algebraic reasoning for probabilisitic concurrent systems. In *Proc. Working Conference on Programming Concepts and Methods*. IFIP TC2, Sea of Gallilee, 1990.

[GSS95] R.J. van Glabbeek, S.A. Smolka, and B. Steffen. Reactive, generative and stratified models of probabilistic processes. *Information and Computation*, 121:59–80, 1995.

[Hen95] T.A. Henzinger. Hybrid automata with finite bisimulations. In Z. Fülöp and F. Gécseg, editors, *Proc. ICALP'95*, pages 324–335. LNCS 944, 1995.

[JL91] B. Jonsson and K.G. Larsen. Specification and refinement of probabilistic processes. In *Proc. LICS'91*, pages 266–277. Amsterdam, 1991.

[JP89] C. Jones and G. Plotkin. A probabilistic powerdomain of evaluations. In *Proc. LICS'89*, pages 186–195. Asilomar, 1989.

[LS91] K.G. Larsen and A. Skou. Bisimulation through probabilistic testing. *Information and Computation*, 94:1–28, 1991.

[RT93] J.J.M.M. Rutten and D. Turi. On the foundations of final semantics: nonstandard sets, metric spaces, partial orders. In J.W. de Bakker, W.-P. de Roever, and G. Rozenberg, editors, *Proc. REX Workshop on Semantics: Foundations and Applications*, pages 477–530. LNCS 666, 1993.

[RT94] J.J.M.M. Rutten and D. Turi. Initial algebra and final coalgebra semantics for concurrency. In J.W. de Bakker, W.-P. de Roever, and G. Rozenberg, editors, *Proc. REX School/Symposium 'A Decade of Concurrency'*, pages 530–582. LNCS 803, 1994.

[Rud66] W. Rudin. *Real and Complex Analysis*. McGraw-Hill, 1966.

[Rut96] J.J.M.M. Rutten. Universal coalgebra: a theory of systems. Report CS-R9652, CWI, 1996. Ftp-available at `ftp.cwi.nl` as pub/CWIreports/-AP/CS-R9652.ps.Z.

Distributed Processes and Location Failures
(Extended Abstract)

James Riely and Matthew Hennessy*

Abstract

Site failure is an essential aspect of distributed systems; nonetheless its effect on programming language semantics remains poorly understood. To model such systems, we define a process calculus in which processes are run at distributed *locations*. The language provides operators to kill locations, to test the status (dead or alive) of locations, and to spawn processes at remote locations. Using a variation of bisimulation, we provide alternative characterizations of strong and weak barbed congruence for this language, based on an operational semantics that uses *configurations* to record the status of locations. We then derive a second, symbolic characterization in which configurations are replaced by logical formulae. In the strong case the formulae come from a standard propositional logic, while in the weak case a temporal logic with past time modalities is required. The symbolic characterization establishes that, in principle, barbed congruence for such languages can be checked efficiently using existing techniques.

1 Introduction

Many semantic theories have been proposed for concurrent processes [18, 16, 6]. Although these theories have been fruitfully applied to the analysis of some distributed systems, for the most part they ignore an essential feature of such systems, namely their *distribution*.

As a simple example consider two implementations of a client-server application in which the client can demand an interactive service provided by the server, such as previewing or updating a document. In one implementation (System A) the server spawns a process to handle the document at its own site, the remote location, and the client previews the document remotely. In the other (System B) the server sends a process, including the document, to the client site, and the client previews the document locally. Using the semantic theories mentioned above it would be difficult to distinguish between these implementations, as the only difference between them is the location at which activity occurs. We aim to develop a useful *extensional* theory of systems which would take this type of property into account.

---

*Research funded by EPSRC project GR/K60701. Authors' address: School of Cognitive and Computing Sciences, Univ. of Sussex, Falmer, Brighton, BN1 9QH, UK, {jamesri,matthewh}@cogs.susx.ac.uk
Acknowledgement: We thank Flavio Corradini and Alan Jeffrey; both made important comments and suggestions in the early stages of this work.

In [8, 20, 10] such theories have been proposed. All of these theories, however, are based on a very strong assumption: that an observer, or user, can determine the location at which every action is performed. Here we start from a weaker premise: that in distributed systems sites are liable to *failure*. The model of failure we have adopted is a *fail stop* model in which failures are independent of each other and the number of failures that can occur is unbounded. Assuming that sites can fail, it is easy to see that Systems A and B, outlined above, are indeed different: if, after the client has begun interaction with the document, a failure occurs at the remote site, then in System A the client deadlocks, while in System B it can continue operation unaffected.

Our work is motivated by the papers [2, 12]. In these papers, distributed languages with location failures are defined and shown to be very expressive. In both of these papers, the semantics is based on *barbed equivalence*, which requires quantification over all program contexts and thus is difficult to use directly. In each of the cited works, the authors provide a translation from their language into a simpler (non-distributed) language and prove that the translations are adequate or fully abstract in some sense. While these translations provide theoretical results about the relative expressiveness of distributed and interleaving calculi, they are sufficiently complicated to make reasoning about examples, even simple ones, very difficult.

By restricting attention to an asynchronous language, Amadio [4] has recently improved on the results of [2], providing simpler translations. Although our work developed independently of [4], the language we study has much in common with the language developed there. The main difference is that our language has no value-passing, allowing us to concentrate on the effects of location failure and simplifying the statement of many of our results. Since the issues raised by failures and value passing are largely independent, this paper may be seen as providing two extensional views of a language similar to Amadio's; the first of these is concrete, as is his translation, the second is more abstract.

In Section 2, we consider a simple language for *located processes* based on pure CCS [18], with which we assume familiarity. For example $(a.p)_\ell$ is a process located at ℓ which, if ℓ is alive, may perform the action a and then behave as $(p)_\ell$. In addition to the usual operators of CCS we have the following new operators: $\text{spawn}(\ell, p)$ which starts process p running at location ℓ; $\text{kill}\,\ell.p$ which, if location ℓ is alive, kills ℓ (with the result that any process located at ℓ is deactivated) and then behaves as p; and if ℓ then p else q which silently evolves to either p or q, depending on whether ℓ is alive or dead when the test is performed.

We give an operational semantics for this language in terms of a labelled transition system. The judgments depend on a set L, of *live* locations, and are of the form $L \triangleright P \overset{\alpha}{\longmapsto} L' \triangleright P'$, where P and P' are located processes and α is either a visible action, which permits synchronization, or the internal action τ. To decide on an appropriate equivalence between process terms we follow the approach advocated in [22]. We define both strong and weak barbed equivalence between processes, $\overset{.}{\sim}$ and $\overset{.}{\approx}$. We then dictate that the required equivalence, which we refer to as *barbed bisimulation equivalence*, is defined (for example in the weak case) as: $P \approx Q$ if and only if for every suitable context $\mathbb{C}[\,]$, $\mathbb{C}[P] \overset{.}{\approx} \mathbb{C}[Q]$. Although this may be reasonable, it is not a very useful definition; the reader is invited to determine whether the following pairs of processes should be equivalent or distinguished.

$$P_1 = [(\alpha)_\ell \mid (\overline{\alpha} + \tau.a)_k] \setminus \alpha \qquad P_2 = [(\text{if } k \text{ then } \alpha \text{ else nil})_\ell \mid (\overline{\alpha}.a)_k] \setminus \alpha$$
$$Q_1 = [(\alpha + \tau)_\ell \mid (\overline{\alpha}.a)_k] \setminus \alpha \qquad Q_2 = (\text{spawn}(k,a))_\ell$$

In Section 3 we define two bisimulation-based relations, strong and weak *Located-Failure equivalence* (*LF-equivalence*) and show that these coincide with the indirectly defined barbed congruences. Since LF-equivalence is defined using bisimulations, the problem of deciding that two systems are semantically congruent can, in principle, be solved using standard proof techniques associated with bisimulation [18]. However, constructing an LF-bisimulation requires that one consider the behavior of the systems under all possible sequences of kills, by both the systems themselves and the environment. The number of states that must be explored may be exponentially larger than the number needed to construct a CCS bisimulation.

In Section 4 we use the ideas of [15] to give alternative *symbolic* characterizations of LF-equivalence that can be decided using a much smaller state space. The idea is to replace the operational judgments $L \triangleright P \xrightarrow{\alpha} L' \triangleright P'$ with judgments of the form $P \xrightarrow{\alpha}_\varphi P'$, where φ is a logical formula that describes the circumstances under which the action α can be performed. In the strong case the required logic is straightforward: a propositional logic that describes the state (dead or alive) of the sites in the system. In the weak case, however, we require a more complicated logic that can express statements of the form *site ℓ was alive at some point in the past*. Using these symbolic transitions, the standard definition of *symbolic bisimulation* [15] requires only minor modification to capture $\simeq$ and $\approx$; hence the symbolic proof techniques and tools of [15] may be used to check the new semantic equivalences proposed in this paper.

In this extended abstract we have omitted several formal definitions and all proofs. The full version [21] includes additional results and examples, including a discussion of basic processes and comparisons with other equivalences.

2 The Language

The syntax of processes is parameterized with respect to several syntactic sets. We assume a set *Loc* of *locations* k, ℓ, m, a set *PConst* of *process constants* A used to define recursive processes, and a set *Act* of *communication actions* a, b, c, such that every action $a \in Act$ has a complement $\overline{a} \in Act$ ($\overline{}$ is a bijection on *Act*). The set $Act_\tau = Act \cup \{\tau\}$ of *actions* α includes also the distinguished *silent action* τ. The formal syntax is as follows. Most of the operators should be familiar from CCS; all of the new constructs have been described in the introduction.

$$p, q \, (\in BProc) ::= \alpha.p \mid \text{spawn}(\ell, p) \mid \text{kill}\,\ell.p \mid \text{if } \ell \text{ then } p \text{ else } q \mid A \mid \Sigma_{i \in I} p_i$$
$$\mid p \mid q \mid p \setminus a \mid p[f]$$
$$P, Q \, (\in LProc) ::= P \mid Q \mid P \setminus a \mid P[f] \mid (p)_\ell$$

We have adopted a two-level syntax which distinguishes between *basic* processes p and *located* processes P. Intuitively, a basic process corresponds to what one normally thinks of as a *process*: a collection of threads of computation that must be run at a single site. A located process, instead, corresponds to a *distribution* of basic processes over several sites. Note that many basic processes may be located at a single site, and a basic process may share a private channel (unknown to other basic processes running at the same site) with a remote process.

The ability of a process to perform an action is dependent on the set of live locations, and consequently the transition relation determining the operational semantics is defined between *configurations*. A *liveset* L is any subset of Loc. A *configuration* $(L \triangleright P)$ is a pair comprising a liveset L and a located process term P. The set of all configurations is $Config$, ranged over by C and D.

In giving the intensional semantics of processes, it will be convenient for later development if we distinguish executions of the operator $kill\,\ell.p$ depending upon whether ℓ is alive or dead at the time of execution. To capture this distinction, we extend the set of actions to the set $KAct = Act \cup \{kill\,\ell \mid \ell \in Loc\}$, which includes the *kill actions* $kill\,\ell$. Unless otherwise specified, μ ranges over $KAct_\tau = KAct \cup \{\tau\}$. In Table 1 we define the transition relation $(\xrightarrow{\mu}) \subseteq Config \times Config$. The definition uses the following simple structural equivalence on processes:

$$(p|q)_\ell \equiv (p)_\ell | (q)_\ell \qquad (p \backslash a)_\ell \equiv (p)_\ell \backslash a \qquad (p[f])_\ell \equiv (p)_\ell [f]$$

While the transition relation $\longrightarrow$ distinguishes effective kill actions from those that have no effect, a basic tenet of our study is that the precise moment of location failure should be unobservable. Thus we extract from $\longrightarrow$ a transition relation $\longmapsto$ in which all kill actions have been replaced with silent actions. It is this derived relation $\longmapsto$ that we take to be fundamental.

Definition 1 ($\longmapsto$). $\qquad C \xmapsto{a} C'$ iff $C \xrightarrow{a} C'$ $\hfill\square$
$\qquad\qquad\qquad\qquad\quad C \xmapsto{\tau} C'$ iff $C \xrightarrow{\tau} C'$ or $\exists k: C \xrightarrow{kill\,k} C'$

Most of the rules in Table 1 are straightforward, being inherited directly from CCS, modulo the constraint that the process $(p)_\ell$ can only move if ℓ is alive. Note that the three new operators — kill, spawn and the conditional — are modeled as τ-transitions; this reflects the fact that in a distributed system the implementation of these operators would involve some computation and thus the passage of some time.

We now discuss the problem of defining an appropriate semantic equivalence for located processes, based on the transition relation $\longmapsto$. An obvious possibility is to adapt the bisimulation equivalences of CCS [18]. (Strong) CCS-*bisimulation* is the largest symmetric relation $\dot\sim^{ccs}$ on configurations such that whenever $C \dot\sim^{ccs} D$ and $C \xmapsto{\alpha} C'$ there exists a D' such that $D \xmapsto{\alpha} D'$ and $C' \dot\sim^{ccs} D'$. A weak version of this relation, $\dot\approx^{ccs}$, can be obtained by adapting this definition to the *weak* transition relation $\Longmapsto$, defined as usual. To see that CCS-bisimulation is not suitable for our language, for example is not a congruence, consider the processes $P_3 = [(\alpha.a)_\ell | (\overline{\alpha})_k] \backslash \alpha$ and $Q_3 = [(\alpha)_\ell | (\overline{\alpha}.a)_k] \backslash \alpha$. $P_3 \dot\sim^{ccs} Q_3$, but P_3 and Q_3 can be distinguished by a context that kills location ℓ, if this kill action is performed after the initial communication on α.

The use of $\dot\approx^{ccs}$ for CCS has been justified in [22] by the fact that it coincides with the congruence obtained from a simple notion of observation called *barbed bisimulation*. Similar results have been obtained for lazy and eager functional languages [1, 14, 7],

Table 1 Transition system with configurations (symmetric rules for | omitted)

Act_c)	$L \triangleright (\alpha.p)_\ell \xrightarrow{\alpha} L \triangleright (p)_\ell$	if	$\ell \in L$
$Spawn_c$)	$L \triangleright (spawn(k, p))_\ell \xrightarrow{\tau} L \triangleright (p)_k$	if	$\ell \in L$
$Kill1_c$)	$L \triangleright (kill\, m.p)_\ell \xrightarrow{kill\, m} L \setminus \{m\} \triangleright (p)_\ell$	if	$\ell \in L, m \in L$
$Kill2_c$)	$L \triangleright (kill\, m.p)_\ell \xrightarrow{\tau} L \triangleright (p)_\ell$	if	$\ell \in L, m \notin L$
$Cond1_c$)	$L \triangleright (if\, m\, then\, p\, else\, q)_\ell \xrightarrow{\tau} L \triangleright (p)_\ell$	if	$\ell \in L, m \in L$
$Cond2_c$)	$L \triangleright (if\, m\, then\, p\, else\, q)_\ell \xrightarrow{\tau} L \triangleright (q)_\ell$	if	$\ell \in L, m \notin L$
Sum_c)	$L \triangleright (\sum_{i \in I} P_i)_\ell \xrightarrow{\mu} L' \triangleright (p'_j)_k$	if	$L \triangleright (p_j)_\ell \xrightarrow{\mu} L' \triangleright (p'_j)_k, j \in I$
Def_c)	$L \triangleright (A)_\ell \xrightarrow{\mu} L' \triangleright (p')_k$	if	$L \triangleright (p)_\ell \xrightarrow{\mu} L' \triangleright (p')_k, A \stackrel{def}{=} p$
Str_c)	$L \triangleright P \xrightarrow{\mu} L' \triangleright Q$	if	$P \equiv P', L \triangleright P' \xrightarrow{\mu} L' \triangleright Q', Q' \equiv Q$
Par_c)	$L \triangleright P \mid Q \xrightarrow{\mu} L' \triangleright P' \mid Q$	if	$L \triangleright P \xrightarrow{\mu} L' \triangleright P'$
$Comm_c$)	$L \triangleright P \mid Q \xrightarrow{\tau} L' \triangleright P' \mid Q'$	if	$L \triangleright P \xrightarrow{a} L' \triangleright P', L \triangleright Q \xrightarrow{\bar{a}} L' \triangleright Q'$
$Restr_c$)	$L \triangleright P \setminus a \xrightarrow{\mu} L' \triangleright P' \setminus a$	if	$L \triangleright P \xrightarrow{\mu} L' \triangleright P', \mu \notin \{a, \bar{a}\}$
Ren_c)	$L \triangleright P[f] \xrightarrow{f(\mu)} L' \triangleright P'[f]$	if	$L \triangleright P \xrightarrow{\mu} L' \triangleright P'$

giving further evidence for the reasonableness of this approach. Roughly, two processes are barbed bisimilar if every silent transition of one can be matched by a silent transition of the other in such a way that the derived states are capable of exactly the same observable actions; in addition, the derived states must also be barbed bisimilar. For our language, the formal definition is as follows.

Definition 2 (Barbed bisimulation). Weak *barbed bisimilarity* ($\dot{\approx}$) is the largest symmetric relation over configurations such that whenever $C \dot{\approx} D$: (a) $C \xrightarrow{\tau} C'$ implies that for some D', $D \stackrel{\varepsilon}{\Longrightarrow} D'$ and $C' \dot{\approx} D'$; and (b) for every a, $C \xrightarrow{a}$ implies $D \stackrel{a}{\Longrightarrow}$. Strong barbed bisimilarity ($\dot{\sim}$) is obtained by replacing $\Longrightarrow$ by $\longmapsto$ everywhere in the definition. □

Barbed bisimulation is a very weak relation; for example, it is not preserved by parallel composition. However, by closing over all contexts we arrive at a reasonable semantic equivalence that by definition enjoys an important property, namely that it is a congruence.

Definition 3 (Barbed equivalence). Located processes P and Q are (weak) *barbed equivalent* ($P \approx Q$) if for every context $\mathbb{C}[\ ]$ such that $\mathbb{C}[P]$ and $\mathbb{C}[Q]$ are configurations, $\mathbb{C}[P] \dot{\approx} \mathbb{C}[Q]$. Strong barbed equivalence ($\sim$) is obtained in the same manner from $\dot{\sim}$. □

Because it requires quantification over all contexts, barbed equivalence is difficult to use directly. For example the processes P_1 and Q_1, given in the introduction, are distinguished by $\approx$ whereas P_2 and Q_2 are identified; it is far from obvious why. Even worse, processes P_5 and Q_5 (given in Section 3) are related, although establishing this fact requires that one prove that P_1 and Q_1 are *related* under the assumption that ℓ is alive at the time P_1 and Q_1 are compared, that is, ℓ is *initially alive*.

We end this section with some additional, simpler examples. The processes $(a)_\ell \mid (b)_k$ and $(b)_\ell \mid (a)_k$ can be distinguished by a context that kills ℓ. The same context can be used to distinguish the basic processes $\mathsf{spawn}(\ell, a)$ and $\mathsf{spawn}(k, a)$, regardless of where they are located. These examples indicate that although the location of an action is not reflected directly in the operational semantics they do impinge on the behavior of processes. The order in which kill actions are executed is also significant. For example kill ℓ.kill k can be distinguished from kill k.kill ℓ using the process $(a)_\ell \mid (b)_k$.

3 Located-Failures Equivalence

In this section and the next, we provide alternative characterizations of barbed equivalence for our language. Note that if $L \triangleright P \xrightarrow{\mu} L' \triangleright P'$, then L' is determined by L and μ. To emphasize this, we adopt the following notation. For each action μ, we define a function "iafter$_\mu$" which reflects the immediate effect of action μ on a liveset. We also define the relations $\xrightarrow{\mu}_L$ and $\xRightarrow{\mu}_L$ on process terms, which capture the capability of action μ under liveset L.

$$\mathsf{iafter}_\mu(L) \stackrel{\text{def}}{=} \begin{cases} L\setminus\{k\}, & \text{if } \mu = \mathit{kill}\,k \\ L, & \text{if } \mu \in \mathit{Act} \cup \{\tau, \varepsilon\} \end{cases} \quad \bigg| \quad \begin{array}{l} P \xrightarrow{\mu}_L P' \stackrel{\text{def}}{\Leftrightarrow} L \triangleright P \xrightarrow{\mu} \mathsf{iafter}_\mu(L) \triangleright P' \\ P \xRightarrow{\mu}_L P' \stackrel{\text{def}}{\Leftrightarrow} L \triangleright P \xRightarrow{\mu} \mathsf{iafter}_\mu(L) \triangleright P' \end{array}$$

For example, $\mathsf{iafter}_\alpha(L) = L$ for any α, and $\mathsf{iafter}_{\mathit{kill}\,\ell}(\{\ell, k\}) = \{k\}$. If $P = (\alpha.a)_\ell \mid (\overline{\alpha})_k$, then $P \xrightarrow{a}_{\{\ell,k\}} \mathsf{nil}$, but P has no a-transition under the liveset $\{k\}$.

We first present the strong case.

Definition 4 (Strong LF-equivalence). Let $\mathcal{S} = \{\mathcal{S}_L\}_{L \subseteq \mathit{Loc}}$ be an indexed family of relations on *LProc*. $\mathcal{S}$ is a *strong LF-bisimulation* if for every L, $\mathcal{S}_L$ is symmetric and whenever $P\,\mathcal{S}_L\,Q$:

(a) $P \xrightarrow{\mu}_L P'$ implies $\exists Q': Q \xrightarrow{\mu}_L Q'$ and $P'\,\mathcal{S}_{\mathsf{iafter}_\mu(L)}\,Q'$

(b) for every $k \in L$ $P\,\mathcal{S}_{L\setminus\{k\}}\,Q$

P and Q are strong LF-*equivalent under* L ($P \simeq_L Q$) if there exists a strong LF-bisimulation $\mathcal{S}$ with $P\,\mathcal{S}_L\,Q$.

P and Q are strong LF-*equivalent* ($P \simeq Q$), if $P \simeq_L Q$ for every subset L of Loc. □

In the full paper, we prove that $\simeq$ and $\sim$ coincide. The alternative characterization of weak barbed equivalence is more complicated: it is not sufficient to change the strong arrows in Definition 4 to weak arrows. To see this, consider the following processes:

$$P_5 = [(b.\beta.\alpha + b.(\alpha + \tau))_\ell \mid (\overline{\beta}.(\overline{\alpha} + \tau.a) + \overline{\alpha}.a)_k] \setminus \alpha \setminus \beta$$
$$Q_5 = \quad [(b.(\alpha + \tau))_\ell \mid \quad\quad\quad (\overline{\alpha}.a)_k] \setminus \alpha$$

If ℓ is initially dead, P_5 and Q_5 are clearly equivalent: both are strong equivalent to nil. If ℓ is initially alive, however, the situation is not so clear. The questionable move is P_5's b-transition to $P_1 \simeq [(\alpha)_\ell \,|\, (\overline{\alpha}+\tau.a)_k] \backslash \alpha$. To match this move Q_5 must perform a weak b-transition to $Q_1 \simeq [(\alpha+\tau)_\ell \,|\, (\overline{\alpha}.a)_k] \backslash \alpha$. But P_1 and Q_1 are not barbed equivalent: if ℓ is dead, then Q_1 is capable of a a transition that P_1 cannot match. This would lead one to believe that P and Q are *not* barbed equivalent; however, they are.

Intuitively this is true because when P_5 reaches P_1, ℓ must be alive; thus P_1 and Q_1 need only be compared under the constraint that ℓ is initially alive. Once this comparison has begun, the environment can distinguish Q_1 from P_1 only by killing ℓ, but it cannot control internal activity on the part of P_1 before ℓ is dead.

Definition 5 (Weak LF-equivalence). For $\mu \in Act_\tau$, define $\hat{\mu}$ such that $\hat{a} = a$ and $\hat{\tau} = \varepsilon$. The definition of $\approx$ is similar to that for $\simeq$, except that when $P \, \mathcal{S}_L \, Q$, we require:

(a) $P \xrightarrow{\mu}_L P'$ implies $\exists Q' : Q \xrightarrow{\hat{\mu}}_L Q'$ and $P' \, \mathcal{S}_L \, Q'$

(b) for every $k \in L$ $\exists Q' : Q \xrightarrow{\varepsilon}_L \cdot \xrightarrow{\varepsilon}_{L\backslash\{k\}} Q'$ and $P \, \mathcal{S}_{L\backslash\{k\}} \, Q'$ □

Whereas the first clause in the definition of weak LF-bisimulation is as one would expect, the second clause is somewhat surprising. It says, in effect, that if the environment kills a location k, then Q must be able to (silently) evolve to a process Q' that matches P; but in reaching Q', Q may exploit the intermediate states of the system (that is, k alive, then k dead).

Theorem 6. *For all located processes $P \approx Q$ if and only if $P \approx Q$.* □

4 Symbolic characterizations

While the LF-equivalences provide a great deal of insight into the meaning of barbed equivalence in distributed process description languages such as ours, they are unwieldy to use in practice. For the most part, this is due to the use of configurations in the operational semantics. In this section, we improve this situation by defining a *symbolic* transition system directly on located process terms, then giving characterizations of strong and weak LF-equivalence using these symbolic transitions. As one should expect, the weak case is quite a bit more subtle than the strong.

We begin by giving the symbolic operational semantics. The symbolic transition relation makes use of propositional formulae π, ρ, which are given a semantics in terms of livesets. Intuitively, a formula indicates a set of constraints on the status of locations (dead or alive) at the time that the transition is enabled. If $P \xrightarrow{\mu}_{\overline{0} \wedge 1} P'$ then if location 0 is dead and 1 is alive, P is capable of making an μ-transition to P'; that is, if $0 \notin L$ and $1 \in L$ then $P \xrightarrow{\mu}_L P'$. In Table 2 we define the transition relation $\xrightarrow{\mu}_\pi \subseteq LProc \times LProc$. The two transition systems are related by the fact that $P \xrightarrow{\mu}_L P'$ if and only if there exists a π such that $P \xrightarrow{\mu}_\pi P'$ and $L \vDash \pi$.

The standard definition of symbolic bisimulation [15] requires that we define entailment between formulae, which we do in the standard way:

$$\pi \Vdash \rho \text{ iff } \forall L: L \vDash \pi \text{ implies } L \vDash \rho$$

Table 2 Symbolic transition system (symmetric rules for | omitted)

Act$_s$)	$(\alpha.p)_\ell \xrightarrow{\alpha}_\ell (p)_\ell$		
Spawn$_s$)	$(\text{spawn}(k,p))_\ell \xrightarrow{\tau}_\ell (p)_k$		
Kill1$_s$)	$(\text{kill}\, m.p)_\ell \xrightarrow{killm}_{\ell \wedge m} (p)_\ell$		
Kill2$_s$)	$(\text{kill}\, m.p)_\ell \xrightarrow{\tau}_{\ell \wedge \overline{m}} (p)_\ell$		
Cond1$_s$)	$(\text{if } m \text{ then } p \text{ else } q)_\ell \xrightarrow{\tau}_{\ell \wedge m} (p)_\ell$		
Cond2$_s$)	$(\text{if } m \text{ then } p \text{ else } q)_\ell \xrightarrow{\tau}_{\ell \wedge \overline{m}} (q)_\ell$		
Sum$_s$)	$(\sum_{i \in I} P_i)_\ell \xrightarrow{\mu}_\pi (p'_j)_\ell$	if	$(p_j)_\ell \xrightarrow{\mu}_\pi (p'_j)_\ell, j \in I$
Def$_s$)	$(A)_\ell \xrightarrow{\mu}_\pi (p')_\ell$	if	$(p)_\ell \xrightarrow{\mu}_\pi (p')_\ell, A \stackrel{def}{=} p$
Str$_s$)	$P \xrightarrow{\mu}_\pi Q$	if	$P \equiv P', P' \xrightarrow{\mu}_\pi Q', Q' \equiv Q$
Par$_s$)	$P\|Q \xrightarrow{\mu}_\pi P'\|Q$	if	$P \xrightarrow{\mu}_\pi P'$
Comm$_s$)	$P\|Q \xrightarrow{\tau}_{\pi \wedge \rho} P'\|Q'$	if	$P \xrightarrow{a}_\pi P', Q \xrightarrow{\overline{a}}_\rho Q'$
Restr$_s$)	$P\backslash a \xrightarrow{\mu}_\pi P'\backslash a$	if	$P \xrightarrow{\mu}_\pi P', \mu \notin \{a, \overline{a}\}$
Ren$_s$)	$P[f] \xrightarrow{f(\mu)}_\pi P'[f]$	if	$P \xrightarrow{\mu}_\pi P'$

Note that entailment is a preorder on formulae. If $\pi \Vdash \rho$ we say that π is *stronger* than ρ. ff is the strongest formula under $\Vdash$, tt the weakest.

We must also identify a set of formulae suitable as parameters in the recursive definition of symbolic equivalence, that is, the analogs of the parameters L in the definition of LF-equivalence. Intuitively, when we say that P and Q are LF-equivalent under L, we are limiting attention to a single possible world, namely that in which exactly the sites in L are alive. The idea of symbolic equivalences, instead, is to treat many possible worlds simultaneously (via entailment). In the case of strong LF-bisimulation, where $P \simeq_L Q$ and $M \subseteq L$ imply $P \simeq_M Q$, this is achieved by restricting attention to *negative formulae* — formulae which contain no positive atoms — in the recursive definition of symbolic equivalence. Finally, we identify a transformation on formulae (indexed by actions) which specifies the conditions under which residual processes are to be compared:

$$M \vDash \text{after}_\alpha(\rho) \text{ iff } \exists L: L \vDash \rho \text{ and } M \subseteq L$$

$$M \vDash \text{after}_{killk}(\rho) \text{ iff } \exists L: L \vDash \rho \text{ and } M \subseteq L\backslash\{k\}$$

Definition 7 (Strong symbolic bisimulation). Let $\mathcal{S}$ be a family of relations on *LProc* indexed by negative formulae ϑ. $\mathcal{S}$ is a *strong symbolic bisimulation* if for every ϑ, $\mathcal{S}_\vartheta$ is symmetric and whenever $P \mathcal{S}_\vartheta Q$ and $P \xrightarrow{\mu}_\pi P'$ then for some π_i, ρ_i, and Q_i:

(a) $\vartheta \wedge \pi \Vdash \bigvee_i \rho_i$, (c) $Q \xrightarrow{\mu}_{\pi_i} Q_i$, and
(b) $\rho_i \Vdash \pi_i$, (d) $P' \mathcal{S}_{\text{after}_\mu(\rho_i)} Q_i$

We write $P \simeq^s_\vartheta Q$ to indicate that there exists a symbolic bisimulation $\mathcal{S}$ with $P \mathcal{S}_\vartheta Q$. □

Theorem 8. $P \simeq_L Q$ iff $\exists \vartheta: P \simeq^s_\vartheta Q$ and $L \vDash \vartheta$. In addition, $(\simeq) = (\simeq^s_{tt})$. □

As a first attempt to define weak symbolic bisimulation, let us try simply replacing the strong transitions in Definition 7 with weak edges defined by conjoining formulae. For example, we would have $P \stackrel{\varepsilon}{\Longrightarrow}_{tt} P$ and $P \stackrel{a}{\Longrightarrow}_{\pi\wedge\rho} P'$ if $P \stackrel{\tau}{\rightarrow}_{\pi} \cdot \stackrel{a}{\rightarrow}_{\rho} P'$. Unfortunately, this definition does not suffice. Consider the processes P_5 and Q_5, previously defined; these have the following symbolic transition graphs (where we have write $\stackrel{\mu}{\rightarrow}_{\pi}$ as $\stackrel{\mu\pi}{\longrightarrow}$):

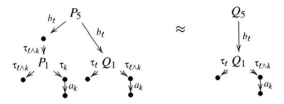

As noted in Section 3, in order to prove these processes equivalent we must compare the processes P_1 and Q_1 under the assumption that ℓ is initially alive, but using our provisional definition we would end up comparing P_1 and Q_1 under the assumption $tt = \text{neg}(\ell \wedge k)$, which is not strong enough to prove that they are related.

As a second attempt, we might simply allow all positive information to carry over into the recursive formula ϑ_i, that is, change the last clause of Definition 7 to $\vartheta_i = \rho_i$. Whereas our first attempt produced an equivalence that was too strong, the revised definition is too weak. For example, the following processes would be identified even though they are not barbed equivalent.

$$P_6 = [(\alpha.a)_\ell \,|\, (\overline{\alpha})_k] \setminus \alpha \qquad Q_6 = [(\alpha)_\ell \,|\, (\overline{\alpha}.a)_k] \setminus \alpha$$
$$\downarrow \tau_{\ell \wedge k} \qquad\qquad \approx \qquad\qquad \downarrow \tau_{\ell \wedge k}$$
$$P_6' \qquad\qquad\qquad\qquad\qquad Q_6'$$
$$\downarrow a_\ell \qquad\qquad\qquad\qquad\qquad \downarrow a_k$$

Here P_6' and Q_6' would be compared under the formula $\ell \wedge k$. This formula, however, says something more than we would like, namely that ℓ and k remain alive until P_6' and Q_6' execute their first action. More complicated examples can be constructed to show that we must be able to express properties such as "ℓ and k must have been alive, then ℓ must have died, and after that k must have died."

Our solution is to define weak symbolic edges using a *past-time temporal logic* [17], interpreted over sequences of livesets. A *live sequence* $\mathcal{L}$ is a finite nonempty sequence of livesets $\langle L_1, \ldots, L_n \rangle$, such that for every i between 1 and $n-1$ there exists a location k such that $L_{i+1} = L_i \setminus \{k\}$. For example, $\langle \{\ell\}, \varnothing \rangle$ is a live sequence, but $\langle \{\ell\}, \{\ell\} \rangle$ and $\langle \{\ell, k\}, \varnothing \rangle$ are not. We write $\mathcal{L}_{(i)}$ for the ith element of $\mathcal{L}$ and, where clear from context, use n to refer to the length of $\mathcal{L}$. Thus, for example, $\mathcal{L}$ models $\overline{\ell}$ if $\ell \notin \mathcal{L}_{(n)}$ and $\mathcal{L}$ models $\Diamond \varphi$ if $\mathcal{L}$ or some prefix of $\mathcal{L}$ models φ. Because live sequences must be strictly decreasing, $\ell \wedge \Diamond \overline{\ell}$ is unsatisfiable; however $\langle \{\ell\}, \varnothing \rangle \models \overline{\ell} \wedge \Diamond \ell$. Weak symbolic transitions are defined as follows:

$$P \stackrel{\varepsilon}{\Longrightarrow}_{tt'} P$$

$$P \stackrel{\alpha}{\Longrightarrow}_{\varphi \wedge \pi} P' \text{ if } P \stackrel{\varepsilon}{\Longrightarrow}_{\varphi} \cdot \stackrel{\alpha}{\rightarrow}_{\pi} P'$$

$$P \stackrel{\varepsilon}{\Longrightarrow}_{\Diamond(\varphi \wedge \pi)} P' \text{ if } P \stackrel{\varepsilon}{\Longrightarrow}_{\varphi} \cdot \stackrel{\tau}{\rightarrow}_{\pi} P' \qquad P \stackrel{killk}{\Longrightarrow}_{\overline{k} \wedge \ominus(\varphi \wedge \pi)} P' \text{ if } P \stackrel{\varepsilon}{\Longrightarrow}_{\varphi} \cdot \stackrel{killk}{\rightarrow}_{\pi} P'$$

$$P \stackrel{\mu}{\Longrightarrow}_{\varphi \wedge \pi} P' \text{ if } P \stackrel{\mu}{\Longrightarrow}_{\varphi} \cdot \stackrel{\tau}{\rightarrow}_{\pi} P'$$

Intuitively $P \stackrel{\mu}{\underset{\varphi}{\Longrightarrow}} P'$ means that P can perform the action μ to become P' in an environment where the change in live sets satisfies the formula φ. For example if $\varphi_1 = (\ell \wedge k) \mathbin{\mathchar"3B} \ell$ and $\varphi_2 = (\ell \wedge k) \mathbin{\mathchar"3B} k$ then P_6 has the symbolic transition $\stackrel{a}{\underset{\varphi_1}{\Longrightarrow}}$ but not $\stackrel{a}{\underset{\varphi_2}{\Longrightarrow}}$, whereas for Q_6 it is the opposite.

As parameters to the weak relation we simply take Boolean formulae, but now interpreted on the initial liveset of a live sequence. Rather than use two logics in the definition or introduce additional operators, we define the function "initially" which converts Boolean formulae into temporal formulae with this interpretation in mind. The transformation function for generating formulae after an action is performed, which we call "finally", must then transform temporal formula into propositional ones. The definitions are as follows: (In the full paper, we show how to calculate these functions.)

$$\mathcal{L} \models \text{initially}(\pi) \quad \text{iff} \quad \mathcal{L}_{(1)} \models \pi$$
$$M \models \text{finally}(\varphi) \quad \text{iff} \quad \exists \mathcal{L}: \ \mathcal{L} \models \varphi \text{ and } M = \mathcal{L}_{(n)}$$

Definition 9 (Weak symbolic bisimulation). Similar to Definition 7, except that when $P\, \mathcal{S}_\pi\, Q$ and $P \stackrel{\hat{\mu}}{\underset{\varphi}{\Longrightarrow}} P'$ we require:

(a) $\text{initially}(\pi) \wedge \varphi \Vdash \bigvee_i \psi_i,$
(b) $\psi_i \Vdash \varphi_i,$
(c) $Q \stackrel{\hat{\mu}}{\underset{\varphi_i}{\Longrightarrow}} Q_i,$ and
(d) $P'\, \mathcal{S}_{\text{finally}(\psi_i)}\, Q_i$ □

Theorem 10. $P \approx_L Q$ iff $\exists \pi: \ L \models \pi$ and $P \approx^s_\pi Q$. In addition, $(\approx) = (\approx^s_{\text{tt}})$. □

5 Conclusions

In this paper we have proposed a new semantic theory for distributed systems which takes into account the possibility of failures at sites. This theory is an adaptation of standard bisimulation-based theories [18] using an operational semantics for *located processes*. The new semantic equivalences are justified in terms of *barbed bisimulations* [22]. We also give *symbolic* characterizations of the new equivalences, which means that they can be investigated using the symbolic methods of [15].

Site failure has also played a role in languages studied in [2, 4, 12]. In these papers abstract languages based on Facile [13] or the pi-calculus [19, 5] are studied. The original motivation for this paper was to provide an alternative characterization of barbed equivalence for languages such as these. Although we have not treated value passing or references, we postulate that our results can be extended in a straightforward way to value-passing languages which retain the assumption that all failures are independent, such as the languages in [2, 4]. More delicate is the extension to languages such as the distributed join-calculus [12] in which the independence assumption is dropped. In this case the logical language used for symbolic bisimulations must be extended to allow statements about the interdependence of locations; we leave this to future work.

A number of location-based equivalences already exist in the literature [8, 9, 20, 10]; however, none of these theories addresses the possible failure of sites. Their emphasis, rather, is to define a measure of the concurrency or distribution of a process: two processes are deemed equivalent only if, informally, they have the same degree of concurrency. In the full paper we give a series of counter-examples which show that $\approx$

is incomparable with all of the equivalences proposed in these papers; we also discuss variations on the language and model of failure.

References

[1] Samson Abramsky. The lazy lambda calculus. In *Research Topics in Functional Programming*, pages 65–117. Addison-Wesley, 1990.

[2] Roberto Amadio and Sanjiva Prasad. Localities and failures. In *FST-TCS*, volume 880 of *LNCS*. Springer, 1994.

[3] Roberto Amadio. From a concurrent λ-calculus to the π-calculus. In *Foundations of Computation Theory*, volume 965 of *LNCS*. Springer, 1995.

[4] Roberto Amadio. An asynchronous model of locality, failure, and process mobility. Technical report, Laboratoire d'Informatique de Marseille, 1997.

[5] Roberto Amadio, Ilaria Castellani, and Davide Sangiorgi. On bisimulations for the asynchronous π-calculus. In *CONCUR96*, volume 1119 of *LNCS*, pages 147–162. Springer, 1996.

[6] J. C. M. Baeten and W. P. Weijland. *Process Algebra*. Cambridge University Press, 1990.

[7] G. Boudol. A lambda calculus for (strict) parallel functions. *Information and Control*, 108:51–127, 1994.

[8] G. Boudol, I. Castellani, M. Hennessy, and A. Kiehn. A theory of processes with localities. *Formal Aspects of Computing*, 6:165–200, 1994.

[9] I. Castellani. Observing distribution in processes: static and dynamic localities. *International Journal of Foundations of Computer Science*, 6(6):353–393, 1995.

[10] Flavio Corradini. *Space, Time and Nondeterminism in Process Algebras*. PhD thesis, Università Degli Studi di Roma "La Sapienza", 1996.

[11] C. Fournet and G. Gonthier. The reflexive CHAM and the join-calculus. In *POPL94*. ACM Press, 1994.

[12] C. Fournet, G. Gonthier, J.J. Levy, L. Marganget, and D. Remy. A calculus of mobile agents. In *CONCUR96*, volume 1119 of *LNCS*, pages 406–421. Springer, 1996.

[13] A. Giacalone, P. Mishra, and S. Prasad. A symmetric integration of concurrent and functional programming. *International Journal of Parallel Programming*, 18(2):121–160, 1989.

[14] Andrew D. Gordon. Bisimilarity as a theory of functional programming. In *MFPS*, volume 1 of *ENTCS*, http://pigeon.elsevier.nl/mcs/tcs/pc/Menu.html. Elsevier, 1995.

[15] M. C. B. Hennessy and H. Lin. Symbolic bisimulations. *Theoretical Computer Science*, 138:353–389, 1995.

[16] C. A. R. Hoare. *Communicating Sequential Processes*. Prentice-Hall, 1985.

[17] Z. Manna and A. Pnueli. *The Temporal Logic of Reactive and Concurrent System: Specification*. Springer, 1992.

[18] Robin Milner. *Communication and concurrency*. Prentice-Hall, 1989.

[19] Robin Milner, Joachim Parrow, and David Walker. A calculus of mobile processes. *Information and Computation*, 100(1), September 1992.

[20] Ugo Montanari and Daniel Yankelovich. Partial order localities. In *ICALP92*, volume 623 of *LNCS*, pages 617–628. Springer, 1992.

[21] James Riely and Matthew Hennessy. Distributed processes and location failures. Technical Report 2/97, University of Sussex, Department of Computer Science, http://www.cogs.susx.ac.uk, 1997.

[22] Davide Sangiorgi. *Expressing Mobility in Process Algebras: First-Order and Higher-Order Paradigms*. PhD thesis, University of Edinburgh, 1992.

Basic Observables for Processes[*]

Michele Boreale[1]   Rocco De Nicola[2]   Rosario Pugliese[2]

[1]Dipartimento di Scienze dell'Informazione, Università di Roma "La Sapienza"
[2]Dipartimento di Sistemi e Informatica, Università di Firenze

Abstract. We propose a general approach to define behavioural pre-orders over process terms by considering the pre–congruences induced by three basic observables. These observables provide information about the initial communication capabilities of processes and about their possibility of engaging in an infinite internal chattering. We show that some of the observables–based pre–congruences do correspond to behavioral pre-orders long studied in the literature. The coincidence proofs shed light on the differences between the *must* preorder of De Nicola and Hennessy and the *fair/should* preorder of Cleaveland and Natarajan and of Brinksma, Rensink and Vogler, and on the rôle played in their definition by tests for internal chattering.

1 Introduction

In the classical theory of functional programming, the point of view is assumed that executing a program corresponds to evaluating it. If we write $M \downarrow v$ to indicate that program M evaluates to value v, the problem of the equivalence of two programs, hence of their semantics, can be stated as follows:

> Two programs M and N are *observationally equivalent* if for every program context C such that both $C[M]$ and $C[N]$ are programs, and for every value v, we have: $C[M] \downarrow v$ if and only if $C[N] \downarrow v$.

An alternative approach, used e.g. for the lazy lambda calculus [1], is that of defining a *simulation* (whose kernel is an equivalence) based on the reduction to normal forms. In general, given a language equipped with a reduction relation, the paradigm for defining equivalence over terms of the language, can be traced back to Morris [16] and can be phrased as follows:

1. Define a set of observables (values, normal forms, ...) to which a program can evaluate by means of successive reductions.
2. Consider the largest (pre–)congruence over the (set of operators of the) language induced by the chosen set of observables.

This paradigm has been the basis for assessing many semantics of sequential languages and is at the heart of the full abstraction problem, see e.g. [18].

Here, we aim at taking advantage of this paradigm also to assess models of concurrent systems and their equivalences. In this case, the choice of the basic observables is less obvious. On one hand, it is well–known that input/output

---
[*] Work partially supported by EEC: HCM project EXPRESS, by CNR project "Specifica ad alto livello e verifica formale di sistemi digitali" and by Istituto di Elaborazione dell'Informazione CNR, Pisa.

relations are not sufficient for describing the semantics of these classes of systems, and thus it would be limitative to use values as observables. On the other hand, studying the evolution to normal forms under all possible contexts is not as inspective as in the case of lambda calculus. Indeed, the interaction between a λ-term and the environment is circumscribed, while that between a process and its environment is less clear.

If we consider the λ-term MN, we know the extent of the influence of N over M, and, in any computation, we know exactly *when* an interaction between M and N occurs, namely when M reduces to a λ-abstraction. Thus by observing M in all possible contexts we can fully understand its behaviour. When considering concurrent systems, the internal evolution of each parallel component is freely intermingled with external communications. Then understanding the semantics of a component via its contextual behaviour turns out to be much less obvious.

Here, we shall consider a simple process description language, TCCS (Tau-less CCS [7]), and will study the impact of three basic observables for concurrent systems on this language. However, our results are easily extensible to general SOS language formats, like GSOS [2].

We shall be interested in testing for the *initial guaranteed* communication capabilities of a system. Indeed, when one is willing to infer the interactive behaviour of a system from its "isolated" behaviour, to know about the system's *possibility* of accepting communications along specific channels is not sufficient: due to the inherent nondeterminism of concurrent computations, it is necessary to know whether the acceptance of the communications is *guaranteed*. This is essential to establish liveness properties, like the absence of deadlock.

Moreover, we shall be interested in the risk a system has of getting involved in an infinite sequence of internal communications (to *diverge*), because this could lead to ignoring all subsequent external stimuli. Finally, with respect to this, it might also be important to know the external communications that can lead to divergent states.

These considerations guide us to introducing three basic observables:

1. $P!\ell$ (P *guarantees* ℓ) asserts that, by internal actions, P can only reach states from which action ℓ can be eventually performed;
2. $P\downarrow$ (P *converges*) asserts that P cannot get involved in an infinite sequence of internal actions;
3. $P\downarrow\ell$ (P *converges along* ℓ) asserts that P converges and does so also after performing ℓ.

For finite process graphs these observables are obviously decidable; in general, they are not, but this is somehow expected since the basic language (TCCS) is Turing powerful.

We shall analyze the impact of the above predicates on the semantics of TCCS. The predicates naturally induce five contextual preorders. These preorders are listed in Table 1; there we represent a contextual preorder using the notation $\preceq^c_{s_1 s_2}$, where s_1 (if present) refers to the used convergence predicate, and s_2 (if present) refers to the guarantees one. The universal relation is denoted by $\mathcal{U}$.

conv./comm.	no req.	$\downarrow$	$\downarrow \ell$
no req.	$\mathcal{U}$	$_\downarrow \precsim^c$	$_{\downarrow\mathcal{L}} \precsim^c$
$!\ell$	$\precsim^c_\mathcal{L}$	$_\downarrow \precsim^c_\mathcal{L}$	$_{\downarrow\mathcal{L}} \precsim^c_\mathcal{L}$

Table 1. Contextual Preorders

conv./comm.	no req.	$\downarrow$	$\downarrow \ell$
no req.	$\mathcal{U}$	$\sqsubseteq_m$	$\sqsubseteq_m$
$!\ell$	$\sqsubseteq_{FT}$	$\sqsubseteq_M$	$\sqsubseteq_S$

Table 2. Main results

Our main results are five full abstraction theorems that make it manifest that our contextual preorders do coincide with well–known and/or intuitive behavioural preorders over processes studied in the literature. Table 2 provides a summary of the claimed results.

More specifically, we will show that:

- $\precsim^c_\mathcal{L}$, the contextual preorder induced by $!\ell$, coincides with $\sqsubseteq_{FT}$, the maximal pre–congruence included in the *fair/should* preorder of [17] and [3]. This pre–congruence can be characterized (see [4]) as the conjunction of the classical trace preorder (called *may* preorder in [6]) with the fair/should preorder;
- $_\downarrow \precsim^c$ and $_{\downarrow\mathcal{L}} \precsim^c$, the contextual preorders induced by $\downarrow$ and $\downarrow \ell$, both coincide with $\sqsubseteq_m$, the (reverse) inclusion of the *convergent traces* preorder, a simple variant of the trace preorder.

Together with the impact of the three observables used in isolation we also study the result of their conjunctions and show that:

- $_\downarrow \precsim^c_\mathcal{L}$, the contextual preorder induced by $\downarrow$ and $!\ell$, coincides with $\sqsubseteq_M$, the original *must* preorder of [6, 10];
- $_{\downarrow\mathcal{L}} \precsim^c_\mathcal{L}$, the contextual preorder induced by $\downarrow \ell$ and $!\ell$, gives rise to a *new* preorder, the *safe–must* preorder $\sqsubseteq_S$, which is supported by a very intuitive testing scenario.

The safe–must preorder has a direct characterization in terms of computations from pairs of observers and processes: a computation is successful if a success state is reached *before* a catastrophic one (this explains the adjective 'safe'). This notion certainly deserves further investigation.

In the rest of the paper, we recall syntax and operational semantics (Sec. 2) and introduce an observational semantics (Sec. 3) for TCCS, then we present our full abstraction results (Sec. 4), compare the semantic preorders (Sec. 5) and briefly discuss related work. Due to space limitations, most proofs have been omitted; they can be found at http://dsi2.dsi.unifi.it/~denicola.

2 Tau–less CCS: TCCS

In this section, we briefly present the syntax and the operational semantics of TCCS, (τ–less CCS [7, 10]). We have preferred to use TCCS rather than CCS because it allows us to avoid the "congruence problems" that arise when the CCS choice operator (+) is used and silent actions are abstracted away. It is worth mentioning that the very same results can be obtained by using CCS and its must pre-congruence obtained from the must preorder by imposing that whenever the "better" process can perform a silent move also the other can do it.

We assume an infinite set of *names* $\mathcal{N}$, ranged over by $a, b, \ldots$, and let $\overline{\mathcal{N}} = \{\overline{a} \mid a \in \mathcal{N}\}$, ranged over by $\overline{a}, \overline{b}, \ldots$, be the set of *co-names*. $\mathcal{N}$ and $\overline{\mathcal{N}}$ are disjoint and are in bijection via the *complementation* function ($\overline{\cdot}$); we define: $\overline{(\overline{a})} = a$. We let $\mathcal{L} = \mathcal{N} \cup \overline{\mathcal{N}}$, ranged over by $\ell, \ell', \ldots$, be the set of *labels*; we shall use B to range over subsets of $\mathcal{L}$ and we define $\overline{B} = \{\overline{\ell} \mid \ell \in B\}$. We also assume a countable set $\mathcal{X}$ of *process variables*, ranged over by $X, Y, \ldots$.

Definition 1. The set of *TCCS terms* is generated by the grammar:

$$E ::= \mathbf{0} \mid \Omega \mid \ell.E \mid E[]F \mid E \oplus F \mid E \mid F \mid E \backslash L \mid E\{f\} \mid X \mid recX.E$$

where $f : \mathcal{L} \to \mathcal{L}$, called *relabelling function* is such that $\{\ell \mid f(\ell) \neq \ell\}$ is finite, $f(a) \in \mathcal{N}$ and $f(\overline{\ell}) = \overline{f(\ell)}$. We let $\mathcal{P}$, ranged over by P, Q, etc., denote the set of *closed* terms or *processes* (i.e. those terms where every occurrence of any agent variable X lies within the scope of some $recX.\_$ operator).

In the following, we often shall write ℓ instead of $\ell.\mathbf{0}$. We write $\_\{\ell'_1/\ell_1, \ldots, \ell'_n/\ell_n\}$ for the relabelling operator $\_\{f\}$ where $f(\ell) = \ell'_i$ if $\ell = \ell_i$, $i \in \{1, \ldots, n\}$, and $f(\ell) = \ell$ otherwise. As usual, we write $E[E_1/X_1, \ldots, E_n/X_n]$ for the term obtained by simultaneously substituting each occurrence of X_i in E with E_i (with renaming of bound process variables possibly involved).

The structural operational semantics of a TCCS term is defined via the two transition relations $\longrightarrow$ and $\rightarrowtail$ induced by the inference rules in Table 3 and in Table 4, respectively. The symmetrical versions of rules **AR4** and **AR5** in Table 3 and of rules **IR5**, **IR6** and **IR7** in Table 4 have been omitted.

AR1 $\ell.P \xrightarrow{\ell} P$		
AR2 $\dfrac{P \xrightarrow{\ell} P'}{P\{f\} \xrightarrow{f(\ell)} P'\{f\}}$	**AR3** $\dfrac{P \xrightarrow{\ell} P'}{P\backslash L \xrightarrow{\ell} P'\backslash L}$ if $\ell \notin L \cup \overline{L}$
AR4 $\dfrac{P \xrightarrow{\ell} P'}{P[]Q \xrightarrow{\ell} P'}$	**AR5** $\dfrac{P \xrightarrow{\ell} P'}{P \mid Q \xrightarrow{\ell} P' \mid Q}$	

Table 3. SOS rules for TCCS: Action Relation

IR1 $\Omega \rightarrowtail \Omega$	**IR2** $recX.E \rightarrowtail E[recX.E/X]$
IR3 $\dfrac{P \rightarrowtail P'}{P\{f\} \rightarrowtail P'\{f\}}$	**IR4** $\dfrac{P \rightarrowtail P'}{P\backslash L \rightarrowtail P'\backslash L}$
IR5 $P \oplus Q \rightarrowtail P$	**IR6** $\dfrac{P \rightarrowtail P'}{P[]Q \rightarrowtail P'[]Q}$
IR7 $\dfrac{P \rightarrowtail P'}{P \mid Q \rightarrowtail P' \mid Q}$	**IR8** $\dfrac{P \xrightarrow{\ell} P', \ Q \xrightarrow{\overline{\ell}} Q'}{P \mid Q \rightarrowtail P' \mid Q'}$

Table 4. SOS rules for TCCS: Internal Relation

As usual, we use $\Longrightarrow$ or $\stackrel{\epsilon}{\Longrightarrow}$ to denote the reflexive and transitive closure of $\rightarrowtail$ and use $\stackrel{s}{\Longrightarrow}$, with $s \in \mathcal{L}^+$, for $\Longrightarrow \stackrel{\ell}{\longrightarrow} \stackrel{s'}{\Longrightarrow}$ when $s = \ell s'$. Moreover, we write $P \stackrel{s}{\Longrightarrow}$ for $\exists P' : P \stackrel{s}{\Longrightarrow} P'$ ($P \stackrel{\ell}{\longrightarrow}$ and $P \rightarrowtail$ will be used similarly). We will call *sort of* P the set $sort(P) = \{\ell \in \mathcal{L} \mid \exists s \in \mathcal{L}^* : P \stackrel{s\ell}{\Longrightarrow}\}$, *successors of* P the set $S(P) = \{\ell \in \mathcal{L} \mid P \stackrel{\ell}{\Longrightarrow}\}$, and *language* generated by P the set $L(P) = \{s \in \mathcal{L}^* \mid P \stackrel{s}{\Longrightarrow}\}$. Note that since we only consider finite relabelling operators, every TCCS process has a finite sort.

A *context* is a TCCS term C with one free occurrence of a process variable, usually denoted by $\_$. If C is a context, we write $C[P]$ instead of $C[P/\_]$. The context *closure* $\mathcal{R}^c$ of a given binary relation $\mathcal{R}$ over processes, is defined as: $P \mathcal{R}^c Q$ iff for each context C, $C[P] \mathcal{R} C[Q]$. $\mathcal{R}^c$ enjoys two important properties: (a) $(\mathcal{R}^c)^c = \mathcal{R}^c$, and (b) $\mathcal{R} \subseteq \mathcal{R}'$ implies $\mathcal{R}^c \subseteq \mathcal{R}'^c$. In the following, we will write $\overline{\mathcal{R}}$ for the complement of $\mathcal{R}$.

3 Observational Semantics

In this section, we introduce different observational semantics for TCCS; we follow two approaches. The first approach takes advantage of basic observables, the second one of the classical *testing* scenario of [6, 10] and variants of it.

3.1 Basic Observables and Observation Preorders

Definition 2. Let P be a process and $\ell \in \mathcal{L}$. We define three basic *observation predicates* over processes as follows:

- $P!\ell$ (P *guarantees* ℓ) iff $\forall P' : P \Longrightarrow P'$ implies $P' \stackrel{\ell}{\Longrightarrow}$;
- $P \downarrow$ (P *converges*) iff there is no infinite sequence of internal transitions $P \rightarrowtail P_1 \rightarrowtail \cdots$ starting from P;
- $P \downarrow \ell$ (P *converges along* ℓ) iff $P \downarrow$ and $\forall P' : P \stackrel{\ell}{\Longrightarrow} P'$ implies $P' \downarrow$.

The above predicates can be combined in five sensible ways and used to define the corresponding basic *observation preorders* over processes, as stated in the following definition.

Definition 3. Let P and Q be processes.

- $P \precsim_{\downarrow} Q$ iff $P \downarrow$ implies $Q \downarrow$;
- $P \precsim_{\downarrow_\mathcal{L}} Q$ iff for each $\ell \in \mathcal{L}$: $P \downarrow \ell$ implies $Q \downarrow \ell$;
- $P \precsim_{\mathcal{L}} Q$ iff for each $\ell \in \mathcal{L}$: $P!\ell$ implies $Q!\ell$;
- $P \precsim_{\downarrow \mathcal{L}} Q$ iff for each $\ell \in \mathcal{L}$: $P \downarrow$ and $P!\ell$ implies $Q \downarrow$ and $Q!\ell$;
- $P \precsim_{\downarrow_\mathcal{L} \mathcal{L}} Q$ iff for each $\ell \in \mathcal{L}$: $P \downarrow \ell$ and $P!\ell$ implies $Q \downarrow \ell$ and $Q!\ell$.

Of course, the basic observation preorders are very coarse. More refined relations can be obtained by closing the above preorders under all TCCS contexts. For each basic observation preorder, say $\precsim$, the *contextual preorder* generated by $\precsim$ is defined as its closure $\precsim^c$.

3.2 Testing Preorders and Alternative Characterizations

Like in the original theory of testing [6, 10], we have that:

- *observers*, ranged over by $O, O', \ldots$, are processes capable of performing an additional distinct "success" action $w \notin \mathcal{L}$;
- *computations* from $P \mid O$ are sequences of internal transitions $P \mid O \rightarrowtail P_1 \mid O_1 \rightarrowtail \cdots$, which are either infinite or such that $P_k \mid O_k \not\rightarrowtail$, $k \geq 0$.

Definition 4. Let P be a process and O be an observer.

1. $P \,\underline{must}_M\, O$ if for each computation from $P \mid O$, say $P \mid O \rightarrowtail P_1 \mid O_1 \rightarrowtail \cdots$, there is some $i \geq 0$ s.t. $O_i \xrightarrow{w}$.
2. $P \,\underline{must}_S\, O$ if for each computation from $P \mid O$, say $P \mid O \rightarrowtail P_1 \mid O_1 \rightarrowtail \cdots$, there is some $i \geq 0$ s.t. $O_i \xrightarrow{w}$ and $P_i \downarrow$.
3. $P \,\underline{must}_F\, O$ if for each computation from $P \mid O$, say $P \mid O \rightarrowtail P_1 \mid O_1 \rightarrowtail \cdots$, it holds that $P_i \mid O_i \xRightarrow{w}$ for each $i \geq 0$.

The first definition of successful computation given above is exactly that of [6]. The second one, considers successful only those computations in which a success state is reached *before* the observed process diverges. The third definition, which is essentially taken from [3], totally ignores the issue of divergence. These three notions allow us to define three preorders: the first one ($\lesssim_M$) is the original *must* preorder of [6, 10], the second one ($\lesssim_S$) is the new *safe-must* preorder and the third one ($\lesssim_F$) is the (reverse of the) *fair/should* preorder of [17] and [3].

Definition 5. Let $i \in \{M, S, F\}$. For all processes P and Q, $P \lesssim_i Q$ iff for every observer O: $P \,\underline{must}_i\, O$ implies $Q \,\underline{must}_i\, O$.

We introduce below alternative characterizations of the preorders *must* and *safe-must*. They support simpler methods for proving (or disproving) that two processes are behaviourally related. We need some additional notation.

Definition 6. Let $s \in \mathcal{L}^*$, $B \subseteq_{\text{fin}} \mathcal{L}$ and S be a set of processes.

- The *convergence* predicate, $\downarrow s$, is defined inductively as follows: $P \downarrow \epsilon$ if $P \downarrow$; $P \downarrow \ell s'$ if $P \downarrow \epsilon$ and $\forall P' : P \xRightarrow{\ell} P'$ implies $P' \downarrow s'$. We write $P \uparrow s$ if $P \downarrow s$ does not hold.
- ($P \,\underline{after}\, s$) denotes the set of processes $\{P' : P \xRightarrow{s} P'\}$.
- We write $P \downarrow B$ if $\forall \ell \in B : P \downarrow \ell$ and $S \downarrow B$ if $\forall P \in S : P \downarrow B$.
- $P \,!\, B$ stands for $\forall P' : P \Longrightarrow P'$ implies $\exists \ell \in B : P' \xRightarrow{\ell}$.
- $S \downarrow!\, B$ stands for $\forall P \in S : P \downarrow B$ and $P \,!\, B$.

Definition 7. For all processes P and Q, we write

- $P \ll_M Q$ if $\forall s \in \mathcal{L}^*$ such that $P \downarrow s$, it holds that:
  (a) $Q \downarrow s$, and (b) for every $B \subseteq_{\text{fin}} \mathcal{L}$: $(P \,\underline{after}\, s) \,!\, B$ implies $(Q \,\underline{after}\, s) \,!\, B$.
- $P \ll_S Q$ is the same as above but predicate $!$ is replaced by $\downarrow!$.

Theorem 8. For all processes P and Q, (1) $P \lesssim_M Q$ iff $P \ll_M Q$ and (2) $P \lesssim_S Q$ iff $P \ll_S Q$.

By taking advantage of the above alternative characterizations it is easy to prove that the must and the safe–must preorders are pre–congruences.

Theorem 9. For all processes P and Q and $i \in \{M, S\}$, $P \lesssim_i Q$ iff $P \sqsubseteq_i^c Q$.

Note that the congruence result does not hold for the fair/should preorder $\lesssim_F$, it is not preserved by the recursion operator. This can be easily seen by considering the following counter–example. Consider the processes $P = a.b[]a.c$ and $Q = a.b$ and the context $C = recX.(\_ | \overline{a}.\overline{b}.X)\backslash\{a,b\}$. It obviously holds that $P \lesssim_F Q$, but $C[P] \not\lesssim_F C[Q]$ (just take $O = \overline{c}.w$); hence $P \not\sqsubseteq_F^c Q$.

An alternative characterization of the closure of the fair/should preorder is given in [4], for a language slightly different from ours.

Definition 10. For all processes P and Q, we write
$$P \lesssim_{FT} Q \text{ if } (P \lesssim_F Q \text{ and } L(P) \subseteq L(Q)).$$

Theorem 11. For all processes P and Q, $P \lesssim_{FT} Q$ iff $P \lesssim_F^c Q$.

4 Full Abstraction Results

From now on, we adopt the following convention: an action declared *fresh* in a statement is supposed to be different from any other name and co–name mentioned in the statement.

4.1 Convergence predicate and convergent traces

In this section, we deal with the first two contextual preorders, $\downarrow \preceq^c$ and $\downarrow_{\mathcal{L}} \preceq^c$, and prove that they have the same distinguishing power and coincide with the reverse inclusion of the convergent traces preorder.

Definition 12. For all processes P and Q, we write $P \lesssim_m Q$ if $\forall s \in \mathcal{L}^*$ such that $P \downarrow s$, it holds that:
a) $Q \downarrow s$, and
b) $s \in L(Q)$ implies $s \in L(P)$.

Theorem 13. For all processes P and Q, $P \lesssim_m Q$ iff $P \sqsubseteq_m^c Q$.

The following special contexts can be used to prove the next theorems. If $s \in \mathcal{L}^*$, say $s = \ell_1 \cdots \ell_n$ ($n \geq 0$), we define
- $C_1^s = \_ | \overline{\ell_1}.\cdots.\overline{\ell_n}.\mathbf{0}$ and
- $C_2^s = \_ | \overline{\ell_1}.\cdots.\overline{\ell_n}.\Omega$.

Theorem 14. For all processes P and Q, $P \lesssim_m Q$ iff $P \downarrow \preceq^c Q$.

Theorem 15. For all processes P and Q, $P \downarrow_{\mathcal{L}} \preceq^c Q$ iff $P \downarrow \preceq^c Q$.

4.2 Guarantees and fair testing

Lemma 16. Let P be a process, O be an observer and let $\ell \in \mathcal{L}$ be a fresh action; (1) $P \underline{must}_F O$ iff $P \mid O\{\ell/w\}\,!\,\ell$, and (2) $P\,!\,\ell$ iff $P \underline{must}_F \bar{\ell}.w$.

Theorem 17. For all processes P and Q, $P \sqsubseteq^c_F Q$ iff $P \preceq^c_{\mathcal{L}} Q$.

PROOF: ($\Longleftarrow$) We prove that $\preceq^c_{\mathcal{L}}$ is contained in $\sqsubseteq_F$, the claimed result follows by closing under contexts. Suppose that $P \preceq^c_{\mathcal{L}} Q$ and that $P \underline{must}_F O$; let ℓ be a fresh action. We have:

$P \underline{must}_F O$ implies (Lemma 16(1))
$P \mid O\{\ell/w\}\,!\,\ell$ implies (hypothesis $P \preceq^c_{\mathcal{L}} Q$, with $C = \_ \mid O\{\ell/w\}$)
$Q \mid O\{\ell/w\}\,!\,\ell$ implies (Lemma 16(1))
$Q \underline{must}_F O$

($\Longrightarrow$) The proof is similar but relies on Lemma 16(2). □

4.3 Guarantees and convergence, and must testing

The next definition introduces two special contexts to be used in the proof of Theorem 20.

Definition 18. Let $s \in \mathcal{L}^*$, say $s = \ell_1 \cdots \ell_n$ ($n \geq 0$), and $B \subseteq_{\text{fin}} \mathcal{L}$. Let f^B denote a function which maps each $\ell \in B$ to a single fresh c. Fix a bijective correspondence among $\ell_1, \ldots, \ell_n$ and n fresh actions $\alpha_1, \ldots, \alpha_n$. We define
- $C_3^s = \_ \mid Q_3^s$ where $Q_3^\epsilon = c$ and $Q_3^{\ell s'} = \bar{\ell}.Q_3^{s'}[]c$, and
- $C_4^{s,B} = (\_ \mid R^s)\{f^B\} \mid Q_4^s$ where $R^s = \bar{\ell}_1.\alpha_1.\cdots\bar{\ell}_n.\alpha_n$, $Q_4^\epsilon = \mathbf{0}$ and $Q_4^{\ell_1 s'} = \bar{\alpha}_1.Q_4^{s'}[]c$.

Lemma 19. Let $s \in \mathcal{L}^*$, $B \subseteq_{\text{fin}} \mathcal{L}$ and c be a fresh action.
a) $P \downarrow s$ iff $C_3^s[P] \downarrow$ iff $C_3^s[P] \downarrow c$.
b) $(P \underline{after}\, s)\,!\,B$ iff $C_4^{s,B}[P]\,!\,c$.

Theorem 20. For all processes P and Q, $P \sqsubseteq^c_M Q$ iff $P \downarrow \preceq^c_{\mathcal{L}} Q$.

PROOF: ($\Longrightarrow$) From the definition, it is easily seen that $\ll_M$ is contained in $\downarrow\preceq^c_{\mathcal{L}}$ (indeed $P\,!\,c$ iff $(P \underline{after}\, \epsilon)\,!\,\{c\}$). From this fact, by closing under contexts and applying Theorem 8, the thesis follows.

($\Longleftarrow$) Here, we show that $\downarrow\preceq^c_{\mathcal{L}}$ is contained in $\ll_M$. From this fact and Theorem 8, the thesis follows. Assume that $P \downarrow\preceq^c_{\mathcal{L}} Q$ and that $P \downarrow s$, for some $s \in \mathcal{L}^*$. We have to show that: (a) $Q \downarrow s$ and (b) $(P \underline{after}\, s)\,!\,B$ implies $(Q \underline{after}\, s)\,!\,B$, for any $B \subseteq_{\text{fin}} \mathcal{L}$. As to part (a), from $P \downarrow s$ and Lemma 19(a), it follows that $C_3^s[P] \downarrow$. Obviously, for every process R, $C_3^s[R]\,!\,c$. From $C_3^s[P] \downarrow$, $C_3^s[P]\,!\,c$ and $P \downarrow\preceq^c_{\mathcal{L}} Q$ it follows that $C_3^s[Q] \downarrow$. By applying again Lemma 19(a), but in the reverse direction, we obtain $Q \downarrow s$. As to part (b), suppose that $(P \underline{after}\, s)\,!\,B$. From this, applying Lemma 19(b), it follows that $C_4^{s,B}[P]\,!\,c$. Moreover, it is easy to see that for every process R, $R \downarrow s$ implies $C_4^{s,B}[R] \downarrow$. From $C_4^{s,B}[P] \downarrow$, $C_4^{s,B}[P]\,!\,c$ and $P \downarrow\preceq^c_{\mathcal{L}} Q$, it follows that $C_4^{s,B}[Q]\,!\,c$. By applying again Lemma 19(b), but in the reverse direction, we obtain $(Q \underline{after}\, s)\,!\,B$. □

4.4 Guarantees and convergence, and safe–must

To prove full abstraction for safe–must, we will use another special context. Again, we assume that $c \in \mathcal{L}$ is always fresh. If $s \in \mathcal{L}^*$, say $s = \ell_1 \cdots \ell_n$ ($n \geq 0$), and $B \subseteq_{\text{fin}} \mathcal{L}$, we define the context

- $C_5^{s,B} = \_ | Q_5^{s,B}$ where $Q_5^{\epsilon,B} = \sum_{\ell \in B} \bar{\ell}.c$ and $Q_5^{\ell s',B} = \bar{\ell}.Q_5^{s',B}[]c$.

The proof of the following theorem is similar to that of Theorem 20, but relies on the context $C_5^{s,B}$ instead of $C_4^{s,B}$.

Theorem 21. For all processes P and Q, $P \lesssim_s Q$ iff $P \downarrow_{\mathcal{L}} \precsim^c Q$.

It is worthwhile to point out why the context $C_5^{s,B}$ cannot be used in place of the context $C_4^{s,B}$ to prove full abstraction for the must preorder (Theorem 20). Indeed, $P \downarrow s$ does not imply that $C_5^{s,B}[P] \downarrow$ (for instance $a.b.\Omega \downarrow a$ but $C_5^{a,\{b\}}[a.b.\Omega] \uparrow$). This would invalidate the proof of the "if" part of Theorem 20.

5 Comparing the preorders

Theorem 22. For all processes P and Q, $P \lesssim_M Q$ implies $P \lesssim_s Q$, but not vice–versa.

PROOF: Paralleling the proof of Theorem 20, part $\Longleftarrow$, it is easy to show that $\downarrow_{\mathcal{L}} \precsim^c$ is contained in $\ll_s$, from which the result will follow by applying Theorems 20 and 8. To show that the vice–versa does not hold, consider $P \stackrel{\text{def}}{=} a.b.\Omega$ and $Q \stackrel{\text{def}}{=} a$. It is easy to see that $P \lesssim_s Q$, but $P \downarrow_{\mathcal{L}} \not\precsim^c Q$ (just consider $\_ | \bar{a}$). □

Theorem 23.

1. $\lesssim_M = \downarrow_{\mathcal{L}} \precsim^c \subset \downarrow_{\mathcal{L}} \precsim^c = \lesssim_s \subset \downarrow_{\mathcal{L}} \precsim^c = \downarrow \precsim^c$.
2. $\precsim^c_{\mathcal{L}} = \lesssim_{FT}$ and $\lesssim_{FT}$ is not comparable with $\lesssim_M$, $\lesssim_s$ and $\downarrow \precsim^c$.

PROOF:
1. The result follows from Theorems 14, 20, 21 and 22. By definition, it is easily seen that $\downarrow_{\mathcal{L}} \precsim^c$ is included in $\downarrow_{\mathcal{L}} \precsim^c$. The inclusion is strict: $a \downarrow_{\mathcal{L}} \precsim^c \mathbf{0}$ but $a \downarrow_{\mathcal{L}} \not\precsim_{\mathcal{L}} \mathbf{0}$.
2. The equality $\precsim^c_{\mathcal{L}} = \lesssim_{FT}$ derives from Theorems 17 and 11. To see that neither of $\lesssim_M$, $\lesssim_s$ and $\downarrow \precsim^c$ is included in $\lesssim_F$ (hence in $\lesssim_{FT}$), consider the processes $P \stackrel{\text{def}}{=} recX.(a.X[]a.b)$ and $Q \stackrel{\text{def}}{=} recX.a.X$. Clearly, $P \lesssim_M Q$, hence $P \lesssim_s Q$ and $P \downarrow_{\mathcal{L}} \precsim^c Q$. However, $P \not\lesssim_F Q$ (because $P \underline{\text{must}}_F O$ and $Q \underline{\text{m}\cancel{\text{us}}\text{t}}_F O$, when $O \stackrel{\text{def}}{=} recX.(\bar{a}.X[]\bar{b}.w)$). To see the converse, observe that $\mathbf{0} \lesssim_{FT} \Omega$, but $\mathbf{0} \downarrow_{\mathcal{L}} \not\precsim \Omega$, hence $\mathbf{0} \not\lesssim_s \Omega$ and $\mathbf{0} \not\lesssim_M \Omega$. □

The mutual relationships among the pre–congruences are simpler if we move to *strongly convergent* processes. We say that a process P is strongly convergent if $P \downarrow s$ for every $s \in \mathcal{L}^*$.

Theorem 24. For strongly convergent processes, it holds that:

$$\precsim^c_{\mathcal{L}} = \lesssim_{FT} \subset \lesssim_M = \downarrow_{\mathcal{L}} \precsim^c = \downarrow_{\mathcal{L}} \precsim^c = \lesssim_s \subset \downarrow_{\mathcal{L}} \precsim^c = \downarrow \precsim^c.$$

6 Conclusions

We have proposed three basic notions of process observables, that, when closed with respect to the contexts of a CCS–like language, induce five pre–congruences that have been proved to coincide with well–known and/or intuitive behavioural relations.

Notions of observables in the same spirit as ours have been proposed in [13], [21], [11], [15], [8] and [12].

In [13], it is shown that the pre–congruence induced by inclusion of maximal traces coincides, both for CCS and CSP, with the must pre–congruence of [6]; another characterization is given by only considering the inclusion of the maximal ϵ–trace, i.e. a sequence of invisible moves leading to a divergent state or to a deadlocked one. The strength of the basic observables (maximal traces are definitely more inspective than our guarantees predicate) prevents from capturing different notions such as fair testing, and hinders the rôle played by the convergence test, which is somehow included in that for maximality.

In [21], two Petri nets are called *d–equivalent* if they both can reach a deadlocked state or if they both cannot do so. Then it is proved that, by closing d–equivalence with respect to parallel composition, the variant of failure semantics [5] that ignores divergence is obtained.

In [11], a series of variants of the testing framework is proposed and results are listed showing that, by changing the expressive power of testers, a number of equivalences ranging from bisimulation to testing can be captured. One of the considered family of observers is that consisting just of agents of the form $\ell.w.\mathbf{0}$, that somehow resemble our $!\ell$ predicates. It is claimed that for strongly convergent processes the pre–congruence induced by this family of observers coincides with the must preorder and the reader is referred to [13] for the proof. However, we could not find the proof in Main's paper.

Milner and Sangiorgi [15] define an equivalence for processes based on elementary observables, namely the *possibility* for a process to synchronize along a specific channel. However, they permit to recursively test for the presence of this observable. The resulting notion of observability (called barbed bisimilarity), when closed under parallel composition, yields bisimulation–based equivalences that are significantly more discriminating than ours.

Ferreira [8] and Lancvc [12] deal with languages significantly different from classical process algebras. In particular, Ferreira uses a predicate which resembles very much the conjunction of our $\downarrow$ and $!\ell$ (based on production of values rather than on communication capabilities) to define a testing preorder for Concurrent ML [20]; this seems to be strongly related to our safe–must preorder. He also conjectures that if one considers pure CCS (and observes communication capabilities instead of value productions) the obtained preorder coincides with the *must* pre–congruence of [6]; here we have proved this conjecture. Laneve discusses the impact of an observables-based testing scenario on the Join Calculus, a language with elaborate synchronization schemata [9].

Acknowledgments

We are grateful to L. Aceto, F. van Breugel, W. Ferreira, A. Rensink and W. Vogler for interesting discussions and suggestions and to F. Focardi for a first debugging of the ideas contained in the paper.

References

1. S. Abramsky. The lazy lambda calculus. *Research Topics in Functional Programming*, David Turner, ed., Addison–Wesley, 1990.
2. B. Bloom, S. Istrail, A.R. Meyer. Bisimulation can't be traced. *Journal of the ACM*, 42(1):232-268, 1995.
3. E. Brinksma, A. Rensink, W. Vogler. Fair Testing. *Proceedings of CONCUR'95*, LNCS 962, pages 313-327, Springer, 1995.
4. E. Brinksma, A. Rensink, W. Vogler. Applications of Fair Testing. In R. Gotzhein and J. Bredereke, ed., Formal Description Techniques IX, Chapman & Hall, 1996.
5. S.D. Brookes, C.A.R. Hoare, A.W. Roscoe. A theory of communicating sequential processes. *Journal of the ACM*, 31(3):560-599, 1984.
6. R. De Nicola, M.C.B. Hennessy. Testing Equivalence for Processes. *Theoretical Computers Science*, 34:83-133, 1984.
7. R. De Nicola, M.C.B. Hennessy. CCS without τ's. *Proceedings of TAPSOFT'87*, LNCS 249, pages 138-152, Springer, 1987.
8. W. Ferreira. *Semantic Theories for Concurrent ML*. Ph.D. Thesis, University of Sussex, 1996.
9. C. Fournet, G. Gonthier, J.-L. Lévy, L. Maranget, D. Rémy. A Calculus of Mobile Agents. *Proceedings of CONCUR'96*, LNCS 1119, 1996.
10. M.C.B. Hennessy. *Algebraic Theory of Processes*. MIT Press, 1988.
11. M.C.B. Hennessy. Observing Processes. In *Linear Time, Branching Time and Partial Order in Logics and Models for Concurrency*, LNCS 354, Springer, 1989.
12. C. Laneve. May and Must Testing in the Join-Calculus. Technical Report UBLCS-96-4, Università di Bologna, Dept. of Computer Science, Bologna, 1996.
13. M.G. Main. Trace, Failure and Testing Equivalences for Communicating Processes. *Int. Journal of Parallel Programming*, 16(5):383-400, 1987.
14. R. Milner. *Communication and Concurrency*. Prentice Hall International, 1989.
15. R. Milner, D. Sangiorgi. Barbed Bisimulation. *Proceedings of ICALP'92*, LNCS 623, Springer, 1992.
16. J.-H. Morris. *Lambda Calculus Models of Programming Languages*. Ph.D. Thesis, MIT, 1968.
17. V. Natarajan, R. Cleaveland. Divergence and Fair Testing. *Proceedings of ICALP'95*, LNCS 944, pages 648-659, Springer, 1995.
18. C.-H.L. Ong. Correspondence between operational and denotational semantics: the full abstraction problem for PCF. *Handbook of Logic in Computer Science*, vol.4, S. Abramsky, D.M. Gabbay and T.S.E. Maibaum, ed., Oxford Science Publ., 1995.
19. G.D. Plotkin. A Structural Approach to Operational Semantics. Technical Report DAIMI FN-19, Aarhus University, Dept. of Computer Science, Aarhus, 1981.
20. J.H. Reppy. Concurrent ML: Design, application and semantics. *Proceedings of Functional Programming, Concurrency, Simulation and Automata Reasoning*, LNCS 693, pages 165-198, Springer, 1993.
21. W. Vogler. Failures Semantics and Deadlocking of Modular Petri Nets. *Acta Informatica*, 26:333-348, 1989.

Constrained Bipartite Edge Coloring with Applications to Wavelength Routing *

Christos Kaklamanis[1] Pino Persiano[2]
Thomas Erlebach[3] Klaus Jansen[4]

[1] Computer Technology Institute, University of Patras, Rio, Greece, kakl@cti.gr
[2] Dipartimento di Informatica ed Appl., Università di Salerno, I-84081 Baronissi, Italy, giuper@dia.unisa.it
[3] Institut für Informatik, TU München, D-80290 München, Germany, erlebach@informatik.tu-muenchen.de
[4] Fachbereich IV – Mathematik, Universität Trier, Postfach 3825, D-54286 Trier, Germany, jansen@dm3.uni-trier.de

Abstract. Motivated by the problem of efficient routing in all-optical networks, we study a constrained version of the bipartite edge coloring problem. We show that if the edges adjacent to a pair of opposite vertices of an L-regular bipartite graph are already colored with αL different colors, then the rest of the edges can be colored using at most $(1+\alpha/2)L$ colors. We also show that this bound is tight by constructing instances in which $(1+\alpha/2)L$ colors are indeed necessary. We also obtain tight bounds on the number of colors that each pair of opposite vertices can see.

Using the above results, we obtain a polynomial time greedy algorithm that assigns proper wavelengths to a set of requests of maximum load L per directed fiber link on a directed fiber tree using at most $5/3L$ wavelengths. This improves previous results of [9, 7, 6, 10].

We also obtain that no greedy algorithm can in general use less than $5/3L$ wavelengths for a set of requests of load L in a directed fiber tree, and thus that our algorithm is optimal in the class of greedy algorithms which includes the algorithms presented in [9, 7, 6, 10].

1 Introduction

In this paper, we study a constrained version of the well-known problem of coloring the edges of an L-regular bipartite graph. It is a classical result from graph theory (see e.g. [3]) that the edges of an L-regular bipartite graphs can be colored using exactly L colors so that edges that share an endpoint are assigned different colors. We call such edge colorings *legal* colorings. The problem does not have any other extra constraint: any given color can be used on any edge provided that no other adjacent edge is colored using that same color. Our constrained version of the bipartite edge coloring problem can be described in the following way.

---

* Partially supported by Progetto MURST 40%, Algoritmi, Modelli di Calcolo e Strutture Informative and by EU Esprit Project GEPPCOM and ALCOM-IT.

We are given an L-regular bipartite graph $G = (\{v_1, \cdots, v_n\}, \{u_1, \cdots, u_n\}, E)$ along with a partial legal coloring of its edges that specifies a color for all edges incident to vertices v_1 and u_1. We denote the total number of constraining colors by αL, where $1 \leq \alpha \leq 2$. We want to color the remaining edges of the graph so as to minimize the total number of colors used and the number of colors used to color the edges touching a pair (u_i, v_i) of opposite vertices.

Our motivation lies in the field of WDM (wavelength division multiplexing) routing in all-optical networks. Optics is emerging as a key technology in state-of-the-art communication networks. A single optical wavelength supports rates of gigabits-per-second (which in turn support multiple channels of voice, data, and video [5] [8]). Multiple laser beams that are propagated over the same fiber on distinct optical wavelengths can increase this capacity much further; this is achieved through WDM (wavelength division multiplexing). We model the underlying fiber network as a directed graph. Communication requests are ordered transmitter-receiver pairs of nodes. WDM technology establishes connectivity by finding transmitter-receiver paths, and assigning a wavelength to each path, so that no two paths going through the same link use the same wavelength. Optical bandwidth is the number of available wavelengths. Bandwidth is a scarce resource: state-of-the-art technology allows for no more than 30-40 optical wavelengths in the laboratory, less than half as many in manufacturing, and there is no anticipation of dramatic progress in the near future [11]. It is thus important to minimize the number of wavelengths used to service a requested communication pattern. Variations of this problem have been studied by several authors [12, 1, 9, 7, 6, 10, 2].

In this paper, we concentrate on tree topologies which are relevant to wide-area networks. In particular we consider directed trees where each edge of the tree consists of two opposite directed fiberlinks. Directedness accurately reflects directed optical amplifiers placed on the fiber as well as asymmetries of the communication requests. Raghavan and Upfal [9] showed that routing requests of maximum load L per link of undirected trees can be satisfied using no more than $3L/2$ optical wavelengths and their arguments extend to give a $2L$ bound for the directed case. Mihail *et al.* [7] were the first to address the directed case. Their main result is a $15L/8$ bound for directed trees. They obtain this bound by reducing the wavelength assignment problem to the constrained bipartite edge coloring problem and obtain a solution specifically for the case $\alpha = 3/2$. This was improved in [6] (and independently in [10]) by solving optimally the constrained bipartite edge coloring problem for the value $\alpha = 3/2$ and yielding a bound of $7/4L$ for directed trees.

1.1 Summary of results

Our results can be summarized in the following theorems. We first present our results on the constrained bipartite edge coloring problem.

Theorem 1. *There exists a polynomial time algorithm that properly colors the uncolored edges of an L-regular bipartite graph constrained by αL colors using* at most $(1 + \alpha/2)L$ colors and so that each pair (v_i, u_i) of opposite vertices sees no more than $\max\{\alpha L, (1 + \alpha/4)L\}$ different colors.

The next lower bound states that the above result is in general tight.

Theorem 2. *For each $1 \leq \alpha \leq 2$ and for each $L > 0$ there exists an L-regular bipartite graph constrained by αL colors for which any legal coloring of the remaining edges requires at least $(1 + \alpha/2)L$ total colors while there exists a pair of opposite vertices that sees at least $\max\{\alpha L, (1 + \alpha/4)L\}$ different colors.*

Next we present our results for wavelength routing on directed trees. We express our results in terms of the maximum load L of a set of requests; i.e., the maximum number of paths between transmitter and receiver that share the same directed fiber link. The proposed algorithm is a greedy algorithm. A greedy algorithm is an algorithm that considers the vertices of the tree one at a time in a DFS manner and, while at vertex v, colors (i.e., assigns a wavelength to) all the requests that touch vertex v (i.e., start at, end at, or go through v) that are still uncolored. Once a request has been colored, a greedy algorithm never recolors it. Greedy algorithms do not require global control and are thus amenable of being implemented in a distributed setting without a "central authority" that has knowledge of the overall request pattern. All known algorithms for the problem of wavelength routing on directed trees are indeed greedy algorithms [7, 6, 10].

Theorem 3. *There exists a greedy polynomial time algorithm that assigns wavelengths to a set of requests of maximum load L on a directed tree using at most $5/3L$ wavelengths.*

Our next theorem shows a lower bound that implies that no greedy algorithm can in general beat the $5/3L$ barrier.

Theorem 4. *For each L, for each $\epsilon > 0$ and for each greedy algorithm G there exists a tree and a pattern of communication requests of maximum load L for which G uses at least $\left(\frac{5}{3} - \epsilon\right) L$ wavelengths.*

Therefore better bounds can only be obtained by non greedy algorithms. The only known general lower bound is $5/4L$ [10].

The rest of our paper is organized as follows.

In Section 2, we prove Theorem 1 by giving an algorithm that solves the constrained bipartite edge coloring problem. Next, in Section 3 we explain the reduction of the wavelength routing problem on directed trees to the constrained bipartite edge coloring problem. This reduction proves Theorem 3. Finally, in Section 4, we present our lower bounds.

2 The algorithm for the constrained bipartite edge coloring problem

In this section we present our algorithm for solving the constrained bipartite edge coloring problem.

The algorithm receives as input an L-regular bipartite graph $G = (\{W_0, ..., W_n\}, \{X_0, \cdots, X_n\}, E)$ where all the edges incident to W_0 and X_0 have been properly colored using αL different colors. We call the edges that are colored *color-forced* edges and a pair (W_i, X_i) of opposite vertices a *line*. We assume without loss of generality that no edge connects two opposite vertices. If a color appears on only one color-forced edge, then we call it a *single* color. If it appears on two color-forced edges, we call it a *double* color; note that one of these two color-forced edges has to be incident to W_0 and the other to X_0. We denote by D and S the number of double and single colors, respectively.

Step 1: Obtaining perfect matchings. We proceed by decomposing the bipartite graph into L perfect matchings which can always be done since it is L-regular. Each such matching includes exactly two color-forced edges: one incident to W_0 and one incident to X_0. A double color is called *separated* if its two color-forced edges appear in different matchings. On the other hand, if they appear in the same matching then the color is said to be *preserved*. We classify the matchings into four types: TT, PP, SS, ST, based on their corresponding color-forced edges. If the two color-forced edges of a matching are colored with separated colors, then the matching is of type TT. If the two color-forced edges are colored with the same preserved color, then the matching is of type PP. If the two color-forced edges are colored with two single colors, then the matching is of type SS. If the two color-forced edges are colored with a single color and with a separated color, then the matching is of type ST.

Step 2: Constructing chains and cycles of matchings. We partition the matchings into groups. Each such group is either a *chain* or a *cycle* of matchings. A chain of matchings is a sequence $M_0, M_1, \cdots, M_{l-1}$ of l matchings such that

1. M_0 and M_{l-1} are matchings of type ST;
2. $M_1, \cdots, M_{l-2}$ are all matchings of type TT;
3. for each $0 \leq i \leq l-2$, matchings M_i and M_{i+1} share exactly one double (separated) color. A chain consists of at least two matchings.

A cycle of matchings is a sequence $\langle M_0, M_1, \cdots, M_{l-1} \rangle$ of l TT matchings such that, for each $0 \leq i \leq l-1$, matchings M_i and $M_{i+1 \bmod l}$ share exactly one double (separated) color.

Step 3: making chains and cycles minimal. A sequence C of matchings (chain or cycle) is minimal if it does not contain any two parallel color-forced edges. A non-minimal sequence of matchings can be split into two shorter sequences in the following way. Consider the sequence $C = \langle M_0, \cdots, M_{l-1} \rangle$ of matchings and suppose that the edge colored c_i of M_i and the edge colored c_j of M_j are parallel. We exchange the two edges thus obtaining two new matchings M_i' and M_j' with color-forced edges colored c_j and c_{i+1} and c_i and c_{j+1} and the two new sequences of matchings $C_1 = \langle M_0, M_1, \cdots M_{i-1}, M_j', M_{j+1}, \cdots, M_{l-1} \rangle$ and $C_2 = \langle M_i', M_{i+1}, \cdots, M_{j-1} \rangle$. The sequence C_1 is of the same type (i.e., a cycle or a chain) as C while C_2 is always a cycle. We repeat this process of splitting one sequence into two new sequences until all sequences are minimal (i.e., they do not contain parallel edges).

Step 4: constructing triplets of matchings. Next we partition all the matchings into groups of three matchings that we call *triplets*. Each such triplet has six color-forced edges; of these, two are colored with single colors and the remaining four with double colors.

We obtain the triplets as follows. First, we consider all the chains of length 3 or greater. From each such chain $C = \langle M_0, M_1, \cdots M_{l-1} \rangle$ we obtain one triplet by stripping off C and grouping together the first two matchings M_0, M_1 and the last matching M_{l-1}. Triplets obtained in this way will consist of two ST matchings (that is M_0 and M_{l-1}) and one TT matching (that is M_1). The color-forced edges are colored with single colors s_0 and s_1 and double colors d_1, d_2, d_{l-1}, with d_1 being the common color of M_0 and M_1. Now we are left with cycles, "stripped chains," chains of length 2, and SS matchings. We consider the even length cycles and stripped chains first and construct triplets each consisting of two consecutive TT matching from the same cycle or stripped chain and one SS matching. We repeat the same process for odd length cycles and stripped chains. However, in this case for each cycle or stripped chain there will be exactly one "leftover" TT matching. We then construct triplets with one SS matching along with a pair of these TT leftover matchings. Finally, if at any time during the construction of the triplets we run out of SS matchings, we continue constructing triplets by grouping together each individual TT matching along with a pair of ST matchings that constitute a chain of length 2. If the total number of old colors is exactly $4/3L$, thus including exactly $2/3L$ single colors and $2/3L$ double colors all matchings can be grouped into such triplets. Instead, if we have less than $4/3L$ old colors, then we are left with some extra TT matchings for which there is no corresponding ST or SS matching. These extra TT matchings will be dealt with separately and we omit from this abstract further details. On the other hand, if the number of old colors exceeds $4/3L$, then we are left with extra SS or ST matchings for which no corresponding TT matching exists. Coloring these matchings is trivial since we can use no new color (we use the single colors to color the uncolored edges) and thus meet the two conditions presented below.

We will color the matchings maintaining the following two conditions which are sufficient to prove Theorem 1.

Condition 1. The number of new colors used is at most $D/2$. This condition will be enforced by using at most one new color per triplet.

Condition 2. Each line sees at most $\max\{(1 + \alpha/4)L, \alpha L\}$ colors.

For values of $\alpha \geq 4/3$ this is enforced by making sure that if a line sees a new color it does not see one of the old colors. Consequently, the number of colors seen by a line does not exceed αL once all edges have been colored.

Lemma 5. *Condition 1 above implies that the total number of colors used is at most $(1 + \alpha/2)L$.*

Proof. Since the number of edges adjacent to W_0 and X_0 is $2L$, we have $2D + S = 2L$ and, since αL colors are used to color these edges, we have that $D + S = \alpha L$.

From these two equalities we get directly that $D = (2-\alpha)L$. Therefore the total number of colors used is at most $D + S + D/2 \leq \alpha L + (2-\alpha)L/2 = (1+\alpha/2)L$.

Step 5: setting the active colors. We color each triplet individually using four of the old colors that appear on the color-forced edges of the triplet and, in some cases, a new color. The four old colors used are called the *active* colors for the triplet and they include the two single colors of the triplet. The remaining two active colors are chosen among the double colors of the triplet so that each double color is active for exactly one triplet.

We continue by determining what the active colors are going to be for each triplet. We have to be careful about consistency among triplets that share double colors; i.e., include TT matchings from the same cycle or chain.

First we fix the active colors of the triplets containing the leftover TT matchings. In order to properly color such a triplet (S, T_1, T_2) while maintaining the properties above we choose the active colors to be the two single colors of matching S along with the color of the color-forced edge touching W_0 in T_1 and the color of the color-forced edge touching X_0 in T_2.

This choice of active colors for such a triplet forces the choice of active colors for the triplets containing TT matchings coming from the same cycle or chain as T_1 and T_2 in the following obvious way. Let (S, T_3, T_4) be a triplet consisting of one SS matching and two consecutive TT matchings from the same cycle or stripped chain as T_1. Then the active colors of such a triplet are the colors of the color-forced edges touching W_0 in T_3 and T_4 along with the two colors of the color-forced edges of S_1. If, instead, T_3 and T_4 belong to the same cycle or stripped chain as T_2, then the active colors are going to be the old single colors appearing in S_1 along with the color of the color-forced edges of T_3 and T_4 that touch X_0.

Finally, we can determine the active colors of the triplets containing two TT matchings belonging to even length cycles or chains (i.e., those cycles or chains that did not give rise to leftover TT matchings) to be for each triplet the two old single colors of the triplet along with the color of the color-forced edges that touch X_0 or W_0, picked arbitrarily as long as we are consistent across each cycle or chain.

Step 6: coloring the triplets. As we mentioned above for each triplet we will use the active colors of the triplet and, sometimes, a new color. If we do use a new color for a triplet, we enforce the property that each line that sees the new color does not see one of the active colors of the triplet. This ensures that the total number of colors that a line will see across all triplets does not exceed the number of old colors and that the total number of new colors introduced for all triplets is at most half the number of old double colors.

There are four general types of triplets:

Type A These are triplets consisting of one SS matching and two leftover TT matchings. A special case of type A triplet occurs when one or both the leftover TT matchings is actually a PP matching that is the leftover matching of a cycle of length 1.

Type B These are triplets consisting of one SS matching and two consecutive TT matchings from the same cycle or chain. A special case of type B triplet occurs when the two TT matchings constitute a cycle of length 2.

Type C These are triplets consisting of the two ST matchings that constitute a chain of length 2 and one TT matching. A special case of type C triplet occurs when the TT matchings is actually a PP matching.

Type D These are triplets that were obtained by stripping off a chain the first two matchings (an ST and a TT matching) and the last matching (an ST matching). A special case of type D triplet occurs when the chain has length exactly 3.

Due to lack of space we next show the coloring algorithm only for triplets of type A. The complete coloring appears in the final version.

Step 6.A: coloring triplets of type A. Consider a triplet $R = (S, T_1, T_2)$ of type A, where $S = (s_1, s_2)$, $T_1 = (x, y)$ and $T_2 = (w, z)$. We note that s_1 and s_2 are single colors and that x, y, w, and z are double colors and let the active colors of R be s_1, s_2, x, and z. Here we concentrate on the case in which the four double colors are distinct separated colors. If $x = y$ or $w = z$, then the corresponding TT matching is actually a PP matching and the coloring is much simpler than what we are going to describe below. If $x = z$ or $y = w$, then R is actually a triplet of type B.

Suppose x, y, w, and z are distinct separated double colors. We consider matchings T_1 and T_2 together as one cycle cover of the bipartite graph. In what follows, for the sake of clarity we assume that the cycle cover of two matchings consists of one single cycle that spans the entire bipartite graph. We remark that all our colorings can be easily adapted if such a cycle cover consists of more than one cycle.

We first check if there exists an uncolored edge whose endpoints are incident to color-forced edges colored with all four active colors. Note that these may include the "fixed" color-forced edges colored with s_1, s_2, x, and z that belong to R as well as the two "free" color-forced edges colored with x and z that belong to other triplets. We denote by e_x and e_z the free color-forced edge colored with x and z, respectively and by e_{s_1} and e_{s_2} the color-forced edges colored with s_1 and s_2, respectively.

Suppose there is no edge restricted by all four active colors. We color the uncolored edges of the cycle cover by starting from one of the color-forced edges of the cycle colored with an active color (i.e., either x or z) and alternating between x and z. When we encounter a vertex v that is incident to a free color-forced edge e, we use color s_2 to color the edge, e', incident to v that would have been colored with the same color as e. Then we color the next edge x and continue alternating between z and x. This is possible unless e' is adjacent to e_{s_2} as well. Note that e_z cannot be incident to the same vertex as e_{s_2}, and, similarly, e_x cannot be incident to the same vertex as e_{s_1}.

Now if e' is restricted by both z and s_2, then we color with s_2 the other edge incident to v, color e' with x and continue alternating z and x (see Figure 1);

we finish by using s_1 to color the edges in the SS matching. This coloring is obviously proper and we do not need to argue about the number of colors seen by a line since we have used no new color.

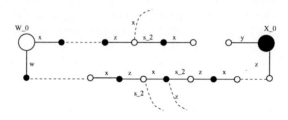

Fig. 1. The case in which an edge is restricted by both z and s_2.

On the other hand, if e_x and e_{s_2} are incident to the same vertex v then we color the conflicting adjacent edge $e = (v, u)$ with s_1 and continue alternating x and z starting with x. The uncolored edges of the SS matching are then colored using s_1 except for the edge $e_u = (u, u^*)$ incident to u. Edge e_u is colored s_2 unless u^* sees the edge colored s_2 used to fix the conflict with e_z in which case e_u is colored with z (see Figure 2).

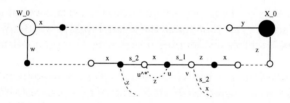

Fig. 2. The case in which an edge is restricted by both x and s_2.

The previous coloring is proper unless e is also adjacent to e_{s_1} in which case a more complex coloring is performed.

Finally we consider the situation where we have an edge (u, v) restricted by all four active colors. Note that such a restricted edge belongs to one of the TT matchings of the triplet, as edges of the SS matching cannot be restricted by s_1 or s_2. We color edge (u, v) with n, the rest of the uncolored edges of the cycle cover by alternating x and z, and the uncolored edges of the SS matching with n. This coloring is obviously proper. No line except the lines containing vertices u and v sees an edge colored with one of the two single colors. Moreover, since u and v cannot be a line as they are adjacent, the line containing u does not see color s_2 and the line containing v does not see color s_1. Therefore, if a line sees n then it does not see at least one of the active colors.

2.1 An alternative coloring approach for $\alpha = 4/3$

In this section, we briefly describe an alternative method for coloring edges of a bipartite graph G for the case $\alpha = 4/3$.

It is possible to show that the L perfect matchings obtained from G can be grouped entirely into triplets such that each triplet can be colored with at most one new color and with ≤ 4 colors per line, thus ensuring Conditions 1 and 2. Due to a result in [10], every triplet $t = (M_1, M_2, M_3)$ with two double colors and one single color incident to each of W_0 and X_0 can be colored in such a way, provided that at least one double color d appears twice in t (call such a triplet a *KS-triplet*) and that t can be partitioned into a gadget (a subgraph where W_0 and X_0 have degree 3 and all other vertices have degree 2) and a matching of all vertices except $\{W_0, X_0\}$. Only the two single colors and the double color d of t as well as one new color are used. Every KS-triplet can be partitioned into gadget and matching unless it contains a PP-matching. Therefore, we assume that G is decomposed into L perfect matchings such that the union of SS-, ST-, and TT-matchings does not contain further PP-matchings.

A PP-matching and a chain of length 2 as well as two PP-matchings and an SS-matching give triplets that can be colored without any new color and 4 colors per line. Chains of odd length and cycles of even length yield triplets of Type B, C, and D, which are KS-triplets. Two chains of even length, one of which has length > 2, yield KS-triplets by combining the first (last) two matchings of the longer chain with the first (last) matching of the shorter chain and producing triplets of Type B or C from the rest. If there is a chain of length 2, a cycle of odd length also yields triplets of Type B and C. Note that there is always a sufficient number of SS-matchings or chains of length 2 to produce KS-triplets, because we have $2/3L$ edges with single colors and $4/3L$ edges with double colors altogether.

After these reductions, we are left with at most one chain of even length > 2, at most one PP-matching, a number of cycles of odd length, and SS-matchings. Two cycles of odd length are handled by choosing an SS-matching and two TT-matchings, one from each cycle, such that the resulting triplet t does not have parallel color-forced edges. If t can be partitioned into gadget and matching, it is colored with reused old colors and one new color using techniques similar to [10], and triplets of Type B are produced from the remainder of the two cycles. Otherwise, the TT-matchings can be reassembled, turning the two given cycles into a single cycle of even length, which is handled as above. A chain of even length > 2 and a cycle of odd length are combined similarly.

For a PP-matching and a cycle of odd length, we choose an arbitrary SS-matching M_1 and a TT-matching M_2 from the cycle such that the cycle cover $M_1 \cup M_2$ does not contain parallel color-forced edges. This cycle cover can be colored with one new color and one of its single colors such that no line sees more than 3 colors. The PP-matching M_3 is colored using its preserved double color, thus ensuring that the coloring for $t = (M_1, M_2, M_3)$ meets the requirements. The remainder of the cycle is combined with SS-matchings into Type B triplets. A PP-matching and a chain of even length > 2 are handled similarly.

3 Reducing the routing problem to a constrained bipartite coloring problem

In this section we reduce the problem of assigning wavelengths to the constrained bipartite edge coloring problem. We do so by giving an algorithm that properly assigns wavelengths by using as a subroutine our algorithm for the constrained bipartite edge coloring problem of the previous section.

Our algorithm for assigning wavelengths is a greedy algorithm as the ones presented in [7, 6, 10]. The algorithm roots the tree at an arbitrary node and computes a depth-first numbering of the nodes of the tree. The algorithm proceeds in phases, one per each node v of the tree. The nodes are considered following their depth first numbering. The phase associated with node v assumes that a partial proper coloring of all paths that touch (i.e., start, end, or go through) nodes with numbers strictly smaller than v's has been computed and extends the partial coloring to one that assigns proper colors to all paths that touch v but have not been colored yet. We stress that the algorithm never recolors paths that have been colored in previous phases.

We now show the reduction of the path coloring problem of a phase associated with node v to an instance of the constrained bipartite edge coloring of a graph G_v. Without loss of generality, we assume to have full load L on each directed link and denote by c_0 v's parent and by $c_1, \cdots, c_k$ the children of v. We construct G_v in the following way. For each vertex c_i, G_v has four vertices W_i, X_i, Y_i, Z_i and the left and right partitions are $\{W_i, Z_i | i = 0, \cdots k\}$ and $\{X_i, Y_i | i = 0, \cdots k\}$. G_v has an edge from W_i to X_j, for each path of the tree directed out of c_i into c_j and an edge from W_i to Y_i, for each path from c_i to v. Finally, for each path from v to c_j, G_v has an edge from Z_i to X_i. See Figure 3. The above edges are called *real*. Notice that no real edge extends across opposite vertices Z_i and Y_i or W_i and X_i and only edges with an endpoint in W_0 or X_0 already have a color as they correspond to requests touching v's parent and have been assigned a color in a previous phase. Notice also that all vertices of type W_i and X_i have degree L whereas vertices of type Z_i and Y_i do not necessarily have degree L. We therefore add *fictitious edges* to the bipartite graph so that all vertices have degree L. Clearly, any proper coloring of the edges of G_v corresponds to a legal assignment of wavelengths to requests that go through vertex v and we compute such a coloring of the edges of G_v by running the algorithm of the previous section on G_v.

4 Lower bound

In this section we present our lower bounds for the wavelength routing problem by showing that any greedy algorithm for assigning paths to requests of load L on a tree cannot use less than $5/3L$ colors even if the tree is binary. The lower bound for the constrained bipartite edge coloring is obtained similarly.

We prove the lower bound inductively. We assume inductively that, for a vertex C there are $\alpha_n/2L$ requests along each link to its parent and that all of these requests are colored using different colors.

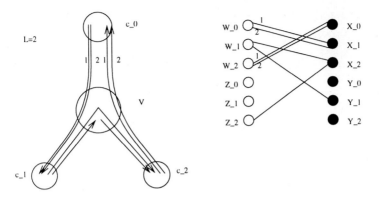

Fig. 3. Requests touching vertex v and the corresponding bipartite graph (only real edges are shown).

Then we assign requests between the two children A and B of C in such a way that $(1+\frac{\alpha_n}{2})L$ colors are used in total and the inductive hypothesis between one of A and B and one of its children is enforced for $\alpha_{n+1} = 1 + \frac{\alpha_n}{4}$. It is easy to see that $\alpha^* = \lim_{n \to \infty} \alpha_n = 4/3$, where $\alpha_1 = 1$ and $\alpha_n = 1 + \frac{\alpha_{n-1}}{4}$ for $n > 1$. Therefore, for any $\epsilon > 0$ and any greedy algorithm G, it is possible to construct a set of communication requests of maximum load L so that G uses at least $(5/3 - \epsilon)L$ colors.

The base of our induction for $\alpha_1 = 1$ is established in the following way. We start with L requests on each direction between the root R and one of its children R'. The greedy algorithm colors these request using at least L colors in each direction. We then choose two sets of $L/2$ request along each direction with each request colored with a different color, propagate them to one of the children of R' and stop at R' the remaining requests.

Let C be a vertex and A and B the two children of its left child. We denote by K_2 the set of colors used along the link (C, A), by K_3 the set of colors used along the link (B, C), by K_1 the set of colors used along the link (B, A) and by K_4 the set of colors used along the link (A, B). We inductively assume that $K_2 \cap K_3 = \emptyset$ and $|K_2| = |K_3| = \alpha/2L$ whence thus $|K_2 \cup K_3| = \alpha L$. We fill the link (B, A) to capacity by assigning $k_1 = L(1 - \alpha/2)$ requests. These requests need to be colored with new colors and thus the total number of colors used increases to $L(1 + \frac{\alpha}{2})$. Next we assign L requests to the link (A, B). The best that any greedy algorithm can do is to color these L requests colored using all the new colors employed for the link (B, A), plus half of the colors of K_3 and half of the colors of K_2. The edge (A, C) thus sees $|K_1 \cup K_2 \cup K_4| = |K_1| + |K_2| + |K_4| - |K_2 \cap K_4| - |K_1 \cap K_4| = (1 + \frac{\alpha}{4})L$. In order to complete the inductive step we have to enforce for A the same situation as in C for $(1 + \frac{\alpha}{4})$. This is achieved in the following way:

1. among the $|K_1| + |K_2| = L$ requests coming from C, we let only the following continue to the left child of A:
   - S_1: $(\frac{1}{2} - \frac{\alpha}{8}) L$ requests from K_1.

- S_2: $\frac{\alpha}{4}L$ requests from K_2.

for a total of $\frac{1}{2}\left(1+\frac{\alpha}{4}\right)$ colors.

2. and the $|K_4| = L$ requests coming up from A to C all originate from A except for the following ones which instead come from the right child of A
   - R_1: $\frac{\alpha}{4}L$ requests that were colored with colors used of K_2 and which were not considered in S_2 above;
   - R_2: $\frac{\alpha}{4}L$ requests that were colored with colors used of K_3;
   - R_3: $\left(\frac{1}{2}-\frac{3}{8}\alpha\right)L$ that were colored with colors of K_1 and which were not considered in S_1 above;

   for a total of $\frac{1}{2}\left(1+\frac{\alpha}{4}\right)$ colors.

Finally, observe that the requests going down to the left child of A and those coming up from the right child of A are colored with different colors (i.e. the sets of colors are disjoint). This completes the proof of Theorem 4.

References

1. A. Aggarwal, A. Bar-Noy, D. Coppersmith, R. Ramaswani, B. Shieber, and M. Sudan, *Efficient Routing and Scheduling Algorithms for Optical Networks*, in Proc. of SODA 93.
2. B. Beauquier, J.-C. Bermond, L. Gargano, P. Hell, S. Perennes, U. Vaccaro, *Graph Problems arising from Wavelength-Routing in All-Optical Networks*, in proc. of Workshop on Optics in Computer Science, 1997.
3. C. Berge, *Graphs*, North-Holland.
4. J.-C. Bermond, L. Gargano, S. Perennes, A. A. Rescigno, and U. Vaccaro, *Efficient Collective Communication in Optical Networks*, in Proc. of ICALP 96.
5. P. E. Green, *Fiber Optic Communication Networks*, Prentice Hall, 1992.
6. C. Kaklamanis and P. Persiano, *Efficient Wavelength Routing on Directed Fiber Trees*, in Proc. of Algorithms – ESA '96, Lecture Notes in Computer Science, 1136, pp. 460–470.
7. M. Mihail, C. Kaklamanis, and S. Rao, *Efficient Access to Optical Bandwidth*, in Proc. of FOCS 1995.
8. D. Minoli, *Telecommunications Technology Handbook*, Artech House, 1991.
9. P. Raghavan and E. Upfal, *Efficient Routing in All-Optical Networks*, in Proc. of STOC 1994.
10. V. Kumar and E. J. Schwabe. Improved access to optical bandwidth in trees. In *Proceedings of SODA '97*, 1997.
11. ONTC-ARPA, Brackett, Acampora, Sweitzer, Tangonan, Smith, Lennon, Wang, Hobbs, A Scalable Multiwavelength Multihop Optical Network: A Proposal for Research in All-Optical Networks, *IEEE J. of Lightwave Technology*, Vol 11 No 5/6, 1993, pp 736-753.
12. R. Pankaj, *Architectures for Linear Lightwave Networks*, Ph.D. Thesis, MIT, 1992.

Colouring Paths in Directed Symmetric Trees with Applications to WDM Routing*

Luisa Gargano[1]   Pavol Hell[2]   Stephane Perennes[3]

[1] Dipartimento di Informatica, Università di Salerno, 84081 Baronissi (SA), Italy
[2] School of Computing Sciences, SFU, Burnaby, B.C. V5A1S6, Canada
[3] Delft University of Technology, The Netherlands.

Abstract. Let T be a symmetric directed tree, i.e., an undirected tree with each edge viewed as two opposite arcs. We prove that the minimum number of colours needed to colour the set of all directed paths in T, so that no two paths of the same colour use the same arc of T, is equal to the maximum number of paths passing through an arc of T. This result is applied to solve the all-to-all communication problem in wavelength–division–multiplexing (WDM) routing in all–optical networks, that is, we give an efficient algorithm to optimally assign wavelengths to the all the paths of a tree network. It is known that the problem of colouring a general subset of all possible paths in a symmetric directed tree is an NP-hard problem. We study conditions for a given set S of paths be coloured efficiently with the minimum possible number of colours/wavelengths.

1 Introduction

Let T be a tree and x, y two vertices of T. The *dipath* $P(x, y)$ in T is the undirected path joining x to y, in which each edge is considered traversed in the direction from x to y. In other words, the dipaths $P(x, y)$ and $P(y, x)$ are different and do not traverse any edge in the same direction. We are interested in colouring the set of dipaths $P(x, y)$, for all ordered pairs x, y of vertices of T, in such a way that two dipaths using the same edge of T in the same direction obtain different colours. Let $c(T)$ denote the minimum number of colours in such a colouring of the dipaths of T. Let $\pi(T)$ denote the maximum number of dipaths $P(x, y)$ which all pass through the same edge of T in the same direction. Clearly $\pi(T) \leq c(T)$ for every tree T. It has been conjectured by Bermond *et al.* [7] that in fact $\pi(T) = c(T)$ holds for every T. Here we prove this conjecture.
Moreover, given a subset S of all the paths on a tree T, we consider conditions for the existence of an efficient algorithm to colour all the paths in S with the minimum possible number of colours; this problem is NP-hard in general.

---

* Work partially supported by the Italian Ministry of the University and of the Scientific Research in the framework of the project: "Efficienza di Algoritmi e Progetto di Strutture Informative" and by Galileo Project.

1.1 Motivations and Related Work

The problem originally arose in the context of all-optical networks. Optical networks are emerging as key technology in communication networks and are expected to dominate many applications, such as video conferencing, scientific visualisation, real-time medical imaging, high–speed super-computing and distributed computing [17, 25, 29]. The books of Green [17] and McAulay [22] offer a comprehensive overview of the physical theory and applications of this emerging technology. All–optical networks exploit photonic technology for the implementation of both switching and transmission functions [16], and maintain the signal in optical form through the transmission, thus allowing for much higher data transmission rates (since there is no prohibitive overhead due to conversions to and from the electronic form). Wavelength–division multiplexing (WDM) [10] partitions the optical bandwidth into a number of channels, and allows multiple data streams to be transferred concurrently along the same optical fiber, on different channels, i.e., different wavelengths. The same wavelength on two input ports of a switch cannot be routed to a same output port, due to electromagnetic interference. There are various switches considered in the literature, with 'generalized switches' being one of the more common variants, [1, 2, 27]. These switches allow different signals to travel on the same communication link into the switch (on different wavelengths), and then exit from it along different links.

All-optical networks are networks where the information, once transmitted as light, reaches its final destination directly without being converted to electronic form in between. Maintaining the signal in optic form allows to reach high speed in these networks since there is no overhead due to conversions to and from the electronic form. Such an approach allows thus the elimination of the "electronic bottleneck" of communications networks with electronic switching.

In an all-optical network one needs to set up a number of communications (paths) between given pairs of nodes, with each path being transmitted on one particular wavelength, and all paths sharing a link having different wavelengths. Specifically, one is given a set of requests $(a_1, b_1), (a_2, b_2), \ldots, (a_k, b_k)$, and is required to connect each a_i to the corresponding b_i by a path P_i and assign wavelengths to each path P_i so that paths of the same wavelength do not share a link. Viewed in this light, the problem has initially been treated in the context of undirected graphs, [2, 1, 27]. However, it has recently become clear that each bidirectional optical link will actually consist of a pair of unidirectional links [25], and hence the new models of the situation tend to represent the network by a symmetric directed graph, or equivalently, view each path as a dipath (as above) [7, 23, 18]. We study the situation in the case of trees. The interest in trees is due to the fact tree-like networks are standard in the telecommunications industry [23]. Furthermore, trees free us from one half of the problem - that of choosing the actual paths for connecting the required nodes (since in a tree these paths are unique). The minimum number of wavelengths corresponds to the minimum number of colours in a colouring of dipaths as detailed above. This parameter is considered of importance in evaluating the competitiveness of the wavelength division multiplexing technology [23].

Thus this general problem becomes one of colouring a given set (or multiset) S of dipaths in a tree T with the minimum number of colours, so that dipaths using one edge in the same direction obtain different colours. We find the above terminology convenient to work with. However, it should be clear to the reader that an equivalent formulation would consider T to be a symmetric directed tree - by replacing each edge of T with the two opposite arcs (optical links) corresponding to it - and then each dipath would simply become a directed path in the usual sense of the word. Conditions on using an edge in one direction then simply translate into conditions on using an arc.

We call *proper* a colouring of S a colouring such that dipaths of S using one edge in the same direction obtain different colours. The minimum number of colours in a proper colouring of a set (multiset) S of dipaths in a tree T will be denoted by $c_S(T)$, and the maximum number of dipaths from S that pass through one edge of T in one direction by $\pi_S(T)$. We clearly must have $\pi_S(T) \leq c_S(T)$.

The problem of colouring a general subset of all possible paths in a symmetric directed tree is an NP-hard problem [12]. Approximation algorithms are given in [27, 23, 18, 19]. The best ratio is obtained in [18] where the authors provide an algorithm that requires at most $5/3\pi_S(T)$ colours, for any set S of paths in a symmetric directed tree T. A recent survey including this topic is given in [6].

1.2 Our Results

In Section 2 we concentrate on the problem of all-to-all communication (or 'gossiping'). In this situation, every node is requesting a connection with every other node. All-to-all communication among the processors is one of the most important issues in multi-processor systems. The need for this kind of communication arises in many problems of parallel and distributed computing including many scientific computations [8, 11, 13] and database management [30]. Due to the considerable practical relevance in parallel and distributed computation and the related interesting theoretical issues, such problems have been extensively studied in the literature (see the surveys [20, 21, 24, 6]). First studies of this problem in the context of optical networks, can be found in [7, 5, 6]

In this paper, we show that the minimum number of colours necessary to establish all-to-all connections in a tree is equal to the maximum number of intersecting dipaths, i.e., we shall prove the following result.

Theorem 1. *Let T be a tree. Then $c(T) = \pi(T)$.*

Above Theorem 1 settles a conjecture by Bermond et al [7].
We stress that our proof also represents an efficient (e.g., polynomial) algorithm for the actual assignment of the colours to the paths.

In Section 3 we study conditions, given a set S of paths on a tree T, for the existence of an efficient algorithm to colour the paths in S with the minimum possible number of colours. We recall that the problem of optimal colouring of paths is NP-hard [12]. We show that $c_S(T) = \pi_S(T)$ for each set of paths S if and only if T is a generalized star, that is, a tree obtained from a star by replacing each edge with a path.

Moreover, given any tree T, we give conditions on the set S assuring that $c_S(T) = \pi_S(T)$ and $c_S(T)$ can be found in polynomial time.

Due to space limitations some proofs are omitted from this extended abstract.

2 Colouring all paths

In this section we consider the problem of all-to-all communication (or 'gossiping'). In this situation, every node is requesting a connection with every other node; thus S consists of the paths $P(x, y)$ for all ordered pairs x, y of vertices of T, and we shall omit the subscripts S and write $c(T), \pi(T)$ instead of $c_S(T), \pi_S(T)$.

We will find it more convenient to prove a weighted version of the theorem. A *weighted tree* is a tree T with positive integer weights $w(x)$ on the vertices x of T. (The intention of the weights is to have a vertex of weight w represent w unweighted vertices). The total weight of a set X of vertices of T is $w(X) = \sum_{x \in X} w(x)$. (In particular $w(T)$ is the weight of the entire tree.)

Let e be an edge of a weighted tree T. The removal of e from T results in two weighted subtrees T_1 and T_2. The *load of e* is the product $w(T_1)w(T_2)$. The *forwarding index* of the weighted tree T, denoted by $\pi(T)$, is the maximum load of any edge in T. It is clear that when all weights are 1 this definition coincides with the previous definition of $\pi(T)$.

In a weighted tree T, we shall consider *the multiset of all dipaths* which consists of $w(a)w(b)$ copies of the dipath from a to b, for every ordered pair of vertices a, b. We denote by $c(T)$ the minimum number of colours in a proper colouring of the multiset of all dipaths. When all weights are 1, the multiset of all dipaths is precisely the set of all dipaths, and so the definition of $c(T)$ also coincides with the one given earlier. For a particular vertex v, we let $In(v)$ (respectively $Out(v)$) consist of those dipaths from the multiset of all dipaths which end (respectively begin) with v. The weighted version of our theorem is as follows:

Theorem 2. *Any weighted tree T satisfies*
$$c(T) = \pi(T)$$
and there exists an efficient algorithm which colours T with $c(T)$ colours.

2.1 Two operations to generate weighted trees

There is a natural way to build all trees from a single edge, by adding and splitting leaves. We will formally define these operations in the context of weighted trees, and then apply them to give an inductive proof of our theorem.

In the following definition we assume that T is a weighted tree of weight $w(T) = W$, x is a leaf of T, f is the parent of x, and finally, that δ is a positive integer $\delta < w(x)$.

Definition 3. The operation $AddLeaf_\delta(x, T)$ modifies T as follows:

- the weight of x is decreased by δ
- a new node y is added with weight δ
- the edge $[y, x]$ is added.

The operation $SplitLeaf_\delta(x, T)$ modifies T as follows:

- the weight of x is decreased by δ
- a new node y is added with weight δ
- the edge $[y, f]$ is added. (Recall that f is the parent of x.)

We say that an operation $AddLeaf_\delta(x, T)$ or $SplitLeaf_\delta xT$ is *legal* if $\delta + w(x) \leq W$ and $w(x) \leq W/2$

We will often abbreviate the notation to simply say that we have performed an operation $AddLeaf$ or $SplitLeaf$ (with respect to the node x and the weight δ if needed). It is easy to see that if an operation $SplitLeaf_\delta(x, T)$ (resp. $AddLeaf_\delta(x, T)$) is legal then in the new tree the load of $[x, f]$ and $[y, f]$ (resp. $[x, y]$) cannot be larger than the load of $[x, f]$ in T. Therefore we have the following property.

Properties 2.1 *If an operation AddLeaf or SplitLeaf is legal then the forwarding index of the new tree does not exceed the forwarding index of T.*

Definition 4. Let T be a weighted tree, and let W denote $W(T)$. T is called W/C-*tree* if the two trees resulting from the removal of an edge of maximum load have weights C and $W - C$, with $C \geq W/2$.

Notice that the above definition is non ambiguous since each edge of maximum load is associated with the same value of C and $\pi(T) = C(W - C)$. In case T is a weighted star then the above definition is equivalent to the fact that the maximum weight of a leaf is $(W - C)C$; we call T a W/C-*star*.

Given a W/C-tree T, we will recursively construct T from some initial W/C-star S by means of a sequence of $AddLeaf$ and $SplitLeaf$ legal operations. By Property 2.1 this will assure that at each step of the construction we have a tree with forwarding index $\pi(S) = C(W - C) = \pi(T)$.

Lemma 5. *T can be generated from some W/C-star T^* by repeated application of legal operations AddLeaf or SplitLeaf.*

Proof (Sketch). We first show that any W/C-tree T contains a vertex u such that the maximum weight of a component of $T \setminus \{u\}$ is $W - C$.

In order to construct our tree T, we start from the W/C–star T^* consisting of the vertex u and all its neighbours in T; for each neighbor v of u we set the weight $w(v)$ of v equal to the weight of the component of $T \setminus \{u\}$ that contains v.

Let t be the number of nodes of T which are not adjacent to u. If $t = 0$ then the tree T is a W/C-star and we don't need to perform any operations. Otherwise we suppose that the result holds for if $t \leq k$, and let $t = k+1$. Let z be a leaf of T of maximum distance from u, and let p be the parent of z. This implies that p has at most one neighbour which is not a leaf.

- If the degree of p is strictly greater than two, then let x be a leaf neighbour of p other than z. Let T' be the weighted tree obtained from T by removing z and increasing the weight of x by $w(z)$. Then T is generated from T' by the operation $SplitLeaf_{w(z)}(x, T')$.
- If the degree of p is two, then let $x = p$. Let T' be the weighted tree obtained from T by removing z and increasing the weight of of x by $w(z)$. Then T is generated from T' by the operation $AddLeaf_{w(z)}(x, T')$.

We then show that in both cases the operations are legal. □

2.2 An inductive colouring

We have seen how an arbitrary W/C-tree T can be constructed from a W/C-star by legal operations, with all intermediate trees being also W/C-trees. We now begin to prove that the multiset of dipaths in each of these trees admits a proper colouring with $W(W - C)$ colours.

Lemma 6. *The multiset of dipaths of any W/C-star T can be efficiently coloured with $W(W - C)$ colours.*

Proof. The crucial observation here is the following: In a star, two dipaths conflict (use some edge in the same direction and hence must obtain different colours) if and only if they have the same beginning or the same end. in other words, two dipaths of the same colour must belong to two different multisets $In(v)$ and to two different multisets $Out(v)$. For each vertex v we have $|In(v)| = |Out(v)|$, but these sizes differ from vertex to vertex. Of course, the maximum $|In(v)| = \pi(T)$. We now add to each $In(v)$ and $Out(v)$, $\pi(T) - |In(v)|$ artificial paths (consisting of the single vertex v), to arrive at a situation where each $In(v)$ and $Out(v)$ has exactly $\pi(v)$ dipaths. Thus the union of any k sets $In(v)$ contains at most k sets $Out(v)$, and, according the the theorem of Hall [28] (Theorem 9.2.1), one can efficiently determine a set of dipaths consisting of exactly one representative from each $In(v)$ and from each $Out(v)$. These dipaths will be coloured by colour 1, and deleted from consideration. Now each $In(v)$ and each $Out(v)$ has $\pi(T) - 1$ dipaths, and so we can continue as above. Clearly, this will produce a proper colouring of the multiset of dipaths in T with $\pi(T) = W(C - W)$ colours. □

We continue, assuming that we have a W/C-tree T with a proper colouring of its multiset of dipaths with $W(W - C)$ colours, and show how to induce a proper colouring of a tree T' obtained by a legal operation.

Let x be a fixed leaf in T (the leaf on which we shall perform the legal operation $AddLeaf$ or $SplitLeaf$). We wish to use again the Theorem of Hall in a fashion similar to the above proof, but treating only the multiset of dipaths starting (and ending) with x. These dipaths are already coloured, and we may have used different colours for $Out(x)$ and $In(x)$. We deal with this complication by introducing the following bijection:

Let Out denote the set of colours used on the dipaths from the multiset $Out(x)$, and let In denote the set of colours used on the dipaths from the multiset $In(x)$. Since all dipaths in $Out(x)$, and in $In(x)$, have different colours, $|Out| = |In|$. Let ϕ be a fixed bijection between Out and In, such that for any $c \in Out \cap In, \phi(c) = c$.

Notice that the $w(x)w(z)$ dipaths between x and any vertex z of $T \setminus \{x\}$ must all obtain different colours, as they all use the unique edge out of x. We now arbitrarily fix (for each vertex z) a partition of these $w(x)w(z)$ colours into $w(z)$ classes of size $w(x)$ denoted by $O_1^z, O_2^z, \cdots, O_{w(z)}^z$. Similarly, we fix (for each z) another partition of the set of $w(x)w(z)$ colours of dipaths from z to x into $I_1^z, I_2^z, \cdots I_{w(z)}^z$, each of size $w(x)$. We shall say that two colours on dipaths starting in x are *I-equivalent* if they belong to the same class I_j^z for some $z \in T \setminus \{x\}, j \in \{1, 2, \cdots, w(z)\}$. Similarly, we shall say that two colours on dipaths ending in x are *O-equivalent* if they belong to the same class O_j^z for some $z \in T \setminus \{x\}, j \in \{1, 2, \cdots, w(z)\}$.

Definition 7. A *supercolour* is a set U of colours such that no colours from U are *I*-equivalent, and no colours from $\phi(U)$ are *O*-equivalent.

Let X be the set of $w(x)(W - w(x))$ colours used by the dipaths starting in x.

Lemma 8. *The set X of colours can be partitioned into $w(x)$ supercolours.*

Proof Omitted. □

The following result allows to complete the proof of Theorem 2.

Proposition 9. *If T is a W/C-tree with a proper colouring of its multiset of all dipaths, and if T' is obtained from T by performing the legal operation $AddLeaf_\delta(x, T)$ or $SplitLeaf_\delta(x, T)$, then T' is a W/C-tree which also admits a proper colouring of its multiset of all dipaths. Such a colouring of T' can be efficiently determined.*

Proof Omitted. □

3 General sets of paths

We have shown that $c(T) = \pi(T)$ for any tree T, even in the general case of weighted trees. Thus the set (multiset) of all dipaths $P(x,y)$ can be coloured with $\pi(T)$ colours. Our proof represents a polynomial algorithm for the actual assignment of the colours. In the more general situation of an arbitrary set S of dipaths, it is known that the problem of optimally colouring the paths in the set S is NP-hard [12], and only approximation algorithms are known [23, 18, 19].

The undirected version of the problem, that is, minimize the number of colours in a colouring of paths of a tree T so that all paths using an edge of T have different colours, is also NP-hard [15, 27].

In this section we make some additional remarks about $c_S(T)$, that is, the minimum number of colours in a colouring, of the paths from a subset S of all the paths on a directed symmetric tree T, such that conflicting paths obtain different colours. We consider situations in which $c_S(T)$ can be efficiently evaluated.

It is easy to see that if T is a path or a star then $\pi_S(T) = c_S(T)$ for every S. In fact, when T is a path $\pi_S(T) = c_S(T)$ is equivalent to the fact for an interval graph the chromatic number is equal to the maximum clique size [14], and when T is a star $\pi_S(T) = c_S(T)$ is equivalent to the fact that for a bipartite graph the edge chromatic index is equal to the maximum degree [9]. These results also imply corresponding polynomial algorithms [9, 14]. We now extend these results (and algorithms) as follows.

Definition 10. The *conflict graph* of set of paths S on a tree T is the undirected graph whose vertices are the dipaths from S, and two dipaths are adjacent if and only if they conflict, i.e., use an edge of T in the same direction.

Definition 11. A *generalized star* is a tree obtained from a star by replacing each edge with a path (the paths may have different lengths).

Notice that a generalized star is a tree in which at most one vertex has degree greater than two, and, conversely, any tree in which at most one vertex has degree greater than two is a generalized star. Also note that stars and paths are generalized stars. We proceed to prove that all conflict graphs in a generalized star T are perfect; this will imply in particular that $\pi_S(T) = c_S(T)$ for all S.

Definition 12. An *odd hole* of an undirected graph is an induced cycle without chords, of odd length greater than three. An *odd antihole* is an induced complement of a cycle without chords, of odd length greater than three.

Lemma 13. *The conflict graph of of any set of paths on a generalized star cannot contain an odd hole or an odd antihole.*

Proof Omitted. □

Since it is not hard to show that the conflict graph of trees satisfies the perfect graph conjecture, the above Lemma 13 implies that conflict graphs in generalized stars are perfect; therefore we have the following result.

Corollary 14. *For any set S of dipaths in a generalized star T we have $c_S(T) = \pi_S(T)$.*

We remark that by combining polynomial algorithms for edge colouring bipartite multigraphs and for vertex colouring interval graphs, we obtain a polynomial algorithm for colouring the dipaths of S in a generalized star with $\pi_S(T)$ colours. It is not difficult to observe that whenever T is a tree other than a generalized star, then there exists a set of dipaths S in T such that $\pi_S(T) \neq c_S(T)$.

Proposition 15. $c_S(T) = \pi_S(T)$ *for all sets S if and only if T is a generalized star.*

We also consider the following condition on S which assures that $\pi_S(T) = c_S(T)$:

Definition 16. A set of paths S is *well distributed* in T if T does not contain an odd number of edges $[v, a_1], [v, a_2], \ldots [v, a_{2k+1}]$ such that some dipath of S contains both edges $[v, a_i]$ and $[v, a_{i+1}]$ (in some direction), for any index $i = 1, \ldots 2k + 1$ (addition on index is taken modulo $2k + 1$).

Proposition 17. *If S is well distributed in T then*

$$c_S(T) = \pi_S(T)$$

and $c_S(T)$ can be found in polynomial time.

Proof (Sketch). We verify that if S is well distributed in T then T admits an orientation such that each dipath in S either uses all edges in the chosen direction, or all in the opposite direction. Since a path which uses edges in the chosen direction cannot conflict with a path which uses edges in the opposite direction, we can colour each set separately. It is easy to see that the conflict graph of each of these sets is chordal and hence $c = \pi$ and π can be found in polynomial time [14].

Acknowledgements

We would like to thank Bruno Beauquier, Jean-Claude Bermond, Huang Jing, Paulraja, Ugo Vaccaro, Joseph Yu, and Zhu Xuding for their interest and advice.

References

1. A. Aggarwal, A. Bar-Noy, D. Coppersmith, R. Ramaswami, B. Schieber, M. Sudan. "Efficient Routing and Scheduling Algorithms for Optical Networks", in *Proceedings of the 5th Annual ACM-SIAM Symposium on Discrete Algorithms (SODA'94)*, (1994), 412–423.

2. Y. Aumann and Y. Rabani. "Improved Bounds for All Optical Routing", *Proceedings of the 6th Annual ACM-SIAM Symposium on Discrete Algorithms (SODA'95)*, (1995), 567–576.
3. B. Awerbuch, Y. Azar, A. Fiat, S. Leonardi, and A. Rosen, "On–Line Competitive Algorithms for Call Admission in Optical Networks", *Proceedings of ESA '96*, LNCS 1136, (1996), 431–444.
4. R. A. Barry and P. A. Humblet, "On the Number of Wavelengths and Switches in All–Optical Networks", in *IEEE Transactions on Communications*, (1993).
5. B. Beauquier. "All–to–All Communication for some Wavelength–Routing All–Optical Networks", *Manuscript*, (1996).
6. B. Beauquier, J-C. Bermond, L. Gargano, P. Hell, S. Perennes, and U. Vaccaro, "Graph Problems arising from Wavelength–Routing in All–Optical Networks", *2nd Workshop on Optics and Computer Science (WOCS)*, April 1997, Geneva, CH.
7. J-C. Bermond, L. Gargano, S. Perennes, A. Rescigno and U. Vaccaro. "Efficient Collective Communications in Optical Networks", *Proc. 23nd ICALP'96, Paderborn, Germany*, (1996).
8. D. P. Bertsekas, and J. N. Tsitsiklis, *Parallel and Distributed Computation: Numerical Methods*, Prentice–Hall, Englewood Cliffs, NJ, 1989.
9. J.A. Bondy and U.S.R. Murty, "Graph Theory and Applications", American Elsevier, N.Y. 1976.
10. N.K. Cheung, K. Nosu and G. Winzer. *IEEE JSAC*: Special Issue on Dense WDM Networks, vol. 8 (1990).
11. J. J. Dongarra and D. W. Walker, "Software Libraries for Linear Algebra Computation on High Performances Computers", *SIAM Review*, vol. 37, (1995), 151–180.
12. T. Erlebach and K. Jansen. "Scheduling of Virtual Connections in Fast Networks", *Proc. of 4th Workshop on Parallel Systems and Algorithms PASA '96*, (1996), 13–32.
13. G. Fox, M. Johnsson, G. Lyzenga, S. Otto, J. Salmon, and D. Walker, *Solving Problems on Concurrent Processors, Volume I*, Prentice Hall, Englewood Cliffs, NJ, 1988.
14. M. C. Golumbic, "Algorithmic Graph Theory and Perfect Graphs", Academic Press, N.Y. 1980.
15. M. C. Golumbic and R. E. Jamison, The edge intersection grpahs of paths in a tree, J. Combinatorial Theory B 38 (1985) 8 -22.
16. P. E. Green. "The Future of Fiber–Optic Computer Networks", *IEEE Computer*, vol. 24, (1991), 78–87.
17. P. E. Green. *Fiber–Optic Communication Networks*, Prentice–Hall, 1992.
18. C. Kaklamanis and P. Persiano, "Efficient Wavelength Routing on Directed Fiber Trees", *Proc. ESA'96*, Springer Verlag, LNCS 1136, (1996), 460–470.
19. C. Kaklamanis and P. Persiano. "Constrained Bipartite Edge Coloring with Applications to Wavelength Routing", *Manuscript* (1996).
20. S. M. Hedetniemi, S. T. Hedetniemi, and A. Liestman, "A Survey of Gossiping and Broadcasting in Communication Networks", *NETWORKS*, 18 (1988), 129–134.
21. J. Hromkovič, R. Klasing, B. Monien, and R. Peine, "Dissemination of Information in Interconnection Networks (Broadcasting and Gossiping)", in: Ding-Zhu Du and D. Frank Hsu (Eds.) *Combinatorial Network Theory*, Kluwer Academic Publishers, 1995, pp. 125-212.
22. A. D. McAulay. *Optical Computer Architectures*, John Wiley, 1991.

23. M. Mihail, C. Kaklamanis, S. Rao, "Efficient Access to Optical Bandwidth", *Proceedings of 36th Annual IEEE Symposium on Foundations of Computer Science (FOCS'95)*, (1995), 548–557.
24. A. Pelc, "Fault Tolerant Broadcasting and Gossiping in Communication Networks", *NETWORKS*, to appear.
25. R. Ramaswami, "Multi-Wavelength Lightwave Networks for Computer Communication", *IEEE Communication Magazine*, vol. 31, (1993), 78–88.
26. Y. Rabani. "Path Coloring on the Meshes", *Proc. of FOCS '96*.
27. P. Raghavan and E. Upfal. "Efficient Routing in All–Optical Networks", *Proceedings of the 26th Annual ACM Symposium on Theory of Computing (STOC'94)*, (1994), 134–143.
28. L. Mirsky, Transversal Theory, Academic Press, New York, London, 1971.
29. R. J. Vetter and D. H. C. Du. "Distributed Computing with High-Speed Optical Networks", *IEEE Computer*, vol. 26, (1993), 8–18.
30. O. Wolfson and A. Segall, "The Communication Complexity of Atomic Commitment and Gossiping", *SIAM J. on Computing*, 20 (1991), 423–450.

On-Line Routing in All-Optical Networks

Yair Bartal[1]   Stefano Leonardi[2]

[1] U.C. Berkeley and International Computer Science Institute (ICSI), Berkeley ***
[2] Dipartimento di Informatica e Sistemistica, Università di Roma "La Sapienza" †

Abstract. The paper deals with *on-line* routing in WDM (wavelength division multiplexing) optical networks. A sequence of requests arrives over time, each is a pair of nodes to be connected by a path. The problem is to assign a *wavelength* and a *path* to each pair, so that no two paths sharing a link are assigned the same wavelength. The goal is to minimize the number of wavelengths used to establish all connections.

We consider trees, trees of rings, and meshes topologies. We give on-line algorithms with competitive ratio $O(\log n)$ for all these topologies. We give a matching $\Omega(\log n)$ lower bound for meshes. We also prove that any algorithm for trees cannot have competitive ratio better than $\Omega(\frac{\log n}{\log \log n})$.

We also consider the problem where every edge is associated with parallel links. While in WDM technology, a fiber link requires different wavelengths for every transmission, SDM (space division multiplexing) technology allows parallel links for a single wavelength, at an additional cost. Thus, it may be beneficial in terms of network economics to combine between the two technologies (this is indeed done in practice). For *arbitrary* networks with $\Omega(\log n)$ parallel links we give an on-line algorithm with competitive ratio $O(\log n)$.

1 Introduction

All-optical networks promise data transmission rates several orders of magnitude higher than current networks. The high speeds in these networks arise from maintaining signals in optical form throughout a transmission thereby avoiding the overhead of conversions to and from electrical form (see [Gr92] for an overview of the topic). *Wavelength division multiplexing* (WDM) supports the propagation of multiple laser beams of distinct wavelengths through an optic fiber. Thus, the high bandwidth of the WDM network is utilized by partitioning it in many "channels", each at a different optical *wavelength*. Intuitively, we may think of wavelengths as light rays of different colors.

A major algorithmic problem for optical networks is that of routing. Each routing request, consists of a pair of nodes in the network, and requires the assignment of a path and a wavelength (color). The key restriction is that two

---

*** Research supported in part by the Rothschild Postdoctoral fellowship and by the National Science Foundation operating grants CCR-9304722 and NCR-9416101. e-mail: yairb@icsi.berkeley.edu

† This work was partially done while the author was a post-doc at the International Computer Science Institute (ICSI), Berkeley. This work is partly supported by EU ESPRIT Long Term Research Project ALCOM-IT under contract n 20244, and by Italian Ministry of Scientific Research Project 40% "Algoritmi, Modelli di Calcolo e Strutture Informative". e-mail: leon@dis.uniroma1.it

requests with equal wavelength cannot be routed through the same link. The main goal is in lowering the number of wavelengths for certain routing requests.

Many of the applications for high speed optical networks are real-time. It is therefore very natural to consider the problem of routing in an on-line setting where routing requests appear over time.

The Path Coloring Problem. The routing problem on a WDM network with generalized switches is referred to as *path coloring*. More formally, let $G = (V, E)$ be a graph representing the network, with $|V| = n$. We are given a sequence of routing requests consisting of pairs $p_i = (s_i, t_i)$ of nodes in G. The algorithm must assign a path connecting s_i and t_i and a color, so that no two paths sharing an edge are assigned with same color. The goal is to minimize the number of colors. The performance measure for an on-line algorithm is the *competitive ratio* [ST85] defined as the worst case ratio over all request sequences between the number of colors used by the on-line algorithm and the optimal number of colors necessary on the same sequence.

While in WDM technology, a fiber link requires different wavelengths for every transmission, SDM (space division multiplexing) technology allows parallel links for a single wavelength, at an additional cost. This can be profitable since only a limited number of wavelengths are available in practice. The two technologies are then combined to find an efficient trade off between the two approaches. This motivates considering a generalization of the path coloring problem where a link of a color is replaced with a number of parallel links. We will alternatively model this case with a bandwidth B available on a link for any color, meaning that B paths of the same color may be routed through a link (that is in the basic path coloring problem $B = 1$).

Related previous work.
The *off-line* path coloring problem has been studied by Raghavan and Upfal [RU94] who give constant approximation algorithms for undirected trees and trees of rings. Further results for trees were given in a sequence of papers [MKR95, KP96, KS97, EJKP97]. Rings have been recently addressed in [GK97] and meshes were studied in [RU94, AR95, KT95]. Kleinberg and Tardos [KT95] give an $O(\log n)$ approximation algorithm for meshes and certain "nearly Eulerian planar graphs". Rabani [Ra96] improves the bound for meshes to $O(\text{poly}(\log \log n))$.

The *on-line* path coloring problem has been studied in the case of a line topology in the context of *interval graph coloring* by Kierstead and Trotter [KT81]. They give an optimal 3-competitive algorithm for the line ([KT81]). Ślusarek [Sl95] proved the same bound for circular arc graphs.

The path coloring problem is closely related to the *virtual circuit routing* problem, motivated by its application to ATM networks. The *load* version of this problem is where every requested pair must be assigned a path as to minimize the maximum number of paths crossing a given edge. Aspnes et al. [AAFPW93] give an $O(\log n)$ competitive algorithm for the load version. Most of the work has concentrated on the *throughput* version of the problem, where every requested pair may be either accepted or rejected. The basic problem, also referred to as *call-control*, is where the paths of all accepted pairs must be edge-disjoint. This can also be generalized to the case where edges may have a given bandwidth B (that can be viewed as having B parallel edges). Awerbuch et al. [AAP93] prove that if $B = \Omega(\log n)$ then there is an $O(\log n)$ competitive algorithm (for throughput). They also give a lower bound of $\Omega(n)$ for deterministic algorithms in the general case. Randomized algorithms have been first studied by Awerbuch et al. [ABFR94, AGLR94] giving an $O(\log \Delta)$ competitive algorithm for trees, where Δ is the diameter of the tree. They also show a matching lower bound. Kleinberg and Tardos [KT95] give $O(\log n)$ competitive algorithm for meshes (and some generalization), improving upon a previous result of [AGLR94].

Bartal et al. [BFL96] prove that for various routing problems including the throughput version of virtual circuit routing and the path-coloring problem there exist networks where the competitive ratio is $\Omega(n^\epsilon)$ (for some fixed ϵ) for any randomized algorithm. Finally, the on-line version of maximizing the throughput in optical networks was addressed in [AAFLR96].

Contributions of this paper. We consider the *on-line* path coloring problem on trees, trees of rings, and meshes topologies:

- We present an $O(\log n)$ competitive deterministic algorithm for path coloring on *meshes*.

- We prove a matching $\Omega(\log n)$ lower bound for the mesh. The lower bound holds for randomized algorithms for the load version of the virtual circuit problem which immediately extends to the path coloring problem.

    We comment that this also provides the first lower bound for the load version of the virtual circuit routing problem in undirected networks with unit edge capacities [AAFPW93].

- We give an $O(\log n)$ competitive algorithm for path coloring on *arbitrary* networks with bandwidth $\Omega(\log n)$ (the actual statement is somewhat more general). This algorithm is also used as a building block for our algorithm for path coloring on meshes. This result can be viewed as a balanced combination of WDM and SDM technologies.

- We give an $O(\log n)$ competitive algorithm for *trees* and *trees of rings*. We also prove that any deterministic algorithm for trees cannot have competitive ratio better than $\Omega(\frac{\log n}{\log \log n})$ (even for trees with $\Delta = O(\log n)$). A logarithmic upper bound and an $\Omega(\sqrt{\log n})$ lower bound for trees have been independently obtained by Borodin, Kleinberg, and Sudan [BKS96].

Paper structure: Section 2 contains the results for path coloring with more bandwidth on arbitrary networks, that are also used in Section 3 for the $O(\log n)$ competitive algorithm for path coloring on meshes. Section 4 contains the lower bound for meshes. Upper and lower bounds for trees are in Section 5. The results and the proofs that are omitted from this abstract can be found in [BL97].

2 Path coloring with more bandwidth

Let $G = (V, E)$ be a network with $|V| = n$ vertices and $|E| = m$ edges. We consider the *path coloring* problem with bandwidth B on the edges. At the j-th step, call j, with endpoints (s_j, t_j), is presented to the algorithm that must assign a color $c(j)$ and a path $P(j)$. The goal of the on-line is to use a set of colors of minimum cardinality C under the constraint that the bandwidth on any edge does not exceed B.

We give an algorithm for general networks for this problem. The algorithm fixes a set $\mathcal{C}$ of C colors that it may choose from, at the beginning, based on an estimate for the optimal performance. The basic algorithm chooses, at every step, one path and one of these colors according to some optimization criteria. This criteria assigns to any edge of any color an exponential function of the current load. Our goal is in proving that the algorithm never exceeds a certain bandwidth on every edge.

A variant for this algorithm proves to be useful (see Section 3) in obtaining an algorithm for path coloring on meshes (with edge bandwidth = 1).

In this variant we restrict the choice of the on-line algorithm for call j to a subset $\mathcal{C}(j)$ of $\mathcal{C}$ (that may be chosen according to some arbitrary rule) whose cardinality is at least αC.

We thus state our results in terms of this parameter α. However, for the scope of this section alone it is enough to set $\alpha = 1$.

Let C^* be the number of colors used by the optimal solution to accommodate the whole set of calls, and let B^* be the bandwidth available by the optimal solution on any edge of any color.

We compare our algorithm to a *stronger adversary* that uses a bandwidth $\Lambda^* \leq B^* C^*$ on a single color, rather than being restricted to using C^* colors and bandwidth B^* for every color.

We assume that the on-line algorithm knows a value Λ such that $\Lambda^* \leq \Lambda \leq 2\Lambda^*$. This is performed by applying a doubling technique (whose description is omitted in this abstract) that results in increasing the competitive ratio at most by a factor of 4.

Let the load on edge e for color c, denoted by $\lambda_e^c(j)$, be the number of calls assigned with color c and a path crossing edge e when call j is presented. Let $a = 2^\beta$. Call j is assigned with a color $c(j)$ and a path $P(j)$ which achieve the minimum, over all the colors in $\mathcal{C}(j)$ and all paths connecting s_j and t_j, of the following "exponential cost":

$$\sum_{e \in P(j)} a^{\lambda_e^{c(j)}(j)}.$$

Theorem 1. *If the number of colors used by the algorithm is $C = 8\Lambda \frac{1}{\alpha}(2^\beta - 1)$ where $\Lambda \geq \Lambda^*$ then the bandwidth is $B \leq 1 + \frac{1}{\beta} \log \frac{2m}{\alpha}$.*

Proof.

Let $\overline{\lambda}$ be the maximum load on any edge for any color in the solution of the on-line algorithm at the end of the sequence. Thus $\overline{\lambda}$ calls are assigned with same color and a path crossing a given edge. When the last such path $P(k)$ is assigned to a call k, its exponential cost is at least $a^{\overline{\lambda}-1}$. By definition of the algorithm, the chosen path is the minimum cost path over all paths and colors $\mathcal{C}(k)$. Therefore, at the time this call arrived, for $\mathcal{C}(k) \geq \alpha C$ colors, any path connecting the same pair of vertices has a cost of at least $a^{\overline{\lambda}-1}$. It follows that the sum of the exponential costs over all edges and all colors at the end of the sequence is

$$Z(f) \geq \alpha C a^{\overline{\lambda}-1}, \qquad (1)$$

where $X(f)$ indicates the value of a function X at the end of the sequence.

Let $l_e^*(j)$ be the number of calls in the adversary solution assigned with path crossing edge e when call j is presented.

We use the following potential function:

$$\Phi(j) = \sum_{c \in \mathcal{C}} \sum_{e \in E} a^{\lambda_e^c(j)} \left(1 - \frac{l_e^*(j)}{2\Lambda}\right).$$

The sum of the exponential costs of the on-line algorithm at the end of the sequence is also bounded by the following:

$$Z(f) \leq 2 \sum_{c \in \mathcal{C}} \sum_{e \in E} a^{\lambda_e^c(f)}(1 - \frac{l_e^*(f)}{2\Lambda}) \leq 2(\Phi(f) - \Phi(0)) + 2mC, \quad (2)$$

In the following we prove that for the claimed choice of C, the potential function does not increase after each step of the algorithm. Therefore $\Phi(f) \leq \Phi(0)$, and thus the equations 1 and 2 can be combined to achieve:

$$B \leq \bar{\lambda} \leq 1 + \frac{1}{\beta} \log \frac{2m}{\alpha}.$$

To complete the proof we prove that if $C = 2\Lambda \frac{1}{\alpha}(2^\beta - 1)$ then for every j, $\Phi(j+1) - \Phi(j) \leq 0$. An extra factor of 4 is due to the application of the doubling technique to estimate the value of Λ. Let $P^*(j)$ be the path assigned by the adversary for call j. The change in the potential function due to call j is:

$$\Phi(j+1) - \Phi(j) \leq \sum_{e \in P(j)} (a^{\lambda_e^{c(j)}(j+1)} - a^{\lambda_e^{c(j)}(j)}) - \frac{1}{2\Lambda} \sum_{c \in \mathcal{C}} \sum_{e \in E} (a^{\lambda_e^c(j+1)} l_e^*(j+1))$$

$$- a^{\lambda_e^c(j)} l_e^*(j) \leq \sum_{e \in P(j)} (a-1) a^{\lambda_e^{c(j)}(j)} - \frac{1}{2\Lambda} \sum_{c \in \mathcal{C}} \sum_{e \in P^*(j)} a^{\lambda_e^c(j)}.$$

Observe that for any color $c \in \mathcal{C}(j)$, the cost of any path P connecting s_j to t_j is not less than the cost of the path $P(j)$ on color $c(j)$ chosen by the on-line algorithm for call j. Therefore we get for any $c \in \mathcal{C}(j)$:

$$\sum_{e \in P} a^{\lambda_e^c(j)} \geq \sum_{e \in P(j)} a^{\lambda_e^{c(j)}(j)}.$$

The above inequality also holds for $P = P^*(j)$, and hence

$$\Phi(j+1) - \Phi(j) \leq ((a-1) - \frac{\alpha C}{2\Lambda}) \sum_{e \in P(j)} a^{\lambda_e^{c(j)}(j)}.$$

Recall that $a = 2^\beta - 1$. Thus, by choosing $C = 2\Lambda \frac{1}{\alpha}(2^\beta - 1)$ we have that the potential function does not increase. ∎

As an application we get the following result for the on-line load balancing problem ([AAFPW93]), in which only one color is available and the goal is to minimize the number of paths assigned to a single edge of the network.

By applying Theorem 1 with $\beta = 1$ and $\alpha = 1$ we get:

Corollary 2. *There exists an algorithm for on-line load balancing that uses $O(\Lambda^*)$ colors with bandwidth $O(\log n)$.*

Note that Corollary 2 gives a stronger result than that of [AAFPW93] that only shows that the on-line load is bounded by $O(\Lambda^* \log n)$.

Finally, going back to the path coloring problem, recall that $\Lambda^* \leq C^* B^*$. For an appropriate choice of β (the proof is omitted), Theorem 1 implies the following:

Corollary 3. *Let δ be such that $B^* = \delta \log \frac{2m}{\alpha}$, and let $\gamma > 0$ be some positive coefficient. The algorithm for on-line path coloring with more bandwidth uses $C \leq 8C^* \frac{\delta}{\alpha}(\log \frac{2m}{\alpha}(2^{\frac{1}{\gamma\delta}} - 1) + 1)$ colors with bandwidth $B \leq \gamma B^*$.*

The above corollary shows that if the bandwidth is $\Omega(\log n)$, then the on-line algorithm does not exceed the bandwidth by using $O(\log n)$ more colors. We thus obtain the result for optical networks with general topology when the technologies WDM and SDM are combined in a network that contains $\Omega(\log n)$ parallel fiber optic links on each connection.

3 Path coloring on meshes

In this section we present an $O(\log n)$ competitive algorithm for path coloring on meshes.

$G = (V, E)$ denotes the $\sqrt{n} \times \sqrt{n}$ two dimensional mesh. We consider $\sqrt{n}$ to be a power of 2. Let $|E| = m$ be the number of edges of the mesh. The vertex of the mesh with row i and column j is denoted with $G[i, j]$. Given two vertices $v = G[i, j], v' = G[i', j']$ we define their distance as the length of the shortest path connecting the two vertices: $d(v, v') = |i - i'| + |j - j'|$.

Let α and σ be parameters that will be fixed later. Calls are divided into *short* calls and *long* calls. A call (s, t) is long if $d(s, t) > 2\sigma \log \frac{2m}{\alpha}$, and short if $d(s, t) \leq 2\sigma \log \frac{2m}{\alpha}$. α and σ will be chosen so that $\sigma \log \frac{2m}{\alpha}$ is a power of two.

We use two different algorithms for long calls and short calls. The algorithm for long calls translates the problem in a mesh, to a problem of coloring with more bandwidth in a simulated network that is also a mesh. Theorem 1 allows a logarithmic competitive ratio with a logarithmic bandwidth on any edge. The route obtained in the simulated network is later translated into a route in the original mesh, satisfying the constraint that paths associated to calls with the same color are disjoint. We describe in Section 3.1 the construction of the simulated network, and in Section 3.2 how a route in the simulated network is transformed into a route in the original mesh.

The algorithm for short calls, whose description is omitted in this abstract, classifies the calls on the basis of their length, and applies a greedy algorithm within each class.

Both algorithms for long and short calls have competitive ratio $O(\log n)$. Therefore, we can state the following theorem.

Theorem 4. *There exists a $O(\log n)$ competitive algorithm for path coloring on meshes.*

3.1 The algorithm for the simulated network

In this section we describe the algorithm for the problem of coloring and routing calls on a simulated network of a mesh of size $\sqrt{n} \times \sqrt{n}$.

The algorithm divides the mesh into $\frac{\sqrt{n}}{\sigma \log \frac{2m}{\alpha}} \times \frac{\sqrt{n}}{\sigma \log \frac{2m}{\alpha}}$ squares of size $\sigma \log \frac{2m}{\alpha} \times \sigma \log \frac{2m}{\alpha}$. Square $S[p, q], p, q = 1, \ldots, \frac{\sqrt{n}}{\sigma \log \frac{2m}{\alpha}}$, is the subgraph of G induced by the set of vertices $\{G[i, j] | i = (p-1)\sigma \log \frac{2m}{\alpha} + 1, \ldots, p\sigma \log \frac{2m}{\alpha}; j = (q-1)\sigma \log \frac{2m}{\alpha} + 1, \ldots, q\sigma \log \frac{2m}{\alpha}\}$.

Note that long calls have their endpoints in different squares, since the distance between the endpoints is bigger than $2\sigma \log \frac{2m}{\alpha}$.

The *simulated network* N of the mesh $G = (V, E)$ is a mesh of size $\frac{\sqrt{n}}{\sigma \log \frac{2m}{\alpha}} \times \frac{\sqrt{n}}{\sigma \log \frac{2m}{\alpha}}$.

Let m' be the number of edges of the simulated network. Every edge of N

is associated with a bandwidth equal to $\sigma \log \frac{2m}{\alpha} = \delta \log \frac{2m'}{\alpha}$. (Observe that $\log m' = \log m - \Theta(\log \log m)$. Hence $\sigma \approx \delta$ for large m.)

This mesh corresponds to the network obtained from the original mesh by contracting every square of G onto a vertex and connecting every pair of vertices representing adjacent squares with an edge. The bandwidth of the edges models the fact that at most $\sigma \log \frac{2m}{\alpha}$ edge-disjoint paths can pass between two adjacent squares.

The basic idea is to color and route long calls in the simulated network using the algorithm of Section 2 for path coloring with more bandwidth, and then translate the assigned paths into an appropriate routing in the original network.

The sequence of long calls in the mesh G is transformed into a sequence of calls in the simulated network in the most natural way: Each long call (s,t) is replaced by a call between the two vertices of N representing the two squares containing s and t.

The path obtained for a call in the simulated network is transformed into a path in the original mesh G respecting the following rule: The path in G will cross between adjacent squares in G where the path in N passes through the edge connecting the corresponding nodes in N.

However we need that the paths with same color crossing any square are edge disjoint. For this purpose we will restrict the set of candidate colors for each call to a constant fraction of the overall number of colors. (Observe that the design of the algorithm of Section 3.2 includes this feature).

For this purpose we distinguish between the two squares that include the endpoints of a call, and the squares that are crossed by the path connecting the endpoints. We say that a call is *internal* to a square if one of its endpoints belongs to the square. A call is called *external* to a square if it is not internal to the square and the path derived by the routing in the simulated network crosses the square.

We furthermore define in any square S of the mesh, three concentric regions: S^1, S^2 and S^3 (see Figure 1). Each region contains $2 \log \frac{2m'}{\alpha}$ concentric rings of the square. S^1 is the most external region, S^2 is internal to S^1 and S^3 is internal to both S^1 and S^2. Finally, the area surrounded by S^3 is called the *central region* of the square.

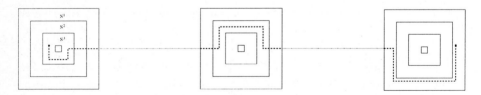

Fig. 1. The routing of a long call.

The set of colors $\mathcal{C}$ used by the on-line algorithm is partitioned into three sets $\mathcal{C}^1, \mathcal{C}^2, \mathcal{C}^3$ of *equal size*. If a call is associated with a color $c \in \mathcal{C}^i$, $i = 1, 2, 3$, then its two endpoints must lie on a region different from S^i, while the path connecting the two endpoints will cross any square of the mesh different from the two squares containing the endpoints using a ring of region S^i.

In Section 3.2 we will show how this requirement allows to avoid intersections between calls with same color crossing a square.

We further impose an additional requirement: for any square, at most one internal call is associated with any color. This requirement is to avoid conflicts between paths assigned to internal calls that leave a square.

Consider the j'th long call (s_j, t_j). Let $S(s_j)$ and $S(t_j)$ be the squares containing s_j and t_j, respectively. The set $\mathcal{C}(j)$ of candidate colors for call j is defined as follows. A color $c \in \mathcal{C}^i$ is in $\mathcal{C}(j)$ if the two following conditions hold:

1. $s_j \notin S^i(s_j)$ and $t_j \notin S^i(t_j)$, e.g. both endpoints are not in region i of their corresponding squares.

2. No call with an endpoint in $S(s_j)$ or $S(t_j)$ has been previously assigned with color c.

The algorithm for path coloring with more bandwidth in the simulated network is run with parameters satisfying: $\alpha \leq \frac{1}{9}$; $\delta \geq 13$; and $\gamma = \frac{1}{\delta}$. The value σ that defines the size of each square is chosen in order to satisfy $\sigma \log \frac{2m}{\alpha} = \delta \log \frac{2m'}{\alpha}$.

The choice of the parameters is such that the adversary bandwidth $B^* = \delta \log \frac{2m'}{\alpha}$ is equal to the maximum number of calls that can be routed through two adjacent squares, and the width $\delta \log \frac{2m'}{\alpha}$ of a square is equal to 13 times the maximum bandwidth $B = \log \frac{2m'}{\alpha}$ used by the on-line algorithm for routing between two adjacent squares.

To apply the result of Corollary 3 we need the following lemma whose proof is omitted.

Lemma 5. *The set of feasible colors $\mathcal{C}(j)$ for a call (s_j, t_j) has size at least αC.*

Therefore, from Corollary 3 we can derive the following corollary, on the number of colors and the bandwidth used by the on-line algorithm for path coloring with more bandwidth in the simulated network:

Corollary 6. *The algorithm for on-line path coloring with more bandwidth in the simulated network N uses $C = 8C^* \frac{\delta}{\alpha}(\log \frac{2m'}{\alpha} + 1)$ colors with bandwidth $B \leq \frac{1}{\delta} \log \frac{2m'}{\alpha}$.*

3.2 Routing of long calls

In this section we describe how to transform a path in the simulated network N into a path in the mesh G, so that the paths associated to calls with same color are mutually edge-disjoint. A path in the simulated network indicates the squares to cross to connect the two endpoints of a call. We are left to describe the route followed by the path within each square.

Given a color $c \in \mathcal{C}^i$, the set of calls accepted with that color have the following property

1. At most one call is internal to each square.

2. Both endpoints of each call are outside region S^i of their squares.

The run of the algorithm for path coloring with more bandwidth ensures that the maximum bandwidth of the on-line algorithm in the simulated network is $B = \frac{\sigma}{13} \log \frac{2m}{\alpha}$. It follows that at most B calls are assigned with paths crossing the boundary between two adjacent squares, and there are at most $2B$ external calls for each square.

We will maintain inductively the following property: A call crosses the boundary between two squares on a row or on a column connecting the central regions of the two squares. The central region of a square has size $B \times B$. Since B is the maximum number of calls routed between adjacent squares, a distinct row or column can be associated with any call.

We first consider external calls. By induction, each external call enters the square on a row or on a column leading to the central area. We route it towards the central area until a free ring of region S^i is reached. This is always the case since there are at most $2B$ external calls and $2B$ available rings in each region S^i. The call then follows the ring until it reaches a free row or a free column connecting the central region of the square to the central region of the adjacent square to which the call is directed. The route follows such row or such column until the adjacent square.

Finally, we consider the routing of the possible single internal call. The endpoint of the internal call is outside the area S^i. If it is originated in the central area, then it can be routed through a path that reaches a free row or column that connects the central area to the central area of the adjacent square to which the internal call is directed. The route goes through such row or column until the adjacent square is reached. If the internal call has the endpoint outside both the central area and the region S^i, it is routed through the ring on which the endpoint lies until it reaches a free row or column connecting the central area of the square to the central area of the adjacent square to which the call is directed, and then follow it until the appropriate adjacent square.

The routing of a call associated with a color of set C^2 is shown in Figure 1. In particular, it is described the route followed in the two squares where the call is internal, and in one square where the call is external.

4 Lower Bounds on Meshes

In this section we give a randomized lower bound of $\Omega(\log n)$ for the path coloring problem on meshes. The lower bound also applies to the load balancing problem ([AAFPW93]) on meshes.

The lower bound is based on an application of Yao's Lemma to on-line algorithms. We construct a distribution over request sequences, such that the number of colors used by an optimal algorithm is always bounded by a constant while the expected on-line load (i.e., the maximum number of paths crossing an edge) of a deterministic algorithm is $\Omega(\log n)$. We recall that the load of a path coloring algorithm is bounded above by the number of colors and thus the lower bound follow.

The distribution over request sequences is defined recursively in $L = \log_4 n$ stages as follows. At the i'th stage of the recursion, $i = 1, 2, \ldots, L$, we define a probability distribution for an $4^{L-i+1} \times 4^{L-i+1}$ square S_i of the mesh. We consider a partition of S_i into 16 subsquares of size $4^{L-i} \times 4^{L-i}$. The internal part of the square S_i is defined as the square I consisting of the 4 internal subsquares in the above partition. $S \setminus I$ is called the external part of the square. Let $I[x, y]$ denote the vertex with row x and column y in the submesh defined by I where $0 \leq x, y \leq 2 \cdot 4^{L-i}$. We now give for each $0 \leq x \leq 2 \cdot 4^{L-i}$ a set of 8 vertical calls from $I[0, x]$ to $I[2 \cdot 4^{L-i}, x]$. Then choose at random one of the 16 subsquares and proceed with the $(i+1)$'st stage of the probability distribution for that subsquare recursively. The $(L+1)$'st stage of the probability distribution contains no requests.

The next two claims give bounds on the optimal and the on-line solutions.

Claim 7 *The number of colors used by an optimal algorithm for the above probability distribution is 8.*

Proof. We prove the claim by induction on i. If the subsquare of size $4^{L-i} \times 4^{L-i}$ chosen in the probability distribution is not in the internal part I, then we route the calls given in the i'th stage through the internal part of the square, and otherwise we route the calls through the external part of the square, so that none of the routes will cross the routes for calls in stages $j > i$. This can be done so that calls with distinct source and destination have disjoint paths and thus the number of colors is 8. ∎

Claim 8 *Let A_i be the expected average load of the on-line algorithm on the edges in the square S_i. Then $A_i \geq i$.*

Proof. We first prove that the average increase in the load of the edges of the square S_i due to the requests given at the i'th stage is at least 1. The number of edges of the mesh S_i is $2 \times 4^{2(L-i+1)}$. The requests given at the ith stage include $8 \times 2 \times 4^{L-i}$ calls between pairs of vertices such that any path between them includes at least $2 \times 4^{L-i}$ edges in S_i (even if the path passes outside the square). Therefore, the increase of the average load on edges of S_i is 1.

We now prove by induction that $A_i \geq i$. For $i = 1$ it follows from the above claim. We assume the claim holds for i and prove it for $i+1$. Since the subsquare for the $(i+1)$'st stage is chosen at random the expected average load of the edges of S_{i+1} is equal to A_i. Since the average increase in the load of the edges of S_{i+1} is at least 1 we have $A_{i+1} \geq i + 1$. ∎

We conclude the following.

Theorem 9. *The competitive ratio of any on-line randomized path coloring algorithm on meshes is $\Omega(\log n)$ against oblivious adversaries. The same lower bound holds for load balancing on meshes.*

5 Path coloring on trees and trees of rings

In this section we consider the on-line path coloring problem on trees and on trees of rings.

An algorithm for trees and trees of rings is obtained by showing that these graphs are $O(C)$-inductive graph, where C is the maximum number of paths that crosses an edge, which is a lower bound on the optimal cost. We omit the proof of this fact in this abstract. The upper bounds follow from a result by Irani [I90] that the greedy on-line coloring algorithm uses $O(d \log n)$ colors on a d-inductive graph of n vertices. We can therefore conclude:

Theorem 10. *There exists a $O(\log n)$-competitive algorithm for on-line path coloring on trees and trees of rings of n vertices.*

We also prove the following lower bound on the competitive ratio of deterministic algorithms for on-line path coloring on trees whose description is omitted in this abstract.

Theorem 11. *Any algorithm for path coloring on trees of n vertices has a competitive ratio of $\Omega(\frac{\log n}{\log \log n})$.*

Acknowledgments: We would like to thank Yossi Azar, Allan Borodin, Amos Fiat, Sandy Irani, Hal Kierstead and Gerhard Woeginger for useful discussions.

References

[ABC+94] A. Aggarwal, A. Bar-Noy, D. Coppersmith, R. Ramaswami, B. Schieber, and M. Sudan. Efficient Routing and Scheduling Algorithms for Optical Networks. *Proc. of SODA '94*, pp. 412–423.

[AAFPW93] J. Aspens, Y. Azar, A. Fiat, S. Plotkin and O. Waarts. On-line Load Balancing with Applications to Machine Scheduling and Virtual Circuit Routing. *Proc. of STOC'93*.

[AAFLR96] B. Awerbuch, Y. Azar, A. Fiat, S. Leonardi, A. Rosén. On-line Competitive Algorithms for Call Admission in Optical Networks. *Proc. of ESA '96*, LNCS 1136, pp. 431-444.

[AAP93] B. Awerbuch, Y. Azar, and S. Plotkin. Throughput Competitive On-line Routing. *Proc. of FOCS'93*.

[ABFR94] B. Awerbuch, Y. Bartal, A. Fiat, and A. Rosén. Competitive Non-Preemptive Call Control. *Proc. of SODA'94*.

[AGLR94] B. Awerbuch, R. Gawlick, F.T. Leighton, and Y. Rabani. On-line Admission Control and Circuit Routing for High Performance Computing and Communication. *Proc. of FOCS '94*.

[AR95] Y. Aumann, Y. Rabani. Improved Bounds for All-Optical Routing. Proc. of SODA'95.

[BFL96] Y. Bartal, A. Fiat, S. Leonardi. Lower Bounds for On-line Graph Problems with Application to On-line Circuit and Optical Routing. *Proc. of STOC'96*.

[BKS96] A. Borodin, J. Kleinberg, and M. Sudan. Personal communication.

[BL97] Y. Bartal and S. Leonardi. On-line Routing in All-Optical Networks. *Tech Rep 02-97, Dipartimento di Informatica e Sistemistica, Università di Roma "La Sapienza"*, 1997.

[Gr92] P.E. Green. *Fiber-Optic Communication Networks*. Prentice Hall, 1992.

[EJKP97] T. Erlebach, K. Jansen, C. Kaklamanis and P. Persiano. Constrained Bipartite Edge Coloring with Applications to Wavelength Routing. *Proc. of ICALP'97* (this proceedings).

[GK97] O. Gerstel and S. Kutten. Dynamic Wavelength Allocation in WDM Ring Networks. To appear in *Proceedings of ICC '97*.

[I90] S. Irani. Coloring inductive graphs on-line. *Proc of FOCS'90*, pp. 470-479.

[KP96] C. Kaklamanis and P. Persiano. Efficient wavelength routing in directed fiber trees. *Proc. of ESA '96*, LNCS 1136, pp. 460-470.

[KS97] V. Kumar and E. Schwabe. Improved Access to Optical Bandwidth in Trees. *Proc. of SODA '97*.

[KT81] H.A. Kierstead and W.T. Trotter. An Extremal Problem in Recursive Combinatorics. *Congress. Numer.*, 33, pp. 143-153, 1981.

[KT95] J. Kleinberg and E. Tardos. Disjoint Paths in Densely Embedded Graphs. *Proc. of FOCS'95*, pp. 52–61.

[MKR95] M. Mihail, C. Kaklamanis, and S. Rao. Efficient Access to Optical Bandwidth. *Proc. of FOCS'95*, pp. 548–557.

[Ra96] Y. Rabani. Path-Coloring on the Mesh. *Proc. of FOCS'96*, pp. 400–409.

[RU94] P. Raghavan and U. Upfal. Efficient Routing in All-Optical Networks. *Proc. of STOC'94*, pp. 133–143.

[Sl95] M. Ślusarek. Optimal Online Coloring of Circular Arc Graphs. Informatique Theoretique et Applications, vol. 29, n. 5, pp. 423-429.

[ST85] D. Sleator, R.E. Tarjan. Amortized Efficiency of List Update and Paging Rules. *Communications of ACM* 28, 1985.

A Complete Characterization of the Path Layout Construction Problem for ATM Networks with Given Hop Count and Load

(Extended Abstract)

Tamar Eilam[1], Michele Flammini[2,3,*] and Shmuel Zaks[1]

[1] Department of Computer Science,
Technion, Haifa 32000, Israel
{eilam,zaks}@cs.technion.ac.il

[2] Dipartimento di Matematica Pura ed Applicata,
University of L'Aquila,
via Vetoio loc. Coppito, I-67100 l'Aquila, Italy
flammini@univaq.it

[3] Project SLOOP I3S - CNRS URA / INRIA / Univ. Nice–Sophia Antipolis,
930 route des Colles, Sophia Antipolis, F-06903 Cedex, France

Abstract. We investigate the time complexity of deciding the existence of layouts of virtual paths in high-speed networks, that enable a connection from one vertex to all others and have maximum hop count h and maximum edge load l, for a stretch factor of one. We prove that the problem of determining the existence of such layouts is NP-complete for every given values of h and l, except for the cases $h = 2, l = 1$ and $h = 1$, any l, for which we give polynomial-time layout constructions.

1 Introduction

1.1 Motivation

Asynchronous Transfer Mode (ATM for short) is widely accepted as the most popular architecture that supports high-speed networks, and is thoroughly described in the literature [14, 13, 16]. ATM is based on relatively small fixed-size packets, that are routed independently, based on two small routing fields at their header (termed *virtual channel index* (VCI) and *virtual path index* (VPI)). At each intermediate switch, these fields serve as indices to two routing tables, and the routing is done in accordance to the predetermined information in the appropriate entries.

Routing in ATM is hierarchical in the sense that the VCI of a cell is ignored as long as its VPI is not null. This algorithm effectively creates two types of

---

* This author has been partly supported by the EU TMR Research Training Grant N. ERBFMBICT960861, by the EU ESPRIT Long Term Research Project ALCOM-IT under contract N. 20244 and by the Italian MURST 40% project "Algoritmi, Modelli di Calcolo e Strutture Informative".

predetermined simple routes in the network - namely routes which are based on VPIs (called *virtual paths* or VPs) and routes based on VCIs and VPIs (called *virtual channels* or VCs). VCs are used for connecting network users, and VPs are used for simplifying network management (routing of VCs in particular). Thus the route of a VC may be viewed as a concatenation of complete VPs.

As far as the mathematical model is concerned, given a communication network, the VPs form a set of simple paths in the network (termed the *virtual path layout* (VPL for short)) on the same vertices. Each VC is thus a concatenation of such virtual paths.

The VP layout must satisfy certain conditions to guarantee important performance aspects of the network (see [1, 12] for technical justification of the model for ATM networks). In particular, there are restrictions on the following parameters:

The hop count: The number of VPs which comprise the path of a VC in the virtual graph. This parameter determines the efficiency of the setup of a VC (see, e.g., [4, 17, 18]).
The load: The number of virtual paths that share any physical edge. This number determines the size of the VP routing tables (see, e.g., [6]).
The stretch factor: The ratio between the length of the path that a VC takes in the physical graph and the shortest possible path between its endpoints. This parameter controls the efficiency of the utilization of the network.

In many works (e.g., [2, 3, 12, 5]), a general routing problem is solved using a simpler sub-problem as a building block; In this sub-problem it is required to enable routing between all vertices and a single vertex (rather than between any pair of vertices). This restricted problem for the ATM VP layout problem is termed the *rooted (or one-to-many) VPL* problem [12] and is the focus of the present work.

1.2 Related Work

A few works have tackled the VP layout problem, some using empirical techniques [1, 15], and some using theoretical analysis [12, 5, 11].

The VP layout problem is closely related to graph-embedding problems since in both cases it is required to embed one graph in another graph. However, while in most embedding problems both graphs are given, here we are given only the physical (host) graph, and we can *choose* the embedded graph (in addition to the choice of the embedding itself).

Most of the performance parameters are also different in both cases:

– While the association between the host graph and the embedded graph is made by the *dilation* parameter in embedding problems, here it is made by the *stretch factor*. In other words, in embedding problems it is important to minimize the length of each individual embedded edge, while in this model it is important to minimize the length of paths.

- The *hop count* parameter is closely related to the *distance* in the virtual graph, however, while the distance depends only on one graph, the hop count also depends on the physical graph (unless the stretch factor is unbounded).
- The *load* parameter is identical to the *congestion* in embedding problems, and the different terminology is due to the loaded meaning of congestion in the communication literature.

The computational complexity of determining the existence of a VP layout for a given network within a given maximum hop count and a given maximum load was investigated in [12], where the authors showed that this problem is NP-complete when there is no limit on the stretch factor. In [12] also some polynomial construction algorithms are given for trees for the stretch factor equal to one, i.e. when the physical routed paths are shortest.

1.3 Summary of Results

In this paper we improve the results of [12], concerning the computational complexity of constructing virtual path layouts from a given node to all other nodes in the network. While in [12] the maximum hop count h and the maximum load l are not constant, here we tightly establish the border between tractability and intractability, by determining the lowest (constant) values of h and l that make the problem computationally hard. Moreover, we give efficient construction algorithms for all the tractable cases.

Specifically, we show that the problem of determining the existence of such layouts is NP-complete for every given values of h and l, except for the cases $h = 2, l = 1$ and $h = 1$, any l, for which we give polynomial-time constructions. All results in this paper concern the stretch factor of one.

The paper is organized as follows: In Section 2 we define the model and the related performance measures. In Section 3 we give the above-mentioned NP-completeness results. In Section 4 we present efficient construction algorithms for the polynomial cases, and in Section 5 we conclude and list some open problems. Some proofs are only briefly sketched in this Extended Abstract.

2 The Model

Following [12] we model the underlying communication network as an undirected graph $G = (V, E)$, where V corresponds to the set of switches and E to the set of physical links between them.

Definition 1. A *rooted virtual path layout (RVPL for short)* Ψ is a collection of simple paths in G, termed *virtual paths* (VPs for short), and a vertex $r \in V$ termed the *root* of the layout (denoted $root(\Psi)$).

Definition 2. The *hop count* $\mathcal{H}(v)$ of a vertex $v \in V$ in a RVPL Ψ is the minimum number of VPs whose concatenation forms a shortest path in G from v to $root(\Psi)$. If no such VPs exist, define $\mathcal{H}(v) \equiv \infty$. (Note that the assumption of stretch factor equal to one is reflected by the requirement of using shortest paths.)

Definition 3. The *maximal hop count* of a RVPL Ψ is defined as $\mathcal{H}_{max}(\Psi) \equiv \max_{v \in V} \{\mathcal{H}(v)\}$.

Definition 4. The *load* $\mathcal{L}(e)$ of an edge $e \in E$ in a RVPL Ψ is the number of VPs $\psi \in \Psi$ that include e.

Definition 5. The *maximal load* $\mathcal{L}_{max}(\Psi)$ of a RVPL Ψ is $\max_{e \in E} \mathcal{L}(e)$. [4]

To minimize the load, one can use a RVPL Ψ which has a VP on each physical link, i.e., $\mathcal{L}_{max}(\Psi) = 1$, however such a layout can have a hop count equal to the diameter of the network. The other extreme is connecting a direct VP from the root to each other vertex, yielding $\mathcal{H}_{max} = 1$ but usually a very high $\mathcal{L}_{max}$. In general, we are interested in the intermediate cases where we trade one parameter for the other. The following decision problem then naturally arises.

Definition 6. $\langle h, l \rangle$-**RVPL Problem**:

INSTANCE: A network $G = (V, E)$ and a given root $r \in V$.
QUESTION: Is there a $\langle h, l \rangle$- RVPL Ψ for G with root r, i.e. a RVPL such that $\mathcal{H}_{max}(\Psi) \leq h$ and $\mathcal{L}_{max}(\Psi) \leq l$?

3 The NP-complete Cases

In this section we tightly establish the values of h and l that make the problem of determining the existence of virtual path layouts NP-complete. Namely, we prove the following theorem.

Theorem 7. *The $\langle h, l \rangle$-RVPL problems are NP-complete for any h and l except for the cases $h = 1$, any l and $h = 2, l = 1$.*

First observe that the $\langle h, l \rangle$-RVPL problems belong to the class NP. In fact, given an RVPL Ψ for $G = (V, E)$ with a given root $r \in V$, one can easily check whether $\mathcal{L}(e) \leq l$ for every edge $e \in E$ and whether $\mathcal{H}_{max}(\Psi) \leq h$. For the latter task we define a weighted graph $G' = (V, E')$, termed *virtual graph*, with an edge of weight l connecting vertices a and b if and only if there is a virtual path of length l between them; then, if d is the (unweighted) distance between r and v in G, by using slight modifications of usual shortest path algorithms we verify that, for every vertex $v \in V - \{r\}$, there is a path from r to v in G' of length d that has at most h edges.

We prove Theorem 7 in the following four lemmas. In Lemma 8 we prove that $\langle 3, 1 \rangle$-RVPL is NP-complete. In Lemmas 9 and 10 we prove that for every l the $\langle 2, l \rangle$-RVPL problems are NP-complete. Finally, we prove in Lemma 11 that for every h and l, if $\langle h, l \rangle$-RVPL is NP-complete then so is $\langle h+1, l \rangle$-RVPL. Thus, the first three lemmas establish the basis of an inductive proof and Lemma 11 is the inductive step.

---
[4] As mentioned above, the load on an edge is identical to its congestion.

Lemma 8. *The $\langle 3,1 \rangle$-RVPL problem is NP-complete.*

Sketch of proof. In order to prove the NP-completeness of the $\langle 3,1 \rangle$-RVPL problem, we provide a polynomial time transformation from the *Dominating Set* problem (DS) (known to be NP-complete; see [10]). In this problem we have a universe set $U = \{u_1, \ldots, u_m\}$ of m elements, a family $\{A_1, \ldots, A_f\}$ of f subsets of U and an integer $k \leq f$; we want to decide if there exist k subsets $A_{j_1}, \ldots, A_{j_k}$ which cover U, i.e. such that $\bigcup_{i=1}^{k} A_{j_i} = U$.

Starting from an instance I_{DS} of DS, we construct a graph G that admits a $\langle 3,1 \rangle$-RVPL if and only if I_{DS} admits a cover.

Let $G = (V, E)$, where $V = \{r\} \cup V_1 \cup \{v\} \cup V_2 \cup V_3$ and $E = E_1 \cup E_2 \cup E_3 \cup E_4$ (see Figure 1), with $V_1 = \{q_a \mid a = 1, \ldots, k+1\}$, $V_2 = \{w_b \mid b = 1, \ldots, f\}$, $V_3 = \{z_c \mid c = 1, \ldots, m\}$, and $E_1 = \{\{r, q_a\} \mid a = 1, \ldots, k+1\}$, $E_2 = \{\{q_a, v\} \mid a = 1, \ldots, k+1\}$, $E_3 = \{\{v, w_b\} \mid b = 1, \ldots, f\}$, $E_4 = \{\{w_b, z_c\} \mid u_c \in A_b\}$.

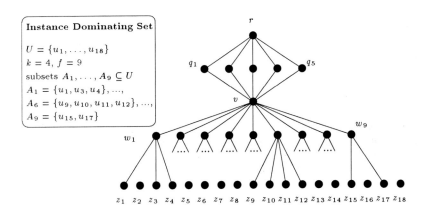

Fig. 1. The reduction graph for $\langle 3,1 \rangle$-RVPL

We show that if there are k dominating sets $A_{j_1}, \ldots, A_{j_k}$, then there exists a $\langle 3,1 \rangle$-RVPL Ψ for G, and that if there are no k dominating sets, then no $\langle 3,1 \rangle$-RVPL Ψ for G exists. The details are omitted in this Extended Abstract. □

Lemma 9. *The $\langle 2,2 \rangle$-RVPL problem is NP-complete.*

Sketch of proof. We prove the claim by providing a polynomial time transformation from the *3-SAT* problem (see [10]). An instance of this problem is constituted by a boolean formula f over m variables $x_1, \ldots, x_m$, where f is in conjunctive normal form, i.e. f is the conjunction of g clauses $c_1, \ldots, c_g$, each of which is the disjunction of three literals. We want to determine whether there exists a truth assignment for $x_1, \ldots, x_m$ which satisfies f.

Starting from an instance of 3-SAT, we construct a graph G that admits a $\langle 2, 2 \rangle$-RVPL if and only if f is satisfiable.

Let $G = (V, E)$, where $V = \{r\} \cup V_1 \cup V_2 \cup V_3 \cup V_4 \cup V_5$, and $E = E_1 \cup E_2 \cup E_3 \cup E_4 \cup E_5 \cup E_6$ (see Figure 2), with $V_1 = \{\overline{u}_a, u_a \mid a = 1, \ldots, m\}$, $V_2 = \{\overline{v}_a, v_a \mid a = 1, \ldots, m\}$, $V_3 = \{q_a \mid a = 1, \ldots, m\}$, $V_4 = \{w_{a,i} \mid a = 1, \ldots, m, i = 1, \ldots, 4\}$, $V_5 = \{z_{b,j} \mid b = 1, \ldots, g, j = 1, \ldots, 4\}$, and $E_1 = \{\{r, \overline{u}_a\}, \{r, u_a\} \mid a = 1, \ldots, m\}$, $E_2 = \{\{\overline{u}_a, \overline{v}_a\}, \{u_a, v_a\} \mid a = 1, \ldots, m\}$, $E_3 = \{\{\overline{u}_a, q_a\}, \{u_a, q_a\} \mid a = 1, \ldots, m\}$, $E_4 = \{\{q_a, w_{a,i}\} \mid a = 1, \ldots, m, i = 1, \ldots, 4\}$, $E_5 = \{\{\overline{v}_a, z_{b,j}\} \mid a = 1, \ldots, m, b = 1, \ldots, g, j = 1, \ldots, 4, \overline{x}_a \in c_b\}$, $E_6 = \{\{v_a, z_{b,j}\} \mid a = 1, \ldots, m, b = 1, \ldots, g, j = 1, \ldots, 4, x_a \in c_b\}$.

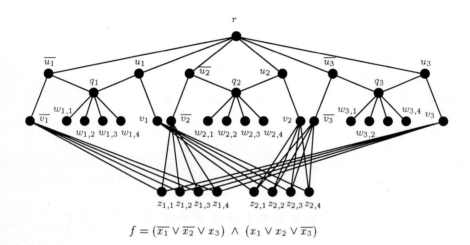

Fig. 2. The reduction graph for $\langle 2, 2 \rangle$-RVPL

Informally, in G we associate to each variable x_a a truth setting component constituted by the subgraph induced by the vertices r, $\overline{u}_a$, u_a, $\overline{v}_a$, v_a, q_a, $w_{a,1}$, $w_{a,2}$, $w_{a,3}$ and $w_{a,4}$. To explain the intuition for our construction, consider any path layout for the graph G. The restriction of this layout to this subgraph can be associated in a natural way to a truth assignment for x_a. In fact, in order for r to reach the four vertices $w_{a,1}, w_{a,2}, w_{a,3}, w_{a,4}$ in at most two hops, the VPs $\langle r, \overline{u}_a, q_a \rangle$ or $\langle r, u_a, q_a \rangle$ must belong to the RVPL.

W.l.o.g. we can then assume that either $\langle r, \overline{u}_a, q_a \rangle$ or $\langle r, u_a, q_a \rangle$ are in the RVPL. In the first case the truth assignment associated to x_a is true, and in the second it is false. If the truth assignment of x_a is true (resp. false), then the RVPL can contain the VP $\langle r, u_a, v_a \rangle$ (resp. $\langle r, \overline{u}_a, \overline{v}_a \rangle$), so that all vertices $z_{b,j}$ corresponding to clauses c_b containing x_a (resp. $\overline{x}_a$) can be reached in at most two hops, as they are directly connected to v_a (resp. $\bar{v}_a$). (See Figure 3.)

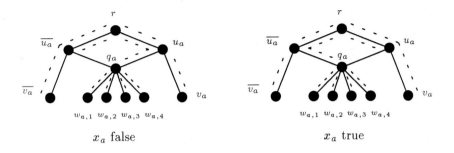

Fig. 3. Path layout and truth assignment in the case of $\langle 2, 2 \rangle$-RVPL

We show (details omitted in this Extended Abstract) that there is a truth assignment satisfying f if and only if there exists a $\langle 2, 2 \rangle$-RVPL Ψ for G. □

Lemma 10. *For every l, the $\langle 2, l \rangle$-RVPL problem is NP-complete.*

Sketch of proof. Given any $l \geq 2$, we will prove that $\langle 2, l \rangle$-RVPL is an NP-complete problem by a polynomial time transformation from the *3-SAT* problem which is a generalization of the transformation presented in the proof of Lemma 9. Let an instance of the 3-SAT problem be as defined in the proof of Lemma 9. Starting from this instance, we construct a graph G that admits a $\langle 2, l \rangle$-RVPL if and only if f is satisfiable.

The idea is to add a construction to each of the vertices u_a and $\bar{u}_a$ which will force an addition of $l - 2$ VPs on each of the edges $\{r, u_a\}$ and $\{r, \bar{u}_a\}$ (in order to reach all vertices in the new construction in 2 hops). In addition, we have to enlarge the number of $w_{a,i}$ vertices (actually we will have $2l$ such vertices for every variable a in f), and the number of $z_{b,j}$ vertices to $3(l-1) + 1$ for each clause b in the formula f. Note that for the special case $l = 2$ we will get exactly the same construction as in the proof of Lemma 9.

Formally, we specify only the additions to the construction of Lemma 9. We add the following sets of vertices $V_6 = \{\bar{s}_{a,i}, s_{a,i} \mid a = 1, \ldots, m, i = 1, \ldots, l - 2\}$, $V_7 = \{\bar{t}_{a,i,j}, t_{a,i,j} \mid a = 1, \ldots, m, i = 1, \ldots, l - 2, j = 1, \ldots, l\}$, and the following sets of edges: $E_7 = \{\{u_a, s_{a,i}\}, \{\bar{u}_a, \bar{s}_{a,i}\} \mid a = 1, \ldots, m, i = 1, \ldots, l - 2\}$, $E_8 = \{\{s_{a,i}, t_{a,i,j}\}, \{\bar{s}_{a,i}, \bar{t}_{a,i,j}\} \mid a = 1, \ldots, m, i = 1, \ldots, l - 2, j = 1, \ldots, l\}$.

We enlarge the number of $w_{a,i}$ and $z_{b,j}$ vertices as follows: $V_4 = \{w_{a,i} \mid a = 1, \ldots, m, i = 1, \ldots, 2l\}$, $V_5 = \{z_{b,j} \mid b = 1, \ldots, g, j = 1, \ldots, 3(l-1) + 1\}$, and we correspondingly enlarge the number of edges in the sets E_4, E_5 and E_6 as follows: $E_4 = \{\{q_a, w_{a,i}\} \mid a = 1, \ldots, m, i = 1, \ldots, 2l\}$, $E_5 = \{\{\bar{v}_a, z_{b,j}\} \mid a =$

$1, \ldots, m$, $b = 1, \ldots, g$, $j = 1, \ldots, 3(l-1)+1$, $\overline{x}_a \in c_b\}$, $E_6 = \{\{v_a, z_{b,j}\} \mid a = 1, \ldots, m$, $b = 1, \ldots, g$, $j = 1, \ldots, 3(l-1)+1$, $x_a \in c_b\}$.

Clearly to reach the $t_{a,i,j}$ and $\bar{t}_{a,i,j}$ vertices in two hops, we must reach each of the $s_{a,i}$ and $\overline{s}_{a,i}$ vertices in one hop, which uses $l-2$ VPs on each of the edges $\{r, u_a\}$ and $\{r, \overline{u}_a\}$.

The rest of the proof is a generalization of the proof of Lemma 9, and is omitted in this Extended Abstract. □

Lemma 11. *For every h and l, if the $\langle h, l \rangle$-RVPL problem is NP-complete then $\langle h+1, l \rangle$-RVPL is also an NP-complete problem.*

Sketch of proof. We assume that $\langle h, l \rangle$-RVPL is NP-complete and prove that $\langle h+1, l \rangle$-RVPL is also NP-complete by a polynomial transformation from $\langle h, l \rangle$-RVPL. Given an instance of $\langle h, l \rangle$-RVPL, a graph $G = (V, E)$ and a vertex $r \in V$, we construct a graph $G' = (V', E')$ and a vertex $r' \in V'$ such that there exists a $\langle h+1, l \rangle$-RVPL for G' if and only if there exists a $\langle h, l \rangle$-RVPL for G. For every vertex in V, let $deg(v)$ be the degree of the vertex (i.e., the number of vertices adjacent to v in G). The graph G' is constructed from G by adding $deg(v) \cdot l$ new vertices to each vertex v in G, and connecting each of them to v. Formally, $V' = V \cup \{w_{v,i} \mid v \in V, i = 1, \ldots, deg(v) \cdot l\}$, $E' = E \cup \{\{v, w_{v,i}\} \mid v \in V, i = 1, \ldots, deg(v) \cdot l\}$.

The root r' of G' is the vertex $r \in V$. We term the vertices and edges of G in G' *original* and the rest of the vertices and edges in G' *new*. Obviously the transformation is polynomial in the size of the input graph G.

Assume that there is an $\langle h, l \rangle$-RVPL Ψ for G with root r. To get an $\langle h+1, l \rangle$-RVPL Ψ' for G' with root r' we add to Ψ the VPs of length 1 from every $v \in V$ to every $w_{v,i}$. It can also be shown (detailed omitted here) that if there is an $\langle h+1, l \rangle$-RVPL Ψ' for G' with root r', then in Ψ' for every original vertex v, $\mathcal{H}(v) \leq h$ and thus Ψ' induces an $\langle h, l \rangle$-RVPL for G in the natural way (remove from Ψ' all VPs with an endpoint which is a new vertex). □

Sketch of proof. [of Theorem 7] We prove that for every h and l except for the cases $h = 1$, any l, and, $h = 2, l = 1$ the $\langle h, l \rangle$-RVPL problem is NP-complete by induction on h. The basis is established in Lemmas 8, 9, and 10, where we prove that the problems $\langle 2, l \rangle$-RVPL for every l, and $\langle 3, 1 \rangle$-RVPL are NP-complete. The induction step is established in Lemma 11, where we prove that for every h and l, the NP-completeness of the $\langle h, l \rangle$-RVPL problem derives the NP-completeness of $\langle h+1, l \rangle$-RVPL. □

4 Polynomial Cases

In this section we show that the above NP-completeness results are strict, by giving polynomial running time algorithms for the $\langle 2, 1 \rangle$-RVPL problem and the

$\langle 1, l \rangle$-RVPL problems for any $l \geq 1$. We do this by applying algorithms to find flow in networks, which are known to be polynomial (e.g., [9, 8, 7]).

Given a directed graph $G = (V, E)$, with capacities $c(e)$- positive integers - for the edges $e \in E$, and two specified vertices s and t, we want to find a flow of maximum total value from s to t. It is well-known that in the case of unit capacities there is a flow of value k from s to t in G iff there are k edge-disjoint paths connecting s and t, and that this holds also in a general network with integral capacities, provided that each edge e is replaced by x parallel edges of unit weight each, where x is the original capacity of e.

Given a graph $G = (V, E)$ and a specified vertex r, to construct a $\langle 1, l \rangle$-RVPL for it, we construct the graph $G' = (V', E')$, as follows. $V' = V \cup \{t\}$. For E' we construct a shortest-path BFS graph, rooted at r; this gives a directed layered graph (whose layers are identical to those constructed by the Dinic's Algorithm; see [8, 7]); The vertices in layer $i, i \geq 0$ are exactly the vertices in V whose distance from r is exactly i. There are no edges within a layer, and all edges are from layer i to layer $i+1$, for some $i \geq 0$. All these edges have a capacity of l. We then add all the edges (v, t) for every vertex v in $V - \{r\}$, with capacity 1. The source and destination of G' are r and t, respectively. (See Figure 4(b)).

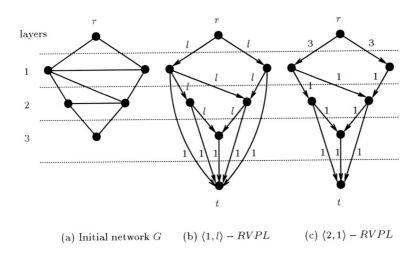

(a) Initial network G (b) $\langle 1, l \rangle - RVPL$ (c) $\langle 2, 1 \rangle - RVPL$

Fig. 4. The flow constructions

We then run any algorithm to determine the maximum flow in this network. By the above, there is a flow of value $|V| - 1$ in G' iff there are paths in G from r to all other vertices in $|V|$, such that no edge is used in more than l of them, which means that there is a solution to the $\langle 1, l \rangle$-RVPL problem iff there is a flow of value at least $|V| - 1$ in G', which thus supplies a polynomial-time algorithm for the $\langle 1, l \rangle$-RVPL problem.

Given a graph $G = (V, E)$ and a specified vertex r, to construct a $\langle 2, 1 \rangle$-RVPL for it, we construct the following graph $G' = (V', E')$. $V' = V \cup \{t\}$, and E' is defined as follows. Let U denote the set of neighbors of r in G. For E' we construct a shortest-path BFS graph, rooted at r (as above). This partitions the vertices of V into layers, such that the vertices in layer $i, i \geq 0$ are exactly the vertices in V whose distance from r is exactly i. As above, there are no edges within a layer, and all edges are from layer i to layer $i + 1$, for some $i \geq 0$. We then add to E' all the edges (v, t) for every vertex v in $V - (U \cup \{r\})$. All the edges in E' have a capacity of 1, except for the edges emanating from r, whose capacity is $|V| - |U| - 1$. The source and destination of G' are r and t, respectively. (See Figure 4(c)). It can be shown that there is a $\langle 2, 1 \rangle$-RVPL iff the maximum flow in the network is equal to $|V| - |U| - 1$, and thus the problem is solved in polynomial time by solving the corresponding flow problem.

Note that, in the case of $h = 1$ and arbitrary l, if we run the network flow algorithm on the original network to which t is added as above, and where each edge is replaced with two anti-parallel edges (rather than using the layered network) and capacities are similarly defined, we can determine whether a layout exists, but with an arbitrary stretch factor.

5 Summary and Open Problems

We have considered a routing problem termed the "rooted VP layout problem" that arises in ATM networks and we have investigated the computational complexity of determining the existence of RVPL fulfilling a maximum hop count h and a maximum load l. We have shown that deciding the existence of such layouts is NP-complete for all values of h and l, except for the cases $h = 2, l = 1$ and $h = 1$, any l, for which we presented polynomial-time layout constructions, based on network flow algorithms.

In classical graph embedding problems vertices are mapped to vertices and edges are mapped to paths connecting the endpoints of their corresponding vertices; this is a very common situation in embeddings within a VLSI networks. In this context, the term *dilation* is used to denote the longest path onto which an edge is embedded. Since in our constructions for the NP-complete results the virtual paths were of length at most two, it follows that the above two problems remain NP-complete for any given bound on the dilation.

An open problem is to extend these results to many-to-many virtual path layouts, where we are interested to connect all pairs of vertices with virtual paths under similar constraints, or to other cases when the pairs to be connected are specified.

A more difficult problem seems to be the one in which not only shortest path layouts are considered, but also layouts that are within a given stretch factor f (that is, one in which the virtual channel between the desired vertices is bounded by f times the shortest path between these vertices). Our polynomial-time algorithms do not apply for a given stretch factor (though, as we noted, we can use simpler algorithms for the case $h = 1$, any l, under the assumption of an arbitrary stretch factor).

Acknowledgment: We thank Shlomo Moran for very helpful comments.

References

1. S. Ahn, R.P. Tsang, S.R. Tong, and D.H.C. Du. Virtual path layout design on ATM networks. In *INFOCOM'94*, pages 192–200, 1994.
2. B. Awerbuch, A. Bar-Noy, N. Linial, and D. Peleg. Compact distributed data structures for adaptive routing. In *21st Symp. on Theory of Computing*, pages 479–489, 1989.
3. B. Awerbuch and D. Peleg. Routing with polynomial communication-space trade-off. *SIAM Journal on Discrete Math*, 5(2):151–162, May 1992.
4. J. Burgin and D. Dorman. Broadband ISDN resource management: The role of virtual paths. *IEEE Communicatons Magazine*, 29, 1991.
5. I. Cidon, O. Gerstel, and S. Zaks. A scalable approach to routing in ATM networks. In G. Tel and P.M.B. Vitányi, editors, *The 8th International Workshop on Distributed Algorithms (LNCS 857)*, pages 209–222, Terschelling, The Netherlands, October 1994. Submitted for publication in IEEE/ACM Trans. on Networking.
6. R. Cohen and A. Segall. Connection management and rerouting in ATM networks. In *INFOCOM'94*, pages 184–191, 1994.
7. E. A. Dinic. Algorithm for solution of a problem of maximum flow in a network with power estimation. *Soviet Math. Dokl., Vol. 11*, pages 1277 – 1280, 1970.
8. S. Even. *Graph Algorithms*. Computer Science Press, 1979.
9. L. R. Ford and D.R. Fulkerson. *Flows in Networks*. Princeton Univ. Press, Princeton, NJ, 1962.
10. M.R. Garey and D.S. Johnson. *Computers and Intractability: A Guide to the Theory of NP-Completeness*. W.H. Freeman and Co., 1979.
11. O. Gerstel, A. Wool, and S. Zaks. Optimal layouts on a chain ATM network. *3rd Annual European Symposium on Algorithms (ESA), (LNCS 979)*, Corfu, Greece, September 1995, pages 508-522. To appear in *Discrete Applied Mathematics*.
12. O. Gerstel and S. Zaks. The virtual path layout problem in fast networks. In *The 13th ACM Symp. on Principles of Distributed Computing*, pages 235–243, Los Angeles, USA, August 1994.
13. R. Händler and M.N. Huber. *Integrated Broadband Networks: an introduction to ATM-based networks*. Addison-Wesley, 1991.
14. ITU recommendation. I series (B-ISDN), Blue Book, November 1990.
15. F.Y.S. Lin and K.T. Cheng. Virtual path assignment and virtual circuit routing in ATM networks. In *GLOBCOM'93*, pages 436–441, 1993.
16. C. Partridge. *Gigabit Networking*. Addison Wesley, 1994.
17. K.I. Sato, S. Ohta, and I. Tokizawa. Broad-band ATM network architecture based on virtual paths. *IEEE Transactions on Communications*, 38(8):1212–1222, August 1990.
18. Y. Sato and K.I. Sato. Virtual path and link capacity design for ATM networks. *IEEE Journal of Selected Areas in Communications*, 9, 1991.

Efficiency of Asynchronous Systems and Read Arcs in Petri Nets

Walter Vogler *, Universität Augsburg, Germany

Abstract

Two solutions to the MUTEX-problem are compared w.r.t. their temporal efficiency. For this, a formerly developed efficiency-testing for asynchronous systems is adapted to nets with so-called read arcs. The close relation between efficiency-testing and fairness is pointed out, and it is shown that read arcs are necessary for any solution to the MUTEX-problem.

1 Introduction

The testing scenario of [DNH84] has been developed further in [Vog95b, JV96] in order to compare the temporal efficiency of asynchronous systems – using Petri nets as system models. This approach is applied here to two solutions of the MUTEX-problem based on token passing. The corresponding nets contain what we call read arcs, and one of our main results is that this is in fact necessary.

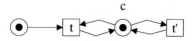

Figure 1

In Petri nets, the check of a side-condition is modelled with a loop as in Figure 1: the occurrence of t removes the condition c and restores it afterwards; hence, t and t' can occur in any order, but not at the same time. This is certainly adequate if e.g. c models the processor that t and t' run on. But if e.g. c is a value from a data base which can be read concurrently, then t and t' *can* occur at the same time. We model such cases with special read arcs instead of loops.

Read arcs have not found so much attention in the past, probably because loops and read arcs are treated just the same if we only look at interleaving semantics. But they do make a difference when we explicitly take into account concurrency. E.g. [CH93] discusses a step semantics and [MR95] defines net-

---

*This work was partially supported by the DFG-project 'Halbordnungstesten'. Author's address: Institut für Informatik, Universität Augsburg, D-86135 Augsburg, Germany. email: vogler@informatik.uni-augsburg.de

processes for nets with read arcs. In both approaches, a net with read arcs can be translated to an equivalent net without, but it is argued in [MR95] that the former is more natural and compact. In clear contrast, read arcs are even better motivated in our setting, since they add relevant expressivity: the MUTEX-problem can only be solved with nets having read arcs; this also holds, if we disregard efficiency and simply take fair behaviour as a basis.

In the testing approach of [DNH84], a system is an implementation if it performs in all environments, i.e. for all users, just as well as the specification. While in the classical setting successful performance only depends on the functionality, i.e. which actions are executed, the testing approach was refined in [Vog95b] to consider also efficiency. The must-version of this efficiency testing (concerned with worst case behaviour) is not so easy to define in the case of asynchronous systems, where the components work with indeterminate relative speeds; most often, this is interpreted as 'each component may work arbitrarily slow'. With this view, the worst case is simply that nothing is done for a long time, hence every test is failed and we do not have a sensible theory of testing.

As a way out, [JV96] assumes that each action is performed within one unit of time (or is disabled within this time). Such an upper time bound is a reasonable basis for judging the efficiency; since actions can also be performed arbitrarily fast, the components work with indeterminate relative speeds even under this assumption, and we have a valid theory for asynchronous systems. It turns out that, for the resulting testing scenario, the implementation preorder is a sensible faster-than relation. Three variants based on dense time are considered and each of them is shown to coincide with a discretely timed version. In the most simple variant, which we will generalize here to nets with read arcs, transitions must fire within time 1 after enabling, but the firing itself is instantaneous.

After defining some basic concepts in Section 2, we define our asynchronous firing rule in Section 3 and present a characterization of the faster-than relation that results from testing; also, the use of loops is discussed. Section 4 shows the close relation between efficiency testing and fairness (in the sense of progress) demonstrating that our efficiency testing is concerned with asynchronous behaviour; it is also described how to determine the fair behaviour of a composed system in a modular fashion. The two MUTEX-solutions with read arcs are studied in Section 5. We view a MUTEX-solution as a scheduler, i.e. an independent component the users have to synchronize with. This view allows a clean formulation of the correctness requirements and fits very well the behaviour notions we have given in Sections 3 and 4; we prove the correctness of one of our solutions and then show that no ordinary net without read arcs can be correct in this sense. Finally, we show that, from the point of view of one user, one solution is more efficient than the other.

Due to lack of space, the proofs had to be omitted; see [Vog96], also for a discussion of the literature on the efficiency of asynchronous systems. I thank Roberto Gorrieri and Lars Jenner for their comments, which helped to improve the presentation of this paper.

2 Basic Notions of Petri Nets with Read Arcs

We use safe nets (extended with read arcs) whose transitions are labelled with *actions* from some infinite alphabet Σ or with the empty word λ, indicating internal, unobservable actions. Σ contains a special action ω, which we will need in our tests to indicate success.

Thus, a *net* $N = (S, T, F, R, l, M_N)$ consists of finite disjoint sets S of *places* and T of *transitions*, the *flow* $F \subseteq S \times T \cup T \times S$ consisting of (ordinary) *arcs*, the set of *read arcs* $R \subseteq S \times T \cup T \times S$, the *labelling* $l : T \to \Sigma \cup \{\lambda\}$, and the *initial marking* $M_N : S \to \{0, 1\}$; R is always symmetric with $R \cap F = \emptyset$. As usual, we draw transitions as boxes, places as circles and arcs as arrows; read arcs are drawn as lines without arrow heads, i.e. we identify the two elements $(x, y), (y, x) \in R$. The net is called *ordinary*, if $R = \emptyset$.

For each $x \in S \cup T$, the *(full) preset* of x is $^\bullet x = \{y \mid (y, x) \in F \cup R\}$ and the *(full) postset* of x is $x^\bullet = \{y \mid (x, y) \in F \cup R\}$; the *reduced preset* of x is $^\circ x = \{y \mid (y, x) \in F\}$ and the *reduced postset* of x is $x^\circ = \{y \mid (x, y) \in F\}$. If $x \in {^\circ y} \cap y^\circ$, then x and y form a *loop*. A *marking* is a function $S \to \mathbb{N}_0$. We sometimes regard sets as characteristic functions, which map the elements of the sets to 1 and are 0 everywhere else; hence, we can e.g. add a marking and a postset of a transition or compare them componentwise.

Our basic firing rule extends the firing rule for ordinary nets by regarding the read arcs as loops, i.e. as ordinary arcs (since R is symmetric). A transition t is *enabled* under a marking M, denoted by $M[t\rangle$, if $^\bullet t \leq M$. If $M[t\rangle$ and $M' = M + t^\bullet - {^\bullet t}$ (which is the same as $M + t^\circ - {^\circ t}$), then we write $M[t\rangle M'$ and say that t can *occur* or *fire* under M yielding the follower marking M'.

Enabling and occurrence is extended to sequences as usual. If $w \in T^*$ is enabled under M_N, it is called a *firing sequence*. We extend the labelling to sequences of transitions as usual, i.e. homomorphically; thus, internal actions are deleted in this *image* of a sequence. With this, we lift the enabledness and firing definitions to the level of actions: a sequence v of actions is *enabled* under a marking M, denoted by $M[v\rangle\rangle$, if $M[w\rangle$ and $l(w) = v$ for some $w \in T^*$. If $M = M_N$, then v is called a *trace*; the set of traces is the *language* of N.

A marking M is called *reachable* if $M_N[w\rangle M$ for some $w \in T^*$. The net is *safe* if $M(s) \leq 1$ for all places s and reachable markings M.

General assumption: All nets considered in this paper are safe and only have transitions t with $^\circ t \neq \emptyset$. (The latter condition is no serious restriction, since it can be satisfied by adding a loop between t and a new marked place, if $^\circ t$ were empty otherwise; this addition does not change the firing sequences.)

We use a TCSP-like *parallel composition* $\|_A$ and write $\|$ for $\|_{\Sigma - \{\omega\}}$. Nets combined with $\|_A$ run in parallel and have to synchronize on actions from A. To construct $N_1 \|_A N_2$, we take the disjoint union of N_1 and N_2, combine each a-labelled transition t_1 of N_1 with each a-labelled transition t_2 from N_2 if $a \in A$ (i.e. introduce a new a-labelled transition (t_1, t_2) that inherits all arcs from t_1 and t_2), and delete all the original a-labelled transitions in N_1 and N_2 if $a \in A$.

3 Timed Behaviour of Asynchronous Systems

We now describe the asynchronous behaviour of a parallel system, taking into account at what times things happen. Hence, the components of the system vary in speed – but we assume that they are guaranteed to perform each enabled action within at most one unit of time; this upper time bound allows the relative speeds of the components to vary arbitrarily, since we have no positive lower time bound. Thus, the behaviour we define is truly asynchronous.

For ordinary nets, [JV96] bases a testing preorder on such an asynchronous firing rule using dense time, shows that one can just as well use discrete time, and gives a characterization of the testing preorder. These results can be generalized to nets with read arcs [Vog96]; here, we immediately define an asynchronous firing rule using discrete time and present the respective characterization.

Due to the time bound 1, a newly enabled transition fires or is disabled within time 0 – or it becomes urgent after one time-unit (denoted by σ), i.e. it has no time left and must fire or must be disabled before the next σ.

The crucial point of read arcs is that they differ from loops w.r.t. disabling. If we have a loop (c,t), (t,c) and an arc or read arc (c,t') for a place c and urgent transitions t and t' (see Figure 1), then firing t removes the token from c and, thus, disables t' momentarily; hence, t' is not urgent any more. If, instead, (c,t) and (t,c) form a read arc, t just checks for the presence of a token without removing it and, thus, t' is not disabled and remains urgent; hence, t and t' will occur faster – and this is what we should expect since t does not block t'.

Definition 3.1 An *instantaneous description* $ID = (M, U)$ consists of a marking M and a set U of *urgent* transitions. The *initial ID* is $ID_N = (M_N, U_N)$ with $U_N = \{t \mid M_N[t\rangle\}$. We write $(M, U)[\varepsilon\rangle(M', U')$ in one of the following cases:
1. $\varepsilon = t \in T$, $M[t\rangle M'$, $U' = U - \{t' \mid {}^\circ t \cap {}^\bullet t' \neq \emptyset\})$
2. $\varepsilon = \sigma$, $M = M'$, $U = \emptyset$, $U' = \{t \mid M[t\rangle\}$

$DFS(N) = \{w \mid ID_N[w\rangle ID\}$ is the set of *discrete(ly timed) firing sequences* of N, $DL(N) = \{l(w) \mid w \in DFS(N)\}$ is the *discrete language* of N containing the *discrete traces* of N, where $l(\sigma) = \sigma$. For $w \in DFS(N)$ or $w \in DL(N)$, $\zeta(w)$ is the number of σ's in w. The behaviour inbetween two σ's is called a *round*.

We call a net *testable*, if none of its transitions is labelled with ω. A testable net N *satisfies* a timed test (O, D), N must (O, D), if each $w \in DL(N\|O)$ with $\zeta(w) \geq D$ contains some ω; we call a net N_1 *faster* than a net N_2, $N_1 \sqsupseteq N_2$, if for all (O, D) we have N_2 must $(O, D) \Rightarrow N_1$ must (O, D). □

Part 1 allows enabled transitions – urgent or not – to fire; hence, $DL(N)$ includes the language of N and describes an asynchronous behaviour. $U = \emptyset$ in Part 2 requires that no urgent transition is delayed over the following σ. Each enabled transition is urgent after σ. Thus, a discrete trace is any ordinary trace subdivided into rounds by σ's such that no transition enabled at (i.e. immediately before) one σ is continuously enabled until after the next σ.

The definitions for testing are standard except for the time bound, where we require that every run of the system embedded in the test environment is successful within time D; hence, we do not consider traces that do not last for time D. We call the implementation N_1 *faster*, since it might satisfy more tests and, in particular, some test nets within a shorter time.

The test-preorder $\sqsupseteq$ formalizes observable difference in efficiency; referring to all possible tests, it is not easy to work with directly. Thus, we now characterize $\sqsupseteq$ by so-called i-refusal traces [JV96]: we replace the σ's in a discrete trace by sets of actions, indicating the time-steps now. Such a set contains actions that are not urgent, i.e. can be refused when the time-step occurs.

Definition 3.2 For discrete instantaneous descriptions (M, U) and (M', U') we write $(M, U)[\varepsilon\rangle_r (M', U')$ if one of the following cases applies:
1. $\varepsilon = t \in T$, $M[t\rangle M'$, $U' = U - \{t' | {}^\circ t \cap {}^\bullet t' \ne \emptyset\})$
2. $\varepsilon = X \subseteq \Sigma$, $M = M'$, $U' = \{t \mid M[t\rangle\}$, $\forall t \in U : l(t) \notin X \cup \{\lambda\}$; X is a *refusal set*.

The corresponding *i-refusal firing sequences* form the set $RFS(N)$. $RT(N) = \{l(w) \mid w \in RFS(N)\}$ is the set of *i-refusal traces* where $l(X) = X$. □

Occurrence of Σ exactly corresponds to that of σ, hence:

Prop. 3.3 *For nets N_1 and N_2, $RT(N_1) \subseteq RT(N_2)$ implies $DL(N_1) \subseteq DL(N_2)$.*

We will show later that read arcs add relevant expressivity; here, we state that ordinary loops are in fact not needed in nets with read arcs.

Prop. 3.4 *For each net N, there is a loopless net N' with $RT(N) = RT(N')$.*

Still, loops are certainly often adequate: if two activities run on the same processor, they cannot occur together; if one takes place, the other has to wait a little – and this is just how we treat two transitions with a common loop-place here. Also, our construction for 3.4 makes nets possibly exponentially larger. Finally, on the level of discrete firing sequences, loops have expressivity of their own, since no net without loops has the same discrete firing sequences as the one shown in Figure 1:

Prop. 3.5 *If N is a loopless net such that $t', tt' \in DFS(N)$, then $t\sigma \notin DFS(N)$.*

To show that RT-semantics induces a congruence for $\|_A$, one defines $\|_A$ for i-refusal traces: actions from A are merged, while others are interleaved; refusal sets are combined as in ordinary failure semantics.

Definition 3.6 Let $u, v \in (\Sigma \cup \mathcal{P}(\Sigma))^*$, $A \subseteq \Sigma$. Then $u \|_A v$ is the set of all $w \in (\Sigma \cup \mathcal{P}(\Sigma))^*$ such that for some n we have $u = u_1 \ldots u_n$, $v = v_1 \ldots v_n$, $w = w_1 \ldots w_n$ and for $i = 1, \ldots, n$ one of the following cases applies:
- $u_i = v_i = w_i \in A$
- $u_i = w_i \in (\Sigma - A)$ and $v_i = \lambda$, or $v_i = w_i \in (\Sigma - A)$ and $u_i = \lambda$
- $u_i, v_i, w_i \subseteq \Sigma$ and $w_i \subseteq ((u_i \cup v_i) \cap A) \cup (u_i \cap v_i)$ □

Theorem 3.7 gives us one half of the characterization in 3.8.

Theorem 3.7 *For $A \subseteq \Sigma$ and nets N_1 and N_2, we have that $RT(N_1 \|_A N_2) = \bigcup \{u \|_A v \mid u \in RT(N_1), v \in RT(N_2)\}$.*

Theorem 3.8 *For testable nets, $N_1 \sqsupseteq N_2$ if and only if $RT(N_1) \subseteq RT(N_2)$.*

Observe that a faster system has less i-refusal traces; such a trace is a witness for slow behaviour, it is something 'bad' due to the refusal information.

Corollary 3.9 *Inclusion of RT-semantics is fully abstract w.r.t. inclusion of DL-semantics and parallel composition, i.e. it is the coarsest precongruence for parallel composition that respects DL-inclusion.*

Theorem 3.8 essentially reduces $\sqsupseteq$ to an inclusion of regular languages, which implies decidability. The testing preorder $\sqsupseteq$ is also compatible with some other interesting operations, namely relabelling, hiding and restriction.

4 Efficiency Testing and Fairness

Now we relate our notion of asynchronous behaviour to (weak) fairness (or progress assumption); at the same time, we study compositionality for fair behaviour. Fairness requires that a continuously enabled activity should eventually occur; in real life, this is automatically true, i.e. it does not have to be implemented. First, we extend the definition of the various firing sequences to infinite sequences taking into account that an infinite run should take infinite time.

Definition 4.1 An infinite sequence is a *(discrete/i-refusal) firing sequence* if all its finite prefixes are (discrete/i-refusal) firing sequences.

A *progressing (i-refusal) firing sequence* is an infinite discrete, i-refusal resp., firing sequence with infinitely many σ's, sets resp. The images of these sequences are the *progressing (refusal) traces*, forming $PL(N)$, $PRT(N)$ resp.

For a progressing (refusal) trace v, $\alpha(v)$ denotes the sequence of actions in v, which remains after removing all σ's, sets resp. □

PRT-(PL-)semantics extends RT-(DL-)semantics to infinite runs, required to take infinite time. Using König's Lemma, one can show that nets have the same PRT-(or PL-)semantics if and only if they have the same RT-(or DL-)semantics.

Classically, an infinite firing sequence $M_N[t_0\rangle M_1[t_1\rangle M_2 \ldots$ would be called fair if we have: if some transition t is enabled under all M_i for $i > j$, then $t = t_i$ for some $i > j$; hence, an infinite sequence of t''s would not be fair in the net of Figure 1, since t is enabled under all states reached, but never occurs. But the sequence should be fair: t is not continuously enabled, since every occurrence of t' disables it momentarily, compare [Rei84, Vog95a]. Thus, we will require that t is enabled also *while* each t_i with $i > j$ is firing. For this, we have to keep in mind that a read arc does not consume a token.

Definition 4.2 For a transition t, a finite firing sequence $M_N[t_0\rangle M_1[t_1\rangle \ldots M_n$ is *t-fair*, if not $M_n[t\rangle$. An infinite firing sequence $M_N[t_0\rangle M_1[t_1\rangle M_2 \ldots$ is *t-fair*, if we have: if t is enabled under all $M_i - {}^\circ t_i$ for i greater than some j, then $t = t_i$ for some $i > j$. A finite or infinite firing sequence is *fair*, if it is t-fair for all transitions t. The *fair language* of N is $Fair(N) = \{v \mid v = l(w)$ for some fair firing sequence $w\}$. □

Now we establish a first relation of our approach to fairness: $PL(N)$, the infinite version of $DL(N)$, describes an asynchronous behaviour just as $Fair(N)$.

Theorem 4.3 *For all nets N, $Fair(N) = \{v \mid \exists u \in PL(N) : v = \alpha(u)\}$*

Next, we determine the coarsest precongruence refining fair-language inclusion, something that is needed when systems are constructed bottom-up with $\|_A$. Theorem 4.5 was first obtained in [Gol88]. We improve the original results by allowing read arcs and loops; also, Gold considered safe nets where always $°t \neq \emptyset$ – as we do –, but allowed unsafe nets with isolated transitions as environments in the proof of 4.5 iii); this is improved, too.

Definition 4.4 For a net N, define the *fair failure semantics* by $\mathcal{FF}(N) = \{(v, X) \mid X \subseteq \Sigma$ and $v = l(w)$ for some, possibly infinite, firing sequence w that is t-fair for all transitions t with $l(t) \in X \cup \{\lambda\}\}$. □

The intuition for $(v, X) \in \mathcal{FF}(N)$ is that all actions in X can be refused when v is performed – in the sense, that fairness does not force additional performance of these actions.

Theorem 4.5 *i) For all nets N, $Fair(N) = \{v \mid (v, \Sigma) \in \mathcal{FF}(N)\}$.*
 ii) For $A \subseteq \Sigma$ and nets N_1 and N_2, $\mathcal{FF}(N_1\|_A N_2) = \{(w, X) \mid \exists (w_i, X_i) \in \mathcal{FF}(N_i), i = 1, 2 : w \in w_1\|_A w_2$ and $X \subseteq ((X_1 \cup X_2) \cap A) \cup (X_1 \cap X_2)\}$.
 iii) Inclusion of $\mathcal{FF}$-semantics is fully abstract w.r.t. fair-language inclusion and parallel composition in the sense of Corollary 3.9.

This result and the following, second relation to our testing approach will also be useful in the next section.

Theorem 4.6 *For a net N, $(v, X) \in \mathcal{FF}(N)$ if and only if there is some $w \in PRT(N)$ such that $v = \alpha(w)$ and, for each $x \in X$, there is some suffix of w where x is in all refusal sets.*

5 Two Token-Passing MUTEX-Processes

In this section we will show how useful, in fact necessary, read arcs are to achieve mutual exclusion. Both our processes pass an access-token around and only the owner of the token may access the critical section, which guarantees mutual exclusion. $MUTEX_1$, shown below, is a modification – using read arcs – of a Petri net solution given in [KW95]. The first user has *priority*, i.e. owns the access-token lying on p_1. He can repeatedly *request* access with r_1, *enter* the critical section with e_1 (marking c_1) and *leave* it with l_1. The second user *misses* the access-token (m_2 is marked); if she requests access, she has to *order* the token by marking o_2, and now the first user might *grant* the token by marking g_2.

For $MUTEX_1$ to work properly, [KW95] assumes fairness in general: e.g., if the internal transition ordering the token is enabled, it has to fire eventually; otherwise the token will never be passed and the requesting user will never enter the critical section. In our solution, it is essential that the upper e_1-transition checks with a read arc that the token has not been ordered. This check does not disable the ordering transition; so, if the latter is enabled and time progresses, then it *will* order the token, which now cannot be used by the owner to enter the critical section again and will be passed eventually.

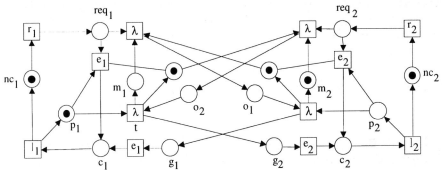

As usual, $MUTEX_1$ is seen in [KW95] as 'code', which has to be inserted into the code of the users; e.g. the r_1-transition *is* the first user requesting access. Since the first user should not be obliged to request, [KW95] has a special class of 'weak' transitions for which fairness is not assumed. This concept is not needed in our view: we see a net such as $MUTEX_1$ as a scheduler guaranteeing mutual exclusion; the user processes are put in parallel with such a MUTEX-process using $\|_{\{r_1,e_1,l_1,r_2,e_2,l_2\}}$, they issue their requests to it and are then allowed to enter the critical section. In this view, the r_1-transition is the MUTEX-process offering the possibility to request; if this offer is not used, then, technically, time can pass in an i-refusal trace with a refusal set not containing r_1.

Our view seems to be very beneficial as a clean way to deal with the question what users do while being noncritical; they may e.g. communicate with each other and even run into deadlocks – it is not completely clear whether this is allowed in the usual view. Here, it obviously is allowed, but we do not have to deal with it explicitly, since such a behaviour is not part of the MUTEX-process. The obligation to prove that a user can indeed request becomes obvious in our view – this obligation is often ignored, see also below.

For the solution of [KW95], fairness is actually not enough; [KW95] therefore requires a restricted form of strong fairness by introducing 'fair arcs'. We will show that, using read arcs, strong fairness is not needed at all.

While in $MUTEX_1$ the token has to be ordered, it is passed automatically in $MUTEX_2$ below if it has been used or is not needed. The check whether the token is needed or not is performed by the read arcs from nc_1 and nc_2.

We will now argue in our setting that $MUTEX_2$ is correct, omitting the similar arguments for $MUTEX_1$. Safety is easy: if one user enters, then he must leave before another enter is possible, since we always have exactly one token on the places c_1, p_1, p_2 and c_2. (This set is an S-invariant, as also used e.g. in [KW95].) Also, $MUTEX_2$ ensures that the users follow the right protocol, i.e.

it allows the actions r_i, e_i and l_i only to be performed cyclically in this order. Liveness – i.e. whenever a user wishes to enter he will be able to do so eventually – is more difficult and requires to assume fairness. First, we have to make sure that a user may always perform a request.

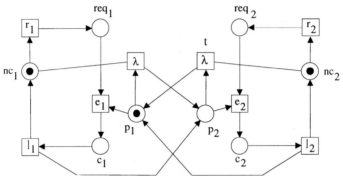

Prop. 5.1 *Let $(w, X) \in \mathcal{FF}(MUTEX_2)$ and $i \in \{1, 2\}$. Then in w r_i occurs and each l_i is followed by another r_i, or $r_i \notin X$.*

This proposition says that if the environment, i.e. the i-th user, tries to request (enables an r_i-transition permanently) at a proper moment (initially or after leaving, i.e. when he is not already requesting or in the critical section), then the request will be performed. If it were not, neither the user (by assumption) nor $MUTEX_2$ (by 5.1) would refuse r_i, hence the combined run according to 4.5 ii) would not refuse r_i, i.e. it would violate fairness according to 4.5 i). By Theorem 4.6, we can formulate 5.1 equivalently as: each $w \in PRT(MUTEX_2)$ contains r_i as in 5.1 or at some stage no following refusal set contains r_i. The proof of 5.1 uses this variant and shows in fact that, after l_i, r_i can be refused at most once before it occurs again. Similar variants are used to prove 5.2 and 5.3, where the former states that a user that enters and then wants to leave will do so. (In fact, he will do so in the present or next round.)

Prop. 5.2 *Let $(w, X) \in \mathcal{FF}(MUTEX_2)$ and $i \in \{1, 2\}$. Then each e_i in w is followed by an l_i, or $l_i \notin X$.*

The most difficult part is to show that a requesting user will eventually enter; here, we must require that a requesting user is indeed willing to enter and also that a user that enters is willing to leave after a while. Since by 5.2, willingness to leave ensures that this happens indeed, we can restrict attention to fair failures where each e_i is followed by l_i; for these we show that each requesting user will enter unless some user has requested but is not willing to enter.

Prop. 5.3 *Let $i \in \{1, 2\}$ and $(w, X) \in \mathcal{FF}(MUTEX_2)$ such that each e_i is followed by l_i. Then either each r_i is followed by e_i or for some $j \in \{1, 2\}$ some r_j is not followed by e_j and $e_j \notin X$.*

We now come to the main result regarding the expressiveness of read arcs.

Theorem 5.4 *Let N be a correct MUTEX-process, i.e. a net that satisfies Propositions 5.1 to 5.3 and guarantees mutual exclusion, namely that e- and l-transitions occur alternatingly. Then N has read arcs.*

Independently, [KW96] have shown a similar result. For correctness, some state-properties are required and a certain net-structure is prescribed there. The latter makes the result quite dependent on Petri nets as system models, whereas our MUTEX-specification in 5.1-5.3 is action-oriented and, thus, fairly model-independent. Also our proof seems to be transferable to other models.

One could also view 5.4 as evidence that a 'simple' progress assumption is not enough to achieve mutual exclusion, as argued in [KW96], who recommend 'fair arcs' as a way to introduce strong fairness in a limited way. Read arcs seem less drastic, but they allow a 'refined' progress assumption, since with read arcs repeated read accesses to one location do not block a write access to this location. This is a restricted form of what [Ray86] calls fairness of hardware.

In fact, the discussion of Dekker's and Knuth's algorithms in [Ray86, p.27/28] might give the impression that the latter does not rely on any fairness of hardware – something that should be false in view of our theorem. And it is: without this fairness, one user-process in Knuth's algorithm can e.g. repeatedly test the variable *turn* in its pre-protocol, thereby preventing the other process from writing *turn* in its post-protocol and in effect from requesting again. Thus, 5.1 treats a realistic possibility for failure that is often ignored.

We conclude the discussion of the MUTEX-problem by comparing the efficiency of $MUTEX_1$ and $MUTEX_2$. Our results are intuitively plausible, hence they demonstrate the feasability of our approach.

The first observation is that both processes have their advantages: if there is no competition, then moving the access-token to the other part of the net is a useless and time consuming effort; on the other hand, if the competition is strong, ordering the token is an additional overhead. This is demonstrated by the following i-refusal traces. If in $MUTEX_2$ the access-token is moved to p_2 immediately before r_1, then t becomes urgent only in the second round, at the end of which e_1 can still be refused; we get $r_1\{e_1\}\{e_1\} \in RT(MUTEX_2) \setminus RT(MUTEX_1)$ showing that sometimes $MUTEX_2$ is slower – namely if the second user is not interested in entering the critical section. Vice versa, $MUTEX_1$ is sometimes slower as witnessed by $r_2\{e_2\}\{e_2\}\{e_2\} \in RT(MUTEX_1) \setminus RT(MUTEX_2)$, where an additional round is needed to order the token.

$RT(MUTEX_i)$ shows how efficiently the respective MUTEX-process serves the environment consisting of both users. Interestingly, we can also use our approach to study a different view: how efficiently are the needs of the first user met by the system, which for him consists of a MUTEX-process and the second user? As second user, we take a standard user who, in the non-critical section, can choose between requesting with r_2 and some other internal activity; if she requests, she is willing to enter the critical section in the next round and to leave it again in the round after. As a net, this user looks like the right hand side of $MUTEX_2$, i.e. has places nc_2, req_2 and c_2 and the transitions between them, plus an internal transition on a loop with nc_2. We compose this user with $MUTEX_i$ via $\|_{\{r_2,e_2,l_2\}}$ and hide the synchronized actions (change them to λ), since from the point of view of the first user they are internal activities of the system. Thus, $MUTEX_1$ and $MUTEX_2$ are transformed to $MUTEX_3$ and $MUTEX_4$. It is plausible that $MUTEX_4$ is more efficient than $MUTEX_3$: we consider the worst case efficiency; naturally, for the first user strong competition is the worst case, and in the case of strong competition $MUTEX_2$ is more efficient since it saves the additional effort of ordering the token.

Theorem 5.5 *i) $MUTEX_4$ is strictly faster than $MUTEX_3$.*
ii) The efficiency of $MUTEX_2$ and that of $MUTEX_1$ are incomparable.

References

[CH93] S. Christensen, N.D. Hansen. Coloured Petri nets extended with place capacities, test arcs, and inhibitor arcs. In M. Ajmone-Marsan, editor, *Applications and Theory of Petri Nets 1993*, LNCS 691, 186–205. Springer, 1993.

[DNH84] R. De Nicola, M.C.B. Hennessy. Testing equivalence for processes. *Theoret. Comput. Sci.*, 34:83–133, 1984.

[Gol88] R. Gold. Verklemmungsfreiheit bei modularer Konstruktion fairer Petrinetze. Diplomarbeit, Techn. Univ. München, 1988.

[JV96] L. Jenner, W. Vogler. Fast asynchronous systems in dense time. In F. Meyer auf der Heide et al., editors, *ICALP'96*, LNCS 1099, 75–86. Springer, 1996.

[KW95] E. Kindler, R. Walter. Message passing mutex. In J. Desel, editor, *Structures in Concurrency Theory*, Worksh. in Computing, 205–219. Springer, 1995.

[KW96] E. Kindler, R. Walter. Mutex needs fairness. To appear, 1996.

[MR95] U. Montanari, F. Rossi. Contextual nets. *Acta Informatica*, 32:545–596, 1995.

[Ray86] M. Raynal. *Algorithms for Mutual Exclusion*. North Oxford Academic, 1986.

[Rei84] W. Reisig. Partial order semantics versus interleaving semantics for CSP-like languages and its impact on fairness. In J. Paredaens, editor, *ICALP'84*, LNCS 172, 403–413. Springer, 1984.

[Vog95a] W. Vogler. Fairness and partial order semantics. *Information Processing Letters*, 55:33–39, 1995.

[Vog95b] W. Vogler. Timed testing of concurrent systems. *Information and Computation*, 121:149–171, 1995.

[Vog96] W. Vogler. Efficiency of asynchronous systems and read arcs in Petri nets. 1996. See http://www.math.uni-augsburg.de/∼vogler/.

Bisimulation Equivalence is Decidable for One-Counter Processes [1]

Petr Jančar
Univ. of Ostrava and Techn. Univ. of Ostrava, Czech Republic
e-mail: jancar@osu.cz

Abstract. It is shown that bisimulation equivalence is decidable for the processes generated by (nondeterministic) pushdown automata where the pushdown behaves like a counter, in fact. Also regularity, i.e. bisimulation equivalence with some finite-state process, is shown to be decidable for the mentioned processes.

1 Introduction

In recent years, growing effort has been devoted to the area of verification of (potentially) infinite-state systems. An important studied question is that of (un)decidability for various (behavioural) equivalences. A prominent role among these equivalences is played by *bisimulation equivalence*, or *bisimilarity*, which is more appropriate for (concurrent, reactive etc.) systems than e.g. the traditional language equivalence (cf. [Mil89]). Roughly speaking, two processes (states of systems) are bisimilar iff for any evolving of one process caused by performing an action labelled a there is an action labelled a which causes evolving of the other process in such a way that the resulting processes (states) are again bisimilar.

Several recent results help to highlight and understand the decidability boundaries for bisimilarity, which are different from those for language equivalence. It is e.g. known that bisimilarity is decidable for Basic Parallel Processes ([CHM93]) while the language equivalence is undecidable for them ([Hir93]). More relevant here are context-free processes (generated by context-free grammars), also called BPA-processes, where the language equivalence is well-known to be undecidable while bisimilarity is decidable ([CHS95]). Pushdown automata (which are in the 'language sense' equivalent to context-free grammars) generate a richer family than that of context-free processes when considering bisimulation equivalence. These *pushdown processes* can be identified with 'state-pushdown' configurations, whose behaviour is determined by the transition rules (not allowing ε-rules). Recently Stirling ([Sti96]) has shown the decidability of bisimilarity for *normed* pushdown processes, while the question remains open for the whole class.

Here we show the decidability of bisimilarity for another subclass of pushdown processes: we will not impose the restriction of normedness but we consider the case when the pushdown behaves like a counter, in fact; i.e. there is only one stack symbol, besides a special bottom symbol which enables to test 'emptiness' of the pushdown. Let us call such processes as *one-counter processes*. The

---

[1] Supported by the Grant Agency of the Czech Republic, Grant No. 201/97/0456, and also by the Univ. of Ostrava grant No. 031/97

decidability result for one-counter processes also confirms the conjecture by the author ([Jan93]) that bisimilarity for labelled Petri nets with one unbounded place is decidable (while two unbounded places suffice for undecidability).

Semidecidability of *nonbisimilarity* of pushdown processes can be derived easily in the standard way applied for image finite systems. Therefore semidecidability of bisimilarity is what matters here. In similar cases, the key point is to show that the *bisimilarity* case has always a finite (or finitely presented) witness whose validity can be checked algorithmically. In our case, at the one-counter processes, the role of such witnesses is played by (descriptions of) semilinear sets; this approach was already used in [Jan93] or [Esp95].

Roughly speaking, the existence of such witnesses (i.e. semilinear bisimulations) for one-counter processes can be anticipated from the intuition that two bisimilar processes have to have the same 'distance' (minimum number of steps) to a 'bottom process' (configuration with only the bottom symbol in the pushdown=counter) when such bottom processes matter at all; it can be guessed that then the counter heights of such processes have to be, in principle, linearly related. The possibility of an algorithmic checking of a witness' validity can be easily observed due to the decidability of Presburger arithmetic (although this deep result is surely not needed in its whole).

Another natural decidability question is that of *regularity* of a given process, which will in our context mean the bisimulation equivalence with some finite-state process. This problem has been shown to be decidable for labelled Petri nets ([JE96]), which include BPP-processes. In [BCS96], the decidability is shown for BPA-processes (where the 'language regularity' is well-known to be undecidable). The question for the whole class of pushdown processes is still open (while for the class of normed pushdown processes is easily seen to be decidable). As an additional result, we demonstrate that regularity is also decidable for one-counter processes.

In fact, one-counter processes can be 'almost' identified with labelled Petri nets with one unbounded place; but unlike Petri nets they can 'test for zero'. Nevertheless the strategy used in the proof of decidability of regularity for labelled Petri nets ([JE96]) applies for them as well.

Section 2 contains definitions and claims the results; the proofs are given in Section 3. Section 4 adds some further comments.

2 Definitions and Results

We begin with recalling some standard notions.

A *labelled transition system*, a *system* for short, is a tuple $\mathcal{T} = (\mathcal{S}, \{\xrightarrow{a}\}_{a \in \mathcal{A}})$ where $\mathcal{S}$ is the set of *states*, $\mathcal{A}$ is the set of *actions* (or *action names*) and each $\xrightarrow{a}$ is a binary *(transition) relation* on $\mathcal{S}$ ($\xrightarrow{a} \subseteq \mathcal{S} \times \mathcal{S}$). By $E \to F$ ($E, F \in \mathcal{S}$) we mean that $E \xrightarrow{a} F$ for some a; $\to^*$ denotes the reflexive and transitive closure of the relation $\to$. By $E \to^* \mathcal{S}'$ ($\mathcal{S}'$ is *reachable* from E), where $\mathcal{S}' \subseteq \mathcal{S}$, we mean $E \to^* F$ for some $F \in \mathcal{S}'$. In the obvious sense, we also use $E \xrightarrow{u} F$ where $u \in \mathcal{A}^*$; $|u|$ denotes the length of the sequence u.

A transition system $\mathcal{T} = (\mathcal{S}, \{\xrightarrow{a}\}_{a\in\mathcal{A}})$ is *finite* iff $\mathcal{S}$ and $\mathcal{A}$ are finite. $\mathcal{T}$ is *image finite* iff $succ(E) = \bigcup_{a\in\mathcal{A}} succ_a(E)$ is finite for any $E \in \mathcal{S}$, where we define $succ_a(E) = \{E' \mid E \xrightarrow{a} E'\}$.

Speaking of a *process* E, we always consider it as (being associated with) a state in a transition system which is clear from the context. When necessary, we denote the relevant transition system by $\mathcal{T}(E)$. Using the term of a *finite*, or rather a *finite-state*, *process* E, we mean that $\mathcal{T}(E)$ is finite; similarly for an *image finite process*.

A binary relation $\mathcal{R}$ between processes is a *bisimulation relation* provided that whenever $(E, F) \in \mathcal{R}$, for each action a

if $E \xrightarrow{a} E'$ then there is F' s.t. $F \xrightarrow{a} F'$ and $(E', F') \in \mathcal{R}$, and
if $F \xrightarrow{a} F'$ then there is E' s.t. $E \xrightarrow{a} E'$ and $(E', F') \in \mathcal{R}$.

Two processes E and F are *bisimulation equivalent*, or *bisimilar*, written $E \sim F$, if there is a bisimulation relation $\mathcal{R}$ relating them.

The family $\{\sim_n \mid n \geq 0\}$ (of relations between processes) is defined inductively:

1/ $E \sim_0 F$ for all processes E, F
2/ $E \sim_{n+1} F$ iff for each a

if $E \xrightarrow{a} E'$ then there is F' s.t. $F \xrightarrow{a} F'$ and $E' \sim_n F'$, and
if $F \xrightarrow{a} F'$ then there is E' s.t. $E \xrightarrow{a} E'$ and $E' \sim_n F'$.

Let us recall some 'folklore' results.

Proposition 2.1 *For image finite processes, $E \sim F$ iff $\forall n \geq 0 : E \sim_n F$.*

Let us call $\mathcal{T} = (\mathcal{S}, \{\xrightarrow{a}\}_{a\in\mathcal{A}})$ an *admissible system* iff the state set $\mathcal{S}$ is finite or countably infinite (identified with a set of sequences over a finite alphabet), the action set $\mathcal{A}$ is finite, $\mathcal{T}$ is image finite, and all the successor functions $succ_a : \mathcal{S} \longrightarrow 2^{\mathcal{S}}$ are effectively computable.

Proposition 2.2 *Considering only admissible transition systems, all the relations $E \sim_n F$ ($n \in \mathcal{N}$) are decidable. Therefore the problem $E \not\sim F$ is semidecidable.*

Now we define the pushdown processes (cf. e.g. [Sti96]); loosely speaking, these are state-pushdown configurations of a given (nondeterministic) pushdown automaton without ε-rules. Then we introduce the 'one-counter case'.

Suppose a given collection (i.e. a pushdown automaton viewed as a 'pushdown process generator') $M = (\mathcal{P}, \Gamma, \mathcal{A}, \mathcal{B})$ where $\mathcal{P} = \{p_1, p_2, \ldots, p_k\}$ is a finite set of *states*, $\Gamma = \{X_1, X_2, \ldots, X_m\}$ is a finite set of *stack symbols*, $\mathcal{A} = \{a_1, a_2, \ldots, a_n\}$ is a finite set of *actions*, and $\mathcal{B}$ is a finite set of basic transitions, each of the form $pX \xrightarrow{a} q\alpha$ where p, q are states, a is an action, X is a stack symbol and α is a sequence of stack symbols (i.e. $\alpha \in \Gamma^*$). The transition system $\mathcal{T}_M$ generated by M has the expressions $p\alpha$ ($p \in \mathcal{P}$, $\alpha \in \Gamma^*$), called *pushdown processes*, as states, $\mathcal{A}$ is its action set, and the transition relations are in the straightforward way determined by the basic transitions together with the following *prefix rule*: if $pX \xrightarrow{a} q\alpha$ then $pX\beta \xrightarrow{a} q\alpha\beta$ (for any $\beta \in \Gamma^*$).

When $\Gamma = \{X, Z\}$ and any basic transition is of the form $pX \xrightarrow{a} q\alpha$ or $pZ \xrightarrow{a} q\alpha Z$ where $\alpha \in \{X\}^*$ (we call $M = (\mathcal{P}, \Gamma, \mathcal{A}, \mathcal{B})$ a *one-counter machine* in such a case), then any $pXX\dots XZ$ is called a *one-counter process*. For convenience, a process $pX^m Z$ will be denoted by $p(m)$ ($m \in \mathcal{N}$, where $\mathcal{N}$ denotes the set of all nonnegative integers).

Notice that any process reachable from a one-counter process is a one-counter process as well. Thus for a one-counter machine M we can safely suppose that $\mathcal{T}_M$ has states of the form $p(m)$ only.

Our main aim here is to show

Theorem 2.3 *Bisimulation equivalence is decidable for one-counter processes.*

More precisely it means that there is an algorithm which inputs (descriptions of) two one-counter processes $p(m)$, $p'(m')$ together with the respective one-counter machines M, M', and after a finite amount of time answers whether or not $p(m) \sim p'(m')$.

An additional result is expressed in the following theorem; here a process E is called *regular* iff there is a finite-state process p s.t. $E \sim p$.

Theorem 2.4 *Regularity (wrt bisimilarity) is decidable for one-counter processes.*

Each of the two decidability results is implied by two semidecision procedures. We can immediately note that semidecidability of *nonbisimilarity* $E \not\sim F$ follows from Proposition 2.2 since one-counter systems (as well as pushdown systems) are obviously admissible.

We finish this section by recalling some known notions and results which are then used in the proofs in Section 3.

Given a transition system $\mathcal{T} = (\mathcal{S}, \{\xrightarrow{a}\}_{a \in \mathcal{A}})$, we define the class of all n-incompatible processes as $INC_n^{\mathcal{T}} = \{E \mid \forall F \in \mathcal{S} : E \not\sim_n F\}$.

More specific variants of the following two propositions were used in [JM95], [JE96].

Proposition 2.5 *For any n, $E \sim F$ implies that $E \sim_n F$ and $E \not\to^* INC_n^{\mathcal{T}(F)}$. In addition, the implication can be reversed for any n s.t. $\sim_{n-1}$ coincides with $\sim_n$ (and hence with $\sim$) on $\mathcal{T}(F)$.*

Corollary 2.6 *Let A be a finite transition system with k states. For any states p, q, it holds that $p \sim_{k-1} q$ iff $p \sim_k q$ (iff $p \sim q$). It yields for any process E and a state p of A: $E \sim p$ iff $E \sim_k p$ and $E \not\to^* INC_k^A$.*

The *distance* of a process E to F, denoted by $Dist(E, F)$, is the length of the shortest sequence u s.t. $E \xrightarrow{u} F$; if F is not reachable from E, we put $Dist(E, F) = \infty$. For a set $\mathcal{F}$ of processes, we define $Dist(E, \mathcal{F}) = \min\{Dist(E, F) \mid F \in \mathcal{F}\}$.

Proposition 2.7 *If $E \sim F$ then $Dist(E, \mathcal{F}) = Dist(F, \mathcal{F})$ for any quotient class $\mathcal{F}$ of $\sim_n$ on the set of all processes.*

We need the notion of *semilinear sets*. An important fact is that they are precisely the *sets expressible in Presburger arithmetic* (cf. [GS66]); we will use it implicitly when arguing that some sets are semilinear.

A set $V \subseteq \mathcal{N}^r$ of vectors ($r \geq 1$) is *linear* if there is a *base vector* $\vec{y}$ and *period vectors* $\vec{x}_1, \vec{x}_2, \ldots, \vec{x}_m$ in $\mathcal{N}^r$ such that $V = \{\vec{y} + \sum_{i=1}^{m} c_i \vec{x}_i \mid c_i \in \mathcal{N}\}$. V is *semilinear* if it is a finite union of linear sets.

In fact, here we are mainly interested in dimensions $r = 1, 2$. The next fact on *one-dimensional* semilinear sets is easily derivable:

Proposition 2.8 *Suppose a set $V \subseteq \mathcal{N}$. Then:*

1/ If there are $c, \delta \in \mathcal{N}$ s.t. $\forall m > c : m \in V \Rightarrow m + \delta \in V$ then V is semilinear.

2/ If V is semilinear then there are constants c and Δ s.t. for any $m > c$, the value $m \bmod \Delta$ determines whether $m \in V$ or $m \notin V$.

3 Proofs

In this section we always (implicitly) suppose a given one-counter machine M with k states (and the stack alphabet $\{X, Z\}$); the states are denoted by p, q (often primed or with subscripts).

Subsection 3.1 proves the crucial fact of this paper (Proposition 3.3) which shows that the set $\{(m, n) \mid p(m) \sim q(n)\}$ is semilinear for any p, q. Subsections 3.2 and 3.3 then prove the theorems.

In the proofs we need the notion of the underlying automaton A_M which behaves like M as long as the bottom of the stack is not reached, and also the notion of processes which are 'Basically Incompatible' with (states of) A_M:

The *underlying finite automaton* A_M (viewed as a finite transition system) has the same set of states as M, and it has the transition $p \xrightarrow{a} q$ iff M has a basic transition $pX \xrightarrow{a} q\alpha$ ($\alpha \in \{X\}^*$).

We define $BInc = \{p(m) \mid p(m) \in INC_k^{A_M}\} = \{p(m) \mid p(m) \not\sim_k q$ for each state $q\}$.

When we observe that $p(m) \sim_k p$ for $m \geq k$, the next lemma is clear:

Lemma 3.1 *If $p(m) \in BInc$ then $m < k$. Therefore $BInc$ is a finite, and effectively computable, set.*

Due to corollary 2.6 we can add (recall that k denotes the number of states of M and hence also of A_M):

Lemma 3.2 *For $m \geq k$ (and any state p), $p(m) \not\sim p$ iff $p(m) \to^* BInc$.*

Notation. By $p(m) \to^*_{\geq r} q(n)$ ($r \in \mathcal{N}$) we mean that there is a path $p(m) = q_1(n_1) \to q_2(n_2) \to \ldots \to q_s(n_s) = q(n)$ s.t. $n_i \geq r$ for $i = 1, 2, \ldots, s$. By $p(m) \to^*_{POS} q(n)$ (POSitive) we mean that $p(m) \to^*_{\geq 1} q(n)$.

Observe the obvious fact (used implicitly in what follows): if $r \geq 1$ then $p(m) \to^*_{\geq r} q(n)$ iff $p(m+\delta) \to^*_{\geq r+\delta} q(n+\delta)$ for any $\delta \in \mathcal{N}$. In particular $p(m) \to^*_{POS} q(n)$ implies $p(m+\delta) \to^*_{POS} q(n+\delta)$.

3.1 Semilinearity Proof

This subsection is devoted to a proof of the next crucial proposition:

Proposition 3.3 *For any one-counter machine and its states p, q, the set $\{(m, n) \mid p(m) \sim q(n)\}$ is semilinear.*

First observe that if $p(m) \to^* BInc$ and $q(n) \not\to^* BInc$ then surely $p(m) \not\sim q(n)$ (cf. Proposition 2.7). Therefore the set $B = \{(m, n) \mid p(m) \sim q(n)\}$ can be written as $B = B_1 \cup B_2$ where

$$B_1 = \{(m, n) \mid p(m) \sim q(n), p(m) \not\to^* BInc, q(n) \not\to^* BInc\},$$

$$B_2 = \{(m, n) \mid p(m) \sim q(n), p(m) \to^* BInc, q(n) \to^* BInc\}.$$

Therefore it suffices to show semilinearity of B_1 and B_2.

The next lemma is a means for proving semilinearity of B_1.

Lemma 3.4 *For any state p (of the one-counter machine M), the set $\{m \mid p(m) \to^* BInc\}$ is semilinear; therefore also $\{m \mid p(m) \not\to^* BInc\}$ is semilinear.*

Proof: Recall that we suppose M with k states; let $\mathcal{P}$ be the state set.

We have to show semilinearity of $R = \{m \mid p(m) \to^* BInc\}$. For any $Q \subseteq \mathcal{P}$ we define the set $R_Q \subseteq R$ as follows: $m \in R_Q$ iff there is a 'witness' path

$$p(m) = q_1(n_1) \to q_2(n_2) \to \ldots \to q_s(n_s) \in BInc \qquad (1)$$

s.t. $q_i \in Q$ for $i = 1, 2, \ldots, s'$ where $s' \leq s$ is the maximum number s.t. $n_i \geq 1$ for $i = 1, 2, \ldots, s'$ (the path goes through states from Q solely while after the first reaching of the stack bottom – if it happens at all – there are no restrictions).

It is clear that $R_\mathcal{P} = R$ and it suffices to show semilinearity of all R_Q. We proceed by induction on $|Q|$.

When $Q = \emptyset$ then R_Q is obviously semilinear ($R_Q = \emptyset$ or $R_Q = \{0\}$).

Now we show semilinearity of R_Q, $|Q| > 0$, while supposing semilinearity for each $R_{Q'}$, $|Q'| < |Q|$. Let some $m \geq 2k$ be in R_Q (otherwise R_Q is finite, hence semilinear) and let (1) be a relevant witness path; recall that $k > n_s$ (Lemma 3.1). We can take the leftmost subsequence $q_{i_1}(m), q_{i_2}(m-1), \ldots, q_{i_{k+1}}(m-k)$; due to the pigeonhole principle, there is $q = q_{i_a} = q_{i_b}$ for $a \neq b$. Therefore $p(m) \to^*_{\geq n'_1} q(n'_1) \to^*_{\geq n'_2} q(n'_2) \to^* q_s(n_s) \in BInc$ where $\delta = n'_1 - n'_2 > 0$, $n'_2 > 0$; hence $q(n + \delta) \to^*_{\geq n} q(n)$ for any $n > 0$.

We can write $R_Q = R_Q^q \cup R_{Q\setminus\{q\}}$ where

$R_Q^q = \{m \in R_Q \mid \text{there is a witness path with } q = q_i \text{ for some } i, 1 \le i \le s'\}$.

Since $m \in R_Q^q$ obviously implies $m + \delta \in R_Q^q$, R_Q^q is semilinear (cf. Proposition 2.8 1/); semilinearity of $R_{Q\setminus\{q\}}$ follows from the induction hypothesis. □

Corollary 3.5 $B_1 = \{(m,n) \mid p(m) \sim q(n), p(m) \not\to^* BInc, q(n) \not\to^* BInc\}$ is semilinear.

Proof: Given $r < k$, consider $B_1(r,-) = \{n \mid (r,n) \in B_1\}$ Note that for any $n \in B_1(r,-)$, $n \ge k$ implies $q(n) \sim q$. Therefore when $B_1(r,-)$ is infinite, it is the union of a finite set and the set $\{n \ge k \mid q(n) \not\to^* BInc\}$; in any case, $B_1(r,-)$ is semilinear. Semilinearity of $B_1(-,r) = \{m \mid (m,r) \in B_1\}$ can be established similarly. B_1 can be written

$$B_1 = \bigcup_{r=0}^{k-1} \{(r,n) \mid n \in B_1(r,-)\} \cup \bigcup_{r=0}^{k-1} \{(m,r) \mid m \in B_1(-,r)\} \cup B_1'$$

where

$B_1' = \{(m,n) \mid m \ge k, n \ge k, p(m) \sim q(n), p(m) \not\to^* BInc, q(n) \not\to^* BInc\}$.

B_1' is either empty (when $p \not\sim q$) or equals to $\{(m,n) \mid m \ge k, n \ge k, p(m) \not\to^* BInc, q(n) \not\to^* BInc\}$ (when $p \sim q$).

Thus semilinearity of B_1 is clear. □

We also need another corollary.

Corollary 3.6 There are constants c and Δ s.t. for any p and any $m > c$, the value $m \bmod \Delta$ determines whether or not $p(m) \to^* BInc$.

Proof: For any state p, we get the relevant c_p, Δ_p due to Proposition 2.8 2/. The constant c desired here can be taken as the maximum of c_p's and Δ can be taken as the product of Δ_p's. □

Our aim now is to show semilinearity of $B_2 = \{(m,n) \mid p(m) \sim q(n), p(m) \to^* BInc, q(n) \to^* BInc\}$.

Notation. $Dist(p(m), BInc)$ will be denoted by $Dist(p(m))$ for short.

Since $Dist(p(m)) = Dist(q(n))$ is a necessary condition for $p(m) \sim q(n)$, we will explore which relation it imposes for m and n. First we show that $Dist(p(m))$ is, in fact, linear (when finite) in m with the provision that the coefficient depends on $m \bmod \Delta$. Here and further, Δ is taken from Corollary 3.6.

Lemma 3.7 There is a constant $d \in \mathcal{N}$, and for any state p and any congruence class $\langle i \rangle_{\bmod \Delta}$ ($0 \le i \le \Delta - 1$) there is a rational constant k' s.t. the following holds for any $m, m \equiv i(\bmod \Delta)$: if $Dist(p(m))$ is finite then

$$Dist(p(m)) \in \langle k'm - d, k'm + d \rangle.$$

Proof: Suppose some p and $\langle i \rangle_{\mathrm{mod}\ \Delta}$. In the proof, for each number denoted by m we implicitly suppose $m \equiv i(\mathrm{mod}\ \Delta)$. We show that there are k' and d' s.t. $Dist(p(m)) \in \langle k'm - d', k'm + d' \rangle$, by which we will be done (the desired d can be taken as the maximum of all relevant constants d').

Observe that $p(m) \to^* BInc$, for a large m, implies a *decreasing cycle*:

$$p(m) \to^*_{POS} q(n+\delta) \to^*_{\geq n} q(n) \to^* BInc \text{ for some } q, n > 0, \delta > 0.$$

Let $Q = \{q \mid p(m) \to^*_{POS} q(n) \to^* BInc \text{ for some } m, n\}$. Now let q' be a state of Q which allows a decreasing cycle $q'(n+\delta_w) \xrightarrow{w}_{\geq n} q'(n)$ (for some w, $\delta_w > 0$, and all $n \geq 1$) with the *best decreasing rate* – i.e. $\delta_w/|w|$ is maximal possible. The existence of such q' can be easily derived (by 'pigeonhole principle reasoning' we could suppose $|w| \leq k$). Moreover we can safely suppose that δ_w is a multiple of Δ (otherwise we take w^Δ which yields the same decreasing rate), and thus $q'(n+\delta_w) \to^* BInc$ iff $q'(n) \to^* BInc$ for $n > c$, c taken from Proposition 3.6.

Let us choose $m > c + \delta_w + k$ s.t. $p(m) \xrightarrow{u}_{POS} q'(n) \to^* BInc$ for some u and $n, c < n \leq c + \delta_w$; denote $\delta_u = m - n$. Note that $p(m+j\Delta) \xrightarrow{u}_{POS} q'(n+j\Delta) \to^* BInc$ for any $j \geq 0$.

Now let $d_0 = |u|$, $d_1 = max\{Dist(p'(c+x)) \mid x \in \{0, 1, \ldots, \delta_w\}$ and $Dist(p'(c+x))$ is finite $\}$. Then it is clear that for any $m > c + \delta_w + k$

$$Dist(p(m)) \leq d_0 + \Big((m - \delta_u - c)/\delta_w\Big)|w| + d_1$$

On the other hand it is easily verifiable that

$$Dist(p(m)) \geq \Big((m - \delta_u - c)/\delta_w - 1\Big)|w|.$$

Calculating the desired k', d' is now a technical routine (d' has to be chosen large enough to 'cover' the finitely many $m \leq c + \delta_w + k$ as well). □

Corollary 3.8 *There is a constant $d \in \mathcal{N}$ s.t. for any p, q and congruence classes $\langle i \rangle_{\mathrm{mod}\ \Delta}$, $\langle j \rangle_{\mathrm{mod}\ \Delta}$, there is a rational constant k' s.t. the following holds for any m, n, $m \equiv i(\mathrm{mod}\ \Delta)$, $n \equiv i(\mathrm{mod}\ \Delta)$: if $Dist(p(m)) = Dist(q(n)) < \infty$ then $n \in \langle k'm - d, k'm + d \rangle$.*

Proof: Because there are constants k_1, k_2 and d' s.t. $Dist(p(m)) \in \langle k_1 m - d', k_1 m + d' \rangle$ and $Dist(q(n)) \in \langle k_2 n - d', k_2 n + d' \rangle$ then it must hold $k_2 n - d' \leq k_1 m + d'$ and $k_2 n + d' \geq k_1 m - d'$. Hence we have $mk_1/k_2 - 2d'/k_2 \leq n \leq mk_1/k_2 + 2d'/k_2$. □

Recall that our aim is to show semilinearity of B_2. We already know that there is $d \in \mathcal{N}$ and a finite set $K = \{k_1, k_2, \ldots, k_r\}$ of rational constants s.t. it suffices, for each $k' \in K$, to show semilinearity of the set

$$B_{k'} = \{(m, n) \mid p(m) \sim q(n), n \in \langle k'm - d, k'm + d \rangle\}.$$

(The union of $B_{k'}$'s consists of B_2 and a subset of B_1 which is obviously semi-linear, i.e. expressible in the Presburger arithmetic).

In fact, we will consider only the subset of $B_{k'}$ where $m > c$ for a sufficiently large c (the rest being finite and therefore causing no problems); c will be chosen so that for any m, n, $m > c$, $|n - k'm| \leq d$, the following holds: for any p', q' and any moves $p'(m) \xrightarrow{a} p''(m')$, $q'(n) \xrightarrow{a} q''(n')$ it is ensured that $|n' - k''m'| > d$ for each $k'' \in K$, $k'' \neq k'$ (a pair of moves cannot lead from '$B_{k'}$-area' into '$B_{k''}$-area').

Given k', let us denote $Cut(m) = \cap_{i=0}^{\infty} Cut_i(m)$ where $Cut_i(m) = \{(p', q', x) \mid x \in \{-d, -d+1, \ldots, d\}, p'(m) \sim_i q'(round(k'm) + x)\}$.

Observe that there surely is an infinite sequence $m_0 < m_1 < m_2 < \ldots$ s.t. for all $i \geq 0$: $k'm_i$ is integer, $m_{i+1} - m_i \equiv 0 \pmod{\Delta}$, $k'm_{i+1} - k'm_i \equiv 0 \pmod{\Delta}$. Since, for any m, $Cut(m)$ is a boundedly finite set, there are surely m, m' satisfying the assumption of the next lemma; and it is easily observable that the lemma demonstrates semilinearity of $B_{k'}$ and thus finishes the proof of Proposition 3.3.

Lemma 3.9 *When $Cut(m) = Cut(m')$ for sufficiently large m where $m < m'$, $k'm, k'm'$ are integers, $m' - m \equiv 0 \pmod{\Delta}$, $k'm' - k'm \equiv 0 \pmod{\Delta}$, then $Cut(m + \delta) = Cut(m' + \delta)$ for any $\delta \geq 0$.*

Proof: We show $Cut(m + \delta) \subseteq Cut(m' + \delta)$ while the other inclusion will be completely symmetric.

In fact, we show by induction on i that $(p, q, x) \in Cut(m + \delta)$ implies $(p, q, x) \in Cut_i(m' + \delta)$ for all i; for $i = 0$ it is trivial as well as for $\delta = 0$.

Induction hypothesis: for any p, q, x, δ, if $(p, q, x) \in Cut(m+\delta)$ then $(p, q, x) \in Cut_i(m' + \delta)$.

Now we consider arbitrary (but fixed) $p, q, x, \delta \geq 1$ s.t. $(p, q, x) \in Cut(m+\delta)$ and we show that $(p, q, x) \in Cut_{i+1}(m' + \delta)$ by which the whole proof will be finished.

In other words, denoting $m_1 = m+\delta$, $n_1 = round(k'(m+\delta))+x$, $m_2 = m'+\delta$, $n_2 = round(k'(m' + \delta)) + x$, we suppose $p(m_1) \sim q(n_1)$ and we have to show $p(m_2) \sim_{i+1} q(n_2)$.

Let $p(m_2) \xrightarrow{a} p'(m_2 + y)$ ($-1 \leq y \leq max$, max depending on the machine M). There is the corresponding move $p(m_1) \xrightarrow{a} p'(m_1+y)$ and there has to be a move $q(n_1) \xrightarrow{a} q'(n_1+z)$ ($-1 \leq z \leq max$) s.t. $p'(m_1+y) \sim q'(n_1+z)$. We claim that the corresponding move $q(n_2) \xrightarrow{a} q'(n_2+z)$ yields $p'(m_2+y) \sim_i q'(n_2+z)$.

When $|k'(m_1 + y) - (n_1 + z)| \leq d$ (hence also $|k'(m_2 + y) - (n_2 + z)| \leq d$), it follows from the inductive hypothesis. Otherwise $Dist(p'(m_1 + y)) = Dist(q'(n_1+z)) = \infty$ and $p' \sim q'$. But then also $Dist(p'(m_2+y)) = Dist(q'(n_2+z)) = \infty$ (recall the property of Δ); therefore $p'(m_2 + y) \sim q'(n_2 + z)$.

The remaining parts of the proof are completely similar. □

3.2 Decidability of Bisimilarity

Now we can provide a proof for Theorem 2.3:

Theorem. *Bisimulation equivalence is decidable for one-counter processes.*

Proof: First notice that we can always consider the bisimilarity problem instance '$p(m) \sim q(n)$?' where $p(m)$, $q(n)$ are associated to the same one-counter machine (which can be achieved by taking the union of two machines – i.e. union of action sets, and disjoint union of state sets and basic transition sets).

Recall that it suffices to show *semi*decidability for '$p(m) \sim q(n)$?' (cf. Proposition 2.2). Now due to Proposition 3.3 it suffices to generate all bisimulation candidates $\mathcal{R}$ s.t. the set $\{(m', n') \mid (p'(m'), q'(n')) \in \mathcal{R}\}$ is semilinear for each pair of states p', q', and for each such candidate to check if $\mathcal{R}$ actually is a bisimulation containing $(p(m), q(n))$. (Descriptions of) such candidate relations can be obviously generated in a systematic way, and the condition to be checked is easily seen to be expressible in Presburger arithmetic, which is decidable (cf. e.g. [Opp78]). □

3.3 Decidability of Regularity

Here we provide a proof for Theorem 2.4:

Theorem. *Regularity (wrt bisimilarity) is decidable for one-counter processes.*

Proof: Semidecidability of regularity of $p(m)$ follows from Theorem 2.3. (We can generate all finite state processes $\mathcal{F}$, viewed as special cases of one-counter processes, and to check for each of them whether $p(m) \sim \mathcal{F}$).

Semidecidability of nonregularity will follow when we show that $p(m)$ is nonregular iff there is a path

$$p(m) \to^* p'(m_1) \to^*_{POS} p'(m_2) \to^*_{POS} q'(n_1) \to^*_{POS} q'(n_2) \to^* BInc$$

where $m_1 < m_2$, $n_1 > n_2$.

The existence of such a path ensures for any $i \geq 0$ that

$$p(m) \to^* p'(m_2 + i(n_1 - n_2)(m_2 - m_1)) \to^* q'(n_1 + i(m_2 - m_1)(n_1 - n_2)) \to^* BInc$$

which implies that there are reachable states with arbitrarily large (but finite) distances to $BInc$ – and this obviously implies nonregularity of $p(m)$. The opposite direction can be also easily established. □

4 Further Comments

The example of a pushdown process used in [Sti96]

$$pX \xrightarrow{a} pXX, pX \xrightarrow{c} q\varepsilon, pX \xrightarrow{b} r\varepsilon, qX \xrightarrow{d} sX, sX \xrightarrow{d} q\varepsilon, rX \xrightarrow{d} r\varepsilon$$

can be easily transformed in a one-counter process with the isomorphic transition system. This process can serve as an example of a one-counter process which is not equivalent to a BPA-process, nor a BPP-process, and when adding a rule $pX \xrightarrow{f} q_{fin}$ we get a one-counter process not equivalent to any normed pushdown process.

References

[BCS96] Burkart O., Caucal D. and Steffen B.: Bisimulation collapse and the process taxonomy; in Proc. CONCUR'96, *Lecture Notes in Computer Science*, Vol. 1119 (Springer, 1996) 247–262

[CHM93] Christensen S., Hirshfeld Y. and Moller F.: Bisimulation equivalence is decidable for all Basic Parallel Processes, in Proc. CONCUR'93, *Lecture Notes in Computer Science*, Vol. 715 (Springer, 1993) 143–157

[CHS95] Christensen S., Hüttel H. and Stirling C.: Bisimulation equivalence is decidable for all context-free processes; *Information and Computation* **121** (1995) 143–148

[Esp95] Esparza J.: Petri nets, commutative context-free grammars, and Basic Parallel Processes; in Proc. Fundamentals of Computation Theory (FCT) 1995, *Lecture Notes in Computer Science*, Vol. 965 (Springer, 1995) 221–232

[GS66] Ginsburg S. and Spanier E.: Semigroups, Presburger formulas, and languages; *Pacific J. of Mathematics* **16** (1966) 285–296

[Hir93] Hirshfeld Y.: Petri nets and the equivalence problem; in Proc. Computer Science Logic (CSL) '93, *Lecture Notes in Computer Science*, Vol. 832 (Springer, 1994) 165–174

[Jan93] Jančar P.: Decidability questions for bisimilarity of Petri nets and some related problems; Techn.rep. ECS-LFCS-93-261, Dept. of comp. sci., Univ. of Edinburgh, UK, April 1993
(cf. also Jančar P.: Undecidability of Bisimilarity for Petri Nets and Related Problems; *Theoretical Computer Science* **148** (1995) 281–301)

[JE96] Jančar P., Esparza J.: Deciding finiteness of Petri nets up to bisimulation; in Proc. ICALP'96, Paderborn, Germany, July 1996, *Lecture Notes in Computer Science*, Vol. 1099 (Springer, 1996) 478–489

[JM95] Jančar P., Moller F.: Checking regular properties of Petri nets; in Proc. CONCUR'95, Philadelphia, U.S.A., August 1995, *Lecture Notes in Computer Science*, Vol. 962 (Springer, 1995) 348–362

[Mil89] Milner R.: *Communication and Concurrency* (Prentice Hall, 1989)

[Opp78] Oppen D.C.: A $2^{2^{2^{pn}}}$ upper bound on the complexity of Presburger Arithmetic, *J. of Comp. and System Sci.* **16** (1978) 323–332

[Sti96] Stirling C.: Decidability of bisimulation equivalence for normed pushdown processes; in Proc. CONCUR'96, *Lecture Notes in Computer Science*, Vol. 1119 (Springer, 1996) 217–232

Symbolic Reachability Analysis of FIFO-Channel Systems with Nonregular Sets of Configurations

(extended abstract)

Ahmed Bouajjani   Peter Habermehl

VERIMAG, Centre Equation, 2 av. de Vignate, 38610 Gieres, France.
Email: Ahmed.Bouajjani@imag.fr, Peter.Habermehl@imag.fr

Abstract. We address the verification problem of FIFO-channel systems by applying the symbolic analysis principle. We represent their sets of states (configurations) using structures called CQDD's combining finite-state automata with linear constraints on number of occurrences of symbols. We show that CQDD's allow forward and backward reachability analysis of systems with nonregular sets of configurations. Moreover, we prove that CQDD's allow to compute the exact effect of the repeated execution of any fixed cycle in the transition graph of a system. We use this fact to define a generic reachability analysis semi-algorithm parametrized by a set of cycles Θ. Given a set of configurations, this semi-algorithm performs a least fixpoint calculation to construct the set of its successors (or predecessors). At each step, this calculation is *accelerated* by considering the cycles in Θ as additional "meta-transitions" in the transition graph, generalizing the approach adopted in [5].

1 Introduction

Analyzing the behaviour of systems relies basically on solving reachability problems in their models, that are in general finite-state automata supplied with (possibly unbounded) data structures (Petri nets, timed or hybrid automata, fifo-channel systems, etc). It is therefore fundamental to compute the set of *all successors* or *all predecessors* of a given set of states S, i.e., the set of states that are reachable from S, or those from which it is possible to reach S.

Let $post(S)$ (resp. $pre(S)$) denote the set of immediate successors (predecessors) of the set S, and let $post^*(S)$ ($pre^*(S)$) denote the set of all its successors (predecessors). Clearly, $post^*(S)$ is the limit of the *infinite* increasing sequence $(X_i)_{i \geq 0}$ with $X_0 = S$ and $X_{i+1} = X_i \cup post(X_i)$ for every $i \geq 0$. Similarly, $pre^*(S)$ is the limit of the infinite sequence obtained by considering pre instead of $post$.

Unfortunately, for any interesting class of infinite-state systems, the sets X_i are in general infinite and the sequence $(X_i)_{i \geq 0}$ is not guaranteed to reach its limit. Hence, the first problem is to find a class of *finite* structures that can represent the infinite sets of states we are interested in. This class of structures should be effectively closed under union and the *post* and *pre* functions such that the X_i's can be calculated. Moreover, to compare two sets and to check whether a given state belongs to an infinite set, the membership and the inclusion problems of the class should be decidable.

For instance, for systems manipulating integer or real valued variables (Petri nets or timed and hybrid automata), representation structures based on polyhedra or sets of linear constraints are used [3, 6, 2, 13]. In systems manipulating sequential data structures like stacks or queues sets of states are vectors of words, and automata-based representation structures can naturally be used.

Another problem is the convergence of the sequence of X_i's. In general this sequence never reaches its limit and an *exact acceleration* of the computation of the limit is considered by defining another increasing sequence $(Y_i)_{i \geq 0}$ such that for every $i \geq 0$, $X_i \subseteq Y_i$, and $Y_i \subseteq \bigcup_{i \geq 0} X_i$. This approach has been used [9, 7] to define model-checking algorithms for pushdown systems using (alternating) finite-state automata to represent sets of stack contents.

In [5], finite-state automata-based structures called QDD's are used to represent queue contents of fifo-channel systems (communicating finite-state machines, CFSM). However, contrary to the case of pushdown systems, the set of reachable states of a CFSM is not regular in general, and hence not QDD representable. Moreover, there is no algorithm allowing to construct the set of reachable states even if we know that it is regular [10, 12, 1]. To face this problem [5] proposes an acceleration technique based on adding to each X_{i+1} the set of states $post^*_\theta(X_i)$ which corresponds to the set of all successors after repeating as much as possible a cycle θ of a special kind (called meta-transitions). The restriction on the nature of θ guarantees that the $post^*_\theta$ image of a regular set is also regular.

In this paper, we also consider CFSM's and propose a generalization of the approach adopted in [5] by allowing an exact acceleration of the fixpoint calculation with the successors by *any* cycle in the transition graph of the system. The difficulty comes from the fact that the set of reachable states by a cycle is in general nonregular. Therefore, we propose a representation structure called CQDD (constrained QDD) allowing the representation of such sets. This structure is based on a combination of (simple) finite-state automata with Presburger arithmetics formulas expressing constraints on the number of occurrences of symbols.

We show that CQDD's satisfy the desirable properties of a representation structure mentioned above. Moreover, and this constitutes our main result, we prove that the class of CQDD representable sets of states is effectively closed under the function $post^*_\theta$ for every cycle θ. We prove also that the class of CQDD reverse representable sets of states (their reverse image is CQDD representable) is effectively closed under the function pre^*_θ for every cycle θ. These results allow to define a generic reachability analysis semi-algorithm which is parametrized by a set of cycles in the transition graph of the system. When it terminates, our algorithm returns the exact set of successors (or predecessors) of a given CQDD representable (or CQDD reverse representable) set of states. Several analysis algorithms can be derived from our algorithm by determining adequate strategies for choosing the set of cycles to be considered to accelerate the fixpoint calculation. The algorithm of [5] can be seen as a particular instance of our algorithm.

Related work: In [16, 11] a model-checking semi-algorithm is proposed for CFSM, based on a finite representation of the state-graph by means of graph grammars. This approach is different from ours since it is based on a finite representation of the state-graph instead of a finite representation of the set of states. There are other existing works on the analysis of CFSM's assuming that the systems have lossy or unreliable channels (queues) [1, 12]. In our work we do not have such assumptions. Other works propose (terminating) algorithms generating an upper approximation of the set of reachable states [15]. This is different from our approach because we construct the exact set of reachable states as a fixpoint calculation and helping the termination of this calculation by *exact* accelerations.

The rest of this paper is organized as follows. In Section 2 we introduce some basic definitions. In Section 3 we define CFSM's and the successors and predecessors functions. In Section 4, we define CQDD's and give basic results. In Section 5, we show how CQDD's can be used to represent nonregular sets of states and give our main results on the class of CQDD representable and reverse representable sets of states. In Section 6, we present our generic forward and backward analysis algorithm. Finally, we conclude in Section 7. Due to lack of space we omit the proofs of the theorems. They can be found in [8].

2 Preliminaries

Presburger arithmetics is the first order logic of natural numbers with addition, subtraction and the usual ordering. We say that f is a Presburger formula *over* a set of variables $X = \{x_1, \ldots, x_n\}$, and we write $f(X)$, if the set of free variables in f is precisely X. The semantics of Presburger formulas is defined in the standard way. Given a formula f with free variables $X = \{x_1, \ldots, x_n\}$, and a valuation $\nu : X \to I\!\!N$, we say that ν *satisfies* f, and write $\nu \models f$, if the evaluation of f under ν is true. We say that a formula f is valid if every valuation satisfies f.

A *simple automaton* over Σ (SA) is a finite-state automaton $A = (Q, q_0, \to, q_m)$ where Q is a finite set of states with $Q = \{q_0, q_1, \ldots, q_m\} \cup \bigcup_{i=0}^{m} P_i$ where $P_i = \{p_i^1, \ldots, p_i^{\ell_i}\}$, q_0 (resp. q_m) is the initial (resp. final) state, $\to \subseteq Q \times \Sigma \times Q$ is a set of transitions (transition relation) defined as the smallest set such that :

1. $\forall i \in \{0, \ldots, m-1\}$. $\exists! a \in \Sigma$. $q_i \xrightarrow{a} q_{i+1}$,
2. $\forall i \in \{0, \ldots, m\}$, if $P_i \neq \emptyset$ then $\exists! a \in \Sigma$. $q_i \xrightarrow{a} p_i^1$, $\exists! a \in \Sigma$. $p_i^{\ell_i} \xrightarrow{a} q_i$
   and $\forall j \in \{1, \ldots, \ell_i - 1\}$. $\exists! a \in \Sigma$. $p_i^j \xrightarrow{a} p_i^{j+1}$,
3. $\forall i \in \{0, \ldots, m-1\}$, if $P_i = \emptyset$ then there is at most one $a \in \Sigma$ with $q_i \xrightarrow{a} q_i$
4. if $P_m = \emptyset$ then $\exists \Sigma' \subseteq \Sigma$. $\forall a \in \Sigma'$. $q_m \xrightarrow{a} q_m$,

A *restricted simple automaton* (RSA) is a simple automaton where point (4) in the definition above is replaced by: (4'). if $P_m = \emptyset$ then q_m has no successors.

Notice that in simple automata, the outdegree of the states q_i's, except maybe q_m, is at most 2, whereas the outdegree of the states in the P_i is always 1. Each state different from q_m belongs to at most one loop which is of the form $q_i \xrightarrow{a_0} p_i^1 \xrightarrow{a_1} \ldots p_i^{\ell_i} \xrightarrow{a_{\ell_i}} q_i$. We say that q_i is the *root* of this loop. The state q_m has a particular status since it may be the root of several loops, but in this case all these loops must be self-loops. In RSA q_m has the same status as the other q_i's. Nondeterministic choices may occur only at the states q_i. A simple automaton is *deterministic* if every state q_i has at most one successor by each symbol in Σ. We write DSA (resp. DRSA) for deterministic SA (resp. RSA).

Given a word $w = a_0 \ldots a_\ell \in \Sigma^*$, a *run* of A over w is a sequence of transitions $\rho = (s_0, a_0, s_1) \ldots (s_\ell, a_\ell, s_{\ell+1}) \in \rightarrow^{\ell+1}$ such that $s_0 = q_0$. The run ρ is *accepting* if $s_{\ell+1} = q_m$. The language accepted by A, denoted by $L(A)$, is the set of words $w \in \Sigma^*$ such that there is an accepting run of A over w.

Notice that RSA's accept languages over Σ which are definable by regular expressions of the form $u_1 v_1^* u_2 v_2^* \cdots u_m v_m^* u_{m+1}$ where the u_i's and the v_i's are words over Σ such that only u_1 and u_{m+1} may be empty. SA's accept words of the form $u_1 v_1^* u_2 v_2^* \cdots u_m (a_1 + \ldots + a_\ell)^*$ where the a_i's are symbols in Σ.

Let $A = (Q, q_0, \rightarrow, q_m)$ be a simple automaton. Let X be the set of variable $\{x_t : t \in \rightarrow\}$. We consider for each run ρ of A the valuation ν_ρ of the variable in X such that, for every $x_t \in X$, $\nu_\rho(x) = |\rho|_t$. Then, we define a Presburger formula $[A]$ over X which characterizes the set of valuations corresponding to all accepting runs of A. For that, let us introduce some notations. We denote by $\mathcal{T}$ the set of transitions $\{t \in \rightarrow \; : \; \exists i \in \{0, \ldots, m-1\}. \; \exists a \in \Sigma. \; t = (q_i, a, q_{i+1})\}$. For each state $q \in Q$, we denote by $In(q)$ (resp. $Out(q)$) the set of transitions of the form (q', a, q) (resp. (q, a, q')) for some $q' \in Q$ and $a \in \Sigma$. Now, let $[A]$ be the formula $(\bigwedge_{t \in \mathcal{T}} x_t = 1) \wedge (\bigwedge_{q \in Q \setminus \{q_0\}} \sum_{t \in In(q)} x_t = \sum_{t \in Out(q)} x_t) \wedge (1 + \sum_{t \in In(q_0)} x_t = \sum_{t \in Out(q_0)} x_t)$. It can be checked that for each valuation ν of the variables in X, ν satisfies $[A]$ if and only if there exists an accepting run ρ of A such that $\nu = \nu_\rho$.

It is well known that every finite-state automaton has a characteristic Presburger formula due to Parikh's theorem [14]. However, the formula we give above is simpler and exploits the particular structure of simple automata.

3 Communicating Finite-State Machines

We consider a generalization of communicating finite-state machines (CFSM) defined in [4]. A CFSM is a finite-state machine which can send and receive messages over a finite set of unbounded FIFO queues. Usually, a transition either appends a message to the end of a queue or removes a message from the head of a queue. We generalize this by allowing simultaneously appending and removing messages from several queues.

Formally, a *Communicating Finite-State Machine* $\mathcal{M}$ is a tuple (S, K, Σ, T) where S is a finite set of *control states*, K is a finite set of unbounded FIFO queues, Σ is a finite set of messages, T is a finite set of transitions. Each transition is of the form (s_1, op, s_2), where s_1 and $s_2 \in S$, and op is a finite set of *queue operations* of the form $\kappa_i ! w$ or $\kappa_i ? w$ with $\kappa_i \in K$ and $w \in \Sigma^*$ such that for each queue κ_i there is at most one label $\kappa_i ! w$ or $\kappa_i ? w$ in op.

A *configuration* of $\mathcal{M}$ is a tuple $\gamma = (s, \vec{w})$ where s is a control state in S, and $\vec{w} = (w_1, \ldots, w_{|K|})$ is a $|K|$-dim multi-word (i.e., a tuple in $(\Sigma^*)^{|K|}$), each w_i being the contents of the queue κ_i, for $i \in \{1, \ldots, |K|\}$. We denote by $Conf$ the set of all configurations of $\mathcal{M}$, i.e., $Conf = S \times (\Sigma^*)^{|K|}$.

We define a *global transition relation* between configurations in the following manner: Let $\gamma = (s, w_1, \ldots, w_{|K|})$ and $\gamma' = (s', w'_1, \ldots, w'_{|K|})$ be two configurations, and let op be a set of queue operations. Then, we have $\gamma \stackrel{op}{\to} \gamma'$ if and only if there exists a transition $(s_1, op, s_2) \in T$ such that, for every $i \in \{1, \ldots, |K|\}$,

- if $\kappa_i?w \in op$ then $ww'_i = w_i$ else if $\kappa_i!w \in op$ then $w'_i = w_iw$,
- otherwise $w'_i = w_i$.

Given a transition $\tau = (s, op, s') \in T$, we say that τ is *executable* at $\gamma = (s, \vec{w})$ if there exists $\gamma' = (s', \vec{w}')$ such that $\gamma \stackrel{op}{\to} \gamma'$. In this case, γ' (resp. γ) is the *immediate successor* (resp. *predecessor*) of γ (resp. γ') by τ. We define the predecessor and successor functions pre_τ and $post_\tau$, both in $2^{Conf} \to 2^{Conf}$, such that, for every set of configurations C, $pre_\tau(C)$ (resp. $post_\tau(C)$) is the set of immediate predecessors (resp. successors) of the configurations in C by τ. The pre (resp. $post$) function is defined as the union of the functions pre_τ (resp. $post_\tau$), for all $\tau \in T$. The notion of *executability* can be generalized to sequences of transitions, in particular to *cycles* in the transition graph T. A sequence θ of transitions in T is called a *cycle* if it is of the form $(s_0, op_0, s_1)(s_1, op_1, s_2) \cdots (s_{n-1}, op_n, s_0)$.

The definitions of pre_τ and $post_\tau$ can also be generalized to sequences of transitions: $pre_{\tau_1 \ldots \tau_n} = pre_{\tau_1} \circ \ldots \circ pre_{\tau_n}$ and $post_{\tau_1 \ldots \tau_n} = post_{\tau_n} \circ \ldots \circ post_{\tau_1}$.

Given a sequence of transitions θ, the functions pre_θ^* and $post_\theta^*$ are the reflexive transitive closures of pre_θ and $post_\theta$, i.e. given a set of configuration C, $pre_\theta^*(C)$ (resp. $post_\theta^*(C)$) is the set of predecessors (resp. successors) of configurations in C obtained by iterating an arbitrary number of times θ.

We define the functions pre^* and $post^*$ as the reflexive transitive closures of pre and $post$. The function pre^* (resp. $post^*$) yields the set of all predecessors (resp. successors) of a given set of configurations.

4 Constrained Queue Description Diagrams

In this section we introduce representation structures for sets of queue contents. These structures consist of a combination of finite-state automata (restricted deterministic simple automata) with linear constraints on the number of times transitions in these automata are taken. This combination allows to represent nonregular sets of queue contents.

4.1 Definition

Constrained Queue Description Diagrams (CQDD's) are a particular case of *constrained simple automata*. For any $n \geq 1$, a n-dim constrained simple automaton (CSA) is a set of accepting components $\mathcal{C} = \{\langle \mathcal{A}_1, f_1 \rangle, \ldots, \langle \mathcal{A}_m, f_m \rangle\}$ where, for every $i \in \{1, \ldots, m\}$, $\mathcal{A}_i$ is a tuple of n simple automata $(A_1^i, \ldots, A_n^i)$ over Σ and f_i is a Presburger formula over a set of variables V_i containing the set $X_i = \{x_t : t \in \mathcal{T}_i\}$, where $\mathcal{T}_i$ is the set of all the transitions of the automata in $\mathcal{A}_i$, i.e., $\mathcal{T}_i = \bigcup_{j=1}^n \to_j^i$.

The CSA $\mathcal{C}$ accepts a n-dim *multi-language*, i.e., a set of tuples of n words. For every $i \in \{1, \ldots, m\}$, the multi-language of the accepting component $\langle \mathcal{A}_i, f_i \rangle$, denoted by $L(\langle \mathcal{A}_i, f_i \rangle)$ is the set of tuple of words $(w_1, \ldots, w_n) \in (\Sigma^*)^n$ for which there are accepting runs $(\rho_1, \ldots, \rho_n)$ of the automata $(A_i^1, \ldots, A_i^n)$ respectively, such that $\exists (V_i \setminus X_i)$. f_i is satisfied by the valuation $(\nu_{\rho_1}, \ldots, \nu_{\rho_n})$ (i.e., the valuation associating with each variable x_t the integer $|\rho_1 \ldots \rho_n|_t$, where $t \in \mathcal{T}_i$). The multi-language of the CSA $\mathcal{C}$, denoted by $L(\mathcal{C})$, is the union $\bigcup_{i=1}^m L(\langle \mathcal{A}_i, f_i \rangle)$.

A n-dim CDSA is a n-dim CSA such that all its automata are deterministic. A n-dim CQDD is a n-dim CDSA such that all its automata are restricted (DRSA's). For every $n \geq 1$, we denote by n-CQDD (resp. n-CDSA) the class of all n-dim CQDD's (resp. n-dim CDSA's). We say that a n-dim multi-language $\mathcal{L}$ is CQDD (resp. CDSA) *definable* if there exists a n-dim CQDD (resp. CDSA) $\mathcal{C}$ such that $L(\mathcal{C}) = \mathcal{L}$. A n-dim multi-language $\mathcal{L}$ is CQDD (resp. CDSA) *reverse definable* if its reverse image, denoted by $\mathcal{L}^R$, is CQDD (resp. CDSA) *definable*.

4.2 Expressiveness

CQDD's allow to define nonregular multi-languages. For instance, consider the context-sensitive language $L_1 = \{a^n b^m a^n b^m : n, m \geq 1\}$. To define L_1, we use the automaton A_1 represented by the following picture:

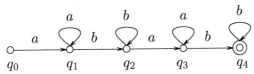

Then, L_1 is defined by the 1-dim CQDD $\{\langle A_1, f_1 \rangle\}$ where f_1 is given by $x_{(q_1,a,q_1)} = x_{(q_3,a,q_3)} \wedge x_{(q_2,b,q_2)} = x_{(q_4,b,q_4)}$. Consider the 2-dim multi-language $L_2 = \{(a^n b^m a^n b^m, c^m d^n a^m) : n, m \geq 1\}$. To define this multi-language, we use two automata, the automaton A_1 above and A_2 given by the following picture:

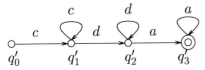

Then, L_2 is defined by the 2-dim CQDD $\{\langle (A_1, A_2), f_2 \rangle\}$ where f_2 is given by $(x_{(q_1,a,q_1)} = x_{(q_3,a,q_3)} = x_{(q_2',d,q_2')} = n) \wedge (x_{(q_2,b,q_2)} = x_{(q_4,b,q_4)} = x_{(q_1',c,q_1')} = x_{(q_3',a,q_3')} = m)$.

These examples show that CQDD's can be used to express nonregular multi-languages involving constraints on number of occurrences of symbols at some positions that may be in a same word (as in L_1), or even in different words (as in L_2). This allows to represent sets of queue contents such that there are counting constraints relating the contents of different queues.

4.3 Basic operations and decision problems

Here we give the main results about boolean operations on CQDD's. We show that they are closed under union, intersection, concatenation and left-derivation, but their complementation yields CDSA's. Concatenation and left-derivation are useful operations in the construction of sets of successors and predecessors (see

Section 5). Moreover, the intersection of a CQDD with a CDSA is a CQDD. Finally, we show that the membership and inclusion are decidable for CQDD's.

Let $\mathcal{L}_1$ and $\mathcal{L}_2$ be two n-dim multi-languages. The *concatenation* of $\mathcal{L}_1$ and $\mathcal{L}_2$, denoted by $\mathcal{L}_1 \cdot \mathcal{L}_2$, is the set $\{\vec{w} \in (\Sigma^*)^n : \exists u \in \mathcal{L}_1. \exists v \in \mathcal{L}_2. \vec{w} = \vec{u}\vec{v}\}$ where $\vec{u}\vec{v}$ is the component-wise concatenation of $\vec{u}$ and $\vec{v}$. The *left-derivative* of $\mathcal{L}_1$ by $\mathcal{L}_2$, denoted by $\mathcal{L}_2^{-1} \cdot \mathcal{L}_1$, is the set $\{\vec{w} \in (\Sigma^*)^n : \exists \vec{w}' \in \mathcal{L}_2. \vec{w}'\vec{w} \in \mathcal{L}_1\}$, i.e., the set of multi-words allowing to extend elements of $\mathcal{L}_2$ to elements of $\mathcal{L}_1$.

Proposition 4.1 *For every $n \geq 1$, n-CQDD is closed under union, intersection, concatenation and left-derivation.*

It can be observed that the product of a DRSA with a DSA is a DSA. Hence:

Proposition 4.2 *For every $n \geq 1$, the intersection of an n-CQDD with an n-CDSA is an n-CQDD.*

Because the simple automata in a CQDD are deterministic we can show:

Proposition 4.3 *For every $n \geq 1$, the complement of a n-CQDD is a n-CDSA.*

Let $\mathcal{C} = \{\langle A, f \rangle\}$ be a CSA. Clearly, $L(\mathcal{C}) \neq \emptyset$ if and only if the Presburger formula $[A] \wedge f$ is satisfiable. Hence:

Proposition 4.4 *The emptiness problem is decidable for CSA's.*

From Propositions 4.2, 4.3, and 4.4 we deduce:

Corollary 4.1 *For every $n \geq 1$, the membership problem as well as the inclusion problem are decidable for n-dim CQDD's.*

5 Representing and manipulating sets of configurations

Let $\mathcal{M} = (S, K, \Sigma, T)$ be a CFSM. Every set of configurations $C \subseteq Conf$ can be written as a union $\bigcup_{s \in S} \{s\} \times \mathcal{L}_s$ where the $\mathcal{L}_s$'s are $|K|$-dim multi-languages. We say that C is CQDD *representable* (resp. *reverse representable*) if for every $s \in S$, the multi-language $\mathcal{L}_s$ is CQDD definable (resp. reverse definable). Let us consider as an example the system $\mathcal{M}$:

$$\{\kappa_1!a\} \overset{\{\kappa_1?a, \kappa_2!b\}}{\underset{\{\kappa_2!a, \kappa_3!a\}}{\rightleftarrows}} s_1$$

The set of configurations of $\mathcal{M}$ reachable from the $(s_0, \epsilon, \epsilon)$ is given by:

$$\{s_0\} \times \{(a^n, (ba)^m, a^m) : n, m \geq 0\} \cup \{s_1\} \times \{(a^n, (ba)^m b, a^m) : n, m \geq 0\} \quad (1)$$

and is clearly CQDD representable.

In the sequel, we present results allowing to manipulate and to reason about sets of configurations that are CQDD representable or reverse representable. First of all, by Propositions 4.1, 4.4 and Corollary 4.1, we deduce:

Theorem 5.1 *The class of CQDD representable (resp. reverse representable) sets of configurations is effectively closed under union and intersection, and has decidable emptiness, membership and inclusion problems.*

The closure property under concatenation and left-derivation of CQDD's (Proposition 4.1) allows us to show:

Theorem 5.2 *For every CQDD representable (resp. reverse representable) set of configurations C, the set of configurations $post(C)$ (resp. $pre(C)$) is CQDD representable (resp. reverse representable) and effectively constructible.*

Now, we give our main results.

Theorem 5.3 *For every CQDD representable set of configurations C, and every cycle θ, the set of configurations $post_\theta^*(C)$ is CQDD representable and effectively constructible.*

We give hereafter a rough scheme of the proof: Let C be a set configurations given by a m-dim CQDD of the form $\{\langle(A_1,\ldots,A_m),f\rangle\}$ (this is not a restriction since $post$ is distributive w.r.t. union). The principle of the construction is to compute the effect of n successive executions of the cycle θ on each queue, n being a parameter.

Then, for every $i \in \{1,\ldots,m\}$, we construct several automata A_i' and constraints g_i. These constraints depend on n (considered as a new free variable), and relate variables corresponding to the transitions of A_i' with those corresponding to the transitions of A_i. The set $post_\theta^*(C)$ is then represented by a union of CQDD's $\{\langle(A_1',\ldots,A_m'),f \wedge \bigwedge_{i=1}^m g_i\rangle\}$. Note that, since all the g_i's depend on the variable n, this expresses the fact that the number of executions of θ must be the same for every queue.

The construction of the automaton A_i' and the constraint g_i is done by identifying the configurations from which the cycle θ can be executed an unbounded number of times and those allowing only a bounded number of executions. Then, we show that in each case, A_i' and g_i are obtained from A_i using basic operations on CQDD's such as concatenation and left-derivation.

We can also prove that the class of CQDD reverse representable sets of configurations is effectively closed under the pre_θ^* function, for every cycle θ.

Theorem 5.4 *For every CQDD reverse representable set of configurations C, and every cycle θ, the set of configurations $pre_\theta^*(C)$ is CQDD reverse representable and effectively constructible.*

6 Forward and backward reachability analysis

The basic (safety) verification problem consists in checking that a *bad* configuration can never be reached from an initial configuration. Thus, given a set of initial configurations I and a set of bad configurations B, this problem can be formulated either as

- (P1) $B \cap post^*(I) = \emptyset$, or
- (P2) $I \cap pre^*(B) = \emptyset$.

The first formulation consists of a forward reachability analysis of the configuration space whereas the second one consists of a backward reachability analysis. Hence, given a set of configurations C, we wish to compute the set of its successors and predecessors, i.e., $post^*(C)$ and $pre^*(C)$. By definition, for $\phi \in \{post, pre\}$, we have

$$\phi^*(C) = \bigcup_{i \geq 0} C_i$$

where

$$C_0 = C$$
$$C_{i+1} = C_i \cup \phi(C_i) \quad \text{for every } i \geq 0.$$

In the case $\phi = post$ (resp. $\phi = pre$), if C is CQDD representable (resp. reverse representable), it can be deduced from Theorems 5.1 and 5.2 that all the C_i's are CQDD representable (resp. reverse representable). Hence, the equations above yield a semi-algorithm for calculating $\phi^*(C)$ based on a iterative calculation of the C_i's. Since the sequence of the C_i's is increasing, the limit is reached if for some i we have $C_i = C_{i+1}$. Then the algorithm stops and returns C_i. We can detect this since the inclusion problem is decidable for CQDD representable (resp. reverse representable) set of configurations (by Theorem 5.1). Then, if the set of initial and bad states is also CQDD representable (resp. reverse representable), the problem P1 (resp. P2) above can be solved by Theorem 5.1.

Of course, since the reachability problem is undecidable for CFSM's, an index i such that $C_i = C_{i+1}$ does not exist in general, and the (naive) algorithm described above may never stop.

We propose to tackle this divergence problem by performing an "*exact acceleration*" of the iterative calculation of the limit $\phi^*(C)$. The idea is as follows: Given a set of cycles in the transition graph of the system, say Θ, add at each step the set of successors (or predecessors) by each of the cycles in Θ. This operation is sound (exact) since all the added configurations belongs to $\phi^*(C)$. So, we compute $\phi^*(C)$ as the limit of another increasing sequence of configurations $(D_i)_{i \geq 0}$ given by:

$$D_0 = C$$
$$D_{i+1} = D_i \cup \phi(D_i) \cup \bigcup_{\theta \in \Theta} \phi_\theta^*(D_i) \quad \text{for every } i \geq 0$$

Clearly, for every $i \geq 0$, we have $C_i \subseteq D_i$. Hence, the chance to reach the limit $\phi^*(C)$ in a finite number of steps is greater (or at least equal) by considering the D_i's instead of the C_i's, and this chance should increase with the size of Θ.

Therefore, using Theorems 5.1, 5.2, 5.3 and 5.4, we obtain a generic reachability analysis semi-algorithm which computes (when it terminates) the exact set of successors (resp. predecessors) of a given CQDD representable (resp. reverse representable) set of configurations. This algorithm is given by:

Reachability (Θ, C):
   $X := C$;
   repeat
      $Y := X$;
      $X := X \cup \phi(X) \cup \bigcup_{\theta \in \Theta} \phi_\theta^*(X)$
   until $X = Y$;
   return(X)
end Reachability

A variety of reachability algorithms can be derived from the generic algorithm above by determining adequate strategies for choosing the set of cycles Θ.

For instance, the forward reachability analysis algorithm given in [5] can be seen as a possible instance of our algorithm[1]. Indeed, in [5] the authors consider the set of cycles that are of one of the following three forms: $(s, \{\kappa!w\}, s)$, $(s, \{\kappa?w\}, s)$, or $(s, \{\kappa_1?w\}, s')(s', \{\kappa_2!w\}, s)$. These kind of cycles do not introduce counting constraints on queue contents. Hence, starting from a regular set of configurations (finite-state automata definable), the set of reachable configurations by these cycles is also regular. Then, a representation structure based only on finite-state automata (QDD's) can be used and allows to analyze some significant systems. But considering QDD's and only cycles of the form specified above does not allow to reason about systems with nonregular sets of configurations like the system $\mathcal{M}$ given in Section 5. However, it is easy to see that our algorithm terminates and computes the exact set of configurations of the system $\mathcal{M}$ (given by 1) if we consider as Θ the set of the two elementary cycles $(s_0, \{\kappa_1!a\}, s_0)$, and $(s_0, \{\kappa_1?a, \kappa_2!b\}, s_1)(s_1, \{\kappa_2!a, \kappa_3!a\}, s_0)\}$.

7 Conclusion

We have applied the symbolic analysis principle to fifo-channel systems (communicating finite state machines). These systems have in general nonregular sets of configurations. We have proposed a representation structure for their sets of configurations combining finite-state automata with counting constraints expressed in Presburger arithmetics. We have shown that this structures allow to compute the exact effect of the repeated execution of any fixed cycle in the transition graph of a system. We have defined a generic reachability analysis semi-algorithm which is parametrized by a set of cycles. This semi-algorithm computes iteratively the set of successors (or predecessors) by considering these cycles as additional "meta-transitions" in the graph, following the approach adopted in [6, 5].

It can be seen that our reachability analysis procedure computes a fixpoint of a function on set of configurations which is of a very particular form. Actually, this procedure can be generalized to a model-checking procedure for any positive fixpoint formula constructed using disjunctions, conjunctions, and the successor (predecessor) function, starting from basic CQDD (reverse) representable sets.

---

[1] In the definition of QDD's, any deterministic finite-state automata can be used. However, it can be checked that, starting from an initial configuration with empty queues, the constructed QDD is a union of DRSA's.

References

1. P. Abdulla and B. Jonsson. Verifying programs with unreliable channels. *Information and Computation*, 127:91–101, 1996.
2. R. Alur, C. Courcoubetis, N. Halbwachs, T. Henzinger, P. Ho, X. Nicollin, A. Olivero, J. Sifakis, and S. Yovine. The Algorithmic Analysis of Hybrid Systems. *TCS*, 138, 1995.
3. R. Alur and D. Dill. A Theory of Timed Automata. *TCS*, 126, 1994.
4. G.V. Bochmann. Finite State Description of Communication Protocols. *Computer Networks*, 2, October 1978.
5. B. Boigelot and P. Godefroid. Symbolic Verification of Communication Protocols with Infinite State Spaces using QDDs. In *CAV'96*. LNCS 1102, 1996.
6. B. Boigelot and P. Wolper. Symbolic Verification with Periodic Sets. In *CAV'94*. LNCS 818, 1994.
7. A. Bouajjani, J. Esparza, and O. Maler. Reachability Analysis of Pushdown Automata: Application to Model Checking. In *CONCUR '97*, 1997.
8. A. Bouajjani and P. Habermehl. Symbolic reachability analysis of fifo-channel systems with nonregular sets of configurations. 1997. full version available at http://www.imag.fr/VERIMAG/PEOPLE/Peter.Habermehl.
9. A. Bouajjani and O. Maler. Reachability Analysis of Pushdown Automata. In *Infinity'96*. tech. rep. MIP-9614, Univ. Passau, 1996.
10. D. Brand and P. Zafiropulo. On communicating finite-state machines. *JACM*, 2(5):323–342, 1983.
11. O. Burkart and Y.M. Quemener. Model-Checking of Infinite Graphs Defined by Graph Grammars. In *Infinity'96*. tech. rep. MIP-9614, Univ. Passau, 1996.
12. Gérard Cécé, Alain Finkel, and S. Iyer. Unreliable Channels Are Easier to Verify Than Perfect Channels. *Information and Computation*, 124(1):20–31, 1996.
13. S. Melzer and J. Esparza. Checking System Properties via Integer Programming. In *ESOP'96*. LNCS 1058, 1996.
14. R.J. Parikh. On Context-Free Languages. *JACM*, 13, 1966.
15. W. Peng and S. Purushotaman Iyer. Data Flow Analysis of Communicating Finite State Machines. *ACM Transactions on Programming Languages and Systems*, 13(3):399–442, July 1991.
16. Y.M. Quemener and T. Jéron. Finitely Representing Infinite Reachability Graphs of CFSMs with Graph Grammars. In *FORTE/PSTV'96*. Chapman and Hall, 1996.

Axiomatizations for the Perpetual Loop in Process Algebra

Wan Fokkink

University of Wales Swansea
Department of Computer Science
Singleton Park, Swansea SA2 8PP, Wales
e-mail: w.j.fokkink@swan.ac.uk

Abstract. Milner proposed an axiomatization for the Kleene star in basic process algebra, in the presence of deadlock and empty process, modulo bisimulation equivalence. In this paper, Milner's axioms are adapted to no-exit iteration x^ω, which executes x infinitely many times in a row, and it is shown that this axiomatization is complete for no-exit iteration in basic process algebra with deadlock and empty process, modulo bisimulation.

1 Introduction

Kleene [15] defined a binary operator x^*y in the context of finite automata, which denotes the iterate of x on y. Intuitively, the expression x^*y can choose to execute either x, after which it evolves into x^*y again, or y, after which it terminates. A feature of the Kleene star is that on the one hand it can express recursion, while on the other hand it can be captured in equational laws. Hence, one does not need meta-principles such as the Recursive Specification Principle [10]. Kleene formulated several equations for his operator, notably the defining equation $x^*y = x(x^*y) + y$. In later years it became more fashionable to consider the unary version x^* of the Kleene star. In the presence of the empty process, the unary and the binary Kleene star are equally expressive.

Salomaa [22] presented a finite complete axiomatization for the Kleene star in language theory, modulo completed trace equivalence, which incorporates one conditional axiom, namely, if $x = y \cdot x + z$, and y cannot terminate immediately, then $x = y^*z$. Salomaa's completeness proof basically consists of two steps: first he shows that the solutions of a guarded recursive specification are all provably equal to the same term, and next he shows that if two terms are completed trace equivalent, then there exists a guarded recursive specification for which both terms are solutions.

Milner [17] was the first to study the (unary) Kleene star modulo bisimulation, and proposed an axiomatization for it, being an adaptation of Salomaa's axiom system. Milner [17, page 461] raised the question whether his axiomatization is complete for the Kleene star in process theory, and remarked that this question may be hard to answer: "The difficulty is that the method [...] of Salomaa's original completeness proof cannot be applied directly, since -in contrast with the case of languages- an arbitrary system of guarded equations [...] cannot in general be solved in star expressions".

In this paper the instantiation $x^*\delta$ of the binary Kleene star is studied, which carries two names: perpetual loop and no-exit iteration. Since the deadlock δ blocks the exits, this construct executes x an infinite number of times in a row. The perpetual loop is closely related to the Kleene star, and shares several of its characteristics. In this paper no-exit iteration, which is denoted by x^ω, is studied in Basic Process Algebra [9] with deadlock and empty process, denoted by $\mathrm{BPA}^\omega_{\delta\varepsilon}(A)$. No-exit iteration can be used to formally describe programs that repeat a certain procedure without end. A significant advantage of iteration over recursion as a means to express infinite processes is that does it not involve a parametric process definition, because the development of process theory is easier if parametrization does not have to be taken as primitive (see e.g. Milner [18, page 212]). Since the syntax of process algebra with iteration has an inductive term structure, it allows simpler axiomatizations than recursion, and it does not need a guardedness restriction to locate the class of meaningful terms. Therefore, the Kleene star is used for example in the specification and verification of Grid protocols [7], which describe parallel computations in a grid-like architecture, and in the ToolBus [8], which enables to link separate tools. In both cases, iteration is used almost exclusively in the form of the perpetual loop. No-exit iteration is also used in the educational vein [21], because it enables to specify and verify infinite processes in a simple and intuitive way.

The three axioms for the unary Kleene star in Milner's axiom system (being Kleene's defining equation, Salomaa's conditional axiom and an equation which describes the interplay of Kleene star and empty process) have obvious counterparts for no-exit iteration. It turns out that these three axioms, together with the standard axioms for $BPA_{\delta\varepsilon}^{\omega}(A)$, make a complete axiomatization for $BPA_{\delta\varepsilon}^{\omega}(A)$ modulo bisimulation. The completeness proof is based on a strategy that originates from [11]. It also uses new techniques, which will hopefully turn out to be applicable in a possible proof of Milner's conjecture (see Section 4 for a discussion on this topic). For a detailed presentation of the completeness proof for $BPA_{\delta\varepsilon}^{\omega}(A)$, and for omitted proofs in this paper, the reader is referred to [12].

This paper focuses on the process algebra $BPA_{\delta}^{\omega}(A)$, in which the empty process is not present. This setting allows a more concise presentation of the ideas that are used in the completeness proof for the perpetual loop in process algebra. We will see that Kleene's defining equation and Salomaa's conditional axiom for the the perpetual loop, together with the standard axioms for $BPA_{\delta}(A)$, are complete for $BPA_{\delta}^{\omega}(A)$ modulo bisimulation.

Sewell [23] proved that there does not exist a complete finite equational axiomatization for the Kleene star in combination with deadlock modulo bisimulation, due to the fact that a^{ω} is bisimilar to $(a^n)^{\omega}$ for $n = 1, 2, \ldots$. Since these equivalences are also present in $BPA_{\delta\varepsilon}^{\omega}$, Sewell's argument can be copied to conclude that there does not exist a complete finite equational axiomatization for $BPA^{\omega}(A)$. Hence, the adaptation of Salomaa's conditional axiom for the perpetual loop is essential for the obtained completeness results.

The requirement 'y cannot terminate immediately' in Salomaa's conditional axiom can be defined inductively on the syntax. According to Kozen [16] this requirement is not algebraic, in the sense that it is not preserved under substitution of terms for actions. He proposed two alternative conditional axioms which do not have this drawback. These axioms, however, are not sound with respect to bisimulation equivalence.

Bergstra, Bethke and Ponse [6] suggested a finite equational axiomatization for BPA^*, i.e, for basic process algebra with the binary Kleene star without the special constants δ and ϵ, modulo bisimulation. Their conjecture that it is complete was solved by Fokkink and Zantema [14]. (In contrast with this result, Aceto, Fokkink and Ingólfsdóttir [3] showed that there does not exist a complete finite equational axiomatization for BPA^* modulo any process semantics in between ready simulation and completed traces.) In [11], a new proof for the completeness result from [14] was presented. This new proof technique was was applied successfully not only in this paper, but also in a paper on a restricted version of iteration called prefix iteration, which is better suited for a setting with prefix multiplication or with communication [2], and in a paper on a more expressive variant of iteration called multi-exit iteration [1].

Acknowledgements. This research was initiated by a question from Alban Ponse. Luca Aceto, Jaco van de Pol, Alban Ponse and an anonymous referee provided useful comments, and Jan Bergstra is thanked for stimulating discussions.

2 The Perpetual Loop in Process Algebra

2.1 Syntax

We assume a non-empty alphabet A of atomic actions, with typical elements a, b, c. We also assume two special constants δ, which represents deadlock, and ε, which represents the empty process, and ξ ranges over $A \cup \{\delta, \varepsilon\}$. Furthermore, we have two binary operators: alternative composition $x + y$, which combines the behaviours of x and y, and sequential composition $x \cdot y$, which puts the behaviours of x and y in sequence. Finally, we have the unary operator x^ω, which executes x infinitely many times in a row. We will refer to this operator both as *perpetual loop* and as *no-exit iteration*. The language $\text{BPA}^\omega_{\delta\varepsilon}(A)$, with typical elements $p, q, ..., w$, consists of all the terms that can be constructed from the atomic actions, the two special constants, the two binary composition operators, and the perpetual loop. That is, the BNF grammar for the collection of process terms is:

$$p ::= a \mid \delta \mid \varepsilon \mid p + p \mid p \cdot p \mid p^\omega.$$

$\text{BPA}^\omega_\delta(A)$ is obtained by deleting the empty process ε, and $\text{BPA}^\omega(A)$ is obtained by deleting the deadlock δ and the empty process ε from the syntax. The sequential composition operator will often be omitted, so pq denotes $p \cdot q$. As binding convention, alternative composition binds weaker than sequential composition and no-exit iteration.

Remark: The presence of the special constant δ in $\text{BPA}^\omega_{\delta\varepsilon}(A)$ is redundant, because it can be expressed in $\text{BPA}^\omega_\varepsilon(A)$ modulo bisimulation: ε^ω is bisimilar with δ, because both processes do not exhibit any behaviour. However, δ is maintained in the syntax as a standard abbreviation.

2.2 Operational Semantics

Table 1 presents an operational semantics for $\text{BPA}^\omega_{\delta\varepsilon}(A)$ in Plotkin style [20], where $x \xrightarrow{a} x'$ represents that process x can evolve into process x' by the execution of action a, and $x \xrightarrow{a} \surd$ denotes that process x can terminate by the execution of action a, and the unary predicate $x \xrightarrow{\varepsilon} \surd$ denotes that process x can terminate immediately.

$$\varepsilon \xrightarrow{\varepsilon} \surd \qquad\qquad a \xrightarrow{a} \surd$$

$$\frac{x \xrightarrow{\xi} \surd}{x + y \xrightarrow{\xi} \surd \quad y + x \xrightarrow{\xi} \surd} \qquad \frac{x \xrightarrow{a} x'}{x + y \xrightarrow{a} x' \quad y + x \xrightarrow{a} x'}$$

$$\frac{x \xrightarrow{\varepsilon} \surd \;\; y \xrightarrow{\xi} \surd}{x \cdot y \xrightarrow{\xi} \surd} \qquad \frac{x \xrightarrow{\varepsilon} \surd \;\; y \xrightarrow{a} y'}{x \cdot y \xrightarrow{a} y'} \qquad \frac{x \xrightarrow{a} \surd}{x \cdot y \xrightarrow{a} y} \qquad \frac{x \xrightarrow{a} x'}{x \cdot y \xrightarrow{a} x' \cdot y}$$

$$\frac{x \xrightarrow{a} \surd}{x^\omega \xrightarrow{a} x^\omega} \qquad\qquad \frac{x \xrightarrow{a} x'}{x^\omega \xrightarrow{a} x' \cdot (x^\omega)}$$

Table 1. Transition rules for $\text{BPA}^\omega_{\delta\varepsilon}(A)$

Definition 1. p' is a *derivative* of p if p can evolve into p' by zero or more transitions. p' is a *proper* derivative of p if p can evolve into p' by one or more transitions.

Note that a process term can be a proper derivative of itself, for example, $a^*b \xrightarrow{a} a^*b$. In the sequel, p' and p'' will denote derivatives of process term p. The following lemma can easily be deduced, using structural induction.

Lemma 2. *Each process term in* $\mathrm{BPA}^\omega_{\delta\varepsilon}(A)$ *has only finitely many derivatives.*

Process terms are considered modulo bisimulation equivalence from Park [19]. Intuitively, two processes are bisimilar if they have the same branching structure.

Definition 3. Two processes p and q are *bisimilar*, denoted by $p \leftrightarrow q$, if there exists a symmetric binary relation B on processes which relates p and q, such that:

- if $r\ B\ s$ and $r \xrightarrow{a} r'$, then there is a transition $s \xrightarrow{a} s'$ such that $r'\ B\ s'$;
- if $r\ B\ s$ and $r \xrightarrow{\xi} \sqrt{}$, then $s \xrightarrow{\xi} \sqrt{}$.

Bisimulation equivalence is a congruence with respect to all the operators, which means that if $p \leftrightarrow p'$ and $q \leftrightarrow q'$, then $p+q \leftrightarrow p'+q'$ and $pq \leftrightarrow p'q'$ and $p^\omega \leftrightarrow (p')^\omega$. Namely, the transition rules in Table 1 are in the 'path' format, which guarantees that the generated bisimulation equivalence is a congruence, see [5, 13].

2.3 Axiomatizations

Table 2 presents the standard axioms A1-9 for $\mathrm{BPA}_{\delta\varepsilon}(A)$. Furthermore, Table 3 contains the defining equation NEI1 together with the conditional axiom RSP^ω for the perpetual loop. The axiomatization A1-7+NEI1+RSP^ω is sound for $\mathrm{BPA}^\omega_\delta(A)$, i.e., if $p = q$ in $\mathrm{BPA}^\omega_\delta(A)$ is provable from these axioms, then $p \leftrightarrow q$. Since bisimulation equivalence is a congruence for $\mathrm{BPA}^\omega_\delta(A)$, soundness can be verified by checking this property for each axiom separately, which is left to the reader.

$$
\begin{array}{ll}
\text{A1} & x + y = y + x \\
\text{A2} & (x + y) + z = x + (y + z) \\
\text{A3} & x + x = x \\
\text{A4} & (x + y) \cdot z = x \cdot z + y \cdot z \\
\text{A5} & (x \cdot y) \cdot z = x \cdot (y \cdot z) \\
\text{A6} & x + \delta = x \\
\text{A7} & \delta \cdot x = \delta \\
\text{A8} & x \cdot \varepsilon = x \\
\text{A9} & \varepsilon \cdot x = x
\end{array}
$$

Table 2. The axioms for $\mathrm{BPA}_{\delta\varepsilon}(A)$

$$
\begin{array}{ll}
\text{NEI1} & x \cdot (x^\omega) = x^\omega \\
\text{RSP}^\omega & x = y \cdot x \implies x = y^\omega
\end{array}
$$

Table 3. The axioms for the perpetual loop in the absence of ε

However, the axiom RSP$^\omega$ is not sound in the presence of the empty process. Namely, due to the axiom A9, $x = \varepsilon x$, it then implies $x = \varepsilon^\omega$, which is clearly unsound. Therefore, in Table 4 an adaptation RSP$^\omega_\varepsilon$ is introduced, where the condition $y \not\downarrow$ expresses that y cannot terminate immediately. This condition, which is similar to the so-called guardedness restriction in the Recursive Specification Principle from Bergstra and Klop [10], can be defined inductively on the syntax:

$$a \not\downarrow$$
$$\delta \not\downarrow$$
$$x \not\downarrow \wedge y \not\downarrow \Rightarrow (x + y) \not\downarrow$$
$$x \not\downarrow \vee y \not\downarrow \Rightarrow (x \cdot y) \not\downarrow$$
$$(x^\omega) \not\downarrow$$

Table 4 contains the defining equation NEI1, and an extra equation NEI2 which describes the interplay of no-exit iteration with the empty process. The axiomatization A1-9+NEI1,2+RSP$^\omega_\varepsilon$ is sound for BPA$^\omega_{\delta\varepsilon}(A)$.

$$\text{NEI1} \quad x \cdot (x^\omega) = x^\omega$$
$$\text{NEI2} \quad (x + \varepsilon)^\omega = x^\omega$$
$$\text{RSP}^\omega_\varepsilon \quad x = y \cdot x \wedge y \not\downarrow \implies x = y^\omega$$

Table 4. The axioms for the perpetual loop in the presence of ε

The purpose of this paper is to present the following three completeness results.

Theorem 4. *The axiomatization A1-9+NEI1,2+RSP$^\omega_\varepsilon$ is complete for BPA$^\omega_{\delta\varepsilon}(A)$ with respect to bisimulation.*

That is, if $p \leftrightarrow q$ for process terms p and q in BPA$^\omega_{\delta\varepsilon}(A)$, then $p = q$ can be derived from the axioms A1-9+NEI1,2+RSP$^\omega_\varepsilon$.

Theorem 5. *The axiomatization A1-7+NEI1+RSP$^\omega$ is complete for BPA$^\omega_\delta(A)$ with respect to bisimulation.*

Theorem 6. *The axiomatization A1-5+NEI1+RSP$^\omega$ is complete for BPA$^\omega(A)$ with respect to bisimulation.*

This paper focuses on the completeness proof for BPA$^\omega_\delta(A)$. The completeness proof for BPA$^\omega(A)$ is closely related to the one for BPA$^\omega_\delta(A)$ (missing only some minor cases for δ in the construction of basic terms in Lemma 17). The completeness proof for BPA$^\omega_{\delta\varepsilon}(A)$ also uses the same proof strategy, but, due to the presence of the empty process, the technical details are considerably more complicated. The reader is referred to [12] for a detailed exposition on the completeness proof for BPA$^\omega_{\delta\varepsilon}(A)$.

3 Proof of the Main Theorem

This section presents preliminaries that are needed in the proof of Theorem 5, together with the completeness proof itself. Many preliminary definitions in this section originate from [11]. For omitted proofs the reader is referred to [12].

3.1 Expansions

From now on, process terms in $\text{BPA}^\omega_\delta(A)$ are considered modulo associativity and commutativity of the $+$, that is, modulo the axioms A1,2. We write $p =_{AC} q$ if p and q can be equated by axioms A1,2. As usual, $\sum_{i=1}^n p_i$ represents the term $p_1 + \ldots + p_n$, and the p_i are called the summands of this term. The empty sum represents δ, where $\sum_{i\in\emptyset} p_i + q$ is not considered empty.

Definition 7. For each process term p, its collection of possible transitions is finite, say $\{p \xrightarrow{a_i} p_i \mid i = 1, \ldots, n\} \cup \{p \xrightarrow{b_j} \sqrt{} \mid j = 1, \ldots, m\}$. The *expansion* of p is

$$\sum_{i=1}^n a_i p_i + \sum_{j=1}^m b_j.$$

Lemma 8. *Each process term p in $\text{BPA}^\omega_\delta(A)$ is provably equal to its expansion, using A4-7+NEI1.*

Proof: By structural induction with respect to p.

3.2 Normed Processes

The following terminology stems from [4].

Definition 9. A process term p is called *normed* if it can terminate in finitely many transitions, that is, $p \xrightarrow{a_1} p_1 \xrightarrow{a_2} \cdots \xrightarrow{a_n} p_n \xrightarrow{b} \sqrt{}$.

The class of normed processes in $\text{BPA}^\omega_\delta(A)$ can be defined inductively as follows:

- $a \in A$ is normed;
- if p or q is normed, then $p + q$ is normed;
- if p and q are normed, then pq is normed.

Lemma 10. *If p is not normed, then $pq = p$ is provable using A4,5,7+NEI1+RSP$^\omega$.*

Proof: By structural induction with respect to p.

3.3 An Ordering on Pairs of Terms

The following weight function on process terms in $\text{BPA}^\omega_\delta(A)$, which represents the maximum nesting of ω's in a term, will be used to formulate an ordering on pairs of terms.

$$\begin{aligned} g(a) &= 0 \\ g(\delta) &= 0 \\ g(p+q) &= \max\{g(p), g(q)\} \\ g(pq) &= \max\{g(p), g(q)\} \\ g(p^\omega) &= g(p) + 1. \end{aligned}$$

Note that g-value is invariant under axioms A1,2. The following lemma can easily be deduced, using structural induction.

Lemma 11. *If p' is a derivative of p, then $g(p') \leq g(p)$.*

We consider pairs of process terms modulo commutativity. The ordering $<$ on pairs of process terms is defined as follows.

Definition 12. The ordering $<$ on pairs of terms is obtained by taking the transitive closure of the union of the three relations below.

1. $(r,s) < (p,q)$ if $g(r) < g(p)$ and $g(s) < g(p)$;
2. $(r,s) < (p,q)$ if $g(r) < g(p)$ and $g(s) \leq g(q)$;
3. $(p',q') < (p,q)$ if p' is a derivative of p, and not vice versa, and q' is a derivative of q.

The proof of the completeness theorem is based on induction with respect to this ordering, so we need to know that it is well-founded.

Lemma 13. *The ordering $<$ on pairs of process terms is well-founded modulo $=_{AC}$.*

Proof: Omitted.

3.4 Basic Terms

We construct a set $\mathbb{B}$ of *basic* process terms, such that each process term is provably equal to a basic term, and the derivatives of basic terms are basic terms. We will prove the completeness theorem by showing that bisimilar basic terms are provably equal.

Definition 14. The set $\mathbb{B}$ of *basic* process terms is defined inductively as follows:

1. if $a_1,...,a_n, b_1,...,b_m \in A$ and $p_1,...,p_n \in \mathbb{B}$, then $\sum_{i=1}^{n} a_i p_i + \sum_{j=1}^{m} b_j \in \mathbb{B}$;
2. if $p \in \mathbb{B}$ then $p^\omega \in \mathbb{B}$;
3. if $p \in \mathbb{B}$ and p' is a proper derivative of p, then $p'(p^\omega) \in \mathbb{B}$.

For notational convenience, we distinguish the following set $\mathbb{C}$ of cycles in $\mathbb{B}$.

Definition 15. $\mathbb{C} = \{p^\omega, p'(p^\omega) \mid p \in \mathbb{B},\ p'\text{ proper derivative of } p\}$.

The following facts for basic terms will be needed in the completeness proof.

Lemma 16.
1. *If $p \in \mathbb{C}$ and $p \xrightarrow{a} p'$, then $p' \in \mathbb{C}$.*
2. *If $p \in \mathbb{B}$ and $p \xrightarrow{a} p'$, then $p' \in \mathbb{B}$.*
3. *If $p \in \mathbb{B}$ and p is a proper derivative of itself, then $p \in \mathbb{C}$.*

Lemma 17. *For each term p there exists a basic term q with $g(q) \leq g(p)$ such that $p = q$ is provable using A4-7+NEI1+RSP$^\omega$.*

3.5 The Auxiliary Function ϕ

Before starting with the completeness proof, first we need to develop some theory. The proposition that will be proved at the end of this section makes an important stepping stone to obtain the desired completeness result for BPA$_\delta^\omega(A)$.

$p'(p^\omega) \leftrightarrow p''(p^\omega)$, with p' and p'' derivatives of p, does not imply $p' \leftrightarrow p''$. For example, clearly $a((aa)^\omega) \leftrightarrow aa((aa)^\omega)$, but $a \not\leftrightarrow aa$. In order to solve this ambiguity, we define an operator ϕ_p on basic terms, where intuitively the term $\phi_p(q)$, for $q \notin \mathbb{C}$, is obtained from the argument q as follows: all proper derivatives q' of q with $q'(p^\omega) \leftrightarrow p^\omega$ are removed in $\phi_p(q)$. We will see that if $p'(p^\omega) \leftrightarrow p''(p^\omega)$ then $\phi_p(p') \leftrightarrow \phi_p(p'')$.

Definition 18. Given $q \in \mathbb{B}$, the term $\phi_p(q)$ is defined as follows, using structural induction. We distinguish two cases: either $q \in \mathbb{C}$ or $q \notin \mathbb{C}$.

- CASE 1: $q \in \mathbb{C}$. Then put
$$\phi_p(q) =_{AC} q.$$

- CASE 2: $q \notin \mathbb{C}$, so that
$$q =_{AC} \sum_{i \in I} a_i q_i + \sum_{j \in J} b_j.$$
Then define $I_0 = \{i \in I \mid q_i(p^\omega) \not\leftrightarrow p^\omega\}$, and put
$$\phi_p(q) =_{AC} \sum_{i \in I_0} a_i \phi_p(q_i) + \sum_{i \in I \setminus I_0} a_i + \sum_{j \in J} b_j.$$

Lemma 19. *For $q \in \mathbb{B}$ we have $g(\phi_p(q)) \leq g(q)$.*

Proof: By structural induction with respect to q.

The proofs of the next two technical lemmas are quite involved, and therefore omitted.

Lemma 20. *Assume that for some natural number N_0:*
A. *for all terms u with $g(u) < N_0$ we have $p^\omega \not\leftrightarrow u$.*
Let $q, r \in \mathbb{B}$ and $g(q+r) < N_0$. If $q(p^\omega) \leftrightarrow r(p^\omega)$ then
$$\phi_p(q) \leftrightarrow \phi_p(r).$$

Proof: Omitted.

Lemma 21. *Assume that for some natural number N_0:*
A. *for all terms u with $g(u) < N_0$ we have $p \not\leftrightarrow u$;*
B. *for all pairs (u, v) of bisimilar terms with $g(u+v) < N_0$ we have $u = v$.*
Let $p, q \in \mathbb{B}$ and $g(p+q) < N_0$. Then
$$q(\phi_p(p)^\omega) = \phi_p(q)(\phi_p(p)^\omega).$$

Proof: Omitted.

Proposition 22. *Assume that for some natural number N_0:*
A. *for all terms u with $g(u) < N_0$ we have $p^\omega \not\leftrightarrow u$;*
B. *for all pairs (u, v) of bisimilar terms with $g(u+v) < N_0$ we have $u = v$.*
Let $g(p+q+r) < N_0$ and $q(p^\omega) \leftrightarrow r(p^\omega)$. Then
$$q(p^\omega) = r(p^\omega).$$

Proof: By Lemma 17
$$p = s \tag{1}$$
with $s \in \mathbb{B}$ and $g(s) \leq g(p) < N_0$. Since conditions A and B hold, Lemma 21 can be applied to derive $s(\phi_s(s)^\omega) = \phi_s(s)(\phi_s(s)^\omega) = \phi_s(s)^\omega$. RSP$^\omega$ then yields
$$s^\omega = \phi_s(s)^\omega. \tag{2}$$
According to Lemma 17 there exist basic terms t and u with $g(t) \leq g(q) < N_0$ and $g(u) \leq g(r) < N_0$ and
$$q = t \tag{3}$$
$$r = u. \tag{4}$$
Since $t(s^\omega) \leftrightarrow q(p^\omega) \leftrightarrow r(p^\omega) \leftrightarrow u(s^\omega)$, and since $g(t+u) < N_0$ and requirement A of Lemma 20 is satisfied, it implies $\phi_s(t) \leftrightarrow \phi_s(u)$. Since $g(\phi_s(t) + \phi_s(u)) < N_0$ (Lemma 19), condition B yields
$$\phi_s(t) = \phi_s(u). \tag{5}$$
Hence,
$$q(p^\omega) \stackrel{(1),(3)}{=} t(s^\omega) \stackrel{(2)}{=} t(\phi_s(s)^\omega) \stackrel{\text{Lem. 21}}{=} \phi_s(t)(\phi_s(s)^\omega)$$
$$\stackrel{(5)}{=} \phi_s(u)(\phi_s(s)^\omega) \stackrel{\text{Lem. 21}}{=} u(\phi_s(s)^\omega) \stackrel{(2)}{=} u(s^\omega) \stackrel{(1),(4)}{=} r(p^\omega). \quad \square$$

3.6 Completeness Proof

Proof of Theorem 5: Assume $p, q \in \mathbb{B}$ with $p \leftrightarrow q$; we show that $p = q$ can be derived from A1-7+NEI1+RSP$^\omega$, by induction on the well-founded ordering $<$ on pairs of terms. So suppose that we have already dealt with pairs of bisimilar basic terms that are smaller than (p, q). By symmetry it is sufficient to consider two cases: either $p \notin \mathbb{C}$ or $p, q \in \mathbb{C}$.

- CASE 1: $p \notin \mathbb{C}$.
  According to Lemma 8 p and q are provably equal to their expansions. Since $p \leftrightarrow q$, these expansions can be adapted, using axiom A3, to obtain:

  $$p = \sum_{i=1}^{n} a_i p_i + \sum_{j=1}^{m} b_j, \quad q = \sum_{i=1}^{n} a_i q_i + \sum_{j=1}^{m} b_j,$$

  where $p_i \leftrightarrow q_i$ for $i = 1, ..., n$. Since $p \notin \mathbb{C}$, Lemma 16.3 says that p is not a derivative of p_i for $i = 1, ..., n$. Since the p_i and the q_i for $i = 1, ..., n$ are derivatives of p and q respectively, it follows that $(p_i, q_i) < (p, q)$ for $i = 1, ..., n$ (by item 3 in Definition 12). So induction yields $p_i = q_i$ for $i = 1, ..., n$. Hence, $p = q$.

- CASE 2: $p, q \in \mathbb{C}$.
  Since $p \in \mathbb{C}$, either $p =_{AC} r^\omega = r(r^\omega)$ or $p =_{AC} r'(r^\omega)$, where $r \in \mathbb{B}$ and r' is a proper derivative of r. In both cases $p = r'(r^\omega)$ with $r \in \mathbb{B}$ and r' a derivative (not necessarily proper) of r. Even so, $q = s'(s^\omega)$ with $s \in \mathbb{B}$ and s' a derivative of s.
  By symmetry, it is sufficient to distinguish two cases: either r' is not normed, or both r' and s' are normed.

  * CASE 2.1: r' is not normed.
    Then by Lemma 10 $r'(r^\omega) = r'$. Since $g(r') \leq g(r) < g(p)$, item 2 in Definition 12 yields $(r', q) < (p, q)$. So, since $r' \leftrightarrow r'(r^\omega) \leftrightarrow q$, induction yields $r' = q$. Hence, $p = r'(r^\omega) = r' = q$.

  * CASE 2.2: Both r' and s' are normed.
    For convenience of notation put $N_0 = \max\{g(p), g(q)\}$. Again, we consider two cases: either there exists or there does not exist a term t with $g(t) < N_0$ and $p \leftrightarrow t$.

    o CASE 2.2.1: There exists a term t with $g(t) < N_0$ and $p \leftrightarrow t$ (and so $q \leftrightarrow t$).
      Since by the assumption at case 2.2 r' is normed, and $r'(r^\omega) \leftrightarrow t$, there exists a derivative t' of t with $r^\omega \leftrightarrow t'$, and so $rt' \leftrightarrow t'$. Furthermore, Lemma 11 implies $g(t') \leq g(t) < N_0$, and so $g(rt' + t') < N_0$. So after using Lemma 17 to reduce rt' and t' to basic form, we can apply induction, by item 1 in Definition 12, to conclude $rt' = t'$. RSP$^\omega$ then yields $r^\omega = t'$, so $p = r't'$. By Lemma 17 $r't' = u$ with $u \in \mathbb{B}$ and $g(u) < N_0$. Thus, $p = u$. Even so, $q = v$ for some basic term v with $g(v) < N_0$. Then $u \leftrightarrow p \leftrightarrow q \leftrightarrow v$, so since $g(u+v) < N_0$, induction yields $u = v$. Hence, $p = u = v = q$.

    o CASE 2.2.2: For each term t, if $g(t) < N_0$ then $p \not\leftrightarrow t$ (and so $q \not\leftrightarrow t$).
      Since $p \leftrightarrow q$, the assumption of this case implies $g(p) = g(q)$.
      Note that the requirements A and B for Proposition 22 are satisfied, by the assumption at case 2.2.2 together with the induction hypothesis (item 1 of Definition 12). So we are allowed to apply Proposition 22 in this case.
      By the assumption at case 2.2 r' is normed, so since $r'(r^\omega) \leftrightarrow s'(s^\omega)$, there exists a derivative s'' of s such that $r^\omega \leftrightarrow s''(s^\omega)$. Even so, $s^\omega \leftrightarrow r''(r^\omega)$ for some derivative r'' of r such that $r''(r^\omega) \leftrightarrow s^\omega$.
      Since $s''r''(r^\omega) \leftrightarrow s''(s^\omega) \leftrightarrow r^\omega \leftrightarrow r(r^\omega)$, and $g(s''r'' + r) < N_0$, Proposition 22 yields $s''r''(r^\omega) = r(r^\omega) \stackrel{NEI1}{=} r^\omega$. RSP$^\omega$ then yields

      $$r^\omega = (s''r'')^\omega. \tag{6}$$

Even so,
$$s^\omega = (r''s'')^\omega. \tag{7}$$
Since $s''((r''s'')^\omega) \stackrel{\text{NEI1}}{=} s''((r''s'')((r''s'')^\omega)) \stackrel{\text{A5}}{=} (s''r'')(s''((r''s'')^\omega))$, RSP$^\omega$ yields
$$s''((r''s'')^\omega) = (s''r'')^\omega. \tag{8}$$
Since $r's''(s^\omega) \underline{\leftrightarrow} r's''((r''s'')^\omega) \underline{\leftrightarrow} r'((s''r'')^\omega) \underline{\leftrightarrow} r'(r^\omega) \underline{\leftrightarrow} s'(s^\omega)$, and $g(r's''+s') < N_0$, Proposition 22 yields
$$r's''(s^\omega) = s'(s^\omega). \tag{9}$$
So finally,
$$p =_{\text{AC}} r'(r^\omega) \stackrel{(6)}{=} r'((s''r'')^\omega) \stackrel{(8)}{=} r's''((r''s'')^\omega) \stackrel{(7)}{=} r's''(s^\omega) \stackrel{(9)}{=} s'(s^\omega) =_{\text{AC}} q. \quad \square$$

3.7 An Example

We give an example as to how the construction in the completeness proof acts on particular pairs of bisimilar basic terms.

Example 1. $(a\delta + b)((c(a\delta + b))^\omega) \underline{\leftrightarrow} (a\delta + bc)^\omega$.

This equivalence belongs with case 2.2.2. It can be derived as follows.
$$(a\delta + b)((c(a\delta + b))^\omega) \stackrel{\text{NEI1}}{=} (a\delta + b)((c(a\delta + b))((c(a\delta + b))^\omega))$$
$$\stackrel{\text{A4,5}}{=} ((a\delta + b)c)((a\delta + b)((c(a\delta + b))^\omega)).$$
Then RSP$^\omega$ yields
$$(a\delta + b)((c(a\delta + b))^\omega) = ((a\delta + b)c)^\omega. \tag{10}$$
So finally,
$$(a\delta + b)((c(a\delta + b))^\omega) \stackrel{(10)}{=} ((a\delta + b)c)^\omega \stackrel{\text{A4,5,7}}{=} (a\delta + bc)^\omega.$$

4 Conclusion

In this paper, Milner's axiomatization for iteration was restricted to the case of no-exit iteration, and it was proved that this yields a complete axiomatization for no-exit iteration in process algebra modulo bisimulation. The main new idea in the proof was to introduce a function ϕ which can help to minimize the argument p of a no-exit iteration term p^ω, in such a way that p does not contain any proper derivatives p' with $p'(p^\omega) \underline{\leftrightarrow} p^\omega$. For example, using this function ϕ, the term $(aa)^\omega$ can be reduced to a^ω.

The completeness result in this paper may be a step forward to a positive answer to the question whether Milner's axiomatization is complete for iteration in process algebra modulo bisimulation. Namely, the main problem in solving this question is to deal with no-exit iteration terms p^ω where p is not minimal. Unfortunately, it is not obvious how to extend the definition of the function ϕ to all terms in process algebra with iteration. For example, consider the term
$$(a((a(ba + a))^*c))^\omega$$
where the argument $a((a(ba + a))^*c)$ of no-exit iteration is not minimal. Minimization of this argument would yield a so-called 'double-exit' term (with exits b and c), which cannot be expressed in process algebra with iteration modulo bisimulation (see [6, 1]). The only way to obtain a no-exit iteration term with a minimal argument in this particular case is to rewrite the term to
$$a((a(ba + a) + ca)^\omega)$$
A minimization strategy for all possible arguments of no-exit iteration would probably be the key to solving Milner's question.

References

1. L. Aceto and W.J. Fokkink. An equational axiomatization for multi-exit iteration. Report RS-96-22, BRICS, Aalborg University, 1996. Accepted for publication in *Information and Computation*.
2. L. Aceto, W.J. Fokkink, R.J. van Glabbeek, and A. Ingólfsdóttir. Axiomatizing prefix iteration with silent steps. *Information and Computation*, 127(1):26–40, 1996.
3. L. Aceto, W.J. Fokkink, and A. Ingólfsdóttir. A menagerie of non-finitely based process semantics over BPA*: from ready simulation to completed traces. Report RS-96-23, BRICS, Aalborg University, 1996.
4. J.C.M. Baeten, J.A. Bergstra, and J.W. Klop. Decidability of bisimulation equivalence for processes generating context-free languages. *Journal of the ACM*, 40(3):653–682, 1993.
5. J.C.M. Baeten and C. Verhoef. A congruence theorem for structured operational semantics with predicates. In *Proceedings CONCUR'93*, LNCS 715, pp. 477–492. Springer, 1993.
6. J.A. Bergstra, I. Bethke, and A. Ponse. Process algebra with iteration and nesting. *The Computer Journal*, 37(4):243–258, 1994.
7. J.A. Bergstra, J.A. Hillebrand and A. Ponse. Grid protocols based on synchronous communication. *Science of Computer Programming*, 1997, To appear.
8. J.A. Bergstra and P. Klint. The discrete time toolbus. In *Proceedings AMAST'96*, LNCS 1101, pp. 286–305. Springer, 1996.
9. J.A. Bergstra and J.W. Klop. Process algebra for synchronous communication. *Information and Control*, 60(1/3):109–137, 1984.
10. J.A. Bergstra and J.W. Klop. Verification of an alternating bit protocol by means of process algebra. In *Proceedings Spring School on Mathematical Methods of Specification and Synthesis of Software Systems*, LNCS 215, pp. 9–23. Springer, 1985.
11. W.J. Fokkink. On the completeness of the equations for the Kleene star in bisimulation. In *Proceedings AMAST'96*, LNCS 1101, pp. 180–194. Springer, 1996.
12. W.J. Fokkink. An axiomatization for the terminal cycle. Logic Group Preprint Series 167, Utrecht University, 1996. Available at http://www.phil.ruu.nl.
13. W.J. Fokkink and R.J. van Glabbeek. Ntyft/ntyxt rules reduce to ntree rules. *Information and Computation*, 126(1):1–10, 1996.
14. W.J. Fokkink and H. Zantema. Basic process algebra with iteration: completeness of its equational axioms. *The Computer Journal*, 37(4):259–267, 1994.
15. S.C. Kleene. Representation of events in nerve nets and finite automata. In *Automata Studies*, pages 3–41. Princeton University Press, 1956.
16. D. Kozen. A completeness theorem for Kleene algebras and the algebra of regular events. *Information and Computation*, 110(2):366–390, 1994.
17. R. Milner. A complete inference system for a class of regular behaviours. *Journal of Computer and System Sciences*, 28:439–466, 1984.
18. R. Milner. The polyadic π-calculus: a tutorial. In *Proceedings Marktoberdorf Summer School '91, Logic and Algebra of Specification*, NATO ASI Series F94, pp. 203–246. Springer, 1993.
19. D.M.R. Park. Concurrency and automata on infinite sequences. In *Proceedings 5th GI Conference*, LNCS 104, pp. 167–183. Springer, 1981.
20. G.D. Plotkin. A structural approach to operational semantics. Report DAIMI FN-19, Aarhus University, 1981.
21. A.Ponse. Personal communication, March 1997. See also the handouts for the course "Concurrency and Distributed Systems", avalailable at http://adam.wins.uva.nl/ alban.
22. A. Salomaa. Two complete axiom systems for the algebra of regular events. *Journal of the ACM*, 13(1):158–169, 1966.
23. P.M. Sewell. Bisimulation is not finitely (first order) equationally axiomatisable. In *Proceedings LICS'94*, pp. 62–70. IEEE Computer Society Press, 1994.

Discrete-Time Control for Rectangular Hybrid Automata*

Thomas A. Henzinger[1]   Peter W. Kopke[2]

[1] University of California, Berkeley, CA. Email: tah@eecs.berkeley.edu
[2] William H. Kopke, Jr. Inc., Lake Success, NY. Email: 75467.2651@compuserve.com

Abstract. Rectangular hybrid automata model digital control programs of analog plant environments. We study rectangular hybrid automata where the plant state evolves continuously in real-numbered time, and the controller samples the plant state and changes the control state discretely, only at the integer points in time. We prove that rectangular hybrid automata have finite bisimilarity quotients when all control transitions happen at integer times, even if the constraints on the derivatives of the variables vary between control states. This is sharply in contrast with the conventional model where control transitions may happen at any real time, and already the reachability problem is undecidable. Based on the finite bisimilarity quotients, we give an exponential algorithm for the symbolic sampling-controller synthesis of rectangular automata. We show our algorithm to be optimal by proving the problem to be EXPTIME-hard. We also show that rectangular automata form a maximal class of systems for which the sampling-controller synthesis problem can be solved algorithmically.

1 Introduction

Hybrid systems are dynamical systems with both discrete and continuous components. A paradigmatic example of a hybrid system is a digital control program for an analog plant environment, like a furnace or an airplane: the controller state moves discretely between control modes, and in each control mode, the plant state evolves continuously according to physical laws. A natural mathematical model for hybrid systems is the *hybrid automaton*, which represents discrete components using finite-state machines and continuous components using real-numbered variables [ACH+95]. A particularly important subclass of hybrid automata are the *rectangular automata*, where in each control mode v, the given n variables follow a nondeterministic differential equation of the form $\frac{d\mathbf{x}}{dt} \in B(v)$, for an n-dimensional rectangle $B(v) \subset \mathbb{R}^n$ [HKPV95]. Rectangular automata are useful as (1) they can be made to approximate, arbitrarily closely, complex continuous behavior using lower and upper bounds on derivatives [HH95], and (2) they can be analyzed automatically using (semi)algorithms based on symbolic execution, such as those implemented in HYTECH [HHW97].

For systems that can be executed symbolically, verification and control yield to a (semi)algorithmic approach even if the state space is infinite [Hen96]. For such systems, a temporal formula can be verified automatically and a controller can be synthesized automatically by computing, using iterative approximation, a fixpoint of an operator on state sets [BCM+92, MPS95]. The fixpoint computation is guaranteed to terminate in the presence of a suitable finite quotient space. For example, symbolically-executable systems with finite bisimilarity quotients allow symbolic LTL and CTL model checking, and symbolic

---

*This research was supported in part by the ONR YIP award N00014-95-1-0520, by the NSF CAREER award CCR-9501708, by the NSF grant CCR-9504469, by the AFOSR contract F49620-93-1-0056, by the ARO MURI contract DAAH-04-96-1-0341, by the ARO contract DAAL03-91-C-0027 through the MSI at Cornell University, by the ARPA grant NAG2-892, and by the SRC contract 95-DC-324.036.

safety controller synthesis. While rectangular automata can be executed symbolically, they do not necessarily have finite bisimilarity quotients, and simple reachability questions are undecidable [HKPV95]. A noted subclass of rectangular automata with finite bisimilarity quotients are *timed automata*, where all variables are clocks with derivative 1 [AD94]. As a consequence, the symbolic model checking and controller synthesis problems have been solved for timed automata [HNSY94, MPS95].

While previous results on timed and hybrid automata allow edge transitions (i.e., control switches) to occur at any real-numbered points in time, this is not necessarily a natural assumption for controller synthesis, as it permits controllers that, in a single time unit, can interact with the plant an unbounded number of times (even infinitely often, if no special care is taken [AH97]). By contrast, we study the control problem under the assumption that while the plant evolves continuously, the controller samples the plant state discretely, at the integer points in time only.[3] This leads to the following formulation of the *sampling-controller synthesis problem* for rectangular automata: given a continuous-time rectangular automaton, is there a discrete-time controller that samples the automaton state at integer times and switches the control mode accordingly so that the resulting closed-loop system satisfies a given invariant?

To solve this problem, we study the *discrete-time transition systems* of timed and rectangular automata, where all time transitions have unit duration. It should be noticed that all variables still evolve continuously, in real-numbered time; only edge transitions are restricted to discrete time. We prove that unlike in the case of dense time, the discrete-time transition system of every rectangular automaton has a finite bisimilarity quotient.[4] As a corollary, we conclude that the standard approaches to symbolic model checking and controller synthesis are guaranteed to terminate when all control switches must occur at integer times. The running times of the verification and control algorithms depend on the number of bisimilarity equivalence classes, which, while exponential in the description of the automaton, is less by a multiplicative exponential factor than the number of region equivalence classes used for the dense-time verification and control of timed automata. Thus, the often more realistic sampling-controller synthesis problem can be solved for a wider class of hybrid systems than dense-time control (rectangular vs. timed), at a smaller cost.

We prove that our sampling-control algorithm is optimal, by giving lower bounds on the control problem for timed and hybrid systems: we show that the safety control decision problem (does there exist a controller that maintains an invariant?) is complete for EXPTIME already in the restricted case of discrete-time timed automata. We also identify the boundary of sampling controllability by proving that several generalizations of rectangular automata lead to an undecidable reachability problem, even in discrete time. The undecidability of dense-time reachability for rectangular automata has led [PV94] to consider the restriction that the flow rectangle $B(v)$ must be the same for each control mode v. For the resulting class of *initialized* rectangular automata, reachability is decidable [HKPV95]. Our work can be viewed as pointing out an orthogonal restriction of rectangularity, namely, that the flow rectangle may change only at integer points in time. Unlike initialization, our restriction guarantees not only a finite language equivalence quotient but a finite bisimilarity quotient on the infinite state space of a rectangular automaton.

---

[3] The sampling rate of the controller may be any rational, but without loss of generality we assume it to be 1.

[4] Under the technical restriction that either the invariant and flow rectangles are positive, or the automaton state stays within a bounded region.

2 Definitions and Previous Results

2.1 Labeled Transition Systems

Definition 2.1 [Transition system] A *transition system* $S = (Q, \Sigma, \rightarrow, Q_I, \Pi, \models)$ consists of a set Q of *states*, a finite set Σ of *events*, a multiset $\rightarrow \subset Q \times \Sigma \times Q$ called the *transition relation*, a set $Q_I \subset Q$ of *initial states*, a set Π of *propositions*, and a *satisfaction relation* $\models \subset Q \times \Pi$. We write $q \xrightarrow{\sigma} q'$ instead of $(q, \sigma, q') \in \rightarrow$, and $q \models \pi$ instead of $(q, \pi) \in \models$. The transition system S is *finite* if Q is finite. We assume for simplicity that S is deadlock-free; that is, for each state $q \in Q$, there exists an event $\sigma \in \Sigma$ and a state $r \in Q$ such that $q \xrightarrow{\sigma} r$. A *region* is a subset of Q. Given a proposition $\pi \in \Pi$, we write $R_\pi = \{q \in Q \mid q \models \pi\}$ for the region of states that satisfy π. ∎

Verification as reachability

Definition 2.2 [Weakest precondition] Let S be a transition system. For each event $\sigma \in \Sigma$, the σ-*predecessor operator* $Pre_\sigma : 2^Q \rightarrow 2^Q$ is defined by $Pre_\sigma(R) = \{q \in Q \mid \exists r \in R . q \xrightarrow{\sigma} r\}$. In particular, $Pre_\sigma(Q)$ is the set of states in which the event σ is enabled. Define $Pre : 2^Q \rightarrow 2^Q$ by $Pre(R) = \bigcup_{\sigma \in \Sigma} Pre_\sigma(R)$. A region $R \subset Q$ is *reachable* in S if $Q_I \cap Pre^k(R) \neq \emptyset$ for some $k \in \mathbb{N}$. ∎

The basic verification problem for transition systems asks whether an unsafe state is unreachable.

Definition 2.3 [Safety verification] Let $\mathcal{C}$ be a class of transition systems. The *safety verification problem* for $\mathcal{C}$ is stated in the following way: given a transition system $S \in \mathcal{C}$ and a proposition $\pi \in \Pi$, determine whether the region R_π is not reachable in S. ∎

For finite transition systems, the safety verification problem is the complement of graph reachability, which can be solved in linear time and is complete for NLOGSPACE. The safety verification problem can be generalized to the safety control problem.

Control as alternating reachability We use the following model for control: for each state q of a transition system, a (memory-free) controller chooses an enabled event σ so that in state q, the controlled system always proceeds via event σ. Since q may have several σ-successors, the controlled system may still be nondeterministic. Alternative models for memory-free control are equivalent.

Definition 2.4 [Control map] Let S be a transition system. A *control map* for S is a function $\kappa : Q \rightarrow \Sigma$ such that for each state $q \in Q$, there exists a state $r \in Q$ with $q \xrightarrow{\kappa(q)} r$. The *closed-loop system* $\kappa(S)$ is the transition system $(Q, \Sigma, \Rightarrow, Q_I, \Pi, \models)$, where $q \xrightarrow{\sigma}_\Rightarrow q'$ iff $q \xrightarrow{\sigma} q'$ and $\kappa(q) = \sigma$. ∎

The basic control problem for transition systems asks whether an unsafe state is avoidable by applying some control map.

Definition 2.5 [Safety control] Let $\mathcal{C}$ be a class of transition systems. The *safety control decision problem* for $\mathcal{C}$ is stated in the following way: given a transition system $S \in \mathcal{C}$ and a proposition $\pi \in \Pi$, determine whether there exists a control map κ such that the region R_π is not reachable in the closed-loop system $\kappa(S)$. If so, then we say π is *avoidable* in S. The *safety controller synthesis problem* requires the construction of a witnessing control map κ when π is avoidable. ∎

For finite transition systems, the safety control decision problem is the complement of AND-OR graph reachability, which can be solved in quadratic time and is complete for PTIME.

Definition 2.6 [Alternating reachability] An *AND-OR graph* $G = (V_A, V_O, V_I, \rightarrow)$ consists of a finite set $V = V_A \cup V_O$ of vertices that is partitioned into a set V_A of AND vertices and a set V_O of OR vertices, a set $V_I \subset V$ of initial vertices, and a multiset $\rightarrow \subset V \times V$ of edges. We assume deadlock freedom, namely, that for each vertex $v \in V$, there exists a vertex $w \in V$ such that $v \rightarrow w$. The *controllable predecessor operator* $CPre: 2^V \rightarrow 2^V$ is defined by $CPre(R) = \{q \in V_O \mid \exists r \in R. q \rightarrow r\} \cup \{q \in V_A \mid \forall r \in V. q \rightarrow r \text{ implies } r \in R\}$. A set $R \subset V$ of vertices is *alternating reachable* in G if $V_I \cap CPre^k(R) \neq \emptyset$ for some $k \in \mathbb{N}$. The *alternating reachability problem* asks whether a given set of vertices is alternating reachable in a given AND-OR graph. ∎

Theorem 2.1 [Imm81] *The alternating reachability problem is complete for PTIME.*

There is a simple correspondence between safety control and alternating reachability. Let S be a finite transition system and let π be a proposition. Define an AND-OR graph G_S as follows: let $V_A = Q$ and $V_O = Q \times \Sigma$ and $V_I = Q_I$; for each vertex $q \in V_A$ and each event $\sigma \in \Sigma$, let $q \rightarrow (q, \sigma)$ in G_S iff $q \in Pre_\sigma(Q)$ in S; and for each vertex $(q, \sigma) \in V_O$, let $(q, \sigma) \rightarrow r$ in G_S iff $q \xrightarrow{\sigma} r$ in S. Then the proposition π is avoidable in S iff the set R_π of AND vertices is not alternating reachable in G_S.

Corollary 2.1 *The safety control decision problem for finite transition systems is complete for PTIME.*

Moreover, a byproduct of a negative alternating reachability computation is a control map that avoids π. Note that for each set $R \subset Q$ of AND vertices, $CPre^2(R) = \bigcap_{\sigma \in \Sigma}(Pre_\sigma(R) \cup (Q \setminus Pre_\sigma(Q)))$. Thus the region $CPre^2(R)$ is the set of all states that no control map can keep out of R at the next transition. Let $R_F = CPre^{2|Q|}(R_\pi)$. Then π is avoidable in S iff $Q_I \cap R_F = \emptyset$. Each application of $CPre^2$ can be computed in linear time, so R_F can be computed in quadratic time. If π is indeed avoidable, then a witnessing control map may be constructed by choosing for each state $q \in Q \setminus R_F$ an event σ such that $q \in Pre_\sigma(Q) \setminus Pre_\sigma(R_F)$.

Theorem 2.2 [RW87] *The safety controller synthesis problem for finite transition systems can be solved in quadratic time.*

Effectively-presented transition systems with finite bisimilarity quotients The safety controller synthesis problem can be solved not only for finite transition systems, but also for effectively-presented transition systems with finite bisimilarity quotients.

Definition 2.7 [Effective presentation] A *symbolic execution theory* for the transition system S consists of a set $\mathcal{F}$ of *formulas*, a formula $\phi_I \in \mathcal{F}$, and a map $[\![\cdot]\!]: \mathcal{F} \rightarrow 2^Q$ such that (1) every proposition $\pi \in \Pi$ is a formula: $[\![\pi]\!] = R_\pi$; (2) for all formulas $\phi_1, \phi_2 \in \mathcal{F}$, the three expressions $\phi_1 \wedge \phi_2$ and $\phi_1 \vee \phi_2$ and $\neg \phi_1$ are formulas: $[\![\phi_1 \wedge \phi_2]\!] = [\![\phi_1]\!] \cap [\![\phi_2]\!]$ and $[\![\phi_1 \vee \phi_2]\!] = [\![\phi_1]\!] \cup [\![\phi_2]\!]$ and $[\![\neg \phi_1]\!] = Q \setminus [\![\phi_1]\!]$; (3) $[\![\phi_I]\!] = Q_I$; (4) the set $\{\phi \in \mathcal{F} \mid [\![\phi]\!] = \emptyset\}$ is recursive; and (5) for each event $\sigma \in \Sigma$, there is a computable map $Pre_\sigma: \mathcal{F} \rightarrow \mathcal{F}$ such that $[\![Pre_\sigma(\phi)]\!] = Pre_\sigma([\![\phi]\!])$ for all formulas $\phi \in \mathcal{F}$. An *effectively-presented transition system* consists of a transition system S together with a symbolic execution theory for S. ∎

Definition 2.8 [Bisimilarity] A *bisimulation* on the transition system S is an equivalence relation $\cong$ on the state set Q such that (1) if $q \cong r$ then for all propositions $\pi \in \Pi$, we have $q \models \pi$ iff $r \models \pi$, and (2) if $q \cong r$ and $q \xrightarrow{\sigma} q'$, then there exists a state $r' \in Q$ such that $r \xrightarrow{\sigma} r'$ and $q' \cong r'$. The largest bisimulation on S is denoted by $\equiv$. The *bisimilarity quotient* $S/_\equiv$ is the transition system $(Q/_\equiv, \Sigma, \rightarrow_\exists, Q_\exists, \Pi, \models_\exists)$, where $R \xrightarrow{\sigma}_\exists R'$ iff there exist two states $q \in R$ and $q' \in R'$ such that $q \xrightarrow{\sigma} q'$, where $R \in Q_\exists$ iff $R \cap Q_I \neq \emptyset$, and where $R \models_\exists \pi$ iff $R \cap R_\pi \neq \emptyset$. ∎

The controllable-predecessor operator $CPre^2$ can be computed on any effectively-presented transition system. When the bisimilarity quotient has $k \in \mathbb{N}$ equivalence classes, the R_F computation converges in at most k iterations of $CPre^2$. Synthesizing a control map is accomplished by first computing the bisimilarity quotient, and then choosing for each state in each equivalence class R disjoint from R_F, an event $\sigma \in \Sigma$ such that $R \cap Pre_\sigma(Q) \neq \emptyset$ and $R \cap Pre_\sigma(R_F) = \emptyset$.

Theorem 2.3 [Hen95] *The safety control decision problem is decidable for effectively-presented transition systems with finite bisimilarity quotients. Moreover, when a proposition is avoidable, a witnessing control map can be computed.*

This result can be generalized to liveness verification such as μ-calculus model checking, and to memory-free liveness control such as control-map synthesis for Rabin chain conditions.

2.2 Rectangular Hybrid Automata

Definition 2.9 [Rectangle] Let $X = \{x_1, \ldots, x_n\}$ be a set of real-valued variables. A *rectangular inequality* over X is a formula of the form $x_i \sim c$, where c is an integer constant, and $\sim$ is one of $<, \leq, >, \geq$. A *rectangular predicate* over X is a conjunction of rectangular inequalities. The rectangular predicate ϕ defines the set of vectors $[\![\phi]\!] = \{\mathbf{y} \in \mathbb{R}^n \mid \phi[X := \mathbf{y}] \text{ is true}\}$. A set of the form $[\![\phi]\!]$, where ϕ is a rectangular predicate, is called a *rectangle*. Given a positive integer $m \in \mathbb{N}_{>0}$, the rectangular predicate ϕ and the rectangle $[\![\phi]\!]$ are *m-definable* if $|c| \leq m$ for every conjunct $x_i \sim c$ of ϕ. The set of all rectangular predicates over X is denoted $Rect(X)$. ∎

Definition 2.10 [Rectangular automaton][HKPV95] A *rectangular automaton* A consists of the following components:

Variables. A finite set $X = \{x_1, \ldots, x_n\}$ of real-valued *variables* representing the continuous component of the system. The number n is the *dimension* of A. We write $\dot{X}$ for the set $\{\dot{x}_i \mid x_i \in X\}$ of dotted variables, and X' for the set $\{x'_i \mid x_i \in X\}$ of primed variables.

Control graph. A finite directed multigraph (V, E) representing the discrete component of the system. The vertices in V are called *control modes*. The edges in E are called *control switches*.

Invariant conditions. A function $inv: V \rightarrow Rect(X)$ mapping each control mode to its *invariant condition*, a rectangular predicate.

Initial conditions. A function $init: V \rightarrow Rect(X)$ mapping each control mode to its *initial condition*, a rectangular predicate.

Jump conditions. A function $jump$ mapping each control switch $e \in E$ to a predicate $jump(e)$ of the form $\phi \wedge \phi' \wedge \bigwedge_{i \notin update(e)}(x'_i = x_i)$, where $\phi \in Rect(X)$ and $\phi' \in Rect(X')$ are rectangular predicates, and $update(e) \subseteq \{1, \ldots, n\}$. The jump condition $jump(e)$ specifies the effect of the change in control mode on the values of the variables:

each unprimed variable x_i refers to a value before the control switch e, and each primed variable x_i' refers to the corresponding value after the control switch.

Flow conditions. A function $flow: V \to Rect(\dot{X})$ mapping each control mode v to a *flow condition*, a rectangular predicate that constrains the behavior of the first derivatives of the variables while time passes in control mode v.

Events. A finite set Σ of *events*, and a function $event: E \to \Sigma$ mapping each control switch to an event.

Thus a rectangular automaton A is a tuple $(X, V, E, inv, init, jump, flow, \Sigma, event)$. The automaton A is *m-definable* if every rectangular predicate in the definition of A is m-definable. The automaton A is *positive* if for every control mode $v \in V$, the invariant rectangle $[\![inv(v)]\!]$ and the flow rectangle $[\![flow(v)]\!]$ are subsets of the positive orthant $\mathbb{R}^n_{\geq 0}$. The automaton A is *bounded* if for every control mode $v \in V$, the invariant rectangle $[\![inv(v)]\!]$ is a bounded set. ∎

The state of a rectangular automaton has two parts: a discrete (or control) part, and a continuous (or plant) part. The discrete state is a control mode. The continuous state is a valuation for the variables.

Definition 2.11 [States of rectangular automata] Let A be a rectangular automaton. A *state* of A is a pair $(v, \mathbf{y})$, where $v \in V$ is a control mode and $\mathbf{y} \in [\![inv(v)]\!]$ is a vector satisfying the invariant condition of v. Thus the set of states is $Q = \{(v, \mathbf{y}) \in V \times \mathbb{R}^n \mid \mathbf{y} \in [\![inv(v)]\!]\}$. A subset of Q is called a *region* of A. A *rectangular state predicate* for A is a function ψ from V to $Rect(X)$. The rectangular state predicate ψ defines the region $[\![\psi]\!] = \{(v, \mathbf{y}) \in Q \mid \mathbf{y} \in [\![\psi(v)]\!]\}$. A region of the form $[\![\psi]\!]$, where ψ is a rectangular state predicate for A, is called a *rectangular region*. The initial condition map defines the rectangular region $Q_I = [\![init]\!]$ of *initial states*. ∎

A rectangular automaton makes two types of transitions: jump (or edge, or control) transitions, and flow (or time, or plant) transitions. Jump transitions are instantaneous. They are characterized by a change in control mode, and are accompanied by discrete modifications to the variables in accordance with the jump condition of the control switch. During flow transitions, while time elapses, the control mode remains fixed and the variables evolve continuously via a trajectory that satisfies the flow condition of the active control mode.

Definition 2.12 [Transitions of rectangular automata] Let A be a rectangular automaton. For each event $\sigma \in \Sigma$, we define the *jump relation* $\xrightarrow{\sigma} \subset Q^2$ by $(v, \mathbf{y}) \xrightarrow{\sigma} (v', \mathbf{y}')$ iff there exists a control switch $e = (v, v') \in E$ such that $event(e) = \sigma$ and $(\mathbf{y}, \mathbf{y}') \in [\![jump(e)]\!]$. For each nonnegative real $\delta \in \mathbb{R}_{\geq 0}$, we define the *flow relation* $\xrightarrow{\delta} \subset Q^2$ by $(v, \mathbf{y}) \xrightarrow{\delta} (v', \mathbf{y}')$ iff (1) $v = v'$, and (2) there exists a differentiable function $f: [0, \delta] \to [\![inv(v)]\!]$ such that $f(0) = \mathbf{y}$ and $f(\delta) = \mathbf{y}'$, and $\dot{f}(\epsilon) \in [\![flow(v)]\!]$ for all reals $\epsilon \in (0, \delta)$, where $\dot{f}$ is the first derivative of f. We say that δ is the *duration* of the flow transition. Since the rectangle $[\![inv(v)]\!]$ is a convex set, it follows that for $\delta > 0$, condition (2) is equivalent to $\frac{\mathbf{y}'-\mathbf{y}}{\delta} \in [\![flow(v)]\!]$; that is, all flows can be thought of as straight lines. ∎

Every rectangular automaton defines two transition systems.

Definition 2.13 [Discrete time and dense time] Let A be a rectangular automaton. Define the binary relation $\xrightarrow{time} \subset Q^2$ by $(v, \mathbf{y}) \xrightarrow{time} (v', \mathbf{y}')$ iff $(v, \mathbf{y}) \xrightarrow{\delta} (v', \mathbf{y}')$ for some duration $\delta \in \mathbb{R}_{\geq 0}$. Define Π to be the set of rectangular state predicates for A, and for all states $(v, \mathbf{y}) \in Q$, define $(v, \mathbf{y}) \models \pi$ iff $(v, \mathbf{y}) \in [\![\pi]\!]$. The *discrete-time transition system* of A

is defined by $S_A^{disc} = (Q, \Sigma \cup \{1\}, \rightarrow, Q_I, \Pi, \models)$. The *dense-time transition system* of A is defined by $S_A^{dense} = (Q, \Sigma \cup \{time\}, \rightarrow, Q_I, \Pi, \models)$. Thus all flow transitions in the discrete-time transition system are required to have duration 1, while flow transitions in the dense-time transition system can have any nonnegative real duration. We refer to the safety verification problem for transition systems of the form S_A^{disc} (resp. S_A^{dense}), for some rectangular automaton A, as the *discrete-time* (resp. *dense-time*) *safety verification problem for rectangular automata*, and similarly for the control decision and controller synthesis problems. ∎

Dense-time undecidability results In dense time, the verification and control of rectangular automata cannot be fully automated.

Theorem 2.4 [ACH+95] *For positive and bounded rectangular automata, the dense-time safety verification problem (and thus the dense-time safety control decision problem) is undecidable.*

Research has therefore concentrated on subclasses of rectangular automata. In [HKPV95] it is shown that for *initialized* rectangular automata, whose flow condition map is a constant function (i.e., all control modes have the same flow condition), the dense-time safety verification problem (in fact, LTL model checking) can be decided. These automata, however, have no finite bisimilarity quotients in dense time [Hen95], and therefore further restrictions are desirable.

Timed automata An important special case of initialized rectangular automata are timed automata. All variables of a timed automaton are clocks, which advance uniformly at rate 1 while time elapses.

Definition 2.14 [Timed automaton][AD94] A *timed automaton* is a positive rectangular automaton A with the restriction that $flow(v) = \bigwedge_{i=1}^{n} (\dot{x}_i = 1)$ for every control mode v. A *triangular inequality* over a set X of variables is a formula of the form $x_i - x_j \sim c$, where $x_i, x_j \in X$ are variables, c is an integer constant, and $\sim$ is one of $<, \leq, >, \geq$. A *triangular predicate* over X is a conjunction of rectangular and triangular inequalities. A *triangular state predicate* for a timed automaton A is a function that maps every control mode of A to a triangular predicate over the variables of A. ∎

The fundamental theorem for timed automata states that the dense-time transition system S_A^{dense} of a timed automaton A has a finite bisimilarity quotient and can be presented effectively using triangular state predicates.

Theorem 2.5 [AD94, HNSY94] *For every m-definable n-dimensional timed automaton A with k control modes, the dense-time transition system S_A^{dense} has a finite bisimilarity quotient with $O(k \cdot (n+1)! \cdot (2m)^n)$ many equivalence classes. Moreover, the boolean combinations of triangular state predicates for A form a symbolic execution theory for S_A^{dense}.*

Corollary 2.2 *For timed automata, the dense-time safety verification problem (in fact, LTL and CTL model checking) can be solved in PSPACE, and the dense-time safety controller synthesis problem can be solved in EXPTIME.*

As for finite transition systems, control is harder than verification. In [AD94] it is shown that the dense-time safety verification problem for timed automata is hard for PSPACE. From Theorem 3.2 below it follows that the dense-time safety control decision problem for timed automata is hard for EXPTIME.

3 Discrete-Time Rectangular Automata

3.1 Finite Bisimilarity Quotients and Effective Presentation

We show that the discrete-time transition system S_A^{disc} of a positive or bounded rectangular automaton A has a finite bisimilarity quotient and can be presented effectively using rectangular state predicates. More precisely, in discrete time, two states of a rectangular automaton are bisimilar if (1) they have the same control mode, (2) corresponding variable values agree on their integer parts, and (3) corresponding variable values agree on whether they are integral. Moreover, if an m-definable rectangular automaton is positive, then it cannot distinguish variable values greater than m. For m-definable bounded rectangular automata, the continuous part of the state is contained in the cube $[-m, m]^n$. It follows that in both the positive and the bounded case, the bisimilarity quotient is finite.

Definition 3.1 Define the equivalence relation $\approx_n$ on $\mathbb{R}^n$ by $\mathbf{y} \approx_n \mathbf{z}$ iff $\lfloor y_i \rfloor = \lfloor z_i \rfloor$ and $\lceil y_i \rceil = \lceil z_i \rceil$ for all $1 \leq i \leq n$. Given $m \in \mathbb{N}_{>0}$, define the equivalence relation $\approx_n^m$ on $\mathbb{R}^n$ by $\mathbf{y} \approx_n^m \mathbf{z}$ iff for each $1 \leq i \leq n$, either $y_i \approx_1 z_i$, or both y_i and z_i are greater than m, or both y_i and z_i are less than $-m$. For an n-dimensional rectangular automaton A, define the equivalence relations $\cong_A$ and $\cong_A^m$ on the states of A by $(v, \mathbf{y}) \cong_A (w, \mathbf{z})$ iff $v = w$ and $\mathbf{y} \approx_n \mathbf{z}$, and $(v, \mathbf{y}) \cong_A^m (w, \mathbf{z})$ iff $v = w$ and $\mathbf{y} \approx_n^m \mathbf{z}$. ∎

Lemma 3.1 *Consider two vectors $\mathbf{y}, \mathbf{z} \in \mathbb{R}^n$. Then $\mathbf{y} \approx_n \mathbf{z}$ iff for every rectangle $B \subset \mathbb{R}^n$, we have $\mathbf{y} \in B$ iff $\mathbf{z} \in B$. Moreover, $\mathbf{y} \approx_n^m \mathbf{z}$ iff for every m-definable rectangle $B \subset \mathbb{R}^n$, we have $\mathbf{y} \in B$ iff $\mathbf{z} \in B$.*

Theorem 3.1 *Let A be an n-dimensional rectangular automaton with k control modes. The equivalence relation $\cong_A$ is a bisimulation on the discrete-time transition system S_A^{disc}. If A is m-definable and either positive or bounded, then $\cong_A^m$ is also a bisimulation on S_A^{disc}. The number of equivalence classes of $\cong_A^m$ is $k \cdot (4m+3)^n$.*

Proof. We argue that $\cong_A^m$ is a bisimulation for positive m-definable A; the other parts of the proof are similar. Suppose that $(v, \mathbf{y}) \cong_A^m (w, \mathbf{z})$ and $(v, \mathbf{y}) \xrightarrow{\sigma} (v', \mathbf{y}')$. We must show that there exists a state $(w', \mathbf{z}')$ such that $(w, \mathbf{z}) \xrightarrow{\sigma} (w', \mathbf{z}')$ and $(v', \mathbf{y}') \cong_A^m (w', \mathbf{z}')$. First, assume that $\sigma \in \Sigma$. In this case there exists a control switch e with source $v = w$ such that $event(e) = \sigma$ and $(\mathbf{y}, \mathbf{y}') \in [\![jump(e)]\!]$, and $y_i = y_i'$ for each $i \notin update(e)$. Define $\mathbf{z}'$ by $z_i' = z_i$ for $i \notin update(e)$, and $z_i' = y_i'$ for $i \in update(e)$. By Lemma 3.1, $(\mathbf{z}, \mathbf{z}') \in [\![jump(e)]\!]$ and $\mathbf{z}' \in [\![inv(v')]\!]$. It follows that $(w, \mathbf{z}) \xrightarrow{\sigma} (v', \mathbf{z}')$.

Second, assume that $\sigma = 1$ (cf. Fig. 1). In this case $v' = v = w$, and $\mathbf{y}' - \mathbf{y} \in [\![flow(v)]\!]$. We must show that there exists a vector $\mathbf{z}'$ such that $\mathbf{z}' - \mathbf{z} \in [\![flow(v)]\!]$ and $\mathbf{y}' \approx_n^m \mathbf{z}'$ (notice that by Lemma 3.1, $\mathbf{y}' \approx_n^m \mathbf{z}'$ implies $\mathbf{z}' \in [\![inv(v)]\!]$). We do this one coordinate at a time. Fix $i \in \{1, \ldots, n\}$. Suppose that $y_i > m$. It follows that $y_i' > m$ and $z_i > m$, because A is positive. Choose any $c \in [\![flow(v)]\!]_i$, and define $z_i' = z_i + c$. Since $c \geq 0$, we have $y_i' \approx_1^m z_i'$. Now suppose that $y_i \leq m$. If $y_i \in \mathbb{N}$ then $z_i = y_i$, because $y_i \approx_1 z_i$. Define $z_i' = y_i'$. Then $z_i' - z_i = y_i' - y_i \in [\![flow(v)]\!]_i$. If $y_i \notin \mathbb{N}$ then $\lfloor y_i \rfloor < y_i, z_i < \lceil y_i \rceil$. The set $[\![flow(v)]\!]_i$ is an interval, say, with endpoints $a, b \in \mathbb{N}$ (it is easy to extend the argument to the case $b = \infty$). Thus $[\![flow(v)]\!]_i$ contains the open interval (a, b), and $y_i' \in [y_i + a, y_i + b]$. We show that there exists a number $c \in (a, b)$ such that $y_i' \approx_1 z_i + c$. Since $a, b \in \mathbb{N}$ and $y_i \approx_1 z_i$, it follows that $y_i + a \approx_1 z_i + a$ and $y_i + b \approx_1 z_i + b$. Thus the closed interval $[z_i + a, z_i + b]$ intersects the same $\approx_1$-equivalence classes as does $[y_i + a, y_i + b]$. Since neither $z_i + a$ nor $z_i + b$ is an integer, the same is true for the open interval $(z_i + a, z_i + b)$. Therefore there exists a number $c \in (a, b)$ such that $y_i' \approx_1 z_i + c$. ∎

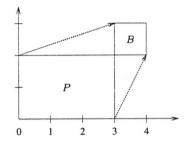

Fig. 1. Given a control mode v, consider the flow condition $flow(v) = (1 \leq \dot{x}_1 \leq 3 \wedge 1 \leq \dot{x}_2 \leq 2)$. Let $B = [\![3 \leq x_1 \leq 4 \wedge 2 \leq x_2 \leq 3]\!]$ and $P = [\![0 \leq x_1 \leq 3 \wedge 0 \leq x_2 \leq 2]\!]$. Then $Pre_1(\{v\} \times B) = \{v\} \times P$.

Corollary 3.1 *For every rectangular automaton A, the boolean combinations of rectangular state predicates for A form a symbolic execution theory for the discrete-time transition system S_A^{disc}.*

Corollary 3.2 *For positive or bounded rectangular automata, the discrete-time safety verification problem (in fact, LTL and CTL model checking) can be solved in PSPACE, and the discrete-time safety controller synthesis problem can be solved in EXPTIME.*

The LTL and CTL parts of the corollary follow from the facts that both model-checking problems can be solved in space logarithmic in the size of the transition system and polynomial in the size of the temporal formula [Kup95]. It should be noted that while in the same complexity class, the actual running times of the discrete-time algorithms for rectangular automata are better by a multiplicative exponential factor than the running times of the corresponding dense-time algorithms for timed automata. This is because there, the number of equivalence classes of the bisimilarity quotient is $\Omega(k \cdot n! \cdot (m+1)^n)$. By providing tight lower bounds, the following theorem shows that our algorithms are optimal. The second part of the theorem follows from Theorem 3.4 below.

Theorem 3.2 *For bounded timed automata, the discrete-time safety verification problem is hard for PSPACE [AD94], and the discrete-time safety control decision problem is hard for EXPTIME.*

3.2 Sampling-Controller Synthesis

The dense-time and discrete-time control problems are not realistic, as a controller may enforce arbitrarily many (even infinitely many) consecutive instantaneous jumps. A more natural control model for hybrid systems involves a controller that samples the plant state once per time unit, and then issues a command based upon its measurement. The command may cause a switch in control mode, after which the plant state evolves continuously for one time unit, before receiving the next command. We call this model "sampling control" to distinguish it from discrete-time control. Moreover, we wish to ensure that a proposition is avoided not only at the sampling points but also between sampling points. Given a rectangular automaton A, we define a third transition system, S_A^{sample}, such that (1) any control map behaves in a sampling manner and (2) the propositional regions are "large enough" so that they cannot be entered and left by a single flow transition of duration 1. For example, if π is a rectangular state predicate that maps each control mode of A to either *true* or *false*, then R_π is large enough. If the region of unsafe states is not large enough, this may be correctable by increasing the sampling rate (i.e., by reducing the unit of time).

Definition 3.2 [Sampling control] Let A be a rectangular automaton. A rectangular state predicate $\pi \in \Pi$ is *large enough* for A if there are no three states $(v, \mathbf{y}), (v, \mathbf{y}') \notin R_\pi$ and $(v, \mathbf{y}'') \in R_\pi$ such that $(v, \mathbf{y}) \stackrel{\delta}{\to} (v, \mathbf{y}'')$ and $(v, \mathbf{y}'') \stackrel{1-\delta}{\to} (v, \mathbf{y}')$ for some real $\delta \in (0, 1)$. Define $\Pi' \subset \Pi$ to be the set of rectangular state predicates that are large enough for A, and define $((v, \mathbf{y}), \lambda) \models' \pi$ iff $(v, \mathbf{y}) \models \pi$. The *sampling-control transition system* of A is defined by $S_A^{sample} = (Q \times \{control, plant\}, \Sigma \cup \{1\}, \Rightarrow, Q_I \times \{control\}, \Pi', \models')$, where the binary relation $\Rightarrow$ is defined by: (1) for each event $\sigma \in \Sigma$, we have $((v, \mathbf{y}), control) \stackrel{\sigma}{\Rightarrow} ((v', \mathbf{y}'), plant)$ iff $(v, \mathbf{y}) \stackrel{\sigma}{\to} (v', \mathbf{y}')$, and (2) $((v, \mathbf{y}), plant) \stackrel{1}{\Rightarrow} ((v', \mathbf{y}'), control)$ iff $(v, \mathbf{y}) \stackrel{1}{\to} (v', \mathbf{y}')$. Thus in the sampling-control transition system the controller and the plant take turns: first the controller specifies a jump transition, then one time unit passes in a flow transition, and so on. We refer to the safety control decision problem for transition systems of the form S_A^{sample}, for some rectangular automaton A, as the *sampling-control decision problem* for rectangular automata, and similarly for the sampling-controller synthesis problem. ∎

Theorem 3.3 *For positive or bounded rectangular automata, the sampling-controller synthesis problem can be solved in EXPTIME.*

Proof. Consider an n-dimensional positive or bounded rectangular automaton A. We reduce the sampling-control problems to discrete-time control problems by constructing a rectangular automaton $Ctrl(A)$ such that S_A^{sample} is isomorphic to $S_{Ctrl(A)}^{disc}$. Moreover, if A is positive, then $Ctrl(A)$ is positive, and if A is bounded, then $Ctrl(A)$ is bounded. Let $X_{Ctrl(A)} = X_A \cup \{x_{n+1}\}$ for a clock $x_{n+1} \notin X_A$. The control graph and events of $Ctrl(A)$ are identical to those of A. Let $inv_{Ctrl(A)}(v) = inv_A(v) \wedge 0 \leq x_{n+1} \leq 1$, let $init_{Ctrl(A)}(v) = init_A(v) \wedge x_{n+1} = 1$, let $jump_{Ctrl(A)}(e) = jump_A(e) \wedge x_{n+1} = 1 \wedge x'_{n+1} = 0$, and let $flow_{Ctrl(A)}(v) = flow_A(v) \wedge \dot{x}_{n+1} = 1$. It follows that in the discrete-time transition system $S_{Ctrl(A)}^{disc}$, jump transitions must alternate with flow transitions (of duration 1). Hence the map $f : Q_{Ctrl(A)} \to Q_A \times \{control, plant\}$, defined by $f(v, \mathbf{y}, 0) = (v, \mathbf{y}, plant)$ and $f(v, \mathbf{y}, 1) = (v, \mathbf{y}, control)$, is an isomorphism between the transition systems $S_{Ctrl(A)}^{disc}$ and S_A^{sample}. If A is m-definable with k control modes, by Theorem 3.1, the bisimilarity quotient of $S_{Ctrl(A)}^{disc}$ has no more than $k \cdot (4m+3)^{n+1}$ equivalence classes, which is singly exponential in the size of A. ∎

Lemma 3.2 *Let $G = (V_A, V_O, V_I, \to)$ be an AND-OR graph, and let R be a set of vertices of G. Define the transition system $S_G = (V_A \cup V_O, \Sigma, \to, V_I, \{\pi\}, \models)$ such that (1) $v \models \pi$ iff $v \in R$, (2) for all OR states $v \in V_O$, if $v \stackrel{\sigma}{\to} w$ and $v \stackrel{\sigma}{\to} w'$, then $\sigma = \sigma'$, and (3) for all AND states $v \in V_A$, if $v \stackrel{\sigma}{\to} w$ and $v \stackrel{\sigma}{\to} w'$ and $w \neq w'$, then $\sigma \neq \sigma'$. Then R is alternating reachable in G iff π is not avoidable in S_G.*

Theorem 3.4 *For bounded timed automata, the sampling-control decision problem is hard for EXPTIME.*

Proof sketch. We reduce the halting problem for alternating Turing machines using polynomial space [CKS81] to the sampling-control decision problem for bounded timed automata. Let M be an alternating Turing Machine with input s so that M uses space $p(|s|)$. Then M accepts s iff the unique final state u_F is alternating reachable in an AND-OR graph whose vertices are configurations of M. The set of configurations of M is $U \times \{1, \ldots, p(|s|)\} \times \Gamma^{p(|s|)}$, where U is the state set of M, the second component of the product gives the position of the tape head, and Γ is the tape alphabet. Without loss of generality, we assume that $\Gamma = \{0, 1, 2\}$, where 0 is the "blank" symbol. We first define a bounded positive rectangular automaton A whose states are configurations of M, and a proposition π_F, large enough for A, that is true exactly in the configurations containing u_F. This is done in a way consistent with Lemma 3.2, so that π_F is not avoidable in S_A^{sample} iff M accepts s. Then we turn A into a bounded timed automaton.

The automaton A uses $p(|s|)$ variables $x_1, \ldots, x_{p(|s|)}$ to store the tape contents. The set of control modes of A is $U \times \{1, \ldots, p(|s|)\}$. The invariant and flow conditions are constant functions: $inv(u, i) = \bigwedge_{j=1}^{p(|s|)}(0 \le x_j \le 2)$ and $flow(u, i) = \bigwedge_{j=1}^{p(|s|)}(\dot{x}_j = 0)$ for all u and i; thus flow transitions have no effect. The initial condition is defined by $init(u, i) = false$ except when u is the initial state u_I of M and $i = 1$; in that case, $init(u_I, 1) = \bigwedge_{j=1}^{|s|}(x_j = s_j) \wedge \bigwedge_{j=|s|+1}^{p(|s|)}(x_i = 0)$. Each transition t of M consists of a source state $u \in U$, a tape symbol $\gamma \in \Gamma$, and a list of triples (u_j, γ_j, d_j), where $u_j \in U$ is a target state, $\gamma_j \in \Gamma$ is written on the current tape cell, and $d_j \in \{-1, 1\}$ gives the direction moved by the tape head (there is exactly one transition for each source state u). For every transition $t = (u, \gamma, (u_j, \gamma_j, d_j)_{j \in J})$ of M, every tape position $1 \le i \le p(|s|)$, and every $j \in J$, we define in A a control switch $e_{t,i,j}$ with source (u, i) and target $(u_j, i + d_j)$. The jump condition $jump(e_{t,i,j})$ is $x_i = \gamma \wedge x_i' = \gamma_j \wedge \bigwedge_{k \ne i}(x_k' = x_k)$. If u is an AND state of M, then $event(e_{t,i,j}) = (u, i, j)$. If u is an OR state of M, then $event(e_{t,i,j}) = 0$. To turn A into a timed automaton, all variables are replaced by clocks, and between any two control switches of A, a sequence of $p(|s|)$ control switches is added, one for each clock, to subtract $p(|s|) + 1$ from each clock value. ∎

4 Beyond Rectangular Automata

Discrete-Time Undecidability Results We show that the pleasant properties of discrete-time rectangular automata (Theorem 3.1) depend on both conditions, (1) positivity or boundedness and (2) rectangularity. If either condition is violated, then already the discrete-time safety verification problem becomes undecidable.

Definition 4.1 [Triangular automaton] A *triangular automaton* A has the same components as a rectangular automaton, except that the predicates defining A may be triangular predicates, and need not necessarily be rectangular. ∎

Theorem 4.1 *The discrete-time safety verification problem (and thus the discrete-time control decision problem) is undecidable for the class of all rectangular automata, and also for the class of bounded positive triangular automata.*

Proof sketch. Both parts use a reduction from the halting problem for two-counter machines. For the first part, the reduction is simple, as counter values can be represented by variable values, as in [KPSY93]. For the second part, counter values must be encoded, so that the counter value c corresponds to the variable value $\frac{1}{2^c}$. For this purpose, the wrapping-clock technique of [HKPV95] can be modified as follows. The set $\{x_1, \ldots, x_n\}$ of dense-time clocks used for encoding counter values is simulated in discrete time by variables with the triangular flow condition $\dot{x}_1 = \cdots = \dot{x}_n$. Then the variables are enforced to represent valid encodings at those integer times when the wrapping clock shows 0. ∎

Generalized Rectangular Automata It is well-known that the pleasant properties of timed automata (Theorem 2.5) are preserved if rectangularity is relaxed to triangularity in invariant, initial, and jump conditions. We conclude with a similar observation for rectangular automata. A *generalized rectangular automaton* is a triangular automaton whose flow conditions are rectangular predicates. It follows from our arguments that for every generalized rectangular automaton A, the boolean combinations of triangular state predicates for A form a symbolic execution theory for the discrete-time transition system S_A^{disc}. Consequently, if A is a *bounded* generalized rectangular automaton, then S_A^{disc} has a finite bisimilarity quotient (which is identical to the region equivalence of timed automata [AD94], and finer by a multiplicative exponential factor than the equivalence of Theorem 3.1). For such automata, we can automatically synthesize sampling controllers that avoid triangular state predicates.

References

[ACH+95] R. Alur, C. Courcoubetis, N. Halbwachs, T.A. Henzinger, P.-H. Ho, X. Nicollin, A. Olivero, J. Sifakis, S. Yovine. The algorithmic analysis of hybrid systems. *Theoretical Computer Science*, 138:3–34, 1995.

[AD94] R. Alur, D.L. Dill. A theory of timed automata. *Theoretical Computer Science*, 126:183–235, 1994.

[AH97] R. Alur, T.A. Henzinger. Modularity for timed and hybrid systems. In *CONCUR: Concurrency Theory*, LNCS. Springer, 1997.

[BCM+92] J.R. Burch, E.M. Clarke, K.L. McMillan, D.L. Dill, L.J. Hwang. Symbolic model checking: 10^{20} states and beyond. *Information and Computation*, 98:142–170, 1992.

[CKS81] A.K. Chandra, D.C. Kozen, L.J. Stockmeyer. Alternation. *J. ACM*, 28:114–133, 1981.

[Hen95] T.A. Henzinger. Hybrid automata with finite bisimulations. In *ICALP: Automata, Languages, and Programming*, LNCS 944, pp. 324–335. Springer, 1995.

[Hen96] T.A. Henzinger. The theory of hybrid automata. In *Proc. 11th Symp. Logic in Computer Science*, pp. 278–292. IEEE, 1996.

[HH95] T.A. Henzinger, P.-H. Ho. Algorithmic analysis of nonlinear hybrid systems. In *CAV: Computer-Aided Verification*, LNCS 939, pp. 225–238. Springer, 1995.

[HHW97] T.A. Henzinger, P.-H. Ho, H. Wong-Toi. HYTECH: a model checker for hybrid systems. In *CAV: Computer-Aided Verification*, LNCS. Springer, 1997.

[HKPV95] T.A. Henzinger, P.W. Kopke, A. Puri, P. Varaiya. What's decidable about hybrid automata? In *Proc. 27th Symp. Theory of Computing*, pp. 373–382. ACM, 1995.

[HNSY94] T.A. Henzinger, X. Nicollin, J. Sifakis, S. Yovine. Symbolic model checking for real-time systems. *Information and Computation*, 111:193–244, 1994.

[Imm81] N. Immerman. Number of quantifiers is better than number of tape cells. *J. Computer and System Sciences*, 22:384–406, 1981.

[KPSY93] Y. Kesten, A. Pnueli, J. Sifakis, S. Yovine. Integration graphs: a class of decidable hybrid systems. In *Hybrid Systems*, LNCS 736, pp. 179–208. Springer, 1993.

[Kup95] O. Kupferman. *Model Checking for Branching-Time Temporal Logics*. PhD thesis, The Technion, Haifa, Israel, 1995.

[MPS95] O. Maler, A. Pnueli, J. Sifakis. On the synthesis of discrete controllers for timed systems. In *STACS: Theoretical Aspects of Computer Science*, LNCS 900, pp. 229–242. Springer, 1995.

[PV94] A. Puri, P. Varaiya. Decidability of hybrid systems with rectangular differential inclusions. In *CAV: Computer-Aided Verification*, LNCS 818, pp. 95–104. Springer, 1994.

[RW87] P.J. Ramadge, W.M. Wonham. Supervisory control of a class of discrete-event processes. *SIAM J. Control and Optimization*, 25:206–230, 1987.

Maintaining Minimum Spanning Trees in Dynamic Graphs

Monika R. Henzinger[1] and Valerie King[2]

[1] Systems Research Center, Digital Equipment Corporation, 130 Lytton Ave, Palo Alto, CA, USA 94301; email: monika@pa.dec.com.
[2] University of Victoria, Dept. of Computer Science, P.O. Box 3055, Victoria, BC., Canada, V8W 3P6; email: val@csr.uvic.ca

Abstract. We present the first fully dynamic algorithm for maintaining a minimum spanning tree in time $o(\sqrt{n})$ per operation. To be precise, the algorithm uses $O(n^{1/3} \log n)$ amortized time per update operation. The algorithm is fairly simple and deterministic. An immediate consequence is the first fully dynamic deterministic algorithm for maintaining connectivity and, bipartiteness in amortized time $O(n^{1/3} \log n)$ per update, with $O(1)$ worst case time per query.

1 Introduction

We consider the problem of maintaining a minimum spanning tree during an arbitrary sequence of edge insertions and deletions. Given an n-vertex graph G with edge weights, the *fully dynamic minimum spanning tree problem* is to maintain a minimum spanning forest F under an arbitrary sequence of the following update operations:

insert(u,v): Add the edge $\{u,v\}$ to G. Add $\{u,v\}$ to F if this reduces the cost of F, and return the edge of F that has been replaced.
delete(u,v): Remove the edge $\{u,v\}$ from G. If $\{u,v\} \in F$, then (a) remove $\{u,v\}$ from F and (b) return the minimum-cost edge e of $G \setminus F$ that reconnects F if e exists or return *null* if e does not exist.

In addition, the data structure permits the following type of query:

connected(u,v): Determine if vertices u and v are connected.

In 1985 [7], Fredrickson introduced a data structure known as *topology trees* for the fully dynamic minimum spanning tree problem with a worst case cost of $O(\sqrt{m})$ per update His data structure permitted connectivity queries to be answered in $O(1)$ time. In 1992, Eppstein et. al. [3, 4] improved the update time to $O(\sqrt{n})$ using the *sparsification technique*. If only edge insertions are allowed, the Sleator-Tarjan dynamic tree data structure [13] maintains the minimum spanning forest in time $O(\log n)$ per insertion or query. If only edge deletions are allowed ("deletions-only"), then no algorithm faster than the $\Omega(\sqrt{n})$ fully dynamic algorithm was known.

Using randomization, it was recently shown that the fully dynamic connectivity problem, i.e., the restricted problem where all edge costs are the same, can be solved in amortized time $O(\log^2 n)$ per update and $O(\log n)$ per connectivity query [9, 10]. However, this approach could not be extended to arbitrary edge weights, leaving the question open as to whether the fully dynamic minimum spanning tree problem can be solved in time $o(\sqrt{n})$.

In this paper we give a positive answer to this question: We present a fully dynamic minimum spanning tree data structure that uses $O(n^{1/3} \log n)$ amortized time per update and $O(1)$ worst case time per query when update time is averaged over any sequence of $\Omega(m_{in})$ updates, for m_{in} the initial size of the graph. Our technique is very different from [7].

The result is achieved in two steps: First, we give a deletions-only minimum spanning tree algorithm that uses $O(m'^{1/3} \log n + n^\epsilon)$ amortized time per update and $O(1)$ worst case time per query when the update time is averaged over any sequence of $\Omega(m_{in})$ updates. Here ϵ is any constant such that $0 < \epsilon < 1/3$, and m' is the number of *nontree* edges at the time of the update.

Then we present a general technique which, given a deletions-only minimum spanning tree data structure with a certain property, generates a fully dynamic data structure with the same running time as the deletions-only data structure. Let $f(m', n)$ be the amortized time per deletion in the deletions-only data structure with m' nontree edges and n vertices. The property required is that, upon inserting into the graph no more than m' edges at the same time (a "batch insertion"), the deletions-only data structure can be modified to reflect these insertions and up to m' subsequent deletions can be performed in a total of $O(m' f(m', n))$ time.

Using this technique, we develop a fully dynamic minimum spanning tree algorithm with *amortized* time per update of $O(m^{1/3} \log n)$, for a sequence of updates of length $\Omega(m_{in})$, where m is the size of G at the time of the update. In other words, letting $m_{(i)}$ denote the size of G (vertices and edges) after update i, the total amount of work for processing a sequence of updates of length l is $O(\sum_{i=0}^{l} m_{(i)}^{1/3} \log n)$. We then apply sparsification [3, 4] to reduce the running time for the sequence to $O(l n^{1/3} \log n)$.

Our result immediately gives faster *deterministic* fully dynamic algorithms for the following problems: connectivity, bipartiteness, k-edge witness, maximal spanning forest decomposition, and Euclidean minimum spanning tree. See [9] for all but the last reduction; see Eppstein [2] for the last reduction. For these problems, the new algorithm achieves an $O(n^{1/6}/\log n)$ factor improvement over the previously best *deterministic* running time. If randomization is allowed, however, much faster times are achievable [9, 10].

Additionally, improvements can be achieved in the following static problems (see [4, 3]): randomly sampling spanning forests of a given graph [6]; finding a color-constrained minimum spanning tree [8].

The paper is structured as follows: In Section 2 we give a deletions-only minimum spanning tree algorithm. In Section 3, we show how to use a sequence of deletions-only data structures to create a fully dynamic data structure.

2 Maintaining a minimum spanning tree–deletions-only

In this section, we give an algorithm which maintains a minimum spanning tree while edges are being deleted. The amortized update time is $O(m^{1/3} \log n)$ and the query time is $O(1)$ for queries of the form "Are vertices i and j connected?". Let $G = (V, E)$ be an undirected graph with edge weights. Without loss of generality, we assume that edge weights are distinct.

Initially, we compute the minimum spanning forest F of G. Let m'_{in} be the number of nontree edges in G initially and $k = {m'_{in}}^{1/3} \log n$. We sort the nontree edges by weight and partition them into m'_{in}/k levels of size k so that the k lightest are in level 0, the next k lightest are in level 1 and so on. The set of edges in a level i is denoted by E_i. In addition, all tree edges of the initial minimum spanning forest F are placed in level 0.

Throughout the algorithm, the level of an edge remains unchanged, and F denotes the minimum spanning forest. For $i = 0, 1, ..., (m'_{in}/k) - 1$, let F_i denote the minimum spanning forest of the graph with vertex set V and edgeset $\cup_{j \leq i} E_j$. (Initially, all $F_i = F$, but in later stages, an edge from any level may become a tree edge. Thus, $F_0 \subseteq F_1 \subseteq \ldots F_{(m'_0/k)-1} = F$.) Let $T_i(x)$ denote the tree in F_i which contains x and let $T(x)$ without the subscript denote the tree in F containing x.

The main idea is the following. If a nontree edge is deleted, then the minimum spanning forest F is unchanged. Suppose a tree edge $\{u, v\}$ in level i is deleted. Then for each F_j, $j \geq i$, the deletion splits the tree in F_j containing u and v into $T_j(u)$ and $T_j(v)$. We search for the minimum weight nontree edge e (called the "replacement edge") that connects $T(u)$ and $T(v)$ by gathering and then testing a set S of candidate edges on level i. If none is found, we repeat the procedure on level $i+1$, etc. until one is found or all levels are exhausted. We now describe the update operations:

delete(u, v): Delete edge $\{u, v\}$ from any data structures in which it occurs. If a tree edge $\{u, v\}$ from level i is deleted, then remove $\{u, v\}$ from F and search for a replacement by calling **Replace**(i, u, v). We refer to i as the level of the call to **Replace**.

In the algorithm below, the subroutine **Search** when applied to a tree in F_i finds all nontree edges in level i which are incident to the tree. A phase consists of the examination of a single edge. (Its exact definition and the details of **Search** are given in Section 2.2 below.)

Replace(i, u, v)

1. Alternating in lockstep, one phase at a time, **Search**$(T_i(u))$ and **Search**$(T_i(v))$ until $k/\log n$ phases are executed (Case A) or one of the searches has stopped (Case B).
   - Case A: Let S be the set of all nontree edges in level i.
   - Case B: Let S be the set of (nontree) edges produced by the **Search** that stopped.

2. Test every edge in S to see if it connects $T(u)$ and $T(v)$.
   - If a connecting edge is found, insert the minimum weight connecting edge into F and the data structures representing the F_j, $j \geq i$.
   - Else if i is not the last level, call **Replace**$(i+1, u, v)$.

2.1 Data Structures

The idea here is to use the ET-tree data structure developed in [9]: (1) to represent and update each tree in F, so that in constant time, we can quickly test if a given edge joins two trees; and (2) to represent each tree in an F_i in such a way that we can quickly retrieve nontree edges in E_i which are incident to the tree. To avoid excessive cost, we explicitly maintain only those F_i where i is a multiple of $m'^{1/3}_{in}/\log n$. An unpleasant consequence of this is that when retrieving nontree edges in E_i, other nontree edges are also retrieved.

Below, we refer to input graph vertices as "vertices" and use "node" to mean a nodes of the B-tree in which we store the "ET-sequences."

ET-trees: An *ET-sequence* is a sequence generated from a tree by listing each vertex each time it is encountered ("an occurrence of the vertex") as a tree is searched depth-first. Each ET-sequence is stored in a B-tree of degree d. This allows us to implement the deletion or insertion of an edge in the forest as follows: we split a tree by deleting an edge or join two trees by inserting an edge in time $O(d \log_d n)$, using a constant number of splits and joins on the corresponding B-trees. Also we can test two vertices of the forest to determine whether they are in the same tree in time $O(\log_d n)$. See for example [1, 11] for operations on B-trees. If $d = n^\alpha$, for α a positive constant, then the join and split operations take time $O(d)$ and the test operation takes time $O(1)$. We refer to the B-trees used to store ET-sequences as ET-trees.

This data structure allows us to keep information about a vertex so that the cumulative information about all vertices in a tree may be maintained. For example, we may keep the number of nontree edges incident to a vertex at one designated occurrence of the vertex. Then each internal node of the ET-tree stores the sum of the numbers of nontree edges kept with designated occurrences in its subtree. In a degree d ET-tree, each split or join operation or each change to the number associated with an occurrence requires the adjustment of $O(\log_d n)$ internal nodes with each adjustment taking $O(d)$ timesteps.

We maintain the following data structures.

- Each edge is labelled by its level and a bit which indicates if it is a tree edge.
- Let $k' = \max\{m'^{1/3}_{in} \log n, n^\epsilon\}$, for any constant $0 < \epsilon \leq 1/3$. Each tree in F is represented as an ET-sequence which is stored in a degree k' B-tree.
- Let $c = m'^{1/3}_{in}/\log n$. We map each level i to the j which is the largest multiple of c no greater than i by the function $f(i) = c\lfloor i/c \rfloor$.
  For each level j such that $c|j$ ("c divides j"):

- we represent each tree in F_j as an ET-sequence which is stored in a binary B-tree;
- for each vertex v, we create a list $L_j(v)$ which contains:
  (i) all nontree edges incident to v which are in any level $i \in f^{-1}(j)$ and;
  (ii) all tree edges incident to v which are in any level $i > j, i \in f^{-1}(j)$.
- We mark each designated occurrence of a vertex v whose list $L_j(v)$ is nonempty. Each internal node of the ET-tree is marked if its subtree contains a marked occurrence.

2.2 The Search routine

Search$(T_i(x))$ returns all nontree edges in level i incident to $T_i(x)$. It begins by searching $T_{f(i)}(x)$ which is a subtree of $T_i(x)$. It proceeds by examining all edges in $L_{f(i)}(v)$ for all vertices v in the tree being searched. Nontree edges in level i are picked out and tree edges in levels $i', f(i) < i' \leq i$ are followed to other trees of $F_{f(i)}$ which are then searched in turn. Note that all such tree edges lead to other trees of $F_{f(i)}$ which are subtrees of $T_i(x)$. A *phase* of the algorithm consists of the examination of one edge e in a list L.

Search$(T_i(u))$

1. $S' \leftarrow \emptyset$;
2. $treelist \leftarrow T_{f(i)}(u)$;
3. Repeat until $treelist$ is empty:
   - Remove an ET-tree from the $treelist$.
   - For each marked vertex u in the ET-tree and for each edge e in each $L_{f(i)}(u)$:
     - If $\{u,v\}$ is a nontree edge on level i, add it to the set of edges to return.
     - Else if $\{u,v\}$ is a tree edge on level l such that $l \leq i$, then add $T_{f(i)}(v)$ to $treelist$.

2.3 Analysis

Initialization: We compute the minimum spanning forest F, create the ET-trees for F_j, for each j such that $c|j$, and partition the nontree edges by weight. Recall that m'_{in} is the number of nontree edges in the initial graph. Let t be the number of edges in the initial minimum spanning forest. The creation of all the lists L takes time proportional to the number of nontree edges m'_{in}. The building of ET-trees for F and all F_j such that $c|j$ and the marking of internal nodes takes time proportional to the size of each forest or $O(((m'_{in}/k)/c)t + m'_{in}) = O(m'^{1/3}_{in}t + m'_{in})$.

Deletions of nontree edges: Deleting a nontree edge on any level may require resetting the bit of an occurrence of a vertex in some ET-tree, which may require resetting bits on all internal nodes on the path to the root in $O(\log n)$ time.

Deletions and insertions of tree edges: Deleting a tree edge takes $O(k')$ time to delete it from the ET-tree of F and $O(\log n)$ time to delete it from the ET-tree of each F_j such that $c|j$, for a total of $O(k' + ((m'_{in}/k)/c)\log n)$ time per edge. Inserting a replacement edge takes the same time.

Finding a replacement edge: We first analyze the cost of **Search**. Let the *weight* $w(T)$ of a tree T of some F_i be $\sum |L_{f(i)}(v)|$ summed over all vertices v in T. It costs $O(\log n)$ to move down the path from the root to a leaf in an ET-tree to find a marked occurrence of a vertex, or to move up a tree from an occurrence to the root. Thus, the cost of **Search**$(T_i(x))$ is $O(\log n)$ times the number of edges examined, or $O(w(T_i(x))\log n)$, if **Search** is carried out until it ends, and $O(k)$ if it is run for $k/\log n$ phases.

In **Replace**(u,v,i), if $w(T_i(u)) \leq w(T_i(v))$, then we refer to $T_i(u)$ as the *smaller component* T_1; otherwise T_1 is $T_i(v)$. The cost of a call to **Replace**(u,v,i) is the cost of the **Search** plus the cost of testing each edge in S. The number of edges in S is $O(\min\{k, w(T_1)\})$. We may use the k'-degree ET-tree representation for F to test each edge at cost $O(1)$. Thus the cost of a call to **Replace** is $O(\min\{k, w(T_1)\log n\})$.

To pay for these costs: We charge the cost of a call to **Replace**(u,v,i) to level i if no replacement edge is found on that level. In that case, a tree of F_i which was split by the deletion remains split. Otherwise, we charge the cost to the deletion. In addition, we charge the cost of modifying F to the deletion so the total cost charged to the deletion is $O(\min\{k, w(T_1)\}\log n\} + ((m'_{in}/k)/c)\log n + k') = O(((m'_{in}/k)/c)\log n + k')$.

Claim 1 $O(\sum w(T_1))$ *summed over all smaller components* T_1 *which split from a tree* T *on any given level during all* **Replace** *operations is* $O(w(T)\log n)$.

The proof of the claim is not hard and follows [5]. The details are omitted here.

There are at most k edges per level (except for level 0, which has at most k nontree edges). Each $L_j(v)$ consists of edges from c levels. Since level 0 tree edges do not belong to any list $L_j(v)$, the maximum weight of a tree $w(T)$ is ck. Thus the total cost charged to a level is $O(ck\log^2 n)$. Summing over all levels we have $O((m'_{in}/k)(ck\log^2 n) = O(m'_{in} c\log^2 n)$, or an amortized cost per deletion of $O(c\log^2 n) = O(m'^{1/3}_{in}\log n)$, if $\Omega(m'_{in})$ edges are deleted.

The cost charged to each deletion is $O((m'_{in}/ck)(\log n) + k')$. For $k' = \max\{m'^{1/3}_{in}\log n, n^\epsilon\}$ and $c = m'^{1/3}_{in}/\log n$, this is $O(m'^{1/3}_{in}\log n + n^\epsilon)$.

To summarize the cost of initialization when amortized over $\Omega(m_{in})$ operations is $O(m'^{1/3}_{in})$ and the cost per deletion of an edge and finding replacement edges, when amortized over $\Omega(m'_{in}))$ operations is $O(m'^{1/3}_{in}\log n + n^\epsilon)$. Thus for a sequence of $\Omega(m_{in})$ operations, the amortized time per update is $O(m'^{1/3}_{in}\log n + n^\epsilon)$.

Finally, we note that the query of the form "Are nodes i and j connected?" may be answered using the ET-tree data structure for F in $O(1)$ time.

3 Allowing insertions

As in the previous section, we assume all edge weights are unique. We refer to the current minimum spanning forest of G as the MST. Let m' be the number of nontree edges in the current graph.

Let $s = \lceil \lg m' \rceil$. Initially, we build and maintain s simultaneous deletions-only data structures $A_s, A_{s-1}, .., A_1$ and a set of edges B. We call this the *composite data structure*. We maintain the MST in a Sleator-Tarjan dynamic tree [13] and also in an ET-tree of degree $\max\{m'^{1/3} \log n, n^\epsilon\}$.

Below, we distinguish between the number of edges inserted into G and the number of edge insertions into B, as an edge of G may be inserted more than once into B even though it has not been deleted and reinserted into G. The minimum spanning forests of the deletions-only data structures are referred to as *local spanning forests*. A *local nontree edge* of an A_i is an edge which is not in A_i's local spanning forest or the MST. We will see that every nontree edge of G is a local nontree edge of some A_i or B, but may also be a tree edge in a local spanning forest of an $A_j, j \neq i$.

Initially A_s is the deletions-only data structure described in the previous section, with $F = MST$ and the set of local nontree edges being all nontree edges of G, and the parameter m'_{in} set to 2^s. The set B is empty and the remaining A_j, $1 \leq j < s$, are initialized ("built") as though the edges of the MST were the only edges in A_j, i.e., they contain no nontree edges. The set B is empty.

For $i = 1, \ldots, s$, let $m_i = 2^i$, $k_i = m_i^{1/3} \log m_i$, and $k'_i = \max\{m_i^{1/3} \log n, n^\epsilon\}$. When an edge is inserted into G, it is placed into B or into the MST.

Let x_i be the number of local nontree edges in $\cup_{j \leq i} A_j \cup B$. Each A_i is built (or rebuilt) when i is the smallest index such that $m_i \geq x_i$ and the number of edges in B has increased to $m'^{1/3}$. At that time, B is emptied and all local nontree edges $\cup_{j \leq i} A_j$ and edges in B are removed from $A_j, j < i$, and B and placed into A_i. Then A_i becomes the deletions-only data structure described in the previous section, which is initialized (or reinitialized) to contain the tree edges of the MST and the local nontree edges previously contained in $\cup_{j \leq i} A_j$ and the edges B. Thus, throughout the algorithm, B contains fewer than $m'^{1/3}$ edges, i.e., the most recent insertions into B, which have not yet been added to some A_j and for $j < \lfloor \lg m'^{1/3} \rfloor$ A_j never contains any nontree edges. These A_j are maintained in the event that they will be used later, if m' is reduced.

To insert edge e into G: Use the dynamic tree to find the maximum weight edge f on the path between e's endpoints in the MST. If e is lighter than f, remove f from the MST, and insert f into B. Else insert e into B.

To delete an edge e from G: (1) Delete e from all data structures in which it appears. (2) For each A_i which contained e in its local spanning forest, update the A_i by determining e's local replacement edge e' (if there is one). Insert e' into A_i's local forest and into B, if it is not already there. (3) If e was in the MST, then for each local replacement edge e' and each edge in B, use the ET-tree representation of the MST to determine which of those edges connect the two subtrees resulting from the deletion of e. Insert the lightest connecting edge into the MST.

3.1 Proof of correctness

Our algorithm maintains the following invariant:

Invariant: Every edge in the local forest of some A_i is (1) in the MST, or (2) is a local nontree edge in some $A_j, j \neq i$, or (3) is in B.

Lemma 2. *The invariant stated above holds throughout the execution of the algorithm.*

The proof of the lemma is straightforward and is omitted here.

The correctness of the algorithm follows easily from the invariant. We use the well-known fact that an edge is in the minimum spanning tree iff it is not the heaviest edge in any cycle ("red rule" [14]). We also note that every edge in the composite data structure is an edge in G.

Let e be an edge of the MST which is deleted. Let e' be the correct replacement edge. Consider the state of the composite data structures right before the deletion of e. By the invariant, since e' was not in the MST, it was a local nontree edge in some A_i or in B. If e' was in B it would be checked in Step 2 above.

If e' was a local nontree edge in A_i, then consider the subgraph G' of G whose edgeset consists of edges in A_i. Since e' is the correct replacement edge for e in the MST then after e's deletion, e' is not the heaviest edge in any cycle of G and therefore is not the heaviest edge of any cycle of G'. Hence, after e's deletion, e' becomes a local forest edge, i.e., e' is a local replacement edge for e in A_i. Recall that e' is the minimum weight edge which connects the two subtrees of the MST resulting from the deletion of e. Thus, e' is the lightest connecting edge from among the edges of B and the set of local replacement edges, and is chosen in Step 2 by the algorithm.

3.2 Implementation and analysis

At the start of the algorithm, A_i for $i \leq s$ are built. After that, A_i for $i \leq s$ may be "rebuilt". Depending of the value of m', $A_{s+1}, A_{s+2},\ldots$ may be built later. We first consider the (one-time) cost of building the A_i's, then the cost of their rebuilding, and finally the cost of maintaining the A_i between rebuilds.

Initialization of the A_i: Let m_{in} be the size of the initial graph (number of vertices plus edges), let m'_{in} be the initial number of nontree edges, and for each operation let m be the size of the current graph. Recall that the total cost of initialization for a deletions-only data structure with m_i nontree edges is $O(m_i^{1/3} n + m_i)$ and that we are given a sequence of $\Omega(m_{in})$ operations.

We will amortize the building of the first $\lceil \lg m_{in} \rceil$ A_i's to the sequence of $\Omega(m_{in})$ operations, even though only s A_i's are built initially. If more than

$\lceil \lg m_{in} \rceil$ A_i's are necessary at some point, we know that at this point $m > m_{in}$ and at least m_{in} insertions happened. Let $A_{s_{max}}$ be the largest deletions-only data structure built during the execution of the algorithm where $2^{s_{max}} > m_{in}$. Then there was a sequence of $\Omega(2^{s_{max}})$ operations during which m was $\Omega((2^{s_{max}}))$ and we can amortize the initialization cost over these operations.

The total cost of initializing the s_{max} A_i's is $O(\sum_{i=1}^{s_{max}} 2^{i/3}n + 2^i) = O(2^{s_{max}/3}n + m_{in})$. The average cost over a sequence of $\Omega(2^{s_{max}})$ operations is thus $O(2^{2s_{max}/3}n)$, which is $O(m^{1/3})$ per operation.

Rebuilding: We create ET-trees for the new A_i by modifying the ET-trees for the previous A_i. For each i, we keep a list of all changes made to each ET-tree of A_i since the last rebuild, and a list of all changes made to the MST. We use this list to first restore all the ET-trees for A_i to their previous state when A_i was last built or rebuilt, MST_{old}, by undoing each change, edge by edge. We then transform each MST_{old} to MST, edge by edge.

The cost of restoring the ET-trees of A_i to MST_{old} is charged to operations on the deletions-only data structure A_i which caused the initial change. This results in only a doubling of cost per operation, as the cost for inserting a tree edge into an ET-tree is the same as for deleting a tree edge.

The cost of transforming MST_{old} to MST is charged to the update operation that causes the change in MST (each update causes $O(1)$ changes in MST) as follows: For each A_i, there are $m_i^{1/3}$ forests of ET-trees represented by binary B-trees and one forest (the ET-tree for the local spanning forest F) represented by a degree-k_i' B-tree. Thus, a single tree edge insertion or deletion costs $O(m_i^{1/3} \log n + k_i')$ for A_i. Note that for each i one change to the MST contributes to the cost of only one rebuild of A_i. The total cost per change over all levels is $O(\sum_{i=1}^{s} m_i^{1/3} \log n + k_i' + \log n) = O(m'^{1/3} \log n + n^\epsilon \log n)$.

Also, when A_i is rebuilt, all local nontree edges from $A_j, j \leq i$ and B are moved from A_j and inserted into A_i. That is, the edges are sorted by weight, assigned to levels in A_i, and put in the appropriate list L. The bits on the internal nodes of ET-trees for $A_j, j \leq i$ are set appropriately. Since each local nontree edge is stored in only one ET-tree on a level, the cost of moving a single local nontree edge is $O(\log n)$. Thus, the total cost is $O(m_i \log n)$. Since A_{i-1} is not rebuilt, $x_{i-1} > m_{i-1} = m_i/2$. We amortize this cost by charging $O(\log n)$ to each edge in $\cup_{j<i} A_j \cup B$, i.e. each edge that is newly added to A_i. We show below (type (3) charges) how to amortize these costs over the update operations.

Maintaining the deletions-only data structures: After a rebuild in A_i there are at most m_i nontree edges in A_i. In Section 2, we have two types of charges: (1) the cost charged to each deletion in a deletions-only data structure which is $O(m_i^{1/3} \log n + n^\epsilon)$ and (2) the cost charged to all the levels which is $O(m_i^{1/3} \log n)$ per nontree edge. Additionally, the rebuilding of an A_i above charged $O(\log n)$ to each nontree edge in A_i. We call these costs type (3) charges.

Type (1) charges: When there is a deletion in G in the fully dynamic data structure, an edge (or one of its copies) may be deleted from each of $A_i, i =$

$1,\ldots s$. We may charge that deletion in G with the (1) charges for all levels for a total cost of $O(\sum_i m_i^{1/3} \log n + n^\epsilon) = O(m'^{1/3} \log n + n^\epsilon \log n)$.

Type (2) and (3) charges: As a special case, the charges incurred by the first deletions-only data structure containing nontree edges A_s must be amortized over the initial sequence of $\Omega(m_{in})$ operations which follow its initialization. Since each A_i, $i < s$ is initialized to contain no nontree edges, there are no type (2) and (3) for these data structures until they are rebuilt.

Note that the A_i, $i > s$ contain nontree edges when initialized. The type(2) and type(3) charges for their building and rebuilding and the rebuilding of the other A_i $i \leq s$ are amortized over the insertions which occurred previous to its building or rebuilding, as analyzed below.

Suppose A_i is rebuilt. Since A_{i-1} was not rebuilt, $x_{i-1} > m_{i-1} = m_i/2$ at the time of the rebuilding of A_i. Thus, $\Omega(m_i)$ insertions into B occurred since the previous rebuild of A_i, and $\Omega(m_i)$ of these occurred when the graph had $\Omega(m_i)$ nontree edges. Thus, we may charge each insertion into B with $O(\sum_i^{s'} m_i^{1/3} \log n + n^\epsilon) = O(m_{s'}^{1/3} \log n + n^\epsilon \log n)$ where $s' = 2^{\lceil \lg m' \rceil}$ where m' is the number of nontree edges in G when the insertion occurred.

To amortize costs over insertions into G, rather than B, we use the following simple but crucial observation: When an edge is inserted into B *that edge may contribute to the type (2) and (3) costs for A_i (when it belongs to A_i) iff it increases x_i.* Note that $m_i \geq x_i \geq x_{i-1} > m_{i-1} = m_i/2$. We charge $O(m_i^{1/3} \log n)$ to each local nontree edge inserted into A_{i+1} to pay for the type (2) and (3) charges while the edges are in A_i.

We examine the types of insertions into B to see how they affect x_i: (a) when an edge is first inserted into B, i.e., when the edge is inserted into G; (b) when an edge is replaced in the MST; (c) when an edge is deleted in G and it is replaced in up to s local spanning forests. The first two cases result in a single insertion into B. The third case may cause up to s' insertions. However, the s insertions do not affect all A_i the same. Each insertion in this case results from a local nontree edge e becoming a local forest edge. Hence if this occurs in some $A_j, j \leq i$, the increase of x_i resulting from the insertion of a copy of e into B is offset by the decrease of x_i caused by the change in status of e from a local nontree edge to a local tree edge. Thus $x_{s'}$ is unchanged by a case-(c) insertion into B, $x_{s'-1}$ is changed by at most 1, and in general, x_i is changed by at most $s' - i$. The type (2) and (3) cost per deletion is $O(\sum_i (s'-i) m_i^{1/3} \log n) = O(\sum i (m_{s'}/2^i)^{1/3} \log n) = O(m'^{1/3} \log n)$.

Thus the deletion cost per update operation is $O(m'^{1/3} \log n + n^\epsilon \log n)$.

Insertion cost: Testing a newly inserted edge to see if it should be added to the MST using the Sleator-Tarjan dynamic trees is an $O(\log n)$ cost operaton. Adding an edge to B can be done in constant time, as B is an unsorted list.

Summary: For rebuilding and maintaining the deletions-only data structures, the algorithm achieves an amortized cost of $O(m'^{1/3} \log n + n^\epsilon)$ per update, where m' is the number of nontree edges in the graph, for processing a sequence of $\Omega(m_{in})$

operations, where m_{in} is the initial size of the graph (vertices plus edges). For the initializations of the deletions-only data structures, the amortized cost per update is $O(m^{1/3} \log n)$, where m is the size of the graph at the time of the update, for a sequence of $\Omega(m_{in})$ operations.

Note: For unweighted graphs, a simpler fully dynamic data structure can be constructed which uses only one deletions-only data structure and adds levels as needed. The details are omitted here.

References

1. T. Corman, C. Leiserson, and Rivest. *Introduction to Algorithms.* MIT Press (1989), p. 381-399.
2. D. Eppstein, "Dynamic Euclidean minimum spanning trees and extrema of binary functions", Disc. Comp. Geom. 13 (1995), 111-122.
3. D. Eppstein, Z. Galil, G. F. Italiano, "Improved Sparsification", Tech. Report 93-20, Department of Information and Computer Science, University of California, Irvine, CA 92717.
4. D. Eppstein, Z. Galil, G. F. Italiano, A. Nissenzweig, "Sparsification - A Technique for Speeding up Dynamic Graph Algorithms" *Proc. 33rd Symp. on Foundations of Computer Science,* 1992, 60–69.
5. S. Even and Y. Shiloach, "An On-Line Edge-Deletion Problem", *J. ACM* 28 (1981), 1–4.
6. T. Feder and M. Mihail, "Balanced matroids", *Proc. 24th ACm Symp. on Theory of Computing,* 1992, 26–38.
7. G. N. Frederickson, "Data Structures for On-line Updating of Minimum Spanning Trees", *SIAM J. Comput.,* 14 (1985), 781–798.
8. G. N. Frederickson and M. A. Srinivas, "Algorithms and data structures for an expanded family of matroid intersection problems", *SIAM J. Comput.* 18 (1989), 112-138.
9. M. R. Henzinger and V. King. Randomized Dynamic Graph Algorithms with Polylogarithmic Time per Operation. *Proc. 27th ACM Symp. on Theory of Computing,* 1995, 519–527.
10. M. R. Henzinger and M. Thorup. Improved Sampling with Applications to Dynamic Graph Algorithms. To appear in *Proc. 23rd International Colloquium on Automata, Languages, and Programming (ICALP),* LNCS 1099, Springer-Verlag, 1996.
11. K. Mehlhorn. "Data Structures and Algorithms 1: Sorting and Searching", Springer-Verlag, 1984.
12. H. Nagamochi and T. Ibaraki, "Linear time algorithms for finding a sparse k-connected spanning subgraph of a k-connected graph", *Algorithmica* 7, 1992, 583–596.
13. D. D. Sleator, R. E. Tarjan, "A data structure for dynamic trees" *J. Comput. System Sci.* 24, 1983, 362–381.
14. R. E. Tarjan, *Data Structures and Network Flow,* SIAM (1983) p. 71.

Efficient Splitting and Merging Algorithms for Order Decomposable Problems

(Extended Abstract)

Roberto Grossi * and Giuseppe F. Italiano **

Abstract. We present a general and novel technique for solving decomposable problems on a set S whose items are sorted with respect to $d > 1$ total orders. We show how to dynamically maintain S in the following time bounds: $O(\log p)$ for the insertion or the deletion of a single item, where p is the number of items currently in S; $O(p^{1-1/d})$ for splits and concatenates along any total order; $O(p^{1-1/d})$ plus an output sensitive cost for rectangular range queries. The space required is $O(p)$. We provide several applications of our technique ranging from two–dimensional priority queues and d–dimensional search trees to concatenable interval trees. This allows us to improve many previously known results on decomposable problems under split and concatenate operations, such as membership query, minimum–weight item, range query, and convex hulls. Our technique is suitable for efficient external memory implementation.

1 Introduction

Let $\mathcal{P}$ be a searching problem defined on an input set S with p items, and let $\mathcal{P}(x, S)$ denote its solution for a query item x. Problem $\mathcal{P}$ is decomposable if we can find an answer to query $\mathcal{P}(x, S)$ by first partitioning set $S = S' \cup S''$ and computing the answers to queries $\mathcal{P}(x, S')$ and $\mathcal{P}(x, S'')$ recursively, and then combining them through a suitable operator $\Diamond$. Formally, $\mathcal{P}$ is said to be $f(p)$–decomposable if and only if $\mathcal{P}(x, S) = \Diamond(\mathcal{P}(x, S'), \mathcal{P}(x, S''))$ for any partition $S = S' \cup S''$ and any query item x, where $\Diamond$ is an operator whose computation requires $O(f(p))$ time. (We assume that function $f(p)$ is smooth, i.e., $f(O(p)) = O(f(p))$, and nondecreasing.) Some examples of $O(1)$–decomposable searching problems include: membership queries (with $\Diamond$ being the logical–or function); closest point queries (with $\Diamond$ the minimal distance); range queries (with $\Diamond$ the list append operation). Convex hull searching is not decomposable as the fact that a point $x \in S$ belongs to the convex hull of S' or S'' does not necessarily imply that x belongs

---

* Dipartimento di Sistemi e Informatica, Università di Firenze, Italy. Email: grossi@dsi2.dsi.unifi.it. Part of this work was done while visiting ICSI, Berkeley.
** Dipartimento di Matematica Applicata ed Informatica, Università "Ca' Foscari" di Venezia, Italy. Email: italiano@dsi.unive.it. Work supported in part by CEE under ESPRIT LTR Project no. 20244 (ALCOM–IT), by the Italian MURST Project "Efficienza di Algoritmi e Progetto di Strutture Informative", and by a Research Grant from University of Venice "Ca' Foscari". Part of this work was done while at University of Salerno and while visiting ICSI, Berkeley.

to the convex hull of $S = S' \cup S''$. The definition of decomposable *search* problems can be extended also to the decomposable *set* problems in which the query item is not specified (e.g., finding the minimum–weight item, where $\Diamond$ is the minimum), and we shall denote a generic solution to a decomposable problem $\mathcal{P}$ by $\mathcal{P}(S)$. Let $d > 1$ total orders $\prec_1, \ldots, \prec_d$ be defined on S, and let $\prec_i$ be a given total order, $1 \leq i \leq d$. A problem $\mathcal{P}$ is $f(p)$-*order* decomposable with respect to total order $\prec_i$ if $\mathcal{P}(S) = \Diamond(\mathcal{P}(S'), \mathcal{P}(S''))$ for any *ordered partition* $S = S' \cup S''$ (i.e., $x' \prec_i x''$ for all $x' \in S'$ and $x'' \in S''$), where operator $\Diamond$ takes $O(f(p))$ time. Problem $\mathcal{P}$ is $f(p)$-*order decomposable* if it is $f(p)$-order decomposable with respect to any total order $\prec_i$, $1 \leq i \leq d$. Convex hull searching is $O(\log p)$-order decomposable. Other examples of order decomposable problems include multidimensional range queries and Voronoi diagrams, and many other decomposable problems in basic data structures, computational geometry, database applications and statistics [7, 17, 21].

In this paper, we present a general technique for maintaining a *dynamic* set S with d total orders, for constant d, under insertions of a single item, deletions of a single item, and re–arrangements of any of the total orders $\prec_1, \ldots, \prec_d$ on S by means of split and concatenate operations. Our queries involve finding the solution $\mathcal{P}(R)$ for only the items in the subset $R \subseteq S$ identified by some ranges in the orders $\prec_1, \ldots, \prec_d$. More formally, we introduce the following *multiordered set splitting and merging problem*:

split$(S, z, \prec_i)$: Split S into S' and S'' according to item z and the specified total order $\prec_i$ ($1 \leq i \leq d$). That is, $x' \prec_i z$ and $z \prec_i x''$ for all $x' \in S'$ and $x'' \in S''$. S is no longer available after this operation.

concatenate$(S', S'', \prec'_i, \prec''_i)$: Combine S' and S'' together according to their respective i-th total orders $\prec'_i$ and $\prec''_i$ ($1 \leq i \leq d$) into a new set $S = S' \cup S''$. The items in the resulting set S undergo the new order $\prec_i$ obtained by concatenating $\prec'_i$ and $\prec''_i$. That is, $x \prec_i y$ in S if and only if either (a) $x \prec'_i y$ and $x, y \in S'$; or (b) $x \prec''_i y$ and $x, y \in S''$; or (c) $x \in S'$ and $y \in S''$. S' and S'' are no longer available after this operation.

insert(z, S): Insert item z into set S according to all orders $\prec_1, \ldots, \prec_d$.

delete(z, S): Delete item z from set S.

range$(\langle a_1, b_1 \rangle, \ldots, \langle a_d, b_d \rangle, S)$: Let $R = \{z \in S : a_i \prec_i z \prec_i b_i, \text{ for } 1 \leq i \leq d\}$. Find the solution $\mathcal{P}(R)$ to problem $\mathcal{P}$ restricted to region R only.

For $d = 1$, the recursive nature of order decomposable problems gives an immediate tree structure, and each of the above operations can be simply implemented in $O(f(p) \log p)$ time by using a 2–3–tree [2]. Maintaining $d > 1$ total orders on the same set S, while splitting or merging each order independently of the others, makes things much more complicated than this simple case. In the case of two or more different orders, indeed, there are some technical difficulties, which are mainly due to the interplay among different orders.

Related Work. Decomposable problems were first introduced by Bentley [6] for dynamizing static data structures, while other dynamization techniques were introduced in [7, 15, 18, 19, 24]. All these techniques rely on two main methods, the *equal block method* [14, 15, 18] and the *logarithmic method* [6, 7, 24], in which a big data structure is decomposed into small data structures, called *blocks*; the number of blocks is properly tuned so as to obtain a good tradeoff between queries and updates. Some lower bounds on the best possible tradeoff were given in [7, 16]. Optimal solutions were obtained by combining the equal block and the logarithmic method by means of the amortized solution in [19] and by the global rebuilding technique yielding worst–case bounds in [23, 25]. The notion of order decomposable problems was first introduced in [20] by generalizing the results of [22] and was independently presented in [10]. Solving an ordered decomposable problem only for the items contained in an input rectangular region can be done by range queries on quad–trees [9] and k-d trees [5], but it is difficult to keep them balanced (e.g., see [26, 27]). Many other elegant data structures for range queries were devised subsequently and we refer the reader to [8] for a comprehensive survey on this topic and a list of references. Among them, [28] and [29] show how to combine decomposable problems and range queries together so as to add some range restrictions to dynamic data structures. Split and concatenate operations were subsequently introduced in [11, 13] for a set of multidimensional points in addition to the standard operations: range queries, insertions and deletions. Specifically, the divided k–d trees [11] for a set of p items supported a range, a split or a concatenate operation in $O(p^{1-1/d} \log^{1/d} p)$ time and an insertion or a deletion in $O(\log p)$ time, with $O(p)$ space. In [13], a general technique, based on the ordered equal block method, was described for solving order decomposable problems and producing efficient concatenable data structures in $O(p)$ space. The following time bounds were obtained for a split or concatenate: $O(\sqrt{p} \log p)$ in concatenable interval trees, $O(p^{1-1/d} \log p)$ in d–dimensional 2-3–trees and $O(\sqrt{p \log p} \log p)$ in a data structure for convex hulls. The bound for insertions and deletions of items is $O(\log p)$ amortized, except for the $O(\log^2 p)$ amortized bound in the data structure for convex hulls. The range query bounds equal the split/concatenate cost plus an output sensitive cost $O(occ)$, where occ is the size of the output reported by the query. Although the range queries in [28] and [29] are faster than the ones in [13], the solutions in [13] support efficient splits and concatenates, require less space and can be used to obtain an efficient dynamic version of static data structures.

Our results. In this paper, we present a novel technique for solving order decomposable problems on S under insertions, deletions, splits, concatenates and range queries, yielding new and efficient concatenable data structures for dimension $d > 1$. All these data structures are based on a new multidimensional data structure, which we call the *cross–tree*. Differently from the approach of [13], our general technique is based more on simple geometric properties rather than on underlying sophisticated data structures, and exploits the fact that some data structures can be built on sorted items more efficiently. By using our technique we maintain a set S of p items in $O(p)$ space with the following *worst–case* time bounds: $O(\log p)$ for the insertion or the deletion of a single item, and $O(p^{1-1/d})$ for splits and concatenates along any order. We use this new technique in a simple way for a wide range of applications to shave some log factors from the best known bounds [11, 13]. We obtain new multidimensional data structures implementing two–dimensional priority queues, two–dimensional search trees, and concatenable interval trees. We achieve the following time bounds for a split or concatenate: $O(\sqrt{p})$ in concatenable interval trees, $O(p^{1-1/d})$ in d–dimensional 2–3–trees (or divided k–d trees) and $O(\sqrt{p \log p})$ in a data structure for the convex hull. We also improve the query bounds because they are equal to the split/concatenate cost plus an $O(occ)$ cost due to the output. Furthermore, we make the bounds for insertions and deletions of a single item worst–case rather than amortized. The new data structures work for many other order decomposable problems under split and concatenate operations. For example, point insertions and deletions in a planar Voronoi diagram of p points take $O(p)$ time in $O(p \log \log p)$ space [21] (a result in [1] is a semi–dynamic algorithm with $O(p)$ deletion time and space). We obtain an $O(p)$ cost also for range, split and concatenate operations in $O(p \log \log p)$ space (the techniques in [13, 28, 29] require more time or space). This solves a problem posed in [1] (i.e., compute the Voronoi diagram for any given subset $R \subseteq S$ of points in less than $\Theta(p \log p)$ time) for the special case in which R is defined by range queries on a dynamic set S. Our technique for order decomposable problems is suitable for efficient external memory algorithms. For the case $d = 1$, B–trees [4] are very popular data structures that can be successfully employed in decomposable search problems analogously to concatenable 2–3–trees. For $d > 1$, no provably good external memory data structures for splitting and concatenating along any dimension were previously known in the literature. In this extended abstract, many details are omitted for lack of space.

2 Splitting and Merging Data Structures

In this section, we describe how to maintain $d = 2$ total orders, which we denote by $\prec_X$ and $\prec_Y$, under split and concatenate operations. Let p be the number of items in S. Each item $z \in S$ can be associated with a dynamic point $(X(z), Y(z))$ in the Cartesian plane, such that $X(z)$ is the rank of z in S with respect to current order $\prec_X$ and $Y(z)$ is the rank of z in S with respect to current order $\prec_Y$. Starting from p items in S, we obtain p points in the Cartesian plane, which can be stored in the form of a $p \times p$ *sparse* and *dynamic* matrix $\mathcal{M}$.

The operations in S can be simulated by a certain number of operations in $\mathcal{M}$. Operation $split(S, z, \prec_X)$ corresponds to splitting matrix $\mathcal{M}$ horizontally at a certain position $X(z)$, which is the rank of z in S with respect to $\prec_X$, while doing the same according to its order $\prec_Y$ is equivalent to handling $\mathcal{M}$ vertically at position $Y(z)$. Concatenating is analogous. Operations $insert(z, S)$ and $delete(z, S)$ require a new operation which sets entry $\mathcal{M}[X(z), Y(z)]$ to item z or to an empty value, respectively. Finally, solving problem $\mathcal{P}$ in the region specified by $range(\langle a_X, b_X \rangle, \langle a_Y, b_Y \rangle, S)$ can be done by solving $\mathcal{P}$ for the points contained in the rectangular part of $\mathcal{M}$ defined by the ranks of a_X, b_X, a_Y, b_Y in their corresponding order. We can state our multiordered set splitting and merging problem by using our sparse matrix $\mathcal{M}$. Formally, for any integers h_1, h_2, v_1, v_2

$(1 \leq h_1 \leq h_2 \leq p, 1 \leq v_1 \leq v_2 \leq p)$, we use $\mathcal{M}[h_1, h_2; v_1, v_2]$ to denote the submatrix of $\mathcal{M}$ that contains entries $\mathcal{M}[i,j]$ with $h_1 \leq i \leq h_2$ and $v_1 \leq j \leq v_2$. We call this submatrix a *region*. We can disassemble and reassemble a single matrix $\mathcal{M}$ in many different ways by using any sequence of the following operations:

h_split$(\mathcal{M}, i)$: Split $\mathcal{M}$ horizontally at row i and obtain two new matrices $\mathcal{M}_1$ and $\mathcal{M}_2$, such that $\mathcal{M}_1 = \mathcal{M}[1, i; 1, p]$ and $\mathcal{M}_2 = \mathcal{M}[i+1, p; 1, p]$. In other words, $\mathcal{M}_1$ is given by the first i rows of $\mathcal{M}$ and $\mathcal{M}_2$ is given by the last $(p - i)$ rows of $\mathcal{M}$. $\mathcal{M}$ is no longer available after the operation.

h_concatenate$(\mathcal{M}_1, \mathcal{M}_2)$: Let $\mathcal{M}_1$ have size $m \times p$ and $\mathcal{M}_2$ have size $n \times p$. We meld $\mathcal{M}_1$ and $\mathcal{M}_2$ horizontally and produce a matrix $\mathcal{M}$ of size $(m+n) \times p$, such that $\mathcal{M}[1, m; 1, p] = \mathcal{M}_1$ and $\mathcal{M}[m+1, m+n; 1, p] = \mathcal{M}_2$. In other words, the first m rows of $\mathcal{M}$ are given by $\mathcal{M}_1$ and the last n rows of $\mathcal{M}$ are given by $\mathcal{M}_2$. This operation assumes that $\mathcal{M}_1$ and $\mathcal{M}_2$ have the same number of columns. $\mathcal{M}_1$ and $\mathcal{M}_2$ are no longer available after the operation.

set$(i, j, w, \mathcal{M})$: Update $\mathcal{M}$ by setting $\mathcal{M}[i, j] = w$. This corresponds either to an insertion (if w is nonempty) or to a deletion (if w is empty).

range$(h_1, h_2, v_1, v_2, \mathcal{M})$: Find the solution $\mathcal{P}(R)$ to problem $\mathcal{P}$ restricted to the nonempty entries contained in region $R = \mathcal{M}[h_1, h_2; v_1, v_2]$.

Operations *v_concatenate*$(\mathcal{M}_1, \mathcal{M}_2)$ and *v_split*$(\mathcal{M}, j)$ are similarly defined. We restrict ourselves to the special case where each row or column of $\mathcal{M}$ contains a constant number of points but our technique works for a general matrix $\mathcal{M}$. We need some preliminary definitions. Let $X = \{x_1, x_2, \ldots, x_q\}$ be a sorted sequence of q elements, according to a total order $\prec$: $x_1 \prec x_2 \prec \cdots \prec x_q$. Let $I_1, \ldots, I_s$ be a partition of X into adjacent intervals, so that for $1 \leq i \leq s-1$ all the elements in I_i precedes all the elements in I_{i+1}. For $1 \leq i \leq s$, let $|I_i|$ denote the size of interval I_i, defined as the number of elements in I_i.

Definition 1. (Size Invariant) Let $k \geq 1$ be a positive integer. The adjacent intervals $I_1, \ldots, I_s$ satisfy the *size invariant of order k* if the following two conditions are met: (a) $|I_i| \leq k$, $1 \leq i \leq s$; and (b) $|I_i| + |I_{i+1}| > k$, $1 \leq i \leq s - 1$.

The size invariant of order k in Definition 1 implies that the number s of intervals is $O(q/k)$. Moreover, the size invariant can be maintained in $O(\log k)$ time when an element is deleted from X or a new element is inserted into X.

We now introduce the *cross-tree*, which is a 2-dimensional data structure supporting efficient split and concatenate operations. Intuitively, a cross-tree describes a balanced decomposition of a 2-dimensional set, and it is based upon a variant of 2-3-tree [2], which we call *1-2-tree*. A 1-2-tree satisfies two conditions: (a) All the leaves are on the same level and each internal node has at most two children. (b) The children of all the internal nodes on the same level satisfy the size invariant of order 2 according to Definition 1. It follows that no two adjacent nodes can have a single child. It can be shown that 1-2-trees are balanced and that a 1-2-tree with n leaves can be modified by means of split, concatenate, insert and delete operations in $O(\log n)$ time per operation, with each operation involving at most $O(1)$ nodes and parent pointers per level.

Definition 2. (Cross–Tree) Let T and S be two 1–2–trees, having the same height. The *cross–tree* $CT(T \times S)$ is the cross product of T and S defined as follows. For each node u in T, there is a node α_{uv} in $CT(T \times S)$ for every node v in S on the same level as u. For each edge $(u, \hat{u})$ in T, there is an edge $(\alpha_{uv}, \alpha_{\hat{u}\hat{v}})$ in $CT(T \times S)$ for every edge $(v, \hat{v})$ in S, such that u and v are on the same level.

A cross–tree has either 1, 2 or 4 children and it is balanced (i.e., its height is logarithmic with respect to the number of its leaves). We can update a cross–tree $CT(T \times S)$ by modifying either T or S (i.e., we can split, concatenate, insert or delete in one of the 1–2–trees) and obtain the corresponding cross–tree efficiently. We can show:

Theorem 3. *We can split a 1–2–tree T into T_1 and T_2 in order to obtain cross-trees $CT(T_1 \times S)$ and $CT(T_2 \times S)$ from cross–tree $CT(T \times S)$ in $O(|S|)$ time. We can concatenate 1–2–trees T_1 and T_2 into T to obtain $CT(T \times S)$ from $CT(T_1 \times S)$ and $CT(T_2 \times S)$ in $O(|S|)$ time.*

2.1 The General Technique

We now treat our splitting and merging problem for a matrix $\mathcal{M}$. We refer to the p nonempty entries of $\mathcal{M}$ as the *points* of $\mathcal{M}$ and let k be a slack parameter, where k is an integer with $1 \leq k \leq p$. We handle the sparse $p \times p$ matrix $\mathcal{M}$ as if it were a dense $\Theta(p/k+k) \times \Theta(p/k+k)$ matrix. We then tune k according to the chosen problem $\mathcal{P}$ and the cost $f(p)$ of operator $\Diamond$. We proceed as follows. We group adjacent rows and columns of matrix $\mathcal{M}$ into respectively horizontal and vertical *stripes*, such that the stripes satisfy the size invariant of order k (Definition 1), where the size of a horizontal (respectively vertical) stripe is given by its number of rows (respectively columns). The size invariant guarantees that each stripe contains at most $O(k)$ points and that the total number of horizontal and vertical stripes is $O(p/k)$. The partition into horizontal and vertical stripes induces a partition of $\mathcal{M}$ into $O(p^2/k^2)$ squares, such that each square intersects no more than k rows and k columns. We call these the *basic squares* in $\mathcal{M}$. We maintain the solution to $\mathcal{P}$ for each such basic square and store these solutions in the leaves of a *cross–tree* $CT(T_H \times T_V)$, which describes recursively the partition of $\mathcal{M}$ into its basic squares. For this purpose, we employ two 1–2–trees, denoted by T_H and T_V, whose leaves are in one-to-one correspondence to the horizontal and vertical stripes, respectively. Trees T_H and T_V have $O(p/k)$ leaves, one for each stripe of $\mathcal{M}$, and a total of $O(p/k)$ nodes. Consequently, cross–tree $CT(T_H \times T_V)$ has height $O(\log(p/k))$ and $O(p^2/k^2)$ leaves, one for each basic square of $\mathcal{M}$, and a total of $O(p^2/k^2)$ nodes. Its leaves corresponding to the nonempty basic squares in either a horizontal or vertical stripe can be retrieved in $O(p/k)$ time, and the points in the stripe can be retrieved in additional $O(k)$ time. We then percolate the solutions from the leaves of the cross–tree towards its internal nodes in a heap–like fashion by means of operator $\Diamond$. If the solutions occupy more than $O(f(p))$ space, we save space whenever $\Diamond$ is invertible: We say that $\Diamond$ is *invertible* if we can keep $O(f(p))$ additional information associated with any solution $\mathcal{P}(R) = \Diamond(\mathcal{P}(R'), \mathcal{P}(R''))$ so that we can compute $\Diamond^{-1}(\mathcal{P}(R)) = \{\mathcal{P}(R'), \mathcal{P}(R'')\}$ in $O(f(p))$ time. For example, if $\mathcal{P}$ is the range query problem and $\Diamond$ is the destructive list append with cost $f(p) = O(1)$, we can simply keep a pointer to the last item in the appended lists to "de–append" them in $O(1)$ time.

Our data structure has the following additional features. For each nonempty basic square of $\mathcal{M}$, we keep its points sorted according to a *total order* $\prec_P$ (not necessarily equal to $\prec_X$ or $\prec_Y$) by means of a *threaded* binary search tree, whose nodes are linked together in symmetrical order. Searching, inserting and deleting a point takes $O(\log k)$ time. Scanning the points in a basic square in their $\prec_P$-order takes constant time per scanned point. We introduce order $\prec_P$ because some data structures can be built more efficiently on a sorted set of points. Each node in cross–tree $CT(T_H \times T_V)$ corresponds to a region R of matrix $\mathcal{M}$. The cross–tree leaves correspond to the basic squares (leaves corresponding to the empty basic squares can be ignored). An internal node ρ corresponds to region $R = \mathcal{M}[h_1, h_2; v_1, v_2]$ and has no more than four children ρ_1, ρ_2, ρ_3, and ρ_4 corresponding to four subregions of R (if a child ρ_i is empty then the corresponding subregion is empty.) We store the solutions to $\mathcal{P}$ in the following way. For each nonempty basic square of $\mathcal{M}$, we store the solution for its points in the corresponding cross–tree leaf. For each internal node ρ of the cross–tree, we use that fact the $\mathcal{P}$ is order decomposable to store $\Diamond(s_1, \ldots, s_j)$ in ρ, where $s_1, \ldots, s_j$ are the solutions stored in its $j \leq 4$ children. This is indeed the solution $\mathcal{P}(R)$ for the points in the region R corresponding to ρ, and is stored in an efficient way depending on the problem $\mathcal{P}$.

We now show how to use our data structure for solving problem $\mathcal{P}$. We denote by $P(k)$ the cost of preprocessing an $O(k)$–point stripe to solve problem $\mathcal{P}$ for every basic square in the stripe. We will exploit the fact that the basic squares are already $\prec_P$–ordered to determine $P(k)$ and we assume that $P(k) \geq k$ is a smooth nondecreasing function. Furthermore, we use $U(k)$ to denote the cost of updating the solution to problem $\mathcal{P}$ for a basic square in an $O(k)$–point stripe after its preprocessing. We assume that $U(k) \geq \log k$, since we have to update at least the threaded search tree in the basic square. Finally, we denote by $S(k)$ the space occupied by an $O(k)$–point stripe. We also assume that $S(k) \geq k$ is a smooth nondecreasing function. In most of our applications, we will have $P(k), S(k) = O(k)$ and $U(k) = O(f(k) \log k)$. In the preprocessing, we put $(k+1)/2$ rows (columns) per stripe except for the last one, which has some dummy rows (columns) added, and build the cross–tree. This takes $O(p \log p + P(p) + f(p) \, p^2/k^2)$ time. We now describe in some details how to perform a $v\_split(\mathcal{M}, j)$. Column j might fall inside a vertical stripe σ, which must necessarily be split. We examine the basic squares of σ. Given a basic square, we scan its points according to their $\prec_P$–order and produce two $\prec_P$–*ordered* lists in linear time: one list contains all the points whose second coordinate is smaller than or equal to j and the other list contains the remaining points, i.e., the points whose second coordinate is larger than j. We split this basic square into two squares and build two threaded search trees for them in linear time by using the two $\prec_P$–ordered lists. Since each stripe consists of $O(p/k)$ basic squares and contains $O(k)$ points, we can examine stripe σ square by square in $O(k+p/k)$ time and split it into new stripes σ_1 and σ_2, such that σ_1 contains all the points of σ before and including column j, and σ_2 contains all the points of σ after column j. This creates $O(p/k)$ smaller squares and costs $O(k+p/k)$ time. We check to see if we can combine σ_1 and σ_2 with their neighbor stripes to maintain the size invariant of order k. For any two such stripes to be merged, we examine their basic squares in pairs (a square per stripe), such that the two squares are on the same horizontal stripe. We take their two $\prec_P$-ordered lists of points and merge them to build a threaded search tree on the resulting list in linear time. Again, this requires $O(k+p/k)$ total time. It is worth noting that splitting and merging stripes preserves the order of their presorted points. Next, we determine the solutions for the basic squares in the $O(1)$ stripes involved at a total cost of $P(k)$ time. It remains to split cross-tree $CT(T_H \times T_V)$ to reflect the split operation on the vertical stripes. We first focus on the cross-tree topology and discuss later on how to maintain the solutions to $\mathcal{P}$ in its nodes. We have to split the 1–2-tree T_V at the leaf w corresponding to stripe σ. We split w into two new leaves w_1 and w_2, corresponding to the split of σ into the new stripes σ_1 and σ_2. If σ_1 or σ_2 are combined with their neighbor stripes, we should do the same on w_1 and w_2 and their neighbor leaves. We check to see if the 1–2-tree T_V satisfies the size invariant of order 2 along a leaf-to-root path and update the corresponding $O(p/k)$ cross-tree leaves. Globally, we create no more than $O(p/k)$ leaves corresponding to the new basic squares in $O(1)$ stripes and we traverse and reorganize their ancestor nodes all the way up to the cross-tree root by Theorem 3 (with $T = T_H$ and $S = T_V$). Consequently, maintaining the cross-tree topology takes $O(p/k)$ time. Next, we recompute the solutions to $\mathcal{P}$ in the traversed cross-tree nodes by applying operator $\Diamond$ to them upwards, in $O(f(p))$ time per node (we show in the full paper how to do this with $\Diamond^{-1}$ if $\Diamond$ is invertible). Since we traverse a total of $O(p/k)$ nodes, it takes $O(f(p)\,p/k)$ time to recompute their solutions. It therefore takes a total of $O(k+f(p)\,p/k+P(k)) = O(f(p)\,p/k+P(k))$ time to execute v_split, as $P(k) \geq k$. The implementation of h_split is completely analogous. We do not discuss here the other operations due to lack of space and refer the reader to the full paper. There, we prove the following main theorem:

Theorem 4. *The splitting and merging problem on p points can be solved with the following time bounds for a parameter k $(1 \leq k \leq p)$ and an operator cost $f(p)$: range, h_split, v_split, h_concatenate, v_concatenate: $O((p/k)f(p) + P(k) + P(p)/p)$, with $P(k) \geq k$; set: $O(\log(p/k)f(p) + U(k) + (p/k^2)f(p) + P(p)/p)$, with $U(k) \geq \log k$. The space required is $O(S(p) + (p^2/k^2)f(p))$ and the preprocessing time is $O(p\log p + P(p) + (p^2/k^2)f(p))$.*

Theorem 4 states the bounds needed for solving a general decomposable problem $\mathcal{P}$ in terms of the parameter k, $1 \leq k \leq p$. In most of our applications, $f(p) = O(p^\epsilon)$ for a non-negative constant $\epsilon < 1$ and the preprocessing cost of a stripe is $P(k) = O(k)$ because we have presorted points. In this case, since $U(k) = O(f(k)\log k)$ and $S(p) = O(k)$ [20, 21], we can tune $k = \lceil \sqrt{p\,f(p)}\,\rceil$:

Theorem 5. *The splitting and merging problem on p items can be solved with the following time bounds whenever the cost of operator $\Diamond$ is $f(p) = O(p^\epsilon)$ for a non-negative constant $\epsilon < 1$: range, h_split, v_split, h_concatenate, v_concatenate in $O\left(\sqrt{p\ f(p)}\right)$; set in $O(\log(p)f(p))$. The space required is $O(p)$ and the preprocessing time is $O(p \log p)$.*

The analysis in Theorem 4 is overly pessimistic when $f(p) = \Theta(p)$. Using weighted balanced B–trees [3] in place of 1–2–trees yields a different analysis and better bounds:

Theorem 6. *The splitting and merging problem on p items can be solved with the following time bounds when $f(p) = \Theta(p)$: set, range, h_split, v_split, h_concatenate, v_concatenate in $O(p)$. The space required is $O(p \log \log p)$ and the preprocessing time is $O(p \log p)$.*

3 Some Applications

In this section we list few applications of Theorems 4–6. The problems in Theorems 7–9 are all $O(1)$–order decomposable; the problem in Theorem 10 is $O(\log p)$–order decomposable while the one in Theorem 11 is $O(p)$–order decomposable. Most of the worst–case bounds reported in this section improve the best previously known bounds for the same problems [11, 13]. The improvement consists of shaving a logarithmic factor from the previous bounds and of making some bounds worst–case rather than amortized. We omit the details.

Theorem 7. *A two–dimensional priority queue for a set of p items can be maintained in the following time bounds: an item insertion or deletion in $O(\log p)$; a split or concatenate of any order in $O(\sqrt{p})$; and a minimum–weight query in a region in $O(\sqrt{p})$. The space required is $O(p)$ and the preprocessing time is $O(p \log p)$.*

Theorem 8. *A two–dimensional 2-3–tree storing p points can be maintained with the following bounds: a point insertion or deletion in $O(\log p)$; a split or concatenate along any coordinate in $O(\sqrt{p})$; a range search in $O(\sqrt{p} + occ)$, where occ is the number of points reported by the search. The space required is $O(p)$ and the preprocessing time is $O(p \log p)$.*

Theorem 9. *An interval tree that stores n (overlapping) intervals from the line can be maintained with the following time bounds: an interval insertion or deletion in $O(\log n)$; a stabbing query (i.e., find all the intervals containing a given point) retrieving only the intervals whose lengths are between two input values $\ell_1 \ldots \ell_2$ in $O(\sqrt{n} + occ)$, where occ is the number of such intervals; a split or concatenate of the intervals according to a perpendicular stabbing line in $O(\sqrt{n})$. The space required is $O(n)$ and the preprocessing time is $O(n \log n)$.*

Theorem 10. *The convex hull for p points in the Cartesian plane can be maintained with the following time bounds: a point insertion or deletion in $O(\log^2 p)$; a split or concatenate along one coordinate in $O(\sqrt{p \log p})$; a query checking if a point is inside or outside the convex hull in $O(\log p)$; a query reporting the convex hull for the points in any input region in $O(\sqrt{p \log p} + h)$, where h is the output size. The space required is $O(p)$ and the preprocessing time is $O(p \log p)$.*

Theorem 11. *The Voronoi diagram for p points in the Cartesian plane can be maintained with the following worst–case time bounds: a point insertion or deletion: $O(p)$; a split or concatenate along one coordinate: $O(p)$; a query reporting the Voronoi diagram for the points in an input region: $O(p)$. The space required is $O(p \log \log p)$ and the preprocessing time is $O(p \log p)$.*

The above results show that our technique is a general paradigm on which we can cast many other split–and–concatenate data structures in some basic problems (e.g., member searching, predecessor, ranking), computational geometry (e.g., neighbor queries, union and intersection queries), database applications (e.g., partial match queries, range queries) and statistics (e.g., maxima queries). We refer the interested reader to [7, 17, 21] for more decomposable problems. We only mention here that our technique can be extended to $d > 2$ total orders $\prec_1, \ldots, \prec_d$ and can be efficiently implemented in external memory. Details will be given in the full paper.

Acknowledgments. We are indebted to Amnon Nissenzweig and to Giuseppe Persiano for many delightful conversations at the beginning of this research. We are grateful to Lars Arge and to Paolo Ferragina for helpful discussions and to Marc van Kreveld and Mark Overmars for sending us a copy of [13].

References

1. A. Aggarwal, L. Guibas, J. Saxe and P.W. Shor, A linear–time algorithm for computing the Voronoi diagram of a convex polygon. *Discrete and Computational Geometry* 4 (1989), 591–604.
2. A.V. Aho, J.E. Hopcroft, and J.D. Ullman. *The Design and Analysis of Computer Algorithms*. Addison–Wesley, Reading, MA, 1974.
3. L. Arge and J.S. Vitter, Optimal dynamic interval management in external memory. 37th IEEE Symp. on Foundations of Computer Science (1996).
4. R. Bayer and C. McCreight, Organization and maintenance of large ordered indexes. *Acta Informatica 1*, 3 (1972), 173–189.
5. J.L. Bentley, Multidimensional binary search trees used for associated searching. *Comm. ACM*, 19 (1975), 509–517.
6. J.L. Bentley, Decomposable Searching Problems. *Information Processing Letters*, 8 (1979), 244–251.
7. J.L. Bentley and J.B. Saxe, Decomposable Searching Problems I. Static–to–Dynamic Transformation. *J. of Algorithms*, 1 (1980), 301–358.
8. Y.-J. Chiang and R. Tamassia, Dynamic Algorithms in Computational Geometry, *Proceedings of the IEEE*, Special issue on Computational Geometry, G. Toussaint, ed., 80 (1992) 1412–1434.

9. R.A. Finkel and J.L. Bentley, Quad–trees: a data structure for retrieval of composite keys. *Acta Inform.*, 4 (1974), 1–9.
10. I.G. Gowda and D.G. Kirkpatrick, Exploiting linear merging and extra storage in the maintenance of fully dynamic geometric data structures. In Proc. 19th Allerton Conference on Communication, Control and Computing (1980), 1–10.
11. M.J. van Kreveld and M.H. Overmars, Divided k–d trees, *Algorithmica*, 6 (1991), 840–858.
12. M.J. van Kreveld and M.H. Overmars, Union–copy structures and dynamic segment trees, *J. ACM*, 40 (1993), 635–652.
13. M.J. van Kreveld and M.H. Overmars, Concatenable structures for decomposable problems, *Information and Computation*, 110 (1994), 130–148.
14. J. van Leeuwen and M.H. Overmars, The art of dynamizing. In Proc. 10th Mathematical Foundations of Computer Science, *LNCS*, 118 (1981), 121–131.
15. J. van Leeuwen and D. Wood, Dynamization of decomposable searching problems. *Information Processing Letters*, 10 (1980), 51–56.
16. K. Mehlhorn, Lowerbounds on the efficiency of transforming static data structures into dynamic structures. *Mathematical System Theory*, 15 (1981), 1–16.
17. K. Mehlhorn, *Multi–Dimensional Searching and Computational Geometry* EATCS Monographs on Theoretical Computer Science, vol. 3, Springer–Verlag, 1984.
18. H.A. Maurer and T.A. Ottmann, Dynamic solutions of decomposable searching problems. In *Discrete Structures and Algorithms*, U. Pape ed., Hanser Verlag, Wien, (1979), 17–24.
19. K. Mehlhorn and M.H. Overmars, Optimal dynamization of decomposable searching problems. *Information Processing Letters*, 12 (1981), 93–98
20. M. H. Overmars, Dynamization of order decomposable set problems. *J. Algorithms*, 2 (1981), 245–260.
21. M.H. Overmars, *The Design of Dynamic Data Structures*, LNCS 156, Springer–Verlag, Berlin/New York, 1983.
22. M.H. Overmars and J. van Leeuwen, Maintenance of configurations in the plane. *Journal of Computer and System Sciences*, 23 (1981), 166–204.
23. M.H. Overmars and J. van Leeuwen, Dynamization of decomposable searching problems yielding good worst-case bounds. In Proc. 5th GI Conference on Theoretical Computer Science, *LNCS*, 104 (1981), 224–233.
24. M.H. Overmars and J. van Leeuwen, Some principles for dynamizing decomposable searching problems. *Information Processing Letters*, 12 (1981), 49–53.
25. M.H. Overmars and J. van Leeuwen, Worst-case optimal insertion and deletion methods for decomposable searching problems. *Information Processing Letters*, 12 (1981), 168–173.
26. M.H. Overmars and J. van Leeuwen, Dynamic Multi–dimensional data structures based on quad– and k-d trees. *Acta Inform.*, 17 (1982), 267–285.
27. H. Samet, Bibliography on quad–trees and related hierarchical data structures. In *Data Structures for Raster Graphics*, L. Kessenaar, F. Peters, and M. van Lierop eds., Springer–Verlag, Berlin, (1986), 181–201.
28. H.W. Scholten and M.H. Overmars, General methods for adding range restrictions to decomposable searching problems, *J. of Symbolic Computation*, 7 (1989), 1–10.
29. D.E. Willard and G.S. Lueker, Adding range restriction capability to dynamic data structures, *J. ACM*, 32 (1985), 597–617.

Efficient Array Partitioning

Sanjeev Khanna[1], S. Muthukrishnan[2] and Steven Skiena[3]

[1] Mathematical Sciences Research Center, Bell Laboratories, Lucent Technologies, 700 Mountain Avenue, Murray Hill, NJ 07974. sanjeev@research.bell-labs.com
[2] Information Sciences Center, Bell Laboratories, Lucent Technologies, 700 Mountain Avenue, Murray Hill, NJ 07974. muthu@research.bell-labs.com
[3] Dept. of Computer Science, State University of New York, Stony Brook, NY 11794-4400. skiena@cs.sunysb.edu. This work is partially supported by ONR award 400x116yip01 and NSF Grant CCR-9625669.

Abstract. We consider the problem of partitioning an array of n items into p intervals so that the maximum weight of the intervals is minimized. The currently best known bound for this problem is $O(n+p^{1+\epsilon})$ [HNC92] for any fixed $\epsilon < 1$. In this paper, we present an algorithm that runs in time $O(n \log n)$; this is the fastest known algorithm for arbitrary p.
We consider the natural generalization of this partitioning to two dimensions, where an $n \times n$ array of items is to be partitioned into p^2 blocks by partitioning the rows and columns into p intervals each and considering the blocks induced by this partition. The problem is to find that partition which minimizes the maximum weight among the resulting blocks. This problem is known to be NP-hard [GM96]. Independently, Charikar et. al. have given a simple proof that shows that the problem is in fact NP-hard to approximate within a factor of two. Here we provide a polynomial time algorithm that determines a solution at most $O(1)$ times the optimum; the previously best approximation ratio was $O(\sqrt{p})$ [HM96]. Both the results above are proved for the case when the weight of an interval or block is the sum of the elements in it. These problems arise in load balancing for parallel machines and data partitioning in parallel languages. Applications in motion estimation by block matching in video and image compression give rise to the dual problem, that of minimizing the number of dividers p so that the maximum weight of a block is at most δ. We give an $O(\log n)$ approximation algorithm for this problem. All our results for two dimensional array partitioning extend to any higher fixed dimension.

1 Introduction

The problem of partitioning a set of items into roughly equal weight subsets is a fundamental one. We study two dual versions of this, namely, (a) given B, partition a given array into at most B blocks so as to minimize the maximum weight of any block in the partition, and (b) given δ, partition a given array into minimum number of blocks such that their individual weight is no larger than δ. The definition of the weight function for a block, the type of partitions allowed, the dimensionality of the arrays, and the relevant version depends upon the application at hand. The problems we consider arise in load balancing for parallel processing, compilers for high-performance parallel languages, and motion estimation in videos by block matching, and hence have been extensively researched in several communities. In this paper, we present algorithms for these problems which are more efficient than the best ones so far, and give improved approximations over those previously known. In what follows, we describe the setting of the problems (Section 1.1), and describe various application scenarios where three such problems arise (Section 1.2). We state our results for such problems in Section 1.3 and present the technical details in sections 2, 3 and 4.

1.1 Problems

We begin with the one dimensional version. Consider an array $A[1\cdots n]$ of nonnegative numbers, and a *weight* function f that maps intervals of A to nonnegative integers. The function f is trivially assumed to be 0 on empty intervals. The *p-partition* of A is a division of A into p intervals, that is, setting dividers $d_0 = 0 < d_1 \leq d_2 \leq \cdots \leq d_{p-1} \leq d_p = n$. Here the ith interval is $[d_{i-1}+1\cdots d_i]$ if $d_{i-1} \neq d_i$ and is denoted empty otherwise. The MAX norm of a partition is $\max_{i=1}^{i=p} f(A[d_{i-1}+1\cdots d_i])$. Two weight functions arise commonly in practice: the additive weight function $F(A[i,j]) = \sum_{k=i}^{k=j} A[k]$ and the Hamming weight function H_c for a given parameter c, relative to another array B of size n, given by $H_c(A[i,j]) = \min_{-c \leq k \leq c} \mathcal{H}(B[i+k, j+k], A[i,j])$ where $\mathcal{H}(X,Y)$ gives the Hamming distance between two segments X and Y of identical length.

The 1D p-partition problem. Given p, find the p-partition that minimizes the MAX norm. □

This notion can be naturally extended to a $p \times p$ *partition* in two dimensions as follows. Consider an $n \times n$ array A. Divide the rows $[1, n]$ into p intervals given by horizontal dividers $h_0 = 0 < h_1 \leq h_2 \leq \cdots \leq h_{p-1} \leq h_p = n$, and the columns $[1, n]$ into p other intervals given by the vertical dividers $v_0 = 0 < v_1 \leq v_2 \leq \cdots \leq v_{p-1} \leq v_p = n$. This induces p^2 *blocks* given by $A[h_{i-1}+1\cdots h_i, v_{j-1}+1\cdots v_j]$ for each i, j. The MAX norm of a partition is $\max_{i=1, j=1}^{i=p, j=p} f(A[h_{i-1}+1\cdots h_i, v_{j-1}+1\cdots v_j])$. Again, the common weight functions on blocks are F and H_c defined analogously as above for intervals.

The 2D $p \times p$-partition problem. Given p, find the $p \times p$ partition that minimizes the MAX norm. □

The 2D δ-weight partition problem. Given δ, find the minimum p for which there exists a $p \times p$ partition of the array with the MAX norm of at most δ. □

Remarks. There are many different ways to partition 2D arrays, as discussed in [GM96, KRW95, MS96, MM+96]. Here we consider only the $p \times p$ partition. These problems can be naturally generalized to higher dimensions. Our solutions for the 2D case extend to higher dimensions in a straightforward way. However, the 1D and 2D cases are fundamentally different, and they will be contrasted later.

1.2 Application Scenarios

Array partitioning problems arise in load balancing, scheduling, data layout, video compression, etc. We focus on three specific array partitioning problems. Here we briefly describe the application context for each; further details of modeling will be discussed in the journal version.

One dimensional case under F. This problem was abstracted for load balancing in pipelined, parallel environments in [B88] and studied in [OM95, AF91, HL92, MS95, M93, CN91, HNC92, N91] etc.

Two dimensional case under F. This problem arises in balanced data distribution as implemented in the Superb environment [ZBG86] and HPF2 [HPF] (High Performance Fortran). See [M93, CM+95] for more applications to particle-in-cell computations and sparse matrix computations.

Two dimensional case under H_c. This arises in motion-compensated video compression by block matching. Roughly this involves compressing a frame in a video sequence by cutting it into rectangles each of which is encoded in terms of a block in the previous frame. See [MM+96] and then references therein for the precise setting.

1.3 Results

We state our results for each of the three problems of our interest.
1D p-partition under F. This problem has been extensively researched. We summarize the previous work and our results in the table below, providing all citations where identical bounds were obtained independently.

Reference	Bound
Bokhari [B88]	$O(n^3 p)$
Anily & Federgruen [AF91]	$O(n^2 p)$
Hansen & Liu [HL92]	$O(n^2 p)$
Manne & Sorevik [MS95]	$O(np \log p)$
Choi & Narahari [CN91]	$O(np)$
Olstad & Manne [OM95]	$O(np)$
Nicol [N91]	$O(n + p^2 \log^2 n)$
Charikar, Chekuri & Motwani [CCM96]	$O(n + p^2 \log^2 n)$
Han, Narahari & Choi [HNC92]	$O(n + p^{1+\epsilon}), \epsilon < 1$
This paper	$O(n \log n)$

Our result relies on a binary search over a space of $O(n^2)$ items. However, at each test, an approximate median among these items is identified in only $O(n)$ (as opposed to $O(n^2)$) time by exploiting the structure in our search space. In particular, we design and use an algorithm that finds an approximate median of the $O(n^2)$ elements which are organized into n sorted lists in only $O(n)$ time.

Throughout we have made no assumptions on the range of F's. However, improved bounds may be obtained if the F's lie in a restricted range; we omit the details here.

The 2D δ-weight problem under H_c. A number of algorithms are known for block matching, and in particular, for the 2D δ-weight problem under H_c. These essentially work by splitting subareas greedily until each subarea has weight at most δ and do not provide any guarantees on the number of blocks used. Building on the result of Grigni and Manne [GM96], this problem can in fact be shown to be NP-hard.

Here we provide an $O(\log n)$ approximate polynomial time algorithm. We obtain our result by a rather simple reduction to the classical set cover problem. Our algorithm works for a general class of metrics including F and H_c.

The 2D $p \times p$ partition problem under F. Grigni and Manne [GM96] showed that the 2D problem is NP-hard even when the given array consists of 0/1 entries. Independently, Charikar et. al. have given a simple proof that this problem is APX-hard, that is, the problem is in fact NP-hard to approximate within a factor of two. While a number of natural heuristic algorithms are known for this problem (See for example [MS96]), most of them can be shown to be bad (typically $\Omega(\sqrt{p})$) approximations. One such heuristic has been recently shown to have a performance guarantee of $O(\sqrt{p})$ by Halldorsson & Manne [HM96]. This is the currently best known approximation for this problem.

Reference	Result
Grigni & Manne [GM96]	NP-Hardness
Charikar et. al. [CCM96]	APX-Hardness
Halldorsson & Manne [HM96]	$O(\sqrt{p})$ approximation
This paper (Section 4)	$O(1)$ approximation

We observe that using our result for the 2D δ-weight problem above, one can easily obtain an $O((\log n)^2)$ approximation algorithm for the 2D $p \times p$-partitioning problem under F. But our main contribution is an $O(1)$-factor approximation for this problem which builds on an inherent connection between "independent" rectangles of large weight within the array and the cost of the optimal solution. Surprisingly, we are able to show that after a suitable preprocessing of the input array, a locally optimal collection of independent rectangles can be used to generate a solution which is at most a constant factor away from the optimal.

2 The One Dimensional Case Under F

We assume for convenience that $F(A[i]) \neq 0$ for any i; this assumption can be easily removed and we omit that detail. Define the Boolean function $M_A(\ell, k, v)$ to be true if and only if there exists a partition of the elements $A[\ell, n]$ into k intervals, such that the MAX norm of these intervals is $< v$. In our analysis below, we count only the complexity of calls to the F oracle; F can be simulated in constant time after linear preprocessing.

Lemma 1. $M_A(l, k, v)$ can be determined using $O(n)$ calls to the F oracle for arbitrary k, l, and v.

Proof. Note that without loss of generality the $(j+1)$st divider can be placed as far to the right of the jth divider such that F value of the elements in that interval is $< v$. By incrementally inserting dividers from left to right so as to prevent the total in any interval from exceeding v, we find the minimum number of dividers required in $O(n)$ time. If this total exceeds k, then $M_A(l, k, v) = false$. Otherwise, $M_A(l, k, v) = true$. □

In the optimal partitioning with k dividers, there will be an interval $A[i, j]$ which will prove the *bottleneck* of the partitioning: an interval is a bottleneck to the partitioning if it is the largest weight interval that results from this partitioning. There are $\binom{n}{2}$ candidates for this bottleneck interval. Performing a binary search on these candidates, using the linear-time oracle of Lemma 1, would yield an $O(n \log n)$ algorithm to search for the k-partition. However, this requires a method to efficiently compute the sequence of (approximate) median candidates to support the binary search. Conventional linear-time median-finding is clearly inadequate, since we have only $O(n)$ time to find the median of $\Theta(n^2)$ elements.

We take advantage of the fact that this collection of $\Theta(n^2)$ elements is not arbitrary, but has rather been derived from interval sums over n elements. We partition the $\binom{n}{2}$ intervals $f(A[i, j])$ into n *columns*, where column c consists of the elements $f(A[i, c])$, $1 \leq i \leq c$. Let $C_c[i] = f(A[i, c])$ denote the ith element of column c. The *subcolumn* $S_c[i, j]$ comprises elements $C_c[x]$, $i \leq x \leq j$. These definitions are illustrated in Figure 1(a).

Lemma 2. $C_c[i] \geq C_c[j]$ *iff* $i \leq j$. *Further, the median element of any subcolumn* $S_c[i, j]$ *can be determined with one call to the F oracle.*

Proof. The first claim follows since the elements of each column are monotonically non-increasing. The second claim follows since the median of $S_c[i, j]$ is $F(A[\lfloor (i+j)/2 \rfloor, c])$. □

Theorem 3. *The 1D p-partition problem under F can be solved in $O(n \log n)$ time.*

Proof. As per the above discussion, we effectively perform a binary search over the set of $\binom{n}{2}$ interval values. For each column, we will maintain one subcolumn containing the range of intervals which might include the optimum. Let U be the set of elements representing the union of the elements in all the active subcolumns. A *splitter* for U is an element m such that the rank of m in U, say r_m, satisfies

$$|U|/c \leq r_m \leq |U|(c-1)/c$$

for some constant c. The following algorithm finds a splitter for the active subcolumns in $O(n)$ time:

1. We find m, the median element of the set of $\leq n$ median elements of the active subcolumns. Using Lemma 2, this set of median elements can be identified in $O(n)$ time. The median of this collection, m, can now be identified in $O(n)$ time using the standard linear-time median finding algorithm.

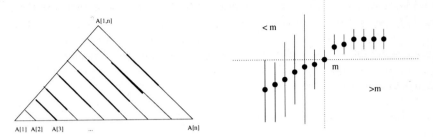

Fig. 1. (a) Columns and subcolumns of A. (b) The median of medians is not necessarily a splitter.

2. We divide the active subcolumns into two sets according to whether their median is $\leq m$ or not. Let C_l (C_r) denote the set of elements in subcolumns whose medians are $\leq m$ ($> m$). If $\min(|C_l|, |C_r|) \geq (|C_l|+|C_r|)/8$, we return m as a good splitter.
3. As illustrated by Figure 1(b), this median of medians is not necessarily a good splitter. If not, we recur on the appropriate set of subcolumns (the ones containing the larger number of elements) for the splitter search. Because the set of subcolumns under consideration is halved on each iteration, the total search time remains linear.

If $M_A(1, p, m) = true$, then m is a lower bound on the optimal partitioning. Half of the elements in each subcolumns in C_l may be eliminated, by replacing subcolumn $C_c[i, j]$ with $C_c[\lfloor(i+j)/2\rfloor, j]$. If $M_A(1, p, m) = false$, then m is an upper bound on the optimal partitioning. Half of the elements in each subcolumns in C_r may be eliminated, by replacing subcolumn $C_c[i, j]$ with $C_c[i, \lfloor(i+j)/2\rfloor]$. In either case, a constant fraction of the elements are eliminated in each linear-time round, and hence the optimal partition is identified in $O(n \log n)$ time. □

3   2D δ-weight partition under H_c

We begin by considering the following geometric problem. We say that a rectangle is *stabbed* by a line if the line passes through the interior of the rectangle.

Stabbing Problem. Given a set of axis-parallel rectangles in the $[1, n] \times [1, n]$ two dimensional integer grid, determine a set R of grid rows and C of grid columns such that each rectangle is stabbed by one of the rows in R or one of the columns in C and furthermore, $s = \max\{|R|, |C|\}$ is minimized. □

Lemma 4. *The stabbing problem is $O(\log n)$-approximable.*

Proof. The proof is by reduction to set cover; the details are deferred to the final version. □

Theorem 5. *There exists a polynomial time $O(\log n)$ factor approximation for the 2D δ-weight partition problem under H_c.*

Proof. We reduce this problem to the stabbing problem above. Consider the collection of *all* possibly overlapping *minimal* rectangles where the F value of each rectangle is $> \delta$; rectangles are minimal in the sense that if two rectangles have F value $> \delta$ and one is contained in the other, we retain the smaller one. Now the 2D δ-weight partition problem is precisely the stabbing problem for which a $O(\log n)$ factor approximation exists. □

4 2D $p \times p$-partition under F

Grigni and Manne [GM96] have shown that the 2D $p \times p$-partition problem under F is NP-Complete. Charikar et al [CCM96] proved that it is NP-Complete even to approximate the solution within a factor of 2. In this section, we present a polynomial time heuristic which provides an $O(1)$ factor approximation.

The following lemma is crucially used in our arguments.

Lemma 6. *Let c and d be two positive integers, $c, d \leq k$. If there exists a $k \times k$ partitioning such that MAX norm of the blocks is B under F, then there exists a $k/c \times k/d$ partitioning with MAX norm $\leq cdB$ under F.*

Proof. Consider a $k \times k$ partitioning with MAX norm B and take every cth row as well as every dth column. The maximum F value of a block of this $k/c \times k/d$ partitioning is at most cdB since each new block contains cd of the previous blocks. □

This lemma can be combined with the observation that Theorem 5 holds for 2D δ-weight partition problem under F as well, to get the following.

Theorem 7. *There exists a polynomial time $O(\log^2 n)$-approximation for the 1D $p \times p$-partition problem under F.*

We omit the proof in this extended abstract.

The main result in this section is a substantially improved approximation algorithm; our algorithm computes an $O(1)$ factor approximation.

Let a $\langle W, \ell \rangle$-partition be a $\ell \times \ell$-partition such that the MAX norm of the blocks is at most W. We will now show that given an input instance for which a $\langle W, \ell \rangle$-partition exists, we can construct in polynomial time a $\langle O(W), \ell \rangle$-partition. The basic idea behind our algorithm is the notion of independent rectangles:

Definition 8. *Two axis-parallel rectangles are said to be independent if their projections are disjoint along both the x-axis and the y-axis.*

Clearly, no single horizontal or vertical line can stab a pair of independent rectangles. So if an array has a $\langle W, \ell \rangle$-partition, then it may contain at most 2ℓ independent rectangles of weight strictly greater than W. As a result, independent rectangles constitute a useful tool in establishing a lower bound on the optimal solution value. The algorithm presented below builds on this idea to construct a partition whose cost is $O(W)$.

4.1 The Algorithm

Let W be the optimal solution value. We assume a knowledge of this value in the presentation below – this value will be determined by performing a binary search over the interval $[0, \sum_{i,j} A[i,j]]$. Observe that $W \geq \max_{ij} A[i,j]$. Our algorithm constitutes of the following five steps:

Step 1. We obtain an $\ell \times \ell$ partition of the array such that each row or column within any block in the partition has weight at most $2W$. □

This can done by performing independent horizontal and vertical scans. During the horizontal scan, we keep a running sum of the weight of each row since the most recent vertical partition and set down the next vertical partition when the weight of any one of the rows exceeds W. Likewise, we set horizontal partitions based on running sums of the weights of columns during the vertical scan. Since each time a new column (row, respectively) is considered, the weight of the rows (columns, respectively) can increase by at most W, it follows that the weight of any row (column, respectively) within any block induced by the vertical and horizontal partitions does not exceed $2W$. Henceforth we consider the array with this $\ell \times \ell$ partition which we refer to as the partition P.

Step 2. We construct the set S of all *minimal* rectangles whose weight exceeds W and which are entirely contained within the blocks induced by the partition from Step 1. A rectangle is *minimal* if there does not exist another rectangle properly contained in it with weight larger than W. □

This can be done by starting from each location within a block and considering rectangles with their top left corner at that location in turn in the order of increasing sides until all minimal rectangles of weight strictly greater than W are discovered.

Step 3. We determine a *local 3-optimal* set $M \subseteq S$ of independent rectangles. M is a local 3-optimal set if there does not exist $i \in \{1, 2, 3\}$ independent rectangles in $S - M$ which can be added to M by removing at most $(i-1)$ rectangles from M without violating the independence condition. □

Such a set can be easily constructed in polynomial time by repeatedly performing swaps which increase the size of the current independent collection. Each swap takes polynomial time and the procedure terminates in polynomial time since any independent collection can have at most $O(n)$ rectangles.

Step 4. We now introduce another partition based on M. For each rectangle in M, we set two *straddling* horizontal and two *straddling* vertical partitions so as to induce that rectangle. In all, this introduces at most $2M$ horizontal and

$2M$ vertical partitions. The partition P from Step 1 together with this partition induced by rectangles in M is our new partition now. □

Step 5. We now have a partition of the input array which uses $h \leq 2M + \ell$ horizontal lines and $v \leq 2M + \ell$ vertical lines. To get a $\ell \times \ell$ partition from this, we simply retain only every $\lceil h/l \rceil$th horizontal line and only every $\lceil v/l \rceil$th vertical line. By Lemma 6, this increases the maximum block weight by at most a factor of $\lceil h/l \rceil \lceil v/l \rceil$.

4.2 Analysis: Approximation Guarantee and Correctness

We need to establish two properties of the above algorithm: (a) given a choice W for which the input array has a $\langle W, \ell \rangle$-partition, the weight of any block in the partition constructed by the above algorithm is $O(W)$, and (b) the smallest value W for which the analysis of the algorithm holds, identified via binary search, is upper bounded by the optimum solution value. We begin by establishing the first property above; the following lemma is central to the analysis here.

Lemma 9. *Let b be a block contained in some block of the partition P constructed in Step 1 above. Then if the weight of block b is at least $27W$, it can be partitioned into 3 independent rectangles, each with weight strictly exceeding W.*

Proof. Given a block of weight at least $27W$, we construct three independent rectangles of weight exceeding W as follows. First we perform a vertical scan, placing a horizontal cut as soon as the weight of the slab seen thus far exceeds $7W$; we place two horizontal cuts in all. This gives us three slabs each of weight strictly greater than $7W$. Now we perform a horizontal scan from right to left placing the first vertical cut as soon as one of the horizontal slabs exceeds weight W. Without loss of generality assume that it is the top slab. Then the top right block has weight greater than W but does not exceed $3W$, and the two lower horizontal slabs to the left of that vertical cut have weight greater than $4W$ each. Now in a similar manner we place a second vertical cut to obtain two independent blocks of weight exceeding W from these two horizontal slabs. Thus we get three independent rectangles of weight greater than W each. □

Lemma 10. *The weight of any block in the partition constructed at the end of Step 4 is $O(W)$.*

Proof. We begin by observing the following easily verifiable properties of the solution: (a) each block of the solution is completely contained in some block of the partition P, and (b) given a block $b \in M$ and another block $b' \notin M$, their projections on the x-axis or the y-axis have either completely disjoint or have a perfect overlap.

Now consider a block b in the solution; using the preceding observations, it is readily seen to fall into one of the following categories: (1) the block b belongs to M, or (2) the block b does not belong to M but has a perfect overlap along one of the axes with a block $b' \in M$, or (3) the block b does not belong to M but has a perfect overlap along the x-axis with a block $b' \in M$ and a perfect overlap along the x-axis with a block $b'' \in M$.

In Case 1, the weight of b is $O(W)$ since the set S as defined in Step 2 has the property that any rectangle r in it has weight at most $3W$. This is because otherwise, we can always remove either a row or a column (of weight at most $2W$) from r to obtain a rectangle r' of weight greater than W, contained in r, which violates the minimality of the rectangles in S.

In Cases 2 and 3, each block has weight at most $27W$; this follows from an application of Lemma 9. We observe that at most two blocks in M, say b' and b'', may not be independent of a block which falls into these two cases. So if b has weight greater than $27W$, we can replace b' and b'' with at least three independent rectangles which are constructible from b (and are contained in S). But this contradicts the local 3-optimality of the collection M constructed in Step 3. Hence b must weight at most $27W$. □

Lemma 11. *The number of rectangles in M is 2ℓ for any choice W for which there exists a $\langle W, \ell \rangle$-partition of the input array.*

Proof. If M had x rectangles, then each of those rectangles must be stabbed in the optimal solution since the optimal solution value is bounded by W and every rectangle in M has weight strictly greater than W. Stabbing x rectangles requires at least $x/2$ horizontal or vertical partitions and hence x must be at most 2ℓ. □

Lemma 12. *The weight of any block in the final solution returned in* Step 5 *is at most $O(W)$ for any choice W for which there exists a $\langle W, \ell \rangle$-partition of the input array.*

Proof. Lemma 11 tells us that the number of horizontal and vertical partitions at the end of Step 4 is $O(\ell)$ each. This fact, along with an application of Lemma 6, allows us to conclude that the weight of every resulting block in the $\ell \times \ell$ partition is $O(W)$. □

This completes the proof of the first property of our algorithm that it gives a solution of weight $O(W)$ whenever a $\langle W, \ell \rangle$-partition exists. To conclude, we observe that the least value W for which the algorithm either fails to construct the partition P in Step 1 or yields a collection M in Step 3 with more than 2ℓ rectangles, must exceed the optimum. Thus the binary search procedure works to identify a suitable W.

Theorem 13. *There exists a polynomial time algorithm that computes an $O(1)$-factor approximation to the two dimensional block partitioning problem.*

5 Acknowledgements

Sincere thanks to the referees for bringing [HNC92] to our attention, and to the authors of [CCM96] for giving us a copy of their manuscript. The third author thanks Estie Arkin and Joe Mitchell for helpful discussions.

References

[AF91] S. Anily and A. Federgruen. Structured partitioning problems. *Operations Research*, 13, 130–149, 1991.

[B88] S. Bokhari. Partitioning problems in parallel, pipelined, and distributed computing. *IEEE Transactions on Computers*, 37, 38–57, 1988.

[CM+95] B. Chapman, P. Mehrotra, and H. Zima. High performance Fortran languages: Advanced applications and their implementation. *Future Generation Computer Systems*, 401 – 407, 1995.

[CCM96] M. Charikar, C. Chekuri, and R. Motwani. *Personal Communication*, 1996.

[CN91] H.-A. Choi and B. Narahari. Algorithms for mapping and partitioning chain structured parallel computations. *Proc. Intl. Conf. on Parallel Processing*, Vol I, 625-628, 1991.

[HPF] High Performance Fortran Forum Home Page. *http://www.crpc.rice.edu/HPFF/home.html*.

[GM96] M. Grigni and F. Manne. On the complexity of the generalized block distribution. Proc. of 3rd international workshop on parallel algorithms for irregularly structured problems (IRREGULAR '96), Lecture notes in computer science 1117, Springer, 319-326, 1996.

[HNC92] Y. Han, B. Narahari and H.-A. Choi. Mapping a chain task to chained processors. *Information Processing Letters* 44, 141-148, 1992.

[HM96] M. Halldorsson and F. Manne. *Manuscript*, 1996.

[HL92] P. Hansen and K. Liu Improved algorithms for Partitioning problems in parallel, pipelined, and distributed computing. *IEEE Trans. Computers*, 41, 769–771, 1992.

[KRW95] M. Kaddoura, S. Ranka and A. Wang. Array decomposition for nonuniform computational environments. *Technical Report*, Syracuse University, 1995.

[M93] F. Manne. *Load Balancing in Parallel Sparse Matrix Computations.* Ph.d. thesis, Department of Informatics, University of Bergen, Norway, 1993.

[MM+96] I. Rhee, G. Martin, S. Muthukrishnan, and R. Packwood. Fast algorithms for variable size block matching motion estimation with minimal error. *Manuscript*, 1996.

[MS95] F. Manne and T. Sorevik. Optimal partitioning of sequences. *Journal of Algorithms*, 19, 235 – 249, 1995.

[MS96] F. Manne and T. Sorevik. Partitioning an array onto a mesh of processors. Proc. of *Workshop on Applied Parallel Computing in Industrial Problems*. 1996.

[N91] D. Nicol. Rectilinear partitioning of irregular data parallel computations. *ICASE* Report 91-55, 1991. *J. Parallel and Distributed Computing*, 23, 119-134, 1994.

[OM95] B. Olstad and F. Manne. Efficient partitioning of sequences. *IEEE Transactions on Computers*, 44, 1322 – 1325, 1995.

[ZBG86] H. Zima, H. Bast and M. Gerndt. Superb: A tool for semi-automatic MIMD/AIMD parallelization. *Parallel Computing*, 1–18, 1986.

Constructive Linear Time Algorithms for Branchwidth*

Hans L. Bodlaender   Dimitrios M. Thilikos

Department of Computer Science, Utrecht University,
P.O. Box 80.089, 3508 TB Utrecht, the Netherlands
E-mail: {hansb,sedthilk}@cs.ruu.nl

Abstract. Let $\mathcal{G}_k$ be the class of graphs with branchwidth at most k. In this paper we prove that one can construct, for any k, a linear time algorithm that checks if a graph belongs to $\mathcal{G}_k$ and, if so, outputs a branch decomposition of minimum width. Moreover, we find the obstruction set for $\mathcal{G}_3$ and, for the same class, we give a safe and complete set of reduction rules. Our results lead to a *practical* linear time algorithm that checks if a graph has branchwidth ≤ 3 and, if so, outputs a branch decomposition of minimum width.

1 Introduction

This paper considers the problem to find branch decompositions of graphs with small branchwidth. The notion of branchwidth has a close relationship to the more well-known notion of treewidth, a notion that has come to play a large role in many recent investigations in algorithmic graph theory. (See Section 2 for definitions of treewidth and branchwidth.) One reason for the interest in this notion is that many graph problems can be solved by linear time algorithms, when the inputs are restricted to graphs with some uniform upper bound on their treewidth. Most of these algorithms first try to find a tree decomposition of small width, and then utilise the advantages of the tree structure of the decomposition.

The branchwidth of a graph differs from its treewidth by at most a multiplicative constant factor (see Theorem 1.) As branchwidth is also reflecting some optimal tree structure arrangement, it is possible to have algorithmic applications analogous to those of treewidth. Hence, instead of using tree decompositions, one also can use branch decompositions as starting point for the linear time algorithms for problems restricted to graphs with bounded treewidth (and hence also bounded branchwidth.) In fact, in some cases, it appears that branchwidth is more convenient to use, and seems to give better constant factors in the implementation of the algorithms; for instance, Cook used branch decompositions as an important ingredient in a practical approximation algorithm for the Travelling

---

* The secont author was supported by the Training and Mobility of Researchers (TMR) Program, (EU contract no ERBFMBICT950198).

Salesman Problem [11], and remarked that branchwidth was the more natural notion (instead of treewidth) to use for that problem [10]: where tree decompositions primarily are concerned with vertices, branch decompositions deal more with edges (in a loose sense.) We also mention that the branchwidth of planar graphs can be computed in polynomial time (see [20]). As both treewidth and branchwidth are NP-complete parameters (see [2, 20]), it appears an interesting task to find algorithms solving the following problems (k is assumed to be a fixed constant).

$\Pi_k^d(B)$ ($\Pi_k^d(T)$): Check if for some input graph has branchwidth (treewidth) $\leq k$.

$\Pi_k^c(B)$ ($\Pi_k^c(T)$): Given a graph with branchwidth (treewidth) at most k, output a minimum width branch (tree) decomposition.

According to the results of Robertson and Seymour, for any minor closed class of graphs there exist a finite set of graphs, its *obstruction set*, such that a graph G belongs to the class iff no element of the obstruction set is a minor of G. It is also known that for, any k, the class of graphs where treewidth (or branchwidth) is bounded by a fixed k is minor closed (see also Theorem 1). An immediate consequence of this fact (using results from Robertson and Seymour and the algorithm from [6]) is the existence of a linear time algorithm solving $\Pi_k^d(B)$ or $\Pi_k^d(T)$. Unfortunately, in this way, we only get a non-constructive proof of the existence of such an algorithm, but in order to construct the algorithm, we must know the corresponding obstruction set. Additionally, we would like to have an algorithm that does not only decides on branchwidth, but also constructs corresponding branch decompositions.

Much research has been done towards the construction of linear time algorithms solving $\Pi_k^d(T)$ and $\Pi_k^c(T)$. In [6], a linear (on the size of the input) time algorithm for treewidth was constructed. As this algorithm appears to be heavily exponential on k (and thus impractical, at least without considerably optimisations in the implementation), more practical algorithms have been presented for small values of k: (treewidth 1 and 2 [14, 22], treewidth 3 [4, 12, 14], treewidth 4 [18].) Also, the obstruction sets for treewidth 1, 2, and 3 are known [5, 19, 22]. In this paper, we find analogous results to those of [4, 5, 6, 19, 12, 14, 19] for the parameter of branchwidth. Namely, for any fixed k, one can construct:

• A linear time algorithm that solves $\Pi_k^d(B)$ and $\Pi_k^c(B)$.

• A parallel algorithm that solves $\Pi_k^d(B)$ in $O(\log n \log^* n)$ time on a EREW PRAM or $O(\log n)$ time on a CRCW PRAM and needs $O(n)$ operations.

• A sentence in monadic second order logic expressing whether a graph has branchwidth at most k or not.

• The obstruction set of the graphs of branchwidth at most k.

As, (similarly to the case of treewidth) the algorithms above appears to be non-practical we provide special results for the case where $k \leq 3$. More specifically, for the class of graphs with branchwidth ≤ 3, we identify the obstruction set and we give a set of safe and complete reduction rules enabling the construction of a practical linear time algorithm that checks if a graph has branchwidth ≤ 3 and, if so, outputs an minimum width branch decomposition.

The paper is organised as follows. In Section 2 the basic definition and preliminary results are presented. In Section 3 we give several graph theoretic results on $\mathcal{G}_3$. These results concern the obstruction set of $\mathcal{G}_3$ and the identification of a complete and safe set of reduction rules for $\mathcal{G}_3$ leading to the construction of a practical linear time algorithm solving $\Pi_3^d(B)$ and $\Pi_3^c(B)$. In Section 4 we present a general (for any fixed value of k) solution for $\Pi_k^d(B)$ and $\Pi_k^c(B)$.

2 Definitions and Preliminary Results

We consider undirected graphs without parallel edges or self-loops. (It is easy to extend the results to graphs with parallel edges and/or self-loops.) Given a graph $G = (V, E)$ we denote its vertex set V and edge set E with $V(G)$ and $E(G)$ respectively. For any vertex $v \in V(G)$, we define as $N_G(v)$ the set of vertices in $V(G)$ adjacent with v. Also, given a set $S \subseteq V(G)$ we denote as $G[S]$ the graph induced by S. We also denote as K_r the complete graph with r vertices.

Given two graphs G, H we say that H is a *minor* of G (denoted by $H \leq G$) if H can be obtained by a series of vertex/edge deletions and/or edge contractions (a contraction of an edge $\{u, v\}$ in G is the operation that replaces u and v by a new vertex whose neighbours are the vertices that were adjacent to u and/or v). Let $\mathcal{G}$ be a class of graphs. We say that $\mathcal{G}$ is *closed under taking of minors* when all minors of any graph in $\mathcal{G}$ belong also to $\mathcal{G}$. Robertson and Seymour proved (see e.g. [16]) that any class of graphs $\mathcal{G}$ contains a finite set of minor minimal elements. We call such a set the *obstruction set* of $\mathcal{G}$. It follows that if $\mathcal{G}$ is closed under taking of minors, then, for any graph H, $G \in \mathcal{G}$ iff there is no graph in the obstruction set of $\mathcal{G}$ such that $H \leq G$.

A *tree decomposition* of a graph G is a pair $(\{X_i \mid i \in I\}, T = (I, F))$, where $\{X_i \mid i \in I\}$ is a collection of subsets of V and T is a tree, such that

- $\bigcup_{i \in I} X_i = V(G)$,
- for each edge $\{v, w\} \in E(G)$, there is an $i \in I$ such that $v, w \in X_i$, and
- for each $v \in V$ the set of nodes $\{i \mid v \in X_i\}$ forms a subtree of T.

The *width* of a tree decomposition $(\{X_i \mid i \in I\}, T = (I, F))$ equals $\max_{i \in I}\{|X_i| - 1\}$. The *treewidth* of a graph G is the minimum width over all tree decompositions of G.

A *branch decomposition* of a graph G is a pair (T, τ), where T is a tree with vertices of degree 1 or 3 and τ is a bijection from the set of leaves of T to $E(G)$. The *order* of an edge e in T is the number of vertices $v \in V(G)$ such that there are leaves t_1, t_2 in T in different components of $T(V(T), E(T) - e)$ with $\tau(t_1)$ and $\tau(t_2)$ both incident with v (we also say: v belongs to e.) The *width* of (T, τ) is the maximum order over all edges of T, and the *branchwidth* of G is the minimum width over all branch decompositions of G (in case where $|E(G)| \leq 1$, then we define the branchwidth to be 0; if $|E(G)| = 0$, then G has no branch decomposition; if $|E(G)| = 1$, then G has a branch decomposition consisting of a tree with one vertex – the width of this branch decomposition is considered to be 0).

Instead, we can use different types of functions τ. If τ is a surjective function that maps every leaf of T to an edge $e \in E(G)$, then we have an *amplified branch decomposition*: for each edge $e \in E(G)$ there exist at least one leave v of T with $\tau(v) = e$. If, instead, we have a partial function τ, mapping only some leaves to an edge, but that is injective (every edge has a unique leaf), then we have an *extended branch decomposition*.

In what follows we denote as $\mathcal{B}_k$ ($\mathcal{T}_k$) the obstruction set of the graphs with branchwidth (treewidth) at most k.

Theorem 1 ([17]) *The following statements hold. (a) The class of graphs with bounded branchwidth is closed under taking of minors. (b) branchwidth$(G) \leq$ treewidth$(G) + 1 \leq \lfloor \frac{3}{2}$branchwidth$(G)\rfloor$. (c) A graph has branchwidth 0 (≤ 1) iff each connected component contains at most one edge (vertex of degree ≥ 2). (d) $\mathcal{B}_2 = \{K_4\}$.*

The results from [17] give algorithms for $\Pi_k^d(B)$ and $\Pi_k^c(B)$ for $k = 0, 1, 2$; for instance, graphs have branchwidth 2 if and only if they have treewidth 2, and a tree decomposition of width 2 can be transformed into a branch decomposition of width 2 in linear time. The following lemma is easy to show.

Lemma 1 *There exist an algorithm that given an amplified branch decomposition (T, τ) of a graph G with width ≤ 3, outputs a branch decomposition of G with width ≤ 3, in $O(|V(T)|)$ time. Moreover, there exist an algorithm that given a branch decomposition (T, τ) of a graph G with width ≤ 3, outputs a branch decomposition of any subgraph of G with width ≤ 3 in $O(|V(G)|)$ time.*

A *reduction* R is a triple (H, S, f), where H is a graph $S \subseteq V(H), S \neq \emptyset$ and $f : V(H) \rightarrow \omega + 1$ is a labelling of vertices in H by ordinals (finite ones and ω), such that $\forall v \in S$ $f(v) = 0$. We say that a reduction $R = (H, S, f)$ *occurs* in G if H is a subgraph of G and for any $v \in V(H)$ the degree of v in $G[V(G) - V(H) \cup \{v\}]$ is at most $f(v)$. The *result of applying R on G* is the graph arising from G if we remove the vertices in S and connect as a clique in G all vertices in $V(H) - S$. Given a graph class $\mathcal{G}$, we say that a set $\mathcal{R}$ of reductions is *safe* if, for any $R \in \mathcal{R}$ and for any G such that R occurs in G, the result of applying R on G is a graph in $\mathcal{G}$ if and only if $G \in \mathcal{G}$. Also, $\mathcal{R}$ is called *complete* for $\mathcal{G}$, if for every non-empty graph $G \in \mathcal{G}$, there is a reduction in $\mathcal{R}$ occurring in G. Clearly, if a set $\mathcal{R}$ of reduction rules is safe and complete for a graph class $\mathcal{G}$, then, for any graph G, it holds that $G \in \mathcal{R}$ if and only if there exist a sequence of reduction rules in $\mathcal{R}$ that, when successively applied, can reduce G to the empty graph.

We denote as $\mathcal{R}_{t \leq 3}$ the set of reduction rules shown in Figure 1. For any $R = (H, S, f) \in \mathcal{R}_{t \leq 3}$, S is represented by the white cycles and the values of f are shown only when they are not ω and correspond to vertices not in S.

Theorem 2 ([4, 12, 15]) $\mathcal{R}_{t \leq 3}$ *is a safe and complete set of reduction rules for the class of graphs with treewidth ≤ 3.*

We call a graph G *chordal* when it does not contain any induced cycle of length ≥ 4. We call a vertex $v \in V(G)$ *simplicial* if $G[N_G(v)]$ is a clique. Let k be an integer. A k-*tree* is a graph which is defined recursively as follows. A clique with $k + 1$ vertices is a k-tree. Given a k-tree G with n vertices, a k-tree with $n + 1$

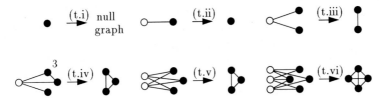

Fig. 1. The reduction rules for the class of graphs with treewidth ≤ 3.

vertices can be constructed by making a new vertex adjacent to the vertices of a k-clique in G. A graph is a partial k-tree if either it has at most k vertices or it is a subgraph of a k-tree G with the same vertex set as G. k-Trees are chordal graphs with $\omega(G) = k+1$ ($\omega(G)$ is the size of the maximum clique in a graph G). It can be easily proved that a graph has treewidth $\leq k$ iff it is a partial k-tree (see e.g. [21]). Also, if G is a k-tree, then $|E(G)| = O(k|V(G)|)$. A set $S \subseteq V(G)$ is an *s-t-separator* in G ($s, t \in V$), if s and t belong to different connected components of $G[V - S]$. S is a *minimal s-t-separator*, if it does not contain another s-t-separator as a proper subgraph. S is a *minimal separator*, if there exist vertices $s, t \in V$ for which S is a minimal s-t-separator. It is known that any minimal separator of a chordal graph induces a clique. We call a graph G' a *triangulation* of G if G' is chordal and $V(G) = V(G')$. We call a triangulation of G with a minimum number of edges *minimal triangulation*.

Theorem 3 ([9]) *Let G' be a minimal triangulation of a graph G. Then any minimal separator in G' is also a minimal separator in G.*

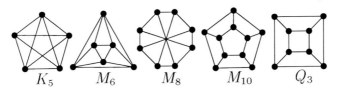

Fig. 2. The graphs K_5, M_6, M_8, M_{10}, and Q_3

Theorem 4 ([5, 19]) $\mathcal{T}_3 = \{K_5, M_6, M_8, M_{10}\}$ *(graphs K_5, M_6, M_8, and M_{10} are shown in Figure 2).*

The following can be proved using Theorems 1.b and 4.

Lemma 2 *The following three statements hold.*
a. There are no graphs in $\mathcal{B}_3$ with treewidth ≤ 2.
b. $Q_3 \in \mathcal{B}_3$ and treewidth$(Q_3) = 3$ (graph Q_3 is shown in Figure 2).
c. The set $\{K_5, M_6, M_8\}$ contains all the graphs of $\mathcal{B}_3$ that have treewidth ≥ 4.

Let G be a graph and $S \subseteq V(G), |S| = 4$. We call $S = \{v_1, v_2, v_3, v_4\}$ a *cross* if the sets $S_i = S - \{v_i\}, 1 \leq i \leq 4$ are all minimal separators of G. We also define as $att(G, S_i)$ the set of all the vertices of the connected components of $G[V(G) - S_i]$ that do not contain the single vertex in $S - S_i$. If a graph does not contain any cross then we call it *crossless*.

Using Theorem 3 we can easily prove the following.

Lemma 3 *Let G be a crossless graph of treewidth at most 3 and G' be a minimal triangulation of G. Then, G' is a crossless chordal graph with $\omega(G) \leq 4$.*

Let G be a 3-tree G. A tree T_G is the *clique tree* of G if each vertex in $V(T_G)$ represents a 4-clique in G and where two vertices $\mathbf{v} = \{v_1, v_2, v_3, v_4\}, \mathbf{u} = \{u_1, u_2, u_3, u_4\} \in V(T_G)$ are connected by an edge $\{\mathbf{v}, \mathbf{u}\}$ in T_G iff $|\mathbf{v} \cap \mathbf{u}| = 3$, i.e., they have exactly 3 vertices in common (notice that each such triple of vertices is a minimal separator of G). Given an edge $e = \{\mathbf{v}, \mathbf{u}\} \in E(T_G)$ we define the *separation set* of e as $\mathsf{sep}(e) = \mathbf{v} \cap \mathbf{u}$.

We will need the following results which we present without proof.

Lemma 4 *There exist an algorithm that given a 3-tree G constructs the clique tree of G in $O(|V(G)|)$ time.*

Lemma 5 *There exists an algorithm that given a crossless chordal graph G with $\omega(G) \leq 4$, outputs, in $O(|V(G)|)$ time, a crossless 3-tree G' such that G' is a subgraph of G where $V(G) = V(G')$.*

3 Graphs with branchwidth at most 3

In this section we will identify the set $\mathcal{B}_3$ and find a complete and safe set of reduction rules for the class of graphs with branchwidth ≤ 3. Our results lead to the construction of a linear time algorithm testing whether a graph has branchwidth ≤ 3 and, if so, computes a branch decomposition of minimum width. According to Theorem 1.c, it is trivial to check in linear time if G has branchwidth ≤ 1 and, if so, to construct a branch decomposition of minimum width. Also, from Theorem 1.d, we can check in linear time if a graph has branchwidth ≥ 3 or not. In what follows, we examine the non trivial case where the input is a graph with branchwidth ≥ 3. We omit the case where we are given a graph with branchwidth 2 as it is a very similar (and much easier) version of the non trivial case.

The following lemma defines the notion of the labelled clique tree of a crossless 3-tree (the proof is omitted).

Lemma 6 *Let T_G be the clique tree of a crossless 3-tree G. Let also, for any $\mathbf{v} \in V(T_G)$, $E_\mathbf{v} = \{e \in E(T_G) : \mathbf{v} \text{ is incident to } e\}$. Then, for each $\mathbf{v} \in V(T_G)$, $|\{\mathsf{sep}(e) : e \in E_\mathbf{v}\}| \leq 3$. Moreover, it is possible in $O(n)$ time to compute a labelling function $l : |E(T_G)| \to \{1, 2, 3\}$ such that $\forall \mathbf{v} \in V(T_G) \; \forall e_1, e_2 \in E_\mathbf{v} \; (\mathsf{sep}(e_1) = \mathsf{sep}(e_2) \text{ iff } l(e_1) = l(e_2))$, i.e. edges in $E_\mathbf{v}$ with the same separation set have the same label. We call such a clique tree 3-labelled and we denote it as (T_G, l).*

Given a labelled clique tree (T_G, l) we define the *span degree* of a vertex $\mathbf{v}$ to be equal to $|\{l(e) : e \in E_\mathbf{v}\}|$. We also call a leaf $\mathbf{u}$ of T_G that is adjacent to a vertex $\mathbf{v}$ *simple* if $|\{\mathbf{e} \in E_\mathbf{v} : l(\mathbf{e}) = l(\{\mathbf{u}, \mathbf{v}\})\}| = 1$.

The following can be easily proved by induction on $|V(T_G)|$.

Lemma 7 *Let (T_G, l) be a labelled tree with more than 3 vertices. Then one of the following holds: (i) There exist no simple leaves. (ii) There exist a simple leaf $\mathbf{u}$ in T_G adjacent to a vertex $\mathbf{v}$ of span-degree 2. (iii) There exist two simple leaves $\mathbf{u}_1$ and $\mathbf{u}_2$ in T_G adjacent to a vertex $\mathbf{v}$ of span-degree 3.*

Using now Lemma 1 we have a proof of the following Lemma which provides the basic algorithm of this section (the proof is long and is omitted due to space limitations).

Lemma 8 *There exist a linear time algorithm that, given a 3-labelled clique tree of a crossless 3-tree G constructs a branch width decomposition of G of width 3.*

Combining Lemmas 1, 4, 5, 6, and 8 gives the following result.

Theorem 5 *Any crossless chordal graph with $\omega(G) \leq 3$ has a branch decomposition of width 3. Moreover it is possible to construct an algorithm that finds such a branch decomposition in $O(|V(G)|)$ time.*

Using Theorems 1 and 5 and Lemma 2, we can now proof the following.

Theorem 6 *The following two propositions hold: (i) branchwidth$(G) \leq 3 \Leftrightarrow$ treewidth$(G) \leq 3 \wedge Q_3 \not\leq G \Leftrightarrow$ treewidth$(G) \leq 3 \wedge G$ is crossless. (ii) The obstruction set of the graphs of branchwidth three, $\mathcal{B}_3$ equals $\{K_5, M_6, M_8, Q_3\}$.*

We denote as $\mathcal{R}_{b \leq 3}$ the set of reduction rules shown in Figure 3.

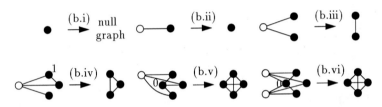

Fig. 3. The reduction rules for the class of graphs with branchwidth ≤ 3.

Using Theorems 1.a, 2, and 6, we can prove the following.

Lemma 9 $\mathcal{R}_{b \leq 3}$ *is a safe set of reduction rules for the class of graphs with bounded branchwidth.*

Also, using the case analysis of Lemma 7, we can prove the following.

Lemma 10 *The following two propositions hold: (i) If G is a crossless 3-tree then, there exists one reduction rule in $\mathcal{R}_{b \leq 3}$ occurring in G. (ii) If there exist some reduction rule in $\mathcal{R}_{b \leq 3}$ occurring in a graph G, then, for any subgraph G' of G where $V(G) = V(G')$, there exist also some rule in $\mathcal{R}_{b \leq 3}$ occurring in G'.*

Using now Lemmas 3, 5, 9, and 10, we can prove the following.

Theorem 7 $\mathcal{R}_{b \leq 3}$ *is a safe and complete set of rules for rewriting graphs of branchwidth≤ 3.*

Using now Theorems 7 and 5, we can prove the following.

Theorem 8 *One can construct an algorithm that tests if a given graph has branchwidth at most 3 and, if so, outputs a branchwidth decomposition of minimum width, and that uses $O(n)$ time.*

4 A linear time algorithm for graphs with branchwidth $\leq k$

In this section, we will show the following theorem.

Theorem 9 *For every k, one can construct an algorithm, that given a graph $G = (V, E)$, decides whether the branchwidth of G is at most k, and if so, constructs a branch decomposition of G of minimum width, and that uses $O(|V|)$ time.*

(Note that if the branchwidth of a graph is bounded by a constant, then $|E| = O(|V|)$.) While the theorem generalises the result of the previous Section, it should be noted that the algorithm here has a large (exponential) constant factor, making it (at least without considerably optimisations in the implementation) not practical, whereas the algorithm in the previous section for the case that $k = 3$ is a practical and efficient algorithm.

Note that a non-constructive version of Theorem 9 can almost directly be obtained from the results in [16] and [6] as, for every k, the class of graphs with branchwidth at most k is closed under taking of minors. Also, note from the result in [6], that it is sufficient to prove the following result:

Lemma 11 *For every k, l, one can construct an algorithm, that given a graph $G = (V, E)$, with a tree decomposition of G of width at most l, decides whether the branchwidth of G is at most k, and if so, constructs a branch decomposition of G of minimum width, and that uses linear time.*

A terminal graph is a triple $G = (V, E, X)$ with $G = (V, E)$ a graph, and X an ordered set of vertices from V. If $|X| = k$, we call (V, E, X) a k-terminal graph. Given an extended branch decomposition, we can build a branch decomposition of the same graph with the same width as follows: repeatedly remove leaves that have no edge associated with them and contract over nodes of degree 2 in the tree. We call the obtained branch decomposition the *shrunken form* of the extended branch decomposition.

Let $G = (V, E)$ be a graph, and let $H = (V', E')$ be a subgraph of G. A branch decomposition (T, τ) of G is an *extension* of a branch decomposition (T', τ') of H, if (T', τ') can be obtained as follows: let τ'' be the restriction of τ to those leaves that map to an edge in E'. (T, τ'') is an extended branch decomposition of H. Now let (T', τ') be the shrunken form of (T, τ'').

The *full model* of a branch decomposition (T, τ) of a terminal graph $G = (V, E, X)$ is the 4-tuple (T, τ, β, γ), where β is a function, that maps each edge e of T to the set of vertices in X that belong to e, and γ is a function that maps each edge in T to its order.

A *model* of a branch decomposition (T, τ) of terminal graph $G = (V, E, X)$ is a 4-tuple $(T', \tau', \beta, \gamma)$, that is obtained by the full model of (T, τ) by applying 0 or more of the following operations:

- Suppose $\{v, w\}$ and $\{v, x\}$ are edges in T, w, x leaves in T, and $\beta(\{v, w\}) \subseteq \beta(\{v, y\})$, $\beta(\{v, x\}) \subseteq \beta(\{v, y\})$; y the third neighbor of v in T. Then remove edges $\{v, w\}$, and $\{v, x\}$ from T, and restrict τ, β, and γ to the edges in the smaller tree.

- Suppose $v_1, v_2, \ldots, v_r$ form a path in T, with all vertices $v_2, \ldots v_{r-1}$ adjacent to a leaf ($\neq v_1, v_r$) in T. Suppose that $\beta(\{v_1, v_2\}) = \beta(\{v_2, v_3\}) = \cdots \beta(\{v_{r-1}, v_r\})$. Suppose also that $\gamma(\{v_1, v_2\}) \leq \min\{\gamma(\{v_2, v_3\}, \ldots, \gamma(\{v_{r-2}, v_{r-1}\})$ and that $\gamma(\{v_{r-1}, v_r)\} \leq \max\{\gamma(\{v_2, v_3\}), \ldots, \gamma(\{v_{r-2}, v_{r-1}\})$. Then, identify v_2 and v_{r-1}, remove vertices $v_3, \ldots, v_{r-2}$, their adjacent edges, their leaf-neighbors, and the leaf-neighbor ($\neq v_1$) of v_2.

The *characteristic* of a branch decomposition (T, τ) of terminal graph $G = (V, E, X)$ is the model of the characteristic that cannot be reduced by applying one of these two operations. One can show that the characteristic of a branch decomposition is unique.

Lemma 12 *Let k be fixed. There are functions f_1, such that each characteristic of a branch decomposition of a terminal graph $G = (V, E, X)$ of width at most k has at most $f_1(|X|)$ nodes on its tree, and each terminal graph $G = (V, E, X)$ has at most $f_2(|X|)$ different characteristics of branch decompositions of G of width at most k.*

Lemma 13 *Let $G_1 = (V_1, E_1, X)$, $G_2 = (V_2, E_2, X)$ be terminal graphs with the same set of terminals. If $H = (V_3, E_3, X)$ is a terminal graph, and (T_1, τ_1), (T_2, τ_2) are branch decompositions, respectively of G_1 and G_2, with the same characteristic, then there is an extension of (T_1, τ_1) that is a branch decomposition of $G_1 \oplus H$ of width $\leq k$, if and only if there is an extension of (T_2, τ_2) that is a branch decomposition of $G_2 \oplus H$ of width $\leq k$.*

The two lemmas above imply together that the property that a graph has branchwidth at most k is *finite state*, or *regular* (see e.g. [1]). (A class of graphs $\mathcal{G}$ is finite state, if the equivalence relation on l-terminal graphs $G \sim H \Leftrightarrow (\forall K : G \oplus K \in \mathcal{G} \Leftrightarrow H \oplus K \in \mathcal{G})$ has a finite number of equivalence classes, for each fixed l.)

Combining the above results, and results and techniques from [1, 3, 6, 7, 13], we obtain the following result. (The results show that each of these is computable, although real practical efficiency and doabily is not guaranteed.)

Theorem 10 *One can construct, for each k,*
- *linear time algorithms that decide whether a graph has branchwidth at most k (these algorithms can either use a tree- or branch decomposition and dynamic programming, or use graph reduction),*
- *parallel algorithms that decide whether a graph has branchwidth at most k, that use $O(\log n \log^* n)$ time on a EREW PRAM or $O(\log n)$ time on a CRCW PRAM and $O(n)$ operations,*
- *a sentence in monadic second order logic expressing whether a graph has branchwidth at most k or not,*
- *the obstruction set of the graphs of branchwidth at most k.*

However, we do not only want to decide whether the branchwidth is at most k, but also to build a corresponding branch decomposition.

A model $(T_1, \tau_1, \beta_1, \gamma_1)$ is *dominated* by a model $(T_2, \tau_2, \beta_2, \gamma_2)$, if $(T_2, \tau_2, \beta_2, \gamma_2)$ can be obtained from $(T_1, \tau_1, \beta_1, \gamma_1)$ by 0 or more of the following operations:

- Contract some edges in T_1.
- Remove vertices from sets $\beta_1(e)$, for some edges e in T.
- Decrease numbers $\beta_1(e)$ for some edges e in T.

Lemma 14 *Let $G_1 = (V_1, E_1, X)$, $G_2 = (V_2, E_2, X)$ be terminal graphs with the same set of terminals. If $H = (V_3, E_3, X)$ is a terminal graph, and if the characteristic of branch decomposition (T_1, τ_1) of G_1 is dominated by the characteristic of branch decomposition (T_2, τ_2) of G_2, then if there is an extension of (T_1, τ_1) that is a branch decomposition of $G_1 \oplus H$ of width at most k, then there is an extension of (T_2, τ_2) that is a branch decomposition of $G_2 \oplus H$ of width at most k.*

A *full set of characteristics (for branchwidth = k)* of a terminal graph $G = (V, E, X)$ is a set of characteristics S of branch decompositions of G of width at most k, such that each characteristic of a branch decomposition of G of width at most k is dominated by an element of S.

Suppose we have a tree decomposition $(\{X_i \mid i \in I\}, T = (I, F))$ of a graph G, with T a rooted binary tree. To each $i \in I$, we associate the terminal graph $G_i = (V_i, E_i, X_i)$, with $V_i = \cup_j$ is a descendant of i or $j = i X_i$, and $E_i = E(G[V_i])$.

The following lemma (with a proof that resembles a technique from [8]) gives us a method to compute full sets of characteristics.

Lemma 15 *Let k, l be constants. Let $(\{X_i \mid i \in I\}, T = (I, F))$ be a tree decomposition of G of width at most l. For any node $i \in I$ with at most 2 children in T, we can compute a full set of characteristics for branchwidth = k of G_i, given full sets of characteristics for branchwidth = k of all terminal graphs associated with the children of i.*

Using the algorithm of the lemma above, we can compute full sets of characteristics for all graphs associated with nodes in the given tree decomposition of the input graph, in time, linear in the size of T (process nodes in a bottom-up order). When the full set of the root node is non-empty, the branchwidth of the input graph is at most k, otherwise not. Finally, by extra bookkeeping, we can build a branch decomposition of width at most k (when existing), in linear time. As the details are cumbersome, they are omitted here.

Acknowledgements

We thank Babette de Fluiter and Koichi Yamazaki for discussions on this research.

References

1. K. R. Abrahamson and M. R. Fellows. Finite automata, bounded treewidth and well-quasiordering. In N. Robertson and P. Seymour, editors, *Proceedings of the AMS Summer Workshop on Graph Minors, Graph Structure Theory*, Contemporary Mathematics vol. 147, pages 539–564. American Mathematical Society, 1993.

2. S. Arnborg, D. G. Corneil, and A. Proskurowski. Complexity of finding embeddings in a k-tree. *SIAM J. Alg. Disc. Meth.*, 8:277–284, 1987.
3. S. Arnborg, B. Courcelle, A. Proskurowski, and D. Seese. An algebraic theory of graph reduction. *J. ACM*, 40:1134–1164, 1993.
4. S. Arnborg and A. Proskurowski. Characterization and recognition of partial 3-trees. *SIAM J. Alg. Disc. Meth.*, 7:305–314, 1986.
5. S. Arnborg, A. Proskurowski, and D. G. Corneil. Forbidden minors characterization of partial 3-trees. *Disc. Math.*, 80:1–19, 1990.
6. H. L. Bodlaender. A linear time algorithm for finding tree-decompositions of small treewidth. *SIAM J. Comput.*, 25:1305–1317, 1996.
7. H. L. Bodlaender and T. Hagerup. Parallel algorithms with optimal speedup for bounded treewidth. In Z. Fülöp and F. Gécseg, editors, *Proceedings 22nd International Colloquium on Automata, Languages and Programming*, pages 268–279, Berlin, 1995. Springer-Verlag, Lecture Notes in Computer Science 944.
8. H. L. Bodlaender and T. Kloks. Efficient and constructive algorithms for the pathwidth and treewidth of graphs. *J. Algorithms*, 21:358–402, 1996.
9. H. L. Bodlaender, T. Kloks, and D. Kratsch. Treewidth and pathwidth of permutation graphs. In *Proceedings 20th International Colloquium on Automata, Languages and Programming*, pages 114–125, Berlin, 1993. Springer Verlag, Lecture Notes in Computer Science, vol. 700.
10. W. Cook, 1996. Personal communication.
11. W. Cook and P. D. Seymour. An algorithm for the ring-routing problem. Bellcore technical memorandum, Bellcore, 1993.
12. Y. Kajitani, A. Ishizuka, and S. Ueno. A characterization of the partial k-tree in terms of certain substructures. *Graphs and Combinatorics*, 2:233–246, 1986.
13. J. Lagergren and S. Arnborg. Finding minimal forbidden minors using a finite congruence. In *Proceedings of the 18th International Colloquium on Automata, Languages and Programming*, pages 532–543. Springer Verlag, Lecture Notes in Computer Science, vol. 510, 1991.
14. J. Matoušek and R. Thomas. Algorithms finding tree-decompositions of graphs. *J. Algorithms*, 12:1–22, 1991.
15. J. Matoušek and R. Thomas. On the complexity of finding iso- and other morphisms for partial k-trees. *Disc. Math.*, 108:343–364, 1992.
16. N. Robertson and P. D. Seymour. Graph minors — a survey. In I. Anderson, editor, *Surveys in Combinatorics*, pages 153–171. Cambridge Univ. Press, 1985.
17. N. Robertson and P. D. Seymour. Graph minors. X. Obstructions to tree-decomposition. *J. Comb. Theory Series B*, 52:153–190, 1991.
18. D. P. Sanders. On linear recognition of tree-width at most four. *SIAM J. Disc. Meth.*, 9(1):101–117, 1996.
19. A. Satyanarayana and L. Tung. A characterization of partial 3-trees. *Networks*, 20:299–322, 1990.
20. P. D. Seymour and R. Thomas. Call routing and the ratcatcher. *Combinatorica*, 14(2):217–241, 1994.
21. J. van Leeuwen. Graph algorithms. In *Handbook of Theoretical Computer Science, A: Algorithms and Complexity Theory*, pages 527–631, Amsterdam, 1990. North Holland Publ. Comp.
22. J. A. Wald and C. J. Colbourn. Steiner trees, partial 2-trees, and minimum IFI networks. *Networks*, 13:159–167, 1983.

The Word Matching Problem Is Undecidable for Finite Special String-Rewriting Systems That Are Confluent

Paliath Narendran[1]* and Friedrich Otto[2]

[1] Institute of Programming and Logics, Department of Computer Science
State University of New York at Albany, Albany, NY 12222, U.S.A.
E-mail: dran@cs.albany.edu

[2] Fachbereich Mathematik/Informatik
Universität Kassel, 34109 Kassel, Germany
E-mail: otto@theory.informatik.uni-kassel.de

Abstract. We present a finite, special, and confluent string-rewriting system for which the word matching problem is undecidable. Since the word matching problem is the non-symmetric restriction of the word unification problem, this presents a non-trivial improvement of the recent result that for this type of string-rewriting systems, the word unification problem is undecidable (Otto 1995). In fact, we show that our undecidability result remains valid even when we only consider very restricted instances of the word matching problem.

Keywords: matching, unification, equational theory, string-rewriting systems

1 Introduction and basic definitions

Equational unification and matching have generated a lot of interest recently, mainly due to their importance in term-rewriting systems and equational reasoning. Historically, one of the earliest equational unification problems that have been studied extensively is *word unification*, which is the problem of solving word equations[3]. The general question of whether the solvability of a word equation is decidable or not remained open for a long time, until it was finally settled positively by Makanin [Mak77].

Since Makanin's paper appeared, his algorithm has been the subject of many research activities. The objectives have been to simplify the proof of the termination and correctness of his algorithm [Pec81, Sch93], to develop simpler algorithms for deciding the solvability of word equations [Jaf90, Sch90], and to compute a description for the set of all solutions of a solvable word equation [MaAb94]. Observe that a word equation can have a minimal complete set of

---
* Partially supported by the NSF grants CCR-9404930 and INT-9401087.
[3] This is also known as Markov's problem or Löb's problem.

most general unifiers that is infinite, that is, the theory of associativity is of unification type infinitary.

Makanin also extended his result further by showing that the word unification problem is decidable for finitely generated free groups [Mak83, Mak85]. Since finitely generated free groups can be specified by finite, special, and confluent string-rewriting systems, this leads naturally to the question of whether the solvability of word equations modulo finite, special, and confluent string-rewriting systems is decidable in general. Here a string-rewriting system is called *special*, if it only contains rules of the form $\ell \to \lambda$, where λ denotes the empty string. Obviously, rewriting modulo such a system is particularly simple, since it simply amounts to the deletion of substrings. A special system R is called *confluent*, if each string has a unique irreducible descendant with respect to the reduction relation induced by R. In this situation the set IRR(R) of irreducible strings modulo R forms a set of unique representatives for the Thue congruence induced by R (see, e.g., [BoOt93]). But here the answer to the solvability question turned out to be negative as shown by Otto [Ott95], who presents a particular finite, special, and confluent string-rewriting system for which the word unification problem is undecidable.

Now where exactly is the borderline between the decidable and the undecidable cases of the problem of deciding the solvability of word equations? On the one hand, one could try to restrict the finite string-rewriting systems considered even further. A reasonable candidate would be the class of finite, special, and confluent string-rewriting systems that present groups. Is the solvability of word equations in general decidable or undecidable for this class of string-rewriting systems? (This question is still open.)

Here we follow a different approach. Instead of restricting the class of string-rewriting systems considered even further, we put an additional restriction on the form of the word equations that we admit. While a typical instance of the word unification problem consists of a pair of strings (u, v), where both u and v contain variables that must be instantiated in order to get instances $\theta(u)$ and $\theta(v)$ that are congruent modulo the system R considered, we look at word equations of the form (u, v), where only one side, say u, contains variables. Hence, such a word equation has a solution modulo R if and only if there exists an instantiation θ such that the strings $\theta(u)$ and v are congruent modulo R. This restricted version of the word unification problem is known as the *word matching problem*.

Here we strengthen the above-mentioned undecidability result by showing that there is a finite, special, and confluent string-rewriting system for which the word matching problem is undecidable. In fact, we consider rather restricted instances of the word matching problem, since we look at word equations of the form (u, λ). Recall that λ is used to denote the empty string. Such a word equation has a solution modulo R if and only if there exists an instantiation θ of the variables occurring in u such that the string $\theta(u)$ is congruent to the empty string λ modulo R. We present a finite, special, and confluent string-rewriting system for which this restricted variant of the word matching problem is undecidable.

This paper is organized as follows. In Section 2 we present a finite, special, and confluent string-rewriting system $\mathcal{R}$ for which the word matching problem is undecidable, thus establishing a weak version of the intended undecidability result. In the following section we extend $\mathcal{R}$ to a finite, special, and confluent system $\mathcal{R}_1$ for which even the above-mentioned restricted variant of the word matching problem is undecidable.

One may ask why we actually give both these proofs, since the latter result is clearly stronger than the former. However, the proof of the former is simpler and therefore, it illustrates the key ideas used in the reductions more clearly. Further, the technical results on the system $\mathcal{R}$ established in Section 2 are needed anyway in proving the stronger result in Section 3.

We close this section by providing the main definitions and notation necessary to follow our arguments. For a more detailed treatment of the basics and for additional information on the notions introduced, we refer the interested reader to the following surveys – [BaSi94, JoKi91] for unification, [DeJo90] for term-rewriting, and [BoOt93] for string-rewriting.

A *string-rewriting system* (often called a 'Thue system') on an alphabet Σ is a finite set of pairs of strings $R \subset \Sigma^* \times \Sigma^*$. In this note we will only be dealing with string-rewriting systems that are *length-reducing*, that is, we assume that $|\ell| > |r|$ for each pair $(\ell, r) \in R$. These pairs are often referred to as *rewrite rules* and are sometimes represented as $\ell \to r$. A string-rewriting system R is said to be *special* if $r = \lambda$ for each pair (ℓ, r) in R. As mentioned before, λ denotes the empty string.

By $\to_R$ we denote the *single-step reduction relation* that is defined by the string-rewriting system R: $\to_R := \{(u\ell v, urv) \,|\, (\ell, r) \in R, u, v \in \Sigma^*\}$. Its reflexive and transitive closure $\to_R^*$ is the *reduction relation* induced by R. The relation $\leftrightarrow_R^* := (\to_R \cup \to_R^{-1})^*$ is called the *Thue congruence* generated by R. By IRR(R) we denote the set of irreducible strings modulo R, that is, $u \in \text{IRR}(R)$ if and only if $u \to_R v$ does not hold for any string v.

Let V be a set of variables that range over Σ^*. A *string equation* or *word equation* is an equation of the form $u = v$ where u and v are strings over $(\Sigma \cup V)^*$. An assignment or substitution $\theta : V \to \Sigma^*$ is a *solution* of the equation $u = v$ modulo R if and only if $\theta(u) \leftrightarrow_R^* \theta(v)$. Here θ is extended to a morphism $\theta : (V \cup \Sigma)^* \to \Sigma^*$ in the obvious way.

The *word matching problem* for a string-rewriting system R is the following decision problem:

INSTANCE : A string $u \in (V \cup \Sigma)^*$ and a string $v \in \Sigma^*$.
QUESTION : Is there a substitution θ satisfying $\theta(u) \leftrightarrow_R^* v$?

2 Undecidability of the word matching problem

The announced undecidability result will be proved by a reduction from the following undecidable problem.

Theorem 1. [NaOt90]
There exists a set of pairs of non-empty strings $P = \{(x_i, y_i) \mid i = 1, \ldots, n\} \subset \{a,b\}^+ \times \{a,b\}^+$ such that the following problem is undecidable:

INSTANCE : Two non-empty strings $x_0, y_0 \in \{a,b\}^+$.
QUESTION : Do there exist indices $i_1, \ldots, i_k$, each i_j from $\{1, \ldots, n\}$, such that $x_0 x_{i_1} x_{i_2} \cdots x_{i_k} = y_0 y_{i_1} y_{i_2} \cdots y_{i_k}$?

We reduce this problem, which is a specialized form of the well-known *Modified Post Correspondence Problem* (MPCP), to the word matching problem modulo a particular finite, special, and confluent string-rewriting system. The system we construct is on the alphabet consisting of the letters a and b, symbols for each of the numbers 1 to n, and special symbols S, $\#$, and $\$$. In other words, let

$$\Gamma := \{a,b\} \cup \{1, \ldots, n\} \cup \{S, \#, \$\}.$$

The rules of our system are divided into three classes. The first class corresponds to the x_i's, the first components of the pairs in P, the second to the y_i's, and the third class is to ensure that a string is indeed from $\{a,b\}^*$.

The rules from Class I are

$$x_i S i \to \lambda, \quad i \in \{1, \ldots, n\}.$$

Class II consists of

$$y_i i S \to \lambda, \quad i \in \{1, \ldots, n\}.$$

Class III consists of the rules

$$a\#\# \to \lambda, \quad b\#\# \to \lambda, \quad a\$\$ \to \lambda, \quad b\$\$ \to \lambda.$$

Observe that the string-rewriting system $\mathcal{R}$ that consists of the above three classes of rules is a finite special system that is in addition confluent. In fact, there are no non-trivial critical pairs at all for this system (see, e.g., [BoOt93]).

We will show that the *simultaneous* variant of the word matching problem is undecidable for this system $\mathcal{R}$, which is the following decision problem:

INSTANCE : A finite sequence $(u_1, v_1), \ldots, (u_m, v_m)$ of pairs of strings from $(V \cup \Gamma)^* \times \Gamma^*$.
QUESTION : Is there a substitution θ satisfying $\theta(u_i) \leftrightarrow^*_{\mathcal{R}} v_i$ simultaneously for all $i = 1, \ldots, m$?

Since this simultaneous variant of the word matching problem is reducible to the (*single*) word matching problem by introducing a new letter, say $\cent$, this will give our intended undecidability result.

Lemma 2. *Let X be an irreducible string from Γ^*. Then*

$$X\#Y_1 \to^*_{\mathcal{R}} \lambda, \quad XY_1\# \to^*_{\mathcal{R}} \lambda, \quad X\$Y_2 \to^*_{\mathcal{R}} \lambda, \quad \text{and } XY_2\$ \to^*_{\mathcal{R}} \lambda,$$

for some $Y_1, Y_2 \in \Gamma^*$ if and only if $X \in \{a,b\}^*$.

Proof: The 'if'-part is trivial. The proof of the 'only if'-part is by contradiction. Let Z be a shortest counterexample, that is, a shortest irreducible string such that

$$Z\#Y_1 \to_{\mathcal{R}}^* \lambda, \ ZY_1\# \to_{\mathcal{R}}^* \lambda, \ Z\$Y_2 \to_{\mathcal{R}}^* \lambda, \text{ and } ZY_2\$ \to_{\mathcal{R}}^* \lambda,$$

for some Y_1, Y_2, where $Z \notin \{a,b\}^*$. Clearly Z cannot end in a $\$$ or a $\#$, for if $Z = Z'\$$, then $Z'\$\#Y_1$ will not be reducible. Thus Z has to end in either an a or a b.

Let $Z = Z'\alpha$, where $\alpha \in \{a,b\}$. For $Z'\alpha\#Y_1$ to be reducible, the leftmost symbol of Y_1 must be a $\#$. Because of the second condition $ZY_1\# \to_{\mathcal{R}}^* \lambda$, the next (that is, second from the left) symbol in Y_1 must also be a $\#$. In other words, $Y_1 = \#\#Y_3$ for some Y_3. By similar reasoning, $Y_2 = \$\$Y_4$ for some Y_4. Now,

$$\begin{aligned} Z'\alpha\#\#Y_3 &\to_{\mathcal{R}} Z'\#Y_3 \to_{\mathcal{R}}^* \lambda, \\ Z'\alpha\#\#Y_3\# &\to_{\mathcal{R}} Z'Y_3\# \to_{\mathcal{R}}^* \lambda, \\ Z'\alpha\$\$Y_4 &\to_{\mathcal{R}} Z'\$Y_4 \to_{\mathcal{R}}^* \lambda, \text{ and} \\ Z'\alpha\$\$Y_4\$ &\to_{\mathcal{R}} Z'Y_4\$ \to_{\mathcal{R}}^* \lambda, \end{aligned}$$

which shows that Z' is a shorter counterexample. This contradicts the choice of Z, and hence, we conclude that there is no such counterexample. □

Lemma 3. Let $x_0, y_0 \in \{a,b\}^+$. Then, for $X_1, X_2 \in \{a,b\}^*$,

$$X_1 SY \to_{\mathcal{R}}^* x_0 S \text{ and } X_2 Y \to_{\mathcal{R}}^* y_0$$

for some Y if and only if there exist indices $j_1, \ldots, j_\ell \in \{1, \ldots, n\}$ such that

$$X_1 = x_0 x_{j_1} \cdots x_{j_\ell} \text{ and } X_2 = y_0 y_{j_1} \cdots y_{j_\ell}.$$

Proof: The 'if' part is straightforward. By examining the rules from Classes I and II, we can easily see that by taking $Y := j_\ell S \cdots j_1 S$ we can reduce $X_1 SY$ by Class I rules to $x_0 S$, and $X_2 Y$ by Class II rules to y_0.

We prove the 'only if' part by contradiction. Let $U_1, U_2 \in \{a,b\}^*$ be minimal counterexamples in terms of their combined length. Obviously, we may assume without loss of generality that Y is irreducible. Clearly $Y \neq \lambda$: if Y were λ, then $U_1 S = x_0 S$ and $U_2 = y_0$.

If $Y \neq \lambda$, then the only rules that are applicable to $U_1 SY$ are the rules from Class I. Thus there must be a rule $x_p Sp \to \lambda$ and strings U_1' and Y' such that $U_1 = U_1' x_p$, $Y = pY'$, and $U_1 SY \to_{\mathcal{R}} U_1' Y' \to_{\mathcal{R}}^* x_0 S$.

Then $U_2 Y = U_2 pY'$ must also be reducible, and it can be seen that only the rule $y_p pS \to \lambda$ from Class II can apply. In other words, $U_2 = U_2' y_p$, $Y' = SY''$ for some U_2', Y'', and $U_2 pY' \to_{\mathcal{R}} U_2' Y'' \to_{\mathcal{R}}^* y_0$.

Thus we get

$$U_1' SY'' \to_{\mathcal{R}}^* x_0 S \text{ and } U_2' Y'' \to_{\mathcal{R}}^* y_0,$$

and this contradicts the minimality of U_1 and U_2. □

The following result is now immediate.

Lemma 4. Let $x_0, y_0 \in \{a,b\}^+$. Then there is a string $X \in \{a,b\}^*$ satisfying

$$XSY \to_{\mathcal{R}}^* x_0 S \quad \text{and} \quad XY \to_{\mathcal{R}}^* y_0$$

for some Y if and only if the instance $\{(x_0, y_0)\}$ of the MPCP has a solution.

Combining the technical results obtained we arrive at the following result.

Theorem 5. For all $x_0, y_0 \in \{a,b\}^+$, the following two statements are equivalent:

(a.) the MPCP has a solution for $\{(x_0, y_0)\}$;
(b.) there exist strings $X, Y, Z_1, Z_2 \in \Gamma^*$ such that the following congruences are satisfied simultaneously:

1. $XSY \leftrightarrow_{\mathcal{R}}^* x_0 S$,
2. $XY \leftrightarrow_{\mathcal{R}}^* y_0$,
3. $X\#Z_1 \leftrightarrow_{\mathcal{R}}^* \lambda$,
4. $XZ_1\# \leftrightarrow_{\mathcal{R}}^* \lambda$,
5. $X\$Z_2 \leftrightarrow_{\mathcal{R}}^* \lambda$, and
6. $XZ_2\$ \leftrightarrow_{\mathcal{R}}^* \lambda$.

Thus, we have the following undecidability result.

Corollary 6. The simultaneous variant of the word matching problem is undecidable for the finite, special, and confluent string-rewriting system $\mathcal{R}$.

Now consider the extended alphabet $\Delta := \Gamma \cup \{\mathfrak{c}\}$, and the following two strings $u \in (\Delta \cup V)^*$ and $v \in \Delta^*$:

$$u := v_1 S v_2 \mathfrak{c} v_1 v_2 \mathfrak{c} v_1 \# v_3 \mathfrak{c} v_1 v_3 \# \mathfrak{c} v_1 \$ v_4 \mathfrak{c} v_1 v_4 \$,$$
$$v := x_0 S \mathfrak{c} y_0 \mathfrak{c} \mathfrak{c} \mathfrak{c} \mathfrak{c},$$

where $v_1, \ldots, v_4 \in V$.

Then there exists a substitution $\theta : \{v_1, \ldots, v_4\} \to \Delta^*$ satisfying $\theta(u) \leftrightarrow_{\mathcal{R}}^* v$ if and only if the following congruences are satisfied simultaneously, where $X := \theta(v_1), Y := \theta(v_2), Z_1 := \theta(v_3),$ and $Z_2 := \theta(v_4)$:

1. $XSY \leftrightarrow_{\mathcal{R}}^* x_0 S$,
2. $XY \leftrightarrow_{\mathcal{R}}^* y_0$,
3. $X\#Z_1 \leftrightarrow_{\mathcal{R}}^* \lambda$,
4. $XZ_1\# \leftrightarrow_{\mathcal{R}}^* \lambda$,
5. $X\$Z_2 \leftrightarrow_{\mathcal{R}}^* \lambda$, and
6. $XZ_2\$ \leftrightarrow_{\mathcal{R}}^* \lambda$.

By the theorem above this means that the instance (u, v) of the word matching problem for $\mathcal{R}$ has a solution over Δ if and only if the MPCP has a solution for $\{(x_0, y_0)\}$. Hence, we obtain our first main result.

Corollary 7. Over the alphabet Δ the word matching problem is undecidable for the finite, special, and confluent string-rewriting system $\mathcal{R}$.

3 A restricted variant of the word matching problem

As described in the introduction we are interested in the following restricted variant of the word matching problem, which we call the *special word matching problem*:

INSTANCE: A string $u \in (V \cup \Sigma)^*$.
QUESTION: Is there a substitution θ satisfying $\theta(u) \leftrightarrow_R^* \lambda$?

Here we present a finite, special, and confluent string-rewriting system $\mathcal{R}_1$ for which this problem is undecidable. We obtain $\mathcal{R}_1$ from the system $\mathcal{R}$ constructed in the previous section by adding three new rules.

Let $a', b', @$, and $\cent$ be four additional symbols, let $\Gamma_0 := \Gamma \cup \{a', b'\}$ $(= \{a, b, a', b'\} \cup \{1, \ldots, n\} \cup \{S, \#, \$\})$, let $\Delta_0 := \Gamma_0 \cup \{@, \cent\}$, and let $\mathcal{R}_1$ be the following string-rewriting system on Δ_0:

$$\mathcal{R}_1 := \mathcal{R} \cup \{aa' \to \lambda, bb' \to \lambda, @\cent \to \lambda\}.$$

It is easily seen that $\mathcal{R}_1$ is a finite, special string-rewriting system that is confluent.

Now, for given strings $x_0, y_0 \in \{a, b\}^+$, we consider the following instance of the special word matching problem for $\mathcal{R}_1$:

$$v_1 v_2 S y_0'^\sim @ v_1 S v_2 x_0'^\sim \cent @ v_1 \# v_3 \cent @ v_1 v_3 \# \cent @ v_1 \$ v_4 \cent @ v_1 v_4 \$ \cent v_1 v_4 \$,$$

where $' : \{a, b\}^* \to \{a', b'\}^*$ denotes the canonical isomorphism induced by $a \mapsto a'$ and $b \mapsto b'$, and $u^\sim$ denotes the reversal of the string u. Here v_1, v_2, v_3, v_4 are variables.

Lemma 8. *If the instance $\{(x_0, y_0)\}$ of the MPCP has a solution, then the above instance of the special word matching problem has a solution for $\mathcal{R}_1$.*

Proof. Let $i_1, \ldots, i_k \in \{1, 2, \ldots, n\}$ such that $x_0 x_{i_1} x_{i_2} \cdots x_{i_k} = y_0 y_{i_1} y_{i_2} \cdots y_{i_k}$. Let $w_1 := x_0 x_{i_1} x_{i_2} \cdots x_{i_k}$, $w_2 := i_k S i_{k-1} S \cdots i_2 S i_1$, $w_3 := \#^{2 \cdot |w_1| - 1}$, and $w_4 := \$^{2 \cdot |w_1| - 1}$. Then we have the following reductions modulo $\mathcal{R}_1$:

$$\begin{aligned}
w_1 w_2 S y_0'^\sim &= y_0 y_{i_1} \cdots y_{i_k} i_k S \cdots i_2 S i_1 S y_0'^\sim \to^* y_0 y_0'^\sim \to^* \lambda, \\
w_1 S w_2 x_0'^\sim &= x_0 x_{i_1} \cdots x_{i_k} S i_k S i_{k-1} \cdots i_2 S i_1 x_0'^\sim \to^* x_0 x_0'^\sim \to^* \lambda, \\
w_1 \# w_3 &= w_1 w_3 \# = x_0 x_{i_1} \cdots x_{i_k} \#^{2 \cdot |w_1|} \to^* \lambda, \text{ and} \\
w_1 \$ w_4 &= w_1 w_4 \$ = x_0 x_{i_1} \cdots x_{i_k} \$^{2 \cdot |w_1|} \to^* \lambda.
\end{aligned}$$

Thus, if φ denotes the morphism defined by $\{v_i \leftarrow w_i \mid i = 1, \ldots, 4\}$, then

$$\varphi(v_1 v_2 S y_0'^\sim @ \cdots \cent v_1 v_4 \$) = w_1 w_2 S y_0'^\sim @ \cdots \cent w_1 w_4 \$ \to_{\mathcal{R}_1}^* (@\cent)^5 \to_{\mathcal{R}_1}^* \lambda.$$

□

We claim that also the converse implication holds. So let φ be a morphism satisfying $\varphi(v_1 v_2 S y_0'^\sim @ \cdots \cent v_1 v_4 \$) \leftrightarrow_{\mathcal{R}_1}^* \lambda$, and let $w_i := \varphi(v_i)$, $i = 1, \ldots, 4$. Without loss of generality we can assume that $w_1, \ldots, w_4$ are irreducible modulo $\mathcal{R}_1$, that is, $w_1, \ldots, w_4 \in \mathrm{IRR}(\mathcal{R}_1)$. Denote $\varphi(v_1 v_2 S y_0'^\sim @ \cdots \cent v_1 v_4 \$)$ simply by w.

Lemma 9. $w_1, \ldots, w_4 \in \Gamma_0^*$.

Proof.

Claim 1. $|w_4|_@ = 0$.

Proof. Assume that $w_4 = g_1@g_2$ for some $g_1 \in \Delta_0^*$ and $g_2 \in (\Delta_0 \smallsetminus \{@\})^*$. Since w ends in $w_4\$$, and since $@\cent \to \lambda$ is the only rule of $\mathcal{R}_1$ containing an occurrence of the symbol $@$, we see that $w \to_{\mathcal{R}_1}^* \lambda$ implies that $|g_2|_\cent > 0$, that is, $g_2 = g_3 \cent g_4$ for some $g_3 \in \Gamma_0^*$. Hence, $w_4 = g_1@g_3 \cent g_4$, and $g_3 \to_{\mathcal{R}_1}^* \lambda$. This, however, contradicts our assumption that $w_4 \in \mathrm{IRR}(\mathcal{R}_1)$. Thus, $|w_4|_@ = 0$. □

Claim 2. $|w_1|_\cent = 0$.

Proof. This follows analogously. □

Claim 3. $|w_1|_@ = 0$.

Proof. Assume that $w_1 = g_1@h_1$ for some $g_1 \in (\Delta_0 \smallsetminus \{\cent\})^*$ and $h_1 \in \Gamma_0^*$. Since w ends in $w_1w_4\$$, this implies that $w_4 = h_2 \cent g_2$ for some $g_2 \in \Gamma_0^*$ satisfying $h_1h_2 \to_{\mathcal{R}_1}^* \lambda$. Since w also contains the substring $w_1\$w_4$, and therewith the substring $@h_1\$h_2\cent$, we see that also $h_1\$h_2 \to_{\mathcal{R}_1}^* \lambda$ must hold. The system $\mathcal{R}_1$ contains only two rules that involve occurrences of the symbol $\$$: $a\$\$ \to \lambda$ and $b\$\$ \to \lambda$. Hence, $h_1\$h_2 \to_{\mathcal{R}_1}^* \lambda$ implies that $|h_1h_2|_\$ \equiv 1 \bmod 2$, while $h_1h_2 \to_{\mathcal{R}_1}^* \lambda$ implies $|h_1h_2|_\$ \equiv 0 \bmod 2$. This contradiction yields $|w_1|_@ = 0$. □

The string w begins with the prefix w_1w_2. Since $|w_1|_@ = 0$, we obtain the following analogously to Claim 1.

Claim 4. $|w_2|_\cent = 0$.

Claim 5. $|w_3|_@ = 0$.

Proof. Assume that $w_3 = g_1@h_1$ for some $g_1 \in \Delta_0^*$ and $h_1 \in (\Delta_0 \smallsetminus \{@\})^*$. Since w_3 is irreducible, we see that $|h_1|_\cent = 0$, that is, $h_1 \in \Gamma_0^*$. Since w contains the substrings $w_3\cent$ and $w_3\#\cent$, we conclude that $h_1 \to_{\mathcal{R}_1}^* \lambda$ and $h_1\# \to_{\mathcal{R}_1}^* \lambda$. But w_3 being irreducible yields $h_1 = \lambda$, which in turn implies that $h_1\# = \# \not\to_{\mathcal{R}_1}^* \lambda$. Thus, $|w_3|_@ = 0$. □

Claim 6. $|w_3|_\cent = 0$.

Proof. Let $w_3 = g_1 \cent h_1$ for some $g_1 \in \Gamma_0^*$ and $h_1 \in (\Delta_0 \smallsetminus \{@\})^*$. Since w contains the substrings $@w_1\#w_3$ and $@w_1w_3$, we see that $w \to_{\mathcal{R}_1}^* \lambda$ and $w_1 \in \Gamma_0^*$ imply that $w_1\#g_1 \to_{\mathcal{R}_1}^* \lambda$ and $w_1g_1 \to_{\mathcal{R}_1}^* \lambda$. The only rules of $\mathcal{R}_1$ that contain occurrences of the symbol $\#$ are the following two: $a\#\# \to \lambda$ and $b\#\# \to \lambda$. Hence, $w_1\#g_1 \to_{\mathcal{R}_1}^* \lambda$ implies that $|w_1g_1|_\# \equiv 1 \bmod 2$, while $w_1g_1 \to_{\mathcal{R}_1}^* \lambda$ implies that $|w_1g_1|_\# \equiv 0 \bmod 2$. Thus, $|w_3|_\cent = 0$. □

Because of $w_1 \in \Gamma_0^*$ and $w \to_{\mathcal{R}_1}^* \lambda$, the fact that w contains the substrings $@w_1\$w_4$ and $@w_1w_4$ implies analogously the following.

Claim 7. $|w_4|_\cent = 0$.

By Claims 2, 4, 6, and 7 $w_1, w_2, w_3, w_4 \in (\Delta_0 \smallsetminus \{\mathsf{\not c}\})^*$. Thus, $|w|_{\mathsf{\not c}} = 5$. Hence, we also have $|w|_@ = 5$, and so $w_1, w_2, w_3, w_4 \in \Gamma_0^*$. This completes the proof of Lemma 9. □

Since $w \leftrightarrow_{\mathcal{R}_1}^* \lambda$, we can conclude the following from Lemma 9:

(1.) $w_1 w_2 S y_0'^{\sim} \to_{\mathcal{R}_1}^* \lambda$,
(2.) $w_1 S w_2 x_0'^{\sim} \to_{\mathcal{R}_1}^* \lambda$,
(3.) $w_1 \# w_3 \to_{\mathcal{R}_1}^* \lambda$,
(4.) $w_1 w_3 \# \to_{\mathcal{R}_1}^* \lambda$,
(5.) $w_1 \$ w_4 \to_{\mathcal{R}_1}^* \lambda$, and
(6.) $w_1 w_4 \$ \to_{\mathcal{R}_1}^* \lambda$.

From Lemma 2 and its proof we see that the reductions (3.) to (6.) imply that $w_1 \in \{a, b\}^*$.

Lemma 10. *Let $x_0, y_0 \in \{a, b\}^+$. Then, for $X_1, X_2 \in \{a, b\}^*$,*

$$X_1 S Y x_0'^{\sim} \to_{\mathcal{R}_1}^* \lambda \quad \text{and} \quad X_2 Y S y_0'^{\sim} \to_{\mathcal{R}_1}^* \lambda$$

for some $Y \in \Gamma_0^*$ if and only if there exist indices $i_1, \ldots, i_\ell \in \{1, \ldots, n\}$ such that

$$X_1 = x_0 x_{i_1} \cdots x_{i_\ell} \quad \text{and} \quad X_2 = y_0 y_{i_1} \cdots y_{i_\ell}.$$

Proof. If $X_1 = x_0 x_{i_1} \cdots x_{i_\ell}$ and $X_2 = y_0 y_{i_1} \cdots y_{i_\ell}$, choose $Y := i_\ell S \cdots i_2 S i_1$. Then

$$X_1 S Y x_0'^{\sim} = x_0 x_{i_1} \cdots x_{i_\ell} S i_\ell S \cdots i_2 S i_1 x_0'^{\sim} \to_{\mathcal{R}}^* x_0 x_0'^{\sim} \to_{\mathcal{R}_1}^* \lambda \text{ and}$$
$$X_2 Y S y_0'^{\sim} = y_0 y_{i_1} \cdots y_{i_\ell} i_\ell S \cdots i_2 S i_1 S y_0'^{\sim} \to_{\mathcal{R}}^* y_0 y_0'^{\sim} \to_{\mathcal{R}_1}^* \lambda.$$

We prove the converse implication by contradiction. Let $U_1, U_2 \in \{a, b\}^*$ be minimal counterexamples in terms of their combined length such that $U_1 S Y x_0'^{\sim} \to_{\mathcal{R}_1}^* \lambda$ and $U_2 Y S y_0'^{\sim} \to_{\mathcal{R}_1}^* \lambda$ for some $Y \in \Gamma_0^*$. Obviously, we may assume without loss of generality that $Y \in \Gamma_0^*$ is irreducible modulo $\mathcal{R}_1$. Clearly, $Y \neq \lambda$, since $U_1 S x_0'^{\sim} \in \{a, b\}^* \cdot S \cdot \{a', b'\}^+ \subseteq \mathrm{IRR}(\mathcal{R}_1)$.

Claim 1. $Y x_0'^{\sim} \in \mathrm{IRR}(\mathcal{R}_1)$.

Proof. If $Y x_0'^{\sim}$ is reducible modulo $\mathcal{R}_1$, then $Y = Y_1 c$ and $x_0'^{\sim} = c' z'$ for some $c \in \{a, b\}$. However, then $U_2 Y S y_0'^{\sim}$ ends in $c S y_0'^{\sim}$, and we see from the form of the rules of $\mathcal{R}_1$ that each descendant of $U_2 Y S y_0'^{\sim}$ then also ends in $c S y_0'^{\sim}$. □

Claim 2. $Y S y_0'^{\sim} \in \mathrm{IRR}(\mathcal{R}_1)$.

Proof. If $Y S y_0'^{\sim}$ is reducible modulo $\mathcal{R}_1$, then $Y = Y_1 y_i i$ for some $i \in \{1, \ldots, n\}$. Hence, $U_1 S Y x_0'^{\sim}$ ends in $y_i i x_0'^{\sim}$, and therewith each descendant of $U_1 S Y x_0'^{\sim}$ also ends in $y_i i x_0'^{\sim}$, since $y_i \in \{a, b\}^+$. Thus, $Y S y_0'^{\sim} \in \mathrm{IRR}(\mathcal{R}_1)$. □

Thus, the only rule that is applicable to $U_1 S Y x_0'^{\sim}$ is of the form $x_p S p \to \lambda$, that is, $U_1 = U_1' x_p$, $Y = p Y'$, and

$$U_1 S Y x_0'^{\sim} \to_{\mathcal{R}} U_1' Y' x_0'^{\sim} \to_{\mathcal{R}_1}^* \lambda.$$

Hence, $U_2 Y S y_0'^\sim = U_2 p Y' S y_0'^\sim$, and since $Y' S y_0'^\sim$ is irreducible by Claim 2, we see that the rule $y_p p S \to \lambda$ must apply, that is, $U_2 = U_2' y_p$, $Y' = SY''$, and

$$U_2 p Y' S y_0'^\sim \to_\mathcal{R} U_2' Y'' S y_0'^\sim \to_{\mathcal{R}_1}^* \lambda.$$

Thus, $U_1', U_2' \in \{a, b\}^*$ satisfy $U_1' Y' x_0'^\sim = U_1' SY'' x_0'^\sim \to_{\mathcal{R}_1}^* \lambda$ and $U_2' Y'' S y_0'^\sim \to_{\mathcal{R}_1}^* \lambda$ for some string $Y'' \in \Gamma_0^*$. This, however, contradicts the minimality of U_1 and U_2. □

Lemma 10 applied to the reductions (1.) and (2.) above implies that there exist indices $i_1, \ldots, i_\ell \in \{1, \ldots, n\}$ such that $w_1 = x_0 x_{i_1} \cdots x_{i_\ell} = y_0 y_{i_1} \cdots y_{i_\ell}$, that is, the instance $\{(x_0, y_0)\}$ of the MPCP has a solution. This observation together with Lemma 8 yields the following equivalence.

Corollary 11. *The instance $\{(x_0, y_0)\}$ of the MPCP has a solution if and only if the above instance of the special word matching problem has a solution for $\mathcal{R}_1$.*

The choice of P thus implies the intended undecidability result.

Theorem 12. *For the finite, special, and confluent string-rewriting system $\mathcal{R}_1$ the special word matching problem is undecidable.*

4 Conclusion and open problems

We have shown that extending Makanin's result to the general case of all finite, special, and confluent string-rewriting systems is even 'more impossible' than we thought. The simplicity of special confluent string-rewriting systems is deceptive; they are powerful enough to even make word matching problems undecidable.

However, there is still one interesting case that remains open, the case of finite, special, and confluent string-rewriting systems *that present groups*. Note that the systems constructed above do certainly not present groups, since symbols like a do not have left-inverses. A helpful factor for attacking this open problem is the fact that the class of groups that are presented by finite, special, and confluent string-rewriting systems can be characterized algebraically: they are exactly those groups that are isomorphic to the free products of a free group of finite rank and finitely many finite cyclic groups [Coc76]. In any case, even if this problem should turn out to be decidable, its complexity is likely to be very high, since Makanin's algorithm for free groups is itself not even primitive recursive [KoPa].

References

[BaSi94] F. Baader and J.S. Siekmann. Unification theory. In: D.M. Gabbay, C.J. Hogger, and J.A. Robinson (eds.), *Handbook of Logic in Artificial Intelligence and Logic Programming*, Oxford University Press, 1994.

[BoOt93] R. Book and F. Otto. *String-Rewriting Systems*. Springer Verlag, New York, 1993.

[Coc76]   Y. Cochet. Church-Rosser congruences on free semigroups. *Colloquia Mathematica Societatis János Bolyai* 20 (1976) 51–60.

[DeJo90]   N. Dershowitz and J.P. Jouannaud. Rewrite systems. In: J. van Leeuwen (ed.), *Handbook of Theoretical Computer Science, Vol. B: Formal Models and Semantics*, Elsevier, Amsterdam, 1990, pages 243–320.

[Jaf90]   J. Jaffar. Minimal and complete word unification. *Journal Association Computing Machinery* 37 (1990) 47–85.

[JoKi91]   J.P. Jouannaud and C. Kirchner. Solving equations in abstract algebras: a rule-based survey of unification. In: J.L. Lassez and G. Plotkin (eds.), *Computational Logic: Essays in Honor of Alan Robinson*, MIT Press, 1991, pages 360-394.

[KoPa]   A. Kościelski and L. Pacholski. Makanin's group algorithm is not primitive recursive. *Theoretical Computer Science*, to appear.

[Mak77]   G.S. Makanin. The problem of solvability of equations in a free semigroup. *Mat. Sbornik* 103 (1977) 147–236.

[Mak83]   G.S. Makanin. Equations in a free group. *Math. USSR Izvestija* 21 (1983) 483–546.

[Mak85]   G.S. Makanin. Decidability of the universal and positive theories of a free group. *Math. USSR Izvestija* 25 (1985) 75–88.

[MaAb94]   G.S. Makanin and H. Abdulrab. On general solution of word equations. In: J. Karhumäki, H. Maurer, and G. Rozenberg (eds.), *Results and Trends in Theoretical Computer Science*, Lecture Notes Computer Science 812, Springer Verlag, Berlin, 1994, pages 251–263.

[NaOt90]   P. Narendran and F. Otto. Some results on equational unification. In: M.E. Stickel (ed.), *Proceedings 10th CADE*, Lecture Notes in Artificial Intelligence 449, Springer Verlag, Berlin, 1990, pages 276–291.

[Ott95]   F. Otto. Solvability of word equations modulo finite special and confluent string-rewriting systems is undecidable in general. *Information Processing Letters* 53 (1995) 237–242.

[Pec81]   J.P. Pecuchet. *Equations avec Constantes et Algorithme de Makanin*. These 3e Cycle, Université de Rouen, France, Dec. 1981.

[Sch90]   K.U. Schulz. Makanin's algorithm for word equations - Two improvements and a generalization. In: K.U. Schulz (ed.), *Word Equations and Related Topics*, Proceedings, Lecture Notes Computer Science 572, Springer Verlag, Berlin, 1990, pages 85–150.

[Sch93]   K.U. Schulz. Word unification and transformation of generalized equations. *Journal of Automated Reasoning* 11 (1993) 149–184.

The Geometry of Orthogonal Reduction Spaces

Zurab Khasidashvili[1] and John Glauert[2]

[1] NTT Basic Research Laboratories, Atsugi, Kanagawa, 243-01, Japan
zurab@theory.brl.ntt.co.jp
[2] School of Information Systems, UEA, Norwich NR4 7TJ England
jrwg@sys.uea.ac.uk ***

Abstract. We investigate mutual *dependencies* of subexpressions of a computable expression, in orthogonal rewrite systems, and identify conditions for their concurrent *independent* computation. To this end, we introduce concepts familiar from ordinary Euclidean Geometry (such as *basis*, *projection*, *distance*, etc.) for reduction spaces. We show how a basis for an expression can be constructed so that any reduction starting from that expression can be decomposed as the sum of its projections on the axes of the basis. To make the concepts more relevant computationally, we relativize them w.r.t. *stable* sets of results, and show that an optimal concurrent computation of an expression w.r.t. S consists of optimal computations of its S-independent subexpressions. All these results are obtained for *Stable Deterministic Residual Structures*, Abstract Reduction Systems with an axiomatized *residual* relation, which model all orthogonal rewrite systems.

1 Introduction

Efficient evaluation of expressions requires concurrent evaluation of subexpressions. In computation in general, it is normal that intermediate results of computation of different subexpressions are used by other subexpressions, and contribute to creation of new computable subexpressions. In concurrent languages like the π-calculus [Mil92] this is expressed explicitly by value-passing, while in sequential languages computations in different subexpressions can only *interact* by joint creation of new redexes. Our aim in this paper is to give a formal numerical characterization of *dependencies* of subexpressions of an expression (or subprograms of a modular program), and in particular to identify conditions for *independent* evaluation of subexpressions. Computation of different independent subexpressions can be conducted in isolation from computations elsewhere in the expression, concurrently, and the results can then be combined to yield the final result.

We restrict our attention to functional languages, and consider their operational model – orthogonal rewrite systems – of which the λ-calculus [Bar84] is the prime example, although we believe that our results can be generalized to the non-orthogonal case and cover concurrent languages as well. To remain as general as possible, and at the same time to avoid syntactic structure of computable expressions (terms, graphs, etc.), which is irrelevant for our purpose, we assume that the rewrite system is given in the form of a *Stable Deterministic Residual Structure*, SDRS [GK96].

*** Part of this work was supported by the Engineering and Physical Sciences Research Council of Great Britain under grant GR/H 41300

SDRSs are Abstract Rewrite Systems with an axiomatized residual relation, which model all orthogonal rewrite systems. Standard important results like the Standardization and Normalization theorems can already be proven in SDRSs [GK96, KG96]. Furthermore, via *Deterministic Family Structures*, DFSs [GK96], which are SDRSs with an axiomatized *family* relation on redexes, one can prove optimality results of Lévy [Lév80], and achieve Prime Event Structure [Win89] style semantics for orthogonal rewrite systems in a uniform way [KG97a].

The idea we want to pursue is very simple and natural, and the concepts we introduce have their counterparts in ordinary Euclidean Geometry, although there will be some differences. For expository purposes, let us assume first that the given SDRS is *linear* – there are no duplication or erasure of redexes. The main analogy is the following. In a Euclidean 3-dimensional space, one can decompose a vector as the sum of its projections on the axes X, Y and Z, which form a Euclidean basis. Similarly, we can construct a basis at any expression t, consisting of *independent* reductions P_i starting from t, such that any reduction P starting from t can be decomposed as the sum of its projections on P_i. Here P_i and P_j are independent if no finite initial parts of them can *interact*, i.e., by joint creation of a new redex. In the basis that we construct, every reduction P_i is a maximal reduction *internal* to U_i, i.e, P_i contracts residuals of redexes in U_i and created redexes; every U_i is *independent*, i.e., no reduction internal to U_i can interact with a reduction internal to the complement $\overline{U}$ of U, which consists of redexes of t not in U; U_i are pairwise non-overlapping, and cover all redexes of t.

Further, the *distance* $|P, Q|$ between co-initial reductions P, Q is the number of their 'different' steps, and characterizes 'how far apart' the reductions have progressed. Here 'different steps' means that they cannot be related by the *zig-zag* relation (which is the transitive and symmetric closure of the residual relation) [Lév80], so they are in different zig-zag families. $|P, Q|$ coincides with the minimal number of reduction steps needed to reach a common reduct from the endpoints of P and Q. This is different from the Euclidean measure of distance. For example, in the simplest case, if two vectors $\vec{P}$ and $\vec{Q}$ are orthogonal (say parallel to axes X and Y respectively), then the distance is $|\vec{P}, \vec{Q}| = \sqrt{|\vec{P}|^2 + |\vec{Q}|^2}$, while the distance between reductions P and Q that contract redexes in different families is $|P, Q| = |P| + |Q|$. However, this is because the Euclidean space is continuous and allows 'shortcuts'. If we were to allow joining of the endpoints of the vectors $\vec{P}$ and $\vec{Q}$ only by moves parallel to X and Y, then we would get the same distance measure as for reductions!

Finally, the *independence degree* of a redex set U of an expression t is the length of a shortest reduction P internal to U such that there is a reduction Q internal to $\overline{U}$ that interacts with P, and is ∞ otherwise. So at least $|P|$ steps can be performed in U independently from the rest of the computation, after which results of computing U and $\overline{U}$ must be combined in order for the computation to proceed 'as concurrently as possible'. Note that if and only if U is independent, its independence degree is ∞.

These concepts can very naturally be explained in terms of *Prime Event Structures* (PES) [Win89], which in the conflict-free case (in which we are interested) are simply *event* sets E partially ordered by a *causal dependency* relation $\leq$, such that every event $e \in E$ can only dominate a finite number of others. Computations in a linear SDRS $\mathcal{R}$ are interpreted as left-closed sets of events (i.e., closed under

$\leq$), called *configurations*, in the PES $\mathcal{E} = (E, \leq)$ whose events correspond to (the zig-zag classes of) redexes in $\mathcal{R}$. Those event sets $X_i \subseteq E$ that are closed under $\geq$ are *independent*, as they correspond to independent reductions in $\mathcal{R}$. Further, if $\{X_i \mid i \in I\}$ are disjoint independent sets covering E, they form a *basis* for E, as for any configuration α, $\alpha = \cup_{i \in I} \alpha \cap X_i$. Here $\alpha \cap X_i$ is the *projection* of α on X_i, and coincides with the *restriction* of α to the set X_i^0 of all initial (i.e., minimal w.r.t. $\leq$) events of X_i. And the set $\{X_i^0\}_{i \in I}$ is an *independent covering* of the set of initial events of E. Further, the *distance* between configurations α and β is defined as the cardinality of $\alpha \cup \beta \setminus \alpha \cap \beta$ (as is usual for sets), and it precisely corresponds to the distance measure for reductions in linear SDRSs – $|P, Q| = |\alpha_P, \alpha_Q|$, where α_P, α_Q are configurations corresponding to P, Q. The *independence degree* of a set α_0 of initial events is the cardinality of the smallest configuration α, whose initial events are in α_0, such that there exists a configuration β not containing elements of α_0 and an event e such that $\alpha \cup \beta \cup \{e\}$ is a configuration, while neither $\alpha \cup \{e\}$ nor $\beta \cup \{e\}$ are (i.e, α and β both contribute to creation, or enabling, of e, and they interact to create e).

Most of the technical difficulties come from the erasure of redexes in SDRSs. To cope with the erasure problems, and to have (most of the) concepts invariant under Lévy-equivalence, we work with *standard* reductions, which in SDRSs are reductions in which later steps 'do not erase' the preceding ones [KG96]. If the SDRS is duplicating, concepts like 'restriction of P to a redex-set U' cannot be defined correctly for arbitrary P – we need P to be a *family-reduction*, that is, a multi-step reduction contacting all members of a (zig-zag) family in parallel, in every multi-step. However, as we have shown in [KG97a], duplicating SDRSs can be interpreted via non-duplicating, also called affine, SDRSs, and the family-reductions in the former become reductions in the latter. Therefore, via that encoding, the results obtained here for affine SDRSs are applied to all SDRSs. (Restriction to family-reductions is inevitable when one studies *adequate* simulation of a duplicating system with an affine one [KKSV94].)

In order to make the introduced concepts more meaningful computationally, we relativize them w.r.t. the semantics one may be interested in. For example, in the λ-calculus, one might be interested in computing normal forms, head-normal forms, weak-head-normal forms, etc. In [GK96], we have characterized all reasonable sets of finite '(partial) results' as *stable* sets S of terms, and have shown that (only) w.r.t. stable sets S, S-needed reductions are S-normalizing. This allows us to ignore S-unneeded redexes, and for example, we can define P, Q to be S-*independent* if there is no joint creation of S-needed redexes. So reductions that interact may be S-independent. This is profitable since redex sets that are not independent may become S-independent, and this allows for finer independent splitting of redex-sets of terms, implying more parallelism in the computation. And indeed, if $\{U_i\}_{i \in I}$ is an S-independent covering of an S-normalizable term t, we show that an optimal S-normalizing reduction is the sum of maximal S-needed reductions internal to U_i.

In Section 2, we recall SDRSs and DFSs. In section 3, we introduce the restriction and projection concepts and prove the *Decomposition* theorem. In section 4, we define the geometry of orthogonal reduction spaces, and prove the *Independent Decomposition* theorem. In section 5, we relativize the geometry w.r.t. stable sets of results S, and show that optimal computation w.r.t. S can be achieved by combining optimal computations of S-independent redex-sets. Conclusions appear in section 6.

2 Deterministic Residual and Family Structures

Let us recall some basic theory for DRSs and DFSs developed in [GK96, KG96, KG97]. DRSs are *Abstract Reduction Systems* (ARSs) with axiomatized notions of *residual*. A definition and a survey of results about ARSs can be found in [Klo92]. Our definition is slightly different, and follows that of Hindley [Hin69].

An ARS is a triple $A = (Ter, Red, \rightarrow)$ where Ter is a set of *terms*, ranged over by t, s, o, e; Red is a set of *redexes* (or *redex occurrences*), ranged over by u, v, w; and $\rightarrow : Red \mapsto (Ter \times Ter)$ is a (total) function such that for any $t \in Ter$ there is only a finite set of $u \in Red$ such that $\rightarrow(u) = (t, s)$, written $t \xrightarrow{u} s$. This set will be known as the redexes of term t, where $u \subseteq t$ denotes that u is a member of the redexes of t and $U \subseteq t$ denotes that U is a subset of the redexes. Note that one can identify u with the triple $t \xrightarrow{u} s$. A *reduction* is a sequence $t \xrightarrow{u_1} t_2 \xrightarrow{u_2} \ldots$.

Notation Reductions are denoted by P, Q, N. We write $P : t \twoheadrightarrow s$ or $t \xrightarrow{P} s$ if P denotes a reduction (sequence) from t to s, write $P : t \twoheadrightarrow$ if P may be infinite, and write $P : t \twoheadrightarrow \infty$ if P is infinite (i.e, of the length ω). $P + Q$ denotes the concatenation of P and Q. u also denotes the reduction that contracts u. The final term of a finite reduction P is denoted by $ft(P)$. If $U \subseteq t$, then $\overline{U}$ will denote the *complement* of U, i.e., the set of redexes in t not in U.

DRSs model orthogonal rewrite systems. They are similar to Stark's *Determinate Concurrent Transition Systems* (DCTSs) [Sta89] and ARSs of Gonthier et al. [GLM92]. Unlike DCTSs, the residual relation in DRSs may be duplicating, and unlike ARSs of [GLM92], we do not have a nesting relation on redexes. Several refined concepts of abstract rewriting are studied in [Oos94, Mel96, Raa96].

Definition 2.1 A DRS is a pair $\mathcal{R} = (A, /)$, where A is an ARS and $/$ is a *residual* relation on redexes relating redexes in the source and target term of every reduction $t \xrightarrow{u} s \in A$, such that for $v \subseteq t$, the set v/u of *residuals of v under u* is a set of redexes of s; a redex in s may be a residual of only one redex in t under u, and $u/u = \emptyset$. If v has more than one u-residual, then u *duplicates* v. If $v/u = \emptyset$, then u *erases* v. A redex of s which is not a residual of any $v \subseteq t$ under u is said to be u-*new* or *created* by u. The set u/P of residuals of u under P is defined by transitivity.

A *development* of $U \subseteq t$ is a reduction $P : t \twoheadrightarrow$ that only contracts residuals of redexes from U; it is *complete* if $U/P = \bigcup_{u \in U} u/P = \emptyset$. Development of $\emptyset$ is identified with the empty reduction. U will also denote a complete development of $U \subseteq t$. The residual relation satisfies the following two axioms:

- [FD] ([GLM92]) All developments are terminating; all complete developments of $U \subseteq t$ end at the same term; and residuals of a redex $v \subseteq t$ under all complete developments of U are the same.
- [weak acyclicity] ([Sta89]) Let $u, v \subseteq t$, $u \neq v$, and $u/v = \emptyset$. Then $v/u \neq \emptyset$.

We call a DRS $\mathcal{R}$ *stable* (SDRS) if:

- [stability] If $u, v \subseteq t$ are different redexes, $t \xrightarrow{u} e$, $t \xrightarrow{v} s$, and u creates a redex $w \subseteq e$, then the redexes in $w/(v/u)$ are not u/v-residuals of redexes of s, i.e., they are created along u/v.

We call a DRS $\mathcal{R}$ *non-duplicating* or *affine* if a redex may have at most one residual under contraction of another redex. Affine SDRSs will be called ASDRSs.

In a DRS $\mathcal{R}$, the residual relation on redexes is extended to all co-initial finite reductions exactly as in syntactic orthogonal rewrite systems [HL91, Lév80, Sta89]: $(P_1 + P_2)/Q = P_1/Q + P_2/(Q/P_1)$ and $P/(Q_1 + Q_2) = (P/Q_1)/Q_2$, and *Lévy-equivalence* or *permutation-equivalence*, $\approx_L$, is defined as the smallest relation on co-initial reductions satisfying: $U + V/U \approx_L V + U/V$ for any $U, V \subseteq t$, and $Q \approx_L Q' \Rightarrow P + Q + N \approx_L P + Q' + N$. Further, one defines $P \trianglelefteq Q$ iff $P/Q = \emptyset$, and can show that $P \approx_L Q$ iff $P \trianglelefteq Q$ and $Q \trianglelefteq P$; and $P \trianglelefteq Q$ iff $Q \approx_L P + N$ for some N. The following *Strong Church-Rosser* property can be proved: for any co-initial finite reductions P, Q, $P \sqcup Q \approx_L Q \sqcup P$, where $P \sqcup Q = P + Q/P$.

The relations $\trianglelefteq, \approx_L$ and / are extended to co-initial possibly infinite reductions N, N' as follows. $N \trianglelefteq N'$, or equivalently, $N/N' = \emptyset$ if, for any redex v contracted in N, say $N = N_1 + v + N_2$, $v/(N'/N_1) = \emptyset$; and $N \approx_L N'$ iff $N \trianglelefteq N'$ and $N' \trianglelefteq N$. Here, for any infinite P, $u/P = \emptyset$ (called *u is erased in P* or *u is P-erased*) if $u/P' = \emptyset$ for some finite initial part P' of P, and P/Q is only defined for finite Q, as the reduction whose initial parts are residuals of initial parts of P under Q.

The essence of stability is better understood by the following lemma, which extends [stability] axiom from one step reductions to any co-initial *external* reductions, that is, reductions that do not contact redexes having common residuals.

Definition 2.2 ([GK96]) • Let $u \in U \subseteq t$ and $P : t \twoheadrightarrow$. We call P *external* to U (resp. u) if P does not contract residuals of redexes in U (resp. residuals of u).

• Let $P : t_0 \xrightarrow{P_i} t_i \xrightarrow{u_i} t_{i+1} \twoheadrightarrow$ and $Q : t_0 = s_0 \xrightarrow{Q_j} s_j \xrightarrow{v_j} s_{j+1} \twoheadrightarrow$. We call P *external* to Q if for any i, j, $u_i/(Q_j/P_i) \cap v_j/(P_i/Q_j) = \emptyset$.

Lemma 2.3 (Stability [GK96]) Let $P : t \twoheadrightarrow s$ be external to $Q : t \twoheadrightarrow e$, in an SDRS, and let P create redexes $W \subseteq s$. Then the residuals $W/(Q/P)$ of redexes in W are created by P/Q, and Q/P is external to W.

Definition 2.4 ([KG96]) • Let $P : t \twoheadrightarrow$ and $u \subseteq t$. We call u *P-needed* if there is no $Q \approx_L P$ that is external to u, and call it *P-unneeded* otherwise.

• Let $Q : t \twoheadrightarrow$, $P : t \xrightarrow{P'} s \twoheadrightarrow$, and $u \subseteq s$. We call u (or more precisely, u with *creation history P'*, denoted by $P'u$) *Q-needed* if u is Q/P'-needed. We call P *Q-needed* if so is every redex contracted in P.

• We call P *self-needed* or *standard* if it is P-needed. We write $Q \approx_S P$ if $Q \approx_L P$ and $Q, P \in STA$, where STA denotes the set of all standard reductions. We call N a *standard variant* of P if $P \approx_L N$ and $N \in STA$.

Note that P-neededness does not depend on the choice of a reduction in the class of reductions Lévy-equivalent to P, but this is not true for the externality concept.

The following is a relativized standardization algorithm for reductions in AS-DRSs. Let $P, Q : t \twoheadrightarrow$. The *canonical P-needed variant of Q*, $ST_P(Q)$, is defined as follows: let $v \subseteq t$ be such that it is P-needed and its residual is contracted in Q first among P-needed residuals of P-needed redexes in t. Then $ST_P(Q) = v + ST_{P/v}(Q/v)$. If there is no such a redex in t, then $ST_P(Q) = \emptyset$. We write $ST(P)$ for $ST_P(P)$.

The *Standardization* theorem [KG96], when restricted to ASDRSs, states that, for co-initial reductions Q, P, finite or infinite, $ST_P(Q)$ is a standard P-needed reduction whose length coincides with the number of P-needed steps in Q, and

$ST_P(Q) \trianglelefteq Q, P$. Further, if Q is finite, then $Q \approx_L ST(Q)$; otherwise, $Q \approx_L ST(Q)$ need not hold.

It has been shown in [KG97] that, in ASDRSs, all standard variants of a finite reduction P can be constructed effectively (as P-neededness is decidable and there are only a finite number of such reductions, all of the same length), and that $\approx_S$ is decidable. So standard reductions can be used as canonical representatives of their Lévy-equivalence classes (which may have an infinite number of elements).

Next we recall an axiomatization of Lévy's concept of *redex-family* for DRSs. All family and sharing concepts for orthogonal reduction systems known to us (such as [Lév80, Mar92, AL94, Oos96]) satisfy our family axioms, which allow for abstract proofs of Relative Normalization and Optimality Theorems [GK96].

Definition 2.5 ([GK96]) A *Deterministic Family Structure* (DFS) is a triple $\mathcal{F} = (\mathcal{R}, \simeq, \hookrightarrow)$, where $\mathcal{R}$ is a DRS; $\simeq$ is an equivalence relation on redexes with *histories*; and $\hookrightarrow$ is the *contribution* relation on co-initial families, defined as follows:

(1) For co-initial reductions P and Q, a redex Qv in the final term of Q (read as v with history Q) is called a *copy* of a redex Pu if $P \trianglelefteq Q$, i.e., $P+Q/P \approx_L Q$, and v is a Q/P-residual of u; the *zig-zag* relation $\simeq_z$ is the symmetric and transitive closure of the copy relation. The *family* relation $\simeq$ is an equivalence relation among redexes with histories containing $\simeq_z$. A *family* is an equivalence class of the family relation; families are ranged over by $\phi, \psi, \ldots$. $Fam(\ )$ denotes the family of its argument.

(2) Further, $\simeq$ and $\hookrightarrow$ satisfy the following axioms:
- [initial] Let $u, v \subseteq t$ and $u \neq v$, in $\mathcal{R}$. Then $Fam(\emptyset_t u) \neq Fam(\emptyset_t v)$, where $\emptyset_t$ is the empty reduction starting from t.
- [contribution] $\phi \hookrightarrow \phi'$ iff for any $Pu \in \phi'$, P contracts at least one redex in ϕ.
- [creation] if $e \xrightarrow{P} t \xrightarrow{u} s$ and u creates $v \subseteq s$, then $Fam(Pu) \hookrightarrow Fam((P+u)v)$.
- [FFD] (*Finite Family Developments*) Any reduction that contracts redexes of a finite number of families is terminating.

It is shown in [GK96] that every DFS is a stable DRS. Further, we have proven in [KG97] that the zig-zag relation $\simeq_z$, as well as the zig-zag contribution relation $\hookrightarrow_z$, are decidable in ASDRSs, and that $\simeq_z$ is a family relation.

Below, $FAM(P)$ (resp. $SFAM(P)$) denotes the set of zig-zag families, or simply families, whose member (resp. P-needed) redexes are contracted in P, in an ASDRS. Further, for any $U \subseteq t$, $FAM_0(U)$ denotes the set of families (relative to t) of redexes in U, and $FAM^+(U)$ will denote the minimal set of families containing $FAM_0(U)$ and closed under the contribution relation $\hookrightarrow_z$.

3 Decomposition of Reductions in ASDRSs

In this section, we introduce *restriction* of a reduction to a redex-set, and its *projection* onto another reduction, study their properties, and use them to decompose reductions as the sum of their restrictions to non-overlapping redex sets.

Let $P : t \twoheadrightarrow$ be a reduction in a DRS, and let $U \subseteq t$ be a set of redexes in t. We call P *internal* to U or a U-*reduction* if it is external to $\overline{U}$, that is, if it contracts residuals of redexes in U and created redexes. We call such redexes U-*redexes*.

Definition 3.1 We call $ST_P(ST(Q))$ the *projection* of Q onto P, written $Q|P$.

Definition 3.2 (1) Let t be a term in an ASDRS $\mathcal{R}$, let $U \subseteq t$, and let $P : t \twoheadrightarrow s$ be standard. The concepts P *respects* U and the *restriction of P to U*, written $P|U$, are defined by induction on $n = |P|$ as follows. If $n = 0$, then P respects U and $P|U = \emptyset$. Now let $P = P' + u$ and let P' respect U. Assume that $P'|U$ and $P'|\overline{U}$ are defined as reductions internal to U and $\overline{U}$, respectively, such that $P' \approx_S P'|U \sqcup P'|\overline{U}$. Then we say that P respects U if either $u = u'/(P'|\overline{U}/P'|U)$ for $u' \subseteq ft(P'|U)$ such that $(P'|U) + u'$ is still internal to U, or $u = u'/(P'|U/P'|\overline{U})$ for $u' \subseteq ft(P'|\overline{U})$ such that $(P'|\overline{U}) + u'$ is still internal to $\overline{U}$. In the first case (depicted on the picture below), we define $P|U = P'|U + u'$ and $P|\overline{U} = P'|\overline{U}$, and define $P|U = P'|U$ and $P|\overline{U} = P'|\overline{U} + u'$ in the second case.

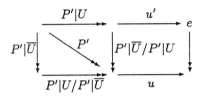

(2) We say that a finite reduction Q respects U if so does $ST(Q)$, and define $Q|U = ST(Q)|U$. We say that Q respects $\Im = \{U_i\}_{i \in I}$ if it respects every U_i.

One can easily show that Definition 3.2 is correct, that is, $ST(P) \approx_S P|U \sqcup P|\overline{U}$. The intuition is that, P respects U iff $ST(P)$ contracts only redexes to which either only redexes in U contribute, or only those in $\overline{U}$, but not redexes in both U and $\overline{U}$. More precisely, if $U \subseteq t$ and $P : t \twoheadrightarrow s$, then P respects U iff $SFAM(P) \subseteq FAM^+(U) \cup FAM^+(\overline{U})$, and in the latter case, $(S)FAM(P|U) = SFAM(P) \cap FAM^+(U)$ and $SFAM(P) = (S)FAM(P|U) \cup (S)FAM(P|\overline{U})$. Further, if P, Q are co-initial, then $(S)FAM(ST_P(Q)) = (S)FAM(Q) \cap SFAM(P)$. It follows that the restriction and projection concepts for finite reductions are invariant under $\approx_L$.

In the above definition, we need to take a standard variant of Q before restricting it to U to ensure that the restriction notion is invariant under Lévy-equivalence. As shown by the following simple example, this is necessary. Let $R = \{f(x) \rightarrow a, g(x) \rightarrow x\}$, let $P : f(g(x)) \xrightarrow{v} f(x) \xrightarrow{u} a$, and let $U = \{v\}$. Then 'direct restriction' of P to U is v, while $P|U = ST(P)|U = \emptyset$, and $v \not\approx_L \emptyset$.

We call a U-reduction P U-*fair* if each U-redex is erased in P, and call *strongly U-cofinal* if, for any U-reduction Q, $Q \trianglelefteq P$. If U is the set of all redexes in t, then U-fair reductions are *fair*, and strongly U-cofinal reductions will be called *strongly cofinal*. One can show that a U-reduction P is strongly U-cofinal iff it is U-fair. (Recall that if P is fair, then it is *cofinal*, but not conversely [Klo80].)

If $Q|P$ is finite, then so is $P|Q$ and $P|Q \approx_S Q|P$. Further, for any $P : t \twoheadrightarrow$ internal to $U \subseteq t$ such that $SFAM(P) = FAM^+(U)$, and any finite $Q : t \twoheadrightarrow s$, $Q|U = Q|P$. If the SDRS is linear, then every U-fair reduction is such.

Let U_i with $i \in I$ be nonempty sets of redexes in t such that $\cup_{i \in I} U_i$ contains each redex of t and $U_i \cap U_j = \emptyset$ when $i \neq j$. Then we call the set $\Im = \{U_i\}_{i \in I}$ a *(redex-)covering* of t.

The restriction concept enjoys nice algebraic properties: If $P : t \twoheadrightarrow s$ respects $U_1, U_2 \subseteq t$, then P respects $U_1 \cup U_2$ and $U_1 \cap U_2$; and $P|U_1 \cup U_2 \approx_S P|U_1 \sqcup P|U_2$ and $P|U_1 \cap U_2 \approx_S (P|U_1)|U_2$. This allows us to prove the following

Theorem 3.3 (Decomposition) Let $\Im = \{U_i\}_{i \in I}$ be a redex-covering of a term t in an ASDRS $\mathcal{R}$.

(1) Let P_i be finite reductions internal to U_i, and let $P = \sqcup_i P_i$. Then P respects $\Im$ and $P|U_i \approx_S P|P_i$.

(2) Let $P : t \twoheadrightarrow s$ respect $\Im$. Then $P \approx_L \sqcup_i P|U_i$.

4 The Geometry of Reduction Spaces

In this section, we introduce the *Reduction Geometry* and prove the *Independent Decomposition* theorem, which reflects the main analogy of orthogonal reduction spaces with the Euclidean Geometry.

Let $P : t \twoheadrightarrow$. We call the *strict domain* of P, written $SDom(P)$, the minimal set of redexes $U \subseteq t$ such that P is internal to U. We call the *domain* of P, written $Dom(P)$, the set $\cup_{Q \approx_L P} SDom(Q)$, i.e., the minimal set of redexes $U \subseteq t$ such that any Q that is Lévy-equivalent to P is internal to U. And we call the *minimal domain* of P, written $MDom(P)$, the set $\cap_{Q \approx_L P} SDom(Q)$.

It is easy to see that $Dom(P)$ is $SDom(P)$ augmented by all P-erased redexes not contracted in P, and $MDom(P)$ is the set of all P-needed redexes in t. Obviously $P \approx_L Q$ implies $Dom(P) = Dom(Q)$ and $MDom(P) = MDom(Q)$, but not $SDom(P) = SDom(Q)$. It follows from the Standardization Theorem that $MDom(P) = SDom(ST(P))$ for any P.

Definition 4.1 (The Reduction Geometry) Let t be a term in an ASDRS $\mathcal{R}$.

- Let $P : t \twoheadrightarrow s$ and $Q : t \twoheadrightarrow e$. We say that P and Q are *independent* or *do not interact*, written $P \perp Q$, if $MDom(P) \cap MDom(Q) = \emptyset$ and any created redex in $ft(P \sqcup Q)$ is a residual of a redex either from $ft(P)$ or from $ft(Q)$.

- We call a set $\Pi = \{P_i\}_{i \in I}$ of reductions starting from t *independent* if $P'_i \perp \sqcup_{i \neq j} P'_j$ for every $i \in I$ and any finite initial parts P'_i of P_i. We call Π a *basis* of $\mathcal{R}$ at t if Π is independent and for any $P : t \twoheadrightarrow s$, $P \trianglelefteq \sqcup P_i$.

- The *distance* $|P, Q|$ between co-initial finite reductions $P, Q : t \twoheadrightarrow$ is the number of families whose *essential* member redexes are contracted either in P or in Q (but not in both). Here a redex $v \subseteq s$ is essential [Kha93] (or Maranget-needed [Mar92]) if in any fair reduction starting from s a residual of v is contracted.

- The *independence degree* of $U \subseteq t$ is the length of a shortest finite P internal to U such that there exists a reduction Q external to U that interacts with P, and is ∞ otherwise.

- We call $U \subseteq t$ *independent* if every pair of finite U- and $\overline{U}$-reductions is so. We call a redex-covering $\Im = \{U_i\}_{i \in I}$ of t an *independent covering* if each U_i is independent.

Example 4.2 (Bases) Consider a term t containing three redexes u, v, w, let $w/(u \sqcup v) = \emptyset$, $w/u \neq \emptyset$, $w/v \neq \emptyset$, and assume no redexes can be created by contraction of these redexes. Then $\Pi_1 = \{u, v\}$, $\Pi_2 = \{u, v, w\}$, $\Pi_3 = \{u, w \sqcup v\}$ and $\Pi_4 = \{u, v \sqcup w\}$ are all bases at t (there are others too), as all reductions are independent, and $u \sqcup v \approx_L u \sqcup v \sqcup w \approx_L u \sqcup (w \sqcup v) \approx_L u \sqcup (v \sqcup w)$ are all normalizing, hence strongly cofinal. For Π_1, the strict domains of the axes do not form a covering of t, while for other bases they do. Note also that for Π_4, u erases the second step of $v \sqcup w - (w/v)/(u/v) = \emptyset$.

Note that, in the definition of $P \perp Q$, a created redex in $ft(P \sqcup Q)$ cannot be a residual of redexes from both $ft(P)$ and $ft(Q)$, as otherwise the same redex would be a residual of redexes from $ft(ST(P))$ and $ft(ST(Q))$, which is impossible by the Stability Lemma (since $MDom(P) \cap MDom(Q) = \emptyset$ implies that $ST(P)$ and $ST(Q)$ are external; the converse implication need not hold). Note also that, if $P \perp Q$, $Dom(P) \cap Dom(Q) = \emptyset$ need not hold. Indeed, consider the modified example from [Lév80]: let $t = (\lambda x.K_a(xY))K_b$, where $K_a = \lambda x.a$, $K_b = \lambda x.b$, and $Y = (\lambda x.f(xx))(\lambda x.f(xx))$, and let $P : t \xrightarrow{t} K_a(K_b Y) \xrightarrow{Y} K_a(K_b(fY)) \xrightarrow{K_b} K_a b$ and $Q : t \xrightarrow{Y} (\lambda x.K_a(x(fY)))K_b \xrightarrow{K_a} (\lambda x.a)K_b$. Then $Y \in Dom(P), Dom(Q)$, but $Y \notin MDom(P), MDom(Q)$, since Y is not needed either in P or in Q.

In the definition of distance between reductions P and Q, one might think that it would be more appropriate to consider $P \sqcup Q$-needed redexes only. The following example shows that the distance would not be a metric. Indeed, take $t = Kxw$, $P : t \xrightarrow{\omega} t \xrightarrow{\omega} t \xrightarrow{\omega} t$, $Q : t \xrightarrow{\omega} t$, and $N : t \xrightarrow{\omega} t \xrightarrow{\omega} t \xrightarrow{\omega} t \xrightarrow{K} x$. Then $|P, Q| = 3$ and $|P, N| = |N, Q| = 1$. It is easy to check that our distance measure on finite co-initial reductions satisfies the triangle inequality. To make it a metric, we define for co-initial finite reductions P, Q, $P \approx_f Q$ iff $FFAM(P) = FFAM(Q)$, where $FFAM(P)$ denotes the set of families of essential redexes in P. Clearly, $\approx_f$ is an equivalence relation, and the (co-initial) reduction space quotiented w.r.t. it is a metric, as $|P, Q| = 0$ implies $P \approx_f Q$. Note that $\approx_L \subseteq \approx_f$, but not conversely.

The independence degree of $U \subseteq t$, if finite, characterizes the minimal amount of work that can be performed in U independently from the rest of the computation.

It follows easily from Definition 3.2 and Definition 4.1 that $U \subseteq t$ is independent iff any finite reduction $P : t \twoheadrightarrow s$ respects it. Now, using Theorem 3.3.(2), we can prove the following

Theorem 4.3 (Independent Decomposition) Let $\Im = \{U_i\}_{i \in I}$ be an independent redex-covering of a term t in an ASDRS $\mathcal{R}$, let $P : t \twoheadrightarrow s$, and let P_i be U_i-fair. Then $P \approx_L \sqcup_i P|U_i$. Further, $B = \{P_i\}_{i \in I}$ is a basis at t, and there are reductions $P'_i \trianglelefteq P_i$ such that $P \approx_L \sqcup_i P'_i$.

We have seen in Example 4.2 that not all bases are of the form described in Theorem 4.3. That is, if $\{P_i\}_{i \in I}$ is a basis at t, P_i need not be an U_i-fair reduction for some independent covering $\Im = \{U_i\}_{i \in I}$ of t, as it is the case for Π_1 (since $w/u \neq \emptyset$ and $w/v \neq \emptyset$). We could exclude this situation, by requiring in the definition of independence of $U \subseteq t$ that for any pair of finite reductions P, Q respectively internal and external to U, Q does not erase any steps of P, that is, $|P| = |P/Q|$. We have chosen not to do so, since also in the relativized bases which we introduce in the next section, axes do not need to be maximal reductions on their strict domains.

Note that every term t in an ASDRS has an independent redex covering – $\{U(t)\}$, where $U(t)$ is the set of all redexes of t, and has an independent basis – a fair reduction starting from t. One can construct finer bases from existing ones, as if $\Im = \{U_i\}_{i \in I}$ and $\Im' = \{U'_j\}_{j \in J}$ are bases, then $\Im \cap \Im' = \{U_i \cap U'_j\}_{(i,j) \in (I,J)}$ is a basis too. It is interesting to note that for any $P : t \twoheadrightarrow s$ and a created redex $u \subseteq s$, any 'smallest' reduction needed to create u, obtainable by *extraction* of Pu [Lév80], which for ASDRSs is defined in [KG97], is internal to some finest independent set of redexes in t.

5 The Optimal Decomposition Theorem

Next we show that an optimal computation of a term, w.r.t. a stable set S of results, can be decomposed into optimal computations of its S-independent redex-sets.

The concepts introduced in Definition 4.1 (independence of reductions and redex sets, covering, basis, etc.) immediately relativize w.r.t. any stable set S, simply by replacing 'independence', 'covering', 'basis', etc. with 'S-independence', 'S-covering', 'S-basis', etc., respectively, and by replacing '(essential) redex' and 'reduction' with 'S-needed redex' and 'S-needed reduction'. Recall that, for any set of terms S, a redex $u \subseteq t$ S-needed iff at least one residual of it is contracted in any reduction from t to a term in S, and S is called *stable* if (a) it is closed under reduction (this condition can be relaxed slightly), and (b) any step entering S is S-needed. The *Relative Normalization* theorem [GK96], for ASDRSs, states that any S-normalizable term $t \notin S$ contains an S-needed redex, any S-needed reduction starting from t is eventually S-normalizing, and is a shortest S-normalizing reduction starting from t.

Let $U \subseteq t$. We call a U-reduction $P : t \twoheadrightarrow$ (U, S)-*fair* if each S-needed U-redex is erased in P (P need not be U-fair). It is not difficult to show that, if $\Im = \{U_i \mid i \in I\}$ is an S-independent covering of an S-normalizable term $t \notin S$, in an ASDRS, then $P : t \twoheadrightarrow s$ is an S-normalizing S-needed reduction iff $P_i = P|U_i : t \twoheadrightarrow s_i$ are (U_i, S)-fair S-needed U_i-reductions; and P_i ia an optimal (U_i, S)-fair U_i-reduction iff it is an S-needed (U_i, S)-fair U_i-reduction. Hence we have from the Relative Normalization theorem that

Theorem 5.1 (Optimal Decomposition) Let S be a stable set of terms in an ASDRS $\mathcal{R}$, let $\Im = \{U_i\}_{i \in I}$ be an S-independent covering of an S-normalizable term t in $\mathcal{R}$, let $\Im' = \{U_j\}_{j \in J \subseteq I}$ contain all U_i that contain at least one S-needed redex of t, and let P_j be internal to U_j. Then P_j are optimal (i.e., shortest) (U_j, S)-fair reductions iff $P = \sqcup_j P_j$ is an optimal S-normalizing reduction starting from t.

6 Conclusions

We have defined concepts similar to those in *Vector Spaces* for orthogonal rewrite systems, and described how these can be used in distributed evaluation of sequential programs. The constructed *Reduction Geometry* is not just a nice piece of mathematics. Obviously, (relative) independence of redex-sets is undecidable in general, as is neededness. However, we hope that decidable approximations for independence can be defined which will yield decidable concepts for large classes of rewrite systems, as is the case for the neededness [HL91]. For example, all the introduced concepts are decidable for Recursive Program Schemes, both in first [Kha93] and higher order [Kha94] cases, but the latter do not have full computational power (as the *if − then − else* operator is only evaluated semantically). Actually, because of a specific simple form of redex-creation in such systems, one has maximal possible independence there − any redex forms an independent redex-set. Further, TRSs in which there is no upwards creation of redexes (such as Klop's TRS which models a Turing machine, in Exercise 2.2.21 of [Klo92]) do have full computational power, and any set consisting of all redexes occurring inside an outermost redex is independent.

Acknowledgements We thank J.R. Kennaway, V. van Oostrom and F.-J. de Vries for useful comments. The diagram was drawn using Paul Taylor's Diagram package.

References

[AL94] Asperti A., Laneve C. Interaction Systems I: The theory of optimal reductions. MSCS 11:1-48, Cambridge University Press, 1993.

[Bar84] Barendregt H. P. The Lambda Calculus, its Syntax and Semantics. North-Holland, 1984.

[GK96] Glauert J.R.W., Khasidashvili Z. Relative normalization in deterministic residual structures. CAAP'96, Springer LNCS, vol. 1059, H. Kirchner, ed. 1996, p. 180-195.

[GLM92] Gonthier G., Lévy J.-J., Melliès P.-A. An abstract Standardisation theorem. In: Proc. of LICS 1992, p. 72-81.

[Hin69] Hindley R.J. An abstract form of the Church-Rosser theorem I. JSL, 34(4):545-560, 1969.

[HL91] Huet G., Lévy J.-J. Computations in Orthogonal Rewriting Systems. In: Computational Logic, Essays in Honor of Alan Robinson, J.-L. Lassez and G. Plotkin, eds. MIT Press, 1991, p. 394-443.

[KKSV94] Kennaway J. R., Klop J. W., Sleep M. R, de Vries F.-J. On the adequacy of Graph Rewriting for simulating Term Rewriting. ACM Transactions on Programming Languages and Systems, 16(3):493-523, 1994.

[Kha93] Khasidashvili Z. Optimal normalization in orthogonal term rewriting systems. In: Proc. of RTA'93, Springer LNCS, vol. 690, C. Kirchner, ed. Montreal, 1993, p. 243-258.

[Kha94] Khasidashvili Z. On higher order recursive program schemes. In: Proc. of CAAP'94, Springer LNCS, vol. 787, S. Tison, ed. Edinburgh, 1994, p. 172-186.

[KG96] Khasidashvili Z., Glauert J. R. W. Discrete normalization and Standardization in Deterministic Residual Structures. In proc. of ALP'96, Springer LNCS, vol. 1139, M. Hanus, M. Rodríguez-Artalejo, eds. 1996, p.135-149.

[KG97] Khasidashvili Z., Glauert J.R.W. Zig-zag, extraction and separable families in non-duplicating stable deterministic residual structures. Technical Report IR-420, Free University, February 1997.

[KG97a] Khasidashvili Z., Glauert J. R. W. Relating conflict-free transition and event models. Submitted.

[Klo80] Klop J. W. Combinatory Reduction Systems. Mathematical Centre Tracts n. 127, Amsterdam, 1980.

[Klo92] Klop J. W. Term Rewriting Systems. In: S. Abramsky, D. Gabbay, and T. Maibaum eds. Handbook of Logic in Computer Science, vol. II, Oxford U. Press, 1992, p. 1-116.

[Lév80] Lévy J.-J. Optimal reductions in the λ-calculus. In: To H. B. Curry: Essays on Combinatory Logic, Lambda-calculus and Formalizm, Hindley J. R., Seldin J. P. eds, Academic Press, 1980, p. 159-192.

[Mar92] Maranget L. La stratégie paresseuse. Thèse de l'Université de Paris VII, 1992.

[Mil92] Milner R. Functions as processes. MSCS 2(2):119–141, 1992.

[Mel96] Melliès P.-A. Description Abstraite des Systèmes de Réécriture. Thèse de l'Université Paris 7, 1996.

[Oos94] Van Oostrom V. Confluence for Abstract and Higher-Order Rewriting. Ph.D. Thesis, Free University, Amsterdam, 1994.

[Oos96] Van Oostrom V. Higher order families. In: Proc. of RTA'96, Springer LNCS, vol. 1103, Ganzinger, H., ed., 1996, p. 392–407.

[Raa96] Van Raamsdonk F. Confluence and normalisation for higher-order rewriting. Ph.D. Thesis, Free University, Amsterdam, 1996.

[Sta89] Stark E. W. Concurrent transition systems. J. TCS, 64(3):221-270, 1989.

[Win89] Winskel G. An introduction to Event Structures. Springer LNCS, vol. 354, 1989, p. 364-397.

The Theory of Vaccines

Massimo Marchiori
Department of Pure and Applied Mathematics, University of Padova
Via Belzoni 7, 35131 Padova, Italy
max@math.unipd.it

Abstract

Despite the major role that modularity occupies in computer science, all the known results on modular analysis only treat particular problems, and there is no general unifying theory. In this paper we provide such a general theory of modularity. First, we study the space of the criteria for modularity (the so-called modularity space), and give results on its complexity. Then, we introduce the notion of vaccine and show how it can be used to completely analyze the modular space. It is also shown how vaccines can be effectively used to solve a variety of other modularity problems, providing the best solutions. As an application, we successfully apply the theory to the study of modularity for term rewriting, giving for the first time optimality results, and show how modularity problems can be completely solved.

1 Introduction

The field of modular analysis is of fundamental importance, and is nowadays attracting increasing interest by the scientific community. In essence, modularity allows to study a complex object by studying his smaller subparts: given a 'big' object composed by smaller subparts (via some composition operator), we want to state that it enjoys a certain property by simply investigating its smaller subcomponents. Hence, modular analysis allows to develop correct complex objects 'bottom-up', just building correct smaller submodules, and even dually to verify the correctness of a complex object by decomposing it into its submodules and verifying them.

Besides for the theoretical relevance, the increasing complexity of nowadays applications has made modularity analysis a task of primary importance from the practical side as well.

At the present moment, the field of modular analysis consists of several results that study the modularity of a particular property for a certain specific paradigm (see e.g. [7, 2, 20, 13, 8, 16, 5, 18]). However, there is no general theory on modular analysis. In this paper, we introduce such a theory.

Given the property to be verified, and the 'composition operator' that builds complex objects from smaller submodules, we analyze the corresponding *modularity space*, that is to say the collection of all the criteria for the modularity of the property w.r.t. the composition operator.

First, a complete description of this space by means of its maximal criteria is provided (roughly speaking, the 'best' results that can be obtained), and its complexity is studied (how many maximal criteria can exist). Next, we introduce the notion of *vaccine*, which is used for analyzing in an effective way the modularity space. Intuitively, a vaccine extracts from a possibly non-modular property a maximal modular sub-property, that is a maximal criterion of the modularity space for that property. Therefore, vaccines provide a convenient way to represent the modularity space. We propose a methodology for finding vaccines (and so the optimal modularity criteria). Moreover, we provide suitable conditions that ensure that the analysis of the modularity space is *completely solved*, i.e., it covers all the optimal criteria, and consequently every possible modularity criterion (being all the others subsumed by the maximal criteria).

Furthermore, it is shown that an analysis which is completely solved, is relevant for the study of the class of the *disjunctive criteria* (cf. [13, 20]), because it provides the *best* disjunctive criterion.

Finally, we consider also the other side of the coin, namely the case when modularity does not hold. We introduce the notion of *counterexample structure*, which is used together with the notion of vaccine for recovering the *best* description of the failure of modularity. The above results are successfully applied to the study of the modularity of important properties of term rewriting systems: termination, completeness and uniqueness of normal forms (the only main properties of TRSs that are not modular). In particular, we show that $\mathcal{C}_\mathcal{E}$-termination (cf. [5, 15]) is a maximal criterion, and provide a formal justification in terms of complexity of the difficulty of the study of the modularity of termination in TRS. Moreover, we completely solve the problem of the modularity of termination for left-linear TRSs, providing the only two optimal criteria. We give analogous results for the other major properties of completeness and uniqueness of normal forms, thus not only improving on all the works on the modularity of these properties, but completely solving the problem of their modular analysis.

The paper is organized as follows. Section 2 starts with some short preliminaries. Soon afterwards, Section 3 presents the notion of modular analysis and of a criterion for modularity. Then, Section 4 introduces the *modularity space* and gives some results on its complexity. In Section 5 the concept of *vaccine* is introduced. Next, Section 6 shows how vaccines can be successfully employed for the study of the modularity space via the notion of *vaccines basis*. Section 7 analyzes another kind of criteria, the so-called disjunctive criteria, and shows how they can be successfully analyzed via vaccines. Section 8 performs the same task for the study of counterexample structures, giving a complete analysis of the failure of modularity. Sections 9 successfully presents practical applications of the theory for the field of term rewriting. Finally, Section 10 ends with some other remarks on the further applications of the theory.

2 Preliminaries

$\mathcal{O}$ denotes the class of generic objects we will consider: every object is understood to be in $\mathcal{O}$. As usual, properties of objects will be identified with the classes of objects that belong to them. So, we will write equivalently $\mathcal{Q}_1 \wedge \mathcal{Q}_2$ or $\mathcal{Q}_1 \cap \mathcal{Q}_2$ to denote the intersection of two properties $\mathcal{Q}_1$ and $\mathcal{Q}_2$. We will also write $\neg \mathcal{Q}$ to indicate the complement property of $\mathcal{Q}$ (i.e. $T \in \neg \mathcal{Q}$ iff $T \notin \mathcal{Q}$).

As far as TRSs are concerned, we only require knowledge of the basic notions (see e.g. [3, 7]). The reader interested in modularity topics of TRSs can find extensive surveys in [14, 16].

3 Modularity

Suppose we want to perform the modular (w.r.t. some composition operator $\odot$) analysis of the property $\mathcal{P}$: given a complex object $T_1 \odot \cdots \odot T_n$ we want to infer it belongs to $\mathcal{P}$ by separately analyzing its smaller submodules $T_1, \ldots, T_n$.

The best case occurs when the property $\mathcal{P}$ is *modular* (w.r.t. a binary composition operator $\odot$): whenever n objects $T_1, \ldots, T_n$ are in $\mathcal{P}$, their composition $T_1 \odot \ldots \odot T_n$ is in $\mathcal{P}$ as well. Thus, to check a complex object $T_1 \odot \ldots T_n$ belongs to $\mathcal{P}$, it just suffices to check its submodules $T_1, \ldots, T_n$ belong to $\mathcal{P}$. In general, however, $\mathcal{P}$ may not be modular, and so we need a more general concept to formalize modular analysis. We so define what is the notion of a criterion for modularity:

Definition 3.1 $\mathcal{Q}$ is a *criterion (for the $\odot$-modularity of $\mathcal{P}$)* if $\mathcal{Q} \neq \emptyset$ and $\forall T_1, \ldots, T_n$. $T_1 \in \mathcal{Q}, \ldots, T_n \in \mathcal{Q} \Rightarrow T_1 \odot \ldots \odot T_n \in \mathcal{P}$. □

In the sequel we will often talk simply of criterion, omitting $\mathcal{P}$ and $\odot$.

So, having a criterion $\mathcal{Q}$ we can perform modular analysis of a complex object $T_1 \odot \ldots \odot T_n$ just by separately checking that every submodule belongs to $\mathcal{Q}$.

3.1 Assumptions

Given the property $\mathcal{P}(\neq \emptyset)$ whose modular behaviour we want to analyze, we call *healthy* the objects in $\mathcal{P}$, and *sick* the others (the reasons for this terminology will become clear when we will introduce vaccines in Section 5). We say that two objects A and B are *compatible* (resp. *uncompatible*) w.r.t. $\mathcal{P}$ and $\odot$, if $A \odot B$ is healthy (resp. sick).

Since the observable of interest is the property $\mathcal{P}$, we introduce the following notion: two objects A and B are said to be $\mathcal{P}$-*equivalent* ($A =_{\mathcal{P}} B$) if $A \in \mathcal{P} \Leftrightarrow B \in \mathcal{P}$.

Recall from algebra that a *groupoid* $(\mathcal{S}, \tau)$ is a set $\mathcal{S}$ equipped with a binary operation τ. Although this is not strictly needed for the development of our theory, for simplicity we suppose that in every groupoid we talk about there is a neutral element (if it is not the case, one can always be added by the standard lifting technique).

We say that a groupoid $(\mathcal{S}, \tau)$ is a $\mathcal{P}$-*semilattice* if for every objects A, B and C in $\mathcal{S}$ we have that $(A \tau B) \tau C =_{\mathcal{P}} A \tau (B \tau C)$, $A \tau B =_{\mathcal{P}} B \tau A$, and $A \tau A =_{\mathcal{P}} A$. That is to say, a $\mathcal{P}$-semilattice is like a semilattice, but for the fact that the equations for associativity, commutativity and idempotence are weakened by considering $=_{\mathcal{P}}$-equivalence in place of equivalence.

Another crucial definition is the following:

Definition 3.2 A groupoid $(\mathcal{S}, \tau)$ is said to be $\mathcal{P}$-*dense* if $\forall T_1, T_2 \in \mathcal{S}$. $T_1 \tau T_2 \in \mathcal{P} \Rightarrow T_1 \in \mathcal{P} \wedge T_2 \in \mathcal{P}$. □

Roughly speaking, density corresponds to the very reasonable assumption that objects constituting a healthy object are themselves healthy.

Now we have all the ingredients to define this main notion:

Definition 3.3 A $\mathcal{P}$-*acid groupoid* (briefly, a $\mathcal{P}$-*acid*), is a groupoid $(\mathcal{S}, \tau)$ that is a $\mathcal{P}$-dense $\mathcal{P}$-semilattice. □

The name "acid" stems from the fact a semilattice can equivalently be seen as an *aci*-groupoid (viz. a groupoid that is *a*ssociative, *c*ommutative and *i*dempotent), and so *acid* stands for *aci* and *d*ense.

Assumption: Throughout the paper, we assume that $(\mathcal{O}, \odot)$ is a $\mathcal{P}$-acid.

We remark that for most of the results all of the above assumptions are not necessary. We take all of them at once to simplify readability (for discussions on the minimal required hypotheses, see e.g. [10, 12]).

4 The Modularity Space

The study of modularity for a given healthiness property is tantamount to the study of the criteria for its modularity. We are so interested in the *modular space* (*m-space*), that is in the collection of all the criteria for modularity. A way to express this information is to consider only the most significant objects in this space. The m-space has a natural partial ordering, namely the set inclusion; the idea is so to consider only the tops of the m-space:

Definition 4.1 The *modular basis* (*m-basis* for short) is the collection of all the maximal criteria. The *modular dimension* (*m-dimension*) is the cardinality of the m-basis. □

The modular basis is a good representative of the modular space, since from it we can build up the whole modular space (the maximal criteria entail all the other criteria):

Theorem 4.2 *Every criterion is contained in a maximal criterion.*

4.1 k-counterexamples

The m-dimension gives an abstract measure of the complexity of the modular space. It is not difficult to see that the m-dimension is one iff $\mathcal{P}$ is modular, and if $\mathcal{P}$ is not modular the m-dimension is at least two. We now give more precise results on the m-dimension, introducing the concept of k-counterexample.

Given an ordinal k, a k-counterexample (to the $\odot$-modularity of $\mathcal{P}$) is a collection $A_1, \ldots, A_k$ of pairwise uncompatible healthy objects.

Usually, a 2-counterexample will be simply called a *counterexample*.

The next two lemmata provide the link between k-counterexamples and the m-dimension. The first result gives a lower bound:

Lemma 4.3 *If there is a k-counterexample ($k<\omega$), then the m-dimension is at least k.*

The second result, dually, gives an upper bound:

Lemma 4.4 *If there is not a k-counterexample ($k < \omega$), then the m-dimension is less than k.*

Combining the above bounds gives the following characterization of the m-dimension in the finite case:

Corollary 4.5 *The m-dimension is k ($k < \omega$) iff there is a k-counterexample but there is no $k+1$-counterexample.*

5 Vaccines

We said the basic notion of the theory is that of vaccine. A vaccine is "*a preparation of living attenuated organisms, or living fully virulent organisms that is administered to produce or artificially increase immunity to a particular disease*" (Webster's 7th Collegiate Dictionary). So, suppose we want to ensure an organism enjoys a particular property. We can inject a specific vaccine for this property to it: if it does not get sick, due to collateral effects, we are sure it is immunized and enjoys that property.

In this paper, we utilize the notion of vaccine in a formal setting to study modularity. Therefore, suppose we want to study the modularity behaviour of the class of objects $\mathcal{P}$. The idea is to consider $\mathcal{P}$ as a 'healthiness condition', and select some representative objects that make things go wrong (i.e. that cause modularity to fail), using them as a vaccines: we can 'inject' one of them, say A, to any other object in $\mathcal{P}$ via the composition operator $\odot$: in case there are no collateral effects, i.e. in case the object is still healthy (belonging to $\mathcal{P}$), it will become 'immunized' to that particular disease that made modularity fail.

More formally, an object A is a vaccine if for the class of its vaccinated objects ($\{T: T\odot A \in \mathcal{P}\}$), $\mathcal{P}$ becomes $\odot$-modular.

The nice fact, as said in the introduction, is that we will show that the criteria defined by vaccines are *optimal* (i.e. maximal). This way, vaccines provide a tool to completely describe the modular space, providing the best criteria.

We now start giving rigorous formal definitions.

Definition 5.1 The class of *objects vaccinated via A with injection operator $\odot$ and healthiness property $\mathcal{P}$* is
$$\mathbf{V}_A^\odot(\mathcal{P}) = \{T: T\odot A \in \mathcal{P}\} \qquad \square$$

That is, we take every object T and inject A to it, obtaining the healthy object $T\odot A$.

The operator $\odot$ and the healthiness property $\mathcal{P}$ will be mostly omitted and considered understood, hence we will also write simply $\mathbf{V}_A$.

Now, we can define what a vaccine for modularity is:

Definition 5.2 A is a *vaccine (for the $\odot$-modularity of $\mathcal{P}$)* if $\mathbf{V}_A$ is a criterion for the $\odot$-modularity of $\mathcal{P}$. $\qquad \square$

That is to say,
$$\emptyset \neq \mathbf{V}_A,\ T_1 \in \mathbf{V}_A, \ldots, T_k \in \mathbf{V}_A \Rightarrow T_1 \odot \ldots \odot T_k \in \mathcal{P}$$

Vaccines can be composed to get new vaccines, as the following results show:

Lemma 5.3 (Composition) *Suppose A is a vaccine for $\mathcal{P}_1$ and B is a vaccine for $\mathcal{P}_2$. If $A\odot B \in \mathcal{P}_1 \wedge \mathcal{P}_2$, then $A\odot B$ is a vaccine for $\mathcal{P}_1 \wedge \mathcal{P}_2$.*

Corollary 5.4 *If A and B are compatible vaccines, then $A\odot B$ is a vaccine.*

Vaccines are only representatives of the corresponding criteria. It is therefore important to ask when different vaccines are representative of the same class. The following lemma gives a neat answer to this question:

Lemma 5.5 *Let A and B be vaccines. Then, $\mathbf{V}_A = \mathbf{V}_B \Leftrightarrow A$ and B are compatible*

6 Vaccines Bases

Every vaccine for modularity defines a criterion for modularity given by the class $\mathbf{V}_A$. The most important reason that makes vaccines attractive to study is that this criterion is *optimal* in the sense that *cannot be improved*.

Theorem 6.1 (Optimality) *If A is a vaccine, then $\mathbf{V}_A$ is a maximal criterion.*

The m-basis is an abstract concept. Anyway, we have just seen that vaccines can conveniently represent the maximal criteria. So, we introduce a new manageable representative of the m-space:

Definition 6.2 A *vaccines basis (v-basis)* is a collection of vaccines $\{A_i\}_{i=1...k}$ (k an ordinal) such that every maximal criterion is represented by exactly one vaccine. □

Hence, $A_1, \ldots, A_k$ is a v-basis iff $\mathbf{V}_{A_1}, \ldots, \mathbf{V}_{A_k}$ is the m-basis.

A v-basis does not only give a complete description of the modular space. It also allows to easily derive that a property is indeed a criterion by proving that it is weaker than an optimal criterion. The precise technique is described in the full paper. This also holds for the other kind of criteria, namely d-criteria (cf. Section 7). Hence not only easy proofs of the previously existing results on modularity can be given, but also investigation of new practical criteria is possible.

6.1 v-Bases versus k-Counterexamples

We now analyze the tight relationships between v-bases and k-counterexamples. First we introduce the notion of partial v-basis, which formalizes the uncomplete knowledge of a v-basis.

Definition 6.3 A *partial vaccines basis* is a collection $A_1, \ldots, A_k$ (k an ordinal) of vaccines giving pairwise different maximal criteria. □

Lemma 6.4 *Every partial vaccines basis* $\{A_1, \ldots, A_k\}$ *is a k-counterexample.*

As a corollary, we get that every v-basis $\{A_1, \ldots, A_k\}$ is a k-counterexample. The next important result shows that also the other direction holds, thus providing a way to find the v-bases:

Theorem 6.5 *If the modular dimension is $k < \omega$, then every k-counterexample is a v-basis.*

Combining these results, we get the following characterization of the v-bases:

Corollary 6.6 (Characterization) *If the modular dimension is $k < \omega$, then the v-bases are exactly the k-counterexamples.*

Therefore, the above results suggest a way to find the optimal criteria: seek for vaccines produced by objects in k-counterexamples.

In fact, Theorem 6.5 says much more: if we know that the m-dimension is $k < \omega$ (e.g. via Corollary 4.5), then a v-basis is automatically provided by a k-counterexample.

Another immediate consequence of Theorem 6.5 is about the existence of v-bases:

Corollary 6.7 *If the modular dimension is $k < \omega$, there is a v-basis.*

In order to effectively find a v-basis, Theorem 6.5 requires the knowledge of the m-dimension, which as said can be computed using Corollary 4.5. Anyway, there is another fundamental result that, starting from a not complete knowledge of it (a partial v-basis), ensures that we have found a v-basis:

Theorem 6.8 (Covering) *Let $A_1, \ldots, A_k$ ($k < \omega$) be a partial v-basis. It is a v-basis iff every healthy object belongs to at least one $\mathbf{V}_{A_i}$: $\cup_{i \in [1,k]} \mathbf{V}_{A_i} = \mathcal{P}$ (i.e. the criteria 'cover' the healthy objects).*

The above theorem thus provides an alternative powerful methodology to find a v-basis: build up a k-counterexample with k as great as possible; prove that its elements are vaccines (Theorem 6.5); check if the criteria cover the healthy objects (Theorem 6.8).

We will later (Section 9) successfully employ this methodology in the applications of the theory to term rewriting.

7 Disjunctive Criteria

The notion of criterion for modularity that we have given in Definition 3.1 is not the only one which has been studied. Another kind of criteria, e.g. studied in [13, 20], requires only one of the objects to be constrained in order to ensure their combination is healthy. So, we introduce this concept:

Definition 7.1 $\mathcal{Q}$ is a *disjunctive criterion (for the $\odot$-modularity of $\mathcal{P}$)*, or *d-criterion* for short, if $\forall T_1, \ldots, T_n. T_1 \in \mathcal{Q} \vee \ldots \vee T_n \in \mathcal{Q} \Rightarrow T_1 \odot \ldots \odot T_n \in \mathcal{P}$. □

The motivation for the adjective 'disjunctive' should be clear from the definition; analogously, the usual criterion of Definition 3.1 could be dubbed 'conjunctive'.

Unlike the standard criteria, the d-criteria space is linearly ordered, since only one object instead of all objects is constrained. The following definition formalizes the top object in this space:

Definition 7.2 The *kernel* $\mathcal{K}$ is the greatest disjunctive criterion, that is $\mathcal{K} = \{T \in \mathcal{P} : \forall T' \in \mathcal{P}. T \odot T' \in \mathcal{P} \ni T' \odot T\}$. □

It is easy to prove that, rather interestingly, the kernel has an important algebraic meaning, since it is just the class of $=_\mathcal{P}$-*neutral elements* (i.e. those elements N such that for every T we have $T \odot N =_\mathcal{P} T =_\mathcal{P} N \odot T$).

Nicely, from a v-basis we can obtain right away the kernel:

Theorem 7.3 *Suppose* $\{A_i\}_{i=1\ldots k}$ *is a vaccines basis. Then the kernel is* $\bigcap_{i=1\ldots k} \mathbf{V}_{A_i}$.

8 Counterexample Structures

In this section we turn our attention to the other side of the coin: when modularity fails. We formally study what happens when two objects give a counterexample to modularity.

Definition 8.1 A couple of classes $\{\mathcal{Q}_1, \mathcal{Q}_2\}$ is a *counterexample structure (c-structure)*, (w.r.t. $\odot$ and $\mathcal{P}$) if in every counterexample one of the two objects belongs to $\mathcal{Q}_1$ and the other to $\mathcal{Q}_2$. □

The canonical ordering on structures is: $\{\mathcal{Q}_1, \mathcal{Q}_2\} \subseteq_{struct} \{\mathcal{Q}'_1, \mathcal{Q}'_2\}$ iff $(\mathcal{Q}_1 \subseteq \mathcal{Q}'_1 \wedge \mathcal{Q}_2 \subseteq \mathcal{Q}'_2) \vee (\mathcal{Q}_1 \subseteq \mathcal{Q}'_2 \wedge \mathcal{Q}_2 \subseteq \mathcal{Q}'_1)$. Then, we say that a structure $\{\mathcal{Q}_1, \mathcal{Q}_2\}$ is *better than* another structure $\{\mathcal{Q}'_1, \mathcal{Q}'_2\}$ if $\{\mathcal{Q}_1, \mathcal{Q}_2\} \subset_{struct} \{\mathcal{Q}'_1, \mathcal{Q}'_2\}$: this means we can provide with $\{\mathcal{Q}_1, \mathcal{Q}_2\}$ a more precise (smaller) description than with $\{\mathcal{Q}'_1, \mathcal{Q}'_2\}$. The best structure is so the minimum w.r.t. $\subseteq_{struct}$.

From a v-basis we can recover the *best* counterexample structure, as the next result shows:

Theorem 8.2 *If* $\{A_1, A_2\}$ *is a vaccines basis, then* $\{\neg \mathbf{V}_{A_1} \wedge \mathcal{P}, \neg \mathbf{V}_{A_2} \wedge \mathcal{P}\}$ *is the best counterexample structure.*

Actually, more can be proved, i.e. that such c-structure is *perfect* in the sense that it provides a characterization of the counterexamples (cf. [12]).

Analogous results can be stated for v-bases of higher dimension.

9 Applications to Term Rewriting

We now provide some applications of the theory to the study of the modularity of termination for Term Rewriting Systems.

So, we let $\mathcal{O}$ =TRSs and consider as usual the combination operator $\odot$ to be the disjoint sum ($\oplus$) of two TRSs: when the signatures overlap the TRSs are renamed to get disjoint signatures, and then their (disjoint) union is taken. The healthiness property is $\mathcal{P}$ =Termination (Termination will be also indicated with the acronym SN, after Strong Normalization). We have that

Lemma 9.1 *(TRSs, $\oplus$) is SN-acid.*

Among the many results on the modularity of termination (see e.g. [14, 8, 16, 18] for a panoramic), the best results so far obtained are the ones in [15] and [9]. We will come back to the result of [9] in the next subsection. In [15] Ohlebusch, generalizing a previous result of Gramlich for finitely branching TRSs ([5]), proved that '$\mathcal{C}_\mathcal{E}$-termination' is modular. It is straightforward to see that the class of $\mathcal{C}_\mathcal{E}$-terminating TRSs coincides with the class of TRSs vaccinated via $\{or(X,Y) \to X, or(X,Y) \to Y\}$. This, a posteriori, implies that the above TRS is a vaccine (for the modularity of termination).

Hence, using Theorem 6.1 we obtain right away:

Theorem 9.2 $\mathcal{C}_\mathcal{E}$-*termination is a maximal criterion.*

That is to say, the result of [15] *cannot* be improved.

But what is the complexity of the modular space for termination? The following result gives a formal confirmation that the topic is quite intricated:

Theorem 9.3 *The m-dimension is at least three.*

The proof of the above result makes use of Lemma 4.3.

Whether the m-dimension is indeed three, is still one of the most important open problems (we conjecture it is).

9.0.1 The Left-Linear Case

As just seen, the situation for termination is quite complicated, since we have proved that the m-dimension is at least three, and only one vaccine has been found so far. In the left-linear case we will be able to *completely solve* the problem, finding a v-basis.

There are two best results on the modularity of termination for left-linear TRSs. The first stems from the one seen above: in the left-linear case, $\{or(X,Y) \to X, or(X,Y) \to Y\}$ is a vaccine.

So, by Theorem 6.1 we can infer that $\mathcal{C}_\mathcal{E}$-termination is a maximal criterion even for left-linear TRSs.

The second is the result proved in [9]. Recall that a TRS is said consistent (with respect to reduction), briefly $CON^\to$, if no term reduces to two different variables. In the aforementioned paper it has been shown that termination is modular for left-linear and consistent TRSs.

We have seen in Section 4 that there are deep relationships between k-counterexamples and v-bases. The most famous counterexample to the modularity of termination has been given by Toyama in [19]: $\{F(0,1,X) \to F(X,X,X)\}$ and $\{or(X,Y) \to X, or(X,Y) \to Y\}$. As seen above, $\{or(X,Y) \to X, or(X,Y) \to Y\}$ is a vaccine. Hence, a stimulating hypothesis is that $\{F(0,1,X) \to F(X,X,X)\}$ is a vaccine as well. Amazingly, this turns out to be true:

Theorem 9.4 *For left-linear TRSs,* $V_{\{F(0,1,X)\to F(X,X,X)\}} = \text{SN} \wedge \text{CON}^{\to}$.

That is to say, the class of left-linear TRSs vaccinated by $\{F(0,1,X) \to F(X,X,X)\}$ is just the criterion found in [9].

Corollary 9.5 *In the left-linear case,* $\{F(0,1,X) \to F(X,X,X)\}$ *is a vaccine.*

Hence, we get

Corollary 9.6 *In the left-linear case,* $\text{SN} \wedge \text{CON}^{\to}$ *is a maximal criterion.*

Thus, the result of [9] *cannot* be improved.

The remarkable thing is that with these two vaccines we have completed the analysis of the modular space, since they form a v-basis:

Theorem 9.7 *The m-dimension for left-linear TRSs is two, and a vaccines basis is given by* $\{F(0,1,X) \to F(X,X,X)\}$, $\{or(X,Y) \to X, or(X,Y) \to Y\}$.

That is to say, the above two optimal criteria completely solve the problem of modularity of termination for left-linear TRSs: there are *no other* optimal criteria and *all the other* criteria are subsumed by one of the two.

Also, being the m-dimension 2, by Corollary 6.6 we have a characterization of the v-bases: they are just the counterexamples.

As far as d-criteria are concerned, Middeldorp in [13] showed that whenever one of two terminating TRSs is both non-collapsing and non-duplicating, then their disjoint sum is terminating; that is to say, he proved that "terminating and non-collapsing and non-duplicating" is a disjunctive criterion. Toyama, Klop and Barendregt showed in [20] that whenever one of two terminating TRSs is confluent and non-collapsing, then their disjoint sum is terminating (hence, they proved that "terminating and confluent and non-collapsing" is a d-criterion).

Using the result on d-criteria (Theorem 7.3), we can properly generalize both of these results in the left-linear case, giving the *best* d-criterion (the kernel):

Theorem 9.8 *For left-linear TRSs,* $\text{CON}^{\to} \wedge C_{\mathcal{E}}$*-termination is the greatest disjunctive criterion for the modularity of termination.*

We now consider c-structures. Ohlebusch in [15] (again, extending a result of Gramlich in [5] for finitely branching TRSs), showed that in every counterexample one of the TRSs is not $C_{\mathcal{E}}$-terminating and the other is collapsing (hence, in our terminology, he showed that { $C_{\mathcal{E}}$-termination, non-collapsibility } is a c-structure). Schmidt-Schauß, Marchiori and Panitz showed in [18] that, in the left-linear case, in every counterexample one of the TRSs is $\text{CON}^{\to}$ and the other is $\neg\text{CON}^{\to}$ (that is, { $\text{CON}^{\to}$, $\neg\text{CON}^{\to}$ } is a c-structure). Both of these results require a not easy proof. Via Theorem 8.2, we can easily not only generalize all of these results in the left-linear case, but also provide the *best* c-structure:

Theorem 9.9 $\{\neg\text{CON}^{\to} \wedge \text{SN}, \neg C_{\mathcal{E}}\text{-termination} \wedge \text{SN}\}$ *is the best counterexample structure.*

The above theorem gives the following result: *in every counterexample to the modularity of termination, one of the TRSs is non consistent and the other is non $C_{\mathcal{E}}$-terminating.*

Other applications, as mentioned in Section 6, include the possibility to give easy proofs of previously existing results on modularity (for example the results in [17] and [13] can be provided, in the left-linear case, with an easy proof).

Finally, the optimality of the v-basis allows to infer right away results on the relative strength of other criteria.

For instance, it has been directly proved with some effort in [5] that Simple Termination implies $\mathcal{C}_\mathcal{E}$-termination, and that termination plus non-duplication imply $\mathcal{C}_\mathcal{E}$-termination. These results immediately follow from Theorem 9.2, once noticed that Simple Termination ([8]) and termination plus non-duplication ([17]) are criteria, and that $\{or(X,Y) \to X, or(X,Y) \to Y\}$ is both simply terminating and non-duplicating.

10 Remarks

In this extended abstract we have sketched the core of the theory of vaccines, and presented as a particular instance some successful applications to modularity in term rewriting. However, so far the theory of vaccines has been employed to obtain a variety of other results. For instance, we have applied it to study the modularity problem for completeness and uniqueness of normal forms w.r.t. reduction ($UN^\to$), finding vaccines for their modularity, and this way improving many existing results so far obtained in the literature. Also, besides many other results which are variations and generalizations of the main results here presented, we have investigated the major topic of *multimodularity*, where other combinations of more than two objects are studied (see [10, 12]). Again, via a v-basis we can obtain precise information on what kind of multimodular behaviour a certain property satisfies.

Currently, we are investigating practical applications of the theory to the study of modularity for other paradigms, like functional or logic programming (cf. [2]). Note that even in the rewriting field there are still many other modularity topics to which the theory of vaccines can be applied, including e.g. more involved combinations of TRSs (like composable ones, cf. [16] for a survey), higher order rewriting in its various forms (see e.g. [7, 6]), conditional rewriting ([7, 14]), combinations with λ-calculus and systems in the λ-cube (cf. e.g. [1]), and so on. For instance, the theory of vaccines can be applied to the criterion developed in [4] for conditional rewriting, showing that it is optimal for finitely branching CTRSs. Also, we have shown that the theory of vaccines nicely interacts with *unraveling theory* (cf. [11]), and shown how one can thus automatically translate a lot of modularity results from term rewriting to conditional rewriting: for instance, we have lifted the result of Theorem 9.7, showing that, for left-linear normal CTRSs, the same two TRSs provide a v-basis for decreasingness.

Acknowledgments

I wish to thank Jan Willem Klop and Aart Middeldorp for scrutinizing a previous version of this paper.

References

[1] F. Barbanera, M. Fernandez, and H. Geuvers. Modularity of strong normalization and confluence in the algebraic λ-cube. In *Proceedings Nineth IEEE Symposium on Logic in Computer Science*, pages 406–415, 1994.

[2] A. Brogi, P. Mancarella, D. Pedreschi, and F. Turini. Modular logic programming. *ACM TOPLAS*, 16(4), 1994.

[3] N. Dershowitz and J.-P. Jouannaud. Rewrite systems. In J. van Leeuwen, editor, *Handbook of Theoretical Computer Science*, volume B, chapter 6, pages 243–320. Elsevier – MIT Press, 1990.

[4] B. Gramlich. Sufficient conditions for modular termination of conditional term rewriting systems. In *3rd Workshop on Conditional Term Rewriting Systems*, vol. 656 of *LNCS*, pp. 128–142. Springer-Verlag, 1993.

[5] B. Gramlich. Generalized sufficient conditions for modular termination of rewriting. *Applicable Algebra in Engineering, Communication and Computing*, 5:131–158, 1994.

[6] J.-P. Jouannaud and M. Okada. Executable higher-order algebraic specification languages. In *Proc. Sixth IEEE Symposium on Logic in Computer Science*, pp. 350–361, 1991.

[7] J.W. Klop. Term rewriting systems. In S. Abramsky, Dov M. Gabbay, and T.S.E. Maibaum, editors, *Handbook of Logic in Computer Science*, volume 2, chapter 1, pages 1–116. Clarendon Press, Oxford, 1992.

[8] M. Kurihara and A. Ohuchi. Modularity of simple termination of term rewriting systems. *Journal of IPS Japan*, 31(5):633–642, 1990.

[9] M. Marchiori. Modularity of completeness revisited. In J. Hsiang, editor, *Proceedings of the Sixth International Conference on Rewriting Techniques and Applications*, volume 914 of *LNCS*, pages 2–10. Springer-Verlag, 1995.

[10] M. Marchiori. The theory of vaccines. Technical Report 27, Dept. of Pure and Applied Mathematics, University of Padova, 1995.

[11] M. Marchiori. Unravelings and ultra-properties. In *Proceedings of the Fifth International Conference on Algebraic and Logic Programming (ALP'96)*, volume 1139 of *LNCS*, pages 107–121. Springer-Verlag, 1996.

[12] M. Marchiori. *Local Analysis and Localizations*. PhD thesis, Dept. of Pure and Applied Mathematics, University of Padova, February 1997. In Italian.

[13] A. Middeldorp. A sufficient condition for the termination of the direct sum of term rewriting systems. In *Proc. Fourth IEEE Symposium on Logic in Computer Science*, pp. 396–401, 1989.

[14] A. Middeldorp. *Modular Properties of Term Rewriting Systems*. PhD thesis, Vrije Universiteit, Amsterdam, November 1990.

[15] E. Ohlebusch. On the modularity of termination of term rewriting systems. *Theoretical Computer Science*, 136(2):333–360, 1994.

[16] E. Ohlebusch. Modular properties of composable term rewriting systems. *Journal of Symbolic Computation*, 20(1):1–41, 1995.

[17] M. Rusinowitch. On termination of the direct sum of term rewriting systems. *Information Processing Letters*, 26:65–70, 1987.

[18] M. Schmidt-Schauß, M. Marchiori, and S.E. Panitz. Modular termination of r-consistent and left-linear term rewriting systems. *Theoretical Computer Science*, 149(2):361–374, 1995.

[19] Y. Toyama. On the Church-Rosser property for the direct sum of term rewriting systems. *Journal of the ACM*, 1(34):128–143, 1987.

[20] Y. Toyama, J.W. Klop, and H.P. Barendregt. Termination for direct sums of left-linear complete term rewriting systems. *Journal of the ACM*, 42(6):1275–1304, November 1995.

The Equivalence Problem for Deterministic Pushdown Automata is Decidable

Géraud Sénizergues

LaBRI
Université de Bordeaux I
351, Cours de la Libération 33405 Talence, France [**]

Abstract. The equivalence problem for deterministic pushdown automata is shown to be decidable. We exhibit a *complete formal system* for deducing equivalent pairs of deterministic rational series on the alphabet associated with a dpda $\mathcal{M}$.

Keywords: deterministic pushdown automata; rational series; finite dimensional vector spaces; matrix semigroups; complete formal systems.

1 Introduction

The so-called "equivalence problem for deterministic pushdown automata" (dpda for short), is the following decision problem:

INSTANCE: two dpda A, B.   QUESTION: $L(A) = L(B)$?

where $L(A)$ (resp. $L(B)$) is the language recognized by A (resp. B). (This problem is often denoted by $Eq(D, D)$, where D stands for the class of all dpda). The question of whether this problem is *decidable* or not is raised in [GG66] and has received much attention since this time. Beside the fact that this question was natural from the point of view of *formal language* theory, it appeared later as Turing-equivalent with other equivalence-problems for different types of recursive *program schemes* (see [Cou90] for a survey). Some other Turing-equivalent problems on *semi-Thue systems* were also found (see [Sén94] for a survey) and formulations in terms of bisimulation equivalence of infinite *graphs* (or *processes*) have been found too (see [Cau95] for a survey).

Among a large number of papers let us only quote [Val74, VP75, Bee76, Rom85, Oya87, Sti96] which proved decidability of $Eq(D', D')$ for subclasses D' of the full class D of dpda. (We refer the reader to the surveys ([Cou90, Cau95, Lis96]) for other results on problems related to $Eq(D, D)$). The work [Mei89, Mei92] is an attempt to solve the general problem. On account of its incompleteness (see for example the comment in [Lis96, p.219]) it does not provide a full solution; nevertheless it introduced a fundamental new idea: the notion of *linear independence* for languages.

We prove here that the equivalence problem for dpda is *decidable* (theorem 9.3).

We obtain this result by providing a *complete* formal system $\mathcal{D}_0$ for equivalence identities between *deterministic rational series* (we use here a type of formal system inspired by [Cou83] and a notion of deterministic series inspired by [HHY79]). The proof of this completeness property leans on three types of arguments:

- in section 3 we develop around the fundamental idea of [Mei89, Mei92] an *algebraic* theory of "d-spaces",
- in sections 5,7 these structure results are turned into a construction of *strategies* for the formal system $\mathcal{D}_0$,
- in section 8 we analyze the *infinite trees* generated by some strategies associated with $\mathcal{D}_0$.

[**] mailing adress:LaBRI and UFR Math-info. Université Bordeaux1
351 Cours de la libération -33405- Talence Cedex.
email:ges@labri.u-bordeaux.fr
fax: 05-56-84-66-69

2 Preliminaries

2.1 Pushdown automata

A *pushdown automaton* on the alphabet X is a 6-tuple $\mathcal{M} = <X, Z, Q, \delta, q_0, z_0>$ where Z is the finite stack-alphabet, Q is the finite set of states, $q_0 \in Q$ is the initial state, z_0 is the initial stack-symbol and $\delta : QZ \times (X \cup \{\epsilon\}) \to \mathcal{P}_f(QZ^*)$, is the transition mapping.

Let $q, q' \in Q, \omega, \omega' \in Z^*, z \in Z, f \in X^*$ and $a \in X \cup \{\epsilon\}$; we note $(qz\omega, af) \longmapsto_{\mathcal{M}} (q'\omega'\omega, f)$ if $q'\omega' \in \delta(qz, a)$. $\stackrel{*}{\longmapsto}_{\mathcal{M}}$ is the reflexive and transitive closure of $\longmapsto_{\mathcal{M}}$. For every $q\omega, q'\omega' \in QZ^*$ and $f \in X^*$, we note $q\omega \stackrel{f}{\longmapsto}_{\mathcal{M}} q'\omega'$ iff $(q\omega, f) \stackrel{*}{\longmapsto}_{\mathcal{M}} (q'\omega', \epsilon)$. $\mathcal{M}$ is said *deterministic* iff, for every $z \in Z, q \in Q$:

$$\text{either } \mathrm{Card}(\delta(qz, \epsilon)) = 1 \text{ and for every } x \in X, \mathrm{Card}(\delta(qz, x)) = 0, \tag{1}$$

$$\text{or } \mathrm{Card}(\delta(qz, \epsilon)) = 0 \text{ and for every } x \in X, \mathrm{Card}(\delta(qz, x)) \leq 1. \tag{2}$$

$\mathcal{M}$ is said *real-time* iff, for every $qz \in QZ$, $\mathrm{Card}(\delta(qz, \epsilon)) = 0$. A dpda $\mathcal{M}$ is said *normalized* iff, for every $qz \in QZ, x \in X$:

$$q'\omega' \in \delta(qz, x) \Rightarrow |\omega'| \leq 2, \text{ and } q'\omega' \in \delta(qz, \epsilon) \Rightarrow |\omega'| = 0 \tag{3}$$

Given some finite set $F \subseteq QZ^*$ of configurations, the *language recognized by $\mathcal{M}$ with final configurations* F is defined by $L(\mathcal{M}, F) = \{w \in X^* \mid \exists c \in F, q_0 z_0 \stackrel{w}{\longmapsto}_{\mathcal{M}} c\}$.

2.2 Deterministic context-free grammars

Let $\mathcal{M}$ be some deterministic pushdown automaton (for sake of simplicity [3], we suppose here that $\mathcal{M}$ is normalized). The *variable* alphabet $V_\mathcal{M}$ associated to $\mathcal{M}$ is defined as: $V_\mathcal{M} = \{[p, z, q] \mid p, q \in Q, z \in Z\}$. The *context-free* grammar $G_\mathcal{M}$ associated to $\mathcal{M}$ is then $G_\mathcal{M} = <X, V, P>$ where $V = V_\mathcal{M}$ and P is the set of all the pairs of one of the following forms:

$$([p, z, q], x[p', z_1, p''][p'', z_2, q]) \text{ or } ([p, z, q], x'[p', z', q]) \text{ or } ([p, z, q], a) \tag{4}$$

where $p, q \in Q, z \in Z, x, x' \in X, a \in X \cup \{\epsilon\}, p'z_1z_2 \in \delta(pz, x), p'z' \in \delta(pz, x'), q \in \delta(pz, a)$. $G_\mathcal{M}$ is a *strict-deterministic* grammar. (A general theory of this class of grammars is exposed in [Har78] and used in [HHY79]). We call *mode* every element of $QZ \cup \{\epsilon\}$. For every $q \in Q, z \in Z$, qz is said ϵ-*bound* (respectively ϵ-*free*) iff condition (1) (resp. condition (2)) in the above definition of deterministic automata is realized. The mode ϵ is said ϵ-free. We define a mapping $\mu : V^* \to QZ \cup \{\epsilon\}$ by

$$\mu(\epsilon) = \epsilon \text{ and } \mu([p, z, q] \cdot \beta) = pz,$$

for every $p, q \in Q, z \in Z, \beta \in V^*$. For every $w \in V^*$ we call $\mu(w)$ the *mode* of the word w.
For technical reasons (which will be made clear in section 7), we suppose that Z contains a special symbol e such that, for every $q \in Q, \delta(qe, \epsilon) = \{q\}$ and $\mathrm{im}(\delta) \subseteq \mathcal{P}_f(Q(Z - \{e\})^*)$.

2.3 Free monoids acting on semi-rings

Semi-ring B $<W>$ Let $(B, +, \cdot, 0, 1)$ where $B = \{0, 1\}$ denote the semi-ring of "booleans". Let W be some alphabet. By $(B<W>, +, \cdot, \emptyset, \epsilon)$ we denote the semi-ring of *boolean series* over W: every boolean series $S \in B<W>$ can be written in a unique way as: $S = \Sigma_{w \in W^*} S_w \cdot w$, where, for every $w \in W^*$, $S_w \in B$. The *support* of S is the language

$$\mathrm{supp}(S) = \{w \in W^* \mid S_w \neq 0\}.$$

In the particular case where the semi-ring of coefficients is B (which is the only case considered in this article) we sometimes identify the series S with its support. We recall that for every $S \in B<W>$, S^* is the series defined by: $S^* = \sum_{0 \leq n} S^n$. Given two alphabets W, W', a map $\psi : B<W> \to B<W'>$ is said σ-*additive* iff it fulfills: for every denumerable family $(S_i)_{i \in \mathbb{N}}$ of elements of $B<W>$, $\psi(\sum_{i \in \mathbb{N}} S_i) = \sum_{i \in \mathbb{N}} \psi(S_i)$. A map $\psi : B<W> \to B<W'>$ which is both a semi-ring homomorphism and a σ-additive map is usually called a *substitution*.

---

[3] but without loss of generality for the equivalence problem

Actions of monoids Given a semi-ring $(S, +, \cdot, 0, 1)$ and a monoid $(M, \cdot, 1_M)$, a map $\circ : S \times M \longrightarrow S$ is called a *right-action* of the monoid M over the semi-ring S iff, for every $S, T \in S, m, m' \in M$:

$$0 \circ m = 0, \quad S \circ 1_M = S, \quad (S + T) \circ m = (S \circ m) + (T \circ m) \quad \text{and} \quad S \circ (m \cdot m') = (S \circ m) \circ m' \qquad (5)$$

In the particular case where $S = B < W >$, $\circ$ is said to be a σ-right-action if it fulfills the additional property that, for every denumerable family $(S_i)_{i \in \mathbb{N}}$ of elements of S and $m \in M$:

$$(\sum_{i \in \mathbb{N}} S_i) \circ m = \sum_{i \in \mathbb{N}} (S_i \circ m). \qquad (6)$$

The action of W^* on $B < W >$ We recall the following classical σ-right-action $\bullet$ of the monoid W^* over the semi-ring $B < W >$: for all $S, S' \in B < W >, u \in W^*$

$$S \bullet u = S' \Leftrightarrow \forall w \in W^*, (S'_w = 1 \text{ iff } S_{u \cdot w} = 1).$$

(i.e. $S \bullet u$ is the *left-quotient* of S by u, or the *residual* of S by u). For every $S \in B < W >$ we denote by $Q(S)$ the set of residuals of S: $Q(S) = \{S \bullet u \mid u \in W^*\}$. We recall that S is said *rational* iff the set $Q(S)$ is *finite*. We define the *norm* of a series $S \in B < W >$, denoted $\|S\|$ by: $\|S\| = \text{Card}(Q(S)) \in \mathbb{N} \cup \{\infty\}$.

The action of X^* on $B < V >$ Let us fix now a deterministic (normalized) pda $\mathcal{M}$ and consider the associated grammar G. We define a σ-right-action $\otimes$ of the monoid $(X \cup \{e\})^*$ over the semi-ring $B < V >$ by: for every $p, q \in Q, A \in Z, H \in V^*, \beta \in V^*, x \in X$

$$[p, A, q] \cdot \beta \otimes x = H \cdot \beta \text{ iff } ([p, A, q], x \cdot H) \in P, \quad [p, A, q] \cdot \beta \otimes e = H \cdot \beta \text{ iff } ([p, A, q], H) \in P \qquad (7)$$

$$\epsilon \otimes x = \emptyset, \quad \epsilon \otimes e = \emptyset. \qquad (8)$$

A series $S \in B < V >$ is said ϵ-*free* iff $\forall w \in V^*, S_w = 1 \Rightarrow \mu(w)$ is ϵ − free. We denote by $B_\epsilon < V >$ the subset of ϵ-free series. We define the map $\rho_\epsilon : B < V > \rightarrow B < V >$ as the unique σ-additive map such that, for every $p \in Q, z \in Z, q \in Q, \beta \in V^*$,

$$\rho_\epsilon([p, z, q] \cdot \beta) = \rho_\epsilon(([p, z, q] \otimes e) \cdot \beta) \text{ if } pz \text{ is } \epsilon - \text{bound}, \quad \rho_\epsilon([p, z, q] \cdot \beta) = [p, z, q] \cdot \beta \text{ if } pz \text{ is } \epsilon - \text{free},$$

and $\rho_\epsilon(\epsilon) = \epsilon$. The above definition is sound because, by hypothesis (3), every $[p, z, q] \otimes e$ is either the unit series ϵ or the empty series $\emptyset$. One can notice that for every $w \in V^*$, $\rho_\epsilon(w) \in V^* \cup \{\emptyset\}$. We call ρ_ϵ the ϵ-reduction map. We then define $\odot$ as the unique right-action of the monoid X^* over the semi-ring $B < V >$ such that: for every $S \in B < V >, x \in X, S \odot x = \rho_\epsilon(\rho_\epsilon(S) \otimes x)$. One can notice that if $u \neq \epsilon$, then $S \odot u$ is ϵ-free. Let us consider the unique substitution $\varphi : B < V > \rightarrow B < X >$ fulfilling: for every $p, q \in Q, z \in Z, \varphi([p, z, q]) = \{u \in X^* \mid [p, z, q] \odot u = \epsilon\}$, (in other words, φ maps every subset $L \subseteq V^*$ on the language generated by the grammar G from the set of axioms L).

Lemma 2.1 φ *is a morphism of right-actions i.e. for every* $S \in B < V >, u \in X^*, \varphi(S \odot u) = \varphi(S) \bullet u$.

We denote by $\equiv$ the kernel of φ i.e.: for every $S, T \in B < V >, S \equiv T \Leftrightarrow \varphi(S) = \varphi(T)$.

3 Series and languages

3.1 Deterministic series and matrices

We introduce here a notion of *deterministic* series which, in the case of the alphabet V associated to a dpda $\mathcal{M}$, generalizes the classical notion of *configuration* of $\mathcal{M}$. The main advantage of this notion is that, unlike for configurations, we shall be able to define *nice algebraic operations* on these series (this is done in section 3.2). Let us consider a pair $(W, \sim)$ where W is an alphabet and $\sim$ is an equivalence relation over W. We call $(W, \sim)$ a *structured* alphabet. The two examples we have in mind are:

- the case where $W = V$, the variable alphabet associated to $\mathcal{M}$ and $[p, A, q] \sim [p', A', q']$ iff $p = p'$ and $A = A'$ (see [Har78])
- the case where $W = X$, the terminal alphabet of $\mathcal{M}$ and $x \sim y$ holds for every $x, y \in X$ (see [Har78]).

Definition 3.1 Let $S \in \mathsf{B} < W >$. S is said left-deterministic iff either (1) $S = \emptyset$ or (2) $S = \epsilon$ or (3) $\forall w, w' \in W^*, S_w = S_{w'} = 1 \Rightarrow \exists A, A' \in W, w_1, w_1' \in W^*, A \sim A', w = A \cdot w_1$ and $w' = A' \cdot w_1'$.

Definition 3.2 Let $S \in \mathsf{B} < W >$. S is said deterministic iff, for every $u \in W^*$, $S \bullet u$ is left-deterministic.

This notion is the straighforward extension to the infinite case of the notion of (finite) *set of associates* defined in [HHY79].

We denote by $\mathsf{DB} < W >$ the subset of deterministic boolean series over W. Let us denote by $\mathsf{B}_{n,m} < W >$ the set of (n, m)-matrices with entries in the semi-ring $\mathsf{B} < W >$.

Definition 3.3 Let $m \in \mathbb{N}, S \in \mathsf{B}_{1,m} < W >: S = (S_1, \cdots, S_m)$. S is said left-deterministic iff either (1) $\forall i \in [1, m], S_i = \emptyset$ or (2) $\exists i_0 \in [1, m], S_{i_0} = \epsilon$ and $\forall i \neq i_0, S_i = \emptyset$ or (3) $\forall w, w' \in W^*, \forall i, j \in [1, m], (S_i)_w = (S_j)_{w'} = 1 \Rightarrow \exists A, A' \in W, w_1, w_1' \in V^*, A \sim A', w = A \cdot w_1$ and $w' = A' \cdot w_1'$.

The right-action $\bullet$ on $\mathsf{B} < W >$ is extended componentwise to $\mathsf{B}_{n,m} < W >$: for every $S = (s_{i,j})$, $u \in W^*$, the matrix $T = S \bullet u$ is defined by $t_{i,j} = s_{i,j} \bullet u$.

Definition 3.4 Let $S \in \mathsf{B}_{1,m} < W >$. S is said deterministic iff, for every $u \in W^*$, $S \bullet u$ is left-deterministic.

We denote by $\mathsf{DB}_{1,m} < W >$ the subset of deterministic row-vectors of dimension m over $\mathsf{B} < W >$.

Definition 3.5 Let $S \in \mathsf{B}_{n,m} < W >$. S is said deterministic iff, for every $i \in [1, n]$, $S_{i,*}$ is a deterministic row-vector.

The following property is crucial for establishing a correct theory of *deterministic spaces* (see §3.2 below).

Lemma 3.6 For every $S \in \mathsf{DB}_{n,m} < W >, T \in \mathsf{DB}_{m,s} < W >, S \cdot T \in \mathsf{DB}_{n,s} < W >$.

W=V Let $(W, \sim)$ be the structured alphabet $(V, \sim)$ associated with $\mathcal{M}$ and let us consider a bijective numbering of the elements of Q: $(q_1, q_2, \ldots, q_{n_Q})$. Some particular "vectorial" notions turn out to be useful:

– we define a *Q-series* to be a family $(S_q)_{q \in Q}$ such that the row-vector $(S_{q_1}, S_{q_2}, \ldots, S_{q_{n_Q}})$ is deterministic
– we define a *Q-form* to be a family $\Phi = (\Phi_q)_{q \in Q}$ of deterministic series.

Given a Q-series S and a Q-form Φ, their Q-product $S * \Phi$ is the deterministic series defined by $S * \Phi = \sum_{q \in Q} S_q \cdot \Phi_q$. If the Q-series $(S_q)_{q \in Q}$ is identified with the row-vector $(S_{q_1}, S_{q_2}, \ldots, S_{q_{n_Q}})$ and the Q-form $(\Phi_q)_{q \in Q}$ with the column-vector $(\Phi_{q_j})_{j \in [1, n_Q]}$, then the Q-product appears to be just the ordinary product of matrices.

Let us define here handful notations for some particular row-vectors or Q-series. Let us use the *Kronecker symbol* $\delta_{i,j}$ meaning ϵ if $i = j$ and $\emptyset$ if $i \neq j$. For every $1 \leq n, 1 \leq i \leq n$, we define the row-vector ϵ_i^n as: $\epsilon_i^n = (\epsilon_{i,j}^n)_{1 \leq j \leq n}$ where $\forall j, \epsilon_{i,j}^n = \delta_{i,j}$. We call *unit row-vector* any vector of the form ϵ_i^n. For every $\omega \in Z^*, p, q \in Q$, $[p\omega q]$ is the deterministic series defined inductively by:

$$[p\epsilon q] = \emptyset \text{ if } p \neq q, [p\epsilon q] = \epsilon \text{ if } p = q,$$

$$[p\omega q] = \sum_{r \in Q} [pAr] \cdot [r\omega' q] \text{ if } \omega = A \cdot \omega' \text{ for some } A \in Z, \omega' \in Z^*.$$

By $[p\omega]$ we denote the Q-series: $[p\omega] = ([p\omega q])_{q \in Q}$. (In particular $[q_i] = \epsilon_i^{n_Q}$). By $[\omega]$ we denote the Q-matrix: $[\omega] = ([p\omega q])_{p \in Q, q \in Q}$. The next lemma relates the right-action $\odot$ with the right-action $\bullet$.

Lemma 3.7 Let $S \in \mathsf{DB} < V >, u \in X^*$. One of the three following cases must occur: (1) $S \odot u = \emptyset$, or (2) $S \odot u = \epsilon$, or (3) $\exists u_1, u_2 \in X^*, v_1 \in V^*, q \in Q, A \in Z, \Phi$ Q-form such that $u = u_1 \cdot u_2, S \odot u_1 = S \bullet v_1 = [qA] * \Phi$ and $S \odot u = ([qA] \odot u_2) * \Phi$.

Corollary 3.8 Let $S \in \mathsf{DB} < V >, u \in X^*$. Then $S \odot u \in \mathsf{DB} < V >$.

The particular letters $[p, e, q]$ for $p, q \in Q$ play a special role in sections 7 and 8: we use them as *marks* in the series (somehow like the ceilings of [Val74]). We define below a map ρ_e which removes the marks in the series. Let us define $\rho_e : \mathsf{DB} < V > \longrightarrow \mathsf{B} < V >$ as the unique substitution such that:

$$\rho_e([p, e, q]) = \epsilon \quad \text{if } p = q, \quad \rho_e([p, e, q]) = \emptyset \quad \text{if } p \neq q.$$

Lemma 3.9 For every $S \in \mathsf{DB} < V >$. $\rho_e(S) \in \mathsf{DB} < V >$ and $\|\rho_e(S)\| \leq \|S\|$.

Rational series, norm Let us generalize the definition of *rationality* of series in $B<W>$ to matrices. Given $M \in B_{n,m}<W>$ we denote by $Q(M)$ the set of *residuals* of M: $Q(M) = \{M \bullet u \mid u \in W^*\}$. Similarly, we denote by $Q_r(M)$ the set of *row-residuals* of M: $Q_r(M) = \bigcup_{1 \leq i \leq n} Q(M_{i,\cdot})$. M is said rational iff the set $Q(M)$ is finite. One can check that it is equivalent to the property that every coefficient $M_{i,j}$ is rational, or to the property that $Q_r(M)$ is finite. We denote by $DRB_{n,m}<W>$ the set of deterministic, rational matrices over $B<W>$. For every $M \in DRB_{n,m}<W>$, we define the norm of M as: $\|M\| = \text{Card}(Q_r(M))$.

Lemma 3.10 *Let $A \in DB_{n,m}<W>, B \in DB_{m,s}<W>$. Then $\|A \cdot B\| \leq \|A\| + \|B\|$.*

3.2 Deterministic spaces

We adapt here the key-idea of [Mei89, Mei92] to series.

Definitions Let $(W, \sim)$ be some structured alphabet and let us consider the set $E = DRB<W>$. A series $U = \sum_{i=1}^{n} \gamma_i \cdot U_i$ where $\gamma \in DRB_{1,n}<W>$, $U_i \in DRB<W>$ is called a *linear combination* of the U_i's. We call *deterministic space* of rational series (d-space for short) any subset V of E which is closed under finite linear combinations. Given any set $\mathcal{G} = \{U_i \mid i \in I\}$, one can check that the set V of all (finite) linear combinations of elements of $\mathcal{G}$ is a d-space (by lemma 3.6) and that it is the smallest d-space containing $\mathcal{G}$. Therefore we call V the d-space *generated* by $\mathcal{G}$ and we call $\mathcal{G}$ a *generating set* of V (we note $V = V(\{U_i \mid i \in I\})$). (Similar definitions can be given for *families* of series).
We let now $W = V$. Following an analogy with classical linear algebra, we develop now a notion corresponding to a kind of *linear independence* of the images by φ of the given series Let us extend the equivalence relation $\equiv$ to d-spaces by: for every d-spaces V_1, V_2, $V_1 \equiv V_2 \Leftrightarrow \forall i, j \in \{1,2\}, \forall S \in V_i, \exists S' \in V_j, S \equiv S'$.

Lemma 3.11 *Let $S_1, \ldots, S_j, \ldots, S_m \in DRB<V>$. The following are equivalent*

1. $\exists \alpha, \beta \in DRB_{1,m}<V>, \alpha \not\equiv \beta$, such that $\sum_{1 \leq j \leq m} \alpha_j \cdot S_j \equiv \sum_{1 \leq j \leq m} \beta_j \cdot S_j$,
2. $\exists j_0 \in [1, m], \exists \gamma \in DRB_{1,m}<V>, \gamma \not\equiv \epsilon_{j_0}^m$, such that $S_{j_0} \equiv \sum_{1 \leq j \leq m} \gamma_j \cdot S_j$,
3. $\exists j_0 \in [1, m], \exists \gamma' \in DRB_{1,m}<V>, \gamma'_{j_0} \equiv \emptyset$, such that $S_{j_0} \equiv \sum_{1 \leq j \leq m} \gamma'_j \cdot S_j$,
4. $\exists j_0 \in [1, m]$, such that $V((S_j)_{1 \leq j \leq m}) \equiv V((S_j)_{1 \leq j \leq m, j \neq j_0})$.

The equivalence between (1),(2) and (3) was first proved in [Mei89, Mei92], in the case where the S_j's are configurations $q_j\omega$, with the same ω.

4 Deduction systems

4.1 General deduction systems

We follow here the general philosophy of [HHY79, Cou83]. Let us call *deduction system* any triple $\mathcal{D} = <\mathcal{A}, H, \vdash>$ where $\mathcal{A}$ is a denumerable set called the *set of assertions*, H, the *cost function* is a mapping $\mathcal{A} \to \mathbb{N} \cup \{\infty\}$ and $\vdash$, the *deduction relation* is a subset of $\mathcal{P}_f(\mathcal{A}) \times \mathcal{A}$; $\mathcal{A}$ is given with a fixed bijection with $\mathbb{N}$ (an "encoding" or "Gödel numbering") so that the notions of recursive subset, recursively enumerable subset, recursive function, ... over $\mathcal{A}, \mathcal{P}_f(\mathcal{A}),...$ are defined, up to this fixed bijection ; we assume that $\mathcal{D}$ satisfies the following axioms:
(A 1) $\vdash$ is recursively enumerable
(A 2) $\forall (P, A) \in \vdash$, $(\min\{H(p), p \in P\} < H(A))$ or $(H(A) = \infty)$. (We let $\min(\emptyset) = \infty$).

In the sequel we use the notation $P \vdash A$ for $(P, A) \in \vdash$. We call *proof* in the system $\mathcal{D}$, any subset $P \subseteq \mathcal{A}$ fulfilling : $\forall p \in P, (\exists Q \subseteq P, Q \vdash p)$. Let us define the total map $\chi : \mathcal{A} \to \{0, 1\}$ and the partial map $\overline{\chi} : \mathcal{A} \to \{0, 1\}$ by :
$\chi(A) = 1$ if $H(A) = \infty$, $\chi(A) = 0$ if $H(A) < \infty$, $\overline{\chi}(A) = 1$ if $H(A) = \infty$, $\overline{\chi}$ is undefined if $H(A) < \infty$. (χ is the 'truth-value function", $\overline{\chi}$ is the "1-value function").

Lemma 4.1 *Let P be a proof and $A \in P$. Then $\chi(A) = 1$.*

In other words : every provable assertion is true. The deduction system $\mathcal{D}$ will be said *complete* iff, conversely, $\forall A \in \mathcal{A}, \chi(A) = 1 \Longrightarrow$ there exists some *finite* proof P such that $A \in P$. (In other words, $\mathcal{D}$ is complete iff every true assertion is "finitely" provable).

Lemma 4.2 : *If $\mathcal{D}$ is complete, $\overline{\chi}$ is a recursive partial map.*

In order to define deduction relations from more elementary ones, we set the following definitions. Let $\vdash \subseteq \mathcal{P}_f(\mathcal{A}) \times \mathcal{A}$. For every $P, Q \in \mathcal{P}_f(\mathcal{A})$ we set:
$P \vdash^{[0]} Q$ iff $P \supseteq Q$; $\quad P \vdash^{[1]} Q$ iff $\forall q \in Q, \exists R \subseteq P, R \vdash q$; $\quad P \vdash^{<0>} Q$ iff $P \vdash^{[0]} Q$; $\quad P \vdash^{<1>} Q$ iff $\forall q \in Q, (\exists R \subseteq P, R \vdash q)$ or $(q \in P)$; $\quad P \vdash^{<n+1>} Q$ iff $\exists R \in \mathcal{P}_f(\mathcal{A}), P \vdash^{<1>} R$ and $R \vdash^{<n>} Q$ (for every $n \geq 1$).;
$\vdash^{<*>} = \bigcup_{n \geq 0} \vdash^{<n>}$.
Given $\vdash_1, \vdash_2 \subseteq \mathcal{P}_f(\mathcal{A}) \times \mathcal{P}_f(\mathcal{A})$, for every $P, Q \in \mathcal{P}_f(\mathcal{A})$ we set : $P(\vdash_1 \circ \vdash_2) Q$ iff $\exists R \subseteq \mathcal{A}, (P \vdash_1 R) \wedge (R \vdash_2 Q)$.

4.2 System $\mathcal{D}_0$

Let us define here a particular deduction system $\mathcal{D}_0$ "Taylored for the equivalence problem for dpda's".

Given a fixed dpda $\mathcal{M}$ over the terminal alphabet X, we consider the variable alphabet V associated to $\mathcal{M}$ (see section 3.1) and the set $\mathsf{DRB} < V >$ (the set of Deterministic Rational Boolean series over V^*). The set of assertions is defined by : $\mathcal{A} = \mathbb{N} \times \mathsf{DRB} < V > \times \mathsf{DRB} < V >$ i.e. an assertion is here a *weighted equation* over $\mathsf{DRB} < V >$.
The "cost-function" $H : \mathcal{A} \to \mathbb{N} \cup \{\infty\}$ is defined by : $H(n, S, S') = n + 2 \cdot \mathrm{Div}(S, S')$, where $\mathrm{Div}(S, S')$, the *divergence* between S and S', is defined by : $\mathrm{Div}(S, S') = \min\{|u| \mid u \in \Delta(\varphi(S), \varphi(S'))\}$. (We recall $\min(\emptyset) = \infty$).
Let us notice that here : $\chi(n, S, S') = 1 \iff S \equiv S'$.
We define a binary relation $\Vdash \subseteq \mathcal{P}_f(\mathcal{A}) \times \mathcal{A}$, the *elementary deduction relation*, as the set of all the pairs having one of the following forms:

(R0) $\{(p, S, T)\}$	$\Vdash (p+1, S, T)$	
(R1) $\{(p, S, T)\}$	$\Vdash (p, T, S)$	
(R2) $\{(p, S, S'), (p, S', S'')\}$	$\Vdash (p, S, S'')$	
(R3) $\emptyset$	$\Vdash (0, S, S)$	
(R'3) $\emptyset$	$\Vdash (0, [qzr], \epsilon)$	(for $q, r \in Q, z \in Z, [qzr] \equiv \epsilon$)
(R4) $\{(p+1, S \odot x, T \odot x) \mid x \in X\}$	$\Vdash (p, S, T)$	(for $S \not\equiv \epsilon \wedge T \not\equiv \epsilon$)
(R5) $\{(p, S, S')\}$	$\Vdash (p+2, S \odot x, S' \odot x)$	(for $x \in X$)
(R6) $\{(p, S \cdot T' + T, T')\}$	$\Vdash (p, S^* \cdot T, T')$	(for $S \not\equiv \varepsilon$)
(R7) $\{(p, S, S')\}$	$\Vdash (p, S + T, S' + T)$	
(R8) $\{(p, S, S')\}$	$\Vdash (p, S \cdot T, S' \cdot T)$	
(R9) $\{(p, S, S')\}$	$\Vdash (p, U \cdot S, U \cdot S')$	

where $p \in \mathbb{N}, S, S', T, T' \in \mathsf{DRB} < V >, U \in \mathsf{RB} < V >$. (By set of "all" these pairs we mean, all the pairs which fulfill both properties "to belong to $\mathcal{P}_f(\mathcal{A}) \times \mathcal{A}$" and "to have one of these 11 possible forms" ; but of course, for example, not all the triples $(p, S + T, S' + T)$ belong to $\mathcal{A}$ because $\mathsf{DRB} < V >$ is not closed under sum).

Lemma 4.3 : *Let $P \in \mathcal{P}_f(\mathcal{A}), A \in \mathcal{A}$ such that $P \Vdash A$. Then $\min\{H(p) \mid p \in P\} \leq H(A)$.*

Let us define $\vdash$ by : for every $P \in \mathcal{P}_f(\mathcal{A}), A \in \mathcal{A}, P \vdash A \iff P \Vdash^{<*>} \circ \Vdash^{[1]}_{0,3,4} \circ \Vdash^{<*>} \{A\}$, where $\Vdash_{0,3,4}$ is the relation defined by R_0, R_3, R'_3, R_4 only. We let $\mathcal{D}_0 = < \mathcal{A}, H, \vdash >$.

Lemma 4.4 : *$\mathcal{D}_0$ is a deduction system.*

The key-statement of this work is that $\mathcal{D}_0$ is complete (theorem 9.2). We prove this completeness result by exhibiting a "strategy" $\mathcal{S}$ which, for every true assertion (n, S, S'), constructs a finite $\mathcal{D}_0$-proof of this assertion. Notice that, by lemma 4.2. we do not need to prove that $\mathcal{S}$ is computable in any sense to establish that $\overline{\chi}$ is partial-recursive.

4.3 Strategies

Let $\mathcal{D} =< \mathcal{A}, H, \vdash >$ be a deduction system. We call a *strategy* for $\mathcal{D}$ any partial map $\mathcal{S} : \mathcal{A}^+ \to \mathcal{A}^*$ such that :
(S1) if $\mathcal{S}(A_1 A_2 \cdots A_n) = B_1 \cdots B_m$ then $\exists Q \subseteq \{A_i \mid 1 \leq i \leq n-1\}$ such that

$$\{B_j \mid 1 \leq j \leq m\} \cup Q \vdash A_n,$$

(S2) if $\mathcal{S}(A_1 A_2 \cdots A_n) = B_1 \cdots B_m$ then

$$\min\{H(A_i) \mid 1 \leq i \leq n\} = \infty \Longrightarrow \min\{H(B_j) \mid 1 \leq j \leq m\} = \infty.$$

Given a strategy $\mathcal{S}$, we define $\mathcal{T}(\mathcal{S}, A)$, the proof-tree associated to the strategy $\mathcal{S}$ and the assertion A as the unique tree t such that :
$\varepsilon \in dom(t)$, $t(\varepsilon) = A$, and, for every path $x_0 x_1, \cdots x_{n-1}$ in t, with labels $t(x_i) = A_{i+1}$ (for $0 \leq i \leq n-1$) if x_{n-1} has m sons $x_{n-1} \cdot 1, \cdots .x_{n-1} \cdot m \in dom(t)$ with labels $t(x_{n-1} \cdot j) = B_j$ (for $1 \leq j \leq m$) then

$$\mathcal{S}(A_1 \cdots A_n) = B_1 \cdots B_m \text{ or } (m = 0 \text{ and } A_1 \cdots A_n \notin dom(\mathcal{S})).$$

Let us say that $\mathcal{S}$ *terminates* iff, $\forall A \in \chi^{-1}(1), \mathcal{T}(\mathcal{S}, A)$ is finite; $\mathcal{S}$ is said *closed* iff, $\forall W \in \mathcal{A}^+, W \in (\chi^{-1}(1))^+ \Longrightarrow W \in dom(\mathcal{S})$ (i.e. $\mathcal{S}$ is defined on every non-empty sequence of true assertions).

Lemma 4.5 : *If S is a closed strategy for $\mathcal{D}$, then, for every true assertion A, the set of labels of $\mathcal{T}(\mathcal{S}, A)$ is a $\mathcal{D}$-proof.*

Lemma 4.6 : *If $\mathcal{D}$ admits some terminating, closed strategy then $\mathcal{D}$ is complete.*

5 Triangulations

Let $S_1, S_2, \cdots, S_d$ be a family of deterministic series over the structured alphabet V (we recall V is the alphabet associated with some dpda $\mathcal{M}$ as defined in section 2.2).
Let us consider a sequence $\mathcal{S}$ of n "weighted" linear equations :

$$(\mathcal{E}_i) : p_i, \sum_{j=1}^{d} \alpha_{i,j} S_j \, , \, \sum_{j=1}^{d} \beta_{i,j} S_j$$

where $p_i \in \mathbb{N}$, and $A = (\alpha_{i,j}), B = (\beta_{i,j})$ are deterministic rational matrices of dimension (n, d), with indices $m \leq i \leq m+n-1, 1 \leq j \leq d$. For any weighted equation, $\mathcal{E} = (p, S, S')$, we recall the "cost" of this equation is : $H(\mathcal{E}) = p + 2 \cdot \text{Div}(\varphi(S), \varphi(S'))$.
We associate to such a system another system of equations, INV($\mathcal{S}$), which "translates the equations of $\mathcal{S}$ into equations over $(\alpha_{i,j}, \beta_{i,j})$ only". This function INV is in some sense an "elaborated version" of the *inverse* systems defined in [Mei89, Mei92]. The general idea of the construction of INV consists in iterating the transformation used in the proof of (1) $\Rightarrow$ (2) $\Rightarrow$ (3) in lemma 3.11, i.e. the classical idea of *triangulating* a system of linear equations. Of course we must deal with the weights and relate the construction with the deduction system $\mathcal{D}_0$. Let us assume here that

$$\forall j \in [1, d], S_j \neq \emptyset. \tag{9}$$

For every $S \in \mathbb{B} < X >$ (resp. $S' \in \mathbb{B}_{1,d} < X >$), we define $\nu(S) = \min\{|u|, u \in \text{supp}(S)\}$ (resp. $\nu(S') = \min\{|u|, u \in \cup_{1 \leq j \leq d} \text{supp}(S'_j)\}$). Let us define INV($\mathcal{S}$), W($\mathcal{S}$) $\in \mathbb{N} \cup \{\bot\}$, D($\mathcal{S}$) $\in \mathbb{N}$ by induction on n. W($\mathcal{S}$) is the *weight* of $\mathcal{S}$. D($\mathcal{S}$) is the *weak codimension* of $\mathcal{S}$.
Case 1 : $\varphi(\alpha_{m,*}) = \varphi(\beta_{m,*})$ or $n = 1$

$$\text{INV}(\mathcal{S}) = ((W(\mathcal{S}), \alpha_{m,j}, \beta_{m,j}))_{1 \leq j \leq d}, \, W(\mathcal{S}) = p_m - 1, \, D(\mathcal{S}) = 0.$$

Case 2 : $\varphi(\alpha_{m,*}) \neq \varphi(\beta_{m,*}), n \geq 2, p_{m+1} - p_m \geq 2 \cdot \nu(\Delta(\varphi(\alpha_{m,*}), \varphi(\beta_{m,*}))) + 1$
Let $u = \min \Delta(\varphi(\alpha_{m,*}), \varphi(\beta_{m,*}))$. Suppose $u \in \Delta(\varphi(\alpha_{m,j_0}), \varphi(\beta_{m,j_0}))$.
Subcase 1 : $\alpha_{m,j_0} \odot u = \varepsilon, \beta_{m,j_0} \odot u = \emptyset$.

Let us consider the equation $(p_m, S_{j_0}, \sum_{j=1}^d (\beta_{m,j} \odot u) S_j)$ and define a new system of weighted equations $\mathcal{S}' = (\mathcal{E}'_i)_{m+1 \leq i \leq m+n-1}$ by :

$$(\mathcal{E}'_i) : p_i, \sum_{j \neq j_0}(\alpha_{i,j} + \alpha_{i,j_0}(\beta_{m,j} \odot u))S_j \ , \ \sum_{j \neq j_0}(\beta_{i,j} + \beta_{i,j_0}(\beta_{m,j} \odot u))S_j$$

where the above equation is seen as as an equation between two linear combinations of the S_i's where the j_0-th coefficient is $\emptyset$ on both sides. We then define :

$$\text{INV}(\mathcal{S}) = \text{INV}(\mathcal{S}'), \text{W}(\mathcal{S}) = \text{W}(\mathcal{S}'), \text{D}(\mathcal{S}) = \text{D}(\mathcal{S}') + 1. \tag{10}$$

Subcase 2 : $\alpha_{m,j_0} \odot u = \varepsilon, \beta_{m,j_0} \odot u \neq \emptyset$.
Let us consider the w-equation $(p_m, S_{j_0}, \sum_{j=1}^d (\beta_{m,j_0} \odot u)^*(\beta_{m,j} \odot u) S_j)$ and define a new system of weighted equations $\mathcal{S}' = (\mathcal{E}'_i)_{m+1 \leq i \leq m+n-1}$ by :

$$(\mathcal{E}'_i) : p_i, \sum_{j \neq j_0}[(\alpha_{i,j} + \alpha_{i,j_0}(\beta_{m,j_0} \odot u)^*(\beta_{m,j} \odot u)]S_j \ , \ \sum_{j \neq j_0}[(\beta_{i,j} + \beta_{i,j_0}(\beta_{m,j_0} \odot u)^*(\beta_{m,j} \odot u)]S_j.$$

We then set the same definitions (10) as above.
Subcase 3 : $\alpha_{m,j_0} \odot u = \emptyset, \beta_{m,j_0} \odot u = \varepsilon$. (Analogous to subcase 1).
Subcase 4 : $\alpha_{m,j_0} \odot u \neq \emptyset, \beta_{m,j_0} \odot u = \varepsilon$. (Analogous to subcase 2).
Case 3 : $\varphi(\alpha_{m,*}) \neq \varphi(\beta_{m,*}), n \geq 2, p_{m+1} - p_m \leq 2 \cdot \nu(\Delta(\varphi(\alpha_{m,*}), (\varphi(\beta_{m,*})))$.
We then define: $\text{INV}(\mathcal{S}) = \bot, \text{W}(\mathcal{S}) = \bot, \text{D}(\mathcal{S}) = 0$, where $\bot$ is a special symbol which can be understood as meaning "undefined".

Lemma 5.1 : *Let $\mathcal{S}$ be a system of linear equations. If $\text{INV}(\mathcal{S}) \neq \bot$ then $\text{INV}(\mathcal{S}) = (\bar{\mathcal{E}}_j)_{1 \leq j \leq d}$ fulfills:*

1. $\forall j \in [1,d], \bar{\mathcal{E}}_j$ is a linear equation with deterministic coefficients,
2. $\{\bar{\mathcal{E}}_j \mid 1 \leq j \leq d\} \cup \{\mathcal{E}_i \mid m \leq i \leq m + \text{D}(\mathcal{S}) - 1\} \vdash \mathcal{E}_{m+\text{D}(\mathcal{S})}$,
   If, in addition, $n \geq d$ then :
3. $\min\{H(\mathcal{E}_i) \mid m \leq i \leq m + \text{D}(\mathcal{S})\} = \infty \Longrightarrow \min\{H(\bar{\mathcal{E}}_j) \mid 1 \leq j \leq d\} = \infty$.

Let us consider the function F defined by :

$$F(n) = max\{\nu(\varphi(A)\Delta\varphi(B)) \mid A, B \in \mathsf{DRB}_{1,d} < V >, \parallel A \parallel \leq n, \parallel B \parallel \leq n, \varphi(A) \neq \varphi(B)\}.$$

For every integer parameters $K_1, K_2, K_3, K_4 \in \mathbb{N} - \{0\}$, we define integer sequences $(\delta_i, \ell_i, L_i, s_i, S_i, \Sigma_i)_{m \leq i \leq m+n-1}$ by :

$$\delta_m = 0, \ell_m = 0, L_m = K_2, s_m = K_3 \cdot K_2 + K_4, S_m = 0, \Sigma_m = 0, \tag{11}$$

and for every $m \leq i \leq m + n - 2$,

$$\delta_{i+1} = 2 \cdot F(s_i + \Sigma_i) + 1, \qquad \ell_{i+1} = 5 \cdot \delta_{i+1} + 14, \qquad L_{i+1} = K_1 \cdot (L_i + \ell_{i+1}) + K_2,$$
$$s_{i+1} = K_3 \cdot L_{i+1} + K_4, \quad S_{i+1} = s_i + \Sigma_i + \mid Q \mid F(s_i + \Sigma_i), \quad \Sigma_{i+1} = \Sigma_i + S_{i+1}. \tag{12}$$

For every weighted, deterministic rational linear equation $\mathcal{E} = (p, \sum_{j=1}^d \alpha_j S_j \ . \ \sum_{j=1}^d \beta_j S_j)$, we define

$$\parallel\mid \mathcal{E} \mid\parallel = max\{\parallel \alpha \parallel, \parallel \beta \parallel\}.$$

Lemma 5.2 *Let $\mathcal{S} = (\mathcal{E}_i)_{m \leq i \leq m+d-1}$ be a system of d weighted linear equations such that :*

(1) $\forall i \in [m, m+d-1], \parallel\mid \mathcal{E}_i \mid\parallel \leq s_i$
(2) $\forall i \in [m, m+d-2], \text{W}(\mathcal{E}_{i+1}) - \text{W}(\mathcal{E}_i) \geq \delta_{i+1}$.

Then $\text{INV}(\mathcal{S}) \neq \bot, \text{D}(\mathcal{S}) \leq d - 1, \forall \mathcal{E} \in \text{INV}(\mathcal{S}), \parallel\mid \mathcal{E} \mid\parallel \leq \Sigma_{m+\text{D}(\mathcal{S})} + s_{m+\text{D}(\mathcal{S})}$.

6 Constants

The following constants will be used in the sequel.

$k_0 = \max\{\nu([pAq]) \mid p,q \in Q, A \in Z, [pAq] \not\equiv \emptyset\}$, $k_1 = \max\{2k_0+1, 3\}$, $k_2 = 4k_1 + 2(k_1)^2 + k_0$,
$D_1 = 4k_0 + 2$, $K_1 = k_1 + 1$, $K_2 = 2(k_1)^3 + 3(k_1)^2 + k_1 + 1$,
$K_3 = k_0|Q|$, $K_4 = k_0|Q|^2 + (k_2+6)|Q|$,
$d_0 = 2 \cdot |Q| \cdot \text{Card}(X^{\leq k_1})$.

We consider now the integer sequences $(\delta_i, \ell_i, L_i, s_i, S_i, \Sigma_i)_{m \leq i \leq m+n-1}$ defined by the relations (11,12) of section 5 where the parameters $K_1, \ldots, K_4$ are chosen to be the above constants, the functions F is associated with $d = d_0$ and $m = 1, n = d_0$.

$$D_2 = \Sigma_{d_0} + s_{d_0}.$$

7 Strategies for $\mathcal{D}_o$

Let us define strategies for the particular system $\mathcal{D}_0$.
We define first auxiliary strategies $T_{cut}, T_\emptyset, T_\varepsilon, T_A, T_B, T_C$ and then derive some closed strategies from them. Let us fix here some total ordering on $X : x_1 < x_2 < \cdots < x_\alpha$ and also some total ordering $\leq$ of type ω on $\mathcal{A}$ (inherited from the usual well-ordering of $\mathbb{N}$ by the fixed encoding). From these orderings one can construct in the usual way an ordering of type ω on the sets $X^*, \mathcal{A}^*$ and $\mathbb{N}^* \times (\text{DRB} < V >)^*$.

Let us adapt the usual notion of *stacking derivation* to derivations of series. For every $u \in X^*$ we define the binary relation $\uparrow (u)$ over $\text{DB} < V >$ by: for every $S, S' \in \text{DB} < V >, S \uparrow (u) S' \Leftrightarrow \exists A \in Z, \omega \in Z^+, p, q \in Q, \Psi \in \text{DB}_{Q,1} < V >$ such that

$$S = [pA] * \Psi, [pA] \odot u = [q\omega], S' = [q\omega] * \Psi.$$

A sequence of deterministic series $S_0, S_1, \ldots, S_n$ is a *derivation* iff there exist $x_1, \ldots, x_n \in X$ such that $S_0 \odot x_1 = S_1, \ldots, S_{n-1} \odot x_n = S_n$. If $u = x_1 \cdot x_2 \cdot \ldots \cdot x_n$ we call $S_0, S_1, \ldots, S_n$ the derivation *associated* with (S, u). A derivation $S_0, S_1, \ldots, S_n$ is said to be *stacking* iff it is the derivation associated to a pair (S, u) such that $S = S_0$ and $S_0 \uparrow (u) S_n$.

T_{cut}: $T_{cut}(A_1 \cdots A_n) = B_1 \cdots B_m$ iff $\exists i \in [1, n-1], \exists S, T$,

$$A_i = (p_i, S, T), A_n = (p_n, S, T), p_i < p_n \text{ and } m = 0$$

$T_\emptyset$: $T_\emptyset(A_1 A_2 \cdots A_n) = B_1 \cdots B_m$ iff $\exists S, T, A_n = (p, S, T), p \geq 0, S \equiv T \equiv \emptyset$ and $m = 0$
T_ε: $T_\varepsilon(A_1 \cdots A_n) = B_1 \cdots B_m$ iff $A_n = (p, S, T), p \geq 0, S \equiv T \equiv \varepsilon$ and $m = 0$
T_A: $T_A(A_1 \cdots A_n) = B_1 \cdots B_m$ iff

$$A_n = (p, S, T), m = |X|, B_1 = (p+1, S \odot x_1, T \odot x_1), \cdots, B_m = (p+1, S \odot x_m, T \odot x_m),$$

where $S \not\equiv \varepsilon, T \not\equiv \varepsilon$
T_B^+: $T_B^+(A_1 \cdots A_n) = B_1 \cdots B_m$ iff $n \geq k_1, A_{n-k_1} = (\pi, \overline{U}, U')$, (where $\overline{U}$ is unmarked)

$$U' = \sum_{q \in Q} [\overline{p}Aq] \cdot V_q \quad (\text{ for some } (\overline{p} \in Q))$$

$A_i = (\pi + k_1 + i - n, U_i, U'_i)$ for $n - k_1 \leq i \leq n$, $(U'i)_{n-k_1 \leq i \leq n}$ is a "stacking derivation" (see the above definition).

$$U'_n = \sum_{q \in Q} [p\tau q] \cdot V_q, \quad \text{for some } p \in Q, \tau \in Z^+,$$

$m = 1, B_1 = (\pi + k_1 - 1, V, V'), V = U_n, V' = \sum_{q \in Q'} [p\tau q] \cdot [qeq] \cdot (\overline{U} \odot u_q)$,
where $Q' = \{q \in Q \mid [\overline{p}Aq] \not\equiv \emptyset\}, \forall q \in Q', u_q = \min(\varphi([\overline{p}Aq]))$.
T_B^-: T_B^- is defined in the same way as T_B^+ by exchanging the left series (S^-) and right (S^+) series in every assertion (p, S^-, S^+).

T_C: $T_C(A_1 \cdots A_n) = B_1 \cdots B_m$ iff there exists $d \in [1, d_0], S_1, S_2, \cdots, S_d \in \mathsf{DRB} < V >, 1 \leq \kappa_1 \leq \kappa_2 < \cdots < \kappa_d = n$, such that,
  (C1) every equation $\mathcal{E}_i = A_{\kappa_i}$ is a weighted equation over $S_1, S_2, \cdots, S_d$,
  (C2) $\mathcal{S} = (\mathcal{E}_i)_{1 \leq i \leq d}$ fulfills the hypothesis of lemma 5.2,
  (C3) $(\kappa_1, \kappa_2, \cdots, \kappa_d, S_1, \cdots, S_d) \in \mathbb{N}^* \times (\mathsf{DRB} < V >)^*$ is the minimal vector satisfying conditions (C1,C2) for the given sequence $(A_1 \cdots A_n)$ and
  (C4) $B_1 \cdots B_m = \rho_e(INV(\mathcal{S}))$ (where ρ_e is the obvious extension of ρ_e to pairs of series and then to sequences of weighted equations; in other words the result of T_C is $INV(\mathcal{S})$ where the marks have been removed).

Let us notice that, by lemma 5.2 and lemma 3.9, for every $j \in [1, m], ||| B_j ||| \leq \Sigma_{m+D(\mathcal{S})} + s_{m+D(\mathcal{S})} \leq \Sigma_{m+d_0} + s_{m+d_0} = D_2$. This inequality is *independant* of the sizes of the series appearing as lefthand sides (or rhs) of the initial equations $A_1 \cdots A_n$.

Lemma 7.1 : $T_{cut}, T_\emptyset, T_\varepsilon, T_A, T_B, T_C$ are $\mathcal{D}_0$ *strategies*.

Let us define the strategy $\mathcal{S}_{AB}$ by : for every $W = A_1 A_2 \cdots A_n$,

(0) if $W \in \text{dom}(T_{cut})$, then $\mathcal{S}_{AB}(W) = T_{cut}(W)$ (1) elsif $W \in \text{dom}(T_\emptyset)$, then $\mathcal{S}_{AB}(W) = T_\emptyset(W)$
(2) elsif $W \in \text{dom}(T_\varepsilon)$, then $\mathcal{S}_{AB}(W) = T_\varepsilon(W)$ (4) elsif $W \in \text{dom}(T_B^+)$, then $\mathcal{S}_{AB}(W) = T_B^+(W)$
(5) elsif $W \in \text{dom}(T_B^-)$, then $\mathcal{S}_{AB}(W) = T_B^-(W)$ (6) elsif $W \in \text{dom}(T_A)$, then $\mathcal{S}_{AB}(W) = T_A(W)$
(7) else $\mathcal{S}_{AB}(W)$ is undefined.

The strategy $\mathcal{S}_{ABC}$ is obtained by inserting "(3) elsif $W \in \text{dom}(T_C)$, then $\mathcal{S}_{ABC}(W) = T_C(W)$" in the above list of cases.

Lemma 7.2 $\mathcal{S}_{ABC}, \mathcal{S}_{AB}$ *are closed*.

8 Tree analysis

This section is devoted to the analysis of the proof-trees τ produced by the strategy $\mathcal{S}_{AB}$ defined in section 7. The main results are [Sén97, lemma 8.14 , 8.15] whose combination asserts that if some path (from a node x to a node y) of τ is such that its origin has a "small norm" and its length is "large enough", then the transformation T_C is defined at some ancestor of y. [4]

9 Completeness of $\mathcal{D}_o$

Lemma 9.1 : $\mathcal{S}_{ABC}$ *is terminating*.

The proof leans on the two delicate lemmas [Sén97, lemma 8.14 , 8.15] mentioned above.

Theorem 9.2 *The system $\mathcal{D}_0$ is complete.*

Proof: By lemma 7.1 $\mathcal{S}_{ABC}$ is a strategy for $\mathcal{D}_0$, by lemma7.2 $\mathcal{S}_{ABC}$ is closed , by lemma 9.1 it is terminating and by lemma 4.6, $\mathcal{D}_0$ is complete. □

Theorem 9.3 *The equivalence problem for deterministic pushdown automata is decidable.*

Proof: Let $\mathcal{M}$ be some dpda. The equivalence relation $\equiv$ on $\mathsf{DRB} < V >$ (where V is the structured alphabet associated to the given $\mathcal{M}$) has a recursively enumerable complement (this is well-known). By theorem 9.2 and lemma 4.2 $\equiv$ is recursively enumerable too. Hence $\equiv$ is recursive. In addition, the system $\mathcal{D}_0$ associated with $\mathcal{M}$ is computable from $\mathcal{M}$, hence the theorem follows. □

---

[4] Technically speaking, this is the most difficult part of the full proof; we cannot sketch it here due to the lack of space.

Acknowledgements I thank: L. Boasson and J.M. Autebert for supervising my first works on Eq(D.D), B. Courcelle for initiating me to the classical equivalence algorithms, to his notion of decision systems and for numerous discussions along the years, M. Oyamaguchi for discussions, M.S. Paterson for his hospitality and interest, J.E.Pin and W. Thomas for pointing to me Meitus' works, D. Caucal and C. Stirling for useful informations, H. Comon and J.P. Jouannaud for stimulating me to read Meitus' work in details, J. Karhumaki for his encouragements , M. Nivat for his support, J. Engelfriet, L.P. Lisovik, Y. Matiyasevich for discussions. I am also indebted to the CNRS (and, obviously, to my collegues in the CNRS comitee) who allowed me to have my *full time for research* during the academic year 1996/1997.

References

[Bee76] C. Beeri. An improvement on Valiant's decision procedure for equivalence of deterministic finite-turn pushdown automata. *TCS 3*, pages 305–320, 1976.

[Cau95] D. Caucal. Bisimulation of context-free grammars and of pushdown automata. *To appear in CSLI, Modal Logic and process algebra, vol. 53, Stanford*, pages 1–20, 1995.

[Cou83] B. Courcelle. An axiomatic approach to the Korenjac-Hopcroft algorithms. *Math. Systems theory*, pages 191–231, 1983.

[Cou90] B. Courcelle. Recursive applicative program schemes. In *Handbook of THeoretical Computer Science, edited by J. Van Leeuwen*, pages 461–490. Elsevier, 1990.

[GG66] S. Ginsburg and S. Greibach. Deterministic context-free languages. *Information and Control*, pages 620–648, 1966.

[Har78] M.A. Harrison. *Introduction to Formal Language Theory*. Addison-Wesley, Reading, Mass., 1978.

[HHY79] M.A. Harrison, I.M. Havel, and A. Yehudai. On equivalence of grammars through transformation trees. *TCS 9*, pages 173–205, 1979.

[Lis96] L.P. Lisovik. Hard sets methods and semilinear reservoir method with applications. In *Proceedings 23rd ICALP*, pages 229–231. Springer, LNCS 1099, 1996.

[Mei89] Y.V. Meitus. The equivalence problem for real-time strict deterministic pushdown automata. *Kibernetika 5 (in russian, english translation in Cybernetics and Systems analysis)*, pages 14–25, 1989.

[Mei92] Y.V. Meitus. Decidability of the equivalence problem for deterministic pushdown automata. *Kibernetika 5 (in russian, english translation in Cybernetics and Systems analysis)*, pages 20–45, 1992.

[Oya87] M. Oyamaguchi. The equivalence problem for real-time d.p.d.a's. *J. assoc. Comput. Mach. 34*, pages 731–760, 1987.

[Rom85] V.Yu. Romanovskii. Equivalence problem for real-time deterministic pushdown automata. *Kibernetika no 2*, pages 13–23, 1985.

[Sén94] G. Sénizergues. Formal languages and word-rewriting. In *Term Rewriting, Advanced Course*, pages 75–94. Springer, LNCS 909, edited by H. Comon and J.P.Jouannaud, 1994.

[Sén97] G. Sénizergues. L(A) = L(B)? Technical report, LaBRI, Université Bordeaux I, report nr1161-97. can be accessed at URL, http://www.labri.u-bordeaux.fr/, 1997.

[Sti96] C. Stirling. Decidability of bisimulation equivalence for normed pushdown processes. In *Proceedings CONCUR 96*, 1996.

[Val74] L.G. Valiant. The equivalence problem for deterministic finite-turn pushdown automata. *Information and Control 25*, pages 123–133, 1974.

[VP75] L.G. Valiant and M.S. Paterson. Deterministic one-counter automata. *Journal of Computer and System Sciences 10*. pages 340–350, 1975.

On Recognizable and Rational Formal Power Series in Partially Commuting Variables[*]

Manfred Droste[1] and Paul Gastin[2]

[1] Institut für Algebra, Technische Universität Dresden, D-01062 Dresden,
droste@math.tu-dresden.de
[2] LITP, Université Paris 7, 2 place Jussieu, F-75251 Paris Cedex 05,
Paul.Gastin@litp.ibp.fr

Abstract. We will describe the recognizable formal power series over arbitrary semirings and in partially commuting variables, i.e. over trace monoids. We prove that the recognizable series are certain rational power series, which can be constructed from the polynomials by using the operations sum, product and a restricted star which is applied only to series for which the elements in the support all have the same connected alphabet. The converse is true if the underlying semi-ring is commutative. Moreover, if in addition the semiring is idempotent then the same result holds with a star restricted to series for which the elements in the support have connected (possibly different) alphabets. It is shown that these assumptions over the semiring are necessary. This provides a joint generalization of Kleene's, Schützenberger's and Ochmański's theorems.

1 Introduction

In the theory of automata and formal languages, Kleene's foundational theorem on the coincidence of regular and rational languages in free monoids has been extended in many ways. Schützenberger [15] investigated formal power series over arbitrary semirings (e.g., like the natural numbers) and the free monoid, i.e. in noncommuting variables, and showed that the recognizable formal power series coincide with the rational ones. This was the starting point for a large amount of work on formal power series, cf. [14,9,2,8] for surveys. The concept of recognizable formal power series has also been defined for arbitrary monoids instead of the free monoid, but it was clear and has been stressed by several authors (cf., e.g. [14]) that in general then the recognizable and the rational series do not coincide.

On the other hand, Mazurkiewicz [10,11] introduced an important mathematical model for the behaviour of concurrent systems: trace monoids (or free partially commutative monoids), see also [3,1,4–6] for their well-developed theory. They are monoids whose generators are partially commutative. Again, their recognizable languages do not coincide with the rational ones, but by Ochmański's

---

[*] This research was partly carried out during a stay of the first author in Paris and another stay of the second author in Dresden.

theorem [12] they coincide with the c-rational languages where the iteration is restricted to connected languages.

It is the aim of this paper to investigate recognizable formal power series over trace monoids, thereby obtaining a generalization of both Schützenberger's and Ochmański's results.

We denote by $K\langle\!\langle \mathbb{M} \rangle\!\rangle$ the set of all formal power series over the semiring K and the free partially commutative monoid $\mathbb{M}$. It is known that in general the recognizable series in $K\langle\!\langle \mathbb{M} \rangle\!\rangle$ form a proper subclass of the rational ones. We therefore define the subclasses of *c-rational* and *mc-rational* series. We say that a series S is *connected*, if each element of its support is connected, and S is *mono-alphabetic*, if all elements of its support have the same set of generators. The c-rational series are obtained from the polynomials by allowing the operations sum, product, and star, but the latter applied only to proper and connected series. The mc-rational series are constructed in the same way, but using star only for series which are proper, mono-alphabetic and connected. In view of Ochmański's result, one might expect that the recognizable series in $K\langle\!\langle \mathbb{M} \rangle\!\rangle$ coincide with the c-rational ones. However, we will show that this fails in general even for the semiring $(\mathbb{N}, +, \times)$. Our main result is the following:

Theorem 1. *Let $\mathbb{M}$ be a trace monoid and K a semiring.*
(a) Each recognizable series in $K\langle\!\langle \mathbb{M} \rangle\!\rangle$ is mc-rational.
(b) If K is commutative, each mc-rational series in $K\langle\!\langle \mathbb{M} \rangle\!\rangle$ is recognizable.
(c) If K is commutative and idempotent, each c-rational series in $K\langle\!\langle \mathbb{M} \rangle\!\rangle$ is recognizable.

The fact that the recognizable series in $K\langle\!\langle \mathbb{M} \rangle\!\rangle$ are closed under the product operation was proved before already by Fliess [7], but only for very specific semirings K (strong Fatou semirings or the Boolean semiring). By Theorem 1(b), this holds for arbitrary commutative semirings, and we show by example that the commutativity of K is needed for this.

Theorem 1(b,c) is proved in section 3. There we also show that if the star S^* of a recognizable proper series S is connected, then it is also recognizable. This gives another closure property of the recognizable series under the star-operation. Part (a) of Theorem 1 is proved in section 4, and in section 5 we give examples and discuss the relationship with Schützenberger's and Ochmański's results. For lack of space, most proofs are not contained in this extended abstract.

It seems a very interesting research road to investigate which other results from the theory of formal power series over non-commuting variables can be extended to series over partially commuting variables, i.e. over trace monoids.

2 Background

Here we recall the necessary notation and background for formal power series and of trace theory. For more details, we refer the reader to [14,2,4,6].

Let M be any monoid and $K = (K, +, \cdot, 0, 1)$ any semiring, i.e., $(K, +, 0)$ is a commutative monoid, $(K, \cdot, 1)$ is a monoid, multiplication distributes over addition, and $0 \cdot x = x \cdot 0 = 0$ for each $x \in K$. If multiplication is commutative, we say that K is *commutative*. If the addition is idempotent, then the semiring is called *idempotent*. For instance, the semiring $(\mathbb{R} \cup \{\infty\}, min, +, \infty, 0)$ is both commutative and idempotent.

Mappings S from M into K are called *formal power series*. They are denoted as formal sums $S = \sum_{m \in M}(S,m).m$ where $(S,m) = S(m) \in K$. The set $supp(S) = \{m \in M \mid (S,m) \neq 0\}$ is called the *support* of S, and if it is finite, then S is called a *polynomial*. The collection of all formal power series is denoted by $K\langle\langle M \rangle\rangle$, and its subset of all polynomials by $K\langle M \rangle$. We consider elements of K also as polynomials in the natural way, having a non-zero entry only at $1 \in M$. If $L \subseteq M$, we define the *characteristic series* of L by $1_L = \sum_{m \in L} 1 \cdot m$.

Let $n \geq 1$ and $[n] = \{1, \ldots, n\}$. We let $K^{n \times n}$ be the monoid of all $(n \times n)$-matrices over K (with matrix multiplication as usual). A series $S \in K\langle\langle M \rangle\rangle$ is called *recognizable*, if there exists an integer $n \geq 1$, a monoid morphism $\mu : M \longrightarrow K^{n \times n}$ and vectors $\lambda \in K^{1 \times n}, \gamma \in K^{n \times 1}$ such that

$$(S,m) = \lambda \cdot (\mu m) \cdot \gamma = \sum_{i,j \in [n]} \lambda_i (\mu m)_{ij} \gamma_j$$

for each $m \in M$. In this case, the triple (λ, μ, γ) is called a *representation* of S, and we often shortly write $S = (\lambda, \mu, \gamma)$ to denote this. If $i, j \in [n]$, we also abbreviate $(\mu m)_{ij} =: \mu m_{ij}$. We let $K^{rec}\langle\langle M \rangle\rangle$ denote the set of all recognizable formal power series.

With componentwise addition, $K\langle\langle M \rangle\rangle$ becomes a commutative monoid. Now, the *(Cauchy) product* of two series S, S' in $K\langle\langle M \rangle\rangle$ is the series defined for $m \in M$ by $(S \cdot S', m) = \sum_{m = m_1 \cdot m_2} (S, m_1) \cdot (S, m_2)$ provided the sum is defined (e.g. when the sum is finite). With this, $K\langle\langle M \rangle\rangle$ is a semiring. The powers $S^n (n \geq 0)$ are defined in the natural way. We call S *proper*, if $(S, 1) = 0$, and then we put, in the natural way, $S^* = \sum_{n \geq 0} S^n$, the *star* (or iteration) of S, and $S^+ = \sum_{n \geq 1} S^n$, provided it is defined. We let $K^{rat}\langle\langle M \rangle\rangle$ denote the smallest subset of $K\langle\langle M \rangle\rangle$ which contains all polynomials and is closed under the operations sum, product and star, where the latter is only applied to proper series. Its elements are called *rational* formal power series. Now Schützenberger's theorem states the following equivalence between recognizable and rational series over the free monoid.

Theorem 2 (Schützenberger, [15]). *Let Σ be any finite set and K any semiring. Then*

$$K^{rec}\langle\langle \Sigma^* \rangle\rangle = K^{rat}\langle\langle \Sigma^* \rangle\rangle.$$

From this, Kleene's theorem on the coincidence of regular and rational languages follows by considering the Boolean semiring $\mathbb{B} = \{0, 1\}$ (with $1 + 1 = 1 \cdot 1 = 1$) and noting that a language $L \subseteq \Sigma^*$ is regular iff its characteristic series $1_L \in \mathbb{B}\langle\langle \Sigma^* \rangle\rangle$ is recognizable, and similarly for rationality.

Later we will also need the *Hadamard product* $S \odot T$ of two series $S, T \in K\langle\langle M \rangle\rangle$. It is defined by $(S \odot T, m) = (S, m) \cdot (T, m)$ for all $m \in M$.

Next we recall basic notions from trace theory. A pair (Σ, I) is called a *trace alphabet*, if Σ is a finite set and I is an irreflexive symmetric binary *independence* relation on Σ. Let $\sim$ denote the smallest congruence on Σ^* containing $\{(ab, ba) : a\ I\ b\}$. The quotient monoid $\mathbb{M} = \mathbb{M}(\Sigma, I) := \Sigma^*/\sim$ is called the *trace monoid* (or free partially commutative monoid) over (Σ, I). If $w \in \Sigma^*$, we let $[w]$ denote the equivalence class of w in $\mathbb{M}$. Also, let $\alpha(w)$ be the set of all letters of Σ occurring in w, called the *alphabet* of w. Since equivalent words have the same alphabet, we may put $\alpha([w]) = \alpha(w)$. If $A, B \subseteq \Sigma$, we write $A\ I\ B$ to denote that $a\ I\ b$ for all $a \in A, b \in B$. We also write $w\ I\ A$ or $[w]\ I\ A$ to abbreviate that $\alpha(w)\ I\ A$, similarly, $w\ I\ w'$ for $\alpha(w)\ I\ \alpha(w')$, etc. A subset $\Delta \subseteq \Sigma$ is called *connected*, if it cannot be split $\Delta = A \cup B$ into two non-empty subsets such that $A\ I\ B$. Again, w and $[w]$ are connected, if $\alpha(w)$ is connected. A language $L \subseteq \mathbb{M}$ or $L \subseteq \Sigma^*$ is called *connected*, if each of its elements is connected, and *mono-alphabetic*, if $\alpha(m) = \alpha(m')$ for all $m, m' \in L$. Then the collection of all *c-rational* languages in $\mathbb{M}$ (respectively, in Σ^*) is defined as the smallest set of languages of $\mathbb{M}$ (respectively, of Σ^*) containing all finite languages and which is closed under the operations union, product and star, where the latter is applied only to connected languages. The following characterizes the recognizable languages of $\mathbb{M}$ (recall that a language $L \subseteq \mathbb{M}$ is recognizable iff it is accepted by some finite $\mathbb{M}$-automaton, or, equivalently, iff its syntactic monoid is finite).

Theorem 3 (Ochmański, [12,4,6]). *Let (Σ, I) be any trace alphabet and $\mathbb{M}$ its trace monoid. Then a language $L \subseteq \mathbb{M}$ is recognizable iff it is c-rational.*

Again, one should note that the Kleene's theorem mentioned above is a special case of Theorem 3 since when the independence relation is empty, the trace monoid $\mathbb{M}(\Sigma, \emptyset)$ is the free monoid Σ^* and in this case all languages are connected, hence rational sets are also c-rational.

The goal of this paper is a common generalization of Theorems 2 and 3, that is, a characterization of the recognizable formal power series in $K\langle\langle\mathbb{M}\rangle\rangle$ where K is a semiring and $\mathbb{M}$ a trace monoid. Let $S \in K\langle\langle\mathbb{M}\rangle\rangle$. We say that S is *connected*, if $supp(S)$ is a connected language in $\mathbb{M}$, and *mono-alphabetic*, if $supp(S)$ is mono-alphabetic. In the latter case, we put $\alpha(S) = \alpha(m)$ if $S \neq 0$ and $m \in supp(S)$. Now let $K^{mc-rat}\langle\langle\mathbb{M}\rangle\rangle$ (mono-alphabetic-connected rational) be the smallest subset of $K\langle\langle\mathbb{M}\rangle\rangle$ which contains all polynomials and is closed under the operations sum, product and star, where the latter gets applied only to proper, mono-alphabetic and connected series. Similarly, we let $K^{c-rat}\langle\langle\mathbb{M}\rangle\rangle$ (connected rational) be the collection of series obtained from the polynomials by allowing the operations sum, product and star, where now star is applied to all proper and connected series. Similarly, we define connected series in $K\langle\langle\Sigma^*\rangle\rangle$ and the collection of *mc-rational* series in $K\langle\langle\Sigma^*\rangle\rangle$.

3 Mc-rational series are recognizable

In this section, let (Σ, I) be a trace alphabet and $\mathbb{M} = \mathbb{M}(\Sigma, I)$ its trace monoid. We will prove Theorem 1(b,c). This will require a more particular notion of representations which we introduce first.

Definition 4. Let $S = (\lambda, \mu, \gamma) \in K\langle\langle\mathbb{M}\rangle\rangle$ be a recognizable series with $\mu : \mathbb{M} \longrightarrow K^{n \times n}$. The representation (λ, μ, γ) is alphabetic, if there exist two functions $\overleftarrow{\alpha}, \overrightarrow{\alpha} : [n] \longrightarrow \mathcal{P}(\Sigma)$ such that for all $u \in \mathbb{M}$, the following three conditions are satisfied:
(1) Whenever $\mu u_{ij} \neq 0$, then $\overleftarrow{\alpha}(j) = \overleftarrow{\alpha}(i) \cup \alpha(u)$ and $\overrightarrow{\alpha}(i) = \overrightarrow{\alpha}(j) \cup \alpha(u)$;
(2) whenever $\lambda_i \neq 0$, then $\overleftarrow{\alpha}(i) = \emptyset$;
(3) whenever $\gamma_j \neq 0$, then $\overrightarrow{\alpha}(j) = \emptyset$.
We call $(\lambda, \mu, \gamma; \overleftarrow{\alpha}, \overrightarrow{\alpha})$ an alphabetic representation of S. Here, $\overleftarrow{\alpha}(k)$ describes the *past alphabet* of k and $\overrightarrow{\alpha}(k)$ the *future alphabet* of k. We say that k is *initial*, if $\overleftarrow{\alpha}(k) = \emptyset$, and k is *final*, if $\overrightarrow{\alpha}(k) = \emptyset$.

We will often use the fact that if (λ, μ, γ) is alphabetic and $\mu u_{ij} \neq 0$, then i initial implies that $\overleftarrow{\alpha}(j) = \alpha(u)$, and j final implies $\overrightarrow{\alpha}(i) = \alpha(u)$. Moreover, if $u \neq 1$, then i initial implies $\mu u_{ki} = 0$, and j final implies $\mu u_{jk} = 0$, for any k.

Proposition 5. *Let $S \in K\langle\langle\mathbb{M}\rangle\rangle$ be a recognizable series. Then there exists an alphabetic representation of S.*

First we want to show that the product of two recognizable series in $K\langle\langle\mathbb{M}\rangle\rangle$ is again recognizable. For more particular semirings K (strong Fatou semirings or the Boolean semiring), the result has been obtained already by Fliess [7, Prop. 2.2.14 and 2.2.15]. Our proof will not use the full notion of alphabetic representation, since it can be based either on the past alphabets (the function $\overleftarrow{\alpha}$) or the future alphabets, only. The full notion of alphabetic representation will come into use when we deal with iteration.

Theorem 6. *Let K be a commutative semiring and let $S_1, S_2 \in K\langle\langle\mathbb{M}\rangle\rangle$ be two recognizable series. Then their product $S = S_1 \cdot S_2$ is also recognizable.*

Proof. Let $(\lambda^1, \mu_1, \gamma^1)$ be a representation of S_1 and let $(\lambda^2, \mu_2, \gamma^2; \overleftarrow{\alpha}, \overrightarrow{\alpha})$ be an alphabetic representation of S_2 (Proposition 5). We assume that $\mu_i : \mathbb{M} \longrightarrow K^{n_i \times n_i}$ for $i = 1, 2$, and let $n = n_1 \cdot n_2$. Subsequently we identify $[n]$ with $[n_1] \times [n_2]$. Next, we define $\mu : \Sigma^* \longrightarrow K^{n \times n}$ by

$$\mu(a)_{(i_1,i_2)(j_1,j_2)} = \delta_{i_2,j_2} I(a,i_2)\mu_1(a)_{i_1,j_1} + \delta_{i_1,j_1}\mu_2(a)_{i_2,j_2}$$

where

$$\delta_{i,j} = \begin{cases} 1 & \text{if } i = j \\ 0 & \text{otherwise} \end{cases} \quad \text{and} \quad I(u,i) = \begin{cases} 1 & \text{if } u\ I\ \overleftarrow{\alpha}(i) \\ 0 & \text{otherwise} \end{cases}$$

Note that $I(a, j_2)\mu_2(a)_{i_2,j_2} = 0$, hence at most one of the two terms is non-zero.

One can prove that $\mu(a) \cdot \mu(b) = \mu(b) \cdot \mu(a)$ for all $(a, b) \in I$. Hence, μ factorizes to a morphism $\mu : \mathbb{M} \longrightarrow K^{n \times n}$. Next we claim that this factorization is given by the explicit formula

$$\mu(w)_{(i_1,i_2)(j_1,j_2)} = \sum_{w=uv} I(u,i_2)\mu_1(u)_{i_1,j_1}\mu_2(v)_{i_2,j_2}$$

Finally, define $\lambda \in K^{1\times n}, \gamma \in K^{n\times 1}$ by $\lambda_{(i_1,i_2)} = \lambda_{i_1}^1 \lambda_{i_2}^2$, $\gamma_{(k_1,k_2)} = \gamma_{k_1}^1 \gamma_{k_2}^2$. We can verify that $S = (\lambda, \mu, \gamma)$ which proves the theorem.

The following result shows that a mono-alphabetic recognizable series has an alphabetic representation $(\lambda, \mu, \gamma; \overleftarrow{\alpha}, \overrightarrow{\alpha})$ with an even more specific form. For this, let $e_1 = (1, 0, \ldots, 0) \in K^{1\times n}$ and $e_n = (0, \ldots, 0, 1)^t \in K^{n\times 1}$.

Proposition 7. *Let $S \in K\langle\!\langle \mathbb{M} \rangle\!\rangle$ be recognizable, proper and mono-alphabetic with $\alpha(S) = A$. Then there exists an alphabetic representation $(e_1, \mu, e_n; \overleftarrow{\alpha}, \overrightarrow{\alpha})$ of S with $\overrightarrow{\alpha}(1) = \overleftarrow{\alpha}(n) = A$.*

We will now prove the following essential closure property of recognizable series. Note that Theorem 1(b) follows easily from Theorems 6 and 8.

Theorem 8. *Let K be a commutative semiring and let $S \in K\langle\!\langle \mathbb{M} \rangle\!\rangle$ be a proper, connected, mono-alphabetic and recognizable series. Then, S^* is recognizable.*

The proof of this theorem is based on a rather involved construction. Let $S \in K\langle\!\langle \mathbb{M} \rangle\!\rangle$ be a proper, recognizable, connected and mono-alphabetic series with $\alpha(S) = A$. Let $S = (e_1, \mu, e_n; \overleftarrow{\alpha}, \overrightarrow{\alpha})$ be an alphabetic representation with $\overrightarrow{\alpha}(1) = \overleftarrow{\alpha}(n) = A$ (Proposition 7). Let $m \geq 1$. We identify $[n^m]$ with the set $[n]^m$ of all m-tuples with entries from $[n]$. We use $\tilde{\imath}$ as abbreviation for such an m-tuple $(i_1, \ldots, i_m)$, similarly $\tilde{\jmath}, \tilde{k}$. Now we define functions $\mu^0, \ldots, \mu^m : \Sigma^* \longrightarrow K^{n^m \times n^m}$ by

$$\mu^0 a_{\tilde{\imath}\tilde{\jmath}} = \begin{cases} \mu a_{i_1 n} & \text{if } \tilde{\jmath} = (i_2, \ldots, i_m, 1) \\ 0 & \text{otherwise} \end{cases}$$

$$\mu^p a_{\tilde{\imath}\tilde{\jmath}} = \begin{cases} \mu a_{i_p j_p} & \text{if } j_l = i_l \text{ for all } l \neq p \\ 0 & \text{otherwise} \end{cases} \quad (p \geq 1)$$

Also, let

$$H_{\tilde{\imath}} = \begin{cases} 1 & \text{if } \overrightarrow{\alpha}(i_p) \cup \overleftarrow{\alpha}(i_p) = A = \alpha(S) \text{ for all } p, \; \overrightarrow{\alpha}(i_1) \neq \emptyset \text{ and} \\ & \overrightarrow{\alpha}(i_p) \; I \; \overleftarrow{\alpha}(i_q) \text{ for all } p < q \\ 0 & \text{otherwise} \end{cases}$$

Let $H \in K^{n^m \times n^m}$ be given by $H_{\tilde{\imath}\tilde{\jmath}} = H_{\tilde{\imath}} \cdot H_{\tilde{\jmath}}$, and define $\mu^* : \Sigma^* \longrightarrow K^{n^m \times n^m}$ by $\mu^* = H \odot (\mu^0 + \cdots + \mu^m)$, where $(H \odot \mu^p)(w)_{\tilde{\imath}\tilde{\jmath}} = H_{\tilde{\imath}\tilde{\jmath}} \cdot \mu^p w_{\tilde{\imath}\tilde{\jmath}}$ for any $w \in \Sigma^*$ and $\tilde{\imath}, \tilde{\jmath} \in [n]^m$.

Theorem 8 results clearly from the following two essential results.

Proposition 9. *Let K be a commutative semiring and assume that $m \geq |A|$. Then $\mu^*(ab) = \mu^*(ba)$ for all $a, b \in \Sigma$ such that $a \; I \; b$.*

Hence μ^* factorizes to a morphism from $\mathbb{M}$ to $K^{n^m \times n^m}$, and we have:

Proposition 10. *Let K be a commutative semiring and assume that $m \geq |A|$. Then $S^* = (\lambda_{\tilde{1}}, \mu^*, \gamma_{\tilde{1}})$ where $\lambda_{\tilde{1}}, \gamma_{\tilde{1}}$ are the row respectively column vectors which have a 1 only at entry $\tilde{1} = (1, \ldots, 1)$, and 0 otherwise.*

Next we wish to derive a further closure properties of $K^{rec}\langle\!\langle \mathbb{M} \rangle\!\rangle$.

Definition 11. *Let $S \in K\langle\!\langle \mathbb{M} \rangle\!\rangle$ or $S \in K\langle\!\langle \Sigma^* \rangle\!\rangle$ and $A \subseteq \Sigma$. Then the restriction of S to A is the series S_A defined by*

$$(S_A, w) = \begin{cases} (S, w) & \text{if } \alpha(w) = A \\ 0 & \text{otherwise} \end{cases}$$

First we show that the restriction preserves both recognizability and mc-rationality of series.

Proposition 12. *Let $S \in K\langle\!\langle \mathbb{M} \rangle\!\rangle$ be recognizable. Then S_A is also recognizable.*

Proposition 13. *Let $S \in K\langle\!\langle \Sigma^* \rangle\!\rangle$ or $S \in K\langle\!\langle \mathbb{M} \rangle\!\rangle$ be mc-rational. Then S_A is also mc-rational.*

The following lemma generalizes a result of Pighizzini [13] for trace languages.

Lemma 14. *Let $S \in K\langle\!\langle \mathbb{M} \rangle\!\rangle$ be proper and $A \subseteq \Sigma$ be nonempty. Then $(S^*)_A = Z^+X$ where $X = \sum_{B \subset A}(S^*)_B$ and $Z = (X \cdot S)_A$.*

Next we derive another sufficient condition which implies that the star of a recognizable series is again recognizable and, also, that the star of an mc-rational series is again mc-rational.

Theorem 15.

1. *Let K be a commutative semiring and $S \in K\langle\!\langle \mathbb{M} \rangle\!\rangle$ be proper and recognizable such that S^* is connected. Then S^* is recognizable.*
2. *Let K be any semiring and $S \in K\langle\!\langle \Sigma^* \rangle\!\rangle$ or $S \in K\langle\!\langle \mathbb{M} \rangle\!\rangle$ be proper and mc-rational such that S^* is connected. Then S^* is mc-rational.*

For positive semirings, the condition S^* connected is stronger than S connected. This latter condition is actually sufficient to obtain the closure properties stated in Theorem 15 when the semiring is commutative and idempotent. This is an easy consequence of Theorem 1(a) and of Theorem 17 for which the following lemma is crucial.

Lemma 16. *Let K be a commutative and idempotent semiring. Let $S \in K\langle\!\langle \mathbb{M} \rangle\!\rangle$ be a connected series and let $B, C \subseteq \Sigma$ be independent subsets of the alphabet. Then, $(S^*)_{B \cup C} = (S^*)_B \cdot (S^*)_C$.*

Theorem 17. *Let K be a commutative and idempotent semiring. A series in $K\langle\!\langle \mathbb{M} \rangle\!\rangle$ is mc-rational iff it is c-rational.*

Proof. One direction is clear and for the converse, it suffices to show that the star of an mc-rational connected series S is still mc-rational. We will first show by induction on the size of $A \subseteq \Sigma$ that if S is an mc-rational connected series then $(S^*)_A$ is mc-rational. The theorem follows directly since $S^* = \sum_{A \subseteq \Sigma} (S^*)_A$.

Clearly, $(S^*)_\emptyset = 1$ is mc-rational. Now, assume $A \neq \emptyset$ and let $A_1, \ldots, A_n$ be the connected components of A: $A = A_1 \cup \cdots \cup A_n$ and $A_i \, I \, A_j$ for $i \neq j$. By Lemma 16, we obtain $(S^*)_A = (S^*)_{A_1} \cdots (S^*)_{A_n}$ and we are reduced to the case A connected. Now, using Lemma 14 we obtain $(S^*)_A = Z^+ X$ where $X = \sum_{B \subsetneq A} (S^*)_B$ and $Z = (X \cdot S)_A$. Then X is mc-rational by induction hypothesis. By Proposition 13, it follows that Z is also mc-rational. Since we have assumed A connected, we deduce that $(S^*)_A = Z \cdot Z^* \cdot X$ is mc-rational.

Note that Theorem 1(c) follows from Theorem 1(b) and Theorem 17.

4 Recognizable series are mc-rational

Throughout this section, let K be an arbitrary (possibly non-commutative) semiring and (Σ, I) a trace alphabet. We will prove that all recognizable series in $K \langle\!\langle \mathbb{M} \rangle\!\rangle$ are mc-rational. This uses the concept of lexicographic normal forms of traces and LNF-representations of series which we introduce first. For this, fix any linear order $\leq$ on Σ. We extend this to the lexicographic linear order, also denoted by $\leq$, on Σ^*. We say that a word w is the lexicographic normal form of $[w]$, if it is the smallest element of $[w]$ with respect to $\leq$. Then LNF is the set of all words which are lexicographic normal forms. Note that LNF is closed under prefixes (and suffixes). Now let $\mathcal{A}_{\mathrm{LNF}} = (Q, \Sigma, \delta, q_0, Q)$ be the minimal (reduced) automaton for LNF.

Definition 18. We will call a morphism $\mu : \Sigma^* \longrightarrow K^{n \times n}$ an LNF-morphism, if there exists a function $\pi : [n] \longrightarrow Q$ such that for all $a \in \Sigma$ and all $i, j \in [n]$, $\mu a_{ij} \neq 0$ implies $\pi(i) \xrightarrow{a} \pi(j)$ in $\mathcal{A}_{\mathrm{LNF}}$. Then any representation (λ, μ, γ) with an LNF-morphism μ of a series $S \in K \langle\!\langle \Sigma^* \rangle\!\rangle$ will be called an LNF-representation of S.

Proposition 19. *Let $S' \in K \langle\!\langle \Sigma^* \rangle\!\rangle$ be recognizable. Then $S = S' \odot 1_{\mathrm{LNF}}$ has an LNF-representation.*

Next we note that for any $n \geq 1$ there is a canonical isomorphism Φ between the semiring of $n \times n$-matrices $K \langle\!\langle \Sigma^* \rangle\!\rangle^{n \times n}$ and the semiring of formal power series $K^{n \times n} \langle\!\langle \Sigma^* \rangle\!\rangle$, given by $(\Phi(A), w) = ((A_{ij}, w))$ if $A = (A_{ij}) \in K \langle\!\langle \Sigma^* \rangle\!\rangle^{n \times n}$. Subsequently, we will often identify A with its image $\Phi(A)$.

We will also use the following result.

Lemma 20 (Ochmański, [12,4]). *Let $w \in \Sigma^*$ be a word such that $w, w^2 \in$ LNF. Then w is connected.*

Proposition 21. *Let $\mu : \Sigma^* \longrightarrow K^{n \times n}$ be an LNF-morphism, and let $M = \sum_{a \in \Sigma} \mu a \cdot a \in K^{n \times n} \langle \Sigma^* \rangle$. Then the entries of M^* are mc-rational series.*

Proof. We first show, by induction on the length of w, that $(M^*, w) = \mu w$ for any word w. Indeed, clearly $(M^*, 1) = 1 = \mu 1$ and $(M^*, wa) = (1 + M^*M, wa) = (M^*M, wa) = (M^*, w)(M, a) = \mu w \cdot \mu a = \mu(wa)$.

By lack of space we only give the proof for $n = 1$, which already shows several connections between all the results. Hence, assume that $n = 1$. Then $M \in K\langle \Sigma^* \rangle$ is proper and mc-rational. Now, let $w \in \Sigma^*$. If $(M^*, w) = \mu w \neq 0$, since μ is an LNF-morphism, we have a path $\pi(1) \xrightarrow{w} \pi(1)$ in $\mathcal{A}_{\text{LNF}}$. Therefore, $w, w^2 \in \text{LNF}$ and by Ochmański's lemma 20, w is connected. Hence M^* is connected and so, by Theorem 15, mc-rational.

Theorem 22. *Let $S \in K\langle\!\langle \Sigma^* \rangle\!\rangle$ be recognizable. Then $S \odot 1_{\text{LNF}}$ is mc-rational.*

Proof. By Proposition 19 we can choose an LNF-representation (λ, μ, γ) of $S' = S \odot 1_{\text{LNF}}$. Let $M = \sum_{a \in \Sigma} \mu a \cdot a$. We have seen in the proof of Proposition 21 that $(M^*, w) = \mu w$ for any word w.

Now, λ and γ are vectors with entries in K, and M^* has only mc-rational series as entries by Proposition 21. Hence $\lambda M^* \gamma \in K\langle\!\langle \Sigma^* \rangle\!\rangle$ is an mc-rational series. Finally, observe that for any word w,

$$(\lambda M^* \gamma, w) = (\sum_{i,j} \lambda_i (M^*)_{ij} \gamma_j, w) = \sum_{i,j} \lambda_i ((M^*)_{ij}, w) \gamma_j)$$

$$= \sum_{i,j} \lambda_i \mu w_{ij} \gamma_j = \lambda \mu w \gamma = (S', w).$$

Therefore $S \odot 1_{\text{LNF}} = S' = \lambda M^* \gamma$ is mc-rational.

Corollary 23. *Let $S \in K\langle\!\langle \Sigma^* \rangle\!\rangle$ be recognizable with $\text{supp}(S) \subseteq \text{LNF}$. Then S is mc-rational.*

Let M, N be two monoids and $h : M \longrightarrow N$ be a morphism. Then $h^{-1} : K\langle\!\langle N \rangle\!\rangle \longrightarrow K\langle\!\langle M \rangle\!\rangle$ given by $(h^{-1}(S), w) = (S, h(w))$ $(w \in N)$ is a semiring morphism. Moreover, if $S = (\lambda, \mu, \gamma) \in K^{rec}\langle\!\langle N \rangle\!\rangle$, then $(h^{-1}(S), w) = (S, h(w)) = \lambda \mu h(w) \gamma$, hence (cf. [14, p.32])

$$h^{-1}(S) = (\lambda, \mu \circ h, \gamma) \in K^{rec}\langle\!\langle M \rangle\!\rangle.$$

Let $\varphi : \Sigma^* \longrightarrow \mathbb{M}$ be the canonical epimorphism. Then φ extends naturally to a mapping, denoted by Φ, from $K\langle\!\langle \Sigma^* \rangle\!\rangle$ to $K\langle\!\langle \mathbb{M} \rangle\!\rangle$ given by

$$\Phi(S) = \sum_{w \in \Sigma^*} (S, w) \varphi(w) = \sum_{t \in \mathbb{M}} \left(\sum_{w \in \varphi^{-1}(t)} (S, w) \right) . t.$$

As is well-known from general results (cf., e.g., [14, pp.13,14]), Φ is a semiring morphism and if S is proper, then $\Phi(S^*) = \Phi(S)^*$. Furthermore, if S is connected (respectively, mono-alphabetic), then $\Phi(S)$ is also connected (respectively, mono-alphabetic). From this, it is clear that if S is mc-rational, then $\Phi(S)$ is also mc-rational. Now we prove Theorem 1(a).

Theorem 24. *Let $S \in K\langle\langle\mathbb{M}\rangle\rangle$ be recognizable. Then S is mc-rational.*

Proof. Let $S = (\lambda, \mu, \gamma) \in K^{rec}\langle\langle\mathbb{M}\rangle\rangle$. As noted before, $\varphi^{-1}(S) \in K^{rec}\langle\langle\Sigma^*\rangle\rangle$. By Theorem 22, $\varphi^{-1}(S) \odot 1_{\text{LNF}}$ is mc-rational. Hence also $\Phi(\varphi^{-1}(S) \odot 1_{\text{LNF}})$ is mc-rational. Now for each $t \in \mathbb{M}$ we have

$$(\Phi(\varphi^{-1}(S) \odot 1_{\text{LNF}}), t) = \sum_{w \in \varphi^{-1}(t)} (\varphi^{-1}(S) \odot 1_{\text{LNF}}, w)$$

$$= \sum_{w \in \varphi^{-1}(t) \cap \text{LNF}} (\varphi^{-1}(S), w) = \sum_{w \in \varphi^{-1}(t) \cap \text{LNF}} (S, \varphi(w)) = (S, t).$$

Therefore, $S = \Phi(\varphi^{-1}(S) \odot 1_{\text{LNF}})$ is mc-rational.

5 Examples and consequences

Here we will give two examples to show that the assumptions in Theorems 6 and 8 (hence, in Theorem 1(b,c)) are necesssary. We also indicate the relationship with the results of Schützenberger and Ochmański. First, we show that in Theorem 6 the commutativity of K is necessary.

Example 25. Consider the trace alphabet (Σ, I) with $\Sigma = \{a, b\}$ and $a\, I\, b$, and let $K = \mathbb{B}\langle\Sigma^*\rangle$. Let $S = \sum_n a^n.a^n, T = \sum_n b^n.b^n \in K\langle\langle\mathbb{M}\rangle\rangle$. Then S and T are recognizable. Indeed, if $\mu : \Sigma^* \longrightarrow K$ is defined by $\mu(a) = a$ and $\mu(b) = 0$ and $\lambda = \gamma = 1$, then $S = (\lambda, \mu, \gamma)$. However, we can show that $S \cdot T \in K\langle\langle\mathbb{M}\rangle\rangle$ is *not* recognizable.

Secondly, we want to show that in general $K^{rec}\langle\langle\mathbb{M}\rangle\rangle$ is properly contained in $K^{c-rat}\langle\langle\mathbb{M}\rangle\rangle$. That is, we show that the star of a connected recognizable series may not be recognizable. (Thus by Theorem 15, the star of this series will not be connected.)

Example 26. Again consider the trace alphabet (Σ, I) with $\Sigma = \{a, b\}$ and $a\, I\, b$, and let $S = a + b \in \mathbb{N}\langle\mathbb{M}\rangle$. Then, obviously, S is a connected polynomial and $(S^*, t) = \binom{|t|_a + |t|_b}{|t|_a}$ for all $t \in \mathbb{M}$. Hence, $S^* = \sum_{n,m \in \mathbb{N}} \binom{n+m}{n} a^n b^m$. We can prove that S^* is not recognizable.

Let Σ be any finite alphabet. If $I = \emptyset$, the trace monoid $\mathbb{M}(\Sigma, I)$ is isomorphic to Σ^*. Hence, by Theorem 24 we have $K^{rec}\langle\langle\Sigma^*\rangle\rangle \subseteq K^{mc-rat}\langle\langle\Sigma^*\rangle\rangle \subseteq K^{rat}\langle\langle\Sigma^*\rangle\rangle$. Now, using one inclusion of Theorem 2, we obtain $K^{rec}\langle\langle\Sigma^*\rangle\rangle = K^{mc-rat}\langle\langle\Sigma^*\rangle\rangle = K^{rat}\langle\langle\Sigma^*\rangle\rangle$ which is in fact a strengthening of Theorem 2.

Now we show how to deduce and actually strengthen Theorem 3 from our results. The following can be proved in the same way as classically for the free monoid (cf. [14,2]).

Proposition 27. *$L \subseteq \mathbb{M}$ is recognizable (resp. rational, c-rational, mc-rational) iff $1_L \in \mathbb{B}\langle\langle\mathbb{M}\rangle\rangle$ is recognizable (resp. rational, c-rational, mc-rational).*

Since the boolean semiring $\mathbb{B}$ is both commutative and idempotent, we deduce from Theorem 1 that a series in $\mathbb{B}\langle\!\langle\mathbb{M}\rangle\!\rangle$ is recognizable iff it is c-rational iff it is mc-rational. Using Proposition 27, we deduce that a trace language $L \subseteq \mathbb{M}$ is recognizable iff it is c-rational iff it is mc-rational. The first equivalence is precisely Ochmański's theorem. The second one is a strengthening of a result by Pighizzini [13] which characterizes the recognizable languages as those languages obtained from finite sets of traces using union, concatenation, *restriction to subalphabet* and star restricted to monoalphabetic and connected languages.

References

1. I.J. Aalbersberg and G. Rozenberg. Theory of traces. *Theoretical Computer Science*, 60:1–82, 1988.
2. J. Berstel and Ch. Reutenauer. *Rational Series and Their Languages*, volume 12 of *EATCS Monographs in Theoretical Computer Science*. Springer Verlag, 1988.
3. Ch. Choffrut. Free partially commutative monoids. Rapport LITP 86.20, Université Paris 7 (France), 1986.
4. V. Diekert. *Combinatorics on Traces*. Number 454 in Lecture Notes in Computer Science. Springer Verlag, 1990.
5. V. Diekert and Y. Métivier. Partial commutation and traces. In G. Rozenberg and A. Salomaa, editors, *Handbook on Formal Languages*, volume III. Springer Verlag. To appear.
6. V. Diekert and G. Rozenberg, editors. *Book of Traces*. World Scientific, Singapore, 1995.
7. M. Fliess. Matrices de Hankel. *J. Math. Pures et Appl.*, 53:197–224, 1974.
8. W. Kuich. Semirings and formal power series: Their relevance to formal languages and automata. In *Handbook on Formal Languages*. Springer Verlag, 1997. To appear.
9. W. Kuich and A. Salomaa. *Semirings, Automata, Languages*, volume 6 of *EATCS Monographs in Theoretical Computer Science*. Springer Verlag, 1986.
10. A. Mazurkiewicz. Concurrent program schemes and their interpretations. Tech. rep. DAIMI PB 78, Aarhus University, 1977.
11. A. Mazurkiewicz. Trace theory. In W. Brauer et al., editors, *Advances in Petri Nets'86*, number 255 in Lecture Notes in Computer Science, pages 279–324. Springer Verlag, 1987.
12. E. Ochmański. Regular behaviour of concurrent systems. *Bulletin of the European Association for Theoretical Computer Science (EATCS)*, 27:56–67, Oct 1985.
13. G. Pighizzini. Synthesis of nondeterministic asynchronous automata. In M. Droste and Y. Gurevich, editors, *Semantics of Programming Languages and Model Theory*, number 5 in Algebra, Logic and Applications, pages 109–126. Gordon and Breach Science Publ., 1993.
14. A. Salomaa and M. Soittola. *Automata-Theoretic Aspects of Formal Power Series*. Texts and Monographs in Computer Science. Springer Verlag, 1978.
15. M.P. Schützenberger. On the definition of a family of automata. *Information and Control*, 4:245–270, 1961.

On a Conjecture of J. Shallit

Julien Cassaigne

Institut de Mathématiques de Luminy,
Case 930, F-13288 Marseille Cedex 9, France
`cassaigne@iml.univ-mrs.fr`

Abstract. We solve a conjecture of J. Shallit related to the automaticity function of a unary language, or equivalently to the first occurrence function in a symbolic sequence. The answer is negative: the conjecture is false, but it can be corrected by changing the constant involved. The proof is based on a study of paths in the Rauzy graphs associated with the sequence.

1 Introduction

In a recent paper [6], Shallit proposed a conjecture on the automaticity function of a unary language, i.e. the size of the minimum finite-state machine that correctly decides membership in the language for words of length at most n. See [9] for more details on the automaticity function and its applications; in short, it measures how close the language is from a regular language. The conjecture arises from a natural question: apart from regular languages (which have bounded automaticity), what is the lowest possible automaticity that a language can have? Shallit rephrased his conjecture in combinatorial terms as follows:

Conjecture 1. *Let* $\mathbf{u} = u_1 u_2 u_3 \ldots$ *be an infinite word over a finite alphabet that is not ultimately periodic. Define $S(n)$ to be the length of the longest suffix of $u_1 u_2 \ldots u_{n+1}$ that is also a factor of $u_1 u_2 \ldots u_n$. Then*

$$\liminf_{n \to \infty} \frac{S(n)}{n} \leq 2 - \varphi = \frac{3 - \sqrt{5}}{2} \simeq .381966$$

where $\varphi = (1 + \sqrt{5})/2 \simeq 1.61803$ *is the golden ratio.*

He also proved that if it is true, then this conjecture is optimal as the value $2 - \varphi$ is attained for the famous Fibonacci word,

0100101001001010010100100101001010010100100101001010...

which is the fixed point of the substitution $0 \mapsto 01$, $1 \mapsto 0$.

Allouche and Bousquet-Mélou [1] noticed a similarity between this conjecture and an older conjecture of Rauzy [7], also involving the golden ratio:

Conjecture 2. *Let* $\mathbf{u}$ *be an infinite word over a finite alphabet that is not ultimately periodic. Let $R(n)$ be the recurrence function of $\mathbf{u}$, i.e. the size of the smallest window containing an occurrence of every factor of $\mathbf{u}$ of length n whatever its position on $\mathbf{u}$, or ∞ if no such window exists. Then*

$$\limsup_{n \to \infty} \frac{R(n)}{n} \geq \varphi + 2 = \frac{5 + \sqrt{5}}{2} \simeq 3.61803 \ .$$

They proposed a modified ("Rauzy-like") conjecture, and proved that it was equivalent to Shallit's conjecture:

Conjecture 3. *Let* **u** *be an infinite word over a finite alphabet that is not ultimately periodic. Let $R'(n)$ be the length of the shortest prefix of* **u** *containing an occurrence of every factor of* **u** *of length n. Then*

$$\limsup_{n \to \infty} \frac{R'(n)}{n} \geq \varphi + 1 = \frac{3 + \sqrt{5}}{2} \simeq 2.61803 \ .$$

Using Rauzy graphs, we have been able to prove Conjecture 2 [3]. We then tried to adapt the proof to Conjecture 3. In principle, Conjecture 3 should have been easier to prove in this way than Conjecture 2, as the constant is smaller and the number of different cases to study is therefore reduced. However we did not succeed in this attempt, and we resolved to first restrict to the case of Sturmian words, which we had previously dismissed as trivial, following Allouche and Bousquet-Mélou: "[...] the case of the Sturmian words [...] can certainly be addressed by adapting the arguments of [5] for the computation of $\limsup R(n)/n$, but we have not written the details." We did not try to use the method of Morse and Hedlund [5] which is specific to Sturmian words, but our general method with (pointed) Rauzy graphs. And it appeared that contrarily to what we expected, the Fibonacci word is not optimal for $R'(n)/n$. Indeed, the infinite word

$$\mathbf{z}_3 = 01001010010010010100100101001001010010001\ldots$$

defined as the fixed point of the substitution $0 \mapsto 01001010$, $1 \mapsto 010$ satisfies

$$\limsup_{n \to \infty} \frac{R'(n)}{n} = \frac{29 - 2\sqrt{10}}{9} \simeq 2.51949 < \varphi + 1 \ .$$

Conjectures 1 and 3 are therefore false. However, we are now able to prove a modified conjecture, with a different constant:

Theorem 1. *Let* **u** *be an infinite word over a finite alphabet that is not ultimately periodic. Let $R'(n)$ be the length of the shortest prefix of* **u** *containing an occurrence of every factor of* **u** *of length n. Then*

$$\limsup_{n \to \infty} \frac{R'(n)}{n} \geq \frac{29 - 2\sqrt{10}}{9} \simeq 2.51949 \ ,$$

and this value is optimal.

Fortunately, Allouche and Bousquet-Mélou [1] proved much more than the equivalence of Conjectures 1 and 3: they proved that the numbers $\liminf S(n)/n$ and $\limsup R'(n)/n$ are inverses of each other. Therefore, we immediately deduce a modified version of Shallit's conjecture, where the constant is optimal for the same Sturmian word $\mathbf{z}_3$ as above:

Corollary 1. *Let* $\mathbf{u} = u_1 u_2 u_3 \ldots$ *be an infinite word over a finite alphabet that is not ultimately periodic. Define $S(n)$ to be the length of the longest suffix of $u_1 u_2 \ldots u_{n+1}$ that is also a factor of $u_1 u_2 \ldots u_n$. Then*

$$\liminf_{n \to \infty} \frac{S(n)}{n} \leq \frac{29 + 2\sqrt{10}}{89} \simeq .396905 \ ,$$

and this value is optimal.

In Section 2, we define precisely the tools that we will use in the proof. We then study in Section 3 the case of Sturmian words, and the word z_3 occurs naturally in this process. Finally, we explain in Section 4 how the general case can be reduced to the Sturmian case.

2 Preliminaries

2.1 Complexity and First Occurrence Functions

Let Σ be a finite alphabet, and Σ^ω the set of one-way infinite sequences over Σ. If $\mathbf{u} = u_1 u_2 u_3 \ldots$ is an element of Σ^ω, and n is a non-negative integer, we denote by $F_n(\mathbf{u})$ the set of factors (also called subwords) of length n of $\mathbf{u}$, i.e. of words of length n consisting of letters occurring consecutively in $\mathbf{u}$: $F_n(\mathbf{u}) = \{u_k u_{k+1} u_{k+2} \ldots u_{k+n-1} \mid k \geq 1\}$, and we denote by $F(\mathbf{u})$ the union of these sets. We also denote by $\mathrm{pref}_n(\mathbf{u})$ the prefix of length n of $\mathbf{u}$, i.e. the word $u_1 u_2 \ldots u_n$.

The complexity function of $\mathbf{u}$ is then defined as the function mapping a non-negative integer n to the number of factors of length n of $\mathbf{u}$: $p_{\mathbf{u}}(n) = \#F_n(\mathbf{u})$. When there is no ambiguity on the sequence $\mathbf{u}$, we shall write $p(n)$ instead of $p_{\mathbf{u}}(n)$. It is clear that for all $n \geq 0$, $1 \leq p(n) \leq (\#\Sigma)^n$; moreover, it is well-known that $p(n) \geq n+1$ when the sequence $\mathbf{u}$ is not ultimately periodic [5].

To study Shallit's conjecture, we will use the *first occurrence function* $\ell_{\mathbf{u}}$ (or simply ℓ) defined as follows. For any word $w \in F(\mathbf{u})$, let $\ell(w)$ be the smallest positive integer m such that $w = u_m u_{m+1} \ldots u_{m+|w|-1}$, so that for instance $\ell(\mathrm{pref}_n(\mathbf{u})) = 1$, and let $\ell(n) = \max\{\ell(w) \mid w \in F_n(\mathbf{u})\}$.

Proposition 1. *The function R' defined in Conjecture 3 satisfies the relation $R'(n) = \ell(n) + n - 1$.*

Proof. The function $R'(n)$ is defined as the length of the shortest prefix of $\mathbf{u}$ containing every factor of length n of $\mathbf{u}$. A factor $w \in F_n(\mathbf{u})$ occurs in $\mathrm{pref}_m(\mathbf{u})$ if and only if $\ell(w) \leq m - (n-1)$, therefore $\mathrm{pref}_m(\mathbf{u})$ contains all factors if and only if $m \geq \ell(n) + n - 1$. □

Defining $\lambda(\mathbf{u}) = \limsup \ell(n)/n$, we get as a corollary that
$$\limsup_{n \to \infty} \frac{R'(n)}{n} = \lambda(\mathbf{u}) + 1 \ .$$

Proposition 2. *The first occurrence and complexity functions satisfy the inequality $\ell(n) \geq p(n)$.*

Proof. For two distinct factors v and w of the same length n, $\ell(v)$ and $\ell(w)$ are two distinct positive integers. The set $\{\ell(w) \mid w \in F_n(\mathbf{u})\}$ contains therefore $p(n)$ distinct positive integers, hence its maximum $\ell(n)$ is at least $p(n)$. □

If the sequence $\mathbf{u}$ is ultimately periodic, then it is easy to see that the function ℓ has a finite limit (it is the minimum value of $|uv|$, where u and v are words such that $\mathbf{u} = uv^\omega$), hence $\lambda(\mathbf{u}) = 0$. Otherwise, the complexity is at least $n+1$ [5], therefore $\ell(n) \geq n+1$ by Proposition 2, and $\lambda(\mathbf{u}) \geq 1$. Theorem 1 says that in fact $\lambda(\mathbf{u}) \geq (20 - 2\sqrt{10})/9 \simeq 1.51949$.

2.2 Rauzy Graphs

To study the structure of the factors of a sequence **u**, it is usually convenient to define a sequence of graphs G_n, called *Rauzy graphs* or *factor graphs*, as follows. For any non negative integer n, let G_n be the directed graph with $p(n)$ vertices labelled with elements of $F_n(\mathbf{u})$, and with an edge from u to v if and only if there exist two letters $x, y \in \Sigma$ such that $uy = xv \in F_{n+1}(\mathbf{u})$. The graph G_n has therefore $p(n+1)$ edges.

Unlike other problems for which only $F(\mathbf{u})$ is important, for Shallit's conjecture we need to know which factors occur first in the sequence. We shall add this information to the Rauzy graphs by singling out one vertex, the one labelled with the prefix of length n of **u**. We will therefore consider the *pointed Rauzy graph* $(G_n, \text{pref}_n(\mathbf{u}))$.

We choose to label edges of G_n with letters, in the following way: if $uy = xv$ with $x, y \in \Sigma$, then the edge (u, v) is labelled with the letter x. We then define the label of a finite path of length k in G_n as the word of length k obtained by concatenating the labels of the edges in the order they are met, and similarly the label of an infinite path as an infinite word.

With this definition, there is a unique infinite path in G_n labelled with **u** and starting in $\text{pref}_n(\mathbf{u})$: it is the path $(w_1, w_2, w_3, \ldots)$ where w_k is the k-th block of length n of **u**, i.e. $w_k = u_k u_{k+1} \ldots u_{k+n-1}$ (in particular, $w_1 = \text{pref}_n(\mathbf{u})$). Knowing this path, we can now read $\ell(n)$ on the graph.

Proposition 3. *Let $(w_1, w_2, w_3, \ldots)$ be the path labelled with **u** in G_n. Then $\ell(n) - 1$ is the length of the shortest prefix of this path that goes through every vertex of G_n, and $\ell(n+1)$ is the length of the shortest prefix of this path that goes through every edge of G_n.*

Proof. For a given $w \in F_n(\mathbf{u})$, we have $\ell(w) = \min\{k \geq 1 \mid w_k = w\}$. Consequently, a prefix $(w_1, w_2, \ldots, w_k)$ of length $k - 1$ of the path labelled with **u** goes through the vertex w if and only if $k \geq \ell(w)$, and it goes through every vertex if and only if $k \geq \ell(n)$. Similarly, for a given edge (u, v) labelled with $x \in \Sigma$, we have $\ell(xv) = \min\{k \geq 1 \mid w_k = u \text{ and } w_{k+1} = v\}$. Consequently, a prefix $(w_1, w_2, \ldots, w_{k+1})$ of length k of the path labelled with **u** goes through the edge (u, v) if and only if $k \geq \ell(xv)$, and it goes through every edge if and only if $k \geq \ell(n+1)$. □

It should be noted that in the graph G_n, every vertex has outdegree at least one (i.e. has at least one outgoing edge), and every vertex except possibly the one labelled with $\text{pref}_n(\mathbf{u})$ has indegree at least one. The sequence is said to be *recurrent* if every factor occurs infinitely often; in this case $\text{pref}_n(\mathbf{u})$ has also indegree at least one. If **u** is not recurrent, then for n large enough the prefix $\text{pref}_n(\mathbf{u})$ occurs only once, and therefore the corresponding vertex in G_n has indegree zero.

A vertex v of G_n is called *bispecial* if both its indegree and its outdegree are greater than 1 (the word v is then a *bispecial factor* of **u** [4]).

2.3 From G_n to G_{n+1}

The reader is warmly encouraged to construct the Rauzy graphs G_n, for small n, for a simple sequence like the Fibonacci word, to get acquainted with the manipulation of these graphs. One crucial point, on which the rest of this article relies heavily, is the relation between G_n and G_{n+1}, which is explained in detail in [2, 4, 8] and summarized below.

Knowing G_n, one constructs its *line graph* $D(G_n)$ as follows: for every edge (u, v) labelled with x in G_n, there is a vertex in $D(G_n)$ labelled with xv; and for every pair of consecutive edges $((u, v), (v, w))$ in G_n labelled with x and y, there is an edge (xv, yw) in $D(G_n)$ labelled with x.

Proposition 4. *The Rauzy graph of order $n+1$ is a subgraph of $D(G_n)$. Namely:*
— *If G_n has no bispecial vertex, then $G_{n+1} = D(G_n)$.*
— *If G_n has bispecial vertices, then some (possibly none) edges (xv, vy), with v bispecial, have to be removed from $D(G_n)$ to obtain G_{n+1}.*

3 The Sturmian Case

In this section, we assume that **u** is a Sturmian sequence, i.e. a sequence with complexity $p(n) = n+1$. As $p(1) = 2$, the alphabet Σ has only two letters. Rauzy graphs of Sturmian sequences are described by the following proposition [2].

Proposition 5. *If **u** is a Sturmian sequence, then the Rauzy graphs are of one of the following two types (vertices with indegree and outdegree 1 are not represented).*

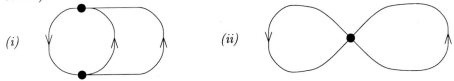

Moreover, both types occur infinitely often.

We shall give a particular importance to graphs of the second type, which we number $G_{n_0} = G_0$, G_{n_1}, G_{n_2}, etc. Adding the initial vertex $\text{pref}_n(\mathbf{u})$ (marked with a black triangle), we get the following pointed graph G_{n_k}.

(1)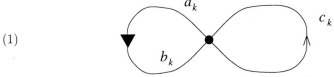

The three branches are labelled with the words a_k, b_k, c_k (in the case where the initial vertex is also the bispecial one, a_k is the empty word and the loop labelled with b_k is the first one used in a path labelled with **u**). They satisfy $p(n_k + 1) = n_k + 2 = |a_k b_k c_k|$.

We are now interested in the evolution of the graphs when n grows from n_k to n_{k+1}.

Proposition 6. *For every k, the transition between G_{n_k} and $G_{n_{k+1}}$ is of one of the following three types:*
transition A: $n_{k+1} = n_k + |a_k b_k|$, $a_{k+1} = a_k$, $b_{k+1} = b_k$, $c_{k+1} = a_k b_k c_k$;
transition B: $n_{k+1} = n_k + |a_k b_k|$, $a_{k+1} = a_k$, $b_{k+1} = b_k c_k$, $c_{k+1} = a_k b_k$;
transition C: $n_{k+1} = n_k + |c_k|$, $a_{k+1} = c_k a_k$, $b_{k+1} = b_k$, $c_{k+1} = c_k$.

Proof. We have to construct the graphs G_n for $n_k < n \leq n_{k+1}$, using Proposition 4 repetitively. Let w denote the bispecial factor of length n_k. Let x, y, z respectively denote the last letters of a_k, b_k, c_k (note that $y \neq z$). The line graph $D(G_{n_k})$ is then

(2)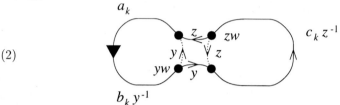

To obtain G_{n_k+1}, which is a graph of type (i), one of the two dotted edges has to be removed from (2). (Note that the other two central edges cannot be removed because the resulting graphs would only have ultimately periodic paths.) Therefore, G_{n_k+1} is either

(3)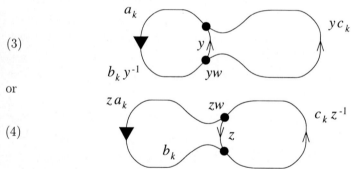

or

(4)

In the first case, the next graphs G_n have the same morphology until $n = n_k + |b_k|$, where we get

(5)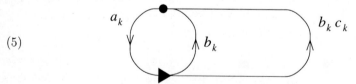

Then the next graph ($n = n_k + |b_k| + 1$) depends on which branch contains the prefix. There are therefore two subcases,

(6)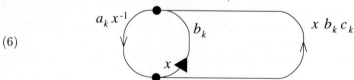

which then evolves to $G_{n_{k+1}}$, at $n_{k+1} = n_k + |a_k b_k|$ (transition A):

(7)

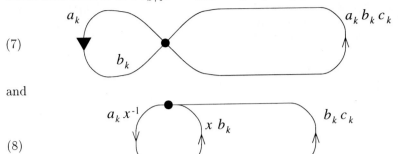

and

(8)

which then evolves to $G_{n_{k+1}}$, also at $n_{k+1} = n_k + |a_k b_k|$ (transition B):

(9)

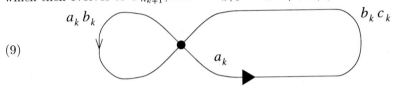

In the second case, the following graphs have the same morphology until $n = n_k + |c_k|$, where we get directly $G_{n_{k+1}}$ (transition C):

(10)

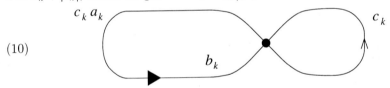

We observe that there are three possible transitions, corresponding to the three transformations A, B, and C. □

Proposition 6 allows us to define a new representation of the sequence **u**. Let Δ be the alphabet $\Delta = \{A, B, C\}$; the *adic representation* of **u** is the sequence $\mathbf{t} = t_1 t_2 t_3 \ldots \in \Delta^\omega$, where t_k indicates which kind of transition occurs between $G_{n_{k-1}}$ and G_{n_k}. The adic representation is related to similar representations studied in [10], and also (in the case of Sturmian words only) to the usual continued fraction expansion of real numbers [5].

Proposition 7. *Given a sequence* $\mathbf{t} \in \Delta^\omega \setminus \Delta^*(A^\omega \cup C^\omega)$, *there exists a unique Sturmian sequence* **u** *(up to renaming of the letters) such that* **t** *is the adic representation of* **u**.

Proof. We first take the graph $G_{n_0} = G_0$ to be the graph with one vertex and two loops of length 1, labelled with the two letters of Σ (b_0 will be the first letter of **u**, and c_0 the other letter), and $a_0 = \varepsilon$ since the starting vertex is the bispecial one. Then a_k, b_k, and c_k are entirely defined by the sequence **t**, using the recurrence relations of Proposition 6. We thus know the labels of the edges of the graphs G_{n_k}, and from this information we can also find the labels of the vertices (the label of a vertex in G_n is the label of any path of length n starting at this vertex). In particular we obtain the words $\text{pref}_{n_k}(\mathbf{u})$, which as a limit give a sequence $\mathbf{u}$.

We still have to check that $\mathbf{u}$ is indeed a Sturmian sequence. Its Rauzy graph of order n_k is a subgraph of G_{n_k}, but it may not be exactly G_{n_k} in the event where the path associated with $\mathbf{u}$ never reaches certain branches of the graph. In this case the sequence would be ultimately periodic, which implies that for n large enough, all graphs G_n have a loop of the same size (equal to the period). This occurs only when $\mathbf{t} \in \Delta^* A^\omega$ or $\mathbf{t} \in \Delta^* C^\omega$ (words in $\Delta^* B^\omega$ define legal Sturmian sequences, for instance B^ω is the adic representation of the Fibonacci sequence). □

We can now turn to the study of $\lambda(\mathbf{u})$. Knowing a few consecutive terms of $\mathbf{t}$, we are able, using the corresponding graphs, to evaluate certain values of $\ell(n)$ as a function of $|a_k|$, $|b_k|$, and $|c_k|$. In some cases, we can prove that it is more than φn. If these terms occur infinitely many times in $\mathbf{t}$, we deduce that $\lambda(\mathbf{u}) \geq \varphi$.

Proposition 8. *If $\mathbf{t}$ contains infinitely many occurrences of the words $BC^m A$ (with $m \geq 0$), ACA, ACC, $CBCB$, $CBCC$, $BBCCB$, $BBCCC$ (i.e. if either $\mathbf{t}$ contains infinitely many occurrences of one word in the list, or if $\mathbf{t}$ contains $BC^m A$ for infinitely many values of m), then $\lambda(\mathbf{u}) \geq \varphi$.*

Proof. We shall study in detail only the case of the word BCA; the other words are dealt with similarly. Suppose that $t_{k+1} = B$, $t_{k+2} = C$, and $t_{k+3} = A$; let $n = n_k$, $a = a_k$, $b = b_k$, and $c = c_k$. Then we have :

i	n_{k+i}	a_{k+i}	b_{k+i}	c_{k+i}
0	n	a	b	c
1	$n + |ab|$	a	bc	ab
2	$n + 2|ab|$	aba	bc	ab
3	$n + 4|ab| + |c|$	aba	bc	$ababcab$

Note that all paths of $G_{n_{k+3}}$ starting at the pointed vertex begin with $bcababcabab$. Thus $bcababcabab$ is a prefix of $\mathbf{u}$. This gives the beginning of the path followed by $\mathbf{u}$ in the graphs $G_{n'}$ for $n' \geq n$. In particular, in G_{n+1} (see (3)), the shortest prefix going through every edge has length $|bcab|$, hence $\ell(n+2) = |bcab|$; in $G_{n+|b|+1}$ (see (8)), this shortest prefix has length $|bcaba|$, hence $\ell(n+|b|+2) = |bcaba|$; and in $G_{n+|ab|+1}$ (see (4)), with k replaced by $k+1$, $\ell(n+|ab|+2) \geq |bcababcab|$. Now let $d_1 = \ell(n+|b|+2) - \varphi(n+|b|+2)$ and $d_2 = \ell(n+|ab|+2) - \varphi(n+|ab|+2)$, and let us compute $d_1 + \varphi d_2$, recalling that $n + 2 = |abc|$, and $\varphi^2 = \varphi + 1$:

$$\begin{aligned} d_1 + \varphi d_2 &= (n+|b|+2) - \varphi(n+|b|+2)) + \varphi(\ell(n+|ab|+2) - \varphi(n+|ab|+2)) \\ &\geq (|bcaba| - \varphi|bcab|) + \varphi(|bcababcab| - \varphi|bcaba|) \\ &\geq (1 + \varphi - \varphi^2)|bcaba| = 0 \end{aligned}$$

This shows that at least one of d_1 and d_2 has to be non-negative, i.e. that $\ell(n') \geq \varphi n'$ for $n' = n + |b| + 2$ or $n' = n + |ab| + 2$.

Similarly, for each occurrence in **t** of a word in the list, there is a length n' for which $\ell(n') \geq \varphi n'$. If there are infinitely many such occurrences, then $\lambda(\mathbf{u}) \geq \varphi$. □

Most sequences **t** satisfy the conditions of Proposition 8; the only words that do not satisfy them are the elements of the set $\Delta^*(CB^2B^*)^\omega \cup \Delta^*B^\omega$. If $\mathbf{t} \in \Delta^*B^\omega$, then **u** is a morphic image of the Fibonacci sequence and it is easy to see that $\lambda(\mathbf{u}) = \varphi$; for the other set however, the method of Proposition 8 does not seem to work.

It is then natural to study the simplest examples of these sequences, for which **t** is periodic with a short period. Namely, take $\mathbf{t} = (CB^m)^\omega$, with $m \geq 2$. The recurrences for a_k, b_k, and c_k can be solved; taking the limit of (b_k), one finds in particular that the associated Sturmian sequence, $\mathbf{z}_m$, is the fixed point of the substitution $f^m \circ g$, where $f(0) = 01$, $f(1) = 0$ (f is the substitution defining the Fibonacci word) and $g(0) = 01$, $g(1) = 1$. Computing ℓ for these sequences, although rather technical, is not very difficult, as the lengths of the paths in the Rauzy graphs can be computed from the lengths of a_k, b_k, and c_k. If only $\lambda(\mathbf{z}_m)$ is of interest, this amounts to computing the eigenvectors of the matrix of the substitution $f^m \circ g$, combining them in several ways, and taking the maximum. Proposition 9 summarizes the results for the first values of m.

Proposition 9. *The sequences $\mathbf{z}_m$, $2 \leq m \leq 5$, yield the following limits:*

$$\lambda(\mathbf{z}_2) = \frac{3 + \sqrt{3}}{3} \simeq 1.57735 \;,$$

$$\lambda(\mathbf{z}_3) = \frac{20 - 2\sqrt{10}}{9} \simeq 1.51949 \;,$$

$$\lambda(\mathbf{z}_4) = \frac{18 + \sqrt{24}}{15} \simeq 1.52660 \;,$$

$$\lambda(\mathbf{z}_5) = \frac{415 + 3\sqrt{65}}{280} \simeq 1.56852 \;.$$

Among these four examples, the sequence $\mathbf{z}_3$ appears to give the lowest value; it is indeed possible to compute explicit formulas for all $\lambda(\mathbf{z}_m)$ and to prove that they are increasing for $m \geq 3$. This observation suggests that $\lambda(\mathbf{z}_3)$ could be the lowest possible value for Sturmian sequences.

To prove this, we proceed as in Proposition 8, loosening the researched inequality by replacing φ with 1.52.

Proposition 10. *If **t** contains infinitely many occurrences of one of the words B^5, B^3CB^4, $B^2CB^2CB^4$ or CB^3CB^2CB, then $\lambda(\mathbf{u}) \geq 1.52 > \lambda(\mathbf{z}_3)$.*

The only sequences **t** satisfying neither Proposition 8 nor Proposition 10 are elements of the set $\Delta^*(CB^2)^\omega \cup \Delta^*(CB^3)^\omega$, i.e. the corresponding Sturmian sequences are morphic images of $\mathbf{z}_2$ or $\mathbf{z}_3$, among which $\mathbf{z}_3$ is optimal according to Proposition 9 (changing a finite prefix of **t** does not change the value of $\lambda(\mathbf{u})$). We have thus finished the proof of Theorem 1 in the Sturmian case.

It should be noted that when $\mathbf{t} \notin \Delta^*(CB^3)^\omega$, then $\lambda(\mathbf{u}) \geq 1.52 > \lambda(\mathbf{z}_3)$: the spectrum of possible values for $\lambda(\mathbf{u})$ for Sturmian sequences is not continuous. We have not tried to find what the next attainable value is, and 1.52 is just a rough minoration.

4 The General Case

Let us now turn to the general case. As noted in [1], sequences with large enough complexity can be easily eliminated.

Proposition 11. *If there is an integer n_0 such that the sequence $\mathbf{u}$ satisfies $p(n+1) - p(n) \geq 2$ for all $n \geq n_0$, then $\lambda(\mathbf{u}) \geq 2$.*

Proof. If this is the case, then $p(n) \geq p(n_0) + 2(n - n_0)$ for $n \geq n_0$, hence there is a constant C such that $p(n) \geq 2n - C$ for all n. According to Proposition 2, $\ell(n) \geq p(n)$, hence

$$\lambda(\mathbf{u}) = \limsup \frac{\ell(n)}{n} \geq \limsup \frac{p(n)}{n} \geq 2 \ .$$

$\square$

We can therefore suppose that $p(n+1) - p(n) = 1$ for infinitely many n, which implies that for infinitely many n, the Rauzy graphs are of the types of Proposition 5, at least if the sequence is recurrent (non-recurrent sequences have slightly different graphs, the initial vertex being connected by an additional branch to the main part of the graph, but they can be handled similarly). As for Sturmian sequences, we can define the sequences n_k, a_k, b_k, and c_k, and study the possible transitions. There are infinitely many possible transitions (including A, B, and C), as the intermediate graphs can be very complicated. However, in most cases we will obtain a sufficiently large minoration for $\lambda(\mathbf{u})$.

The graph G_{n_k+1} may be graph (3) or graph (4), in which case we find the same transitions A, B, and C as with Sturmian words, but it may also be graph (2), the complete line graph $D(G_{n_k})$. In this graph, the shortest paths starting from the pointed vertex and going through every edge are $bccab$ and $babcc$ (for simplicity, we now note $n = n_k$, $a = a_k$, etc.) hence $\ell(n+2) \geq |a|+2|bc|$. What happens next depends on the respective sizes of b and c. If b is shorter, we get the following graph of order $n + |b|$

(11)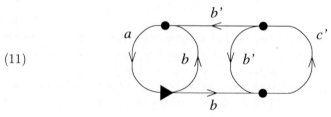

where $c = c'b'$, $|b'| = |b|$. There are then two possibilities for $n + |b| + 1$,

(12)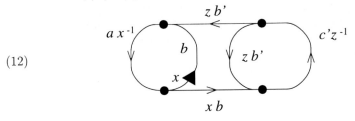

and

(13)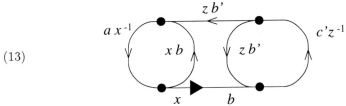

In both cases, a path starting from the pointed vertex and going through every edge has length at least $2|abc|$, i.e. $\ell(n + |b| + 2) \geq 2|abc|$. As in the proof of Proposition 8, let $d_1 = \ell(n+2) - \varphi(n+2)$ and $d_2 = \ell(n+|b|+2) - \varphi(n+|b|+2)$, and let us compute $d_1 + \varphi d_2$, using $|abc| = p(n+1) \geq n+2$:

$$\begin{aligned} d_1 + \varphi d_2 &= (\ell(n+2) - \varphi(n+2)) + \varphi(\ell(n+|b|+2) - \varphi(n+|b|+2)) \\ &\geq (|a| + 2|bc| - \varphi|abc|) + \varphi(2|abc| - \varphi|bcab|) \\ &= (1 + \varphi - \varphi^2)|abc| + |bc| - \varphi|b| = |bc| - \varphi|b| \end{aligned}$$

As $|b| \leq |c|$, this number is positive, hence also one of d_1 and d_2, i.e. $\ell(n') \geq \varphi n'$ for $n' = n+2$ or $n' = n + |b| + 2$. If this transition occurs infinitely often, then $\ell(\mathbf{u}) \geq \varphi$; we can therefore assume that this transition does not occur when k is large enough.

If c is shorter than b or has the same length, several subcases are possible, most of which can be eliminated with the same kind of arguments. The only transitions that remain are those where the loop labelled with c is taken a fixed number of times $j \geq 2$ by every path, with $(j-1)|c| \leq |b|$. We eventually get at order $n + |b|$ the graph

(14)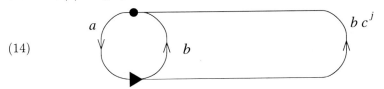

which is essentially the same as graph (5) except that c is repeated j times. This gives rise to transitions A_j and B_j analogous to $A = A_1$ and $B = B_1$.

We can now define the adic representation of a sequence $\mathbf{u}$ that satisfies $\lambda(\mathbf{u}) < \varphi$: it is a sequence $\mathbf{t} = t_{k_0} t_{k_0+1} t_{k_0+2} \ldots$ on the infinite alphabet $\Delta' = \{A_j, B_j \mid j \geq 1\} \cup \{C\}$, where t_k indicates the transition between $G_{n_{k-1}}$ and G_{n_k}. As replacing A and B by A_j and B_j may only increase the values of ℓ, the rest of the proof for Sturmian words works with the general case as well, and we can conclude that Theorem 1 is true for any recurrent binary sequence.

As noted above, the case of non-recurrent sequence uses graphs with a slightly different morphology, but does not cause any additional problem. The case of an arbitrary finite alphabet can be easily reduced to the binary case with a simple projection argument, and this completes the proof of Theorem 1.

Acknowledgments

I would like to warmly thank J.-P. Allouche for introducing me to this problem, W. Plandowski for an interesting discussion that helped me realize that considering pointed graphs was essential, and D. Bernardi for numerically checking the unexpected counterexamples z_m.

References

1. J.-P. ALLOUCHE AND M. BOUSQUET-MÉLOU, On the conjectures of Rauzy and Shallit for infinite words, *Comment. Math. Univ. Carolinae* **36** (1995), 705–711.
2. P. ARNOUX AND G. RAUZY, Représentation géométrique des suites de complexité $2n + 1$, *Bull. Soc. Math. France* **119** (1991), 199–215.
3. J. CASSAIGNE, A proof of Rauzy's conjecture through an exhaustive study of factor graphs. In preparation.
4. J. CASSAIGNE, Facteurs spéciaux et complexité, *Bull. Belg. Math. Soc.* **4** (1997), 67–88. Special issue: Actes des Journées Montoises d'Informatique Théorique 1994.
5. M. MORSE AND G. A. HEDLUND, Symbolic dynamics II: Sturmian trajectories, *Amer. J. Math.* **61** (1940), 1–42.
6. C. POMERANCE, J. M. ROBSON, AND J. SHALLIT, Automaticity II: Descriptional complexity in the unary case, *Theoret. Comput. Sci.* To appear.
7. G. RAUZY, Suites à termes dans un alphabet fini, *Sém. de Théorie des Nombres de Bordeaux*, 1982–1983, 25.01–25.16.
8. G. ROTE, Sequences with subword complexity $2n$, *J. Number Th.* **46** (1994), 196–213.
9. J. SHALLIT AND Y. BREITBART, Automaticity I: Properties of a measure of descriptional complexity, *J. Comput. System Sci.* **53** (1996), 10–25.
10. A. M. VERSHIK, Locally transversal symbolic dynamics, *St. Petersburg Math. J.* **6** (1995), 529–540.

On Characterizations of Escrow Encryption Schemes

Yair Frankel*   Moti Yung

CertCo LLC
frankely,moti@certco.com, moti@cs.columbia.edu

Abstract. Designing escrow encryption schemes is an area of much recent interest. However, the basic design issues, characterizations and difficulties of escrow systems are not fully understood or specified yet. This paper demonstrates that in public-key based escrow, the combination of (1) two different receivers (intended receiver and potentially law enforcement); and (2) on-line verified compliance assurance by the sender which ensures that law enforcement can decrypt ciphertext upon court order, is equivalent to a "chosen ciphertext secure public-key system" (i.e., one secure against an adversary who uses the decryption oracle before trying to decipher a target ciphertext). If we further add measures to ensure that law enforcement is given access to messages only within an authorized context and law enforcement is assured to comply as well (i.e., it cannot frame users), then the escrow system is equivalent to "non-malleable encryption schemes". The characterizations provide a theoretical underpinning for escrow encryption and also lead us to new designs.

1 Introduction

The intent of escrow encryption schemes is to enable strong cryptography for users while protecting society from criminal behavior. Namely, users can send encrypted messages while enabling law enforcement (when and only when allowed by the court) to read their clear messages. The first scheme was the Escrow Encryption Standard (EES) and its Clipper implementation [19], after which many systems have been suggested world-wide [10, 19, 25]. Governments, industry and international organizations are all investigating escrow encryption solutions.

Many of the early and recent designs focused on various specific aspects of escrow encryption, but no rigorous investigations of the technical issues have been done. One of the basic issues that the initial Clipper implementation [10] and the EES gave rise to, is the notion of "compliance assurance and verification" implemented through the use of a LEAF authentication field in Clipper. This was only based on an intuitive understanding drawn from an obvious need, and, in fact, due to design errors and lack of understanding of requirements, some severe flaws were found [6, 20]. Here we attempt a step in the direction of theoretical understanding of escrow systems.

---

* Research performed while at Sandia National Laboratories. This work was performed under U.S. Department of Energy Contract number DE-AC04-76AL85000.

Our Results:

An escrow encryption system can be viewed as a system providing private messages for a regular receiver and a potential additional "shadow receiver" called law-enforcement. We concentrate on "public key based" schemes, i.e. where the sender and receiver do not have to meet. We have collected available requirements and we formally model "escrow systems" based on these basic requirements. We concentrate on a very basic property of "compliance and its verification" in our modeling. This property assures that a message sent is made available for future authorized law enforcement access; it is discussed by several documents and proposed systems. As an example, the Law Enforcement Activation Field (LEAF) in Clipper has the purpose of enforcing availability of "sufficient information" that enables the shadow receiver to read the messages when a proper escrow procedure takes place, connecting the availability to the decoding availability. Another example motivating us to investigate compliance issues is a statement in a NIST (The US National Institute of Science and Technology) document [34] which says: "To meet these criteria, encryption products will need to implement key escrow mechanisms that can not be readily altered or bypassed so as to defeat the purpose of key escrowing".

To model compliance verification, we add a formal entity called a "gateway" G that assures that messages sent into the systems (from a sender to the receiver and the potential "law enforcement") are in compliance; G is less obtrusive than the recently suggested "Trusted Third Party entity" [25]. We then ask: Given the escrow encryption system models with the basic property of compliance, what type of cryptosystems and security notions characterize them? Such characterization helps in understanding the requirements and may also help in future designs. It may potentially allow implementations to exploit available cryptographic knowledge and prevent flaws in future system designs.

We call schemes which provide the sender's compliance checking capability *compliance verifiable escrow encryption systems*. These systems, which have a seemingly necessary ingredient required for full-fledged escrow encryption, are shown to be strongly related to chosen ciphertext secure encryption public-key systems which were first introduce by Naor and Yung [33] and further developed in various works [35, 9, 40, 28, 4, 21]. We concentrate on systems with formal proof of security and we prove that under quite a broad definition of the respective systems (avoiding narrow scenarios and limiting resources, concentrating on the principle security requirements), the following holds:

A compliance verifiable escrow encryption exists iff a chosen ciphertext secure cryptosystem exists

Furthermore, if we require more from the escrow system and also assume that the system has to limit untrusted law enforcement as well, then we get what we call *basic escrow encryption systems*. These systems require compliance verification and in addition they ask for the binding of a message to a proper limitation of context (namely time, sender and receiver identities). Context is required to be checked (by an authority– escrow agents) before messages are opened to law enforcement (as advocated in the recent design in [27] and by formal documents). For secure basic escrow systems we show that:

A basic escrow encryption system exists iff a non-malleable encryption system exists

Non-malleable systems in essence do not allow an attacker to modify chosen ciphertexts to create a new meaningful ciphertext and were introduced by Dolev, Dwork and Naor [16].

Our results demonstrate inherent complexities in implementing systems like the key recovery (by the US government) and trusted third party (by some European governments) on top of a public key infrastructure. They show that the difficulties of assuring compliance in such systems is not only a property of the current ad-hoc designs, but rather they are inherent to any system attempting to build key escrow with compliance in the available infrastructure.

On Reductions between Cryptosystems:
We note that in "one-way function based cryptography," a characterization was completed and its various primitives (one-way functions, digital signatures, random generators, private cipher systems) have been shown to be equivalent (in a long research program which is reviewed in [29]). In "public-key encryption systems" (our subject) the picture is much less clear. A sufficient condition for a secure public key communication (i.e., secrecy without the parties sharing a key) is either "trapdoor function" or "key exchange protocol". These imply the existence of "one way function" since they enable an authentication protocol [23], but there are indications that one-way functions by themselves, cannot easily imply (based on black-box reductions) "public-key cryptography" (since such a construction separates NP from P) [24]. Necessary conditions beyond this are not known, and, therefore, equivalence among various public-key notions is mostly open and intriguing.

Another issue is the quality of the cryptographic reduction. In [29] a reduction is quantified by the amount it reduces the security parameter of a problem when a problem is reduced to another one (an idea attributed to L. Levin). Our reductions are high quality in this sense, they are *linear preserving*.

Related Work:
Various designs have been suggested concentrating on several important aspects and crucial stages of escrow schemes. Most of these aspects are orthogonal to the issue of compliance as investigated here. In [30] the issue of key distribution to trustees was discussed, while a more rigorous approach to a distribution channel with minimization of various potential exposures was given in [26]. In [14] tracing receivers was discussed (this is criticized in [15, 20]), and in [31] the distribution of pseudorandom functions was discussed. The issue of limiting time and context of escrow was discussed in [5, 27]. Also, a few alternatives to escrow based on partial key has been put forth [37, 3, 2]. Opening of ciphertexts based on small (message by message) granularity was put forth in [11]. Characterizing universal escrow trapdoor using public-key systems was presented in [7]. Issues for systems design, based on findings of initial failures were discussed in [20].

2 Compliance-Verification Systems

We first concentrate on a system with minimal requirements for escrow encryption, assuming honest law enforcement. As motivated by [34], the system satisfies the following requirements: **(1) compliance verification:** the messages sent are assured to be open-able by law enforcement when the sender actually employs the system for privacy; and **(2) "limiting surveillance"**: as said in [34]: "information both sent and received by the user can be decrypted without release of keys of other users."

We note that we allow the systems in our definitions to use all available resources such as interaction and perhaps inefficient constructions as long as they are polynomial. We are interested in relations and we do not necessarily limit ourselves to restricted models.

Remark: We note that the sender may use another mechanism to encrypt the message (pre-encryption) or employ a covert channel. Our definition does not attempt to prevent such transmissions; all we argue about is that if a certain message is generated by the system and is sent via the gateway to the receiver–it should be the case that law enforcement gets this message when needed.

The parties: There are four parties which are polynomial-time and each has its respective key: S is the sender and G is the gateway, through which messages are passed, R is the receiver, and L is law enforcement. The sender and gateway may be active at message sending and are probabilistic algorithms. (We explain the motivation/source of the gateway below).

Basic properties and definition

- Based on the above two requirements, in order to satisfy the compliance verification the gateway G is introduced here which does not allow a ciphertext that does not pass the verification (of compliance) to be opened (or received) by the receiver. Such assurance seems to be a minimal requirement in a mandatory escrow process. Physically, this gateway may reside at the sender module, receiver module (as the LEAF checker in Clipper), or anywhere on the communication channel (the network router, the firewall, etc.). The gateway is a checking function of the sender and is similar but less involved than the recently suggested "trusted third party" [25] endorsed by a number of European government and financial institutes.
- To satisfy the minimal surveillance requirement, the law enforcement key must be different than the receiver's key.

Definition 1. Compliance verifiable escrow encryption system (CV-EES): Let k be the security parameter for an encryption system which for any Law enforcement (L) with a randomly chosen public key e_L (and corresponding private key d_L), for any public key e_R (chosen at random) and corresponding private keys d_R of the Receiver (R), and a verification key v_G for a compliance Gateway (G), for any Sender (S), the following holds.

Let α be the encryption of a message m generated by S, namely $S(e_R, e_L, m) = \alpha$, then there is a protocol between S and G, and then R acts on α (to get the message), L may apply to α later. We are assure that:

Certification of Compliance: for any α, let $G(v_G, \alpha)$ the result of a protocol between G and S computed by G, then $G(v_G, \alpha) = 1$ implies that there exists an m such that: $R(d_R, \alpha) = m$ and $L(d_L, \alpha) = m$ with probability. $1 - \frac{1}{k^d}$ for any constant d, for parameter k large enough.

Security: The system is polynomially secure [22], namely for any two messages m_0, m_1 computed by a message finder, for any ciphertext α that encrypts one of the message m_b ($b \in \{0,1\}$ chosen at random), for any message distinguisher that is given $\langle e_R, e_L, v_G, m_0, m_1, \alpha \rangle$ returns $b' = \{0,1\}$, then $b = b'$ with probability less than $\frac{1}{2} + \frac{1}{k^d}$ for any constant d, for k large enough.

Note that we can have a number of variations that do not change the system in a fundamental way: we can assume that the ciphertext generated is performed interactively with the gateway; also, the order of choice and publication of the keys does not matter as a receiver cannot help itself by using e_L in its key generation so e_R is actually drawn at random to be secure.

2.1 Chosen ciphertext security

Let us recall the definition of chosen-ciphertext secure systems [33].

Definition 2. Chosen ciphertext secure encryption system (ccs system): Let k be the security parameter for a public key encryption system which generates public/private key pair $\langle e, d \rangle$ for each user of the system. The adversary attacking a user (A *CC-attacker*) is a sender who is allowed the following **attack:** It generates a history tape h from $1^k, e$ and input/output pairs from (poly in k) ciphertext queries it provides adaptively to a decryption oracle which has d. Then, the following holds:

Security: Two messages m_0, m_1 from the message space are generated from a probabilistic polynomial time called a *message finder* on input 1^k and auxiliary input tape which may include h and e and other public information. Let α be the encryption of m_b with e for some randomly chosen bit b. Lastly, a message distinguisher given $\langle e, m_0, m_1, h, \alpha \rangle$ returns $b' = \{0,1\}$. A system is secure against chosen ciphertext if it is polynomially secure after the attack namely, for any CC-attacker, for any message finder, for any message distinguisher, then $b = b'$ with probability less than $\frac{1}{2} + \frac{1}{k^d}$ for any constant d and k large enough.

Remark: the definition above assumed non-adaptive attacker in the sense that the target ciphertext was not available to it when producing h, we may also allow adaptive attacker (that gets to see the challenge first, but is not allowed to query the oracle on it).

2.2 Equivalence of the systems

Next we compare the systems above: the first, motivated by requirements of an escrow encryption environment, and the second which assures level of security. We prove the following:

Theorem 3. *The following are equivalent: (1) Existence of compliance verifiable escrow system, and (2) Existence of chosen ciphertext secure encryption system.*

To prove the theorem we will show reductions in the next two Lemmas:

Lemma 4. *If there exists a compliance verifiable escrow system CV-EES then there exists chosen ciphertext secure (CCS) encryption system.*

Proof. **(Sketch)** We assume that there exists a CV-EES and build a CCS system. Let G be the gateway and v_G be the verification key for the CV-EES. Let e_R, e_L be the public keys and d_L be the law enforcement key corresponding to e_L for the CV-EES.

In the following we will use a "tinkering argument" that will move keys and components around to have a public-key system based solely on the components available to us from the CV-EES system. We demonstrate that the following is a CCS system in a complete public-key environment where every participant has a key (as in Rackoff and Simon [35]). The following is done

- Each user u publishes a public key as a receiver e_u which is drawn from the family of receivers' public key;
- in addition it publishes a sender key which is from the family of Law enforcement keys e_L^u.

Let V be the sender and U the receiver, we define them as following:
Encryption of m: $\alpha = S(e_U, e_L^V, m)$
Decryption of α: If $G(v_G, \alpha) = 1$ then return $m = L(d_U, \alpha)$ else return NULL.

First note that the system is polynomially secure, this is derived from the security definition of the CV-EES system. To prove that this is secure against a chosen ciphertext attack we use the following argument. When G returns 1 it means that the two decryptions (under d_U and under d_L^V) retrieves the same message with overwhelming probability. Thus, in a similar argument to [35], if $G(v_G, \alpha) = 1$ then the sender must have known the input which generated ciphertext α (by knowing and applying d_L^V to retrieve the message. Observe that this key is the sender private key drawn from the family of law enforcement keys which is corresponding to e_L^V. Since "the sender" already knew the value that is encrypted we are sure that revealing it after the check "the sender" won't learn anything new since "the sender" already must have known this information. Hence, the attacker being the sender was reduced to a "known plaintext attack" which is taken care of by the property of polynomial security. In fact we can show is that by providing the ciphertext queries and getting a corresponding cleartext answer in a CC-attack producing history h, the attacker has no more power than the (message only) attacker that produces by itself (without the help of the oracle) a cleartext message and then produces its ciphertext and produce a history h' of ordered ciphertexts and their corresponding messages.

The above reduction is direct (using the same keys used in the original system) and thus one does not lose in the size of the security parameter when translating the CV-EES to the CCS cryptosystem. This implies that a success ratio in breaking the CCS cryptosystem implies the same ration for the escrow related scheme (CV-EES) which means a linear preserving reduction.

Lemma 5. *If there exists a chosen ciphertext secure (CCS) encryption system then there exists compliance verifiable escrow system.*

Proof. **(Sketch)** First, we have an encryption scheme which is polynomially secure (by definition of CCS-cryptosystem); in fact all we need is a secure encryption for proving the lemma. So L can publish a public key e_L in such a system, and R publishes another public key e_R in this system. Then note that since we have chosen ciphertext secure encryption system we have a one-way function (we can use the encryption function for authentication protocol among two parties, thus by [23] one-way function exist).

We continue by a simple version of a construction of [33] to construct the CV-EES. To send a message m, a sender first generates two encryptions of m one for the receiver under e_R and one for law enforcement under e_L. The verification algorithm of G is done by a zero-knowledge proof of knowledge of the fact that "The sender (i.e., prover) knows a unique message m such that the two ciphertext are encryptions of it under their respective keys". This is an NP statement and can be proven to G interactively in a zero-knowledge fashion by the sender (that knows the preimages) using the availability of one-way functions. This proof assures that the message opened by the receiver using his public key is the same message available to law enforcement if they wish to open it using their key, thus G can allow the two ciphertexts to be transmitted together over the communication line. (In the next version we will formally recall the definition of proof of knowledge [17, 38, 1] and use it to show that the system has the required properties and that with very high probability both security and certification of compliance hold).

Using amplification of one-way functions [39] we can have the probability of extracting any computational advantage in the CV-EES system based on the zero knowledge proof, inverse exponential in the security parameter for any polynomial-time computation. Thus, breaking the system based on breaking the ZK proofs adds a negligible value to the time-success ratio. Now observe that the reduction just uses two CCS encryption systems, and the above zero-knowledge proof (which has the inverse exponential success probability). Breaking the CV-EES encryption means that in most of the time (at least $1/2$ of the cases) we break one of them. Therefore, this reduction is linear preserving.

3 Basic escrow encryption systems and non-malleability

CV-EES systems are not sufficient to protect individuals' privacy rights from unlawful search and seizure as they impose no compliance restriction on opening of ciphertext by law enforcement. For example, it was shown that with Clipper it is possible to modify the ciphertext so that it appears the ciphertext was generated by (or for) a Clipper chip different from the actual participants [20]. Let us review what we want from a "basic escrow encryption".

- First, we want "basic escrow encryption systems" to assure compliance of senders (as in CV-EES), and to be secure (as in CV-EES).
- In addition, it has been concluded by many that we need to have some context associated with a ciphertext which determines if law enforcement has the right to open that message. Then, an authorization body (judge, escrow agents, etc.) can use this context to determine whether to allow law enforcement to open or not to open a ciphertext (This is **context-limited escrow**). This is motivated by various designs [6, 20, 30, 8, 27, 5] and primarily by the correspondences on Clipper [18, 10, 19]. The ciphertext context includes the sender and receiver identity since, formally, "Law enforcement agencies require (1) information from the service provider to verify the association of the intercepted communications with the intercept subject, ..." [18]. This seems a reasonable minimal requirement. Note that "context limitation" is a double-edge sword. Namely, the sender who knows that law enforcement is allowed to escrow based on restricted context, can attach "wrong context" to evade legal escrowing. Thus, the sender's compliance has to be revised and to include also compliance with "a correct context" which is assured by extended compliance assurance which includes (**context certification**).
- Next, from a security point of view, we would need to be able to identify a sender with the message and not enable law enforcement to modify the sender's ID nor the other content and context (opening of messages is allowed only within a context). This makes the system **spoofing-free** (with respect to law enforcement that tries to modify messages or fabricate ones based on past opened messages and even when it can control some of the earlier messages sent in a conversation).

The notion of spoofing-freeness looked to us related to the one of non-malleability (defined in [16]). The later helped the formalization of the above requirements as following:

Definition 6. Basic Escrow encryption system (B-EES): Let k be the security parameter for an encryption system which for any Law enforcement (L) with a randomly chosen public key e_L (and corresponding private key d_L), for any verification key v_G for a compliance Gateway (G), and a legal authority J with authorization key $a_J = a_J(e_L, d_L)$, for any randomly chosen public key e_R and private keys d_R of the Receiver (R) (each of the keys drawn from a corresponding key family with parameter k), then, for any Sender (S):

Let $\kappa\tau$ be the context of a message which at minimum includes the identity of the sender $\kappa\tau_s$ and receiver $\kappa\tau_r$. Let α be the encryption of a message m generated by S using receiver's key e_R, namely $S(e_R, e_L, \kappa\tau, m) = \alpha$ then:

(1) Compliance and Context Certification and Correctness: for any α, let $G(v_G, \alpha)$ the result of a protocol between G and S computed by G, then $G(v_G, \alpha) = 1$ implies that with probability $1 - k^d$ for any d, for k large enough:

- there exists an m such that: $R(d_R, \alpha) = m, \kappa\tau$, and
- the result of $J(a_J, \alpha) = context, key$ and $context = \kappa\tau$ and $L(d_L, \alpha, key, ...) = m, \kappa\tau$. (We may assume that the context is part of the message).

(2) Context-Limited Escrow: For any ciphertext α and an authorized context $\kappa\tau$, Let $J(a_J, \alpha) = context, key$. If $context = \kappa\tau$ then L is activated and $L(d_L, \alpha, key) = m, \kappa\tau$.

(3) Spoofing-freeness: We define poly-time adversary $\mathcal{A}$ which may try to produce a message by modifying another message according to some poly-time relation REL (REL different from the identity relation), thus spoofing the system (and generating a message out of context or with different content that may be opened by the judge and in effect will frame the user); formally:

- The adversary $\mathcal{A}$ first generates a history tape h_1 from $1^k, e, e_L, d_L, e_R$ and input/output pairs from (poly in k) queries it provides to the authorizing authority with authorized contexts; For any ciphertext α_i and an authorized context $\kappa\tau_i$, Let $J(a_J, \alpha_i) = context_i, key_i$. If $context_i = \kappa\tau_i$ then L is activated and $L(d_L, \alpha_i, key_i) = m_i, \kappa\tau_i$. The record $(\alpha_i, \kappa\tau_i, key_i, m_i)$ is put on the history tape.
- Then, $\mathcal{A}$ produces a distribution $\mathcal{M}$ on messages (and contexts).
- Then $\mathcal{A}$ receives the challenge ciphertext $\alpha \in_R S(e, e_L, \kappa\tau, m)$ for $m \in_R \mathcal{M}$ and some knowledge about the message (e.g., its context) called $hint(m)$ which is polynomial time computable from m.
- $\mathcal{A}$ again generates a history tape h_2 from $1^k, e, e_L, d_L, e_R$ and input/output pairs from (poly in k) queries all different from α that it provides to the authorizing authority with authorized contexts; For any ciphertext α_j and an authorized context $\kappa\tau_j$, Let $J(a_J, \alpha_j) = context_j, key_j$. If $context_j = \kappa\tau_j$ then L is activated and $L(d_L, \alpha_j, key_j) = m_j, \kappa\tau_j$. The record $(\alpha_j, \kappa\tau_j, key_j, m_j)$ is put on the history tape.
- $\mathcal{A}$ now produces polynomially many ciphertexts f_i such that f_i is an encryption of β_i. Then $\mathcal{A}$ succeeds if $REL(m, \beta_i)$ holds for some i.

The system is called spoofing-free if for any polynomial $\mathcal{A}$ for any polynomial modification relation REL the probability of success is smaller than $1/k^d$ for any constant d, for k large enough. This concludes the definition.

We note that spoofing-freeness is modeled after non-malleability and implies polynomial security. In fact, we can show more strongly (proof omitted) that following the proof strategy of the Theorem 3 gives:

Theorem 7. *The following are equivalent: (1) Existence of basic escrow encryption systems, and (2) Existence of non-malleable encryption systems.*

Designs:

The characterizations have led us to a number of designs based on secure public key systems (and their relaxations). The designs introduce various ways to implement the compliance verifying gateway.

Based on private key system we can consider a server-based key distribution system (where users do not meet but each user shares a permanent key with a server). We can adapt our results and conclude that by augmenting such a system we can have an escrow system in this model. What we need is the notion of "publicly certified key distribution" where each key given to a user has also a publicly announced version which is encrypted or one-way processed by the trusted server. Now, each key distribution to a pair of users can be on-line verified by a gateway G for compliance. Unlike [25], this design needs only one way functions. We get (proof omitted):

Theorem 8. *Based on a trusted server and the existence of a one-way function (only), there exists a basic escrow system.*

References

1. M. Bellare and O. Goldreich, *On Defining Proofs of Knowledge*, Crypto '92.
2. M. Bellare and O. Goldwasser, *Verifiable Partial Key Escrow*, ACM, 4-th Symp. on Computer and Comm. Security, 1997.
3. M. Bellare and R. Rivest, *Translucent Cryptography – an alternative to key escrow and its implementation via fractional oblivious transfer*, a manuscript.
4. M. Bellare and P. Rogaway, *Random Oracles are Practical: a paradigm for designing efficient protocols*, ACM, 1-st Comp. and Com. Sec. 1993.
5. T. Beth, H.-J. Knobloch, M. Otten, G.J. Simmons and P.Wichmann, *Towards Acceptable Key Escrow Systems*, In the Proceedings of The 2nd ACM Symp. on Comp. and Comm. Security, 1994 51–58.
6. M. Blaze, *Protocol failure in the Escrowed Encryption Standard*, In the Proceedings of The 2nd ACM Symp. on Comp. and Comm. Security, 1994, 59–67.
7. M. Blaze, J. Feigenbaum and T. Leighton, *Master-Key Cryptosystems*, Crypto-95 Rump session.
8. *Building in Big Brothers*: the cryptographic policy debate, ed. L.J. Hoffman, Springer Verlag, 1995.
9. I. Damgård, *Towards practical public key cryptosystems secure against chosen ciphertext attacks*, Crypto '91.
10. D. E. Denning and M. Smid, *Key Escrowing Now*, IEEE Communications Magazine, Sep. 1994, pp. 54-68.
11. A. De Santis, Y. Desmedt, Y. Frankel and M. Yung, *How to Share a Function Securely*, ACM STOC 94.
12. A. De Santis, and G. Persiano, *Non-Interactive Zero-Knowledge Proof of Knowledge*, FOCS 93.
13. Y. Desmedt and Y. Frankel, *Threshold cryptosystems*, Crypto '89.

14. Y. Desmedt, *Securing Traceability of Ciphertexts: Towards a Secure Software Key Escrow Systems*, Eurocrypt 95.
15. L. Knudsen and T. Pedersen, *On the Difficulty of Software Escrowing*, Eurocrypt 96.
16. D. Dolev, C. Dwork and M. Naor, *Non-Malleable Cryptography*, STOC 91.
17. U. Feige, A. Fiat and A. Shamir, *Zero-Knowledge Proofs of Identity*, Journal of Cryptology, vol. 1, 1988, pp. 77–94. (Originally: STOC 87).
18. The FBI, *Law Enforcement requirements for the Surveillance of Electronic Communications*, June 1994.
19. FIPS PUB 185, *Escrowed Encryption Standard* Feb.94. (Dep. of Commerce).
20. Y. Frankel and M. Yung, *Escrow Encryption Visited: Attacks, Analysis and Designs*. Crypto '95.
21. Y. Frankel and M. Yung, *Cryptanalysis of the immunized LL public key systems*. Crypto '95.
22. S. Goldwasser and S. Micali, *Probabilistic Encryption*, J. Com. Sys. Sci. 28 (1984), pp 270-299.
23. R. Impagliazzo and M. Luby, *One-way Functions are Essential for Complexity-Based Cryptography* FOCS 89.
24. R. Impagliazzo and S. Rudich, *Limits on the Provable Consequences of Oneway Permutations*, STOC 89.
25. N. Jefferies, C. Mitchell and M. Walker, *A Proposed Architecture for Trusted Third Party Services*, in *Cryptography: Policy and Algorithms*, Springer Verlag LNCS 1029, 1996. (Also: Royal Holloway, U. of London Report, 95).
26. J. Kilian and F.T. Leighton, *Fair Cryptosystems, Revisited*, Crypto '95.
27. A. Lenstra, P. Winkler and Y. Yacobi, *A key escrow system with warrant bounds*, Crypto '95.
28. C. H. Lim and P. J. Lee, *Another method for attaining security against adaptive chosen ciphertext attacks*, Crypto '93.
29. M. Luby, *Pseudorandomness and its Cryptographic Applications*, Princeton Univ. Press, 1995.
30. S. Micali, *Fair public-key cryptosystems*, Crypto '92.
31. S. Micali and R. Sidney, *A simple method for generating and sharing pseudorandom functions with applications to clipper-like key escrow systems*, Crypto '95.
32. M. Naor and M. Yung, *Universal One-way Hash Functions and their Cryptographic Applications*, STOC 89.
33. M. Naor and M. Yung, *Public-key cryptosystem provably secure against chosen ciphertext attack*, STOC 1990.
34. NIST, *Issues: Export of software key escrow encryption*, August 1995. see: http://csrc.ncsl.nist.gov/keyescrow/
35. C. Rackoff and D. Simon, *Non-Interactive Zero-Knowledge Proof of Knowledge and Chosen Ciphertext Attacks*, Crypto '91.
36. J. Rompel *One-way Functions are Necessary and Sufficient for Secure Signatures*, STOC 90.
37. A. Shamir, *Partial Key Escrow*, Crypto 95 Rump Session.
38. M. Tompa and H. Woll, *Random Self-Reducibility and Zero-Knowledge Interactive Proofs of Possession of Information*, FOCS 87.
39. A. C. Yao, *Theory and Applications of Trapdoor functions*, FOCS 82.
40. Y. Zheng and J. Seberry, *Immunizing public key cryptosystems against chosen ciphertext attacks*, IEEE JSAC 93.

Randomness-Efficient Non-Interactive Zero Knowledge

(Extended Abstract)

Alfredo De Santis,[1] Giovanni Di Crescenzo,[2] Pino Persiano[1]

[1] Dipartimento di Informatica ed Applicazioni
Università di Salerno, 84081 Baronissi (SA), Italy
E-mail: {ads,giuper}@dia.unisa.it

[2] Computer Science and Engineering Department
University of California at San Diego, La Jolla, CA, 92093, USA
E-mail: giovanni@cs.ucsd.edu
(Part of this work was done while at Università di Salerno, Italy)

Abstract. The model of *Non-Interactive Zero-Knowledge* allows to obtain minimal interaction between prover and verifier in a zero-knowledge proof if a public random string is available to both parties. In this paper we investigate upper bounds for the length of the random string for proving one and many statements, obtaining the following results:

- We show how to prove in non-interactive *perfect* zero-knowledge any polynomial number of statements using a random string of fixed length, that is, not depending on the number of statements. Previously, such a result was known only in the case of *computational* zero-knowledge.
- Under the quadratic residuosity assumption, we show how to prove any NP statement in non-interactive zero-knowledge on a random string of length $\Theta(nk)$, where n is the size of the statement and k is the security parameter, which improves the previous best construction by a factor of $\Theta(k)$.

1 Introduction

Zero-knowledge proofs [19, 17] require quite a rich scenario in terms of resources needed and much effort has been devoted to presenting alternative poorer settings in which zero-knowledge proofs were possible.

In [5, 6, 12], the shared-string model for non-interactive zero-knowledge was put forward. Here, the prover and the verifier share a random string and the mechanism of the proof is mono-directional: the prover sends one message to the verifier. Non-interactive zero-knowledge proofs have found several applications in Cryptography (most notably the construction of cryptosystems secure against chosen-cyphertext attacks [24]) and can be employed in any setting in which communication is a precious and scarce resource. Thus, the shared-string model trades the need for interaction with the need for shared randomness. Since non-interactive zero-knowledge proofs from scratch can be obtained only for BPP languages [18], the shared-string model provides a minimal enough setting for non-interactive zero-knowledge.

Randomness has played a major role in several theoretical and applied fields of Computer Science. Several are the examples of computational tasks which are impossible to execute deterministically or whose efficiency is greatly enhanced if a source of random bits is available. Unfortunately, good random sources are difficult to find and this has motivated the study of the minimal amount of randomness needed for certain tasks (e.g., computing the sum in a secure way [7]), of techniques for reducing the number of random bits used by probabilistic algorithms (see for instance [20]) and the construction of pseudorandom generator specific for certain computational tasks: pseudorandom generator for constant-depth circuits [1, 25], space bounded computation [26, 27] and network computation [23] have been presented. The randomness in interactive proof systems has been studied in [2] and [3].

In this paper we consider the shared string model for non-interactive zero knowledge of [5, 6] and study the amount of shared randomness needed for zero-knowledge proofs.

Perfect zero-knowledge on a fixed random string. The first problem we investigate is the possibility of proving many statements using a random string of fixed length, i.e., not depending on the number of statements. This problem has found early solutions for the case of *computational* zero-knowledge in [5], assuming the intractability of quadratic residuosity, and, later, in [15], assuming the existence of certified one-way permutations. The certification requirement for one-way permutations was later removed in [4]. In [15, 13] the case of many provers was solved. Unfortunately, these constructions do not preserve *perfect* zero knowledge and thus cannot be used in our context. Before the current paper, no indication had been given that this problem might have a positive solution in the case of perfect zero-knowledge. The state of this problem was particularly unclear also because not many non-interactive perfect zero-knowledge protocols have been found in the literature (see [11]).

OUR RESULTS. We show how to prove many statements in non-interactive perfect zero-knowledge using a fixed random string. First we give a protocol for the language of quadratic non residuosity. Then we identify a general class of languages, called Simulator-Rankable languages, for which we give a protocol. Finally, we show that all languages known having a non-interactive perfect zero-knowledge proof system are Simulator-Rankable.

Non-interactive zero-knowledge for all NP on a short random string. Another problem we investigate is the possibility of proving any NP statement using a random string of short length. Many non-interactive zero-knowledge proof systems for NP-complete languages have been given in the literature, motivated by attempts both of reducing the complexity assumption necessary and of increasing the efficiency of the proof system. The first proof system for all NP was given in [6], under a specific number-theoretic assumption, and used a random string of length $\Theta(kn^3)$, where by k we denote the security parameter, and by n the size of the input. The proof system in [5, 12] reduced the assumption to the intractability of deciding quadratic residuosity modulo composite integers, and used a string of length $\Theta(kn^3)$. The proof system in [15] reduced the assumption to the intractability of inverting one-way permutations, and used a string of length $\Theta(kn^{5.5})$. Under the same assumption, [21] and [22] obtained proof systems using a string of length $\Theta(k^2 n \log n)$ and $\Theta(k^2 n)$, respectively. Under the quadratic residuosity assumption, [9] and [8] obtained proof systems using a random string of length $\Theta(k^2 n)$. As a result, the best known proof system for all NP before this paper uses a random string of length $\Theta(k^2 n)$.

OUR RESULT. Under the quadratic residuosity assumption, we show how to prove any NP statement in non-interactive zero-knowledge using a random string of length $\Theta(kn)$, thus improving the previous best result by a factor of $\Theta(k)$.

Lower bounding the length of the random string. In order to best estimate the efficiency of our proof systems, we have also looked at the question of finding lower bounds on the length of the random string necessary to obtain a non-interactive zero-knowledge proof. Previously, a result in [18] showed that non-interactive (computational or perfect) zero-knowledge proofs without the random string are possible only for languages in BPP.

OUR RESULT. We can show that non-interactive (computational or perfect) zero-knowledge proofs on a random string of length less than $\max(k, c \log n)$, for any constant c, can be given only for languages in BPP.

Organization of the paper. In Section 2, we review the definitions for non-interactive zero-knowledge proofs. In Section 3, we present our results on proving multiple non-interactive perfect zero-knowledge on a fixed random string. In Section 4, we present our result on proving any NP statement in non-interactive zero-knowledge on a short random string. Formal proofs and descriptions of some protocols are omitted from this extended abstract for lack of space. For the same reason, we follow the notation of [5] without explictly repeating it and advise the reader to refer to [28] or [5] for the necessary number-theoretic background.

2 Non-Interactive Zero-Knowledge

We review the definition of non-interactive zero-knowledge proof systems of [5], referring the reader to the original paper for motivations and discussions. We start with the definition of non-interactive proof systems.

Definition 1. Let P a probabilistic Turing machine and V a deterministic Turing machine that runs in time polynomial in the length of its first input. We say that (P,V) is a *Non-Interactive Proof System* with security parameter $k > 1$ for the language L if there exists a constant c such that the following hold:

1. *Completeness.* $\forall x \in L$, $|x| = n$, and for all sufficiently large n,
$$\mathbf{Pr}(\sigma \leftarrow \{0,1\}^{n^c}; Proof \leftarrow P(\sigma, x) : V(\sigma, x, Proof) = 1) \geq 1 - 2^{-k}.$$

2. *Soundness.* $\forall x \notin L$, $|x| = n$, for all Turing machines P', and for all sufficiently large n,
$$\mathbf{Pr}(\sigma \leftarrow \{0,1\}^{n^c}; Proof \leftarrow P'(\sigma, x) : V(\sigma, x, Proof) = 1) \leq 2^{-k}.$$

We will call the random string σ, input to both P and V, the *reference string*. Now we recall the definitions of non-interactive computational and perfect zero-knowledge proof systems. We will denote by $View(n,x)$ the probability space $View(n,x) = \{\sigma \leftarrow \{0,1\}^{n^c}; Proof \leftarrow P(\sigma,x) : (\sigma, Proof)\}$, where c is a constant.

Definition 2. Let (P,V) be a non-interactive proof system for the language L. We say that (P,V) is *Computational Zero-Knowledge* if there exists an efficient algorithm S, called the *Simulator* such that $\forall x \in L$, $|x| = n$, for all efficient non-uniform (distinguishing) algorithms D_n, $\forall d > 0$, and all sufficiently large n,

$$\left| Pr(s \leftarrow View(n,x) : D_n(s) = 1) - Pr(s \leftarrow S(1^n, x) : D_n(s) = 1) \right| < n^{-d}.$$

Definition 3. Let (P,V) be a non-interactive proof system for the language L. We say that (P,V) is *Perfect Zero-Knowledge* if there exists an efficient algorithm S, called the *Simulator* such that $\forall x \in L$, $|x| = n$, and all sufficiently large n, the two probability spaces $S(1^n, x)$ and $View(n,x)$ are equal.

3 Perfect zero-knowledge on a fixed random string

In this section we show how to prove any polynomial number of statements in non-interactive perfect zero-knowledge using a reference string of fixed length. In Subsection 3.1 we present our technique with respect to the language of quadratic non residuosity. In Subsection 3.2 we give a result that will be useful when proving this result for a more general class of languages: a transformation between any non-interactive zero-knowledge proof system with expected polynomial time simulator to one with strict polynomial time simulator. In Subsection 3.3 we describe a protocol that applies to a more general class of languages, that we call Simulator-Rankable languages.

Some simplifications. For simplicity, in our protocol for quadratic non residuosity we will assume that the modulus x is already known (or has already been proven) to be a Blum integer and, unless explicitly specified, that the reference string is made of integers in Z_x^{+1}, instead than of just n-bit integers. Techniques used, for instance, in [5] and [11], allow to deal with the general cases by losing only a constant factor in the length of the reference string, and preserving perfect zero knowledge.

3.1 Quadratic non residuosity

We present a perfect zero-knowledge proof system (A,B) with security parameter k that uses a reference string of length $\Theta(nk)$ for proving that any polynomial number $m(n)$ of elements $y_1, \ldots, y_{m(n)}$ are quadratic non residues modulo an integer x of length n.

The proof system of [5] for one statement. The non-interactive perfect zero-knowledge proof system of [5] for proving *one* quadratic non residuosity statement of size n uses a reference string of length nk. On input a pair (x, y), where x is a Blum integer and y an element of Z_x^{+1}, the reference string is viewed as the concatenation of k elements $z_1 \circ \cdots \circ z_k$ of Z_x^{+1}. If y is a quadratic non residue, then for each j, exactly one of z_j and $yz_j \bmod x$, call it u_j, is a quadratic residue and the prover gives a random square root of u_j. The soundness of the proof system relies on the fact that if y is a quadratic residue and z_j is a quadratic non residue then neither z_j and $yz_j \bmod x$ is a quadratic residue and thus the prover cannot satisfy the verifier's verifications. Since the z_j's are chosen at random and since exactly half of the elements in Z_x^{+1} are quadratic non residues, the prover has probability 2^{-k} of making the verifier accept when y is a quadratic residue.

Proving many statements. We modify the above described proof system in such a way that the following two properties are satisfied: 1) the prover can generate exactly one proof for each input and each reference string; 2) each proof has the same distribution as the reference string. We use the following definition. Let x be a Blum integer; for $z \in Z_x^{+1}$ and $b \in \{0,1\}$, define $u = sqrt(x, z, b)$ as the integer $u \in Z_x^{+1}$ such that (a) $u^2 = z \bmod x$ and (b) if $b = 0$ then $u \leq x/2$ else $u > x/2$. Now we give a formal description of our proof system (A,B).

Input to A and B:
- A $k(n+1)$-bit reference string $\sigma = z_1 \circ \cdots \circ z_k \circ b_1 \circ \cdots \circ b_k$, where $z_j \in Z_x^{+1}$, $b_j \in \{0,1\}$, for $j = 1, \ldots, k$.
- An $(m+1)$-tuple $(x, y_1, \ldots, y_m)$, where $|x| = n$, $y_i \in Z_x^{+1}$, for $i = 1, \ldots, m$.

Input to A: x's factorization.

Instructions for A.

A.1 Set $u_{1,j} = z_j$, $b_{1,j} = b_j$, for $j = 1, \ldots, k$.
A.2 For $i = 1, \ldots, m$,
 for $j = 1, \ldots, k$,
  if $u_{i,j} \in QR_x$ then
   compute $u_{i+1,j} = sqrt(x, u_{i,j}, b_{i,j})$ and set $b_{i+1,j} = 0$;
  if $u_{i,j} \in NQR_x$ then
   compute $u_{i+1,j} = sqrt(x, y \cdot u_{i,j} \bmod x, b_{i,j})$ and set $b_{i+1,j} = 1$;
  set $Proof_i = (u_{i+1,1}, \ldots, u_{i+1,k}, b_{i+1,1}, \ldots, b_{i+1,k})$.
A.3 Send $(Proof_1, \ldots, Proof_m)$ to B.

Input to B: A sequence of proofs $(Proof_1, \ldots, Proof_m)$, where $Proof_i = (u_{i+1,1}, \ldots, u_{i+1,k}, b_{i+1,1}, \ldots, b_{i+1,k})$, $u_{i+1,j} \in Z_x^{+1}$, $b_{i+1,j} \in \{0,1\}$, for $j = 1, \ldots, k$.

Instructions for B.

B.1 Set $u_{1,j} = z_j$, $b_{1,j} = b_j$, for $j = 1, \ldots, k$.
B.2 For $i = 1, \ldots, m$, and $j = 1, \ldots, k$,
 verify that $u_{i+1,j}^2 = y^{b_{i+1,j}} \cdot u_{i,j} \bmod x$.
B.3 If all verifications are satisfied then output: ACCEPT else output: REJECT.

Completeness, Soundness and Perfect Zero Knowledge: intuition. The completeness property is not hard to check. To prove soundness and perfect zero-knowledge, the following characterization of the distribution of a proof for a quadratic non residue is useful. The i-th proof $Proof_i$ is a string of k integers $u_{i+1,j}$ in Z_x^{+1}, and k bits $b_{i+1,j}$, such that the each $u_{i+1,j}$ is uniformly distributed (and so is its quadratic residuosity) and each bit $b_{i+1,j}$ is also uniformly distributed. The soundness of (A,B) can then be proved by induction on the number m of integers y_i. The base case is simple; for the inductive case, we assume that $y_1, \ldots, y_{i-1}$ are quadratic non residues modulo x, and that y_i is a quadratic residue, and use the above characterization of the distribution for the proof for y_{i-1}, that is also the reference string to be used for proving y_i. The perfect zero-knowledge of (A,B) can be proved by generating the m proofs starting from the last one, using the above characterization. Here the main difficulty consists in simulating the generation of a square root $u_{i+1,j}$ of $y^{b_i+1,j} \cdot u_{i,j} \bmod x$ which is less than $x/2$ or not, according to the value of the random bit $b_{i,j}$ taken from the reference string. The generation is accomplished as follows. The simulator will first choose bit $b_{i,j}$ at random and $u_{i+1,j} \in Z_x^{+1}$ and then compute $u_{i,j}$ such that $u_{i,j}^2 = y^{b_i+1,j} \cdot u_{i+1,j} \bmod x$; now, the value of bit $b_{i,j}$ is then determined depending on whether $u_{i,j}$ is greater or smaller than $x/2$. It is possible to see that if x is a Blum integer and y is a quadratic non residue, then bit $b_{i,j}$ (or in other words, the predicate saying whether $u_{i,j} \leq x/2$ or not) is uniformly distributed, no matter how quadratic residues are distributed in Z_x^{+1}. We obtain the following

Theorem 4. *(A,B) is a non-interactive perfect zero-knowledge proof system with security parameter k that can prove any polynomial number of quadratic non residuosity statements, each of size n and uses a reference string of length $\Theta(kn)$.*

3.2 Expected vs. strict polynomial time simulators

The zero-knowledge requirement in the definition of a non-interactive zero-knowledge proof system requires the simulator associated to the proof system to run in expected polynomial time. We can transform any non-interactive zero-knowledge proof system into one having the additional property that the simulator runs in strict polynomial time. The transformation preserves the kind of zero-knowledge, i.e., computational or perfect. We obtain the following

Theorem 5. *Let L be a language having a non-interactive zero-knowledge proof system. Then L has a non-interactive zero-knowledge proof system such that the simulator associated runs in strict polynomial time.*

3.3 A general class of languages

In this subsection we show a non-interactive perfect zero-knowledge proof system for proving many statements on a fixed reference string, which applies to some general class of languages, not necessarily depending on number-theoretic properties. We start with an informal discussion, and then define a class of languages and give a protocol for all languages in such class.

An informal discussion. Generalizing the proof system of previous section, an idea to construct a randomness-efficient protocol for proving many statements in non-interactive perfect zero-knowledge would be the following: a first statement x_1 is proved on a given reference string σ_1 and then the proof itself is used in order to compute a new reference string for the next statement x_2, and so on. Specifically, instead of using the proof, whose structure is not known in general, we would like to use the randomness needed by the simulator to simulate a proof for x_1 in order to compute a new reference string for the next statement x_2. Notice that because of Theorem 5, we can assume that the amount of randomness needed by the simulator to simulate a proof is a fixed and well defined quantity.

Simulator-Rankable languages. Let L be a language and let (A,B) be a non-interactive perfect zero-knowledge proof system for L; also, denote by M the simulator associated to (A,B), by σ the reference string, by x the common input, and by $S_{M,\sigma,x}$ the set $\{R \mid M(R,x) = (\sigma, Proof)\}$. If $|x| = n$, let $|R| = r(n)$, $|\sigma| = s(n)$ and $|S_{M,\sigma,x}| = 2^{t(n)}$ (we can assume a fixed length $r(n)$ for string R because of Theorem 5). We say that (A,B) is *simulator-rankable* if there exists a polynomial-time computable function $F : \{0,1\}^n \times \{0,1\}^{r(n)} \to \{0,1\}^{t(n)}$ such that if $x \in L$ then $F(x,R)$ is the rank of R in set $S_{M,\sigma,x}$, where σ is such that $M(R,x) = (\sigma, Proof)$. We say that language L is *simulator-rankable* if there exists a non-interactive perfect zero-knowledge proof system (A,B) for L which is simulator-rankable.

A protocol for any simulator-rankable language. Let L be a simulator-rankable language; now we describe a non-interactive perfect zero-knowledge proof system (P,V) for proving any polynomial number $m = m(n)$ of membership statements of size n to L which uses a fixed reference string. By $rank_S(x)$ we denote the rank of element x in set S. Now we give a formal description of (P,V).

Input to P and V: n-bit strings $x_1, \ldots, x_m$, and an $r(n)$-bit string σ.

Instructions for P:

P.1 Set $\tau_1 = \sigma$.
P.2 For $i = 1, \ldots, m$,
   write $\tau_i = \gamma_i \circ ind_i$, where $|\gamma_i| = s(n)$ and $|ind_i| = r(n) - s(n)$;
   compute $R_i \in S_{M,\gamma_i,x_i}$ such that $rank_{S_{M,\gamma_i,x_i}}(R_i) = ind_i$;
   set $\tau_{i+1} = R_i$.
P.3 Send $(\tau_1, \ldots, \tau_{m+1})$ to V.

Input to V: a sequence of $r(n)$-bit strings $(\tau_1, \ldots, \tau_{m+1})$.

Instructions for V:

V.1 Set $\tau_1 = \sigma$.
V.2 For $i = m, \ldots, 1$,
   write $\tau_i = \gamma_i \circ ind_i$, where $|\gamma_i| = s(n)$ and $|ind_i| = r(n) - s(n)$;
   set $R_i = \tau_{i+1}$ and $\sigma_i = M(R_i, x_i)$;
   check that $\sigma_i = \gamma_i$ and $F(x, R_i) = ind_i$.
V.3 If all verifications are successful then output: ACCEPT and halt, else output: REJECT and halt.

We obtain the following

Theorem 6. *Let L be a simulator-rankable language and let (A,B) be a simulator-rankable non-interactive perfect zero-knowledge proof system for L. Then (P,V) is a non-interactive perfect zero-knowledge proof system that can prove any polynomial $m = m(n)$ number of membership statements each of size n and uses a reference string of length $r(n)$, (that is, not depending on m), where $r(n)$ is the length of the random string used by the simulator M associated to (A,B).*

Examples of simulator-rankable languages. A first example of a simulator-rankable language is the language of quadratic non residuosity modulo Blum integers. This can be seen by using the protocol in [5], revised in Section 3.1: for each reference string σ, there exist exactly 2^k random strings R in set $S_{M,\sigma,x}$, since each integer $z_i \in Z_x^{+1}$ might have been generated from two different square roots: r_i and $-r_i$ mod x. This allows to compute the rank of any random string in $S_{M,\sigma,x}$, for any reference string σ. Later, in Section 4.2 we show that the language of all 1-out-of-3 thresholds over quadratic non residuosity is simulator-rankable. Using this fact, we can show the same for the language of k-out-of-m thresholds over quadratic non residuosity [11] and for the language of all secret-sharing based compositions over quadratic non residuosity [10]. Also, it is easy to see that the language of all elements in a family of trapdoor permutations [4] is simulator-rankable. This implies that all known languages having a non-interactive perfect zero-knowledge proof system are simulator-rankable.

4 A randomness-efficient protocol for NP

We start by reviewing the non-interactive zero-knowledge proof system for the NP-complete language 3SAT given in [5]. We will denote by k the security parameter of the proof system, by n the number of variables and by m the number of clauses of the 3-SAT input formula ϕ. Also, we choose the size of the Blum integer used as a modulus to be equal to k.

The protocol in [5] for 3SAT. The non-interactive zero-knowledge proof system for 3SAT given in [5] uses a reference string of length $\Theta(kn^3)$ and can be divided into three steps.

1. *Committing to truth values.* First of all the prover uniformly chooses a k-bit Blum integer x and a quadratic non residue y. Then, using x, y, and a satisfying assignment t for variables in ϕ, the prover assigns an integer $y_i \in Z_x^{+1}$ to each literal l_i in ϕ in such a way that if y is a quadratic non residue modulo x, then the following is true: y_i is a quadratic non residue modulo x if and only if literal l_i is true under the assignment t.
2. *Proving that the commitments are consistent.* Here the prover sends a non-interactive zero-knowledge proof that x is a Blum integer and y is a quadratic non residue modulo x.
3. *Proving that clauses are satisfied.* For each clause $(l_{i1} \lor l_{i2} \lor l_{i3})$ of ϕ, the prover proves that at least one of y_{i1}, y_{i2}, y_{i3} is a quadratic non residue modulo x, where integer y_{ij} was assigned to literal l_{ij}.

Our contribution. We give a significantly different implementation of the first and third step in the above protocol, and obtain the following

Theorem 7. *Under the quadratic residuosity assumption, there exists a non-interactive computational zero-knowledge proof system with security parameter k for 3SAT, using a reference string of length $\Theta(nk)$, where n is the number of variables of the input formula.*

Now we informally describe our implementation of the first and third steps of the above protocol, omitting a formal description. We remark that our protocol satisfies also the requirement of strong soundness, that is, it is sound also if a malicious prover chooses the statement after seeing the reference string.

4.1 Committing to the truth values of the literals

Let t be an assignment for variables $v_1, \ldots, v_n$ in the 3SAT formula ϕ; let x be the input modulus and let $w_1 \circ \cdots \circ w_n$ be a portion of the random string, where each $w_i \in Z_x^{+1}$. Also, denote by q_i the quadratic residuosity of w_i, for each $i = 1, \ldots, n$. Then the prover P commits to each v_j and $\overline{v}_j$ as follows. For each $i = 1, \ldots, n$, P sets $d_i = t(v_i) \oplus q_i$, $tcom_i = y^{d_i} \cdot w_i \bmod x$ and $ncom_i = y \cdot tcom_i \bmod x$. The commitments are then $(v_i, tcom_i), (\overline{v}_i, ncom_i)$, for $i = 1, \ldots, n$. It is easy to check that $tcom_i$ ($ncom_i$) is a quadratic non-residue if and only if variable v_i ($\overline{v}_i$) is true under assignment t. We remark that the above commitments are generated using integers from the reference string, while in [5] they were generated from the prover by using some private randomness. In our analysis, this will decrease significantly the cheating power of a dishonest prover and will allow us to use a shorter reference string in the proof system for proving that the clauses have been correctly constructed.

4.2 Proving that the clauses are satisfied

In order to prove that a single clause is satisfied, we use a non-interactive perfect zero-knowledge proof system for the language 3-OR(NQR$_x$) of triples (y_1, y_2, y_3) such that at least one out of y_1, y_2, y_3 is a quadratic non residue modulo the Blum integer x. We do not yet know whether such language is simulator-rankable, since it is not clear how to use the two protocols given in [14, 11] for this language in order to derive such property. Here we describe a non-interactive perfect zero-knowledge proof system (A,B) for language 3-OR(NQR$_x$), which allows to conclude that such language is simulator-rankable, and thus allows to prove all m clauses of formula ϕ on one fixed random string.

An informal description. We start with some definitions. Let x be a Blum integer and $b_1, b_2, b_3 \in \{0, 1\}$; we say that a triple (z_1, z_2, z_3) of integers in Z_x^{+1} has *quadratic character* (b_1, b_2, b_3), if $\mathcal{Q}_x(z_i) = b_i$, for $i = 1, 2, 3$. Also, we say that two triples (y_1, y_2, y_3) and (z_1, z_2, z_3) of integers in Z_x^{+1} have *different* quadratic characters if the two triples of bits representing the quadratic characters of (y_1, y_2, y_3) and (z_1, z_2, z_3) are different. Finally, we define the *OR-triples* of (y_1, y_2, y_3), for any triple (y_1, y_2, y_3) of integers in Z_x^{+1}, as the 7 triples (y_1, y_2, y_3), $(y_1 y_2 y_3, y_1 y_3, y_1)$, $(y_2, y_3, y_1 y_2)$, $(y_3, y_1 y_2, y_2 y_3)$, $(y_1 y_2, y_2 y_3, y_1 y_2 y_3)$, $(y_2 y_3, y_1 y_2 y_3, y_1 y_3)$, $(y_1 y_3, y_1, y_2)$, where all computations are done modulo x. We will use the following

Fact 1 *Let x be a Blum integer, and let $y_1, y_2, y_3 \in Z_x^{+1}$. Then the OR-triples of (y_1, y_2, y_3) satisfy the following properties:*

1. *If (y_1, y_2, y_3) has quadratic character $(0, 0, 0)$ then all OR-triples of (y_1, y_2, y_3) have quadratic character $(0, 0, 0)$;*
2. *If (y_1, y_2, y_3) has quadratic character different from $(0, 0, 0)$ then*
   - *each OR-triple of (y_1, y_2, y_3) has quadratic character different from $(0, 0, 0)$;*
   - *each two OR-triples of (y_1, y_2, y_3) have different quadratic character.*

The proof system (A,B) uses a reference string of length $\Theta(nk)$, viewed as the concatenation of triples $(z_{i,1}, z_{i,2}, z_{i,3})$ of integers in Z_x^{+1}, for $i = 1, \ldots, \lceil k/3 \rceil$. On input (x, y_1, y_2, y_3), the prover A computes the quadratic character $(d_{i,1}, d_{i,2}, d_{i,3})$ of each triple $(z_{i,1}, z_{i,2}, z_{i,3})$. Now, if $(d_{i,1}, d_{i,2}, d_{i,3}) = (0, 0, 0)$ then A computes and sends to B square roots of $z_{i,1}, z_{i,2}, z_{i,3}$. Instead, if $(d_{i,1}, d_{i,2}, d_{i,3}) \neq (0, 0, 0)$, A computes and sends to B square roots of $z_{i,1} \cdot v_1 \bmod x$, $z_{i,2} \cdot v_2 \bmod x$, and $z_{i,3} \cdot v_3 \bmod x$, where (v_1, v_2, v_3) is the OR-triple with quadratic character $(d_{i,1}, d_{i,2}, d_{i,3})$. The verifier B checks that the square roots are correctly computed. A formal description of (A,B) is omitted. Similarly as done for the language of quadratic non residuosity, we can show that the language 3-OR(NQR$_x$) is simulator-rankable. Using Theorem 6, we obtain the following

Theorem 8. *There exists a non-interactive perfect zero-knowledge proof system with security parameter k for proving any polynomial number $m(n)$ of membership statements for the language 3-OR(NQR$_x$) of size n, which uses a reference string of length $\Theta(kn)$.*

Acknowledgements. We thank Russell Impagliazzo for valuable discussions. Part of the second author's research was supported by NSF YI Award CCR-92-570979 and Sloan Research Fellowship BR-3311.

References

1. M. Ajtai and A. Wigderson, *Deterministic Simulation of Probabilistic Constant Depth Circuits*, in Proceedings of STOC 85.
2. M. Bellare, O. Goldreich, and S. Goldwasser, *Randomness in Interactive Proof Systems*, in Proceedings of FOCS 90.
3. M. Bellare and J. Rompel, *Randomness in Interactive Proof Systems*, in Proceedings of FOCS 94.
4. M. Bellare and M. Yung, *Certifying Cryptographic Tools: the case of Trapdoor Permutations*, in Journal of Cryptology, vol. 9, n. 1, pp. 149–166.

5. M. Blum, A. De Santis, S. Micali, and G. Persiano, *Non-Interactive Zero-Knowledge*, SIAM Journal of Computing, vol. 20, no. 6, Dec 1991, pp. 1084–1118.
6. M. Blum, P. Feldman, and S. Micali, *Non-Interactive Zero-Knowledge and Applications*, in Proceedings of STOC 88.
7. C. Blundo, A. De Santis, G. Persiano, and U. Vaccaro, *On the number of random bits in totally private computations*, in Proceedings of ICALP 95.
8. J. Boyar and R. Peralta, *Short Discreet Proofs*, in Proc. of EUROCRYPT 96.
9. I. Damgaard, *Non-interactive circuit-based proofs and non-interactive perfect zero-knowledge with preprocessing*, in Proceedings of EUROCRYPT 92.
10. A. De Santis, G. Di Crescenzo, and G. Persiano, *Secret Sharing and Perfect Zero-Knowledge*, in Proceedings of CRYPTO 93.
11. A. De Santis, G. Di Crescenzo, and G. Persiano, *The Knowledge Complexity of Quadratic Residuosity Languages*, in Theor. Comp. Sc., Vol. 132, pp. 291–317.
12. A. De Santis, S. Micali, and G. Persiano, *Non-Interactive Zero-Knowledge Proof Systems*, in Proceedings of CRYPTO 87.
13. A. De Santis and M. Yung, *Cryptographic Applications of the Metaproof and Many-Prover Systems*, in Proceedings of CRYPTO 90.
14. G. Di Crescenzo, *Recycling Random Bits for Composed Perfect Zero-Knowledge*, in Proceedings of EUROCRYPT 95.
15. U. Feige, D. Lapidot, and A. Shamir, *Multiple Non-Interactive Zero-Knowledge Proofs Based on a Single Random String*, in Proceedings of FOCS 90.
16. L. Fortnow, *The Complexity of Perfect Zero-Knowledge*, in Proc. of STOC 87.
17. O. Goldreich, S. Micali, and A. Wigderson, *Proofs that Yield Nothing but their Validity or All Languages in NP Have Zero-Knowledge Proof Systems*, Journal of the ACM, vol. 38, n. 1, 1991, pp. 691–729.
18. O. Goldreich and Y. Oren, *Definitions and Properties of Zero-Knowledge Proof Systems*, Journal of Cryptology, vol. 7, 1994, pp. 1–32.
19. S. Goldwasser, S. Micali, and C. Rackoff, *The Knowledge Complexity of Interactive Proof-Systems*, SIAM Journal on Computing, vol. 18, n. 1, February 1989.
20. R. Impagliazzo and D. Zuckerman, *How to Recycle Random Bits*, in Proceedings of FOCS 89.
21. J. Kilian, *On the Complexity of Bounded-interaction and Non-interactive Zero-knowledge Proofs*, in Proceedings of FOCS 94.
22. J. Kilian and E. Petrank, *An Efficient Zero-knowledge Proof System for NP under General Assumptions*, in Electronic Colloquium on Computational Complexity, Technical Report no. TR95-038.
23. R. Impagliazzo, N. Nisan, and A. Wigderson, *Pseudorandomness for Network Algorithms*, in Proceedings of STOC 94.
24. M. Naor and M. Yung, *Public-Key Cryptosystems Provably Secure against Chosen Ciphertext Attack*, in Proceedings of STOC 90.
25. N. Nisan, *Pseudorandom Bits for Constant Depth Circuits*, Combinatorica, 11, pp. 63-70, 1991.
26. N. Nisan, *Pseudorandom Sequences for Space Bounded Computations*, Combinatorica, 12, pp. 449–461, 1992.
27. N. Nisan and D. Zuckerman, *More deterministic simulation in LOGSPACE*, in Proceedings of STOC 93.
28. I. Niven and H. S. Zuckerman, *An Introduction to the Theory of Numbers*, John Wiley and Sons, 1960, New York.

Approximation Results
for the Optimum Cost Chromatic Partition Problem

Klaus Jansen[1]

Fachbereich IV - Mathematik, Universität Trier, 54 286 Trier, Germany, email:
jansen@dm3.uni-trier.de

Abstract. In this paper, we study the optimum cost chromatic partition (OCCP) problem for several graph classes. The OCCP problem is the problem of coloring the vertices of a graph such that adjacent vertices get different colors and that the total coloring costs are minimum.
We prove that there exists no polynomial approximation algorithm with ratio $O(|V|^{0.5-\epsilon})$ for the OCCP problem restricted to bipartite and interval graphs, unless $P = NP$.
Furthermore, we propose approximation algorithms with ratio $O(|V|^{0.5})$ for bipartite, interval and unimodular graphs. Finally, we prove that there exists no polynomial approximation algorithm with ratio $O(|V|^{1-\epsilon})$ for the OCCP problem restricted to split, chordal, permutation and comparability graphs, unless $P = NP$.

1 Introduction

In this paper, we study the optimum cost chromatic partition (OCCP) problem for several graph classes. The graph classes used in this paper are defined e.g. in [5]. The OCCP problem can be described as follows: Given a graph $G = (V, E)$ with n vertices and a sequence of coloring costs $(k_1, \ldots, k_n)$, find a feasible coloring $f(v)$ for each vertex $v \in V$ such that the total coloring costs $\sum_{v \in V} k_{f(v)}$ are minimum. A coloring $f : V \to \{1, \ldots, n\}$ is feasible if adjacent vertices have different colors. Alternatively, the OCCP problem can be formulated as follows: Given a graph $G = (V, E)$ with n vertices and a sequence of coloring costs $(k_1, \ldots, k_n)$, find a partition into independent sets $U_1, \ldots, U_s$ such that $\sum_{c=1}^{s} k_c \cdot |U_c|$ is minimum. We may assume that $k_c \leq k_d$ whenever $c < d$.

A VLSI layout problem introduced by Supowit [11] with terminals on a circle or on two opposite parallel lines corresponds to the OCCP problem restricted to circle or permuation graphs. Another application is given by Kroon et al. [9]. The OCCP problem for interval graphs is equivalent to the Fixed Interval Scheduling Problem (FISP) with machine dependent processing costs. It is not difficult to see that the OCCP problem is NP-complete for arbitrary graphs. Sen et al. [10] proved that the OCCP problem for circle graphs is NP-complete.

Kroon et al. [9] studied the OCCP problem for interval graphs and trees. They showed that the problem restricted to trees can be solved in linear time and that the problem restricted to interval graphs is NP-complete even if there are only four different values for the coloring costs. If there are only two different values for the coloring costs, then the OCCP problem is equivalent to the maximum q-colorable subgraph problem. Suppose that the first q costs are equal and that the last $n - q$ costs are equal ($k_1 = \ldots = k_q < k_{q+1} = \ldots = k_n$). Then, we get an optimum solution if the maximum q-colorable subgraph is colored with the colors $1, \ldots, q$ and if the other vertices are colored with the remaining colors. The maximum q-colorable subgraph problem has been studied extensively by Frank [3], Gavril [4], Yannakakis and Gavril [12], Jansen et al. [6] and Chang et al. [2]. Further complexity results for the OCCP problem can be found in [7].

We give several approximation results for the OCCP problem restricted to bipartite, chordal, comparability, interval, permutation, unimodular and split graphs. We prove that there exists no polynomial approximation algorithm with ratio $O(|V|^{0.5-\epsilon})$ for the OCCP problem restricted to bipartite and interval graphs, unless $P = NP$. Furthermore, we propose approximation algorithms with ratio $O(|V|^{0.5})$ for both graph classes and for unimodular graphs. Finally, we prove that there exists no polynomial approximation algorithm with ratio $O(|V|^{1-\epsilon})$ for the OCCP problem restricted to split, chordal, permutation and comparability graphs, unless $P = NP$.

2 Bipartite graphs

In this section we prove that OCCP is hard to approximate $O(|V|^{0.5-\epsilon})$ for bipartite graphs. After that, we propose an approximation algorithm with ratio $O(|V|^{0.5})$.

2.1 Non-Approximability result

We use the precoloring extension problem that is NP-complete for bipartite graphs proved by Bodlaender, Jansen and Woeginger [1]. Given a bipartite graph $G = (V, E)$ with vertex set $V = A \cup B$, edge set $E \subset \{\{v, w\} | v \in A, w \in B\}$ and three specified vertices $a_1, a_2, a_3 \in A$, the 1-PrExt problem is to decide whether there exists a 3-coloring of G with $f(a_1) = 1$, $f(a_2) = 2$ and $f(a_3) = 3$.

First, we show the NP-completeness of the OCCP problem using an integer parameter K. Later, we specify the parameter K to achieve our non - approximability result.

Theorem 1. *The OCCP problem for bipartite graphs is NP-complete if there are at least four different cost values.*

Proof. The theorem is proved by a reduction from 1-PrExt restricted to bipartite graphs. We may assume that $G = (A \cup B, E)$ contains three further vertices $b_1, b_2, b_3 \in B$ with $\{a_i, b_j\} \in E$ for $1 \leq i \neq j \leq 3$. Let n be the number of vertices in G.

Let I be an instance of 1-PrExt containing the bipartite graph $G = (A \cup B, E)$ with $a_1, a_2, a_3 \in A$ and $b_1, b_2, b_3 \in B$ as described above. Let K be a positive integer with $K \geq 1$. An instance I' of the OCCP problem is constructed as follows. First, we define a bipartite graph $G' = (V', E')$ with vertex set

$$V' = \{v_{1,j}, v_{2,j} | 1 \leq j \leq 2000K^2n\} \cup \\ \{v_{3,j'}, v_{4,j'} | 1 \leq j' \leq 100Kn\} \cup \{v_5, v_6\}$$

and edge set

$$E' = \{\{v_{1,j}, v_{3,j'}\}, \{v_{2,j}, v_{4,j'}\} | 1 \leq j \leq 2000K^2n, 1 \leq j' \leq 100Kn\} \cup \\ \{\{v_5, v_{3,j'}\}, \{v_6, v_{4,j'}\} | 1 \leq j' \leq 100Kn\} \cup \\ \{v_5, v_6\}.$$

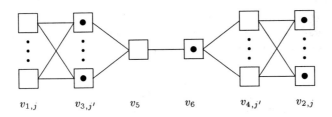

Fig. 1. The constructed graph G' and a feasible 2-coloring of G'.

The bipartite graph G' illustrated in Figure 1 contains $4000K^2n + 200Kn + 2$ vertices. Then, we connect G and G' using the following edges:

$$\bar{E} = \{\{a_1, v_{3,j'}\}, \{b_1, v_{4,j'}\}, \{b_1, v_5\}, \{a_2, v_{2,j}\}, \\ \{b_2, v_{1,j}\}, \{b_2, v_5\}, \{a_3, v_{3,j'}\}, \{a_3, v_{2,j}\}, \\ \{b_3, v_{1,j}\}, \{b_3, v_{4,j'}\} \mid 1 \leq j \leq 2000K^2n, 1 \leq j' \leq 100Kn\}\}.$$

In total, the bipartite graph $\bar{G}$ for I' is given by

$$\bar{G} = (A \cup B \cup V', E \cup E' \cup \bar{E}).$$

The cost values are $k_1 = 1$, $k_2 = 10K$, $k_3 = 100K^2$ and $k_4 = 15000K^3n$. A cheap coloring of $\bar{G}$ has to use only three colors; otherwise the costs would be more than $15000K^3n$.

We can prove the following statements: I is a yes instance of 1-PrExt if and only if the minimum total costs of coloring all vertices in I' don't exceed $6100K^2n + 100K^2 + 1$.

If there is no solution of the 1-PrExt problem, then we have either four colors in $\bar{G}$ with coloring costs of at least $k_4 = 15000K^3n$ or a 3-coloring with coloring costs of at least $10000K^3n$. □

Theorem 2. *For each $\epsilon < \frac{1}{2}$, there exists no polynomial approximation algorithm with ratio $O(|V|^{0.5-\epsilon})$ for the OCCP problem restricted to bipartite graphs, unless $P = NP$.*

Proof. Let H be an approximation algorithm for the OCCP problem that computes a coloring with costs $H(I') \leq c|V|^{0.5-\epsilon}OPT(I')$, where c is a constant and $OPT(I')$ are the minimum costs of a solution I'.

We construct for an instance I of the 1-PrExt problem an instance I' of the OCCP problem as described in the proof above. We obtain a graph with at most $4300K^2n$ vertices. If there exists a solution of the 1-PrExt problem, the optimum solution of the OCCP instance I' has costs of at most $6200K^2n$. In this case, our approximation algorithm produces the value $H(I') \leq 6200K^2cn|V|^{0.5-\epsilon}$. Since the number of vertices in I' is at most $4300K^2n$, we have

$$c|V|^{0.5-\epsilon} \leq (4300)^{0.5}K^{1-2\epsilon}cn^{0.5}.$$

If there exists no solution of the 1-PrExt instance, then $OPT(I') > 10000K^3n$ and, therefore, algorithm H generates a solution with costs greater than $10000K^3n$. Next, we consider the inequality

$$10000K^3n > (4300)^{0.5}6200K^{3-2\epsilon}cn^{1.5}.$$

This inequality is satisfied if and only if

$$K^{2\epsilon} > \frac{c(4300)^{0.5}6200}{10000}n^{0.5}.$$

We define

$$K = \left\lceil \sqrt[2\epsilon]{\frac{c(4300)^{0.5}6200}{10000}n^{0.5}} \right\rceil + 1.$$

Since c and ϵ are constant, K is a polynomial in n and, therefore, the instance I' can be constructed in polynomial time. If there exists no solution of the 1-PrExt problem, then H generates a solution with costs of at least

$$10000K^3n > (4300)^{0.5}6200K^{3-2\epsilon}cn^{1.5} \geq 6200K^2cn|V|^{0.5-\epsilon}.$$

Therefore, by using the polynomial time approximation algorithm H, we could decide the existence of a solution for the 1-PrExt problem, which would imply $P = NP$. □

2.2 Approximability result

The key idea of the approximation algorithm is to compute two colorings for the problem and to choose the cheaper one.

Algorithm A

given: Instance I of the OCCP problem containing a bipartite graph $G = (V, E)$ and cost vector $(k_1, \ldots, k_{|V|})$.

(1) Compute a 2-coloring of G with n_1 vertices colored with color 1 and $|V| - n_1$ vertices colored with color 2 such that n_1 is maximum and, therefore, $n_1 \geq \frac{|V|}{2}$. The costs of the first coloring are $A_1(I) = n_1k_1 + (|V| - n_1)k_2$.

(2) Compute a maximum independent set U in G with $\alpha(G)$ vertices and color the vertices in U with color 1. Then, compute a 2-coloring of $G[V \setminus U]$ with n'_1 vertices colored with color 2 and $|V| - \alpha(G) - n'_1$ vertices colored with color 3 such that $n'_1 \geq \frac{|V|-\alpha(G)}{2}$. The costs of the second coloring are $A_2(I) = \alpha(G)k_1 + n'_1 k_2 + (|V| - \alpha(G) - n'_1)k_3$.

(3) Choose the cheaper coloring among the two colorings.

We note that the costs of the second coloring are bounded by

$$\alpha(G)k_1 + \frac{|V| - \alpha(G)}{2}(k_2 + k_3) \leq \alpha(G)k_1 + (|V| - \alpha(G))k_3.$$

Theorem 3. *Algorithm A computes a solution of the OCCP problem restricted to bipartite graphs with approximation ratio $\leq |V|^{0.5}$.*

Proof. Let I be an instance of the OCCP problem containing a bipartite graph $G = (V, E)$ and cost vector $(k_1, \ldots, k_{|V|})$. Then, we have two lower bounds for the optimum value $OPT(I)$:

(1) $OPT(I) \geq |V|k_1$,
(2) $OPT(I) \geq \alpha(G)k_1 + (|V| - \alpha(G))k_2$.

We consider two cases $k_3 \leq |V|^{0.5}k_2$ and $k_3 > |V|^{0.5}k_2$ and can prove that $A(I) \leq |V|^{0.5}OPT(I)$. □

3 Interval graphs

In this section we prove that the OCCP problem restricted to interval graphs is hard to approximate with ratio $O(|V|^{0.5-\epsilon})$. Furthermore, we propose an approximation algorithm with ratio $O(|V|^{0.5})$ for interval graphs and also for unimodular graphs.

3.1 Non-Approximability result

The NP-completeness proof uses a reduction from Numerical Three Dimensional Matching (N3DM) and is a modification of the pure NP-completeness proof of the OCCP problem given by Kroon et al. [9].

Theorem 4. *For each $\epsilon < \frac{1}{2}$, there exists no polynomial approximation algorithm with ratio $O(|V|^{0.5-\epsilon})$ for the OCCP problem restricted to interval graphs, unless $P = NP$.*

Proof. First, we give a reduction from N3DM with variable parameter $K \in \mathbb{N}$ and, later, we specify the parameter $K \in \mathbb{N}$ to achieve our non-approximability result. Let I_1 be an instance of N3DM with integer t and rational numbers $0 < a_i, b_i, c_i < 1$ for $1 \leq i \leq t$ with $\sum_{i=1}^{t}(a_i + b_i + c_i) = t$. The N3DM problem is to decide whether there exist permutations ρ and δ of $\{1, \ldots, t\}$ such that $a_i + b_{\rho(i)} + c_{\delta(i)} = 1$ for $1 \leq i \leq t$.

We choose further rational numbers A_i, B_j and X_{ij} such that all these numbers are different and that $4 < A_i < 5 < B_j < 6$ and $7 < X_{ij} < 9$ for $1 \leq i,j \leq t$. Next, we construct an instance I_2 of the OCCP problem. We use the intervals given in Table 1 for the interval graph.

interval	interval	numbers
$(0, B_j]$	$(11 - c_k, 13]$	$1 \leq j \leq t$ or $1 \leq k \leq t$
$(1, 2]$	$(2, A_i]$	t times or $1 \leq i \leq t$
$(A_i, X_{ij}]$	$(X_{ij}, 10 + a_i + b_j]$	$1 \leq i, j \leq t$
$(B_j, X_{ij}]$	$(X_{ij}, 14]$	$1 \leq i, j \leq t$
$(3, B_j]$	$(0, A_i]$	$t-1$ times and $1 \leq j \leq t$ or $1 \leq i \leq t$
$(\frac{100Kt^2}{2000K^2t^4}, 3]$	$(12, 14 - \frac{100Kt^2}{2000K^2t^4}]$	$t^2 - t$ times
$(\frac{\ell}{2000K^2t^4}, \frac{\ell+1}{2000K^2t^4}]$	$(13 + \frac{\ell}{2000K^2t^4}, 13 + \frac{\ell+1}{2000K^2t^4}]$	t times, $0 \leq \ell < 2000K^2t^4$
$(\frac{\ell}{2000K^2t^4}, \frac{\ell+1}{2000K^2t^4}]$	$(14 - \frac{\ell+1}{2000K^2t^4}, 14 - \frac{\ell}{2000K^2t^4}]$	$t^2 - t$ times, $0 \leq \ell < 100Kt^2$

Table 1. The intervals in the interval graph

Furthermore, there are t colors with costs 1, $t^2 - t$ colors with costs $10Kt^2$, t^2 colors with costs $100K^2t^4$ and all other colors have costs $20000K^3t^6$.

The **first claim** (see also [8]) is to prove the following statement: I_1 is a yes instance of N3DM if and only if the minimum total costs of coloring all intervals of I_2 do not exceed

$$costs(K) := 2000K^2t^5 + 5t + 2300K^2t^6 + 50Kt^4 - 50Kt^3.$$

If I_1 is a no instance of N3DM, then the total costs of coloring all intervals of I_2 are greater than $10000K^3t^6$. We notice that the value $costs(K)$ is bounded by $4355K^2t^6$.

The **second part** of the proof is the specification of the parameter K to achieve our non-approximability result. We define

$$K = \left\lceil \sqrt[2\epsilon]{\frac{(4208)^{0.5}4355}{10000} t^{2.5}} \right\rceil + 1$$

and get our non-approximability result (see also [8]). □

3.2 Approximability result

Next, we propose an approximation algorithm A with ratio $O(|V|^{0.5})$ for the OCCP problem restricted to interval graphs or to unimodular graphs. The key idea is to analyse the structure of the optimum solution and to solve a special coloring problem.

Suppose that the optimum solution consists of $b_{opt} \geq \chi(G)$ colors. Furthermore, we assume that the colors $a_{opt}, a_{opt} + 1, \ldots, b_{opt}$ cover at least $\lceil \sqrt{|V|} \rceil$

vertices and that the colors $a_{opt}+1, \ldots, b_{opt}$ cover less than $\lceil\sqrt{|V|}\rceil$ vertices of G. This implies that $a_{opt} \in \{1, \ldots, b_{opt}\}$. Let $n_{a_{opt}, b_{opt}-a_{opt}}$ be the number of vertices colored with colors $1, \ldots, a_{opt}$ and let $\bar{n}_{a_{opt}, b_{opt}-a_{opt}}$ be the number of vertices colored with the other colors $a_{opt}+1, \ldots, b_{opt}$. Therefore, $\bar{n}_{a_{opt}, b_{opt}-a_{opt}}$ is bounded by $\sqrt{|V|}$.

Using these assumptions, we obtain the following lower bounds for the minimum costs $OPT(I)$ of a coloring:

(1) $OPT(I) \geq \lceil\sqrt{|V|}\rceil \cdot k_{a_{opt}}$,
(2) $OPT(I) \geq k_{b_{opt}}$.

The first inequality is satisfied, since $\lceil\sqrt{|V|}\rceil$ vertices are colored with the colors $a_{opt}, a_{opt}+1, \ldots, b_{opt}$ and since $k_{a_{opt}} \leq k_{a_{opt}+1} \leq \ldots \leq k_{b_{opt}}$. The second inequality follows from the fact that color b_{opt} occurs at least once in the optimum coloring.

For our approximation algorithm we have to solve the following graph theoretical problem (called *maximum $(a, b-a)$-colorable subgraph problem*).

Maximum $(a, b-a)$-colorable subgraph
Given: A graph $G=(V,E)$, and numbers $a,b \in \mathbb{N}$ with $a \leq b$ and $b \geq \chi(G)$.
Question: Compute a partition $(V', V \setminus V')$ of V such that V' has maximum cardinality and can be colored with a colors and $V \setminus V'$ can be colored with $b-a$ colors.

Let H be an optimum algorithm to solve the maximum $(a, b-a)$-colorable subgraph problem. A call of this algorithm with parameters a and b is denoted by $H(a, b-a)$. Note, that the maximum $(a, b-a)$-colorable subgraph problem is harder as the maximum q-colorable subgraph problem. This implies that the decision problem corresponding to the maximum $(a, b-a)$-colorable subgraph problem is NP-complete for e.g. split graphs, undirected path graphs and their complements and for k-trees with unbounded k.

We have proved the following results:

Theorem 5. (1) *The maximum $(a, b-a)$-colorable subgraph problem for interval graphs is solvable in polynomial time using a mincost flow algorithm.*
(2) *The maximum $(a, b-a)$-colorable subgraph problem for unimodular graphs is solvable in polynomial time using a linear program.*

We denote by $\alpha_{a,b-a}(G)$ the maximum cardinality of such a subset V' and with $\bar{\alpha}_{a,b-a}(G)$ the number of vertices in $V \setminus V'$. Clearly, $\alpha_{a_{opt},b_{opt}-a_{opt}}(G) \geq n_{a_{opt},b_{opt}-a_{opt}}$ and $\bar{\alpha}_{a_{opt},b_{opt}-a_{opt}}(G) \leq \bar{n}_{a_{opt},b_{opt}-a_{opt}}$. Given a solution with sets V' (and $V \setminus V'$), a coloring with at most a (and $b-a$) colors can be computed with an optimum coloring algorithm for several classes of graphs (e.g. interval or unimodular graphs). Since the colors $a_{opt}+1, \ldots, b_{opt}$ cover less than $\sqrt{|V|}$ vertices in the optimum solution and since $\bar{\alpha}_{a_{opt},b_{opt}-a_{opt}}(G) \leq \bar{n}_{a_{opt},b_{opt}-a_{opt}}$, the value $\bar{\alpha}_{a_{opt},b_{opt}-a_{opt}}(G)$ is bounded by $\sqrt{|V|}$.

Let $C_H(a_{opt}, b_{opt} - a_{opt})$ be the costs of a coloring computed by a call of the algorithm $H(a_{opt}, b_{opt} - a_{opt})$ and a corresponding coloring algorithm. Then, we can bound the costs $C_H(a_{opt}, b_{opt} - a_{opt})$ using the lower bounds (1) and (2) as follows:

$$\begin{aligned}C_H(a_{opt}, b_{opt} - a_{opt}) &\leq \alpha_{a_{opt}, b_{opt} - a_{opt}}(G) \cdot k_{a_{opt}} + \bar{\alpha}_{a_{opt}, b_{opt} - a_{opt}}(G) \cdot k_{b_{opt}} \\ &\leq |V| \cdot k_{a_{opt}} + \sqrt{|V|} \cdot k_{b_{opt}} \\ &\leq \sqrt{|V|}\sqrt{|V|} k_{a_{opt}} + \sqrt{|V|} \cdot k_{b_{opt}} \\ &\leq 2\sqrt{|V|} OPT(I).\end{aligned}$$

For the approximation algorithm for the OCCP problem the values a and $b - a$ can be bounded by $\chi(G)$. If $a_{opt} \geq \chi(G)$, the optimum costs $OPT(I)$ are greater than $\lceil\sqrt{|V|}\rceil k_{\chi(G)}$. In this case, we get an approximate solution with $a = \chi(G)$ and $b = a$ using

$$C_H(\chi(G), 0) \leq \alpha_{\chi(G), 0}(G) k_{\chi(G)} \leq \sqrt{|V|}\sqrt{|V|} k_{\chi(G)} \leq \sqrt{|V|} OPT(I).$$

If $b_{opt} > \chi(G) + a_{opt}$ and $a_{opt} \leq \chi(G)$, then the optimum costs $OPT(I) \geq k_{\chi(G)+a_{opt}}$. In this case, we get an approximate solution with $a = a_{opt}$ and $b = a + \chi(G)$ using

$$C_H(a_{opt}, \chi(G)) \leq |V| k_{a_{opt}} + \sqrt{|V|} k_{\chi(G)+a_{opt}} \leq 2\sqrt{|V|} OPT(I).$$

These arguments imply that at most $O(\chi(G)^2)$ calls of the maximum $(a, b-a)$ - colorable subgraph are sufficient for our approximation algorithm. In the next part of this section we improve this bound. We show that at most $O(\log \chi(G))$ calls of the maximum $(a, b - a)$ colorable subgraph algorithm H are needed.

For each $a \in \{1, \ldots, \chi(G) - 1\}$, let x be the smallest integer with $x \in \{1, \ldots, \chi(G)\}$ (if possible) such that $\bar{\alpha}_{a,x}(G) < \lceil\sqrt{|V|}\rceil$. We notice that $\bar{\alpha}_{a,\chi(G)}(G)$ can be greater than $\lceil\sqrt{|V|}\rceil$ and that

$$\bar{\alpha}_{a,1}(G) \geq \bar{\alpha}_{a,2}(G) \geq \ldots \geq \bar{\alpha}_{a,\chi(G)}(G).$$

For $a \in \{1, \ldots, \chi(G) - 1\}$ we define

$$first(a) = \begin{cases} x & \text{if } \bar{\alpha}_{a,\chi(G)}(G) < \lceil\sqrt{|V|}\rceil \\ \infty & \text{otherwise} \end{cases}$$

Since G can be colored with $\chi(G)$ colors, we define $first(\chi(G)) = 0$.

Lemma 6. *If $a < a'$ and $first(a), first(a') \neq \infty$ then $first(a') + a' \leq first(a) + a$.*

The smallest $\bar{a}$ with $first(\bar{a}) < \infty$ can be found using binary search with calls $H(a, \chi(G))$. Therefore, $\bar{a}$ can be found with $O(\log \chi(G))$ calls of the maximum a-colorable subgraph algorithm. We notice that $first(a') < \infty$ for each $a' \in \{\bar{a}, \ldots, \chi(G)\}$. This implies that the mapping $\varphi : a \to first(a) + a$ is non-increasing.

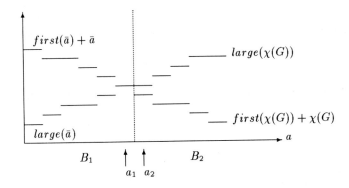

Fig. 2. The mappings $\varphi : a \to first(a) + a$ and $\psi : a \to large(a)$

For each $a \in \{1, \ldots, \chi(G)\}$, let x be the smallest number in $\{a+1, \ldots, |V|\}$ (if existent) with $k_x > \sqrt{|V|k_a}$. Notice that $k_{|V|}$ can be smaller than $\sqrt{|V|k_a}$. We define

$$large(a) = \begin{cases} x & \text{if } k_{|V|} > \sqrt{|V|k_a} \\ \infty & \text{otherwise} \end{cases}$$

For $\chi(G) = |V|$, we define $large(|V|) = \infty$.

Lemma 7. *If $a < a'$ then $large(a) \leq large(a')$.*

This Lemma implies that the mapping $\psi : a \to large(a)$ is non-decreasing. Next, we define two regions $B_1 = \{a \in \{\bar{a}, \ldots, \chi(G)\} | large(a) \leq first(a) + a\}$ and $B_2 = \{a \in \{\bar{a}, \ldots, \chi(G)\} | large(a) > first(a) + a\}$. We define $a_1 = max(B_1)$ and $a_2 = min(B_2)$ (if the corresponding sets are non-empty). In Figure 2, we have illustrated the mappings $\psi : a \to large(a)$ and $\varphi : a \to first(a) + a$. Since the mappings φ and ψ are non-increasing and non-decreasing, for each pair $a \in B_1$, $a' \in B_2$ we have $a < a'$.

Consider an optimum solution with parameters (a_{opt}, b_{opt}) where $a_{opt} \leq \chi(G)$. We prove that is sufficient to compute at most two solutions using a $(a_1, first(a_1))$ and a $(a_2, first(a_2))$ maximum colorable subgraph.

Lemma 8. *Let $(V', V \setminus V')$ be an optimum solution with parameters (a_{opt}, b_{opt}) such that $a_{opt} \leq \chi(G)$. If $B_1 \neq \emptyset$ and $B_2 \neq \emptyset$ then*

$$min(C_H(a_1, first(a_1)), C_H(a_2, first(a_2))) \leq 2\sqrt{|V|}OPT(I).$$

If $B_1 = \emptyset$, then $a_2 = \bar{a}$ and

$$C_H(\bar{a}, first(\bar{a})) \leq 2\sqrt{|V|}OPT(I).$$

Proof. We analyse the case with $B_1 \neq \emptyset$ and $B_2 \neq \emptyset$; the other case with $B_1 = \emptyset$ follows then directly. Clearly, we have $a_2 = a_1 + 1$. We have to consider three cases.

Case 1: $large(a_{opt}) \leq first(a_{opt}) + a_{opt}$ and $a_{opt} \geq \bar{a}$. In this case, the optimum solution lies in the first region B_1. Notice that $first(a_{opt}) + a_{opt} \leq b_{opt}$. Since the mapping $\varphi : a \rightarrow first(a) + a$ is non-increasing and $a_1 \geq a_{opt}$, we have $first(a_1) + a_1 \leq first(a_{opt}) + a_{opt}$. Moreover, it holds that $large(a_1) \leq first(a_1) + a_1$ and that $k_{large(a_1)} > \sqrt{|V|} k_{a_1}$.

The costs $C_H(a_1, first(a_1)) \leq |V| k_{a_1} + \sqrt{|V|} k_{first(a_1)+a_1}$. Using $\sqrt{|V|} k_{a_1} < k_{large(a_1)}$ and $large(a_1) \leq first(a_1) + a_1 \leq first(a_{opt}) + a_{opt} \leq b_{opt}$, we obtain

$$C_H(a_1, first(a_1)) \leq 2\sqrt{|V|} k_{b_{opt}} \leq 2\sqrt{|V|} OPT(I).$$

Case 2: $large(a_{opt}) > first(a_{opt}) + a_{opt}$ and $a_{opt} \geq \bar{a}$. In this case, the optimum solution lies in region B_2 and similar as above we get $C_H(a_2, first(a_2)) \leq 2\sqrt{|V|} OPT(I)$.

Case 3: $a_{opt} < \bar{a}$. In this case, we get a contradiction. □

The next Lemma implies how the values a_1 and a_2 can be computed using binary search and calls $H(a, large(a) - a)$ and $H(a, large(a) - a - 1)$.

Lemma 9. *Let $a \in \{\bar{a}, \ldots, \chi(G)\}$. Using calls $H(a, large(a) - a)$ and $H(a, large(a) - a - 1)$ we can decide whether $a \in B_1$ or $a \in B_2$.*

Now, we are ready for our **approximation algorithm**.

Algorithm B
(1) compute $large(a)$ for each $a \in \{1, \ldots, \chi(G)\}$ (using preprocessing in $O(\chi(G) + |V|)$ time),
(2) compute a first solution with the call $H(\chi(G), 0)$ (this is an arbitrary coloring for the case $a_{opt} > \chi(G)$),
(3) find the smallest $\bar{a}$ with $first(\bar{a}) < \infty$ using binary search with calls $H(a, \chi(G))$ (these are maximum a - colorable subgraphs),
(4) compute a_1 (if existent) and a_2 using binary search with calls $H(a, large(a) - a)$ and $H(a, large(a) - a - 1)$ (these are maximum (a, x)-colorable subgraphs),
(5) compute $first(a_1)$ (if a_1 exists) and $first(a_2)$ using binary search with calls $H(a_i, x)$ (these are maximum (a_i, x)-colorable subgraphs),
(6) choose the cheapest solution among the solutions with costs $C_H(a_1, first(a_1))$ (if a_1 exists), $C_H(a_2, first(a_2))$ and $C_H(\chi(G), 0)$.

Using the calculations above, we obtain the following result.

Theorem 10. *The approximation algorithm B above computes a coloring of the OCCP problem restricted to interval graphs (and unimodular graphs) with approximation ratio $O(|V|^{0.5})$.*

The time complexity of this algorithm for interval graphs is given by $O(\log \chi(G))$ calls of a minimum cost flow algorithm. For unimodular graphs, we need at most $O(\log \chi(G))$ calls of a linear programming algorithm.

4 Other perfect graphs

In [8], we have proved the following further results for the OCCP problem restricted to chordal, comparability, permutation and split graphs.

Theorem 11. *For each $\epsilon < 1$, there exists no polynomial approximation algorithm with ratio $O(|V|^{1-\epsilon})$ for the OCCP problem restricted to permutation graphs (and to comparability graphs), unless $P = NP$.*

Theorem 12. *For each $\epsilon < 1$, there exists no polynomial approximation algorithm with ratio $O(|V|^{1-\epsilon})$ for the OCCP problem restricted to split graphs (and to chordal graphs), unless $P = NP$.*

Acknowledgement. I thank Thomas Erlebach (TU München) for his helpful comments and many fruitful discussions.

References

1. Bodlaender, H.L., Jansen, K., Woeginger, G.: Scheduling with incompatible jobs. Graph Theoretic Concepts in Computer Science, LNCS **657** (1992) 37–49
2. Chang, M.S., Chen, Y.H., Chang, G.J., Yan J.H.: Algorithmic aspects of the generalized clique-transversal problem on chordal graphs. Discrete Applied Mathematics **66** (1996) 189–203
3. Frank, A.: On chain and antichain families of a partially ordered set. Journal Combinatorial Theory **(B)** 29 (1980) 176–184
4. Gavril, F.: Algorithms for maximum k-colorings and k-coverings of transitive graphs. Networks **17** (1987) 465–470.
5. Golumbic, M.C.: Algorithmic graph theory and perfect graphs, Academic Press, New York (1980)
6. Jansen, K., Scheffler, P., Woeginger, G.J.: Maximum covering with D cliques. Fundamentals of Computation Theory, LNCS **710** (1993) 319–328
7. Jansen, K.: The optimum cost chromatic partition problem. Algorithms and Complexity, LNCS **1203** (1997) 25–36
8. Jansen, K.: Approximation results for the optimum cost chromatic partition problem. Universität Trier, Forschungsbericht 1 (1997)
9. Kroon, L.G., Sen, A., Deng, H., Roy, A.: The optimal cost chromatic partition problem for trees and interval graphs. Graph Theoretical Concepts in Computer Science, LNCS (1996)
10. Sen, A., Deng, H., Guha, S.: On a graph partition problem with an application to VLSI layout. Information Processing Letters **43** (1992) 87–94
11. Supowit, K.J.: Finding a maximum planar subset of a set of nets in a channel. IEEE Transactions on Computer Aided Design CAD **6, 1** (1987) 93–94
12. Yannakakis, M., Gavril, F.: The maximum k-colorable subgraph problem for chordal graphs. Information Processing Letters **24** (1987) 133–137

The Minimum Color Sum of Bipartite Graphs[*]

Amotz Bar-Noy[**]   Guy Kortsarz[***]

Abstract. The problem of *minimum color sum* of a graph is to color the vertices of the graph such that the sum (average) of all assigned colors is minimum. Recently, in [BBH+96], it was shown that in general graphs this problem cannot be approximated within $n^{1-\epsilon}$, for any $\epsilon > 0$, unless $NP = ZPP$. In the same paper, a 9/8-approximation algorithm was presented for bipartite graphs. The hardness question for this problem on bipartite graphs was left open. In this paper we show that the minimum color sum problem for bipartite graphs admits no polynomial approximation scheme, unless $P = NP$. The proof is by L-reducing the problem of finding the maximum independent set in a graph whose maximum degree is four to this problem. This result indicates clearly that the minimum color sum problem is much harder than the traditional coloring problem which is trivially solvable in bipartite graphs. As for the approximation ratio, we make a further step towards finding the precise threshold. We present a polynomial 10/9-approximation algorithm. Our algorithm uses a flow procedure in addition to the maximum independent set procedure used in previous results.

1 Introduction

One of the most fundamental problems in *scheduling* theory is scheduling efficiently (under some optimization goals) dependent tasks on a single machine. At any given time, the machine is capable to perform (serve) any number of tasks as long as these tasks are independent. When the serving time of each task is the same, this problem is identical to the well known coloring problem of graphs. The vertices of the graph represent the tasks and an edge in the graph between vertices v and u represents the dependency between the two corresponding tasks. That is, the machine

---

[*] The full version of this extended abstract can be found in URL http://www.eng.tau.ac.il/ amotz/publications.html.
[**] Department of Electrical Engineering, Tel-Aviv University, Tel-Aviv 69978, Israel. E-mail: amotz@eng.tau.ac.il.
[***] Department of Computer Science, The Open University of Israel, Ramat Aviv, Israel. E-mail: guyk@tavor.openu.ac.il.

cannot perform the tasks corresponding to vertices u and v concurrently. Another important application arises in the context of distributed resource allocation. Here, the vertices represent processors each has one job to execute. An edge between two vertices indicates that the jobs belonging to the corresponding processors cannot be executed concurrently since they require the usage of the same common resource. This problem is known in the literature as the dining (drinking) philosophers problem ([LYN81, CM84]).

More formally, the coloring problem can be defined as follows. let $G = (V, E)$ be an undirected simple graph with n vertices. where V denotes the set of n vertices and E denotes the set of edges. A coloring of the vertices of G is a mapping into the set of positive integers, $f : V \mapsto \mathcal{Z}^+$, such that adjacent vertices are assigned different colors. We refer to $f(v)$ as the *color* of v.

The traditional optimization goal is to minimize the number of different assigned colors. We call this problem the *minimum coloring* (MC) problem. In the setting of tasks system, this is equivalent to finding a schedule in which the machine finishes performing all the tasks as early as possible. In the setting of resource allocation, this is equivalent to finding a schedule in which the last processor finishes executing its job the earliest. This is an optimization goal that favors the system. However, from the point of view of the tasks (or processors) themselves, we might wish to find the best coloring such that the average waiting time to be served (or to execute the job) is minimized.

Clearly, minimizing the average waiting time is equivalent to minimizing the sum of all assigned colors. The *minimum color sum* (MCS) problem is defined as follows. Let $G = (V, E)$ be an undirected simple graph with n vertices. We are looking for a coloring in which the sum of the assigned colors of all the vertices of G is minimized. That is, the value of $\sum_{v \in V} f(v)$ is minimized.

The minimum color sum problem was introduced by Kubicka in [K89]. In [KS89] it was shown that computing the MCS of a given graph is NP-hard. A polynomial time algorithm was given for the case where G is a tree. In [KKK89] it was shown that approximating the MCS problem within an additive constant factor is NP-hard. In a recent paper, [BBH+96], it was proven that the MCS problem cannot be approximated within $n^{1-\epsilon}$, for any $\epsilon > 0$, unless $NP = ZPP$. On the other hand, this paper showed that an algorithm based on finding iteratively a maximum independent set is a 4-approximation to the MCS problem. This bound yields a 4ρ-approximation polynomial algorithm for the MCS problem for classes of graphs for which the maximum independent set problem can be polynomially approximated within a factor of ρ.

A special and important sub-class of graphs is the class of *bipartite graphs*. In a bipartite graph the set of vertices V is partitioned into two disjoint sets V_l and V_r such that both sets are independent. That is, all the edges of E are between vertices of V_l and V_r. Coloring V_l by 1 and V_r by 2 yields a 2-coloring of any bipartite graph. Obviously this is the best possible solution for the MC problem. However, for the MCS problem the answer is not straightforward. Denote by MBCS the MCS problem on bipartite graphs.

Coloring the largest set between V_l and V_r by 1 and the other set by 2 yields a solution to the MBCS problem the value of which is at most $3n/2$. Obviously the value of the optimal solution is at least n, and therefore this solution is at least a 3/2-approximation to the optimal solution. The paper [BBH+96] presents a better approximation of 9/8 using as a sub-procedure the algorithm for finding a maximum independent set. In bipartite graphs, finding maximum independent set can be done in polynomial time. Therefore, their approximation algorithm is also polynomial.

New results: The contributions of this paper are the following two results:

- We prove the first hardness result for MBCS. We show that the MBCS problem admits no polynomial approximation scheme, unless $P = NP$. The proof is by L-reducing the problem of finding the maximum independent set in a graph whose maximum degree is four to the MBCS problem which implies that MBCS is MAXSNP-hard [PY88]. This result indicates clearly that the MCS problem is much harder than the traditional coloring problem.

- We improve the approximation ratio for the MBCS problem by presenting a 10/9-approximation algorithm. Our algorithm introduces a new technique. It employs a flow procedure in addition to the maximum independent set procedure used in [BBH+96].

Max-type vs. sum-type problems: Our impossibility result raises the general question of the connection between "max-type" and "sum-type" problems. The MC problem is a max-type problem whereas the MCS problem is a sum-type problem. The input and the feasible solutions for both problems are the same, the difference lies in the optimization goal. In the full version of this paper ([BK97]) we examine the "max-type" and the "sum-type" of the *Traveling Salesperson problem* (TSP). The discussion there raises the interesting question of classifying problems according to the relationship between their "max-type" version with the "sum-type" version. The coloring problem and the traveling salesperson problem each belongs to a different class.

2 Preliminaries

Given a graph $G(V, E)$ we use the following notations. Let $\text{MIS}(G)$ denote the largest independent set in G. For any set $S \subseteq V$, let $N(S)$ be the set of neighbors of S and $\text{MIS}(S)$ denotes the maximum independent set in the graph *induced by* S. We also use the term S to denote the size of S. Given any coloring f of a graph, we denote by $\text{SC}(f)$ the sum of colors in f, i.e., $\text{SC}(f) = \sum_{v \in V} f(v)$. When all the vertices in a set $S \subseteq V$ are colored by the same color c, we say that S is colored by c.

We say that problem P admits a polynomial approximation scheme, if for any $\epsilon > 0$ there exists a polynomial time approximation algorithm for P, whose approximation ratio is bounded by $(1 + \epsilon)$.

L–reduction The L-reduction ([PY88]) is a tool that helps proving hardness results. Unlike the usual NP-hardness reductions, it "preserves" approximation ratios. In order to define L-reduction we need the following notations. Let P be an optimization (either minimization or maximization) problem. Denote by $I(P)$ the set of instances for problem P, by $sol(P)$ the set of feasible solutions of problem P, and by $c_P(s)$ the cost function of any feasible solution s for P. Suppose now that P and Q are two optimization problems. In order to construct an L−reduction we need to define two (polynomially computable) functions $\mathcal{R} : I(P) \mapsto I(Q)$ and $\mathcal{S} : sol(Q) \mapsto sol(P)$. For any instance $x \in I(P)$ let $c_{OPT}(x)$ be the value of the optimal solution for x and let $c_{OPT}(\mathcal{R}(x))$ be the value of the optimal solution for $\mathcal{R}(x)$. The two functions $\mathcal{R}$ and $\mathcal{S}$ are an L−reduction from problem P to problem Q, if there exist two constants α and β such that the two following properties hold:

1. $c_{OPT}(\mathcal{R}(x)) \leq \alpha \cdot c_{OPT}(x)$.

2. For any feasible solution $s \in sol(Q)$ of $\mathcal{R}(x)$, $\mathcal{S}(s)$ is a feasible solution for x and $|c_{OPT}(x) - c_P(\mathcal{S}(s))| \leq \beta \cdot |c_{OPT}(\mathcal{R}(x)) - c_Q(s)|$.

Theorem 1 [PY88]. *Suppose that Problem P admits no polynomial approximation scheme and that Problem P can be L−reduced to problem Q. Then Problem Q admits no polynomial approximation scheme.*

The MIS *and* 4-MIS *problems* The *Maximum Independent Set* (MIS) problem is the following. Given an undirected graph $G(V, E)$ with n vertices, the goal is to find a maximum independent set. I.e., a maximum sized set $S \subseteq V$ such that no two vertices of S share an edge. The 4-MIS problem is the MIS problem restricted to graphs with maximum degree 4.

Theorem 2 [ALM+92]. *There exists some $\epsilon > 0$ such that the* 4-MIS *admits no $(1 + \epsilon)$-approximation algorithm, unless $P = NP$ (and hence* 4-MIS *admits no polynomial approximation scheme).*

Known algorithms for the MBCS *problem* We recall the approximation algorithm presented in [BBH+96]. For a given bipartite graph G, denote by I_1 the maximum independent set in G, by I_2 the maximum independent set in $G \setminus I_1$, by I_3 the maximum independent set in $G \setminus (I_1 \cup I_2)$, and so on. The algorithm of [BBH+96] is best explained by the definition of a sequence of (roughly) $\log n$ possible algorithms. Let A(2) be the algorithm that colors the vertices of G with two colors, the larger side of V by 1 and the smaller side by 2. Let A(3) be the following algorithm: color the vertices of I_1 by 1, and then color the vertices of $G \setminus I_1$ by 2 and 3 (i.e., color the larger side in the remaining graph by 2 and the smaller side by 3). In general, for $i \geq 3$ and for $1 \leq j \leq i - 2$, algorithm A(i) colors the sets I_j with color j, and then colors the larger side of the remaining graph by $i - 1$ and the smaller side by i. All together, algorithm A(i) uses i colors. Note that we have defined at most $\lfloor \log n \rfloor$ algorithms, because the maximum independent set in any bipartite graph with n vertices contains at least $n/2$ vertices. Let A' be the last possible algorithm in this family of algorithms. Since G is a bipartite graph, it follows that $I_1 \geq n/2$. Therefore, algorithm A(2) is a 3/2-approximation algorithm. Consider now the following algorithm, denoted by $\mathcal{B}$, that runs algorithms A(2) and A(3) and picks the best solution.

Theorem 3 [BBH+96]. *Algorithm $\mathcal{B}$ is a 9/8-approximation algorithm to the* MBCS *problem.*

An algorithmic tool We now describe the new tool used in our approximation algorithm. Define the 2-Neighborhood problem as follows. Given a bipartite graph $G(V_l, V_r, E)$ we look for a set $S \subseteq V_l$ such that $d_S = 2S - N(S)$ is maximum. We note that the order in which V_l and V_r are specified in the problem-presentation is important, that is the solution S is a subset of V_l. Polynomial time solutions for problems of this nature are known (see, e.g., [GGT89]).

3 A hardness result for the MBCS problem

In this section, we prove that (unless $P = NP$) the MBCS problem has no polynomial approximation scheme. We do that by proving an L-reduction from the 4-MIS problem to the MBCS problem (hence showing that the MBCS problem is MAXSNP-hard). By Theorems 1 and 2 the hardness result is implied.

3.1 The construction – the function $\mathcal{R}$

Let $G(V, E)$ be an instance of the 4-MIS problem. The $\mathcal{R}$ function should map G into a graph $\tilde{G}$ which is an instance of the MBCS problem. First, $\tilde{G}$ contains a vertex corresponding to each vertex in V. In $\tilde{G}$, V is an independent set. We assume an order on the vertices of G. Whenever we consider an edge $(x, y) \in E$ we assume that $x < y$. The construction involves adding a *gadget* for each edge $e = (x, y) \in E$. Each gadget is composed of twelve independent sets of vertices containing no internal edges (edges only cross from one different set to the other). The sets of vertices corresponding to different edges are *disjoint*.

Before describing the sets of vertices and the edges of any gadget we need some definitions. We say that two (independent) sets A and B are *cliqued*, if every vertex in A is connected to every vertex in B that is, the sets A and B induce a complete bipartite graph. We say that the two sets are *matched* if $|A| = |B|$ and every vertex x in A has a *single* neighbor $m(x)$ in B, that is, the sets A and B induce a perfect matching. The sets and edges in the gadget corresponding to the edge $e = (x, y)$ are as follows.

Main and matched sets:

1. A set XYX of 3 vertices and a *matched* set $m(XYX)$ of 3 vertices.

2. A set XYY of 3 vertices and a matched set $m(XYY)$ of 3 vertices.

3. A set XY of 6 vertices and a matched set $m(XY)$ of 6 vertices.

Imposing sets:

1. A set $I_1(XYX)$ of 18 vertices and a *cliqued* set $I_2(XYX)$ of 9 vertices.

2. A set $I_1(m(XYX))$ of 6 vertices and a *cliqued* set $I_2(m(XYX))$ of 3 vertices.

3. Two sets $I_1(XY)$ of 24 vertices and $I_1(m(XY))$ of 12 vertices.

Additional edges between the sets:

1. The vertex x (y) is connected to all 3 vertices of XYX (XYY).
2. The sets XYX and XYY each is *cliqued* with XY.
3. The sets XYX ($m(XYX)$) and $I_2(XYX)$ ($I_2(m(XYX))$) are *cliqued*.
4. The sets XY ($m(XY)$) and $I_1(XY)$ ($I_1(m(XY))$) are *cliqued*.

This completes the description of the gadget corresponding to each edge $e = (x, y)$ and the description of the $\mathcal{R}$-function. The above sets depend on e, that is, there is such a gadget for every edge $e \in E$. We avoid adding e as a subscript in these sets, for the simplicity of notation. In order for the $\mathcal{R}$ function to be valid we demonstrate a 2 coloring for $\tilde{G}$ proving that the graph $\tilde{G}$ is a bipartite graph.

Lemma 4. *The graph $\tilde{G}$ is bipartite.*

The intuition behind the construction: The goal of the construction is to enable us to define the right function $\mathcal{S}$. The role of the imposing sets is to force a situation in which some sets cannot be colored by a specific color. For example, it will be shown that in an optimal coloring the imposing set $I_2(XYX)$ is colored by 2. Consequently, the set XYX cannot be colored by 2. In general, in an optimal solution, all the sets of type I_1 are colored by 1 and all the sets of type I_2 are colored by 2. The role of the matched sets is to assure that the sum coloring of two matched sets is fixed in any optimal coloring. For example, if a vertex in XYX is colored by 1, then its matched vertex is colored by 3, and vice versa (recalling that these two sets can not be colored by 2). Thus every pair in XYX and $m(XYX)$ adds exactly 4 to the sum coloring in an optimal coloring and the contribution of XYX and $m(XYX)$ is fixed. Now let us explain the main idea in the construction. Let x and y be two vertices adjacent in G (i.e., $(x, y) \in E$). We will show that we lose in the sum coloring if both x and y are colored by 1. Indeed, say that both x and y are colored 1, and consider the colors of XY, XYX, XYY. In the best coloring XYX is colored by 3 and XYY by 2. Therefore, since the set $I_1(XY)$ is colored by 1, it follows that XY is colored by at least 4. On the other hand, if one of x and y is not colored by 1, we may gain by assigning XY a color less then 4. This follows since XYX and XYY will "waste" only one of the colors 2 and 3. Hence, it is possible to color XY with either 2 or 3. Therefore, a "good" sum coloring colors as large as possible independent set in G by 1. Thus, a "good" approximation for the MBCS problem implies a "good" approximation for the 4-MIS problem.

3.2 The function $\mathcal{S}$

A coloring $\tilde{f}$ of the vertices in $\tilde{G}$ is *proper*, if the two following properties hold for every edge.

Imposing properties: The sets $I_1(XYX)$, $I_1(m(XYX))$, $I_1(XY)$, and $I_1(m(XY))$ are colored by 1. The sets $I_2(XYX)$ and $I_2(m(XYX))$ are colored by 2.

Independence property: All the vertices of G that are colored by 1 in $\tilde{f}$ form an independent set in G.

The process of constructing $\mathcal{S}$ is as follows. We start with any feasible coloring f of $\tilde{G}$. We then show in five stages that f can be transformed to a *proper* coloring $\tilde{f}$ such that the sum of colors in $\tilde{f}$ is no larger than the sum of colors in f ($\mathsf{SC}(\tilde{f}) \leq \mathsf{SC}(f)$). The mapping $\mathcal{S}$ is now defined by choosing the set of vertices in G that are colored by 1 by $\tilde{f}$ denoted by $I_1(\tilde{f})$. Note, that by the independence property, $I_1(\tilde{f})$ is also an independent set in G.

In the first stage we transform f into f_1 such that all the vertices in any independent set in any gadget are colored by the same color. In the second stage, we transform f_1 into a coloring f_2 that is locally minimal, that is a coloring such that each set in the gadget is colored by no more than $k+1$ where k is the number of neighboring sets to this set. In the third stage, we show how to transform f_2 into a coloring f_3 such that the imposing properties hold. In the forth stage, we transform f_3 into a coloring f_4 in which all the sets XYX and XYY in all the gadgets are colored by no more than 3. Finally, in the fifth stage we transform f_3 into the desired coloring $\tilde{f}$ by showing how to achieve the independence property. In all five stages the new coloring has no worse sum coloring then the previous one. The full proof appears in [BK97].

3.3 The L-reduction properties

We now turn to prove the two L-reduction properties. Let OPT be the minimum sum coloring in $\tilde{G}$ and let $\mathsf{MIC} = \mathsf{SC}(\mathsf{OPT})$. The next lemma proves the first property of the L-reduction.

Lemma 5. *There exists a constant α such that $\mathsf{MIC} \leq \alpha \cdot \mathsf{MIS}(G)$.*

For the second property of the L-reduction, we need to show the existence of a constant β such that for any legal coloring f of $\tilde{G}$ the following holds: $\mathsf{MIS}(G) - \mathcal{S}(f) \leq \beta(\mathsf{SC}(f) - \mathsf{MIC})$. We prove this inequality with $\beta = 1$. The proof uses the following two lemmas. Let I_1 be the maximum independent set in G.

Lemma 6. MIC $\leq 135 \cdot E + 2n - I_1$.

Now let f be an arbitrary coloring of $\tilde{G}$ and let $\tilde{f}$ be its corresponding proper coloring. Let $I_1(\tilde{f})$ be the set of vertices colored by 1 in $\tilde{f}$, and thus $\mathcal{S}(f) = I_1(\tilde{f})$.

Lemma 7. $\mathrm{SC}(\tilde{f}) \geq 135 \cdot E + 2n - I_1(\tilde{f})$.

The following lemma states the second property of the L-reduction.

Lemma 8. $\mathrm{MIS}(G) - \mathcal{S}(f) \leq \mathrm{SC}(f) - \mathrm{MIC}$.

We completed constructing a valid L-reduction from the 4-MIS problem to the MBCS problem. The following theorem follows from Theorems 1 and 2.

Theorem 9. *There exists an $\epsilon > 0$ such that there is no $(1+\epsilon)$-ratio approximation algorithm for the MBCS problem unless $P = NP$.*

4 Improved approximation algorithm for MBCS

In the previous section we have shown that there exists some $\epsilon > 0$ such that the MBCS problem has no $(1+\epsilon)$-approximation algorithm. However, the precise threshold for the approximation is yet to be determined. We take a further step in this direction. In this section, we present a new algorithm $\mathcal{C}$ that utilizes a new procedure Neig. We prove that this procedure, combined with algorithms A(2), A(3), and A(4) yield a 10/9-approximation algorithm for the MBCS problem.

4.1 Procedure Neig and Algorithm $\mathcal{C}$

Procedure Neig utilizes the solution to the 2-Neighborhood problem. It uses the following subsets and subgraphs of G.

1. I_1 – the maximum independent set in G. $I_1^l = I_1 \cap V_l$ and $I_1^r = I_1 \cap V_r$.
2. Z – the larger side of $G \backslash I_1$ and W – the smaller side of $G \backslash I_1$. Without loss of generality, assume that $Z \subset V_l$ and $W \subset V_r$.
3. $G_Z = (Z, I_1^r, E_Z)$ – the (bipartite) subgraph induced by Z and I_1^r.
   $G_W = (W, I_1^l, E_W)$ – the (bipartite) subgraph induced by W and I_1^l.
4. S_Z – the set maximizing $d_{S_Z} = 2S_Z - N(S_Z)$ in G_Z.
   S_W – the set maximizing $d_{S_W} = 2S_W - N(S_W)$ in G_W.
5. $N_1(S_Z) = N(S_Z) \cap I_1^r$ and $N_1(S_W) = N(S_W) \cap I_1^l$.

Procedure Neig*:*

If $d_{S_Z} \geq d_{S_W}$ then color:
1. $I_1^l \cup S_Z \cup (I_1^r \setminus N_1(S_Z))$ by 1.
2. $W \cup N_1(S_Z)$ by 2.
3. $Z \setminus S_Z$ by 3.

If $d_{S_Z} < d_{S_W}$ then color:
1. $I_1^l \cup S_W \cup (I_1^r \setminus N_1(S_W))$ by 1.
2. $Z \cup N_1(S_W)$ by 2.
3. $W \setminus S_W$ by 3.

For the case $d_{S_Z} \geq d_{S_W}$, procedure Neig can be described as follows. Start with the initial coloring of A(3), that is I_1 is colored by 1, Z (the larger of the two remaining sides) is colored by 2 and W by 3. Thus $\mathtt{SC}(\mathtt{A(3)}) = I_1^l + I_1^r + 2Z + 3W$. Next, re-color Z by 3 and W by 2, losing $Z - W$ in the sum coloring. Next, change the color of S_Z from 3 to 1 gaining $2S_Z$ in the sum coloring. This forces all the neighbors of Z in I_1, $N_1(S_Z)$, to be colored by a color different than 1, thus color them by 2. Here we lose $N_1(S_Z)$ in the sum coloring. The net profit in the sum coloring is therefore $2S_Z - N_1(S_Z) + W - Z = d_{S_Z} + W - Z$. Similarly, it can be shown that for the case $d_{S_W} > d_{S_Z}$, the net profit is d_{S_W}. (This case is better for us since we do not need to switch the colors of Z and W, loosing $Z - W$.) Thus, we proved the following proposition.

Proposition 10.
(1). If $d_{S_Z} \geq d_{S_W}$ then $\mathtt{SC}(\mathtt{Neig}) = \mathtt{SC}(\mathtt{A(3)}) - d_{S_Z} + (Z - W)$.
(2). If $d_{S_W} > d_{S_Z}$ then $\mathtt{SC}(\mathtt{Neig}) = \mathtt{SC}(\mathtt{A(3)}) - d_{S_W}$.

We conclude this subsection with the description of algorithm $\mathcal{C}$. It clearly follows that the algorithm has a polynomial running time.

Algorithm $\mathcal{C}$

- Run algorithms A(2), A(3), A(4), and Procedure Neig.

- Pick the solution whose sum coloring is the minimum among the four coloring solutions.

4.2 Analysis

All through the analysis, let $Z = (n-I_1)/2 + \epsilon_d n$ and $W = (n-I_1)/2 - \epsilon_d n$. The term $\epsilon_d n$ quantifies the extent in which the graph induced by $Z \cup W$ is unbalanced. This is the graph resulting once the maximum independent set I_1 is deleted from G.

Outline of the analysis: If $Z - W = 2\epsilon_d n$ is "large" enough, then the 10/9-ratio is already yielded by $\min\{\text{SC}(\text{A(2)}), \text{SC}(\text{A(3)})\}$. Otherwise, $Z - W$ is not too "large". If I_2 is "large" enough, then this time already $\min\{\text{SC}(\text{A(2)}), \text{SC}(\text{A(4)})\}$ yields the 10/9-ratio. Otherwise, W is almost as "large" as Z and I_2 is not too "large". If W is "small" enough and therefore Z is also "small" and I_1 is "large" enough, then $\text{SC}(\text{A(3)})$ alone yields the 10/9-ratio. Otherwise $Z - W$ and I_2 are not too "large" and W is not too "small". If the optimal algorithm does not deviate much from algorithm A(3), then again $\min\{\text{SC}(\text{A(2)}), \text{SC}(\text{A(3)})\}$ yields the 10/9-ratio. Finally, if all the previous conditions do not hold, we use the new procedure Neig and show that $\min\{\text{SC}(\text{A(2)}), \text{SC}(\text{Neig})\}$ yields the 10/9-ratio. The analysis is partitioned into the above five cases. The complete analysis appears in [BK97].

Theorem 11. *Algorithm $\mathcal{C}$ is a polynomial 10/9-approximation algorithm for the MBCS problem.*

References

[ALM+92] S. Arora, C. Lund, R. Motwani, M Sudan, and M. Szegedy. Proof verification and intractability of approximation problems. In *Proc. of the 33'rd IEEE Symp. on the Foundations of Computer Science*, pages 14–23, 1992.

[BK97] A. Bar-Noy and G. Kortsarz. The Minimum Color Sum of Bipartite Graphs. In URL: http://www.eng.tau.ac.il/ amotz/publications.html.

[BBH+96] A. Bar-Noy, M. Bellare, M. M. Halldórsson, H. Shachnai, and T. Tamir. On chromatic sums and distributed resource allocation. In *Proc. of the fourth Israel Symp. on Theory and Computing and Systems*, pages 119–128, 1996. (Also in URL: http://www.eng.tau.ac.il/ amotz/publications.html.)

[CM84] K. Chandy and J. Misra. The Drinking Philosophers Problem. *ACM Trans. Programming Languages and Systems*, 6:632–646, 1984.

[Chr76] N. Christofides. Worst case analysis of a new heuristic for the traveling salesman problem. Technical report GSIA, Carnegie-Mellon Univ., 1976.

[GGT89] G. Gallo, M.D. Grigoriadis, and R.E. Tarjan. A fast parametric maximum flow algorithm and applications. *SIAM J. on Comput.*, 18:30–55, 1989.

[K89] E. Kubicka. The Chromatic Sum of a Graph. PhD thesis, Western Michigan University, 1989.

[KKK89] E. Kubicka, G. Kubicki, and D. Kountanis. Approximation Algorithms for the Chromatic Sum. In *Proc. of the First Great Lakes Computer Science Conf.*, Springer LNCS 507, pages 15–21, 1989.

[KS89] E. Kubicka and A. J. Schwenk. An Introduction to Chromatic Sums. In *Proc. of the ACM Computer Science Conf.*, pages 39–45, 1989.

[LYN81] N. Lynch. Upper Bounds for Static Resource Allocation in a Distributed System. *J. of Computer and System Sciences*, 23:254–278, 1981.

[PY88] C. H. Papadimitriou and M. Yannakakis. Optimization approximation and complexity classes. In *Proc. of the 20'th IEEE Symp. on The Theory of Computing*, pages 229–234, 1988.

A Primal–Dual Approach to Approximation of Node–Deletion Problems for Matroidal Properties *
(Extended Abstract)

Toshihiro Fujito[1]

Dept. of Electrical Engineering, Faculty of Engineering
Hiroshima University
1-4-1 Kagamiyama, Higashi–Hiroshima 739 JAPAN
e-mail: fujito@huis.hiroshima-u.ac.jp

Abstract. This paper is concerned with the polynomial time approximability of node–deletion problems for hereditary properties.

We will focus on such graph properties that are derived from matroids definable on the edge set of any graph. It will be shown first that all the node–deletion problem for such properties can be uniformly formulated by a simple but non–standard form of the integer program. A primal-dual approximation algorithm based on this and the dual of its linear relaxation is then presented.

When a property has infinitely many minimal forbidden graphs no constant factor approximation for the corresponding node–deletion problem has been known except for the case of the Feedback Vertex Set (FVS) problem in undirected graphs. It will be shown next that FVS is not the sole exceptional case and that there exist infinitely many graph (hereditary) properties with an infinite number of minimal forbidden graphs, for which the node–deletion problems are efficiently approximable to within a factor of 2. Such properties are derived from the notion of matroidal families of graphs and relaxing the definitions for them.

1 Introduction

This paper is concerned with the polynomial time approximability of node–deletion problems for hereditary properties. The *node–deletion problem for a graph property* π (denoted ND(π) throughout the paper) is a typical graph optimization problem; that is, given a node–weighted graph G, find a node set of the minimum weight sum s.t. deletion of it (along with all the incident edges) from G leaves a subgraph satisfying the property π. A graph property π is *hereditary* if every subgraph of a graph satisfying π also satisfies π. A number of well–studied graph properties are hereditary such as independent set, planar,

---

* This work is partially supported by a grant from the Okawa Foundation for Information and Telecommunications.

bipartite, degree–constrained, circular–arc, circle graph, chordal, comparability, permutation, perfect. Consequently, many well known graph problems fall into this class of problems when desired graph properties are specified appropriately. Lewis and Yannakakis proved, however, that whenever π is nontrivial and hereditary on induced subgraphs $ND(\pi)$ is NP–hard [LY80]. When this general NP–hardness result was established in 1980, almost nothing was known about the approximability of $ND(\pi)$'s except for good approximation algorithms for the Vertex Cover (VC) problem (i.e., $\pi =$ "independent set"). Moreover, their generic reductions from VC to other $ND(\pi)$'s are approximation preserving, and as such, no $ND(\pi)$ can be approximated better than VC can be. One question posed therein was thus concerned with the other direction of approximability: Can other node–deletion problems be approximated as good as VC can be ?

It has been long known that VC can be approximated with ratio 2 (achievable by a simple maximal matching heuristic [Gav74] for the unweighted case) and a better approximation has been a subject of extensive research over the years. Yet the best constant bound has remained the same at 2 while the best known heuristics can accomplish only slightly better ($2 - \frac{\log \log n}{2 \log n}$ of [BE85, MS85]). On the other hand very few other $ND(\pi)$'s have been shown to be approximable within a factor of c, for *any* constant c, not to mention a constant of 2. As observed in [LY93] whenever hereditary π has only a finite number of minimal forbidden graphs $ND(\pi)$ can be efficiently approximated to within some constant factor of the optimum. It was in fact conjectured therein that those with finitely many minimal forbidden graphs are the only hereditary properties which yield constant factor approximable node–deletion problems (see also [Yan94]). It was found later, however, that this conjecture does not hold as is when the (unweighted) Feedback Vertex Set (FVS) problem (i.e., $\pi =$ "acyclic") in undirected graphs was shown to be approximable to within a factor of 4 [BGNR94] (Note: every simple cycle of each length is a minimal forbidden graph for this π). Until now this problem has been the only known exception to the Lund–Yannakakis' conjecture.

1.1 Our results

In this paper we will show that there exist infinitely many $ND(\pi)$'s for π with an infinite number of minimal forbidden graphs, each of which approximable to a factor of 2. For that purpose we shall concentrate on such hereditary properties that can be derived from (independent sets of) matroids definable on the edge set of any graph (details given later). The class of $ND(\pi)$'s for such properties includes VC, FVS, and many others. It will be shown first that all $ND(\pi)$'s in this class can be uniformly formulated by a simple but non–standard form of the integer program using matroid rank functions. A primal–dual approximation algorithm for such $ND(\pi)$'s is then designed based on this formulation and the dual of its linear programming relaxation, which is simpler than those algorithms for FVS given in [BBF95, BG94, CGHW96]. In particular our algorithm does not look into nor modify explicitly, unlike the previous algorithms for FVS, any special structure in graphs under consideration. An analysis of this algorithm reveals that its performance ratio can be reduced to the combinatorial bound arising from the underlying structures of the problems.

It will be shown next, as an application of the current primal–dual approach, that FVS is not the sole exceptional case; i.e., there exist other (hereditary) properties π's with an infinite number of minimal forbidden graphs, s.t. ND(π)'s are efficiently approximable to within a factor of 2, the best constant factor known for either VC or FVS. In fact, we will show, there are infinitely many of them (at least countably many). Such properties are derived from the notion of matroidal families of graphs and relaxing the definitions for them (details later). The infinite sequence of these properties will be constructed having those for both VC and FVS at its basis and thus providing a proper generalization of them. It is also worth pointing out that our formulation for these ND(π)'s introduces the integrality gap of at most 2 unlike the more natural "covering" formulations for them.

1.2 Other related work

Every ND(π) for nontrivial hereditary π is *MAX SNP*–hard, as pointed out in [LY93], due to the reductions of [LY80] and the result of [PY91]. Thus, no polynomial time algorithm can approximate ND(π) to within a factor of $1+\epsilon$ for some positive ϵ unless $P = NP$ [ALM+92]. Yet a better lower bound is provided by the one in approximation of VC as it serves as a lower bound for every ND(π) for hereditary π. Such a bound for VC has been continuously improved in the last few years, and currently it is known to be as large as $\frac{7}{6}$ [Hås97].

The approximation ratio of [BGNR94] for the unweighted FVS was subsequently extended to the one for the weighted FVS and was further improved to 2 in [BBF95, BG94], matching the best constant factor known for VC. Recently Chudak et al. [CGHW96] gave a primal–dual interpretation of these 2-approximation algorithms of [BBF95, BG94]. They also provided a new primal–dual algorithm for FVS, which has the same performance ratio but is slightly simpler than the previous two.

2 Preliminaries

2.1 Notation and Definitions

For any graph G let $V(G)$ and $E(G)$ denote the vertex set and the edge set, respectively, of G. The subgraph of $G = (V, E)$ induced by $X \subseteq V$ is denoted by $G[X]$. Let $E[X]$ denote the set of edges induced by $X \subseteq V$, and conversely, let $V[F]$ for $F \subseteq E$ denote the set of vertices incident to some edge in F. $E[X, Y]$ is the set of edges with one end in X and the other in Y. The set of edges incident to some node of X is denoted $\delta(X)$ and when those edges are restricted to the ones in a subgraph $G[Y]$ we denote it by $\delta_Y(X) (= \delta(X) \cap E[Y])$. Let $\delta(u)$ ($\delta_Y(u)$, resp.) be a shortening of $\delta(\{u\})$ ($\delta_Y(\{u\})$, resp.).

A graph property π is *nontrivial* if infinitely many graphs satisfy π and infinitely many graphs fail to satisfy it. It is *hereditary (on induced subgraphs)* if, in any graph satisfying π, every (node–induced, resp.) subgraph also satisfies π. For a hereditary property π any graph which does not satisfy π is called a *forbidden* graph for π, and it is a *minimal* one if, additionally, every "proper" (induced, resp.) subgraph of it satisfies π. Any hereditary property π is equivalently characterized by the set of all minimal forbidden graphs for π.

It is customary to measure the quality of an approximation algorithm by its *performance ratio*, which is the worst case ratio of the optimal solution value to the value of an approximate solution returned by the algorithm.

2.2 Matroidal Properties

One way to represent a *matroid* M is by a pair of a *ground set* E and a *rank function* r defined on 2^E. A set $F \subseteq E$ is called

- *independent* if $r(F) = |F|$ (and conversely, $r(F')$ is the cardinality of a largest independent subset of F' for an arbitrary $F' \subseteq E$),
- *dependent* if $r(F) < |F|$,
- a *base* if it is a maximal (and hence, maximum in any matroid) independent set, and a *circuit* if it is a minimal dependent set,
- *spanning* if $r(F) = r(E)$.

For any matroid $M = (E, r)$ there is the *dual matroid* $M^d = (E, r^d)$ defined on the same ground set E. The rank functions r and r^d are related s.t.

$$r^d(E - F) = (|E| - r(E)) - (|F| - r(F))$$

for any $F \subseteq E$ (For more on matroid theory see, for instance, [Wel76]).

Let M be a matroid which can be defined on the edge set of any graph (called an *edge set matroid*) and denote by $M(G)$ the matroid defined by M on the edge set of G. To avoid any possible anomaly we stipulate that for any subgraph H of G, $M(H)$ is the *restriction* of $M(G)$ onto $E(H)$. This means that the rank function of $M(H)$ is that of $M(G)$, but its domain restricted to subsets of $E(H)$.

We say that a graph property π is *matroidal* if for some edge set matroid M a (sub)graph G satisfies π iff its edge set is independent in $M(G)$ (Such a property is said to be *derived from* the matroid M). Such a property is hereditary on induced subgraphs because a subset of an independent set is independent in any matroid. Therefore, node–deletion problems for any nontrivial matroidal properties are *NP*–hard and *MAX SNP*–hard according to the results of [LY80] and [LY93]. Also note that the family of minimal forbidden graphs for such a property π corresponds to the family of circuits of the corresponding matroid $M(G)$ for all possible G.

2.3 Matroidal Families of Graphs

A *matroidal family of graphs* is a non–empty collection P of finite, *connected* graphs with the following property: given an arbitrary graph G, the edge sets of the subgraphs of G that are isomorphic to some member of P are the circuits of a matroid on $E(G)$. The matroid defined this way by the matroidal family P on the edge set of graph G will be denoted by $P(G)$.

The following four matroidal families, P_0, P_1, P_2, and P_3, are those that were discovered first [Sim72, Sim73]. The family P_0 consists of one graph only, namely two nodes with one edge in between. This is also the only finite matroidal family. The family P_1 consists of all the cycles; thus, $P_1(G)$ is the cycle matroid defined on $E(G)$. The family P_2 consists of all the bicycles, where a bicycle is a graph formed by minimally connecting two independent cycles. These two cycles can be joined together by either (1) sharing only a single node, (2) sharing only a connected path, or (3) having a simple path attached only at each end of it. The family P_3 consists of all the even cycles (i.e. cycles of even length) and the bicycles with no even cycle. The matroidal properties derived from these families thus correspond, respectively, to "a graph has no edge" (P_0), "a graph contains no cycle" (P_1), "every connected component contains at most one cycle" (P_2), and "every connected component contains at most one odd cycle and no even cycle" (P_3). Therefore, ND(π) is actually the VC (FVS, respectively) problem when π is the matroidal property derived from P_0 (P_1, respectively).

It has been known that in fact there exist infinitely many (uncountably many) matroidal families of graphs, and the first description of them (countably many matroidal families) was obtained by Andreae:

Proposition 1 [And78]. *Let s and t be integers, $s \geq 0$ and $-2s + 1 \leq t \leq 1$. Let $P_{s,t}$ be the set of all graphs G s.t.*

(i) $s|V(G)| + t = |E(G)|$, *and*
(ii) G *is minimal with respect to property (i); i.e., no graph isomorphic to a proper subgraph of G satisfies property (i).*

Then $P_{s,t}$ is a matroidal family.

It is not so hard to verify that $P_1 = P_{1,0}$, $P_2 = P_{1,1}$, and $P_0 = P_{s,-2s+1}$ (P_3 is not of the form $P_{s,t}$).

3 Primal–Dual Approximation for Matroidal Properties

One of the most natural integer program formulations of ND(π), presented here for the sake of comparison, is the one for a "covering problem":

$$\text{Min} \sum_{u \in V} w_u x_u$$

subject to:
$$\sum_{u \in V(H)} x_u \geq 1 \qquad H \in \Omega_G(\pi)$$
$$x_u \in \{0,1\} \qquad u \in V$$

where $\Omega_G(\pi)$ is the set of minimal forbidden graphs of ND(π) contained as subgraphs in G. It was indicated in [ENSZ96] that in case of the FVS problem the linear relaxation of the formulation above introduces the *integrality gap* (i.e., the ratio between the integer and fractional optima) of size as large as $\Omega(\log |V|)$. We shall show later that there exists another formulation of which integrality gap is bounded by 2 for many $\mathrm{ND}(\pi)$'s including FVS (see Corollary 9). Chudak et al. gave new primal–dual formulations and the algorithms based on them for the FVS problem in undirected graphs [CGHW96]. These algorithms are not new ones but actually are primal–dual "interpretations" of the algorithms previously known from [BBF95, BG94]. We shall show below that in fact every $\mathrm{ND}(\pi)$ with matroidal π has a simple and identical primal–dual formulation as well as an algorithm based on it. Chudak et al. also gave a new algorithm for the FVS problem which is a slight simplification of the previous algorithms cited above. Our algorithm for $\mathrm{ND}(\pi)$ is even simpler than theirs.

We claim that $\mathrm{ND}(\pi)$ on graph $G = (V, E)$ can be formulated by the following integer program when π is a matroidal property derived from $M = (E(G), r)$:

(IP)
$$\text{Min} \sum_{u \in V} w_u x_u$$
subject to:
$$\sum_{u \in S} r^d(\delta_S(u)) x_u \geq r^d(E[S]) \qquad S \subseteq V$$
$$x_u \in \{0, 1\} \qquad u \in V$$

Theorem 2. *When π is a matroidal property, F is a solution of $\mathrm{ND}(\pi)$ iff $x^F \in \{0,1\}^V$ (incidence vector of F) is a feasible solution to (IP).*

Consider now the dual of the linear programming relaxation of (IP):

(D)
$$\text{Max} \sum_{S \subseteq V} r^d(E[S]) y_S$$
subject to:
$$\sum_{S: u \in S} r^d(\delta_S(u)) y_S \leq w_u \qquad u \in V$$
$$y_S \geq 0 \qquad S \subseteq V$$

The primal–dual approximation algorithm, based on (IP) and (D) above, for $\mathrm{ND}(\pi)$ with matroidal π is presented in Fig. 1. We elaborate more on it. The algorithm starts with $F = \emptyset$, the original graph $G[S'] = (V, E)$ and the dual feasible solution $y = 0$. Given F, if it is not yet a solution of $\mathrm{ND}(\pi)$ there must exist some set $S \subseteq V$ corresponding to a violated constraint of (IP). In particular the set of all the remaining nodes $S'(= V - F)$ must be always such a set, and thus we can always choose S' as a "violated set". The algorithm then increases the dual variable $y_{S'}$ as much as possible until for some node u in S' the dual constraint for u becomes tight; i.e., $\sum_{S: u \in S} r^d(\delta_S(u)) y_S = w_u$. Notice that $y_{S'}$ here can be indeed increased because S' is the collection of all those nodes whose corresponding dual constraints were not yet tight. The algorithm adds u into a solution set F and at the same time removes it from S'. Clearly F eventually becomes a solution of $\mathrm{ND}(\pi)$ (and to (IP)) while y is kept feasible to (D). Lastly,

```
Initialize F = ∅, S' = V, y = 0, l = 0.
While F is not a solution of ND(π) do
    l ← l + 1.
    Increase y_S' until for some u ∈ S' the dual constraint corresponding to u
        becomes tight.
    Let u_l ← u.
    Add u_l into F and remove u from S'.
For j = l downto 1 do
    If F − {u_j} is a solution of ND(π) in G then remove u_j from F.
Output F.
```

Fig. 1. Primal–Dual Approximation Algorithm for ND(π)

the nodes in F are examined one by one, in the reverse order of their inclusion to F, and whenever any of them is found to be extraneous it is thrown out of F.

The algorithm clearly constructs a feasible solution F of ND(π) and a solution y feasible for (D). These two solutions are related such that

$$\sum_{u \in F} w_u = \sum_{u \in F} \sum_{S:u \in S} r^d(\delta_S(u)) y_S = \sum_{S \subseteq V} \left(\sum_{u \in S \cap F} r^d(\delta_S(u)) \right) y_S \quad (1)$$

An analysis of this algorithm reduces its performance ratio to the following combinatorial bound.

Theorem 3. *Let π be a matroidal property derived from $M = (E(G), r)$. Then the performance ratio of the primal-dual algorithm is bounded by*

$$\max \left\{ \frac{\sum_{u \in X} r^d(\delta(u))}{r^d(E(G))} \right\}$$

where max *is taken over any minimal solution X of ND(π) in any graph G.*

4 Uniformly Sparse Graph Properties

It was shown in [Fuj96] that when π is derived either from P_0 or P_1 (i.e., the VC or FVS property) (an essentially same algorithm as) the primal–dual algorithm delivers a solution with approximation ratio of 2. We add here one more to this list:

Theorem 4. *When π is derived from P_3 the primal-dual algorithm for ND(π) has performance ratio of 2.*

The case of $P_2 \equiv P_{1,1}$ will be subsumed by the general result given below.

We now turn our attention to a "relaxation" of the matroidal families of graphs, dropping the connectivity requirement on graphs in the families. Recall the countably many matroidal families $P_{s,t}$ ($s \geq 0, -2s+1 \leq t \leq 1$) of graphs from Proposition 1. Fix s to 1, let t be any integer $\geq -2s+1 = -1$, and consider the sets of graphs, that are no longer necessary to be connected, using the same set of the definitions for $P_{s,t}$; i.e., $P_{1,t}$ is the set of all graphs G s.t.

(i) $|V(G)| + t = |E(G)|$, and
(ii) G is minimal with respect to property (i); i.e., no graph isomorphic to a proper subgraph of G satisfies property (i).

Let $Q_k \stackrel{\text{def}}{=} P_{1,k+1}$ for $k \geq -2$.

Proposition 5. *Q_k defines the set of circuits of a matroid on any (edge set of) graph for all k.*

It is useful to observe here what graph properties are actually derived from Q_k's. A graph $G = (V, E)$ satisfies the property iff for every $F \subseteq E$, $|F| - |V[F]| \leq k$, and thus we may call a graph with such a property *uniformly k-sparse*. We should also note:

Proposition 6. *Q_k consists of an infinite number of distinct graphs for all $k \geq -1$.*

The next is a key lemma of the present paper, proof of which is postponed till Sec. 5.

Lemma 7. *Let π be a property derived from $Q_k = (E(G), r)$. Suppose $X \subseteq V(G)$ is any minimal solution of $ND(\pi)$ in any G. Then,*

$$\sum_{u \in X} r^d(\delta(u)) \leq 2 \cdot r^d(E(G)).$$

Finally, observe that given $G = (V, E)$ we can compute efficiently the rank $r(F)$ of any $F \subseteq E$ (and thus $r^d(\delta(u))$ for each $u \in V$) under $Q_k(G)$ (for instance, using the formula (2)). Therefore, our primal–dual algorithm runs in polynomial time for every Q_k. Now from Lemma 7 and Theorem 3 it easily follows that

Theorem 8. *When π is the property derived from Q_k for any fixed k the primal–dual algorithm computes a solution of $ND(\pi)$ in polynomial time; its performance ratio is bounded above by 2.*

And hence, there exist at least countable many nontrivial hereditary properties with an infinite number of minimal forbidden graphs, for which the node–deletion problems are efficiently approximable to within a factor of 2.

We also deduce from Lemma 7, (1), and the fact that $F \cap S$ is a minimal solution in $G[S]$ whenever y_S is nonzero, that the integrality gap in our formulation is at most 2 when π is derived from Q_k. Let Z_{IP}^* and Z_D^* be the optimal values of (IP) and (D), respectively. And then, for any F and y computed by the primal–dual algorithm,

$$Z_{IP}^* \leq \sum_{u \in V} w_u x_u^F = \sum_{u \in F} w_u = \sum_{S \subseteq V} (\sum_{u \in S \cap F} r^d(\delta_S(u))) y_S \leq \sum_{S \subseteq V} 2 r^d(E[S]) y_S$$
$$\leq 2 Z_D^*.$$

Corollary 9. *When r is the rank function of Q_k for any k, $\frac{Z_{IP}^*}{Z_D^*} \leq 2$.*

5 Proof of Lemma 7

Definitions. Let C be a connected component. Define the *surplus* of C by $sp(C) \stackrel{\text{def}}{=} |E(C)| - |V(C)|$ and the *bounded surplus* $\bar{sp}$ of C by $\bar{sp}(C) = \min\{k, sp(C)\}$. Let $C^+(F)$ ($C^-(F)$) denote the set of components, induced by an edge set F, with a positive bounded surplus (with a negative bounded surplus, respectively). When E' is an edge subset of E define $sp(E')$ to be the surplus of the graph induced by E'.

Notice that $C^-(F)$ consists of all the acyclic components, each with a (bounded) surplus of -1, induced by F. Also notice that for any component C and for any $E' \subseteq E(C)$, $sp(E') \leq sp(C)$. The rank function of the matroid $Q_k(G)$ defined on $G = (V, E)$ can be given by

$$r(F) = |V[F]| + \min\{k, \sum_{C \in C^+(F)} \bar{sp}(C)\} - |C^-(F)| \quad (2)$$

for any $F \subseteq E$.

Assume throughout that $k \geq 0$ (the case of $k \leq -1$ is no harder). Consider first the edge set $E[V - X]$, which must be an independent set of $Q_k(G)$, since X is a solution of ND(π). Using (2) we have

$$|E[V - X]| = r(E[V - X])$$
$$= |V - X| - (\# \text{ of acyclic components in } G[V - X]) + l \quad (3)$$

for some $0 \leq l \leq k$. We shall use the following auxiliary lemma in proving Lemma 7.

Lemma 10. *Assume (3). If X is a minimal solution of ND(π) then*

$$|E[X, V - X]| \geq (k - l + 1)|X| + \sum_{u \in X} (\# \text{ of acyclic components, in } G[V - X], \text{ adjacent to } u) \quad (4)$$

Suppose G contains an acyclic component T. Then since X contains no node of T, due to its minimality, we can restrict ourselves w.l.o.g. to G without T. So assume that G contains no acyclic component. Now suppose $r(E) < |V| + k$. Then E must be independent in $Q_k(G)$ and G satisfies π. But then a solution X minimal in G must be empty and the inequality in question trivially holds.

So assume that $r(E) = |V| + k$, and using (3) we can write

$$r^d(E) = |E| - r(E)$$
$$= |E[X]| + |E[X, V - X]| - (r(E) - |E[V - X]|)$$
$$= |E[X]| + |E[X, V - X]|$$
$$\quad - (|X| + k - l + (\# \text{ of acyclic components in } G[V - X])) \quad (5)$$

Assume that $|X| \geq 2$ (the case of $|X| \leq 1$ is more straightforward and omitted here). Call such a component in $G[V - X]$ that is adjacent to a single node in X as a *leaf* component. Recall that the dual rank of any $E' \subseteq E$ under a matroid M can be equivalently defined by

$$r^d(E') = \max\{|E' - B| : B \text{ is a base of } M\}$$

Take any node u of X. To estimate the value of $r^d(\delta(u))$ we observe how many edges incident to u must belong to a base of $Q_k(G)$. Let $I, J \subseteq E$ be mutually disjoint sets s.t. I is an independent set and $sp(J) \leq 0$. And then, $I \cup J$ in general is an independent set of $Q_k(G)$. This observation allows us to argue that, for every acyclic leaf component T which is adjacent only to u, any base B of $Q_k(G)$ must use all the edges in T and (at least) one edge connecting u and T. Besides them B must use at least one more edge from $\delta(u)$. To see why notice first that if no other edges of $\delta(u)$ belong to B the component of B containing u is a tree at best. As observed above, however, it is always possible to extend this component by one more edge incident to it *if it exists*. So it remains to see that at least one more edge is incident to u, and this is easy to do for, otherwise, u belongs to an acyclic component of G, which we've excluded at the beginning of the current analysis. Therefore, we can write

$$r^d(\delta(u)) \leq |\delta(u)| - ((\# \text{ of acyclic leaf components adjacent to } u) + 1)$$

and hence,

$$\sum_{u \in X} r^d(\delta(u)) \leq 2|E[X]| + |E[X, V - X]| - |X|$$
$$-(\# \text{ of acyclic leaf components}) \qquad (6)$$

Notice that, since there is no isolated acyclic component in G, we can reduce (4) to

$$|E[X, V - X]| \geq (k - l + 1)|X| + 2(\# \text{ of acyclic non-leaf components})$$
$$+ (\# \text{ of acyclic leaf components}) \qquad (7)$$

Combining (5), (6) and (7),

$$2r^d(E) - \sum_{u \in X} r^d(\delta(u))$$

$$\geq |E[X, V - X]| - 2(|X| + k - l + (\# \text{ of acyclic components in } G[V - X]))$$
$$+ |X| + (\# \text{ of acyclic leaf components in } G[V - X])$$
$$= |E[X, V - X]| - (|X| + 2(k - l))$$
$$- (2(\# \text{ of acyclic non-leaf components}) + (\# \text{ of acyclic leaf components}))$$
$$\geq (k - l)(|X| - 2)$$
$$\geq 0$$

References

[ALM+92] S. Arora, C. Lund, R. Motwani, M. Sudan, and M. Szegedy. Proof verification and hardness of approximation problems. In *33rd FOCS*, pages 14–23, 1992.

[And78] T. Andreae. Matroidal families of finite connected nonhomeomorphic graphs exist. *J. of Graph Theory*, 2:149–153, 1978.

[BBF95] V. Bafna, P. Berman, and T. Fujito. Constant ratio approximations of the weighted feedback vertex set problem for undirected graphs. In *ISAAC '95*, pages 142–151, 1995.

[BE85] R. Bar-Yehuda and S. Even. A local–ratio theorem for approximating the weighted vertex cover problem. In *Annals of Discrete Mathematics*, volume 25, pages 27–46. North–Holland, 1985.

[BG94] A. Becker and D. Geiger. Approximation algorithms for the loop cutset problem. In *Proc. of the 10th conference on Uncertainty in Artificial Intelligence*, pages 60–68, 1994.

[BGNR94] R. Bar-Yehuda, D. Geiger, J. Naor, and R. M. Roth. Approximation algorithms for the vertex feedback set problem with applications to constraint satisfaction and bayesian inference. In *5th SODA*, pages 344–354, 1994.

[CGHW96] F.A. Chudak, M.X. Goemans, D.S. Hochbaum, and D.P. Williamson. A primal–dual interpretation of recent 2-approximation algorithms for the feedback vertex set problem in undirected graphs. Manuscript, 1996.

[ENSZ96] G. Even, J. Naor, B. Schiever, and L. Zosin. Approximating minimum subset feedback sets in undirected graphs with applications. In *4th ISTCS*, pages 78–88, 1996.

[Fuj96] T. Fujito. A unified local ratio approximation of node–deletion problems. In *ESA '96*, pages 167–178, 1996.

[Gav74] F. Gavril, 1974. cited in [GJ79, page 134].

[GJ79] M. R. Garey and D. S. Johnson. *Computers and Intractability: A Guide to the Theory of NP-Completeness*. W. H. Freeman and co., New York, 1979.

[Hås97] J. Håstad. Some optimal in–approximability results. In *29th STOC*, to appear, 1997.

[LY80] J.M. Lewis and M. Yannakakis. The node–deletion problem for hereditary properties is NP–complete. *JCSS*, 20:219–230, 1980.

[LY93] C. Lund and M. Yannakakis. The approximation of maximum subgraph problems. In *20th ICALP*, pages 40–51, 1993.

[MS85] B. Monien and E. Speckenmeyer. Ramsey numbers and an approximation algorithm for the vertex cover problem. *Acta Inform.*, 22:115–123, 1985.

[PY91] C. Papadimitriou and M. Yannakakis. Optimization, approximation and complexity classes. *JCSS*, 43:425–440, 1991.

[Sim72] J.M.S. Simões-Pereira. On subgraphs as matroid cells. *Math. Z.*, 127:315–322, 1972.

[Sim73] J.M.S. Simões-Pereira. On matroids on edge sets of graphs with connected subgraphs as circuits. In *Proc. Amer. Math. Soc.*, volume 38, pages 503–506, 1973.

[Wel76] D.J.A. Welsh. *Matroid Theory*. Academic Press, London, 1976.

[Yan94] M. Yannakakis. Some open problems in approximation. In *CIAC '94*, pages 33–39, 1994.

Independent Sets in Asteroidal Triple-Free Graphs

Hajo Broersma[1] Ton Kloks[1] Dieter Kratsch[2] Haiko Müller[2]

[1] University of Twente
Faculty of Applied Mathematics
P.O. Box 217
7500 AE Enschede, the Netherlands
{H.J.Broersma,A.J.J.Kloks}@math.utwente.nl
[2] Fakultät für Mathematik und Informatik
Friedrich-Schiller-Universität
07740 Jena, Germany
{kratsch,hm}@minet.uni-jena.de

Abstract. An asteroidal triple is a set of three vertices such that there is a path between any pair of them avoiding the closed neighborhood of the third. A graph is called AT-free if it does not have an asteroidal triple. We show that there is an $O(n^2 \cdot (\overline{m} + 1))$ time algorithm to compute the maximum cardinality of an independent set for AT-free graphs, where n is the number of vertices and $\overline{m}$ is the number of non edges of the input graph. Furthermore we obtain $O(n^2 \cdot (\overline{m} + 1))$ time algorithms to solve the INDEPENDENT DOMINATING SET and the INDEPENDENT PERFECT DOMINATING SET problem on AT-free graphs. We also show how to adapt these algorithms such that they solve the corresponding problem for graphs with bounded asteroidal number in polynomial time. Finally we observe that the problems CLIQUE and PARTITION INTO CLIQUES remain NP-complete when restricted to AT-free graphs.

1 Introduction

Asteroidal triples were introduced in 1962 to characterize interval graphs as those chordal graphs that do not contain an asteroidal triple (short AT) [20]. Graphs not containing an AT are called asteroidal triple-free graphs (short AT-free graphs). They form a large class of graphs containing interval, permutation, trapezoid and cocomparability graphs. Since 1989 AT-free graphs have been studied extensively by Corneil, Olariu and Stewart. They have published a collection of papers presenting many structural and algorithmic properties of AT-free graphs (see e.g. [6, 7]). Further results on AT-free graphs were obtained in [18, 23].

Up to now the knowledge on the algorithmic complexity of NP-complete graph problems when restricted to AT-free graphs was relatively small compared to other graph classes. The problems TREEWIDTH, PATHWIDTH and MINIMUM FILL-IN remain NP-complete on AT-free graphs [1, 25]. On the other hand, domination-type problems like CONNECTED DOMINATING SET [7], DOMINATING SET [19] and TOTAL DOMINATING SET [19] can be solved by polynomial time algorithms for AT-free graphs. However there is a collection of classical NP-complete graph problems for which the algorithmic complexity when restricted to AT-free graphs was not known. Prominent representatives are INDEPENDENT SET, CLIQUE, GRAPH k-COLORABILITY, PARTITION INTO CLIQUES, HAMILTONIAN CIRCUIT and HAMILTONIAN PATH.

A crucial reason for the lack of progress in designing efficient algorithms for NP-complete problems on AT-free graphs seems to be that none of the typical representations, that are useful for the design of efficient algorithms on special graph classes, is known to exist for AT-free graphs. Contrary to well-known graph classes such as chordal, permutation and circular-arc graphs, AT-free graphs do not seem to have a representation by a geometric intersection model, an elimination scheme of vertices or edges, small separators, a small number of minimal separators etc. However it turns out that the design of all our algorithms is supported by a structural property of AT-free graphs, that can be obtained from the definition of AT-free graphs rather easily.

Our approach in this paper is similar to the one used to design algorithms for problems such as TREEWIDTH [14, 17] MINIMUM FILL-IN [17] and VERTEX RANKING [18] on AT-free graphs. However these algorithms have polynomial running time only under the additional constraint that the number of minimal separators is bounded by a polynomial in the number of vertices of the graph. (Notice that all three problems are NP-complete on AT-free graphs.) Technically, for the three different independent set problems in this paper, we are able to replace the set of all minimal separators, used in [14, 17, 18] – which might be 'too large' in size – by the 'small' set of all closed neighborhoods of the vertices of the graph.

Finding out the algorithmic complexity of INDEPENDENT SET on AT-free graphs is a challenging task. Besides the fact that INDEPENDENT SET is a classical and well-studied NP-complete problem, the problem is also interesting since, contrary to well-known subclasses of AT-free graphs such as cocomparability graphs, not all AT-free graphs are perfect. Thus the polynomial time algorithm for perfect graphs of Grötschel, Lovász and Schrijver [11] solving the INDEPENDENT SET problem does not apply to AT-free graphs.

We present the first polynomial time algorithm solving the NP-complete problem IN-DEPENDENT SET, when restricted to AT-free graphs. More precisely, our main result is the $O(n^2 \cdot (\overline{m} + 1))$ algorithm to compute the maximum cardinality of an independent set in an AT-free graph. Furthermore we present an $O(n^2 \cdot (\overline{m} + 1))$ time algorithm to solve the problem INDEPENDENT DOMINATING SET. A similar algorithm solves the problem INDEPENDENT PERFECT DOMINATING SET in time $O(n^2 \cdot (\overline{m} + 1))$ [3]. We also observe that the problems CLIQUE and PARTITION INTO CLIQUES remain NP-complete when restricted to AT-free graphs.

A natural generalization of asteroidal triples are the so-called asteroidal sets. Structural results for asteroidal sets and algorithms for graphs with bounded asteroidal number were obtained in [15, 21]. Computing the asteroidal number (i.e., the maximum cardinality of an asteroidal set) turns out to be NP-complete in general, but solvable in polynomial time for many graph classes [16]. Furthermore the results for problems as TREEWIDTH and MINIMUM FILL-IN on AT-free graphs can be generalized to graphs with bounded asteroidal number [15]. We show how to adapt our algorithms to obtain polynomial time algorithms for graphs with bounded asteroidal number solving the problems INDEPENDENT SET, INDEPENDENT DOMINATING SET and INDEPENDENT PERFECT DOMINATING SET.

2 Preliminaries

For a graph $G = (V, E)$ we denote $|V|$ by n, $|E|$ by m and the number of edges of the complement of G, which is equal to the number of non edges of G, by $\overline{m}$.

Recall that an independent set in a graph G is a set of pairwise nonadjacent vertices. The independence number of a graph G denoted by $\alpha(G)$ is the maximum cardinality of an independent set in G.

For a graph $G = (V, E)$ and $W \subseteq V$, $G[W]$ denotes the subgraph of G induced by the vertices of W; we write $\alpha(W)$ for $\alpha(G[W])$. For convenience, for a vertex x of G we write $G - x$ instead of $G[V \setminus \{x\}]$. Analogously, for a subset $X \subseteq V$ we write $G - X$ instead of $G[V \setminus X]$. We consider components of a graph as (maximal connected) subgraphs as well as vertex subsets. For a vertex x of $G = (V, E)$, $N(x) = \{y \in V : \{x,y\} \in E\}$ is the neighborhood of x and $N[x] = N(x) \cup \{x\}$ is the closed neighborhood of x. For $W \subseteq V$, $N[W] = \bigcup_{x \in W} N[x]$.

A set $S \subseteq V$ is a separator of the graph $G = (V, E)$ if $G - S$ is disconnected.

Definition 1. Let $G = (V, E)$ be a graph. A set $\Omega \subseteq V$ is an *asteroidal set* if for every $x \in \Omega$ the set $\Omega \setminus \{x\}$ is contained in one component of $G - N[x]$. An asteroidal set with three vertices is called an *asteroidal triple* (short AT).

Notice that every asteroidal set is an independent set.

Remark. A triple $\{x, y, z\}$ of vertices of G is an asteroidal triple if and only if for every two of these vertices there is a path between them avoiding the closed neighborhood of the third.

Definition 2. A graph $G = (V, E)$ is called *asteroidal triple-free* (short AT-free) if G has no asteroidal triple.

It is well-known that the INDEPENDENT SET problem 'Given a graph G and a positive integer k, decide whether $\alpha(G) \geq k$', is NP-complete [9]. The problem remains NP-complete, even when restricted to cubic planar graphs [13]. Moreover the independence number is hard to approximate within a factor of $n^{1-\epsilon}$ for any constant $\epsilon > 0$ [12]. Despite this discouraging recent result on the complexity of approximation, the independence number can be computed in polynomial time on many special classes of graphs (see [13]). For example, the best known algorithm to compute the independence number of a cocomparability graph has running time $O(n + m)$ [24].

The main result of this paper is an $O(n^2 \cdot (\overline{m} + 1))$ algorithm to compute the maximum cardinality of an independent set in a given AT-free graph. The structural properties enabling the design of our algorithms are given in the next three sections. In this extended abstract, we restrict ourselves to the cardinality case of the problems. Nevertheless our algorithms can be extended in a straightforward manner such that they solve the corresponding problems on graphs with real vertex weights (see [3]).

3 Intervals

Let $G = (V, E)$ be an AT-free graph, and let x and y be two distinct nonadjacent vertices of G. Throughout the paper we use $C^x(y)$ to denote the component of $G - N[x]$ containing y, and $r(x)$ to denote the number of components of $G - N[x]$.

Definition 3. A vertex $z \in V \setminus \{x, y\}$ is *between* x and y if x and z are in one component of $G - N[y]$ and y and z are in one component of $G - N[x]$.

Equivalently, z is between x and y in G if there is an x, z-path avoiding $N[y]$ and there is a y, z-path avoiding $N[x]$.

Definition 4. The *interval* $I = I(x, y)$ of G is the set of all vertices of G that are between x and y.

Thus $I(x, y) = C^x(y) \cap C^y(x)$.

4 Splitting intervals

Let $G = (V, E)$ be an AT-free graph, let $I = I(x, y)$ be a nonempty interval of G and let $s \in I$. Let $I_1 = I(x, s)$ and $I_2 = I(s, y)$.

Lemma 5. *The vertices x and y are in different components of $G - N[s]$.*

Proof. Assume x and y would be in the same component of $G - N[s]$. Then there is an x, y-path avoiding $N[s]$. However $s \in I$ implies that there is an s, y-path avoiding $N[x]$ and an s, x-path avoiding $N[y]$. Thus $\{s, x, y\}$ is an AT of G, a contradiction. □

Corollary 6. $I_1 \cap I_2 = \emptyset$.

Proof. Assume $z \in I_1 \cap I_2$. Then $z \in I_1$ implies that there is a component C^s of $G - N[s]$ containing both x and z. Furthermore $z \in I_2$ implies that also $y \in C^s$, contradicting Lemma 5. □

Lemma 7. $I_1 \subseteq I$ and $I_2 \subseteq I$.

Proof. Let $z \in I_1$. Clearly $s \in I$ implies $s \in C^x(y)$. Thus $z \in I_1$ implies $z \in C^x(y)$. Clearly $z \in C^s(x)$ since $z \in I_1$. By Lemma 5, $C^s(x)$ is contained in a component of $G - N[y]$ and obviously this component contains x. This proves $z \in I$. Consequently $I_1 \subseteq I$.

$I_2 \subseteq I$ can be shown analogously. □

Theorem 8. *There exist components $C_1^s, C_2^s, \ldots, C_t^s$ of $G - N[s]$ such that*

$$I \setminus N[s] = I_1 \cup I_2 \cup \bigcup_{i=1}^{t} C_i^s.$$

Proof. By Lemma 7, we have $I_1 \subseteq I \setminus N[s]$ and $I_2 \subseteq I \setminus N[s]$. By Lemma 5, x and y belong to different components $C^s(x)$ and $C^s(y)$ of $G - N[s]$. Let $z \in I \setminus N[s]$.

Assume $z \in C^s(x)$. There is a z, y-path avoiding $N[x]$. This path must contain a vertex of $N[s]$, showing the existence of a z, s-path avoiding $N[x]$. Hence $z \in I_1$.

Similarly $z \in C^s(y)$ implies $z \in I_2$.

Assume $z \notin C^s(x)$ and $z \notin C^s(y)$. Since $z \notin N[s]$, z belongs to the component $C^s(z)$ of $G - N[s]$. For any vertex $p \in C^s(z)$, there is a p, z-path avoiding $N[x]$, since $C^s(z) \neq C^s(x)$. Since $z \in I$, there is a z, y-path avoiding $N[x]$. Hence there is also a p, y-path avoiding $N[x]$. This shows $C^s(z) \subseteq I \setminus N[s]$. □

Corollary 9. *Every component of $G[I \setminus (N[s] \cup I_1 \cup I_2)]$ is a component of $G - N[s]$.*

5 Splitting components

Let $G = (V, E)$ be an AT-free graph. Let C^x be a component of $G - N[x]$ and let y be a vertex of C^x. We study the components of the graph $C^x - N[y]$.

Theorem 10. *Let D be a component of $C^x - N[y]$. Then $N[D] \cap (N[x] \setminus N[y]) = \emptyset$ if and only if D is a component of $G - N[y]$.*

Proof. Let D be a component of $C^x - N[y]$ with $N[D] \cap (N[x] \setminus N[y]) = \emptyset$. Since no vertex of D has a neighbor in $N[x] \setminus N[y]$, D is a component of $G - N[y]$.

Now let $D \subseteq C^x$ be a component of $G - N[y]$. Then $N[D] \cap N[x] \subseteq N[y]$. □

Corollary 11. *Let B be a component of $C^x - N[y]$. Then $N[B] \cap (N[x] \setminus N[y]) \neq \emptyset$ if and only if $B \subseteq C^y(x)$.*

Theorem 12. *Let $B_1, \ldots, B_\ell$ denote the components of $C^x - N[y]$ that are contained in $C^y(x)$. Then $I(x, y) = \bigcup_{i=1}^{\ell} B_i$.*

Proof. Let $I = I(x, y)$. First we show that $B_i \subseteq I$ for every $i \in \{1, \ldots, \ell\}$. Let $z \in B_i$. There is an x, z-path avoiding $N[y]$, since some vertex in B_i has a neighbor in $N[x] \setminus N[y]$. Clearly, there is also a z, y-path avoiding $N[x]$, since z and y are both in C^x. This shows that $z \in I$. Consequently $\bigcup_{i=1}^{\ell} B_i \subseteq I$.

Suppose $z \in I \setminus \bigcup_{i=1}^{\ell} B_i$. Since $z \notin \bigcup_{i=1}^{\ell} B_i$, the component D of $C^x - N[y]$ containing z does not contain a vertex with a neighbor in $N[x] \setminus N[y]$. Thus $z \notin C^y(x)$, implying $z \notin I$, a contradiction. □

6 Computing the independence number

In this section we describe our algorithm to compute the independence number of an AT-free graph. The algorithm we propose uses dynamic programming on intervals and components. All intervals and all components are sorted according to nondecreasing number of vertices. Following this order, the algorithm determines the independence number of each component and of each interval using the formulas given in Lemmas 13, 14 and 15.

We start with an obvious lemma.

Lemma 13. *Let $G = (V, E)$ be any graph. Then*

$$\alpha(G) = 1 + \max_{x \in V} \left(\sum_{i=1}^{r(x)} \alpha(C_i^x) \right),$$

where $C_1^x, C_2^x, \ldots, C_{r(x)}^x$ are the components of $G - N[x]$.

Applying Lemma 13 to the decomposition given by Theorems 10 and 12, we obtain the following lemma.

Lemma 14. *Let $G = (V, E)$ be an AT-free graph. Let $x \in V$ and let C^x be a component of $G - N[x]$. Then*

$$\alpha(C^x) = 1 + \max_{y \in C^x} \left(\alpha(I(x, y)) + \sum_i \alpha(D_i^y) \right),$$

where the D_i^y's are the components of $G - N[y]$ contained in C^x.

Applying Lemma 13 to the decomposition given by Theorem 8, we obtain the following lemma.

Lemma 15. *Let $G = (V, E)$ be an AT-free graph. Let $I = I(x, y)$ be an interval of G. If $I = \emptyset$ then $\alpha(I) = 0$. Otherwise*

$$\alpha(I) = 1 + \max_{s \in I} \left(\alpha(I(x, s)) + \alpha(I(s, y)) + \sum_i \alpha(C_i^s) \right),$$

where the C_i^s's are the components of $G - N[s]$ contained in $I(x, y)$.

Remark. Notice that the components D_i^y and C_i^s as well as the intervals $I(x, s)$ and $I(s, y)$ on the right-hand side of the formulas in Lemma 14 and Lemma 15 are proper subsets of C^x and I, respectively. Hence $\alpha(C^x)$ (resp. $\alpha(I)$) can be computed by table look-up to components and intervals with a smaller number of vertices.

Consequently we obtain the following algorithm to compute the independence number $\alpha(G)$ for a given AT-free graph $G = (V, E)$, which is based on dynamic programming.

Step 1 For every $x \in V$ compute all components $C_1^x, C_2^x, \ldots, C_{r(x)}^x$ of $G - N[x]$.
Step 2 For every pair of nonadjacent vertices x and y compute the interval $I(x, y)$.
Step 3 Sort all the components and intervals according to nondecreasing number of vertices.
Step 4 Compute $\alpha(C)$ and $\alpha(I)$ for each component C and each interval I in the order of Step 3.
Step 5 Compute $\alpha(G)$.

Theorem 16. *There is an $O(n^2 \cdot (\overline{m} + 1))$ time algorithm to compute the independence number of a given AT-free graph.*

Proof. The correctness of our algorithm follows from the formulas of Lemmas 13, 14 and 15 as well as the order of the dynamic programming.

We show how to obtain the stated time complexity. Clearly, Step 1 can be implemented such that it takes $O(n(n+m))$ time using a linear time algorithm to compute the components of the graph $G - N[x]$ for each vertex x of G. For each component of $G - N[x]$, a sorted linked list of all its vertices and its number of vertices is stored. For all nonadjacent vertices x and y there is a pointer $P(x, y)$ to the list of $C^x(y)$. Thus in Step 2, an interval $I(x, y)$ can be computed using the fact that $I(x, y) = C^x(y) \cap C^y(x)$. Hence a sorted vertex list of $I(x, y)$ can be computed in time $O(n)$ for each interval. Consequently the overall time bound for Step 2 is $O(n \cdot (\overline{m} + 1))$. There are at most n^2 components and at most n^2 intervals and each has at most n vertices. Thus using the linear time sorting algorithm bucket sort, Step 3 can be done in time $O(n^2)$.

The bottleneck for the time complexity of our algorithm is Step 4. First consider a component C^x of $G - N[x]$ and a vertex $y \in C^x$. We need to compute the components of $G - N[y]$ that are contained in C^x. Each component D of $G - N[y]$ except $C^y(x)$ is contained in C^x if and only if $D \cap C^x \neq \emptyset$. Thus the components D of $G - N[y]$ with $D \subseteq C^x$ are exactly those components of $G - N[y]$ addressed by $P(y, z)$ for some $z \in C^x$. Thus all such components can be found in time $O(|C^x|)$ for fixed vertices x and $y \in C^x$. Hence the computation of $\alpha(C)$ for all components C takes time $\sum_{\{x,y\} \notin E} O(|C^x(y)|) = O(n \cdot (\overline{m} + 1))$.

Now consider an interval $I = I(x, y)$, and a vertex $s \in I$. We need to add up the independence numbers of the components C_i^s of $G - N[s]$ that are contained in I. The components of $G - N[y]$ that are contained in I are exactly those components addressed by $P(y,z)$ for some $z \in I$, except $C^s(x)$ and $C^s(y)$. Thus all such components can be found in time $O(|I(x,y)|)$ for a fixed interval $I(x,y)$ and $s \in I(x,y)$. Hence the computation of $\alpha(I)$ for all intervals I takes time $\sum_{\{x,y\}\notin E} \sum_{s \in I(x,y)} O(|I(x,y)|) = O(n^2 \cdot (\overline{m}+1))$.

Clearly Step 5 can be done in $O(n^2)$ time. Thus the running time of our algorithm is $O(n^2 \cdot (\overline{m}+1))$. □

7 Independent domination

The approach used to design the presented polynomial time algorithm to compute the independence number for AT-free graphs can also be used to obtain a polynomial time algorithm solving the INDEPENDENT DOMINATING SET problem on AT-free graphs. The best known algorithm to solve the weighted version of the problem on cocomparability graphs has running time $O(n^{2.376})$ [4].

Definition 17. Let $G = (V, E)$ be a graph. Then $S \subseteq V$ is a *dominating set* of G if every vertex of $V \setminus S$ has a neighbor in S. A dominating set $S \subseteq V$ is an *independent dominating set* of G if S is an independent set.

We denote by $\gamma_i(G)$ the minimum cardinality of an independent dominating set of the graph G. Given an AT-free graph G, our next algorithm computes $\gamma_i(G)$. It works very similar to the algorithm of the previous section.

We present only the formulas used in Step 4 and 5 of the algorithm (which are similar to those in Lemma 13, Lemma 14 and Lemma 15).

Lemma 18. *Let $G = (V, E)$ be a graph. Then*

$$\gamma_i(G) = 1 + \min_{x \in V} \Big(\sum_{j=1}^{r(x)} \gamma_i(C_j^x) \Big),$$

where $C_1^x, C_2^x, \ldots, C_{r(x)}^x$ are the components of $G - N[x]$.

Lemma 19. *Let $G = (V, E)$ be an AT-free graph. Let $x \in V$ and let C^x be a component of $G - N[x]$. Then*

$$\gamma_i(C^x) = 1 + \min_{y \in C^x} \Big(\gamma_i(I(x,y)) + \sum_j \gamma_i(D_j^y) \Big),$$

where the D_j^y's are the components of $G - N[y]$ contained in C^x.

Lemma 20. *Let $G = (V, E)$ be an AT-free graph. Let $I = I(x,y)$ be an interval. If $I = \emptyset$ then $\gamma_i(I) = 0$. Otherwise*

$$\gamma_i(I) = 1 + \min_{s \in I} \Big(\gamma_i(I(x,s)) + \gamma_i(I(s,y)) + \sum_j \gamma_i(C_j^s) \Big),$$

where the C_j^s's are the components of $G - N[s]$ contained in $I(x,y)$.

Design and analysis of the algorithm is done similar to the previous section. We obtain the following theorem.

Theorem 21. *There exists an $O(n^2 \cdot (\overline{m}+1))$ time algorithm to compute the independence domination number γ_i of a given AT-free graph.*

In the full version [3] we also show how to obtain an $O(n^2 \cdot (\overline{m}+1))$ algorithm to compute a minimum cardinality independent perfect dominating set for AT-free graphs.

8 Bounded asteroidal number

In this section we show that the independence number of graphs with bounded asteroidal number can be computed in polynomial time.

Definition 22. *The* asteroidal number *of a graph G is the maximum cardinality of an asteroidal set in G.*

Hence a graph is AT-free if and only if its asteroidal number is at most two. Furthermore the asteroidal number of a graph G is bounded by $\alpha(G)$, since every asteroidal set is an independent set.

Definition 23. *Let Ω be an asteroidal set of G. The* lump *$L(\Omega)$ is the set of vertices v such that for all $x \in \Omega$ there is a component of $G - N[x]$ containing v and $\Omega \setminus \{x\}$.*

Let $\Omega = \{x_1, \ldots, x_\kappa\}$ be an asteroidal set of cardinality $\kappa \geq 2$ and consider the lump $L = L(\Omega)$.

Let s be an arbitrary vertex in L. In this section we show how $N[s]$ splits the lump analogous to Theorem 8.

Consider the components of $G - N[s]$. These components partition Ω into sets $\Omega_1, \ldots, \Omega_\tau$, where each Ω_i is a maximal subset of Ω contained in a component of $G - N[s]$.

Lemma 24. *For each $i = 1, \ldots, \tau$, the set $\Omega_i^* = \Omega_i \cup \{s\}$ is an asteroidal set in G.*

Proof. Consider $x \in \Omega_i$. Then, by definition, $\Omega \setminus \{x\}$ and s are contained in one component of $G - N[x]$. Hence, $\Omega_i^* \setminus \{x\}$ is contained in one component of $G - N[x]$. This proves the claim. □

Lemma 25. *Let $z \in L$ be in some component C^* of $G - N[s]$ that contains no vertices of Ω. Then $C^* \subseteq L$.*

Proof. Let $p \in C^* \setminus \{z\}$. There is a p, z-path avoiding $N[x]$ for any vertex $x \in \Omega$. This proves the claim. □

First we consider the case where $\tau = 1$, i.e., where Ω is in one component of $G - N[s]$. Then $\Omega \cup \{s\}$ is an asteroidal set.

Lemma 26. *If Ω is contained in one component C of $G - N[s]$, then $L(\Omega \cup \{s\}) = L \cap C$.*

Proof. Clearly $L(\Omega \cup \{s\}) \subseteq L \cap C$. Let $z \in L \cap C$ and consider a vertex $x \in \Omega$. Clearly, there is an x, z-path avoiding $N[s]$, since z and x are in the component C of $G - N[s]$. Hence z is in the component of Ω of $G - N[s]$. Consider any other vertex $y \in \Omega$. (Such vertices exist since $|\Omega| \geq 2$). There exists a z, y-path avoiding $N[x]$ since $z \in L$. But also, there exists a y, s-path avoiding $N[x]$ since $\Omega \cup \{s\}$ is an asteroidal set. Hence z is in the component of $(\Omega \cup \{s\}) \setminus \{x\}$ of $G - N[x]$. □

Now we consider the case where $\tau > 1$. Let $L_i = L(\Omega_i \cup \{s\})$ for $i = 1, \ldots, \tau$. Clearly, $L_i \cap L_j = \emptyset$ for every $i \neq j$.

Lemma 27. *Assume $\tau > 1$ and let C be the component of $G - N[s]$ containing Ω_i. Then $L_i = L \cap C$.*

Proof. First let $z \in L \cap C$. Then for all x and y in Ω_i there is a z, x-path avoiding $N[s]$ since $z \in C$ (showing that z and Ω_i are in one component of $G - N[s]$), and there is a z, x-path avoiding $N[y]$ since $z \in L$. For $y' \in \Omega_j$ for any $j \neq i$ there is a z, y'-path avoiding $N[x]$, since $z \in L$. Such a path contains a vertex of $N[s]$, and consequently there is a z, s-path avoiding $N[x]$. This shows that z, s and $\Omega_i \setminus \{x\}$ are in one component of $G - N[x]$ and hence $L \cap C \subseteq L_i$.

Now let $z \in L_i$. This clearly implies $z \in C$. For a vertex $y \in \Omega_j$, $j \neq i$, s and the set $\Omega \setminus \{y\}$ are in one component of $G - N[y]$ since $s \in L$. There is an s, z-path avoiding $N[y]$ since y and z belong to different components of $G - N[s]$. Consequently, z and $\Omega \setminus \{y\}$ are in one component of $G - N[y]$.

For a vertex $x \in \Omega_i$, there is a component of $G - N[x]$ containing s and $\Omega \setminus \{x\}$, since $s \in L$. Since $z \in L_i$, there is an s, z-path avoiding $N[x]$. Hence also z is in this component of $G - N[x]$ and therefore $L_i \subseteq L \cap C$. □

Theorem 28. *There exist components $C_1, \ldots, C_t$ of $G - N[s]$ which contain no vertex of Ω such that*

$$L \setminus N[s] = \bigcup_{i=1}^{t} C_i \cup \bigcup_{j=1}^{\tau} L_j.$$

Proof. Let $C_1, \ldots, C_t$ be the components of $G - N[s]$ which contain a vertex of L but no vertex of Ω. Then by Lemma 25 we have $\bigcup_{i=1}^{t} C_i \subseteq L \setminus N[s]$, and by Lemmas 26 and 27 we have $\bigcup_{j=1}^{\tau} L_j \subseteq L \setminus N[s]$.

Now let $l \in L \setminus N[s]$. If l is in a component containing Ω_i, $1 \leq i \leq \tau$, then $l \in L_i$ by Lemma 26 or 27. Otherwise there is an index i, $1 \leq i \leq t$ such that $l \in C_i$. This completes the proof. □

Theorem 28 enables us to generalize Lemmas 15 and 20 in the following way.

Lemma 29. *Let $L = L(\Omega)$ be a lump of G. If $L = \emptyset$ then $\alpha(L) = \gamma_i(L) = 0$. Otherwise*

$$\alpha(L) = 1 + \max_{s \in L} \Big(\sum_{j=1}^{t} \alpha(C_j) + \sum_{i=1}^{\tau} \alpha(L_i) \Big),$$

$$\gamma_i(L) = 1 + \min_{s \in L} \Big(\sum_{j=1}^{t} \gamma_i(C_j) + \sum_{k=1}^{\tau} \gamma_i(L_k) \Big),$$

where $C_1, \ldots, C_t$ are the components of $G - N[s]$ which contain no vertex of Ω, $L_1, \ldots, L_\tau$ are the lumps $L(\Omega_i + s)$ as used in Lemma 24.

Together with Lemmas 13 and 14, 18 and 19, the formulas of Lemma 29 lead to recursive algorithms computing $\alpha(G)$ and $\gamma_i(G)$ for a graph G. For any positive integer k, these algorithms can be implemented to run in time $O(n^{k+2})$ for all graphs with asteroidal number at most k. Analogously to the proof of Theorem 16, the time complexity is now dominated by the term $\sum_\Omega \sum_{s \in L(\Omega)} O(|L(\Omega)|) = O(n^{k+2})$, where the sum is taken over all asteroidal sets Ω of G and all $s \in L(\Omega)$.

As before, our algorithms for graphs with a bounded asteroidal number can be extended to the weighted cases of the problems and the corresponding algorithms have the same timebounds.

9 Conclusions

In this paper we have shown that the independence number as well as the independence domination number of an AT-free graph can be computed in time $O(n^2 \cdot (\overline{m} + 1))$. The same approach can be used to obtain an $O(n^2 \cdot (\overline{m} + 1))$ algorithm to solve the INDEPENDENT PERFECT DOMINATING SET problem on AT-free graphs. We have shown how to adapt the algorithm computing the independence number in such a way that the new algorithm computes the independence number of a graph with a bounded asteroidal number in polynomial time.

In the full version [3] we show how to extend our algorithms for the problems INDEPENDENT SET and INDEPENDENT DOMINATING SET to AT-free graphs with real vertex weights. Both algorithms run in time $O(n^2 \cdot (\overline{m} + 1))$. Furthermore our algorithms can also be modified such that they compute a maximum weight independent set and a minimum weight independent dominating set in time $O(n^2 \cdot (\overline{m} + 1))$.

Contrary to the independent set problems considered so far, the NP-complete graph problems CLIQUE and PARTITION INTO CLIQUES, that are closely related to INDEPENDENT SET, both remain NP-complete when restricted to the class of AT-free graphs. Concerning CLIQUE recall that Poljak has shown that INDEPENDENT SET remains NP-complete on triangle-free graphs [9]. Consequently CLIQUE remains NP-complete on graphs with independence number at most two, and thus on AT-free graphs. Similarly, it follows from a recent result due to Maffray and Preissman (showing that GRAPH k-COLORABILITY remains NP-complete when restricted to triangle-free graphs [22]), that the problem PARTITION INTO CLIQUES remains NP-complete on AT-free graphs.

Consequently CLIQUE and PARTITION INTO CLIQUES are the first NP-complete graph problems (known to us) which are NP-complete on AT-free graphs, but solvable in polynomial time on the class of cocomparability graphs. The latter graph class is the largest well-studied subclass of AT-free graphs which is also a class of perfect graphs.

It would be interesting to find out the algorithmic complexity of the following well-known NP-complete graph problems when restricted to AT-free graphs: GRAPH k-COLORABILITY, HAMILTONIAN CIRCUIT, HAMILTONIAN PATH. These three problems are all known to have polynomial time algorithms for cocomparability graphs [8, 10].

References

1. Arnborg, S., D. G. Corneil and A. Proskurowski, Complexity of finding embeddings in a k-tree, *SIAM J. Alg. Disc. Meth.* **8** (1987), pp. 277–284.
2. Brandstädt, A., Special graph classes – A survey, Schriftenreihe des Fachbereichs Mathematik, SM-DU-199, Universität Duisburg Gesamthochschule, 1991.
3. Broersma, H. J., T. Kloks, D. Kratsch and H. Müller, Independent sets in asteroidal triple-free graphs, Memorandum No. 1359, Faculty of Applied Mathematics, University of Twente, Enschede, The Netherlands, 1996.
4. Breu, H. and D. G. Kirkpatrick, Algorithms for domination and Steiner tree problems in cocomparability graphs, Manuscript 1993.
5. Chang, M. S., Weighted domination on cocomparability graphs, *Proceedings of ISAAC'95*, Springer-Verlag, LNCS 1004, 1996, pp. 122–131.
6. Corneil, D. G., S. Olariu and L. Stewart, The linear structure of graphs: Asteroidal triple-free graphs, *Proceedings of WG'93*, Springer-Verlag, LNCS 790, 1994, pp. 211–224.
7. Corneil, D. G., S. Olariu and L. Stewart, A linear time algorithm to compute dominating pairs in asteroidal triple-free graphs, *Proceedings of ICALP'95*, Springer-Verlag, LNCS 944, 1995, pp. 292–302.
8. Deogun, J. S. and G. Steiner, Polynomial algorithms for hamiltonian cycle in cocomparability graphs, *SIAM J. Comput.* **23** (1994), pp. 520–552.
9. Garey, M. R. and D. S. Johnson, *Computers and Intractability: A guide to the theory of NP-completeness*, Freeman, San Francisco, 1979.
10. Golumbic, M. C., *Algorithmic graph theory and perfect graphs*, Academic Press, New York, 1980.
11. Grötschel, M., L. Lovász and A. Schrijver, Polynomial algorithms for perfect graphs, *Annals of Discrete Mathematics* **21** (1984), pp. 325–356.
12. Hastad, J., Clique is hard to approximate within $n^{1-\epsilon}$, to appear in the *Proceedings of FOCS'96*.
13. Johnson, D. S., The NP-completeness column: An ongoing guide, *J. Algorithms* **6** (1985), pp. 434–451.
14. Kloks, T., *Treewidth – Computations and Approximations*, Springer-Verlag, LNCS 842, 1994.
15. Kloks, T., D. Kratsch and H. Müller, A generalization of AT-free graphs and some algorithmic results, Manuscript 1996.
16. Kloks, T., D. Kratsch and H. Müller, Asteroidal sets in graphs, Memorandum No. 1347, Faculty of Applied Mathematics, University of Twente, Enschede, The Netherlands, 1996.
17. Kloks, T., D. Kratsch and J. Spinrad, On treewidth and minimum fill-in of asteroidal triple-free graphs, to appear in *Theoretical Computer Science* **175** (1997).
18. Kloks, T., H. Müller and C. K. Wong, Vertex ranking of asteroidal triple-free graphs, *Proceedings of ISAAC'96*, Springer-Verlag, LNCS 1178, 1996, pp. 174-182.
19. Kratsch, D., Domination and total domination on asteroidal triple-free graphs, Forschungsergebnisse Math/Inf/96/25, FSU Jena, Germany, 1996.
20. Lekkerkerker, C. G. and J. Ch. Boland, Representation of a finite graph by a set of intervals on the real line, *Fund. Math.* **51** (1962), pp. 45–64.
21. Lin, I. J., T. A. McKee and D. B. West, Leafage of chordal graphs, Manuscript 1994.
22. Maffray, F. and M. Preissman, On the NP-completeness of the k-colorability problem for triangle-free graphs, *Discrete Mathematics* **162** (1996), pp. 313–317.
23. Möhring, R. H., Triangulating graphs without asteroidal triples, *Discrete Applied Mathematics*, **64** (1996), pp. 281–287.
24. McConnell, R. M. and J. P. Spinrad, Modular decomposition and transitive orientation, Manuscript 1995.
25. Yannakakis, M., Computing the minimum fill-in is NP-complete, *SIAM J. Alg. Disc. Meth.* **2** (1981), pp. 77–79.

Refining and Compressing Abstract Domains

Roberto Giacobazzi*   Francesco Ranzato**

*Dipartimento di Informatica, Università di Pisa
Corso Italia 40, 56125 Pisa, Italy
`giaco@di.unipi.it`

**Dipartimento di Matematica Pura ed Applicata, Università di Padova
Via Belzoni 7, 35131 Padova, Italy
`franz@math.unipd.it`

Abstract. In the context of Cousot and Cousot's abstract interpretation theory, we present a general framework to define, study and handle operators modifying abstract domains. In particular, we introduce the notions of operators of refinement and compression of abstract domains: A refinement enhances the precision of an abstract domain; a compression operator (compressor) can exist relatively to a given refinement, and it simplifies as much as possible a domain of input for that refinement. The adequateness of our framework is shown by the fact that most of the existing operators on abstract domains fall in it. A precise relationship of adjunction between refinements and compressors is also given, justifying why compressors can be understood as inverses of refinements.

1 Introduction

It is well known that abstract domains play a fundamental rôle in abstract interpretation [5, 6], since the precision of an abstract interpretation-based program analysis strongly depends on the expressive power of the chosen abstract domain. Much work has been therefore devoted to define systematic operators for enhancing the precision of representation of abstract domains. Relevant examples are Cousot and Cousot's reduced product, disjunctive completion and reduced cardinal power [6], Nielson's tensor product [18], Giacobazzi and Ranzato's dependencies and dual-Moore-set completion [13], the open product and pattern completion of Cortesi et al. [4], to cite the most known ones. The basic idea is that richer abstract domains can be obtained by combining simpler ones or by lifting them by adding new information. These operators on abstract domains provide high level facilities to tune the analysis in accuracy and cost, and some of them have been included as tools for abstract domain design aid in modern systems for program analysis, like for instance in System Z [22] and in PLAI [1].

We carry on this idea of operators enhancing the precision of abstract domains and we present in Sect. 3 a general and precise framework to handle these operators, which encompasses and improves the ideas sketched in [9]. The central notion is that of *abstract domain refinement*, that intuitively is any operator performing an action of refinement on abstract domains, with respect to their standard ordering relation of precision. There exists a strong link between refinements and *closure operators*, and many lattice-theoretic properties of closures are inherited by refinements. We introduce a generic pattern of definition for domain refinements, which allows to recover most of the important refinements listed above. Moreover, as an instance of this scheme, we present a new *refinement of completeness*. Roughly speaking, an abstract domain D is *complete* for a semantic function f defined on the concrete domain when no loss of precision is introduced by approximating f in the best possible way (i.e. by considering its best correct approximation, cf. [5, 6]) with respect to D. Thus, for a domain D, our refinement of completeness provides the most abstract domain which is more precise than D and complete for a given continuous concrete semantic function.

Recently, also operators of simplification of abstract domains have been defined and studied, like the operations of complementation in [3] and least disjunctive basis in [14]. As well as refinements, we show in Sect. 4 that these operators can be expressed in a formal and precise way in our framework. Actually, these operators are instances of our notion of *operator of compression* (or *compressor*). Roughly speaking, for a given abstract domain refinement $\Re$, its relative compressor simplifies a domain D of input for $\Re$, by returning the domain (if this exists) which contains the least amount of information required as input by $\Re$ to reach the same enhancement obtainable from D. This is somehow similar to the operation of compression on files – hence our terminology. In more precise terms, if $\Re$ is a unary refinement and D is an abstract domain, then an abstract domain A is the *optimal basis* of D for $\Re$, if A is the most abstract solution to the equation $\Re(X) = \Re(D)$. Obviously, if an optimal basis exists then it is necessarily unique. We say that $\Re$ is *invertible* on a given class of abstract domains if there exists the optimal basis of any domain D in the class. In this case, the compressor $\Re^-$ relative to $\Re$ (also called the *inverse* of $\Re$) provides the optimal basis $\Re^-(D)$ of D for $\Re$. The problem of inverting a refinement is often hard to solve in a satisfactory way, and, in general, not all domain refinements admit a corresponding compressor defined for a significant class of abstract domains. We show that complementation and least disjunctive basis give rise, respectively, to the compressors relative to reduced product and disjunctive completion refinements, and we give a generic scheme for defining invertible refinements. Moreover, we show that invertible refinements provide solutions to the problem of decomposing abstract domains into simpler factors. If $\Re$ is an n-ary refinement and $D = \Re(D_1, \ldots, D_n)$, then the tuple $\langle D_1, \ldots, D_n \rangle$ can be considered as a decomposition of D relative to $\Re$. We then present a general iterative method which starting from any decomposition relative to an invertible refinement provides minimal decompositions, i.e. decompositions involving the most abstract factors.

It is important to note that our notion of inversion of a refinement does not correspond to the more customary inversion in the sense of adjunctions – on the contrary, we observe that, in general, this is not possible. However, we show in Sect. 5 that this asymmetry can be overcome by considering a modified ordering relation between abstract domains, that is induced in a natural way by the refinement itself. We prove that for this lifted order on abstract domains, an invertible refinement and its compressor do constitute an adjunction. This provides a firm mathematical relationship between refinements and compressors, and gives a more precise justification to the use of the term "inverse".

2 Preliminaries

The structure $\langle uco(C), \sqsubseteq, \sqcup, \sqcap, \lambda x.\top, \lambda x.x \rangle$ denotes the complete lattice of all *upper closure operators* (shortly closures) on a complete lattice $\langle C, \leq, \vee, \wedge, \top, \bot \rangle$, where $\rho \sqsubseteq \eta$ iff $\forall x \in C.\ \rho(x) \leq \eta(x)$. The complete lattice of all *lower closure operators* on C is denoted by $lco(C)$ and is dual-isomorphic to $uco(C)$. Recall that each closure operator $\rho \in uco(C)$ is uniquely determined by the set of its fixpoints, which is its image, i.e. $\rho(C) = \{x \in C \mid \rho(x) = x\}$, that $\rho \sqsubseteq \eta$ iff $\eta(C) \subseteq \rho(C)$, and that a subset $X \subseteq C$ is the set of fixpoints of a closure iff $X = \{\wedge Y \mid Y \subseteq X\}$ (note that $\top \in X$). $\langle \rho(C), \leq \rangle$ is a complete meet subsemilattice of C but, in general, it is not a complete sublattice of C.

In the standard Cousot and Cousot abstract interpretation theory, abstract domains can be equivalently specified either by Galois connections or by closure operators [6]. In the first case, concrete and abstract domains are related by a pair of adjoint functions. This provides a way to relate domains containing objects having different representation. In the second case instead, an abstract domain is specified as (the set of fixpoints of) an upper closure on the concrete domain. Thus, the closure operator approach is particularly convenient when reasoning about properties of abstract domains independently from the representation of their objects, as in our case. Hence, we will identify $uco(C)$ with the complete lattice of all possible abstract domains of the concrete domain (i.e. any complete lattice) C. The ordering on $uco(C)$ corresponds precisely to the standard order used in abstract interpretation to compare abstract domains with regard to their precision: D_1 is *more precise* than D_2 iff $D_1 \sqsubseteq D_2$ in $uco(C)$ ($\sqsubset$ denotes strict ordering). The lub and glb on $uco(C)$ have therefore the following meaning as operators on domains. Suppose $\{D_i\}_{i \in I} \subseteq uco(C)$: (i) $\sqcup_{i \in I} D_i$ is the most concrete among the domains which are abstractions of all the D_i's, i.e. it is their least common abstraction; (ii) $\sqcap_{i \in I} D_i$ is (isomorphic to) the well-known reduced product of all the D_i's, and, equivalently, it is the most abstract among the domains (abstracting C) which are more concrete than every D_i. Whenever C is a meet-continuous complete lattice (i.e., for any chain $Y \subseteq C$ and $x \in C$: $x \wedge (\vee Y) = \vee_{y \in Y}(x \wedge y)$), $uco(C)$ enjoys the lattice-theoretic property of *pseudocomplementedness* (cf. [12]). This property allowed to define the operation of *complementation* of abstract domains (cf. [3]), namely an operation which, starting from any two domains $C \sqsubseteq D$, where C is meet-continuous, gives as result the most abstract domain $C \sim D$, such that $(C \sim D) \sqcap D = C$.

3 Abstract Domain Refinements

Intuitively, an abstract domain refinement is an operator that, for any tuple $\langle D_i \rangle_{1 \leq i \leq n}$ of domains of input (ranging on a given domain of definition), provides as output a domain more precise than each D_i. It is also very reasonable to expect that such an operator is monotone. These observations naturally lead to the definition below. In the following, a generic tuple of objects is denoted by $\mathbf{O}$, $\pi_i(\mathbf{O})$ denotes its i-th component, and $\mathbf{O}[X/i]$ denotes the tuple obtained from $\mathbf{O}$ by replacing $\pi_i(\mathbf{O})$ with X. Also, C is a complete lattice acting as the concrete domain and $\mathbf{U} \subseteq uco(C)^n$, $n \geq 1$, is a given tuple of sets of domains abstracting C (for simplicity, we only consider refinements of finite arity

– actually those having a practical meaning – although a generalization would be straightforward). When $n = 1$ we denote $\mathbf{U}$ as the set $U \subseteq uco(C)$. We extend on tuples the glb of $uco(C)$: For any tuple of domains $\mathbf{D}$, $\sqcap \mathbf{D} = \sqcap_{1 \leq i \leq n} \pi_i(\mathbf{D})$.

Definition 3.1 A map $\Re : \mathbf{U} \to uco(C)$ is a (*n-ary abstract domain*) *refinement* if: (i) $\Re$ is monotone; (ii) $\Re$ is reductive: $\forall \mathbf{D} \in \mathbf{U}. \Re(\mathbf{D}) \sqsubseteq \sqcap \mathbf{D}$. □

The kernel of definition of any refinement $\Re : \mathbf{U} \to uco(C)$ is given by $\mathbb{K}_{\mathbf{U}} = \cap_{1 \leq i \leq n} \pi_i(\mathbf{U})$. Often, refinements are defined on any tuple of abstract domains, i.e., $\Re : uco(C)^n \to uco(C)$, as in the case of reduced product and disjunctive completion, later considered. We will call them *full* refinements, in order to distinguish them from generic ones as allowed by Definition 3.1. Any n-ary refinement $\Re : \mathbf{U} \to uco(C)$ induces a family of refinements of lower arity obtained by fixing some of the domains of input. For instance, by fixing $n - 1$ domains, we get the unary refinements $\lambda X. \Re(\mathbf{D}[X/i]) : \pi_i(\mathbf{U}) \to uco(C)$. Also, $\Re$ induces the canonical unary *self-refinement* $\Re_1 : \mathbb{K}_{\mathbf{U}} \to uco(C)$ defined as $\Re_1(D) = \Re(D, ..., D)$. Conversely, any n-uple $\mathbf{R} = \langle \Re_i \rangle_{1 \leq i \leq n}$ of unary refinements $\Re_i : U_i \to uco(C)$ induces an n-ary refinement $\Re_{\mathbf{R}} : U_1 \times ... \times U_n \to uco(C)$ defined as $\Re_{\mathbf{R}}(\mathbf{D}) = \sqcap_{1 \leq i \leq n} \Re_i(\pi_i(\mathbf{D}))$, and called *attribute independent*.

It is important to remark that Definition 3.1 lacks of any requirement of idempotence. For instance, for a unary refinement $\Re : U \to uco(C)$ may well happen that a refined domain $\Re(D) \in U$ can still be object of further refinement, i.e. $\Re(\Re(D)) \sqsubset \Re(D)$. Due to lack of space, in the paper we will only consider examples of idempotent refinements, although a relevant example of nonidempotent refinement can be given by the dependencies between abstract domains of [13]. However, it is worth noting that, by monotonicity, any refinement can be lifted to an idempotent one as the limit of a possibly transfinite Kleene fixpoint iteration sequence. It is therefore reasonable requiring idempotence for refinements, i.e. that a refinement upgrades abstract domains all at once.

Definition 3.2 An n-ary refinement $\Re : \mathbf{U} \to uco(C)$ is *idempotent* if for any $i \in [1, n]$ and $\mathbf{D} \in \mathbf{U}$ such that $\Re(\mathbf{D}) \in \mathbb{K}_{\mathbf{U}}$, $\Re(\mathbf{D}) = \Re(\mathbf{D}[\Re(\mathbf{D})/i])$. □

Proposition 3.3 *For any $\Re : \mathbf{U} \to uco(C)$, the following are equivalent:*
(a) *$\Re$ is idempotent;*
(b) *For any $\mathbf{D} \in \mathbf{U}$ such that $\Re(\mathbf{D}) \in \mathbb{K}_{\mathbf{U}}$, $\Re(\mathbf{D}) = \Re_1(\Re(\mathbf{D}))$.*
If $\mathbb{K}_{\mathbf{U}}$ is a (finitely) meet subsemilattice of $uco(C)$ then (a) is equivalent to:
(c) *$\Re_1$ is idempotent and for any $\mathbf{D} \in \mathbf{U}$ such that $\Re(\mathbf{D}) \in \mathbb{K}_{\mathbf{U}}$, $\Re(\mathbf{D}) = \Re_1(\sqcap \mathbf{D})$.*

The following example yields a generic and useful pattern of definition for full idempotent refinements.

Example 3.4 Consider any property P of abstract domains, i.e. a subset of the lattice of abstract interpretations $P \subseteq uco(C)$. For any fixed $n \in \mathbb{N}$, define the operator $\Re_P : uco(C)^n \to uco(C)$ as $\Re_P = \lambda \mathbf{D}. \sqcup \{A \in uco(C) \mid A \in P, A \sqsubseteq \sqcap \mathbf{D}\}$. Thus, $\Re_P(\mathbf{D})$ is the least common abstraction of all domains that satisfy P and are more concrete (viz. precise) than every $\pi_i(\mathbf{D})$ for $i \in [1, n]$. It is immediate to observe that $\Re_P$ is monotone and reductive. Also, it is easily seen that $\Re_P$ satisfies the condition (c) of Proposition 3.3. Thus, $\Re_P$ always defines a full idempotent refinement. However, in general, $\Re_P(\mathbf{D})$ may not satisfy P. On the other hand, the following characterization holds.

Proposition 3.5 $\forall \mathbf{D}.\ \Re_P(\mathbf{D}) \in P \Leftrightarrow P \in lco(uco(C)) \Rightarrow \Re_P = \lambda \mathbf{D}.P(\sqcap \mathbf{D})$.

Thus, for a property P which is a lower closure, $\Re_P(\mathbf{D})$ is the most abstract domain which satisfies P and is more concrete than every $\pi_i(\mathbf{D})$, or, equivalently, $\Re_P(\mathbf{D})$ is the least extension of $\sqcap \mathbf{D}$ that satisfies P. It is also worth noting that $\Re_P(\mathbf{D})$ is the greatest fixpoint of the equation $X = P(X) \sqcap (\sqcap \mathbf{D})$ in $uco(C)$. □

Note that any unary idempotent refinement $\Re : U \to uco(C)$ such that $\Re(U) \subseteq U$ (we say in this case that $\Re$ is *well-defined* on U) actually is a *lower closure operator* on the poset $\langle U, \sqsubseteq \rangle$, with the order inherited from $uco(C)$, i.e. $\Re \in lco(U)$. In particular, any unary full idempotent refinement $\Re$ is a lower closure on $uco(C)$, i.e. $\Re \in lco(uco(C))$, a case already considered in [9]. Also, for any n-ary full idempotent refinement $\Re : uco(C)^n \to uco(C)$, we have that any unary refinement $\lambda X.\Re(\mathbf{D}[X/i])$ ($i \in [1, n]$) induced by $\Re$ is a lower closure operator on $uco(C)$, as well as the self-refinement $\Re_1$. It would be straightforward, although notationally tedious, to generalize this latter observation to generic n-ary (possibly nonfull) idempotent refinements that satisfy a suitably generalized condition of well-definedness. These observations are fairly important, since unary idempotent refinements inherit all the lattice-theoretic properties of lower closures (see [23] for a few of them). For instance, whenever the domain of definition $\langle U, \sqsubseteq \rangle$ is a complete lattice, we get that these refinements well-defined on U form a complete lattice $\langle lco(U), \sqsubseteq \rangle$ (by a slight abuse of notation, we always use the ordering symbol $\sqsubseteq$ for any kind of closures), where $\Re_1 \sqsubseteq \Re_2$ iff for any $A \in U$, $\Re_1(A) \sqsubseteq \Re_2(A)$ iff the set of abstract domains refined by $\Re_1$ is contained in the set of those refined by $\Re_2$. Thus, analogously to the case of abstract domains, the complete ordering $\sqsubseteq$ between idempotent refinements can be interpreted as a relation of precision among refinement operators, where $\Re_1$ is more precise than $\Re_2$ iff $\Re_1 \sqsubseteq \Re_2$. Moreover, any unary idempotent refinement well-defined on a complete subsemilattice U of $uco(C)$ enjoys the following properties of *compositionality w.r.t. the reduced product and least common abstraction*.

Proposition 3.6 *If U is a complete meet (join) subsemilattice of $uco(C)$, $\Re : U \to U$ is an idempotent refinement, and $\{D_i\}_{i \in I} \subseteq \wp(U)$, then $\Re(\sqcap_{i \in I} D_i) = \Re(\sqcap_{i \in I} \Re(D_i))$ ($\Re(\sqcup_{i \in I} \Re(D_i)) = \sqcup_{i \in I} \Re(D_i)$).*

Reduced Product Refinement. The simplest and probably most familiar example of abstract domain refinement is the reduced product [6], which is the glb in the lattice of abstractions. For simplicity, we consider it as a binary refinement. For any fixed concrete domain C (i.e., any complete lattice), reduced product is obviously an idempotent full refinement $\Re_\sqcap : uco(C) \times uco(C) \to uco(C)$. Thus, the unary refinement induced by $\Re_\sqcap$, i.e. $\lambda X.\ A \sqcap X$ is a lower closure, and for it the properties discussed above hold. It is worth noting that $\Re_\sqcap$ is the simplest instance of the family of refinements defined in Example 3.4, since $\Re_\sqcap$ is $\Re_P$ for the trivial property $P = uco(C)$. Also, $\Re_\sqcap$ is the attribute independent combination of the trivial identity refinements. Reduced product has been successfully applied as a domain refinement in program analysis e.g. in [1, 16, 21].

Disjunctive Completion Refinement. The disjunctive completion [6] enhances an abstract domain so that its disjunction operation (i.e. lub) becomes *precise* (as that of the concrete domain). Abstract domains with a precise disjunction (also called disjunctive abstract domains) correspond to additive closure operators. Disjunctive completion can be given as an instance of the general scheme of Example 3.4, where the property P is given by additivity: $P = uco^a(C)$, the subset of $uco(C)$ of additive closures. Hence, the disjunctive completion $\Re_\vee : uco(C) \to uco(C)$ is defined as $\Re_\vee(D) = \sqcup\{A \in uco^a(C) \mid A \sqsubseteq D\}$. Thus, $\Re_\vee$ is an idempotent full refinement. It is easy to observe that $uco^a(C)$ defines a lower closure on $uco(C)$. Then, by Proposition 3.5, $\Re_\vee(D)$ is the most abstract disjunctive domain that is more concrete than D. The disjunctive completion refinement has been applied in program analysis e.g. in [8, 15, 10].

Negative Completion Refinement. Assume the concrete domain C be a complete Boolean algebra. It is easy to verify that if $\rho \in uco^a(C)$ then $\neg\rho = \{\neg x \in C \mid x \in \rho\} \in uco^a(C)$. The negative completion refinement is then defined on disjunctive abstract domains, $\Re_\neg : uco^a(C) \to uco(C)$, as follows: $\Re_\neg(A) = A \sqcap \neg A$. Thus, $\Re_\neg$ lifts a given disjunctive abstract domain A to the reduced product of A with its negative abstract domain, namely to the most abstract domain containing both A and $\neg A$. It is now simple to check that $\Re_\neg : uco^a(C) \to uco(C)$ is an idempotent refinement. It is worth noting that, in general, $\Re_\neg(A)$ may not be disjunctive (i.e., $\Re_\neg$ is not well-defined on $uco^a(C)$).

The Refinement of Completeness. Abstract interpretation is intended to create *sound* approximations of the concrete semantics of programs. If the program semantics is specified as the least fixpoint of a monotone semantic operation $f : C \to C$ on a complete lattice C, then, in the closure operator approach, the soundness criterion for an abstract domain given by $\rho \in uco(C)$ and for an abstract monotone semantic operation $f^\sharp : \rho(C) \to \rho(C)$, is $\forall c \in C. \rho(f(c)) \leq f^\sharp(\rho(c))$. This ensures the global soundness of the abstract semantics, i.e. $\rho(\textit{lfp}(f)) \leq \textit{lfp}(f^\sharp)$ (cf. [5]). *Completeness* is the dual relation $\forall c \in C. f^\sharp(\rho(c)) \leq \rho(f(c))$. Because soundness is always required in abstract interpretation, in the following we abuse terminology and say that $f^\sharp$ is complete for f if $\rho \circ f = f^\sharp \circ \rho$. In this case $\rho(\textit{lfp}(f)) = \textit{lfp}(f^\sharp)$. Completeness in abstract interpretation is a quite rare ideal situation, where for a given abstract domain no loss of precision is introduced by abstract semantic operations. Completeness is especially recurrent between (concrete) semantics of programming languages (cf. [2, 7, 11]). Issues of completeness and related notions have also been studied in [17, 19, 20]. Completeness can be made a property of abstract domains, by making this notion independent on the choice for $f^\sharp$. Recall that the best correct approximation of f w.r.t. ρ is given by $\rho \circ f : \rho(C) \to \rho(C)$. Thus, we consider completeness of the best correct approximation: $\rho \in uco(C)$ is *complete* for f if $\rho \circ f = \rho \circ f \circ \rho$. For example, let us consider the canonical 4-point abstract domain $Sign = \{\emptyset, \mathbb{Z}_{<0}, \mathbb{Z}_{>0}, \mathbb{Z}\}$, which is an obvious abstraction of $\langle \wp(\mathbb{Z}), \subseteq \rangle$. It is simple to show that $Sign$ is complete for the monotone operation of integer multiplication $\lambda X.n \cdot X : \wp(\mathbb{Z}) \to \wp(\mathbb{Z})$ (where $n \in \mathbb{Z}$ and $n \cdot X = \{n \cdot m \mid m \in X\}$). On the other hand, $\rho = \{\mathbb{Z}_{>0}, \mathbb{Z}\}$ (with $Sign \sqsubseteq \rho$) is not complete for $\lambda X. n \cdot X$ with $n < 0$: In fact, e.g., $\rho(n \cdot \{-3\}) = \mathbb{Z}_{>0}$, but, because $\rho(\{-3\}) = \mathbb{Z}$, $\rho(n \cdot \rho(\{-3\})) = \rho(\mathbb{Z}) = \mathbb{Z}$. The property of completeness for a semantic function f is therefore given by $\Gamma(f) = \{\rho \in uco(C) \mid \rho \circ f = \rho \circ f \circ \rho\}$. Following the scheme of Example 3.4, we can define an idempotent full refinement of completeness $\Re_{\Gamma(f)} : uco(C) \to uco(C)$ as $\Re_{\Gamma(f)}(\rho) = \sqcup\{\eta \in uco(C) \mid \eta \in \Gamma(f), \eta \sqsubseteq \rho\}$.

Theorem 3.7 *If f is continuous then $\Gamma(f) \in lco(uco(C))$.*

Thus, by Proposition 3.5, we have that for a continuous f, $\Re_{\Gamma(f)}(D)$ actually is the (unique) most abstract domain which includes D and is complete for f. For instance, it is possible to check that for $n < 0$, $\Re_{\Gamma(\lambda X.n\cdot X)}(\{\mathbb{Z}_{>0}, \mathbb{Z}\}) = Sign$.

4 Abstract Domain Compressors

We have introduced the notion of abstract domain refinement as a formalization (and generalization) of many existing operators devoted to enhance the expressiveness of abstract domains. However, no operator performing a dual action of simplification on abstract domains has been proposed up till now. We now formalize the idea of a simplifying operator that gives as input to a fixed refinement the simplest domains (i.e. most abstract) which can be object of that refinement. Let $\Re : \mathbf{U} \to uco(C)$ be a (possibly nonidempotent) refinement. Define $\Re_k^- : \mathbf{U} \to uco(C)$, $k \in [1,n]$, as $\Re_k^- = \lambda \mathbf{D}. \sqcup \{A \in \pi_k(\mathbf{U}) \mid \Re(\mathbf{D}[A/k]) = \Re(\mathbf{D})\}$. For $\mathbf{D} \in \mathbf{U}$, $\Re_k^-(\mathbf{D})$ is the least common abstraction of all domains in $\pi_k(\mathbf{U})$ that, when substituted to $\pi_k(\mathbf{D})$ as k-th input for $\Re$, do not change the output.

Definition 4.1 $\Re_k^-(\mathbf{D})$ *is the k-th optimal basis of $\mathbf{D} \in \mathbf{U}$ for $\Re$ if $\Re_k^-(\mathbf{D}) \in \pi_k(\mathbf{U})$ and $\Re(\mathbf{D}) = \Re(\mathbf{D}[\Re_k^-(\mathbf{D})/k])$. The refinement $\Re$ is k-invertible (or admits the k-th inverse) on $V \subseteq \pi_k(\mathbf{U})$ if for all $\mathbf{D} \in \mathbf{U}[V/k]$, $\Re_k^-(\mathbf{D})$ is the k-th optimal basis of $\mathbf{D}$ for $\Re$. When $\Re$ is k-invertible, the map $\Re_k^- : \mathbf{U}[V/k] \to \pi_k(\mathbf{U})$ is called the k-th compressor for $\Re$.* □

Note that if the domain of definition $V \subseteq \pi_k(\mathbf{U})$ of the k-th compressor $\Re_k^-$ is a complete join subsemilattice of $uco(C)$, then the condition $\Re_k^-(\mathbf{D}) \in \pi_k(\mathbf{U})$ in the above definition can be omitted. For $K \subseteq [1,n]$, we say that $\Re$ is K-invertible on a $|K|$-tuple $\mathbf{V}$, where $\forall i \in K$. $\pi_i(\mathbf{V}) \subseteq \pi_i(\mathbf{U})$, if it is k-invertible on $\pi_k(\mathbf{V})$, for any $k \in K$. In particular, $\Re$ is *fully invertible* on $\mathbf{V} \subseteq \mathbf{U}$ if it is $[1,n]$-invertible on $\mathbf{V}$. For the simpler case of a unary refinement $\Re : U \to uco(C)$, we have that $\Re^- : U \to uco(C)$ is defined as $\Re^-(D) = \sqcup\{A \in U \mid \Re(A) = \Re(D)\}$, and $\Re$ is invertible on $V \subseteq U$ iff for any $D \in V$, $\Re(\Re^-(D)) = \Re(D)$. It is simple to observe that the above definition of k-invertibility can be formulated by using the unary refinements induced by a (n-ary) refinement. More precisely, if $\Re : \mathbf{U} \to uco(C)$ is a refinement, then we have already seen that for any $k \in [1,n]$ and $\pi_i(\mathbf{D}) \in \pi_i(\mathbf{U})$ ($i \neq k$), $\lambda X.\Re(\mathbf{D}[X/k]) : \pi_k(\mathbf{U}) \to uco(C)$ is a unary refinement. It is then easily seen that $\Re$ is k-invertible in $V \subseteq \pi_k(\mathbf{U})$ iff $\lambda X.\Re(\mathbf{D}[X/k])$ is (1-)invertible on $V \subseteq \pi_k(\mathbf{U})$. In this case, for the compressor $(\lambda X.\Re(\mathbf{D}[X/k]))^- : V \to \pi_k(\mathbf{U})$ and the k-th compressor $\Re_k^-$ of $\Re$, the following mutual equality result holds: $\forall D \in V$. $(\lambda X.\Re(\mathbf{D}[X/k]))^-(D) = \Re_k^-(\mathbf{D})$.

Not all domain refinements are invertible in a satisfactory way. An example is provided by the negative completion refinement $\Re_\neg$ of Sect. 3. In fact, as observed in [9], the optimal basis of the domain $Sign$ (in Sect. 3) for $\Re_\neg$ does not exist.

Since *Sign* enjoys all most important lattice-theoretic properties, this means that $\Re_\neg$ is not invertible on any really significant class of abstract domains.

As the following result says, compressors relative to idempotent refinements are extensive and idempotent.

Proposition 4.2 *If $\Re : \mathbf{U} \to uco(C)$ is idempotent and k-invertible in V, then the compressor $\Re_k^- : \mathbf{U}[V/k] \to \pi_k(\mathbf{U})$ is extensive (i.e. $\pi_k(\mathbf{D}) \sqsubseteq \Re_k^-(\mathbf{D})$) and idempotent (i.e. $\Re_k^-(\mathbf{D}) \in V \Rightarrow \Re_k^-(\mathbf{D}[\Re_k^-(\mathbf{D})/k]) = \Re_k^-(\mathbf{D})$).*

In general, compressors are neither monotone nor antimonotone: [14] proves that the least disjunctive basis operator is neither monotone nor antimonotone, and later we will show that the least disjunctive basis is the compressor relative to the disjunctive completion refinement. On the other hand, as expected, a compressor applied to a refined domain performs no further simplification.

Proposition 4.3 *If $\Re : \mathbf{U} \to uco(C)$ is idempotent and k-invertible in V then for any $\mathbf{D} \in \mathbf{U}[V/k]$ such that $\Re(\mathbf{D}) \in V$, $\Re_k^-(\mathbf{D}[\Re(\mathbf{D})/k]) = \Re_k^-(\mathbf{D})$.*

An n-ary refinement $\Re : \mathbf{U} \to uco(C)$ is *commutative* if for any permutation τ of $\{1,\ldots,n\}$, $\Re(\pi_{\tau(1)}(\mathbf{D}),\ldots,\pi_{\tau(n)}(\mathbf{D})) = \Re(\mathbf{D})$ holds. For instance, the reduced product refinement $\Re_\sqcap$ is obviously commutative as well as any attribute independent refinement. For commutative refinements, the following result holds (this result admits a straightforward, although notationally tedious, generalization for generic K-commutativity and invertibility).

Proposition 4.4 *If $\Re : \mathbf{U} \to uco(C)$ is a (possibly nonidempotent) commutative refinement and $k \in [1,n]$ then, $\Re$ is fully invertible on $\mathbf{V} \subseteq \mathbf{U}$ iff $\Re$ is k-invertible on $\pi_k(\mathbf{V})$ iff for all $\mathbf{D} \in \mathbf{V}$, $\lambda X.\Re(\mathbf{D}[X/k]) : \pi_k(\mathbf{U}) \to uco(C)$ is (1-)invertible on $\pi_k(\mathbf{V})$.*

Not all refinements are commutative. Examples of noncommutative refinements are reduced power [6], dependencies [13], and tensor product [18]. Due to lack of space, we do not formalize these operators as refinements. In the following, we show how the results in [3, 12, 14] on complementation and least disjunctive basis of abstract domains, actually permit to define the compressors relative to reduced product and disjunctive completion respectively. These results also suggest a generalization towards a general pattern of invertible refinements.

The Inverse of Reduced Product. Since $\Re_\sqcap$ is commutative, by Proposition 4.4, $\Re_\sqcap$ is fully invertible on some $V \times V \subseteq uco(C)^2$ iff for any $D \in V$, $\lambda X.(D \sqcap X)$ is invertible on V. As recalled in Sect. 2, for any meet-continuous complete lattice C and $D \in uco(C)$, one can define the complement abstract domain $C \sim D$. Moreover, it is immediate to note that, for any complete lattice C, if $D_1, D_2 \in uco(C)$ satisfy the ascending chain condition (to be ACC, for short; DCC is dual) then $D_1 \sqcap D_2$ is ACC as well, and hence meet-continuous. These observations directly imply that we can invert the reduced product on the ACC abstractions of any concrete domain C (i.e. a plain complete lattice). Let us define $ACC(C) = \{D \in uco(C) \mid D \text{ is ACC}\}$, for any complete lattice C.

Theorem 4.5 *If $D \in ACC(C)$ then $\lambda X.D \sqcap X$ is invertible on $ACC(C)$, and the corresponding compressor $(\lambda X.D \sqcap X)^- : ACC(C) \to uco(C)$ is defined as $(\lambda X.D \sqcap X)^-(E) = (D \sqcap E) \sim D$.*

Thus, $\Re_\sqcap$ is fully invertible on $ACC(C) \times ACC(C)$. For instance, if $D_1, D_2 \in ACC(C)$, we have that the first compressor is $(\Re_\sqcap)_1^-(D_1, D_2) = (D_1 \sqcap D_2) \sim D_2$.

The Inverse of Disjunctive Completion. Giacobazzi and Ranzato defined and studied in [14] the operator of least disjunctive basis on abstract domains, that corresponds exactly to the compressor for the disjunctive completion refinement. Hence, the results in [14] can be reformulated as follows.

Theorem 4.6
(i) If C is co-algebraic completely distributive then $\Re_\vee$ is invertible on all $uco(C)$.
(ii) If C is distributive then $\Re_\vee$ is invertible on $\{A \in uco(C) \mid A \text{ is finite}\}$.

Compressing Lower and Upper Refinements. Define an *upper (lower) improvement* on C as any map $\mathcal{I} : \wp(C) \to \wp(C)$ such that $\forall S \in \wp(C).\forall s \in S.\forall s' \in \mathcal{I}(S)$. $s \leq s'$ ($s' \leq s$). Glb and lub are obvious examples of lower and upper improvements. We prove that upper and lower improvements induce invertible refinements in a natural way. This provides a general pattern for defining new invertible refinements. For an upper (lower) improvement $\mathcal{I}$ on C, define the corresponding *upper (lower) set-refinement* $\Re^\mathcal{I} : \wp(C) \to \wp(C)$ as: $\Re^\mathcal{I}(X) = X \cup (\cup_{S \subseteq X} \mathcal{I}(S))$. It turns out that $\Re^\mathcal{I}$ is a lower closure on $\langle \wp(C), \supseteq \rangle$. However, in general, for a closure $\rho \in uco(C)$, $\Re^\mathcal{I}(\rho)$ may not be in $uco(C)$. But, when a unary full idempotent refinement $\Re \in lco(uco(C))$ is the restriction on $uco(C)$ of an upper (lower) set-refinement, i.e. there exists an upper (lower) improvement $\mathcal{I}$ on C such that $\Re = \Re^\mathcal{I}_{\lceil uco(C)}$ (in this case, we call $\Re$ an *upper (lower) refinement*), the following general theorem of inversion for $\Re$ holds.

Theorem 4.7 *If C is a complete lattice satisfying the DCC (ACC), then any upper (lower) refinement $\Re \in lco(uco(C))$ is invertible on all $uco(C)$.*

For instance, if C is distributive and $\mathcal{I} = \vee$, we get for free the inversion of disjunctive completion of Theorem 4.6 (ii). By Proposition 4.4, the attribute independent refinement induced by a family of upper or lower refinements is invertible under suitable hypotheses derived by Theorem 4.7. By this last observation, it would be possible (but we omit the details) to derive as a consequence of Theorem 4.7 the result of inversion for the reduced product of Theorem 4.5.

Minimal $\Re$-decompositions. For a given refinement $\Re : \mathbf{U} \to uco(C)$ of arity $n > 1$, we say that $\mathbf{D} \in \mathbf{U}$ is a $\Re$-*decomposition* of $D \in uco(C)$, if $D = \Re(\mathbf{D})$. If $\mathbf{D}, \mathbf{E} \in \mathbf{U}$ are two $\Re$-decompositions of D then $\mathbf{D}$ is *better* than $\mathbf{E}$ if $\mathbf{E} \sqsubseteq \mathbf{D}$ componentwise.[1] The intended meaning is that $\mathbf{D}$ is better than $\mathbf{E}$ because it is a less costly decomposition (in particular, $\sum_{i=1}^n |\pi_i(\mathbf{D})| \leq \sum_{i=1}^n |\pi_i(\mathbf{E})|$). Obviously, this relation induces a partial ordering between $\Re$-decompositions of D, but, in general, optimal (i.e. least) $\Re$-decompositions for this order do not exist. For instance, $\langle D, \{\top\} \rangle$ and $\langle \{\top\}, D \rangle$ are uncomparable minimal $\Re_\sqcap$-decompositions of D. It is easy to see that, if $\Re$ is idempotent and fully invertible on $\mathbf{V} \subseteq \mathbf{U}$ and $\mathbf{D}$ is a $\Re$-decomposition of D, then for any $k \in [1, n]$ the tuple $\mathbf{D}\left[\Re_k^-(\mathbf{D})/k\right]$ ($*$) is still a $\Re$-decomposition of D which is better than $\mathbf{D}$, and

---

[1] For commutative refinements, both this definition and the successive development would identify decompositions up to permutation – however, we omit the details.

that $\mathbf{D}$ is a minimal $\Re$-decomposition of D iff $\forall k \in [1,n]$. $\pi_k(\mathbf{D}) = \Re_k^-(\mathbf{D})$. Thus, each $\Re$-decomposition can be improved by iterating the above step $(*)$ as shown in the following nondeterministic function $\Re$-min, where *choose* selects an arbitrary element from its input set.

```
fun ℜ-min (D:array[1,n] of domains)
J := {1,...,n};
repeat
    k := choose(J);
    J := J \ {k};
    D := D[ℜ_k^-(D)/k]
until J = ∅
output D
```

Theorem 4.8 *Let $\Re$ be an idempotent and fully invertible refinement on $\mathbf{V}$. If, for any $k \in [1,n]$, $\Re_k^-$ is anti-monotone then for any $\mathbf{D} \in \mathbf{V}$, $\Re$-min$(\mathbf{D})$ is a minimal $\Re$-decomposition of $\Re(\mathbf{D})$.*

Note that, for a $\Re$-decomposition $\mathbf{D}$ of D, we can get at most $n!$ different minimal $\Re$-decompositions of D. For instance, if $\Re(D_1, D_2) = D$ then $\langle \Re_1^-(D_1, \Re_2^-(D_1, D_2)), \Re_2^-(D_1, D_2) \rangle$ is a minimal $\Re$-decomposition of D. Theorem 4.8 generalizes the results of [3, Sect. 4], since the compressor relative to reduced product is anti-monotone (cf. [3]).

5 A Relation of Adjunction between Refinements and Compressors

Assume that $\Re$ is an idempotent n-ary refinement $\Re : \mathbf{U} \to uco(C)$ that is k-invertible on $V \subseteq \pi_k(\mathbf{U}) \subseteq uco(C)$, for some $k \in [1,n]$. We saw in Sect. 4 that any (k-th) unary refinement $\lambda X. \Re(\mathbf{D}[X/k]) : \pi_k(\mathbf{U}) \to uco(C)$ induced by $\Re$ is invertible in V, and the corresponding compressor (of type $V \to \pi_k(\mathbf{U})$) is defined as $(\lambda X. \Re(\mathbf{D}[X/k]))^- = \lambda X \in V. \Re_k^-(\mathbf{D}[X/k])$. In general, the refinement $\lambda X. \Re(\mathbf{D}[X/k])$ and the relative compressor $\lambda X. \Re_k^-(\mathbf{D}[X/k])$ do not constitute an adjunction on the poset of domains $\langle V, \sqsubseteq \rangle$ of invertibility, i.e. for all $A \in \pi_k(\mathbf{U})$ and $B \in V$, $\Re(\mathbf{D}[A/k]) \sqsubseteq B \Leftrightarrow A \sqsubseteq \Re_k^-(\mathbf{D}[B/k]))$ may not hold. This is due to the fact that compressors, in general, are not monotone, as observed after Proposition 4.2. Since, by Proposition 4.2, compressors are idempotent and extensive, this also implies that compressors $\lambda X. \Re_k^-(\mathbf{D}[X/k])$, well-defined on V, are not upper closures on $\langle V, \sqsubseteq \rangle$, as instead we would expect by viewing compressors as inverses of refinements.

We solve this asymmetry between abstract domain refinements and compressors by modifying the standard ordering $\sqsubseteq$ of precision between domains, so as to keep into account the rôle of $\Re$. We maintain the above scenario and also suppose that the refinement $\lambda X. \Re(\mathbf{D}[X/k])$ is well-defined in $\pi_k(\mathbf{U})$, namely for any $D \in \pi_k(\mathbf{U})$, $\Re(\mathbf{D}[D/k]) \in \pi_k(\mathbf{U})$, and that $\langle \pi_k(\mathbf{U}), \sqsubseteq \rangle$ is a complete sublattice of $\langle uco(C), \sqsubseteq \rangle$. These hypotheses imply that $\lambda X. \Re(\mathbf{D}[X/k])$ is a lower closure on the complete lattice $\langle \pi_k(\mathbf{U}), \sqsubseteq \rangle$. Then, we define the following relation $\sqsubseteq^\Re$ (that actually depends also on the fixed arguments $\pi_i(\mathbf{D})$, $i \neq k$) on $\pi_k(\mathbf{U})$:

$A \sqsubseteq^\Re B$ iff $\Re(\mathbf{D}[A/k]) \sqsubseteq \Re(\mathbf{D}[B/k])$ & $(\Re(\mathbf{D}[B/k]) \sqsubseteq \Re(\mathbf{D}[A/k]) \Rightarrow A \sqsubseteq B)$.

Theorem 5.1 $\langle \pi_k(\mathbf{U}), \sqsubseteq^\Re \rangle$ *is a complete lattice.*

Note that $A \sqsubseteq B \Rightarrow A \sqsubseteq^\Re B$. Thus, we call $\sqsubseteq^\Re$ the *lifting of* $\sqsubseteq$ *via* $\Re$. This lifted complete partial order reflects precisely the relative precision of domains with respect to the refinement $\lambda X. \Re(\mathbf{D}[X/k])$: A is more precise than B in the lifted order if the refinement of A is more precise than the refinement of B in the standard sense and, when they are the same (i.e. $(\Re(\mathbf{D}[B/k]) = \Re(\mathbf{D}[A/k]))$, then A contains more information. For this ordering $\sqsubseteq^{\Re}$, we get back a relation of adjunction between the invertible refinement and its compressor.

Theorem 5.2 $\forall A \in \pi_k(\mathbf{U}), B \in V.\ \Re(\mathbf{D}[A/k]) \sqsubseteq^{\Re} B \Leftrightarrow A \sqsubseteq^{\Re} \Re_k^-(\mathbf{D}[B/k]))$.
As a consequence, $\lambda X.\ \Re_k^-(\mathbf{D}[X/k])$ is an upper closure operator on $\langle V, \sqsubseteq^{\Re} \rangle$ (provided it is well-defined on V). For example, we get an adjunction between reduced product and complementation w.r.t. the lifted order. For any complete lattice C and $D \in uco(C)$, the lifted order on $uco(C)$ is defined as follows: For all $A, B \in uco(C)$, $A \sqsubseteq^{\sqcap} B$ iff $D \sqcap A \sqsubseteq D \sqcap B\ \&\ (D \sqcap B \sqsubseteq D \sqcap A \Rightarrow A \sqsubseteq B)$. Hence, the adjunction between refinement (reduced product) and compressor (complementation) is the following: For any $A \in uco(C)$ and $B \in ACC(C)$, $D \sqcap A \sqsubseteq^{\sqcap} B \Leftrightarrow A \sqsubseteq^{\sqcap} (D \sqcap B) \sim D$.

Acknowledgments. We are grateful to Francesca Scozzari for her contribution to Theorem 3.7 and to one anonymous referee for many helpful suggestions.

References

1. M. Codish, A. Mulkers, M. Bruynooghe, M. García de la Banda, and M. Hermenegildo. Improving abstract interpretations by combining domains. *ACM TOPLAS*, 17(1):28–44, 1995.
2. M. Comini and G. Levi. An algebraic theory of observables. In *Proc. ILPS'94*, pp. 172–186, 1994.
3. A. Cortesi, G. Filé, R. Giacobazzi, C. Palamidessi, and F. Ranzato. Complementation in abstract interpretation. *ACM TOPLAS*, 19(1):7–47, 1997.
4. A. Cortesi, B. Le Charlier, and P. Van Hentenryck. Combinations of abstract domains for logic programming. In *Proc. POPL'94*, pp. 227–239, 1994.
5. P. Cousot and R. Cousot. Abstract interpretation: a unified lattice model for static analysis of programs by construction or approximation of fixpoints. In *Proc. POPL'77*, pp. 238-252, 1977.
6. P. Cousot and R. Cousot. Systematic design of program analysis frameworks. In *Proc. POPL'79*, pp. 269–282, 1979.
7. P. Cousot and R. Cousot. Inductive definitions, semantics and abstract interpretation. In *Proc. POPL'92*, pp. 83–94, 1992.
8. P. Cousot and R. Cousot. Higher-order abstract interpretation (and application to comportment analysis generalizing strictness, termination, projection and PER analysis of functional languages). In *Proc. IEEE ICCL'94*, pp. 95–112, 1994.
9. G. Filé, R. Giacobazzi, and F. Ranzato. A unifying view of abstract domain design. *ACM Comput. Surv.*, 28(2):333–336, 1996.
10. G. Filé and F. Ranzato. Improving abstract interpretations by systematic lifting to the powerset. In *Proc. ILPS'94*, pp. 655–669, 1994.
11. R. Giacobazzi. "Optimal" collecting semantics for analysis in a hierarchy of logic program semantics. In *Proc. STACS'96*, LNCS 1046, pp. 503-514, 1996.
12. R. Giacobazzi, C. Palamidessi, and F. Ranzato. Weak relative pseudo-complements of closure operators. *Algebra Universalis*, 36(3):405-412, 1996.
13. R. Giacobazzi and F. Ranzato. Functional dependencies and Moore-set completions of abstract interpretations and semantics. In *Proc. ILPS'95*, pp. 321–335, 1995.
14. R. Giacobazzi and F. Ranzato. Optimal domains for disjunctive abstract interpretation. To appear in *Sci. Comput. Program.* Preliminary version in LNCS 1058, pp. 141–155, 1996.
15. T.P. Jensen. Disjunctive strictness analysis. In *Proc. LICS'92*, pp. 174–185. 1992.
16. K. Muthukumar and M. Hermenegildo. Combined determination of sharing and freeness of program variables through abstract interpretation. In *Proc. ICLP'91*, pp. 49–63, 1991.
17. A. Mycroft. Completeness and predicate-based abstract interpretation. In *Proc. PEPM'93*.
18. F. Nielson. Tensor products generalize the relational data flow analysis method. In *Proc. 4th Hungarian Comput. Sci. Conf.*, pp. 211–225, 1985.
19. U. Reddy and S. Kamin. On the power of abstract interpretation. In *Proc. IEEE ICCL'92*, 1992.
20. R.C. Sekar, P. Mishra, and I.V. Ramakrishnan. On the power and limitation of strictness analysis. To appear in *J. ACM*. Preliminary version in *Proc. POPL'91*, pp. 37–48, 1991.
21. R. Sundararajan and J. Conery. An abstract interpretation scheme for groundness, freeness, and sharing analysis of logic programs. In *Proc. FST&TCS'92*, LNCS 652, pp. 203–216, 1992.
22. K. Yi and W.L. Harrison. Automatic generation and management of interprocedural program analyses. In *Proc. POPL'93*, pp. 246–259, 1993.
23. M. Ward. The closure operators of a lattice. *Ann. Math.*, 43(2):191–196, 1942.

Labelled Reductions, Runtime Errors, and Operational Subsumption

Laurent Dami

Centre Universitaire d'Informatique, Université de Genève
24, rue Général-Dufour, CH-1211 Genève 4, Switzerland
http://cuiwww.unige.ch/~dami

1 Introduction

Consider the "name-switching" function $F \stackrel{\text{def}}{=} \lambda x.\{l_1 = x.l_2, l_2 = x.l_1\}$ in a λ-calculus with records. Most type systems would reject program $(F\{l_1 = 3\}).l_2$ because the type of F is $\{l_1 : X, l_2 : Y\} \to \{l_2 : Y, l_1 : X\}$ and $\{l_1 : X, l_2 : Y\}$ cannot be unified with $\{l_1 : \mathbf{Int}\}$, the type of the record argument. However this program reduces to 3 without error. This shows that the common notion of "erroneous" terms, as implemented in most typed languages, is sometimes overrestrictive. Here we propose a general framework for studying the semantics of programs containing "uncatchable" errors, and a language-independent classification of error propagation properties; this is then applied to a comparison of various λ-calculi. In this approach, errors (written ε) can be passed around as any other value, sometimes in a lazy way, and therefore an error occurring inside a term is not necessarily propagated to the top level; a term is considered "erroneous" if and only if it *always* generates ε. We define an operational ordering of terms, called "subsumption", which gives a formal foundation for the notion of "substitutability" or "safe replacement" often used informally in the object-oriented literature: a term subsumes another iff it generates fewer errors in all program contexts. Subsumption often implies and sometimes equals the usual approximation ordering (Theorems 21, 26); its main interest is to directly interpret subtyping in a term model, which is simpler than the partial equivalence relations (PERs) of [6] or the coercion functions of [5]. Since we require that errors are "absorbing" (any attempt to interact with an error yields an error again), ε is the *top* element. Therefore the semantic structure is a lattice, like in the original work of Scott [19].

For the technical development below we make heavy use *labelled reductions*, an old idea used in the λ-calculus to restrict the interaction behaviour of a term to a finite number of steps. Here this is generalised in an abstract way to other reduction systems. Labelled reductions allow us to classify both terms and contexts according to the number of interaction steps they can perform, and therefore introduce an operational notion of finite approximation. This in turn can be used as an alternative to the contractive maps of [15] or the embedding-projection pairs of [7] for solving recursive type equations.

2 Basic definitions: error generation and preservation

This section defines a number of abstract notions, independent of any particular language. However, since some concepts need illustrations, informal examples will be drawn from the standard λ-calculus extended with constants and records. Precise definitions for this calculus and other calculi will be given later in Section 5. Prior knowledge of the λ-calculus and the notions of call-by-name (CBN), call-by-value (CBV) and lazy evaluation is assumed; standard references are [3,17,1]. As a reminder, common abbreviations for λ-terms are $\mathbf{I} \stackrel{\text{def}}{=} \lambda x.x, \mathbf{K} \stackrel{\text{def}}{=} \lambda xy.x, \Delta \stackrel{\text{def}}{=} \lambda x.xx, \Omega \stackrel{\text{def}}{=} \Delta\Delta, \mathbf{Y} \stackrel{\text{def}}{=} \lambda f.(\lambda x.f(xx))(\lambda x.f(xx))$; furthermore $\mu x.a$ abbreviates $\mathbf{Y}(\lambda x.a)$.

Notation. We consider languages of the form $(\mathcal{T}, \mathcal{V}, \to)$ where $\mathcal{T}$ is a set of *terms*, $\mathcal{V} \subset \mathcal{T}$ is the set of *values*, and $\to$ is a binary relation on terms (*one-step reduction*) satisfying $\forall v \in \mathcal{V}, v \to v' \implies v' \in \mathcal{V}$. The letters a, b, c range over arbitrary terms, v, u range over values. We assume a set $\mathcal{X} \subset \mathcal{T}$ of *variables* and standard notions of bound and free variables; the function $FV : \mathcal{T} \to 2^{\mathcal{X}}$ gives the free variables of a term; letters x, y, z range over $\mathcal{X}$. $\mathcal{T}^C$ and $\mathcal{V}^C$ denote the sets of *closed* terms and values, i.e. those for which FV returns the empty set. The substitution of b for free occurrences of x in a is written $a[x := b]$. *Contexts* are terms possibly containing occurrences of a "hole" $[-]$; if $C[-]$ is a context, then $C[a]$ is the term obtained by filling the hole in $C[-]$ with a, possibly capturing variables. The set of contexts is written $\mathcal{T}[-]$; since there is no restriction on the number of holes, we have $\mathcal{T} \subset \mathcal{T}[-]$. A *subterm* of a is a term a' such that $a \equiv C[a']$ for some $C[-]$. The reflexive, transitive closure of $\to$ is written $\stackrel{*}{\to}$ and $\stackrel{*}{=}$ is its symmetric closure; $(a \to)$ is an abbreviation for $\exists b, a \to b$. Finally, if $\sqsubseteq_\theta$ is one of the operational ordering relations defined below, with θ representing any collection of subscripts/superscripts, then $\cong_\theta$ is its symmetric closure and $\sqsubset_\theta$ is its strict restriction, i.e. the relation $\sqsubseteq_\theta \setminus \cong_\theta$.

Definition 1 (Reduction properties). For a language $\mathcal{L} \stackrel{\text{def}}{=} (\mathcal{T}, \mathcal{V}, \to)$ we say that

- a is *stuck* iff $a \notin \mathcal{V}$ and $\neg(a \to)$
- a *diverges* (written $a \Uparrow$) iff, for each b such that $a \stackrel{*}{\to} b$, we have $((b \notin \mathcal{V}) \land (b \to))$. Conversely, a *converges* $(a \Downarrow)$ iff $\exists v \in \mathcal{V}, a \stackrel{*}{\to} v$.
- $\to$ is *Church-Rosser* (CR) iff $((a \stackrel{*}{\to} b) \land (a \stackrel{*}{\to} c)) \implies \exists d.((b \stackrel{*}{\to} d) \land (c \stackrel{*}{\to} d))$
- $\to$ is *compatible* iff $a \to b \implies C[a] \stackrel{*}{\to} C[b]$ for any context $C[-]$

Definition 2 (Relevant contexts). A context $C[-]$ is *relevant* iff $a \Uparrow \implies C[a] \Uparrow$ and there is a term b such that $C[b] \Downarrow$.

Example 3. Contexts $[-], ([-]ab), ((\lambda x.[-])ab), [-].l$ are relevant. The context $(\mathbf{K}[-]a)$ is relevant with CBV evaluation, but not with CBN. The context $\lambda x.[-]$ is relevant with both CBV and CBN, but not with lazy evaluation.

Definition 4 (Solvable terms). A term a is *solvable* iff, for every term b, there is a relevant context $C[-]$ such that $C[a] \xrightarrow{*} b$.

Definition 5 (Language properties). A language $(\mathcal{T}, \mathcal{V}, \rightarrow)$

- *has divergence* iff there is at least a term $\Omega \in \mathcal{T} \setminus \mathcal{V}$ such that $\Omega \Uparrow$.
- is *stuck-free* iff $\mathcal{T}$ contains no stuck terms.
- *has errors* iff there is a nonempty subset $\mathcal{E} \subset \mathcal{V}$ of *error values* satisfying $v \in \mathcal{E} \implies \neg(v \rightarrow)$. Most often we will consider a singleton set and write ε to denote the single error value. We write $a\dagger^0$ if $a \xrightarrow{*} v \in \mathcal{E}$.
- is *error-generating* iff there is an $a \in \mathcal{T}$ such that $a\dagger^0$ and for every subterm a' of a, $a' \notin \mathcal{E}$.
- is *error-complete* iff, for every value $v \in \mathcal{V}^C$, there is a relevant context $C[-]$ such that $C[v]\dagger^0$.
- is *error-preserving* iff there are no relevant context $C[-]$ and error value $v \in \mathcal{E}$ such that $C[v] \Uparrow$.

Some comments are of order. Absence of stuck terms is easily obtained by adding an error term ε and completing the reduction relation so that stuck terms explicitly reduce to ε. In that case the language is also error-generating. Error-completeness is a closely related, but different property: we will show examples of languages which are error-generating but not error-complete, or vice-versa. Finally, error-preservation ensures that errors are not observable internally; in other words, there is no "catch" construct to recover from errors.

Example 6. The pure λ-calculus with an added error constant ε has stuck terms: (εa) does not reduce and is not a value. With an added reduction rule $\forall a, \varepsilon a \rightarrow \varepsilon$ the language becomes stuck-free; however it is not error-generating. Error-completeness varies with the evaluation strategy: with CBN evaluation, all values are solvable, and therefore can become errors in some context. By contrast, lazy evaluation admits values which are unsolvable, so then the language is not error-complete: there is no relevant context which can turn $\lambda x.\Omega$ into an error.

Example 7. The λ-calculus with integers and integer operators is error-complete, independently of the evaluation strategy: this is because there are contexts such as $([-][-])$ and $([-] + [-])$ which discriminate between functional values and integer values, even if they are unsolvable.

Example 8. A language like the one in [16], containing constructs **isnat**, **islam**, **ispr**, ... for identifying various syntactic classes of values such as numbers, λ-abstractions or pairs, is not error-preserving: for example the context

$$\textbf{if } (\textbf{islam}([-])) \textbf{ then } + 1 \textbf{ else } - 1$$

returns -1 for all terms which are not λ-abstractions, including ε. By contrast, the approach of [2], who discriminate between syntactic classes through a single construct

$$\textbf{cases } a \textbf{ nat} : a_1 \textbf{ fun} : a_2 \textbf{ pair} : a_3 \ldots \textbf{end}$$

is error-preserving, provided of course that the **cases** construct has no "default" clause and no clause to recognize errors.

Example 9. The λ-calculus extended with ε, with records $\{l_1 = a_1 \ldots l_n = a_n\}$ and with a field selection construct $a.l$, together with the obvious reduction and error generation rules, is stuck-free, error-generating, error-complete and error-preserving.

Following [1,13,16], we can define approximation in an operational way:

Definition 10 (Contextual approximation). *Contextual approximation* $\sqsubseteq_\Downarrow$ is defined as:
$$(a \sqsubseteq_\Downarrow b) \iff (\forall C[-], C[a] \Downarrow \implies C[b] \Downarrow)$$

In error-preserving languages, since ε always converges, then $\Omega \sqsubseteq_\Downarrow a \sqsubseteq_\Downarrow \varepsilon$ for any a.

3 Labelled reduction

This section borrows from Chapter 14 of [3] the idea of *labelled reductions*. Labelled terms are obtained from usual terms by decorating subterms with natural numbers which limit the number of reduction steps they can perform. For example
$$(((\lambda x.x^4)^{3^1})(\lambda yz.y^3))^{2^0}$$
is a labelled λ-term. Subterms without any label are implicitly labelled with ∞. We write $a_\ell, b_\ell, \ldots, C_\ell[-], D_\ell[-], \ldots$ for labelled terms and contexts, and $\mathcal{T}_\ell$ for the set of labelled terms. Given a set $\mathcal{V}$ of values, we define $\mathcal{V}_\ell$ as the set of labelled values satisfying
$$v_\ell \equiv C_\ell[(a_\ell)^0] \iff C[\Omega] \in \mathcal{V}$$

In other words, labelled values can contain 0 labels only in places where the corresponding subterm, replaced by a divergent term, still yields a value in the original language: this is typically the case in lazy computation systems [12], in which the outermost term constructor is enough to determine whether a term is a value or not.

For defining labelled reduction we assume that the original reduction relation $\rightarrow$ is given by a set of rules ($lhs \rightarrow rhs$) in some form of rewrite system (possibly dealing with bound variables, as in [14,12,20]). Operators (function symbols) in the left-hand side of a rule which are not at the outermost level are called *internal*. Given a left-hand side lhs of a rule, a labelling $\ell_N(lhs)$ is obtained by decorating internal operators in lhs with labels in N. Each original rule ($lhs \rightarrow rhs$) generates labelled rules of shape
$$\ell_{\{n+1 \mid n \in N\}}(lhs) \rightarrow_\ell rhs^{\min(N)}$$

Labelled reduction is the relation on $\mathcal{T}_\ell$ given by all such labelled rules, together with the label elimination rules

$$(lab1)\ (a^m)^n \to_\ell a^{min(m,n)}$$
$$(lab2)\quad a^0 \to_\ell \Omega$$

Example 11. β-reduction on λ-terms is expressed in [14] as $@(\lambda([x]Z(x)), Z') \to Z(Z')$. The only internal operator is λ, so the corresponding rule for labelled β-reduction is $@(\lambda^{n+1}([x]Z(x)), Z') \to (Z(Z'))^n$, which in more familiar notation is written

$$(\lambda x.a)^{n+1} b \to (a[x := b])^n$$

This is not exactly like the definition of [3], which reads:

$$(\lambda x.a)^{n+1} b \to (a[x := b^n])^n$$

so our labelled reductions are not strongly normalizing, because b could be a divergent term. Nevertheless for the current purpose this is not a problem: labelled reductions still introduce an appropriate notion of finite approximation, as will be shown below. Hence these are intended as a general, abstract mechanism to replace the language-dependent finite projection functions of [2,16,1].

Example 12. In a record calculus, the field extraction rule $\overline{\{l_i = a_i\}}.l_k \to a_k$ has corresponding labelled rule $\overline{\{l_i = a_i\}}^{n+1}.l_k \to a_k^n$

Proposition 13. *If* $(\mathcal{T}, \mathcal{V}, \to)$ *is stuck-free, with compatible and Church-Rosser reduction, then so is its labelled extension* $(\mathcal{T}_\ell, \mathcal{V}_\ell, \to_\ell)$.

Note that $\mathcal{L}_\ell$ is never error-preserving, as can be seen easily by a context like $([-]^1 \mathbf{I})$ which diverges when filled with ε.

Definition 14 (k-relevant contexts). 1. A context $C[-]$ is *k-relevant* iff ($a \Uparrow \implies C[a] \Uparrow$) and there is a term b such that $C[b^{k+1}] \Downarrow$.
2. The *relevance index* for $C[-]$, written $RI(C[-])$, is the smallest k such that $C[-]$ is k-relevant, or undefined if there is no such k.
3. $\mathcal{C}^k$ denotes the set $\{C[-] \in \mathcal{T}[-] | RI(C[-]) = k\}$.

The notion of k-relevance captures the number of interaction steps between a context and the term filling it. 0-relevant contexts are contexts which only carry the hole around without interacting with it, like $[-], (\mathbf{I}[-])$ or $(\{l = [-]\}.l)$; 1-relevant contexts include the 0-relevant ones, but in addition also include contexts like $([-]\mathbf{I})$ or $([-].l)$ which perform one single interaction step with the hole. More generally, we have:

Lemma 15. *1. Any k-relevant context is also $(k+1)$-relevant.*
2. A context is relevant iff it is k-relevant for some $k > 0$.

Lemma 16 (context decomposition).

$$C[-] \in \mathcal{C}^{k+1} \implies \exists C_1[C_2[-]] \stackrel{*}{=} C[-], C_1[-] \in \mathcal{C}^1 \wedge C_2[-] \in \mathcal{C}^k$$

Proof. If $k = 0$, there is an easy solution $C_1[-] \equiv C[-], C_2[-] \equiv [-]$. If $k > 0$, we know i) $\exists a, v, C[a^{k+2}] \stackrel{*}{\to} v$ and ii) $\forall b, C[b^{k+1}] \Uparrow$. Suppose $v \equiv C'[a'^{k+2}]$, with $C[-] \stackrel{*}{\to} C'[-], a \stackrel{*}{\to} a'$. Then by definition $C'[a'^{k+1}]$ must be a value, contradicting ii). So necessarily

$$C[a^{k+2}] \stackrel{*}{\to} D_1[D_2[a'^{k+2}]] \to D_1[b^{k+1}] \stackrel{*}{\to} v$$

where $D_2[a'^{k+2}] \to b^{k+1}$ is an instance of a labelled reduction rule. Now by rule (lab1), $D_1[(D_2[a'^{k+2}])^{k+1}] \stackrel{*}{\to} v$, so $D_1[-] \in \mathcal{C}^k$; moreover $D_2[a'^2] \to b^1 \Downarrow$, which implies $D_2 \in \mathcal{C}^1$. □

Now we can use relevance indices of contexts to measure the interactivity of terms; intuitively, a term is k-interactive if it can performe k interaction steps.

Definition 17 (k-interactivity). 1. every term is 0-*interactive*
2. a is $(k+1)$-*interactive* iff $\exists C[-] \in \mathcal{C}^k, C[a] \Downarrow$.
3. the *interactivity index* of a term a, written $II(a)$, is the biggest k such that a is k-interactive, or ∞ if a is k-interactive for every k.
4. $\mathcal{T}^k$ denotes the set $\{a \in \mathcal{T} | II(a) \leq k\}$.

Example 18. – In the lazy λ-calculus [1] all λ-abstractions are values, so the term $\lambda x.\Omega$ is 1-interactive, as well as $(\lambda x.a)^1$ for any function $\lambda x.a$.
– In the standard call-by-name λ-calculus, the term $\lambda x.x\Omega$ is 1-interactive.

As demonstrated by these examples, the notion of k-interactivity not only applies to labelled terms, but also to unlabelled ones. Labels are used as an auxiliary study tool, but then the results can be extracted and give information about the unlabelled language.

4 Erroneous Terms and Subsumption

We want to allow some errors to occur inside terms, because of the assumption that these will not necessarily be propagated to the top level. However, if a term contains *only* errors, then it is observationally not different from an error itself. For example, the term $\lambda x.\varepsilon$ is not β-equal to ε, but only yields errors in any context . By contrast, lazy systems admit unsolvable values like $\mu x.\lambda y.x, \mu x.\{l = x\}$ which can interact without ever generating errors. Hence we come to define the erroneous terms are those which always yield errors after a finite number of interaction steps:

Definition 19 (Erroneous terms). A term a is k-*erroneous*, written $a\dagger^k$, iff $C[a] \stackrel{*}{\to} \varepsilon$ for every context $C[-] \in \mathcal{C}^k$. A term a is *erroneous*, written $a\dagger$, iff it is k-erroneous for some k.

Clearly 0-erroneous terms must belong to the class $\{a | a \stackrel{*}{\to} \varepsilon\}$. Examples of 1-erroneous terms are $\lambda x.\varepsilon$ or $\{l = \varepsilon\}$. .

Definition 20 (Subsumption). A term a *subsumes* another term b, written $a \sqsubseteq_\varepsilon b$, iff it generates fewer errors in all program contexts:

$$a \sqsubseteq_\varepsilon b \iff \forall C[-], C[a]\dagger \implies C[b]\dagger$$

As for $\sqsubseteq_\Downarrow$, we have $\Omega \sqsubseteq_\varepsilon a \sqsubseteq_\varepsilon \varepsilon$ for any a in error-preserving languages. The obvious question then is how the two orderings relate. This in general depends on the language properties, as shown through several examples in the next section. Nevertheless, a general result can be stated already:

Theorem 21. *In an error-complete language,* $a \sqsubseteq_\varepsilon b \implies a \sqsubseteq_\Downarrow b$.

Proof. We will show $(a \sqsubseteq_\varepsilon b) \implies (\forall C[-], C[b] \Uparrow \implies C[a] \Uparrow)$, from which $(a \sqsubseteq_\Downarrow b)$ directly follows by definition. Suppose $a \sqsubseteq_\varepsilon b$. For any context $C[-]$, furthermore suppose $C[b] \Uparrow$ and $C[a] \Downarrow$. If the language is error-complete, then there exists a relevant context $D[-]$ with $D[C[a]]\dagger^0$; but since $D[-]$ is relevant, $D[C[b]] \Uparrow$, contradicting $a \sqsubseteq_\varepsilon b$. Hence $C[a]$ must diverge. □

5 Comparing various lambda calculi

We will now apply our abstract framework to several languages, all related to the λ-calculus, but with various kinds of extensions, and with two different notions of values: *head normal forms* (terms withouth a head redex) or *lazy values* (terms with an outermost abstraction construct). These are described by fairly standard rules, given in the appendix. Head and lazy versions are distinguished by the superscripts ^{H} and ^{L}.

For the pure λ-calculus Λ the relation $\cong_\varepsilon$ clearly is inconsistent since there are no errors. By contrast, $\sqsubseteq_\Downarrow$ on Λ^H is the usual approximation relation, and its reflexive closure $\cong_\Downarrow$ is the sensible theory of [3], equating all unsolvable terms; $\cong_\Downarrow$ on Λ^L is the semi-sensible, *lazy* theory of [1], which equates unsolvable terms of the same order. So in Λ^H we have $\Omega \cong_\Downarrow \mathbf{YK} \sqsubseteq_\Downarrow a$ for every a, while in Λ^L we have $\Omega \sqsubseteq_\Downarrow a \sqsubseteq_\Downarrow \mathbf{YK}$. A detailed discussion of these different relations can be found in [1].

Lemma 22. *1.* $\lambda x.a \sqsubseteq_\Downarrow \lambda x.b \iff a \sqsubseteq_\Downarrow b$
2. $\lambda x.a \sqsubseteq_\Downarrow b \implies (b \xrightarrow{*} \lambda x.b') \wedge (a \sqsubseteq_\Downarrow b')$

5.1 Standard λ-calculus with ε

Λ_ε is the pure λ-calculus with an added constant ε and corresponding reduction rule $\varepsilon a \to \varepsilon$.

Lemma 23. *In* Λ_ε, $a\dagger \iff a \xrightarrow{*} \lambda x_1 \ldots x_n.\varepsilon$

Proof. ($\Longleftarrow$): easy, $\lambda x_1 \ldots x_n.\varepsilon$ is n-erroneous. ($\Longrightarrow$): a must be k-erroneous for some k, so we can use induction on k. □

Lemma 24. 1. Λ_ε is not error-generating, but is error-preserving.
2. $\Lambda_\varepsilon^L \models \mathbf{YK} \cong_\Downarrow \varepsilon$.
3. Λ_ε^H is error-complete, but not Λ_ε^L.

Proof. 1: Easy by inspection of rules β and ε. 2: Both are ever-convergent. 3: Values in Λ_ε^H are λ-terms in head normal form, or ε. Since HNFs are solvable, for every v there is always a context $C[-]$ such that $C[v]\dagger^0$. By contrast, value $\lambda x.\Omega$ in Λ_ε^L never reduces to an error. $\square$

Lemma 25. $\Lambda_\varepsilon^H \models a \sqsubseteq_\varepsilon b \iff \Lambda_\varepsilon^L \models a \sqsubseteq_\varepsilon b$.

Proof. By the Lemma 23 the error terms in both calculi are the same. $\square$

Theorem 26. 1. In both Λ_ε^H and Λ_ε^L, $a \sqsubseteq_\Downarrow b \implies a \sqsubseteq_\varepsilon b$
2. In Λ_ε^H, $a \sqsubseteq_\Downarrow b \iff a \sqsubseteq_\varepsilon b$

Proof. 1: suppose $a \sqsubseteq_\Downarrow b$. By Lemma 23, for any context $C[-]$, if $C[a]\dagger$ then $C[a] \xrightarrow{*} \lambda x_1 \ldots x_n.\varepsilon$. Therefore by Lemma 22 $C[b] \xrightarrow{*} \lambda x_1 \ldots x_n.b'$ with $\varepsilon \sqsubseteq_\Downarrow b'$, so $C[b]\dagger$. 2: ($\implies$):preceding part of the theorem. ($\impliedby$): from Theorem 21, knowing that Λ_ε^H is error-complete. $\square$

5.2 λ-calculus with records

The λ-calculus is now extended with records, i.e. collections of bindings from *names* to terms. As usual, these are written with curly braces; we use the vector notation $\{\overline{l_i = a_i}\}$ to denote the record with finite list of fields $l_1 = a_1, \ldots, l_n = a_n$, with all l_i distinct. The expression $(\overline{l_i = a_i} \setminus l)$ denotes removal of field l (if present) in a collection of bindings. Here all records are considered as values, which is perhaps a debatable choice, but conforms to an often similar choice in calculi with tuples [16].

Lemma 27. 1. $\Lambda_{\{\}}$ is error-generating, error-complete and error-preserving for both the head and the lazy calculus.
2. $\Lambda_{\{\}}^H \models a \sqsubseteq_\varepsilon b \iff \Lambda_{\{\}}^L \models a \sqsubseteq_\varepsilon b$.

Proof. 1: Error-generating: obvious. Error-complete: each closed value is either of record shape or of functional shape. In each case there is a context $([-]a)$ or $[-].l$ which generates an error. Error-preserving: easy by inspection of the reduction rules. 2: As for Λ_ε (Lemma 25): the error terms are the same (although the proof here is slightly more complex, as error terms may also be of record shape). $\square$

Since now even the lazy calculus is error-complete, the "ogre" **YK** has a different status than in Λ_ε:

Proposition 28. In $\Lambda_{\{\}}, \neg(\mathbf{YK} \cong_\Downarrow \varepsilon)$

Proof. Because $\Lambda^L_{\{\}}$ is error-complete and because of Theorem 21, it suffices to show $\neg(\mathbf{YK} \cong_\varepsilon \varepsilon)$. In the empty context $[-]$, there is no k such that $\mathbf{YK}$ is k-erroneous, because it can consume an infinite number of arguments without yielding an error. □

On the other hand there is a new term which is erroneous, namely the empty record:

Proposition 29. *In* $\Lambda_{\{\}}, \{\} \cong_\varepsilon \varepsilon$

Proof. By inspection of the reduction rules, $\{\}$ cannot interact without yielding an error, so it is 1-erroneous. □

However if the calculus is augmented with a record extension construct $a\Leftarrow l = b$ (like in [18,21]) then the empty record becomes solvable: for any value v there is a relevant context $([-]\Leftarrow l = v).l$ yielding that value, so in that case $\{\}$ is not equal to ε.

6   Types

This section illustrates the usefulness of both subsumption and labelled reductions for the semantics of types : subsumption is a natural foundation for interpreting subtyping, and labelled terms are a natural foundation for interpreting recursive types, following the approach of [7]. This is just an appetizer, as lack of space prevents us from going through full technical developments. Nevertheless the general approach borrows well-known techniques and therefore should be easy to follow.

Types are interpreted as non-empty, downward-closed subsets of terms in the $\sqsubseteq_\varepsilon$ ordering. Let **Tset** denote the set of such subsets. For any $t \in \mathbf{Tset}$, t^n denotes the set $\{a^n | a \in t\}$ (finite projection). A *type environment* η is a mapping from **Tvar** to **Tset**. Given a type environment, a type interpretation function $\mathbf{Ti}[-]$ maps types to members of **Tset**. We will illustrate this approach on the $\Lambda_{\{\}}$ calculus of the previous section, considering types of the following syntax.

$$T, U ::= \top \mid X \mid T \to U \mid \{\overline{l_i : T_i}\} \mid \mu X.T$$

Type assignment rules and subtyping rules are not displayed here: standard rules are assumed (see for example [8]). We also assume a rule *(top)* assigning type $\top$ to *any term*. Figure 1 gives the type interpretation. A well-known difficulty associated with recursive types is the fact that arrow types are contravariant on the left. The ideal model of [15] solves the problem through contractive maps on ideals in the semantic domain; this requires some conditions on the syntax of type expressions to enforce contractiveness. By contrast we follow here the idea of [7], using a family of indexed type interpretations, where the index denotes finite approximations. In this approach non-contractive type expressions are naturally mapped to the bottom type (the one containing only divergent terms), without any syntactic constraints. With labelled terms this can be done in an operational way, without needing to resort to denotational semantics.

$$\begin{aligned}
\mathbf{Ti}[T]_\eta^0 &= \{a | a \sqsubseteq_\varepsilon \Omega\} \\
\mathbf{Ti}[\top]_\eta^{n+1} &= \mathcal{T}^{n+1} \\
\mathbf{Ti}[X]_\eta^{n+1} &= \eta(X)^{n+1} \\
\mathbf{Ti}[T \to U]_\eta^{n+1} &= \{a \in \mathcal{T}^{n+1} | b \in \mathbf{Ti}[T]_\eta^n \implies a(b) \in \mathbf{Ti}[U]_\eta^n\} \\
\mathbf{Ti}[\{\overline{l_i : T_i}\}]_\eta^{n+1} &= \{a \in \mathcal{T}^{n+1} | \forall i, a.l_i \in \mathbf{Ti}[T_i]_\eta^n\} \\
\mathbf{Ti}[\mu X.T]_\eta^{n+1} &= \mathbf{Ti}[T]_{\eta[X \mapsto \mathbf{Ti}[\mu X.T]_\eta^n]}^{n+1} \\
\mathbf{Ti}[T]_\eta &= \{a | \forall n \in \omega, a^n \in \mathbf{Ti}[T]_\eta^n\}
\end{aligned}$$

Fig. 1. Type interpretation for functions and records

Lemma 30. $\forall T, \eta, \mathbf{Ti}[T]_\eta \in \mathbf{Tset}$.

Lemma 31. $T \leq U \implies \mathbf{Ti}[T]_\eta \subseteq \mathbf{Ti}[U]_\eta$.

Definition 32. A closing substitution σ *satisfies* a basis Γ, written $\sigma \models \Gamma$, iff, $\forall \eta, \forall x \in dom(\Gamma), \sigma(x) \in \mathbf{Ti}[\Gamma(x)]_\eta$.

Theorem 33. $\Gamma \vdash a : T \implies (\forall \sigma \models \Gamma, a\sigma \in \mathbf{Ti}[T])$.

Definition 34 (Trivial types). The set **Triv** of trivial types is defined inductively as:

$$\mathbf{Triv} = \top \cup \{T \to U | U \in \mathbf{Triv}\} \cup \{\{\overline{l_i : T_i}\} | \forall i, T_i \in \mathbf{Triv}\} \cup \{\mu(X)T | T \in \mathbf{Triv}\}$$

Lemma 35. *In any non-trivial type environment, non-trivial types do not contain erroneous terms. (η is non-trivial iff $\varepsilon \notin \eta(X)$ for each type variable X in $dom(\eta)$)*

Theorem 36. *If $\Gamma \vdash a : T$ and $T \notin \mathbf{Triv}$, then $\forall \sigma \models \Gamma, \neg(a\sigma\dagger)$.*

Proof. Consequence of the preceding lemma and of subject reduction, shown using standard techniques. □

Lemma 37. *The following equality between record types is sound:*

$$\{l : \top, \overline{l_i : T_i}\} = \{\overline{l_i : T_i}\}$$

Proof. Since $\varepsilon \in \mathbf{Ti}[\top]$, the condition $a.l_i \in \mathbf{Ti}[T_i]$ on field l is always satisfied, even for records where field l is absent. □

Example 38. The example of the introduction

$$(\lambda x.\{l_1 = x.l_2, l_2 = x.l_1\})\{l_1 = 3\}$$

has type $\{l_1 : \top, l_2 : \mathbf{Int}\}$, which is equal to $\{l_2 : \mathbf{Int}\}$ and is non-trivial.

References

1. Samson Abramsky and C.-H. Luke Ong. Full Abstraction in the Lazy Lambda Calculus. *Information and Computation*, 105:159-267, 1993.
2. Martin Abadi, Benjamin Pierce and Gordon Plotkin. Faithful Ideal Models for Recursive Polymorphic Types. *Int. J. of Foundations for Computer Science*, 2(1):1-21, 1991.
3. Henk Barendregt. *The Lambda-Calculus, its Syntax and Semantics*. Studies in Logic and the Foundations of Mathematics, North-Holland, 1984.
4. Baard Bloom. Can LCF Be Topped? Flat Lattice Models of Typed λ-calculus. *Information and Computation* 87:264-301, 1990.
5. Val Breazu-Tannen, Thierry Coquand, Carl A. Gunter, and Andre Scedrov. Inheritance as Implicit Coercion. *Information and Computation* 93:172-221, 1991. Also in [11], pp 197-245.
6. A Modest Model of Records, Inheritance, and Bounded Quantification. *Information and Computation* 87:196-240, 1990. Also in [11], pp 151-195.
7. Felice Cardone and Mario Coppo. Two extensions of Curry's Type Inference System. In *Logic and Computer Science*, P. Odifreddi(ed), pp 19-75. Academic Press, 1990.
8. Luca Cardelli and John Mitchell. Operations on Records. In [11], pp 295-350. First appeared in *Math. Structures in Comp. Sc.*, 1991, pp 3-48.
9. Laurent Dami. A Lambda-Calculus for Dynamic Binding. To appear in *Theoretical Comp. Sc.*, special issue on Coordination, 1997.
10. Laurent Dami. Labelled Reductions, Runtime Errors, and Operational Subsumption. Technical Report, U. of Geneva, 1997. Currently available at http://cuiwww.unige.ch/~dami.
11. Carl A. Gunter and John C. Mitchell, eds. *Theoretical aspects of object-oriented programming: types, semantics, and language design*. MIT Press, Foundations of computing series, 1994.
12. D. J. Howe. Equality in lazy computation systems. In *Proc. 4th IEEE Symp. on Logic in Comp. Sc.*, pp 198-203, 1989.
13. Trevor Jim and Albert R. Meyer. Full Abstraction and the Context Lemma. *SIAM J. on Computing* 25(3):663-696, June 1996.
14. Jan W. Klop, Vincent van Oostrom and Femke van Raamsdonk. Combinatory reduction systems: introduction and survey. *Theoretical Computer Science*, 121:279-308, 1993.
15. David MacQueen, Gordon Plotkin and Ravi Sethi. An Ideal Model for Recursive Polymorphic Types. *Information and Control*, 71:95-130, 1986.
16. Ian A. Mason, Scott F. Smith and Carolyn L. Talcott. From Operational Semantics to Domain Theory. In *Information and Computation*, 128:26-47, 1996.
17. Gordon Plotkin. Call-by-name, call-by-value and the λ-calculus. *Theoretical Computer Science*, 1:125-159, 1975.
18. Didier Rémy. Typechecking records and variants in a natural extension of ML. In *Proceedings ACM POPL'89*, pp 242-249. Also in [11], pp 67-96.
19. Dana Scott. Data types as lattices. *SIAM J. of Computing*, 5:522-587, 1976.
20. C. Talcott, A Theory of Binding Structures and Applications to Rewriting, *Theoretical Computer Science*, 112:99-143, 1993.
21. Mitchell Wand. Type Inference for Record Concatenation and Multiple Inheritance. *Information and Computation*, 93(1):1-15, 1991.

A Language Rules

A.1 Standard λ-calculus with ε

Syntax	$(\xi)\dfrac{x \in \mathcal{X}}{x \in \mathcal{T}}$ $\quad$ $(\lambda)\dfrac{x \in \mathcal{X}, a \in \mathcal{T}}{\lambda x.a \in \mathcal{T}}$ $\quad$ $(\beta)\dfrac{a, b \in \mathcal{T}}{(ab) \in \mathcal{T}}$ $\quad$ $(\varepsilon)\dfrac{}{\varepsilon \in \mathcal{T}}$
Red. Rules	$(\beta)\dfrac{}{(\lambda x.a)b \to a[x := b]}$ $\quad$ $(|\lambda|)\dfrac{a \to b}{\lambda x.a \to \lambda x.b}$ $\quad$ $(\varepsilon)\dfrac{}{(\varepsilon a) \to \varepsilon}$ $(|\beta 1|)\dfrac{a \to b}{(ac) \to (bc)}$ $\quad$ $(|\beta 2|)\dfrac{a \to b}{(ca) \to (cb)}$ $\quad$ $(\eta)\dfrac{x \in \mathcal{X}, a \in \mathcal{T}, x \notin FV(a)}{\lambda x.ax \to a}$
Values	$(\xi)\dfrac{}{x \in \mathcal{H}}$ $\quad$ $(\beta)\dfrac{v \in \mathcal{H}, a \in \mathcal{T}}{(va) \in \mathcal{H}}$ $\quad$ $(\chi)\dfrac{v \in \mathcal{H}}{v \in \mathcal{V}}$ $(\lambda^H)\dfrac{v \in \mathcal{V}}{\lambda x.v \in \mathcal{V}}$ $\quad$ $(\lambda^L)\dfrac{a \in \mathcal{T}}{\lambda x.a \in \mathcal{V}}$ $\quad$ $(\varepsilon)\dfrac{}{\varepsilon \in \mathcal{V}}$

A.2 λ-calculus with records

Syntax	$(\rho)\dfrac{\forall i, a_i \in \mathcal{T}}{\{l_i = a_i\} \in \mathcal{T}}$ $\quad$ $(\sigma)\dfrac{a \in \mathcal{T}}{a.l \in \mathcal{T}}$
Red. Rules	$(\sigma_\rho)\dfrac{\exists j, l \equiv l_j}{\{l_i = a_i\}.l \to a_j}$ $\quad\quad$ $(\sigma_\varepsilon)\dfrac{\forall j, l \not\equiv l_j}{\{l_i = a_i\}.l \to \varepsilon}$ $(\sigma_\lambda)\dfrac{}{(\lambda x.a).l \to \varepsilon}$ $\quad\quad$ $(\beta_\rho)\dfrac{}{(\{l_i = a_i\}\, b) \to \varepsilon}$ $(|\rho|)\dfrac{a_i \to a'_i}{\{\ldots, l_i = a_i, \ldots\} \to \{\ldots, l_i = a'_i, \ldots\}}$ $\quad$ $(|\sigma|)\dfrac{a \to a'}{a.l \to a'.l}$ $(\varepsilon_\sigma)\dfrac{}{\varepsilon.l \to \varepsilon}$
Values	$(\rho 1)\dfrac{\forall j, a_j \in \mathcal{T}}{\{l_i = a_i\} \in \mathcal{R}}$ $\quad$ $(\rho 2)\dfrac{a \in \mathcal{R}}{a \in \mathcal{V}}$ $\quad$ $(\sigma)\dfrac{a \in \mathcal{H}}{a.l \in \mathcal{H}}$

A Complete and Efficiently Computable Topological Classification of D-dimensional Linear Cellular Automata over $\mathbf{Z}_m$

Giovanni Manzini[1,2], Luciano Margara[3]

[1] Dipartimento di Scienze e Tecnologie Avanzate, Università di Torino, Via Cavour 84, 15100 Alessandria, Italy.
[2] Istituto di Matematica Computazionale, Via S. Maria, 46, 56126 Pisa, Italy.
[3] Dipartimento di Scienze dell'Informazione, Università di Bologna, Mura Anteo Zamboni 7, 40127 Bologna, Italy.

Abstract. We study the dynamical behavior of D-dimensional linear cellular automata over $\mathbf{Z}_m$. We provide easy-to-check necessary and sufficient conditions for a D-dimensional linear cellular automata over $\mathbf{Z}_m$ to be *sensitive to initial conditions, expansive, strongly transitive*, and *equicontinuous*.

1 Introduction

Cellular Automata (CA) are dynamical systems consisting of a regular lattice of variables which can take a finite number of discrete values. The global state of the CA, specified by the values of all the variables at a given time, evolves in synchronous discrete time steps according to a given *local rule* which acts on the value of each single variable. CA have been widely studied in a number of disciplines (e.g., computer science, physics, mathematics, biology, chemistry) with different purposes (e.g., simulation of natural phenomena, pseudo-random number generation, image processing, analysis of universal model of computations, cryptography). For an introduction to the CA theory and an extensive and up-to-date bibliography see [7].

CA can display a rich and complex temporal evolution whose exact determination is in general very hard, if not impossible. In particular, some properties of the temporal evolution of general CA are undecidable [3, 4, 10]. Despite their simplicity that makes it possible a detailed algebraic analysis, linear CA over $\mathbf{Z}_m$ (CA based on a linear local rule) exhibit many of the complex features of general CA. Several important properties of linear CA have been studied during the last few years [1, 5, 8, 9, 12, 13] and in some cases exact characterizations have been obtained. As an example, in [9] the authors present criteria for surjectivity and injectivity of linear CA, while in [2] the authors present criteria for topological transitivity and ergodicity.

In this paper we investigate the topological behavior of linear D-dimensional CA over $\mathbf{Z}_m$. We focus our attention on a number of topological properties which are widely recognized as fundamental in the determination of the qualitative behavior of any discrete time dynamical system, namely *sensitivity to initial condi-*

Property	Characterization	Reference
Surjectivity	$\gcd(m, \lambda_1, \ldots, \lambda_s) = 1$	[9]
Injectivity	$(\forall p \in \mathcal{P})\,(\exists!\lambda_i)\colon p \nmid \lambda_i$	[9]
Transitivity	$\gcd(m, \lambda_2, \ldots, \lambda_s) = 1$	[2]
Sensitivity	$(\exists p \in \mathcal{P})\colon p \nmid \gcd(\lambda_2, \ldots, \lambda_s)$	This paper
Expansivity	$\gcd(m, a_1, \ldots, a_r) = \gcd(m, a_{-1}, \ldots, a_{-r}) = 1$	This paper
Equicontinuity	$(\forall p \in \mathcal{P})\; p \mid \gcd(\lambda_2, \ldots, \lambda_s)$	This paper
Strong Trans.	$(\forall p \in \mathcal{P})\,(\exists \lambda_i, \lambda_j)\colon p \nmid \lambda_i \wedge p \nmid \lambda_j$	This paper

Fig. 1. Characterization of set theoretic and topological properties of linear CA over $\mathbf{Z}_m$ in terms of the coefficients λ_i's (for D-dimensional CA) or a_i's (for 1-dimensional CA). $\mathcal{P}$ denotes the set of prime factors of m.

tions, expansivity, equicontinuity, and *strong transitivity*. The main contribution of this paper consists in efficiently computable criteria for deciding whether a linear CA satisfies one of the above four properties. Our criteria are reported in Fig. 1 and are given in terms of the coefficients of the linear local map associated to the CA. Note that, using our criteria, one can easily construct a linear CA which satisfies any combination of the above properties. The criteria we propose require only gcd computations and can be checked in polynomial time in the number of coefficients and in the logarithm of the cardinality of the alphabet. The dimension of the lattice does not explicitly affect the computational cost of our criteria. The results of this paper hold for every dimension $D \geq 1$ and for every $m \geq 2$. Our results show that linear CA over $\mathbf{Z}_m$ have dynamical aspects that linear CA over finite fields, such as $\mathbf{Z}_p$ with p prime, cannot have.

2 Basic definitions

Let $\mathbf{Z}_m$, $m \geq 2$, denote the ring of integers modulo m. We consider the *space of configurations*

$$\mathcal{C}_m^D = \{c \mid c\colon \mathbf{Z}^D \to \mathbf{Z}_m\}.$$

which consists of all functions from $\mathbf{Z}^D$ into $\mathbf{Z}_m$. Each element of $\mathcal{C}_m^D$ can be visualized as an infinite D-dimensional lattice in which each cell contains an element of $\mathbf{Z}_m$. A special configuration is the *null* configuration $\mathbf{0}$ which has the property that $\mathbf{0}(\mathbf{v}) = 0$ for all $\mathbf{v} \in \mathbf{Z}^D$.

Let $s \geq 1$. A *neighborhood frame* of size s is an ordered set of distinct vectors $\mathbf{u}_1, \mathbf{u}_2, \ldots, \mathbf{u}_s \in \mathbf{Z}^D$. Given any function $f\colon \mathbf{Z}_m^s \to \mathbf{Z}_m$, a D-dimensional CA based on the *local rule* f is the pair $(\mathcal{C}_m^D, F)$, where $F\colon \mathcal{C}_m^D \to \mathcal{C}_m^D$, is the *global transition map* defined as follows. For every $c \in \mathcal{C}_m^D$ the configuration $F(c)$ is such that for every $\mathbf{v} \in \mathbf{Z}^D$

$$[F(c)](\mathbf{v}) = f\left(c(\mathbf{v} + \mathbf{u}_1), \ldots, c(\mathbf{v} + \mathbf{u}_s)\right), \tag{1}$$

In other words, the content of cell $\mathbf{v}$ in the configuration $F(c)$ is a function of the content of the cells $\mathbf{v} + \mathbf{u}_1, \ldots, \mathbf{v} + \mathbf{u}_s$ in the configuration c. Note that the local rule f and the neighborhood frame completely determine F.

A map $f \colon \mathbf{Z}_m^s \to \mathbf{Z}_m$, is linear if and only if there exist $\lambda_1, \ldots, \lambda_s \in \mathbf{Z}_m$ such that $f(x_1, \ldots, x_s) \equiv \sum_{i=1}^{s} \lambda_i x_i \pmod{m}$. From now on, we say that a CA defined over $\mathbf{Z}_m$ is linear if the local rule on which it is based is linear over $\mathbf{Z}_m$. Note that for a linear D-dimensional CA, equation (1) becomes

$$[F(c)](\mathbf{v}) = \sum_{i=1}^{s} \lambda_i c(\mathbf{v} + \mathbf{u}_i) \mod m.$$

We define the *radius* of the linear CA $(\mathcal{C}_m^D, F)$ as

$$\rho(F) = \max\{\|\mathbf{u}_i\|_\infty, \ 1 \le i \le s\}, \tag{2}$$

where the maximum is restricted to the indices i such that $\lambda_i \not\equiv 0 \pmod{m}$. As usual, $\|\mathbf{v}\|_\infty$ denotes the maximum of the absolute value of the components of $\mathbf{v}$. For linear 1-dimensional CA we use a simplified notation. A local rule of radius r is written as $f(x_{-r}, \ldots, x_r) = \sum_{i=-r}^{r} a_i x_i \mod m$, where at least one between a_{-r} and a_r is nonzero. Using this notation, the global map F of a 1-dimensional CA with $\rho(F) = r$ becomes

$$[F(c)](i) = \sum_{j=-r}^{r} a_j c(i+j) \mod m, \ c \in \mathcal{C}_m^1, \ i \in \mathbf{Z}.$$

In order to study the topological properties of D-dimensional CA, we introduce a distance over the space of the configurations. Let $\Delta \colon \mathbf{Z}_m \times \mathbf{Z}_m \to \{0, 1\}$ defined by $\Delta(i,j) = 0$ if $i = j$ and $\Delta(i,j) = 1$ otherwise. Given $a, b \in \mathcal{C}_m^D$ the Tychonoff distance $d(a, b)$ is given by

$$d(a,b) = \sum_{\mathbf{v} \in \mathbf{Z}^D} \frac{\Delta(a(\mathbf{v}), b(\mathbf{v}))}{2^{\|\mathbf{v}\|_\infty}}. \tag{3}$$

It is easy to verify that d is a metric on $\mathcal{C}_m^D$ and that the topology induced by d coincides with the product topology induced by the discrete topology of $\mathbf{Z}_m$.

2.1 Topological Properties

In this section we recall the definitions of some topological properties which determine the qualitative behavior of any general discrete time dynamical system. Here, we assume that the space of configurations X is equipped with a distance d and that the map F is continuous on X according to the topology induced by d (for CA, Tychonoff distance satisfies this property). We denote by $\mathcal{B}(x, \epsilon)$ the (open) set $\{y \in X \colon d(x, y) < \epsilon\}$.

Definition 1 (Sensitivity). A dynamical system (X, F) is sensitive to initial conditions if and only if there exists $\delta > 0$ such that for any $x \in X$ and for any $\epsilon > 0$, there exists $y \in \mathcal{B}(x, \epsilon)$ and $n \geq 0$, such that $d(F^n(x), F^n(y)) > \delta$. The value δ is called the sensitivity constant. □

Intuitively, a map is sensitive to initial conditions, or simply sensitive, if there exist points arbitrarily close to x which eventually separate from x by at least δ under iteration of F. Note that not all points near x need eventually separate from x under iteration, but there must be at least one such point in every neighborhood of x.

A property stronger than sensitivity is expansivity. Expansivity differs from sensitivity in that all nearby points must eventually separate by at least δ. It is easy to verify that expansive CA are sensitive to initial conditions.

Definition 2 (Expansivity). A dynamical system (X, F) is expansive if and only if there exists $\delta > 0$ such that for every $x, y \in X$ there exists $n \geq 0$ such that $d(F^n(x), F^n(y)) > \delta$. The value δ is called the expansivity constant. □

If a dynamical system is sensitive to initial conditions or, even worse, expansive, then its dynamics defies numerical approximation. As an example, round-off errors may become magnified upon iterations of F and the results of the numerical computation of an orbit, no matter how accurate, may be completely different from the real orbit.

Definition 3 (Equicontinuity at x). A dynamical system (X, F) is equicontinuous at $x \in X$ if and only if for any $\delta > 0$ there exists $\epsilon > 0$ such that for any $y \in \mathcal{B}(x, \epsilon)$ and $n \geq 0$ we have $d(F^n(x), F^n(y)) < \delta$. □

Definition 4 (Equicontinuity). A dynamical system (X, F) is equicontinuous if and only if it is equicontinuous at every $x \in X$. □

The notions of sensitivity and equicontinuity are related. In fact, by comparing the definitions one can easily see that

$$F \text{ is not sensitive} \iff \exists x: F \text{ is equicontinuous at } x. \quad (4)$$

Definition 5 (Strong transitivity). A dynamical system (X, F) is strongly transitive iff for all nonempty open set $U \subseteq X$ we have $\bigcup_{n=0}^{+\infty} F^n(U) = X$. □

A strongly transitive map F has points which, under iteration of F, move from one arbitrarily small neighborhood to all the space of configurations X. A weaker notion is *transitivity*: a map F is transitive iff for all nonempty open set U the set $\bigcup_{n=0}^{+\infty} F^n(U)$ is a dense subset of X. Clearly, strongly transitive maps are transitive, and in view of [2, Theorem 6] ergodic with respect to the normalized Haar measure.

3 Statement of the new results

In this section we state the main results of this paper. The same results are summarized in Fig. 1.

Theorem 6. *Let F denote the global transition map of a linear D-dimensional CA over $\mathbf{Z}_m$ defined by*

$$[F(c)](\mathbf{v}) = \sum_{i=1}^{s} \lambda_i c(\mathbf{v} + \mathbf{u}_i) \bmod m. \tag{5}$$

Assume $\mathbf{u}_1 = \mathbf{0}$, that is, λ_1 is the coefficient associated to the null displacement. The global transition map F is sensitive if and only if there exists a prime p such that

$$p|m \quad \text{and} \quad p \nmid \gcd(\lambda_2, \lambda_3, \ldots, \lambda_s). \tag{6}$$

In other words, F is sensitive unless every prime which divides m divides also all the coefficients λ_i's with $i \neq 1$. □

Note that we can check the above condition without knowing the factorization of m. In fact, (6) holds if and only if $\gcd(\lambda_2, \lambda_3, \ldots, \lambda_s)$ does not contain all the prime factors of m. Since each prime appears in m with a power at most $\lfloor \log_2 m \rfloor$, F is sensitive if and only if $[\gcd(\lambda_2, \lambda_3, \ldots, \lambda_s)]^{\lfloor \log_2 m \rfloor} \not\equiv 0 \pmod{m}$.

Theorem 7. *Let F denote the global transition map of a linear 1-dimensional CA over $\mathbf{Z}_m$ with local rule $f(x_{-r}, \ldots, x_r) = \sum_{i=-r}^{r} a_i x_i \bmod m$. The global transition map F is expansive if and only if*

$$\gcd(m, a_{-r}, \ldots, a_{-1}) = 1 \quad \text{and} \quad \gcd(m, a_1, \ldots, a_r) = 1. \tag{7}$$

□

Note that by Theorem 5.3 in [6] we know that expansive CA, whether linear or not, do not exist in any dimension $D \geq 2$.

Theorem 8. *Let F denote the global transition map of the linear D-dimensional CA over $\mathbf{Z}_m$ defined by (5). The following statements are equivalent: (i) F is equicontinuous in at least one point, (ii) F is equicontinuous at every point, and (iii) for each prime p such that $p|m$ we have $p| \gcd(\lambda_2, \lambda_3, \ldots, \lambda_s)$.* □

By Theorem 8 and (4), a linear CA is either sensitive or equicontinuous. Hence, F is equicontinuous if and only if $[\gcd(\lambda_2, \lambda_3, \ldots, \lambda_s)]^{\lfloor \log_2 m \rfloor} \equiv 0 \pmod{m}$.

Theorem 9. *Let F denote the global transition map of a linear D-dimensional CA over $\mathbf{Z}_m$ defined by (5). The global transition map F is strongly transitive if and only if for each prime p such that $p|m$, there exist at least two coefficients λ_i, λ_j such that $p \nmid \lambda_i$ and $p \nmid \lambda_j$.* □

We can check whether F is strongly transitive without knowing the factorization of m. In fact, the above condition is equivalent to $\gcd(m, \lambda_1, \lambda_2, \ldots, \lambda_{s-1}) = \gcd(m, \lambda_1, \lambda_2, \ldots, \lambda_{s-2}, \lambda_s) = \cdots = \gcd(m, \lambda_2, \lambda_3, \ldots, \lambda_s) = 1$.

4 Proof of the main theorems

We now prove the results stated in Sect. 3. Due to limited space the proof of Theorem 8 is reported in [11]. In our proofs we make use of the *formal power series* (fps) representation of the configuration space $\mathcal{C}_m^D$ (see [9, Sec. 3] for details). For $D = 1$, to each configuration $c \in \mathcal{C}_m^1$ we associate the fps $P_c(X) = \sum_{i \in \mathbf{Z}} c(i) X^i$. The advantage of this representation is that the computation of a linear map is equivalent to power series multiplication. Let $F: \mathcal{C}_m^1 \to \mathcal{C}_m^1$ be a linear map with local rule $f(x_{-r}, \ldots, x_r) = \sum_{i=-r}^{r} a_i x_i$. We associate to F the finite fps $A_f(X) = \sum_{i=-r}^{r} a_i X^{-i}$. Then, for any $c \in \mathcal{C}_m^1$ we have

$$P_{F(c)}(X) \equiv P_c(X) A_f(X) \pmod{m}. \tag{8}$$

Note that each coefficient of $P_{F(c)}(X)$ is well defined since $A_f(X)$ has only finitely many nonzero coefficients. Note also that the finite fps associated to F^n is $A_f^n(X)$. More in general, to each configuration $c \in \mathcal{C}_m^D$ we associate the formal power series

$$P_c(X_1, \ldots, X_D) = \sum_{i_1, \ldots, i_D \in \mathbf{Z}} c(i_1, \ldots, i_D) X_1^{i_1} \cdots X_D^{i_D}.$$

The computation of a linear map F over $\mathcal{C}_m^D$ is equivalent to the multiplication by a finite fps $A(X_1, \ldots, X_D)$ which can be easily obtained by the local rule f and the neighborhood frame $\mathbf{u}_1, \ldots, \mathbf{u}_s$. The finite fps associated to the map F defined by (5) is $A(X_1, \ldots, X_D) = \sum_{i=1}^{s} \lambda_i X_1^{-\mathbf{u}_i(1)} \cdots X_D^{-\mathbf{u}_i(D)}$ where $\mathbf{u}_i(j)$ denotes the j-th component of vector $\mathbf{u}_i$.

Throughout the paper, given a fps $H(X)$ and $i \in \mathbf{Z}$, we use $\langle H(X) \rangle_i$ to denote the coefficient of X^i in $H(X)$.

4.1 Sensitivity

In this section we characterize sensitive linear CA. We prove our results only in the 2-dimensional case, since the proofs for the other dimensions are similar.

Let $F: \mathcal{C}_m^2 \to \mathcal{C}_m^2$ denote the global transition map of a 2-dimensional CA. For any integer $k > 0$, let $\mathcal{V}_k$ denote the set of configurations $c \in \mathcal{C}_m^2$ such that $c(\mathbf{v}) = 0$ for $\|\mathbf{v}\|_\infty < k$. It is straightforward to verify that F is sensitive if and only if there exists $\delta > 0$ such that for any configuration $c \in \mathcal{C}_m^2$ we have

$$\forall k \quad \exists c' \in \mathcal{V}_k: \quad d(F^n(c + c'), F^n(c)) > \delta \quad \text{for some } n \geq 0. \tag{9}$$

In fact, (9) implies that we can find a configuration, arbitrarily close to c, whose distance from c exceeds δ after a sufficiently large number of iterations.

If F is linear we can get rid of the initial configuration c. In fact, we have

$$d(F^n(c + c'), F^n(c)) = d(F^n(c) + F^n(c'), F^n(c)) = d(F^n(c'), 0).$$

Hence, F is sensitive if and only if

$$\forall k \quad \exists c' \in \mathcal{V}_k: \quad d(F^n(c'), 0) > \delta \quad \text{for some } n \geq 0. \tag{10}$$

This observation leads to the following lemma.

Lemma 10. *Let F denote the global transition map of a linear D-dimensional CA over $\mathbf{Z}_m$. F is sensitive if and only if*

$$\limsup_{n \to \infty} \rho(F^n) = \infty; \qquad (11)$$

(the radius ρ of a CA is defined by (2)).

Proof. We prove the result for $D = 2$. If (11) does not hold, there exists M such that $\rho(F^n) < M$ for all n. Thus, if $k > M$, for all $c \in \mathcal{V}_k$ we have $F^n(c) \in \mathcal{V}_{k-M}$. Elementary calculus shows that $c \in \mathcal{V}_t \implies d(c, \mathbf{0}) \leq \frac{8(t+2)}{2^t}$. Hence, for any δ, if k is large enough $c \in \mathcal{V}_k$ implies $d(F^n(c), \mathbf{0}) \leq \delta$ for all n, and F cannot be sensitive.

Assume now (11) holds. Then, for every k we can find n such that $\rho(F^n) = z > k$. Let $\lambda_i^{(n)}$, $\mathbf{u}_i^{(n)}$ denote the coefficients and the displacements of the local map associated to F^n. $\rho(F^n) = z$ implies that there exists j, such that $\lambda_j^{(n)} \neq 0$ and $\|\mathbf{u}_j^{(n)}\|_\infty = z$. Let c be such that $c(-\mathbf{u}_j^{(n)}) = 1$, and $c(\mathbf{v}) = 0$ for $\mathbf{v} \neq -\mathbf{u}_j^{(n)}$. Clearly, $c \in \mathcal{V}_k$ and $[F^n(c)](\mathbf{0}) = \lambda_j^{(n)} \neq 0$ which implies (10). □

Proof of Theorem 6 Let F denote the global transition map of a linear 2-dimensional CA, and let

$$A(X, Y) = \sum_{\substack{v \leq i \leq w \\ y \leq j \leq z}} a_{i,j} X^i Y^j,$$

denote the finite fps associated to F. Assume (6) holds. Then, there exist a prime p and a coefficient $a_{s,u}$ such that $p|m$, $p \nmid a_{s,u}$ and at least one between s and u is nonzero. We now prove that, as a consequence, $\limsup \rho(F^n) = \infty$. Without loss of generality, we can assume $s \neq 0$, and that for $i < s$ we have $p|a_{i,j}$. Let $\tilde{A}(X,Y) = A(X,Y) \bmod p$. By our assumptions, $\tilde{A}(X,Y)$ can be written as $X^s G(Y) + \sum_{s < i \leq w} X^i H_i(Y)$, with $G(Y) \neq 0$. Hence,

$$(A^n(X,Y) \bmod p) = \tilde{A}^n(X,Y) = X^{ns} G^n(Y) + \sum_{ns < i \leq nw} X^i H'_i(Y).$$

Since $\mathbf{Z}_p$ is an integral domain, we have $G^n(Y) \neq 0$ which implies $\rho(F^n) \geq n|s|$.

Assume now $p|m \implies p|\lambda_i$ for all $i \neq 1$. Let $m = p_1^{k_1} \cdots p_n^{k_n}$ denote the factorization of m, and let $k = \max_i k_i$. We prove that $\rho(F^n) \leq \rho(F)(k-1)$. Let $b_{i,j}$ denote the coefficients of the fps associated to F^n. We have

$$b_{i,j} = \sum_{\substack{i_1 + \cdots + i_n = i \\ j_1 + \cdots + j_n = j}} a_{i_1,j_1} a_{i_2,j_2} \cdots a_{i_n,j_n}. \qquad (12)$$

If $\max(|i|, |j|) > \rho(F)(k-1)$, each term $a_{i_1,j_1} a_{i_2,j_2} \cdots a_{i_n,j_n}$ contains at least k coefficients a_{i_h,j_h} with $\max(|i_h|, |j_h|) \neq 0$. Hence, $p|m \implies p^k | a_{i_1,j_1} \cdots a_{i_n,j_n}$, and each term in the sum (12) is a multiple of m. Hence, $\rho(F^n) \leq \rho(F)(k-1)$ and by Lemma 10 F is not sensitive. □

4.2 Expansivity

In this section we characterize expansive linear CA. Since expansive CA do not exist in dimension $D \geq 2$ (see [6, Theorem 5.3]) we can restrict ourselves to the 1-dimensional case.

Let $F: \mathcal{C}_m^1 \to \mathcal{C}_m^1$ denote the global transition map of a 1-dimensional CA. It is straightforward to verify that F is expansive if and only if there exists $\delta > 0$ such for any configuration $c \in \mathcal{C}_m^1$ we have

$$\forall c' \in \mathcal{C}_m^1 \quad \exists n \geq 0: \quad d(F^n(c+c'), F^n(c)) > \delta.$$

Reasoning as in Sect. 4.1, if F is linear we can get rid of the particular configuration c. We have

$$d(F^n(c+c'), F^n(c)) = d(F^n(c) + F^n(c'), F^n(c)) = d(F^n(c'), \mathbf{0}).$$

Hence, F is expansive if and only if for any $c' \in \mathcal{C}_m^1$ we have $d(F^n(c'), \mathbf{0}) > \delta$ for a sufficiently large n. Clearly, this is equivalent to assuming that there exists $M > 0$ such that

$$\forall c' \in \mathcal{C}_m^1 \quad \exists n \geq 0: \quad [F^n(c')](i) \neq 0 \text{ for some } i \text{ with } |i| < M.$$

For any integer $k > 0$, let $\mathcal{W}_k$ denote the set of configurations $c \in \mathcal{C}_m^1$ such that $c(i) = 0$ for $|i| < k$ and at least one between $c(k)$ and $c(-k)$ is different from zero. Since δ can be chosen arbitrarily, we have that F is expansive iff $\exists \tilde{k}$ such that for all $k > \tilde{k}$

$$\forall c' \in \mathcal{W}_k \quad \exists n \geq 0: \quad [F^n(c')](i) \neq 0 \text{ for some } i \text{ with } |i| < M. \tag{13}$$

If we visualize each configuration as a biinfinite array, (13) tells us that the essential feature of expansive maps is that *any* pattern of nonzero values can "propagate" from positions arbitrarily away from 0 up to a position i with $|i| < M$. Informally, we say that any nonzero pattern can propagate for an arbitrarily large distance. For a comparison, sensitive 1-dimensional linear CA can be seen as those CA in which for each $t > 0$ there exists a nonzero pattern which propagates by at least t positions.

Proof of Theorem 7 (sketch) First we prove that (7) is a necessary condition for expansivity. Assume for example $\gcd(a_1, \ldots, a_r) = q_1 > 1$, and let $q_2 = m/q_1$. For any integer $k > 0$ let $c_k \in \mathcal{W}_k$ denote the configuration defined by $c_k(i) = q_2$ if $i = k$ and $c_k(i) = 0$ otherwise. We show that for every $n > 0$ and $i < k$ we have $[F^n(c_k)](i) = 0$ which implies that F is not expansive. Let $A(X) = \sum_{i=-r}^{r} a_{-i} X^i$ be the finite fps associated to f. Since the fps associated to c_k is $q_2 X^k$, we have

$$[F^n(c_k)](i) = \langle q_2 X^k A^n(X) \rangle_i = q_2 \langle A^n(X) \rangle_{i-k}.$$

By hypothesis, for $j < 0$, $\langle A(X) \rangle_j$ is a multiple of q_1. Since the same is true for $A^n(X)$, for $i < k$ we have $[F^n(c_k)](i) \equiv 0 \pmod{m}$ as claimed.

Now we prove that condition (7) implies expansivity. Let $c \in \mathcal{C}_m^1$ such that $c(v) \neq 0$ and $c(i) = 0$ for $i > v$. We show that $\gcd(m, a_{-1}, \ldots, a_{-r}) = 1$ implies that for any integer w there exists n such that $[F^n(c)](i) \neq 0$ for some $i > w$. This proves that any one-sided nonzero pattern can propagate arbitrarily far away to the right. Similarly, $\gcd(m, a_1, \ldots, a_r) = 1$ implies that any one-sided nonzero pattern can propagate arbitrarily far away to the left. Combining these two facts we get (13) (the details will be given in the full paper).

Let $c \in \mathcal{C}_m^1$ such that $c(v) \neq 0$ and $c(i) = 0$ for $i > v$, and let $C(X) = \sum_{i \leq v} c_i X^i$ be the associated fps. Since $m \nmid c_v$, there exists a prime p and an integer k such that $p^k | m$ and $p^k \nmid c_v$. Let $A(X) = \sum_{i=-r}^{r} a_{-i} X^i$ denote the finite fps associated to f. Since $\gcd(m, a_{-1}, \ldots, a_{-r}) = 1$, we can find t, $0 < t \leq r$, such that
$$p \nmid a_{-t} \quad \text{and} \quad p | a_{-i} \text{ for } t < i \leq r. \tag{14}$$
Under these assumptions we show that if n is a multiple of $p^k(k-1)!$ then
$$[F^n(c)](v + nt) = \langle C(X) A^n(X) \rangle_{v+nt} \not\equiv 0 \pmod{m}.$$
Clearly, this proves our claim that every one-sided nonzero pattern propagates arbitrarily far away to the right. Let $\tilde{A}(X) = A(X) \bmod p^k$. By (14) we know that $\tilde{A}(X)$ satisfies the hypothesis of Lemma A.4 of [11]. Hence, if n is a multiple of $p^k(k-1)!$, we have $\tilde{A}^n(X) = \sum_{i=-nr}^{nt} \tilde{a}_i X^i$ with $\gcd(\tilde{a}_{nt}, p^k) = 1$. We have
$$[F^n(c)](v + nt) \equiv \left\langle \tilde{A}^n(X) C(X) \right\rangle_{v+nt} \pmod{p^k}$$
$$\equiv \left\langle \left(\sum_{i=-nr}^{nt} \tilde{a}_i X^i\right) \left(\sum_{i \leq v} c_i X^i\right) \right\rangle_{v+nt} \pmod{p^k}$$
$$\equiv \tilde{a}_{nt} c_v \pmod{p^k}.$$
Since $p^k \nmid c_v$ and $p \nmid \tilde{a}_{nt}$, $[F^n(c)](v+nt)$ is not a multiple of p^k. We conclude that $[F^n(c)](v+nt) \not\equiv 0 \pmod{m}$ as claimed. $\square$

4.3 Strong transitivity

In this section we give a characterization of strongly transitive linear CA. The proof is quite complex and we will need some preliminary lemmas. To simplify the notation we consider only the 1-dimensional case; the proof for dimensions $D > 1$ is analogous and will be given in the full paper. Let $\mathcal{V}_k = \{x \in \mathcal{C}_m^1 \mid x(i) = 0 \text{ for } |i| < k\}$. For any $x \in \mathcal{C}_m^1$ let
$$\mathcal{D}(x, k) = x + \mathcal{V}_k = \{y \in \mathcal{C}_m^1 \mid y = x + z, \ z \in \mathcal{V}_k\}.$$
For any nonempty open subset $U \subseteq \mathcal{C}_m^1$ we can find $x \in X$ and $\epsilon > 0$ such that $\mathcal{B}(x, \epsilon) \subseteq U$. Elementary calculus shows that
$$\mathcal{D}(x, 3 + \lceil \log(1/\epsilon) \rceil) \subseteq \mathcal{B}(x, \epsilon) \subseteq U,$$
hence F is strongly transitive if and only if
$$\forall x \forall k \quad \bigcup_{n=0}^{+\infty} F^n(\mathcal{D}(x, k)) = \mathcal{C}_m^1. \tag{15}$$
We are now ready to establish a simple condition which, for linear maps, implies strong transitivity.

Lemma 11. *Let F be a linear 1-dimensional map over $\mathbf{Z}_m$. If, for all k, there exists n_k such that $F^{n_k}(\mathcal{V}_k) = \mathcal{C}_m^1$, then F is strongly transitive.*

Proof. For all $x \in \mathcal{C}_m^1$ and $k > 0$ we have

$$\bigcup_{n=0}^{+\infty} F^n(\mathcal{D}(x,k)) \supseteq F^{n_k}(x + \mathcal{V}_k) = F^{n_k}(x) + F^{n_k}(\mathcal{V}_k) = \mathcal{C}_m^1.$$

□

We prove the "if" part of Theorem 9 using Lemma 11 and the power series representation of CA. Lemma 12 establishes the result for the special case in which the cardinality of $\mathbf{Z}_m$ is a prime power, while Lemma 13 proves the result in the general case.

Lemma 12. *Let $A(X) = \sum_{-r \le i \le r} a_i X^i$ denote a finite fps over $\mathbf{Z}_{p^k}$ (p prime). Suppose there exist two coefficients a_i, a_j such that $\gcd(p, a_i) = \gcd(p, a_j) = 1$, and let n be any multiple of $p^k(k-1)!$. Then, for each fps $C(X)$ we can find $B(X) = \sum_{i \in \mathbf{Z}} b_i X^i$ such that $B(X)A^n(X) \equiv C(X) \pmod{p^k}$ and*

$$b_{-\lfloor n/2 \rfloor} = b_{-\lfloor n/2 \rfloor + 1} = \cdots = b_{\lfloor n/2 \rfloor - 2} = b_{\lfloor n/2 \rfloor - 1} = 0.$$

Due to limited space we do not report the proof of Lemma 12 here (see [11]).

Lemma 13. *Let $A(X) = \sum_{-r \le i \le r} a_i X^i$ denote a finite fps over $\mathbf{Z}_m$. Suppose that for each prime p which divides m there exist two coefficients a_i, a_j such that $\gcd(p, a_i) = \gcd(p, a_j) = 1$. Then, for any integer $z > 0$ there exists n such that for each fps $C(X) = \sum_{i \in \mathbf{Z}} c_i X^i$ we can find a fps $B(X) = \sum_{i \in \mathbf{Z}} b_i X^i$ such that*

$$b_{-z+1} = \cdots = b_{z-2} = b_{z-1} = 0, \quad \text{and} \quad B(X)A^n(X) \equiv C(X) \pmod{m}. \tag{16}$$

Proof. Let $m = p_1^{k_1} p_2^{k_2} \cdots p_h^{k_h}$, $q_i = p_i^{k_i}$, and $k = \max_i k_i$. Let n denote a multiple of $m(k-1)!$ such that $n > 2z$. Clearly n is a multiple of $q_i(k_i - 1)!$ for $i = 1, \ldots, h$. By Lemma 12 we know that given $C(X)$ we can find $B_i(X) = \sum_{j \in \mathbf{Z}} b_j^{(i)} X^j$ such that

$$b_{-z+1}^{(i)} = \cdots = b_{z-2}^{(i)} = b_{z-1}^{(i)} = 0, \quad \text{and} \quad B_i(X)A^n(X) \equiv C(X) \pmod{q_i}$$

Since $\gcd(q_i, m/q_i) = 1$, we can find β_i such that $\beta_i(m/q_i) \equiv 1 \pmod{q_i}$. Let

$$B(X) = \sum_{i=1}^{h} \beta_i \frac{m}{q_i} B_i(X).$$

For $i = 1, \ldots, h$, we have $B(X) \equiv B_i(X) \pmod{q_i}$. Hence, $B(X)A^n(X) \equiv C(X) \pmod{q_i}$ for all i, which implies (16). □

Proof of Theorem 9 The "if" part follows directly from Lemmas 11 and 13. To prove the "only if" part we use again the power series representation. Let $A(X) = \sum_{-r \leq i \leq r} a_i X^i$ denote the finite fps associated to the map F, and assume there exist a prime p and an integer j such that $p|m$ and $p|a_i$ for all $i \neq j$. Let $a_i^{(n)}$, $-rn \leq i \leq rn$, denote the coefficients of $A^n(X)$. It is straightforward to verify that, for $i \neq jn$, we have that $p|a_i^{(n)}$. Consider now any configuration $b \in \mathcal{V}_1$. The corresponding fps $B(X) = \sum_{i \in \mathbf{Z}} b_i X^i$ is such that $b_0 = 0$. We have

$$[F^n(b)](nj) = \langle A^n(X)B(X)\rangle_{nj} = \sum_{i=-rn}^{rn} a_i^{(n)} b_{nj-i}.$$

Since $b_0 = 0$, all terms in the summation are multiple of p and $p|[F^n(b)](nj)$. Hence, the configuration c such that $c(i) = 1$ for all $i \in \mathbf{Z}$ clearly does not belong to $F^n(\mathcal{V}_1)$, and by (15) F cannot be strongly transitive. □

References

1. H. Aso and N. Honda. Dynamical characteristics of linear cellular automata. *Journal of Computer and System Sciences*, 30:291–317, 1985.
2. G. Cattaneo, E. Formenti, G. Manzini, and L. Margara. On ergodic linear cellular automata over Z_m. In *14th Annual Symposium on Theoretical Aspects of Computer Science (STACS '97)*, volume 1200 of *LNCS*, pages 427–438. Springer Verlag, 1997.
3. K. Culik, J. Pachl, and S. Yu. On the limit sets of cellular automata. *SIAM Journal of Computing*, 18:831–842, 1989.
4. K. Culik and S. Yu. Undecidability of CA classification schemes. *Complex Systems*, 2:177–190, 1988.
5. P. Favati, G. Lotti, and L. Margara. One dimensional additive cellular automata are chaotic according to Devaney's definition of chaos. *Theoretical Computer Science*, 174:157–170, 1997.
6. M. Finelli, G. Manzini, and L. Margara. Lyapunov exponents vs expansivity and sensitivity in cellular automata. *Journal of Complexity*. To appear.
7. M. Garzon. *Models of Massive Parallelism*. EATCS Texts in Theoretical Computer Science. Springer Verlag, 1995.
8. P. Guan and Y. He. Exacts results for deterministic cellular automata with additive rules. *Jour. Stat. Physics*, 43:463–478, 1986.
9. M. Ito, N. Osato, and M. Nasu. Linear cellular automata over Z_m. *Journal of Computer and System Sciences*, 27:125–140, 1983.
10. J. Kari. Rice's theorem for the limit set of cellular automata. *Theoretical Computer Science*, 127(2):229–254, 1994.
11. G. Manzini and L. Margara. A complete and efficiently computable topological classification of D-dimensional linear cellular automata over Z_m. Technical Report B4-96-18, Istituto di Matematica Computazionale, CNR, Pisa, Italy, 1996.
12. T. Sato. Group structured linear cellular automata over Z_m. *Journal of Computer and System Sciences*, 49(1):18–23, 1994.
13. S. Takahashi. Self-similarity of linear cellular automata. *Journal of Computer and System Sciences*, 44(1):114–140, 1992.

Recognizability Equals Definability for Partial k-Paths*

Valentine Kabanets

School of Computing Science, Simon Fraser University, Vancouver, Canada

Abstract. We prove that every recognizable family of partial k-paths is definable in a counting monadic second-order logic. We also show the obstruction set of the class of partial k-paths computable for every k.

1 Introduction

In 1960, Büchi [1] showed that a language is regular iff it is definable by some formula in a monadic second-order logic, MS. Here, MS is the extension of the first-order logic that allows quantification over set variables. A set of objects is definable by an MS-formula if the formula is true exactly on the members of the set. Thus Büchi established that recognizability is equivalent to MS-definability for words. Doner [7] then extended this result to ranked trees.

Graphs are algebraic objects since any graph can be constructed from smaller graphs using certain graph operations. They are also logical structures since any graph is completely determined by the set of its vertices and the adjacency relation on this set. Thus the notions of recognizability and definability can be extended to finite graphs. Courcelle [2] proved that every MS-definable set of finite graphs is recognizable, but not conversely. However, he was able to extend the result of Doner to unordered unbounded trees using a counting monadic second-order logic, CMS, an extension of MS that allows modular counting.

The question remained whether there was a sufficiently large class of graphs for which recognizability would imply CMS-definability. In their study of graph minors, Robertson and Seymour [10] introduced the notion of the tree-width of a graph. A graph of tree-width k exhibits certain tree-like structure. Such a graph can be decomposed into subgraphs of size $k+1$ arranged as nodes of a tree (tree-decomposition) so that the nodes containing a given vertex form a subtree.

The class of graphs of tree-width at most k coincides with that of partial k-trees. Among other classes of graphs of bounded tree-width are trees and forests (tree-width ≤ 1), series-parallel graphs and outerplanar graphs (≤ 2), and Halin graphs (≤ 3).

The class of graphs of bounded tree-width plays an important role for another reason. Courcelle showed in [2] that the MS-theory of the class of partial k-trees

---

* This research was done while the author was at Simon Fraser University [8]. The author's present address is Department of Computer Science, University of Toronto, Toronto, ON, Canada M5S 3G4; kabanets@cs.utoronto.ca.

is decidable. Seese [11] proved that if the MS-theory of a class of finite graphs $\mathcal{G}$ is decidable, then the graphs in $\mathcal{G}$ have uniformly bounded tree-width. Thus, tree-width "characterizes" classes of finite graphs having decidable MS-theories.

Strictly speaking, the above results hold for so-called MS_2 logic, where MS_2 denotes the monadic second-order language using quantification over both vertex sets and edge sets of graphs; MS_1 is the language that uses quantification over vertex sets only (see [5, 6]). In this paper, we are using MS_2 and CMS_2.

For graphs of tree-width at most k, recognizability is defined using a tree automaton working on the corresponding tree-decompositions: A set $\mathcal{G}$ of partial k-trees G is recognizable if there is a tree automaton that accepts any tree-decomposition of each graph $G \in \mathcal{G}$, and rejects tree-decompositions of graphs not in $\mathcal{G}$. Courcelle [3] showed that a recognizable set of partial k-trees is CMS-definable for $k = 1$ and $k = 2$, and conjectured that recognizability implies CMS-definability of partial k-trees for every k. Kaller [9] proved the case of $k = 3$ and the case of k-connected partial k-trees.

We establish that every recognizable set of partial k-paths is CMS-definable, thereby proving a special case of Courcelle's conjecture. A partial k-path, or graph of bounded path-width, is a partial k-tree for which the corresponding tree-decomposition is a path-decomposition. Partial k-paths are recognized by finite automata working on the corresponding path-decompositions.

Our second result deals with computing the obstruction sets of minor-closed graph families. The class of partial k-trees (k-paths) is minor-closed and its obstruction set can be determined from the MS-formula defining that class [4]. We describe how to construct the MS-formula defining the class of partial k-paths for every given k. As a consequence, the obstruction sets of the classes of partial k-paths are computable for each k.

The remainder of this article is organized as follows: In Sect. 2, we give the necessary definitions. In Sect. 3, we show that recognizability implies CMS-definability for a generalization of the class of k-connected partial k-paths, the class of $(k, 1)$-paths. This is a base case of our solution for arbitrary partial k-paths which is outlined in Sect. 4.

2 Preliminaries

2.1 Partial k-Paths

We consider finite and simple graphs $G = (V, E)$, where V is the vertex-set and E is the edge-set of G. A *path-decomposition* (or *decomposition*) of G is a sequence $B = \langle B_1, \ldots, B_m \rangle$ of vertex-subsets, called *bags*, such that

1. every vertex $v \in V$ belongs to some bag B_i ($1 \leq i \leq m$),
2. for each edge $e \in E$, there is a B_i ($1 \leq i \leq m$) containing both ends of e,
3. for any $i, l, j \in \{1, \ldots, m\}$ such that $i \leq l \leq j$, $B_i \cap B_j \subseteq B_l$.

The *path-width of a decomposition* $B = \langle B_1, \ldots, B_m \rangle$ is $\max_{1 \leq i \leq m}\{|B_i|\} - 1$. A decomposition of path-width at most k will be called a k-*decomposition*. The path-width of a graph G is the minimum path-width over all decompositions of G. A *partial k-path* is a graph of path-width at most k.

Example 1. Graphs G_1 (Fig. 1) and G_2 (Fig. 2) are partial 1-path and 2-path, respectively, with possible decompositions: $B(G_1) = \langle \{1,2\}, \{2,3\}, \{3,4\}, \{3,5\}, \{3,6\} \rangle$ and $B(G_2) = \langle \{1, 1', 2\}, \{1, 2, 3\}, \{2, 3, 4\}, \{2, 3, 5\}, \{2, 3, 6\} \rangle$.

Fig. 1. A partial 1-path G_1.   **Fig. 2.** A partial 2-path G_2.

For a partial k-path $G = (V, E)$ with a decomposition $B = \langle B_1, \ldots, B_m \rangle$, first$(v)$ is the number of the bag where a vertex $v \in V$ appears for the first time, i.e., first$(v) = \min_{1 \leq l \leq m}\{l | v \in B_l\}$, new$(B_i)$ $(i \in \{1, \ldots, m\})$ is the set of vertices in B_i that appear in the decomposition for the first time, i.e., new$(B_i) = \{u \in B_i | \text{first}(u) = i\}$, and old$(B_i)$ is the set of vertices in B_i that also appear in some earlier bag, i.e., old$(B_i) = B_i \setminus \text{new}(B_i)$.

For G and B as above, a vertex $u \in B_r$ $(1 \leq r \leq m)$ is called a *drop vertex* of B_r iff for every $w \in V \setminus \cup_{i=1}^{r} B_i$, $\{u, w\} \notin E$. The set of all drop vertices of B_r $(1 \leq r \leq m)$ is denoted by drop(B_r). The remaining vertices of B_r are called *non-drop vertices* of B_r, the set of which is denoted by non-drop(B_r).

2.2 CMS-Definability

A graph $G = (V, E)$ can be viewed as a relational structure $(V \cup E, \{\mathbf{p}_v, \mathbf{p}_e, \mathbf{Inc}\})$, where $\mathbf{p}_v$ and $\mathbf{p}_e$ are unary predicates that define the vertex-set and the edge-set, respectively, and $\mathbf{Inc}$ is the ternary incidence predicate, i.e., for any $e \in E$ and $u, v \in V$, $\mathbf{Inc}(e, u, v) = \mathbf{True}$ iff $e = \{u, v\}$.

The language of *counting monadic second-order logic* corresponding to graphs G has the usual logical connectives: $\neg$ ("not"), $\wedge$ ("and"), $\vee$ ("or"), $\Rightarrow$ ("if-then"), and $\Leftrightarrow$ ("if and only if"), universal ($\forall$) and existential ($\exists$) quantifiers, equality symbol $=$, a sequence $\mathbf{u}, \mathbf{v}, \mathbf{w}, \ldots$, of individual variables, a sequence $\mathbf{U}, \mathbf{V}, \mathbf{W}, \ldots$, of set variables, the membership symbol $\in$, the unary predicate symbols $\mathbf{mod}_{p,q}$, $p < q$ are non-negative integers, and the predicate symbols $\mathbf{p}_v$, $\mathbf{p}_e$, and $\mathbf{Inc}$. In our interpretation, $\mathbf{mod}_{p,q}(\mathbf{V}) = \mathbf{True}$ iff $|S| = p \bmod q$, where S is the set denoted by the set variable $\mathbf{V}$.

A graph property P is called *CMS-definable* over a class of graphs $\mathcal{G}$ iff there is a CMS-formula Φ such that for each $G \in \mathcal{G}$, G satisfies P iff Φ is true on G.

Example 2. Connectedness of a graph G is an MS-definable property:
Connected $\equiv \forall \mathbf{V}_1 \forall \mathbf{V}_2 \ (\mathbf{V}_1 \neq \emptyset \wedge \mathbf{V}_2 \neq \emptyset \wedge \mathbf{V}_1 \cup \mathbf{V}_2 = V) \Rightarrow \text{Adj}(\mathbf{V}_1, \mathbf{V}_2),$

$\text{Adj}(\mathbf{V}_1, \mathbf{V}_2) \equiv \exists \mathbf{v}_1 \, \exists \mathbf{v}_2 \;\; \mathbf{v}_1 \in \mathbf{V}_1 \wedge \mathbf{v}_2 \in \mathbf{V}_2 \wedge \text{adj}(\mathbf{v}_1, \mathbf{v}_2),$
$\text{adj}(\mathbf{v}_1, \mathbf{v}_2) \equiv \exists \mathbf{e} \;\; \text{Inc}(\mathbf{e}, \mathbf{v}_1, \mathbf{v}_2),$
where $(\mathbf{V}_i \neq \emptyset) \equiv \exists \mathbf{v} \;\; \mathbf{p}_v(\mathbf{v}) \wedge \mathbf{v} \in \mathbf{V}_i \;\; (i = 1, 2)$ and
$(\mathbf{V}_1 \cup \mathbf{V}_2 = V) \equiv \forall \mathbf{v} \;\; \mathbf{p}_v(\mathbf{v}) \Rightarrow (\mathbf{v} \in \mathbf{V}_1 \vee \mathbf{v} \in \mathbf{V}_2).$

Using $\mathbf{mod}_{0,2}$, we can express in CMS the property that a given vertex subset of a graph has even cardinality. This cannot be done in MS alone [2].

2.3 Recognizability

We define the notion of recognizability of partial k-paths in terms of deterministic finite automata $A = (\Sigma, Q, \delta, q_0, F)$ working on extended decompositions. A decomposition $\bar{B} = \langle B_1, B_1^-, \ldots, B_m, B_m^- \rangle$ is called *extended* iff dropping old vertices and adding new vertices occur separately, i.e., $B_i^- = \text{non-drop}(B_i)$, $1 \leq i \leq m$.

Example 3. Here is an extended 1-decomposition of the graph G_1: $\bar{B}(G_1) = \langle \{1, 2\}, \{2\}, \{2, 3\}, \{3\}, \{3, 4\}, \{3\}, \{3, 5\}, \{3\}, \{3, 6\}, \{\} \rangle.$

Let $G = (V, E)$ be a partial k-path with an extended k-decomposition $B = \langle B_1, \ldots, B_m \rangle$. Let $\beta : V \to \{1, \ldots, k+1\}$ be a labeling function such that any two distinct vertices in the same bag or in two consecutive bags have different labels. We call such labeling functions *admissible* by B. It is not difficult to see that $k + 1$ labels always suffice in the case of *extended* decompositions. For the labeling function β and any set of vertices $W \subseteq V$, $\beta(W) = \bigcup_{w \in W} \beta(w)$.

For B and β described above, we define the following string $\sigma_\beta(B)$ of colored undirected graphs on at most $k + 1$ vertices: $\sigma_\beta(B) = \langle \sigma_\beta(B_1), \ldots, \sigma_\beta(B_m) \rangle$, where for a bag B_i $(1 \leq i \leq m)$, $\sigma_\beta(B_i) = (V_\beta(B_i), E_\beta(B_i))$ such that $V_\beta(B_i) = \beta(B_i)$, and for every $u, u' \in B_i$, $\{\beta(u), \beta(u')\} \in E_\beta(B_i)$ iff $\{u, u'\} \in E$. Let Σ_g be the set of all colored (with colors $1, \ldots, k+1$) undirected graphs on at most $k+1$ vertices. Clearly, $|\Sigma_g|$ is bounded by a function of k.

A family $\mathcal{G}$ of partial k-paths G is called *recognizable* iff there is an automaton A with the input alphabet Σ_g such that for any G, $G \in \mathcal{G}$ iff $\sigma_\beta(B) \in L(A)$ for any extended k-decomposition B of G and any labeling function β admissible by B, and $G \notin \mathcal{G}$ iff $\sigma_\beta(B) \notin L(A)$ for any B and β as above. Here $L(A)$ denotes the language accepted by A.

3 The Case of $(k, 1)$-Paths

3.1 $(k, 1)$-Paths and k-Generative Orders

A connected partial k-path is called a $(k, 1)$-path if it allows a k-decomposition $B = \langle B_1, \ldots, B_m \rangle$ satisfying the following conditions:

1. $\text{old}(B_i) = \text{non-drop}(B_{i-1})$ for every $i \in \{2, \ldots, m\}$,
2. $\text{drop}(B_i) \neq \emptyset$ for every $i \in \{1, \ldots, m\}$,

3. $|\text{new}(B_i)| = 1$ for every $i \in \{2, \ldots, m\}$.

Here (1) says that vertices are dropped from a bag as soon as possible, (2) that each bag contains at least one drop vertex, and (3) that exactly one new vertex is added to form the next bag. Note that every k-connected partial k-path is a $(k, 1)$-path.

Example 4. The graphs G_1 and G_2 described earlier are $(k, 1)$-paths.

To show that a recognizable family $\mathcal{G}$ of $(k, 1)$-paths G is CMS-definable, it suffices to define in CMS some extended decomposition for every G and then use Büchi's result for sets of words. A decomposition of G can be defined if some linear order on V is known. Let $\leq$ be an arbitrary linear order on V, and let $\langle v_1, \ldots, v_n \rangle$ be the sequence of vertices in V ordered according to $\leq$. We define the sequence $B_\leq = \langle B_1, \ldots, B_n \rangle$, where $B_i = \{v_i\} \cup \{v_j | j < i \text{ and there is } j' \geq i \text{ s.t. } \{v_j, v_{j'}\} \in E\}$. Clearly, $B_\leq$ is a decomposition of G. For a partial k-path G, a linear order $\leq$ on V is called k-*generative* if $B_\leq$ is a k-decomposition. Conversely, from a $(k, 1)$-decomposition B of G, one can define a k-generative linear order on G by setting u to be less than v iff first$(u) <$ first(v), $u, v \in V$, and ordering the vertices in B_1 arbitrarily.

Thus, to show that recognizability implies CMS-definability for $(k, 1)$-paths, it would suffice to define in CMS a k-generative linear order for every given $(k, 1)$-path. However, there are $(k, 1)$-paths for which no linear order can be defined in CMS. Consider the family of $G_n = (\{0, 1, \ldots, n\}, E_n)$, where $E_n = \{\{0, j\} | 1 \leq j \leq n\}$. No linear orders can be CMS-defined on G_n, since these graphs have nontrivial automorphisms, and the size of G_n can be arbitrary large. So, in general, we cannot CMS-define a k-decomposition of a partial k-path.

For a partial k-path G, a partial order on V is called k-*generative* if every completion to a linear order on V is k-generative. We will describe a certain k-generative partial order, which is MS-definable over a suitably colored $(k, 1)$-path G^c. Given such a partial order, one can MS-define a *tree-decomposition* of G of a special form. Since we cannot MS-define a path-decomposition but only a tree-decomposition, we need CMS to get the formula for recognizability of G^c, using an extension of Büchi's theorem. To convert the corresponding CMS-formula into a formula for the underlying uncolored $(k, 1)$-paths G, we "guess" some coloring of G using a constant number of $\exists$ quantifiers, check in MS if it induces the required structure, and apply our CMS-formula to the colored graph.

To MS-define a k-generative partial order on a $(k, 1)$-path G with a $(k, 1)$-decomposition $B = \langle B_1, \ldots, B_m \rangle$, we convert G into the directed graph $G_B^d = (V, E^d)$ using the following algorithm. For a bag $B_r = \text{old}(B_r) \cup \text{new}(B_r)$ $(1 < r \leq m)$, where $\text{old}(B_r) = \{u_1, \ldots, u_s\}$ and $\text{new}(B_r) = \{v\}$, if $\{v, u_j\} \in E$, then $(v, u_j) \in E^d$. That is, we direct the edges from new to old vertices. To simplify the notation, we will often omit the superscript in E^d and the subscript in G_B^d.

Now we label G^d as follows. For $v \in \text{new}(B_r)$ and every $u \in \text{old}(B_r) \cap \text{drop}(B_r)$ $(1 < r \leq m)$, we color the arc $v \to u$ with some new color. This colored arc will be denoted as a double arrow $v \Rightarrow u$, and the set of them as $E_\Rightarrow$.

If $\{v\} = \text{new}(B_r) = \text{drop}(B_r)$, we color v with some new color, the same color for all such vertices; v will be denoted by having a loop arrow.

Example 5. For G_2 defined earlier, the $(k,1)$-decomposition $B(G_2)$ induces the labeled digraph G_2^d (Fig. 3).

Fig. 3. The labeled digraph G_2^d, with double arrows shown as thick single arrows.

3.2 A k-Generative Partial Order

Given the digraph G^d induced by a $(k,1)$-decomposition B of a $(k,1)$-path G, we define the following binary relation of *strong precedence*, denoted by $\stackrel{s}{\prec}$, on the set V: for any $u, v \in V$, $u \stackrel{s}{\prec} v$ iff either $(v, u) \in E$ or there is some $w \in V$ such that $(u, w) \in E$ and $(v, w) \in E_\Rightarrow$. The reflexive and transitive closure of $\stackrel{s}{\prec}$, denoted by $\preceq$, is called *precedence*. Semantically, $u \prec v$ means that $\text{first}(u) < \text{first}(v)$. We extend $\preceq$ so that for any two vertices $u \in B_1$ and $v \notin B_1$ incomparable with respect to $\preceq$, u is less than v. Let $\preceq^1$ denote the transitive closure of that extension. Obviously, $\preceq^1$ is a k-generative partial order on G.

To define the required CMS-formula for recognizability of $(k,1)$-paths, we need a certain refinement of $\preceq^1$. We color G^d so that the precedence relation $\preceq$ is completed to a linear order on the set non-drop(B_1). We do so by coloring the non-drop vertices of B_1 with colors $1, \ldots, k$ so that no two vertices are colored the same. We denote this new colored digraph by G^{d1}.

Using G^{d1} enables us to define the following k sets $P_1, \ldots, P_k$. For any $v \in V$, $v \in P_i$ $(1 \leq i \leq k)$ iff i is the minimum over the labels of the vertices $u \in \text{non-drop}(B_1)$ such that there is a path of double arrows in the digraph G^{d1} from v to u. The set N of *nodes* is defined as $N = \cup_{i=1}^k P_i$, the set L of *leaves* is defined as $L = V \setminus (N \cup B_1)$.

Example 6. The digraph G_2^d from Example 5 can be viewed as G_2^{d1} with the two sets of nodes $P_1 = \{1, 3, 6\}$ and $P_2 = \{2\}$, and the set of leaves $L = \{4, 5\}$.

Since no vertex in G^d can have more than one *incoming* double arrow, each set P_i, $1 \leq i \leq k$, induces a path of double arrows in G^{d1}. Therefore, each P_i is linearly ordered by $\preceq$. Using this fact, we can MS-define a k-generative partial order on G that is a linear order on the set of nodes N. We denote this partial order by $\preceq^n$. Note that we could MS-define a tree-decomposition of G using $\preceq^n$.

We need to order the leaves that are incomparable with respect to $\preceq^n$. By the definition of a $(k,1)$-decomposition, each leaf $w \in L$ has at most k outgoing single arrows pointing to some nodes from *different* sets $P_1, \ldots, P_k$. For a leaf $w \in L$, $P(w)$ denotes the set of nodes to which there are arrows from w, i.e., $P(w) = \{v \in N | (w,v) \in E\}$. We associate with each leaf $w \in L$ its *characteristic vector* $\chi(w) = (\chi_1(w), \ldots, \chi_k(w))$, where for each $1 \leq i \leq k$, $\chi_i(w) = 1$ if $P(w) \cap P_i \neq \emptyset$, and $\chi_i(w) = 0$ otherwise. We extend $\preceq^n$ to a new partial order on V, denoted by $\preceq^{nl}$, by ordering the leaves incomparable with respect to $\preceq^n$ lexicographically according to their characteristic vectors.

For two vertices $w_1, w_2 \in V$, we say that w_1 and w_2 are p-equivalent, denoted by $w_1 \overset{p}{\sim} w_2$, iff $w_1, w_2 \in L$ and $P(w_1) = P(w_2)$. For the quotient graph $G_p = G/\overset{p}{\sim} = (V_p, E_p)$ we extend $\preceq^{nl}$ to the set V_p in the standard way. Clearly, $\preceq^{nl}$ is a linear order on the set $(N \cup L)/\overset{p}{\sim}$. Ordering the drop vertices of B_1 arbitrarily yields a k-generative linear order on G_p, denoted by $\leq_p$. We will denote the digraph G^{d1} with ordered drop vertices of B_1 by $G^{d1'}$.

Example 7. For G_2, the $(k,1)$-decomposition of the corresponding quotient graph is $B'_p = \langle \{[1], [1'], [2]\}, \{[1], [2], [3]\}, \{[2], [3], [4]\}, \{[2], [3], [6]\} \rangle$, where $[u]$ denotes the set of vertices p-equivalent to u, $u \in V$.

3.3 A CMS-Formula

Let $B'_p = \langle B'_1, \ldots, B'_m \rangle$ be the $(k,1)$-decomposition of the graph G_p induced by $\leq_p$. We can construct a $(k,1)$-decomposition of the original graph G as follows. In the sequence B'_p, replace B'_1 with B_1. For every $i \in \{1, \ldots, m\}$, replace $B'_i = \{[u_1]_p, \ldots, [u_{s_i}]_p, [w]_p\}$, where $[w]_p$ is the new vertex of B'_i such that $[w]_p = \{w_1, \ldots, w_{t_i}\}$ ($t_i \geq 1$), with the sequence of bags $B(w_1) = \{u_1, \ldots, u_{s_i}, w_1\}, \ldots, B(w_{t_i}) = \{u_1, \ldots, u_{s_i}, w_{t_i}\}$. Let B' denote thus constructed decomposition of G.

Example 8. For G_2, two decompositions B' are possible: $\langle \{1, 1', 2\}, \{1, 2, 3\}, \{2, 3, 4\}, \{2, 3, 5\}, \{2, 3, 6\} \rangle$ or $\langle \{1, 1', 2\}, \{1, 2, 3\}, \{2, 3, 5\}, \{2, 3, 4\}, \{2, 3, 6\} \rangle$.

Let us convert B'_p into the extended decomposition $\bar{B}'_p$ and color G_p with some labeling function $\beta_p : V_p \to \{1, \ldots, k+1\}$ admissible by $\bar{B}'_p$. Let us also convert the decomposition B' of G into the extended decomposition $\bar{B}'$ and color the graph G with the labeling function $\beta : V \to \{1, \ldots, k+1\}$ such that, for every $v \in V$, $\beta(v) = \beta_p([v]_p)$. The labeling function β is admissible by $\bar{B}'$ since no leaf appears in two consecutive bags. Note that the symbols in the alphabet Σ_g that correspond to the bags $\bar{B}'(w_1)$ and $\bar{B}'(w_2)$, for any two $\overset{p}{\sim}$-leaves w_1 and w_2, are identical. Let $\sigma_{\beta_p}(\bar{B}'_p) = \langle \sigma_1, \sigma_{1'}, \ldots, \sigma_m, \sigma_{m'} \rangle$. Then $\sigma_\beta(\bar{B}')$ can be obtained from $\sigma_{\beta_p}(\bar{B}'_p)$ by repeating every subsequence $\langle \sigma_i, \sigma_{i'} \rangle$ ($2 \leq i \leq m$) $|[w]_p|$ times, where $\text{new}(B'_i) = \{[w]_p\}$. It can be shown that $\sigma_{\beta_p}(\bar{B}'_p)$ is MS-definable.

Let $A = (\Sigma_g, Q, \delta, q_0, F)$ be the automaton recognizing a family $\mathcal{G}$ of $(k,1)$-paths G. To obtain the required CMS-formula for recognizability of $\mathcal{G}$, we use an extension of Büchi's result to words that are defined as sequences of substrings given with their multiplicities (in our case, the sequences $\sigma_{\beta_p}(\bar{B}'_p)$ with the cardinalities of the corresponding p-equivalence classes). By finiteness of A, to determine the behavior of A on a substring ω repeated t times, it suffices to know $t \bmod a$ for some constant a dependent on A. Therefore, every recognizable family of colored $(k,1)$-paths $G^{d1'}$ is CMS-definable.

Let Φ be the CMS-formula checking the recognizability of suitably colored $(k,1)$-paths. We state without proof that there is an MS-formula Φ_{adm} verifying that a given coloring c of a $(k,1)$-path G is such that G is recognized by A iff Φ holds for G colored by c. Then the required CMS-formula for uncolored $(k,1)$-paths G is the following: $\exists$ "coloring c of G" $\Phi_{\mathrm{adm}}(c) \wedge \Phi(G^c)$.

Theorem 1. *Every recognizable family of $(k,1)$-paths is CMS-definable.*

4 The General Case

4.1 Nice Decompositions

In general, a partial k-path is not necessarily a $(k,1)$-path; consider the partial 2-path G_2 from Example 1 with the new edge connecting vertices 4 and 5. We generalize our definition of $(k,1)$-decomposition as follows. A decomposition $B = \langle B_1, \ldots, B_m \rangle$ of G is called *nice* iff all of the following conditions hold:

1. $\mathrm{old}(B_i) = \mathrm{non\text{-}drop}(B_{i-1})$ for every $i \in \{2, \ldots, m\}$,
2. $\mathrm{drop}(B_i) \neq \emptyset$ for every $i \in \{1, \ldots, m\}$,
3. for any $i \in \{2, \ldots, m\}$, if $|\mathrm{new}(B_i)| > 1$, then
   (a) for any $v \in \cup_{j=i}^{m} \mathrm{new}(B_j)$, each decomposition $\langle B_1, \ldots, B_{i-1}, \mathrm{old}(B_i) \cup \{v\}, C_1, \ldots, C_s \rangle$ of G is such that $\mathrm{drop}(\mathrm{old}(B_i) \cup \{v\}) = \emptyset$, and
   (b) for any subset $S \subset \mathrm{new}(B_i)$, each decomposition $\langle B_1, \ldots, B_{i-1}, \mathrm{old}(B_i) \cup S, C_1, \ldots, C_s \rangle$ of G is such that $\mathrm{drop}(\mathrm{old}(B_i) \cup S) = \emptyset$.

Here (1) and (2) are as those for $(k,1)$-decompositions, and (3) says that if more than one new vertex is added to form B_i, then both (a) there was no single non-added vertex to choose instead of the set $\mathrm{new}(B_i)$ so that B_i contained a drop vertex and (b) $\mathrm{new}(B_i)$ is a minimal set with respect to set inclusion such that B_i contains a drop vertex.

It is not difficult to show that every k-decomposition can be converted into a nice k-decomposition. We call a nice k-decomposition $B = \langle B_1, \ldots, B_m \rangle$ a (k,p)-*decomposition* for some $1 \leq p \leq k$ iff $|\mathrm{new}(B_i)| \leq p$ for all $1 < i \leq m$. A partial k-path allowing a (k,p)-decomposition will be called a (k,p)-*path*.

Let $B = \langle B_1, \ldots, B_m \rangle$ be a nice k-decomposition of a partial k-path G. The family of sets $\mathrm{new}(B_i)$ $(1 \leq i \leq m)$ forms a partitioning of the vertex-set V of G. We call the corresponding equivalence on V the *1-equivalence*, denoted by $\overset{1}{\sim}$. The decomposition B also induces a linear order on the quotient set $V/\overset{1}{\sim}$, denoted by $\leq_1$. Clearly, given the pair $(\overset{1}{\sim}, \leq_1)$, we can reconstruct the decomposition B of G. Although we can MS-define the 1-equivalence when G is suitably colored, it is impossible to MS-define $\leq_1$.

We will divide a k-decomposition of a partial k-path G into a sequence of monotonic pieces whose structure resembles that of $(k,1)$-decompositions. Formally, a contiguous subsequence $\langle B_i, \ldots, B_{i+l} \rangle$ $(1 \leq i, i+l \leq m)$ of a decomposition $B = \langle B_1, \ldots, B_m \rangle$ is called *monotonic* iff $|\text{new}(B_i)| > 1$ and $|\text{new}(B_r)| = 1$ for each $i < r \leq i+l$. The nice decomposition B can then be viewed as a sequence of monotonic pieces $\langle M_1, \ldots, M_d \rangle$, where $M_s = \langle B_{i_s}, \ldots, B_{j_s} \rangle$ for each $1 \leq s \leq d$. Note that a nice decomposition is defined so that it is monotonic as long as possible, then there is a "jump" — more than one new vertex is added to a bag — which starts a new monotonic piece, and so on.

We define the sets $\text{new}(M_s) = \bigcup_{r=i_s}^{j_s} \text{new}(B_r)$ $(1 \leq s \leq d)$ the family of which forms a partitioning of the vertex-set V of G. The corresponding equivalence on V is called *2-equivalence* and denoted by $\overset{2}{\sim}$. This sequence of monotonic pieces also induces a linear order on the quotient set $V/\overset{2}{\sim}$, denoted by $\leq_2$. Some k-decomposition of G (possibly different from B) can be constructed given $\overset{1}{\sim}, \overset{2}{\sim}$, and $\leq_2$. Again, we can MS-define the 2-equivalence on a suitably colored graph, but not $\leq_2$.

4.2 k-Generative Structures

For a partial k-path G, a triple $(\overset{1'}{\sim}, \overset{2'}{\sim}, \leq'_2)$, where $\overset{1'}{\sim}$ and $\overset{2'}{\sim}$ are equivalences on V and $\leq'_2$ is a linear order on $V/\overset{2'}{\sim}$, is called a *linear k-generative structure on G* iff there exists some nice k-decomposition B of G such that $\overset{1'}{\sim}$ and $\overset{2'}{\sim}$ are the 1-equivalence and 2-equivalence, respectively, induced by B, and $\leq'_2$ is the linear order on 2-equivalence classes induced by B. For a partial k-path G, a triple $(\overset{1'}{\sim}, \overset{2'}{\sim}, \preceq'_2)$, where $\overset{1'}{\sim}$ and $\overset{2'}{\sim}$ are equivalences on V and $\preceq'_2$ is a partial order on $V/\overset{2'}{\sim}$, is called a *partial k-generative structure on G* iff any completion of $\preceq'_2$ to a linear order yields a linear k-generative structure on G.

Let $\overset{1}{\sim}$ and $\overset{2}{\sim}$ be the 1-equivalence and 2-equivalence, respectively, induced by some nice k-decomposition of a partial k-path G. Let $\preceq$ be the precedence relation defined similarly to the case of $(k,1)$-paths, and let $\overset{2}{\preceq}$ be the extension of $\preceq$ to the quotient set $V/\overset{2}{\sim}$ in the standard way. The triple $(\overset{1}{\sim}, \overset{2}{\sim}, \overset{2}{\preceq})$ is not necessarily a partial k-generative structure on G. One reason is that each $\overset{2}{\sim}$-class $[u]_{\overset{2}{\sim}}$ ($u \in V$) contains several vertices all of which must be put in the same bag. The other reason is that $[u]_{\overset{2}{\sim}}$ can "contribute" more non-drop vertices than drop vertices. We did not have the latter problem in the case of $(k,1)$-paths, because there adding a new vertex always produced at least one drop vertex.

To get around these problems, we put consecutive monotonic pieces of the k-decomposition B of G into sequences of minimal length such that the number of non-drop vertices produced by each sequence, except the first one, is at most that of drop vertices. More formally, let $\mu = \langle M_s, \ldots, M_t \rangle$ be a contiguous subsequence of a nice k-decomposition B that corresponds to the sequence of bags $\langle B_{i_s}, \ldots, B_{j_t} \rangle$. We define the *balance* of μ, $\text{bal}(\mu)$, as $\text{bal}(\mu) =$

$|\text{non-drop}(B_{j_t})| - |\text{old}(B_{i_s})|$. A contiguous subsequence μ of monotonic pieces is called *balanced* if $\text{bal}(\mu) \leq 0$ and no proper non-empty prefix of μ is of non-positive balance.

Let $B = \langle M_1, \ldots, M_d \rangle$, where M_s, $1 \leq s \leq d$, is a monotonic piece. We divide B into disjoint subsequences of monotonic pieces $\mu_1, \ldots, \mu_r$ such that $B = \mu_1 \ldots \mu_r$, $\mu_1 = \langle M_1 \rangle$, and each μ_i, $2 \leq i \leq r$, is balanced. It can be shown that every μ_i, $2 \leq i \leq r$, corresponds to a $(k, k-1)$-subdecomposition of G. The sets $\text{new}(\mu_i)$, $1 \leq i \leq r$, defined in an obvious way induce a partitioning of V. The corresponding equivalence is called 3_1-equivalence and is denoted by $\overset{3_1}{\sim}$. Recursively, we partition each μ_i, $1 \leq i \leq r$, into $\mu_1^i, \ldots, \mu_s^i$ and define 3_2-equivalence classes. Each μ_j^i, $2 \leq j \leq s$, corresponds to a $(k, k-2)$-subdecomposition of G. We stop after k steps when every (not necessarily balanced) sequence μ consists of a single monotonic piece and corresponds to a $(k, 1)$-subdecomposition of G; also note that 3_k-equivalence coincides with 2-equivalence.

Then we define partial orders on these 3_i-equivalence classes, denoted by $\overset{3_i}{\preceq}$, $1 \leq i \leq k$, satisfying the following condition: for any completions of $\overset{3_i}{\preceq}$ to linear orders $\leq^i$, $1 \leq i \leq k$, such that $\leq^j$ is a refinement of $\leq^i$ for every $j > i$ (i.e., the restriction of $\leq^j$ to $V/\overset{3_i}{\sim}$ coincides with $\leq^i$), the triple $(\overset{1}{\sim}, \overset{2}{\sim}, \leq^k)$ is a linear k-generative structure on G. These partial orders as well as 3_i-equivalences can be MS-defined for suitably colored connected partial k-paths thanks to the properties of nice decompositions.

4.3 Defining a CMS-Formula

We partition our set of 3_i-equivalence classes into the sets of 3_i-nodes and 3_i-leaves, $1 \leq i \leq k$. Then we refine each partial order $\overset{3_i}{\preceq}$, $1 \leq i \leq k$, to a linear order on the set of 3_i-nodes within each 3_{i-1}-equivalence class; every two vertices of G are 3_0-equivalent. However, we cannot order leaves in the same way as we did in the case of $(k, 1)$-paths, because now they are not necessarily single vertices but instead correspond to sequences of bags, and hence to words over Σ_g.

Let $A = (\Sigma_g, Q, \delta, q_0, F)$ be an automaton recognizing our family of partial k-paths. We call two incomparable 3_i-leaves within the same 3_{i-1}-equivalence class, $1 \leq i \leq k$, δ_i-*equivalent* if the corresponding words ω_1 and ω_2 over Σ_g are such that for each $q \in Q$, $\delta^*(q, \omega_1) = \delta^*(q, \omega_2)$, where δ^* is the extended transition function of A. To determine if two leaves are δ_i-equivalent, we need to know the behavior of A on the sequences of bags corresponding to those leaves.

The above discussion suggests the following "bottom-up" procedure which can be encoded in CMS. We define the sequence of bags corresponding to each 3_k-equivalence class as in the case of $(k, 1)$-paths, since each 3_k-equivalence class is the set of new vertices of a *monotonic* piece. Then we convert this sequence into the word ω over Σ_g and compute the behavior of A on ω. This behavior is a map from Q to Q, which can be presented as a *state-vector* $q(\omega)$ of length $|Q|$. For each 3_{k-1}-equivalence class C, two 3_k-leaves C' and C'' in $C/\overset{3_k}{\sim}$ are δ_k-equivalent iff $q(C') = q(C'')$. We extend the partial order on the set $C/\overset{3_k}{\sim}$ to a linear order on $C_\delta = (C/\overset{3_k}{\sim})/\overset{\delta_k}{\sim}$ by ordering incomparable leaves lexicographically according to their state-vectors. Let $\langle C_1, \ldots, C_s \rangle$ be thus ordered sequence of elements of C_δ. The behavior of A on C is defined as $q(C) = q(C_1)^{t_1} \circ \cdots \circ q(C_s)^{t_s}$, where $t_i = |C_i|$, $1 \leq i \leq s$, and $\circ$ is the composition. By finiteness of Q, $q(C)$ can be defined in CMS. Continuing in this manner will give us, after k steps, the vector $q(G)$ describing the behavior of A on the entire k-decomposition of G. The graph G is recognized by A iff $q(G)$ maps q_0 to some final state of A.

Thus, we can define a CMS-formula for recognizability of suitably colored connected partial k-paths. As in the case of $(k, 1)$-paths, there is an MS-formula Φ'_{adm} so that recognizability implies CMS-definability for connected partial k-paths. Note that the formula $\exists$ "coloring c of G" $\Phi'_{\text{adm}}(c)$ is true on G iff G is a partial k-path, so the obstruction set of the class of partial k-paths is computable.

For a disconnected partial k-path G, we compute the state-vectors for its connected components, order these vectors lexicographically, and compute their composition in CMS. Together with Courcelle's result this yields our main claim.

Theorem 2. *Recognizability equals definability for partial k-paths.*

Acknowledgements. I am indebted to my supervisor Arvind Gupta at SFU for suggesting this topic and for his encouragement and support. I want to thank David Mould for his assistance in preparing this paper. I am also grateful to the anonymous referees for their comments.

References

1. J. Büchi. Weak second-order arithmetic and finite automata. *Zeitschr. j. math. Logik und Grundlagen d. Math.*, 6:66–92, 1960.
2. B. Courcelle. The monadic second-order logic of graphs. I. Recognizable sets of finite graphs. *Information and Computation*, 85:12–75, 1990.
3. B. Courcelle. The monadic second-order logic of graphs. V. On closing the gap between definability and recognizability. *Theoret. Comput. Sci.*, 80:153–202, 1991.
4. B. Courcelle. The monadic second-order logic of graphs. III. Tree-decompositions, minors and complexity issues. *Informatique théorique et Appl.*, 26:257–286, 1992.
5. B. Courcelle. The monadic second-order logic of graphs. VI. On several representations of graphs by relational structures. *Discr. Appl. Math.*, 54:117–149, 1994.
6. B. Courcelle. The monadic second-order logic of graphs. VIII. Orientations. *Ann. Pure Appl. Logic*, 72:103–143, 1995.
7. J. Doner. Tree acceptors and some of their applications. *J. Computer and System Sciences*, 4:406–451, 1970.
8. V. Kabanets. Recognizability equals definability for partial k-paths. Master's thesis, Simon Fraser University, June 1996.
9. D. Kaller. Definability equals recognizability of partial 3-trees, 1996. Workshop on Graph-Theoretic Concepts in Computer Science (WG '96).
10. N. Robertson and P. Seymour. Graph minors. II. Algorithmic aspects of tree-width. *J. Algorithms*, 7:309–322, 1986.
11. D. Seese. The structure of the models of decidable monadic theories of graphs. *Ann. Pure Appl. Logic*, 53:169–195, 1991.

Molecular Computing, Bounded Nondeterminism, and Efficient Recursion

Richard Beigel[1]* and Bin Fu[2]**

[1] Yale University, University of Maryland, and Lehigh University
[2] Yale University and University of Maryland

Abstract. The maximum number of strands used is an important measure of a molecular algorithm's complexity. This measure is also called the *space* used by the algorithm. We show that every NP problem that can be solved with $b(n)$ bits of nondeterminism can be solved by molecular computation in a polynomial number of steps, with four test tubes, in space $2^{b(n)}$. In addition, we identify a large class of recursive algorithms that can be implemented using bounded nondeterminism. This yields improved molecular algorithms for important problems like 3-SAT, independent set, and 3-colorability.

1. A model of molecular computing

Molecular computation was first studied in [1, 17]. The models we define were inspired as well by the work of [3, 23]. A molecular sequence is a string over an alphabet Σ (we can use any alphabet we like, encoding characters of Σ by finite sequences of base pairs). A test tube is a multi-set of molecular sequences. We describe the allowable operations below. Where set notation is applied to multi-sets, multiplicities are respected. In the definitions T_1, T_2, and T_3 denote distinct test tubes, c denotes a character, and i denotes a positive integer.

Separate(T_1, c, i, T_2, T_3)
 $T_2 :=$ the multi-set of all strings in T_1 whose ith character is c;
 $T_3 :=$ the multi-set of all strings in T_1 whose ith character is not c;
 $T_1 := \emptyset$.
Pour(T_1, T_2)
 $T_2 := T_1$;
 $T_1 := \emptyset$.
Append(T, c)
 $T := \{xc : x \in T\}$.

---

* Address: Dept. of Computer Science, University of Maryland at College Park, College Park, MD 20742-3251, USA. Research supported in part by the National Science Foundation under grants CCR-8958528 and CCR-9415410 and by NASA under grant NAG 52895. On sabbatical from Yale University. Email: beigel@cs.umd.edu
** Address: Dept. of Computer Science, P.O. Box 208285, New Haven, CT 06520-8285, USA. Research supported in part by the National Science Foundation under grants CCR-8958528 and CCR-9415410. Email: fu-bin@cs.yale.edu

Merge(T_1, T_2, T_3)
$\quad T_3 := T_1 \cup T_2;$
$\quad T_1 := \emptyset;$
$\quad T_2 := \emptyset.$

Others have proposed a variant of operation separate, which we will call Sep. It checks whether a string contains the character c anywhere. If we represent the ith symbol z_i of a string z by the symbol $\langle i, z_i \rangle$ instead, then the standard Sep operation can simulate our Separate operation with no additional overhead. The use of polynomial-size alphabets is standard practice in molecular computing. We prefer the Separate operation for convenience in programming.

The running time for a molecular algorithm is proportional to the number of operations on test tubes. An important complexity measure is the *solution space size* (also called simply *space*), i.e., the maximum number of strings in all test tubes at any time, counting multiplicities. Adleman [2] has speculated that molecular computation with a solution space of size 2^{70} (about 0.002 moles) might be possible. Recent papers [3, 19] attempt to optimize solution space size for particular combinatorial problems.

Problem instances are associated with a parameter n called their size. In complexity theory, n is the length of a suitable encoding of the instance. However, in analysis of algorithms, n is usually a more natural representation-independent parameter, such as the number of vertices in a graph or number of variables in a formula. Although the n's of complexity theory and the n's of analysis of algorithms are usually polynomially related, it can make a phenomenal difference when n appears in the exponent. For that reason we take n to be a problem-dependent but representation-independent notion of size through this paper. We write $|x|$ to denote the *size* of a problem instance x rather than its length, and we usually identify n with $|x|$.

We consider a highly restricted model of $t(n)$-time, $s(n)$-space molecular computation, which we think has a good chance of eventually being practical. On input x, one test tube T_0 is initialized to hold encodings of the numbers $1, \ldots, s(|x|)$. A sequence of molecular operations $o_1, \ldots, o_{t(|x|)} = f(x)$ is then performed, where f is a conventional polynomial-time computable function (that is, the program is uniform in a weak but appropriate sense). The computation accepts if T_0 is nonempty after the last operation is performed. MOL$(s(n))$ is the class of languages accepted by such a computation where the running time $t(n)$ is polynomial bounded.

We give the most space-efficient molecular algorithms known for several problems. See Table 1.

2. Bounded Nondeterminism

NP computation with a limited amount of nondeterminism was introduced in [14, 15, 16] and studied further in [10, 11, 20, 9, 12, 25, 13, 7]. The class NPbits$(b(n))$ consists of all languages recognized by an NP machine that make at most $b(n)$ binary nondeterministic choices on each computation path on inputs of size n. (Actually, prior treatments allowed $O(b(n))$ binary choices, but

Problem	Previously Space	Previously Limited Model	Reference	In This Paper Space	In This Paper Limited Model
Hamiltonian Path	$n!$	✓	[1]		
SAT	2^n	✓	[17]		
QBF	2^n	××	[23]		
3-SAT	1.62^n	×	[19]	1.50^n	✓
3-Colorability	1.89^n	✓	[3]	1.35^n	✓
Independent Set	1.51^n	✓	[3]	1.23^n	✓
$(3,2)$-system				1.39^n	✓

Table 1. Results for particular problems

the constant factor turns out to be very important in connection with molecular computation.) We define a refinement of these classes: $\mathrm{NPinit}(s(n))$ consists of all languages recognized by NP machines that nondeterministically choose a number between 1 and $s(n)$ on inputs of size n and then behave deterministically. Clearly, $\mathrm{NPbits}(b(n)) = \mathrm{NPinit}(2^{b(n)})$.

3. $\mathrm{NPinit}(s(n)) \subseteq \mathrm{MOL}(s(n))$

In this section we show how to simulate bounded nondeterministic computation via bounded-space molecular computation. Results of this type appear in [4, 23, 24, 29], but they assume models of molecular computation with more powerful operations, such as Amplify, that may be harder to implement in practice. Independently, Boneh et al. [8] obtained a result similar to ours.

Lemma 1. *Let π be a circuit with m gates. Given a tube T_0, a molecular algorithm using only the operations Pour, Append, and Merge, running in time $O(m)$, and using only four test tubes can create tubes T_1 and T_2 such that T_1 contains all strings z from tube T_0 that satisfy $\pi(z) = 1$ and T_2 contains all strings z from tube T_0 that satisfy $\pi(z) = 0$.*

Proof. Let π's input gates be $g_1, \ldots, g_n$ and internal gates be $g_{n+1}, \ldots, g_m$ in topological order; in particular g_m is the output gate. We will use four tubes T_0, T_1, T_2, T_3. For each i, let g_i compute $f_i(g_{j(i)}, g_{k(i)})$ where $j(i) < i$, $k(i) < i$, and f_i is a binary function. We perform the following algorithm:

    for $i := n+1$ to m do
        Separate$(T_0, 0, j(i), T_1, T_2)$
        Separate$(T_1, 0, k(i), T_0, T_3)$
        Append$(T_0, f_i(0,0))$
        Append$(T_3, f_i(0,1))$
        Merge(T_0, T_3, T_1)
        Separate$(T_2, 0, k(i), T_0, T_3)$
        Append$(T_0, f_i(1,0))$

      Append($T_3, f_i(1,1)$)
      Merge(T_0, T_3, T_2)
      Merge(T_1, T_2, T_0)
Separate($T_0, 0, m, T_1, T_2$)

At completion, T_1 contains all strings that satisfy π and T_2 contains all strings that do not satisfy π. ∎

Theorem 2. NPinit($s(n)$) $\subseteq$ MOL($s(n)$).

Proof. Let L be accepted by an NPinit($s(n)$) machine M. Construct a deterministic machine M' that takes as inputs a string x and a positive integer $z \leq s(n)$ and accepts iff M accepts input x with nondeterministic guess z. Obtain M'_x by fixing the input x, so the only input to M'_x is the number z. Construct a circuit π equivalent to M'_x in the usual way (see [21]). Apply Lemma 1 to π to see that L is in MOL($s(n)$). ∎

4. Implementing Recursion with Bounded Nondeterminism

In this section we show how to enumerate search spaces using bounded nondeterminism. In many nondeterministic searches, some paths are longer than others, which can be inefficient. However, if we can compute the size of subtrees, then we can balance nondeterministic search trees, which reduces the amount of nondeterminism needed.

Recursive algorithms for NP problems usually take the form of d-self-reductions ("d" for disjunctive). Self-reductions were defined in [27] and d-self-reductions were defined in [28].

Definition 3. Let $|y|$ denote the size of the problem instance y. A partial order $\prec$ is polynomial well-founded if there exists a polynomial-bounded function p such that

- $y_m \prec \cdots \prec y_1 \Rightarrow m \leq p(|y_1|)$
- $y_m \prec \cdots \prec y_1 \Rightarrow |y_m| \leq p(|y_1|)$

For technical simplicity we will consider only languages L containing the emptystring, Λ.

Definition 4. A d-self-reduction for a language L consists of a polynomial time computable function $h(x) = \{x_1, \ldots, x_m\}$ and a polynomial-well-founded partial order $\prec$ on problem instances such that

- Λ is the only minimal element under $\prec$
- for all $x \neq \Lambda$, $x \in L \iff h(x) \cap L \neq \emptyset$
- for all x, $x_i \in h(x) \Rightarrow x_i \prec x$

Definition 5. Let $\langle h, \prec \rangle$ be a d-self-reduction and let x be a problem instance.

- $T_{h,\prec}(x)$ is the unordered rooted tree that satisfies the following rules: (1) the root is x; (2) for each y, the set of children of y is $h(y)$.
- $|T_{h,\prec}(x)|$ is the number of leaves in $T_{h,\prec}(x)$.

If $\langle h, \prec \rangle$ is a self-reduction for L, then the corresponding recursive algorithm for L runs in time $|x|^{O(1)} |T_{h,\prec}(x)|$. The analysis of such an algorithm usually provides a bound on $|T_{h,\prec}(x)|$ that is suitable for use in constructing a molecular algorithm for L. We formalize this below:

Definition 6. Let T be a polynomial-time computable function. A language L is in REC($T(x)$) if there is a d-self-reduction $\langle h, \prec \rangle$ for L such that for all x

(1) $|T_{h,\prec}(x)| \leq T(x)$, and
(2) $T(x) \geq \sum_{x_i \in h(x)} T(x_i)$.

Lest conditions (1) and (2) above seem restrictive, we argue that they are quite natural. We consider the typical analysis of a recursive algorithm. One introduces a function T and proves by induction on $|x|$ that $|T_{h,\prec}(x)| \leq T(x)$, which is (1). The inductive hypothesis is that $|T_{h,\prec}(w)| \leq T(w)$ if $|w| < |x|$. Inspection of the algorithm yields

$$|T_{h,\prec}(x)| = \sum_{x_i \in h(x)} T_{h,\prec}(x_i)$$
$$\leq \sum_{x_i \in h(x)} T(x_i) \quad \text{by the inductive hypothesis}$$

The last step in the induction consists of showing that T satisfies $\sum_{x_i \in h(x)} T(x_i) \leq T(x)$, which is (2). The only other requirement on T is that T be polynomial-time computable. We will deal with that later in this section.

The function T above depends on problem instances rather than their size because the analysis of the algorithm may depend on two or more parameters. We will need an analogous variant of NPinit().

Definition 7. NPinit'($S(x)$) consists of languages recognized by NP machines that nondeterministically choose a number between 1 and $S(x)$ on input x and then behave deterministically.

Clearly, if $S(x) \leq s(|x|)$ then NPinit'($S(x)$) $\subseteq$ NPinit($s(n)$).

Theorem 8. REC($T(x)$) $\subseteq$ NPinit'($T(x)$).

Proof. Let $L \in$ REC($T(x)$) via $\langle h, \prec \rangle$. We will define a deterministic polynomial-time computable function path(i, x) taking values in $\{0, 1, \Lambda\}$ such that path($1, x$) $\cdots$ path($T(x), x$) is equal to the sequence of values at the leaves of $T_{h,\prec}(x)$ in canonical order. The proof is completed by having the ith path of an NPinit'($T(x)$) machine compute path(i, x); clearly that machine accepts L. The function path(i, x) will be computed via tail recursion.

```
function path(i, x)
if x = Λ then return true
else if h(x) = ∅ then return false
else
    {x_1, ..., x_m} := h(x)
    for j := 1 to m do
        if i ≤ T(x_j) then return path(i, x_j)
        else i := i - T(x_j)
    return Λ
```
∎

Now we give sufficient conditions for T to be polynomial-time computable.

Definition 9. We say that a partial order $\prec$ on problem instances is *parameterizable* if there are a function m from problem instances to a set M, a partial order $\prec'$ on M, and a polynomial p such that

- $m(x)$ is computable in time polynomial in $|x|$, and
- $x \prec y \Rightarrow m(x) \prec' m(y)$, and
- $||\{i : i \prec' m(x)\}|| \leq p(|x|)$.

In many examples we will take $m(x) = |x|$ and $\prec'$ to be the standard linear order on natural numbers. In other examples, $m(x)$ will be a tuple of parameters (such as the number of 2-clauses and the number of 3-clauses in a Boolean formula); in many (but not all) of these examples we use the partial order $(a_1, \ldots, a_k) \prec' (b_1, \ldots, b_k)$ if $(\forall i)[a_i \leq b_i]$ and $(\exists i)[a_i < b_i]$.

Definition 10. Given h and m, define

- $\text{mh}(x) = $ the multi-set $\{m(x_i) : x_i \in h(x)\}$
- $\text{MH}(x) = $ the set $\{\text{mh}(y) : m(y) = m(x)\}$

Definition 11. A d-self-reduction $\langle h, \prec \rangle$ is *by cases* if $\prec$ is parameterizable via $\langle m, \prec' \rangle$ in such a way that $\text{MH}(x)$ is computable in time polynomial in $|x|$.

Lemma 12. *Let $\langle h, \prec \rangle$ be a d-self-reduction by cases with parameter function $m()$. Let T_0 be the least function T such that*

(1) $|T_{h,\prec}(x)| \leq T(x)$
(2) $T(x) \geq \sum_{x_i \in h(x)} T(x_i)$
(3) $T(x)$ is a function of $m(x)$

Then T_0 exists and $T_0(x)$ is computable in time polynomial in $|x|$.

Proof. Let $\langle h, \prec \rangle$ have a parameterization $\langle m, \prec' \rangle$, where $m(x)$ and $\text{MH}(x)$ are computable in time polynomial in $|x|$. Define a partial function t from M to natural numbers recursively:

$$t(\mu) = \begin{cases} 1 & \text{if } \mu \text{ is a minimal element under } \prec' \\ \max_{m(y)=\mu} \sum_{y_i \in h(y)} t(m(y_i)) & \text{otherwise} \end{cases}$$

If μ is in the range of m, then $t(\mu)$ is defined because $\text{MH}(x)$ is a finite set for every x. Now it is easy to see that $t \circ m$ is the least function satisfying (1,2,3).

By Definition 9, $|\{i : i \prec' m(x)\}| \leq p(|x|)$. If we compute $t(m(x))$ by the obvious recursion, at most $p(|x|)$ different subproblems will arise. If we use a table to avoid recomputation, the recursion will run in polynomial time. ∎

4.1. 3-SAT

In this section, we apply our results to the classic 3-SAT algorithm of Monien and Speckenmeyer [18] and a recent unverified 3-SAT algorithm of Schiermeyer [26]. The former yields a simple $\text{MOL}(1.62^n)$ algorithm, and the latter (assuming that Schiermeyer's paper is correct), yields a $\text{MOL}(1.497^n)$ algorithm.

Monien and Speckenmeyer's Algorithm The size of a satisfiability instance is the number of variables. Consider the 3-SAT algorithm of Monien and Speckenmeyer. Let $f|_\ell$ denote the formulas obtained by replacing in f the literal ℓ by true and $\overline{\ell}$ by false. A k-clause is a disjunction of k literals. The function 3SAT takes a formula f consisting of some 3-clauses and at least one 1-clause or 2-clause.

function 3SAT(f)
if f is the empty set of clauses then return true
else if f contains an empty clause then return false
else if some variable v appears only in positive literals then return $3SAT(f|_v)$
else if some variable v appears only in negative literals then return $3SAT(f|_{\overline{v}})$
else if f contains a clause C consisting of a single literal ℓ then return $3SAT(f|_\ell)$
else if f contains a clause C consisting of two literals ℓ_1, ℓ_2 then
    return $3SAT(f|_{\ell_1}) \vee 3SAT(f|_{\overline{\ell_1}}|_{\ell_2})$
else
    let v be the first variable to appear in f
    return $3SAT(f|_v) \vee 3SAT(f|_{\overline{v}})$

The last case in the recursion is ostensibly the worst, yielding two subproblems of size $n-1$, but it only occurs on the first call or immediately after eliminating a single variable, which yields a single subproblem of size $n-1$; unrolling the recursion, we see that the last case gives two subproblems of size $n-2$. The worst case is the second to the last, which yields subproblems of size $n-1$ and $n-2$. Thus the number of leaves in the self-reduction is at most $2f(n)$ where $f(n)$ is given by the recurrence $f(n) = f(n-1) + f(n-2)$; in particular $2f(n) \leq 1.62^n$ for almost all n.

The algorithm above is clearly a d-self-reduction for 3-SAT. The value function h for a formula is the set of subformulas generated by the recursive algorithm. Let $m(x) = n$, where n is the number of variables in the formula x. $\leq$ is the normal order for the integers. From the analysis above we know $\text{mh}(x)$ is either $\{n-1\}, \{n-2, n-2\}$ or $\{n-2, n-1\}$. $\text{MH}(x)$ is $\{\{n-1\}, \{n-2, n-2\}, \{n-2, n-1\}\}$ that is clearly polynomial time computable. Let $t(n) = 1.62^n$. $2f(n) \leq t(n)$. Hence $t(n)$ is an upper bound of the number of leaves of computation tree for the recursive algorithm. It is easy to see that

$t(n) \geq t(n-2) + t(n-1) \geq t(n-2) + t(n-2)$. Hence, $T_{3S}(x) = t(m(x))$ satisfies the conditions of Lemma 12 and 3-SAT is in REC($T_0(F)$) for some $T_0 \leq T_{3S}$. By Theorem 8 and Theorem 2, 3SAT $\in$ NPinit($t_{3S}(n)$), so 3-SAT is in MOL(1.62^n). The same space bound for 3-SAT was obtained previously by Ogihara [19], but in a model that allows more powerful operations like Polymerization, which can implement the Amplify operation.

Schiermeyer's Algorithm Schiermeyer [26] reports a 1.497^n time algorithm for 3-SAT problem. His algorithm is a d-self-reduction for the 3-SAT problem. We will prove that 3SAT $\in$ REC($T(F)$), where the function $T(F) \leq 1.497^n$ and will be defined below. We follow [26] to define F_3 and F'_3. For a formula F with n variables, let p be the maximum number of 1-clauses and 2-clauses (with preference of 1-clauses) such that no variable occurs more than twice. Let q be the number of remaining 2-clauses and define $m = p + \min(2, q)$. Let $F_3(n) = c\beta^n \cdot \frac{\beta}{\alpha^4}$ and $F'_3(n) = c\beta\alpha^{-m}$, where $\beta = 1.4963$, $\alpha = 1.04855$ and c is a sufficiently large constant.

$$T(F) = \begin{cases} F_3(n) \text{ if } F \text{ has no 1-clauses or 2-clauses} \\ F'_3(n) \text{ otherwise} \end{cases}$$

Schiermeyer states that $F'_3(n) \geq |T_{h,\prec}(F)|$ if F has at least one 1-clause or 2 clause, and that $F_3(n) \geq |T_{h,\prec}(F)|$ for all F. Hence, $T(F) \geq |T_{h,\prec}|$, since $|T_{h,\prec}| \leq$ the number of recursive calls. The inequalities that Schiermeyer gives in the proofs of his Lemma 4.3 and Lemma 4.4 imply that our $T(F)$ satisfies the conditions of Definition 6. Hence, 3SAT $\in$ REC($T(F)$) $\subseteq$ MOL(1.497^n).

4.2. 3-Coloring and (3, 2)-System

Beigel and Eppstein [6] give algorithms for $(3, 2)$-system and 3-coloring. In the (a, b)-system problem, we are given a collection of n vertices, each of which can be given one of a different colors. However certain color combinations are disallowed: we are also given a set of constraints, each of which forbids one coloring of some b-tuple of variables. $(3, 2)$-system generalizes 3-coloring, 3-SAT and 3-edge-coloring.

(3, 2)-System Algorithm The size of a $(3, 2)$ system is the number of variables in it. Beigel and Eppstein's [6] $(3, 2)$-system algorithm can be sketched as follows:

> function 32SYS(F)
> if $|F| \leq 5$ then return brute-force(F)
> else
> $\langle F_1, \ldots, F_k \rangle = h_{32}(F)$
>   return $\bigvee_{i=1}^{m}$ 32SYS(F_i)

In the algorithm above, brute-force(F) means "use the brute force method to solve the $(3, 2)$-system F;" $k \leq 3$; h_{32} is polynomial-time computable; and $|F_i| < |F|$. Let $h = h_{32}$ and let $\prec$ be the standard linear ordering on the natural numbers. Then $\langle h, \prec \rangle$ is a d-self-reduction for $(3, 2)$-system. Define $m(F) = |F|$. In case 1, mh(F) = $\{n - (4 + i), n - 1\}$, where $i \geq 0$.

In cases 2a, 2c, and 3, $\mathrm{mh}(F) = \{n - (3+i), n-2\}$, where $i \geq 0$.
In cases 2b, 2d, 6, 8c and 9, $\mathrm{mh}(F) = \{n'\}$, where $n' < n$.
In case 4, $\mathrm{mh}(F) = \{n-5, n-3, n-3\}$.
In case 5, $\mathrm{mh}(F) = \{n-4, n-4\}$.
In case 7, $\mathrm{mh}(F) = \{n-3, n-3\}$.

$\mathrm{MH}(F)$ is polynomial-time computable by the case analysis above. Let $t(n) = 1.38028^n$. It is easy to see that for every input x with $\{n_1, \ldots, n_k\} = \mathrm{mh}(x)$ and $m(F) = n$, $t(n) \geq t(n_1) + \cdots + t(n_k)$. Define $T(F) = t(m(F))$. By Lemma 12, there is a polynomial-time computable function T_0 such that $T_0(F) \leq T(F)$ and T_0, h, and $\prec$ satisfy the conditions of Definition 6. Thus, $(3,2)$-system is in $\mathrm{REC}(T_0(F))$. So,

$$
\begin{aligned}
(3,2)\text{-system} &\in \mathrm{NPinit}'(T_0(F)) && \text{by Theorem 8} \\
&\subseteq \mathrm{NPinit}'(T(F)) && \text{because } T_0(F) \leq T(F) \\
&= \mathrm{NPinit}(t(n)) && \text{because } T(F) = t(|F|) \\
&= \mathrm{NPinit}(1.38028^n) && \text{because } t(n) = 138028^n \\
&\subseteq \mathrm{MOL}(1.38028^n) && \text{by Theorem 2.}
\end{aligned}
$$

3-Coloring Algorithm There are two parts to Beigel and Eppstein's algorithm. The first part runs in polynomial-time and finds an independent set S with a lot of neighbors. Let $\Gamma(S)$ denote the set of vertices in G that are not in S but are adjacent to an element of S. The second part 3-colors S in all possible ways. Each of these $3^{|S|}$ partially-colored graphs is transformed into an equivalent $(3,2)$-system with $n - |S| - |\Gamma(S)|$ variables, which is solved by calling 32SYS. Their algorithm runs in time $3^{|S|} 1.38028^{n-|S|-|\Gamma(S)|}$, which is less than 1.345^n for sufficiently large n. Thus we have the following $\mathrm{NPinit}(1.345^n)$ algorithm:

> choose a natural number $m < 1.345^n$
> construct Beigel and Eppstein's set S
> let $c = m \bmod 3^{|S|}$
> color S with the cth 3-coloring in the lexicographical ordering
> form the corresponding $(3,2)$-system F
> let $b = \lfloor m/3^{|S|} \rfloor$
> run 32SYS(F) using the nondeterministic choices dictated by b

Therefore 3-coloring is in $\mathrm{MOL}(1.345^n)$.

4.3. Independent Set

For a graph G, an independent set S is a subset of G's nodes such that there is no edge between any two nodes in S. The independent set problem is "given a graph G and a number k, does G contain an independent set of cardinality at lest k?"

Tarjan's Algorithm Consider the following simple algorithm due to Tarjan [30]. ($d(v)$) denotes the degree of v, and $N(v)$ denotes the neighbor set of v. $\max(S,T)$ denotes the larger of the two sets S and T, with ties resolved arbitrarily.)

>   function MIS(G)
>   pick any vertex v in G
>   if $d(v) \leq 1$ then return $\{v\} \cup$ MIS($G - v - N(v)$)
>   else return $\max(\text{MIS}(G - v), \{v\} \cup \text{MIS}(G - v - N(v)))$

This is a self-reduction with at most $T(n)$ leaves where $T(n)$ satisfies $T(n) = T(n-1) + T(n-3)$ where $T(n)$. The recurrence can be solved in polynomial time by an explicit formula or by dynamic programming so the independent set problem is in MOL(1.47^n), which is better than prior results [3]. Because the algorithm is particularly simple, the molecular algorithm can even be made to run in linear time.

Robson's Algorithm The best published purely recursive algorithm for the independent set problem is due to Robson [22] and runs in time 1.229^n for sufficiently large n. A d-self-reduction with 1.229^n leaves is evident from Robson's paper, so we have we have a MOL(1.229^n) algorithm for the independent set problem. Details will be given in the full version of this paper.

Robson has a faster dynamic programming algorithm for independent set, but we see no way to adapt it to molecular computing. Molecular computing may motivate the search for efficient recursive algorithms that do not use dynamic programming. Towards that end we have found a recursive 1.223^n time (for sufficiently large n) algorithm for independent set [5] that is based on a d-self-reduction and hence is directly adaptable to molecular computing.

5. Acknowledgments

We are grateful to William Gasarch for his patience in reading this paper as well as his suggestions in improving the presentation. We are also grateful to Tirza Hirst for helpful discussions and to Ingo Schiermeyer for sharing with us a preliminary draft of [26].

References

1. L. Adleman. Molecular computation of solutions to combinatorial problems. *Science*, 266:1021–1024, Nov. 1994.
2. L. Adleman. On constructing a molecular computer. In *1st DIMACS workshop on DNA Computing*, 1995.
3. E. Bach, A. Condon, E. Glaser, and C. Tanguay. DNA models and algorithms for NP-complete problems. In *Proc. 11th Ann. Conf. Structure in Complexity Theory*, pp. 290–299, 1996.
4. D. Beaver. A universal molecular computer. CSE 95-001, Penn. State Univ., 1995.
5. R. Beigel. Maximum independent set algorithms. Manuscript, 1996.
6. R. Beigel and D. Eppstein. 3-coloring in time $O(1.3446^n)$: a no-MIS algorithm. In *Proc. 36th IEEE FOCS*, pp. 444–452, 1995.

7. R. Beigel and J. Goldsmith. Downward separation fails catastrophically for limited nondeterminism classes. In *Proc. 9th Ann. Conf. Structure in Complexity Theory*, pp. 134–138, 1994.
8. D. Boneh, C. Dunworth, R. J. Lipton, and J. Sgall. On the computational power of DNA. Manuscript, 1996.
9. J. F. Buss and J. Goldsmith. Nondeterminism within P. *SICOMP*, 22:560–572, 1993.
10. J. D. C. Àlvarez and J. Torán. Complexity classes with complete problems between P and NP-complete. In *Foundations of Computation Theory*, pp. 13–24. Springer-Verlag, 1989. LNCS 380.
11. J. Díaz and J. Torán. Classes of bounded nondeterminism. *MST*, 23:21–32, 1990.
12. J. Goldsmith, M. Levy, and M. Mundhenk. Limited nondeterminism. *SIGACT News*, pp. 20–29, June 1996.
13. L. Hemachandra and S. Jha. Defying upward and downward separation. In *Proc. 10th STACS*, pp. 185–195. Springer-Verlag, 1993. LNCS 665.
14. C. M. R. Kintala. *Computations with a restricted number of nondeterministic steps*. PhD thesis, Penn. State Univ., University Park, PA, 1977.
15. C. M. R. Kintala and P. C. Fischer. Computations with a restricted number of nondeterministic steps. In *Proc. 9th ACM STOC*, pp. 178–185, 1977.
16. C. M. R. Kintala and P. C. Fischer. Refining nondeterminism in relativized polynomial-time bounded computations. *SICOMP*, 9(1):46–53, Feb. 1980.
17. R. Lipton. Using DNA to solve NP-complete problems. *Science*, 268:542–545, Apr. 1995.
18. B. Monien and E. Speckenmeyer. Solving satisfiability in less than 2^n steps. *Discrete Appl. Math.*, 10:287–295, 1985.
19. M. Ogihara. Breadth first search 3SAT algorithms for DNA computers. TR 629, U. Rochester, July 1996.
20. C. H. Papadimitriou and M. Yannakakis. On limited nondeterminism and the complexity of the V–C dimension. In *Proc. 8th Ann. Conf. Structure in Complexity Theory*, pp. 12–18, 1993.
21. N. Pippenger and M. Fischer. Relations among complexity measures. *J. ACM*, 26, 1979.
22. J. Robson. Algorithms for maximum independent sets. *J. Algorithms*, 7:425–440, 1986.
23. D. Roos and K. Wagner. On the power of bio-computers. TR, U. of Wurzburg, Feb. 1995. ftp://haegar.informatik.uni-wuerzburg.de/pub/TRs/ro-wa95.ps.gz.
24. P. Rothemund. A DNA and restriction enzyme implementation of Turing machines. http://www.ugcs.caltech.edu/tt~pwkr/oett.html.
25. L. Sanchis. Constructing language instances based on partial information. *International Jour. Found. Comp. Sci.*, 5(2):209–229, 1994.
26. I. Schiermeyer. Pure literal lookahead: An $O(1,497^n)$ 3-satisfiability algorithm. Manuscript, August 14, 1996.
27. C. P. Schnorr. Optimal algorithms for self-reducible problems. In *Proc. 3rd ICALP*, pp. 322–337, 1976.
28. A. L. Selman. Natural self-reducible sets. TR, Northeastern Univ., 1986.
29. W. Smith and A. Schweitzer. DNA computers in vitro and vivo. TR, NEC, 1995.
30. R. Tarjan. Finding a maximum clique. TR 72-123, Cornell Univ., 1972.

Constructing Big Trees from Short Sequences

Péter L. Erdős
Michael A. Steel
László A. Székely
and Tandy J. Warnow

[1] Mathematical Institute of the Hungarian Academy of Sciences. E-mail:
elp@math-inst.hu
[2] Biomathematics Research Centre, University of Canterbury. E-mail:
m.steel@math.canterbury.ac.nz
[3] Department of Mathematics, University of South Carolina. E-mail:
laszlo@math.sc.edu
[4] Department of Computer and Information Science, University of Pennsylvania.
E-mail: tandy@central.cis.upenn.edu.

Abstract. The construction of evolutionary trees is a fundamental problem in biology, and yet methods for reconstructing evolutionary trees are not reliable when it comes to inferring accurate topologies of large divergent evolutionary trees from realistic length sequences. We address this problem and present a new polynomial time algorithm for reconstructing evolutionary trees called the *Short Quartets Method* which is consistent and which has greater statistical power than other polynomial time methods, such as Neighbor-Joining and the 3-approximation algorithm by Agarwala *et al.* (and the "Double Pivot" variant of the Agarwala *et al.* algorithm by Cohen and Farach) for the L_∞-nearest tree problem. Our study indicates that our method will produce the correct topology from shorter sequences than can be guaranteed using these other methods.

1 Introduction

Evolutionary trees indicate how species evolved from a common ancestor and are of fundamental concern to biologists. There are many methods for reconstructing trees from biomolecular sequences, and all potentially competitive methods are evaluated according to their accuracy for topology prediction [11]. However, reconstructing this topology is a difficult task for at least two reasons. First, all accepted optimization problems in this area are NP-hard, so that methods which are efficient typically do not provide good performance on large sets of sequences. More importantly, even if we could solve some of the NP-hard optimization problems in this domain, the sequence length required in order to be able to guarantee an accurate topology estimation can be beyond what is available or even possible. A polynomial time algorithm that can only be guaranteed to be accurate on unavailable sequence lengths is simply not reliable, and it must either not be used, or if used its output must not be believed. On the other hand, a method which is accurate on realistic length sequences *can* be used *even if* it requires more computational resources. We may simply need to use more machines, wait longer, employ more sophisticated techniques to implement the same basic objective, etc. Thus, the sequence length needed by a method imposes a significantly more severe limitation than its computational requirements. The importance to biologists of this measure of accuracy (called *efficiency* or *power* in the systematic biology literature [14]) is reflected in the extensive performance analysis literature in systematic biology in which methods are analyzed according to their performance on model tree reconstruction under various stochastic models of evolution [12]. Initially these studies focused on *consistency* [7], i.e. the question of whether a method would be guaranteed to produce the correct topology given long enough sequences. Since the discovery around 1970 [13] of *consistent distance transformations* (which produce *"corrected distances"*), it has been clear that all reasonable distance-based methods can recover the true tree with high probability given long enough sequences when applied to corrected distances computed on sequences generated by binary trees. All this is well-understood in the systematic biology community. What is not so well-understood is the sequence length needed to obtain an accurate topology with high probability using a given method on a given model tree. Unfortunately, sequence lengths are limited, and especially so when the tree to be reconstructed is large and contains widely divergent sequences.

This paper contains several results:

- We present a probabilistic analysis of the *depth* and *diameter* of random trees under two distributions.
- We describe a framework based upon *topology-invariant neighborhoods* which permits the comparison of the statistical power of different distance-based tree reconstruction methods.
- We develop a new consistent polynomial time method, the *Short Quartet Method* for reconstructing evolutionary trees, and provide an analytical study of its convergence rate for inferring trees under the Cavender-Farris model. (This analysis extends to a large class of r-state Markov models.) We show that this method has superior statistical power to Neighbor-Joining, the most popular distance-based method of phylogenetic tree reconstruction, and to new results from the theoretical computer science community by Agarwala *et al.* (STOC 1996) [1] and Cohen and Farach (SODA 1997 and RECOMB 1997) [5].

Due to space constraints, we cannot give proofs in this extended abstract.

2 Basics

We begin by describing a simple model of sequence evolution, called the *Cavender-Felsenstein* model, or sometimes the *Cavender-Farris* model. The Cavender-Felsenstein model of evolution for binary sequences associates to every edge e in a model tree T a *mutation probability* p_e with $0 < p_e < .5$, and the mutations on each edge are independent. The sites (i.e. positions within the sequences) are assumed to evolve identically and independently, with the state at the root selected according to some distribution (usually uniform). If k sites evolve under this model, then the tree generates a set of sequences of length k at the leaves. We allow the input to our method to be any symmetric zero-diagonal non-negative matrix, and we will abuse the notation and call such matrices *distance matrices*.

Definition 1. A distance matrix D is *additive* if and only if there exists a tree T with non-negative edge weighting w such that for all leaves i, j, $D_{ij} = \sum_{e \in P_{ij}} w(e)$, where P_{ij} is the path between i and j in T. The $\mathbf{L}_\infty$ distance between two distance matrices A and B is defined by $L_\infty(A, B) = max_{ij}|A_{ij} - B_{ij}|$. The $\mathbf{L}_\infty$-*nearest tree* problem takes as input a distance matrix d and returns an additive distance matrix D minimizing $L_\infty(d, D)$. The δ-*neighborhood* around d, denoted $N(d, \delta)$, is the set of all distance matrices d' such that $L_\infty(d, d') < \delta$. A *distance-based method* M for phylogeny construction is a mapping from $n \times n$ distance matrices to $n \times n$ additive distance matrices. A tree T_1 is said to *refine* a tree T if T can be obtained from T_1 by contracting some of the edges in T_1. A method M is said to be *combinatorially consistent* if $M(D) = D$ for all additive distance matrices D, and *continuous at* D if for every $\epsilon > 0$ there exists a $\delta > 0$ such that if $d \in N(D, \delta)$ then $M(d) \in N(M(D), \epsilon)$. We will say that a distance-based method is *reasonable* if it is both combinatorially consistent and continuous at every additive distance matrix defining a binary tree.

An interesting characterization of additive matrices D is the following:

Theorem 2. *Four Point Condition, from [4]: A distance matrix D is an additive matrix if and only if for all i, j, k, l, of the three pairwise sums $D_{ij} + D_{kl}, D_{ik} + D_{jl}, D_{il} + D_{jk}$, the largest two are identical.*

The proof of the theorem shows that the ordering on the three pairwise sums indicates the topology induced by the quartet. Thus, if $D_{ij} + D_{kl}$ is strictly smaller than the other two sums, then the topology induced by the quartet i, j, k, l is a resolved binary tree; otherwise all three sums are identical, and the topology induced by i, j, k, l is a star. Since we assume that T is binary, all such quartets induce resolved subtrees. We will denote this topology by $ij|kl$ when the pairs that are separated by an internal edge are ij and kl.

We now present a characterization of additive distance matrices which define the same topology.

Theorem 3. *Two additive distance matrices D and D' define the same topology if and only if for all quartets, the relative orders of the pairwise sums for that quartet are identical in the two matrices. Therefore, for every reasonable distance-based method M and for every binary tree T defining additive distance matrix D, there will be a $\delta > 0$ such that M is guaranteed to reconstruct the topology of T when applied to any $d \in N(D, \delta)$. Consequently, any reasonable distance-based method M will be consistent on every binary tree when applied to corrected distances. However, for every edge-weighted tree T with minimum* edge weight x, there is a tree T' with a different leaf-labelled topology such that $L_\infty(D, D') = x/2$, where D is the additive distance matrix for T and D' the additive distance matrix for T'.

We will now describe a method we call the *Naive Method*, based on Buneman's Four-Point Condition. For each quartet of species i, j, k, l, compute the topology on that quartet by computing the three pairwise sums (this is called the *four-point method* (FPM) for reconstructing a tree on a single quartet.) If the three sums are distinct and the minimum is attained at $D_{ij} + D_{kl}$, then set the topology on i, j, k, l to be $ij|kl$. If the minimum sum is not unique, constrain the topology to be a star. Construct the tree (if it exists) consistent with all the constraints on the topologies of quartets. If no tree exists consistent with all the constraints, output a star tree. (A similar procedure was described by Fitch in [9].) Constructing a tree consistent with all quartet topologies is easily done in polynomial time through a variety of techniques, hence this is a polynomial time method.

We now present a comparison of various distance based methods based upon topology invariant neighborhoods.

Theorem 4. *Let D be an additive $n \times n$ distance matrix defining a binary tree T, d be a fixed distance matrix, and let $\delta = L_\infty(d, D)$. Assume that x is the minimum weight of internal edges of T in the edge weighting corresponding to D.*
(i) *A hypothetical exact algorithm for the L_∞-nearest tree is guaranteed to return the topology of T from d if $\delta < x/4$.*
(ii) (a) *The 3-approximation algorithm for the L_∞-nearest tree is guaranteed to return the topology of T from d if $\delta < x/8$.* (b) *For all n there exists at least one d with $\delta = x/6$ for which the method can err.* (c) *If $\delta \geq x/4$, the algorithm can err for every such d.*
(iii) *The Naive Method is guaranteed to return the topology of T from d if $\delta < x/2$, and there exists a d for any $\delta > x/2$ for which the method can err.*

In other words, *given any matrix d of corrected distances, if an exact algorithm for the L_∞-nearest tree can be guaranteed to correctly reconstruct the topology of the model tree, then so can the Naive Method.* Thus, an exact algorithm for the L_∞-nearest tree *can err* on longer sequences than the Naive Method, when applied to corrected distances, for *any* model tree T. This suggests an inherent limitation of the L_∞-nearest tree approach to reconstructing evolutionary tree topologies.

3 The Short Quartet Method

The Short Quartet Method is similar in spirit to the Naive Method, in that it is based upon reconstructing trees for quartets, and then combining these trees if possible. However, the essential difference is that we attempt to avoid reconstructing the trees for the difficult quartets. Instead, we attempt to construct topologies only on those quartets that are close within the tree; these are called the *short quartets*. The reconstruction of the tree from these short quartets involves solving a special case of a problem which is in its general form NP-complete [15]. The method we use to reconstruct the topology on each quartet is not specified; if we can afford the time, we may elect to use maximum likelihood which has great statistical power, but which is computationally too expensive to use for all but small trees. However we do not know *apriori* which quartets are short quartets. Thus, the method we actually employ is a greedy method, which surprisingly can be shown to have high probability of accurate reconstruction of the topology provided that the sequence length is adequate, even if we reconstruct topologies on quartets using the same (simple and not particularly statistically powerful) method used by the Naive Method!

3.1 Short Quartet Consistency

We begin by defining the notion of an *edi*-subtree.

Definition 5. The *topological distance* between two leaves i and j in a tree T is the number of edges on the path between i and j, and the *topological length* of a path P is the number of edges on P. Consider the subtrees of a binary T obtained by deleting a single edge e in T but not the endpoints of e; call such subtrees *edi*-subtrees (for *edge-deletion-induced*). Each such edi-subtree can be considered a rooted tree, by rooting it at the endpoint of e to which it was originally attached. Given an *edi*-subtree t, **rep(t)** denotes a leaf in t closest to the root of t. Two *edi*-subtrees which are disjoint and whose roots are distance 2 apart are said to be *sibling edi*-subtrees. In order to simplify the discussion, we may abuse the notation and let t also denote the leaf set of the *edi*-subtree t.

We give some more definitions.

Definition 6. Let the *depth* of an edi-subtree in T be the number of edges on the path from e to the nearest leaf, and let the *depth* of T (denoted by $\mathbf{d(T)}$) be the maximum depth of any edi-subtree in T. We say that a path P in the tree T is *short* if its length is at most $2d(T) + 2$. The quartet i, j, k, l is said to be a *short quartet* if it induces a subtree which contains a single edge connected to four disjoint *short* paths.

Thus, the depth of a complete binary tree of n leaves is $\log_2 n - 1$ but the depth of a caterpillar (a tree consisting of a long path with leaves hanging off the path) is just 1. Consequently, *every* quartet in a complete binary tree on n leaves is a short quartet, but there are only $O(n)$ short quartets in a caterpillar.

We now proceed with the description of the algorithm which we will use to construct binary model trees from a set of topologies on quartets. Our algorithm operates by determining siblinghood, first of leaves, and then of larger and larger rooted *edi*-subtrees, until the tree is constructed from the leaves inward. The determination of siblinghood of *edi*-subtrees is based upon detecting witnesses and anti-witnesses among the quartets, which we now define.

Definition 7. Given a quartet $\{i, j, k, l\}$ of leaves, we will denote by $ij|kl$ the induced topology on i, j, k, l in which i and j are separated in T from k and l via a path. Let t_1 and t_2 be two *edi*-subtrees. A *witness to the siblinghood of t_1 and t_2* is a short quartet $\{u, v, w, x\}$ with topology $uv|wx$ such that $u \in t_1$, $v \in t_2$, and $\{w, x\} \cap (t_1 \cup t_2) = \emptyset$. We call such quartets *witnesses*. An *anti-witness to the siblinghood of t_1 and t_2* is a short quartet $\{p, q, r, s\}$ with topology $pq|rs$, such that $p \in t_1$, $r \in t_2$, and $\{q, s\} \cap (t_1 \cup t_2) = \emptyset$. We will call these *anti-witnesses*.

We now present the property upon which the algorithm is based:

Axiom 1 *Let t_1 and t_2 be disjoint edi-subtrees of T and assume $T - t_1 - t_2$ has at least two leaves. Then t_1 and t_2 are siblings if and only if the following two conditions hold:*

1. *There are leaves y and z such that the quartet $\{rep(t_1), rep(t_2), y, z\}$ is a witness to the siblinghood of t_1 and t_2, and*
2. *If there is an antiwitness to the siblinghood of t_1 and t_2, then there is a witness for it as well.*

This axiom provides the basis for determining if there is at least one tree consistent with the constraints in the set of quartets, but may not be enough to verify that there are not two such trees. Verifying uniqueness of the solution turns out to be easy, fortunately, but it is also necessary due to the way in which we selectively apply the short quartet consistency algorithm.

In each *edi*-subtree, there may be more than one leaf that is closest to the root of the subtree (in terms of the number of edges on the path from the leaf to the root). However, among all such closest leaves in each *edi*-subtree, there is a unique leaf which has a smallest label, if the species are labelled by $1, 2, ..., n$. We call this leaf the **smallest representative** of the edi-subtree. This allows us to define a special set of short quartets, which we call the **representative quartets**, as follows. Each short quartet is composed of a single edge $e = (a, b)$, so that if we delete both a and b from T we create four *edi*-subtrees. We will say that a short quartet is a **representative quartet** if its leaves are the smallest representatives of the four *edi*-subtrees created in this manner. Then the following can be shown:

Theorem 8. *If a binary tree T is consistent with a set Q of quartet topologies such that Q contains all representative quartets, then T is uniqely consistent with Q.*

This observation and the axiom above suggests the following algorithm:

- Start with every leaf of T (i.e. the taxa) defining an *edi*-subtree.
- While the graph has more than three *edi*-subtrees, do:
    - Form the graph on vertex set given by the edi-subtrees, and with edge set defined by siblinghood; i.e., (x, y) is an edge if and only if edi-subtrees x and y satisfy the conditions of Axiom 1 for siblinghood.

* Make a sibling pair out of each connected component, and make the roots of the *edi*-subtrees in that connected component children of a common root r, and replace the pair of *edi*-subtrees by one *edi*-subtree.
    * If no new sibling pairs are found, then return *fail*.
  * If there are at most three *edi*-subtrees left, connect their roots each to one internal node, and call the resultant tree T.
- Verify that T satisfies all the constraints given in the input, and that Q contains the *representative* quartet for every edge in T. If so, return T, and else return *fail*.

The correctness of this algorithm follows from the discussion above, and the runtime of this algorithm depends upon how the two *edi*-subtrees are found that can be siblings. It is obvious that this can be achieved in polynomial time, but the details of the implementation are omitted due to space constraints.

Theorem 9. *Given a set Q containing all short quartets of a tree T and satisfying Axiom 1, we can determine T in $O(|Q|\log n + n^2 \log n)$ time.*

3.2 The entire method

We now describe how we use the short quartet consistency algorithm to construct the tree. One issue we address is how we select the set of quartets to consider. As it turns out, this is done in a greedy fashion, which we now describe:

Definition 10. We define the **similarity** between sequences i and j to be $s(i,j) = 1 - 2H(i,j)/k$, where k is the sequence length, and $H(i,j)$ is the Hamming distance of sequences i and j. Let Q be the set of all possible quartets on $[n]$, and let Q_w be those quartets a,b,c,d such that $\min\{s(a,b), s(a,c), s(a,d), s(b,c), s(b,d), s(c,d)\} \geq w$.

On a given set Q_w, the result of applying the Short Quartet Consistency algorithm will either be a binary tree that is uniquely consistent with all the topology constraints in Q_w, or *fail*. This permits us to define our method as follows. The structure of the method is to do a *"halving"* search among the w by applying the Short Quartet Consistency algorithm to Q_w. starting with $w = 1/2, 1/4$, etc., until we either find a tree that is uniquely consistent with the Short Quartet consistency algorithm or realize that no such tree can be found (this evidence of failure occurs when $w < 1/k$). We can show that with high probability, given adequate sequence length this search will examine a set Q_w which contains all short quartets and which also satisfies Axiom 1. Consequently, in polynomial time we will reconstruct the tree topology.

Theorem 11. *The Short Quartets Method takes $O(n^4 \log n \log k + n^2 k)$ time in the worst case. On any input d of distances derived from sequences generated on a model tree T, if the Naive Method accurately reconstructs the topology of T from d then SQM will also accurately reconstruct the topology of T from d.*

A more realistic analysis of the running time of the Short Quartet Method is based upon analyzing *typical* trees can be obtained by using Theorem 13. Typical trees under both the uniform and Yule-Harding distributions have $O(\log \log n)$ depths. If the p_e probabilities on the edges of a tree of depth $O(\log \log n)$ are equal or almost equal, then certain Q_w's with $|Q_w| = O(n \, polylogn)$ will yield a tree through the consistency algorithm, and the halving search will hit such a w, with probability $1 - o(1)$. Consequently, for typical tree shapes and for mutation probabilities that just slightly vary, applying the Short Quartet Method is likely to take only $O(n^2 k + n^2 \log n)$ time.

We now state our main result:

Theorem 12. *Suppose k sites evolve under the Cavender-Farris model on a binary tree T, so that for all edges e, $p_e \in [f, g]$, where we allow f, g to be functions of n. Assume that g is separated from $1/2$. The Short Quartet Method returns the tree T with probability $1 - o(1)$, if*

$$k > \frac{c \cdot \log n}{(1 - \sqrt{1 - 2f})^2 (1 - 2g)^{4depth(T)}} \tag{1}$$

where c is a fixed constant.

4 Depth vs. Diameter of Random Trees

We have shown that the sequence length needed by our method depends exponentially upon the minimum of the depth or the diameter of the tree it attempts to reconstruct. We study these topological quantities in this section.

Two simple models for describing semi-labelled binary trees are the *uniform* model, in which each tree has the same probability, and the *Yule-Harding* model, studied in [2, 3, 10]. This distribution is based upon a simple model of speciation, and results in "bushier" trees than the uniform model.

The following results are needed to analyse the performance of phylogeny reconstruction algorithms on random binary trees. Recall the definitions of depth and diameter from Section 3.

Theorem 13. *a) For a random semilabelled binary tree T with n leaves under the uniform model, $d(T) \leq (2 + o(1)) \log_2 \log_2 (2n)$ with probability $1 - o(1)$, and $diam(T) > \epsilon \sqrt{n}$ with probability $1 - O(\epsilon^2)$.*

b) For a random semilabelled binary tree T with n leaves under the Yule-Harding distribution, $d(T) = O(\log \log n)$ and $diam(T) = \Theta(\log n)$, with probability $1 - o(1)$

4.1 Analysis of the Short Quartet Method

In [6], Farach and Kannan proposed a method (FK) for reconstructing Cavender-Farris trees based upon applying the 3-approximation of Agarwala et al (discussed in Section 2) for the L_∞-nearest tree problem to corrected distances. They proved that the method converged quickly for the *variational distance* (a related but different concern than the topology estimation), but did not analyze the convergence to the topology of the model tree. Recently, Kannan extended the analysis (personal communication) and obtained the following counterpart to (1): If T is a model tree with mutation probabilities in the range $[f, g]$, and if sequences of length k' are generated on this tree, where

$$k' > \frac{c' \cdot \log n}{f^2(1-2g)^{2diam(T)}}, \qquad (2)$$

and c' is some constant, then with high probability the result of applying Agarwala et al to Cavender-Farris distances will be a tree with the same topology as T.

We now compare the sequence length requirements for the Short Quartet method as compared to the 3-approximation algorithm for the nearest L_∞-tree. Comparing this formula to (1), we note that the the comparison of depth and diameter is the most important issue. We always have $diam(T) \geq 2depth(T)+1$. The constants do not affect the comparison unless the depth and the diameter are close to each other, which in general they are not (from our earlier results, for almost all trees, the depth is $O(\log \log n)$ while the diameter is $\Omega(\sqrt{n})$, under the uniform distribution, while for the Yule-Harding distribution, the depth is still $O(\log \log n)$ and the diameter is $\Omega(\log n)$. Consequently, the Short Quartet Method requires much shorter sequence lengths than the Agarwala et al algorithm for almost all binary trees.

We summarize these results in the following table.

		range of mutation probabilities on edges:	
		$[f,g]$ f,g are constants	$\left[\frac{1}{\log n}, \frac{\log \log n}{\log n}\right]$
binary trees worst-case	SQM FK	polynomial superpolynomial	polylog superpolynomial
random binary trees (uniform model)	SQM FK	polylog superpolynomial	polylog superpolynomial
random binary trees (Yule-Harding)	SQM FK	polylog polynomial	polylog polylog

This comparison establishes that our method requires significantly shorter sequences in order to ensure accuracy of the topology estimation than the algorithm of Agarwala et al, for almost all trees under both probability distributions. The trees for which the two methods need comparable length sequences are those in which the diameter and the depth are as close as possible – such as complete binary trees. In these cases, the previous analysis given in Section 3 indicates that SQM will nevertheless need shorter sequences than Agarwala et al will need to obtain the topology with high probability.

Although their running time is likely to be faster than ours on most data sets, our method is fast enough to be useful for all data sets that we might wish to analyze (even up to several thousand sequences). The real advantage of this method is its increase in accuracy on sequences of realistic length.

However, both algorithms are fast enough to make real-time computation of evolutionary trees feasible even for very large ($n = 500$ to 1000) data sets. This means that the issue of accuracy realistically is the most important issue, and needs to be the focus of the study.

5 Lower bounds

A careful analysis of the table above concerning the sequence length needed by the short quartet method reveals that for almost all trees under either distribution, the required sequence length grows polylogarithmically in the number of taxa for each fixed range of mutation probabilities. In this section, we show that this is a polynomial of the minimum possible sequence length for *any* method, whether deterministic or randomized.

We will henceforth assume that all trees we consider are binary trees bijectively leaf-labelled by the elements of $\{1, 2, \ldots, n\} = [n]$; we will call these *semi-labelled binary trees*. Since the number of semi-labelled binary trees on n leaves is $(2n-5)!!$, encoding deterministically all such trees by binary sequences at the leaves requires that the sequence length, k, satisfy $(2n-5)!! \leq 2^{nk}$, i.e. $k = \Omega(\log n)$. We now show that this information-theoretic argument can be extended for *arbitrary* models of evolution and *arbitrary* deterministic or even randomized algorithms for tree reconstruction. For each semi-labelled binary tree, T, and for each algorithm A, whether deterministic or randomized, we will assume that T is equipped with a mechanism for generating sequences, which allows the algorithm A to reconstruct the topology of the underlying tree T from the shortest possible sequences with constant probability.

Theorem 14. *Let T be a tree with n leaves labelled by sequences of $\{0,1\}^k$, and let A be an arbitrary algorithm, deterministic or randomized. For A to be able to reconstruct the topology of T from the sequences at the leaves with probability greater than $1/2$ (respectively greater than ϵ), it must hold that $(2n-5)!! \leq 2^{nk}$ (respectively, $(2n-5)!!\epsilon \leq 2^{nk}$), and so $k = \Omega(\log n)$.*

The Theorem above shows that model and algorithm have to be a very good match, if not much more than $\log n$ length sequences suffice for tree reconstruction with high probability for each trees. In view of the very mild conditions, it is amazing, that this bound basically can be attained by our SQM, applied to the Cavender-Farris model!

6 Acknowledgements

Thanks to Ken Rice for carefully reading the manuscript for biological accuracy and Scott Nettles for advice about data structures. This research was supported by an NSF Young Investigator Award CCR-9457800, a David and Lucille Packard Foundation fellowship, and generous research support from the Penn Research Foundation and Paul Angello to the fourth author. The second author was supported by the New Zealand Marsden Fund. The first and third authors were supported in part by the Hungarian National Science Fund contracts T 016

358, T 019 367, and European Communities (Cooperation in Science and Technology with Central and Eastern European Countries) contract ERBCIPACT 930 113. This research started when the authors enjoyed the hospitality of DIMACS during the Special Year for Mathematical Support to Molecular Biology in 1995.

References

1. R. Agarwala, V. Bafna, M. Farach, B. Narayanan, M. Paterson, and M. Thorup. On the approximability of numerical taxonomy: fitting distances by tree metrics. *Proceedings of the 7th Annual ACM-SIAM Symposium on Discrete Algorithms*, 1996.
2. D. J. Aldous, Probability distributions on cladograms, in: *Discrete Random Structures*, eds. D. J. Aldous and R. Permantle, Springer-Verlag, IMA Vol. in Mathematics and its Applications. Vol. 76, 1-18, 1995.
3. J. K. M. Brown, Probabilities of evolutionary trees, *Syst. Biol.* **43**(1), 78–91, (1994).
4. P. Buneman, The recovery of trees from measures of dissimilarity, in *Mathematics in the Archaeological and Historical Sciences*, F. R. Hodson, D. G. Kendall, P. Tautu, eds.; Edinburgh University Press, Edinburgh, 1971, pp. 387–395.
5. J. Cohen and M. Farach, Numerical Taxonomy on Data: Experimental Results. SODA '97 and RECOMB '97.
6. M. Farach, and S. Kannan, Efficient algorithms for inverting evolution, *Proceedings of the ACM Symposium on the Foundations of Computer Science*, 230–236, (1996).
7. J. Felsenstein, Cases in which parsimony or compatibility methods will be positively misleading, *Syst. Zool.*, **27**, 401–410 (1978).
8. J. Felsenstein, Numerical methods for inferring evolutionary trees, *Quarterly Review of Biology*, 57 (1982), pp. 379-404.
9. W. Fitch, A non-sequential method for constructing trees and hierarchical classifications. *J. Mol. Evol.*, (18):30-37, 1981.
10. E. F. Harding, The probabilities of rooted tree shapes generated by random bifurcation, *Adv. Appl. Probab.* **3**, 44–77, (1971).
11. D. Hillis, Approaches for assessing phylogenetic accuracy. *Syst. Biol.* **44**(1):3-16, 1995.
12. D. Hillis, J. Huelsenbeck, and D. Swofford, Hobgoblin of phylogenetics? Nature, Vol. 369, 1994, pp. 363-364.
13. J. Neyman, Molecular studies of evolution: a source of novel statistical problems. Pages 1-27 of Gupta, S.S. and J. Yackel (eds), *Statistical Decision Theory and Related Topics*. New York: Academic Press, 1971.
14. D. Penny, M. Hendy, and M. Steel, Progress with methods for constructing evolutionary trees. *Trends Ecol. Evol.* (7): 73-79, 1992.
15. M. A. Steel, The complexity of reconstructing trees from qualitative characters and subtrees, *J. Classification*, **9**, 91–116 (1992).

Termination of Constraint Logic Programs

Salvatore Ruggieri

Dipartimento di Informatica, Università di Pisa
Corso Italia 40, 56125 Pisa, Italy
e-mail: ruggieri@di.unipi.it

Abstract. In this paper, we introduce a method for proving universal termination of constraint logic programs by strictly extending the approach of Apt and Pedreschi [1]. Taking into account a generic constraint domain instead of the standard Herbrand univers, acceptable (CLP) programs are defined. We prove correctness and completeness of the method w.r.t. the leftmost selection rule for the class of *ideal* constraint systems, including CLP($\mathcal{R}_{lin}$), CLP($\mathcal{RT}$), and CLP($\mathcal{FT}$) among the others. Moreover, we investigate the problems arising in extending those results to non-ideal constraint system, by specifically designing sufficient conditions for termination of CLP($\mathcal{R}$) programs.

1 Introduction

Motivations for the termination analysis of logic programs are related to several topics, including systematic program development, control generation, non-monotonic reasoning, decidability issues, applications to abstract interpretation, program transformation and testing.

There are many contributions in the literature on termination of logic and Prolog programs (see [9] for a recent survey). However, research has been mainly focused on Prolog programs. Only recently other logic programming (LP) paradigms have been considered, including logic programs with delay declarations, and constraint logic programming (CLP).

Jaffar and Maher claim in their survey [6], that "the CLP Scheme provides a framework in which the lifting of results from logic programming to CLP is almost trivial". As shown in [7], that statement is certainly true for many results, including the equivalence of declarative, functional and operational semantics. However, we will show that a well-known declarative proof method for termination of logic programs can be easily extended only to a restricted class of systems, namely *ideal* constraint systems. In those systems, the consistency test is correct and complete, in the sense that a computation proceeds iff the accumulated constraints are satisfiable.

Although the class of ideal constraint systems includes CLP($\mathcal{R}_{lin}$), CLP($\mathcal{RT}$), RISC-CLP(Real) and CLP($\mathcal{FT}$) among the others, several real systems are not ideal. As the most representative example, in CLP($\mathcal{R}$) [5] non-linear constraints are *delayed* until some variables in these constraints get unique values during the further computation process so that the constraints become linear. If a computation stops with some delayed non-linear constraints, the system generates a "maybe" answer, i.e. the test cannot ensure consistency of all the answer constraints since the test has been performed only on the linear ones. The delaying of passive constraints is a mechanism for bounding the computational complexity of the constraint solver. Unfortunately, this prevents an early failure detection and may be the cause of infinite derivations.

In this paper, we introduce a method for proving universal termination of constraint logic programs with respect to a leftmost selection rule. We extend the approach of Apt and Pedreschi [1], which declaratively characterize the class of logic programs such that every LD-derivation starting with a ground query is finite, namely *acceptable* logic programs. On the one hand, we lift their results to ideal constraint systems, by taking into account a generic constraint domain instead of the standard Herbrand univers. On the other hand, we improve the method by providing a stronger completeness theorem even in the case of pure logic programming.

Concerning non-ideal constraint systems, we study termination of CLP($\mathcal{R}$) programs by specifically designing two sufficient conditions. Both of them are aimed at preventing the involvement of non-linear constraints in the termination analysis, either by removing them from the analysis, or by imposing a notion of well-modedness which ensures that non-linear constraints become linear at runtime.

Preliminaries We will use throughout the paper the terminology of Jaffar and Maher [6]. By a program we mean a constraint logic program, i.e. a set of clauses of the form $A \leftarrow B_1, \ldots, B_n$ where A is an atom and each $B_i, i \in [1, n]$, is either an atom or a constraint. A *flat* program is a program in which every atom has the form $p(X_1, \ldots, X_n)$, where $X_1, \ldots, X_n$ are (not necessarily distinct) variables.

A constraint domain $\mathcal{D}$ is a first order structure on the signature Σ of the constraints. We denote with D the domain of $\mathcal{D}$. A $\mathcal{D}$-interpretation of a program P is an interpretation of P with the same domain as $\mathcal{D}$ and the same interpretation for the symbols in Σ as $\mathcal{D}$. It can be represented as a subset of $B_\mathcal{D}^P$, where $B_\mathcal{D}^P$ is the set of atoms of the form $p(a_1, \ldots, a_n)$, with $a_i \in D$ for $i \in [1, n]$, and p n-ary predicate symbol appearing in P. When P is clear from the context, we write $B_\mathcal{D}$. A $\mathcal{D}$-model of P is a $\mathcal{D}$-interpretation of P which is also a model of it.

We write $\mathcal{D} \models c\vartheta$ when the constraint c is true in $\mathcal{D}$ w.r.t. the valuation ϑ. Given an atom $p(t_1, \ldots, t_n)$ and a valuation ϑ, $p(t_1, \ldots, t_n)\vartheta$ stands for $p(t_1\vartheta, \ldots, t_n\vartheta)$, where $t_i\vartheta$ is the value of t_i in the valuation ϑ. Analogously for queries and clauses. A $\mathcal{D}$-ground instance of a clause C is then any $C\vartheta$, where ϑ is a valuation. For a $\mathcal{D}$-interpretation I and a $\mathcal{D}$-ground atom A, we write $I \models A$ iff $A \in I$. For a $\mathcal{D}$-ground constraint c, we write $I \models c$ iff $\mathcal{D} \models c$.

Those definitions easily extend to a many-sorted language.

The operational semantics of a constraint system is characterized by a transition relation $\rightarrow$ defined in terms of the relations $\rightarrow_r$, $\rightarrow_c$, $\rightarrow_i$, $\rightarrow_s$ and of the functions $infer$ and $consistent$, as described in [6]. $infer$ is required to satisfy $infer(C, S) = (C', S') \Rightarrow \mathcal{D} \models C \wedge S \leftrightarrow C' \wedge S'$.

consistent is required to satisfy $consistent(C) \Rightarrow \mathcal{D} \models \exists C$.

N is the set of natural numbers. N^∞ is $N \cup \{\infty\}$. The *list-length* function ll is defined as follows: $ll(f(t_1, \ldots, t_n))$ is 0 if $f \neq [.|.]$ and $ll(t_2) + 1$ if $f(t_1, \ldots, t_n) = [t_1|t_2]$. In particular, the length of an infinite list is ∞. $size(t)$ is the number of symbols occurring in a term t. For a pair (C, S), we define the projection on the first element $(C, S)_1 = C$.

2 Termination in LP

A largely acknowledged termination proof method for logic programs was proposed by Apt and Pedreschi in [1], where the class of acceptable logic programs was introduced. First of all, we recall the basic notions of level mappings and ground instances of logic programs.

Definition 1. Given a logic program P

- a *level mapping for* P is a function $|\ |: B_P \to N$ of ground atoms to natural numbers. $|A|$ is called the level of A.
- $ground(P)$ denotes the set of ground instances of clauses from P. □

Intuitively, a program is acceptable if every time a clause is used in a LD-derivation, the level of the head of any of its ground instances is greater than the level of each atom in the body which might be selected further.

Definition 2. Let P be a logic program, and $I \subseteq B_P$ a Herbrand interpretation.

- P is *acceptable by* $|\ |: B_P \to N$ *and* I iff I is a model of P, and for every $A \leftarrow B_1, \ldots, B_n$ in $ground(P)$: for $i \in [1, n]$

$$I \models B_1, \ldots, B_{i-1} \quad implies \quad |A| > |B_i|$$

- A query Q is *acceptable by* $|\ |$ *and* I iff there exists $k \in N$ such that for every ground instance $A_1, \ldots, A_n$ of it: for $i \in [1, n]$

$$I \models A_1, \ldots, A_{i-1} \quad implies \quad k > |A_i| \qquad \square$$

We summarize the main termination properties of acceptable programs in the following Theorem (see [1] for a proof).

Theorem 3. *Every LD-derivation for a logic program P and query Q both acceptable by $|\ |$ and I is finite.*

Conversely, if every LD-derivation for P and Q and for P and every ground query is finite then P and Q are acceptable by some $|\ |$ and I. □

Intuitively, a generalization of acceptability to the CLP Scheme has to consider $\mathcal{D}$-ground instances of clauses, in order to involve the constraint domain to the proof level. As an example, **MEMBER**

```
member(X, [X| Xs]).
member(X, [Y| Xs]) ← member(X, Xs).
```

and the query `Xs = [a| Xs], member(b, Xs)` show different termination behaviors when considering finite trees or rational trees as the underlying constraint domain.

Definition 4. Given a program P defined on a constraint system $\mathcal{D}$,

- a *level mapping for P* is a function $|\ |: B_\mathcal{D} \to N^\infty$ of $\mathcal{D}$-ground atoms to natural numbers plus infinitum. $|A|$ is called the level of A.
- $ground_\mathcal{D}(P)$ denotes the set of $\mathcal{D}$-ground instances of clauses from P. □

Though it is clear why we consider now $B_\mathcal{D}$, it is less obvious why we include ∞ in the codomain of level mappings. The underlying objective is to be able to partly reason on termination of programs and a restricted class of queries. In the case of MEMBER, for instance, it is still legitimate to consider queries of the form `member(2, t)` where t is a *finite* list, since non-termination arises only for infinite lists. To this end, we extend the $>$ order on natural numbers to the relation $\gg$, defined as follows:

$$n \gg m \quad \text{iff} \quad n = \infty \text{ or } n > m$$

Therefore, $\infty \gg \alpha$ for every $\alpha \in N^\infty$, and for $n \in N$, $n \gg m$ iff $m \in N$ and $n > m$. It is worth noting that although $\gg$ is not an ordering relation, there is no infinite descending chain $n_1 \gg n_2 \gg \ldots$ when $n_1 \in N$.

3 From LP to ideal CLP

Acceptability extends to constraint logic programs by replacing the Herbrand univers with the constraint domain, and the ordering $>$ with the relation $\gg$.

Definition 5. Let P be a program on the constraint system $\mathcal{D}$, $I \subseteq B_\mathcal{D}$ a $\mathcal{D}$-interpretation and $|\ |$ a level mapping for P.

- P is *acceptable by* $|\ |$ *and* I iff I is a $\mathcal{D}$-model of P, and for every $A \leftarrow B_1, \ldots, B_n$ in $ground_\mathcal{D}(P)$: for $i \in [1, n]$, if B_i is an atom then

$$I \models B_1, \ldots, B_{i-1} \quad \text{implies} \quad |A| \gg |B_i|$$

- A query Q is *acceptable by* $|\ |$ *and* I iff there exists $k \in N$ such that for every $\mathcal{D}$-ground instance $A_1, \ldots, A_n$ of it: for $i \in [1, n]$, if A_i is an atom then

$$I \models A_1, \ldots, A_{i-1} \quad \text{implies} \quad k \gg |A_i|$$

□

The definition above is quite similar to Definition 2, except for the fact that now we consider atoms whose level is infinitum, and do not require the decreasing of the level mapping from the head of a $\mathcal{D}$-ground clause to the constraints in the body. The latter choice is only a matter of convenience, since the most natural level of a constraint should be always 0.

Relation $>$ plays two roles. On the one hand, it prevents us from reasoning about *badly-typed* clauses, i.e. those for which the level of the head is infinitum. In fact, if the level $|A|$ of the head of a $\mathcal{D}$-ground clause is infinitum, the requirement $|A| > |B_i|$ in Definition 5 is trivially satisfied for every i. On the other hand, $>$ plays the same role of the $>$ order on naturals when the level of the head is finite.

We recall from [6] the definition of *ideal* constraint systems. We denote with (C, S) a pair of sets of active and passive (i.e., delayed) constraints.

Definition 6. A constraint system with operational semantics defined by $\rightarrow$, *consistent* and *infer* is called *ideal* if

(i) $\rightarrow = \rightarrow_{ris} + \rightarrow_{cis}$,
(ii) for every (C, S), $\quad infer(C, S) = (C \cup S, \emptyset)$
(iii) for every C, $\quad consistent(C) \Leftrightarrow \mathcal{D} \models \exists C$. □

Therefore, the operational semantics of ideal constraint systems is defined in terms of $\rightarrow_{ris}$ and $\rightarrow_{cis}$ transitions, the inferred active constraint set $C \cup S$ gathers all the information of the pair (C, S), and the consistency test is complete. CLP($\mathcal{R}_{lin}$), CLP($\mathcal{RT}$), CLP($\mathcal{FT}$), RISC-CLP($\mathcal{R}$) fall in this class. On the contrary, full CLP($\mathcal{R}$) [5] is not ideal, since non-linear constraints are delayed until they become linear.

As an example, let us consider the clp($\mathcal{RT}$) (alias Prolog without occur check) program CURRY, which implements the rules of a simple Curry's type system. The query type(E,M,T) is intended to calculate the type T of a term M in the environment E. Since the elements of the domain are rational trees, *recursive polymorphic* types are allowed, such as the solution of the equation $\alpha = \alpha \rightarrow \beta$. The answer constraint for the query

$$\text{type}([], \text{lambda}(\text{x}, \text{apply}(\text{var}(\text{x}), \text{var}(\text{x}))), \text{T}). \tag{1}$$

binds T to the type α.

```
type(E,var(X),T)       ← in(E,X,T).
type(E,apply(M,N),T)   ← type(E,M,arrow(S,T)), type(E,N,S).
type(E,lambda(X,M),arrow(S,T)) ← type([(X,S)|E],M,T).

in([(X,T)|E],X,T).
in([(Y,T1)|E],X,T) ← X ≠ Y, in(E,X,T).
```

CURRY and the query (1) are both acceptable by $|\ |$ and $B_{\mathcal{RT}}$, where

$$|\text{type}(E,\ M,\ T)| = ll(E) + size(M)$$

$$|\text{in}(E,\ X,\ T)| = ll(E)$$

On the other hand CURRY and a query such as

$$M = \text{lambda}(x, M), \text{type}([], \text{lambda}(x, M), T)$$

are not acceptable by a same level mapping and interpretation. In fact, they have an infinite LD-tree. In general, CURRY and a query type(E, M, T) may not terminate when M is an *infinite* term. The use of ∞ in the codomain of level mappings covers the situations in which we are interested to reason on termination of a restricted class of queries. As another example, consider the well-known *test & generate* programming technique:

$$\text{program}(X,\ Y)\ \leftarrow \text{test}(X,\ Y),\ \text{generate}(X,\ Y).$$

test creates a network of constraints between the variables, whilst generate instantiates the variables. When reasoning on termination, we have to show the decreasing of the level mapping from the head to the generate atom in the body only for those $\mathcal{D}$-ground instances that pass the constraint network. Thus, we should not be worried about the possible divergence arising for generate atoms that do not satisfy the test constraints.

The following theorem states termination of acceptable programs and queries. It extends the first part of Theorem 3 to ideal constraint systems.

Theorem 7. *(Termination Correctness) Consider an ideal constraint system, and a program P and a query Q both acceptable by | | and I. Then every LD-derivation for P and Q is finite.* □

Consider again CURRY. By the theorem, we conclude that the LD-tree of the query (1) is finite.

Focusing on termination completeness, we present a result that extends the second part of Theorem 3. It is even more general, since we relax the hypothesis that the LD-tree of the program and every ground query is finite. In other words, our notion of acceptability is a correct and complete characterization of universal termination with respect to leftmost selection rules.

Theorem 8. *(Termination Completeness) Consider an ideal constraint system, a program P and a query Q such that every LD-derivation for P and Q is finite. Then there exist | | and I such that P and Q are both acceptable by | | and I.* □

4 From ideal CLP to CLP($\mathcal{R}$)

Let us consider now the following program FACT for computing factorial numbers:

```
fact(0, 1).
fact(1, 1).
fact(N, N * F)  ←F >= 1, N >= 2, fact(N-1, F).
```

A query such as `fact(4,F)` is intended to compute the 4th factorial number, i.e. 24. Moreover, the same program can be used to check whether a number is factorial, by means of a query such as $Q = $ `fact(N, 24)`. We point out that FACT and Q are both acceptable by $|\ |$ and $B_\mathcal{R}$ where

$$|\text{fact}(n, f)| = int(f)$$

where $int(f)$ is the integer part of a real f. From Definition 5, the only proof obligation we have to show is that

$$int(n \cdot f) > int(f)$$

when $f \geq 1, n \geq 2$.

Running the program and the query Q on a RISC-CLP(Real) system, the resulting LD-tree is finite. In fact, as RISC-CLP(Real) is ideal, termination is a consequence of Theorem 7. On the contrary, the LD-tree built by the CLP($\mathcal{R}$) system is infinite, since the system eventually runs into an infinite loop by applying the third clause again and again. As CLP($\mathcal{R}$) delays the non-linear constraints, their unsatisfiability is never checked.

As often it happens, real programming language implementations deviate from theoretically desirable properties. They often sacrifice completeness of the *consistency* test for efficiency reasons. The consistency test on *passive* constraints is delayed until they are sufficiently instantiated. This is the case, for example, of non-linear constraints in CLP($\mathcal{R}$). As a consequence, the computation may proceed even in the case that the accumulated constraints are unsatisfiable.

A simple extension of our approach to generic systems is then to prevent the use of any declarative reading of programs in the termination proofs.

Definition 9. A program P is *recurrent* by $|\ |$ iff for every $A \leftarrow B_1, \ldots, B_n$ in $ground_\mathcal{D}(P)$: for $i \in [1, n]$, if B_i is an atom then $|\ A\ | > |\ B_i\ |$. □

The definition of recurrent queries is derived accordingly. It can be easily shown that any derivation is finite with respect to *any selection rule*, when considering programs and queries both recurrent by a same level mapping. Recurrent programs extends recurrent logic programs introduced by Bezem [2]. As an example, consider the program MAP, defined in CLP($\mathcal{R}$).

```
map([], []).
map([X|Xs], [Y|Ys])  ←  Y = X * X, map(Xs, Ys).
```

It is easy to see that it is recurrent by defining $|\text{map}(Ls, Rs)| = ll(Ls)$. However, if we rewrite MAP in a *flat* form, namely the following MAPFLAT

```
map(A, B)  ←  A = [], B = [].
map(A, B)  ←  A = [X|Xs], B = [Y|Ys], Y = X * X, map(Xs, Ys).
```

we obtain a program that is not recurrent.

In the rest of this section, we give some sufficient conditions specially designed for termination of CLP($\mathcal{R}$) programs together with a generalization of the underlying insights to other non-ideal constraint systems.

A first idea is to exclude non-linear constraints from the termination analysis. Next theorem states that if a program and a query with their non-linear constraints removed terminate, then the original program and query do terminate.

Theorem 10. *Consider the CLP($\mathcal{R}$) system. The LD-tree for a program P and a query Q is finite if P' and Q' are both acceptable by $|\ |$ and I, where P' (resp., Q') is obtained by deleting all non-linear constraints from P (resp., Q).* □

Intuitively, the conclusion follows since adding constraints to a clause implies having shorter derivations.

Consider again the MAPFLAT program. It is immediate to observe that the non-linear constraint Y = X * X does not play a relevant role in termination of a query such as map([X,3,5],Z). In fact, termination is given by the decreasing of the length of the list in the first argument of map. By deleting Y = X * X we get the program MAPFLAT'

```
map(A, B)  ←  A = [], B = [].
map(A, B)  ←  A = [X|Xs], B = [Y|Ys], map(Xs, Ys).
```

which is acceptable by $|\ |$ and $B_\mathcal{R}$, where $|$ map(Ls, Rs) $| = ll(Ls)$. Therefore, we conclude that the LD-tree for MAPFLAT and map([X,3,5],Z) is finite.

In general, we have a stronger result for a large class of constraint systems.

Definition 11. *A constraint system with operational semantics defined by $\rightarrow$, consistent and infer is called incremental if $\rightarrow\ =\ \rightarrow_{ris} +\ \rightarrow_{cis}$, and*

[M] *for every $S' \subseteq S$, consistent(infer($\emptyset, S$)$_1$) $\Rightarrow$ consistent(infer($\emptyset, S'$)$_1$)*
[I] *for S, S' sets of constraints, and C set of active constraints*

$$consistent(infer(infer(C, S) \cup (\emptyset, S'))_1) \Leftrightarrow consistent(infer(C, S \cup S')_1)$$ □

As an example, ideal constraint systems and CLP($\mathcal{R}$) are incremental. Basically, [M] requires *monotonicity* of *consistent* and *infer* – a condition naturally satisfied in all practical systems.

[I] is an *incrementality* requirement. Starting from a pair (C, S), if applying *infer* first, then adding the constraints in S' and then re-applying *infer* we obtain a consistent state, then the state obtained by applying *infer* only once to $(C, S \cup S')$ should be consistent as well, and vice-versa.

Theorem 12. *Consider an incremental constraint system. The LD-tree for a program P and a query Q is finite if the LD-tree for P' and Q' is finite, where P' (resp., Q') is obtained by deleting some constraints from P (resp., Q).* □

However, this approach is not sufficient to prove termination when it depends on non-linear constraints. Consider the program SQRT for computing square roots of naturals.

```
srqt(X, R)    ←  A = 0, sqrt2(X, A, R).
sqrt2(X, A, A) ←  (A+1)*(A+1) > X.
sqrt2(X, A, B) ←  (A+1)*(A+1) ≤ X, A1 = A + 1, sqrt2(X, A1, B).
```

If we remove the non-linear constraints, we get a program that has an infinite LD-derivation for any query by applying the third rule again and again. In addition, the non-linear constraints become linear at run-time iff sqrt2 is called with the second argument ground.

To properly reason on programs containing non-linear constraints that become linear at run-time, we introduce a notion of *moding*. Without any loss of generality, we restrict to consider *flat* programs.

Definition 13.

- Consider an n-ary predicate symbol p. A *mode* for p is a function d_p from $\{1,\ldots,n\}$ in $\{+,-,\flat\}$. If $d_p(i) =' +'$ we call i an *input* position. If $d_p(i) =' -'$ then i is called an *output* position. If $d_p(i) =' \flat'$ then i is called a *blank* position (with respect to d_p.) We write d_p in the form $p(d_p(1),\ldots,d_p(n))$.
- A *mode* for a constraint $c(X_1,\ldots,X_n)$ whose *variables* are $X_1,\ldots,X_n$ is a function d_p from $\{X_1,\ldots,X_n\}$ in $\{+,-,\flat\}$. We write d_p in the form $c(X_1 d_p(1),\ldots,X_n d_p(n))$.
- For an atom or a constraint A, we write $A(\mathbf{X},\mathbf{Y},\mathbf{Z})$ to denote that $\mathbf{X}$ are the variables occurring in input positions, $\mathbf{Y}$ are those occurring in output positions, and $\mathbf{Z}$ are those occurring in blank positions.
- We say that a *flat* program P is well-moded iff for every clause

$$A_0(\mathbf{Y}_0, \mathbf{X}_{n+1}, \mathbf{Z}_0) \leftarrow A_1(\mathbf{X}_1, \mathbf{Y}_1, \mathbf{Z}_1),\ldots,A_n(\mathbf{X}_n, \mathbf{Y}_n, \mathbf{Z}_n)$$

of P, for $i \in [1, n+1]$ $\mathbf{X}_i \subseteq \cup_{k<i} \mathbf{Y}_k$.
- We say that a *flat* query $A_1(\mathbf{X}_1, \mathbf{Y}_1, \mathbf{Z}_1),\ldots,A_n(\mathbf{X}_n, \mathbf{Y}_n, \mathbf{Z}_n)$ is well-moded iff for $i \in [1, n]$ $\mathbf{X}_i \subseteq \cup_{k<i} \mathbf{Y}_k$. □

The intuition underlying this definition is to force the input variables in an atom or a constraint selected along a LD-derivation to be grounded by the active constraints. Variables not involved in the input-output relation are marked as blank.

Suppose now that the moding of the constraints is consistent with the operational semantics, i.e. if a constraint $c(\mathbf{X},\mathbf{Y},\mathbf{Z})$ is selected and the active constraints imply $\mathbf{X} = \mathbf{a}$ for some tuple $\mathbf{a}$ of elements of the domain, then the active constraints of the resolvent (if exists) imply $\mathbf{Y} = \mathbf{b}$ for some tuple $\mathbf{b}$. Under this assumption, when a non-linear constraint is selected then the input variables are grounded by the active constraints. We can exploit this fact to impose that non-linear constraints become linear at run-time.

Definition 14. A moding for a program P (resp., a query Q) is consistent w.r.t. CLP($\mathcal{R}$) if for every constraint $c(\mathbf{X}, \mathbf{Y}, \mathbf{Z})$ in P (resp., Q) either

(i) $\mathbf{Y}$ is an empty tuple and $c(\mathbf{X}, \mathbf{Y}, \mathbf{Z})$ is linear in $\mathbf{Z}$, or
(ii) $\mathbf{Y}$ is a tuple of only one variable, $\mathbf{Z}$ is an empty tuple and $c(\mathbf{X}, \mathbf{Y}, \mathbf{Z})$ is an equation linear in $\mathbf{Y}$. □

It is worth noting that both well-modedness and consistency w.r.t. CLP($\mathcal{R}$) are syntactic notions. Consider again the program SQRT. It is immediate to see that it is well-moded with the moding

$$\text{sqrt}(\emptyset, \emptyset), \text{sqrt2}(\emptyset, +, \emptyset), \text{A}- = 0$$
$$(\text{A}++1)*(\text{A}++1) > \text{X}\emptyset, (\text{A}++1)*(\text{A}++1) \leq \text{X}\emptyset, \text{A1}- = \text{A}+ + 1.$$

Moreover, the moding for the constraints is consistent w.r.t. CLP($\mathcal{R}$). Next theorem relates modings, acceptability and termination by providing a sufficient condition for termination of well-moded acceptable CLP($\mathcal{R}$) programs.

Theorem 15. *Consider the CLP($\mathcal{R}$) system. Let P and Q be well-moded flat program and query and let the moding be consistent w.r.t. CLP($\mathcal{R}$). Suppose P and Q are both acceptable by I and $|\ |$. Then every LD-derivation for P and Q is finite.* □

The program SQRT and the query sqrt(n, R) for $n \in N$ are acceptable by $B_{\mathcal{R}}$ and $|\ |$, where

$$|\text{sqrt2}(x, a, b)| = \begin{cases} max(x-a, 0) & \text{if } x, a \in N \\ \infty & \text{otherwise} \end{cases}$$

$$|\text{sqrt}(x, r)| = \begin{cases} x+1 & \text{if } x \in N \\ \infty & \text{otherwise} \end{cases}$$

Therefore, Theorem 15 allows us to state that the LD-tree for SQRT and sqrt(n, R) is finite when $n \in N$.

Theorem 15 can be used together with Theorem 12 in order to prove termination of programs P and queries Q defined on CLP($\mathcal{R}$), by means of the following strategy:

(i) delete some (non-linear) constraints from P and Q, and
(ii) show that the resulting program and query are well-moded and acceptable by the same model and level mapping.

Finally, we point out that this approach is extendible to a generic non-ideal constraint system by appropriately defining a notion of *consistency* of constraint moding w.r.t. the system.

5 Conclusions

There is still little work on the extension of termination approaches to constraint logic programming. The only papers we are aware of are [3] and [8]. [8] provides *sufficient* conditions based on approximation techniques, with the aim of automatizing the termination proof. [3] presents a necessary and sufficient condition for termination based on a radically different approach from ours, which is inspired by the works of Floyd on termination of flowchart programs. We also cite [4], where a class of programs is characterized with no delayed constraints at the end of successful computations. Also, that method is able to discover possible sources of non-termination due to delaying of non-linear constraints.

We presented an extension to the CLP Scheme of a largely acknowledged approach to termination of logic programs. For a large class of constraint systems, namely *ideal constraint systems*, we extend and improve on the results of [1], showing stronger forms of correctness and completeness even in the case of pure logic programming. In the second part of the paper, we investigated termination specifically for the CLP($\mathcal{R}$) system, by proposing two sufficient conditions.

References

1. K.R. Apt and D. Pedreschi. Studies in Pure Prolog: Termination. In J. W. Lloyd, editor, *Symposium on Computational Logic*, pages 150–176. Springer-Verlag, Berlin, 1990.
2. M. Bezem. Characterizing termination of logic programs with level mappings. In E. L. Lusk and R. A. Overbeek, editors, *Proceedings of the North American Conference on Logic Programming*, pages 69–80. The MIT Press, 1989.
3. L. Colussi, E. Marchiori, and M. Marchiori. On Termination of Constraint Logic Programs. In M. Bruynooghe and J. Penjam, editors, *Proc. of PPCP'95*, number 976 in Lectures Notes in Computer Science, 1995.
4. M. Hanus. Analysis of nonlinear constraints in CLP(R). In D. S. Warren, editor, *Proceedings of the 1993 International Conference on Logic Programming*, pages 83–99. MIT Press, 1993.
5. P. Stuckey J. Jaffar, S. Michaylov and R. Yap. The CLP(R) Language and System. *ACM Toplas*, 14(3):339–395, 1992.
6. J. Jaffar and M.J. Maher. Constraint logic programming: A survey. *Journal of Logic Programming*, 19,20:503–581, 1994.
7. M.J. Maher. A Logic Programming View of CLP. In D. S. Warren, editor, *Proceedings of the 1993 International Conference on Logic Programming*. The MIT Press, 1993.
8. F. Mesnard. Inferring Left-terminating Classes of Queries for Constraint Logic Programs. In M. Maher, editor, *Proceedings of the 1996 Joint International Conference and Symposium on Logic Programming*, pages 7–21. The MIT Press, 1996.
9. D. De Schreye and S. Decorte. Termination of logic programs: the never-ending story. *Journal of Logic Programming*, 19-20:199–260, 1994.

The Expressive Power of Unique Total Stable Model Semantics

Francesco Buccafurri,[1] Sergio Greco[2] and Domenico Saccà[2]

[1] ISI-CNR, 87030 Rende, Italy
bucca@si.deis.unical.it
[2] DEIS, Univ. della Calabria, 87030 Rende, Italy
{ greco, sacca }@si.deis.unical.it

Abstract. This paper investigates the expressive power of DATALOG¬ queries under *unique* T-stable model semantics, i.e., a query on a given database yields an answer if and only if there exists a unique T-stable model. Under this semantics DATALOG¬ queries are shown to express exactly all decision problems with unique solutions. Obviously, unique T-stable model semantics is the 'natural' semantics for queries with at most one T-stable model or with exactly one T-stable model for every database. The expressive powers of of these two classes of queries are investigated as well but it turns out that any practical language for such queries cannot get to an expressive power higher than DATALOG with stratified negation.

1 Introduction

Total stable models (*T-stable models*) [9] provide a simple, yet powerful semantics to DATALOG¬, i.e., logic programming with negation but without function symbols. One of the properties of stable models is their multiplicity: a program may have from 0 to n T-stable models, where n can grows exponentially with the size of the universe.

Multiplicity has been recognized by some authors as an important opportunity for either expressing non-determinism or for increasing the expressive power while preserving determinism (e.g., by taking the union or the intersection of all models). On the other hand, multiplicity has been strongly criticized by many other authors mainly because the canonical meaning of a logic program is traditionally based on a unique model. This criticism explains the great deal of interest for special classes of DATALOG¬ programs with 'unique' T-stable models such as stratified model [3] or total well-founded model [23], notwithstanding their reduced expressive power (indeed, only a proper subset of polynomial problems are expressible by such programs).

An interesting question is the following: is there any class of DATALOG¬ queries which preserves the T-stable model uniqueness property but it has an expressive power higher than stratified DATALOG¬? To anwser this question, we investigate the classes $\mathbf{Q}_{0,1}$ and $\mathbf{Q}_1$ of DATALOG¬ queries admitting, respectively, at most one T-stable model and exactly one T-stable model for every input database.

We show that the expressive powers of these two classes is bound by $\mathcal{NP} \cap \mathcal{UW}$ and $\mathcal{NP} \cap co\mathcal{NP}$ respectively. But $\mathcal{NP} \cap \mathcal{UW}$ and $\mathcal{NP} \cap co\mathcal{NP}$ as well as all their meaningful subclasses from $\mathcal{P}$ over are not know to be expressible by a (recursively enumerable) query language [13]. Moreover, although total well-founded semantics is capable to express all fixpoint queries as recently shown in [8], no language is known which has the same power of fixpoint queries and only generates queries having total well-founded models for every database. Thus, it appears that any practical language for queries with unique T-stable model is not more expressive than stratified `DATALOG¬`!

To get more expressive power using a semantics based on a unique total stable model, it probably remains to take the whole class of `DATALOG¬` queries and to check uniqueness a-posteriori. To this end, we introduce the *unique T-stable model semantics*: a ground literal is true if both it is in a T-stable model and there exists no other T-stable model — informally multiplicity corresponds to a negative answer. We show that the class of all `DATALOG¬` queries under unique T-stable model semantics is able to express all the decision problems that can be defined using an existential second-order formula of the form $(\exists!\mathbf{S})\Phi(\mathbf{S})$ with *unique witnesses* for the second-order quantifiers, i.e., there are unique relations $s_1, \ldots, s_m$ in $\mathbf{S}$ satisfying the first-order formula $\Phi(\mathbf{S})$ — we call this class $\mathcal{UW}$. This is an interesting class which consists of most of all decision problems with unique solutions. Observe that T-stable models under a popular version of T-stable model semantics, *certain semantics*, capture $co\mathcal{NP}$; so, as $co\mathcal{NP} \subseteq \mathcal{UW}$, unique T-stable model semantics turns out to be more expressive than certain semantics.

The paper is organized as follows. Background and basic definitions on T-stable model semantics for `DATALOG¬` queries are given in Section 2. The expressive power of unique T-stable model semantics for the class of all `DATALOG¬` queries is investigated in Section 3. The analysis of the subclasses $\mathbf{Q}_{0,1}$ and $\mathbf{Q}_1$ as well as the conclusion are presented in Section 4.

2 Total Stable Models and DATALOG Queries

Let us start by recalling basic concepts and notation of the `DATALOG¬` language, that is logic programming with negative goals in the rules but without function symbols [1, 21].

A *rule* r is a formula of the language of the form $Q \leftarrow Q_1, \ldots, Q_m$, where Q is a atom (*head* of the rule) and $Q_1, \ldots, Q_m$ are literals (*goals* of the rule). A ground rule with no goals is called a *fact*; a rule without negative goals is called *positive*. A `DATALOG¬` *program* is a finite set of function-free rules and it is called *positive* (or, simply, `DATALOG`) when all its rules are positive.

Given a `DATALOG¬` program $\mathcal{LP}$, some of the predicate symbols (*EDB predicates*) do not occur in the rule heads as they are defined by a number of facts stored into a database — the other predicate symbols are called *IDB predicates*. EDB predicate symbols form a relational database scheme $\mathcal{DS}_{\mathcal{LP}}$, thus they are also seen as relation symbols. A database D on $\mathcal{DS}_{\mathcal{LP}}$ is a set of finite relations

$D(r)$ on a countable domain U, one for each r in $\mathcal{DS}_{\mathcal{LP}}$. Given a database D on $\mathcal{DS}_{\mathcal{LP}}$, $\mathcal{LP}_D$ denotes the program obtained from $\mathcal{LP}$ by adding the facts corresponding to the relation tuples in D. Observe that the *Herbrand universe* and the *Herbrand Base* for $\mathcal{LP}_D$ (denoted by $U_{\mathcal{LP}_D}$ and $B_{\mathcal{LP}_D}$, respectively) are both finite; moreover, $U_{\mathcal{LP}_D}$ is a finite subset of U as possible constants in $\mathcal{LP}$ are also taken from the domain U. Any subset of $B_{\mathcal{LP}_D}$ is called an *interpretation*.

Let M be an interpretation of the program $\mathcal{LP}_D$. Let $pos(\mathcal{LP}_D, M)$ be the positive program obtained from the ground instantiation of $\mathcal{LP}_D$ by deleting (a) each rule that has a negative goal $\neg A$ for which $A \in M$, and (b) all negative goals from the remaining rules. Then M is *total stable* (*T-stable*) *model* [9] if and only if $\mathbf{T}^\infty_{pos(\mathcal{LP}_D, M)}(\emptyset) = M$, where the operator $\mathbf{T}$ is the classical *immediate consequence transformation*. The existence of a T-stable model for any program is not guaranteed.

Fact 1 [9, 17] *Given a* DATALOG¬ *program* $\mathcal{LP}$, *a database* D *on* $\mathcal{DS}_{\mathcal{LP}}$, *and an interpretation* M *for* $\mathcal{LP}_D$, *then*

1. *deciding whether M is a T-stable model for $\mathcal{LP}_D$ is in $\mathcal{P}$;*
2. *deciding whether there exists a T-stable model for $\mathcal{LP}_D$ is $\mathcal{NP}$-complete.* □

Three versions of deterministic semantics for T-stable models are known in the literature: the *possible* (or *credulous* or *brave*) semantics [2, 20, 6], the *certain* (or *skeptical* or *cautious*) semantics [9, 2, 20, 6], and the *definite* semantics [19]. We now introduce a fourth version: the *unique T-stable model* semantics.

Definition 1. Given a DATALOG¬ program $\mathcal{LP}$, a database D on $\mathcal{DS}_{\mathcal{LP}}$ and a ground literal A, then

1. A is a $\mathcal{TS}^\exists$ (*possible*) inference of $\mathcal{LP}_D$ if A is true in some T-stable model of $\mathcal{LP}_D$;
2. A is a $\mathcal{TS}^\forall$ (*certain*) inference of $\mathcal{LP}_D$ if A is true in each of the T-stable models of $\mathcal{LP}_D$;
3. A is a $\mathcal{TS}^{\forall!}$ (*definite*) inference of $\mathcal{LP}$ if $\mathcal{LP}_D$ ha at least one T-stable model and A is in each of these models;
4. A is a $\mathcal{TS}^1$ (*unique*) inference of $\mathcal{LP}_D$ if $\mathcal{LP}_D$ ha exactly one T-stable model and A is true in this model. □

The above version of T-stable model semantics will be denoted by $\mathcal{TS}^v$, where v is $\exists$, $\forall$, $\forall!$, or 1.

Definition 2. A (*bound* DATALOG¬) *query* Q is a pair $\langle \mathcal{LP}, G \rangle$, where $\mathcal{LP}$ is a DATALOG¬ program and G is a ground literal (the *query goal*) — possible constants in G are in U as well. The set of all queries is denoted by $\mathbf{Q}$.

Given any T-stable model semantics $\mathcal{TS}^v$, the *database set* of Q under $\mathcal{TS}^v$, denoted by $\mathcal{EXP}_{\mathcal{TS}^v}(Q)$, is the set of all databases D on $\mathcal{DS}_{\mathcal{LP}}$ for which G is a $\mathcal{TS}^v$ inference of $\mathcal{LP}_D$. Moreover, the *expressive power* of the $\mathcal{TS}^v$ semantics is measured by the family of the database sets of all possible queries and is denoted by $\mathcal{EXP}_{\mathcal{TS}^v}[\mathbf{Q}] = \{\mathcal{EXP}_{\mathcal{TS}^v}(Q) | Q \in \mathbf{Q}\}$. □

It is well known that for each query Q and for each T-stable model semantics TS^v, $\mathcal{EXP}_{TS^v}(Q)$ is indeed a *generic* database set [5, 1], i.e., it is closed under renaming of constants in $(U-C)$, where C is the set of constants occurring in $\mathcal{LP}$ and in G — thus the constants not in C are not interpreted and relationships among them are only those explicitly provided by the databases. From now on any generic set of databases on the same scheme will be called a *database collection*.

The expressive power of any T-stable model semantics will be measured w.r.t. classes of database collections defined as follows. Given a (not necessarily Turing machine) complexity class C of decision problems and a database collection $\mathbf{D}$, $\mathbf{D}$ is *C-recognizable* if the problem of deciding whether a database D is in $\mathbf{D}$ is in C. The *database complexity class DB-C* is the family of all C-recognizable database collections — for instance, *DB-$\mathcal{P}$* is the family of all database collections that are recognizable in polynomial time. Observe that any two database collections in a database complexity class do not in general share the same database scheme.

We stress that our expressive power measure follows the *data complexity* approach of [5, 24] for which the query is assumed to be a constant whereas the database is the input variable. The following results are known in the literature:

Fact 2 *Given a* DATALOG¬ *program $\mathcal{LP}$, a database D on $\mathcal{DS}_{\mathcal{LP}}$, and an interpretation M for $\mathcal{LP}_D$, then*

1. $\mathcal{EXP}_{TS^{\exists}} = DB\text{-}\mathcal{NP}$ [17];
2. $\mathcal{EXP}_{TS^{\forall}} = DB\text{-}co\mathcal{NP}$ [20];
3. $\mathcal{EXP}_{TS^{\forall!}} = DB\text{-}\mathcal{D}^p$ [19]. □

Example 1. Let $\mathcal{DS}_K = \{v, e\}$ be a database scheme defining directed graphs and $\mathbf{D}^K$ be the set of all databases on $\mathcal{DS}_K$ corresponding to graphs with a kernel — recall that a kernel of a graph G is a subset V_1 of V such that (a) for any two $x, y \in V_1$, the edge (x, y) is not in E, and (b) for any $y \in V_2 = V - V_1$ there is an $x \in V_1$ such that $(x, y) \in E$. Consider the following DATALOG¬ program K:

```
v1(X)              ← v(X), ¬v2(X).
v2(X)              ← v(X), ¬v1(X).
joined_to_V1(X)    ← v1(Y), e(Y,X).
no_condition_a     ← v1(X), joined_to_V1(X).
no_condition_b     ← v2(X), ¬joined_to_V1(X).
kernel             ← ¬no_condition_a, ¬no_condition_b.
T_constraint       ← ¬kernel, ¬T_constraint.
```

Given any database D on $\mathcal{DS}_K$, say corresponding to the graph G, any possible T-stable model M of K_D must make T_constraint false because of the last rule (otherwise, T_constraint would be undefined); then M must make true kernel, i.e., the vertices selected for V_1 by M through the first rule form a kernel for G. Hence, K_D has exactly one T-stable model for each kernel of the graph. Given the query $Q^K = \langle K, kernel \rangle$, $\mathcal{EXP}_{TS^{\exists}}(Q^K) = \mathcal{EXP}_{TS^{\forall!}}(Q^K) = \mathbf{D}^K$, i.e., under both possible and definite T-stable model semantics Q^K defines the $\mathcal{NP}$-complete problem of whether a graph has a kernel. Moreover, $\mathcal{EXP}_{TS^1}(Q^K) = $

$\mathbf{D}^{1K}$, that is the set of all graphs with exactly one kernel; i.e., under unique T-stable model semantics, Q^K defines the problem of whether a graph has exactly one kernel. On the other hand, as *kernel* is a $\mathcal{TS}^\forall$ inference also when there is no T-stable model, the database set of Q^K under $\mathcal{TS}^\forall$ semantics consists of all graphs, i.e., the query is meaningless under this semantics.

Let K' be obtained from K by removing the last rule. Now, there are T-stable models for K'_D also when D corresponds to a graph without kernel. Consider now the query $Q^{K'} = \langle K', \neg kernel \rangle$. We have that $\mathcal{EXP}_{\mathcal{TS}^\forall}(Q^{K'}) = \mathcal{EXP}_{\mathcal{TS}^{\forall!}}(Q^{K'}) = \overline{\mathbf{D}}^K$, that is the set of all graphs without kernel; i.e., under both certain and definite T-stable model semantics, $Q^{K'}$ defines the $co\mathcal{NP}$-complete problem of whether a graph has no kernel. This query is meaningless under possible and unique T-stable model semantics. □

From Fact 2 it follows that, as far as the expressive powers are concerned, definite semantics subsumes the other two semantics which, in turn, are incomparable with each other (unless $\mathcal{NP} = co\mathcal{NP}$). In the next section we characterize the expressive power of unique T-stable model semantics.

3 Expressive Power of Unique Stable Model Semantics

In this section we prove that unique T-stable model semantics captures the whole class *DB-UW*, consisting of all database collections **D** that can be defined using an existential second-order formula of the form $(\exists!\mathbf{T})\Phi(\mathbf{T})$ with *unique witnesses* for the second-order quantifiers, i.e., there are unique relations in **T** satisfying the first-order formula $\Phi(\mathbf{T})$ on a finite structure $\mathcal{DS}$. Obviously every problem in *UW* is also in *US* (the class of problems with *unique solution* [4]); however, not every problem in *US* can be written in the above logic form [15]. The class *UW* includes $co\mathcal{NP}$ whereas it is not known whether it also includes $\mathcal{NP}$; the latter question is equivalent to the question of whether $\mathcal{D}^p$ equals to *UW* (and to *US* as well).

The formula $(\exists!\mathbf{T})\Phi(\mathbf{T})$ is in *Skolem normal form* if the first-order formula $\Phi(\mathbf{T})$ is in the following format:

$$\Phi(\mathbf{T}) = (\forall \mathbf{x})(\exists \mathbf{y})(\Theta_1(\mathbf{T}, \mathbf{x}, \mathbf{y}) \vee \ldots \vee \Theta_k(\mathbf{T}, \mathbf{x}, \mathbf{y})).$$

Next we show that any existential second-order formula with unique witnesses can be brought into Skolem normal form as it happens for formulas with multiple witnesses.

Lemma 3. *Given a second order formula* $\Gamma = (\exists!\mathbf{T})\Phi(\mathbf{T})$, *there is a a Skolem normal form formula which is equivalent to* Γ.

Proof. We first bring $\Phi(\mathbf{T})$ in prenex normal form and then apply repeatedly the equivalence

$$(\forall \mathbf{u})(\exists \mathbf{v})\Theta(\mathbf{u},\mathbf{v}) \Leftrightarrow (\exists!S)\{(\forall \mathbf{u})(\forall \mathbf{v})[S(\mathbf{u},\mathbf{v}) \leftrightarrow \Theta(\mathbf{u},\mathbf{v})] \wedge (\forall \mathbf{u})(\exists \mathbf{v})S(\mathbf{u},\mathbf{v})\}$$

Observe that our "Skolemization" differs from the classical one for existential second order formulas with multiple witnesses [7, 16] essentially because $S(\mathbf{u}, \mathbf{v}) \rightarrow \Theta(\mathbf{u}, \mathbf{v})$ is replaced by $S(\mathbf{u}, \mathbf{v}) \leftrightarrow \Theta(\mathbf{u}, \mathbf{v})$. Thus, we require that the chosen relation for S be maximal, i.e., it exactly contains all the tuples (u, v) satisfying Θ in addition to have at least one of such tuples for every u as in the classical Skolemization. Therefore, as the maximal relation for S is obviously unique, also $\exists S$ of classical Skolemization can be replaced by $\exists ! S$. Note that our procedure of Skolemization in general requires more steps than the classical one because the implication $S(\mathbf{u}, \mathbf{v}) \leftarrow \Theta(\mathbf{u}, \mathbf{v})$ corresponds to $S(\mathbf{u}, \mathbf{v}) \vee \neg \Theta(\mathbf{u}, \mathbf{v})$ so that negation must be suitably propagated inside Θ by inverting quantifiers and logical connectives. □

Theorem 4. $\mathcal{EXP}_{\mathcal{TS}^1}[\mathbf{Q}] = DB\text{-}\mathcal{UW}$.

Proof. [*Proof of* $\mathcal{EXP}_{\mathcal{TS}^1}[\mathbf{Q}] \subseteq DB\text{-}\mathcal{UW}$.] Take any query $Q = \langle \mathcal{LP}, G \rangle$; without loss of generality assume that G is a zero-arity atom g. Given $\mathbf{D} = \mathcal{EXP}_{\mathcal{TS}^1}(Q)$, we have to show that $\mathbf{D}$ is in $DB\text{-}\mathcal{UW}$, i.e., there exists an existential second-order formula defining $\mathbf{D}$ of the format $(\exists ! \mathbf{S}) \Phi(\mathbf{S})$, where $\Phi(\mathbf{S})$ is a first order formula. By the definition of unique T-stable model semantics, a database D on $\mathcal{DS}_{\mathcal{LP}}$ is in $\mathbf{D}$ if and only if the following conditions hold: (i) there exists exactly one T-stable model for $\mathcal{LP}_D$ and (ii) g is in exactly one T-stable model for $\mathcal{LP}_D$. To complete the proof, it is sufficient to show that each of the above two conditions is in $\mathcal{UW}$. Observe that Condition (i) is not subsumed by Condition (ii); in fact, the latter condition does not forbid to have other T-stable models containing $\neg g$.

Condition (i) can be expressed by the second-order formula $(\exists ! \mathbf{S}) \Gamma(\mathbf{S})$ over the database scheme $\mathcal{DS}_{\mathcal{LP}}$ as follows. $\mathbf{S}$ has a relation symbol for each IDB predicate symbol of $\mathcal{LP}$ and selecting relations $\mathbf{s}$ for $\mathbf{S}$ defines a set $M(\mathbf{s})$ of ground literals $\{s(t) |\ s \in \mathbf{S}$ and t is a tuple in the relation of $\mathbf{s}$ corresponding to $s\}$. We define Γ in such a way that, for each database D on $\mathcal{DS}_{\mathcal{LP}}$, $\Gamma(\mathbf{s})$ is true if and only if $M(\mathbf{s})$ is a T-stable model of $\mathcal{LP}_D$; therefore, the formula $(\exists ! \mathbf{S})\ \Gamma(\mathbf{S})$ is satisfied if there exists a unique T-stable model for $\mathcal{LP}_D$. But testing T-stability is in $\mathcal{P}$ by part 1 of Fact 1. So, as $\mathcal{P} \subseteq co\mathcal{NP} \subseteq \mathcal{UW}$, $\Gamma(\mathbf{s})$ can be expressed by a second-order formula $(\exists ! \mathbf{S}_2) \Omega(\mathbf{s}_1, \mathbf{S}_2)$ where $\Omega(\mathbf{s}_1, \mathbf{S}_2)$ is a first order formula. Hence, Condition (i) is defined by the formula: $(\exists ! \mathbf{S}_1, \mathbf{S}_2)\ \Omega(\mathbf{S}_1, \mathbf{S}_2)$.

It is now easy to see that also Condition (ii) is in $\mathcal{UW}$. Indeed, take the above formula $(\exists ! \mathbf{S}) \Gamma(\mathbf{S})$ with the following extended condition: for each database D on $\mathcal{DS}_{\mathcal{LP}}$, $\Gamma(\mathbf{s})$ is true if and only if both (i) $M(\mathbf{s})$ is a T-stable model of $\mathcal{LP}_D$ and (ii) g is in $M(\mathbf{s})$. Let s be the relation symbol in $\mathbf{s}$ corresponding to g. Then $\Gamma(\mathbf{s})$ can be now expressed by a second-order formula $(\exists ! \mathbf{S}_2)\ (\Omega(\mathbf{s}, \mathbf{S}_2) \wedge s())$, where Ω, defined as above, tests T-stability of $M(\mathbf{s})$ and s checks membership of g to $M(\mathbf{s})$.

[*Proof of* $DB\text{-}\mathcal{UW} \subseteq \mathcal{EXP}_{\mathcal{TS}^1}[\mathbf{Q}]$.] Take any database collection $\mathbf{D}$ on a database scheme $\mathcal{DS}$ whose recognition is in $\mathcal{UW}$. Then, by Lemma 3 $\mathbf{D}$ can be defined by a Skolem normal form second order formula, say:

$$(\exists ! \mathbf{S})(\forall \mathbf{x})(\exists \mathbf{y})(\Theta_1(\mathbf{S}, \mathbf{x}, \mathbf{y}) \vee \ldots \vee \Theta_k(\mathbf{S}, \mathbf{x}, \mathbf{y})).$$

It is now easy to prove that **D** is the database set of a query under unique T-stable model semantics. Indeed, consider the query $Q = \langle \mathcal{LP}, \neg g \rangle$, where $\mathcal{LP}$ is:

$r_1: s_j(\mathbf{W}_j) \leftarrow \neg \hat{s}_j(\mathbf{W}_j). \ (1 \leq j \leq m)$ $\qquad r_4: g \leftarrow \neg q(\mathbf{X}).$
$r_2: \hat{s}_j(\mathbf{W}_j) \leftarrow \neg s_j(\mathbf{W}_j). \ (1 \leq j \leq m)$ $\qquad r_5: p \leftarrow g, \neg p$
$r_3: q(\mathbf{X}) \quad \leftarrow \Theta_i(\mathbf{X}, \mathbf{Y}). \ (1 \leq i \leq k)$

Let D be a database on $\mathcal{DS} = \mathcal{DS}_{\mathcal{LP}}$. We construct a T-stable model for $\mathcal{LP}_D$ as follows. For each tuple $\mathbf{w}_j$, the first two groups of rules make true either $s_j(\mathbf{w}_j)$ or $\hat{s}_j(\mathbf{w}_j)$; using these rules, we perform a non-deterministic selection of relations for **S**. For each **x**, rules 3 makes $q(\mathbf{x})$ true if there exists some **y** for which one of Θ_i is satisfied. By rules 4, g is false if and only if the selected relations for **S** are witnesses for $\Phi(\mathbf{S})$ (i.e., for each **x**, $q(\mathbf{x})$ is true). By rule 5, p is not undefined if and only if g is made false; so the role of this rule is to invalidate any selection for **S** that does not make g false. Therefore, the program $\mathcal{LP}_D$ admits a number of T-stable models, one for every witness for $\Phi(\mathbf{S})$. Hence, if $D \in \mathbf{D}$ then there is a unique witness for $\Phi(\mathbf{S})$ and, therefore, a unique T-stable model of $\mathcal{LP}_D$, say M; since $\neg g \in M$, $D \in \mathcal{EXP}_{TS^1}(Q)$ as well. On the other hand, if $D \notin \mathbf{D}$ then $\mathcal{LP}_D$ admits either no T-stable model or multiple T-stable models, so $D \notin \mathcal{EXP}_{TS^1}(Q)$. It turns out that $\mathbf{D} = \mathcal{EXP}_{TS^1}(Q)$; therefore, $DB\text{-}\mathcal{UN} \subseteq \mathcal{EXP}_{TS^1}[\mathbf{Q}]$. □

We point out that this is not the first time that a relationship between **DATALOG**⁻ and the class $\mathcal{UN}$ is discovered: **DATALOG**⁻ programs with unique fixpoint are characterized in terms of $\mathcal{UN}$ in [16].

As $co\mathcal{NP} \subseteq \mathcal{UN} \subseteq \mathcal{D}^p$, from Theorem 4 and Fact 2 we derive that, measured in terms of expressive powers, unique semantics subsumes certain semantics and, in turn, it is subsumed by definite semantics. The relationships among the various versions of T-stable model semantics is depicted in Fig. 1.

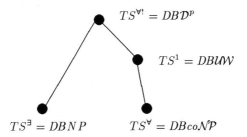

Figure 1: *Relationships among T-stable semantics*

Example 2. In Example 1, we have shown that, under the unique T-stable model semantics, the query $Q^K = \langle K, kernel \rangle$ defines the problem of whether a graph has exactly one kernel — this problem is a typical problem in $\mathcal{UN}$. Since $co\mathcal{NP} \subseteq \mathcal{UN}$, according to Theorem 4 unique T-stable model semantics is also able to express the $co\mathcal{NP}$-complete problem of whether a graph has no kernel. We next show how to modify the query Q^K to formulate this problem.

Let K'' be the program obtained from K by modifying the first two rules into:

$$\text{v1(X)} \leftarrow \text{b, v(X)}, \neg\text{v2(X)}. \qquad \text{v2(X)} \leftarrow \text{b, v(X)}, \neg\text{v1(X)}.$$

and by adding the following three rules:

$$\text{a} \leftarrow \neg\text{b}. \qquad \text{b} \leftarrow \neg\text{a}. \qquad \text{kernel} \leftarrow \text{a}.$$

Under the unique T-stable model semantics, the query $Q^{K''} = \langle K'', kernel \rangle$ expresses the problem of whether a graph has no kernel. In fact, the interpretation $M = \{a, kernel\} \cup D^3$ is a T-stable model for K''_D for any database D. Moreover, the program has an additional T-stable model for every kernel in the graph G corresponding to D; for such models a is false and b is true. Therefore, K''_D has a unique T-stable model (that is, M) if and only if G has no kernel. □

4 Subclasses of Queries with Unique T-stable Model

So far we have analyzed various types of deterministic semantics for the class **Q** of all DATALOG¬ queries. In this subsection we consider two interesting subclasses of **Q** for which the unique T-stable model semantics is the natural semantics:

- $\mathbf{Q}_1 = \{Q = \langle \mathcal{LP}, G \rangle \mid \forall D \text{ on } \mathcal{DS}_{\mathcal{LP}}, \mathcal{LP}_D \text{ admits a unique T-stable model}\}$
- $\mathbf{Q}_{0,1} = \{Q = \langle \mathcal{LP}, G \rangle \mid \forall D \text{ on } \mathcal{DS}_{\mathcal{LP}}, \mathcal{LP}_D \text{ admits at most one T-stable model}\}$

Obviously, $\mathbf{Q}_1 \subset \mathbf{Q}_{0,1} \subset \mathbf{Q}$. Note that, while **Q** is a recursive query language, the two sub-classes are not recursively enumerable as it is not in general decidable whether a DATALOG¬ program has a unique T-stable model for every database. Therefore, $\mathbf{Q}_1$ and $\mathbf{Q}_{0,1}$ are not query languages in the sense of [13].

The two subclasses blur the differences among the various T-stable model semantics.

Proposition 5.
1. *For each* $Q \in \mathbf{Q}_1$, $\mathcal{EXP}_{TS^\exists}(Q) = \mathcal{EXP}_{TS^\forall}(Q) = \mathcal{EXP}_{TS^{\forall !}}(Q) = \mathcal{EXP}_{TS^1}(Q)$;
2. *For each* $Q \in \mathbf{Q}_{0,1}$, $\mathcal{EXP}_{TS^\exists}(Q) = \mathcal{EXP}_{TS^{\forall !}}(Q) = \mathcal{EXP}_{TS^1}(Q)$.

Proof. . Let $Q = \langle \mathcal{LP}, G \rangle$ and D be a database on $\mathcal{DS}_{\mathcal{LP}}$. If Q is in $\mathbf{Q}_1$ then $\mathcal{LP}_D$ has exactly one T-stable model: so all semantics coincide. Suppose now that $Q \in \mathbf{Q}_{0,1}$: if $\mathcal{LP}_D$ has no T-stable model then only certain semantics behaves differently from the other semantics. □

Next we characterize the expressive power of T-stable model semantics for the two subclasses of queries. To this end we need to consider further database classes, first introduced in [15]:

---
[3] The database D is seen as a set of ground atoms, one for each tuple.

1. $DB\text{-}\mathcal{U}\Sigma_1^1$ denotes the subset of $DB\text{-}\mathcal{UW}$ consisting of each database collection $\mathbf{D}$ which is defined by a formula $(\exists!\mathbf{T})\Phi(\mathbf{T})$ such that for each $D \notin \mathbf{D}$, $(\exists!\mathbf{T})\Phi(\mathbf{T})$ is false, i.e., for each D, either the formula is satisfied by exactly one witness (and, then, $D \in \mathbf{D}$) or it is not;
2. $DB\text{-}\mathcal{U}\Delta_1^1$ denotes the subset of $DB\text{-}\mathcal{U}\Sigma_1^1$ of all database collections $\mathbf{D}$ for which the complementary database collection $\mathbf{D}'$ is in $DB\text{-}\mathcal{U}\Sigma_1^1$ as well.

As discussed in [15], $\mathcal{U}\Sigma_1^1$ is related to the complexity class $\mathcal{UP}$ that has been introduced by Valiant [22] and consists of all *unambiguous* computations. Indeed $\mathcal{U}\Sigma_1^1$ captures $\mathcal{UP}$ if an order on the universe is available; in this case, $\mathcal{U}\Delta_1^1$ captures $\mathcal{UP} \cap co\mathcal{UP}$.

Lemma 6. *Given a second order formula* $\Gamma = (\exists \mathbf{T})\Phi(\mathbf{T})$ *in* $\mathcal{U}\Sigma_1^1$, *there is a Skolem normal form formula in* $\mathcal{U}\Sigma_1^1$ *which is equivalent to* Γ.

Proof. We first bring $\Phi(\mathbf{T})$ in prenex normal form and then apply repeatedly the Skolemization introduced in the proof of Lemma 3. It is easy to see that every relation symbol added by the skolemization admits at most one witness. □.

Theorem 7.
1. $DB\text{-}\mathcal{U}\Delta_1^1 \subseteq \mathcal{EXP}_{TS^1}[\mathbf{Q}_1] \subseteq DB\text{-}\mathcal{NP} \cap DB\text{-}co\mathcal{NP}$;
2. $DB\text{-}\mathcal{U}\Delta_1^1 \subseteq \mathcal{EXP}_{TS^1}[\mathbf{Q}_{0,1}] \subseteq DB\text{-}\mathcal{NP} \cap DB\text{-}\mathcal{UW}$.

Proof. (1) Since $\mathcal{EXP}_{TS^\exists}[\mathbf{Q}] = DB\text{-}\mathcal{NP}$ by Fact 2 and $\mathbf{Q}_1 \subseteq \mathbf{Q}$, $\mathcal{EXP}_{TS^\exists}[\mathbf{Q}_1] \subseteq DB\text{-}\mathcal{NP}$; so, as $\mathcal{EXP}_{TS^\exists}[\mathbf{Q}_1] = \mathcal{EXP}_{TS^1}[\mathbf{Q}_1]$ by Proposition 5, $\mathcal{EXP}_{TS^1}[\mathbf{Q}_1] \subseteq DB\text{-}\mathcal{NP}$. By replacing $\exists$ with $\forall$ and $\mathcal{NP}$ with $co\mathcal{NP}$ and repeating the previous argument we obtain $\mathcal{EXP}_{TS^1}[\mathbf{Q}_1] \subseteq DB\text{-}co\mathcal{NP}$. Hence, $\mathcal{EXP}_{TS^1}[\mathbf{Q}_1] \subseteq DB\text{-}\mathcal{NP} \cap DB\text{-}co\mathcal{NP}$. Let us now prove the other relationship. Let $\mathbf{D}$ be a database collection in $DB\text{-}\mathcal{U}\Delta_1^1$, say with database scheme $\mathcal{DS}$. Let $\mathbf{D}'$ be the complementary database collection of $\mathbf{D}$. Then, by definition of $\mathcal{U}\Delta_1^1$, $\mathbf{D}$ and $\mathbf{D}'$ are defined by two formulas in $\mathcal{U}\Sigma_1^1$, say $(\exists \mathbf{S})\phi(\mathbf{S})$ and $(\exists \mathbf{S}')\phi'(\mathbf{S}')$, respectively. By Lemma 6, we can assume that both formulas are in Skolem format say:

$$\phi(\mathbf{S}) = (\forall \mathbf{x})(\exists \mathbf{y})(\Theta_1(\mathbf{S},\mathbf{x},\mathbf{y}) \vee \ldots \vee \Theta_k(\mathbf{S},\mathbf{x},\mathbf{y})),$$
$$\phi'(\mathbf{S}') = (\forall \mathbf{x}')(\exists \mathbf{y}')(\Theta_1'(\mathbf{S}',\mathbf{x}',\mathbf{y}') \vee \ldots \vee \Theta_{k'}'(\mathbf{S}',\mathbf{x}',\mathbf{y}')).$$

Consider the program $\mathcal{LP}''$:

$r_1: a \leftarrow \neg b.$ $\qquad r_2: b \leftarrow \neg a.$

$r_3: s_j(\mathbf{W}_j) \leftarrow a, \neg \hat{s}_j(\mathbf{W}_j). \ (1 \leq j \leq m)$
$r_4: \hat{s}_j(\mathbf{W}_j) \leftarrow a, \neg s_j(\mathbf{W}_j). \ (1 \leq j \leq m)$
$r_5: q(\mathbf{X}) \quad \leftarrow \Theta_i(\mathbf{X},\mathbf{Y}). \ (1 \leq i \leq k)$
$r_6: g \quad \leftarrow \neg q(\mathbf{X}).$

$r_7: s_j'(\mathbf{W}_j') \leftarrow b, \neg \hat{s}_j'(\mathbf{W}_j'). \ (1 \leq j \leq m')$
$r_8: \hat{s}_j'(\mathbf{W}_j') \leftarrow b, \neg s_j'(\mathbf{W}_j'). \ (1 \leq j \leq m')$
$r_9: q'(\mathbf{X}') \quad \leftarrow \Theta_i'(\mathbf{X}',\mathbf{Y}'). \ (1 \leq i \leq k')$
$r_{10}: g' \quad \leftarrow \neg q'(\mathbf{X}').$

$r_{11}: \ g'' \leftarrow \neg g.$ $\qquad r_{12}: \ g'' \leftarrow \neg g'.$ $\qquad r_{13}: \ p \leftarrow \neg g'', \neg p$

The program $\mathcal{LP}''$ consists of two subprograms: $\mathcal{LP}$ (rules 3–6) and $\mathcal{LP}'$ (rules 7–10) plus the first two rules which enable one of the two subprograms plus the rules 11–13 which make p undefined iff neither g nor g' is false. Observe that, under the $TS^\exists$ semantics, the queries $Q = \langle \mathcal{LP}, \neg g \rangle$ and $Q' = \langle \mathcal{LP}', \neg g' \rangle$ defines $\mathbf{D}$ and $\mathbf{D}'$, respectively; moreover for each D, if $D \in \mathbf{D}$ (resp. $\mathbf{D}'$) then there exists exactly one T-stable model M for $\mathcal{LP}_D$ (resp., $\mathcal{LP}'_D$) such that $\neg g \in M$. It is then easy to see that for each D, $\mathcal{LP}''_D$ has exactly one T-stable model. Therefore, the query $Q'' = \langle \mathcal{LP}'', \neg g \rangle$ is in $\mathbf{Q}_1$ and $\mathcal{EXP}_{TS^1}(Q'') = \mathbf{D}$; so $DB\text{-}\mathcal{U}\Delta_1^1 \subseteq \mathcal{EXP}_{TS^1}[\mathbf{Q}_1]$.

(2) As $\mathbf{Q}_1 \subset \mathbf{Q}_{0,1}$, by Part (1), $DB\text{-}\mathcal{U}\Delta_1^1 \subseteq \mathcal{EXP}_{TS^1}[\mathbf{Q}_{0,1}]$. Concerning the second relationship, we have that $\mathcal{EXP}_{TS^\exists}[\mathbf{Q}] = DB\text{-}\mathcal{NP}$ and $\mathcal{EXP}_{TS^1}[\mathbf{Q}] = DB\text{-}\mathcal{UW}$ by Fact 2 and Theorem 4. Therefore, as $\mathcal{EXP}_{TS^\exists}[\mathbf{Q}_{0,1}] = \mathcal{EXP}_{TS^1}[\mathbf{Q}_{0,1}]$ by Proposition 5, we derive that $\mathcal{EXP}_{TS^1}[\mathbf{Q}_{0,1}] \subseteq DB\text{-}\mathcal{NP} \cap DB\text{-}\mathcal{UW}$. □

Note that classes of queries whose expressive power is bounded by $\mathcal{NP} \cap co\mathcal{NP}$ have been studied in [12, 11, 10] and that also such classes are characterized by similar uniqueness conditions.

The above results are rather negative with respect to the possibility to single out a subclass of $\mathbf{Q}_1$ or $\mathbf{Q}_{0,1}$ which can be expressed by a query language more powerful than stratified **DATALOG**¬. In fact, as the classes $\mathcal{NP} \cap \mathcal{UW}$ and $\mathcal{NP} \cap co\mathcal{NP}$ as well as any known subclass of them over $\mathcal{P}$ are not syntactic unless something surprising is true (e.g., $\mathcal{NP} \subseteq \mathcal{UW}$, $\mathcal{NP} = co\mathcal{NP}$ or $\mathcal{NP} \cap co\mathcal{NP} = \mathcal{P}$) [13], it turns out that any query language in $\mathbf{Q}_{TS_1}$ or $\mathbf{Q}_{TS_{0,1}}$ cannot express more than $\mathcal{P}$. But it is not know either whether $\mathcal{P}$ is expressible by a query language and whether there exists a language for total well-founded semantics preserving the capability of expressing all fixpoint queries [1, 10]. Flum et al. have recently shown in [8] that total well-founded semantics has the same expressive power as 'partial' well-founded semantics. However, this result refers to database equivalence in the sense that a 'partial' query on a database can be replaced by the same query on a different database yielding a total model. Thus they have not proved the existence of a language L with the power of fixpoint queries which only generates queries whose well-founded models are total for every database. So follows our conjecture that any practical language for **DATALOG**¬ queries with unique T-stable model is not more expressive than stratified **DATALOG**¬:

Conjecture 1 *Given any subset* $\mathbf{Q}'$ *of* $\mathbf{Q}_{TS_1}$, *if* $\mathbf{Q}'$ *is recursively enumerable then* $\mathcal{EXP}_{TS^1}[\mathbf{Q}'] \subseteq \mathcal{EXP}_{TS^1}[\mathbf{Q}'']$, *where* $\mathbf{Q}''$ *is the class of all* **DATALOG**¬ *queries with stratified negation.* □

ACKNOWLEDGEMENTS: Work partially supported by the EC-US033 project "DEUS EX MACHINA: non-determinism in deductive databases", and by a MURST grant under the project "Sistemi formali e strumenti per basi di dati evolute". The third author's work is also supported by ISI-CNR. The authors would like to thank a anonymous reader of a preliminary draft of this paper for many stimulating suggestions and criticisms.

References

1. Abiteboul S., Hull R., V. Vianu, *Foundations of Databases*, Addison-Wesley, 1994.
2. Abiteboul S., Simon E., and V. Vianu, Non-deterministic languages to express deterministic transformations, *Proc. ACM PODS Symp.*,1990, pp. 218-229.
3. Apt K., Blair H. and A. Walker, Towards a theory of declarative knowledge, in *Foundations of Deductive Databases and Logic Programming* (J. Minker ed.), Morgan Kauffman, 1988, 89-142.
4. Blass A. and Y. Gurevich, On the Unique Satisfiability Problem, *Inform. and Control*, 1982, pp. 80-88.
5. Chandra A., and D. Harel, Structure and Complexity of Relational Queries, *Journal of Computer and System Sciences 25*, 1, 1982, pp. 99-128.
6. Eiter T., Gottlob G. and H. Manila, Expressive Power and Complexity of Disjunctive DATALOG, *Proc. ACM PODS Symp.*, Minneapolis, USA, May 1994.
7. Fagin R., Generalized First-Order Spectra and Polynomial-Time Recognizable Sets, in *Complexity of Computation*, SIAM-AMS Proc., Vol. 7, 1974, pp. 43-73.
8. Flum J., Kubierschky M., and B. Ludascher, Total and partial well-founded Datalog coincide, in *Proc. Int. Conf. on Database Theory (ICDT)*, 1997, pp. 113-124.
9. Gelfond M., and V. Lifschitz, The Stable Model Semantics for Logic Programmin, *Proc. 5th Int. Conf. on Logic Programming*, 1988, pp. 1070-1080.
10. Greco S. and Saccà D., 'Possible is Certain' is desirable and can be expressive, *Annals of Mathematics and Artificial Intelligence*, 1997.
11. Grumbach S. and Z. Lacroix, On non-determinism in machines and languages, *Annals of Mathematics and Artificial Intelligence*, 1997.
12. Grumbach S., Lacroix Z. and S. Lindell, Implicit Definitions on Finite Structures, in *Proc. of the Conf. on Computer Science Logic*, 1995.
13. Gurevich Y., Logic and the Challenge of Computer Science, in E. Borger (ed.), *Trends in Theoretical Computer Science*, Computer Science Press, 1988.
14. Johnson D.S., A Catalog of Complexity Classes, in J. van Leewen (ed.), *Handbook of Theoretical Computer Science*, Vol. 1, North-Holland, 1990.
15. Kolaitis P.G., Implicit definability on finite structures and unambiguous computations, *Proc. 5th IEEE Symp. on Logic in Computer Science*, 1990, pp. 168-180.
16. Kolaitis P.G. and C.H. Papadimitriou, Why not Negation by Fixpoint?, *Journal of Computer and System Sciences 43*, 1991, pp. 125-144.
17. Marek W., M. Truszcynski, Autoepistemic Logic, *J. ACM 38*, 3, 1991, pp. 588-619.
18. Papadimitriou C., *Computational Complexity*, Addison-Wesley, 1994.
19. Saccà D., Multiple Total Stable Models are Definitely Needed to Solve Unique Solution Problems, *Information Processing Letters 58*, 5, 1996, pp. 249-254
20. Schlipf J.S., The Expressive Powers of the Logic Programming Semantics, *Proc. ACM PODS Symp.*, 1990, pp. 196-204.
21. Ullman J.D., *Principles of Database and Knowledge Base Systems*, Computer Science Press, 1989.
22. Valiant L., "Relative complexity of checking and evaluating", *Information Processing Letters 5*, 1976, pp. 20-23.
23. Van Gelder A., Ross K. and J.S. Schlipf, The Well-Founded Semantics for General Logic Program, *Journal of the ACM 38*, 3, 1991, pp. 620-650.
24. Vardi M.Y., The Complexity of Relational Query Languages, *Proc. ACM Symp. on Theory of Computing*, 1982, pp. 137-146.

Author Index

Farid Ablayev, 195
Stephen Alstrup, 270
Andris Ambainis, 401
Alexander E. Andreev, 177
Krzysztof R. Apt, 36
Andrea Asperti, 259
Christel Baier, 430
Amotz Bar-Noy, 738
Yair Bartal, 516
Frédérique Bassino, 76
Marie-Pierre Béal, 76
Martin Beaudry, 110
Richard Beigel, 816
Marco Bernardo, 358
Hans L. Bodlaender, 627
Michele Boreale, 482
Ahmed Bouajjani, 560
Olivier Bournez, 143
Hajo Broersma, 760
Véronique Bruyère, 87
Francesco Buccafurri, 849
Harry Buhrman, 188
Olaf Burkart, 419
Edson Cáceres, 390
Olivier Carton, 17
Julien Cassaigne, 693
Edmund M. Clarke, 430
Andrea E. F. Clementi, 177
Bruno Codenotti, 203
Laurent Dami, 782
Frank Dehne, 390
Rocco De Nicola, 482
Alfredo De Santis, 716
Roberto Di Cosmo, 237
Giovanni Di Crescenzo, 716
Volker Diekert, 336
Pietro Di Gianantonio, 121
Manfred Droste, 682
Bruno Durand, 65
Tamar Eilam, 527
Péter L. Erdős, 827
Funda Ergün, 203
Thomas Erlebach, 493
Stephen Fenner, 188

Alfonso Ferreira, 390
Michele Flammini, 527
Paola Flocchini, 390
Wan Fokkink, 571
Lance Fortnow, 188
Yair Frankel, 705
Bin Fu, 816
Yuxi Fu, 325
Toshihiro Fujito, 749
Luisa Gargano, 505
Paul Gastin, 682
Peter S. Gemmell, 203
Neil Ghani, 237
Roberto Giacobazzi, 771
John Glauert, 649
Sergio Greco, 849
Roberto Grossi, 605
Yuri Gurevich, 154
Peter Habermehl, 560
Torben Hagerup, 292
David Harel, 408
Vasiliki Hartonas-Garmhausen, 430
Pavol Hell, 505
Edith Hemaspaandra, 214
Lane A. Hemaspaandra, 214
Matthew Hennessy, 471
Thomas A. Henzinger, 582
Jacob Holm, 270
Kohei Honda, 225
Giuseppe F. Italiano, 605
Petr Jančar, 549
Klaus Jansen, 493, 727
Valentine Kabanets, 805
Christos Kaklamanis, 493
Juhani Karhumäki, 98
Sanjeev Khanna, 616
Zurab Khasidashvili, 649
Valerie King, 594
Ton Kloks, 760
Peter W. Kopke, 582
Guy Kortsarz, 738
Dieter Kratsch, 760
Sven O. Krumke, 281
Marta Kwiatkowska, 430

Laura F. Landweber, 56
Cosimo Laneve, 259
James I. Lathrop, 132
François Lemieux, 110
Stefano Leonardi, 516
Kristian de Lichtenberg, 270
Richard J. Lipton, 56
Jack H. Lutz, 132
Giovanni Manzini, 794
Madhav V. Marathe, 281
Massimo Marchiori, 660
Luciano Margara, 794
Ian A. Mason, 369
Yuri Matiyasevich, 336
Kurt Mehlhorn, 7
Filippo Mignosi, 98
Robin Milner, 1
Haiko Müller, 760
Anca Muscholl, 336
S. Muthukrishnan, 616
Stefan Näher, 7
Paliath Narendran, 638
Hartmut Noltemeier, 281
Friedrich Otto, 638
Valeria de Paiva, 248
Christos H. Papadimitriou, 2
Stephane Perennes, 505
Dominique Perrin, 17, 76
Pino Persiano, 493, 716
Anna Philippou, 314
Wojciech Plandowski, 98
Rosario Pugliese, 482
Francesco Ranzato, 771
Monika Rauch Henzinger, 594
Ramamurthy Ravi, 281
S.S. Ravi, 281
S. Ravi Kumar, 203
James Riely, 471
Ingo Rieping, 390

Eike Ritter, 248
John M. Robson, 441
José D. P. Rolim, 177
Alessandro Roncato, 390
Jörg Rothe, 214
Salvador Roura, 449
Salvatore Ruggieri, 838
Jan J. M. M. Rutten, 460
Mark Ryan, 430
Domenico Saccà, 849
Davide Sangiorgi, 303
Nicola Santoro, 390
Uwe Schwiegelshohn, 379
Géraud Sénizergues, 671
Eli Singerman, 408
Steven Skiena, 616
Siang W. Song, 390
Michael A. Steel, 827
Bernhard Steffen, 419
Ravi Sundaram, 281
László A. Székely, 827
Carolyn L. Talcott, 369
Denis Thérien, 110
Lothar Thiele, 379
Dimitrios M. Thilikos, 627
Mikkel Thorup, 270
Christian Uhrig, 7
Erik de Vink, 460
Walter Vogler, 538
Andrei Voronkov, 154
David Walker, 314
Tandy J. Warnow, 827
Klaus Weihrauch, 166
Thomas Wilke, 347
Hans-Christoph Wirth, 281
Nobuko Yoshida, 225
Moti Yung, 705
Shmuel Zaks, 527

Lecture Notes in Computer Science

For information about Vols. 1–1179

please contact your bookseller or Springer-Verlag

Vol. 1180: V. Chandru, V. Vinay (Eds.), Foundations of Software Technology and Theoretical Computer Science. Proceedings, 1996. XI, 387 pages. 1996.

Vol. 1181: D. Bjørner, M. Broy, I.V. Pottosin (Eds.), Perspectives of System Informatics. Proceedings, 1996. XVII, 447 pages. 1996.

Vol. 1182: W. Hasan, Optimization of SQL Queries for Parallel Machines. XVIII, 133 pages. 1996.

Vol. 1183: A. Wierse, G.G. Grinstein, U. Lang (Eds.), Database Issues for Data Visualization. Proceedings, 1995. XIV, 219 pages. 1996.

Vol. 1184: J. Waśniewski, J. Dongarra, K. Madsen, D. Olesen (Eds.), Applied Parallel Computing. Proceedings, 1996. XIII, 722 pages. 1996.

Vol. 1185: G. Ventre, J. Domingo-Pascual, A. Danthine (Eds.), Multimedia Telecommunications and Applications. Proceedings, 1996. XII, 267 pages. 1996.

Vol. 1186: F. Afrati, P. Kolaitis (Eds.), Database Theory - ICDT'97. Proceedings, 1997. XIII, 477 pages. 1997.

Vol. 1187: K. Schlechta, Nonmonotonic Logics. IX, 243 pages. 1997. (Subseries LNAI).

Vol. 1188: T. Martin, A.L. Ralescu (Eds.), Fuzzy Logic in Artificial Intelligence. Proceedings, 1995. VIII, 272 pages. 1997. (Subseries LNAI).

Vol. 1189: M. Lomas (Ed.), Security Protocols. Proceedings, 1996. VIII, 203 pages. 1997.

Vol. 1190: S. North (Ed.), Graph Drawing. Proceedings, 1996. XI, 409 pages. 1997.

Vol. 1191: V. Gaede, A. Brodsky, O. Günther, D. Srivastava, V. Vianu, M. Wallace (Eds.), Constraint Databases and Applications. Proceedings, 1996. X, 345 pages. 1996.

Vol. 1192: M. Dam (Ed.), Analysis and Verification of Multiple-Agent Languages. Proceedings, 1996. VIII, 435 pages. 1997.

Vol. 1193: J.P. Müller, M.J. Wooldridge, N.R. Jennings (Eds.), Intelligent Agents III. XV, 401 pages. 1997. (Subseries LNAI).

Vol. 1194: M. Sipper, Evolution of Parallel Cellular Machines. XIII, 199 pages. 1997.

Vol. 1195: R. Trappl, P. Petta (Eds.), Creating Personalities for Synthetic Actors. VII, 251 pages. 1997. (Subseries LNAI).

Vol. 1196: L. Vulkov, J. Waśniewski, P. Yalamov (Eds.), Numerical Analysis and Its Applications. Proceedings, 1996. XIII, 608 pages. 1997.

Vol. 1197: F. d'Amore, P.G. Franciosa, A. Marchetti-Spaccamela (Eds.), Graph-Theoretic Concepts in Computer Science. Proceedings, 1996. XI, 410 pages. 1997.

Vol. 1198: H.S. Nwana, N. Azarmi (Eds.), Software Agents and Soft Computing: Towards Enhancing Machine Intelligence. XIV, 298 pages. 1997. (Subseries LNAI).

Vol. 1199: D.K. Panda, C.B. Stunkel (Eds.), Communication and Architectural Support for Network-Based Parallel Computing. Proceedings, 1997. X, 269 pages. 1997.

Vol. 1200: R. Reischuk, M. Morvan (Eds.), STACS 97. Proceedings, 1997. XIII, 614 pages. 1997.

Vol. 1201: O. Maler (Ed.), Hybrid and Real-Time Systems. Proceedings, 1997. IX, 417 pages. 1997.

Vol. 1203: G. Bongiovanni, D.P. Bovet, G. Di Battista (Eds.), Algorithms and Complexity. Proceedings, 1997. VIII, 311 pages. 1997.

Vol. 1204: H. Mössenböck (Ed.), Modular Programming Languages. Proceedings, 1997. X, 379 pages. 1997.

Vol. 1205: J. Troccaz, E. Grimson, R. Mösges (Eds.), CVRMed-MRCAS'97. Proceedings, 1997. XIX, 834 pages. 1997.

Vol. 1206: J. Bigün, G. Chollet, G. Borgefors (Eds.), Audio- and Video-based Biometric Person Authentication. Proceedings, 1997. XII, 450 pages. 1997.

Vol. 1207: J. Gallagher (Ed.), Logic Program Synthesis and Transformation. Proceedings, 1996. VII, 325 pages. 1997.

Vol. 1208: S. Ben-David (Ed.), Computational Learning Theory. Proceedings, 1997. VIII, 331 pages. 1997. (Subseries LNAI).

Vol. 1209: L. Cavedon, A. Rao, W. Wobcke (Eds.), Intelligent Agent Systems. Proceedings, 1996. IX, 188 pages. 1997. (Subseries LNAI).

Vol. 1210: P. de Groote, J.R. Hindley (Eds.), Typed Lambda Calculi and Applications. Proceedings, 1997. VIII, 405 pages. 1997.

Vol. 1211: E. Keravnou, C. Garbay, R. Baud, J. Wyatt (Eds.), Artificial Intelligence in Medicine. Proceedings, 1997. XIII, 526 pages. 1997. (Subseries LNAI).

Vol. 1212: J. P. Bowen, M.G. Hinchey, D. Till (Eds.), ZUM '97: The Z Formal Specification Notation. Proceedings, 1997. X, 435 pages. 1997.

Vol. 1213: P. J. Angeline, R. G. Reynolds, J. R. McDonnell, R. Eberhart (Eds.), Evolutionary Programming VI. Proceedings, 1997. X, 457 pages. 1997.

Vol. 1214: M. Bidoit, M. Dauchet (Eds.), TAPSOFT '97: Theory and Practice of Software Development. Proceedings, 1997. XV, 884 pages. 1997.

Vol. 1215: J. M. L. M. Palma, J. Dongarra (Eds.), Vector and Parallel Processing – VECPAR'96. Proceedings, 1996. XI, 471 pages. 1997.

Vol. 1216: J. Dix, L. Moniz Pereira, T.C. Przymusinski (Eds.), Non-Monotonic Extensions of Logic Programming. Proceedings, 1996. XI, 224 pages. 1997. (Subseries LNAI).

Vol. 1217: E. Brinksma (Ed.), Tools and Algorithms for the Construction and Analysis of Systems. Proceedings, 1997. X, 433 pages. 1997.

Vol. 1218: G. Păun, A. Salomaa (Eds.), New Trends in Formal Languages. IX, 465 pages. 1997.

Vol. 1219: K. Rothermel, R. Popescu-Zeletin (Eds.), Mobile Agents. Proceedings, 1997. VIII, 223 pages. 1997.

Vol. 1220: P. Brezany, Input/Output Intensive Massively Parallel Computing. XIV, 288 pages. 1997.

Vol. 1221: G. Weiß (Ed.), Distributed Artificial Intelligence Meets Machine Learning. Proceedings, 1996. X, 294 pages. 1997. (Subseries LNAI).

Vol. 1222: J. Vitek, C. Tschudin (Eds.), Mobile Object Systems. Proceedings, 1996. X, 319 pages. 1997.

Vol. 1223: M. Pelillo, E.R. Hancock (Eds.), Energy Minimization Methods in Computer Vision and Pattern Recognition. Proceedings, 1997. XII, 549 pages. 1997.

Vol. 1224: M. van Someren, G. Widmer (Eds.), Machine Learning: ECML-97. Proceedings, 1997. XI, 361 pages. 1997. (Subseries LNAI).

Vol. 1225: B. Hertzberger, P. Sloot (Eds.), High-Performance Computing and Networking. Proceedings, 1997. XXI, 1066 pages. 1997.

Vol. 1226: B. Reusch (Ed.), Computational Intelligence. Proceedings, 1997. XIII, 609 pages. 1997.

Vol. 1227: D. Galmiche (Ed.), Automated Reasoning with Analytic Tableaux and Related Methods. Proceedings, 1997. XI, 373 pages. 1997. (Subseries LNAI).

Vol. 1228: S.-H. Nienhuys-Cheng, R. de Wolf, Foundations of Inductive Logic Programming. XVII, 404 pages. 1997. (Subseries LNAI).

Vol. 1230: J. Duncan, G. Gindi (Eds.), Information Processing in Medical Imaging. Proceedings, 1997. XVI, 557 pages. 1997.

Vol. 1231: M. Bertran, T. Rus (Eds.), Transformation-Based Reactive Systems Development. Proceedings, 1997. XI, 431 pages. 1997.

Vol. 1232: H. Comon (Ed.), Rewriting Techniques and Applications. Proceedings, 1997. XI, 339 pages. 1997.

Vol. 1233: W. Fumy (Ed.), Advances in Cryptology — EUROCRYPT '97. Proceedings, 1997. XI, 509 pages. 1997.

Vol 1234: S. Adian, A. Nerode (Eds.), Logical Foundations of Computer Science. Proceedings, 1997. IX, 431 pages. 1997.

Vol. 1235: R. Conradi (Ed.), Software Configuration Management. Proceedings, 1997. VIII, 234 pages. 1997.

Vol. 1236: E. Maier, M. Mast, S. LuperFoy (Eds.), Dialogue Processing in Spoken Language Systems. Proceedings, 1996. VIII, 220 pages. 1997. (Subseries LNAI).

Vol. 1238: A. Mullery, M. Besson, M. Campolargo, R. Gobbi, R. Reed (Eds.), Intelligence in Services and Networks: Technology for Cooperative Competition. Proceedings, 1997. XII, 480 pages. 1997.

Vol. 1239: D. Sehr, U. Banerjee, D. Gelernter, A. Nicolau, D. Padua (Eds.), Languages and Compilers for Parallel Computing. Proceedings, 1996. XIII, 612 pages. 1997.

Vol. 1240: J. Mira, R. Moreno-Díaz, J. Cabestany (Eds.), Biological and Artificial Computation: From Neuroscience to Technology. Proceedings, 1997. XXI, 1401 pages. 1997.

Vol. 1241: M. Akşit, S. Matsuoka (Eds.), ECOOP'97 – Object-Oriented Programming. Proceedings, 1997. XI, 531 pages. 1997.

Vol. 1242: S. Fdida, M. Morganti (Eds.), Multimedia Applications, Services and Techniques – ECMAST '97. Proceedings, 1997. XIV, 772 pages. 1997.

Vol. 1243: A. Mazurkiewicz, J. Winkowski (Eds.), CONCUR'97: Concurrency Theory. Proceedings, 1997. VIII, 421 pages. 1997.

Vol. 1244: D. M. Gabbay, R. Kruse, A. Nonnengart, H.J. Ohlbach (Eds.), Qualitative and Quantitative Practical Reasoning. Proceedings, 1997. X, 621 pages. 1997. (Subseries LNAI).

Vol. 1245: M. Calzarossa, R. Marie, B. Plateau, G. Rubino (Eds.), Computer Performance Evaluation. Proceedings, 1997. VIII, 231 pages. 1997.

Vol. 1246: S. Tucker Taft, R. A. Duff (Eds.), Ada 95 Reference Manual. XXII, 526 pages. 1997.

Vol. 1247: J. Barnes (Ed.), Ada 95 Rationale. XVI, 458 pages. 1997.

Vol. 1248: P. Azéma, G. Balbo (Eds.), Application and Theory of Petri Nets 1997. Proceedings, 1997. VIII, 467 pages. 1997.

Vol. 1249: W. McCune (Ed.), Automated Deduction – CADE-14. Proceedings, 1997. XIV, 462 pages. 1997. (Subseries LNAI).

Vol. 1250: A. Olivé, J.A. Pastor (Eds.), Advanced Information Systems Engineering. Proceedings, 1997. XI, 451 pages. 1997.

Vol. 1251: K. Hardy, J. Briggs (Eds.), Reliable Software Technologies – Ada-Europe '97. Proceedings, 1997. VIII, 293 pages. 1997.

Vol. 1252: B. ter Haar Romeny, L. Florack, J. Koenderink, M. Viergever (Eds.), Scale-Space Theory in Computer Vision. Proceedings, 1997. IX, 365 pages. 1997.

Vol. 1253: G. Bilardi, A. Ferreira, R. Lüling, J. Rolim (Eds.), Solving Irregularly Structured Problems in Parallel. Proceedings, 1997. X, 287 pages. 1997.

Vol. 1254: O. Grumberg (Ed.), Computer Aided Verification. Proceedings, 1997. XI, 486 pages. 1997.

Vol. 1255: T. Mora, H. Mattson (Eds.), Applied Algebra, Algebraic Algorithms and Error-Correcting Codes. Proceedings, 1997. X, 353 pages. 1997.

Vol. 1256: P. Degano, R. Gorrieri, A. Marchetti-Spaccamela (Eds.), Automata, Languages and Programming. Proceedings, 1997. XIV, 862 pages. 1997.

Vol. 1258: D. van Dalen, M. Bezem (Eds.), Computer Science Logic. Proceedings, 1996. VIII, 473 pages. 1997.

Vol. 1259: T. Higuchi, I. Masaya, W. Liu (Eds.), Evolvable Systems: From Biology to Hardware. Proceedings, 1996. XI, 484 pages. 1997.

Vol. 1260: D. Raymond, D. Wood, S. Yu (Eds.), Automata Implementation. Proceedings, 1996. VIII, 189 pages. 1997.